JACARANDA MATHS QUEST

SPECIALIST MATHEMATICS 12

VCE UNITS 3 AND 4 | SECOND EDITION

JACARANDA MATHS QUEST

SPECIALIST MATHEMATICS 12

VCE UNITS 3 AND 4 | SECOND EDITION

RAYMOND ROZEN

SUE MICHELL

jacaranda
A Wiley Brand

Second edition published 2023 by
John Wiley & Sons Australia, Ltd
155 Cremorne Street, Cremorne, Vic 3121

First edition published 2016

Typeset in 10.5/13 pt TimesLTStd

ISBN: 978-1-119-87674-8

The covers of the *Jacaranda Maths Quest VCE Mathematics* series are the work of Victorian artist Lydia Bachimova.

Lydia is an experienced, innovative and creative artist with over 10 years of professional experience, including five years of animation work with Walt Disney Studio in Sydney. She has a passion for hand drawing, painting and graphic design.

Illustrated by diacriTech and Wiley Composition Services

Typeset in India by diacriTech

A catalogue record for this book is available from the National Library of Australia

Printed in Singapore
M121041_150822

Contents

About this resource

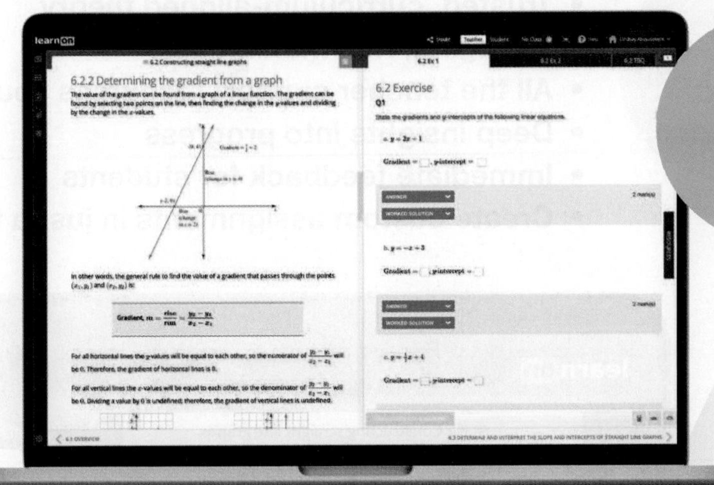

Everything you need for your students to succeed

JACARANDA MATHS QUEST
SPECIALIST MATHEMATICS 12 VCE UNITS 3 AND 4 | SECOND EDITION

Developed by expert Victorian teachers for VCE students

Tried, tested and trusted. The NEW Jacaranda VCE Mathematics series continues to deliver curriculum-aligned material that caters to students of all abilities.

Completely aligned to the VCE Mathematics Study Design

Our expert author team of practising teachers and assessors ensures 100 per cent coverage of the new VCE Mathematics Study Design (2023–2027).

Everything you need for your students to succeed, including:

- NEW! Access targeted question sets including exam-style questions and all relevant past VCAA exam questions since 2013. Ensure assessment preparedness with practice school-assessed coursework.
- NEW! Be confident your students can get unstuck and progress, in class or at home. For every question online they receive immediate feedback and fully worked solutions.
- NEW! Teacher-led videos to unpack concepts, plus VCAA exam questions and exam-style questions to fill learning gaps after COVID-19 disruptions.

Learn online with Australia's most

Everything you need for each of your lessons in one simple view

- **Trusted, curriculum-aligned theory**
- **Engaging, rich multimedia**
- **All the teacher support resources you need**
- **Deep insights into progress**
- **Immediate feedback for students**
- **Create custom assignments in just a few clicks.**

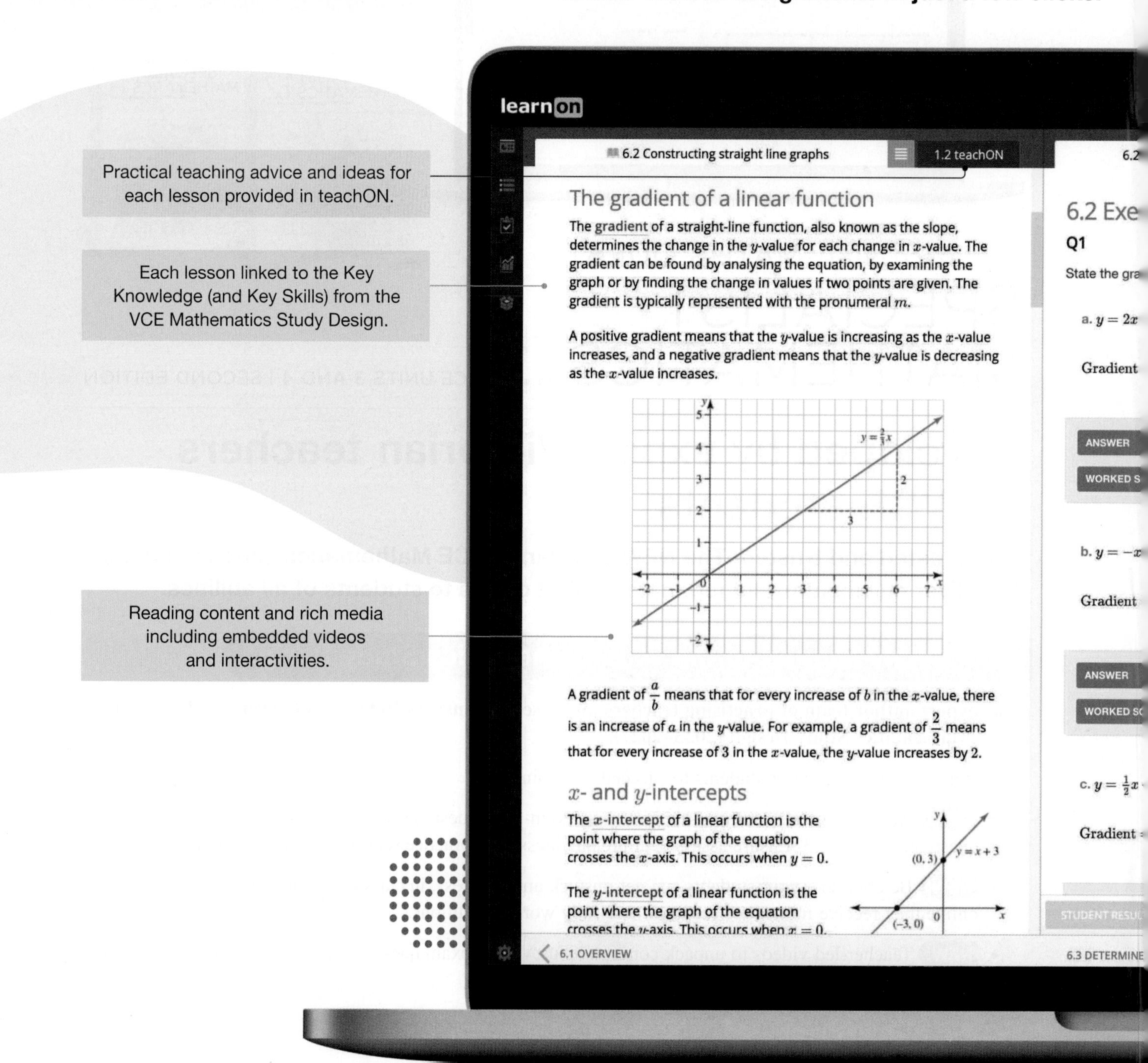

powerful learning tool, learnON

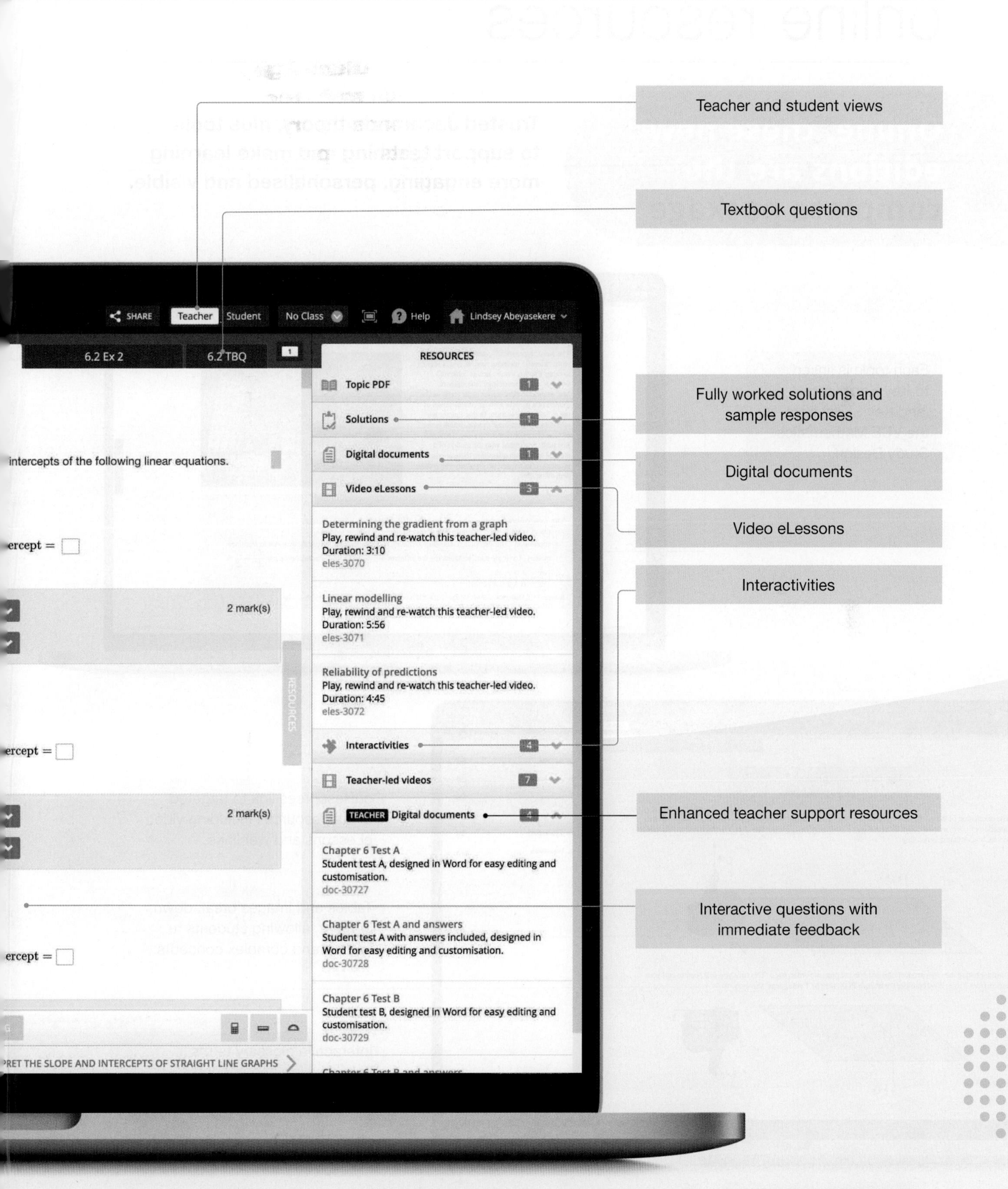

Get the most from your online resources

Online, these new editions are the complete package

Trusted Jacaranda theory, plus tools to support teaching and make learning more engaging, personalised and visible.

Each topic is linked to Key Knowledge (and Key Skills) from the VCE Mathematics Study Design.

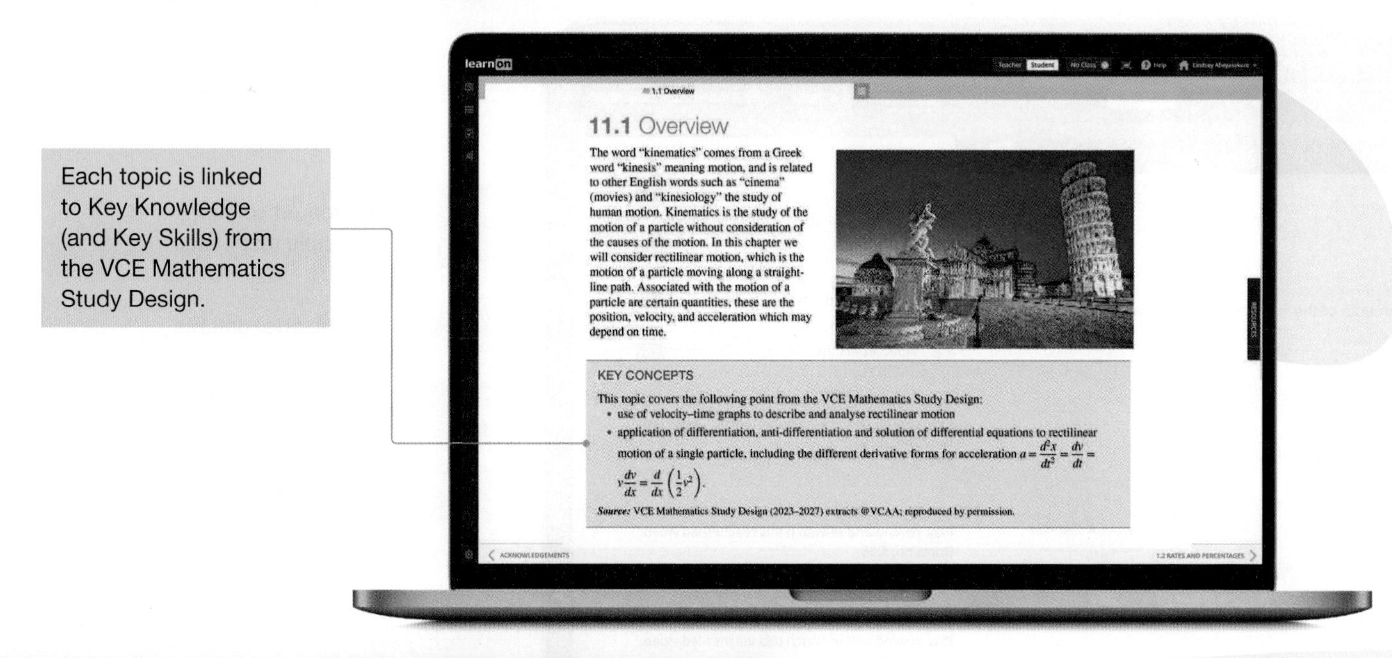

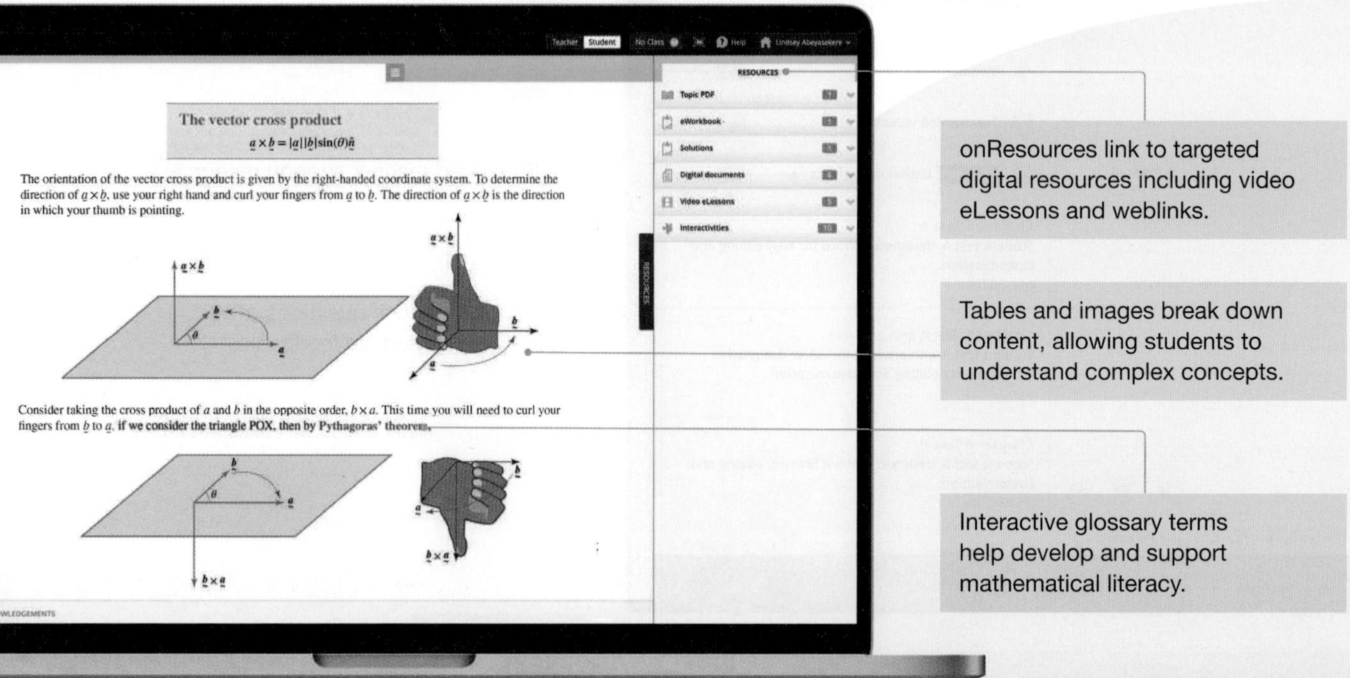

onResources link to targeted digital resources including video eLessons and weblinks.

Tables and images break down content, allowing students to understand complex concepts.

Interactive glossary terms help develop and support mathematical literacy.

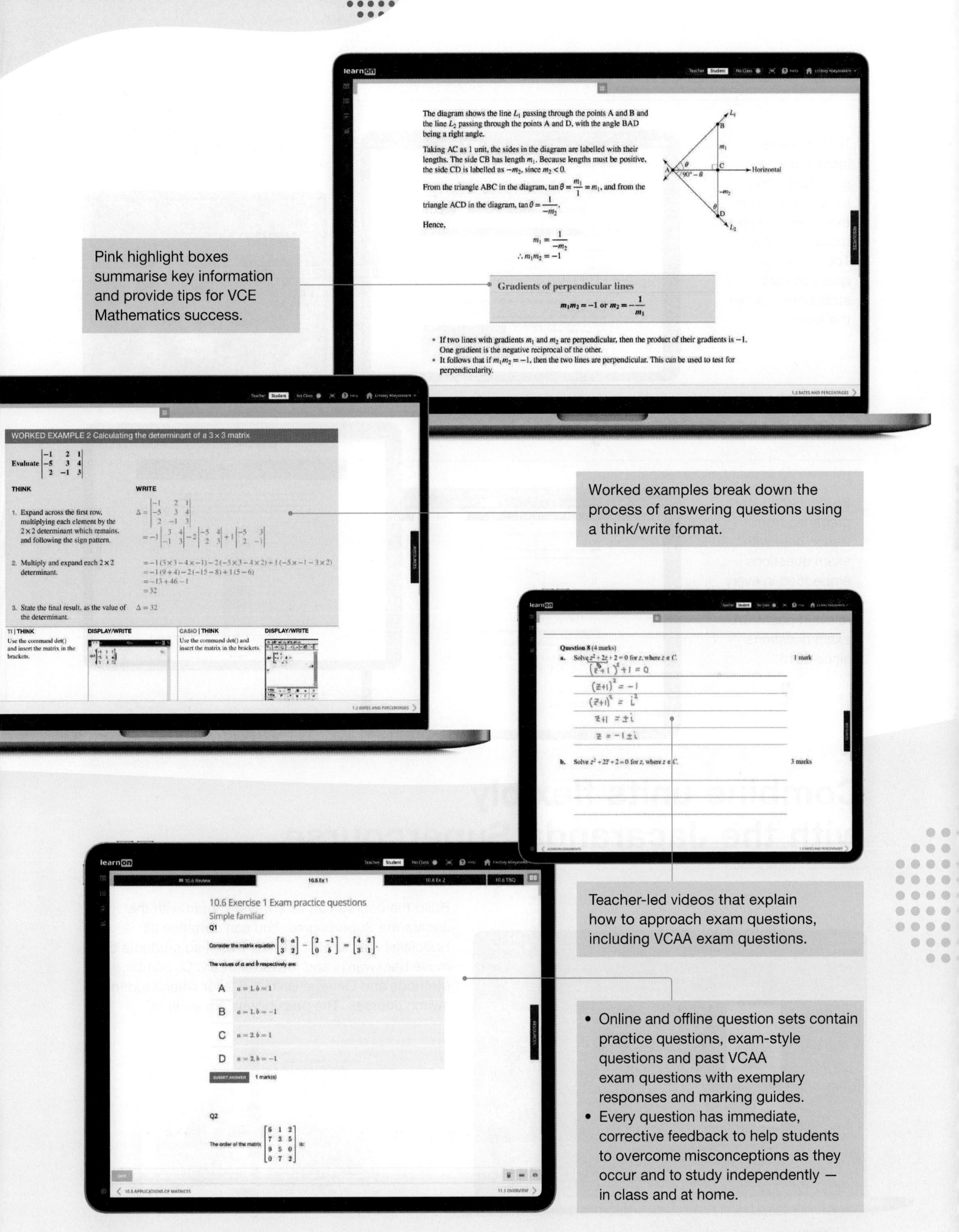

Pink highlight boxes summarise key information and provide tips for VCE Mathematics success.
Gradients of perpendicular lines
Worked examples break down the process of answering questions using a think/write format.
WORKED EXAMPLE 2 Calculating the determinant of a 3 × 3 matrix
THINK
WRITE
Teacher-led videos that explain how to approach exam questions, including VCAA exam questions.
10.6 Exercise 1 Exam practice questions
Simple familiar
Online and offline question sets contain practice questions, exam-style questions and past VCAA exam questions with exemplary responses and marking guides.
Every question has immediate, corrective feedback to help students to overcome misconceptions as they occur and to study independently — in class and at home.

Topic reviews

Topic reviews include online summaries and topic-level review exercises that cover multiple concepts. Topic-level exam questions are structured just like the exams.

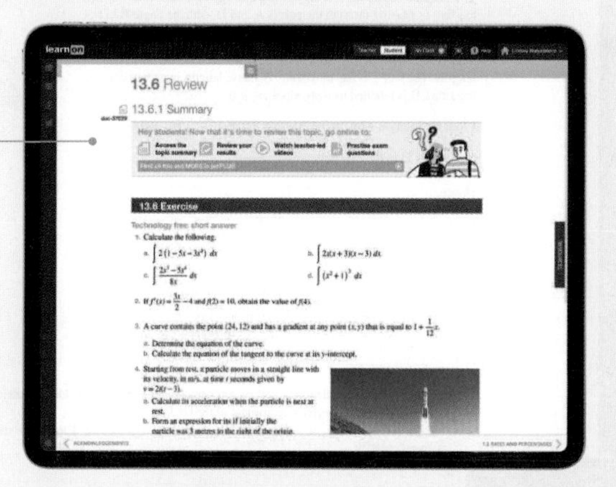

End-of-topic exam questions include relevant past VCE exam questions and are supported by teacher-led videos.

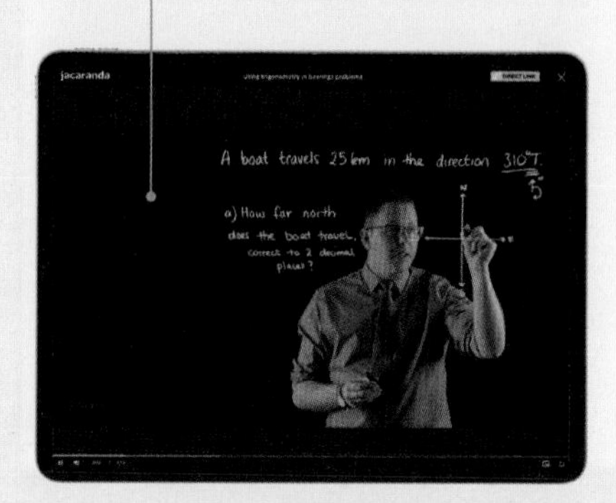

Get exam-ready!

Students can start preparing from lesson one, with exam questions embedded in every lesson — with relevant past VCAA exam questions since 2013.

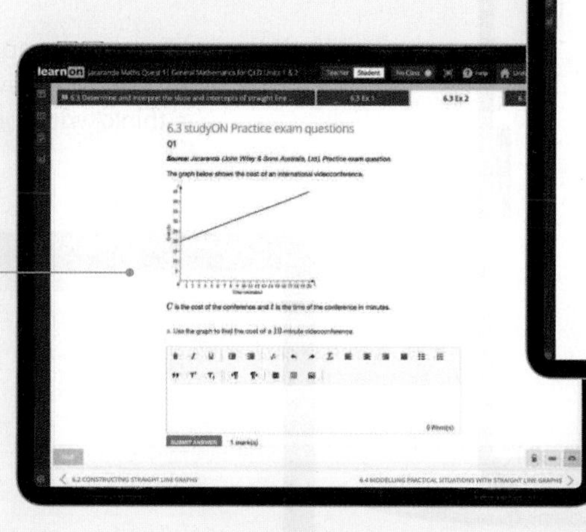

Customisable practice SACs available to build student competence and confidence.

Combine units flexibly with the Jacaranda Supercourse

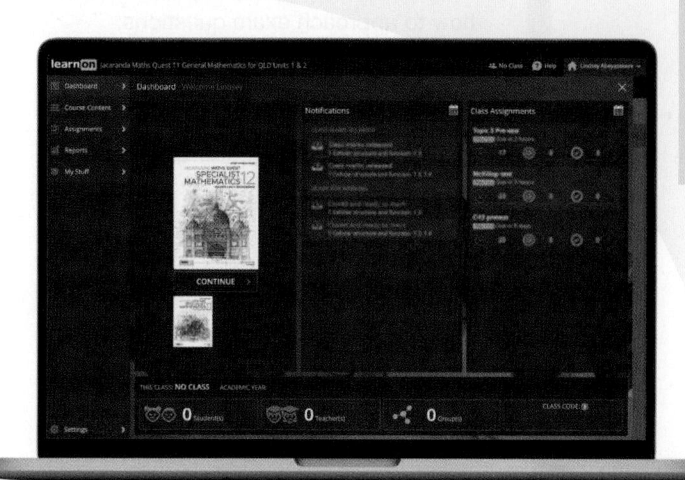

Build the course you've always wanted with the Jacaranda Supercourse. You can combine all Specialist Mathematics Units 1 to 4, so students can move backwards and forwards freely. Or combine Methods and General Units 1 & 2 for when students switch courses. The possibilities are endless!

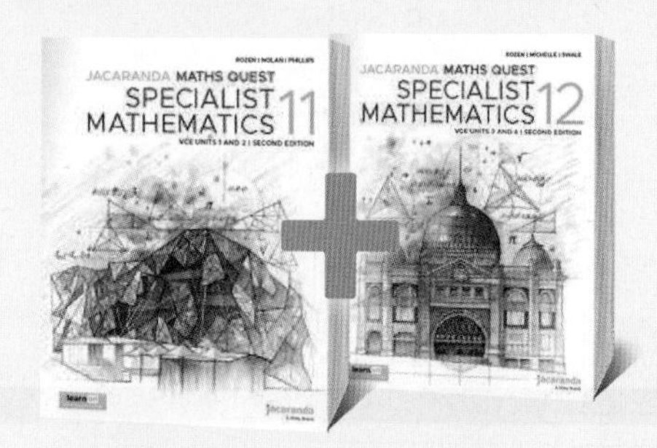

A wealth of teacher resources

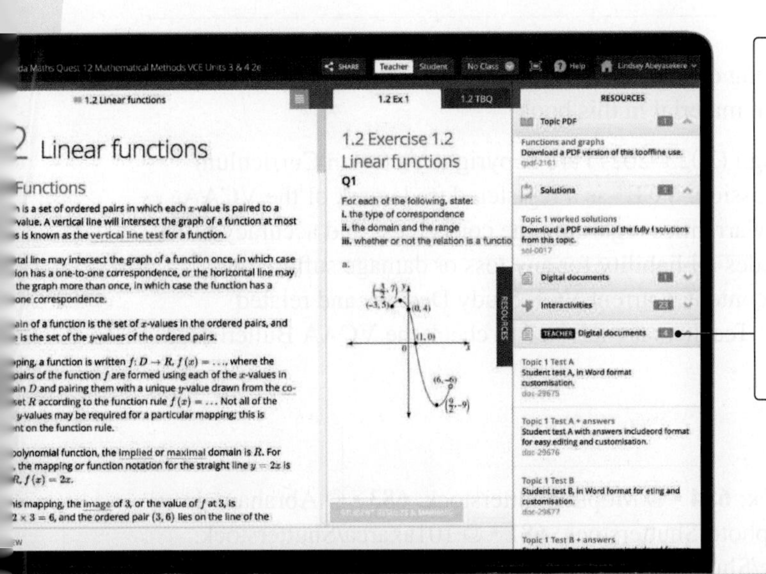

Enhanced teacher support resources, including:

- work programs and curriculum grids
- teaching advice and additional activities
- quarantined topic tests (with solutions)
- quarantined SACs (with worked solutions and marking rubrics)

Customise and assign

A testmaker enables you to create custom tests from the complete bank of thousands of questions (including past VCAA exam questions).

Reports and results

Data analytics and instant reports provide data-driven insights into performance across the entire course.

Show students (and their parents or carers) their own assessment data in fine detail. You can filter their results to identify areas of strength and weakness.

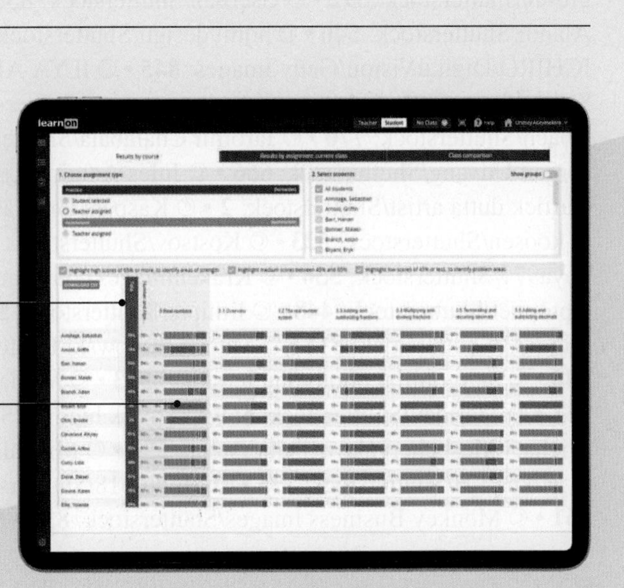

Acknowledgements

The authors and publisher would like to thank the following copyright holders, organisations and individuals for their assistance and for permission to reproduce copyright material in this book.

Selected extracts from the VCE Mathematics Study Design (2023–2027) are copyright Victorian Curriculum and Assessment Authority (VCAA), reproduced by permission. VCE® is a registered trademark of the VCAA. The VCAA does not endorse this product and makes no warranties regarding the correctness and accuracy of its content. To the extent permitted by law, the VCAA excludes all liability for any loss or damage suffered or incurred as a result of accessing, using or relying on the content. Current VCE Study Designs and related content can be accessed directly at www.vcaa.vic.edu.au. Teachers are advised to check the VCAA Bulletin for updates.

Images

• © Alamy Stock Photo: **687** • © Daria Nipot/Shutterstock: **684** • © Mopic/Shutterstock: **683** • © Abraham de moivre/Wikimedia Commons Public Domain: **68** • © 06photo/Shutterstock: **681** • © 101akarca/Shutterstock: **773** • © alexkatkov/Shutterstock: **604** • © andrea crisante/Shutterstock: **559** • © Andreas Gradin/Shutterstock: **614** • © Andrew Lam/Shutterstock: **576** • © Andrew Robins Photography/Shutterstock: **795** • © Andrey_Popov/Shutterstock: **591, 829** • © Ann Worthy/Shutterstock: **292** • © Archvadze Paata/Shutterstock: **553** • © Art Collection 2/Alamy Stock Photo: Shutterstock • © Aspen Photo/**38**: **774** • © assistant/Shutterstock: **640** • © Astronira/Shutterstock: **46** • © azur13/Shutterstock: **186** • © ben bryant/Shutterstock: **577** • © bentaj hicham/Shutterstock: **710** • © Berents/Shutterstock: **524** • © bezikus/Shutterstock: **668** • © bioraven/ Shutterstock: **577** • © Bplanet/Shutterstock: **670** • © By Euclid - http://www.math.ubc.ca/ cass/Euclid/ papyrus/tha.jpg: Public Domain, https://commons.wikimedia.org/w/index.php?curid=1259734, **2** • © ChameleonsEye/Shutterstock: **680** • © Chantal de Bruijne/Shutterstock: **496** • © Chinaview/Shutterstock: **735** • © Christoph Bernhard Francke/Wikipedia: **232** • © Computer Earth/Shutterstock: **174** • © comzeal images/ Shutterstock: **524** • © cristiano barni/Shutterstock: **655** • © damedias/Adobe Stock: **579** • © Dancestrokes/ Shutterstock: **251** • © David Smart/Shutterstock: **832** • © Derek R. Audette/Shutterstock: **642** • © domonabike/ Alamy Stock Photo: **775** • © Dreef/Getty Images: **701** • © englishinbsas/Shutterstock: **778** • © EpicStockMedia/ Shutterstock: **721** • © Everett Historical/Shutterstock: **577** • © Fotofermer/Shutterstock: **552** • © FotografiaBasica/Getty Images: **/E+846** • © Fotokostic/Shutterstock: **701** • © Francesca Baldassari/Wikimedia Commons/**547**Public Domain: • © Georgios Kollidas/Shutterstock: **616** • © Germanskydiver/Shutterstock: **700** • © Gints Ivuskans/Shutterstock: **766** • © Godfrey Kneller/Wikipedia: **232** • © GoodStudio/Shutterstock: **809** • © gresei/Shutterstock: **592** • © Gserban/Shutterstock: **835** • © H0zell/Shutterstock: **303** • © Hanna Alandi/Shutterstock: **546** • © homydesign/Shutterstock: **255** • © hxdyl/Shutterstock: **147** • © ICHIRO/DigitalVision/Getty Images: **845** • © ILYA AKINSHIN/Shutterstock: **521** • © IOIO IMAGES/Alamy Stock Photo: **739** • © Italianvideophotoagency/Shutterstock: **727** • © Jacqui Martin/Shutterstock: **640** • © Jamie Roach/Shutterstock: **770** • © Jaromir Chalabala/Shutterstock: **673** • © JIANG HONGYAN/Shutterstock: **584** • © Jorgen Udvang/Shutterstock: **666** • © Jules Antoine Lissajous/Wikimedia Commons/ Public Domain: **746** • © Kartick dutta artist/Shutterstock: **2** • © Kaspars Grinvalds/Shutterstock: **674** • © katatonia82/Shutterstock: **255** • © koosen/Shutterstock: **523** • © Kostsov/Shutterstock: **285** • © Kovaleva_Ka/Shutterstock: **806** • © koya979/Shutterstock: **536** • © Krakenimages.com/Shutterstock: **3, 4** • © Kzenon/Shutterstock: **696** • © laptopnet/Shutterstock: **448** • © limipix/Shutterstock: **546** • © LStockStudio/Shutterstock: **818** • © luiggi33/Shutterstock: **811** • © Maciej Rogowski Photo/Shutterstock: **775** • © Makushin Alexey/Shutterstock: **146** • © margouillat photo/Shutterstock: **523** • © Maria Gaetana Agnesi/Wikimedia Commons/ Public Domain: **751** • © Maridav/Shutterstock: **669** • © mark higgins/Shutterstock: **585** • © mezzotint/Shutterstock: **290** • © Michael Meshcheryakov/Shutterstock: **590** • © Michal Vitek/Shutterstock: **773** • © Mikbiz/Shutterstock: **829** • © Mikhail H/Shutterstock: **125** • © Modulo18/Shutterstock: **185** • © Monkey Business Images/Shutterstock: **761** • © Monkey Business Images/Shutterstock: **831** • © monticello/Shutterstock.com: **803** • © MUNGKHOOD

STUDIO/Shutterstock: **68919** • © Nadya Ershova/Shutterstock: • © Nataly Studio/Shutterstock: **819** • © Nathan Till/Shutterstock: **641** • © Nelson Charette Photo/Shutterstock: **45** • © NH/Shutterstock: **231** • © Nils Versemann/Shutterstock: • © nito**688**/Shutterstock: **727** • © niwat chaiyawoot/Shutterstock: **288** • © nokwalai/Shutterstock: Shutterstock**796** • © Oksana Shufrych/: **808** • © oksana.perkins/Shutterstock: **682** • © Oksana2010/Shutterstock: **562** • © OlegDoroshinShutterstock: **289** • © Olga Danylenko/Shutterstock: **147** • © Patryk Kosmider/Shutterstock: **656** • © Peter Bernik/Shutterstock: **711** • © Picsfive/Shutterstock: **846** • © Pisit Rapitpunt/Shutterstock: **284** • © Prapat Aowsakorn/Shutterstock: **447** • © r.classen/Shutterstock: **525** • © Rido/Shutterstock: **4, 5** • © Risto Raunio/Shutterstock: **304** • © Robert Ranson/Shutterstock: **289** • © rodimov/Shutterstock: **453** • © Romolo Tavani/Shutterstock: **288** • © Ryan Fletcher/Shutterstock: **676** • © Sakarin Sawasdinaka/Shutterstock: **27** • © Sashkin/Shutterstock: **659** • © Seksun Guntanid/Shutterstock: **569** • © sharpner/Shutterstock: **206** • © soda aha/Shutterstock: **495** • © sommthink/Shutterstock: **555** • © Thanaphat Kingkaew/Shutterstock: **838** • © Tobias Helbig/Shutterstock: **680** • © tomocam/Shutterstock: **582** • © Tono Balaguer/Shutterstock: **699** • © TORWAISTUDIO/Shutterstock: **710** • © TravelerFL/Shutterstock: **1** • © Tupungato/Shutterstock: **664** • © Undrey/Shutterstock: **363, 364** • © Valentyn Volkov/Shutterstock: **602** • © Valery Lisin/Shutterstock: **570** • © VanderWolf Images/Shutterstock: **585** • © vanillaechoes/Shutterstock: **642** • © vichie81/Shutterstock: **290** • © Vipavlenkoff/Shutterstock: **662** • © wavebreakmedia/Shutterstock: **722** • © wellphoto/Shutterstock: **486** • © Xiebiyun/Shutterstock: **816** • © Yusnizam YusofShutterstock/: **21**

Every effort has been made to trace the ownership of copyright material. Information that will enable the publisher to rectify any error or omission in subsequent reprints will be welcome. In such cases, please contact the Permissions Section of John Wiley & Sons Australia, Ltd.

1 Logic and proof

LEARNING SEQUENCE

Fully worked solutions for this topic are available online.

1.1 Overview

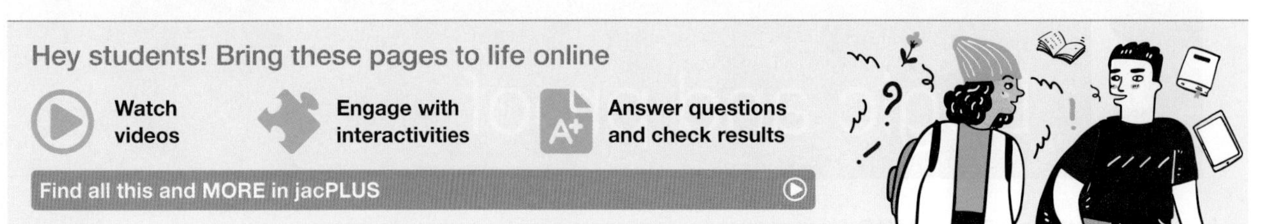

1.1.1 Introduction

The word 'proof' comes from the Latin word 'probare' (to test).

Proofs first started in ancient Greece around the 5th century BCE, with many famous mathematicians and philosophers of that time, including Thales (624–546 BCE) and Hippocrates of Chios (c. 470–410 BCE) who gave some proofs in geometry, Eudoxus (408–355 BCE) and Theaetetus (417–369 BCE) who formulated theorems but did not prove them, Aristotle (384–322 BCE) and Euclid (325−263 BCE) who introduced axioms.

An axiom is a statement that is considered to be true, based on logic and does not need to be proven. This is also called a postulate, or an assumption. Euclid's book called 'The Elements' is a series of 13 books that is still used today in the development of logic. The image shows an old fragment from Euclid's book.

KEY CONCEPTS

This topic covers the following key concepts from the VCE Mathematics Study Design:

- conjecture – making a statement to be proved or disproved
- implications, equivalences and if and only if statements (necessary and sufficient conditions)
- natural deduction and proof techniques: direct proofs using a sequence of direct implications, proof by cases, proof by contradiction, and proof by contrapositive
- quantifiers 'for all' and 'there exists', examples and counter-examples
- proof by mathematical induction.

Source: VCE Mathematics Study Design (2023–2027) extracts © VCAA; reproduced by permission.

1.2 Logic

LEARNING INTENTION

At the end of this subtopic you should be able to:
- apply principles of proof to prove simple tautologies
- prove statements involving quantifiers over different sets of numbers.

Logic comprises both propositional and predicate logic, which are extremely useful and used in computer science. First we revise and define some basic terminology.

1.2.1 Number systems

The set of natural numbers, or counting numbers is the set $N = \{1, 2, 3, ...\}$. If we add or multiply any two natural numbers, we always get a natural number, so we say that the set of natural numbers is closed under the operations of addition and multiplication. However if we subtract two natural numbers, we don't necessarily get a natural number, so we extend the set of natural numbers to the set of integers. The set of integers consists of both positive and negative integers and zero. $Z = \{... -3, -2, -1, 0, 1, 2, 3, ...\}$.

The subsets of positive integers are the set of natural numbers $Z^+ = N = \{1, 2, 3, ...\}$ and the set of negative integers $Z^- = \{-1, -2, -3, ...\}$, so that $Z = Z^+ \cup Z^- \cup \{0\}$. If we add, subtract or multiply any two integers, we always get an integer, so we say that the set of integers is closed under the operations of addition, subtraction and multiplication. However, when we divide two integers we do not necessarily get an integer, so we extend the set of integers to the set of rational numbers. The set of rational numbers are numbers of the form $Q = \left\{\frac{a}{b}\right\}$, $a \in Z, b \in Z, b \neq 0$, where a and b do not have any common factor except for 1. Rational numbers either have a recurring pattern in their decimal expansion, or its decimal expansion terminates.

Rational numbers are closed under the operations of addition, subtraction, multiplication and division. However there are equations such as $x^2 = 2$, $x = \pm\sqrt{2}$, which have no rational solutions, and numbers such as $\sqrt{2}, \pi, e, ...$ are not rational. That is, there are numbers which cannot be expressed as a ratio of two integers and their decimal expansion never terminates, nor has any pattern ever been detected in the decimal expansion.

So we extend the set to include real numbers R.

We have met complex numbers in Year 11, and will extend your knowledge of complex numbers in Topic 2, however at this stage we know there are equations such as $x^2 = -1$ which have no real solutions, so we define $i^2 = -1$ and extend the number system to include the set of complex numbers, $C = \{a + bi\}, a, b \in R$.

$N \subseteq Z \subseteq Q \subseteq R \subseteq C$, where the symbol $\subseteq$ means is a subset, so every natural number is an integer, every integer is rational, every rational number is real and every real number is a complex number (with a zero imaginary part).

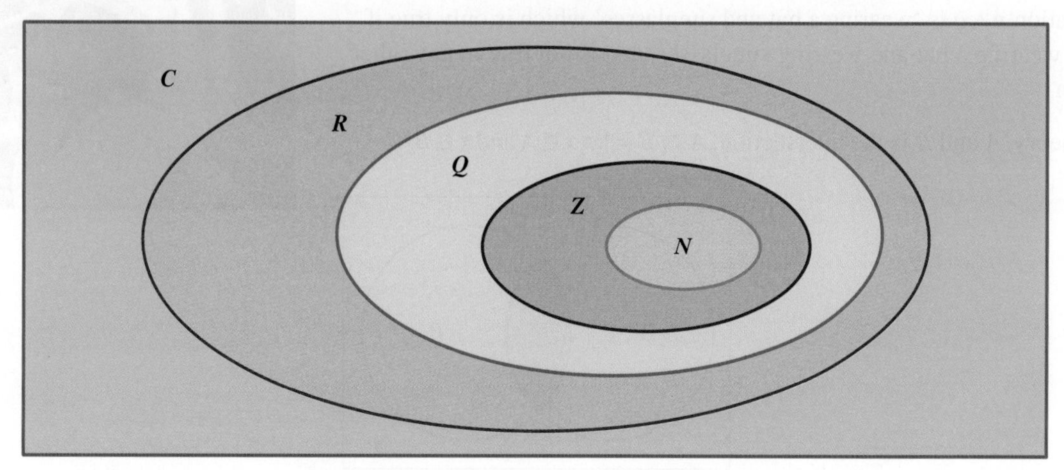

1.2.2 Proposition logic

Propositional logic is concerned with mathematical statements called **propositions**, which are either true (T) or false (F), note that we usually denote propositions by lower case letters, p, q, r.

A couple of examples of propositions are the following.

p: Preston is a suburb of Melbourne. This statement is true.

q: Queenstown is the capital city of Victoria. This statement is false.

We can use a truth table to display all possible true and false combinations.

Negation

When we negate a statement, we insert the word 'not' into it to change its truth status. This means the **negation** of a true statement will be false and vice-versa. The negation of a proposition p is denoted by $\neg p$. As an example, the negation of the proposition p from above is $\neg p$: Preston is NOT a suburb of Melbourne. This statement is now false.

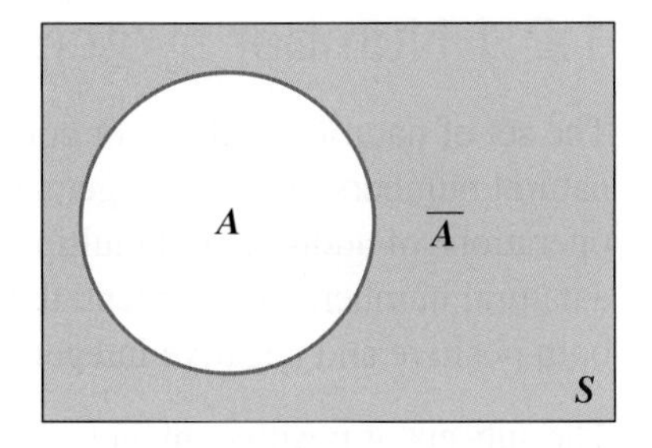

p	$\neg p$
T	F
F	T

In set theory, the negation is equivalent to the complement and can be represented in Venn diagrams as $\overline{A} = \{x : x \notin A\}$.

1.2.3 Combining two propositions

We can combine two or more propositions using connectives. We use a truth table to display all possible true and false combinations.

And

Using 'and' to combine two propositions p and q is also written as $p \wedge q$. $p \wedge q$ is only true when p and q are both true, as shown in the truth table below.

p	q	$p \wedge q$
T	T	T
T	F	F
F	T	F
F	F	F

For example, if p is 'wearing a hat' and q is 'wearing sunglasses', the combination $p \wedge q$ is 'wearing a hat and sunglasses' which is only true if you are wearing a hat and wearing sunglasses, and is not true in any other situation.

In set theory, A and B is the intersection, $A \cap B = \{x : x \in A \text{ and } x \in B\}$.

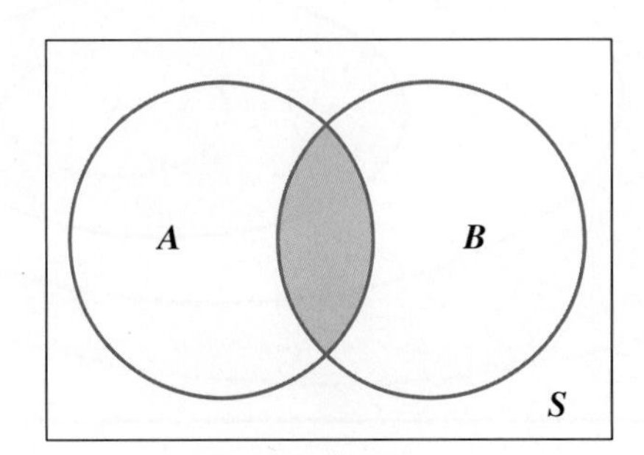

Or

Using 'or' to combine two propositions p or q is also written as $p \vee q$. $p \vee q$ is true when either p or q are true, or both are true.

p	q	$p \vee q$
T	T	T
T	F	T
F	T	T
F	F	F

Continuing the example from above, $p \vee q$ is 'wearing a hat or sunglasses' which is true if you are wearing a hat or wearing sunglasses, or both. $p \vee q$ is false if and only if p and q are both false.

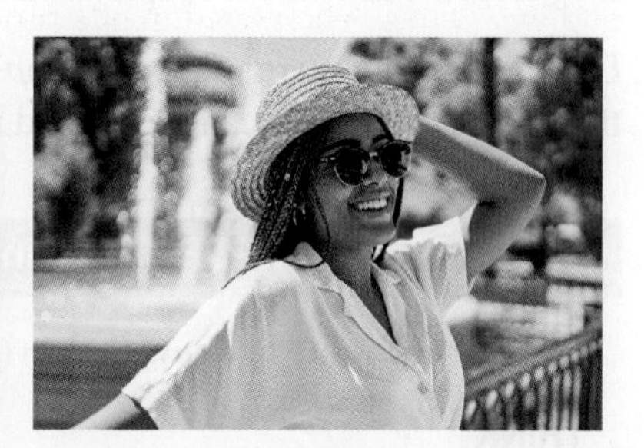

In set theory, A or B is the union $A \cup B = \{x: x \in A \text{ or } x \in B\}$.

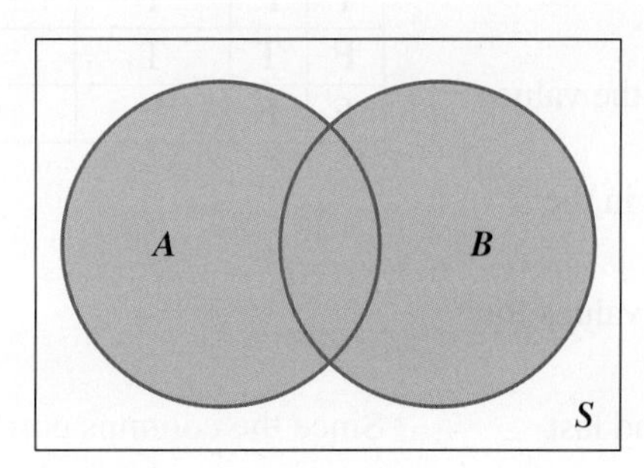

Implication

Implications are statements of the form 'if p then q', alternatively 'p implies q'. This is written as $p \rightarrow q$. The only time $p \rightarrow q$ is false, is when p is true and q is false.

p	q	$p \rightarrow q$
T	T	T
T	F	F
F	T	T
F	F	T

Note that when p is false, the implication statement $p \rightarrow q$ is defined as true by default. This situation is known as a vacuous truth as it essentially holds no meaning or information. Any implication statement that starts with a false 'if' statement is vacuously true. For example, 'If pigs could fly then...' is a vacuously true statement regardless of what comes after the word 'then' because pigs can't fly!

Equivalence

Statements of equivalence are of the form 'p if and only if q', alternatively 'p iff q' (iff is a shorthand for if and only if); this is written as $p \leftrightarrow q$. $p \leftrightarrow q$ is true when both p and q are true and when p and q are both false.

Note that $p \leftrightarrow q$ is also equivalent to $p \rightarrow q$ and $q \rightarrow p$.

p	q	$p \leftrightarrow q$
T	T	T
T	F	F
F	T	F
F	F	T

De Morgan's laws are a couple of famous equivalence statements which are very useful in many proofs.

De Morgan's laws are $\neg(p \vee q) \leftrightarrow \neg p \wedge \neg q$ and $\neg(p \wedge q) \leftrightarrow \neg p \vee \neg q$. The following Worked example shows how the first of these laws can be verified using truth tables.

WORKED EXAMPLE 1 Proving De Morgan's law

Use a truth table to show that $\neg(p \vee q) \leftrightarrow \neg p \wedge \neg q$.

THINK

1. Set up the truth table. Since there are two propositions, we need four rows.
2. In the third column complete the values for $p \vee q$.
3. In the fourth column, negate all the values in the third column.
4. In the fifth column negate p and in the sixth column negate q.
5. In the last column complete the values for $\neg p \wedge \neg q$.
6. Notice that the fourth column and last columns are the same.

WRITE

p	q	$p \vee q$	$\neg(p \vee q)$	$\neg p$	$\neg q$	$\neg p \wedge \neg q$
T	T	T	F	F	F	F
T	F	T	F	F	T	F
F	T	T	F	T	F	F
F	F	F	T	T	T	T

Since the columns corresponding to $\neg(p \vee q)$ and $\neg p \wedge \neg q$ are identical it follows that $\neg(p \vee q) \leftrightarrow \neg p \wedge \neg q$.

1.2.4 Tautologies

In mathematical logic, a **tautology** is a formula or assertion involving propositions that is true in every possible interpretation.

The De Morgan's law $\neg(p \vee q) \leftrightarrow \neg p \wedge \neg q$ is an example of a tautology.

In algebra, multiplication takes precedence over addition and subtraction, so in $3 + (4 \times 5)$ the brackets are implied and not needed; this expression can therefore be expressed as $3 + 4 \times 5$.

Similarly, in logic $\vee$ and $\wedge$ take precedence over implication and equivalence, so $\neg(p \vee q) \leftrightarrow (\neg p \wedge \neg q)$ can simply be written as $\neg(p \vee q) \leftrightarrow \neg p \wedge \neg q$.

WORKED EXAMPLE 2 Proving a simple tautology

Use a truth table to show that $\neg(\neg p) \leftrightarrow p$, is a tautology.

THINK

1. Set up the truth table. Since there is only one proposition, we need only two rows.
2. In the second column negate p.
3. In the third column negate the second column.
5. Notice that the first and third columns are the same.

WRITE

p	$\neg p$	$\neg(\neg p)$
T	F	T
F	T	F

Since the columns corresponding to $\neg p$ and $\neg(\neg p)$ are identical it follows that $\neg(\neg p) \leftrightarrow p$.

1.2.5 Contrapositive, inverse and converse

Consider the conditional implication statement $p \to q$. The **contrapositive** is $\neg q \to \neg p$, and it will be shown in Worked example 3 that these two statements are equivalent since $(p \to q) \leftrightarrow (\neg q \to \neg p)$ is a tautology.

Consider the conditional implication statement, $p \to q$. The **inverse** is $\neg p \to \neg q$ and the inverse's truth value is not dependent on whether the original implication is true.

Consider the conditional implication statement, $p \to q$. The **converse** is $q \to p$ and is actually the contrapositive of the inverse, so it always has the same truth value as the inverse and therefore $(q \to p) \leftrightarrow (\neg p \to \neg q)$ is also a tautology.

Consider the conditional implication statement, $p \to q$. The negation is $\neg(p \to q)$ and if the negation is true, then the original implication and the contrapositive are false. Note that $\neg(p \to q) \leftrightarrow (p \wedge \neg q)$ is a tautology, which we leave as an exercise.

Variations of implication statements

Name	Logic form	Description
Implication	**If p then q, $p \to q$**	**First statement implies truth of the second**
Contrapositive	**If not q then not p, $\neg q \to \neg p$**	**Reversal and negation of both statements**
Inverse	**If not p then not q, $\neg p \to \neg q$**	**Negation of both statements**
Converse	**If q then p, $q \to p$**	**Reversal of both statements**
Negation	**p and not q, $\neg(p \to q) \leftrightarrow (p \wedge \neg q)$**	**Contradicts the implication**

WORKED EXAMPLE 3 Proving the contrapositive law

Use a truth table to show that $(p \to q) \leftrightarrow (\neg q \to \neg p)$ is a tautology.

THINK

1. Set up the truth table. Since there are two propositions, we need four rows.
2. Complete the third column for $p \to q$.
3. In the fourth and fifth columns negate q and p respectively.
4. Complete the sixth column for $\neg q \to \neg p$.
5. Notice that the third and last columns are the same.

WRITE

p	q	$p \to q$	$\neg q$	$\neg p$	$\neg q \to \neg p$
T	T	T	F	F	T
T	F	F	T	F	F
F	T	T	F	T	T
F	F	T	T	T	T

Since the columns corresponding to $p \to q$ and $\neg q \to \neg p$ are identical it follows that $(p \to q) \leftrightarrow (\neg q \to \neg p)$.

WORKED EXAMPLE 4 Contrapositive, inverse, converse and negation statements

Let p and q be the propositions, 'it is cold' and 'I wear a coat,' respectively. Write the implication, $p \rightarrow q$, the contrapositive, inverse, converse and negation statements in words.

THINK	WRITE
1. Implication $p \rightarrow q$	If it is cold then I wear a coat.
2. Contrapositive $\neg q \rightarrow \neg p$	If I do not wear a coat then it is not cold.
3. Inverse $\neg p \rightarrow \neg q$	If it is not cold then I will not wear a coat.
4. Converse $q \rightarrow p$	If I wear a coat then it is cold.
5. Negation $\neg (p \rightarrow q)$	It is not the case that if it is cold then I wear a coat. Equivalently It is cold and I do not wear a coat.

Extension to three propositions

WORKED EXAMPLE 5 Proving a tautology involving three propositions

Use a truth table to show that $[(p \vee q) \rightarrow r] \leftrightarrow [(p \rightarrow r) \wedge (q \rightarrow r)]$ is a tautology.

THINK

1. Set up the truth table. Since there are three propositions, we require $2^3 = 8$ rows.
2. Complete the fourth column for $p \vee q$.
3. Complete the fifth column for $(p \vee q) \rightarrow r$.
4. Complete the sixth column for $p \rightarrow r$.
5. Complete the seventh column for $q \rightarrow r$.

WRITE

p	q	r	$p \vee q$	$(p \vee q) \rightarrow r$	$p \rightarrow r$	$q \rightarrow r$	$(p \rightarrow r) \wedge (q \rightarrow r)$
T	T	T	T	T	T	T	T
T	T	F	T	F	F	F	F
T	F	T	T	T	T	T	T
T	F	F	T	F	F	T	F
F	T	T	T	T	T	T	T
F	T	F	T	F	T	F	F
F	F	T	F	T	T	T	T
F	F	F	F	T	T	T	T

6. Notice that the fifth and last columns are the same.

Since the columns corresponding to $(p \vee q) \rightarrow r$ and $(p \rightarrow r) \wedge (q \rightarrow r)$ are identical it follows that $\left[(p \vee q) \rightarrow r\right] \leftrightarrow \left[(p \rightarrow r) \wedge (q \rightarrow r)\right]$.

1.2.6 Predicate logic

Predicate logic is a form of propositional logic involving quantifiers or propositions which are in terms of a function of a variable x.

The proposition $5 > 3$ is a true statement, while $3 > 5$ is a false statement. The statement $x > 3$ can be true or false for certain values of x. A propositional statement which is a proposition regarding a variable, x, is written as $P(x)$.

The universal quantifier

The universal quantifier 'for all' (or 'for every') is represented by the symbol $\forall$.

As an example, $\forall x \in R$, $\sin^2(x) + \cos^2(x) = 1$, alternatively, $\sin^2(x) + \cos^2(x) = 1$ is true for all real numbers, x.

Proof by counter-example

If we need to show that a statement $\forall x \in D,\ P(x)$ is false, then it is sufficient to determine just **one** value within D for which it is false, no matter how many cases it is true for. D is called the universe of discourse and is simply the set that is specified in the statement.

WORKED EXAMPLE 6 Evaluating propositions for all real numbers

Write the following propositions in words and state whether they are true or false.

a. $\forall x \in R,\ x^2 + 4 > 3$

b. $\forall x \in Z,\ 2x > x$

THINK	WRITE
a. 1. The symbol $\forall$ means for all.	For all real numbers $x^2 + 4 > 3$.
2. For all real numbers $x^2 \geq 0$, therefore $x^2 + 4 \geq 4$. Since $4 > 3$, $x^2 + 4 > 3$.	The proposition $\forall x \in R,\ x^2 + 4 > 3$ is true.
b. 1. The symbol $\forall$ means for all.	For all integers x, $2x > x$.
2. The proposition is true for all positive integers, but it is not true for negative integers. Show an example which disproves the proposition.	When $x = -2$, $2x = -4 < -2$.
3. State the conclusion.	The proposition $\forall x \in Z,\ 2x > x$ is false.

The existential quantifier

The existential quantifier 'there exists' (or 'for some') is represented by the symbol $\exists$. As an example, $\exists x \in R,\ \sin^2(x) = 1$. Alternatively, there exist some real number(s) x, such that $\sin^2(x) = 1$. Note that we are not saying there is only one unique value for which this is true, there may be more than one, or in fact an infinite number, which make the proposition or functional statement true.

Quantifiers

Statement	True when	False when
$\forall x \in D, P(x)$	$P(x)$ is true for all $x \in D$	There is an $x \in D$ for which $P(x)$ is false
$\exists x \in D, P(x)$	There is an $x \in D$ for which $P(x)$ is true	$P(x)$ is false for all $x \in D$

WORKED EXAMPLE 7 Evaluating propositions involving the existential quantifier

Write the following propositions in words and state whether they are true or false.

a. $\exists x \in Q,\ x^2 + 4 = 6$

b. $\exists x \in C,\ x^2 + 4 = 3$

THINK	WRITE
a. 1. The symbol $\exists$ means there exists.	There exists a rational solution to $x^2 + 4 = 6$.
2. $x^2 + 4 = 6$ is equivalent to $x^2 = 2$.	The only solutions to $x^2 = 2$ are $x = \pm\sqrt{2}$, which are not rational. The proposition $\exists x \in Q,\ x^2 + 4 = 6$ is therefore false.

b. 1. The symbol $\exists$ means there exists.	There exists a solution involving complex numbers to $x^2+4=3$.
2. $x^2+4=3$ is equivalent to $x^2=-1$.	$x^2=-1$ $x=\pm\sqrt{-1}=\pm i$ The proposition $\exists x \in C,\ x^2+4=3$ is true.

Negating quantifiers

The negation of the statement $\forall x \in D,\ P(x)$ is the statement $\exists x \in D,\ \neg P(x)$.

The negation of the statement $\exists x \in D,\ P(x)$ is the statement $\forall x \in D,\ \neg P(x)$.

For example, take the statement, 'All quadrilaterals have four sides' or equivalently,

'If a polygon is a quadrilateral, then it has four sides.'

The contrapositive, 'If a polygon does not have four sides, then it is not a quadrilateral.'

We see that the truth of the original statement, is equivalent to the contrapositive.

The inverse, 'If a polygon is not a quadrilateral, then it does not have four sides.'

The converse, 'If a polygon has four sides, then it is a quadrilateral.'

We see that the inverse and converse are also equivalent.

The negation, 'There is at least one quadrilateral that does not have four sides.'

This is obviously false.

Negating quantifiers

$$\neg(\forall x \in D,\ P(x)) \leftrightarrow \exists x \in D,\ \neg P(x)$$
$$\neg(\exists x \in D,\ P(x)) \leftrightarrow \forall x \in D,\ \neg P(x)$$

Combining quantifiers

When a proposition has two variables, x and y, quantifiers are needed for both variables, for example in $\forall x \in D_1$, $\forall y \in D_2, P(x, y)$ and $\exists x \in D_1,\ \exists y \in D_2,\ P(x, y)$, the order of the quantifiers is irrelevant. If the sets are the same, we can write $\forall x, y \in D,\ P(x, y)$ and $\exists x, y \in D,\ P(x, y)$.

WORKED EXAMPLE 8 Evaluating propositions

Write the following propositions in words and state whether they are true or false.

a. $\forall x \in Z^+, \forall y \in Z^-, x > y$

b. $\exists x \in Z, \exists y \in N, x + y = 0$

THINK	WRITE
a. 1. The symbol $\forall$ means for all.	All positive integers are greater than all negative integers.
2. The smallest positive integer is 1, which is greater than all negative integers.	The proposition $\forall x \in Z^+,\ \forall y \in Z^-, x > y$ is true.

b. 1. The symbol ∃ means there exists.	There are integers x and natural numbers y, such that $x+y=0$
2. Take the integer $x=-2$ and the natural number $y=2$.	$x+y=-2+2=0$ The proposition $\exists x \in Z,\ \exists y \in N,\ x+y=0$ is true.

Mixing quantifiers

The statements $\forall x \in D_1,\ \exists y \in D_2,\ P(x,y)$ and $\exists x \in D_1,\ \forall y \in D_2,\ P(x,y)$ have *different* meanings as they are read left to right and the order is extremely important.

Mixing quantifiers

Statement	True when	False when
$\forall x, \forall y, P(x,y)$	**$P(x,y)$ is true for every x and y.**	**There is at least one x and y for which $P(x,y)$ is false.**
$\exists x, \exists y, P(x,y)$	**There is at least one x and y for which $P(x,y)$ is true.**	**$P(x,y)$ is false for every x and y.**
$\forall x, \exists y, P(x,y)$	**For every x, there is a y for which $P(x,y)$ is true.**	**There is an x, for which $P(x,y)$ is false for every y.**
$\exists x, \forall y, P(x,y)$	**There is an x for which $P(x,y)$ is true for every y.**	**For every x, there is a y for which $P(x,y)$ is false.**

WORKED EXAMPLE 9 Evaluating propositions with mixed quantifiers

Write the following propositions in words and state whether they are true or false.

a. $\exists x \in Z^+, \forall y \in Q, xy=1$ **b.** $\forall x \in Z^+, \exists y \in Q, xy=1$

THINK	WRITE
a. 1. The symbols ∃ means there exists and ∀ means for all. Reading from left to right.	There are positive integers x, for every rational number y, such that $xy=1$.
2. Rearrange to make x the subject.	$xy=1 \Leftrightarrow x=\frac{1}{y}$
3. Determine an example which disproves the proposition.	When $y=\frac{2}{3}, x=\frac{1}{y}=\frac{3}{2}$. This x is rational, not an integer.
4. State the conclusion.	The proposition $\exists x \in Z^+,\ \forall y \in Q,\ xy=1$ is false.
b. 1. Read from left to right.	For every positive integer x, there is a rational number y, such that $xy=1$.
2. Rearrange to make y the subject.	$xy=1 \Leftrightarrow y=\frac{1}{x}$
3. For all positive integers x, the value of y is a rational number $\frac{1}{x}$.	The proposition $\forall x \in Z^+,\ \exists y \in Q,\ xy=1$ is true.

1.2 Exercise

Students, these questions are even better in jacPLUS

Receive immediate feedback and access sample responses

Access additional questions

Track your results and progress

Find all this and MORE in jacPLUS

Technology free

1. **WE1** Use a truth table to show that $\neg(p \wedge q) \leftrightarrow \neg p \vee \neg q$.

2. **WE2** Use a truth table to show that $p \vee \neg p$ is a tautology.

3. Use a truth table to show that $p \leftrightarrow (p \vee p)$ is a tautology.

4. **WE3** Use a truth table to show that $(p \to q) \leftrightarrow (\neg p \vee q)$ is a tautology.

5. Use a truth table to show that $(p \leftrightarrow q) \leftrightarrow (\neg p \leftrightarrow \neg q)$ is a tautology.

6. **WE4** Let p and q be the propositions, 'it is raining' and 'I take an umbrella,' respectively. Write the implication, $p \to q$, the contrapositive, inverse, converse and negation statements in words.

7. Using a truth table, show that the converse of $p \to q$ and the inverse of $p \to q$ are tautologies.

8. **WE5** Use a truth table to show the following is a tautology: $(p \wedge (q \wedge r)) \leftrightarrow ((p \wedge q) \wedge r)$. This is one of the associative laws.

9. Use a truth table to show the following is a tautology: $(p \wedge (q \vee r)) \leftrightarrow (p \wedge q) \vee (p \wedge r)$. This is one of the distributive laws.

10. Use a truth table to show the following is a tautology: $(p \vee (q \wedge r)) \leftrightarrow (p \vee q) \wedge (p \vee r)$.

11. **WE6** Write the following propositions in words and state whether they are true or false.
    a. $\forall x \in N,\ x+4>3$
    b. $\forall x \in N,\ x>0$

12. **WE7** Write the following propositions in words and state whether they are true or false.
    a. $\exists x \in N,\ x+4<3$
    b. $\exists x \in Q,\ 3x=2$

13. Write the following propositions in words and state whether they are true or false.
    a. $\exists x \in Z,\ 3x+6=0$
    b. $\exists x \in Z, x^2<1$

14. **WE8** Write the following propositions in words and state whether they are true or false.
    a. $\forall x \in N,\ \forall y \in N,\ x+y>0$
    b. $\forall x \in R,\ \forall y \in R,\ x^2+y^2>0$

15. **WE9** Write the following propositions in words and state whether they are true or false.
    a. $\forall x \in Z^+, \exists y \in Z^-, x+y=0$
    b. $\forall x \in Z, \exists y \in Z,\ y>x$

For questions 16 to 20, write the following propositions in words and state whether they are true or false.

16. a. $\exists x \in Q,\ \forall y \in Q, xy=1$
    b. $\forall x \in Q,\ \exists y \in Q, y=x^2$
    c. $\exists x \in R,\ \forall y \in Q,\ x\sqrt{y}=1$

17. a. $\forall x \in Z, \exists y \in Z, y=\frac{x}{2}$
    b. $\forall x \in Z, \exists y \in Q, y=\frac{x}{2}$

18. $\exists x \in Z, \forall y \in Z, x + y = 0$

19. $\forall x \in Q\backslash\{0\}, \exists y \in Q, xy = 1$

20. $\forall a \in R\backslash\{0\}, \forall b \in R, \forall c \in R, \exists x \in Q, ax^2 + bc + c = 0$

1.2 Exam questions

Question 1 (2 marks) TECH-FREE
Show that the negation of $p \to q$ is equivalent to p and not q, that is show that $\neg(p \to q) \leftrightarrow p \wedge \neg q$ is a tautology.

Question 2 (3 marks) TECH-FREE
Use a truth table to show that $\left((p \to q) \wedge (q \to r)\right) \to (p \to r)$ is a tautology.

Question 3 (2 marks) TECH-FREE
Write the following proposition in words and state whether it is true or false.
$\forall\{a, b, c\} \in R \wedge b^2 > 4ac, \exists x \in R, ax^2 + bc + c = 0$

More exam questions are available online.

1.3 Direct proofs

LEARNING INTENTION

At the end of this subtopic you should be able to:
- prove statements using a sequence of direct implications.

1.3.1 Terminology

There are many types of mathematical statements, in logic and proofs.
- **Conjecture**: A statement that seems to be true, but needs to be proven.
- **Axiom**: A statement that is considered to be true, or a fact, based on logic and does not need to be proven. Also called a postulate, or an assumption.
- **Theorem**: A statement that can be shown to be true using a proof technique. Once proven, theorems can be used in future reasoning and proofs.
- **Propositions**: Statements which are either true or false.
- **Corollary**: A theorem which follows almost immediately from another (related) theorem.
- **Lemma**: A theorem that is used to help prove other theorems. Lemmas are usually not of much interest in isolation, but are important due to the role that they play in the proof of a more important theorem.
- **Definition**: A statement that gives some mathematical meaning to a word (or words) that is then consistently and precisely understood and agreed upon.

At the beginning of this topic we defined the meaning of the sets N, Z, Q and R which we accept as definitions.

We will assume as an axiom or a fact, that the sum, difference and product of integers are integers. That is, let $A(x, y) = x + y$, $D(x, y) = x - y$, $P(x, y) = xy$, then $\forall x \in Z, \forall y \in Z, A(x, y) \in Z$, $\forall x \in Z, \forall y \in Z, D(x, y) \in Z$ and $\forall x \in Z, \forall y \in Z, P(x, y) \in Z$ are all true, and will be used extensively in further proofs.

1.3.2 Number theory

Number theory is a branch of mathematics devoted primarily to the study of integers. French mathematician Pierre De Fermat (1601–1655) was one of the first to explore number theory and he communicated with letters to friends about his findings and theorems, often with little or no proof. He is famous for what is known as Fermat's last theorem which was scribbled in the margin on a page. It states that no three positive integers a, b and c satisfy the equation $a^n + b^n = c^n$ for any integer value of n greater than two. This seemingly simple conjecture has proved to be one of the world's hardest mathematical problems to prove and several mathematicians have devoted their lives in attempting a proof.

Definition of a prime number

A natural number is prime if it has exactly two positive divisors, 1 and itself.

So 1 is not prime, as it has only one divisor, not two. 2 is prime as both itself and 1 are divisors.

Definition of an even number

We say that an integer n is even, if there exists an integer k such that $n = 2k$.

For example, 12 is even because $12 = 2 \times 6$.

An integer n is even, iff $\exists k \in Z,\ n = 2k$.

Definition of an odd number

We say that n is odd, if there exists an integer k such that $n = 2k + 1$.

For example, 17 is odd because $17 = 2 \times 8 + 1$. Note that 17 is not even because there is no integer k for which $17 = 2k$.

We could also define odd numbers as integers which are not even.

WORKED EXAMPLE 10 Identifying sets of integers

Let S be the set of natural numbers less than 20, P be the set of prime numbers in S, E be the set of even numbers in S and F be the set of $x \in S$ such that x is divisible by 4. List the sets

a. P **b.** E **c.** F **d.** $P \cap E$ **e.** $P \cap F$

THINK	WRITE
List the elements in the set S.	$S = \left\{ \begin{matrix} 1, 2, 3, 4, 5, 6, 7, 8, 9, 10, 11, \\ 12, 13, 14, 15, 16, 17, 18, 19 \end{matrix} \right\}$
a. List the elements in the set P.	$P = \{2, 3, 5, 7, 11, 13, 17, 19\}$
b. List the elements in the set E.	$E = \{2, 4, 6, 8, 10, 12, 14, 16, 18\}$
c. List the elements in the set F.	$F = \{4, 8, 12, 16\}$
d. List the elements which are in both set P and set E.	The only number which is in both set P and set E is 2. $P \cap E = \{2\}$
e. List the elements which are in both set P and set F.	The are no prime numbers which are divisible by 4, this is an empty set. $P \cap F = \phi$

1.3.3 Direct proof

Having these definitions in place we are ready to begin writing proofs. A proof is a verification which includes an understanding of the words, phrases and symbols in the proof.

In a direct proof, the conclusion is established by logically combining the axioms, definitions, and earlier theorems.

You might think that for a direct proof, we merely substitute in a couple of values and test these and since they work, the proof is done. However this does not prove the general case and in this case is **not** a valid method of proof.

A direct proof is one common method of proof. It is used to prove statements of the form 'if p then q', or 'p implies q', which is also written as $p \rightarrow q$, as in the truth table.

Row	p	q	$p \rightarrow q$
1	T	T	T
2	T	F	F
3	F	T	T
4	F	F	T

The method of a direct proof is to take a statement p, which we assume to be true, and use it directly to show that q is true. Now since p is true, only rows 1 and 2 of the table concern us, so for $p \rightarrow q$ to be true, all we then have to do then is to show that q is true (row 1).

WORKED EXAMPLE 11 Proving that the sum of two odd integers is even

Prove that the sum of two odd integers is an even integer.

THINK	WRITE
1. Write an equivalent statement to be proved. p: a and b are odd integers. q: their sum $a+b$ is even.	Proof: If a and b are odd integers, then their sum $a+b$ is even.
2. Assume that p is true, that is a and b are two odd integers. Note that it would be incorrect to state $b=2j+1$ as that would imply $a=b$.	Let $a=2j+1$ and $b=2k+1$ where $j, k \in Z$.
3. Evaluate the sum. Recall that the sum of any integers is an integer.	$a+b=(2j+1)+(2k+1)$ $=2(j+k+1)$ $=2i$ Where $i=j+k+1$, therefore $i \in Z$.
4. Since q is true, state the conclusion.	Since $a+b=2i$, it follows from the definition that the sum of two odd integers is an even integer.

At the start of each proof, we need to write the word Proof. Sometimes you might see QED at the end of a proof. It comes from the Latin phrase 'quod erat demonstrandum', meaning 'which was to be demonstrated'. Literally then it states 'what was required to be shown'.

Outline for a direct proof

Proof:

If p then q.

Suppose p is true.

some steps of working

Therefore, q is true.

WORKED EXAMPLE 12 Proving that a function of even integers is odd

Prove the following.
If x is an even integer, then $x^2 - 8x + 19$ is odd.

THINK	WRITE
1. Suppose x is even.	Proof: Let $x = 2j$ for some $j \in Z$.
2. Consider $x^2 - 8x + 19$, substitute for x.	$x^2 - 8x + 19 = (2j)^2 - 8(2j) + 19$
3. Expand and simplify	$= 4j^2 - 16j + 19$ $= 2(2j^2 - 8j + 9) + 1$ $= 2i + 1$ where $i = 2j^2 - 8j + 9$. The sum, product and difference of any integers are integers (these are axioms). Therefore $i \in Z$.
4. State the conclusion.	$x^2 - 8x + 19$ is odd.

Modus ponens

Modus ponendo is Latin for 'putting in place of'. It is just an application of using a theorem or earlier known true proposition, to show some other proof. The use of modus ponens is: if the statement for p has occurred in a proof, and p is true and also if $p \rightarrow q$ is true, then it follows that q is true.

Modus ponens

Given that $p \rightarrow q$ is true and p is true, then q must also be true.

1.3.4 Divisibility

If a and b are non-zero integers, then a divides b, which we write as $a|b$, if $b = ac$ for some $c \in Z$. In other words, a is a factor of b, as it divides into b with zero remainder.

For example, 7 divides 28, because $28 = 4 \times 7$, so $7|28$ is a true statement. $5|19$ is a false statement as there is no integer c for which $19 = 5c$.

Note that $a|b$ is a statement which is either true or false, while $a/b = \frac{a}{b}$ is a number.

The least common multiple of two non-zero integers, a and b is denoted by $\text{lcm}(a, b)$ and is the smallest positive integer that is a multiple of both a and b.

For example $\text{lcm}(6, 15) = 30$ and $\text{lcm}(9, 45) = 45$. The order is not important here since $\text{lcm}(a, b) = \text{lcm}(b, a)$.

The greatest common divisor of two non-zero integers, a and b is denoted by $\text{gcd}(a, b)$ and is the largest integer that divides both a and b.

For example, $\text{gcd}(4, 10) = 2$ and also $\text{gcd}(10, 4) = \text{gcd}(-4, 10) = \text{gcd}(-10, 4) = 2$. The order is not important here since $\text{gcd}(a, b) = \text{gcd}(b, a)$. Now without loss of generality (wlog), another mathematical abbreviation, we will consider $\text{gcd}(a, b)$ and $\text{lcm}(a, b)$ when a and b are positive integers, that is $a, b \in Z^+$.

The functions lcm and gcd are available on many calculators.

WORKED EXAMPLE 13 Proving the transitive property of divisibility

Let a, b and c be positive integers. If $a|b$ and $b|c$ then prove that $a|c$. This is also known as a transitive property.

THINK	WRITE
1. $a\|b$ is true.	There is an integer, d_1 such that $b = ad_1$.
2. $b\|c$ is true.	There is an integer, d_2 such that $c = bd_2$.
3. Substitute for b.	$c = ad_1d_2 = ad_3$ where $d_3 = d_1d_2$ is an integer, since we know that the product of two integers is an integer.
4. State the conclusion.	Therefore it follows that $a\|c$.

1.3.5 Proof by cases

Sometimes a statement can be broken down into a number of simpler cases that need investigation separately. The validity of proof by cases is from the tautology $[(p \vee q) \rightarrow r] \leftrightarrow [(p \rightarrow r) \wedge (q \rightarrow r)]$ which was proved in Worked example 5. We can generalise this tautology to n cases as shown below.

$$[(p_1 \vee ... \vee p_n) \rightarrow r] \leftrightarrow [(p_1 \rightarrow r) \wedge (p_2 \rightarrow r) \wedge ... \wedge (p_n \rightarrow r)]$$

WORKED EXAMPLE 14 Proof by cases

If x is a real number such that $\frac{x+3}{x^2-4} > 0$ then either $x > 2$ or $-3 < x < -2$.

THINK	WRITE
1. Factorise the denominator.	Proof: $\frac{x+3}{x^2-4} = \frac{x+3}{(x+2)(x-2)} > 0$
2. Since a negative times a negative is positive, for a fraction $\frac{a}{bc}$ to be positive, there are 4 cases.	Consider $a = x+3, \quad b = x+2, \quad c = x-2$

3. Case 1. $a > 0,\ b > 0,\ c > 0$ Solve the inequalities.	$x+3>0 \quad x-2>0 \quad x-2>0$ $x>-3, \quad x>-2, \quad x>2$ These imply $x>2$.
4. Case 2. $a > 0,\ b < 0,\ c < 0$ Solve the inequalities.	$x+3>0, \quad x+2<0 \quad x-2<0$ $x>-3, \quad x<-2, \quad x<2$ These imply $-3<x<-2$.
5. Case 3. $a < 0,\ b < 0,\ c > 0$ Solve the inequalities.	$x+3<0, \quad x+2<0, \quad x-2>0$ $x<-3, \quad x<-2, \quad x>2$ There are no values of x which satisfy all of these three inequalities simultaneously.
6. Case 4. $a < 0,\ b > 0,\ c < 0$ Solve the inequalities.	$x+3<0, \quad x+2>0, \quad x-2<0$ $x<-3, \quad x>-2, \quad x<2$ There are no values of x which satisfy all of these three inequalities simultaneously.
7. State the conclusion.	Thus only Case 1 or Case 2 apply, so that $x>2$ (Case 1) or $-3<x<-2$ (Case 2).

1.3 Exercise

Students, these questions are even better in jacPLUS

Receive immediate feedback and access sample responses

Access additional questions

Track your results and progress

Find all this and MORE in jacPLUS

Technology free

1. **WE10** Let S be the set of natural numbers less than 20. Let P be the set of prime numbers x such that $x \in S$. Let O be the set of odd numbers x such that $x \in S$ and T be the set x such that x is divisible by 3 and $x \in S$. List the following sets:

   a. P b. O c. T d. $P \cap O$ e. $P \cap T$

2. **WE11** Prove that the sum of two even integers is an even integer.
3. Prove that if x is an even integer, then $4x - 7$ is odd.
4. Prove that if x is an odd integer, then $3x + 11$ is even.
5. Prove that the square of an even number is even.
6. Prove that the cube of an odd number is odd.
7. **WE12** Prove that if x is an even integer, then $x^2 + 4x + 5$ is odd.
8. If x is an odd integer, prove that $x^2 - 5x + 8$ is even.
9. Prove that if x is an odd integer, then $x^2 - 7x + 18$ is even.
10. Prove that if x is an odd integer, then $x^2 - 1$ is divisible by 4.
11. Prove that if $10^n - 1$ is prime, then n is odd.

12. a. Prove that for $x, y \in R$, $x^2 + y^2 \geq 2xy$.
 b. By expanding $(x-y)^2 + (y-z)^2 + (z-x)^2$ prove that for $x, y, z \in R$ $x^2 + y^2 + z^2 \geq xy + yz + xz$.

13. **WE13** If a and b are positive integers and if $a|b$, prove that $a^2|b^2$.

14. a. Let a, b and c be positive integers. If $a|b$ and $a|c$ then prove that $a|\,(b+c)$.
 b. Let a, b, c and d be positive integers. Prove or disprove if $a|\,(b+c)$ and $a|\,(b+d)$ then $a|\,(c+d)$.

15. Let a, b, c and d be positive integers. If $a|b$ and $c|d$ then prove that $ac|bd$.

16. a. If a, b and c are integers, prove or disprove the following. If $a|b$ and $c|b$ then $a\,c|b$.
 b. If a and b are integers, prove or disprove the following. If $5|\left(a^2+b^2\right)$ then $5|a^2$ and $5|b^2$.

17. Prove that if $5|n$ then $5|n^2$.

18. Evaluate each of the following.
 a. $\gcd(18, 27)$ b. $\gcd(12, 60)$ c. $\gcd(7, 11)$ d. $\text{lcm}(18, 27)$ e. $\text{lcm}(12, 60)$ f. $\text{lcm}(7, 11)$

19. **WE14** If $x \in R$ such that $\dfrac{x+1}{x^2-4} > 0$, prove that either $x > 2$ or $-2 < x < -1$.

20. If $x \in R$ such that $\dfrac{x^2-16}{x+7} > 0$, prove that either $x > 4$ or $-7 < x < -4$.

21. If $x \in R$ such that $\left(36-x^2\right)(x+4) \geq 0$, prove that either $x \leq -6$ or $-4 \leq x \leq 6$.

22. Prove that if x is an integer, then $x^2 + 5x + 8$ is even.

Technology active

23. Evaluate each of the following.
 a. $\gcd(996, 524)$ b. $\gcd(417, 819)$ c. $\gcd(1025, 450)$
 d. $\text{lcm}(996, 524)$ e. $\text{lcm}(417, 819)$ f. $\text{lcm}(1025, 450)$

24. Peter used the sieve of Eratosthenes to list and count all the prime numbers between 1 and 100 inclusive, he found there was 25. The sieve eliminates numbers divisible by 2, 3, 5, 7 etc. He then listed all the prime numbers between 101 and 200 inclusive and found there was 21. He then listed all the prime numbers between 201 and 300 inclusive and found there was 16. He stated that as the set of natural numbers gets larger, there are less and less prime numbers and therefore there will be a finite set of prime numbers. Write a program or use a calculator to count the number of prime numbers within a certain interval.

1	2	3	4	5	6	7	8	9	10
11	12	13	14	15	16	17	18	19	20
21	22	23	24	25	26	27	28	29	30
31	32	33	34	35	36	37	38	39	40
41	42	43	44	45	46	47	48	49	50
51	52	53	54	55	56	57	58	59	60
61	62	63	64	65	66	67	68	69	70
71	72	73	74	75	76	77	78	79	80
81	82	83	84	85	86	87	88	89	90
91	92	93	94	95	96	97	98	99	100

Numbers that are multiples of 2 are shaded blue
Numbers that are multiples of 3 are shaded pink
Numbers that are multiples of 5 are shaded green
Numbers that are multiples of 7 are shaded purple

eratosthenes

Do you agree with Peter's assertion?
Complete the following table to decide on your assertion.

Natural numbers	Number of prime numbers
$1-100$	25
$101-200$	21
$201-300$	16
$301-400$	
$401-500$	
$501-600$	
$601-700$	
$701-800$	
$801-900$	
$901-1000$	
$1-1000$	
$1001-2000$	

1.3 Exam questions

Question 1 (2 marks) TECH-FREE
Prove that every odd integer can be expressed as a difference of two squares.

For example, $5=3^2-2^2,\ 7=4^2-3^2,\ 9=5^2-4^2...$

Question 2 (2 marks) TECH-FREE
Let a, b and c be positive integers. Prove or disprove if $a|b$ and $c|b$ then $(a+c)\,|b$.

Question 3 (3 marks) TECH-FREE
Prove that if x is an integer, then $3x^2+7x+11$ is odd.

More exam questions are available online.

1.4 Indirect proofs

LEARNING INTENTION

At the end of this subtopic you should be able to:
- prove statements using contrapositive and contradiction.

1.4.1 Indirect proofs

Indirect proofs are a roundabout way of trying to show a proof, often by doing things in reverse.

When attempting to prove an assertion we first try a direct proof, that is, we assume the hypothesis that is given is true and try to show directly that the conclusion follows.

The method of a direct proof involves showing $p \to q$ is true by taking a statement p, which we assume to be true, and using it directly to show that q is true.

1.4.2 Proof by contrapositive

When a direct proof can't be used, we use an indirect proof, which is the proof of the contrapositive. Recall that since $p \rightarrow q \leftrightarrow (\neg q \rightarrow \neg p)$ is a tautology, we can indirectly prove $p \rightarrow q$ by proving the equivalent contrapositive $\neg q \rightarrow \neg p$ is true.

This is a very useful fact, as it is often easier to prove the contrapositive statement, than it is to prove the implication directly.

> **Proof using the contrapositive**
>
> **If you can prove that the contrapositive of a statement is true, then the original statement must also be true (and vice versa).**

WORKED EXAMPLE 15 Proof by contrapositive (1)

Prove that if $4^n - 1$ is prime, then n is odd.

THINK	WRITE
1. To prove $p \rightarrow q$ p: '$4^n - 1$ is prime' q: 'n is odd'	To prove: If $4^n - 1$ is prime, then n is odd. A direct proof cannot be established.
2. The contrapositive is $\neg q \rightarrow \neg p$. Not an odd integer is an even integer.	Proof: We prove the contrapositive. If n is an even integer then $4^n - 1$ is not prime.
3. Use the definition of an even integer.	Let $n = 2k$, where $k \in Z$
4. Factorise the expression, using the difference of two squares.	$4^{2k} - 1 = \left(4^k\right)^2 - 1$ $= (4^k + 1)(4^k - 1)$
5. State the conclusion.	Since $4^{2k} - 1$ has factors, it is not prime, and the contrapositive is true, therefore the original statement If $4^n - 1$ is prime, then n is odd, is true.

WORKED EXAMPLE 16 Proof by contrapositive (2)

Prove that for $x \in R$ if $x^3 + 4x > x^2 + 1$ then $x > 0$.

THINK

1. To prove $p \to q$
   p: '$x^3 + 4x > x^2 + 1$'
   q: '$x > 0$'
2. Not positive is negative.
   $\neg q \to \neg p$
3. A negative number to an odd power is negative, and four times a negative number is negative.
4. A negative number to an even power is positive.
5. A positive number is greater than a negative number.
6. State the conclusion.

WRITE

To prove:
$x \in R$ if $x^3 + 4x > x^2 + 1$ then $x > 0$

Proof: We prove the contrapositive.
If $x < 0$ then $x^3 + 4x < x^2 + 1$.
$\neg q: x < 0$,
$\neg p: x^3 + 4x < x^2 + 1$

Since $x < 0$, $x^3 + 4x < 0$

Since $x < 0$, $x^2 + 1 > 0$

$x^3 + 4x < x^2 + 1$

As the contrapositive, $\neg q \to \neg p$ has been shown to be true, the original statement, $p \to q$, for $x \in R$ if $x^3 + 4x > x^2 + 1$ then $x > 0$ has been proved.

WORKED EXAMPLE 17 Proof by contrapositive (3)

Prove that if the product of two integers is even, then one of the integers is even.

THINK

1. Use the contrapositive and De Morgan's Law.
2. Write an equivalent statement to be proved.
   p: 'The product of two integers a and b is even'
   r: 'a is even'
   s: 'b is even'
   Not an even integer is an odd integer.
3. Assume that a and b are two odd integers.
4. Form the product.
5. State the conclusion.

WRITE

The contrapositive of $p \to (r \vee s)$ is $\neg (r \vee s) \to \neg p$ which is equivalent to $(\neg r \wedge \neg s) \to \neg p$.

To prove:
If the product of two integers a and b is even, then a is even or b is even.
Proof: We prove the contrapositive.
If a and b are odd integers, then their product ab is odd.

Let $a = 2j + 1$ and $b = 2k + 1$ where $j, k \in Z$.

$$\begin{aligned} ab &= (2j+1)(2k+1) \\ &= 4jk + 2k + 2j + 1 \\ &= 2(2jk + k + j) + 1 \\ &= 2i + 1 \end{aligned}$$

Where $i = 2jk + k + j$.
Recall that the product and sum of any integers is an integer, so $i \in Z$.

Since $ab = 2i + 1$, it follows that ab is odd, and therefore from the equivalence of the contrapositive the original statement is true.

WORKED EXAMPLE 18 Proof of divisibility of n and n^2

a. Prove that if $n \in N$ and $3|n$ then $3|n^2$.
b. Prove that if $n \in N$ and $3|n^2$ then $3|n$.
***Note:* Proving both parts *a* and *b* is equivalent to proving that, for $n \in N$, n is divisible by 3 iff n^2 is divisible by 3.**

THINK	WRITE
a. 1. Define the propositions	p: $3\|n$, 3 divides into n. q: $3\|n^2$, 3 divides into n^2.
2. Use a direct proof.	To prove: $p \to q$
3. Assume that p is true.	Since $3\|n$ then $n = 3k$, $k \in Z$
4. To show that q is true.	$n^2 = (3k)^2 = 9k^2 = 3\left(3k^2\right) = 3j, \; j \in Z$, so that $3\|n^2$ is true.
b. 1. Use an indirect proof.	To prove: $q \to p$, prove the equivalent contrapositive, $\neg p \to \neg q$.
2. Assume that $\neg p$ is true.	$3 \nmid n$, meaning n is not divisible by 3. Therefore when n is divided by 3, it must have a remainder of either 1 or 2.
3. Consider cases.	Case 1. When n is divided by 3, it has a remainder of 1, then $n = 3k + 1$, $k \in Z$
4. Consider n^2.	$n^2 = (3k+1)^2$ $n^2 = 9k^2 + 6k + 1$ $n^2 = 3(3k^2 + 2k) + 1$ $n^2 = 3j + 1, j \in Z$
5. State the conclusion.	So n^2 has a remainder of 1 when divided by 3. Therefore $3 \nmid n^2$.
6. Consider the other case.	Case 2. When n is divided by 3, it has a remainder of 2, then $n = 3k + 2$, $k \in Z$
7. Consider n^2.	$n^2 = (3k+2)^2$ $n^2 = 9k^2 + 12k + 4$ $n^2 = 3(3k^2 + 4k + 1) + 1$ $n^2 = 3j + 1, j \in Z$
8. State the conclusion.	So n^2 has a remainder of 1 when divided by 3. Therefore $3 \nmid n^2$.
9. Combining the cases.	We have shown that when $3 \nmid n$ then $3 \nmid n^2$, that is $\neg p \to \neg q$. Since the contrapositive is true, the original statement $q \to p$ is true.

1.4.3 Proof by contradiction

Proof by contradiction is another indirect method of proof. Suppose we want to believe in the truth of a mathematical statement or some proposition. With a proof by contradiction, we set out to prove the statement is false (which is easier than proving it to be true) and then as the proof develops, we run into something that does not make any sense, that is, a contradiction. By coming across a contradiction we show that the original statement cannot possibly be false.

The steps to follow when using a proof by contradiction are:

- assume your statement to be false
- proceed as you would when using a direct proof
- come up with a contradiction
- because of the contradiction it can't be the case that the statement is false, so the assumption that the original statement was false is incorrect. Therefore the original statement must be true.

True or false?

A statement cannot be true and false at the same time.

- If a statement can be proven true then it cannot be false.
- If a statement can be proven false then it cannot be true.
- If the negation of a statement can be proven true then the original statement is false.
- If the negation of a statement can be proven false then the original statement is true.

The last statement is the one that is used when doing a proof by contradiction.

Assume that the negation of what you wish to prove is true and show that this implies two contradictory outcomes. In logical form, $\neg p \rightarrow (r \wedge \neg r)$ whose conclusion is a contradiction, that is a false statement. An example of an implication of this form is 'therefore n is odd and n is even'. This is obviously a contradiction.

Proof by contradiction

Step 1. Assume that the negation (opposite) of what you would like to prove is true.
Step 2. Use mathematical reasoning to show that the consequences of this premise/assumption are false/impossible.
Step 3. Since it has been shown that the negation is false, the original statement must be true.

WORKED EXAMPLE 19 Proof by contradiction (1)

Prove the following statement. No integers a and b exist for which $3a + 12b = 23$.

THINK	WRITE
1. We could spend hours trying to find integers a and b for which $3a + 12b = 23$. For example: $a = 4,\ b = 1,\ 3a + 12b = 24$ is close but not a proof. Instead of trying to find a counter-example, assume that the original statement is false.	Proof: Assume that the original statement is false. $\exists a, b \in Z,\ 3a + 12b = 23$
2. Assume a and b are integers and factorise.	$3a + 12b = 23$ $3(a + 4b) = 23$
3. Divide both sides by 3.	$a + 4b = \dfrac{23}{3}$
4. When we add two integers, the result is an integer, not a fraction.	We have a contradiction, the sum of two integers, is always an integer, not a rational number.
5. State the conclusion.	This contradiction means that the statement cannot be proven false, therefore the original statement must be true. No integers a and b exist for which $3a + 12b = 23$.

WORKED EXAMPLE 20 Proof by contradiction (2)

Prove that $\sqrt{3}$ is irrational.

THINK	WRITE
1. A rational number is of the form $\frac{a}{b}$, where a and b are integers which have no common factors (other than 1).	Proof by contradiction: Assume $\sqrt{3}$ is rational. Therefore $\sqrt{3} = \frac{a}{b}$, where $a, b \in Z$ and a and b share no common factors other than 1.
2. Square both sides.	$\sqrt{3} = \frac{a}{b}$ $3 = \frac{a^2}{b^2}$
3. Multiply both sides by b^2.	$3b^2 = a^2$
4. a^2 is a product of 3 and another integer.	Since b^2 is an integer, $a^2 = 3n$ where n is an integer. We have proved that if a^2 is a multiple of 3, then a is a multiple of 3 (from Worked example 18). Now let $a = 3k,\ k \in Z$.
5. Substitute for a. Now b^2 is a product of 3 and another integer.	$a^2 = 9k^2 = 3b^2$ $b^2 = 3k^2$ $b^2 = 3m$ Where m is an integer.
6. b is a product of 3 and another integer.	Since b^2 is a multiple of 3, b is also a multiple of 3.
7. We have a contradiction.	Both a and b are multiples of 3, so they share a common factor of 3. This contradicts the assumption that a and b have no common factors other than 1.
8. State the conclusion.	Therefore $\sqrt{3}$ is irrational.

1.4 Exercise

Technology free

1. WE15 Prove the following statement using proof by contrapositive. If $2^n - 1$ is prime, then n is odd.

2. Prove the following statement using proof by contrapositive. If n is an integer and $3n + 7$ is even, then n is odd.

3. Prove the following statement using proof by contrapositive. If n is an integer and n^2 is even, then n is even.

4. Prove the following statement using proof by contrapositive. If n is an integer and n^2 is not divisible by 4, then n is odd.

5. Prove the following statement using proof by contrapositive. If n^3 is an odd integer, then n is odd.

6. WE16 Prove that for $x \in R$ if $x^3 + 2x > 4x^2 + 9$ then $x > 0$.

7. Prove that for $x \in R$ if $x^5 + 2x^3 + x > x^6 + 3x^4 + x^2 + 8$ then $x > 0$.

8. Prove that for $x, y \in R$ if $y^3 + yx^2 < x^3 + xy^2$ then $y < x$.

9. WE17 Prove that if the product of two integers is odd, then both of the integers are odd.

10. WE19 Prove that no integers a and b exist for which $4a + 8b = 10$.

11. Prove that no integers a and b exist for which $2a - 8b = 21$.

12. Suppose $a, b \in Z$. Prove that if $a + b \geq 19$ then $a \geq 10$ or $b \geq 10$.

13. WE18 a. Prove that if $2|n$ then $2|n^2$.

    b. Prove that if $2|n^2$ then $2|n$.

14. WE20 Prove that $\sqrt{2}$ is irrational.

15. If $n^2 = 10$ then n is not a rational number.

16. Prove that $\log_3(8)$ is irrational.

17. Prove using the contrapositive that if n is an integer and $n^3 + 5$ is odd, then n is even.

18. For all $a, b \in Z^+$, prove that $\frac{a}{b} + \frac{b}{a} \geq 2$.

1.4 Exam questions

Question 1 (3 marks) TECH-FREE

Suppose $x, y \in R$. If $x + y > 20$ prove that $x > 10$ or $y > 10$.

Question 2 (3 marks) TECH-FREE

Prove that $\log_2(5)$ is irrational.

Question 3 (3 marks) TECH-FREE

Prove using a contradiction that if n is an integer and $n^3 + 5$ is odd, then n is even.

More exam questions are available online.

1.5 Proof by mathematical induction

LEARNING INTENTION

At the end of this subtopic you should be able to:
- prove statements using mathematical induction.

1.5.1 What is mathematical induction?

Mathematical induction is a method of proof involving inductive reasoning. It is a step by step process for proving propositional statements involving natural numbers $n \in N$. The first step involves verifying that a propositional statement $P(n)$ is true for the base case (usually when $n = 1$). The next step is called the inductive step and involves assuming that the statement is true when $n = k$ and then showing that this assumption implies that it is true when $n = k + 1$, we can then conclude that $P(n)$ is true for all natural numbers n greater than or equal to the value used in the base case.

It is a bit like climbing a set of stairs, first we need to get to the first step, then we need to be able to get from one step to the next. If you can do both of those things you can climb the stairs, no matter how many there are.

Proof by mathematical induction

Let $P(n)$ be a propositional statement regarding natural numbers, n.

The steps which can be used to prove $P(n)$ by mathematical induction are:

Step 1 (Base step): **Prove that $P(n)$ is true for the first allowable value of n (usually $n = 1$).**
Step 2 (Inductive step): **Assume that $P(k)$ is true and then use this assumption to prove that $P(k + 1)$ is also true.**

You can now conclude that the statement $P(n)$ is true $\forall n \in N$ which are greater than or equal to the value used in the base step.

Summation notation

The symbol $\sum$ is called sigma and is used in mathematics to represent 'the sum of the following'. In general $\sum_{i=1}^{n} f(i)$ means $f(1) + f(2) + ... + f(n)$. The variable i is called a dummy variable and could be replaced with any other letter.

For example: $$\sum_{j=1}^{4} \frac{(-1)^j x^j}{j^2} = \sum_{r=1}^{4} \frac{(-1)^r x^r}{r^2} = \frac{(-1)^1 x^1}{1^2} + \frac{(-1)^2 x^2}{2^2} + \frac{(-1)^3 x^3}{3^2} + \frac{(-1)^4 x^4}{4^2}$$
$$= -x + \frac{x^2}{4} - \frac{x^3}{9} + \frac{x^4}{16}$$

The symbol $\prod$ is called product and is used in mathematics to represent 'the product of the following'. In general $\prod_{i=1}^{n} f(i)$ means $f(1)f(2) \ldots f(n)$.

For example: $\prod_{j=3}^{5}\left(1-\frac{x^j}{j!}\right)=\left(1-\frac{x^3}{3!}\right)\left(1-\frac{x^4}{4!}\right)\left(1-\frac{x^5}{5!}\right)$

1.5.2 Proofs using mathematical induction

Induction proofs using sigma notation

Many proofs using mathematical induction, are logically set out and can be simplified using sigma notation, the expression, $\sum_{r=1}^{k+1} f(r)=\sum_{r=1}^{k} f(r)+f(k+1)$ will be used extensively.

WORKED EXAMPLE 21 Proof by mathematical induction (1)

Use mathematical induction to prove that $1^2+2^2+3^2+\ldots+n^2=\frac{1}{6}n(n+1)(2n+1), \forall n \in N$.

THINK	WRITE
1. State the propositional statement $P(n)$ which we are looking to prove.	Proof: $P(n)$ is the propositional statement: $1^2+2^2+3^2+\ldots+n^2=\frac{1}{6}n(n+1)(2n+1), \forall n \in N.$
2. Verify that $P(1)$ is true.	LHS of $P(1)=1$ RHS of $P(1)=\frac{1}{6}\times 1\times 2\times 3=1$ $P(1)$ is true.
3. Assume that $P(k)$ is true.	$1^2+2^2+3^2+\ldots+k^2=\frac{1}{6}k(k+1)(2k+1)$
4. Write a statement for $P(k+1)$.	$1^2+2^2+3^2+\ldots+(k+1)^2=\frac{1}{6}(k+1)\left((k+1)+1\right)\left(2(k+1)+1\right)$
5. Consider the LHS of $P(k+1)$.	$1^2+2^2+3^2+\ldots+(k+1)^2=\left(1^2+2^2+3^2+\ldots+k^2\right)+(k+1)^2$
6. Use the fact that $P(k)$ is true.	$=\frac{1}{6}k(k+1)(2k+1)+(k+1)^2$
7. Factorise the expression until it looks like the RHS of $P(k+1)$.	$=\frac{1}{6}(k+1)\left[k(2k+1)+6(k+1)\right]$ $=\frac{1}{6}(k+1)(2k^2+7k+6)$ $=\frac{1}{6}(k+1)(k+2)(2k+3)$ $=\frac{1}{6}(k+1)\left((k+1)+1\right)\left(2(k+1)+1\right)$ $=$ RHS of $P(k+1)$
8. State what has been demonstrated and concluded.	We have shown that if $P(k)$ is true, $P(k+1)$ is also true. $P(1)$ is true, therefore by the principle of mathematical induction $P(n)$: $1^2+2^2+3^2+\ldots+n^2=\frac{1}{6}n(n+1)(2n+1)$ is true $\forall n \in N$.

WORKED EXAMPLE 22 Proof by mathematical induction (2)

Use mathematical induction to prove that

$$\frac{1}{1\times 3}+\frac{1}{3\times 5}+\frac{1}{5\times 7}+...+\frac{1}{(2n-1)(2n+1)}=\frac{n}{2n+1}, \forall n \in N.$$

THINK	WRITE
1. State the propositional statement $P(n)$ which we are looking to prove.	Proof: Using sigma notation $\frac{1}{1\times 3}+\frac{1}{3\times 5}+\frac{1}{5\times 7}+...+\frac{1}{(2n-1)(2n+1)}=\sum_{r=1}^{n}\frac{1}{(2r-1)(2r+1)}$ Let $P(n)$ be the propositional statement: $\sum_{r=1}^{n}\frac{1}{(2r-1)(2r+1)}=\frac{n}{2n+1}$
2. Verify that $P(1)$ is true.	LHS of $P(1)=\frac{1}{1\times 3}=\frac{1}{3}$ RHS of $P(1)=\frac{1}{2\times 1+1}=\frac{1}{3}$ $P(1)$ is true.
3. Assume that $P(k)$ is true.	$\sum_{r=1}^{k}\frac{1}{(2r-1)(2r+1)}=\frac{k}{2k+1}$
4. Write a statement for $P(k+1)$.	$\sum_{r=1}^{k+1}\frac{1}{(2r-1)(2r+1)}=\frac{k+1}{2(k+1)+1}$
5. Consider the LHS of $P(k+1)$.	$\sum_{r=1}^{k+1}\frac{1}{(2r-1)(2r+1)}=\sum_{r=1}^{k}\frac{1}{(2r-1)(2r+1)}+\frac{1}{(2(k+1)-1)(2(k+1)+1)}$
6. Use the fact that $P(k)$ is true.	$=\frac{k}{2k+1}+\frac{1}{(2k+1)(2k+3)}$
7. Add the fraction, forming the lowest common denominator.	$=\frac{k(2k+3)+1}{(2k+1)(2k+3)}$
8. Expand the numerator.	$=\frac{2k^2+3k+1}{(2k+1)(2k+3)}$
9. Factorise the numerator and simplify by cancelling the common factor.	$=\frac{(2k+3)(k+1)}{(2k+1)(2k+3)}$ $=\frac{k+1}{2k+3}$
10. Write the expression so that it is clearly the same as the RHS of $P(k+1)$.	$=\frac{k+1}{2(k+1)+1}$
11. State what has been demonstrated and concluded.	We have shown that if $P(k)$ is true, $P(k+1)$ is also true. $P(1)$ is true, therefore by the principle of mathematical induction, $P(n)$: $\sum_{r=1}^{n}\frac{1}{(2r-1)(2r+1)}=\frac{n}{2n+1}$ is true $\forall n \in N$.

Mathematical induction proofs involving powers

WORKED EXAMPLE 23 Proof by mathematical induction (powers)

Use mathematical induction to prove that $4+4^2+4^3+...+4^n=\frac{1}{3}\left(4^{n+1}-4\right), \forall n \in N$.

THINK	WRITE
1. State the propositional statement $P(n)$ which we are looking to prove.	Proof: Using sigma notation $4+4^2+4^3+...+4^n=\sum_{r=1}^{n} 4^r$ Let $P(n)$ be the propositional statement: $\sum_{r=1}^{n} 4^r=\frac{1}{3}\left(4^{n+1}-4\right)$
2. Verify that $P(1)$ is true.	LHS of $P(1)=4$ RHS of $P(1)=\frac{1}{3}\left(4^2-4\right)=4$ $P(1)$ is true.
3. Assume that $P(k)$ is true.	$\sum_{r=1}^{k} 4^r=\frac{1}{3}\left(4^{k+1}-4\right)$
4. Write a statement for $P(k+1)$.	$\sum_{r=1}^{k+1} 4^r=\frac{1}{3}\left(4^{(k+1)+1}-4\right)$
5. Consider the LHS of $P(k+1)$.	$\sum_{r=1}^{k+1} 4^r=\sum_{r=1}^{k} 4^r+4^{k+1}$
6. Use the fact that $P(k)$ is true.	$=\frac{1}{3}\left(4^{k+1}-4\right)+4^{k+1}$
7. Take out the common factor and simplify.	$=\frac{1}{3}\left(\left(4^{k+1}-4\right)+3\times 4^{k+1}\right)$ $=\frac{1}{3}\left(4\times 4^{k+1}-4\right)$
8. Simplify using index laws.	$=\frac{1}{3}\left(4^{k+2}-4\right)$
9. Write the expression so that it is clearly the same as the RHS of $P(k+1)$.	$=\frac{1}{3}\left(4^{(k+1)+1}-4\right)$
10. State what has been demonstrated and concluded.	We have shown that if $P(k)$ is true, $P(k+1)$ is also true. $P(1)$ is true, therefore by the principle of mathematical induction, $P(n)$: $\sum_{r=1}^{n} 4^r=\frac{1}{3}\left(4^{n+1}-4\right)$ is true $\forall n \in N$.

Mathematical induction proofs involving divisibility

Mathematical induction proofs involving divisibility are common and can be quite tricky; the proof involves manipulating the expression for $P(k+1)$ so that it contains the expression for $P(k)$ plus some other terms which are also divisible by the required number.

WORKED EXAMPLE 24 Proof by mathematical induction (divisibility)

a. Prove that $4^n - 1$ is divisible by 3, $\forall n \in N$.
b. Prove that $5^n + 2 \times 11^n$ is divisible by 3, $\forall n \in N$.
c. Prove that $n^3 + 2n$ is divisible by 3, $\forall n \in N$.

THINK	WRITE
a. 1. State the propositional statement $P(n)$ which we are looking to prove.	a. Proof: Let $P(n)$ be the propositional statement: $4^n - 1$ is divisible by 3, $\forall n \in N$.
2. Verify that $P(1)$ is true.	$P(1)$: $4^1 - 1 = 3$ is divisible by 3, $P(1)$ is true.
3. Assume that $P(k)$ is true.	$4^k - 1$ is divisible by 3.
4. Write a statement for $P(k+1)$.	$4^{k+1} - 1$ is divisible by 3.
5. Manipulate the expression from $P(k+1)$ so that it contains the expression from $P(k)$.	$4^{k+1} - 1 = 4 \times 4^k - 1$ $= (3+1) \times 4^k - 1$ $= 3 \times 4^k + \left(4^k - 1\right)$
6. State what has been demonstrated.	Since we have assumed that $4^k - 1$ is divisible by 3, and $3 \times \left(4^k\right)$ is also divisible by 3, $4^{k+1} - 1 = 3 \times 4^k + \left(4^k - 1\right)$ also has 3 as a factor and is divisible by 3.
7. State what has been concluded.	We have shown that if $P(k)$ is true, $P(k+1)$ is also true. $P(1)$ is true. therefore by the principle of mathematical induction, $P(n)$: $4^n - 1$ is divisible by 3, is true $\forall n \in N$.
b. 1. State the propositional statement $P(n)$ which we are looking to prove.	b. Proof: Let $P(n)$ be the propositional statement: $5^n + 2 \times 11^n$ is divisible by 3, $\forall n \in N$.
2. Verify that $P(1)$ is true.	$P(1)$: $5^1 + 2 \times 11^1 = 5 + 22 = 27$ is divisible by 3, $P(1)$ is true.
3. Assume that $P(k)$ is true.	$5^k + 2 \times 11^k$ is divisible by 3.
4. Write a statement for $P(k+1)$.	$5^{k+1} + 2 \times 11^{k+1}$ is divisible by 3.
5. Manipulate the expression from $P(k+1)$ so that it contains the expression from $P(k)$.	$5^{k+1} + 2 \times 11^{k+1} = 5 \times 5^k + 2 \times 11 \times 11^k$ $= 5 \times 5^k + 22 \times 11^k$ $= 5\left(5^k + 2 \times 11^k\right) + 12 \times 11^k$ $= 5\left(5^k + 2 \times 11^k\right) + 3 \times \left(4 \times 11^k\right)$
6. State what has been demonstrated.	Since we have assumed that $5^k + 2 \times 11^k$ is divisible by 3, and $3 \times \left(4 \times 11^k\right)$ is also divisible by 3, $5^{k+1} + 2 \times 11^{k+1} = 5\left(5^k + 2 \times 11^k\right) + 3 \times \left(4 \times 11^k\right)$ also has 3 as a factor and is divisible by 3.
7. State what has been concluded.	We have shown that if $P(k)$ is true. $P(k+1)$ is also true. $P(1)$ is true, therefore by the principle of mathematical induction, $P(n)$: $5^n + 2 \times 11^n$ is divisible by 3, is true $\forall n \in N$.

c. 1. State the propositional statement $P(n)$ which we are looking to prove.	**c.** Proof: Let $P(n)$ be the propositional statement: $n^3 + 2n$ is divisible by 3, $\forall n \in N$.
2. Verify that $P(1)$ is true.	$P(1)$: $1 + 2 = 3$ is divisible by 3, $P(1)$ is true.
3. Assume that $P(k)$ is true.	$k^3 + 2k$ is divisible by 3.
4. Write a statement for $P(k+1)$.	$(k+1)^3 + 2(k+1)$ is divisible by 3.
5. Manipulate the expression from $P(k+1)$ so that it contains the expression from $P(k)$.	$(k+1)^3 + 2(k+1) = k^3 + 3k^2 + 3k + 1 + 2k + 2$ $= k^3 + 3k^2 + 5k + 3$ $= \left(k^3 + 2k\right) + 3\left(k^2 + k + 1\right)$
6. State what has been demonstrated.	Since we have assumed that $k^3 + 2k$ is divisible by 3, and $3\left(k^2 + k + 1\right)$ is also divisible by 3, $(k+1)^3 + 2(k+1) = \left(k^3 + 2k\right) + 3\left(k^2 + k + 1\right)$ also has 3 as a factor and is divisible by 3.
7. State what has been concluded.	We have shown that if $P(k)$ is true. $P(k+1)$ is also true. $P(1)$ is true, therefore by the principle of mathematical induction, $P(n)$: $n^3 + 2n$ is divisible by 3, is true $\forall n \in N$.

Mathematical induction proofs involving factorial and inequalities

Recall some properties of factorials and combinations:

$$n! = n(n-1)(n-2)\ldots 3 \times 2 \times 1$$
$$n! = n(n-1)!$$
$$0! = 1$$
$$\binom{n}{r} = \frac{n!}{r!\,(n-r)!} = \binom{n}{n-r}$$

These properties can be very useful in proofs by mathematical induction.

WORKED EXAMPLE 25 Proof by mathematical induction (factorials)

Use mathematical induction to prove that $n! > 2^n$ for natural numbers $n \geq 4$.

THINK	WRITE
1. State the propositional statement $P(n)$ which we are looking to prove.	Proof: Let $P(n)$ be the propositional statement: $n! > 2^n$ for natural numbers $n \geq 4$.
2. Verify that for the base step this time when $n = 4$, $P(4)$ is true.	LHS of $P(4) = 4! = 24$, RHS of $P(4) = 2^4 = 16$, so $P(4)$ is true.
3. Assume that $P(k)$ is true.	$k! > 2^k$ for natural numbers $k \geq 4$.
4. Write a statement for $P(k+1)$.	$(k+1)! > 2^{k+1}$ for natural numbers $k \geq 4$.
5. Consider the LHS of the propositional statement $P(k+1)$.	$(k+1)! = (k+1)\,k!$
6. Use the fact that $P(k)$ is true.	$(k+1)\,k! > (k+1) \times 2^k$
7. Now when $k \geq 4$, $k+1 > 2$.	$(k+1)! > 2 \times 2^k$

8. Simplify using index laws. $(k+1)! > 2^{k+1}$

9. State what has been demonstrated. $(k+1)! > 2^{k+1}$

10. State what has been concluded. We have shown that if $P(k)$ is true, $P(k+1)$ is also true. $P(4)$ is true, therefore by the principle of mathematical induction, $P(n)$: $n! > 2^n$ is true for natural numbers $n \geq 4$.

Mathematical induction proofs involving sequences

WORKED EXAMPLE 26 Proof by mathematical induction (Fibonacci numbers)

The Fibonacci numbers are defined by $f_1 = f_2 = 1$ and $f_{n+2} = f_{n+1} + f_n$ for all natural numbers $n \geq 1$.

Use mathematical induction to prove that $\sum_{r=1}^{n} f_r^2 = f_n f_{n+1}$.

THINK	WRITE
1. State the propositional statement $P(n)$ which we are looking to prove.	Proof: Let $P(n)$ be the propositional statement: $\sum_{r=1}^{n} f_r^2 = f_n f_{n+1}$ where f_n are the Fibonacci numbers and $n \geq 1$.
2. Verify that $P(1)$ is true.	LHS of $P(1) = f_1^2 = 1^2 = 1$ RHS of $P(1) = f_1 f_2 = 1 \times 1 = 1$ $P(1)$ is true.
3. Assume that $P(k)$ is true.	$\sum_{r=1}^{k} f_r^2 = f_k f_{k+1}$
4. Write a statement for $P(k+1)$.	$\sum_{r=1}^{k+1} f_r^2 = f_{k+1} f_{k+1+1}$
5. Consider the LHS of the propositional statement $P(k+1)$.	$\sum_{r=1}^{k+1} f_r^2$
6. Determine the sum of the first $k+1$ terms by adding in the next term.	$\sum_{r=1}^{k+1} f_r^2 = \sum_{r=1}^{k} f_r^2 + f_{k+1}^2$
7. Use the fact that $P(k)$ is true.	$\sum_{r=1}^{k+1} f_r^2 = f_k f_{k+1} + f_{k+1}^2$
8. Considering the RHS, take out the common factor and use the Fibonacci property, $f_{n+2} = f_{n+1} + f_n$.	$\sum_{r=1}^{k+1} f_r^2 = f_{k+1}\left(f_k + f_{k+1}\right)$ $\sum_{r=1}^{k+1} f_r^2 = f_{k+1} f_{k+2}$
9. Write the expression so that it is clearly the same when $n = k+1$.	$\sum_{r=1}^{k+1} f_r^2 = f_{k+1} f_{k+1+1}$

10. State what has been concluded.	We have shown that if $P(k)$ is true, $P(k+1)$ is also true. $P(1)$ is true, therefore by the principle of mathematical induction, $P(n)$: $\sum_{r=1}^{n} f_r^2 = f_n f_{n+1}$, $n \geq 1$.

Mathematical induction proofs involving matrices

WORKED EXAMPLE 27 Proof by mathematical induction (matrices)

Let $A = \begin{bmatrix} 3 & -1 \\ 0 & 2 \end{bmatrix}$. Prove by mathematical induction that $A^n = \begin{bmatrix} 3^n & 2^n - 3^n \\ 0 & 2^n \end{bmatrix}$, $\forall n \in N$.

THINK	WRITE
1. State the propositional statement $P(n)$ which we are looking to prove.	Proof: If $A = \begin{bmatrix} 3 & -1 \\ 0 & 2 \end{bmatrix}$ Let $P(n)$ be the propositional statement: $A^n = \begin{bmatrix} 3^n & 2^n - 3^n \\ 0 & 2^n \end{bmatrix}$
2. Verify that $P(1)$ is true.	$A^1 = \begin{bmatrix} 3^1 & 2^1 - 3^1 \\ 0 & 2^1 \end{bmatrix} = \begin{bmatrix} 3 & -1 \\ 0 & 2 \end{bmatrix} = A$ $P(1)$ is true.
3. Assume that $P(k)$ is true.	$A^k = \begin{bmatrix} 3^k & 2^k - 3^k \\ 0 & 2^k \end{bmatrix}$
4. Write a statement for $P(k+1)$.	$A^{k+1} = \begin{bmatrix} 3^{k+1} & 2^{k+1} - 3^{k+1} \\ 0 & 2^{k+1} \end{bmatrix}$
5. Consider the matrix A^{k+1}.	$A^{k+1} = A^k A$
6. Substitute for the matrices on the RHS assuming that $P(k)$ is true.	$A^{k+1} = \begin{bmatrix} 3^k & 2^k - 3^k \\ 0 & 2^k \end{bmatrix} \begin{bmatrix} 3 & -1 \\ 0 & 2 \end{bmatrix}$
7. Perform the matrix multiplication.	$A^{k+1} = \begin{bmatrix} 3^k \times 3 + (2^k - 3^k) \times 0 & 3^k \times (-1) + (2^k - 3^k) \times 2 \\ 0 \times 3 + 2^k \times 0 & 0 \times (-1) + 2^k \times 2 \end{bmatrix}$ $A^{k+1} = \begin{bmatrix} 3 \times 3^k & 2 \times 2^k - 3 \times 3^k \\ 0 & 2 \times 2^k \end{bmatrix}$
8. Use index laws to write the expression so that it is clearly the same as the RHS of $P(k+1)$.	$A^{k+1} = \begin{bmatrix} 3^{k+1} & 2^{k+1} - 3^{k+1} \\ 0 & 2^{k+1} \end{bmatrix}$

9. State what has been demonstrated and concluded.	We have shown that if $P(k)$ is true, $P(k+1)$ is also true. $P(1)$ is true, therefore by the principle of mathematical induction, $P(n)$: If $A = \begin{bmatrix} 3 & -1 \\ 0 & 2 \end{bmatrix}$ then $A^n = \begin{bmatrix} 3^n & 2^n - 3^n \\ 0 & 2^n \end{bmatrix}$ is true $\forall n \in N$.

Induction proofs involving trigonometry

Before looking at trigonometric proofs, we must recall some trigonometric identities, which you have seen in Year 11.

Compound angle formulas

$$\sin(A+B) = \sin(A)\cos(B) + \cos(A)\sin(B)$$

$$\sin(A-B) = \sin(A)\cos(B) - \cos(A)\sin(B)$$

$$\cos(A+B) = \cos(A)\cos(B) - \sin(A)\sin(B)$$

$$\cos(A-B) = \cos(A)\cos(B) + \sin(A)\sin(B)$$

$$\tan(A+B) = \frac{\tan(A) + \tan(B)}{1 - \tan(A)\tan(B)}$$

$$\tan(A-B) = \frac{\tan(A) - \tan(B)}{1 + \tan(A)\tan(B)}$$

Double angle formulas

$$\sin(2A) = 2\sin(A)\cos(A)$$

$$\cos(2A) = \cos^2(A) - \sin^2(A)$$

$$\cos(2A) = 1 - 2\sin^2(A)$$

$$\cos(2A) = 2\cos^2(A) - 1$$

$$\tan(2A) = \frac{2\tan(A)}{1 - \tan^2(A)}$$

WORKED EXAMPLE 28 Proof by mathematical induction (trigonometric)

Prove by mathematical induction that $\sin\left(\frac{(2n+1)\pi}{2}\right) = (-1)^n, \forall n \in Z^+ \cup \{0\}$.

THINK	WRITE
1 State the propositional statement $P(n)$ which we are looking to prove.	Proof: Let $P(n)$ be the propositional statement: $\sin\left(\frac{(2n+1)\pi}{2}\right) = (-1)^n$
2 Verify that for the base step this time when $n = 0$, $P(0)$ is true.	LHS of $P(0) = \sin\left(\frac{\pi}{2}\right) = 1$ RHS of $P(0) = (-1)^0 = 1$ $P(0)$ is true.
3 Assume that $P(k)$ is true.	$\sin\left(\frac{(2k+1)\pi}{2}\right) = (-1)^k$
4 Write a statement for $P(k+1)$.	$\sin\left(\frac{(2(k+1)+1)\pi}{2}\right) = (-1)^{k+1}$
5 Consider the LHS of $P(k+1)$.	$\sin\left(\frac{(2(k+1)+1)\pi}{2}\right)$ $= \sin\left(\frac{(2k+3)\pi}{2}\right)$
6 Simplify.	$= \sin\left(\frac{(2k+1+2)\pi}{2}\right)$ $= \sin\left(\frac{(2k+1)\pi}{2} + \pi\right)$
7 Use the compound angle formula with $A = \frac{(2k+1)\pi}{2}$, $B = \pi$ and using $\cos(\pi) = -1$, $\sin(\pi) = 0$.	$= \sin\left(\frac{(2k+1)\pi}{2}\right)\cos(\pi) + \cos\left(\frac{(2k+1)\pi}{2}\right)\sin(\pi)$ $= (-1) \times \sin\left(\frac{(2k+1)\pi}{2}\right) + 0 \times \cos\left(\frac{(2k+1)\pi}{2}\right)$ $= -\sin\left(\frac{(2k+1)\pi}{2}\right)$
8 Use the fact that $P(k)$ is true and index laws.	$= -(-1)^k = (-1)^{k+1}$
9 State what has been demonstrated and concluded.	We have shown that if $P(k)$ is true, $P(k+1)$ is also true. $P(0)$ is true, therefore by the principle of mathematical induction, it is true $\forall n \in Z^+ \cup \{0\}$.

1.5 Exercise

Students, these questions are even better in jacPLUS

Receive immediate feedback and access sample responses

Access additional questions

Track your results and progress

Find all this and MORE in jacPLUS

Technology free

1. WE21 Use mathematical induction to prove that $1+2+3+...+n=\frac{1}{2}n(n+1),\ \forall n \in N$.

2. Use mathematical induction to prove that $1\times 2+2\times 3+3\times 4+...+n(n+1)=\frac{1}{3}n(n+1)(n+2),\ \forall n \in N$.

3. WE22 Use mathematical induction to prove that
$$\frac{1}{2\times 5}+\frac{1}{5\times 8}+\frac{1}{8\times 11}+...+\frac{1}{(3n-1)(3n+2)}=\frac{n}{6n+4},\ \forall n \in N.$$

4. WE23 Use mathematical induction to prove that
$$1\times 4+2\times 4^2+3\times 4^3+...+n\times 4^n=\frac{1}{9}(4(3n-1)\times 4^n+4),\ \forall n \in N.$$

5. WE24 Prove that each of the following are all divisible by 5 for all $n \in N$.
   a. 6^n-1
   b. 6^n+4
   c. $2^{3n}-3^n$

6. Prove that each of the following are all divisible by 6 for all $n \in N$.
   a. $17n^3+103n$
   b. $n(n+1)(2n+1)$

7. Prove that each of the following are all divisible by 7 for all $n \in N$.
   a. $2^{n+2}+3^{2n+1}$
   b. $8^{2n+1}+6^{2n-1}$

8. Use mathematical induction to prove that $1+3+5+...+(2n-1)=n^2\ \forall n \in N$.

9. Prove by mathematical induction that $3^{2n-1}+1$ is divisible by 4, $\forall n \in N$.

10. Use mathematical induction to prove that $a+a^2+a^3+...+a^n=\frac{1}{a-1}(a^{n+1}-a),\ \forall n \in N$, and for $a>1$.

11. WE25 Use mathematical induction to prove that $1\times 1!+2\times 2!+3\times 3!+...+n\times n!=(n+1)!-1,\ \forall n \in N$.

12. Use mathematical induction to prove that $(1+x)^n \geq 1+nx$ for $x>-1$ and $\forall n \in N$.

13. Use mathematical induction to prove that $\left(1-\frac{2}{3}\right)\left(1-\frac{2}{4}\right)\left(1-\frac{2}{5}\right)...\left(1-\frac{2}{n}\right)=\frac{2}{n(n-1)}$ for natural numbers $n \geq 3$.

14. Use mathematical induction to prove that $\left(1-\frac{1}{4}\right)\left(1-\frac{1}{9}\right)...\left(1-\frac{1}{n^2}\right)=\frac{n+1}{2n}$ for natural numbers $n \geq 2$.

15. WE26 The Fibonacci numbers are defined by $f_1=f_2=1$ and $f_{n+2}=f_{n+1}+f_n,\ \forall n \in N$. Use mathematical induction to prove that $f_1+f_3+f_5+...+f_{2n-1}=f_{2n},\ \forall n \in N$.

16. Use mathematical induction to prove that $f_2+f_4+f_6+...+f_{2n}=f_{2n+1}-1,\ \forall n \in N$.

17. Use mathematical induction to prove that if $a = \dfrac{1+\sqrt{5}}{2}$ and $b = \dfrac{1-\sqrt{5}}{2}$ then $f_n = \dfrac{a^n - b^n}{\sqrt{5}}$, $\forall\, n \in N$.

18. The Fibonacci numbers are defined by $f_1 = f_2 = 1$ and $f_{n+2} = f_{n+1} + f_n$, $\forall n \in N$. Prove the following by mathematical induction.

$$\sum_{r=1}^{n} f_r = f_{n+2} - 1$$

19. **WE27** Let $A = \begin{bmatrix} 4 & 0 \\ -1 & 5 \end{bmatrix}$. Prove by mathematical induction that $A^n = \begin{bmatrix} 4^n & 0 \\ 4^n - 5^n & 5^n \end{bmatrix}$, $\forall n \in N$.

20. Let $A = \begin{bmatrix} 2 & 1 \\ -1 & 0 \end{bmatrix}$. Prove by mathematical induction that $A^n = \begin{bmatrix} n+1 & n \\ -n & 1-n \end{bmatrix}$, $\forall n \in N$.

21. Let $B = \begin{bmatrix} 2 & b \\ 0 & 1 \end{bmatrix}$. Prove by mathematical induction that $B^n = \begin{bmatrix} 2^n & b(2^n - 1) \\ 0 & 1 \end{bmatrix}$, $\forall n \in N$.

22. a. Prove by mathematical induction that the sum of three consecutive natural numbers is divisible by 3.
    b. Prove by mathematical induction that the product of three consecutive natural numbers is divisible by 3

23. **WE28** Prove by mathematical induction that $\cos(n\pi) = (-1)^n$, $\forall n \in Z^+ \cup \{0\}$.

24. Use mathematical induction to prove that $\tan(x + n\pi) = \tan(x)$, $\forall n \in Z^+ \cup \{0\}$ and $x \in R$.

25. Use mathematical induction to prove that $\cos(x + 2n\pi) = \cos(x)$, $\forall n \in Z^+ \cup \{0\}$ and $x \in R$.

26. Use mathematical induction to prove that $\sin(x + 2n\pi) = \sin(x)$, $\forall n \in Z^+ \cup \{0\}$ and $x \in R$.

27. Use mathematical induction to prove that $\cos(x + n\pi) = (-1)^n \cos(x)$, $\forall n \in Z^+ \cup \{0\}$ and $x \in R$.

28. Use mathematical induction to prove that $\sin(x + n\pi) = (-1)^n \sin(x)$, $\forall n \in Z^+ \cup \{0\}$ and $x \in R$.

29. a. Show that $\tan\left(x + \dfrac{\pi}{2}\right) = \dfrac{-1}{\tan(x)}$.
    b. Hence prove by mathematical induction that $\tan\left(\dfrac{(2n+1)\pi}{4}\right) = (-1)^n$, $\forall n \in Z^+ \cup \{0\}$ and $x \in R$.

30. Chebyshev (1821–1894) was a famous Russian mathematician. Although known more for his work in the fields of probability, statistics, number theory and differential equations, Chebyshev also devised recurrence relations for trigonometric multiple angles.
    (1) $\cos(nx) = 2\cos(x)\cos((n-1)x) - \cos((n-2)x)$
    (2) $\sin(nx) = 2\cos(x)\sin((n-1)x) - \sin((n-2)x)$
    a. Assuming (1) is true use mathematical induction to show that (2) is true $\forall n \in N,\ n \geq 2$.
    b. Assuming (2) is true use mathematical induction to show that (1) is true $\forall n \in N,\ n \geq 2$.

1.5 Exam questions

Question 1 (4 marks) TECH-FREE

Use mathematical induction to prove that

$$1 \times 2 \times 3 + 2 \times 3 \times 4 + 3 \times 4 \times 5 + ... + n(n+1)(n+2) = \frac{1}{4}n(n+1)(n+2)(n+3), \ \forall n \in N.$$

Question 2 (4 marks) TECH-FREE

Use mathematical induction to prove that $5^{2n+1} + 2^{2n+1}$ is divisible by 7, $\forall n \in N$.

Question 3 (4 marks) TECH-FREE

Use mathematical induction to prove that $\frac{1}{2!} + \frac{2}{3!} + \frac{3}{4!} + ... + \frac{n}{(n+1)!} = 1 - \frac{1}{(n+1)!}, \ \forall n \in N$.

More exam questions are available online.

1.6 Review

1.6.1 Summary

doc-37055

Hey students! Now that it's time to revise this topic, go online to:

Access the topic summary

Review your results

Watch teacher-led videos

Practise VCAA exam questions

Find all this and MORE in jacPLUS

1.6 Exercise

Technology free: short answer

1. a. Prove that the following is a tautology.
   $((p \vee q) \wedge \neg p) \rightarrow q$
   b. Prove that the following is a tautology.
   $(p \leftrightarrow q) \leftrightarrow (p \rightarrow q) \wedge (q \rightarrow p)$

2. Let p and q be the propositions, 'I will study hard' and 'I pass the test,' respectively. Write the implication, $p \rightarrow q$, the contrapositive, inverse, converse and negation statements in words.

3. Write the following propositions in words and state whether they are true or false.
   a. $\exists x \in Z,\ 3x = 2$
   b. $\forall x \in Z,\ x^2 \geq 0$
   c. $\forall x\,(x \in N \rightarrow x \in Z)$

4. Consider the propositions, $F(x)$: 'x is my friend' and $S(x)$: 'x is sincere'.
   Write each of the following in words.
   a. $\forall x\,(F(x) \wedge S(x))$
   b. $\exists x\,(F(x) \wedge S(x))$
   c. $\exists x\,(\neg F(x) \wedge S(x))$
   d. $\exists x\,(F(x) \wedge \neg S(x))$
   e. $\exists x\,(\neg F(x) \wedge \neg S(x))$
   f. $\exists \neg x\,(F(x) \wedge S(x))$

5. a. Prove that the sum of an even and an odd integer is an odd integer.
   b. Prove that the product of an even and an odd integer is an even integer.

6. Use proof by contradiction to show that $2\sqrt{x}\sqrt{y} \leq x + y,\ \forall x, y \in R^+$.

7. Prove the following by mathematical induction. $\forall n \in N,\ 1^3 + 2^3 + 3^3 + \ldots + n^3 = \frac{1}{4}n^2(n+1)^2$

8. Prove by mathematical induction that the sum of three consecutive even numbers is divisible by 6.

Technology active: multiple choice

9. **MC** The propositional statement $(p \wedge \neg p) \wedge q$ is
   A. a tautology.
   B. sometimes true and sometimes false.
   C. always false.
   D. used to prove a contradiction by cases.
   E. used to prove mathematical induction by cases.

10. **MC** The propositional statement $((p \vee \neg p) \rightarrow r) \leftrightarrow ((p \rightarrow r) \vee (q \rightarrow r))$ is
   A. a tautology.
   B. sometimes true and sometimes false.
   C. always false.
   D. used to prove a contradiction by cases.
   E. used to prove mathematical induction by cases.

11. **MC** The statement below which is true is

A. $\forall x \in Q, \exists y \in Q, y = x^2$
B. $\exists y \in Q, \forall x \in Q, y = x^2$
C. $\exists y \in Q, \forall x \in Q, y = \sqrt{x}$
D. $\forall x \in Q, \exists y \in Q, y = \sqrt{x}$
E. $\forall y \in Z, \exists x \in Q, x = \sqrt{y}$

12. **MC** The negation of the converse of $p \to q$ is

A. $\neg q \to p$
B. $\neg q \to \neg p$
C. $q \wedge \neg p$
D. $\neg q \wedge p$
E. $q \vee \neg p$

13. **MC** The statement below which is false is

A. The sum of two even numbers is even.
B. The product of two even numbers is even.
C. The sum of two odd numbers is even.
D. The sum of an even and an odd number is even.
E. The product of an even and an odd number is even.

14. **MC** The statement below which is true is

A. There is a finite number of prime numbers.
B. All prime numbers are odd.
C. If n is a prime number then $n^2 + 1$ is also a prime number.
D. If n is a prime number then $2^n - 1$ is also a prime number.
E. No function $p: N \to N$ to determine the nth prime number exists.

15. **MC** To prove $p \to q$ using a direct proof, we

A. assume p is true and show that q is true.
B. assume p is true and show that q is false.
C. assume $\neg p$ is true and show that $\neg q$ is true.
D. assume $\neg q$ is true and show that $\neg p$ is true.
E. assume $\neg p$ is false and show that $\neg q$ is false.

16. **MC** To prove $p \to q$ using a proof by contraposition, we

A. assume p is true and show that q is true.
B. assume p is true and show that q is false.
C. assume $\neg p$ is true and show that $\neg q$ is true.
D. assume $\neg q$ is true and show that $\neg p$ is true.
E. assume $\neg p$ is false and show that $\neg q$ is false.

17. **MC** To prove $p \to q$ using a proof by contradiction, we

A. assume $\neg p$ is true and show that this implies $\neg q$.
B. assume $\neg q$ is true and show that this implies $\neg p$.
C. assume $\neg q$ is true and show that this implies $r \wedge \neg r$.
D. assume $\neg p \wedge q$ is true and show that this implies $r \wedge \neg r$.
E. assume $p \wedge \neg q$ is true and show that this implies $r \wedge \neg r$.

18. **MC** If $n \in N$, to prove a proof by mathematical induction about $P(n)$ we show a base case and that

A. $P(n)$ and $P(n+1)$ are both true.
B. $\neg P(n)$ and $\neg P(n+1)$ are both false.
C. $P(n) \vee P(n+1)$ is true.
D. $P(n) \to P(n+1)$.
E. $P(n+1) \to P(n)$.

Technology active: extended response

19. Use mathematical induction to prove the following $\forall n \in N$.

a. $1 \times 3 + 2 \times 3^2 + 3 \times 3^3 + ... + n \times 3^n = \frac{3}{4}[(2n-1) \times 3^n + 1]$

b. $\frac{1}{1 \times 2} + \frac{1}{2 \times 3} + \frac{1}{3 \times 4} + ... + \frac{1}{n(n+1)} = \frac{n}{n+1}$

20. a. Peter claimed that the function $p: N \to N,\ p(n) = \frac{n^5}{20} - \frac{5n^4}{8} + \frac{17n^3}{6} - \frac{43n^2}{8} + \frac{307n}{60}$ generates a formula for prime numbers. Determine $p(1), p(2), p(3), p(4) ...$
Prove or disprove Peter's assertion.

b. Quentin claimed that the function $q: N \to N,\ q(n) = n^2 - n + 41$ generates prime numbers. Determine $q(1), q(2), q(3), q(4) ...$
Prove or disprove Quentin's assertion.

1.6 Exam questions

Question 1 (2 marks) TECH-FREE

If x is an even integer, prove that $x^2 + 5x - 11$ is odd.

Question 2 (3 marks) TECH-FREE

Prove that $\sqrt{5}$ is irrational.

Question 3 (4 marks) TECH-FREE

The Fibonacci numbers are defined by $f_1 = f_2 = 1$ and $f_{n+2} = f_{n+1} + f_n,\ \forall n \in N$. Prove the following by mathematical induction.

$$f_n \geq \left(\frac{3}{2}\right)^{n-2}$$

Question 4 (4 marks) TECH-FREE

Prove by mathematical induction that $4^{n+1} + 5^{2n-1}$ is divisible by 21, $\forall n \in N$.

Question 5 (3 marks) TECH-FREE

Prove that the difference between the squares of any two consecutive odd numbers is divisible by 8.

More exam questions are available online.

Answers

Topic 1 Logic and proof

1.2 Logic

1.2 Exercise

1–5. Sample responses can be found in the worked solutions in the online resources.

6. Implication $p \rightarrow q$. If it is raining then I take an umbrella.
Contrapositive $\neg q \rightarrow \neg p$. If I do not take an umbrella then it is not raining.
Inverse $\neg p \rightarrow \neg q$. If it is not raining then I will not take an umbrella.
Converse $q \rightarrow p$. If a take an umbrella then it is raining.
Negation$\neg (p \rightarrow q) \leftrightarrow p \wedge \neg q$. It is raining and I do not take an umbrella.

7–10. Sample responses can be found in the worked solutions in the online resources.

11. a. All natural numbers satisfy $x > -1$, True.
b. All natural numbers are positive, True.

12. a. There are natural numbers that satisfy $x < -1$, False.
b. There is a rational number which satisfies $3x = 2$, True, $x = \frac{2}{3}$.

13. a. There is an integer which satisfies $3x + 6 = 0$, True, $x = -2$.
b. There is an integer whose square is less than 1, True, $x = 0$.

14. a. The sum of every two natural numbers is positive, True.
b. The sum of the squares of every two real numbers is positive, False, $x = 0, y = 0$.

15. a. For every integer, there is a negative integer, such that their sum is zero, True.
b. For every integer, there is an integer greater than it, True.

16. a. The multiplicative reciprocal of every rational number is rational. False, not true when $y = 0$.
b. The square of every rational number is rational. True.
c. There is a real number which is the reciprocal of the square root of a rational number. True.

17. a. For every integer, there is an integer which is half of it, False.
b. For every integer, there is a rational number which is half of it, True.

18. There is an integer such that its sum with all integers is zero, False.

19. For every non-zero rational number there is a multiplicative reciprocal, True.

20. Every quadratic has rational solutions, False.

1.2 Exam questions

Note: Mark allocations are available with the fully worked solutions online.

1. Sample responses can be found in the worked solutions in the online resources.
2. Sample responses can be found in the worked solutions in the online resources.
3. Every quadratic with a positive discriminant and real coefficients has a real solution, True.

1.3 Direct proofs

1.3 Exercise

1. a. $\{2, 3, 5, 7, 11, 13, 17, 19\}$
b. $\{1, 3, 5, 7, 9, 11, 13, 15, 17, 19\}$
c. $\{3, 6, 9, 12, 15, 18\}$
d. $\{3, 5, 7, 11, 13, 17, 19\}$
e. $\{3\}$

2–13. Sample responses can be found in the worked solutions in the online resources.

14. a. Sample responses can be found in the worked solutions in the online resources.
b. The statement is false.

15. Sample responses can be found in the worked solutions in the online resources.

16. a. $4|12$, $6|12$ are both True but $24|12$ is False.
b. Sample responses can be found in the worked solutions in the online resources.

17. Sample responses can be found in the worked solutions in the online resources.

18. a. 9 b. 12 c. 1
d. 54 e. 60 f. 77

19–22. Sample responses can be found in the worked solutions in the online resources.

23. a. 4 b. 3 c. 25
d. 130 476 e. 113 841 f. 18 450

24. Peter's assertion is wrong.

Natural numbers	Number of prime numbers
1 − 100	25
101 − 200	21
201 − 300	16
301 − 400	16
401 − 500	17
501 − 600	14
601 − 700	16
701 − 800	14
801 − 900	15
901 − 1000	14
1 − 1000	168
1001 − 2000	135

1.3 Exam questions

Note: Mark allocations are available with the fully worked solutions online.

1. Sample responses can be found in the worked solutions in the online resources.
2. The statement is false.
3. Sample responses can be found in the worked solutions in the online resources.

1.4 Indirect proofs

1.4 Exercise

Sample responses can be found in the worked solutions in the online resources.

1.4 Exam questions

Note: Mark allocations are available with the fully worked solutions online.

Sample responses can be found in the worked solutions in the online resources.

1.5 Proof by mathematical induction

1.5 Exercise

Sample responses can be found in the worked solutions in the online resources.

1.5 Exam questions

Note: Mark allocations are available with the fully worked solutions online.

Sample responses can be found in the worked solutions in the online resources.

1.6 Review

1.6 Exercise

Technology free: short answer

1. Sample responses can be found in the worked solutions in the online resources.
2. Implication $p \rightarrow q$. If I study hard then I will pass the test.
   Contrapositive $\neg q \rightarrow \neg p$. If I do not pass the test then I did not study hard.
   Inverse $\neg p \rightarrow \neg q$. If I do not study hard then I will not pass the test.
   Converse $q \rightarrow p$. If I pass the test then I study hard.
   Negation $\neg(p \rightarrow q) \leftrightarrow p \wedge \neg q$. I study hard and I do not pass the test.
3. a. There is an integer x which satisfies $3x = 2$, False.
   b. The square of all integers is positive or zero, True.
   c. All natural numbers are integers, True.
4. a. All of my friends are sincere.
   b. Some of my friends are sincere.
   c. There are some people who are not my friends who are sincere.
   d. Some of my friends are not sincere.
   e. There are some people who are not my friends and who are not sincere.
   f. There isn't any person who is my friend and sincere.

5–8. Sample responses can be found in the worked solutions in the online resources.

Technology active: multiple choice

9. C
10. B
11. E
12. C
13. D
14. E
15. A
16. D
17. E
18. D

Technology active: extended response

19. Sample responses can be found in the worked solutions in the online resources.
20. a. $p(6) = 28$ is not prime.
    b. $p(41) = 1681 = 41^2$ is not prime.

1.6 Exam questions

Note: Mark allocations are available with the fully worked solutions online.

Sample responses can be found in the worked solutions in the online resources.

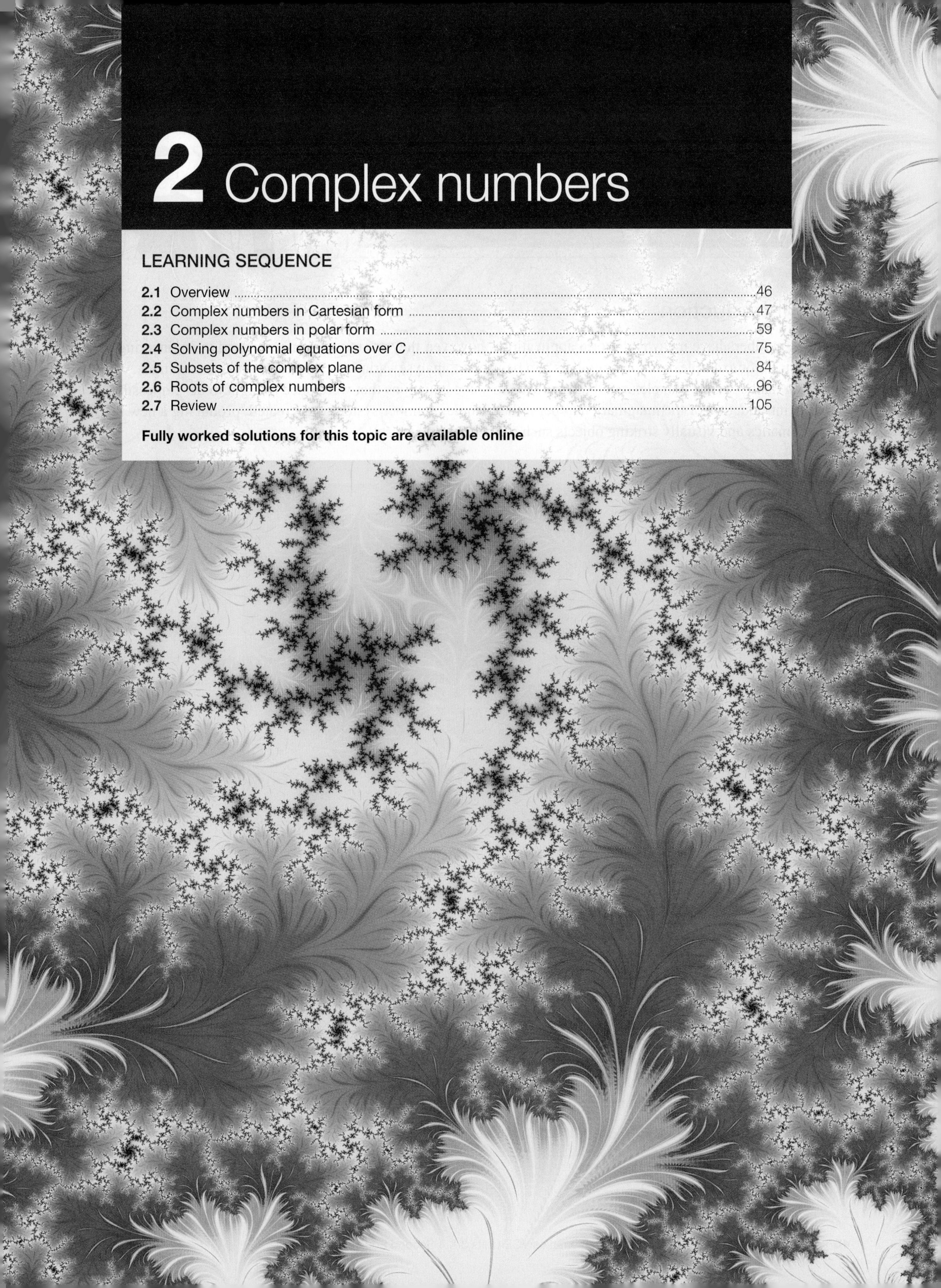

2 Complex numbers

LEARNING SEQUENCE

Fully worked solutions for this topic are available online

2.1 Overview

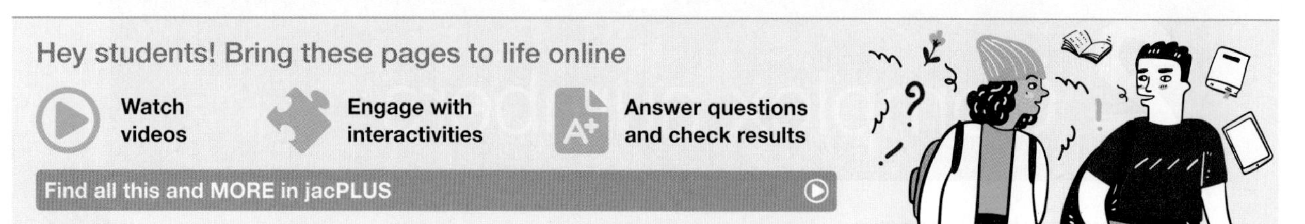

2.1.1 Introduction

Complex numbers have many real-world applications, however, they are most commonly used in engineering and physics. For example, complex numbers can be used to analyse variations in voltage and current in an AC circuit. In mathematics, complex numbers are used to solve equations that were once thought to be 'impossible'. With the introduction of the imaginary number i, the complex number system was created, producing a new field of mathematics and visually striking objects such as the Mandelbrot set.

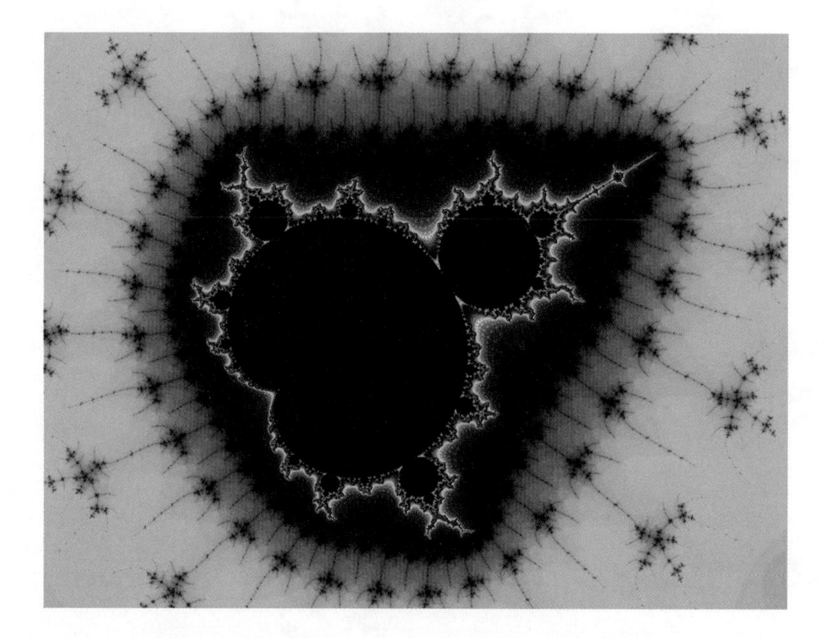

KEY CONCEPTS

This topic covers the following points from the VCE Mathematics Study Design:

- De Moivre's theorem, proof for integral powers, powers and roots of complex numbers in polar form, and their geometric representation and interpretation
- the nth roots of unity and other complex numbers and their location in the complex plane
- factors over C, of polynomials; and introduction to the fundamental theorem of algebra, including its application to factorisation of polynomial functions of a single variable over C, for example, z^8+1, z^2-i or $z^3-(2-i)z^2+z-2+i$
- solution over C of polynomial equations by completing the square, use of the quadratic factorisation and the conjugate root theorem.

2.2 Complex numbers in Cartesian form

LEARNING INTENTION

At the end of this subtopic you should be able to:
- perform operations on complex numbers in Cartesian form.

2.2.1 The definition of a complex number

The **complex number system** is an extension of the real number system. Complex numbers are numbers that include the number i, known as the imaginary unit.

The imaginary unit

The imaginary unit i is defined as:

$$i^2 = -1$$

This means that solutions to equations such as $x^2 = -1$ can now be found in terms of i.

The general form of a complex number is represented by z and defined as $z = x + yi$, where x and $y \in R$, and $z \in C$, where C is used to denote the set of complex numbers (in the same way that R denotes the set of real numbers). Note that $z = x + yi$ is one single number but is composed of two parts: a real part and an imaginary part. The real part is written as $\text{Re}(z) = x$ and the imaginary part is written as $\text{Im}(z) = y$.

A complex number in the form $z = x + yi$, where both x and y are real numbers, is called the Cartesian form or rectangular form or standard form of a complex number. Throughout this chapter, it is assumed that all equations are solved over C.

WORKED EXAMPLE 1 Evaluating real and imaginary components of complex numbers

Evaluate each of the following.

a. **$\text{Re}(6 - 12i)$**

b. **$\text{Im}\left(3 - i + i^3\right)$**

THINK	WRITE
a. The real part of the complex number $6 - 12i$ is 6.	a. $\text{Re}(6 - 12i) = 6$
b. 1. Simplify $3 - i + i^3$ first by recalling that $i^2 = -1$.	b. $\text{Im}\left(3 - i + i^3\right) = \text{Im}\left(3 - i + i^2 \times i\right)$ $= \text{Im}(3 - i - i)$ $= \text{Im}(3 - 2i)$
2. Look for the coefficient of i to determine the imaginary part of any complex number.	$\text{Im}(3 - 2i) = -2$

2.2.2 Operations on complex numbers

The Argand plane

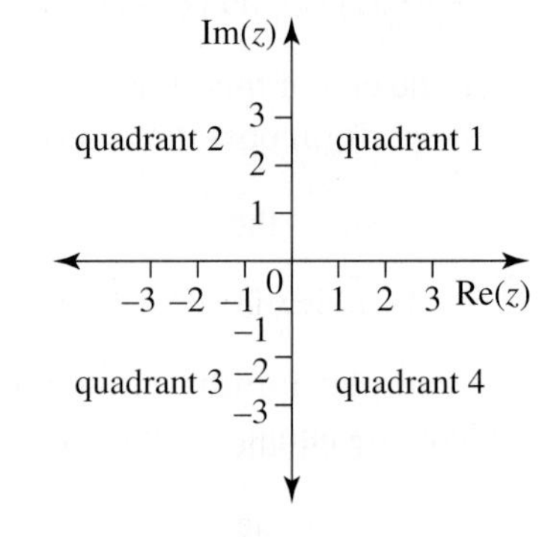

Complex numbers cannot be represented on a traditional Cartesian diagram because of their imaginary part. However, a similar plane was created by the Swiss mathematician Jean-Robert Argand (1768–1822). It is called an **Argand plane** or Argand diagram, and it allows complex numbers to be represented visually.

Because a complex number has two parts, a real part and an imaginary part, the horizontal axis is called the real axis and the vertical axis is called the imaginary axis. A complex number $z = x + yi$ is represented by the equivalent point (x, y) in a Cartesian coordinate system. Note that the imaginary axis is labelled 1, 2, 3 etc., not i, $2i$, $3i$ etc.

Scalar multiplication of complex numbers

If $z = x + yi$, then $kz = kx + kyi$, where $k \in R$. The diagram shows the situation for $x > 0$, $y > 0$ and $k > 1$. Notice that since both the real and imaginary parts are scaled by the same factor, k, the complex numbers z and kz lie on the same line from the origin.

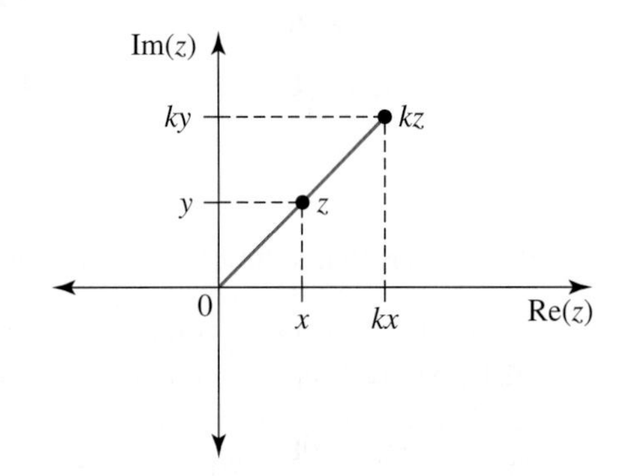

Addition of complex numbers

If $z_1 = x_1 + y_1 i$ and $z_2 = x_2 + y_2 i$, then $z_1 + z_2 = (x_1 + x_2) + (y_1 + y_2)i$. The addition of two complex numbers can be achieved using the same procedure as adding two vectors. Note that the complex numbers are represented by the dots. The directed lines from the origin are there to help demonstrate the similarity between adding complex numbers and adding vectors in 2-dimensions.

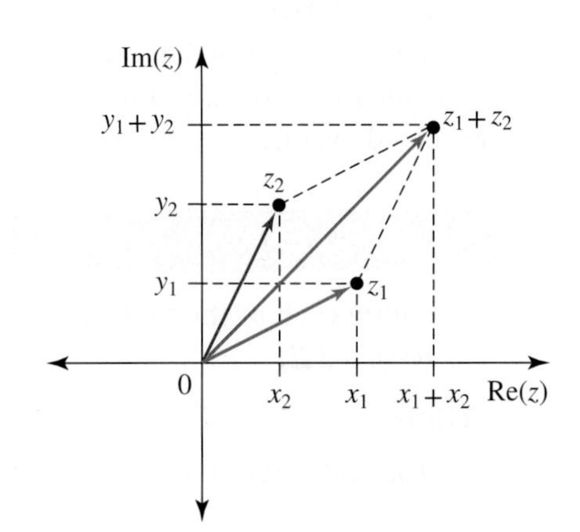

Subtraction of complex numbers

If $z_1 = x_1 + y_1 i$ and $z_2 = x_2 + y_2 i$, then

$$\begin{aligned} z_1 - z_2 &= z_1 + (-z_2) \\ &= (x_1 + y_1 i) - (x_2 + y_2 i) \\ &= (x_1 - x_2) + (y_1 - y_2)i \end{aligned}$$

The subtraction of two complex numbers can be achieved using the same procedure as subtracting two vectors.

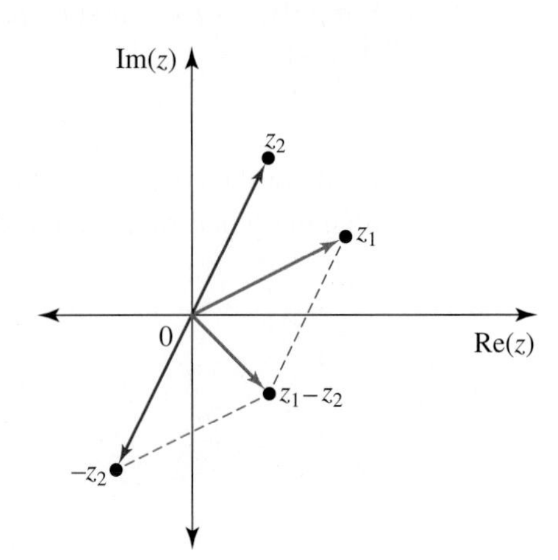

WORKED EXAMPLE 2 Adding and subtracting complex numbers

Given the complex numbers $u = 2 - 5i$ and $v = -3 + 2i$, calculate the complex numbers:

a. $u + v$ b. $u - v$ c. $2u - 3v$

THINK	WRITE
a. 1. Substitute for u and v.	a. $u + v = (2 - 5i) + (-3 + 2i)$
2. Group the real and imaginary parts.	$= (2 - 3) + i(-5 + 2)$
3. Using the rules, state the final result.	$= -1 - 3i$
b. 1. Substitute for u and v.	b. $u - v = (2 - 5i) - (-3 + 2i)$
2. Group the real and imaginary parts.	$= (2 + 3) + i(-5 - 2)$
3. Using the rules, state the final result.	$= 5 - 7i$
c. 1. Substitute for u and v.	c. $2u - 3v = 2(2 - 5i) - 3(-3 + 2i)$
2. Expand by multiplying by the constants.	$= (4 - 10i) - (-9 + 6i)$
3. Group the real and imaginary parts.	$= (4 + 9) + i(-10 - 6)$
4. Using the rules, state the final result.	$= 13 - 16i$

Multiplication of complex numbers

To multiply complex numbers in Cartesian form, proceed as in conventional algebra and replace i^2 with -1 when it appears.

In general, if $z_1 = a + bi$ and $z_2 = c + di$ where a, b, c and $d \in R$, then:

$$\begin{aligned} z_1 z_2 &= (a + bi)(c + di) \\ &= ac + bci + adi + bdi^2 \\ &= ac + (ad + bc)i - bd \\ &= ac - bd + (ad + bc)i \end{aligned}$$

WORKED EXAMPLE 3 Multiplying complex numbers

Given the complex numbers $u = 2 - 5i$ and $v = -3 + 2i$, calculate the complex numbers:

a. uv b. u^2

THINK	WRITE
a. 1. Substitute for u and v.	a. $uv = (2 - 5i)(-3 + 2i)$
2. Expand the brackets using the distributive law.	$= -6 + 15i + 4i - 10i^2$
3. Replace i^2 with -1, group the real and imaginary parts and simplify.	$= -6 + 10 + i(4 + 15)$ $= 4 + 19i$
b. 1. Substitute for u.	b. $u^2 = (2 - 5i)^2$
2. Expand.	$= 4 - 20i + 25i^2$
3. Replace i^2 with -1 and simplify	$= 4 - 20i - 25$ $= -21 - 20i$

TI | THINK	DISPLAY/WRITE	CASIO | THINK	DISPLAY/WRITE
On a Calculator page define the complex numbers u and v as shown.	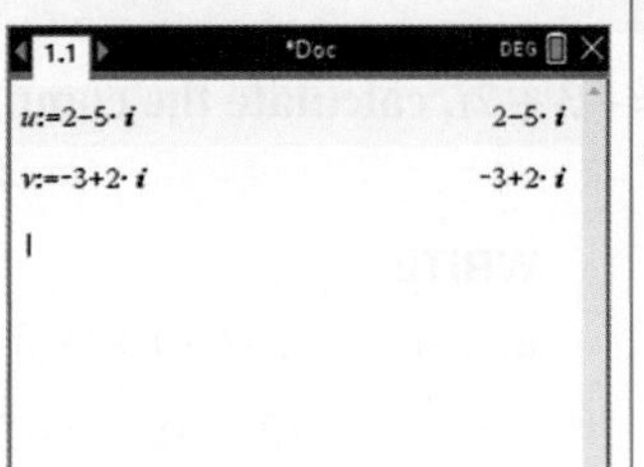	On a Main screen define the complex numbers u and v as shown.	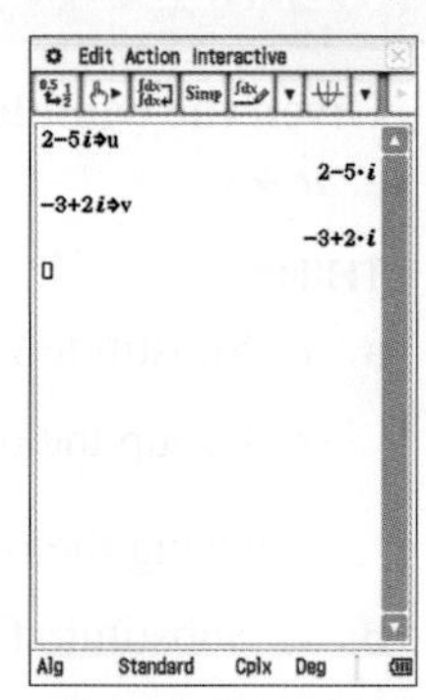
a. Type $u \times v$ and press 'enter'.	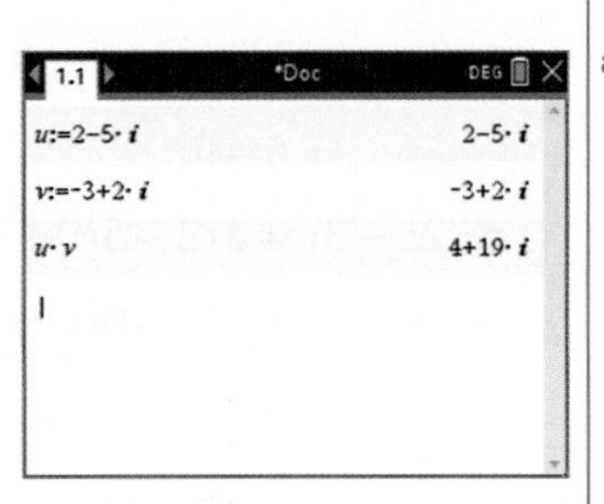	a. Type $u \times v$ and press 'EXE'.	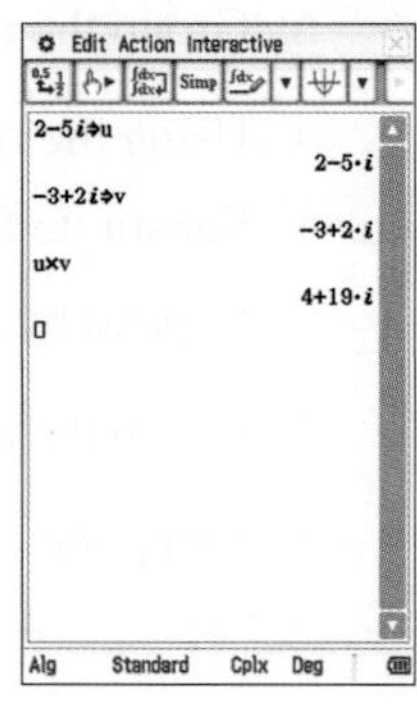
b. Type u^2 and press 'enter'.	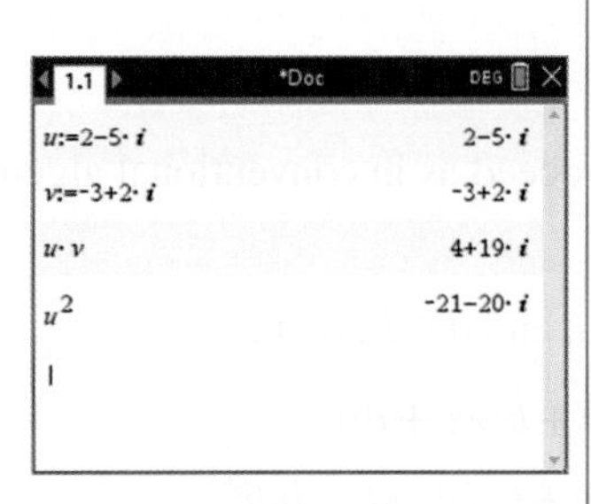	b. Type u^2 and press 'EXE'.	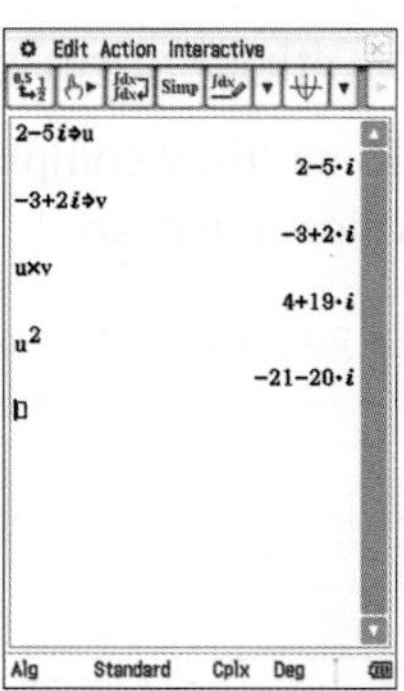

WORKED EXAMPLE 4 Identifying real and imaginary components of complex products

Given the complex numbers $u = 2 - 5i$ and $v = 2 + 5i$, determine:

a. **Re(uv)**

b. **Im(uv)**

THINK	WRITE
1. Substitute for u and v.	$uv = (2 - 5i)(2 + 5i)$
2. Expand the brackets.	$= 4 - 10i + 10i - 25i^2$
3. Simplify and replace i^2 by -1.	$= 29$
a. State the real part.	a. $\text{Re}(uv) = 29$
b. State the imaginary part.	b. $\text{Im}(uv) = 0$

Complex conjugates

In Worked example 4, the complex numbers u and v have the property that the imaginary part of their products is zero. Such numbers are called complex conjugates of each other.

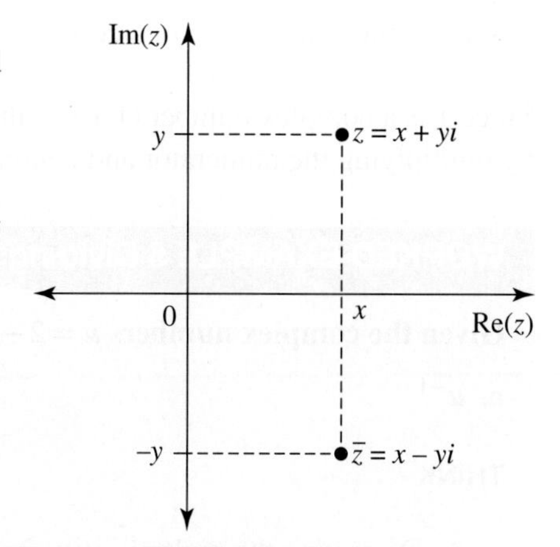

If $z = x + yi$, the conjugate of z is denoted by $\bar{z}$ (read as z bar), and $\bar{z} = x - yi$. That is, the complex conjugate of a number is simply obtained by changing the sign of the imaginary part.

$$\begin{aligned} z\bar{z} &= (x + yi)(x - yi) \\ &= x^2 - xyi + xyi - y^2i^2 \\ &= x^2 + y^2 \end{aligned}$$

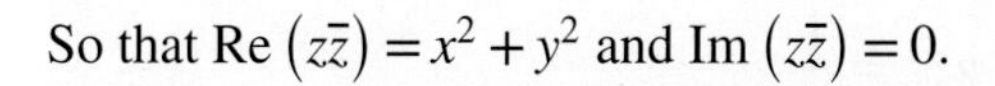

So that $\text{Re}\left(z\bar{z}\right) = x^2 + y^2$ and $\text{Im}\left(z\bar{z}\right) = 0$.

From the diagram it can be seen that $\bar{z}$ is the reflection of the complex number z in the real axis.

The complex conjugate

The conjugate of a complex number $z = x + yi$ is defined as:

$$\bar{z} = x - yi$$

Division of complex numbers

The conjugate is useful in division of complex numbers, because both the numerator and denominator can be multiplied by the conjugate of the denominator. Hence, the complex number can be replaced with a real number in the denominator. This process is similar to rationalising the denominator to remove surds.

In general, if $z_1 = a + bi$ and $z_2 = c + di$ where a, b, c and $d \in R$, then:

$$\begin{aligned} \frac{z_1}{z_2} &= \frac{z_1}{z_2} \times \frac{\bar{z}_2}{\bar{z}_2} \\ &= \frac{a + bi}{c + di} \times \frac{c - di}{c - di} \\ &= \frac{ac + bci - adi - bdi^2}{c^2 - d^2i^2} \\ &= \frac{ac + bd + (bc - ad)i}{c^2 + d^2} \\ &= \frac{ac + bd}{c^2 + d^2} + \left(\frac{bc - ad}{c^2 + d^2}\right)i \end{aligned}$$

Dividing complex numbers

To divide complex numbers, multiply both the numerator and denominator by the conjugate of the denominator.

$$\frac{z}{w} = \frac{z}{w} \times \frac{\bar{w}}{\bar{w}}$$

The multiplicative inverse

The multiplicative inverse of a complex number $z = x + yi$ is $z^{-1} = \frac{1}{z} = \frac{1}{x + yi}$

Since 1 is a complex number $(1 + 0i)$, the multiplicative inverse can be realised in the same way as shown above, by multiplying the numerator and denominator by the conjugate of the denominator.

WORKED EXAMPLE 5 Dividing by complex numbers

Given the complex numbers $u = 2 - 5i$ and $v = -3 + 2i$, evaluate the following:

a. u^{-1}

b. $\frac{u}{v}$

THINK	WRITE
a. 1. Determine the multiplicative inverse and substitute for u.	a. $u^{-1} = \frac{1}{u}$ $= \frac{1}{2 - 5i}$
2. Multiply both the numerator and the denominator by the conjugate of the denominator.	$= \frac{1}{2 - 5i} \times \frac{2 + 5i}{2 + 5i}$
3. Simplify the denominator.	$= \frac{2 + 5i}{4 - 25i^2}$
4. Replace i^2 with -1.	$= \frac{2 + 5i}{29}$
5. State the final answer in $x + yi$ form.	$= \frac{2}{29} + \frac{5}{29}i$
b. 1. Substitute for u and v.	b. $\frac{u}{v} = \frac{2 - 5i}{-3 + 2i}$
2. Multiply both the numerator and the denominator by the conjugate of the denominator.	$= \frac{2 - 5i}{-3 + 2i} \times \frac{-3 - 2i}{-3 - 2i}$
3. Expand the expression in both the numerator and the denominator.	$= \frac{-6 + 15i - 4i + 10i^2}{9 - 4i^2}$
4. Simplify and replace i^2 with -1.	$= \frac{-6 + 11i - 10}{9 + 4}$
5. State the final answer in $x + yi$ form.	$= -\frac{16}{13} + \frac{11}{13}i$

TI | THINK — **DISPLAY/WRITE**

On a Calculator page define the complex numbers u and v as shown.

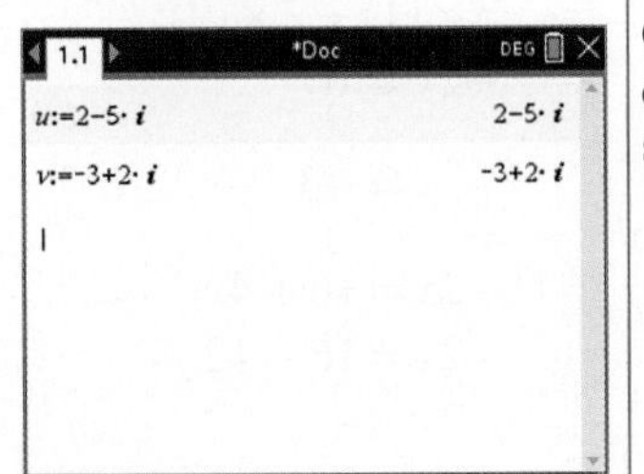

a. Type u^{-1} and press 'enter'.

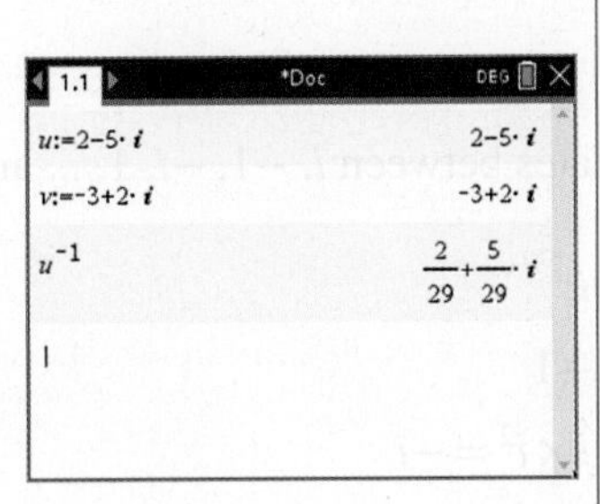

b. Type $\frac{u}{v}$ and press 'enter'.

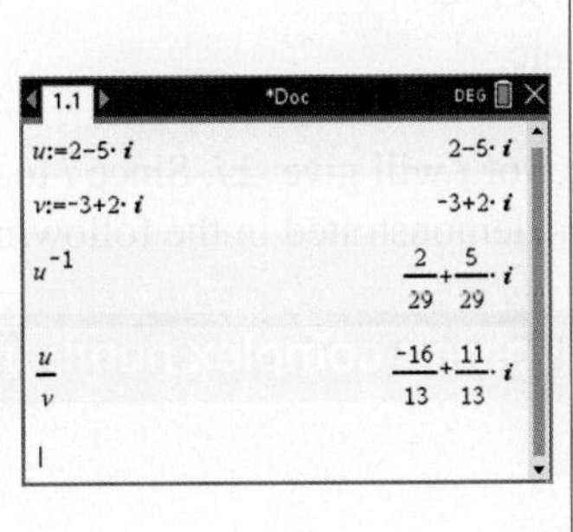

CASIO | THINK — **DISPLAY/WRITE**

On a Main screen define the complex numbers u and v as shown.

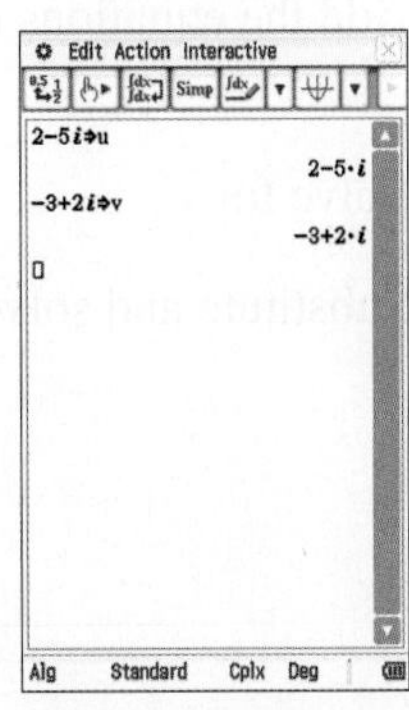

a. Type u^{-1} and press 'EXE'.

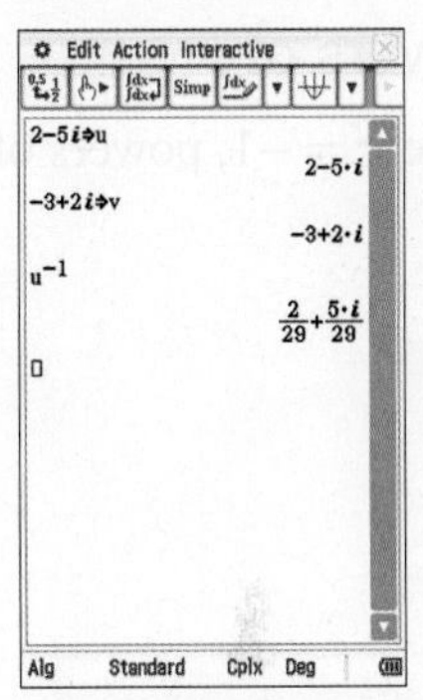

b. Type $\frac{u}{v}$ and press 'EXE'.

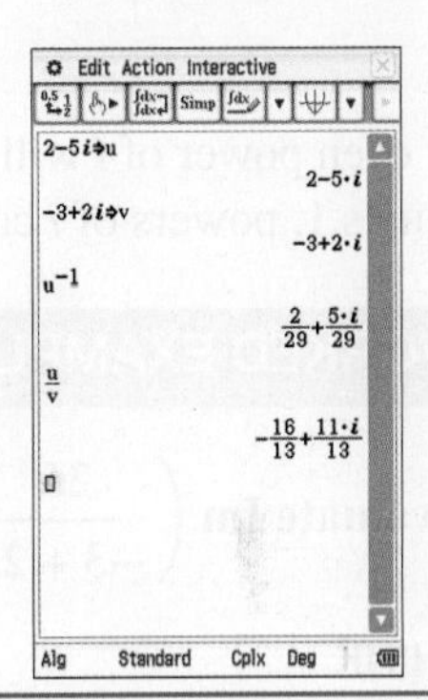

Equality of complex numbers

Two complex numbers are equal if and only if their real parts and their imaginary parts are both equal. For example, if $5+yi=x-3i$, then from equating the real part, we determine that $x=5$, and from equating the imaginary part, we determine that $y=-3$.

WORKED EXAMPLE 6 Equating real and imaginary components

Determine the values of x and y if $2x+5iy-3ix-4y=16-21i$.

THINK	WRITE
1. Group the real and imaginary parts.	$2x+5iy-3ix-4y=16-21i$ $2x-4y+i(5y-3x)=16-21i$
2. Equate the real and imaginary components.	$2x-4y=16 \quad (1)$ $5y-3x=-21 \quad (2)$
3. Solve the simultaneous equations by elimination.	$6x-12y=48 \quad 3\times(1)$ $10y-6x=-42 \quad 2\times(2)$

4. Add the equations to eliminate x.

$$3\times(1)+2\times(2):$$
$$-2y=6$$

5. Solve for y.

$$y=-3$$

6. Substitute and solve for x.

$$\begin{aligned}2x&=16+4y\\2x&=16-12\\2x&=4\\x&=2\end{aligned}$$

Powers of i

Since $i^2=-1$, powers of i form a pattern which alternates between i, -1, $-i$, 1, ... and so on:

$$\begin{aligned}i^1&=i\\i^2&=-1\\i^3&=i\times i^2=-i\\i^4&=(i^2)^2=1\\i^5&=i\times i^4=i\\&\text{etc.}\end{aligned}$$

Any even power of i will give ±1, while any odd power of i will give $\pm i$. Since i to the power of a multiple of 4 equals 1, powers of i can easily be simplified. This is demonstrated in the following worked example.

WORKED EXAMPLE 7 Solving equations involving complex numbers

Evaluate $\text{Im}\left(\dfrac{26}{-3+2i}+i^{69}\right)$.

THINK

WRITE

1. Multiply both the numerator and denominator by the conjugate of the denominator and group the power of i as multiples of i^4 using index laws.

$$\text{Im}\left(\frac{26}{-3+2i}+i^{69}\right)=\text{Im}\left(\frac{26}{-3+2i}\times\frac{-3-2i}{-3-2i}+i^{17\times4+1}\right)$$

2. Expand the expression in the denominator and use index laws on the power of i.

$$=\text{Im}\left(\frac{-26(3+2i)}{9-4i^2}+(i^4)^{17}i\right)$$

3. Simplify and replace i^2 with -1 and i^4 with 1.

$$=\text{Im}\left(\frac{-26(3+2i)}{13}+(1)^{17}i\right)$$

4. Simplify.

$$=\text{Im}(-2(3+2i)+i)$$
$$=\text{Im}(-6-4i+i)$$

5. The imaginary part is the coefficient of the i term. State the final result.

$$=\text{Im}(-6-3i)$$
$$=-3$$

TI | THINK	DISPLAY/WRITE	CASIO | THINK	DISPLAY/WRITE
On a Calculator page type in the command 'imag', place the expression in brackets and press 'enter'.	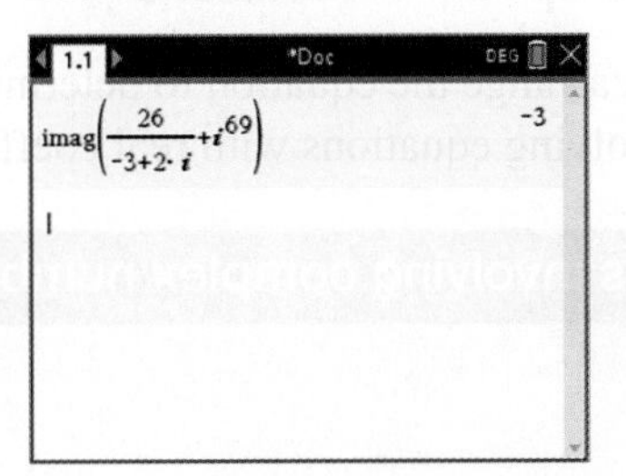	On a Main screen type 'Im', place the expression in brackets and press 'EXE'.	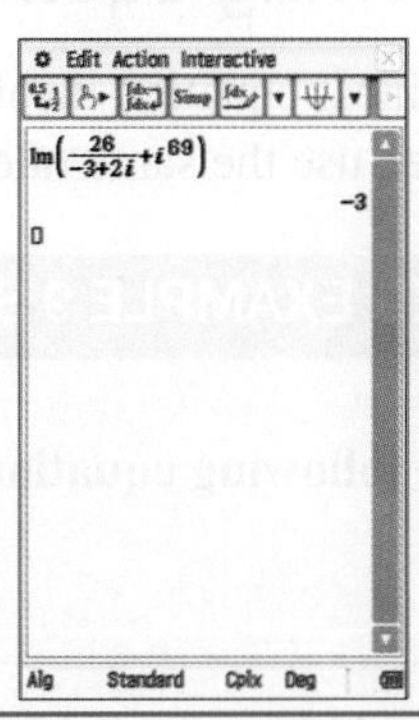

Multiplication by i

If $z = x + yi$, then iz is given by

$$\begin{aligned} iz &= i(x + yi) \\ &= ix + i^2y \\ &= -y + xi \end{aligned}$$

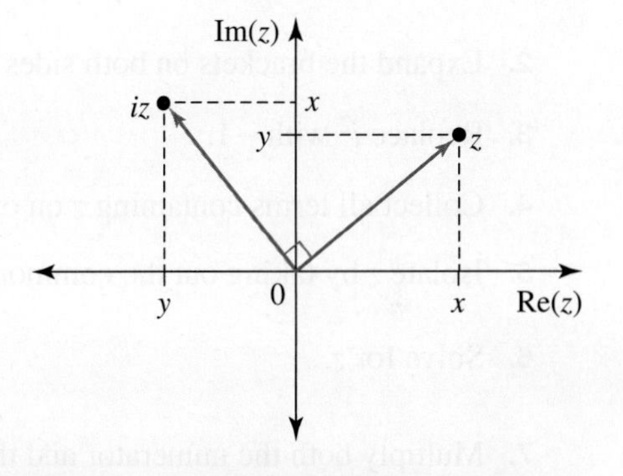

The complex number iz is a rotation of z by 90° anticlockwise.

WORKED EXAMPLE 8 Representing complex numbers on the Argand plane

Given the complex number $z = 2 + 3i$, represent the complex numbers z, $2z$, $\bar{z}$ and iz on one Argand diagram. Comment on their relative positions.

THINK

Determine the values of the complex numbers z, $2z$, $\bar{z}$ and iz in Cartesian form.

$$z = 2 + 3i$$

$$2z = 4 + 6i$$

$$\bar{z} = 2 - 3i$$

$$\begin{aligned} iz &= i(2 + 3i) \\ &= 2i + 3i^2 \\ &= -3 + 2i \end{aligned}$$

Plot each of these complex numbers on the same Argand diagram.

WRITE/DRAW

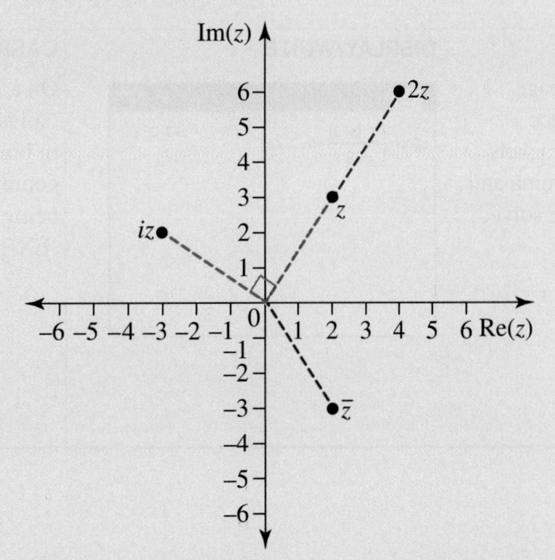

The complex number $2z$ is twice the length of z, the complex conjugate $\bar{z}$ is the reflection of the complex number z in the real axis, and the complex number iz is a rotation of 90° anticlockwise from z.

2.2.3 Solving equations involving complex numbers

To solve a linear equation involving a complex number, rearrange the equation to determine the unknown quantity, then use the same rules and strategies as when solving equations with real coefficients.

WORKED EXAMPLE 9 Solving linear equations involving complex numbers

Solve the following equation for z: $\dfrac{3(z+2)}{z+2i} = 5 - 2i$.

THINK	WRITE
1. Multiply both sides by the expression in the denominator.	$\dfrac{3(z+2)}{z+2i} = 5-2i$ $3(z+2) = (5-2i)(z+2i)$
2. Expand the brackets on both sides of the equation.	$3z+6 = 5z+10i-2iz-4i^2$
3. Replace i^2 with -1.	$3z+6 = 5z+10i-2iz+4$
4. Collect all terms containing z on one side of the equation.	$2-10i = 2z-2iz$
5. Isolate z by taking out the common factors.	$2(1-5i) = 2z(1-i)$
6. Solve for z.	$z = \dfrac{1-5i}{1-i}$
7. Multiply both the numerator and the denominator by the conjugate of the denominator.	$z = \dfrac{1-5i}{1-i} \times \dfrac{1+i}{1+i}$
8. Expand the numerator and denominator.	$z = \dfrac{1+i-5i-5i^2}{1-i^2}$
9. Replace i^2 with -1.	$z = \dfrac{6-4i}{2}$
10. Express the complex number in $x+yi$ form.	$z = 3-2i$

TI | THINK

On a Calculator page type 'cSolve', place the equation in brackets followed by a comma and the variable being solved for. Press 'enter'.

DISPLAY/WRITE

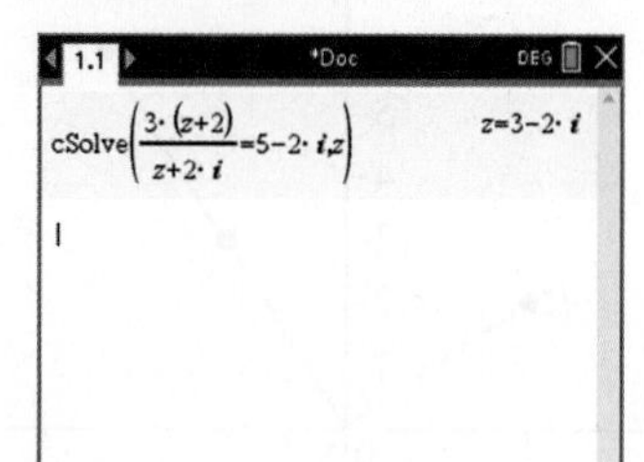

CASIO | THINK

On a Main screen type 'solve', place the equation in brackets followed by a comma and the variable being solved for. Press 'EXE'.

DISPLAY/WRITE

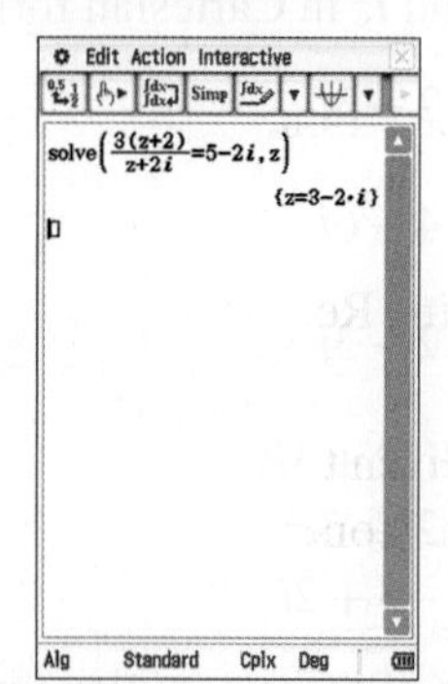

2.2 Exercise

Students, these questions are even better in jacPLUS

Receive immediate feedback and access sample responses

Access additional questions

Track your results and progress

Find all this and MORE in jacPLUS

Technology free

1. **WE1** Evaluate each of the following.

   a. $\text{Re}(8-9i)$
   b. $\text{Im}\left(12-i^2+2i^5\right)$
   c. $\text{Im}(17-i)$
   d. $\text{Im}\left(7-3i^2+i^4-2i\right)$

2. **WE2** Given the complex numbers $u=3-i$ and $v=4-3i$, calculate the complex numbers:

   a. $u+v$
   b. $u-v$
   c. $3u-2v$

3. **WE3** Given the complex numbers $u=3-i$ and $v=4-3i$, calculate the complex numbers:

   a. uv
   b. v^2

4. Given the complex numbers $u=1+3i$ and $v=3+4i$, calculate the complex numbers:

   a. $(u-v)^2$
   b. $(3u-2v)^2$

5. **WE4** Given the complex numbers $u=3-i$ and $v=3+i$, determine:

   a. $\text{Re}(uv)$
   b. $\text{Im}(uv)$

6. **WE5** Given the complex numbers $u=3-i$ and $v=4-3i$, calculate the complex numbers:

   a. u^{-1}
   b. $\frac{u}{v}$.

7. Given the complex numbers $u=1+3i$ and $v=3+4i$, evaluate the complex number $\frac{u+v}{u-v}$.

8. **WE6** Determine the values of x and y if $4x-2iy+3ix-4y=-6-i$.

9. Determine the values of x and y if $(x+yi)(3-2i)=6-i$.

10. **WE7** Determine $\text{Im}\left(\frac{25}{4-3i}+i^{77}\right)$.

11. Determine $\text{Re}\left(\frac{10}{1+3i}+i^{96}\right)$.

12. **WE8** Given the complex number $z=-2-i$, represent the complex numbers z, $3z$, $\bar{z}$ and iz on one Argand diagram. Comment on their relative positions.

13. Given the complex numbers $u=1-2i$ and $v=2+i$, represent the complex numbers $u+v$ and $u-v$ on one Argand diagram.

14. **WE9** Determine the complex number z if $\frac{z-i}{z+i}=2+i$.

15. Determine the complex number z if $\frac{5(z+2i)}{z-2}=11-2i$.

16. a. Draw an Argand diagram and mark on it the complex numbers: $u = 2 + i$, $v = 1 + 2i$ and $w = -2i$.
    b. On the same diagram mark the points: iu, iv and iw.
    c. Describe, geometrically, the relationship between the complex number z and iz on the Argand diagram.
17. a. Draw an Argand diagram and mark on it the complex numbers: $u = 2 + i$, $v = 1 + 2i$ and $w = -2i$.
    b. On the same diagram mark the points: $-u$, $-v$ and $-w$.
    c. Describe, geometrically, the relationship between the complex numbers z and $-z$ on the Argand diagram.
18. a. Draw an Argand diagram and mark on it the numbers: $u = 2 + i$, $v = 1 + 2i$ and $w = -2i$.
    b. On the same diagram mark the points: $-iu$, $-iv$ and $-iw$.
    c. Describe, geometrically, the relationship between the complex numbers z and $-iz$ on the Argand diagram.
19. Let $z = x + yi$. Show that:
    a. $z + \bar{z} = 2\,\text{Re}(z)$
    b. $z - \bar{z} = 2i \times \text{Im}(z)$
20. Let $z_1 = a + bi$ and $z_2 = c + di$. Show that:
    a. $\bar{z}_1 + \bar{z}_2 = \overline{z_1 + z_2}$
    b. $\bar{z}_1 \times \bar{z}_2 = \overline{z_1 \times z_2}$

2.2 Exam questions

Question 1 (1 mark) TECH-ACTIVE

Source: *VCE 2020 Specialist Mathematics Exam 2, Section A, Q5; © VCAA.*

MC Given the complex number $z = a + bi$, where $a \in R\backslash\{0\}$ and $b \in R$, $\dfrac{4z\bar{z}}{(z+\bar{z})^2}$ is equivalent to

A. $1 + \left(\dfrac{\text{Im}(z)}{\text{Re}(z)}\right)^2$

B. $4\,[\text{Re}(z) \times \text{Im}(z)]$

C. $4\left([\text{Re}(z)]^2 \times [\text{Im}(z)]^2\right)$

D. $4\left[1 + (\text{Re}(z) + \text{Im}(z))^2\right]$

E. $\dfrac{2 \times \text{Im}(z)}{[\text{Re}(z)]^2}$

Question 2 (1 mark) TECH-ACTIVE

Source: *VCE 2019 Specialist Mathematics Exam 2, Section A, Q4; © VCAA.*

MC The expression $i^{1!} + i^{2!} + i^{3!} + \ldots + i^{100!}$ is equal to

A. 0
B. 96
C. $95 + i$
D. $94 + 2i$
E. $98 + 2i$

Question 3 (1 mark) TECH-ACTIVE

MC $\sum_{n=1}^{100} ni^n$ is equal to

A. $50 - 50i$
B. 50
C. $-50i$
D. 100
E. 5050

More exam questions are available online.

2.3 Complex numbers in polar form

LEARNING INTENTION

At the end of this subtopic you should be able to:
- convert between the polar and Cartesian forms of complex numbers
- perform operations on complex numbers in polar form.

2.3.1 Review of complex numbers in polar form

In the previous subtopic, we expressed complex numbers in Cartesian form. Another way in which complex numbers can be represented is **polar form**. This form has two parts: the modulus and the argument.

To demonstrate this, consider the complex number $z = x + yi$ as pictured in the Argand diagram.

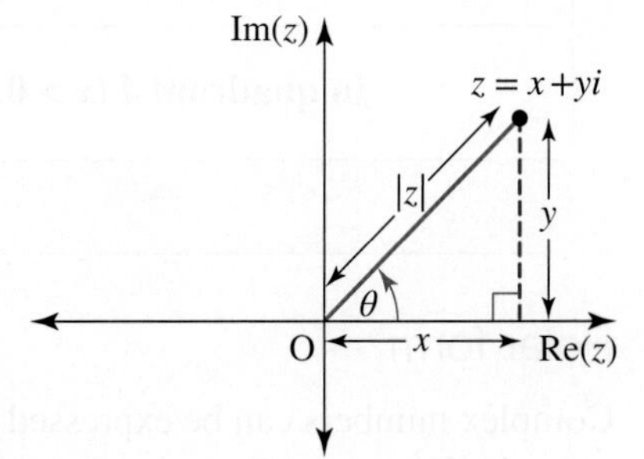

The length, magnitude or modulus of the complex number is the distance of the complex number from the origin. This distance is represented by r or $|z|$. Note that this distance is always a positive real number.

The modulus of a complex number

For a complex number $z = x + yi$, the modulus of z is defined as:

$$r = |z| = \sqrt{x^2 + y^2}$$

The angle θ that the line segment makes with the positive real axis is called the **argument of** z, and is denoted by $\arg(z)$.

Usually this angle is given in radians, as multiples of π, although it can also be given in degrees. Angles measured anticlockwise from the positive real axis are positive angles, and angles measured clockwise from the positive real axis are negative angles. Because any integer multiple of 2π radians (or 360°) can be added or subtracted to any angle to get the same result on the Argand plane, all complex numbers can be represented using an infinite number of arguments.

To overcome this problem, we often refer to the **principal value** of the argument of z. The principal value is defined to be the angle $\text{Arg}(z)$ where $\text{Arg}(z) \in (-\pi, \pi]$. (Note the use of the capital letter A to denote the principal value.)

Recall that $x = r\cos(\theta)$ and $y = r\sin(\theta)$. Therefore:

$$\frac{y}{x} = \frac{r\sin(\theta)}{r\cos(\theta)}$$

$$\tan(\theta) = \frac{y}{x}$$

$$\theta = \text{Arg}(z) = \tan^{-1}\left(\frac{y}{x}\right)$$

Note that since the range of the inverse tan function is $\left(-\frac{\pi}{2}, \frac{\pi}{2}\right)$ this result gives angles in the first and fourth quadrants only. If z lies in quadrant 2, $\text{Arg}(z) = \tan^{-1}\left(\frac{y}{x}\right) + \pi$ and if z lies in quadrant 3, $\text{Arg}(z) = \tan^{-1}\left(\frac{y}{x}\right) - \pi$.

Principal argument of a complex number $z = x + yi$

Location of z	Principal argument
in quadrant 1 ($x > 0,\ y > 0$)	$\text{Arg}(z) = \tan^{-1}\left(\frac{y}{x}\right)$
in quadrant 2 ($x < 0,\ y > 0$)	$\text{Arg}(z) = \tan^{-1}\left(\frac{y}{x}\right) + \pi$
in quadrant 3 ($x < 0,\ y < 0$)	$\text{Arg}(z) = \tan^{-1}\left(\frac{y}{x}\right) - \pi$
in quadrant 4 ($x > 0,\ y < 0$)	$\text{Arg}(z) = \tan^{-1}\left(\frac{y}{x}\right)$

Polar form

Complex numbers can be expressed using the modulus and the argument. By substituting $x = r\cos(\theta)$ and $y = r\sin(\theta)$ into the Cartesian form $z = x + yi$ we get the polar form which is:

$$\begin{aligned} z &= r\cos(\theta) + ir\sin(\theta) \\ &= r\big(\cos(\theta) + i\sin(\theta)\big) \\ &= r\,\text{cis}(\theta) \end{aligned}$$

where 'cis' is just a shorthand for '$\cos + i\sin$'. Note that the polar form of a complex number can also be written as $z = |z|\,\text{cis}(\theta)$, since $r = |z|$.

Polar form of a complex number

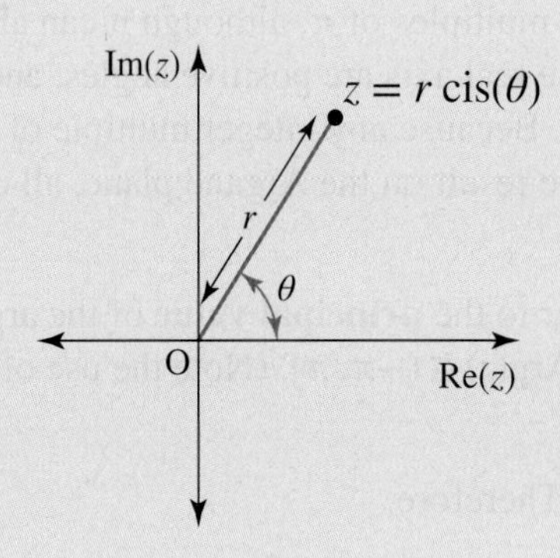

2.3.2 Converting between polar and Cartesian form

Cartesian form to polar form

A complex number in Cartesian form, $z = x + yi$, can be converted into polar form by determining the corresponding values of r and θ.

WORKED EXAMPLE 10 Converting from Cartesian form to polar form

Convert each of the following complex numbers to polar form.

a. $1+i$ b. $-\sqrt{3}+i$ c. $-4-4i$ d. $1-\sqrt{3}i$ e. -5 f. $3i$

THINK	WRITE
a. 1. Draw the complex number on an Argand diagram.	a. $z = 1 + i$ This complex number is in the first quadrant.

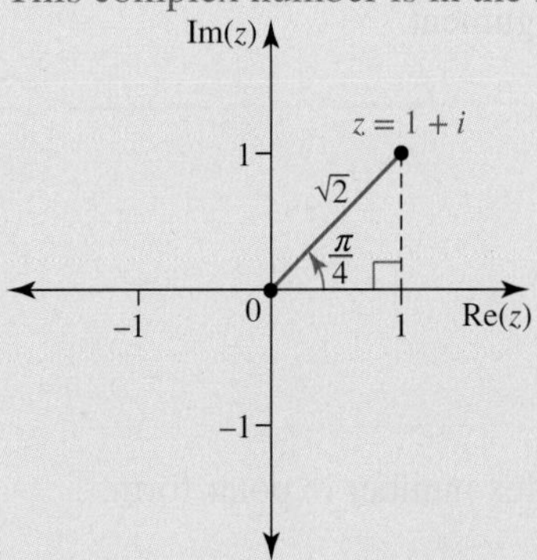

THINK	WRITE
2. Identify the real and imaginary parts.	$x = \text{Re}(z) = 1$ and $y = \text{Im}(z) = 1$
3. Evaluate the modulus.	$\lvert z \rvert = \sqrt{x^2+y^2} = \sqrt{1^2+1^2} = \sqrt{2}$
4. Evaluate the argument.	$\theta = \text{Arg}(z) = \tan^{-1}\left(\frac{y}{x}\right) = \tan^{-1}(1) = \frac{\pi}{4}$
5. State the complex number in polar form.	$z = 1+i = \sqrt{2}\,\text{cis}\left(\frac{\pi}{4}\right)$
b. 1. Draw the complex number on an Argand diagram.	b. $z = -\sqrt{3}+i$ This complex number is in the second quadrant.

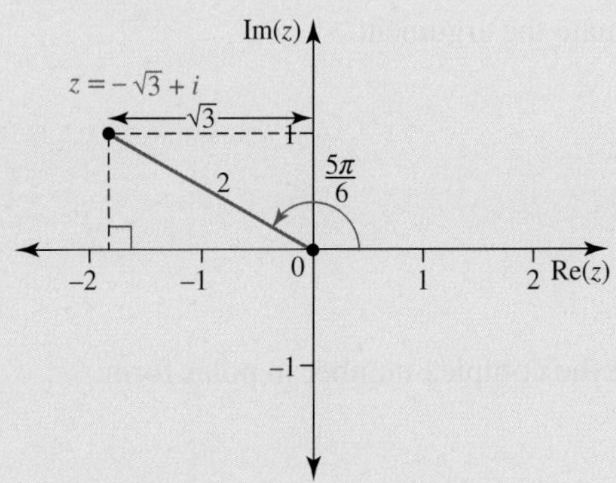

2. Identify the real and imaginary parts. $x = \text{Re}(z) = -\sqrt{3}$ and $y = \text{Im}(z) = 1$

3. Evaluate the modulus.

$$|z| = \sqrt{x^2 + y^2}$$
$$= \sqrt{\left(-\sqrt{3}\right)^2 + 1^2}$$
$$= 2$$

4. Evaluate the argument.

$$\theta = \text{Arg}(z)$$
$$= \pi + \tan^{-1}\left(-\frac{1}{\sqrt{3}}\right)$$
$$= \pi - \frac{\pi}{6}$$
$$= \frac{5\pi}{6}$$

5. State the complex number in polar form. $z = 2\text{cis}\left(\frac{5\pi}{6}\right)$

c. 1. Draw the complex number on an Argand diagram.

c. $z = -4 - 4i$

This complex number is in the third quadrant.

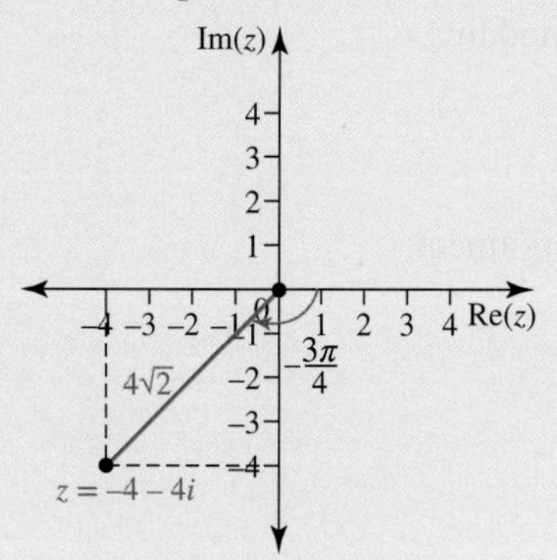

2. Identify the real and imaginary parts. $x = \text{Re}(z) = -4$ and $y = \text{Im}(z) = -4$

3. Evaluate the modulus.

$$|z| = \sqrt{x^2 + y^2}$$
$$= \sqrt{(-4)^2 + (-4)^2}$$
$$= 4\sqrt{2}$$

4. Evaluate the argument.

$$\theta = \text{Arg}(z)$$
$$= -\pi + \tan^{-1}(1)$$
$$= -\pi + \frac{\pi}{4}$$
$$= -\frac{3\pi}{4}$$

5. State the complex number in polar form. $z = 4\sqrt{2}\,\text{cis}\left(-\frac{3\pi}{4}\right)$

d. 1. Draw the complex number on an Argand diagram.

d. $z = 1 - \sqrt{3}i$

This complex number is in the fourth quadrant.

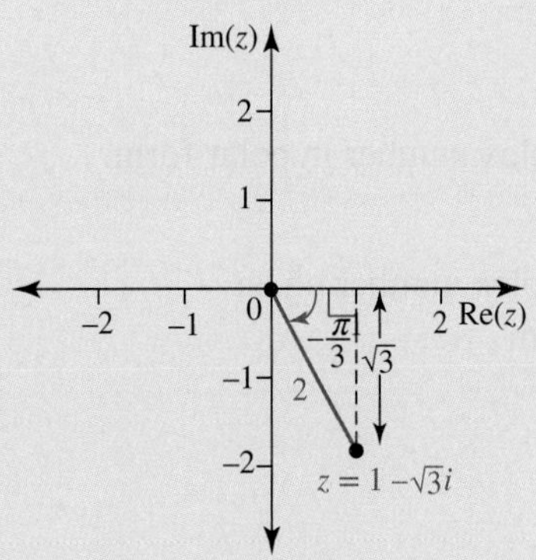

2. Identify the real and imaginary parts.

$x = \text{Re}(z) = 1$ and $y = \text{Im}(z) = -\sqrt{3}$

3. Evaluate the modulus.

$$\begin{aligned}|z| &= \sqrt{x^2 + y^2}\\ &= \sqrt{1^2 + (-\sqrt{3})^2}\\ &= 2\end{aligned}$$

4. Evaluate the argument.

$$\begin{aligned}\theta &= \text{Arg}(z)\\ &= \tan^{-1}\left(\frac{y}{x}\right)\\ &= \tan^{-1}(-\sqrt{3})\\ &= -\frac{\pi}{3}\end{aligned}$$

5. State the complex number in polar form.

$$\begin{aligned}z &= 1 - \sqrt{3}i\\ &= 2\,\text{cis}\left(-\frac{\pi}{3}\right)\end{aligned}$$

e. 1. Draw the complex number on an Argand diagram.

e. $z = -5$

This complex number is actually a real number and lies on the real axis.

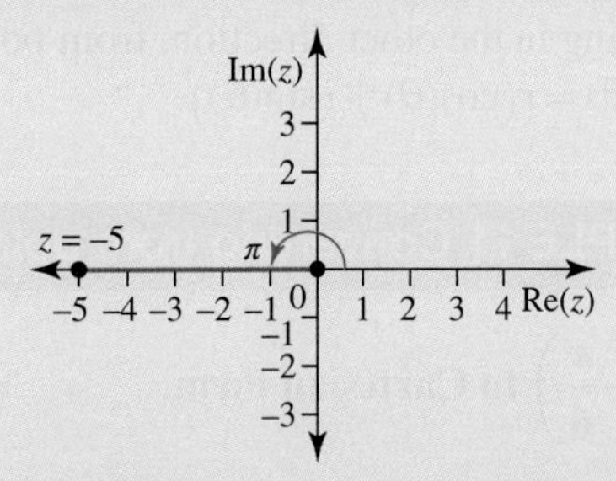

2. Identify the real and imaginary parts.

$x = \text{Re}(z) = -5$ and $y = \text{Im}(z) = 0$

3. Evaluate the modulus.

$$\begin{aligned}|z| &= \sqrt{x^2 + y^2}\\ &= \sqrt{(-5)^2 + 0^2}\\ &= 5\end{aligned}$$

4. Evaluate the argument.	$\theta = \text{Arg}(z)$ $= \pi$ Note that $\theta = \text{Arg}(z) = -\pi$ is not correct, since $-\pi < \text{Arg}(z) \le \pi$.
5. State the complex number in polar form.	$z = -5$ $= 5\,\text{cis}(\pi)$
f. 1. Draw the complex number on an Argand diagram.	**f.** $z = 3i$ This complex number lies on the imaginary axis.

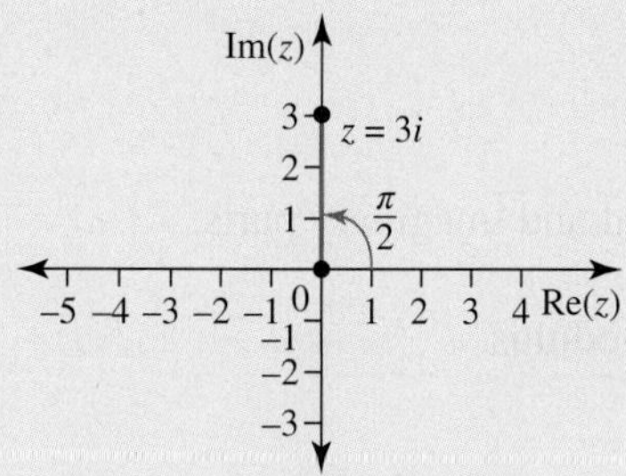

2. Identify the real and imaginary parts.	$x = \text{Re}(z) = 0$ and $y = \text{Im}(z) = 3$
3. Evaluate the modulus.	$\|z\| = \sqrt{x^2 + y^2}$ $= \sqrt{0^2 + 3^2}$ $= 3$
4. Evaluate the argument.	$\theta = \dfrac{\pi}{2}$
5. State the complex number in polar form.	$z = 3i$ $= 3\,\text{cis}\left(\dfrac{\pi}{2}\right)$

Polar form to Cartesian form

Now consider converting in the other direction, from polar form to Cartesian form. To determine the values of x and y recall that $r\,\text{cis}(\theta) = r\big(\cos(\theta) + i\sin(\theta)\big)$.

WORKED EXAMPLE 11 Converting from polar form to Cartesian form

a. Convert $8\,\text{cis}\left(-\dfrac{\pi}{6}\right)$ to Cartesian form. **b. Convert $16\,\text{cis}\left(\dfrac{2\pi}{3}\right)$ to Cartesian form.**

THINK	**WRITE**
a. 1. Expand.	**a.** $8\,\text{cis}\left(-\dfrac{\pi}{6}\right) = 8\left(\cos\left(-\dfrac{\pi}{6}\right) + i\sin\left(-\dfrac{\pi}{6}\right)\right)$
2. Use trigonometric results for functions of negative angles.	$\cos(-\theta) = \cos(\theta)$ $\sin(-\theta) = -\sin(\theta)$ $8\,\text{cis}\left(-\dfrac{\pi}{6}\right) = 8\left(\cos\left(\dfrac{\pi}{6}\right) - i\sin\left(\dfrac{\pi}{6}\right)\right)$

3. Substitute for the exact trigonometric values. Note that the complex number is in the fourth quadrant.

$$=8\left(\frac{\sqrt{3}}{2}-i\times\frac{1}{2}\right)$$

4. Simplify and write in $x+yi$ form.

$$=4\sqrt{3}-4i$$

b. 1. Expand.

b. $$16\operatorname{cis}\left(\frac{2\pi}{3}\right)=16\left(\cos\left(\frac{2\pi}{3}\right)+i\sin\left(\frac{2\pi}{3}\right)\right)$$

2. Substitute for the exact trigonometric values. Note that the complex number is in the second quadrant.

$$=16\left(-\frac{1}{2}+i\times\frac{\sqrt{3}}{2}\right)$$

3. Simplify and write in $x+yi$ form.

$$=-8+8\sqrt{3}i$$

2.3.3 Complex conjugates and multiplicative inverses in polar form

Conjugates in polar form

If $z=x+iy=r\operatorname{cis}(\theta)$, then the conjugate of z is given by $\bar{z}=x-iy=r\operatorname{cis}(-\theta)$.

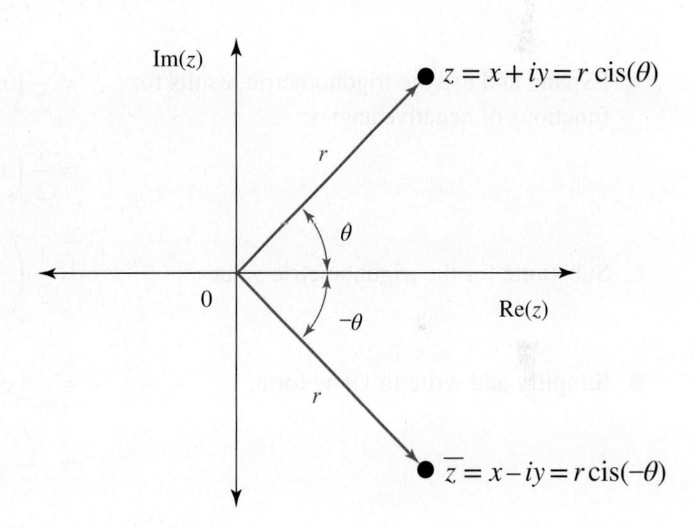

Multiplicative inverses in polar form

If $z=x+iy=r\operatorname{cis}(\theta)$, then the multiplicative inverse or reciprocal of z is given by

$$\begin{aligned} z^{-1} &= \frac{1}{z} \\ &= \frac{1}{x+yi} \\ &= \frac{x-yi}{x^2+y^2} \\ &= \frac{1}{r}\operatorname{cis}(-\theta) \end{aligned}$$

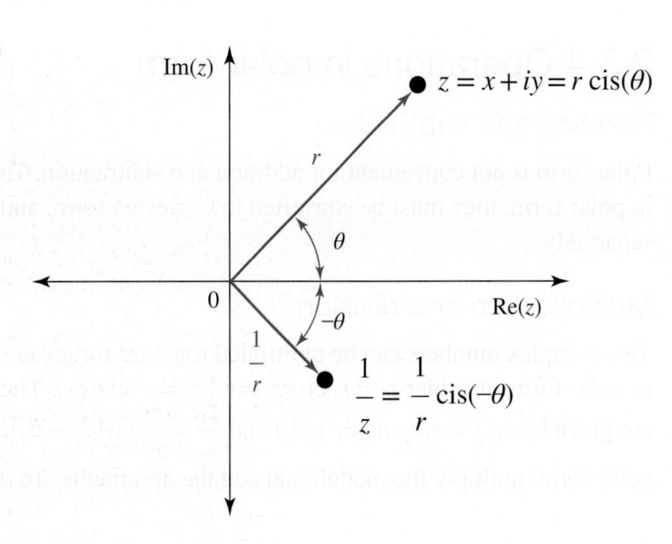

WORKED EXAMPLE 12 Evaluating conjugates and inverses in polar form

If $u = 2\,\text{cis}\left(-\frac{\pi}{6}\right)$, evaluate $\overline{u}^{-1}$ giving your answer in Cartesian form.

THINK	WRITE
1. Use the conjugate rule.	$u = 2\,\text{cis}\left(-\frac{\pi}{6}\right)$ $\overline{u} = 2\,\text{cis}\left(\frac{\pi}{6}\right)$
2. Determine the multiplicative inverse.	$\overline{u}^{-1} = \frac{1}{\overline{u}}$ $= \frac{1}{2\,\text{cis}\left(\frac{\pi}{6}\right)}$
3. Use the results.	$= \frac{1}{2}\,\text{cis}\left(-\frac{\pi}{6}\right)$
4. Expand and use the trigonometric results for functions of negative angles.	$= \frac{1}{2}\left(\cos\left(-\frac{\pi}{6}\right) + i\sin\left(-\frac{\pi}{6}\right)\right)$ $= \frac{1}{2}\left(\cos\left(\frac{\pi}{6}\right) - i\sin\left(\frac{\pi}{6}\right)\right)$
5. Substitute for the trigonometric values.	$= \frac{1}{2}\left(\frac{\sqrt{3}}{2} - i\frac{1}{2}\right)$
6. Simplify and write in $x + yi$ form.	$= \frac{1}{4}\left(\sqrt{3} - i\right)$ $= \frac{\sqrt{3}}{4} - \frac{1}{4}i$

2.3.4 Operations in polar form

Addition and subtraction

Polar form is not convenient for addition and subtraction. To add or subtract complex numbers that are given in polar form, they must be converted to Cartesian form, and the real and imaginary parts added and subtracted separately.

Multiplication and division

Two complex numbers can be multiplied together if they are both given in Cartesian form. If they are both given in polar form, consider $z_1 = r_1\,\text{cis}(\theta_1)$ and $z_2 = r_2\,\text{cis}(\theta_2)$. The rules for multiplication and division in polar form are given by $z_1 z_2 = r_1 r_2\,\text{cis}(\theta_1 + \theta_2)$ and $\frac{z_1}{z_2} = \frac{r_1}{r_2}\,\text{cis}(\theta_1 - \theta_2)$. Therefore, to multiply two complex numbers in polar form, multiply the moduli and add the arguments. To divide two complex numbers in polar form, divide the moduli and subtract the arguments. The proof of these results is as follows.

$$\begin{aligned}
z_1z_2 &= \left(r_1\operatorname{cis}(\theta_1)\right)\left(r_2\operatorname{cis}(\theta_2)\right)\\
&= r_1r_2\left(\cos(\theta_1)+i\sin(\theta_1)\right)\left(\cos(\theta_2)+i\sin(\theta_2)\right)\\
&= r_1r_2\left(\left(\left(\cos(\theta_1)\cos(\theta_2)-\sin(\theta_1)\sin(\theta_2)\right)+i\left(\sin(\theta_1)\cos(\theta_2)+\sin(\theta_2)\cos(\theta_1)\right)\right)\right)\\
&= r_1r_2\left(\cos(\theta_1+\theta_2)+i\sin(\theta_1+\theta_2)\right) \quad \text{(by compound-angle formulas)}\\
&= r_1r_2\operatorname{cis}(\theta_1+\theta_2)
\end{aligned}$$

and

$$\begin{aligned}
\frac{z_1}{z_2} &= z_1z_2^{-1}\\
&= r_1\operatorname{cis}(\theta_1)\times\frac{1}{r_2\operatorname{cis}(\theta_2)}\\
&= \frac{r_1}{r_2}\operatorname{cis}(\theta_1)\operatorname{cis}(-\theta_2)\\
&= \frac{r_1}{r_2}\operatorname{cis}(\theta_1-\theta_2)
\end{aligned}$$

The diagrams demonstrate this geometrically.

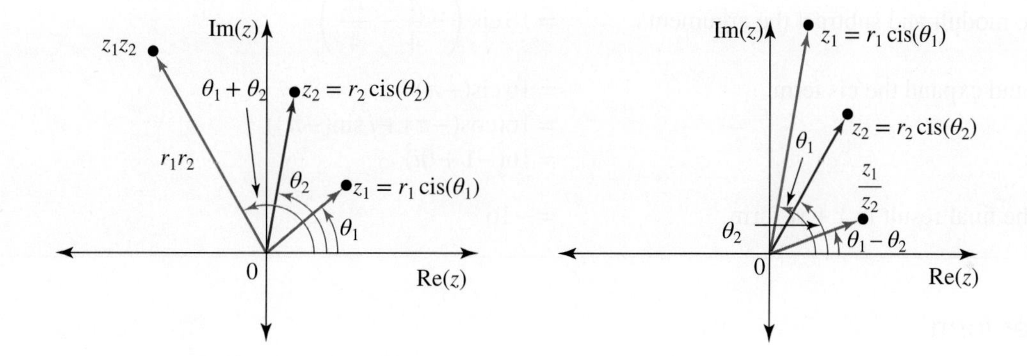

Note that if two complex numbers are given with one in polar form and one in Cartesian form, they cannot be multiplied or divided until they are both in the same form.

Multiplication and division in polar form

If $z_1 = r_1\operatorname{cis}(\theta_1)$ and $z_2 = r_2\operatorname{cis}(\theta_2)$, then:

$$z_1\times z_2 = r_1r_2\operatorname{cis}(\theta_1+\theta_2)$$

and

$$\frac{z_1}{z_2} = \frac{r_1}{r_2}\operatorname{cis}(\theta_1-\theta_2)$$

WORKED EXAMPLE 13 Multiplying and dividing complex numbers in polar form

If $u = 8\,\text{cis}\left(-\frac{\pi}{4}\right)$ and $v = \frac{1}{2}\text{cis}\left(\frac{3\pi}{4}\right)$, evaluate each of the following, giving your answers in Cartesian form.

a. uv

b. $\frac{u}{v}$

THINK	WRITE
a. 1. Substitute for u and v.	a. $uv = 8\,\text{cis}\left(-\frac{\pi}{4}\right) \times \frac{1}{2}\text{cis}\left(\frac{3\pi}{4}\right)$
2. Multiply the moduli and add the arguments.	$= 8 \times \frac{1}{2}\,\text{cis}\left(-\frac{\pi}{4} + \frac{3\pi}{4}\right)$
3. Simplify.	$= 4\,\text{cis}\left(\frac{\pi}{2}\right)$
4. Express in $x + yi$ form.	$= 4i$
b. 1. Substitute for u and v.	b. $\frac{u}{v} = \frac{8\,\text{cis}\left(-\frac{\pi}{4}\right)}{\frac{1}{2}\text{cis}\left(\frac{3\pi}{4}\right)}$
2. Divide the moduli and subtract the arguments.	$= 16\,\text{cis}\left(-\frac{\pi}{4} - \frac{3\pi}{4}\right)$
3. Simplify and expand the cis term.	$= 16\,\text{cis}(-\pi)$ $= 16(\cos(-\pi) + i\sin(-\pi))$ $= 16(-1 + 0i)$
4. Express the final result in $x + yi$ form.	$= -16$

Powers in polar form

If $z = r\,\text{cis}(\theta)$, then:

$$z^2 = r\,\text{cis}(\theta) \times r\,\text{cis}(\theta)$$
$$= r^2\,\text{cis}(2\theta)$$
$$z^3 = r^3\,\text{cis}(3\theta) \ldots$$

In general, $z^n = r^n\,\text{cis}(n\theta)$, for $n \in N$

This result can be proved using mathematical induction and is known as **de Moivre's theorem**.

De Moivre's theorem

If $z = r\,\text{cis}(\theta)$, then:

$$z^n = r^n\,\text{cis}(n\theta),\ n \in N$$

WORKED EXAMPLE 14 Applying de Moivre's theorem

If $u = -\sqrt{3} + i$, determine:

a. **$\text{Arg}(u^{12})$**

b. **u^{12}, giving your answer in Cartesian form.**

THINK	WRITE
a. 1. Convert to polar form.	a. $u = -\sqrt{3} + i$ $= 2\,\text{cis}\left(\frac{5\pi}{6}\right)$
2. Use de Moivre's theorem.	$u^{12} = 2^{12}\text{cis}\left(12 \times \frac{5\pi}{6}\right)$
3. Simplify.	$u^{12} = 4096\,\text{cis}(10\pi)$
4. $-\pi < \text{Arg}(z) \le \pi$ and is unique, but $\arg(z)$ is not unique.	$\arg(u^{12}) = 10\pi$ but $\text{Arg}(u^{12}) \ne 10\pi$
5. Add or subtract an appropriate multiple of 2π to the angle.	$\text{Arg}(u^{12}) = 10\pi - 10\pi$ $= 0$
6. State the answer.	$\text{Arg}(u^{12}) = 0$
b. 1. Expand the cis term.	b. $u^{12} = 4096\,\text{cis}(0)$ $= 4096(\cos(0) + i\sin(0))$ $= 4096(1 + 0i)$
2. State the answer.	$u^{12} = 4096$

2.3.5 Applications of complex numbers in polar form

Using trigonometric compound-angle formulas

Recall the trigonometric compound-angle formulas from Unit 2.

Compound-angle formulas

$$\sin(A+B) = \sin(A)\cos(B) + \cos(A)\sin(B)$$

$$\sin(A-B) = \sin(A)\cos(B) - \cos(A)\sin(B)$$

$$\cos(A+B) = \cos(A)\cos(B) - \sin(A)\sin(B)$$

$$\cos(A-B) = \cos(A)\cos(B) + \sin(A)\sin(B)$$

$$\tan(A+B) = \frac{\tan(A) + \tan(B)}{1 - \tan(A)\tan(B)}$$

$$\tan(A-B) = \frac{\tan(A) - \tan(B)}{1 + \tan(A)\tan(B)}$$

These identities can be used in problems involving complex numbers to obtain or check certain results.

WORKED EXAMPLE 15 Applying compound-angle formulas to complex numbers (1)

a. **Show that $\tan\left(\frac{\pi}{12}\right) = 2 - \sqrt{3}$.**

b. **Given $u = 1 + \left(2 - \sqrt{3}\right)i$, determine the value of iu and hence evaluate $\text{Arg}\left(\sqrt{3} - 2 + i\right)$.**

THINK	WRITE
a. 1. Rewrite the argument as a sum or difference of fractions.	a. $\frac{\pi}{4} - \frac{\pi}{6} = \frac{\pi}{12}$, or in degrees, $45° - 30° = 15°$. $\tan\left(\frac{\pi}{12}\right) = \tan\left(\frac{\pi}{4} - \frac{\pi}{6}\right)$
2. State and use an appropriate identity. $\tan(A - B) = \frac{\tan(A) - \tan(B)}{1 + \tan(A)\tan(B)}$	Let $A = \frac{\pi}{4}$ and $B = \frac{\pi}{6}$. $\tan\left(\frac{\pi}{4} - \frac{\pi}{6}\right) = \frac{\tan\left(\frac{\pi}{4}\right) - \tan\left(\frac{\pi}{6}\right)}{1 + \tan\left(\frac{\pi}{4}\right)\tan\left(\frac{\pi}{6}\right)}$
3. Simplify and use exact values.	Substitute $\tan\left(\frac{\pi}{4}\right) = 1$ and $\tan\left(\frac{\pi}{6}\right) = \frac{\sqrt{3}}{3}$: $\tan\left(\frac{\pi}{12}\right) = \frac{1 - \frac{\sqrt{3}}{3}}{1 + \frac{\sqrt{3}}{3}}$
4. Form common denominators in both the numerator and denominator, and cancel the factors.	$\tan\left(\frac{\pi}{12}\right) = \frac{\frac{3-\sqrt{3}}{3}}{\frac{3+\sqrt{3}}{3}}$ $= \frac{3 - \sqrt{3}}{3 + \sqrt{3}}$
5. To rationalise, multiply both the numerator and denominator by the conjugate surd in the denominator.	$\tan\left(\frac{\pi}{12}\right) = \frac{3 - \sqrt{3}}{3 + \sqrt{3}} \times \frac{3 - \sqrt{3}}{3 - \sqrt{3}}$
6. Expand and simplify.	$\tan\left(\frac{\pi}{12}\right) = \frac{9 - 6\sqrt{3} + 3}{9 - 3}$ $= \frac{12 - 6\sqrt{3}}{6}$ $= \frac{6(2 - \sqrt{3})}{6}$
7. Simplify and state the final answer.	$\tan\left(\frac{\pi}{12}\right) = 2 - \sqrt{3}$
b. 1. State the complex number and its argument, as it is in the first quadrant.	b. $u = 1 + \left(2 - \sqrt{3}\right)i$ $\text{Arg}(u) = \tan^{-1}\left(2 - \sqrt{3}\right)$ $= \frac{\pi}{12}$

2. Determine the value of the complex number iu, which is in the second quadrant. The complex number iu is a rotation of u by 90° anticlockwise.

$$iu = i + \left(2-\sqrt{3}\right)i^2$$
$$= \sqrt{3}-2+i$$
$$\text{Arg}(iu) = \frac{\pi}{12}+\frac{\pi}{2}$$

3. State the final result.

$$\text{Arg}\left(\sqrt{3}-2+i\right) = \frac{7\pi}{12}$$

Applications of de Moivre's theorem

De Moivre's theorem can be used to solve problems involving powers of complex numbers.

WORKED EXAMPLE 16 Applying compound-angle formulas to complex numbers (2)

Determine all values of n such that $\left(-\sqrt{3}+i\right)^n + \left(-\sqrt{3}-i\right)^n = 0$.

THINK	WRITE
1. Express the complex number $-\sqrt{3}+i$ in polar form.	$-\sqrt{3}+i = 2\,\text{cis}\left(\frac{5\pi}{6}\right)$
2. The complex number $-\sqrt{3}-i$ is the conjugate. Express $-\sqrt{3}-i$ in polar form.	$-\sqrt{3}-i = 2\,\text{cis}\left(-\frac{5\pi}{6}\right)$
3. Express the equation in polar form.	$\left(-\sqrt{3}+i\right)^n + \left(-\sqrt{3}-i\right)^n = 0$ $\left(2\,\text{cis}\left(\frac{5\pi}{6}\right)\right)^n + \left(2\,\text{cis}\left(-\frac{5\pi}{6}\right)\right)^n = 0$
4. Use de Moivre's theorem.	$2^n\,\text{cis}\left(\frac{5\pi n}{6}\right) + 2^n\,\text{cis}\left(-\frac{5\pi n}{6}\right) = 0$
5. Take out the common factor, expand cis(θ), and apply the Null Factor Theorem.	$2^n\left(\text{cis}\left(\frac{5\pi n}{6}\right) + \text{cis}\left(-\frac{5\pi n}{6}\right)\right) = 0$ $\cos\left(\frac{5\pi n}{6}\right) + i\sin\left(\frac{5\pi n}{6}\right) + \cos\left(-\frac{5\pi n}{6}\right) + i\sin\left(-\frac{5\pi n}{6}\right) = 0$
6. Use the trigonometric results for functions of negative angles and simplify.	Since $\cos(-\theta) = \cos(\theta)$ and $\sin(-\theta) = -\sin(\theta)$, $2\cos\left(\frac{5\pi n}{6}\right) = 0$
7. Use the formula for the general solutions of trigonometric equations.	$\cos\left(\frac{5\pi n}{6}\right) = 0$ $\frac{5\pi n}{6} = \frac{(2k+1)\pi}{2}$ where $k \in Z$
8. Solve for n and state the final answer.	$n = \frac{3(2k+1)}{5}$ where $k \in Z$

2.3 Exercise

Students, these questions are even better in jacPLUS

Receive immediate feedback and access sample responses

Access additional questions

Track your results and progress

Find all this and MORE in jacPLUS

Technology free

1. **WE10** Convert each of the following complex numbers to polar form.

   a. $1+\sqrt{3}i$ b. $-1+i$ c. $-2-2\sqrt{3}i$ d. $\sqrt{3}-i$ e. 4 f. $-2i$

2. Convert each of the following complex numbers to polar form.

   a. $\sqrt{3}+i$ b. $-1+\sqrt{3}i$ c. $-\sqrt{3}-i$ d. $2-2i$ e. -7 f. $5i$

3. **WE11** a. Convert $4\operatorname{cis}\left(-\frac{\pi}{3}\right)$ to Cartesian form.

   b. Convert $8\operatorname{cis}\left(-\frac{\pi}{2}\right)$ to Cartesian form.

4. Convert $6\sqrt{2}\operatorname{cis}(-135°)$ into $x+yi$ form.

5. **WE12** If $u=6\operatorname{cis}\left(-\frac{\pi}{3}\right)$, evaluate $\bar{u}^{-1}$, giving your answer in Cartesian form.

6. If $u=\frac{\sqrt{2}}{4}\operatorname{cis}\left(\frac{3\pi}{4}\right)$, evaluate $\frac{1}{u}$, giving your answer in Cartesian form.

7. **WE13** If $u=4\sqrt{2}\operatorname{cis}\left(-\frac{3\pi}{4}\right)$ and $v=\sqrt{2}\operatorname{cis}\left(\frac{\pi}{4}\right)$, evaluate each of the following, giving your answers in Cartesian form.

   a. uv b. $\frac{u}{v}$

8. If $u=4\operatorname{cis}\left(\frac{\pi}{3}\right)$ and $v=\frac{1}{2}\operatorname{cis}\left(-\frac{2\pi}{3}\right)$ evaluate each of the following, giving your answers in Cartesian form.

   a. uv b. $\frac{u}{v}$

9. **WE14** If $u=-1-i$, determine:

   a. $\operatorname{Arg}\left(u^{10}\right)$

   b. u^{10}, giving your answer in Cartesian form.

10. Simplify $\frac{(-1+i)^6}{\left(\sqrt{3}-i\right)^4}$, giving your answer in Cartesian form.

11. **WE15** a. Show that $\tan\left(\frac{5\pi}{12}\right) = \sqrt{3}+2$.

b. Given $u = 1 + \left(\sqrt{3}+2\right)i$, determine iu and hence evaluate $\operatorname{Arg}\left(-\sqrt{3}-2+i\right)$.

12. Show that $\tan\left(\frac{11\pi}{12}\right) = \sqrt{3}-2$ and hence evaluate $\operatorname{Arg}\left(1+\left(\sqrt{3}-2\right)i\right)$.

13. **WE16** Determine all values of n such that $\left(1+\sqrt{3}i\right)^n - \left(1-\sqrt{3}i\right)^n = 0$.

14. Determine all values of n such that $\left(1+\sqrt{3}i\right)^n + \left(1-\sqrt{3}i\right)^n = 0$.

15. a. If $z = 2+2i$, calculate z^8.

b. If $z = -3\sqrt{3}+3i$, calculate z^6.

c. If $z = -\frac{5}{2} - \frac{5\sqrt{3}}{2}i$, calculate z^9.

d. If $z = 2\sqrt{3}-2i$, calculate z^7.

16. Let $u = \frac{1}{2}\left(\sqrt{3}-i\right)$.

a. Evaluate $\overline{u}$, $\frac{1}{u}$ and u^6, giving all answers in Cartesian form.

b. Evaluate $\operatorname{Arg}(\overline{u})$, $\operatorname{Arg}\left(\frac{1}{u}\right)$ and $\operatorname{Arg}(u^6)$.

c. Determine if $\operatorname{Arg}(\overline{u})$ is equal to $-\operatorname{Arg}(u)$.

d. Determine if $\operatorname{Arg}\left(\frac{1}{u}\right)$ is equal to $-\operatorname{Arg}(u)$.

e. Determine if $\operatorname{Arg}(u^6)$ is equal to $6\operatorname{Arg}(u)$.

17. Let $u = -1+\sqrt{3}i$ and $v = -2-2i$.

a. Evaluate $\operatorname{Arg}(u)$.

b. Evaluate $\operatorname{Arg}(v)$.

c. Evaluate $\operatorname{Arg}(uv)$.

d. Evaluate $\operatorname{Arg}\left(\frac{u}{v}\right)$.

e. Determine is $\operatorname{Arg}(uv)$ is equal to $\operatorname{Arg}(u)+\operatorname{Arg}(v)$.

f. Determine is $\operatorname{Arg}\left(\frac{u}{v}\right)$ is equal to $\operatorname{Arg}(u)-\operatorname{Arg}(v)$.

18. Let $u = \sqrt{2}(1-i)$ and $v = 2\operatorname{cis}\left(\frac{2\pi}{3}\right)$.

a. Evaluate uv, working with both numbers in Cartesian form and giving your answer in Cartesian form.

b. Evaluate uv, working with both numbers in polar form and giving your answer in polar form.

c. Hence, deduce the exact value of $\sin\left(\frac{5\pi}{12}\right)$.

d. Using the formula $\sin(x-y)$, verify your exact value for $\sin\left(\frac{5\pi}{12}\right)$.

19. Let $u = -4-4\sqrt{3}i$ and $v = \sqrt{2}\operatorname{cis}\left(-\frac{3\pi}{4}\right)$.

a. Evaluate $\frac{u}{v}$, working with both numbers in Cartesian form and giving your answer in Cartesian form.

b. Evaluate $\frac{u}{v}$, working with both numbers in polar form and giving your answer in polar form.

c. Hence, deduce the exact value of $\cos\left(\frac{\pi}{12}\right)$.

d. Using the formula $\cos(x-y)$, verify your exact value for $\cos\left(\frac{\pi}{12}\right)$.

20. Let $u=-1-\sqrt{3}i$ and $v=\sqrt{2}\,\text{cis}\left(\frac{3\pi}{4}\right)$.

a. Evaluate $\frac{u}{v}$, working with both numbers in Cartesian form and giving your answer in Cartesian form.

b. Evaluate $\frac{u}{v}$, working with both numbers in polar form and giving your answer in polar form.

c. Hence, deduce the exact value of $\cos\left(\frac{7\pi}{12}\right)$.

d. Using the formula $\cos(x-y)$, verify your exact value for $\cos\left(\frac{7\pi}{12}\right)$.

21. a. Show that $\tan\left(\frac{\pi}{8}\right)=\sqrt{2}-1$.

b. Let $u=1+\left(\sqrt{2}-1\right)i$ and hence evaluate $\text{Arg}(u)$.

c. Determine the value of iu and hence evaluate $\text{Arg}\left(\left(1-\sqrt{2}\right)+i\right)$.

22. a. Show that $\tan\left(\frac{7\pi}{12}\right)=-\left(\sqrt{3}+2\right)$.

b. Hence, evaluate $\text{Arg}\left(-1+\left(\sqrt{3}+2\right)i\right)$.

c. Hence, evaluate $\text{Arg}\left(1-\left(\sqrt{3}+2\right)i\right)$.

d. Hence, evaluate $\text{Arg}\left(\left(\sqrt{3}+2\right)+i\right)$.

23. Determine all values of n such that:

a. $(1+i)^n+(1-i)^n=0$

b. $(1+i)^n-(1-i)^n=0$

24. If $z=\text{cis}(\theta)$, show that:

a. $|z+1|=2\cos\left(\frac{\theta}{2}\right)$

b. $\text{Arg}(1+z)=\frac{\theta}{2}$

25. Use polar arithmetic to prove the following identities.

a. $|z_1z_2|=|z_1||z_2|$

b. $\arg(z_1z_2)=\arg(z_1)+\arg(z_2)$

26. Prove the following identities using a polar approach.

a. $\left|\frac{z_1}{z_2}\right|=\frac{|z_1|}{|z_2|}$

b. $\arg\left(\frac{z_1}{z_2}\right)=\arg(z_1)-\arg(z_2)$

27. Prove the following identities using polar arithmetic.

a. $|z^n|=|z|^n$

b. $\arg(z^n)=n\arg(z)$

28. Given $z=r\,\text{cis}(\theta)$, prove the following identities.

a. $\left|\frac{1}{z^n}\right|=\frac{1}{|z|^n},\ z\neq 0$

b. $\arg\left(\frac{1}{z^n}\right)=-n\arg(z),\ z\neq 0$

29. Let $z_1=r_1\text{cis}(\theta_1)$ and $z_2=r_2\text{cis}(\theta_2)$. Show that $\frac{z_1}{z_2}=\frac{r_1}{r_2}\text{cis}(\theta_1-\theta_2)$.

30. Let $z=r\text{cis}(\theta)$. Use mathematical induction to show that $z^n=r^n\text{cis}(n\theta)$ for $n\in Z^+$.

2.3 Exam questions

Question 1 (1 mark) TECH-ACTIVE

Source: VCE 2021 Specialist Mathematics Exam 2, Section A, Q4; © VCAA.

MC For $z \in C$, if $\text{Im}(z) > 0$, then $\text{Arg}\left(\frac{z\bar{z}}{z-\bar{z}}\right)$ is

A. $-\frac{\pi}{2}$ **B.** 0 **C.** $\frac{\pi}{4}$ **D.** $\frac{\pi}{2}$ **E.** π

Question 2 (1 mark) TECH-ACTIVE

Source: VCE 2021 Specialist Mathematics Exam 2, Section A, Q6; © VCAA.

MC If $z \in C$, $z \neq 0$ and $z^2 \in R$, then the possible values of $\arg(z)$ are

A. $\frac{k\pi}{2}, k \in Z$ **B.** $k\pi, k \in Z$ **C.** $\frac{(2k+1)\pi}{2}, k \in Z$

D. $\frac{(4k+1)\pi}{2}, k \in Z$ **E.** $\frac{(4k-1)\pi}{2}, k \in Z$

Question 3 (5 marks) TECH-FREE

Source: VCE 2019 Specialist Mathematics Exam 1, Q7; © VCAA.

a. Show that $3-\sqrt{3}i = 2\sqrt{3}\,\text{cis}\left(-\frac{\pi}{6}\right)$. **(1 mark)**

b. Find $\left(3-\sqrt{3}i\right)^3$, expressing your answer in the form $x+iy$, where $x, y \in R$. **(2 marks)**

c. Find the integer values of n for which $\left(3-\sqrt{3}i\right)^n$ is real. **(1 mark)**

d. Find the integer values of n for which $\left(3-\sqrt{3}i\right)^n = ai$, where a is a real number. **(1 mark)**

More exam questions are available online.

2.4 Solving polynomial equations over C

LEARNING INTENTION

At the end of this subtopic you should be able to:

- solve polynomial equations using the fundamental theorem of algebra and the conjugate root theorem.

2.4.1 The fundamental theorem of algebra

A polynomial in z is an expression of the form $P(z) = a_n z^n + a_{n-1}z^{n-1} + a_{n-2}z^{n-2} + \ldots + a_1 z + a_0$, where $n \in N$ is the degree (highest power) of $P(z)$ and a_i (with $a_n \neq 0$).

If $a_i \in R$, that is, all the coefficients are real, then $P(z)$ is said to be a *polynomial over R*. Similarly, if at least one of the a_i is complex, $P(z)$ is said to be a *polynomial over C*.

For example, $P(z) = 3z^4 - 5z^2 + 6$ is a polynomial of degree 4 over R and $P(z) = 2iz^3 + 3z^2 - 8i$ is a polynomial of degree 3 over C.

If a polynomial $P(x)$ has $(x-a)$ as a factor, then $P(x)$ can be written $P(x) = (x-a)\,Q(x)$. From this it is clear that $P(a) = (a-a)\,Q(a) = 0$. The converse can also be shown to be true; that is, if $P(a) = 0$, then $(x-a)$ is a factor of $P(x)$. This result is known as the *factor theorem*.

Factor theorem

If $P(a) = 0$, then $(x - a)$ is a factor of $P(x)$.

The fundamental theorem of algebra

Every polynomial over C has at least one solution that is a complex number.

That is, if $P_n(z)$ is a polynomial of degree n over C, then there exists a $z_1 \in C$ such that $P_n(z_1) = 0$. This important result can be used to show that a polynomial of degree n, with $n \in N$, has n solutions.

The proof relies on a repeated application of the fundamental theorem of algebra and the factor theorem.

Firstly, the fundamental theorem of algebra guarantees that there is a $z_1 \in C$ such that $P_n(z_1) = 0$. The factor theorem states that if $P_n(z_1) = 0$ for some z_1 then $(z - z_1)$ is a factor of $P_n(z)$ so that $P_n(z) = (z - z_1)P_{n-1}(z)$, where $P_{n-1}(z)$ is a polynomial of degree $n - 1$.

Now by applying the fundamental theorem of algebra to $P_{n-1}(z)$ there is a $z_2 \in C$ such that $P_{n-1}(z_2) = 0$ and the factor theorem ensures that $P_{n-1}(z) = (z - z_2)P_{n-2}(z)$.

Hence $P_n(z) = (z - z_1)(z - z_2)P_{n-2}(z)$. By applying this method to each successive polynomial we can state the following:

Implications of the fundamental theorem of algebra

A polynomial of degree n will always have exactly n linear factors. That is;

$$P_n(z) = (z - z_1)(z - z_2)(z - z_3)\ldots(z - z_n)P_0(z)$$

where $z_1, z_2, z_3, \ldots, z_n \in C$ and $P_0(z)$ is a constant.

Note: Although n solutions are obtained, the fundamental theorem of algebra does not prescribe that they are necessarily distinct.

2.4.2 Solving quadratic equations

Recall the quadratic equation $az^2 + bz + c = 0$. If the coefficients a, b and c are all real, then the roots depend upon the discriminant, $\Delta = b^2 - 4ac$.

If $\Delta > 0$, then there are two distinct real roots.

If $\Delta = 0$, then there is one real root.

If $\Delta < 0$, then there is one pair of complex conjugate roots.

Relationship between the roots and coefficients

Given a quadratic equation with real coefficients, if the discriminant is negative, then the roots occur in complex conjugate pairs. A relationship can be formed between the roots and the coefficients.

Given a quadratic $az^2 + bz + c = 0$, if $a \neq 0$, then

$$z^2 + \frac{b}{a}z + \frac{c}{a} = 0.$$

Let the roots be α and β, so the factors are

$$(z - \alpha)(z - \beta).$$

Expanding gives

$$z^2 - (\alpha + \beta)z + \alpha\beta = 0$$

or

$$z^2 - (\text{sum of the roots}) + \text{product of the roots} = 0$$

so that

$$\alpha + \beta = -\frac{b}{a} \text{ and } \alpha\beta = \frac{c}{a}.$$

This provides a relationship between the roots and coefficients.

Rather than solving a quadratic equation, consider the reverse problem of forming a quadratic equation with real coefficients, given one of the roots.

WORKED EXAMPLE 17 Solving quadratic equations

Solve $z^2 + 6z + 13 = 0$ over C using the completing the square method.

THINK	WRITE
1. Use the 'complete the square' method.	$0 = z^2 + 6z + 13$ $0 = z^2 + 6z + 9 - 9 + 13$ $0 = (z+3)^2 + 4$
2. Express the RHS as the difference of two squares by converting 4 into complex form $(i^2 = -1)$.	$0 = (z+3)^2 - 4i^2$ $0 = (z+3)^2 - (2i)^2$
Write the expression in brackets as a quadratic.	$0 = (z+3+2i)(z+3-2i)$
3. Apply the Null Factor Theorem to state the solutions.	$(z+3+2i) = 0$ *or* $(z+3-2i) = 0$ The roots occur as a pair of complex conjugates: $z = -3-2i, -3+2i$.

WORKED EXAMPLE 18 Determining the equation of the quadratic from its roots

Determine the equation of the quadratic $P(z)$ with real coefficients given that $P(-11+2i) = 0$.

THINK	WRITE
1. State the given root.	Let $\alpha = -11 + 2i$.
2. The conjugate is also a root.	Let $\beta = -11 - 2i$.
3. State the linear factors.	$P(z) = (z-\alpha)(z-\beta)$ $= (z+11-2i)(z+11+2i)$
4. Expand the linear factors and state the quadratic equation.	$P(z) = (z+11)^2 - (2i)^2$ $= z^2 + 22z + 121 - 4i^2$ $= z^2 + 22z + 125$

A quadratic equation in the form $P(z) = az^2 + bz + c = 0$ can also be solved by using the quadratic formula as an alternative method. Given a, b and $c \in R$ the equation can be solved using $z = \dfrac{-b \pm \sqrt{b^2 - 4ac}}{2a}$.

2.4.3 Solving cubic equations

Cubic equations with real coefficients

A cubic polynomial equation of the form $az^3 + bz^2 + cz + d = 0$ with $z \in C$, and $a \neq 0$ but with all the coefficients real, will have three linear factors. These may be repeated, but the cubic must have at least one real factor. When solving $az^3 + bz^2 + cz + d = 0$, the roots can be three real roots, not necessarily all distinct, or they can be one real root and one pair of complex conjugate roots.

WORKED EXAMPLE 19 Solving cubic equations

Determine the roots of $z^3 + 6z^2 + 61z + 106 = 0$.

THINK	WRITE
1. Determine the real root using trial and error.	$P(-1) = (-1)^3 + 6(-1)^2 + 61(-1) + 106$ $= 50$ $P(-2) = (-2)^3 + 6(-2)^2 + 61(-2) + 106$ $= 0$
2. Apply the factor theorem.	Therefore, $(z+2)$ is a factor.
3. We can determine the quadratic factor using algebraic methods.	$P(z) = z^3 + 6z^2 + 61z + 106$ $= (z+2)(z^2 + bz + c)$ $= z^3 + bz^2 + cz + 2z^2 + 2bz + 2c$
4. Compare the two expressions for $P(z)$ to calculate c.	$106 = 2c \quad [1]$ $c = 53$
5. Compare the two expressions for $P(z)$ and use equation [1] to calculate b.	$61z = 2bz + cz \quad [2]$ $61z = 2bz + 53z$ $b = \dfrac{61z - 53z}{2z}$ $b = 4$
6. State $P(z)$ as the product of a linear and quadratic factor.	$P(z) = z^3 + 6z^2 + 61z + 106$ $= (z+2)(z^2 + 4z + 53)$
7. Factorise the quadratic equation by completing the square. Apply the Null Factor Theorem.	$0 = (z+2)(z^2 + 4z + 4 - 4 + 53)$ $0 = (z+2)((z+2)^2 + 49)$ $0 = (z+2)((z+2)^2 - (7i)^2)$ $0 = (z+2)(z+2+7i)(z+2-7i)$
8. State the three roots.	The roots are one real and one pair of complex conjugates: $z = -2, -2-7i, -2+7i$.

The conjugate root theorem

The preceding results are true not only for quadratic and cubic equations, but for any nth degree polynomial. In general, provided that all the coefficients of the polynomial are real, if the roots are complex, they must occur in conjugate pairs.

Note that if one or more of the coefficients is a complex number, then the roots do not occur in conjugate pairs.

Rather than formulating a problem such as solving a cubic equation, consider the reverse problem: determine some of the coefficients of a cubic equation with real coefficients, given one of the roots.

The conjugate root theorem

For a polynomial $P(z)$ with all real coefficients, complex roots occur in conjugate pairs. That is:

if $z_1 = a + bi$ is a root of $P(z)$, then $z_2 = a - bi$ is also a root of $P(z)$.

WORKED EXAMPLE 20 Determining the roots of a cubic with real coefficients

If $P(z) = z^3 + bz^2 + cz - 75 = 0$ where b and c are real and $P(-4 + 3i) = 0$, determine the values of b and c and state all the roots of $P(z) = 0$.

THINK	WRITE
1. Apply the conjugate root theorem.	Let $\alpha = -4 + 3i$ and $\beta = -4 - 3i$.
2. Evaluate the sum of the roots.	$\alpha + \beta = -8$
3. Evaluate the product of the roots.	$\alpha\beta = 16 - 9i^2$ $= 25$
4. Determine the quadratic factor.	$z^2 + 8z + 25$
5. Use short division.	$P(z) = z^3 + bz^2 + cz - 75 = 0$ $P(z) = (z^2 + 8z + 25)(z - 3) = 0$
6. Expand the brackets.	$P(z) = z^3 + 5z^2 + z - 75 = 0$
7. Equate coefficients.	From the z^2: $b = 5$ and from the coefficient of z: $c = 25 - 8 \times 3 = 1$.
8. State all the roots and their nature.	The roots are one real root and one pair of complex conjugates: $z = 3$ and $z = -4 \pm 3i$.

Cubic equations with complex coefficients

We can use the grouping technique to solve certain types of cubic equations with complex coefficients.

WORKED EXAMPLE 21 Solving cubic equations with complex coefficients using grouping

Solve for z if $z^3 + iz^2 + 5z + 5i = 0$.

THINK	WRITE
1. This cubic can be solved by grouping terms together.	$z^3 + iz^2 + 5z + 5i = 0$ $z^2(z + i) + 5(z + i) = 0$
2. Factorise.	$(z^2 + 5)(z + i) = 0$

3. Express the quadratic factor as the difference of two squares using $i^2 = -1$. $\quad (z^2 - 5i^2)(z + i) = 0$

4. State the linear factors. $\quad (z + \sqrt{5}i)(z - \sqrt{5}i)(z + i) = 0$

5. Apply the Null Factor Theorem and state all the roots. $\quad z = \pm\sqrt{5}i$ and $z = -i$.

General cubic equations with complex coefficients

If one of the roots of a cubic equation is given, the remaining roots can be determined. Note that if one of the coefficients in the cubic equation is a complex number, the roots do not all occur in conjugate pairs.

WORKED EXAMPLE 22 Solving cubic equations with complex coefficients

Show that $z = 5 - 2i$ is a root of the equation $z^3 + (-5 + 2i)z^2 + 4z + 8i - 20 = 0$, and hence evaluate all the roots.

THINK	WRITE
1. Substitute $5 - 2i$ for z.	$P(5-2i) = (5-2i)^3 + (-5+2i)(5-2i)^2 + 4(5-2i) + 8i - 20$
2. Simplify.	$P(5-2i) = (5-2i)^3 - (5-2i)^3 + 20 - 8i + 8i - 20 = 0$ Therefore, $z = 5 - 2i$ is a root of the cubic equation.
3. Since $z = 5 - 2i$ is a root, $(z - 5 + 2i)$ is a factor. Use short division.	$(z - 5 + 2i)(z^2 + 4) = 0$
4. Express as the difference of two squares using $i^2 = -1$.	$(z - 5 + 2i)(z^2 - 4i^2) = 0$
5. State the linear factors.	$(z - 5 + 2i)(z + 2i)(z - 2i) = 0$
6. Apply the Null Factor Theorem and state all the roots.	$z = 5 - 2i$ and $z = \pm 2i$.

2.4.4 Solving quartic equations

A quartic of the form $P(z) = az^4 + bz^3 + cz^2 + dz + e$ with all real coefficients can have four linear factors. The roots can be either all real roots; two real roots and one pair of complex conjugate roots; or two pairs of complex conjugate roots.

WORKED EXAMPLE 23 Solving quartic equations which can be reduced to quadratics

Solve for z if $z^4 + 3z^2 - 28 = 0$.

THINK	WRITE
1. Use a suitable substitution to reduce the quartic equation to a quadratic equation.	Let $u = z^2$, then $u^2 = z^4$. $z^4 + 3z^2 - 28 = 0$ $u^2 + 3u - 28 = 0$
2. Factorise the expression.	$(u + 7)(u - 4) = 0$
3. Substitute z^2 for u.	$(z^2 + 7)(z^2 - 4) = 0$
4. Express as the difference of two squares using $i^2 = -1$.	$(z^2 - 7i^2)(z^2 - 4) = 0$

5. State the linear factors.

$$\left(z+\sqrt{7}i\right)\left(z-\sqrt{7}i\right)(z+2)(z-2)=0$$

6. Apply the Null Factor Theorem and state all the four roots and their nature.

The equation $z^4+3z^2-28=0$ has two real roots and one pair of complex conjugate roots.
$z=\pm\sqrt{7}i$ and $z=\pm 2$.

TI | THINK

On a Calculator page type 'cSolve', place the equation in brackets followed by a comma and the variable being solved for. Press 'enter'.

DISPLAY/WRITE

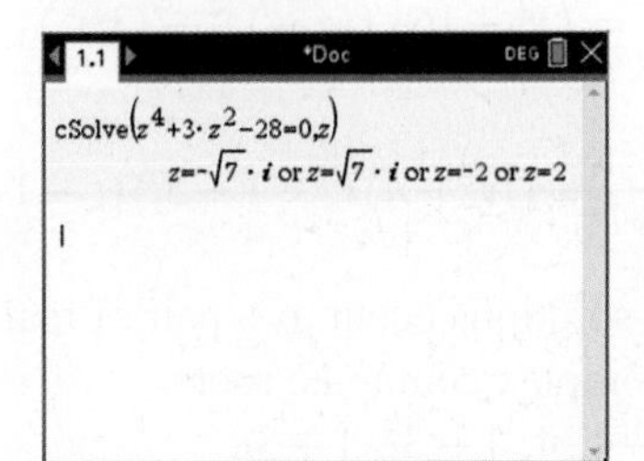

CASIO | THINK

On a Main screen type 'solve' (found in the Math3 keyboard), place the equation in brackets followed by a comma and the variable being solved for. Press 'EXE'.

DISPLAY/WRITE

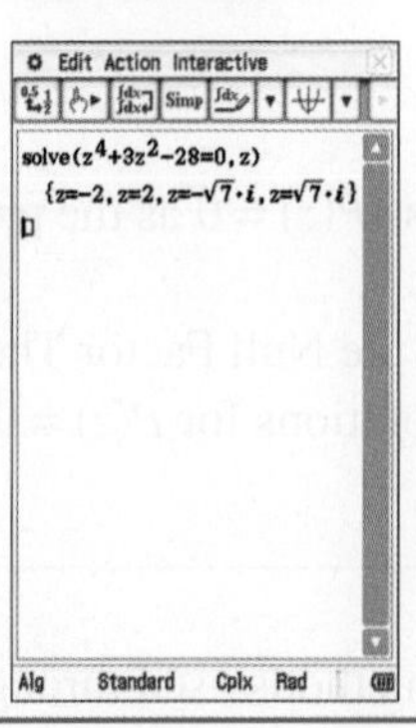

In the previous worked examples, algebraic methods were used to determine unknown coefficients for quadratic factors. Polynomial long division is an alternative method and is demonstrated in the following worked example.

WORKED EXAMPLE 24 Solving quartic equations

Given that $P(4)=0$ where $P(z)=z^4-2z^3-6z^2+32z-160$, use polynomial long division to calculate all the roots for $P(z)$.

THINK

1. State the given real root.

WRITE

Let $z_1=4$.

2. Since one root is real determine another real root by testing $P(-4)$.

$$P(-4)=(-4)^4-2(-4)^3-6(-4)^2+32(-4)-160$$
$$=0$$

Thus $z_2=-4$ is another root.

3. Express $P(z)$ as the product of two quadratic factors.

$$P(z)=(z-z_1)(z-z_2)(z-z_3)(z-z_4)$$
$$=(z-4)(z+4)(z-z_3)(z-z_4)$$
$$=\left(z^2-16\right)\left(az^2+bz+c\right)$$

4. Complete polynomial long division to calculate the unknown quadratic factor.

$$\begin{array}{r l}
 & z^2-2z+10 \\
z^2-16 \,\big) & z^4-2z^3-6z^2+32z-160 \\
 & -(z^4 \quad \downarrow \quad -16z^2 \quad \downarrow \quad \downarrow) \\
\hline
 & -2z^3+10z^2+32z-160 \\
 & -(\;-2z^3 \quad \downarrow \quad +32z \quad \downarrow) \\
\hline
 & 10z^2 \quad -160 \\
 & -(\;10z^2 \quad -160) \\
\hline
 & 0
\end{array}$$

5. Express $P(z)$ as the product of two quadratic factors by using the result from the long division operation.	$P(z) = (z^2 - 16)(z^2 - 2z + 10)$
6. Factorise the quadratic equation by completing the square.	$P(z) = (z^2 - 16)(z^2 - 2z + 1 - 1 + 10)$ $= (z^2 - 16)((z-1)^2 + 9)$ $= (z^2 - 16)((z-1)^2 - (3i)^2)$
7. Express $P(z) = 0$ as the product of four linear factors.	$0 = (z+4)(z-4)(z-1+3i)(z-1-3i)$
8. Apply the Null Factor Theorem to state the four solutions for $P(z) = 0$.	The solutions occur as a pair of real roots and a pair of complex conjugate roots. $z = -4, 4, 1-3i, 1+3i$.

Note: The methods used throughout this subtopic can be extended to determine the roots of higher order polynomials which have real coefficients.

2.4 Exercise

Technology free

1. WE17 Solve $z^2 - 14z + 74 = 0$ over C using the completing the square method.
2. Solve $4z^2 + 4z + 82 = 0$ over C using the completing the square method.
3. Solve $z(z+4) = -29$ for z using the quadratic formula.
4. WE18 Determine the equation of the quadratic $P(z)$ with real coefficients given that $P(-1+13i) = 0$.
5. Determine the equation of the quadratic $P(z)$ with real coefficients given that $P(6-5i) = 0$.
6. WE19 Determine the roots of $z^3 + 9z^2 + 24z - 34 = 0$.
7. Solve the equation $z^3 + 13z^2 + 97z + 85 = 0$ over C.
8. Determine the roots of $z^3 + 4z^2 - 2z - 20 = 0$.
9. WE20 If $P(z) = z^3 + bz^2 + cz - 39 = 0$ where b and c are real, and $P(-2+3i) = 0$, determine the values of b and c, and state all the roots of $P(z) = 0$.
10. If $P(z) = z^3 + bz^2 + cz - 50 = 0$ where b and c are real, and $P(5i) = 0$, determine the values of b and c, and state all the roots of $P(z) = 0$.
11. WE21 Solve for z if $z^3 - 2iz^2 + 4z - 8i = 0$.
12. Determine the linear factors of $z^3 + 3iz^2 + 7z + 21i$.

13. WE22 Show that $z=2-3i$ is a root of the equation $z^3+(-2+3i)z^2+4z+12i-8=0$, and hence determine all the roots.

14. Show that $z=\frac{3}{2}+2i$ is a root of the equation $2z^3-(4i+3)z^2+10z-20i-15=0$, and hence determine all the roots.

15. WE23 Solve for z if $z^4-z^2-20=0$.

16. Solve for z if $2z^4-3z^2-9=0$.

17. Solve each of the following for z.

a. $z^4+5z^2-36=0$

b. $z^4+4z^2-21=0$

18. Solve each of the following for z.

a. $z^4-3z^2-40=0$

b. $z^4+9z^2+18=0$

19. Given $P(z)=z^4+az^3+34z^2-54z+225$ and $P(3i)=0$, determine the value of the real constant a and evaluate all the roots.

20. Given $P(z)=z^4+6z^3+29z^2+bz+100=0$ and $P(-3-4i)=0$, determine the value of the real constant b and evaluate all the roots.

21. WE24 Given that $P(3)=0$ where $P(z)=z^4+10z^3+25z^2-90z-306$, use polynomial long division to calculate all the roots for $P(z)$.

22. If $z=4$ is a zero of $P(z)=z^3+8z^2+69z-468$, use polynomial long division to calculate all the roots for $P(z)$.

23. Given $P(z)=z^4-2z^3-6z^2+32z-160=(z-1-3i)(z-z_2)(z-z_3)(z-z_4)$, use polynomial long division to calculate all the roots for $P(z)$.

Technology active

24. Given that $z=ai$ is a root of the equation $z^4+6z^3+41z^2+96z+400=0$, determine the value of the real constant a and all the roots.

25. Given $P(z)=z^4+bz^3+18z^2+32z+32$ and $P(4i)=0$, calculate the value of the real constant b and determine all the roots.

26. A quartic is given by the equation, $P(z)=\left(az^2+bz+c\right)\left(z^2-10z+41\right)+d$, where $d\in R$. If $P(z)=z^4-8z^3+23z^2+62z+81$ in expanded form, determine the real value d.

2.4 Exam questions

Question 1 (4 marks) TECH-FREE

Source: VCE 2021 Specialist Mathematics Exam 1, Q8; © VCAA.

a. Solve $z^2+2z+2=0$ for z, where $z\in C$. **(1 mark)**

b. Solve $z^2+2\bar{z}+2=0$ for z, where $z\in C$. **(3 marks)**

Question 2 (1 mark) TECH-ACTIVE

Source: VCE 2020 Specialist Mathematics Exam 2, Section A, Q6; © VCAA.

MC For the complex polynomial $P(z)=z^3+az^2+bz+c$ with real coefficients a, b and c, $P(-2)=0$ and $P(3i)=0$.

The values of a, b and c are respectively

A. $-2, 9, -18$

B. $3, 4, 12$

C. $2, 9, 18$

D. $-3, -4, 12$

E. $2, -9, -18$

Question 3 (1 mark) TECH-ACTIVE

Source: *VCE 2017 Specialist Mathematics Exam 2, Section A, Q3; © VCAA.*

MC The number of distinct roots of the equation $(z^4-1)(z^2+3iz-2)=0$, where $z \in C$, is

A. 2 **B.** 3 **C.** 4 **D.** 5 **E.** 6

More exam questions are available online.

2.5 Subsets of the complex plane

LEARNING INTENTION

At the end of this subtopic you should be able to:

- identify and sketch subsets of the complex plane, such as circles, lines and rays.

In previous sections, complex numbers have been used to represent points on the Argand plane. If we consider z as a complex variable, we can define and sketch subsets of the Argand plane such as lines, rays, circles and ellipses.

2.5.1 Lines

If $z=x+yi$, then $\text{Re}(z)=x$ and $\text{Im}(z)=y$. The equation $a\,\text{Re}(z)+b\,\text{Im}(z)=c$ where a, b and $c \in R$ represents the line $ax+by=c$.

WORKED EXAMPLE 25 Sketching lines in the complex plane (1)

Determine the Cartesian equation and sketch the graph defined by $\{z: 3\text{Re}(z)-2\text{Im}(z)=6\}$.

THINK	WRITE/DRAW
1. Consider the equation.	$3\text{Re}(z)-2\text{Im}(z)=6$ As $z=x+yi$, then $\text{Re}(z)=x$ and $\text{Im}(z)=y$. This is a straight line with the Cartesian equation $3x-2y=6$.
2. Evaluate the axial intercepts.	When $y=0$, $3x=6 \Rightarrow x=2$. $(2,0)$ is the intercept with the real axis. When $x=0$, $-2y=6 \Rightarrow y=-3$. $(0,-3)$ is the intercept with the imaginary axis.
3. Identify and sketch the equation.	The equation represents the line $3x-2y=6$.

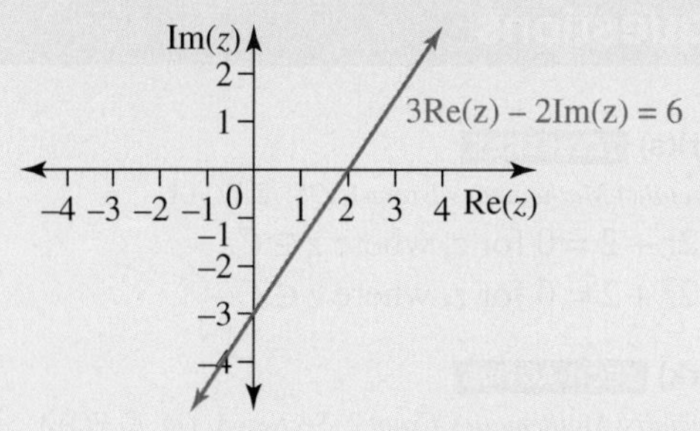

Lines in the complex plane can also be represented as a set of points that are equidistant from two other fixed points. The equations of a line in the complex plane can thus have multiple representations.

WORKED EXAMPLE 26 Sketching lines in the complex plane (2)

Determine the Cartesian equation and sketch the graph defined by $\{z: |z-3| = |z+3i|\}$.

THINK	WRITE/DRAW
1. Consider the equation as a set of points.	$\lvert z-3\rvert = \lvert z+3i\rvert$ Substitute $z = x + yi$: $\lvert x+yi-3\rvert = \lvert x+yi+3i\rvert$
2. Group the real and imaginary parts together.	$\lvert (x-3)+yi\rvert = \lvert x+(y+3)i\rvert$
3. Use the definition of the modulus.	$\sqrt{(x-3)^2+y^2} = \sqrt{x^2+(y+3)^2}$
4. Expand both sides and cancel like terms.	$x^2-6x+9+y^2 = x^2+y^2+6y+9$ $-6x = 6y$
5. Identify the required line.	$y = -x$
6. Identify the line geometrically.	The line is the set of points that is equidistant from the two points $(3, 0)$ and $(0, -3)$.
7. Sketch the required line.	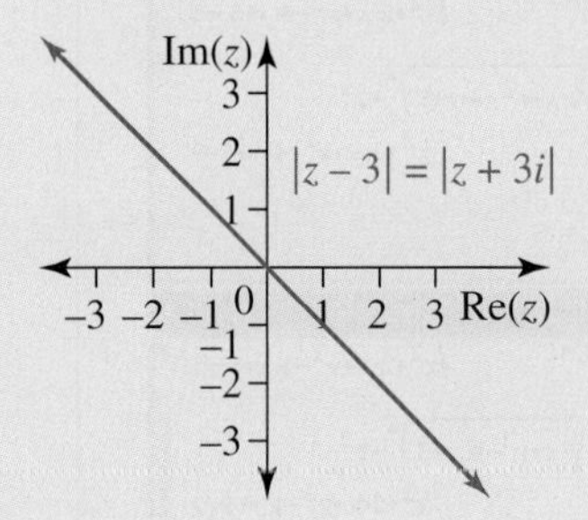

2.5.2 Circles

The equation $|z| = r$ where $z = x + yi$ is given by $|z| = \sqrt{x^2+y^2} = r$. Expanding this produces $x^2 + y^2 = r^2$. This represents a circle with centre at the origin and radius r. Geometrically, $|z| = r$ represents the set of points, or what is called the locus of points, in the Argand plane that are at r units from the origin.

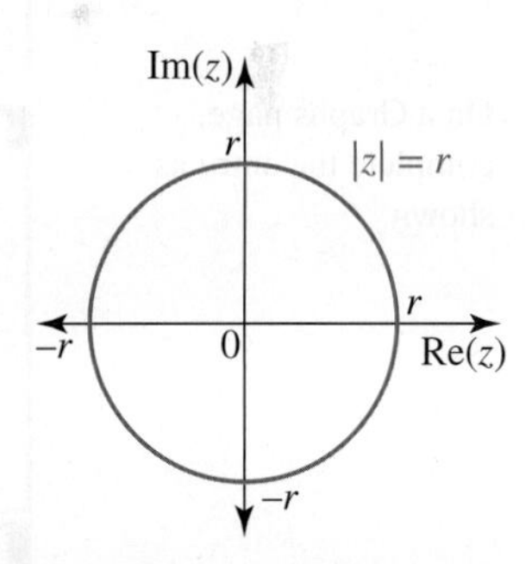

WORKED EXAMPLE 27 Sketching circles in the complex plane

Determine the Cartesian equation and sketch the graph of $\{z: |z+1-2i| = 3\}$.

THINK	WRITE/DRAW
1. Consider the equation.	$\lvert z+1-2i\rvert = 3$ Substitute $z = x + yi$: $\lvert x+yi+1-2i\rvert = 3$
2. Group the real and imaginary parts.	$\lvert (x+1)+i(y-2)\rvert = 3$
3. Use the definition of the modulus.	$\sqrt{(x+1)^2+(y-2)^2} = 3$

4. Square both sides.	$(x+1)^2 + (y-2)^2 = 9$ The equation represents a circle with centre at $(-1, 2)$ and radius 3.
5. Sketch the graph on the Argand plane.	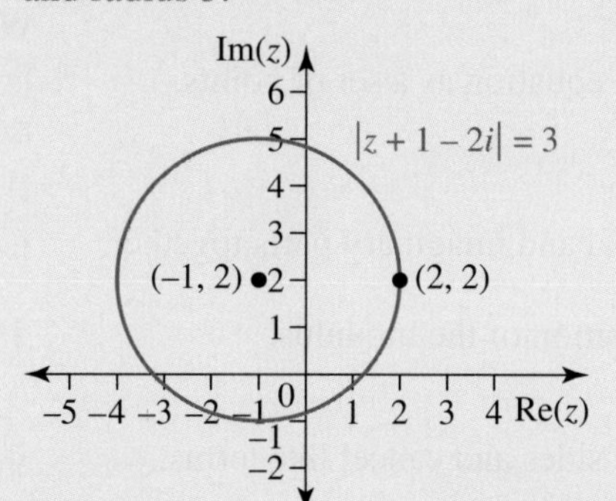

TI | THINK	DISPLAY/WRITE	CASIO | THINK	DISPLAY/WRITE
On a Calculator page, complete the entry as shown.	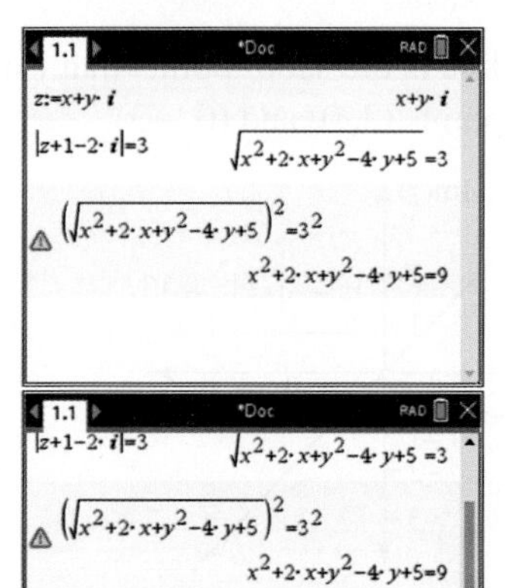	On a Main screen, complete the entry as shown.	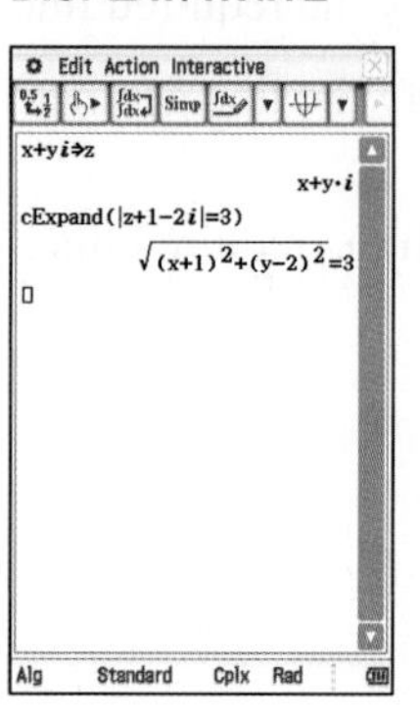
On a Graphs page, complete the entry as shown.	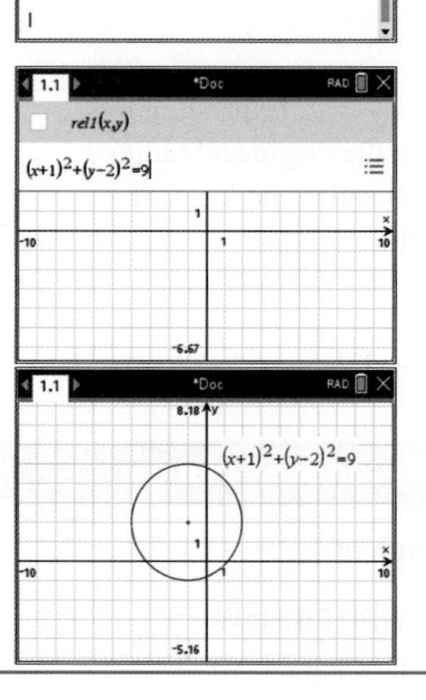	On a Conics screen, complete the entry as shown.	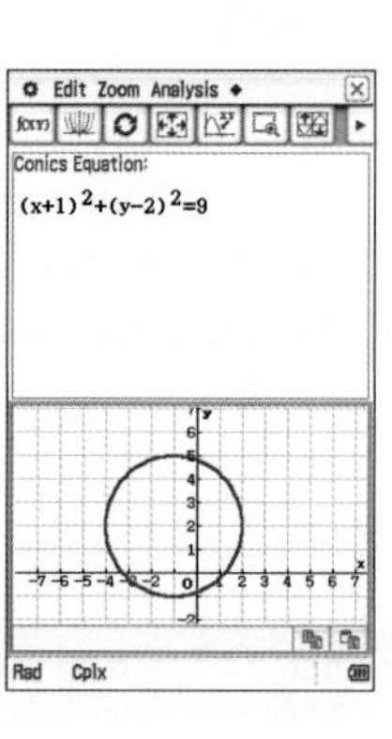

2.5.3 Ellipses

Recall that the general form of an ellipse is $\frac{x^2}{a^2}+\frac{y^2}{b^2}=1$. This ellipse has semi-major axes of a and semi-minor axis of b.

An alternative definition of an ellipse is that for all points on the curve the sum of the distances to two fixed points is a constant.

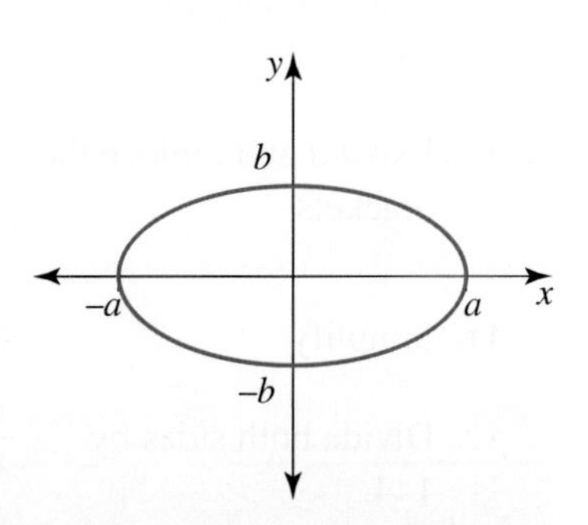

Worked Example 28 Sketching ellipses in the complex plane

Determine the Cartesian equation and sketch the graph of the set of points in the complex plane defined by $\{z: \left|z-2\sqrt{5}\right|+\left|z+2\sqrt{5}\right|=12\}$.

THINK	WRITE/DRAW
1. Substitute for z.	$\left\|z-2\sqrt{5}\right\|+\left\|z+2\sqrt{5}\right\|=12$ Let $z=x+yi$. $\left\|\left(x-2\sqrt{5}\right)+yi\right\|+\left\|\left(x+2\sqrt{5}\right)+yi\right\|=12$
2. Use the definition of the modulus.	$\sqrt{\left(x-2\sqrt{5}\right)^2+y^2}+\sqrt{\left(x+2\sqrt{5}\right)^2+y^2}=12$
3. Transfer one of the square roots to the other side of the equation.	$\sqrt{\left(x-2\sqrt{5}\right)^2+y^2}=12-\sqrt{\left(x+2\sqrt{5}\right)^2+y^2}$
4. Square both sides.	$\left(x-2\sqrt{5}\right)^2+y^2=144+\left(x+2\sqrt{5}\right)^2+y^2-24\sqrt{\left(x+2\sqrt{5}\right)^2+y^2}$
5. Transfer the square root to the other side of the equation.	$24\sqrt{\left(x+2\sqrt{5}\right)^2+y^2}=144+\left(x+2\sqrt{5}\right)^2+y^2-\left[\left(x-2\sqrt{5}\right)^2+y^2\right]$
6. Expand the RHS.	$24\sqrt{\left(x+2\sqrt{5}\right)^2+y^2}=144+x^2+4\sqrt{5}x+20+y^2-\left[x^2-4\sqrt{5}x+20+y^2\right]$
7. Simplify the RHS, by cancelling.	$24\sqrt{\left(x+2\sqrt{5}\right)^2+y^2}=144+8\sqrt{5}x$ $=8\left(18+\sqrt{5}x\right)$
8. Cancel a common factor. Since the LHS is always positive, RHS must also be positive.	$3\sqrt{\left(x+2\sqrt{5}\right)^2+y^2}=18+\sqrt{5}x$ We require $18+\sqrt{5}x>0 \Rightarrow x>-\frac{18}{\sqrt{5}}$

9. Square both sides again. $9\left(\left(x+2\sqrt{5}\right)^2+y^2\right)=\left(18+\sqrt{5}x\right)^2$

10. Expand and remove the brackets. $9\left(x^2+4\sqrt{5}x+20+y^2\right)=324+36\sqrt{5}x+5x^2$

$9x^2+36\sqrt{5}x+180+9y^2=324+36\sqrt{5}x+5x^2$

11. Simplify. $4x^2+9y^2=144$

12. Divide both sides by 144. $\frac{x^2}{36}+\frac{y^2}{16}=1$

13. Describe the equation. The graph is the ellipse with semi-major and semi-minor axes 6 and 4 respectively, since $x>-\frac{18}{\sqrt{5}}>-8$ is satisfied.

14. Sketch the graph of the ellipse.

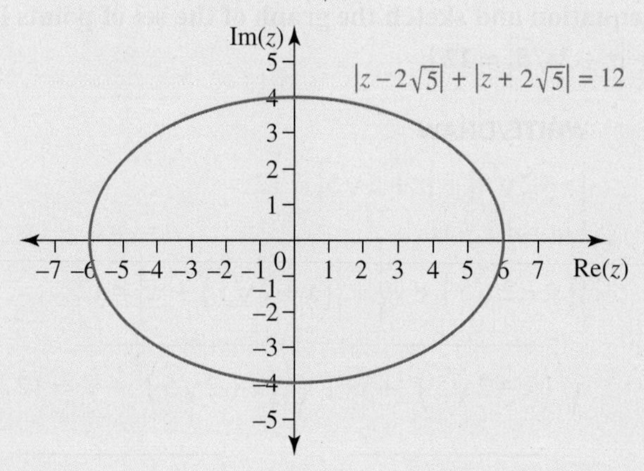

2.5.4 Rays

A ray is a half-line, which can be defined in terms of the argument of the complex number. $\text{Arg}(z)=\theta$ represents the set of all points on the half-line or ray that has one end at the origin and makes an angle of θ with the positive real axis. Note that the endpoint, in this case the origin, is not included in the set since $x>0$ and $y>0$. We indicate this by placing a small open circle at this point.

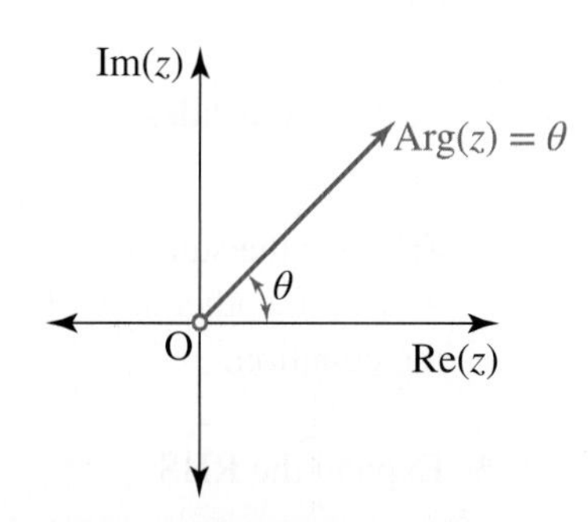

WORKED EXAMPLE 29 Sketching rays in the complex plane

Determine the Cartesian equation and sketch the graph defined by $\left\{z: \text{Arg}(z+3-2i)=-\frac{2\pi}{3}\right\}$.

THINK

1. Determine the Cartesian equation of the ray.

WRITE/DRAW

$\text{Arg}(z+3-2i)=-\frac{2\pi}{3}$

Substitute $z=x+yi$:

$\text{Arg}(x+yi+3-2i)=-\frac{2\pi}{3}$

2. Group the real and imaginary parts. $\operatorname{Arg}\left((x+3)+(y-2)i\right) = -\dfrac{2\pi}{3}$

3. Use the definition of the argument. $\tan^{-1}\left(\dfrac{y-2}{x+3}\right) = -\dfrac{2\pi}{3}$ for $x<-3$

4. Simplify.

$$\frac{y-2}{x+3} = \tan\left(-\frac{2\pi}{3}\right)$$
$$= \sqrt{3} \text{ for } x<-3$$
$$y-2 = \sqrt{3}(x+3) \text{ for } x<-3$$

5. State the Cartesian equation of the ray. $y=\sqrt{3}x+3\sqrt{3}+2$ for $x<-3$.

6. Identify the point from which the ray starts. The ray starts from the point $(-3, 2)$.

7. Determine the angle the ray makes. The ray makes an angle of $-\dfrac{2\pi}{3}$ with the positive real axis.

8. Describe the ray. The point $(-3, 2)$ is not included. Alternatively, consider the ray from the origin making an angle of $-\dfrac{2\pi}{3}$ with the positive real axis to have been translated 3 units to the left parallel to the real axis and 2 units up parallel to the imaginary axis.

9. Sketch the required ray.

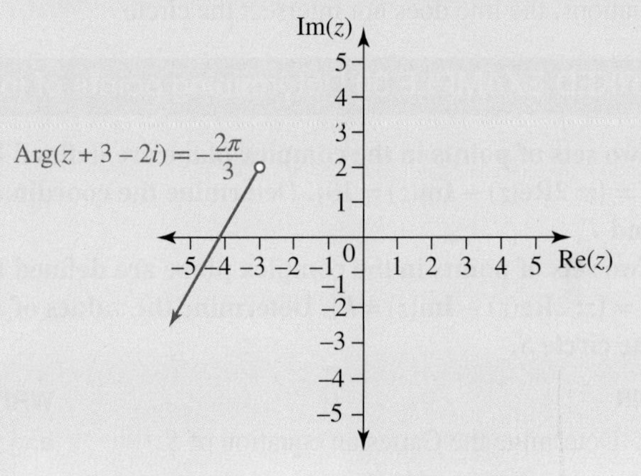

TI | THINK	DISPLAY/WRITE	CASIO | THINK	DISPLAY/WRITE
On a Calculator page, complete the entry as shown.	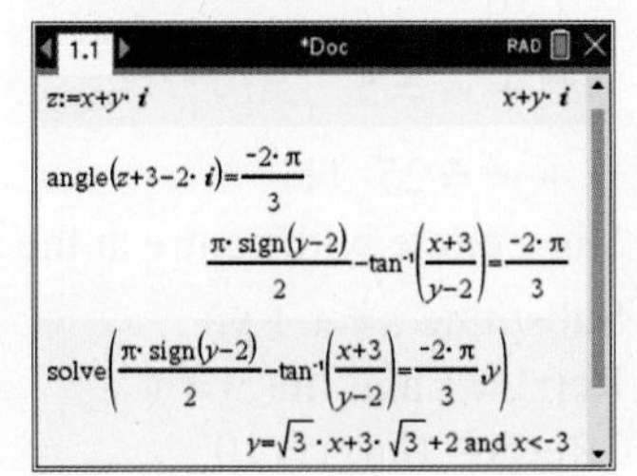	On a Main screen, complete the entry as shown.	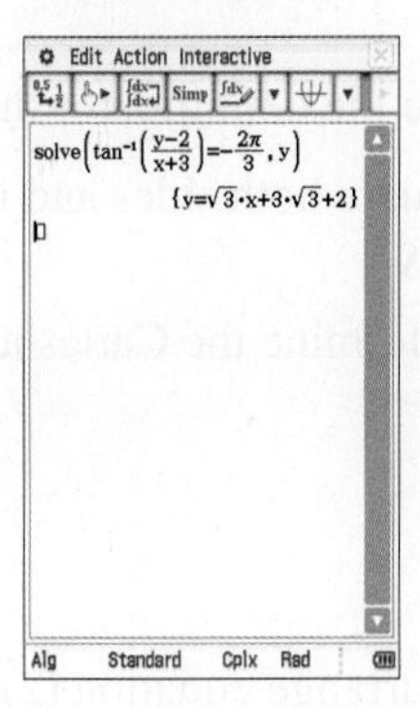

On a Graphs page, complete the entry as shown.

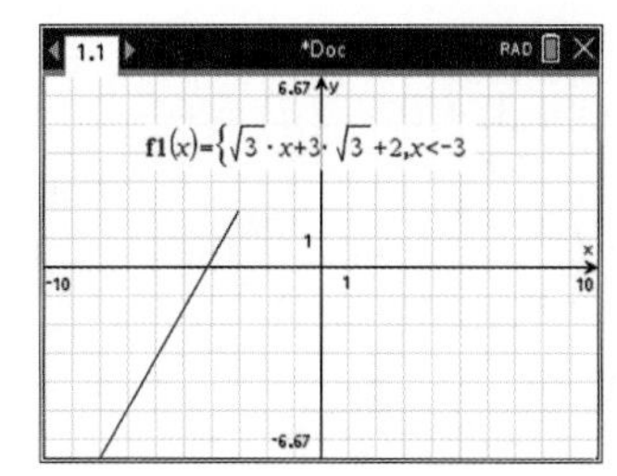

On a Conics screen, complete the entry as shown.

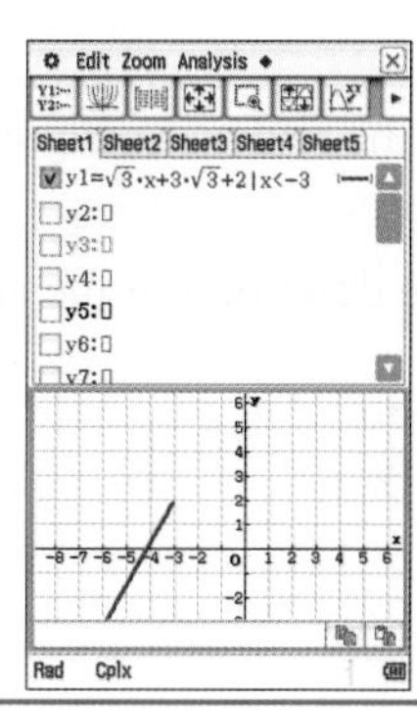

2.5.5 Intersection of subsets of the complex plane

The coordinates of the points of intersection between subsets of the complex plane can be determined algebraically by solving the system of equations. The number of solutions reflects the number of points of intersection. For example, consider the intersection of a circle and a line. If there are two solutions to the equations, the line intersects the circle at two points. If there is one solution to the equation, the line and the circle touch at one point, and the line is a tangent to the circle at the point of contact. If there are no solutions to the equations, the line does not intersect the circle.

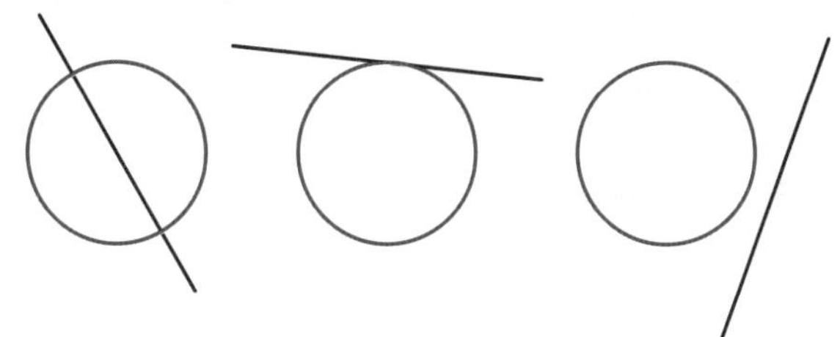

WORKED EXAMPLE 30 Determining points of intersection

a. Two sets of points in the complex plane are defined by $S = \{z: |z| = 5\}$ and $T = \{z: 2\text{Re}(z) - \text{Im}(z) = 10\}$. Determine the coordinates of the points of intersection between S and T.

b. Two sets of points in the complex plane are defined by $S = \{z: |z| = 3\}$ and $T = \{z: 2\text{Re}(z) - \text{Im}(z) = k\}$. Determine the values of k for which the line through T is a tangent to the circle S.

THINK	WRITE
a. 1. Determine the Cartesian equation of S.	a. $\|z\| = 5$ Substitute $z = x + yi$: $\|x + yi\| = 5$
2. Use the definition of the modulus.	$\sqrt{x^2 + y^2} = 5$
3. Square both sides and identify the boundary of S.	$x^2 + y^2 = 25 \quad (1)$ S is a circle with centre at the origin and radius 5.
4. Determine the Cartesian equation of T.	Substitute $z = x + yi$: $\text{Re}(z) = x$ and $\text{Im}(z) = y$ $2\text{Re}(z) - \text{Im}(z) = 10$ $2x - y = 10 \quad (2)$ T is a straight line.
5. Rearrange equation (2) and solve equations (1) and (2) for x and y by substitution.	$y = 2x - 10$ $x^2 + (2x - 10)^2 = 25$

6. Expand and simplify.

$$x^2 + 4x^2 - 40x + 100 = 25$$
$$5x^2 - 40x + 75 = 0$$
$$5(x^2 - 8x + 15) = 0$$

7. Solve for x.

$$x^2 - 8x + 15 = 0$$
$$(x-5)(x-3) = 0$$
$$x = 5 \text{ or } x = 3$$

8. Calculate the corresponding y-values.

From (2) $y = 2x - 10$,
When $x = 5 \Rightarrow y = 0$
and when $x = 3 \Rightarrow y = -4$.

9. State the coordinates of the two points of intersection.

The points of intersection are (5, 0) and (3, −4).

b. 1. Determine the Cartesian equation of S.

b. $|z| = 3$
Substitute $z = x + yi$:
$|x + yi| = 3$

2. Use the definition of the modulus.

$$\sqrt{x^2 + y^2} = 3$$

3. Square both sides and identify the boundary of S.

$x^2 + y^2 = 9 \quad (1)$
S is a circle with centre at the origin and radius 3.

4. Determine the Cartesian equation of T.

Substitute $z = x + yi$:
$\text{Re}(z) = x$ and $\text{Im}(z) = y$
$2\text{Re}(z) - \text{Im}(z) = k$
$2x - y = k \quad (2)$
T is a straight line.

5. Rearrange equation (2) and solve equations (1) and (2) for x and y by substitution.

$$y = 2x - k$$
$$x^2 + (2x - k)^2 = 9$$

6. Expand and simplify.

$$x^2 + 4x^2 - 4kx + k^2 = 9$$
$$5x^2 - 4kx + k^2 - 9 = 0$$

7. If the line through T is a tangent to the circle S, there will be only one solution for x.

The discriminant $\Delta = b^2 - 4ac = 0$, where $a = 5, b = -4k$ and $c = k^2 - 9$.

$$\Delta = (-4k)^2 - 4 \times 5 \times (k^2 - 9)$$
$$= 16k^2 - 20(k^2 - 9)$$
$$= -4k^2 + 180$$
$$= 4(45 - k^2)$$

8. Solve the discriminant equal to zero for k.

$$45 - k^2 = 0$$
$$k = \pm\sqrt{45}$$

9. State the value of k for which the line through T is a tangent to the circle S.

$$k = \pm 3\sqrt{5}$$

2.5 Exercise

Students, these questions are even better in jacPLUS

Receive immediate feedback and access sample responses

Access additional questions

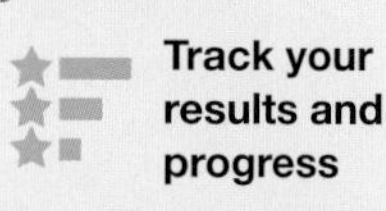

Track your results and progress

Find all this and MORE in jacPLUS

Technology free

1. WE25 Sketch and describe the region of the complex plane defined by $\{z: 4\text{Re}(z) + 3\text{Im}(z) = 12\}$.

2. The region of the complex plane shown can be described by $\{z: a\text{Re}(z) + b\text{Im}(z) = 8\}$. Determine the values of a and b.

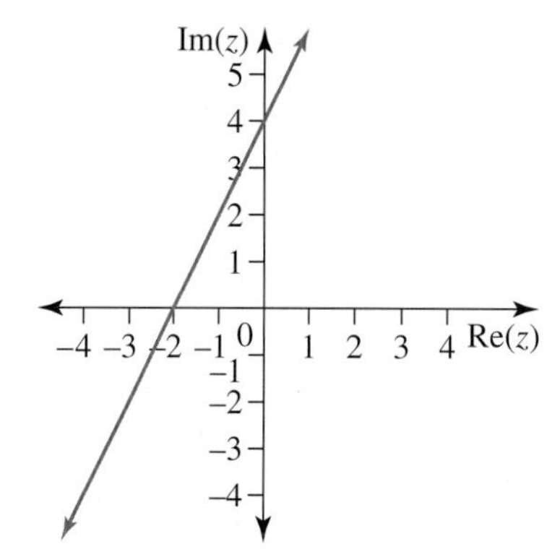

3. Sketch and describe the region of the complex plane defined by:

   a. $\{z: 3\text{Re}(z) + 2\text{Im}(z) = 6\}$

   b. $\{z: 2\text{Re}(z) - \text{Im}(z) = 6\}$

4. WE26 Sketch and describe the region of the complex plane defined by $\{z: |z + 3i| = |z - 3|\}$.

5. Sketch and describe each of the following sets, clearly indicating which boundaries are included.

   a. $\{z: |z - 2| = |z - 4|\}$

   b. $\{z: |z + 4i| = |z - 4|\}$

6. Sketch and describe each of the following sets, clearly indicating which boundaries are included.

   a. $\{z: |z + 4| = |z - 2i|\}$

   b. $\{z: |z + 2 - 3i| = |z - 2 + 3i|\}$

7. WE27 Sketch and describe the region of the complex plane defined by $\{z: |z - 3 + 2i| = 4\}$.

8. The region of the complex plane shown can be described by $\{z: |z - (a + bi)| = r\}$. Determine the values of a, b and r.

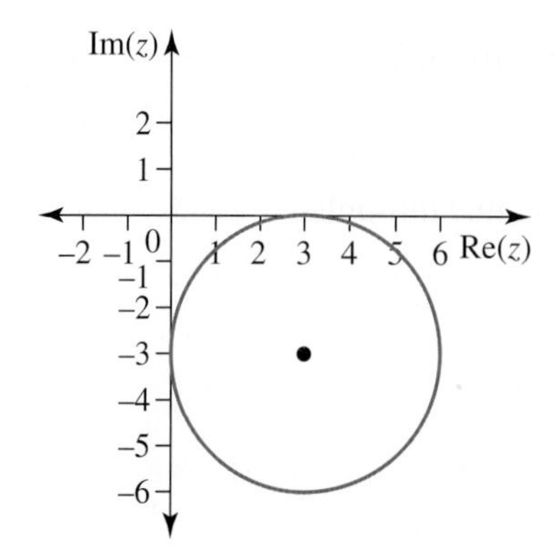

9. For each of the following, sketch and determine the Cartesian equation of the set, and describe the region.
   a. $\{z: |z+2-3i|=2\}$
   b. $\{z: |z-3+i|=3\}$

10. a. Show that the complex equation $\left\{z: \text{Im}\left(\frac{z-2i}{z-3}\right)=0\right\}$ represents a straight line and determine its equation.
    b. Show that the complex equation $\left\{z: \text{Re}\left(\frac{z-2i}{z-3}\right)=0\right\}$ represents a circle and determine its centre and radius.

11. a. Determine the Cartesian equation of $\{z: |z-3|=2|z+3i|\}$.
    b. Determine the locus of the set of points in the complex plane given by $\{z: |z+3|=2|z+6i|\}$.

12 **WE28** Determine the Cartesian equation and sketch the graph of the set of points in the complex plane defined by $\{z : |z-4|+|z+4|=10\}$.

13. **WE29** Determine the Cartesian equation and sketch the graph defined by $\left\{z: \text{Arg}(z-2)=\frac{\pi}{6}\right\}$.

14. Determine the Cartesian equation and sketch the graph defined by $\left\{z: \text{Arg}(z+3i)=-\frac{\pi}{2}\right\}$.

15. Sketch and describe the following subsets of the complex plane.
    a. $\left\{z: \text{Arg}(z)=\frac{\pi}{6}\right\}$
    b. $\left\{z: \text{Arg}(z+i)=\frac{\pi}{4}\right\}$

16. **WE30** a. Two sets of points in the complex plane are defined by $S=\{z: |z|=3\}$ and $T=\{z: 3\text{Re}(z)-4\text{Im}(z)=12\}$. Determine the coordinates of the points of intersection between S and T.
    b. Two sets of points in the complex plane are defined by $S=\{z: |z|=4\}$ and $T=\{z: 4\text{Re}(z)-2\text{Im}(z)=k\}$. Determine the values of k for which the line through T is a tangent to the circle S.

17. a. Let $S=\{z: |z-6|=2|z-3i|\}$ and $T=\{z: |z-(a+bi)|=r\}$. Given that $S=T$, determine the values of a, b and r.
    b. Let $\{z: |z+3|=2|z-3i|\}$ and $T=\{z: |z-(a+bi)|=r\}$. Given that $S=T$, state the values of a, b and r.

18. a. Two sets of points in the complex plane are defined by $S=\left\{z: |z|=\sqrt{29}\right\}$ and $T=\{z: 3\text{Re}(z)-\text{Im}(z)=1\}$. Determine the coordinates of the points of intersection between S and T.
    b. Two sets of points in the complex plane are defined by $S=\{z: |z|=5\}$ and $T=\{z: 2\text{Re}(z)-3\text{Im}(z)=k\}$. Determine the values of k for which the line through T is a tangent to the circle S.

19. Let $S=\{z: |z|=3\}$ and $T=\left\{z: \text{Arg}(z)=-\frac{\pi}{4}\right\}$. Sketch the sets S and T on the same Argand diagram and determine $z: S=T$.

20. Sets of points in the complex plane are defined by $S=\{z: |z+3+i|=5\}$ and $R=\left\{z: \text{Arg}(z+3)=-\frac{3\pi}{4}\right\}$.
    a. Determine the Cartesian equation of S.
    b. Determine the Cartesian equation of R.
    c. If $u \in C$, determine u where $S=R$.

21. Two sets of points in the complex plane are defined by $S=\{z: |z|=3\}$ and $T=\{z: 3\text{Re}(z)+4\text{Im}(z)=15\}$. Show that the line T is a tangent to the circle S and determine the coordinates of the point of contact.

22. Two sets of points in the complex plane are defined by $S=\{z: |z|=6\}$ and $T=\{z: 3\text{Re}(z)-4\text{Im}(z)=k\}$. Determine the values of k for which the line through T is a tangent to the circle S.

23. Show that the complex equation $|z-a|^2-|z-bi|^2=a^2+b^2$, where a and b are real and $b \neq 0$, represents a line.

24. Show that the complex equation $3z\bar{z}+6z+6\bar{z}+2=0$ represents a circle, and determine its centre and radius.

25. a. Let $w = \text{cis}(\theta)$ and $z = \frac{1}{2}\left(9w - \overline{w}\right)$.

i. Express z in terms of θ.

ii. Show that z lies on the ellipse $\frac{x^2}{16} + \frac{y^2}{25} = 1$.

iii. Show that $|z - 3i|^2 = (5 - 3\sin(\theta))^2$ and $|z + 3i|^2 = (5 + 3\sin(\theta))^2$.

iv. Hence, show that $|z + 3i| + |z - 3i| = 10$.

b. Show by substituting $z = x + yi$ that the equation $\{z : |z - 3i| + |z + 3i| = 10\}$ represents the ellipse $\frac{x^2}{16} + \frac{y^2}{25} = 1$ in the Argand plane.

26. Show that the equation $\{z : |z - c| + |z + c| = 2a\}$ represents the ellipse $\frac{x^2}{a^2} + \frac{y^2}{a^2 - c^2} = 1$ provided that $a^2 > c^2$.

27. a. Show that the complex equation $z\overline{z} + (3 + 2i)z + (3 - 2i)\overline{z} + 4 = 0$ represents a circle, and determine its centre and radius.

b. Consider the complex equation $az\overline{z} + \overline{b}z + b\overline{z} + c = 0$ where a and c are real and $b = \alpha + \beta i$ is complex. Show that the equation represents a circle provided $b\overline{b} > ac$ and $a \neq 0$, and determine the circle's centre and radius.

28. a. Show that the complex equation $\left\{z\text{: Im}\left(\frac{z - ai}{z - b}\right) = 0\right\}$ where a and b are real represents a straight line if $ab \neq 0$.

b. Show that the complex equation $\left\{z\text{: Re}\left(\frac{z - ai}{z - b}\right) = 0\right\}$ where a and b are real represents a circle if $ab \neq 0$. State the circle's centre and radius.

29. Given that $c = a + bi$ where a and b are real, show that the complex equation $(z - c)(\overline{z} - \overline{c}) = r^2$ represents a circle, and determine its centre and radius.

30. Given that $c = a + bi$ where a and b are real, show that the complex equation $|z - c| = |2z - \overline{c}|$ represents a circle, and determine its centre and radius.

2.5 Exam questions

Question 1 (1 mark) TECH-ACTIVE

Source: VCE 2021 Specialist Mathematics Exam 2, Section A, Q5; © VCAA.

MC The graph of the circle given by $\left|z - 2 - \sqrt{3}\, i\right| = 1$, where $z \in C$, is shown.

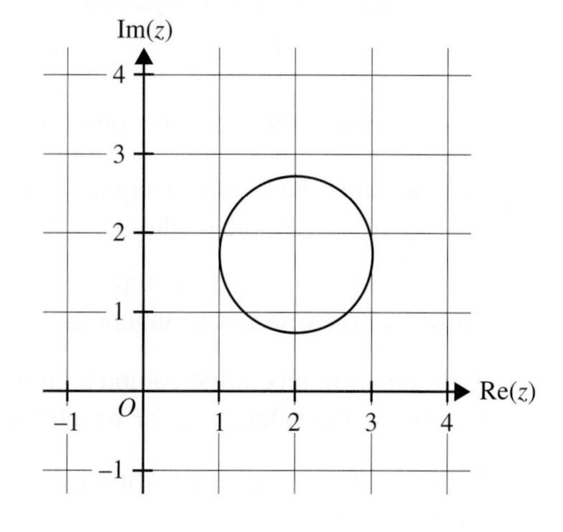

For points on this circle, the maximum value of $|z|$ is

A. $\sqrt{3} + 1$

B. 3

C. $\sqrt{13}$

D. $\sqrt{7} + 1$

E. 8

Question 2 (11 marks) TECH-ACTIVE

Source: *VCE 2020 Specialist Mathematics Exam 2, Section B, Q2; © VCAA.*

Two complex numbers, u and v, are defined as $u = -2 - i$ and $v = -4 - 3i$.

a. Express the relation $|z - u| = |z - v|$ in the cartesian form $y = mx + c$, where $m, c \in R$. **(3 marks)**

b. Plot the points that represent u and v and the relation $|z - u| = |z - v|$ on the Argand diagram below. **(2 marks)**

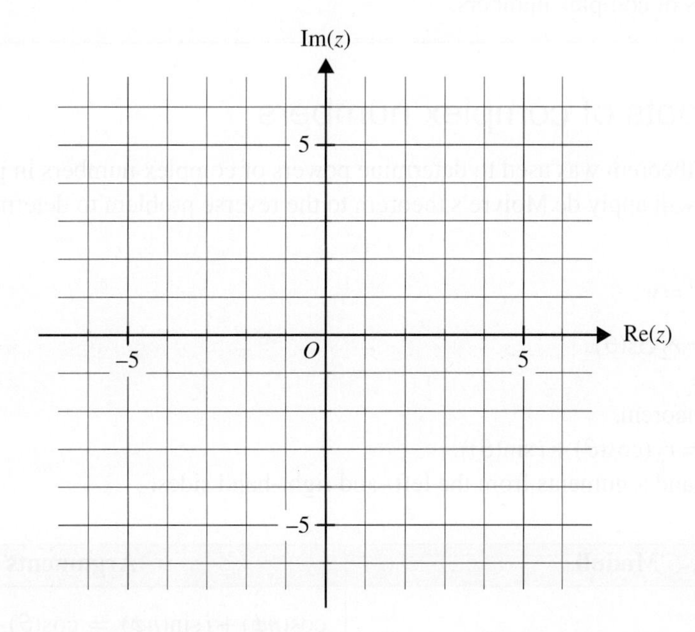

c. State a geometrical interpretation of the graph of $|z - u| = |z - v|$ in the points that represent u and v. **(1 mark)**

d. i. Sketch the ray given by $\text{Arg}(z - u) = \frac{\pi}{4}$ on the Argand diagram in part **b**. **(1 mark)**

ii. Write down the function that describes the ray $\text{Arg}(z - u) = \frac{\pi}{4}$, giving the rule in cartesian form. **(1 mark)**

e. The points representing u and v and $-5i$ lie on the circle given by $|z - z_c| = r$, where z_c is the centre of the circle and r is the radius.

Find z_c in the form $a + ib$, where $a, b \in R$, and find the radius r. **(3 marks)**

Question 3 (1 mark) TECH-ACTIVE

Source: *VCE 2019 Specialist Mathematics Exam 2, Section A, Q5; © VCAA.*

MC Let $z = x + yi$, where $x, y \in R$. The rays $\text{Arg}(z - 2) = \frac{\pi}{4}$ and $\text{Arg}(z - (5 + i)) = \frac{5\pi}{6}$, where $z \in C$, intersect on the complex plane at a point (a, b).

The value of b is

A. $-\sqrt{3}$

B. $2 - \sqrt{3}$

C. 0

D. $\sqrt{3}$

E. $2 + \sqrt{3}$

More exam questions are available online.

2.6 Roots of complex numbers

LEARNING INTENTION

At the end of this subtopic you should be able to:

- calculate the roots of complex numbers.

2.6.1 The *n*th roots of complex numbers

Recall that de Moivre's theorem was used to determine powers of complex numbers in polar form, $z^n = r^n \text{cis}(n\theta)$. Now we will apply de Moivre's theorem to the reverse problem to determine the roots of complex numbers.

Consider the equation $z^n = w$.

Let $z = r_1 \text{ cis}(\phi)$ and $w = r_2 \text{ cis}(\theta)$.
$\therefore (r_1 \text{cis}(\phi))^n = r_2 \text{cis}(\theta)$
Applying de Moivre's theorem,
$r_1^n (\cos(n\phi) + i\sin(n\phi)) = r_2 (\cos(\theta) + i\sin(\theta))$
Now equate the moduli and arguments from the left- and right-hand sides.

Moduli	Arguments
$\begin{aligned} r_1^n &= r_2 \\ \therefore r_1 &= \sqrt[n]{r_2} \\ &= (r_2)^{\frac{1}{n}} \end{aligned}$	$\begin{aligned} \cos(n\phi) + i\sin(n\phi) &= \cos(\theta) + i\sin(\theta) \\ \text{cis}(n\phi) &= \text{cis}(\theta) \\ n\phi &= \theta + 2k\pi \text{ where } k \in Z \\ \phi &= \frac{\theta + 2k\pi}{n} \text{ where } k \in Z \end{aligned}$

The investigation above illustrates that if $z^n = w$ where $z, w \in C$ and $n \in N$, the equation has n distinct solutions, in which z is termed an nth root of w.

Because the solutions start repeating, we keep using k until n solutions are found. The roots are generally denoted as $z_1, z_2, \ldots z_n$ where z_1 represents the first root. The first root is commonly known as the principal nth root of the complex number.

WORKED EXAMPLE 31 Determining the square roots of complex numbers in polar form

Solve $z^2 = 4 \text{ cis}\left(\frac{\pi}{3}\right)$ for z.

THINK	WRITE
1. Use the results from the table above.	$z = \sqrt[2]{4} \text{ cis}\left(\frac{\frac{\pi}{3} + 2k\pi}{2}\right)$
2. Simplify the modulus and argument.	$z = 2 \text{ cis}\left(\frac{\pi}{6} + k\pi\right)$

3. Let $k=0$ to calculate the first root.

$$\text{Let } k=0,\ z_1 = 2\,\text{cis}\left(\frac{\pi}{6}+0\times\pi\right) = 2\,\text{cis}\left(\frac{\pi}{6}\right)$$

4. Let $k=1$ to calculate the second root.

$$\text{Let } k=1,\ z_2 = 2\,\text{cis}\left(\frac{\pi}{6}+1\times\pi\right) = 2\,\text{cis}\left(\frac{7\pi}{6}\right) = 2\,\text{cis}\left(-\frac{5\pi}{6}\right)$$

5. Since the polynomial is quadratic, there are 2 roots, so we have our answer.

$$z_1 = 2\,\text{cis}\left(\frac{\pi}{6}\right),\ z_2 = 2\,\text{cis}\left(-\frac{5\pi}{6}\right)$$

By plotting the solutions on an Argand diagram, it can be seen that the roots to the equation lie on a circle with a radius of 2. The two solutions are separated around the circle by π (or 180°). Although there are two roots, they are not complex conjugates of one another.

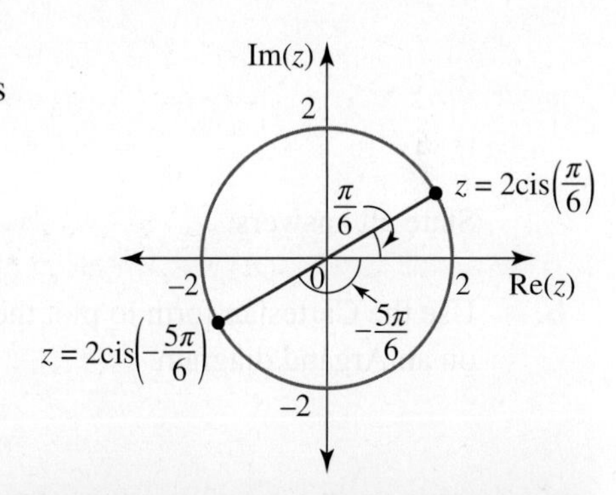

WORKED EXAMPLE 32 Determining the cube roots of complex numbers in polar form

a. If $z^3 = 4 - 4\sqrt{3}i$, solve the equation for z using a polar method.
b. Plot the solutions on an Argand diagram and comment on their location.

THINK | **WRITE**

a. 1. Express z^3 in polar form $\left(z^3 = r\,\text{cis}(\theta)\right)$.

a. $|z^3| = \sqrt{(4)^2 + \left(-4\sqrt{3}\right)^2}$

$|z^3| = 8$

$\theta = \tan^{-1}\left(\frac{-4\sqrt{3}}{4}\right)$

$\theta = \frac{-\pi}{3}$

$\therefore z^3 = 4 - 4\sqrt{3}i = 8\,\text{cis}\left(\frac{-\pi}{3}\right)$

$z^3 = 8\,\text{cis}\left(-\frac{\pi}{3}\right)$

2. Use the results from the table above.

$$z = \sqrt[3]{8}\,\text{cis}\left(\frac{-\frac{\pi}{3}+2k\pi}{3}\right)$$

3. Simplify the modulus and argument.

$$z = 2\operatorname{cis}\left(\frac{-\pi}{9} + \frac{2k\pi}{3}\right)$$

4. Let $k = 0$ to calculate the first root.

$$\text{Let } k = 0,\ z_1 = 2\operatorname{cis}\left(\frac{-\pi}{9} + \frac{2 \times 0 \times \pi}{3}\right) = 2\operatorname{cis}\left(\frac{-\pi}{9}\right)$$

5. Let $k = 1$ to calculate the second root.

$$\text{Let } k = 1,\ z_2 = 2\operatorname{cis}\left(\frac{-\pi}{9} + \frac{2 \times 1 \times \pi}{3}\right) = 2\operatorname{cis}\left(\frac{5\pi}{9}\right)$$

6. Let $k = 2$ to calculate the third root.

$$\text{Let } k = 2,\ z_3 = 2\operatorname{cis}\left(\frac{-\pi}{9} + \frac{2 \times 2 \times \pi}{3}\right) = 2\operatorname{cis}\left(\frac{-7\pi}{9}\right)$$

State all answers.

$$z = 2\operatorname{cis}\left(-\frac{\pi}{9}\right),\ 2\operatorname{cis}\left(\frac{5\pi}{9}\right) \text{ and } 2\operatorname{cis}\left(-\frac{7\pi}{9}\right)$$

b. Use the Cartesian form to plot the solutions on an Argand diagram.

b.

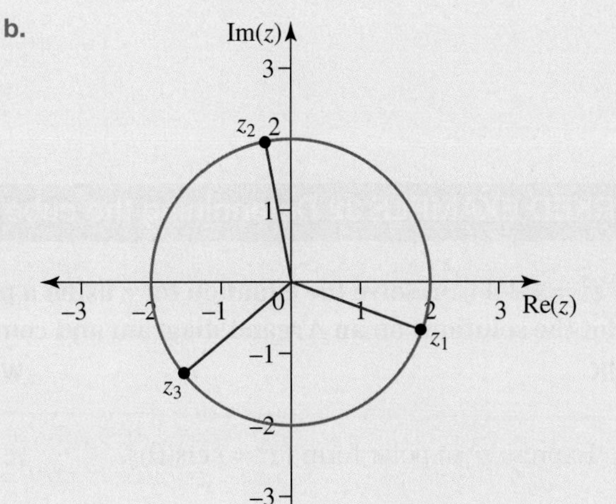

The three roots are equally spaced around a circle of radius 2.

The angle between each solution is $\frac{2\pi}{3}$ (120°).

In the previous worked example there were 3 solutions that were evenly spaced by $\frac{2\pi}{n} = \frac{2\pi}{3}$. Consider the problem below in which the fourth roots of a complex number are determined.

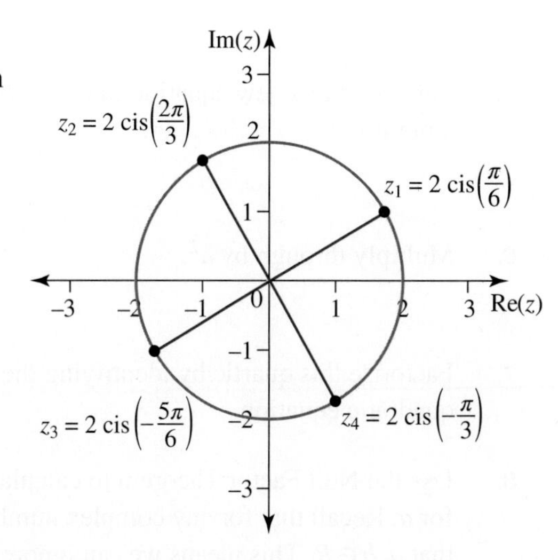

$$z^4 = 16\text{ cis}\left(\frac{2\pi}{3}\right)$$

$$\text{Solutions}\begin{cases} z_1 = 2\text{ cis}\left(\frac{\pi}{6}\right) \\ z_2 = 2\text{ cis}\left(\frac{2\pi}{3}\right) \\ z_3 = 2\text{ cis}\left(-\frac{5\pi}{6}\right) \\ z_4 = 2\text{ cis}\left(-\frac{\pi}{3}\right) \end{cases}$$

In the above case there are 4 solutions that were evenly spaced by $\frac{2\pi}{n} = \frac{2\pi}{4} = \frac{\pi}{2}$. The general properties of solutions to $z^n = w$ are summarised below.

Solutions of $z^n = w$

Solutions to $z^n = w$ where $z, w \in C$ and $n \in N$ are in the form:

$$z = \sqrt[n]{r}\text{ cis}\left(\frac{\theta + 2k\pi}{n}\right), k = 0, 1, 2, \ldots, n-1.$$

The solutions are equidistant from the origin and lie on a circle of radius $\sqrt[n]{r}$.

The solutions are evenly spaced around the circle with the arguments differing by $\frac{2\pi}{n}$.

The following example demonstrates an alternative method for solving equations of the form $z^n = w$ where $w \in C$.

WORKED EXAMPLE 33 Determining the square roots in Cartesian form

Solve $z^2 = 3 - 4i$ using a Cartesian approach.

THINK	WRITE
1. Express z in Cartesian form and substitute this into the equation.	$z = a + bi$ $(a + bi)^2 = 3 - 4i$
2. Expand the brackets using $(a + b)^2 = a^2 + 2ab + b^2$ and use $i^2 = -1$ to simplify the equation.	$a^2 + 2abi + b^2i^2 = 3 - 4i$ $a^2 - b^2 + 2abi = 3 - 4i$
3. Equate the coefficients of the real parts and the imaginary parts.	$a^2 - b^2 = 3$ and $2ab = -4$
4. Rearrange $2ab = -4$ for b.	$2ab = -4 \Rightarrow b = -\frac{2}{a}$

5. Substitute this new equation into $a^2 - b^2 = 3$ and simplify.

$$a^2 - \left(-\frac{2}{a}\right)^2 = 3$$
$$a^2 - \frac{4}{a^2} = 3$$

6. Multiply through by a^2.

$$a^4 - 4 = 3a^2$$
$$a^4 - 3a^2 - 4 = 0$$

7. Factorise this quartic by identifying the hidden quadratic equation.

$$\left(a^2\right)^2 - 3\left(a^2\right) - 4 = 0$$
$$\left(a^2 - 4\right)\left(a^2 + 1\right) = 0$$

8. Use the Null Factor Theorem to calculate the answers for a. Recall that for any complex number $z = a + bi$ that $a, b \in R$. This means we can ignore any solutions to $(a^2 + 1) = 0$.

$a^2 - 4 = 0$ or $a^2 + 1 = 0$
$(a - 2)(a + 2) = 0$ since $a, b \in R$
$\therefore a = 2$ or $a = -2$

9. Determine the corresponding values of b for each value of a.

When $a = 2$, $b = -1$ and when $a = -2$, $b = 1$.

10. State the solutions to the equation.

The solutions to the equation $z^2 = 3 - 4i$ are $z = 2 - i$ and $z = -2 + i$.

2.6.2 The *n*th roots of unity

Since $1 = 1 + 0i = \text{cis}(0)$, de Moivre's theorem is often used to solve equations in the form $z^n = 1$. The solutions to this form of an equation and are known as the *n*th roots of 1 (unity). When interpreting the solutions geometrically, the roots represent the n vertices of a regular polygon of n sides inscribed inside a circle of radius one. This circle is referred to as the unit circle defined by the equation $|z| = 1$.

Using our knowledge of solving equations of the form $z^n = w$ where $w \in C$, we can deduce that the solutions to the equation $z^8 = 1$ where $z \in C$ will lie on a circle of radius 1 and be evenly spaced around the circle by $\frac{\pi}{4}$. The eight 8th roots of 1 can be interpreted geometrically using an Argand diagram.

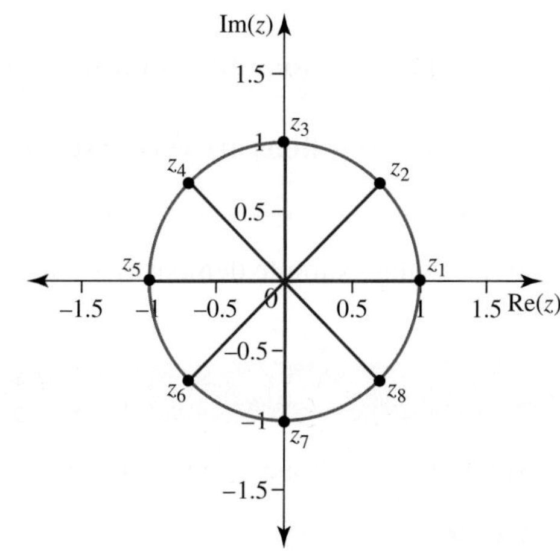

The solution set is

$$z \in \left\{\text{cis}(0), \text{cis}\left(\frac{\pi}{4}\right), \text{cis}\left(\frac{\pi}{2}\right), \text{cis}\left(\frac{3\pi}{4}\right), \text{cis}(\pi), \text{cis}\left(-\frac{3\pi}{4}\right), \text{cis}\left(-\frac{\pi}{2}\right), \text{cis}\left(-\frac{\pi}{4}\right)\right\}.$$

Note that all non-real roots occur as conjugate pairs. In this case, n is even, so there are two real roots $\{1, -1\}$. If n is odd, the only real root will be 1 and the other roots will be imaginary, occurring as conjugate pairs.

*n*th roots of unity

If $z^n = 1$, where $z \in C$ and $n \in N$, then the *n*th roots of 1 (unity) are given by

$$z = \text{cis}\left(\frac{2k\pi}{n}\right),\ k = 0, 1, 2, \ldots, n - 1$$

WORKED EXAMPLE 34 Determining *n*th roots of unity

a. Determine all solutions to the equation, $z^3 = 1$, over C. Express the solutions in polar form.
b. Represent these solutions using a polar grid.

THINK	WRITE
a. 1. Express 1 in polar form.	**a.** $1 = \text{cis}(0)$
2. Use the results from the table above.	$z^3 = \text{cis}(0)$ $z = \sqrt[3]{1}\,\text{cis}\left(\frac{0+2k\pi}{3}\right)$
3. Simplify the modulus and argument.	$z = 1\,\text{cis}\left(\frac{2k\pi}{3}\right)$
4. Let $k=0$ to calculate the first root.	Let $k=0$, $z_1 = \text{cis}\left(\frac{2\times 0\times \pi}{3}\right)$ $= \text{cis}(0)$ $= 1$
5. Let $k=1$ to calculate the second root.	Let $k=1$, $z_2 = \text{cis}\left(\frac{2\times 1\times \pi}{3}\right)$ $= \text{cis}\left(\frac{2\pi}{3}\right)$
6. Let $k=2$ to calculate the third root.	Let $k=2$, $z_3 = \text{cis}\left(\frac{2\times 2\times \pi}{3}\right)$ $= \text{cis}\left(\frac{-2\pi}{3}\right)$
b. Plot the three roots on a polar grid. The circle has a centre (0, 0) and radius, $r=1$. The three solutions are spaced by $\frac{2\pi}{3}$.	**b.** 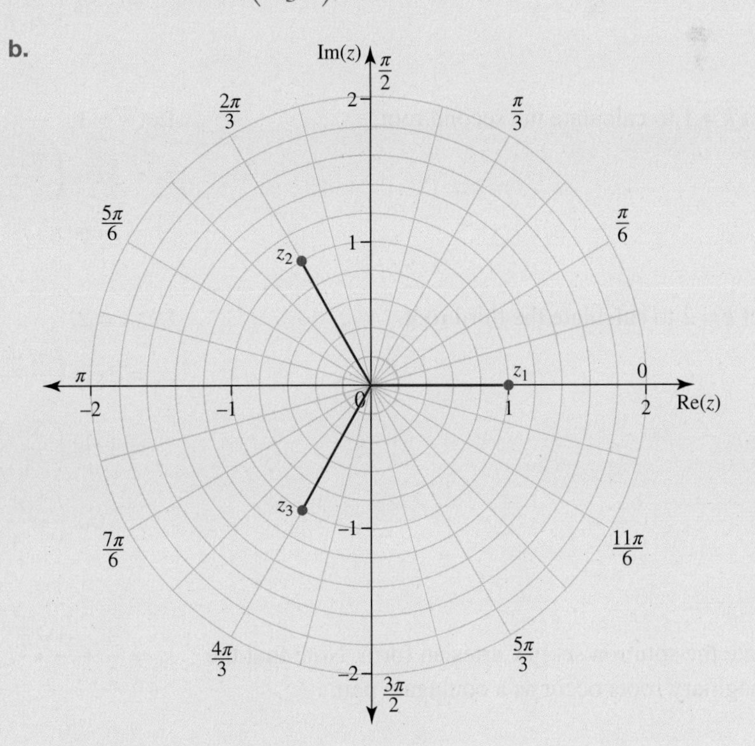

Note: If you were to construct lines between the three solutions, an equilateral triangle would be formed.

2.6.3 Solving $z^n = a$ where $a \in R$

Equations of the form $z^n = a$ where $a \in R$, can be solved in the same way in which equations of the form $z^n = w$ where $w \in C$ were solved. That is, if

$z^n = a, \quad a \in R$ and $n \in N$,

then $z = |a|^{\frac{1}{n}} \text{cis}\left(\frac{\theta + 2k\pi}{n}\right), \; k = 0, 1, 2, \ldots, n-1.$

WORKED EXAMPLE 35 Solving equations of the form $z^n = a$

Determine the solutions to $\{z : z^3 = -27\}$, stating the solution set in Cartesian form.

THINK	WRITE
1. Express −27 in polar form.	$-27 = \lvert -27 \rvert \text{cis}(\pi)$ $= 27 \text{cis}(\pi)$
2. Use the results from the table above.	$z^3 = 27 \text{cis}(\pi)$ $z = \sqrt[3]{27} \text{cis}\left(\frac{\pi + 2k\pi}{3}\right)$ $z = 3 \text{cis}\left(\frac{\pi + 2k\pi}{3}\right)$
3. Let $k = 0$ to calculate the first root.	Let $k = 0$, $z_1 = 3 \text{cis}\left(\frac{\pi + 2 \times 0 \times \pi}{3}\right)$ $= 3 \text{cis}\left(\frac{\pi}{3}\right)$
4. Let $k = 1$ to calculate the second root.	Let $k = 1$, $z_2 = 3 \text{cis}\left(\frac{\pi + 2 \times 1 \times \pi}{3}\right)$ $= 3 \text{cis}(\pi)$
5. Let $k = 2$ to calculate the third root.	Let $k = 2$, $z_3 = 3 \text{cis}\left(\frac{\pi + 2 \times 2 \times \pi}{3}\right)$ $= 3 \text{cis}\left(\frac{5\pi}{3}\right)$ $= 3 \text{cis}\left(-\frac{\pi}{3}\right)$
6. State the solution set in Cartesian form. Note that the imaginary roots occur as a conjugate pair.	$z_1 = \frac{3}{2} + \frac{3\sqrt{3}}{2} i, \; z_2 = -3, \; z_3 = \frac{3}{2} - \frac{3\sqrt{3}}{2} i$

2.6 Exercise

Students, these questions are even better in jacPLUS

Receive immediate feedback and access sample responses

Access additional questions

Track your results and progress

Find all this and MORE in jacPLUS

Technology free

1. WE31 Solve $z^2 = 5\,\text{cis}\left(\frac{\pi}{6}\right)$ for z.

2. Solve $z^3 = 27\,\text{cis}\left(\frac{2\pi}{3}\right)$ for z.

3. Solve $z^4 = 16\,\text{cis}\left(\frac{-3\pi}{4}\right)$ for z.

4. a. WE32 If $z^3 = -1 - \sqrt{3}i$, solve the equation for z giving answers in polar form.
   b. Plot the solutions on an Argand diagram and comment on their location.

5. a. If $z^4 = -5 - 5i$, solve the equation for z giving answers in polar form.
   b. Plot the solutions on an Argand diagram and comment on their location.

6. a. If $z^6 = 3 + \sqrt{3}i$, solve the equation for z giving answers in polar form.
   b. Plot the solutions on an Argand diagram and comment on their location.

7. Solve for z in the following cases.
   a. $z^3 = i$
   b. $z^4 = 64i$

8. WE33 Solve $z^2 = 2 + 2\sqrt{3}i$ using a Cartesian approach.

9. Use the solutions from Worked example 34 to verify that the product of the three cube roots of 1 is 1.

10. WE34 a. Determine all solutions to the equation, $z^4 = 1$, over C. Express the solutions in polar form.
    b. Represent these solutions using a polar grid.

11. Use the solutions from Question **10** to verify that the sum of the four fourth roots of 1 is 0.

12. a. Determine all solutions to the equation, $z^5 = 1$, over C. Express the solutions in polar form.
    b. Represent these solutions using a polar grid.

13. The solutions from Question **12** can be defined as ω_0, ω_1, ω_2, ω_3 and ω_4. Describe how these solutions would compare to the solutions to the equation, $z^5 = 10$, over C.

14. WE35 Determine the solutions to the following equations, stating the solution set in Cartesian form.
    a. $z^3 = 8$
    b. $z^4 = 256$

15. Plot the solutions to $z^{10} = 1$ on a polar grid without determining the solutions.

16. If 1, u and v represent the three cube roots of unity, show that:
    a. $v = \overline{u}$
    b. $u^2 = v$
    c. $1 + u + v = 0$.

17. Let $u = 2 - 2i$.
   a. Determine $\text{Arg}(u^4)$ and hence determine u^4.
   b. Determine all the solutions of $z^4 + 64 = 0$, giving your answers in both Cartesian and polar form.
   c. Plot all the solutions of $z^4 + 64 = 0$ on one Argand diagram and comment on their relative positions.

18. Determine all the solutions for each of the following, giving your answers in both polar and Cartesian form.
   a. $z^6 - 64 = 0$
   b. $z^6 + 64 = 0$

19. Solve $z^8 - 16 = 0$, giving your answers in both polar and Cartesian form.

20. Determine all the roots of the equation $z^{12} - 4096 = 0$, giving your answers in both polar and Cartesian form.

2.6 Exam questions

Question 1 (3 marks) TECH-FREE

Source: *VCE 2020 Specialist Mathematics Exam 1, Q3; © VCAA.*

Find the cube roots of $\frac{1}{\sqrt{2}} - \frac{1}{\sqrt{2}}i$. Express your answer in polar form using principal values of the argument.

Question 2 (1 mark) TECH-ACTIVE

Source: *VCE 2017 Specialist Mathematics Exam 2, Section A, Q4; © VCAA.*

MC The solution to $z^n = 1 + i, n \in Z^+$ are given by

A. $2^{\frac{1}{2n}} \text{cis}\left(\frac{\pi}{4n} + \frac{2\pi k}{n}\right), k \in R$

B. $2^{\frac{1}{n}} \text{cis}\left(\frac{\pi}{4n} + 2\pi k\right), k \in Z$

C. $2^{\frac{1}{2n}} \text{cis}\left(\frac{\pi}{4} + \frac{2\pi k}{n}\right), k \in R$

D. $2^{\frac{1}{n}} \text{cis}\left(\frac{\pi}{4n} + \frac{2\pi k}{n}\right), k \in Z$

E. $2^{\frac{1}{2n}} \text{cis}\left(\frac{\pi}{4n} + \frac{2\pi k}{n}\right), k \in Z$

Question 3 (4 marks) TECH-ACTIVE

Source: *VCE 2015 Specialist Mathematics Exam 1, Q4; © VCAA.*

a. Find all solution of $z^3 = 8i, z \in C$ in cartesian form. **(3 marks)**

b. Find all solution of $(z - 2i)^3 = 8i, z \in C$ in cartesian form. **(1 mark)**

More exam questions are available online.

2.7 Review

2.7.1 Summary

doc-37056

Hey students! Now that it's time to revise this topic, go online to:

 Access the topic summary  Review your results  Watch teacher-led videos  Practise exam questions

Find all this and MORE in jacPLUS

2.7 Exercise

Technology free: short answer

1. a. Calculate $\text{Re}\left(\dfrac{5+2i}{5-2i}\right)+\text{Im}\left(\dfrac{5-2i}{5+2i}\right)$.
   b. Determine the value of d if $\text{Im}\left(\dfrac{d-2i}{3-7i}\right)=0$.
2. Express each of the following in Cartesian form.
   a. $\dfrac{(-1+i)^4}{\left(-\sqrt{3}+i\right)^6}$
   b. $(-1-i)^7\left(\sqrt{3}-i\right)^6$
3. Solve each of the following for z.
   a. $z^3-2iz^2+10z-20i=0$
   b. $2z^4-z^2-45=0$
4. Solve each of the following for n.
   a. $(5-12\,i)^n-(5+12\,i)^n=0$
   b. $(3+4\,i)^n+(3-4\,i)^n=0$
5. For each of the following subsets of the complex plane, determine and describe the Cartesian equation.
   a. $\{z\colon |z-4|=2\,|z-1|\}$
   b. $\{z\colon |z-4|=|z-1|\}$
   c. $\{z\colon |z-4i|=\text{Im}(z)-2\}$
   d. $\{z\colon |z-4|=2\,(\text{Re}(z)-1)\}$
6. Solve each of the following for z, giving your answer in Cartesian $a+bi$ form.
   a. $z^2+i=0$
   b. $z^3+27=0$
   c. $z^3+27i=0$
   d. $z^4+1024=0$
7. a. Sets of points in the complex plane are defined by $S=\{z\colon |z+1-2i|=5\}$ and $T=\{z\colon \text{Re}(z)+2\text{Im}(z)=8\}$. Determine the coordinates of the points of intersection between S and T.
   b. Sets of points in the complex plane are defined by $S=\{z\colon |z|=2\}$ and $T=\{z\colon a\text{Re}(z)-2\text{Im}(z)=5\}$. Determine the values of a for which the line T is a tangent to the circle S.

Technology active: multiple choice

8. MC Which of the following is equal to -1?
   A. i^{20}
   B. $\text{cis}\left(-\dfrac{\pi}{2}\right)$
   C. $\text{cis}(\pi)$
   D. $\text{cis}\left(-\dfrac{\pi}{2}\right)+\text{cis}\left(\dfrac{\pi}{2}\right)$
   E. $\text{cis}\left(-\dfrac{3\pi}{4}\right)+\text{cis}\left(\dfrac{3\pi}{4}\right)$

9. MC If $z = \frac{1}{a} + bi$ where a and b are non-zero real numbers, then $\frac{1}{z}$ is equal to

A. $a - \frac{i}{b}$
B. $a + \frac{i}{b}$
C. $\frac{a}{1+a^2b^2} + \frac{a^2bi}{1+a^2b^2}$
D. $\frac{a}{1+a^2b^2} - \frac{a^2bi}{1+a^2b^2}$
E. $\frac{a}{1-a^2b^2} - \frac{a^2bi}{1-a^2b^2}$

10. MC If $z = a + bi$ where a and b are non-zero real numbers, which of the following statements is **false**?

A. $z^2 = a^2 - b^2 + 2abi$
B. $\text{Arg}(z) = \tan^{-1}\left(\frac{b}{a}\right)$
C. $|z| = |\bar{z}| = \sqrt{a^2 + b^2}$
D. $\text{Im}(z - \bar{z}) = 2b$
E. $\text{Re}(z + \bar{z}) = 2a$

11. MC If $z = 7\text{ cis}\left(\frac{\pi}{11}\right)$ then $(\bar{z})^{-1}$ is equal to

A. $\frac{1}{7}\text{cis}\left(\frac{\pi}{11}\right)$
B. $\frac{1}{7}\text{cis}\left(-\frac{\pi}{11}\right)$
C. $\frac{1}{7}\text{cis}\left(\frac{11}{\pi}\right)$
D. $-7\text{cis}\left(-\frac{11}{\pi}\right)$
E. $-7\text{cis}\left(\frac{11}{\pi}\right)$

12. MC If $z = -1 - i$ then $\text{Arg}(z^7)$ is equal to

A. $\left(-\frac{3\pi}{4}\right)^7$
B. $-\frac{21\pi}{4}$
C. $-\frac{3\pi}{4}$
D. $-\frac{\pi}{4}$
E. $\frac{3\pi}{4}$

13. MC The diagram shows a circle of radius 4 on an Argand diagram. The points shown u, v and w are equally spaced around the circle and are the solutions of the equation $P(z) = 0$. Then

A. $P(z) = z^3 + 64i$
B. $P(z) = z^3 - 64i$
C. $P(z) = z^3 - 44$
D. $P(z) = z^3 - 4i$
E. $P(z) = z^3 + 4i$

Im(z)
u
Re(z)
v
w

14. MC $P(z)$ is a polynomial in z of degree 4 with real coefficients.
Several students stated some information regarding the roots.
Andrew stated that, $P(z) = 0$ can have two real roots and one pair of complex conjugate roots.
Betty stated that, $P(z) = 0$ can have three real roots and one complex root.
Colin stated that, $P(z) = 0$ can have one real root and three complex roots.
Daisy stated that, $P(z) = 0$ can have two pairs of complex conjugate roots.
Edward stated that, $P(z) = 0$ can have four real roots.
Determine which of the following statements is correct.

A. Andrew, Betty, Colin, Daisy and Edward are all correct.
B. only Betty, Colin and Daisy are correct.
C. only Andrew and Edward are correct.
D. only Daisy and Edward are correct.
E. only Andrew, Daisy and Edward are correct.

15. MC The set of points in the complex plane described by $\{z: |z + 8i| = 2|z + 2i|\}$ represents

A. a straight line.
B. a circle.
C. an ellipse.
D. a hyperbola.
E. a parabola.

16. **MC** If $z=-a-ai$, $a\in R^{+}$, then the complex number z^{n}, $n\in R$ is real for:

A. $n=\dfrac{4k}{3}$ where $k\in Z$.

B. $n=2(2k+1)$ where $k\in Z$.

C. $n=3k$ where $k\in Z$.

D. $n=\dfrac{2(2k+1)}{3}$ where $k\in Z$.

E. no values of n.

17. **MC** If $z^2=5-12i$ then z is equal to

A. $\sqrt{5}-2\sqrt{3}\,i$ only

B. $\pm\left(\sqrt{5}-2\sqrt{3}\,i\right)$

C. $\sqrt{5}-2\sqrt{3}\,i$ or $\sqrt{5}+2\sqrt{3}\,i$

D. $3-2i$ or $-3+2i$

E. $3-2i$ only

Technology active: extended response

18. **a.** Given that $\cos\left(\dfrac{\pi}{10}\right)=\dfrac{1}{4}\sqrt{2\left(5+\sqrt{5}\right)}$, show that $\sin\left(\dfrac{\pi}{10}\right)=\dfrac{1}{4}\left(\sqrt{5}-1\right)$.

b. Let $u=\dfrac{1}{4}\sqrt{2\left(5+\sqrt{5}\right)}+\dfrac{1}{4}\left(\sqrt{5}-1\right)i$. Express u in polar form.

c. Evaluate u^{20}, u^{30} and u^{45} giving your answers in simplest rectangular form.

d. Determine the value(s) of n such that $\text{Re}\left(u^{n}\right)=0$.

19. In the complex plane C is the circle $|z-8|=4\sqrt{3}$ and L is the line $\text{Im}\left(\dfrac{z+i\sqrt{3}}{z-1}\right)+\dfrac{\sqrt{3}}{|z-1|^{2}}=0$.

a. Determine the Cartesian equation of the circle C.

b. Show that the Cartesian equation of the line L is given by $y=\sqrt{3}x$.

The part of the line L in the first quadrant can be expressed as $\text{Arg}(z)=\alpha$.

c. State the value of α.

d. Determine the point(s) of intersection between the line L and the circle C in Cartesian form.

e. Sketch the line L and the circle C on the Argand diagram below.

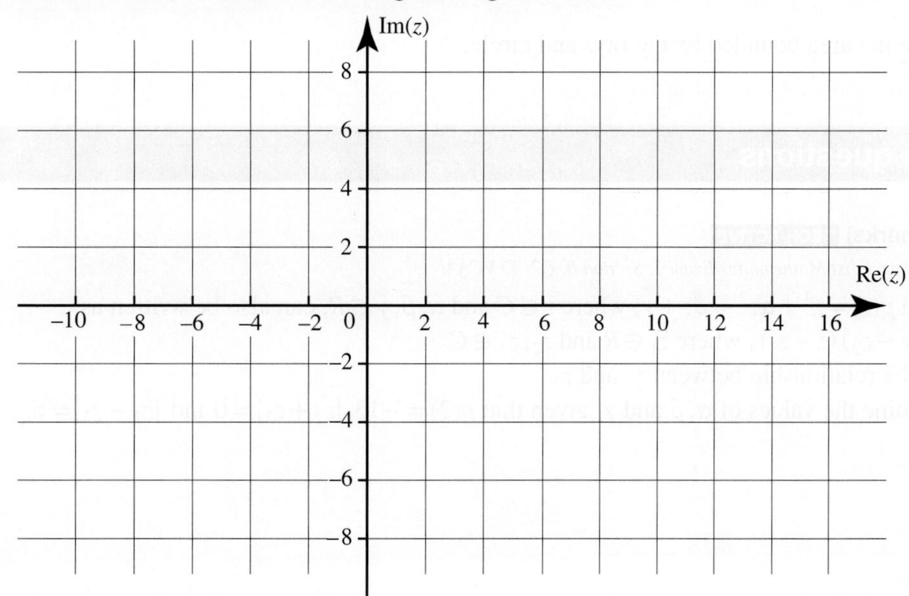

f. Calculate the area in the first quadrant enclosed by the graph of the line L and the circle C and the real axis.

The straight line L can also be written in the form $|z - k| = |z - i|$, where $k \in C$.

g. Express k in the form $r\,\text{cis}(\theta)$, where θ is the principal argument of k.

h. Determine the set of values of β for which $\text{Arg}(z) = \beta$ and the circle C do not intersect.

20. Let $R_1 = \left\{z: \text{Arg}(z) = \dfrac{\pi}{12}\right\}$ and $R_2 = \left\{z: \text{Arg}(z) = \dfrac{5\pi}{12}\right\}$ be two rays and $\left\{z: |z - 2 - 2i| = \sqrt{2}\right\}$ where $z = x + yi,\ z \in C$ be a circle in the complex plane.

a. Express both rays and the circle in Cartesian form.

b. Determine the coordinates of the point(s) of intersection between each ray and the circle.

c. Sketch both rays and the circle on the Argand diagram below.

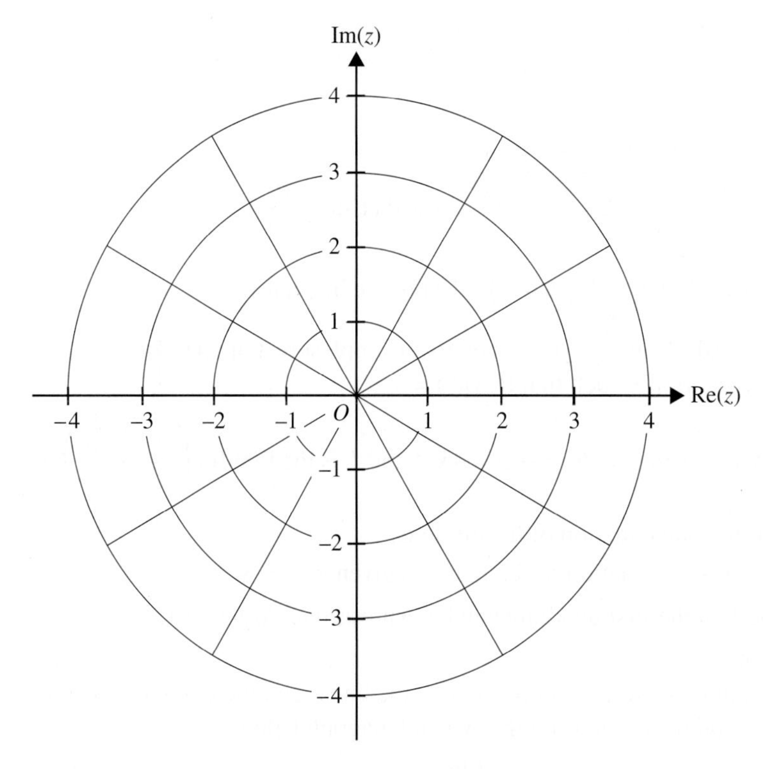

d. Calculate the area bounded by the rays and circle.

2.7 Exam questions

Question 1 (9 marks) TECH-ACTIVE

Source: *VCE 2021 Specialist Mathematics Exam 2, Section B, Q2; © VCAA.*

The polynomial $p(z) = z^3 + \alpha z^2 + \beta z + \gamma$, where $z \in C$ and $\alpha, \beta, \gamma \in R$, can also be written as $p(z) = (z - z_1)(z - z_2)(z - z_3)$, where $z_1 \in R$ and $z_2, z_3 \in C$.

a. **i.** State the relationship between z_2 and z_3. **(1 mark)**

ii. Determine the values of α, β and γ, given that $p(2) = -13$, $|z_2 + z_3| = 0$ and $|z_2 - z_3| = 6$. **(3 marks)**

Consider the point $z_4 = \sqrt{3} + i$.

b. Sketch the ray given by $\text{Arg}(z - z_4) = \dfrac{5\pi}{6}$ on the Argand diagram below. **(2 marks)**

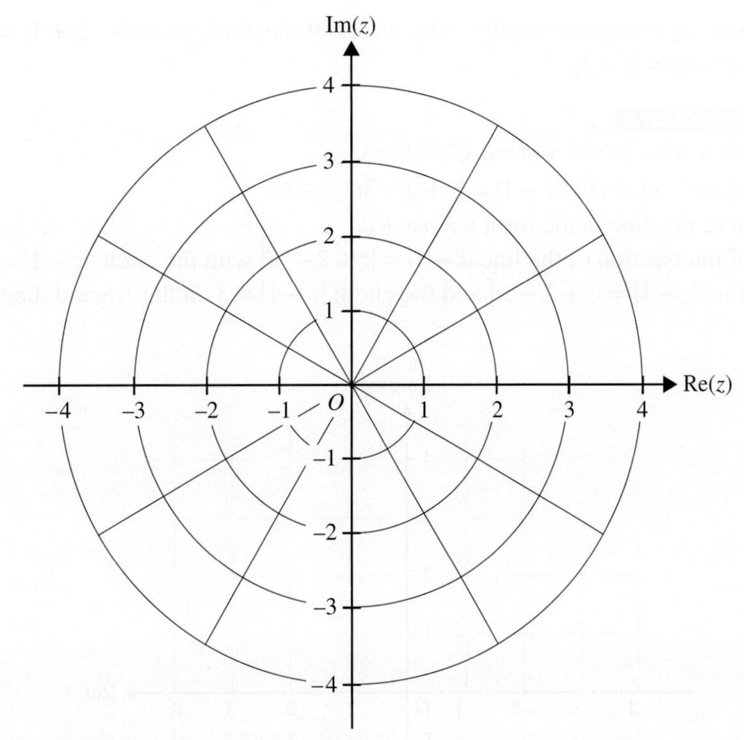

c. The ray $\text{Arg}(z - z_4) = \dfrac{5\pi}{6}$ intersects the circle $|z - 3i| = 1$, dividing it into a major and a minor segment.

i. Sketch the circle $|z - 3i| = 1$ on the Argand diagram in part **b**. **(1 mark)**

ii. Find the area of the minor segment. **(2 marks)**

Question 2 (10 marks) TECH-ACTIVE

Source: *VCE 2018 Specialist Mathematics Exam 2, Section B, Q2; © VCAA.*

a. State the center in the form (x, y), where $x, y \in R$, and the radius of the circle given by $|z - (1 + 2i)| = 2$, where $z \in C$. **(1 mark)**

b. By expressing the circle given by $|z + 1| = \sqrt{2}\,|z - i|$ in cartesian form, show that this circle has the same centre and radius as the circle given by $|z - (1 + 2i)| = 2$. **(2 marks)**

c. Graph the circle given by $|z + 1| = \sqrt{2}\,|z - i|$ on the Argand diagram below, labelling the intercepts with the vertical axis. **(2 marks)**

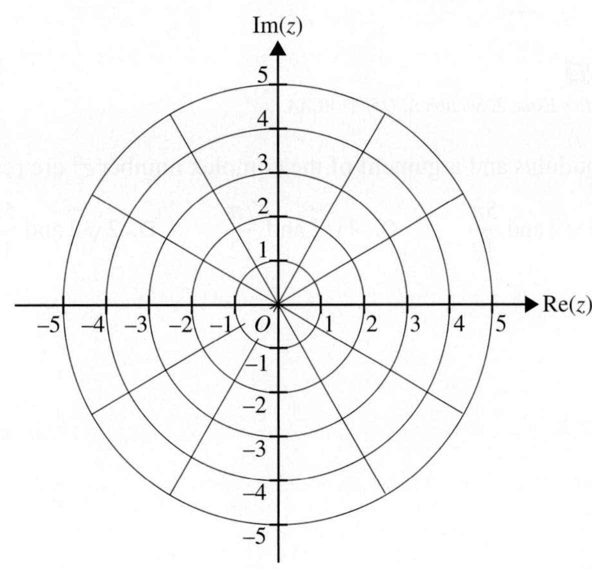

The line given by $|z-1|=|z-3|$ intersects the circle given by $|z+1|=\sqrt{2}\,|z-i|$ in two places.

d. Draw the line given by $|z-1|=|z-3|$ on the Argand diagram in part **c**. Label the points of intersection with their coordinates. **(2 marks)**

e. Find the area of the minor segment enclosed by an arc of the circle given by $|z+1|=\sqrt{2}\,|z-i|$ and part of the line given by $|z-1|=|z-3|$. **(3 marks)**

Question 3 (11 marks) TECH-ACTIVE

Source: *VCE 2016 Specialist Mathematics Exam 2, Section B, Q2; © VCAA.*

A line in the complex plane is given by $|z-1|=|z+2-3i|$, $z \in C$.

a. Find the equation of this line in the form $y=mx+c$. **(2 marks)**

b. Find the points of intersection of the line $|z-1|=|z+2-3i|$ with the circle $|z-1|=3$. **(2 marks)**

c. Sketch both the line $|z-1|=|z+2-3i|$ and the circle $|z-1|=3$ on the Argand diagram below. **(2 marks)**

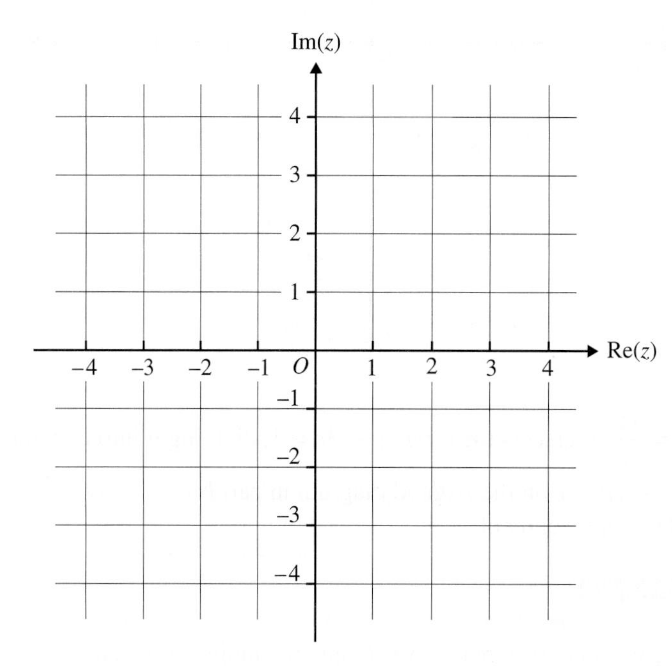

d. The line $|z-1|=|z+2-3i|$ cuts the circle $|z-1|=3$ into two segments. Find the area of the major segment. **(2 marks)**

e. Sketch the ray given by $\operatorname{Arg}(z)=-\dfrac{3\pi}{4}$ on the Argand diagram in part **c**. **(1 mark)**

f. Write down the range of values of α, $\alpha \in R$, for which a ray with equation $\operatorname{Arg}(z)=\alpha\pi$ intersects the line $|z-1|=|z+2-3i|$. **(2 marks)**

Question 4 (1 mark) TECH-ACTIVE

Source: *VCE 2015 Specialist Mathematics Exam 2, Section A, Q5; © VCAA.*

MC Given $z=\dfrac{1+i\sqrt{3}}{1+i}$, the modulus and argument of the complex number z^5 are respectively

A. $2\sqrt{2}$ and $\dfrac{5\pi}{6}$ **B.** $4\sqrt{2}$ and $\dfrac{5\pi}{12}$ **C.** $4\sqrt{2}$ and $\dfrac{7\pi}{12}$ **D.** $2\sqrt{2}$ and $\dfrac{5\pi}{12}$ **E.** $4\sqrt{2}$ and $-\dfrac{\pi}{12}$

Question 5 (12 marks) TECH-ACTIVE

Source: *VCE 2015 Specialist Mathematics Exam 2, Section 2, Q2; © VCAA.*

a. i. On the Argand diagram below, plot and label the point $0 + 0i$ and $1 + i\sqrt{3}$. **(2 marks)**

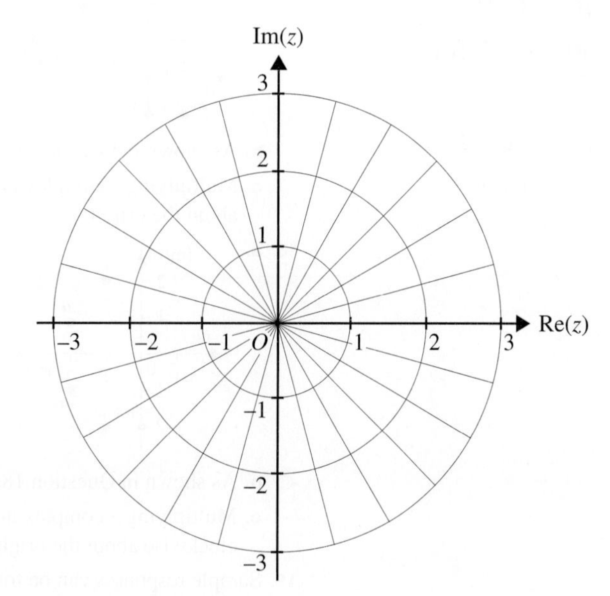

ii. On the same Argand diagram above, sketch the line $\left|z - \left(1 + i\sqrt{3}\right)\right| = |z|$ and the circle $|z - 2| = 1$. **(2 marks)**

iii. Use the fact that the line $\left|z - \left(1 + i\sqrt{3}\right)\right| = |z|$ passes through the point $z = 2$, or otherwise, to find the equation of this line in cartesian form. **(1 mark)**

iv. Find the points of intersection of the line and the circle, expressing your answer in the form $a + ib$. **(3 marks)**

b. i. Consider the equation $z^2 - 4\cos(\alpha)z + 4 = 0$, where α is a real constant and $0 < \alpha < \frac{\pi}{2}$.

Find the roots z_1 and z_2 of this equation, in terms of α, expressing your answers in polar form. **(3 marks)**

ii. Find the values of α for which $\left|\text{Arg}\left(\frac{z_1}{z_2}\right)\right| = \frac{5\pi}{6}$. **(1 mark)**

More exam questions are available online.

Answers

Topic 2 Complex numbers

2.2 Complex numbers in Cartesian form

2.2 Exercise

1. a. 8 b. 2 c. -1 d. -2
2. a. $7-4i$ b. $-1+2i$ c. $1+3i$
3. a. $9-13i$ b. $7-24i$
4. a. $3+4i$ b. $8-6i$
5. a. 10 b. 0
6. a. $\frac{3}{10}+\frac{1}{10}i$ b. $\frac{3}{5}+\frac{1}{5}i$
7. $-3-2i$
8. $x=2,\ y=\frac{7}{2}$
9. $x=\frac{20}{13}, y=\frac{9}{13}$
10. 4
11. 2
12. $3z$ has a length of three times z, $\bar{z}=-2+i$ is the reflection in the real axis, and $iz=1-2i$ is a rotation of 90° anticlockwise from z.

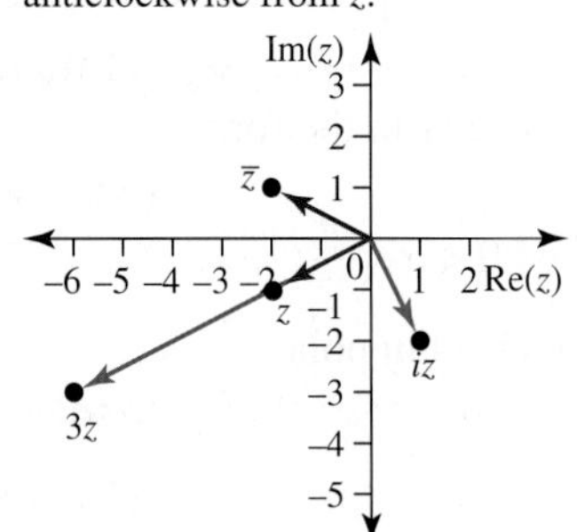

13. $u+v=3-i,\ u-v=-1-3i$

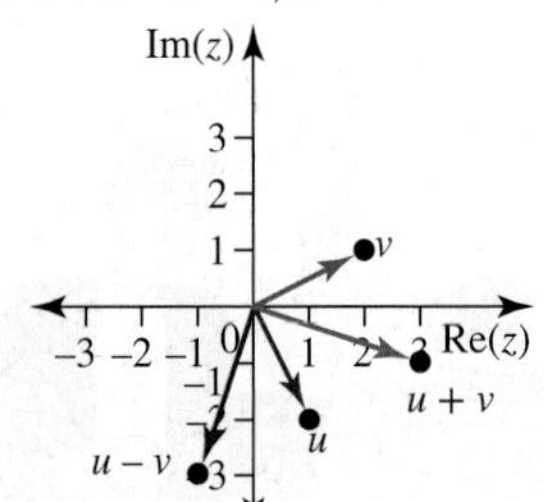

14. $-1-2i$
15. $3+2i$
16. a.

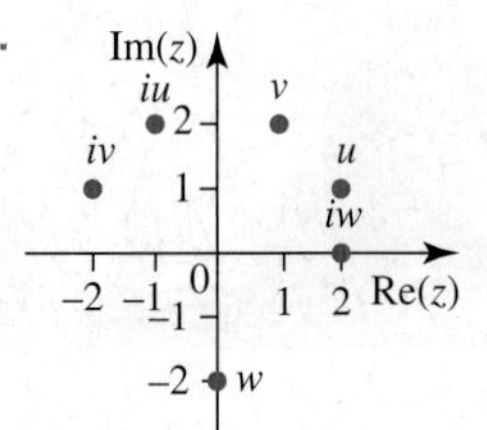

b. As shown in Question 16a.

c. Multiplying a complex number z by i rotates z 90° anticlockwise about the origin.

17. a.

b. As shown in Question 17a.

c. Multiplying a complex number z by -1 rotates z by 180° about the origin.

18. a.

b. As shown in Question 18a.

c. Multiplying a complex number z by $-i$ rotates z 90° clockwise about the origin.

19. Sample responses can be found in the worked solutions in the online resources.
20. Sample responses can be found in the worked solutions in the online resources.

2.2 Exam questions

Note: Mark allocations are available with the fully worked solutions online.

1. A
2. C
3. A

2.3 Complex numbers in polar form

2.3 Exercise

1. a. $2\operatorname{cis}\left(\frac{\pi}{3}\right)$ b. $\sqrt{2}\operatorname{cis}\left(\frac{3\pi}{4}\right)$ c. $4\operatorname{cis}\left(-\frac{2\pi}{3}\right)$
   d. $2\operatorname{cis}\left(-\frac{\pi}{6}\right)$ e. $4\operatorname{cis}(0)$ f. $2\operatorname{cis}\left(-\frac{\pi}{2}\right)$
2. a. $2\operatorname{cis}\left(\frac{\pi}{6}\right)$ b. $2\operatorname{cis}\left(\frac{2\pi}{3}\right)$
   c. $2\operatorname{cis}\left(-\frac{5\pi}{6}\right)$ d. $2\sqrt{2}\operatorname{cis}\left(-\frac{\pi}{4}\right)$
   e. $7\operatorname{cis}(\pi)$ f. $5\operatorname{cis}\left(\frac{\pi}{2}\right)$
3. a. $2-2\sqrt{3}i$ b. $-8i$
4. a. $-6-6i$ b. $-3+4i$
5. $\frac{1}{12}-\frac{\sqrt{3}}{12}i$
6. $-2+2i$
7. a. $-8i$ b. -4
8. a. $1-\sqrt{3}i$ b. -8

9. a. $\frac{\pi}{2}$ b. $32i$

10. $-\frac{\sqrt{3}}{4} - \frac{1}{4}i$

11. a. Sample responses can be found in the worked solutions in the online resources.

b. $-\sqrt{3} - 2 + i, \frac{11\pi}{12}$

12. $-\frac{\pi}{12}$

13. $n = 3k$ where $k \in Z$

14. $n = 3k + \frac{3}{2}$ where $k \in Z$

15. a. 4096 b. $-46\,656$

c. 1 953 125 d. $-8192\sqrt{3} + 8192i$

16. a. $\frac{1}{2}\left(\sqrt{3}+i\right), \frac{1}{2}\left(\sqrt{3}+i\right), -1$

b. $\frac{\pi}{6}, \frac{\pi}{6}, \pi$

c. Yes

d. Yes

e. No

17. a. $\frac{2\pi}{3}$

b. $-\frac{3\pi}{4}$

c. $-\frac{\pi}{12}$

d. $-\frac{7\pi}{12}$

e. In this case yes but not in general

f. No

18. a. $\left(\sqrt{6} - \sqrt{2}\right) + \left(\sqrt{6} + \sqrt{2}\right)i$

b. $4 \text{ cis}\left(\frac{5\pi}{12}\right)$

c. $\frac{1}{4}\left(\sqrt{6} + \sqrt{2}\right)$

d. Sample responses can be found in the worked solutions in the online resources.

19. a. $2\left(\sqrt{3}+1\right) + 2\left(\sqrt{3}-1\right)i$

b. $4\sqrt{2} \text{ cis}\left(\frac{\pi}{12}\right)$

c. $\frac{1}{4}\left(\sqrt{6} + \sqrt{2}\right)$

d. Sample responses can be found in the worked solutions in the online resources.

20. a. $\frac{1}{2}\left(1 - \sqrt{3}\right) + \frac{1}{2}\left(\sqrt{3}+1\right)i$

b. $\sqrt{2} \text{ cis}\left(\frac{7\pi}{12}\right)$

c. $\frac{1}{4}\left(\sqrt{2} - \sqrt{6}\right)$

d. Sample responses can be found in the worked solutions in the online resources.

21. a. Sample responses can be found in the worked solutions in the online resources.

b. $\frac{\pi}{8}$

c. $\frac{5\pi}{8}$

22. a. Sample responses can be found in the worked solutions in the online resources.

b. $\frac{7\pi}{12}$

c. $-\frac{5\pi}{12}$

d. $\frac{\pi}{12}$

23. a. $n = 2(2k+1)$ where $k \in Z$

b. $n = 4k$ where $k \in Z$

24-30. Sample responses can be found in the worked solutions in the online resources.

2.3 Exam questions

Note: Mark allocations are available with the fully worked solutions online.

1. A
2. A
3. a. Sample responses can be found in the worked solutions in the online resources.

b. $-24\sqrt{3}i$

c. $n = 6k, k \in Z$

d. $n = 6k + 3, k \in Z$

2.4 Solving polynomial equations over *C*

2.4 Exercise

1. $z = 7 + 5i, 7 - 5i$
2. $z = -\frac{1}{2} + \frac{9}{2}i, -\frac{1}{2} - \frac{9}{2}i$
3. $z = -2 + 5i, -2 - 5i$
4. $P(z) = z^2 + 2z + 170$
5. $P(z) = z^2 - 12z + 61$
6. $z = 1, -5 - 3i, -5 + 3i$
7. $z = -1, -6 - 7i, -6 + 7i$
8. $z = -2, -3 - i, -3 + i$
9. $-2 \pm 3i, 3, b = 1, c = 1$
10. $\pm 5i, 2, b = -2, c = 25$
11. $\pm 2i$
12. $(z + \sqrt{7}i)(z - \sqrt{7}i)(z + 3i)$
13. $\pm 2i, 2 - 3i$
14. $\frac{3}{2} + 2i, \pm\sqrt{5}i$
15. $\pm 2i, \pm\sqrt{5}$
16. $\pm\frac{\sqrt{6}i}{2}, \pm\sqrt{3}$

17. a. $\pm 3i, \pm 2$ b. $\pm \sqrt{7}i, \pm \sqrt{3}$

18. a. $\pm \sqrt{5}i, \pm 2\sqrt{2}$ b. $\pm \sqrt{6}i, \pm \sqrt{3}i$

19. $a = -6, \pm 3i, 3 \pm 4i$

20. $b = 24, -3 \pm 4i, \pm 2i$

21. $z = -3, 3, -5 - 3i, -5 + 3i$

22. $z = 4, -6 - 9i, -6 + 9i$

23. $z = 4, -4, 1 + 3i, 1 - 3i$

24. $a = \pm 4, \pm 4i, -3 \pm 4i$

25. $b = 2, z = \pm 4i, z = -1 \pm i$

26. $d = -1$

2.4 Exam questions

Note: Mark allocations are available with the fully worked solutions online.

1. a. $z = -1 \pm i$ b. $z = 1 \pm \sqrt{5}i$

2. C

3. D

2.5 Subsets of the complex plane

2.5 Exercise

1. The line $4x + 3y = 12$

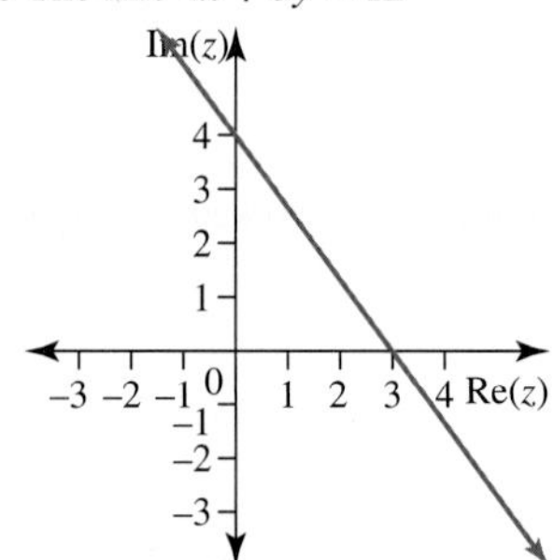

2. $a = -4, b = 2$

3. a. $3x + 2y = 6$

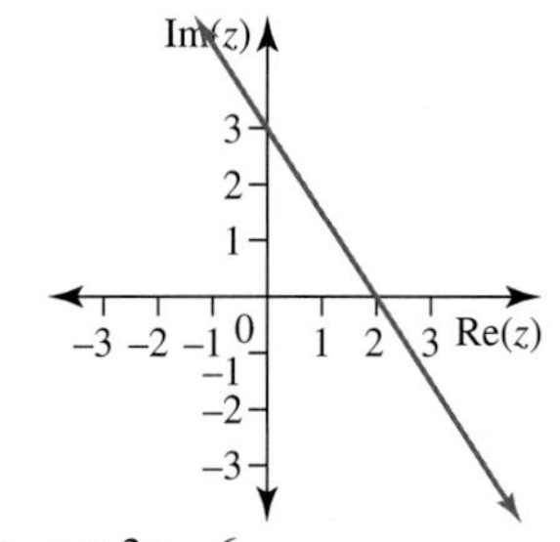

b. $y = 2x - 6$

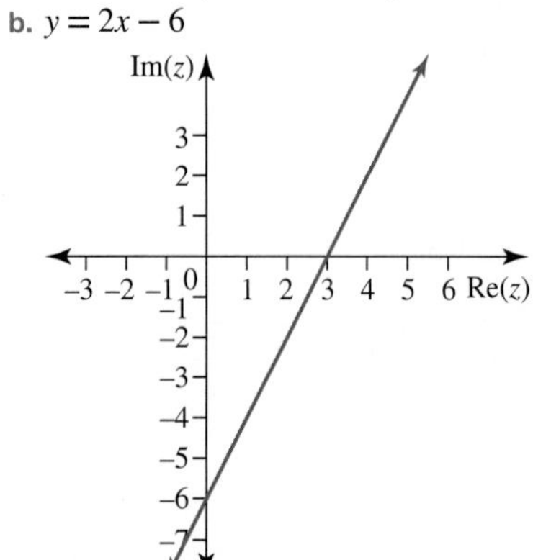

4. Line $y = -x$; the set of points equidistant from $(0, -3)$ and $(3, 0)$

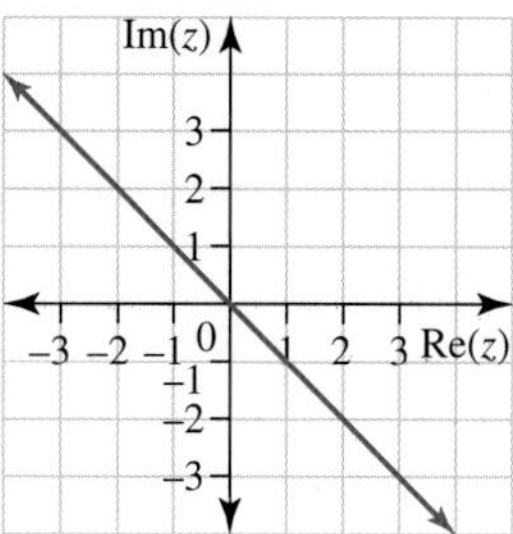

5. a. $x = 3$; line

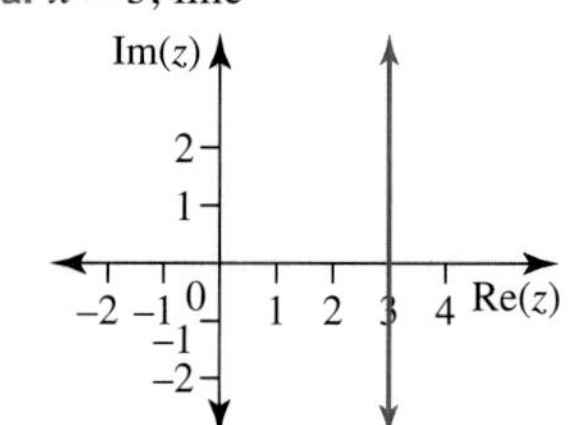

b. $y = -x$

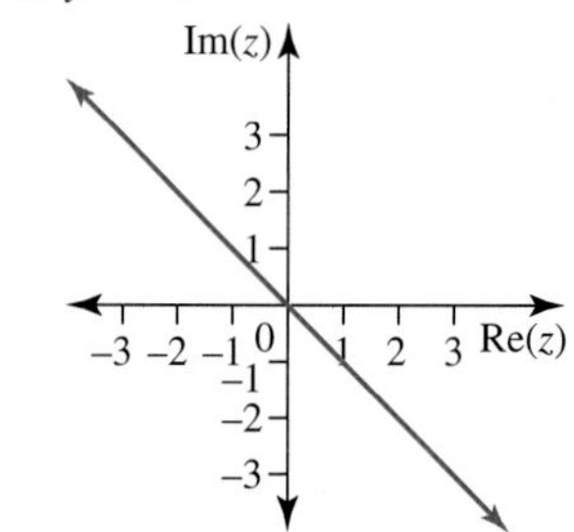

6. a. $y = -2x - 3$

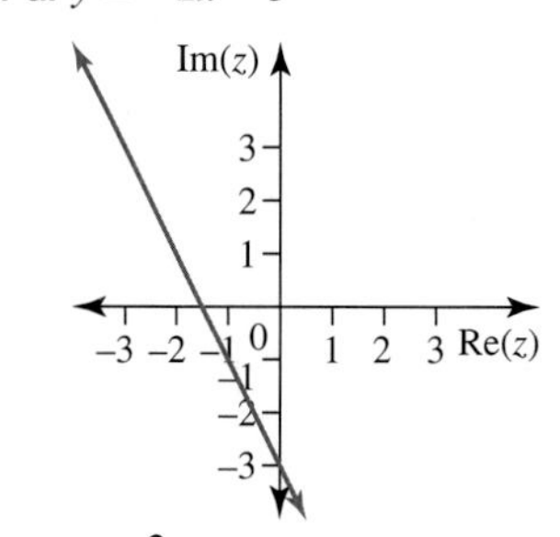

b. $y = \frac{2x}{3}$

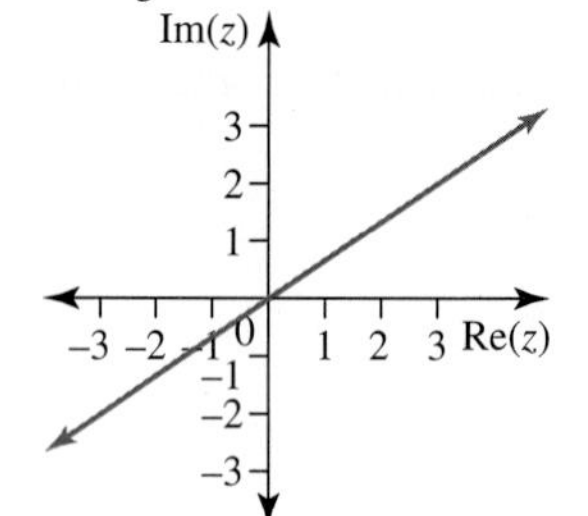

7. Circle $(x-3)^2+(y+2)^2=16$, with centre $(3,-2)$ and radius 4

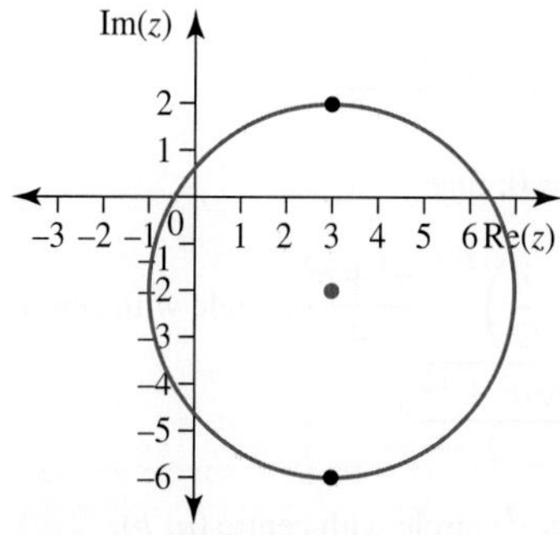

8. $a=3, b=-3, r=3$

9. a. $(x+2)^2+(y-3)^2=4$; circle with centre $(-2,3)$ and radius 2

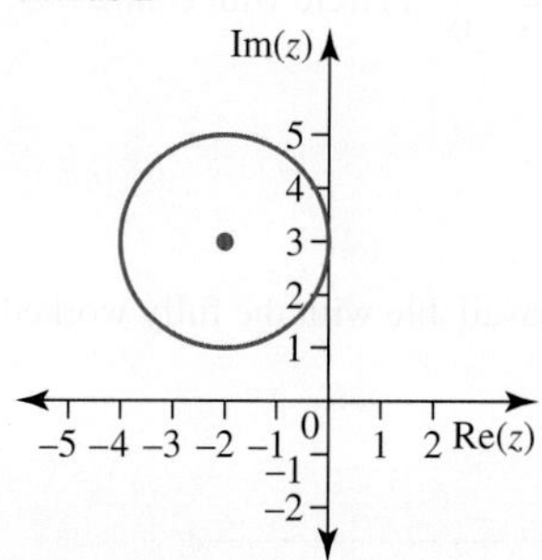

b. $(x-3)^2+(y+1)^2=9$; circle with centre $(3,-1)$ and radius 3

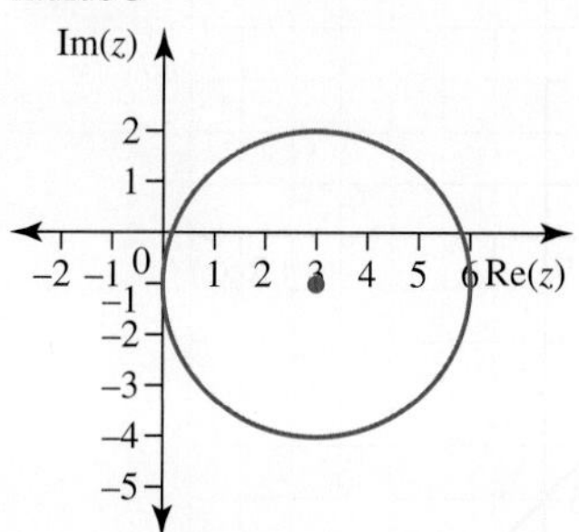

10. a. $y=-\dfrac{2x}{3}+2$; line

b. $\left(x-\dfrac{3}{2}\right)^2+(y-1)^2=\dfrac{13}{4}$; circle with centre $\left(\dfrac{3}{2},1\right)$, radius $\dfrac{\sqrt{13}}{2}$

11. a. $(x+1)^2+(y+4)^2=8$; circle with centre $(-1,-4)$, radius $2\sqrt{2}$

b. $(x-1)^2+(y+8)^2=20$; circle with centre $(1,-8)$, radius $2\sqrt{5}$

12. $\dfrac{x^2}{25}+\dfrac{y^2}{9}=1$

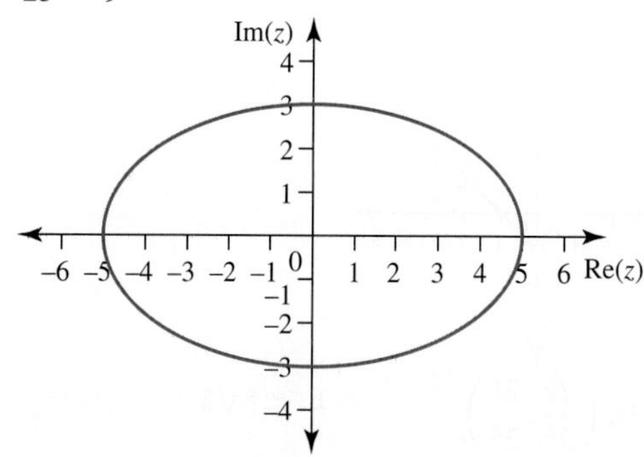

13. A ray from $(2,0)$ making an angle of $\dfrac{\pi}{6}$ or 30° with the real axis

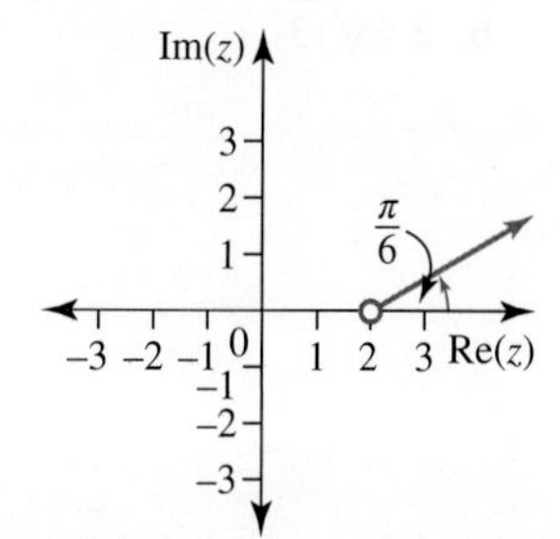

14. A ray from $(0,-3)$ making an angle of $-\dfrac{\pi}{2}$ or 90° with the real axis

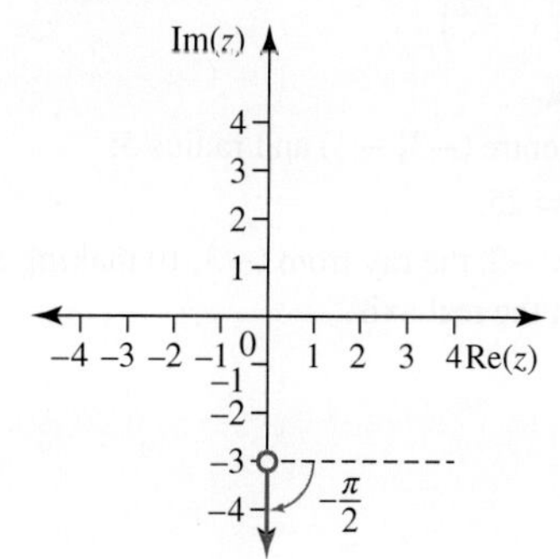

15. a. $y=\dfrac{x}{\sqrt{3}}$ for $x>0$; a ray from $(0, 0)$ making an angle of 30° with the real axis

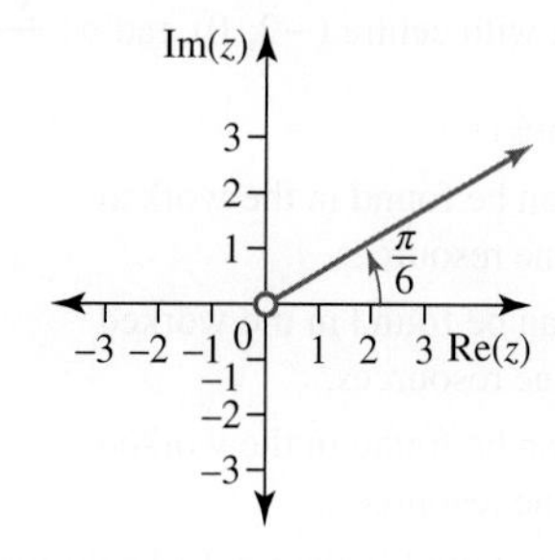

b. $y = x - 1$ for $x > 0$; a ray from $(0, -1)$ making an angle of 45° with the real axis

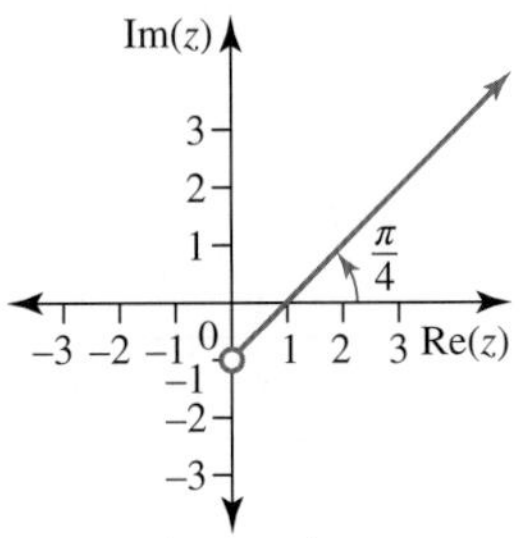

16. a. $(0, 3), \left(\frac{72}{25}, \frac{21}{25}\right)$ b. $\pm 8\sqrt{5}$

17. a. $a = -2, b = 4, r = 2\sqrt{5}$
 b. $a = 1, b = 4, r = 2\sqrt{2}$

18. a. $(2, 5), \left(-\frac{7}{5}, -\frac{26}{5}\right)$ b. $\pm 5\sqrt{13}$

19. $\left(\frac{3\sqrt{2}}{2}. -\frac{3\sqrt{2}}{2}\right)$

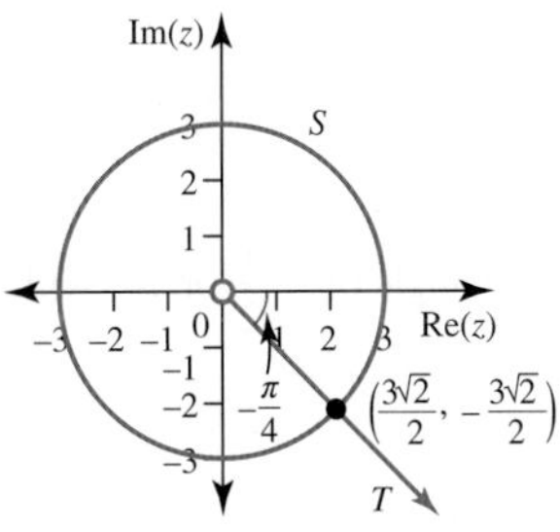

20. a. S is the circle with centre $(-3, -1)$ and radius 5: $(x+3)^2 + (y+1)^2 = 25$.
 b. T is $y = x + 3$ for $x < -3$, the ray from $(-3, 0)$ making an angle of $-135°$ with the real axis.
 c. $u = -7 - 4i$

21. $\left(\frac{9}{5}, \frac{12}{5}\right)$

22. ±30

23. Sample responses can be found in the worked solutions in the online resources.

24. $(x+2)^2 + y^2 = \frac{10}{3}$; circle with centre $(-2, 0)$, radius $\frac{\sqrt{30}}{3}$

25. a. i. $z = 4\cos(\theta) + 5i\sin(\theta)$
 ii. Sample responses can be found in the worked solutions in the online resources.
 iii. Sample responses can be found in the worked solutions in the online resources.
 iv. Sample responses can be found in the worked solutions in the online resources.
 b. Sample responses can be found in the worked solutions in the online resources.

26. Sample responses can be found in the worked solutions in the online resources.

27. a. $(x+3)^2 + (y-2)^2 = 9$; circle with centre $(-3, 2)$, radius 3
 b. $\left(x + \frac{\alpha}{a}\right)^2 + \left(y + \frac{\beta}{a}\right)^2 = \frac{b\bar{b} - ac}{a^2}$; circle with centre $\left(-\frac{\alpha}{a}, -\frac{\beta}{a}\right)$, radius $\frac{\sqrt{b\bar{b} - ac}}{a}$

28. a. $y = -\frac{ax}{b} + a$, $ab \neq 0$; line
 b. $\left(x - \frac{b}{2}\right)^2 + \left(y - \frac{a}{2}\right)^2 = \frac{a^2 + b^2}{4}$; circle with centre $\left(\frac{b}{2}, \frac{a}{2}\right)$, radius $\frac{\sqrt{a^2+b^2}}{2}$

29. $(x-a)^2 + (y-b)^2 = r^2$; circle with centre (a, b), radius r

30. $(x-a)^2 + \left(y + \frac{5b}{3}\right)^2 = \frac{16b^2}{9}$; circle with centre $\left(a, \frac{-5b}{3}\right)$, radius $\frac{4b}{3}$

2.5 Exam questions

Note: Mark allocations are available with the fully worked solutions online.

1. D

2. a. $y = -x - 5$
 b.

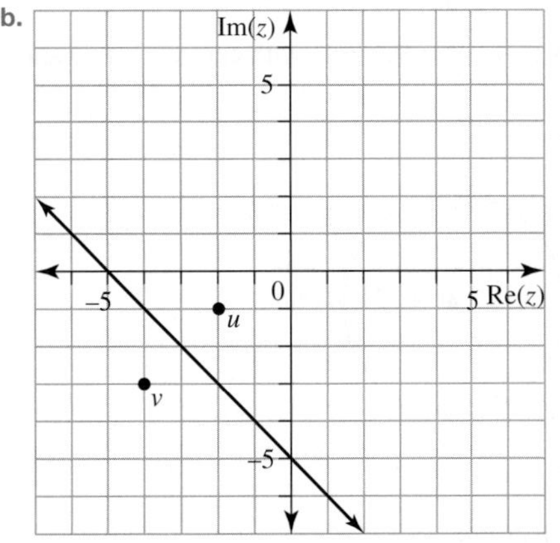

 c. The line is the perpendicular bisector of the line segment joining the points represented by u and v.
 d. i.

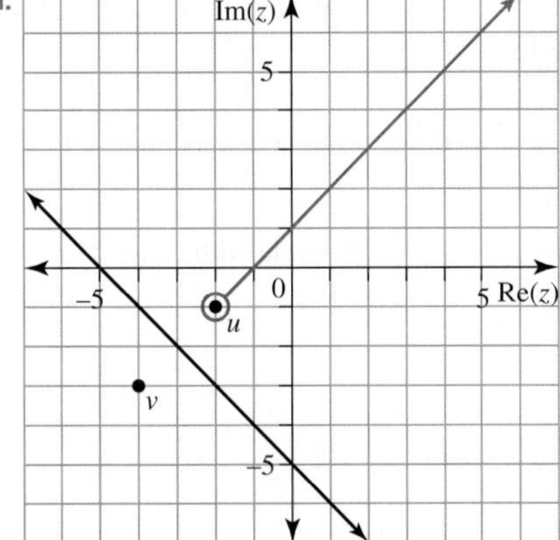

 ii. $y = x + 1, x > -2$

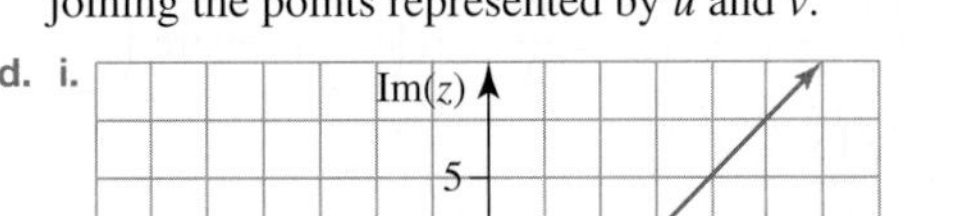

 e. $z_c = -\frac{5}{3} - \frac{10i}{3}$, radius $= \frac{5\sqrt{2}}{3}$.

3. D

2.6 Roots of complex numbers

2.6 Exercise

1. $z = \sqrt{5}\operatorname{cis}\left(\frac{\pi}{12}\right), \sqrt{5}\operatorname{cis}\left(-\frac{11\pi}{12}\right)$

2. $z = 3\operatorname{cis}\left(\frac{2\pi}{9}\right), 3\operatorname{cis}\left(\frac{8\pi}{9}\right), 3\operatorname{cis}\left(-\frac{4\pi}{9}\right)$

3. $z = 2\operatorname{cis}\left(-\frac{3\pi}{16}\right), 2\operatorname{cis}\left(\frac{5\pi}{16}\right), 2\operatorname{cis}\left(\frac{13\pi}{16}\right),$

$2\operatorname{cis}\left(-\frac{11\pi}{16}\right)$

4. a. $z = \sqrt[3]{2}\operatorname{cis}\left(-\frac{2\pi}{9}\right), \sqrt[3]{2}\operatorname{cis}\left(\frac{4\pi}{9}\right), \sqrt[3]{2}\operatorname{cis}\left(-\frac{8\pi}{9}\right)$

b.

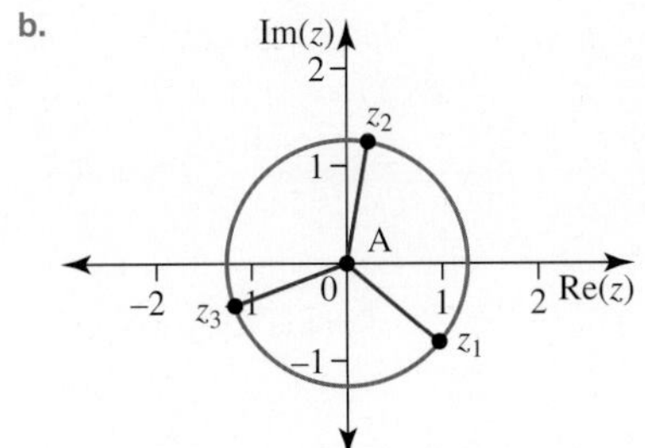

The three roots are equally spaced around a circle of radius $\sqrt[3]{2}$. The angle between each solution is $\frac{2\pi}{3}$.

5. a. $z = \sqrt[4]{5\sqrt{2}}\operatorname{cis}\left(-\frac{3\pi}{16}\right), \sqrt[4]{5\sqrt{2}}\operatorname{cis}\left(\frac{5\pi}{16}\right),$

$\sqrt[4]{5\sqrt{2}}\operatorname{cis}\left(\frac{13\pi}{16}\right), \sqrt[4]{5\sqrt{2}}\operatorname{cis}\left(-\frac{11\pi}{16}\right)$

b.

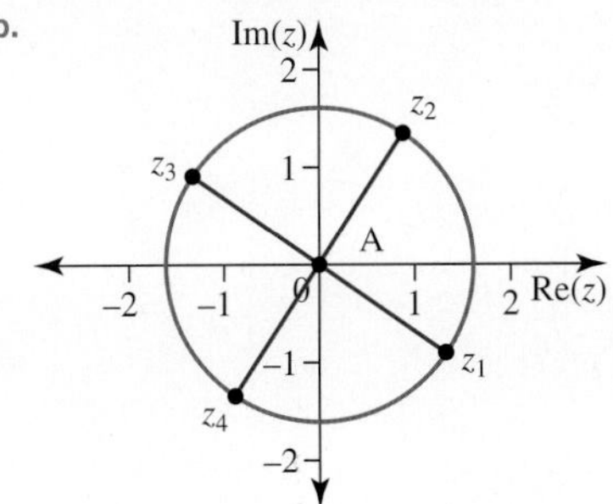

The three roots are equally spaced around a circle of radius $\sqrt[4]{5\sqrt{2}} \approx 1.631$. The angle between each solution is $\frac{\pi}{2}$.

6. a. $z = \sqrt[6]{2\sqrt{3}}\operatorname{cis}\left(\frac{\pi}{36}\right), \sqrt[6]{2\sqrt{3}}\operatorname{cis}\left(\frac{13\pi}{36}\right), \sqrt[6]{2\sqrt{3}}\operatorname{cis}\left(\frac{25\pi}{36}\right), \sqrt[6]{2\sqrt{3}}\operatorname{cis}\left(-\frac{35\pi}{36}\right),$

$\sqrt[6]{2\sqrt{3}}\operatorname{cis}\left(-\frac{23\pi}{36}\right), \sqrt[6]{2\sqrt{3}}\operatorname{cis}\left(-\frac{11\pi}{36}\right)$

b.

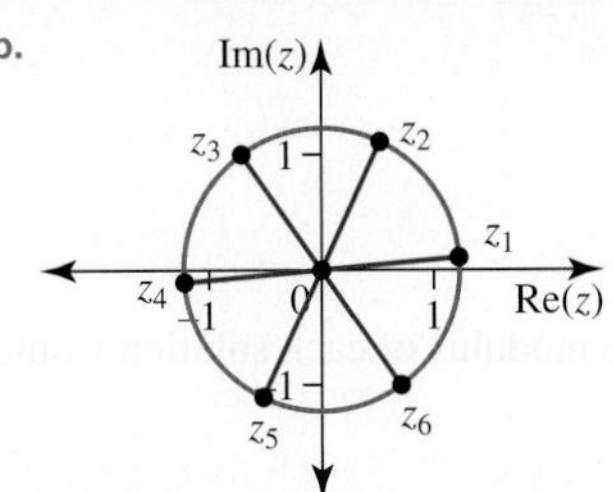

The three roots are equally spaced around a circle of radius $\sqrt[6]{2\sqrt{3}} \approx 1.230$. The angle between each solution is $\frac{\pi}{3}$.

7. a. $z = \text{cis}\left(\frac{\pi}{6}\right), \text{cis}\left(\frac{5\pi}{6}\right), \text{cis}\left(-\frac{\pi}{2}\right)$

b. $z = 2\sqrt{2}\,\text{cis}\left(\frac{\pi}{8}\right), 2\sqrt{2}\,\text{cis}\left(\frac{5\pi}{8}\right), 2\sqrt{2}\,\text{cis}\left(-\frac{7\pi}{8}\right), 2\sqrt{2}\,\text{cis}\left(-\frac{3\pi}{8}\right)$

8. $\sqrt{3} + i, -\sqrt{3} - i$

9. Sample responses can be found in the worked solutions in the online resources.

10. a. $z = 1, \text{cis}\left(\frac{\pi}{2}\right), \text{cis}(\pi), \text{cis}\left(\frac{-\pi}{2}\right)$

b.

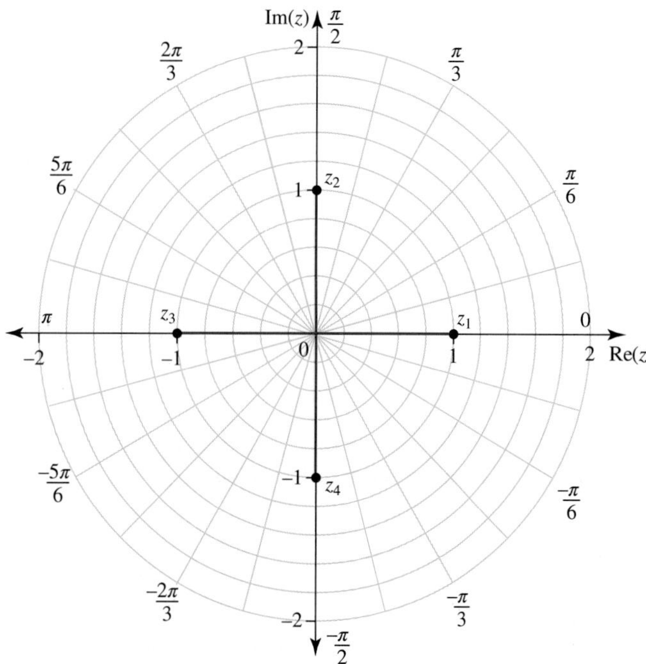

11. Sample responses can be found in the worked solutions in the online resources.

12. a. $z = 1, \text{cis}\left(\frac{2\pi}{5}\right), \text{cis}\left(\frac{4\pi}{5}\right), \text{cis}\left(\frac{-4\pi}{5}\right), \text{cis}\left(\frac{-2\pi}{5}\right)$

b.

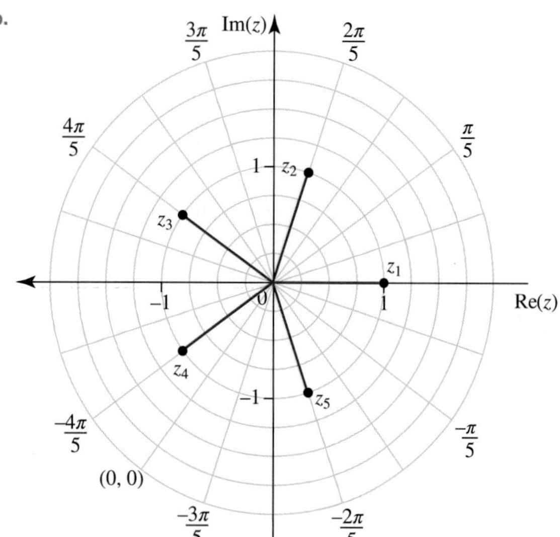

13. The solutions would be equally spaced by $\frac{2\pi}{5}$ radians and lie on a circle of radius $10^{\frac{1}{5}}$. The modulus of each solution would change from $|z| = 1$ to $|z| = 10^{\frac{1}{5}}$.

14. a. $z=-1-\sqrt{3}i,\ z=-1+\sqrt{3}i,\ z=2$

b. $z=4i,\ z=-4i,\ z=-4,\ z=4$

15.

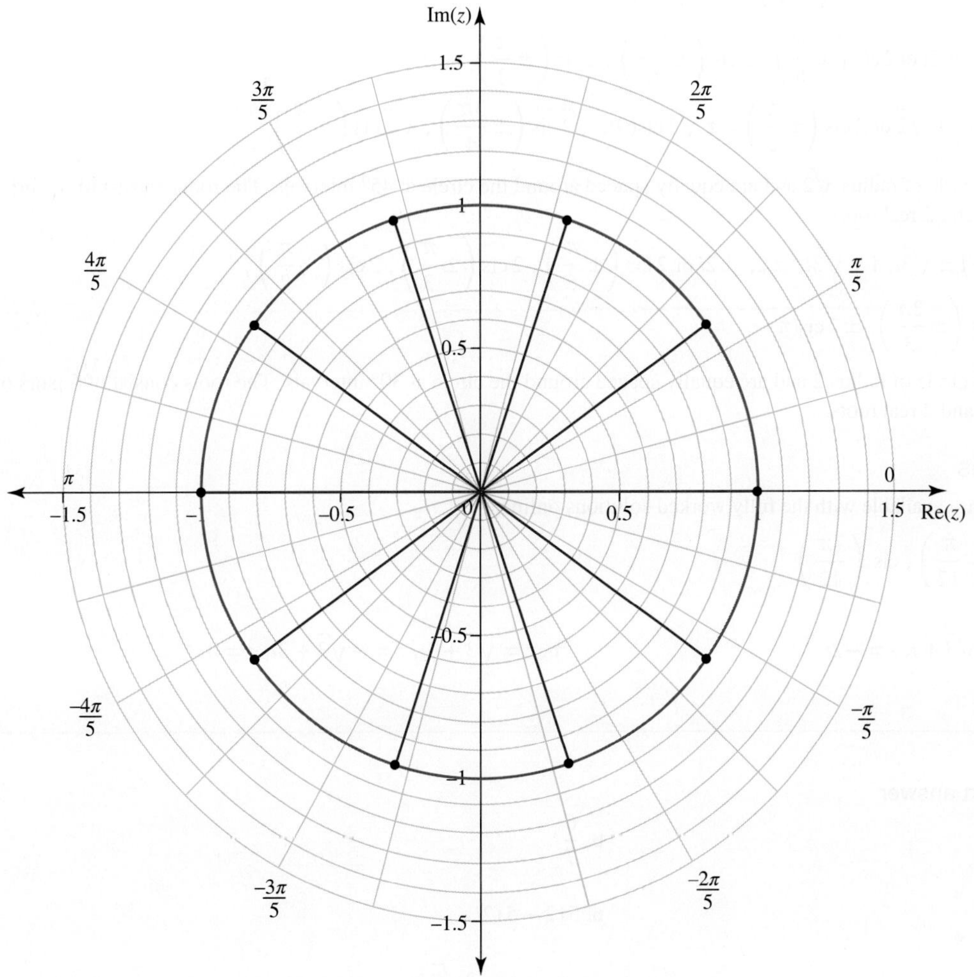

16. Sample responses can be found in the worked solutions in the online resources.

17. a. $-64,\ \pi$

b. $\pm 2(1+i),\ \pm 2(1-i)$ or $2\sqrt{2}\,\text{cis}\left(\pm\frac{\pi}{4}\right),\ 2\sqrt{2}\,\text{cis}\left(\pm\frac{3\pi}{4}\right)$

c. All 4 roots are on a circle of radius $2\sqrt{2}$ and are equally spaced around the circle at 90° intervals. The roots consist of 2 pairs of complex conjugates.

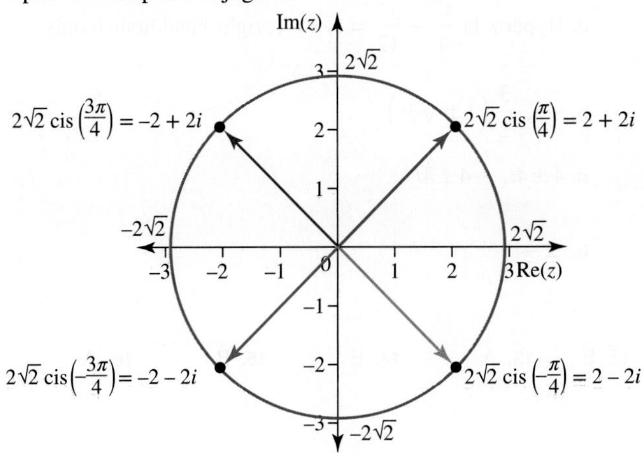

18. a. $1\pm\sqrt{3}i,\ -1\pm\sqrt{3}i,\ \pm 2$ or $2\,\text{cis}\left(\pm\frac{\pi}{3}\right),\ 2\,\text{cis}\left(\pm\frac{2\pi}{3}\right),\ \pm 2\,\text{cis}(\pi)$

b. $-\sqrt{3}\pm i,\ \sqrt{3}\pm i,\ \pm 2i$ or $2\,\text{cis}\left(\pm\frac{\pi}{6}\right),\ 2\,\text{cis}\left(\pm\frac{5\pi}{6}\right),\ 2\,\text{cis}\left(\pm\frac{\pi}{2}\right)$

19. $1\pm i,\ -1\pm i,\ \pm\sqrt{2}i,\ \pm\sqrt{2}$ or $2\,\text{cis}\left(\pm\frac{\pi}{2}\right),\ \pm\sqrt{2}\,\text{cis}(0),\ \sqrt{2}\,\text{cis}\left(\pm\frac{3\pi}{4}\right),\ \sqrt{2}\,\text{cis}\left(\pm\frac{\pi}{4}\right)$

All 8 roots are on a circle of radius $\sqrt{2}$ and are equally spaced around the circle at 45° intervals. The roots consist of 3 pairs of complex conjugates and 2 real roots.

20. $-\sqrt{3}\pm i,\ \sqrt{3}\pm i,\ -1\pm\sqrt{3}i,\ 1\pm\sqrt{3}i,\ \pm 2,\ \pm 2i$ or $2\,\text{cis}\left(\pm\frac{\pi}{6}\right),\ 2\,\text{cis}\left(\pm\frac{\pi}{3}\right),\ 2\,\text{cis}\left(\pm\frac{\pi}{2}\right),$

$2\,\text{cis}\left(\pm\frac{5\pi}{6}\right),\ 2\,\text{cis}\left(\pm\frac{2\pi}{3}\right),\ \pm 2\,\text{cis}(\pi)$

All 12 roots are on a circle of radius 2 and are equally spaced around the circle at 30° intervals. The roots consist of 5 pairs of complex conjugates and 2 real roots.

2.6 Exam questions

Note: Mark allocations are available with the fully worked solutions online.

1. $\text{cis}\left(-\frac{3\pi}{4}\right),\ \text{cis}\left(-\frac{\pi}{12}\right),\ \text{cis}\left(\frac{7\pi}{12}\right)$

2. E

3. a. $z=\sqrt{3}+i,\ z=-\sqrt{3}+i,\ z=-2i$

b. $z=\sqrt{3}+3i,\ z=-\sqrt{3}+3i,\ z=0$

2.7 Review

2.7 Exercise

Technology free: short answer

1. a. $\frac{1}{29}$

b. $\frac{6}{7}$

2. a. $\frac{1}{16}$

b. $512-512i$

3. a. $\pm\sqrt{10}i,\ 2i$

b. $\pm\sqrt{5},\ \pm\frac{3\sqrt{2}i}{2}$

4. a. $\frac{k\pi}{\tan^{-1}\left(\frac{12}{5}\right)},\ k\in Z$

b. $\frac{(2k+1)\pi}{2\tan^{-1}\left(\frac{4}{3}\right)},\ k\in Z$

5. a. Circle $x^2+y^2=4$

b. Line $x=\frac{5}{2}$

c. Parabola $y=\frac{x^2}{4}+3$

d. Hyperbola $\frac{x^2}{4}-\frac{y^2}{12}=1,\ x>1$, right-hand branch only

6. a. $\frac{\sqrt{2}}{2}(1-i),\ \frac{\sqrt{2}}{2}(-1+i)$

b. $-3,\ \frac{3}{2}\left(1\pm\sqrt{3}i\right)$

c. $3i,\ \frac{3}{2}\left(\sqrt{3}-i\right),\ -\frac{3}{2}\left(\sqrt{3}+i\right)$

d. $4\pm 4i,\ -4\pm 4i$

7. a. $(-4,6),\ (4,2)$

b. $\pm\frac{3}{2}$

Technology active: multiple choice

8. C 9. D 10. B 11. A 12. E 13. A 14. E 15. B 16. A

Technology active: extended response

17. D

18. a. Sample responses can be found in the worked solutions in the online resources.

b. $\text{cis}\left(\frac{\pi}{10}\right)$

c. $1,\ -1,\ i$

d. $5(2k+1),\ k \in Z$

19. a. $(x-8)^2 + y^2 = 48$

b. Sample responses can be found in the worked solutions in the online resources.

c. $\alpha = \dfrac{\pi}{3}$

d. $\left(2\ ,\ 2\sqrt{3}\right)$

e.

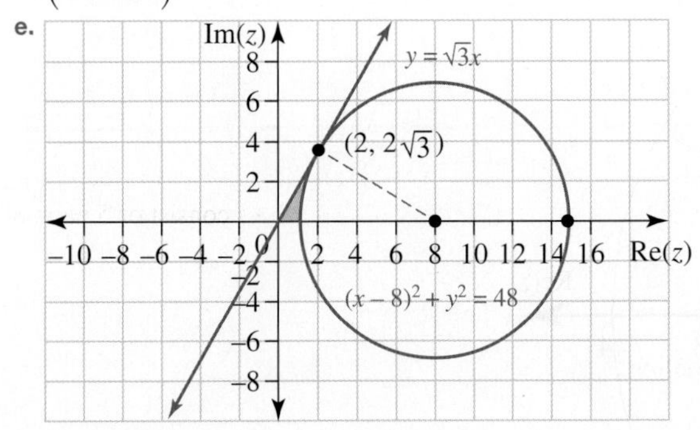

f. $A = 8\sqrt{3} - 4\pi$

g. $k = \text{cis}\left(\dfrac{\pi}{6}\right)$

h. $\beta \in \left(\dfrac{\pi}{3},\ \pi\right] \cup \left(-\pi,\ -\dfrac{\pi}{3}\right)$

20. a. $R_1 : y = \left(2 - \sqrt{3}\right)x,\quad x > 0 \qquad R_2 : y = \left(2 + \sqrt{3}\right)x, x > 0$

Circle: $(x-2)^2 + (y-2)^2 = 2$

b. Ray 1 and the circle: $\left(\dfrac{3+\sqrt{3}}{2},\ \dfrac{3-\sqrt{3}}{2}\right)$

Ray 2 and the circle: $\left(\dfrac{3-\sqrt{3}}{2},\ \dfrac{3+\sqrt{3}}{2}\right)$

c.

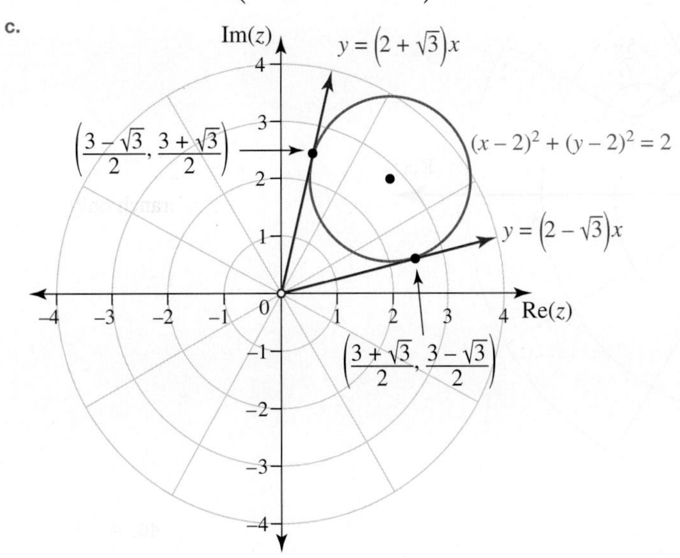

d. $2\sqrt{3} - \dfrac{2\pi}{3}$

2.7 Exam questions

Note: Mark allocations are available with the fully worked solutions online.

1. a. i. z_2 and z_3 are conjugate pairs.

ii. $\alpha = -3, \beta = 9, \gamma = -27$

b.

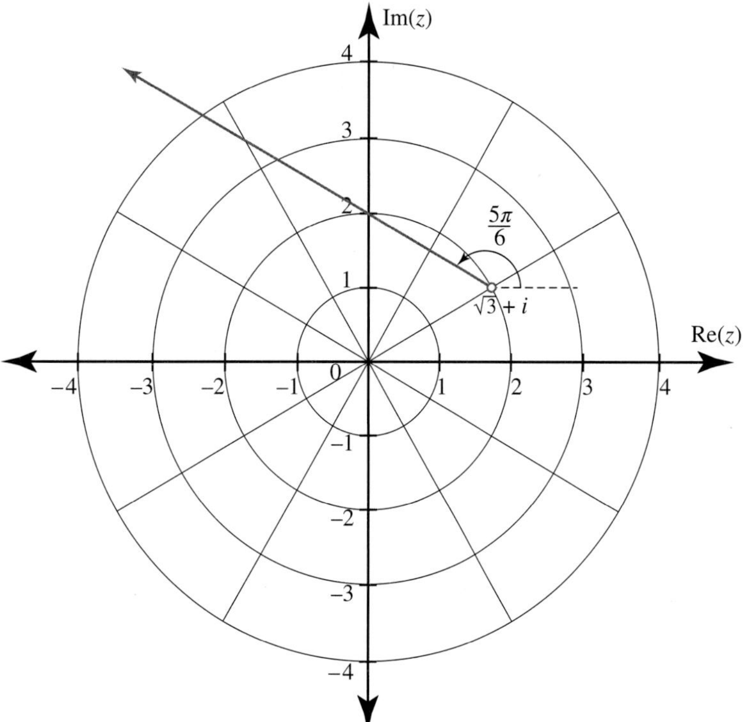

c. i.

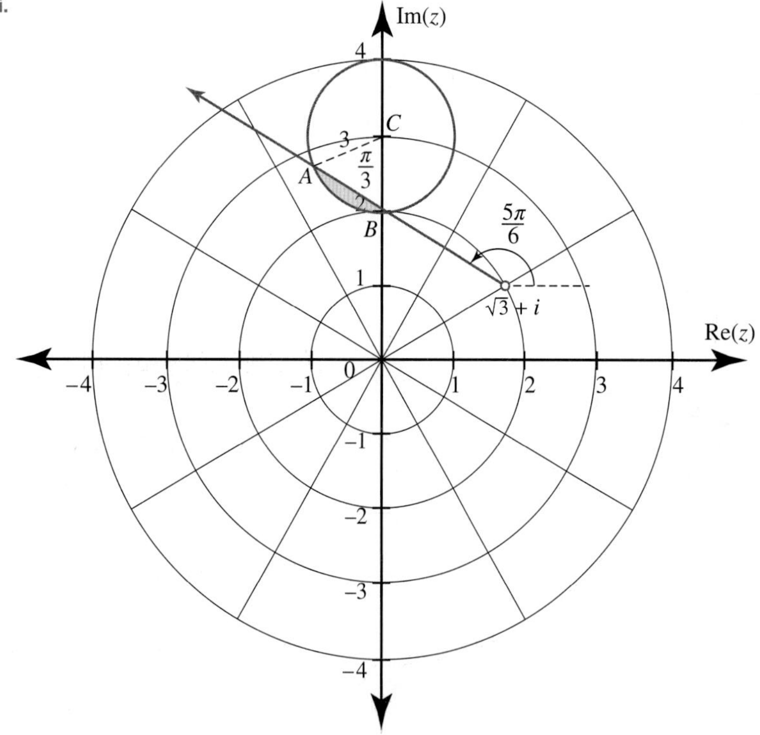

ii. $\frac{1}{12}\left(2\pi - 3\sqrt{3}\right)$

2. a. Centre: $(1, 2)$, radius 2

b. Sample responses can be found in the worked solutions in the online resources.

c. 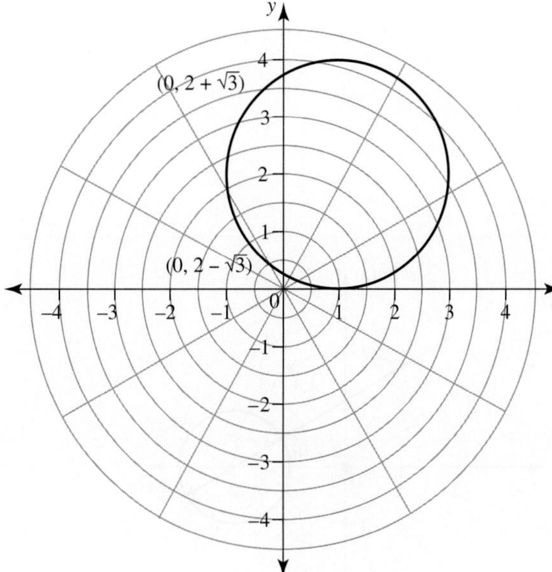

d. 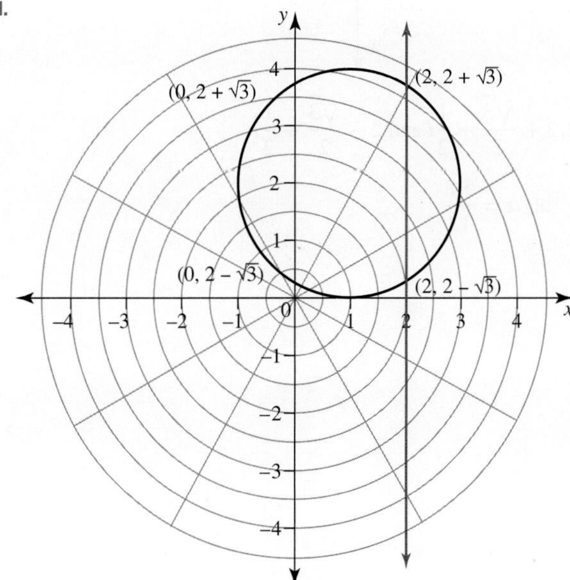

e. $\dfrac{4\pi}{3} - \sqrt{3}$

3. a. $y = x + 2$

b. $(-2, 0)$ and $(1, 3)$

c. 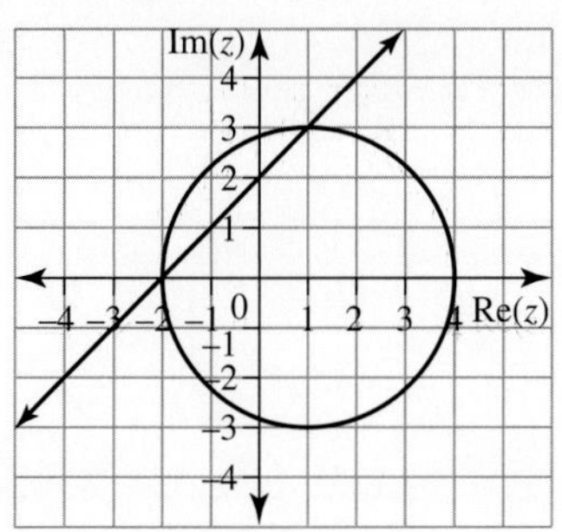

d. Area $= \dfrac{27\pi}{4} + \dfrac{9}{2}$

e. 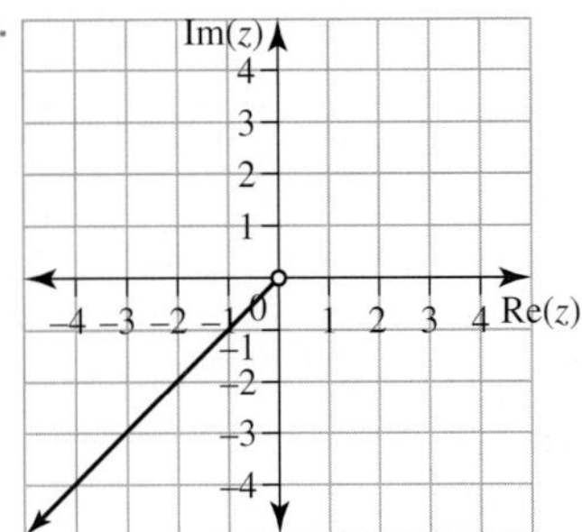

f. $\left(-1, -\frac{3}{4}\right) \cup \left(\frac{1}{4}, 1\right]$

4. B

5. a. i. 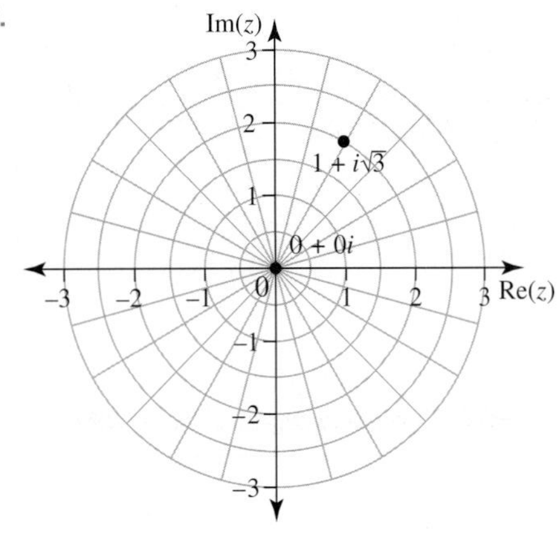

ii. 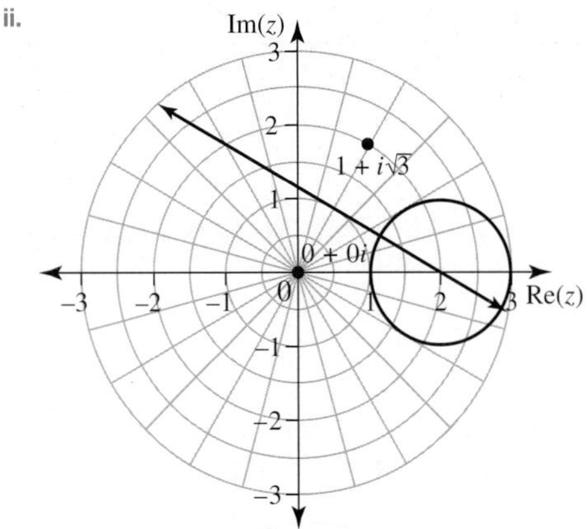

iii. $y = -\dfrac{1}{\sqrt{3}}x + \dfrac{2}{\sqrt{3}}$

iv. $2 + \dfrac{\sqrt{3}}{2} - \dfrac{1}{2}i$ and $2 - \dfrac{\sqrt{3}}{2} + \dfrac{1}{2}i$

b. i. $z_1 = 2\,\text{cis}(\alpha)$, $z_2 = 2\,\text{cis}(-\alpha)$

ii. $\alpha = \dfrac{5\pi}{12}$

3 Vectors

LEARNING SEQUENCE

Fully worked solutions for this topic are available online.

3.1 Overview

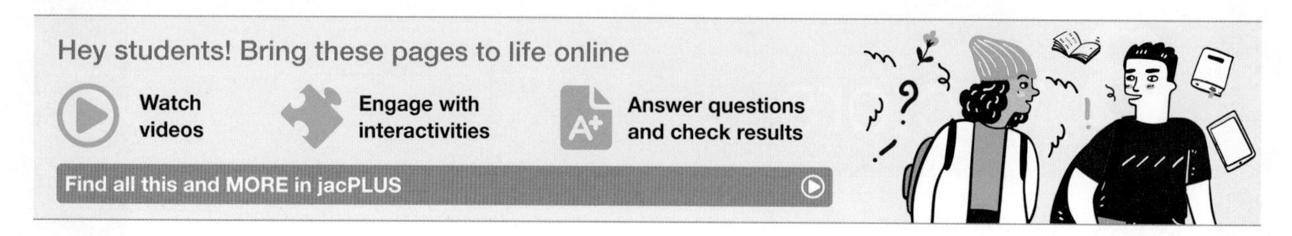

3.1.1 Introduction

Until this point, the study of vectors has been restrained to only two dimensions. By extending our knowledge to three dimensions, the list of applications is endless.

Parametric equations link our conventional understanding of equations of two variables with a third, known as a parameter. The famous Brachistochrone problem incorporates three variables and was posed by the Mathematician James Bernoulli (1654–1705). Bernoulli was determined to identify the path (x, y) in which a particle could slide from two points (P and Q) in the least possible time (t). Challenging fellow scholars such as Newton and L'Hopital, Bernoulli found that if the particle followed a path in the shape of an inverted cycloid, it would reach its destination in the shortest possible time.

KEY CONCEPTS

This topic covers the following key concepts from the VCE Mathematics Study Design:

- addition and subtraction of vectors and their multiplication by a scalar, position vectors
- linear dependence and independence of a set of vectors and geometric interpretation
- magnitude of a vector, unit vector, the orthogonal unit vectors $\underset{\sim}{i}$, $\underset{\sim}{j}$ and $\underset{\sim}{k}$
- resolution of a vector into rectangular components
- scalar (dot) product of two vectors, deduction of dot product for the $\underset{\sim}{i}$, $\underset{\sim}{j}$ and $\underset{\sim}{k}$ vector system and its use to find scalar resolute and vector resolute
- parallel and perpendicular vectors
- vector proofs of simple geometric results, such as 'the diagonals of a rhombus are perpendicular', 'the medians of a triangle are concurrent' and 'the angle subtended by a diameter in a circle is a right angle'
- vector equations and parametric equations of curves in two or three dimensions involving a parameter (and the corresponding Cartesian equation in the two-dimensional case).

3.2 Vectors in two dimensions

LEARNING INTENTION

At the end of this subtopic you should be able to:
- perform basic operations on vectors
- show that two vectors are collinear.

3.2.1 Review of vectors in two dimensions

Scalar and vector quantities

A **scalar** quantity is a real number only, it has magnitude, but no direction. Scalar quantities include temperature, mass and time.

A **vector** is a quantity that has **magnitude** and **direction**. Vectors include weight in a particular direction and displacement with a specific direction.

Vectors are directed line segments with a tail (start) and head (end). In the figure below, the head of the vector is at point B (indicated with an arrow), while the tail is at point A. In this instance, the vector could be expressed in a variety of forms.

Various forms of vector notation

$\overrightarrow{AB} = \underset{\sim}{u} = \boldsymbol{u}$

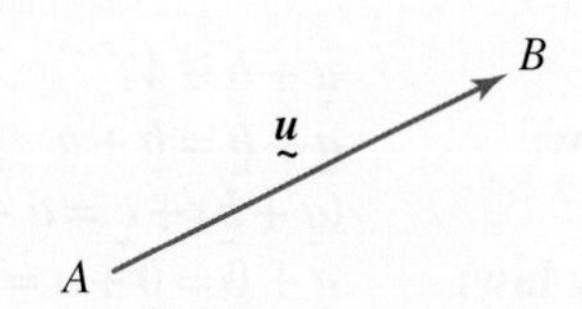

The *tilde* symbol (~) is commonly used because drawing bold letters is often difficult.

Equality of vectors

Two vectors are equal if *both* their magnitude and direction are equal.

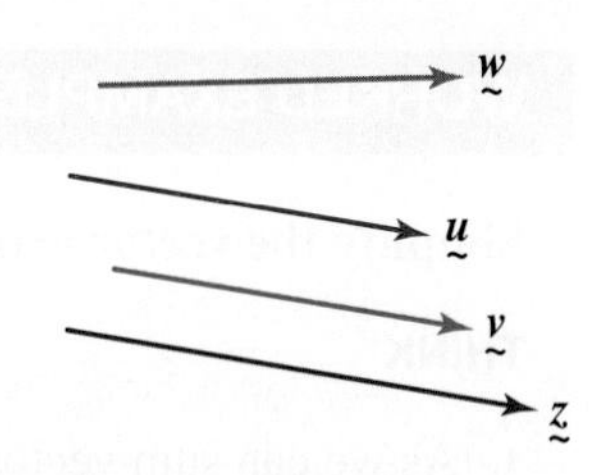

In this figure, the following statements can be made:

$\underset{\sim}{u} = \underset{\sim}{v}$ (direction and magnitude equal)

$\underset{\sim}{u} \neq \underset{\sim}{w}$ (directions are not equal)

$\underset{\sim}{u} \neq \underset{\sim}{z}$ (magnitudes are not equal)

Addition and subtraction of vectors

The triangle rule of vectors can be applied in both the addition and subtraction of vectors.

To add two vectors, take the tail of one vector and join it to the head of another. The result of this addition is the vector from the tail of the first vector to the head of the second vector.

If $\underset{\sim}{u}$ is the vector from A to B, then $-\underset{\sim}{u}$ is the vector from B to A.

To subtract two vectors, add the negative of the second vector to the first vector. For example, $\underset{\sim}{u} - \underset{\sim}{v} = \underset{\sim}{u} + (-\underset{\sim}{v})$.

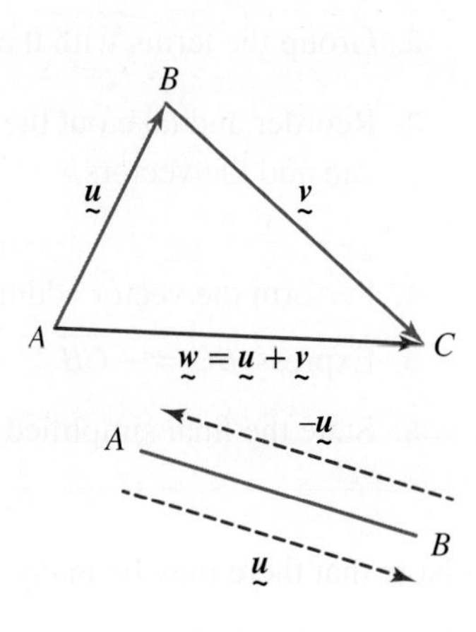

Multiplying a vector by a scalar

Multiplying a vector by a *positive* number (scalar) affects only the *magnitude* of the vector, not the direction.

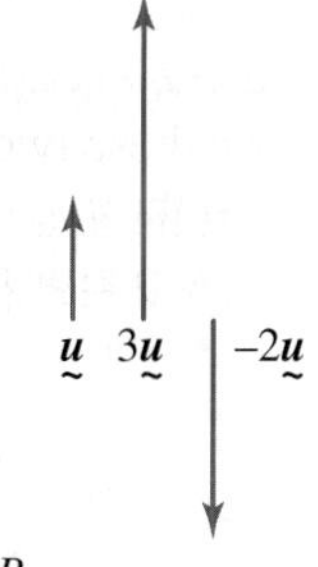

If the scalar is negative, then the direction is reversed.

The zero or null vector

Note that $\underset{\sim}{a} + (-\underset{\sim}{a}) = \underset{\sim}{0}$. This is a vector of no magnitude and no direction. In fact, if all the sides of the triangle are added, then

$$\begin{aligned}\overrightarrow{OA} + \overrightarrow{AB} + \overrightarrow{BO} &= \left(\overrightarrow{OA} + \overrightarrow{AB}\right) + \overrightarrow{BO} \\ &= \overrightarrow{OB} + \overrightarrow{BO} \\ &= \overrightarrow{OB} - \overrightarrow{OB} \\ &= \underset{\sim}{0}\end{aligned}$$

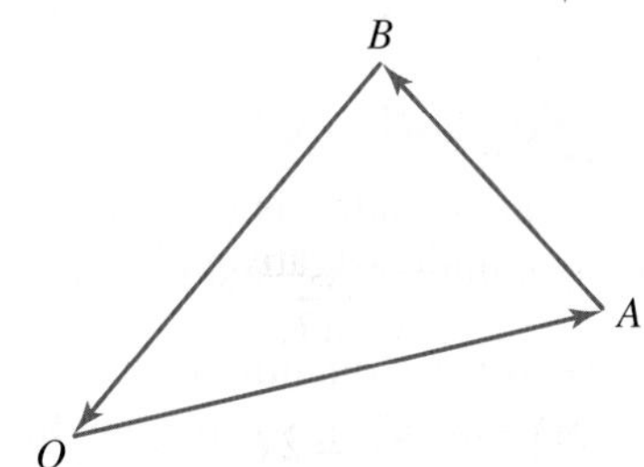

The algebra of vectors

Vectors satisfy the field laws using vector addition. Let $\underset{\sim}{a}, \underset{\sim}{b} \in V$, where V is the set of vectors; then under the operation of vector addition, these five field laws are also satisfied.

The field laws for vectors

1. **Closure:** $\underset{\sim}{a} + \underset{\sim}{b} \in V$
2. **Commutative law:** $\underset{\sim}{a} + \underset{\sim}{b} = \underset{\sim}{b} + \underset{\sim}{a}$
3. **Associative law:** $(\underset{\sim}{a} + \underset{\sim}{b}) + \underset{\sim}{c} = \underset{\sim}{a} + (\underset{\sim}{b} + \underset{\sim}{c})$
4. **Additive identity law:** $\underset{\sim}{a} + \underset{\sim}{0} = \underset{\sim}{0} + \underset{\sim}{a} = \underset{\sim}{a}$
5. **Inverse law:** $\underset{\sim}{a} + (-\underset{\sim}{a}) = (-\underset{\sim}{a}) + \underset{\sim}{a} = \underset{\sim}{0}$

Using these laws, vector expressions can be simplified.

WORKED EXAMPLE 1 Simplifying vector expressions

Simplify the vector expression $4\overrightarrow{AB} - \overrightarrow{CB} - 4\overrightarrow{AC}$.

THINK	WRITE
1. As we can sum vectors, express the negatives as sums of vectors.	$4\overrightarrow{AB} - \overrightarrow{CB} - 4\overrightarrow{AC}$ $= 4\overrightarrow{AB} + \overrightarrow{BC} + 4\overrightarrow{CA}$
2. Group the terms with the common factor.	$= 4\overrightarrow{AB} + 4\overrightarrow{CA} + \overrightarrow{BC}$
3. Reorder and take out the common factor so we can add the vectors.	$= 4\overrightarrow{CA} + 4\overrightarrow{AB} + \overrightarrow{BC}$ $= 4\left(\overrightarrow{CA} + \overrightarrow{AB}\right) + \overrightarrow{BC}$
4. Perform the vector addition.	$= 4\overrightarrow{CB} + \overrightarrow{BC}$
5. Express $\overrightarrow{BC} = -\overrightarrow{CB}$	$= 4\overrightarrow{CB} - \overrightarrow{CB}$
6. State the final simplified vector expression.	$= 3\overrightarrow{CB}$

Note that there may be many ways to arrive at this simplified vector expression.

Collinear points

Three points are said to be collinear if they all lie on the same straight line. In vector terms, if the vector $\overrightarrow{AB}$ is parallel to the vector $\overrightarrow{BC}$, then these two parallel vectors have the point B in common, and so they must lie in a straight line. Hence A, B and C are collinear if $\overrightarrow{AB} = \lambda \overrightarrow{BC}$ where $\lambda \in R \backslash \{0\}$.

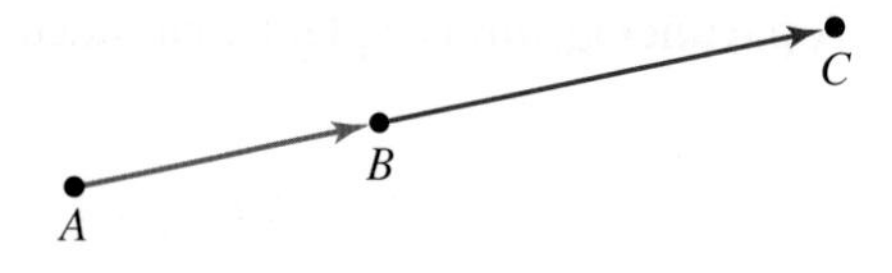

WORKED EXAMPLE 2 Showing that three points are collinear

If $3\overrightarrow{OA} - 2\overrightarrow{OB} - \overrightarrow{OC} = \underset{\sim}{0}$, show that the points A, B and C are collinear.

THINK	WRITE
1. Use the negative of a vector, in the middle term, $-\overrightarrow{OB} = \overrightarrow{BO}$.	$3\overrightarrow{OA} - 2\overrightarrow{OB} - \overrightarrow{OC} = \underset{\sim}{0}$ $3\overrightarrow{OA} + 2\overrightarrow{BO} - \overrightarrow{OC} = \underset{\sim}{0}$
2. Let $3\overrightarrow{OA} = 2\overrightarrow{OA} + \overrightarrow{OA}$.	$2\overrightarrow{OA} + \overrightarrow{OA} + 2\overrightarrow{BO} - \overrightarrow{OC} = \underset{\sim}{0}$
3. Reorder the terms.	$2\overrightarrow{BO} + 2\overrightarrow{OA} + \overrightarrow{OA} - \overrightarrow{OC} = \underset{\sim}{0}$
4. Take out the common factor in the first two terms and use vector addition.	$2\left(\overrightarrow{BO} + \overrightarrow{OA}\right) + \overrightarrow{OA} - \overrightarrow{OC} = \underset{\sim}{0}$ $2\overrightarrow{BA} + \overrightarrow{OA} - \overrightarrow{OC} = \underset{\sim}{0}$
5. Take the last two terms across to the right-hand side.	$2\overrightarrow{BA} = \overrightarrow{OC} - \overrightarrow{OA}$
6. Use the negative of a vector. This statement shows that A, B and C are collinear, since the point A is common.	$2\overrightarrow{BA} = \overrightarrow{AO} + \overrightarrow{OC}$ $2\overrightarrow{BA} = \overrightarrow{AC}$

3.2.2 Geometrical shapes

A general **quadrilateral** is a plane four-sided figure with no two sides necessarily parallel nor equal in length.

Trapeziums

A **trapezium** is a plane four-sided figure with one pair of sides parallel, but not equal.

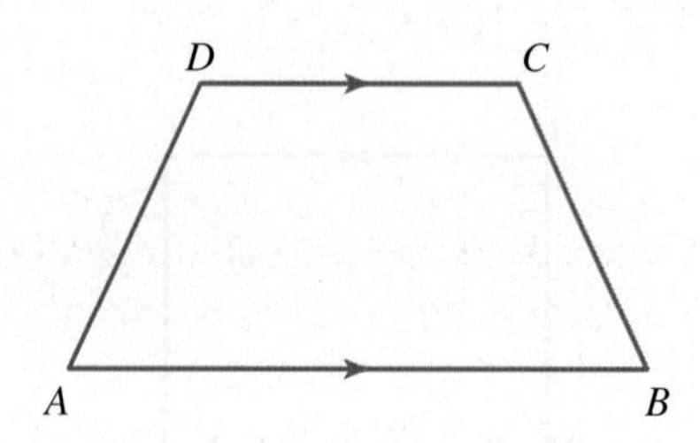

In the trapezium $ABCD$, since AB is parallel to DC, $\overrightarrow{AB} = \lambda \overrightarrow{DC}$.

Parallelograms

A **parallelogram** is a plane four-sided figure with two sets of sides parallel and equal in length.

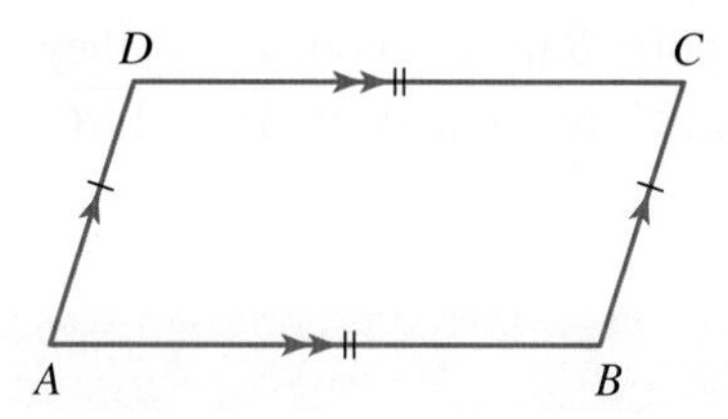

In the parallelogram $ABCD$, $\overrightarrow{AB} = \overrightarrow{DC}$ and $\overrightarrow{AD} = \overrightarrow{BC}$.

Rectangles

A **rectangle** is a parallelogram with all angles being 90°.

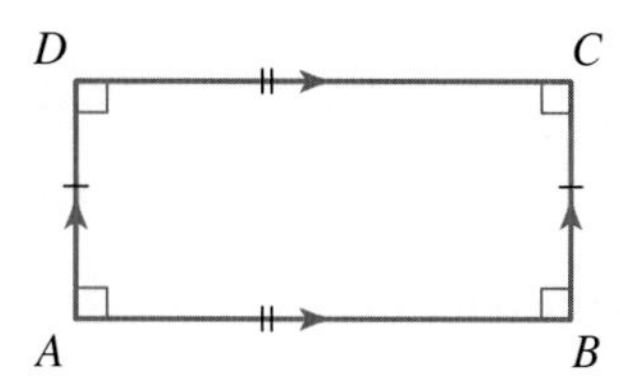

Rhombuses

A **rhombus** is a parallelogram with all sides equal in length.

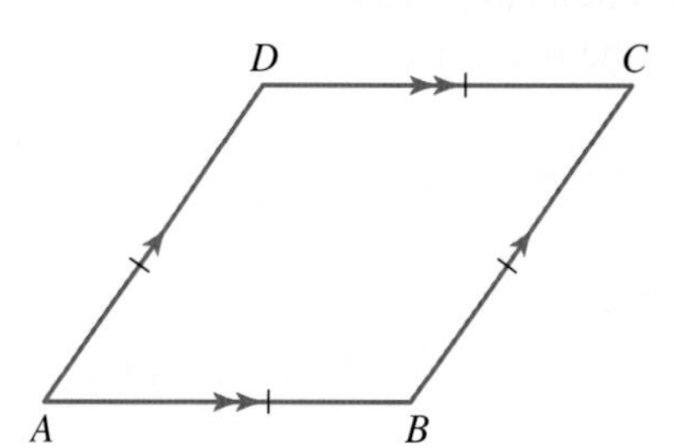

In the rhombus $ABCD$, $\overrightarrow{AB} = \overrightarrow{DC}$ and $\overrightarrow{AD} = \overrightarrow{BC}$; also $\left|\overrightarrow{AB}\right| = \left|\overrightarrow{BC}\right| = \left|\overrightarrow{DC}\right| = \left|\overrightarrow{AD}\right|$, since all sides are equal in length.

Squares

A **square** is a rhombus with all angles 90°.

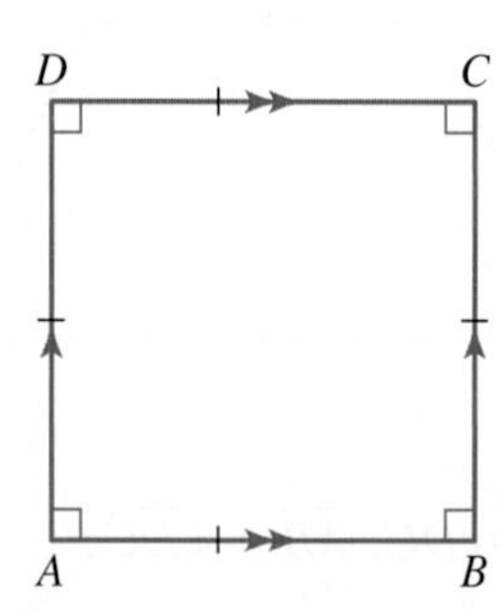

In the square $ABCD$, $\overrightarrow{AB} = \overrightarrow{DC}$, $\overrightarrow{AD} = \overrightarrow{BC}$ and thus $\overrightarrow{AB} \cdot \overrightarrow{BC} = 0$, $\overrightarrow{BC} \cdot \overrightarrow{CD} = 0$, $\overrightarrow{CD} \cdot \overrightarrow{DA} = 0$, $\overrightarrow{DA} \cdot \overrightarrow{AB} = 0$ and $\left|\overrightarrow{AB}\right| = \left|\overrightarrow{BC}\right| = \left|\overrightarrow{DC}\right| = \left|\overrightarrow{AD}\right|$.

Triangles

A median of a **triangle** is the line segment from a vertex to the midpoint of the opposite side. The centroid of a triangle is the point of intersection of the three medians. If G is the centroid of the triangle ABC and O is the origin, then it can be shown that $\overrightarrow{OG} = \frac{1}{3}\left(\overrightarrow{OA} + \overrightarrow{OB} + \overrightarrow{OC}\right)$.

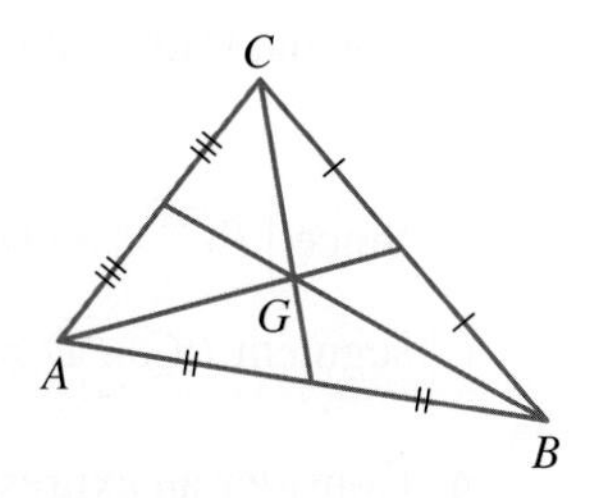

Using vectors to prove geometrical theorems

The properties of vectors can be used to prove many geometrical theorems.

The following statements are useful in proving geometrical theorems.

1. **If O is the origin and A and B are points, the midpoint M of the line segment AB is given by**

$$\begin{aligned}\overrightarrow{OM} &= \overrightarrow{OA} + \overrightarrow{AM}\\ &= \overrightarrow{OA} + \frac{1}{2}\overrightarrow{AB}\\ &= \overrightarrow{OA} + \frac{1}{2}\left(\overrightarrow{OB} - \overrightarrow{OA}\right)\\ &= \frac{1}{2}\left(\overrightarrow{OA} + \overrightarrow{OB}\right)\end{aligned}$$

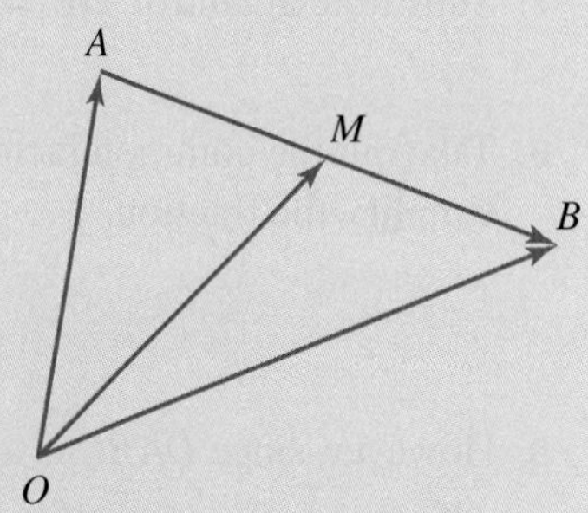

2. **If two vectors $\overrightarrow{AB}$ and $\overrightarrow{CD}$ are parallel, then $\overrightarrow{AB} = \lambda\overrightarrow{CD}$ where $\lambda \in R$ is a scalar.**
3. **If two vectors $\overrightarrow{AB}$ and $\overrightarrow{CD}$ are perpendicular, then $\overrightarrow{AB} \cdot \overrightarrow{CD} = 0$.**
4. **If two vectors $\overrightarrow{AB}$ and $\overrightarrow{CD}$ are equal, then $\overrightarrow{AB}$ is parallel to $\overrightarrow{CD}$; furthermore, these two vectors are equal in length, so that $\overrightarrow{AB} = \overrightarrow{CD} \Rightarrow \left|\overrightarrow{AB}\right| = \left|\overrightarrow{CD}\right|$.**
5. **If $\overrightarrow{AB} = \lambda\overrightarrow{BC}$, then the points A, B and C are collinear; that is, A, B and C all lie on a straight line.**

WORKED EXAMPLE 3 Showing that three points are collinear (2)

***OABC* is a parallelogram. *P* is the midpoint of *OA*, and the point *D* divides *PC* in the ratio 1 : 2. Prove that *O*, *D* and *B* are collinear.**

THINK

1. Draw a parallelogram and label it as $OABC$. Mark in the point P as the midpoint of OA. Mark in the point D dividing PC into thirds, with the point D closer to P on PC since $\left|\overrightarrow{PC}\right| = 3\left|\overrightarrow{DP}\right|$.

WRITE/DRAW

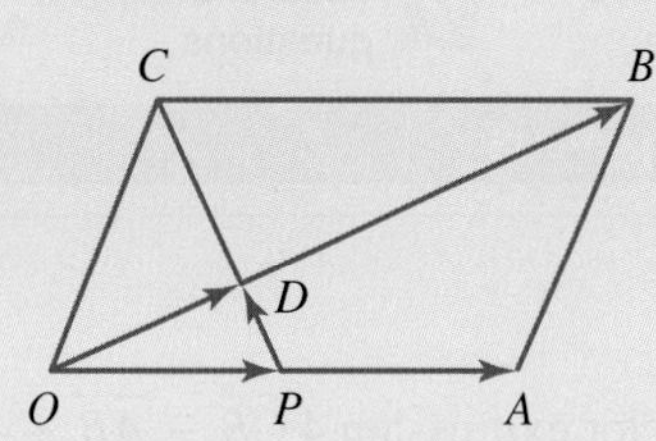

2. P is the midpoint of $\overrightarrow{OA}$.	$\overrightarrow{OP} = \overrightarrow{PA} = \frac{1}{2}\overrightarrow{OA}$
3. Since $\left\|\overrightarrow{PC}\right\| = 3\left\|\overrightarrow{DP}\right\|$, and D is on the line segment PC, $\overrightarrow{PD} = \frac{1}{3}\overrightarrow{PC}$.	$\overrightarrow{PD} = \frac{1}{3}\overrightarrow{PC}$
4. Consider an expression for $\overrightarrow{OD}$.	$\overrightarrow{OD} = \overrightarrow{OP} + \overrightarrow{PD}$
5. Substitute for the expression above.	$\overrightarrow{OD} = \frac{1}{2}\overrightarrow{OA} + \frac{1}{3}\overrightarrow{PC}$
6. Use $\overrightarrow{PC} = \overrightarrow{PO} + \overrightarrow{OC}$, or $\overrightarrow{PC} = \overrightarrow{OC} - \overrightarrow{OP}$	$\overrightarrow{OD} = \frac{1}{2}\overrightarrow{OA} + \frac{1}{3}\left(\overrightarrow{PO} + \overrightarrow{OC}\right)$ $= \frac{1}{2}\overrightarrow{OA} + \frac{1}{3}\left(\overrightarrow{OC} - \overrightarrow{OP}\right)$
7. Substitute again for $\overrightarrow{OP} = \frac{1}{2}\overrightarrow{OA}$.	$\overrightarrow{OD} = \frac{1}{2}\overrightarrow{OA} + \frac{1}{3}\left(\overrightarrow{OC} - \frac{1}{2}\overrightarrow{OA}\right)$
8. Take out the common factor of $\overrightarrow{OA}$ and simplify the fraction.	$\overrightarrow{OD} = \left(\frac{1}{2} - \frac{1}{6}\right)\overrightarrow{OA} + \frac{1}{3}\overrightarrow{OC}$ $= \frac{1}{3}\overrightarrow{OA} + \frac{1}{3}\overrightarrow{OC}$
9. However, since $OABC$ is a parallelogram, $\overrightarrow{OC} = \overrightarrow{AB}$.	$\overrightarrow{OD} = \frac{1}{3}\overrightarrow{OA} + \frac{1}{3}\overrightarrow{AB}$ $= \frac{1}{3}\left(\overrightarrow{OA} + \overrightarrow{AB}\right)$
10. As $\overrightarrow{OD} = \frac{1}{3}\overrightarrow{OB}$, this vector equation implies that the points O, D and B are collinear.	$\overrightarrow{OD} = \frac{1}{3}\overrightarrow{OB}$ O, D and B are collinear.

3.2 Exercise

Students, these questions are even better in jacPLUS

Receive immediate feedback and access sample responses

Access additional questions

Track your results and progress

Find all this and MORE in jacPLUS

Technology free

1. **WE1** Simplify the vector expression $4\overrightarrow{CB} - \overrightarrow{AB} + 4\overrightarrow{AC}$.

2. Simplify the vector expression $2\overrightarrow{BO} - 5\overrightarrow{AO} - 2\overrightarrow{BA}$.

3. Simplify each of the following vector expressions.
   a. $2\overrightarrow{AC} - \overrightarrow{CB} + \overrightarrow{AB}$
   b. $5\overrightarrow{CA} + \overrightarrow{BC} + 4\overrightarrow{OC} - \overrightarrow{BO}$

4. a. $\overrightarrow{OC} + 6\overrightarrow{AB} + \overrightarrow{CA} + 5\overrightarrow{OA}$
   b. $3\overrightarrow{OB} + \overrightarrow{AB} - 3\overrightarrow{AC} + 4\overrightarrow{BC} - \overrightarrow{OC}$

5. **WE2** If $\overrightarrow{AO} + \overrightarrow{OB} - 2\overrightarrow{BO} - 2\overrightarrow{OC} = \underset{\sim}{0}$, show that the points A, B and C are collinear.

6. If $\overrightarrow{PO} - 4\overrightarrow{RO} + 3\overrightarrow{QO} = \underset{\sim}{0}$, show that the points P, Q and R are collinear.

7. For each of the following vector expressions, show that the points A, B and C are collinear.
   a. $\overrightarrow{AO} + \overrightarrow{OB} = \overrightarrow{BO} + \overrightarrow{OC}$
   b. $3\overrightarrow{OA} - 2\overrightarrow{OB} = \overrightarrow{OC}$

8. a. $\overrightarrow{BO} + 4\overrightarrow{AO} - 5\overrightarrow{CO} = \underset{\sim}{0}$
   b. $3\overrightarrow{BO} - 5\overrightarrow{CO} + 2\overrightarrow{AO} = \underset{\sim}{0}$

9. **WE3** $OABC$ is a parallelogram in which P is the midpoint of CB, and D is a point on AP such that $\left|\overrightarrow{AD}\right| = \frac{2}{3}\left|\overrightarrow{AP}\right|$. Prove that $\overrightarrow{OD} = \frac{2}{3}\overrightarrow{OB}$ and that O, D and B are collinear.

10. $OABC$ is a parallelogram in which the point P divides OA in the ratio 1:2, and the point D divides PC in the ratio 1:3. Prove that $\overrightarrow{OD} = \frac{1}{4}\overrightarrow{OB}$ and that O, D and B are collinear.

11. $OABCGDEF$ is a cuboid with $\overrightarrow{OA} = \underset{\sim}{a}$, $\overrightarrow{OC} = \underset{\sim}{c}$ and $\overrightarrow{OD} = \underset{\sim}{d}$.
    a. List all the vectors equal to $\overrightarrow{OA} = \underset{\sim}{a}$.
    b. List all the vectors equal to $\overrightarrow{OC} = \underset{\sim}{c}$.
    c. List all the vectors equal to $\overrightarrow{OD} = \underset{\sim}{d}$.
    d. Express each of the following in terms of $\underset{\sim}{a}$, $\underset{\sim}{c}$ and $\underset{\sim}{d}$.
       i. $\overrightarrow{DF}$　　ii. $\overrightarrow{EB}$
       iii. $\overrightarrow{FO}$　　iv. $\overrightarrow{DB}$

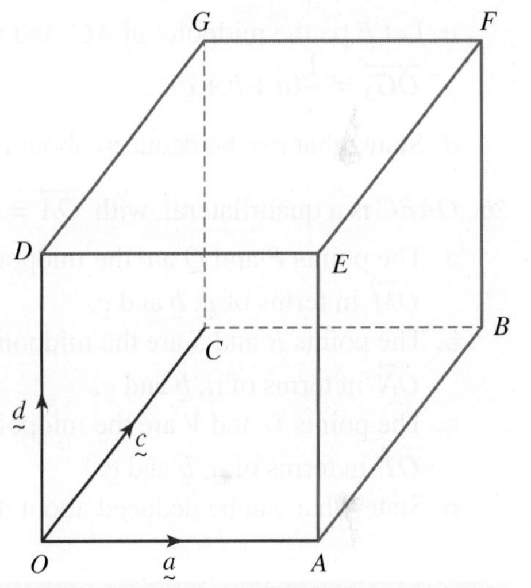

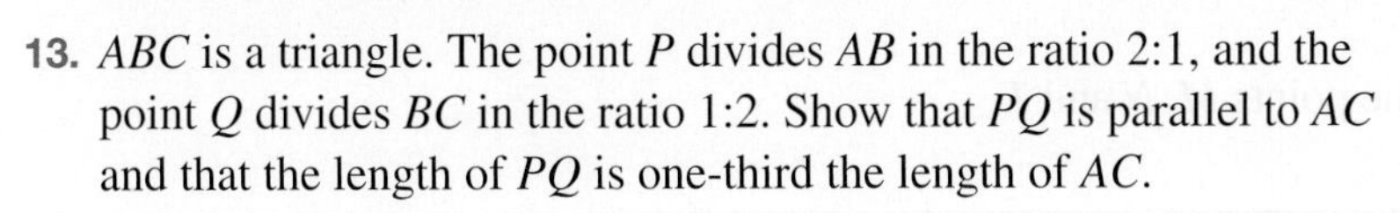

12. ABC is a triangle. The points P and Q are the midpoints of AB and BC respectively. Show that PQ is parallel to AC and that the length of PQ is half the length of AC.

13. ABC is a triangle. The point P divides AB in the ratio 2:1, and the point Q divides BC in the ratio 1:2. Show that PQ is parallel to AC and that the length of PQ is one-third the length of AC.

14. OAB is a triangle with $\overrightarrow{OA} = \underset{\sim}{a}$ and $\overrightarrow{OB} = \underset{\sim}{b}$. Let M be the midpoint of AB. Show that $\overrightarrow{OM} = \frac{1}{2}(\underset{\sim}{a} + \underset{\sim}{b})$.

15. OPQ is a triangle. The point M divides PQ in the ratio 1:2. Show that $\overrightarrow{OM} = \frac{1}{3}\left(2\overrightarrow{OP} + \overrightarrow{OQ}\right)$.

16. OAB is a triangle with $\overrightarrow{OA} = \underset{\sim}{a}$ and $\overrightarrow{OB} = \underset{\sim}{b}$. The point M divides AB in the ratio 1:3. Show that $\overrightarrow{OM} = \frac{1}{4}(3\underset{\sim}{a} + \underset{\sim}{b})$.

17. $ABCD$ is a quadrilateral. The points P and Q are the midpoints of AB and BC respectively. Show that $\overrightarrow{PQ} = \overrightarrow{AP} + \overrightarrow{QC}$.

18. $ABCD$ is a quadrilateral. P is a point on DB such that $\overrightarrow{AP} + \overrightarrow{PB} + \overrightarrow{PD} = \overrightarrow{PC}$. Show that $ABCD$ is a parallelogram.

19. $ABCD$ is a quadrilateral. The points M and N are the midpoints of AB and CD respectively. Show that $2\overrightarrow{MN} = \overrightarrow{BC} + \overrightarrow{AD}$.

20. $ABCD$ is a trapezium with sides AB and DC parallel. P and Q are the midpoints of the sides AD and BC respectively. Show that PQ is parallel to both AB and DC and that the distance PQ is one half the sum of the distances AB and DC.

21. ABC is a triangle. The points P, Q and R are the midpoints of the sides AB, BC and CA respectively. If O is any other point, show that $\overrightarrow{OP} + \overrightarrow{OQ} + \overrightarrow{OR} = \overrightarrow{OA} + \overrightarrow{OB} + \overrightarrow{OC}$.

22. $ABCD$ is a rectangle. The points P, Q, R and S are the midpoints of the sides AB, BC, CD and DA respectively. Show that $PQRS$ is a parallelogram.

23. $ABCD$ is a parallelogram. The points M and N are the midpoints of the diagonals AC and DB respectively. Show that M and N are coincident.

24. $ABCD$ is a quadrilateral. The points M and N are the midpoints of AC and BD respectively. Show that $\overrightarrow{AB} + \overrightarrow{AD} + \overrightarrow{CB} + \overrightarrow{CD} = 4\overrightarrow{MN}$.

25. ABC is a triangle with $\overrightarrow{OA} = \underset{\sim}{a}$, $\overrightarrow{OB} = \underset{\sim}{b}$ and $\overrightarrow{OC} = \underset{\sim}{c}$.
   a. Let D be the midpoint of AB and G_1 be the point which divides CD in the ratio $2:1$. Show that $\overrightarrow{OG_1} = \frac{1}{3}(\underset{\sim}{a} + \underset{\sim}{b} + \underset{\sim}{c})$.
   b. Let E be the midpoint of BC, and G_2 be the point which divides AE in the ratio $2:1$. Show that $\overrightarrow{OG_2} = \frac{1}{3}(\underset{\sim}{a} + \underset{\sim}{b} + \underset{\sim}{c})$.
   c. Let F be the midpoint of AC, and G_3 be the point which divides BF in the ratio $2:1$. Show that $\overrightarrow{OG_3} = \frac{1}{3}(\underset{\sim}{a} + \underset{\sim}{b} + \underset{\sim}{c})$.
   d. State what can be deduced about the points G_1, G_2 and G_3.

26. $OABC$ is a quadrilateral, with $\overrightarrow{OA} = \underset{\sim}{a}$, $\overrightarrow{OB} = \underset{\sim}{b}$ and $\overrightarrow{OC} = \underset{\sim}{c}$.
   a. The points P and Q are the midpoints of OA and BC respectively. Let M be the midpoint of PQ. Express $\overrightarrow{OM}$ in terms of $\underset{\sim}{a}$, $\underset{\sim}{b}$ and $\underset{\sim}{c}$.
   b. The points R and S are the midpoints of AB and OC respectively. Let N be the midpoint of SR. Express $\overrightarrow{ON}$ in terms of $\underset{\sim}{a}$, $\underset{\sim}{b}$ and $\underset{\sim}{c}$.
   c. The points U and V are the midpoints of AC and OB respectively. Let T be the midpoint of UV. Express $\overrightarrow{OT}$ in terms of $\underset{\sim}{a}$, $\underset{\sim}{b}$ and $\underset{\sim}{c}$.
   d. State what can be deduced about the points M, N and T.

3.2 Exam questions

Question 1 (4 marks) TECH-ACTIVE

Source: *VCE 2013 Specialist Mathematics Exam 2, Section B, Q4e; © VCAA.*

For the triangle ABC shown below, the midpoints of the sides are the points M, N and P.

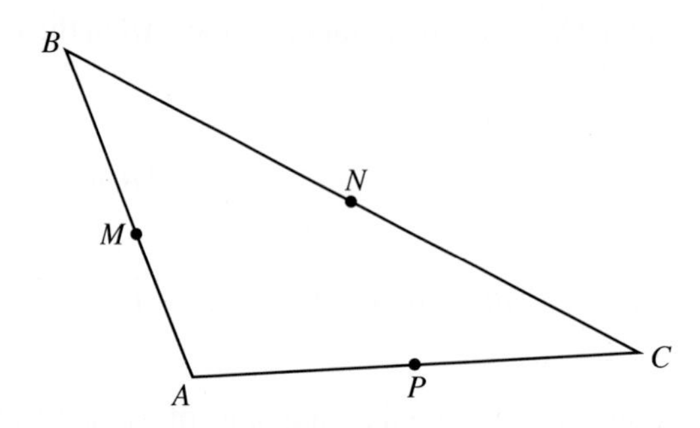

Let $\overrightarrow{AC} = \underset{\sim}{u}$ and $\overrightarrow{CB} = \underset{\sim}{v}$.

   i. Express $\overrightarrow{AN}$ in terms of $\underset{\sim}{u}$ and $\underset{\sim}{v}$. **(1 mark)**
   ii. Express $\overrightarrow{CM}$ and $\overrightarrow{BP}$ in terms of $\underset{\sim}{u}$ and $\underset{\sim}{v}$. **(2 marks)**
   iii. Hence simplify the expression $\overrightarrow{AN} + \overrightarrow{CM} + \overrightarrow{BP}$. **(1 mark)**

Question 2 (3 marks) TECH-FREE

The diagram shows a triangle with vertices Q, R and S. O is the origin, and the vectors $\overrightarrow{OQ} = \underset{\sim}{q}$ and $\overrightarrow{OS} = \underset{\sim}{s}$.

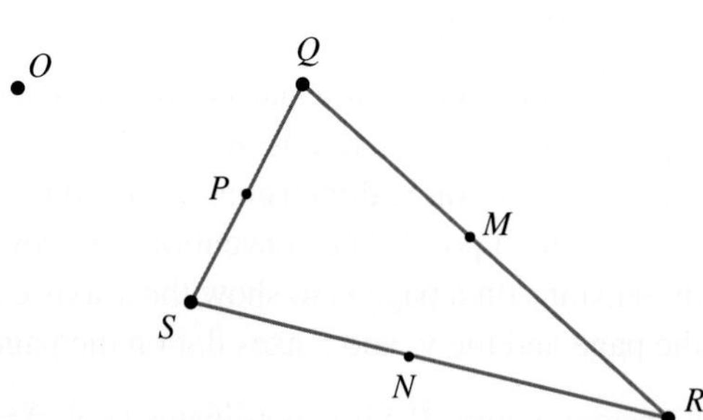

Given that M is the midpoint of the line segment QR, N is the midpoint of the line segment RS and P is the midpoint of the line segment QS, prove that the quadrilateral $MNPQ$ is a parallelogram.

Question 3 (1 mark) TECH-ACTIVE

MC The point M cuts the line segment PQ in the ratio of $3:4$ with M being closer to P.

If the position vectors of P and Q are $\underset{\sim}{p}$ and $\underset{\sim}{q}$ respectively, then the position vector of M is:

A. $\frac{1}{7}(3\underset{\sim}{p} + 4\underset{\sim}{q})$ **B.** $\frac{1}{7}(3\underset{\sim}{p} - 4\underset{\sim}{q})$ **C.** $\frac{1}{7}(4\underset{\sim}{q} - 3\underset{\sim}{p})$ **D.** $\frac{1}{7}(4\underset{\sim}{p} + 3\underset{\sim}{q})$ **E.** $\frac{1}{7}(4\underset{\sim}{p} - 3\underset{\sim}{q})$

More exam questions are available online.

3.3 Vectors in three dimensions

LEARNING INTENTION

At the end of this subtopic you should be able to:
- represent points in 3-dimensional space by position vectors
- determine whether a set of three vectors are linearly dependent or independent
- determine the angle between two vectors.

3.3.1 Position vectors in three dimensions

A unit vector is a vector that has a magnitude of 1. Unit vectors are direction givers.

A hat or circumflex above the vector, is used to indicate that it is a unit vector, for example $\hat{\underset{\sim}{a}}$.

If $\underset{\sim}{a}$ is a vector, then the length or magnitude of the vector is denoted by $|\underset{\sim}{a}|$. Dividing the vector by its length makes it a unit vector, so that $\hat{\underset{\sim}{a}} = \frac{\underset{\sim}{a}}{|\underset{\sim}{a}|}$.

A unit vector in the positive direction parallel to the x-axis is denoted by $\underset{\sim}{i}$.

A unit vector in the positive direction parallel to the y-axis is denoted by $\underset{\sim}{j}$.

A unit vector in the positive direction parallel to the z-axis is denoted by $\underset{\sim}{k}$.

Since we understand that $\underset{\sim}{i}$, $\underset{\sim}{j}$ and $\underset{\sim}{k}$ are unit vectors, we do not need to place the hat or circumflex above these vectors, although some other notations do use this notation and write $\hat{\underset{\sim}{i}}$, $\hat{\underset{\sim}{j}}$ and $\hat{\underset{\sim}{k}}$.

If a vector is expressed in terms of $\underset{\sim}{i}$, then the coefficient of $\underset{\sim}{i}$ represents the magnitude parallel to the x-axis and the $\underset{\sim}{i}$ indicates that this vector is parallel to the x-axis.

If a vector is expressed in terms of $\underset{\sim}{j}$, then the coefficient of $\underset{\sim}{j}$ represents the magnitude parallel to the y-axis and the $\underset{\sim}{j}$ indicates that this vector is parallel to the y-axis.

If a vector is expressed in terms of $\underset{\sim}{k}$, then the coefficient of $\underset{\sim}{k}$ represents the magnitude parallel to the z-axis and the $\underset{\sim}{k}$ indicates that this vector is parallel to the z-axis.

$\underset{\sim}{i}\,\underset{\sim}{j}\,\underset{\sim}{k}$ vectors

The vectors $\underset{\sim}{i}$, $\underset{\sim}{j}$ and $\underset{\sim}{k}$ are unit vectors in the directions of the x, y and z-axes respectively.

Position vectors

In two-dimensional Cartesian coordinates, a point P is represented by (x, y); that is, we need two coordinates to represent it. In three dimensions we need three coordinates to represent a point. The convention for showing three dimensions on a page is to show the x-axis coming out of the page and the y- and z-axes flat on the page.

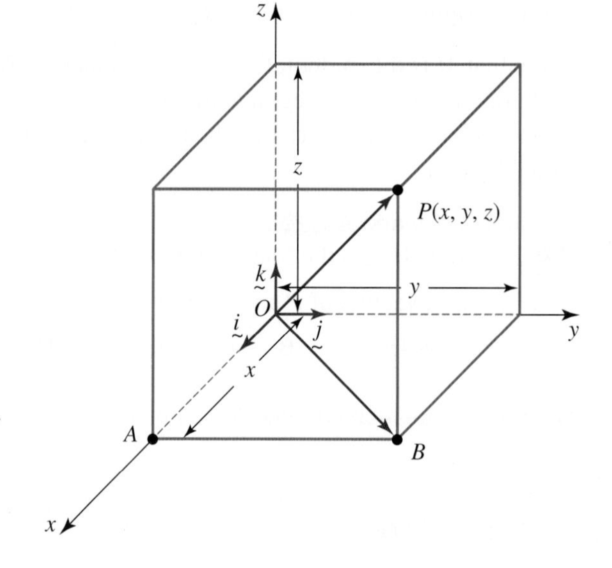

Consider a point P with coordinates (x, y, z) relative to the origin, $O(0, 0, 0)$. Use $\underset{\sim}{r}$ as a notation for the position vector. The vector $\underset{\sim}{r} = \overrightarrow{OP}$ can be expressed in terms of three other vectors: one parallel to the x-axis, one parallel to the y-axis and one parallel to the z-axis. This method of splitting a vector up into its components is called resolution of vectors.

$\overrightarrow{OP} = \overrightarrow{OA} + \overrightarrow{AB} + \overrightarrow{BP}$

but $\left|\overrightarrow{OA}\right| = x$, $\left|\overrightarrow{AB}\right| = y$ and $\left|\overrightarrow{BP}\right| = z$,

so the vectors $\overrightarrow{OA} = x\underset{\sim}{i}$, $\overrightarrow{AB} = y\underset{\sim}{j}$ and $\overrightarrow{BP} = z\underset{\sim}{k}$.

Therefore, $\underset{\sim}{r} = \overrightarrow{OP} = x\underset{\sim}{i} + y\underset{\sim}{j} + z\underset{\sim}{k}$.

A vector expressed in $\underset{\sim}{i}\,\underset{\sim}{j}\,\underset{\sim}{k}$ form is said to be in component form. We can now give answers to questions in terms of $\underset{\sim}{i}$, $\underset{\sim}{j}$ and $\underset{\sim}{k}$.

Magnitude of a vector

The distance between the origin O and the point P is the length or magnitude of the vector. By Pythagoras' theorem using the triangle in the xy plane, $\left|\overrightarrow{OB}\right|^2 = \left|\overrightarrow{OA}\right|^2 + \left|\overrightarrow{AB}\right|^2 = x^2 + y^2$ and using triangle OPB, $\left|\overrightarrow{OP}\right|^2 = \left|\overrightarrow{OB}\right|^2 + \left|\overrightarrow{BP}\right|^2 = x^2 + y^2 + z^2$.

The magnitude of a vector

The magnitude of the position vector $\underset{\sim}{r} = x\underset{\sim}{i} + y\underset{\sim}{j} + z\underset{\sim}{k}$ is given by:

$$|\underset{\sim}{r}| = \left|\overrightarrow{OP}\right| = \sqrt{x^2 + y^2 + z^2}$$

WORKED EXAMPLE 4 Determining the position vector and a unit vector

If the point P has coordinates $(3, 2, -4)$, determine:

a. the vector $\overrightarrow{OP}$

b. a unit vector parallel to $\overrightarrow{OP}$.

THINK	WRITE
a. There is no need to draw a three-dimensional diagram. The vector $\overrightarrow{OP}$ has the x-coordinate of P as the $\underset{\sim}{i}$ component, the y-coordinate of P as the $\underset{\sim}{j}$ component and the z-coordinate of P as the $\underset{\sim}{k}$ component.	a. The point $(3, 2, -4)$ has coordinates $x = 3$, $y = 2$ and $z = -4$. $\overrightarrow{OP} = 3\underset{\sim}{i} + 2\underset{\sim}{j} - 4\underset{\sim}{k}$

b. 1. First determine the magnitude or length of the vector using $\left|\overrightarrow{OP}\right| = \sqrt{x^2+y^2+z^2}$. Leave the answer as an exact answer, that is as a surd.

b. $\left|\overrightarrow{OP}\right| = \sqrt{3^2+2^2+(-4)^2}$
$= \sqrt{9+4+16}$
$= \sqrt{29}$

2. A unit vector parallel to $\overrightarrow{OP}$ is $\widehat{OP} = \dfrac{\overrightarrow{OP}}{\left|\overrightarrow{OP}\right|}$.

There is no need to rationalise the denominator.

$\widehat{OP} = \dfrac{1}{\sqrt{29}}(3\underset{\sim}{i}+2\underset{\sim}{j}-4\underset{\sim}{k})$

3.3.2 Operations on vectors in three dimensions

Addition and subtraction of vectors in three dimensions

If A is the point (x_1, y_1, z_1) and B is the point (x_2, y_2, z_2) relative to the origin O, then $\overrightarrow{OA} = \underset{\sim}{a} = x_1\underset{\sim}{i} + y_1\underset{\sim}{j} + z_1\underset{\sim}{k}$ and $\overrightarrow{OB} = \underset{\sim}{b} = x_2\underset{\sim}{i} + y_2\underset{\sim}{j} + z_2\underset{\sim}{k}$.

To add two vectors in component form, add the $\underset{\sim}{i}$, $\underset{\sim}{j}$ and $\underset{\sim}{k}$ components separately. So $\overrightarrow{OA} + \overrightarrow{OB} = \underset{\sim}{a} + \underset{\sim}{b} = (x_1+x_2)\underset{\sim}{i} + (y_1+y_2)\underset{\sim}{j} + (z_1+z_2)\underset{\sim}{k}$.

To subtract two vectors in component form, subtract the $\underset{\sim}{i}$, $\underset{\sim}{j}$ and $\underset{\sim}{k}$ components separately. So $\overrightarrow{AB} = \overrightarrow{OB} - \overrightarrow{OA} = \underset{\sim}{b} - \underset{\sim}{a} = (x_2-x_1)\underset{\sim}{i} + (y_2-y_1)\underset{\sim}{j} + (z_2-z_1)\underset{\sim}{k}$.

The vector $\overrightarrow{AB}$ represents the position vector of B relative to A, that is B as seen from A.

The distance between the points A and B, is the magnitude of the vector $\overrightarrow{AB}$:

$$\left|\overrightarrow{AB}\right| = \sqrt{(x_2-x_1)^2 + (y_2-y_1)^2 + (z_2-z_1)^2}$$

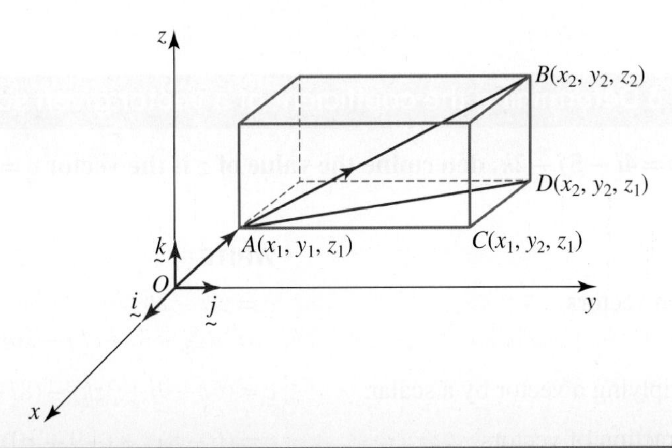

WORKED EXAMPLE 5 Determining a unit vector parallel to a the line between two points

Two points, A and B, have the coordinates (1, −2, 1) and (3, 4, −2) respectively. Determine a unit vector parallel to $\overrightarrow{AB}$.

THINK	WRITE
1. Write the vector $\overrightarrow{OA}$.	$\overrightarrow{OA} = \underset{\sim}{i} - 2\underset{\sim}{j} + \underset{\sim}{k}$
2. Write the vector $\overrightarrow{OB}$.	$\overrightarrow{OB} = 3\underset{\sim}{i} + 4\underset{\sim}{j} - 2\underset{\sim}{k}$

3. The vector $\overrightarrow{AB}$ is obtained by subtracting the two vectors $\overrightarrow{OB}$ and $\overrightarrow{OA}$.	Substitute for the two vectors. $\overrightarrow{AB} = \overrightarrow{OB} - \overrightarrow{OA}$ $= (3\underset{\sim}{i} + 4\underset{\sim}{j} - 2\underset{\sim}{k}) - (\underset{\sim}{i} - 2\underset{\sim}{j} + \underset{\sim}{k})$
4. Use the rules for subtraction of vectors.	$\overrightarrow{AB} = (3-1)\underset{\sim}{i} + (4-(-2))\underset{\sim}{j} + (-2-1)\underset{\sim}{k}$ $= 2\underset{\sim}{i} + 6\underset{\sim}{j} - 3\underset{\sim}{k}$
5. Determine the magnitude of the vector $\overrightarrow{AB}$.	$\left\lvert\overrightarrow{AB}\right\rvert = \sqrt{2^2 + 6^2 + (-3)^2}$ $= \sqrt{4+36+9}$ $= \sqrt{49}$ $= 7$
6. Write the unit vector parallel to $\overrightarrow{AB}$, that is $\widehat{AB} = \dfrac{\overrightarrow{AB}}{\left\lvert\overrightarrow{AB}\right\rvert}$.	$\widehat{AB} = \frac{1}{7}(2\underset{\sim}{i} + 6\underset{\sim}{j} - 3\underset{\sim}{k})$

Equality of two vectors

Given the vectors $\overrightarrow{OA} = \underset{\sim}{a} = x_1\underset{\sim}{i} + y_1\underset{\sim}{j} + z_1\underset{\sim}{k}$ and $\overrightarrow{OB} = \underset{\sim}{b} = x_2\underset{\sim}{i} + y_2\underset{\sim}{j} + z_2\underset{\sim}{k}$, the two vectors are equal, $\underset{\sim}{a} = \underset{\sim}{b}$, if and only if $x_1 = x_2$, $y_1 = y_2$ and $z_1 = z_2$.

Scalar multiplication of vectors

For scalar multiplication of vectors, $\lambda\underset{\sim}{a} = \lambda x_1\underset{\sim}{i} + \lambda y_1\underset{\sim}{j} + \lambda z_1\underset{\sim}{k}$. That is, each coefficient is multiplied by the scalar.

For example, if $\underset{\sim}{a} = \underset{\sim}{i} + 2\underset{\sim}{j} - 3\underset{\sim}{k}$, then $2\underset{\sim}{a} = 2\underset{\sim}{i} + 4\underset{\sim}{j} - 6\underset{\sim}{k}$ and $-\underset{\sim}{a} = -\underset{\sim}{i} - 2\underset{\sim}{j} + 3\underset{\sim}{k}$.

The following worked examples further illustrate scalar multiplication and equality of vectors in component forms.

WORKED EXAMPLE 6 Determining the coefficient of a vector given some information

If $\underset{\sim}{a} = 2\underset{\sim}{i} - 3\underset{\sim}{j} + z\underset{\sim}{k}$ and $\underset{\sim}{b} = 4\underset{\sim}{i} - 5\underset{\sim}{j} - 2\underset{\sim}{k}$, determine the value of z if the vector $\underset{\sim}{c} = 3\underset{\sim}{a} - 2\underset{\sim}{b}$ is parallel to the xy plane.

THINK	WRITE
1. Substitute for the given vectors.	$\underset{\sim}{c} = 3\underset{\sim}{a} - 2\underset{\sim}{b}$ $= 3(2\underset{\sim}{i} - 3\underset{\sim}{j} + z\underset{\sim}{k}) - 2(4\underset{\sim}{i} - 5\underset{\sim}{j} - 2\underset{\sim}{k})$
2. Use the rules for multiplying a vector by a scalar.	$\underset{\sim}{c} = (6\underset{\sim}{i} - 9\underset{\sim}{j} + 3z\underset{\sim}{k}) - (8\underset{\sim}{i} - 10\underset{\sim}{j} - 4\underset{\sim}{k})$
3. Use the rules for subtraction of vectors.	$\underset{\sim}{c} = (6-8)\underset{\sim}{i} + (-9+10)\underset{\sim}{j} + (3z+4)\underset{\sim}{k}$ $\underset{\sim}{c} = -2\underset{\sim}{i} + \underset{\sim}{j} + (3z+4)\underset{\sim}{k}$
4. If this vector is parallel to the xy plane, then its $\underset{\sim}{k}$ component must be zero.	$3z + 4 = 0$
5. Solve for the unknown value in this case.	$3z = -4$ $z = -\dfrac{4}{3}$

Parallel vectors

When solving vector problems, it is sometimes necessary to recall some of the other properties of vectors.

In the diagram, the vectors $\underset{\sim}{a}$, $\underset{\sim}{b}$ and $\underset{\sim}{c}$ are all parallel, as they are pointing in the same direction; however, they have different lengths. Two vectors are parallel if one is a scalar multiple of the other. That is, $\underset{\sim}{a}$ is parallel to $\underset{\sim}{b}$ if $\underset{\sim}{a} = \lambda \underset{\sim}{b}$ where $\lambda \in R$.

WORKED EXAMPLE 7 Determining the coefficient of a vector given some information (2)

Given the vectors $\underset{\sim}{r} = \underset{\sim}{i} - 3\underset{\sim}{j} + z\underset{\sim}{k}$ and $\underset{\sim}{s} = -2\underset{\sim}{i} + 6\underset{\sim}{j} - 7\underset{\sim}{k}$, determine the value of z in each case if:

a. **the length of the vector $\underset{\sim}{r}$ is 8**

b. **the vector $\underset{\sim}{r}$ is parallel to the vector $\underset{\sim}{s}$.**

THINK	WRITE
a. 1. Determine the magnitude of the vector in terms of the unknown value.	a. $\underset{\sim}{r} = \underset{\sim}{i} - 3\underset{\sim}{j} + z\underset{\sim}{k}$ $\lvert\underset{\sim}{r}\rvert = \sqrt{1^2 + (-3)^2 + z^2}$ $= \sqrt{1 + 9 + z^2}$ $= \sqrt{10 + z^2}$
2. Equate the length of the vector to the given value.	Since $\lvert\underset{\sim}{r}\rvert = 8$, $\sqrt{10 + z^2} = 8$
3. Square both sides.	$10 + z^2 = 64$
4. Solve for the unknown value. Both answers are acceptable values.	$z^2 = 54$ $z = \pm\sqrt{54}$ $= \pm\sqrt{9 \times 6}$ $= \pm 3\sqrt{6}$
b. 1. If two vectors are parallel, then one is a scalar multiple of the other. Substitute for the given vectors and expand.	b. $\underset{\sim}{r} = \lambda \underset{\sim}{s}$ $\underset{\sim}{i} - 3\underset{\sim}{j} + z\underset{\sim}{k} = \lambda(-2\underset{\sim}{i} + 6\underset{\sim}{j} - 7\underset{\sim}{k})$ $= -2\lambda\underset{\sim}{i} + 6\lambda\underset{\sim}{j} - 7\lambda\underset{\sim}{k}$
2. For the two vectors to be equal, all components must be equal.	From the $\underset{\sim}{i}$ component, $-2\lambda = 1$, and from the $\underset{\sim}{j}$ component, $6\lambda = -3$, so $\lambda = -\frac{1}{2}$. From the $\underset{\sim}{k}$ component, $z = -7\lambda$.
3. Solve for the unknown value in this case.	$z = -7\lambda$ and $\lambda = -\frac{1}{2}$, so $z = \frac{7}{2}$.

TI | THINK	DISPLAY/WRITE	CASIO | THINK	DISPLAY/WRITE
On a Calculator page, complete the entry as shown.	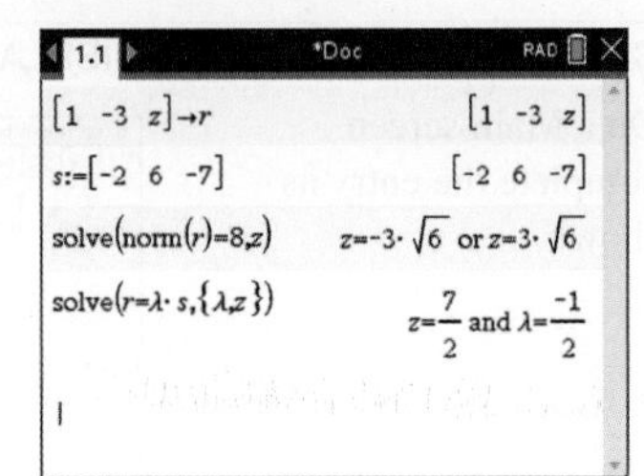	On a Main screen, complete the entry as shown.	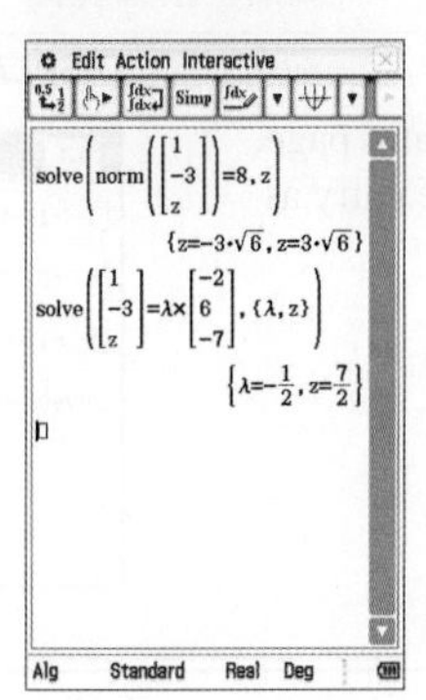

Linear dependence

Consider the three non-zero vectors $\underset{\sim}{a}$, $\underset{\sim}{b}$ and $\underset{\sim}{c}$. The vectors are said to be **linearly dependent** if there exist non-zero scalars α, β and γ such that $\alpha\underset{\sim}{a}+\beta\underset{\sim}{b}+\gamma\underset{\sim}{c}=\underset{\sim}{0}$.

If $\alpha\underset{\sim}{a}+\beta\underset{\sim}{b}+\gamma\underset{\sim}{c}=\underset{\sim}{0}$, then we can write $\underset{\sim}{c}=m\underset{\sim}{a}+n\underset{\sim}{b}$. That is, one of the vectors is a linear combination of the other two, where $m=-\dfrac{\alpha}{\gamma}$ and $n=-\dfrac{\beta}{\gamma}$; since $\alpha\neq 0$, $\beta\neq 0$ and $\gamma\neq 0$, it follows that $m\neq 0$ and $n\neq 0$.

The set of non-zero vectors $\underset{\sim}{a}$, $\underset{\sim}{b}$ and $\underset{\sim}{c}$ are said to be **linearly independent** if $\alpha\underset{\sim}{a}+\beta\underset{\sim}{b}+\gamma\underset{\sim}{c}=\underset{\sim}{0}$, only if $\alpha=0$, $\beta=0$ and $\gamma=0$.

For example, the two vectors $\underset{\sim}{a}=4\underset{\sim}{i}-8\underset{\sim}{j}+12\underset{\sim}{k}$ and $\underset{\sim}{b}=-3\underset{\sim}{i}+6\underset{\sim}{j}-9\underset{\sim}{k}$ are linearly dependent, since $3\underset{\sim}{a}+4\underset{\sim}{b}=\underset{\sim}{0}$ or alternatively $\underset{\sim}{a}=-\dfrac{4}{3}\underset{\sim}{b}$; in fact, the vectors are parallel.

WORKED EXAMPLE 8 Showing that two vectors are linearly dependent

Show that the vectors $\underset{\sim}{a}=\underset{\sim}{i}-\underset{\sim}{j}+4\underset{\sim}{k}$, $\underset{\sim}{b}=4\underset{\sim}{i}-2\underset{\sim}{j}+3\underset{\sim}{k}$ and $\underset{\sim}{c}=5\underset{\sim}{i}-\underset{\sim}{j}+z\underset{\sim}{k}$ are linearly dependent, and determine the value of z.

THINK	WRITE
1. Since the vectors are linearly dependent, they can be written as a linear combination.	$\underset{\sim}{c}=m\underset{\sim}{a}+n\underset{\sim}{b}$
2. Substitute for the given vectors.	$5\underset{\sim}{i}-\underset{\sim}{j}+z\underset{\sim}{k}=m(\underset{\sim}{i}-\underset{\sim}{j}+4\underset{\sim}{k})+n(4\underset{\sim}{i}-2\underset{\sim}{j}+3\underset{\sim}{k})$
3. Use the rules for scalar multiplication.	$5\underset{\sim}{i}-\underset{\sim}{j}+z\underset{\sim}{k}=(m\underset{\sim}{i}-m\underset{\sim}{j}+4m\underset{\sim}{k})+(4n\underset{\sim}{i}-2n\underset{\sim}{j}+3n\underset{\sim}{k})$
4. Use the rules for addition of vectors.	$5\underset{\sim}{i}-\underset{\sim}{j}+z\underset{\sim}{k}=(m+4n)\underset{\sim}{i}-(m+2n)\underset{\sim}{j}+(4m+3n)\underset{\sim}{k}$
5. Since the two vectors are equal, their components are equal.	$\underset{\sim}{i}$: $m+4n=5$ (1) $\underset{\sim}{j}$: $-m-2n=-1$ (2) $\underset{\sim}{k}$: $4m+3n=z$ (3)
6. Solve equations (1) and (2) for m and n.	(1) + (2): $2n=4$ $n=2$
7. Substitute and determine the value of m.	$m=5-4n$ $=5-8$ $=-3$
8. Use the values of m and n to solve for z.	Substitute $n=2$ and $m=-3$ into (3): $z=-12+6$ $=-6$
9. State the result.	The vectors are linearly dependent since $\underset{\sim}{c}=2\underset{\sim}{b}-3\underset{\sim}{a}$.

TI | THINK	DISPLAY/WRITE	CASIO | THINK	DISPLAY/WRITE
On a Calculator page, complete the entry as shown.	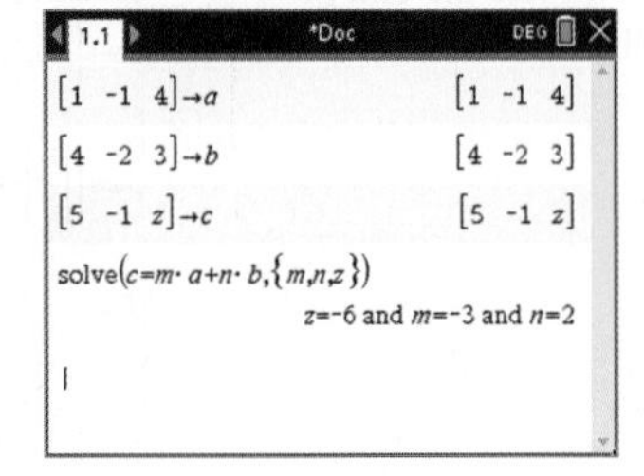	On a Main screen, complete the entry as shown.	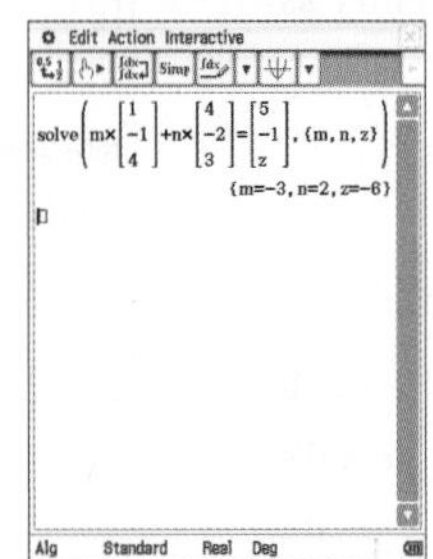

Direction cosines

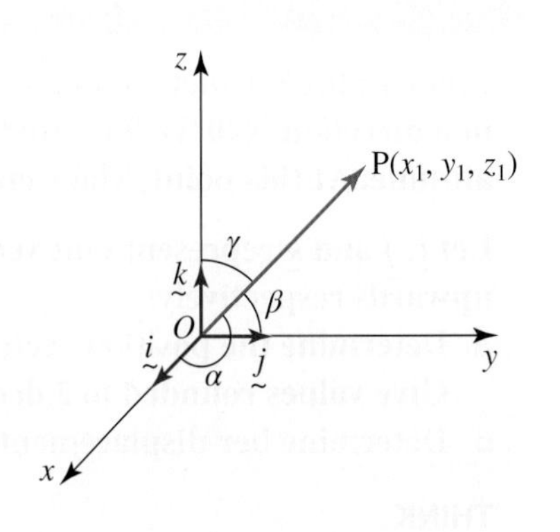

Let α be the angle that the vector $\overrightarrow{OP} = \underset{\sim}{r} = x_1\underset{\sim}{i} + y_1\underset{\sim}{j} + z_1\underset{\sim}{k}$ makes with the positive x-axis, let β be the angle that the vector $\overrightarrow{OP}$ makes with the positive y-axis, and let γ be the angle $\overrightarrow{OP}$ makes with the positive z-axis.

Generalising from the two-dimensional case,

$$\cos(\alpha) = \frac{x}{\sqrt{x^2+y^2+z^2}} = \frac{x}{|\underset{\sim}{r}|},$$

$$\cos(\beta) = \frac{y}{\sqrt{x^2+y^2+z^2}} = \frac{y}{|\underset{\sim}{r}|} \text{ and}$$

$$\cos(\gamma) = \frac{z}{\sqrt{x^2+y^2+z^2}} = \frac{z}{|\underset{\sim}{r}|}.$$

$\cos(\alpha)$, $\cos(\beta)$ and $\cos(\gamma)$ are called the direction cosines. Also,
$\hat{\underset{\sim}{r}} = \cos(\alpha)\underset{\sim}{i} + \cos(\beta)\underset{\sim}{j} + \cos(\gamma)\underset{\sim}{k}$ so $\cos^2(\alpha) + \cos^2(\beta) + \cos^2(\gamma) = 1$.

Often we need to calculate angles to the nearest degree, which involves calculating inverse trigonometric ratios. Throughout this topic we will allow the use of a scientific calculator to calculate approximate values of these inverse trigonometric ratios.

WORKED EXAMPLE 9 Calculating the angle between a vector and an axis

Calculate the angle to the nearest degree that the vector $2\underset{\sim}{i} - 3\underset{\sim}{j} - 4\underset{\sim}{k}$ makes with the z-axis.

THINK	WRITE
1. Give the vector a name.	Let $\underset{\sim}{a} = 2\underset{\sim}{i} - 3\underset{\sim}{j} - 4\underset{\sim}{k}$.
2. Determine the magnitude of the vector.	$\lvert\underset{\sim}{a}\rvert = \sqrt{2^2 + (-3)^2 + (-4)^2}$ $= \sqrt{4+9+16}$ $= \sqrt{29}$
3. The angle that the vector makes with the z-axis is given by $\cos(\gamma) = \frac{z}{\sqrt{x^2+y^2+z^2}} = \frac{z}{\lvert\underset{\sim}{a}\rvert}$	$\cos(\gamma) = \frac{-4}{\sqrt{29}}$
4. Determine the angle using a calculator, making sure the angle mode is set to degrees format.	$\gamma = \cos^{-1}\left(\frac{-4}{\sqrt{29}}\right)$ $= 137.97°$
5. State the result, giving the answer in decimal degrees.	The vector makes an angle of 138° with the z-axis.

Application problems

As in the two-dimensional case, we can determine the position vector of moving objects in terms of $\underset{\sim}{i}$, $\underset{\sim}{j}$ and $\underset{\sim}{k}$. Usually $\underset{\sim}{i}$ is a unit vector in the east direction, $\underset{\sim}{j}$ is a unit vector in the north direction and $\underset{\sim}{k}$ is a unit vector vertically upwards.

WORKED EXAMPLE 10 Application of vectors in three dimensions

Mary walks 500 metres due south, turns and moves 400 metres due west and then turns again to move in a direction N40°W for a further 200 metres. In all three of those movements she is at the same altitude. At this point, Mary enters a building and travels 20 metres vertically upwards in a lift.

Let $\underset{\sim}{i}$, $\underset{\sim}{j}$ and $\underset{\sim}{k}$ represent unit vectors of length 1 metre in the directions of east, north and vertically upwards respectively.

a. **Determine the position vector of Mary when she leaves the lift, relative to her initial position. Give values rounded to 2 decimal places where necessary.**

b. **Determine her displacement correct to 1 decimal place in metres from her initial point.**

THINK	WRITE/DRAW
a. 1. Consider the two-dimensional north–east situation. First Mary walks 500 metres due south, from the initial point O to a point A.	a. $\overrightarrow{OA} = -500\underset{\sim}{j}$
2. Next Mary walks 400 metres due west, from A to B.	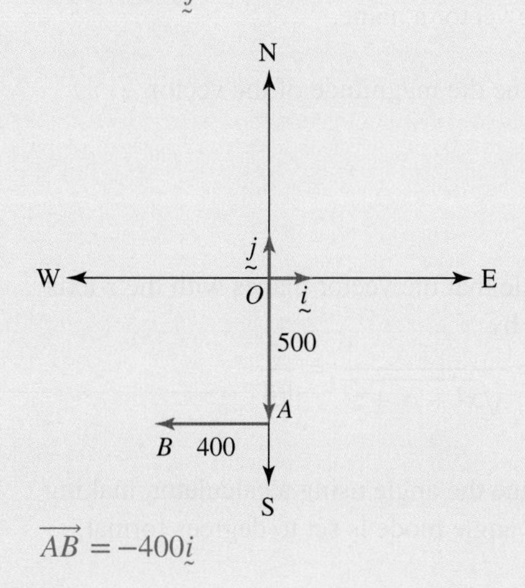 $\overrightarrow{AB} = -400\underset{\sim}{i}$

3. Mary now walks in a direction N40°W for a further 200 metres, from B to a point C.

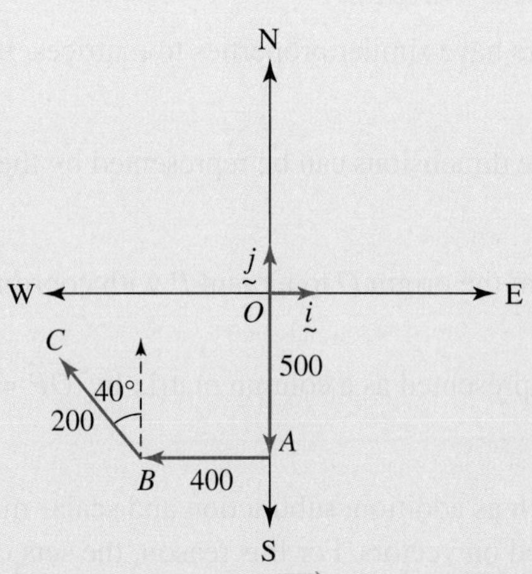

Resolve the vector $\overrightarrow{BC}$ into components:
$\overrightarrow{BC} = -200\sin(40°)\underset{\sim}{i} + 200\cos(40°)\underset{\sim}{j}$

4. Now Mary goes up in the lift from C to her final point, M.

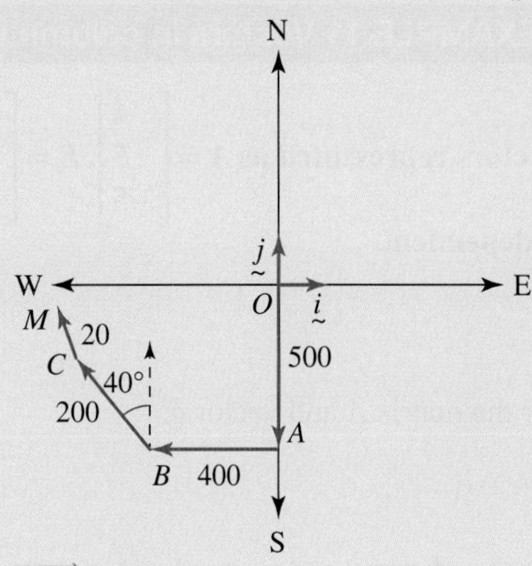

$\overrightarrow{CM} = 20\underset{\sim}{k}$

5. To determine the position vector of Mary, sum the vectors.

$$\begin{aligned}\overrightarrow{OM} &= \overrightarrow{OA} + \overrightarrow{AB} + \overrightarrow{BC} + \overrightarrow{CM}\\ &= -500\underset{\sim}{j} - 400\underset{\sim}{i}\\ &\quad + \left(-200\sin(40°)\underset{\sim}{i} + 200\cos(40°)\underset{\sim}{j}\right) + 20\underset{\sim}{k}\end{aligned}$$

6. Add and subtract the $\underset{\sim}{i}$, $\underset{\sim}{j}$ and $\underset{\sim}{k}$ components.

$$\begin{aligned}\overrightarrow{OM} &= -\left(400 + 200\sin(40°)\right)\underset{\sim}{i}\\ &\quad + \left(200\cos(40°) - 500\right)\underset{\sim}{j} + 20\underset{\sim}{k}\end{aligned}$$

7. Use a calculator, giving the results to 2 decimal places.

The position vector of Mary is
$\overrightarrow{OM} = -528.56\underset{\sim}{i} - 346.79\underset{\sim}{j} + 20\underset{\sim}{k}$

b. Her total displacement is the magnitude of the vector.

b. $$\begin{aligned}\left|\overrightarrow{OM}\right| &= \sqrt{(-528.56)^2 + (-346.79)^2 + 20^2}\\ &= 632.5 \text{ metres}\end{aligned}$$

Column vector notation

Because vectors have similar properties to matrices, it is common to represent vectors as column matrices.

Vectors in three dimensions can be represented by the unit vectors $\underset{\sim}{i} = \begin{bmatrix} 1 \\ 0 \\ 0 \end{bmatrix}$, $\underset{\sim}{j} = \begin{bmatrix} 0 \\ 1 \\ 0 \end{bmatrix}$ and $\underset{\sim}{k} = \begin{bmatrix} 0 \\ 0 \\ 1 \end{bmatrix}$.

The vector from the origin O to a point P with coordinates (x_1, y_1, z_1) can be expressed as $\overrightarrow{OP} = x_1\underset{\sim}{i} + y_1\underset{\sim}{j} + z_1\underset{\sim}{k}$.

Thus, this is represented as a column matrix by $\overrightarrow{OP} = x_1\begin{bmatrix} 1 \\ 0 \\ 0 \end{bmatrix} + y_1\begin{bmatrix} 0 \\ 1 \\ 0 \end{bmatrix} + z_1\begin{bmatrix} 0 \\ 0 \\ 1 \end{bmatrix} = \begin{bmatrix} x_1 \\ y_1 \\ z_1 \end{bmatrix}$.

Operations such as addition, subtraction and scalar multiplication are performed on matrices in similar ways to those performed on vectors. For this reason, the sets of vectors and column matrices are called isomorphic, a Greek word meaning having the same structure.

WORKED EXAMPLE 11 Vectors in column notation

Given the vectors represented as $A = \begin{bmatrix} 4 \\ 5 \\ -5 \end{bmatrix}$, $B = \begin{bmatrix} 5 \\ 3 \\ -2 \end{bmatrix}$ and $C = \begin{bmatrix} 3 \\ 7 \\ -8 \end{bmatrix}$, show that the points A, B and C are linearly dependent.

THINK	WRITE
1. Consider the matrix A and vector $\underset{\sim}{a}$.	$A = \begin{bmatrix} 4 \\ 5 \\ -5 \end{bmatrix}$, $\underset{\sim}{a} = 4\underset{\sim}{i} + 5\underset{\sim}{j} - 5\underset{\sim}{k}$
2. Consider the matrix B and vector $\underset{\sim}{b}$.	$B = \begin{bmatrix} 5 \\ 3 \\ -2 \end{bmatrix}$, $\underset{\sim}{b} = 5\underset{\sim}{i} + 3\underset{\sim}{j} - 2\underset{\sim}{k}$
3. Consider the matrix C and vector $\underset{\sim}{c}$.	$C = \begin{bmatrix} 3 \\ 7 \\ -8 \end{bmatrix}$, $\underset{\sim}{c} = 3\underset{\sim}{i} + 7\underset{\sim}{j} - 8\underset{\sim}{k}$
4. State the matrix C as a linear combination of A and B.	$C = \alpha A + \beta B$
5. Write the equation in matrix form.	$\begin{bmatrix} 3 \\ 7 \\ -8 \end{bmatrix} = \alpha\begin{bmatrix} 4 \\ 5 \\ -5 \end{bmatrix} + \beta\begin{bmatrix} 5 \\ 3 \\ -2 \end{bmatrix}$
6. Use the properties of scalar multiplication and addition of matrices.	$\begin{bmatrix} 3 \\ 7 \\ -8 \end{bmatrix} = \begin{bmatrix} 4\alpha \\ 5\alpha \\ -5\alpha \end{bmatrix} + \begin{bmatrix} 5\beta \\ 3\beta \\ -2\beta \end{bmatrix} = \begin{bmatrix} 4\alpha + 5\beta \\ 5\alpha + 3\beta \\ -5\alpha - 2\beta \end{bmatrix}$
7. Use the properties of equality of matrices.	$3 = 4\alpha + 5\beta \quad (1)$ $7 = 5\alpha + 3\beta \quad (2)$ $-8 = -5\alpha - 2\beta \quad (3)$
8. Solve the simultaneous equations using elimination.	Add (2) and (3) to eliminate α so that $\beta = -1$.

9. Substitute the value of β into (2) to determine the value of α.	$5\alpha = 7 - 3\beta$ $5\alpha = 10$ $\alpha = 2$
10. Since we have not used (1), we must check that this equation is valid.	$3 = 4\alpha + 5\beta$ (1) Substitute $\alpha = 2, \beta = -1$: RHS $= 8 - 5$ $=$ LHS
11. Write the equation relating the matrices.	$C = 2A - B$ or $\underset{\sim}{c} = 2\underset{\sim}{a} - \underset{\sim}{b}$
12. State the conclusion.	The points A, B and C are linearly dependent.

3.3 Exercise

Students, these questions are even better in jacPLUS

Receive immediate feedback and access sample responses

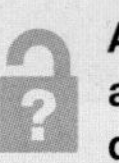
Access additional questions

Track your results and progress

Find all this and MORE in jacPLUS

Technology free

1. **WE4** If the point P has coordinates $(2, -2, -1)$, determine:
   a. the vector $\overrightarrow{OP}$
   b. a unit vector parallel to $\overrightarrow{OP}$.
2. If $\underset{\sim}{a} = 4\underset{\sim}{i} - 8\underset{\sim}{j} - 2\underset{\sim}{k}$, determine $\hat{\underset{\sim}{a}}$.
3. **WE5** Two points are given by $P(2, 1, -3)$ and $Q(4, -1, 2)$ respectively. Calculate a unit vector parallel to $\overrightarrow{PQ}$.
4. Two points are given by $A(-1, 2, -4)$ and $B(2, 6, 8)$. Calculate the distance between the points A and B.
5. **WE6** If $\underset{\sim}{a} = \underset{\sim}{i} + 2\underset{\sim}{j} - z\underset{\sim}{k}$ and $\underset{\sim}{b} = 4\underset{\sim}{i} - 5\underset{\sim}{j} - 2\underset{\sim}{k}$, determine the value of z if the vector $\underset{\sim}{c} = \underset{\sim}{a} - 2\underset{\sim}{b}$ is parallel to the xy plane.
6. Determine the values of x, y and z, given the points $C(x, -2, 4)$ and $D(2, y, -3)$ and the vector $\overrightarrow{CD} = 3\underset{\sim}{i} + 4\underset{\sim}{j} + z\underset{\sim}{k}$.
7. **WE7** Given the vectors $\underset{\sim}{r} = 2\underset{\sim}{i} - \underset{\sim}{j} + z\underset{\sim}{k}$ and $\underset{\sim}{s} = -4\underset{\sim}{i} + 2\underset{\sim}{j} + 5\underset{\sim}{k}$, determine the value of z if:
   a. the length of the vector $\underset{\sim}{r}$ is 5
   b. the vector $\underset{\sim}{r}$ is parallel to the vector $\underset{\sim}{s}$.
8. Given the vectors $\underset{\sim}{a} = 3\underset{\sim}{i} + y\underset{\sim}{j} - 4\underset{\sim}{k}$ and $\underset{\sim}{b} = -6\underset{\sim}{i} + 3\underset{\sim}{j} + 8\underset{\sim}{k}$, determine the value of y if:
   a. the vectors $\underset{\sim}{a}$ and $\underset{\sim}{b}$ are equal in length
   b. the vector $\underset{\sim}{a}$ is parallel to the vector $\underset{\sim}{b}$.
9. **WE8** Show that the vectors $\underset{\sim}{a} = 2\underset{\sim}{i} - \underset{\sim}{j} + 3\underset{\sim}{k}$, $\underset{\sim}{b} = -2\underset{\sim}{i} + 2\underset{\sim}{j} - \underset{\sim}{k}$ and $\underset{\sim}{c} = 2\underset{\sim}{i} + \underset{\sim}{j} + z\underset{\sim}{k}$ are linearly dependent, and determine the value of z.
10. Given the vectors $\underset{\sim}{a} = 3\underset{\sim}{i} - 2\underset{\sim}{j} + 4\underset{\sim}{k}$, $\underset{\sim}{b} = 2\underset{\sim}{i} - 3\underset{\sim}{j} + 5\underset{\sim}{k}$ and $\underset{\sim}{c} = x\underset{\sim}{i} + 2\underset{\sim}{j}$, show that they are linearly dependent and determine the value of x.

Technology active

11. **WE10** Peter is training for a fun run. First he runs 900 metres in a direction S25°E, then a further 800 metres due south and finally 300 metres due east. In all three of these movements he maintains the same altitude. Finally Peter enters a building and runs up the stairs, climbing to a height of 150 metres vertically upwards above ground level.

    Let $\underset{\sim}{i}$, $\underset{\sim}{j}$ and $\underset{\sim}{k}$ represent unit vectors of length 1 metre in the directions of east, north and vertically upwards respectively.

    a. Determine the position vector of Peter at the top of the stairs relative to his initial position, rounding to 2 decimal places where appropriate.
    b. Determine his displacement correct to 1 decimal place in metres from his initial position.

12. A helicopter takes off from a helipad and moves 800 metres vertically upwards. It then turns and moves 2 km due west, and finally it turns again to move in a direction S20°W for a further 2 km. In both of the latter two movements it maintains the same altitude. Let $\underset{\sim}{i}$, $\underset{\sim}{j}$ and $\underset{\sim}{k}$ represent unit vectors of 1 metre in the directions of east, north and vertically upwards respectively.

    a. Determine the position vector of the helicopter relative to its initial position, rounding to 1 decimal place where appropriate.
    b. Determine the displacement to the nearest metre of the helicopter from its initial position.

13. **WE11** Given the vectors represented as $A = \begin{bmatrix} 2 \\ -5 \\ 4 \end{bmatrix}$, $B = \begin{bmatrix} 3 \\ -6 \\ 4 \end{bmatrix}$ and $C = \begin{bmatrix} 1 \\ -7 \\ 8 \end{bmatrix}$, show that A, B and C are linearly dependent.

14. Given the vectors represented as $A = \begin{bmatrix} 2 \\ -4 \\ 3 \end{bmatrix}$, $B = \begin{bmatrix} -1 \\ -3 \\ 2 \end{bmatrix}$ and $C = \begin{bmatrix} 7 \\ 1 \\ -1 \end{bmatrix}$, show that A, B and C are linearly independent.

15. a. Determine a vector of magnitude 6 parallel to the vector $\underset{\sim}{i} - 2\underset{\sim}{j} + 2\underset{\sim}{k}$.
    b. Determine a vector of magnitude 26 in the opposite direction to the vector $-3\underset{\sim}{i} + 4\underset{\sim}{j} + 12\underset{\sim}{k}$.
    c. Determine a vector of magnitude $10\sqrt{2}$ in the opposite direction to the vector $-5\underset{\sim}{i} - 3\underset{\sim}{j} + 4\underset{\sim}{k}$.

16. a. Given the vectors $\underset{\sim}{a} = 2\underset{\sim}{i} + 3\underset{\sim}{j} - \underset{\sim}{k}$ and $\underset{\sim}{b} = -3\underset{\sim}{i} + \underset{\sim}{j} + 2\underset{\sim}{k}$, show that the vector $2\underset{\sim}{a} + \underset{\sim}{b}$ is parallel to the xy plane.
    b. Given the vectors $\underset{\sim}{p} = 3\underset{\sim}{i} + 2\underset{\sim}{j} - 5\underset{\sim}{k}$ and $\underset{\sim}{q} = 2\underset{\sim}{i} + \underset{\sim}{j} - 2\underset{\sim}{k}$, show that the vector $2\underset{\sim}{p} - 3\underset{\sim}{q}$ is parallel to the yz plane.
    c. Given the vectors $\underset{\sim}{r} = 2\underset{\sim}{i} - 3\underset{\sim}{j} + 5\underset{\sim}{k}$ and $\underset{\sim}{s} = \underset{\sim}{i} + y\underset{\sim}{j} - 2\underset{\sim}{k}$, determine the value of y if the vector $4\underset{\sim}{r} + 3\underset{\sim}{s}$ is parallel to the xz plane.

17. a. Show that the points $A(2, -1, 3)$, $B(8, -7, 15)$ and $C(4, -3, 7)$ are collinear.
    b. Show that the points $P(2, 1, 4)$, $Q(1, -2, 3)$ and $R(-1, -8, 1)$ are collinear.
    c. Determine the values of x and y if the points $A(x, 1, 2)$, $B(2, y, -1)$ and $C(3, -4, 5)$ are collinear.

18. a. Given the points $A(3, 1, -2)$ and $B(5, 3, 4)$, determine the position vector of P where P is the midpoint of AB.
    b. Given the points $C(-2, 4, 1)$ and $D(-5, 1, 4)$, determine the position vector of P where P divides CD in the ratio $1:2$.
    c. Given the points $R(1, -2, -3)$ and $S(-3, 2, 5)$, determine the position vector of P where P divides RS in the ratio $3:1$.

19. a. Show that the vectors $\underset{\sim}{a} = 2\underset{\sim}{i} - 4\underset{\sim}{j} - 6\underset{\sim}{k}$, $\underset{\sim}{b} = 3\underset{\sim}{i} + 6\underset{\sim}{j} - 9\underset{\sim}{k}$ and $\underset{\sim}{c} = 3\underset{\sim}{i} + 10\underset{\sim}{j} - 9\underset{\sim}{k}$ are linearly dependent.
    b. Determine the value of y if the vectors $\underset{\sim}{p} = 2\underset{\sim}{i} - 3\underset{\sim}{j} + 4\underset{\sim}{k}$, $\underset{\sim}{q} = 2\underset{\sim}{i} - \underset{\sim}{j} + 5\underset{\sim}{k}$ and $\underset{\sim}{r} = 2\underset{\sim}{i} + y\underset{\sim}{j} + 8\underset{\sim}{k}$ are linearly dependent.
    c. Determine the value of z if the vectors $\underset{\sim}{a} = 2\underset{\sim}{i} - 3\underset{\sim}{j} + 4\underset{\sim}{k}$, $\underset{\sim}{b} = 3\underset{\sim}{i} - 4\underset{\sim}{j} + 2\underset{\sim}{k}$ and $\underset{\sim}{c} = -7\underset{\sim}{i} + 8\underset{\sim}{j} + z\underset{\sim}{k}$ are linearly dependent.

20. A mountain climber walks 2 km due west to a point A, then 1 km due south to a point B. At this point he ascends a vertical cliff to a point C, which is 500 metres above ground level. If $\underset{\sim}{i}$, $\underset{\sim}{j}$ and $\underset{\sim}{k}$ represent unit vectors of 1 kilometre in the directions of east, north and vertically upwards, determine the position vector and displacement of the mountain climber from his initial position. Give values rounded to 2 decimal places where necessary.

21. A student walks on level ground in a north-westerly direction a distance of 200 m to a point P. She then walks 50 m due north to a point R. At this point she turns southward and ascends a set of stairs inclined at an angle of 50° to the horizontal, moving 5 m along the stairs to a point R. If $\underset{\sim}{i}$, $\underset{\sim}{j}$ and $\underset{\sim}{k}$ represent unit vectors of 1 metre in the directions of east, north and vertically upwards, determine the position vector and displacement of the student from her initial position, writing your answers correct to 2 decimal places.

22. A plane takes off and flies upwards facing east for a distance of 30 km at an angle of elevation of 35°. It then moves horizontally east at 300 km/h. If $\underset{\sim}{i}$, $\underset{\sim}{j}$ and $\underset{\sim}{k}$ represent unit vectors of 1 kilometre in the directions of east, north and vertically upwards, determine the position vector, correct to 3 decimal places, and displacement, correct to 2 decimal places, of the plane after it has flown horizontally for 5 minutes.

23. **WE9** Determine the angle in degrees, rounded to 1 decimal place, that the vector $\underset{\sim}{i} + 2\underset{\sim}{j} - 3\underset{\sim}{k}$ makes with the z-axis.

24. The vector $\underset{\sim}{i} + z\underset{\sim}{k}$ makes an angle of 150° with the z-axis. Determine the value of z.

25. Given the point $A(-2, 4, 1)$, determine:
    a. a unit vector parallel to OA
    b. the angle that the vector $\overrightarrow{OA}$ makes with the x-axis, rounded to 1 decimal place.

26. Given the point $B(3, 5, -2)$, determine:
    a. a unit vector parallel to OB
    b. the angle that the vector $\overrightarrow{OB}$ makes with the y-axis, rounded to 1 decimal place.

27. Given the point $C(4, 6, -8)$, determine:
    a. a unit vector parallel to OC
    b. the angle that the vector $\overrightarrow{OC}$ makes with the z-axis, rounded to 1 decimal place.

28. a. Given the points $A(3, 5, -2)$ and $B(2, -1, 3)$, calculate the distance between the points A and B.
    b. Given the points $P(-2, 4, 1)$ and $Q(3, -5, 2)$, determine the angle the vector $\overrightarrow{PQ}$ makes with the y-axis, rounded to 1 decimal place.
    c. Given the points $R(4, 3, -1)$ and $S(6, 1, -7)$, determine a unit vector parallel to $\overrightarrow{SR}$.

29. Given the vectors $\underset{\sim}{a} = \underset{\sim}{i} - 2\underset{\sim}{j} + 4\underset{\sim}{k}$ and $\underset{\sim}{b} = x\underset{\sim}{i} + 6\underset{\sim}{j} - 12\underset{\sim}{k}$, determine the value of x if:
   a. $\underset{\sim}{a}$ is parallel to $\underset{\sim}{b}$
   b. the length of the vector $\underset{\sim}{b}$ is $10\sqrt{2}$.

30. Given the vector $\underset{\sim}{r} = 3\underset{\sim}{i} + y\underset{\sim}{j} + \underset{\sim}{k}$, determine the value of y if:
   a. the length of the vector $\underset{\sim}{r}$ is 10
   b. the vector $\underset{\sim}{r}$ makes an angle of $\cos^{-1}\left(-\frac{1}{3}\right)$ with the y-axis.

31. Given the vector $\underset{\sim}{u} = 2\sqrt{2}\underset{\sim}{i} + 2\underset{\sim}{j} + z\underset{\sim}{k}$, determine the value of z if:
   a. the length of the vector $\underset{\sim}{u}$ is 6
   b. the vector $\underset{\sim}{u}$ makes an angle of 120° with the z-axis.

32. A helicopter moves 600 metres vertically upwards and then moves S60°W for 500 m parallel to the ground. It then moves south-west, moving upwards at an angle of elevation of 50° and a speed of 120 km/h for 1 minute. If $\underset{\sim}{i}$, $\underset{\sim}{j}$ and $\underset{\sim}{k}$ represent unit vectors of one metre in the directions of east, north and vertically upwards, determine the position vector and displacement of the helicopter from its initial position. Give values rounded to 2 decimal places where necessary.

33. a. Determine the value of m if the length of the vector $(m-1)\underset{\sim}{i} + (m+1)\underset{\sim}{j} + m\underset{\sim}{k}$ is $\sqrt{17}$.
   b. A unit vector makes an angle of 45° with the x-axis and an angle of 120° with the y-axis. Determine the angle it makes with the z-axis if it is known that this angle is acute.
   c. A unit vector makes an angle of 60° with the x-axis and an angle of 120° with the z-axis. Determine the angle it makes with the y-axis if it is known that this angle is obtuse.

34. a. Given the vectors $\underset{\sim}{a} = 4\underset{\sim}{i} - 3\underset{\sim}{j} + 2\underset{\sim}{k}$, $\underset{\sim}{b} = -\underset{\sim}{i} + 2\underset{\sim}{j} - 3\underset{\sim}{k}$, $\underset{\sim}{c} = 4\underset{\sim}{i} - \underset{\sim}{j} + 2\underset{\sim}{k}$ and $\underset{\sim}{d} = 7\underset{\sim}{i} + \underset{\sim}{j} + 11\underset{\sim}{k}$, determine the values of the scalars p, q and r if $\underset{\sim}{d} = p\underset{\sim}{a} + q\underset{\sim}{b} + r\underset{\sim}{c}$.
   b. Determine the values of x and y if the two vectors $x\underset{\sim}{i} + 2y\underset{\sim}{j} + 3\underset{\sim}{k}$ and $2x\underset{\sim}{i} + y\underset{\sim}{j} + 4\underset{\sim}{k}$ both have a length of 5.

3.3 Exam questions

Question 1 (4 marks) TECH-FREE

Source: VCE 2021 Specialist Mathematics Exam 1, Q6; © VCAA.

Consider the three vectors $\underset{\sim}{a} = -\underset{\sim}{i} + 6\underset{\sim}{j} - 3\underset{\sim}{k}$, $\underset{\sim}{b} = 2\underset{\sim}{i} - 8\underset{\sim}{j} + 5\underset{\sim}{k}$ and $\underset{\sim}{c} = 3\underset{\sim}{i} + 2\underset{\sim}{j} + |1 - p^2|\,\underset{\sim}{k}$, where p is a real constant.

Find the values of p for which the three vectors are linearly independent.

Question 2 (1 mark) TECH-ACTIVE

Source: VCE 2020 Specialist Mathematics Exam 2, Section A, Q13; © VCAA.

MC The vectors $\underset{\sim}{a} = \underset{\sim}{i} + 2\underset{\sim}{j} - \underset{\sim}{k}$, $\underset{\sim}{b} = \lambda\underset{\sim}{i} + 3\underset{\sim}{j} + 2\underset{\sim}{k}$ and $\underset{\sim}{c} = \underset{\sim}{i} + \underset{\sim}{k}$ will be **linearly dependent** when the value of λ is

A. 1 **B.** 2 **C.** 3 **D.** 4 **E.** 5

Question 3 (2 marks) TECH-FREE

Source: VCE 2016 Specialist Mathematics Exam 1, Q5b; © VCAA.

Consider the vectors $\underset{\sim}{a} = 3\underset{\sim}{i} + 5\underset{\sim}{j} - 2\underset{\sim}{k}$, $\underset{\sim}{b} = \underset{\sim}{i} - 2\underset{\sim}{j} + 3\underset{\sim}{k}$ and $\underset{\sim}{c} = \underset{\sim}{i} + d\underset{\sim}{k}$ where d is a real constant.

Find the value of d if the vectors are **linearly dependent**.

More exam questions are available online.

3.4 Scalar product and applications

LEARNING INTENTION

At the end of this subtopic you should be able to:

- calculate the scalar product of two vectors
- determine the angle between two vectors using the scalar product
- calculate the scalar resolute of one vector in the direction of another vector
- calculate the vector resolutes of a vector parallel to and perpendicular to another vector.

3.4.1 Multiplying vectors

When a vector is multiplied by a scalar, the result is a vector. What happens when two vectors are multiplied together? Is the result a vector or a scalar? What are the applications of multiplying two vectors together? In fact, vectors can be multiplied together in two different ways: either by using the scalar product (also known as the dot product) and obtain a scalar as the result, or by using the vector product and obtain a vector as the result. The vector product is covered in the next topic.

Definition of the scalar product

The scalar or dot product of two vectors $\underset{\sim}{a}$ and $\underset{\sim}{b}$ is defined by

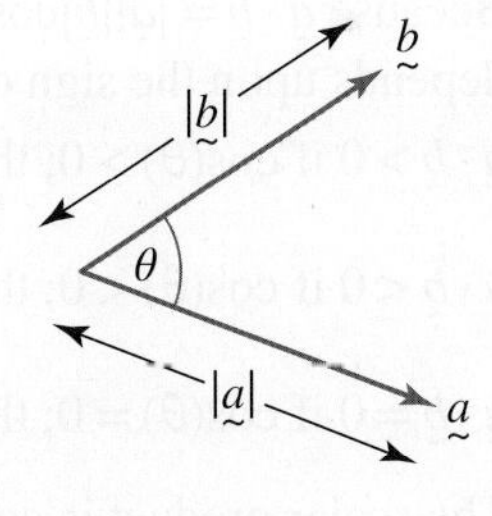

$$\underset{\sim}{a} \cdot \underset{\sim}{b} = |\underset{\sim}{a}||\underset{\sim}{b}|\cos(\theta).$$

This is read as $\underset{\sim}{a}$ dot $\underset{\sim}{b}$, where θ is the angle between the vectors $\underset{\sim}{a}$ and $\underset{\sim}{b}$.

Note that the angle between two vectors must be the angle between the tails of the vectors. The scalar product or dot product is also known as the inner product of two vectors.

The scalar product

The scalar or dot product of vectors $\underset{\sim}{a}$ and $\underset{\sim}{b}$ is:

$$\underset{\sim}{a} \cdot \underset{\sim}{b} = |\underset{\sim}{a}||\underset{\sim}{b}|\cos(\theta)$$

where θ is the angle between $\underset{\sim}{a}$ and $\underset{\sim}{b}$.

WORKED EXAMPLE 12 Calculating the scalar product of two vectors

Given the diagram below, calculate $\underset{\sim}{a} \cdot \underset{\sim}{b}$.

THINK	WRITE
1. Write the magnitudes or lengths of the two vectors.	$\lvert\underset{\sim}{a}\rvert = 6$ and $\lvert\underset{\sim}{b}\rvert = 4\sqrt{3}$
2. Write the angle between the two vectors.	$\theta = 150°$
3. Apply the formula from the definition to calculate the value of the scalar product.	$\underset{\sim}{a} \cdot \underset{\sim}{b} = \lvert\underset{\sim}{a}\rvert\lvert\underset{\sim}{b}\rvert\cos(\theta)$ $= 6 \times 4\sqrt{3}\cos(150°)$ $= 24\sqrt{3} \times \left(\frac{-\sqrt{3}}{2}\right)$
4. State the final result, which is negative because $\cos(150°) < 0$.	$\underset{\sim}{a} \cdot \underset{\sim}{b} = -36$

Properties of the scalar product

- The scalar product always gives a number as the result, hence its name. This number can be positive, negative or zero.
- Because $\underset{\sim}{a} \cdot \underset{\sim}{b} = \lvert\underset{\sim}{a}\rvert\lvert\underset{\sim}{b}\rvert\cos(\theta)$ for non-zero vectors $\underset{\sim}{a}$ and $\underset{\sim}{b}$, both $\lvert\underset{\sim}{a}\rvert > 0$ and $\lvert\underset{\sim}{b}\rvert > 0$, so the sign of $\underset{\sim}{a} \cdot \underset{\sim}{b}$ depends upon the sign of $\cos(\theta)$. Hence, it follows that:
  $\underset{\sim}{a} \cdot \underset{\sim}{b} > 0$ if $\cos(\theta) > 0$; that is, θ is an acute angle or $0 < \theta < \frac{\pi}{2}$.
  $\underset{\sim}{a} \cdot \underset{\sim}{b} < 0$ if $\cos(\theta) < 0$; that is, θ is an obtuse angle or $\frac{\pi}{2} < \theta < \pi$.
  $\underset{\sim}{a} \cdot \underset{\sim}{b} = 0$ if $\cos(\theta) = 0$; that is, $\theta = \frac{\pi}{2}$. This means that $\underset{\sim}{a}$ is perpendicular to $\underset{\sim}{b}$ unless either $\underset{\sim}{a} = \underset{\sim}{0}$ or $\underset{\sim}{b} = \underset{\sim}{0}$.
- The scalar product is commutative. That is, $\underset{\sim}{a} \cdot \underset{\sim}{b} = \underset{\sim}{b} \cdot \underset{\sim}{a}$.
- This follows because the angle between $\underset{\sim}{b}$ and $\underset{\sim}{a}$ is $2\pi - \theta$, and $\cos(2\pi - \theta) = \cos(\theta)$.
- The scalar product of a vector with itself is the square of the magnitude of the vector. That is, $\underset{\sim}{a} \cdot \underset{\sim}{a} = \lvert\underset{\sim}{a}\rvert^2$, as $\theta = 0$ and $\cos(0) = 1$.
- Scalars or common factors in a vector are merely multiples. That is, if $\lambda \in R$, then $\underset{\sim}{a} \cdot (\lambda\underset{\sim}{b}) = (\lambda\underset{\sim}{a}) \cdot \underset{\sim}{b} = \lambda(\underset{\sim}{a} \cdot \underset{\sim}{b})$.

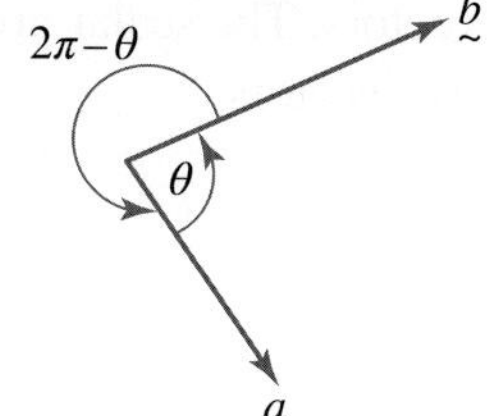

Component forms

The vectors $\underset{\sim}{i}$, $\underset{\sim}{j}$ and $\underset{\sim}{k}$ are all unit vectors; that is, $\lvert\underset{\sim}{i}\rvert = 1$, $\lvert\underset{\sim}{j}\rvert = 1$ and $\lvert\underset{\sim}{k}\rvert = 1$. The angle between the vectors $\underset{\sim}{i}$ and $\underset{\sim}{i}$ is zero, as $\cos(0) = 1$. It follows from the definition and properties of the scalar product that $\underset{\sim}{i} \cdot \underset{\sim}{i} = 1$. Similarly, $\underset{\sim}{j} \cdot \underset{\sim}{j} = 1$ and $\underset{\sim}{k} \cdot \underset{\sim}{k} = 1$.

The unit vectors are mutually perpendicular; that is, the angles between $\underset{\sim}{i}$ and $\underset{\sim}{j}$, $\underset{\sim}{i}$ and $\underset{\sim}{k}$, and $\underset{\sim}{j}$ and $\underset{\sim}{k}$ are all 90°, as $\cos(90°) = 0$. From that fact and the commutative law, it follows that $\underset{\sim}{i} \cdot \underset{\sim}{j} = \underset{\sim}{j} \cdot \underset{\sim}{i} = 0$, $\underset{\sim}{i} \cdot \underset{\sim}{k} = \underset{\sim}{k} \cdot \underset{\sim}{i} = 0$ and $\underset{\sim}{j} \cdot \underset{\sim}{k} = \underset{\sim}{k} \cdot \underset{\sim}{j} = 0$.

In general, if $\underset{\sim}{a} = x_1\underset{\sim}{i} + y_1\underset{\sim}{j} + z_1\underset{\sim}{k}$ and $\underset{\sim}{b} = x_2\underset{\sim}{i} + y_2\underset{\sim}{j} + z_2\underset{\sim}{k}$, then

$$\begin{aligned}\underset{\sim}{a} \cdot \underset{\sim}{b} &= (x_1\underset{\sim}{i} + y_1\underset{\sim}{j} + z_1\underset{\sim}{k}) \cdot (x_2\underset{\sim}{i} + y_2\underset{\sim}{j} + z_2\underset{\sim}{k}) \\ &= x_1x_2\underset{\sim}{i} \cdot \underset{\sim}{i} + y_1y_2\underset{\sim}{j} \cdot \underset{\sim}{j} + z_1z_2\underset{\sim}{k} \cdot \underset{\sim}{k} \\ &= x_1x_2 + y_1y_2 + z_1z_2\end{aligned}$$

The scalar product of vectors in component form

If $\underset{\sim}{a} = x_1\underset{\sim}{i} + y_1\underset{\sim}{j} + z_1\underset{\sim}{k}$ and $\underset{\sim}{b} = x_2\underset{\sim}{i} + y_2\underset{\sim}{j} + z_2\underset{\sim}{k}$, then

$$\underset{\sim}{a} \cdot \underset{\sim}{b} = x_1x_2 + y_1y_2 + z_1z_2$$

WORKED EXAMPLE 13 Calculating the scalar product of two vectors in component form

If $\underset{\sim}{a} = 2\underset{\sim}{i} + 3\underset{\sim}{j} - 5\underset{\sim}{k}$ and $\underset{\sim}{b} = 4\underset{\sim}{i} - 5\underset{\sim}{j} - 2\underset{\sim}{k}$, calculate $\underset{\sim}{a} \cdot \underset{\sim}{b}$.

THINK	WRITE
1. Substitute for the given vectors.	$\underset{\sim}{a} \cdot \underset{\sim}{b} = (2\underset{\sim}{i} + 3\underset{\sim}{j} - 5\underset{\sim}{k}) \cdot (4\underset{\sim}{i} - 5\underset{\sim}{j} - 2\underset{\sim}{k})$
2. Use the result for multiplying vectors in component form.	$\underset{\sim}{a} \cdot \underset{\sim}{b} = 2 \times 4 + 3 \times (-5) + (-5) \times (-2)$ $= 8 - 15 + 10$
3. State the final result.	$\underset{\sim}{a} \cdot \underset{\sim}{b} = 3$

Orthogonal vectors

Two vectors are said to be orthogonal or perpendicular if the angle between them is 90°. In this case, the dot product of the two vectors is zero.

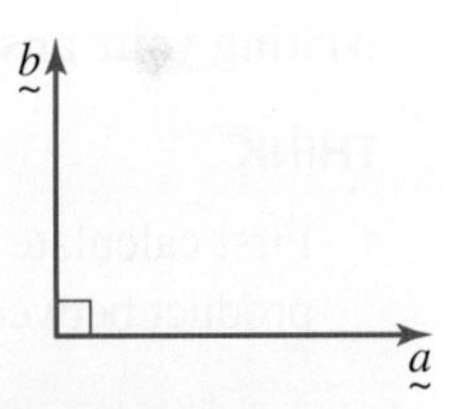

WORKED EXAMPLE 14 Determining a value for which two vectors are orthogonal

If the two vectors $\underset{\sim}{a} = 3\underset{\sim}{i} - 2\underset{\sim}{j} - 4\underset{\sim}{k}$ and $\underset{\sim}{b} = 2\underset{\sim}{i} - 5\underset{\sim}{j} + z\underset{\sim}{k}$ are orthogonal, determine the value of z.

THINK	WRITE
1. First calculate the value of the scalar product between the two vectors.	$\underset{\sim}{a} \cdot \underset{\sim}{b} = (3\underset{\sim}{i} - 2\underset{\sim}{j} - 4\underset{\sim}{k}) \cdot (2\underset{\sim}{i} - 5\underset{\sim}{j} + z\underset{\sim}{k})$ $= 3 \times 2 + (-2) \times (-5) + (-4) \times z$ $= 6 + 10 - 4z$ $= 16 - 4z$
2. The vectors are orthogonal. Equate their dot product to zero.	$\underset{\sim}{a} \cdot \underset{\sim}{b} = 16 - 4z$ $= 0$
3. Solve for z.	$4z = 16$ $z = 4$

TI | THINK	DISPLAY/WRITE	CASIO | THINK	DISPLAY/WRITE
On a Calculator page, complete the entry as shown.	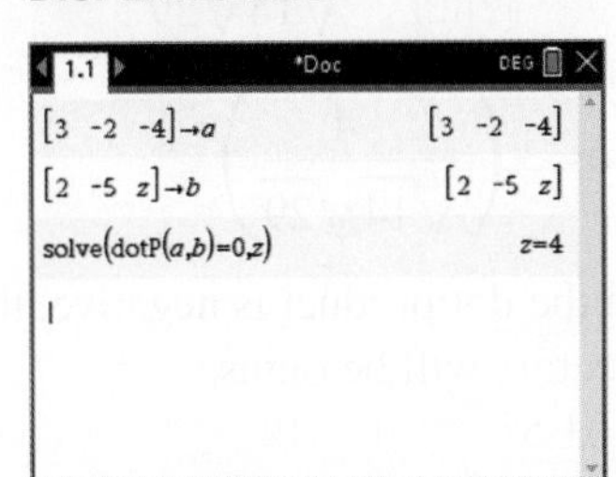	On a Main screen, complete the entry as shown.	

Previously, the angle that a single vector makes with the x-, y- or z-axis was found using direction cosines. Now, by using the scalar product, the angle between two vectors can be calculated.

Angle between two vectors

The formula $\underset{\sim}{a} \cdot \underset{\sim}{b} = |\underset{\sim}{a}||\underset{\sim}{b}|\cos(\theta)$ can be rearranged to determine the angle θ between the two vectors, giving $\cos(\theta) = \dfrac{\underset{\sim}{a} \cdot \underset{\sim}{b}}{|\underset{\sim}{a}||\underset{\sim}{b}|}$ or $\theta = \cos^{-1}\left(\dfrac{\underset{\sim}{a} \cdot \underset{\sim}{b}}{|\underset{\sim}{a}||\underset{\sim}{b}|}\right)$.

The angle between two vectors

The angle between two vectors $\underset{\sim}{a}$ and $\underset{\sim}{b}$ is:

$$\theta = \cos^{-1}\left(\frac{\underset{\sim}{a} \cdot \underset{\sim}{b}}{|\underset{\sim}{a}||\underset{\sim}{b}|}\right)$$

WORKED EXAMPLE 15 Determining the angle between two vectors

Given the vectors $\underset{\sim}{a} = \underset{\sim}{i} - 2\underset{\sim}{j} + 3\underset{\sim}{k}$ and $\underset{\sim}{b} = 2\underset{\sim}{i} - 3\underset{\sim}{j} - 4\underset{\sim}{k}$ determine the angle between the vectors $\underset{\sim}{a}$ and $\underset{\sim}{b}$, writing your answer correct to 1 decimal place.

THINK	WRITE
1. First calculate the value of the scalar product between the two vectors.	$\underset{\sim}{a} \cdot \underset{\sim}{b} = (\underset{\sim}{i} - 2\underset{\sim}{j} + 3\underset{\sim}{k}) \cdot (2\underset{\sim}{i} - 3\underset{\sim}{j} - 4\underset{\sim}{k})$ $= 1 \times 2 + (-2) \times (-3) + 3 \times (-4)$ $= 2 + 6 - 12$ $= -4$
2. Determine the magnitude of the first vector.	$\underset{\sim}{a} = \underset{\sim}{i} - 2\underset{\sim}{j} + 3\underset{\sim}{k}$ $\lvert\underset{\sim}{a}\rvert = \sqrt{1^2 + (-2)^2 + 3^2}$ $= \sqrt{1 + 4 + 9}$ $= \sqrt{14}$
3. Determine the magnitude of the second vector.	$\underset{\sim}{b} = 2\underset{\sim}{i} - 3\underset{\sim}{j} - 4\underset{\sim}{k}$ $\lvert\underset{\sim}{b}\rvert = \sqrt{2^2 + (-3)^2 + (-4)^2}$ $= \sqrt{4 + 9 + 16}$ $= \sqrt{29}$
4. Apply the formula to calculate the angle between the vectors.	$\cos(\theta) = \dfrac{\underset{\sim}{a} \cdot \underset{\sim}{b}}{\lvert\underset{\sim}{a}\rvert\lvert\underset{\sim}{b}\rvert} = \dfrac{-4}{\sqrt{14}\sqrt{29}}$
5. Using a scientific calculator, determine the angle between the two vectors, rounding the answer to 1 decimal place.	$\theta = \cos^{-1}\left(\dfrac{-4}{\sqrt{14}\sqrt{29}}\right)$ Since the dot product is negative, the angle between the vectors will be obtuse. $\theta = 101.5°$

Determining magnitudes of vectors

The magnitude and the sum or difference of vectors can be calculated using the properties of the scalar product: $\underset{\sim}{a} \cdot \underset{\sim}{a} = |\underset{\sim}{a}|^2$ and $\underset{\sim}{a} \cdot \underset{\sim}{b} = \underset{\sim}{b} \cdot \underset{\sim}{a}$.

WORKED EXAMPLE 16 Determining the magnitude of the sum of two vectors

If $|\underset{\sim}{a}| = 3$, $|\underset{\sim}{b}| = 5$ and $\underset{\sim}{a} \cdot \underset{\sim}{b} = -4$, determine $|\underset{\sim}{a} + \underset{\sim}{b}|$.

THINK	WRITE
1. The magnitude of a vector squared is obtained by calculating the dot product of the vector with itself.	$\lvert\underset{\sim}{a}+\underset{\sim}{b}\rvert^2 = (\underset{\sim}{a}+\underset{\sim}{b}) \cdot (\underset{\sim}{a}+\underset{\sim}{b})$
2. Expand the brackets.	$\lvert\underset{\sim}{a}+\underset{\sim}{b}\rvert^2 = \underset{\sim}{a}\cdot\underset{\sim}{a} + \underset{\sim}{a}\cdot\underset{\sim}{b} + \underset{\sim}{b}\cdot\underset{\sim}{a} + \underset{\sim}{b}\cdot\underset{\sim}{b}$
3. Using the properties of the scalar product, $\underset{\sim}{a}\cdot\underset{\sim}{a} = \lvert\underset{\sim}{a}\rvert^2$, $\underset{\sim}{b}\cdot\underset{\sim}{b} = \lvert\underset{\sim}{b}\rvert^2$ and $\underset{\sim}{a}\cdot\underset{\sim}{b} = \underset{\sim}{b}\cdot\underset{\sim}{a}$.	$\lvert\underset{\sim}{a}+\underset{\sim}{b}\rvert^2 = \lvert\underset{\sim}{a}\rvert^2 + 2\underset{\sim}{a}\cdot\underset{\sim}{b} + \lvert\underset{\sim}{b}\rvert^2$
4. Substitute for the given values.	$\lvert\underset{\sim}{a}+\underset{\sim}{b}\rvert^2 = 3^2 + 2\times(-4) + 5^2$ $= 9 - 8 + 25$ $= 26$
5. State the answer.	$\lvert\underset{\sim}{a}+\underset{\sim}{b}\rvert = \sqrt{26}$

3.4.2 Scalar and vector resolutes

As seen previously, a vector can be resolved parallel and perpendicular to the *x*- or *y*-axis. In this section, a generalisation of this process will be considered in which one vector is resolved parallel and perpendicular to another vector.

Scalar resolute

The projection of the vector $\underset{\sim}{a} = \overrightarrow{OA}$ onto the vector $\underset{\sim}{b} = \overrightarrow{OB}$ is defined by dropping the perpendicular from the end of $\underset{\sim}{a}$ onto $\underset{\sim}{b}$, that is at the point *C*. The projection is defined as this distance along $\underset{\sim}{b}$ in the direction of $\underset{\sim}{b}$. This distance, *OC*, is called the scalar resolute of $\underset{\sim}{a}$ onto $\underset{\sim}{b}$, or the scalar resolute of $\underset{\sim}{a}$ parallel to $\underset{\sim}{b}$.

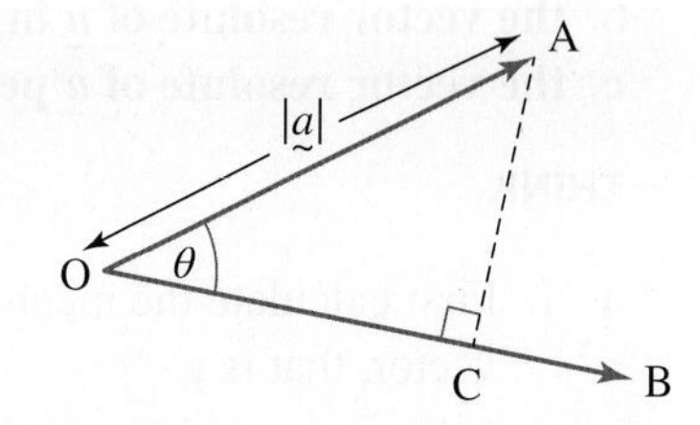

Because $\cos(\theta) = \dfrac{\left|\overrightarrow{OC}\right|}{|\underset{\sim}{a}|}$, it follows that $\left|\overrightarrow{OC}\right| = |\underset{\sim}{a}|\cos(\theta)$.

But $\cos(\theta) = \dfrac{\underset{\sim}{a}\cdot\underset{\sim}{b}}{|\underset{\sim}{a}||\underset{\sim}{b}|}$, so $\left|\overrightarrow{OC}\right| = \dfrac{\underset{\sim}{a}\cdot\underset{\sim}{b}}{|\underset{\sim}{b}|}$

$= \underset{\sim}{a}\cdot\hat{\underset{\sim}{b}}$.

The scalar resolute

The scalar resolute of $\underset{\sim}{a}$ in the direction of $\underset{\sim}{b}$ is given by $\dfrac{\underset{\sim}{a}\cdot\underset{\sim}{b}}{|\underset{\sim}{b}|} = \underset{\sim}{a}\cdot\hat{\underset{\sim}{b}}$.

Parallel vector resolute

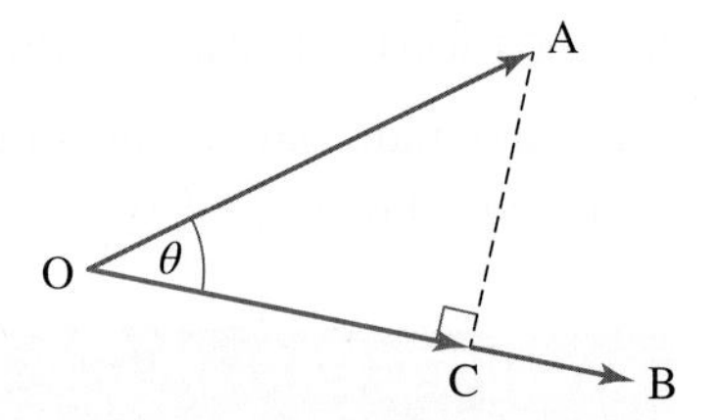

Consider now the vector joining O to C (shown in green). It's magnitude is the scalar resolute $\underset{\sim}{a} \cdot \hat{\underset{\sim}{b}}$, while its direction is the same as $\underset{\sim}{b}$, that is $\hat{\underset{\sim}{b}}$. This is known as the vector resolute of $\underset{\sim}{a}$ parallel to $\underset{\sim}{b}$ and is denoted by the symbol $\underset{\sim}{a}_{\parallel}$.

The vector resolute of $\underset{\sim}{a}$ parallel to $\underset{\sim}{b}$

The vector resolute of $\underset{\sim}{a}$ parallel to $\underset{\sim}{b}$ is given by $\underset{\sim}{a}_{\parallel} = \left(\underset{\sim}{a} \cdot \hat{\underset{\sim}{b}}\right) \hat{\underset{\sim}{b}}$.

Perpendicular vector resolute

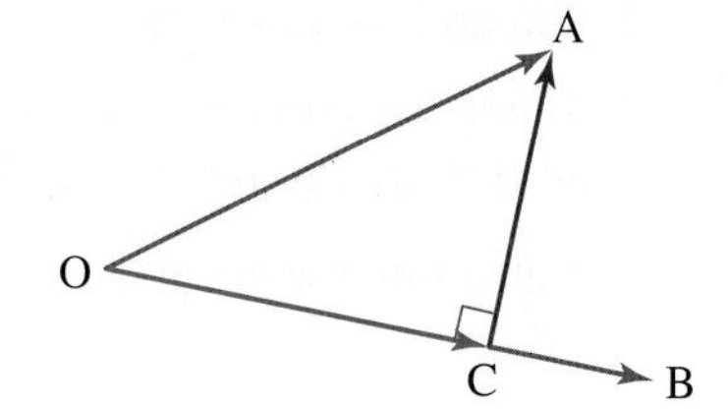

Now consider the vector joining C to A (shown in pink). This is known as the vector resolute of $\underset{\sim}{a}$ perpendicular to $\underset{\sim}{b}$, denoted $\underset{\sim}{a}_{\perp}$, and it is equal to $\underset{\sim}{a} - \underset{\sim}{a}_{\parallel}$.

The vector resolute of $\underset{\sim}{a}$ perpendicular to $\underset{\sim}{b}$

The vector resolute of $\underset{\sim}{a}$ perpendicular to $\underset{\sim}{b}$ is given by $\underset{\sim}{a}_{\perp} = \underset{\sim}{a} - \underset{\sim}{a}_{\parallel} = \underset{\sim}{a} - \left(\underset{\sim}{a} \cdot \hat{\underset{\sim}{b}}\right) \hat{\underset{\sim}{b}}$.

WORKED EXAMPLE 17 Determining scalar and vector resolutes

Given the vectors $\underset{\sim}{u} = 3\underset{\sim}{i} - \underset{\sim}{j} + 2\underset{\sim}{k}$ and $\underset{\sim}{v} = 2\underset{\sim}{i} - 3\underset{\sim}{j} - \underset{\sim}{k}$, determine:

a. **the scalar resolute of $\underset{\sim}{u}$ in the direction of $\underset{\sim}{v}$**
b. **the vector resolute of $\underset{\sim}{u}$ in the direction of $\underset{\sim}{v}$**
c. **the vector resolute of $\underset{\sim}{u}$ perpendicular to $\underset{\sim}{v}$.**

THINK	WRITE
a. 1. First calculate the magnitude of the second vector, that is $\underset{\sim}{v}$.	a. $\underset{\sim}{v} = 2\underset{\sim}{i} - 3\underset{\sim}{j} - \underset{\sim}{k}$ $\lvert\underset{\sim}{v}\rvert = \sqrt{2^2 + (-3)^2 + (-1)^2}$ $= \sqrt{4+9+1}$ $= \sqrt{14}$
2. Write a unit vector parallel to the second given vector, that is $\hat{\underset{\sim}{v}}$.	$\hat{\underset{\sim}{v}} = \dfrac{1}{\sqrt{14}}(2\underset{\sim}{i} - 3\underset{\sim}{j} - \underset{\sim}{k})$
3. Determine the scalar product of the two vectors, $\underset{\sim}{u} \cdot \underset{\sim}{v}$.	$\underset{\sim}{u} \cdot \underset{\sim}{v} = (3\underset{\sim}{i} - \underset{\sim}{j} + 2\underset{\sim}{k}) \cdot (2\underset{\sim}{i} - 3\underset{\sim}{j} - \underset{\sim}{k})$ $= 6 + 3 - 2$ $= 7$

4. The scalar resolute of $\underset{\sim}{u}$ in the direction of $\underset{\sim}{v}$ is given by $\underset{\sim}{u} \cdot \hat{\underset{\sim}{v}}$.

$$\underset{\sim}{u} \cdot \hat{\underset{\sim}{v}} = \frac{\underset{\sim}{u} \cdot \underset{\sim}{v}}{|\underset{\sim}{v}|} = \frac{7}{\sqrt{14}}$$

5. Rationalise the denominator.

$$\underset{\sim}{u} \cdot \hat{\underset{\sim}{v}} = \frac{7}{\sqrt{14}} \times \frac{\sqrt{14}}{\sqrt{14}} = \frac{\sqrt{14}}{2}$$

b. The vector resolute of $\underset{\sim}{u}$ in the direction of $\underset{\sim}{v}$ is given by $(\underset{\sim}{u} \cdot \hat{\underset{\sim}{v}})\hat{\underset{\sim}{v}}$.
Substitute for the scalar resolute $\underset{\sim}{u} \cdot \hat{\underset{\sim}{v}}$ and the unit vector $\hat{\underset{\sim}{v}}$.

b. $$(\underset{\sim}{u} \cdot \hat{\underset{\sim}{v}})\hat{\underset{\sim}{v}} = \frac{\sqrt{14}}{2}\hat{\underset{\sim}{v}} = \frac{\sqrt{14}}{2} \times \frac{1}{\sqrt{14}}(2\underset{\sim}{i} - 3\underset{\sim}{j} - \underset{\sim}{k}) = \frac{1}{2}(2\underset{\sim}{i} - 3\underset{\sim}{j} - \underset{\sim}{k})$$

c. 1. The vector resolute of $\underset{\sim}{u}$ perpendicular to $\underset{\sim}{v}$ is given by $\underset{\sim}{u} - (\underset{\sim}{u} \cdot \hat{\underset{\sim}{v}})\hat{\underset{\sim}{v}}$. Substitute for the given vectors.

c. $$\underset{\sim}{u} - (\underset{\sim}{u} \cdot \hat{\underset{\sim}{v}})\hat{\underset{\sim}{v}} = (3\underset{\sim}{i} - \underset{\sim}{j} + 2\underset{\sim}{k}) - \left(\frac{1}{2}(2\underset{\sim}{i} - 3\underset{\sim}{j} - \underset{\sim}{k})\right)$$

2. Form a common denominator to subtract the vectors.

$$= \frac{1}{2}\left[2(3\underset{\sim}{i} - \underset{\sim}{j} + 2\underset{\sim}{k}) - (2\underset{\sim}{i} - 3\underset{\sim}{j} - \underset{\sim}{k})\right] = \frac{1}{2}\left[(6\underset{\sim}{i} - 2\underset{\sim}{j} + 4\underset{\sim}{k}) - (2\underset{\sim}{i} - 3\underset{\sim}{j} - \underset{\sim}{k})\right]$$

3. State the final result.

$$= \frac{1}{2}(4\underset{\sim}{i} + \underset{\sim}{j} + 5\underset{\sim}{k})$$

TI | THINK

On a Calculator page, complete the entry as shown.

DISPLAY/WRITE

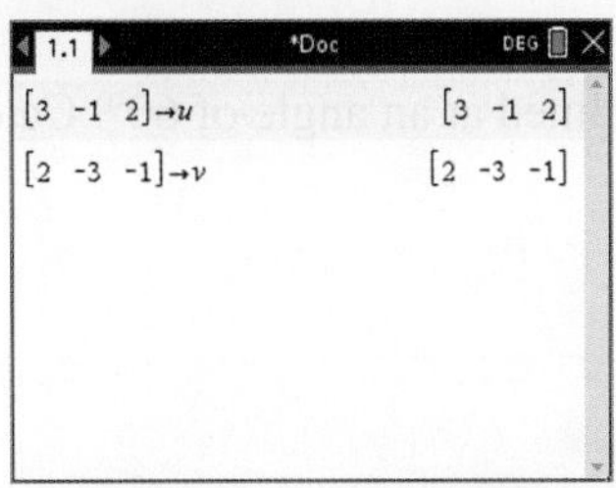

dotP(u,v)/norm(v) → √14/2
dotP(u,v)/norm(v)·unitV(v) → [1 -3/2 -1/2]
u−dotP(u,v)/norm(v)·unitV(v) → [2 1/2 5/2]

CASIO | THINK

On a Main screen, complete the entry as shown.

DISPLAY/WRITE

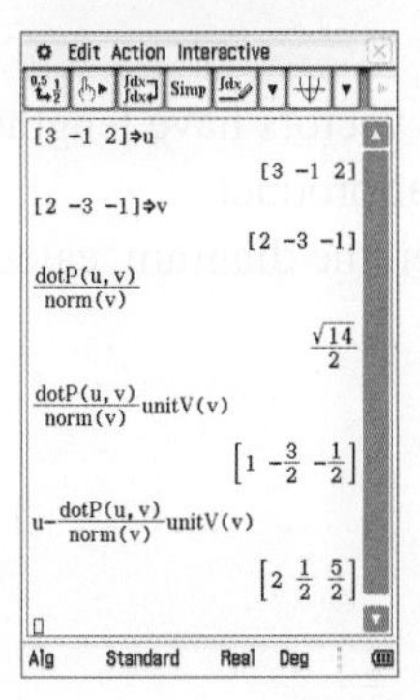

3.4 Exercise

Students, these questions are even better in jacPLUS

Receive immediate feedback and access sample responses

Access additional questions

Track your results and progress

Find all this and MORE in jacPLUS

Technology free

1. WE12 Given the diagram, calculate $\underset{\sim}{a} \cdot \underset{\sim}{b}$.

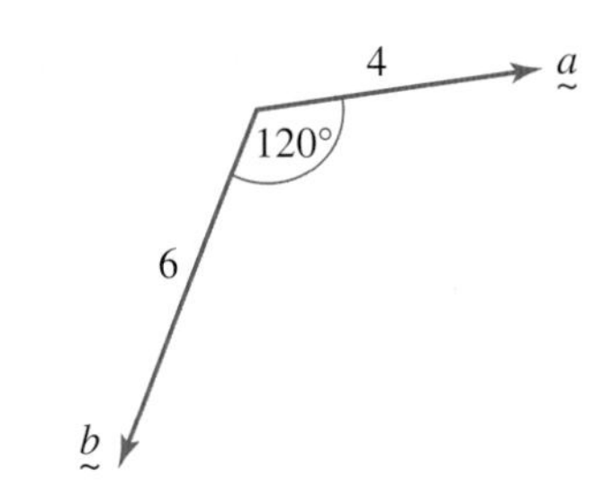

2. Given the diagram, calculate $\underset{\sim}{r} \cdot \underset{\sim}{s}$.

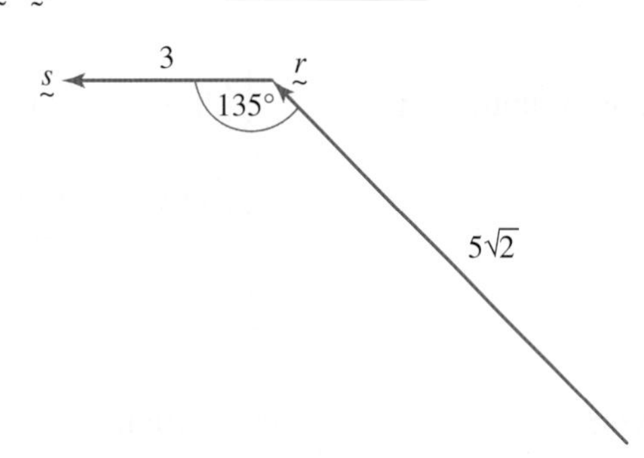

3. a. Two vectors have lengths of 8 and 3 units and are inclined at an angle of 60°. Calculate the value of their scalar product.
   b. Given the diagram, calculate the value of $\underset{\sim}{a} \cdot \underset{\sim}{b}$.

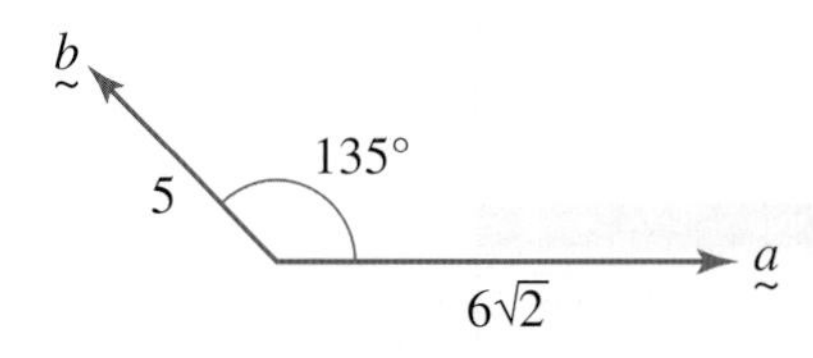

   c. Two vectors have lengths of 7 and 3 units. The vectors are parallel but point in opposite directions. Calculate the value of their scalar product.

4. WE13 Given the vectors $\underset{\sim}{a} = 2\underset{\sim}{i} - 3\underset{\sim}{j} + 5\underset{\sim}{k}$ and $\underset{\sim}{b} = -\underset{\sim}{i} + 3\underset{\sim}{j} - 2\underset{\sim}{k}$, calculate $\underset{\sim}{a} \cdot \underset{\sim}{b}$.

5. Given the vectors $\underset{\sim}{r} = \underset{\sim}{i} + y\underset{\sim}{j} + 4\underset{\sim}{k}$ and $\underset{\sim}{s} = -2\underset{\sim}{i} + 6\underset{\sim}{j} - 7\underset{\sim}{k}$, determine the value of y if $\underset{\sim}{r} \cdot \underset{\sim}{s} = 12$.

6. a. Given the vectors $\underset{\sim}{u} = 2\underset{\sim}{i} - \underset{\sim}{j} + 4\underset{\sim}{k}$ and $\underset{\sim}{v} = -\underset{\sim}{i} + 2\underset{\sim}{j} - 3\underset{\sim}{k}$, calculate the value of $\underset{\sim}{u} \cdot \underset{\sim}{v}$.
   b. Determine the value of $(\underset{\sim}{i} + \underset{\sim}{j} - 3\underset{\sim}{k}) \cdot (2\underset{\sim}{i} + \underset{\sim}{j})$.
   c. $\underset{\sim}{r} = 2\underset{\sim}{i} - 3\underset{\sim}{j}$ and $\underset{\sim}{s} = 3\underset{\sim}{i} + 2\underset{\sim}{j} + 4\underset{\sim}{k}$, determine the value of $\underset{\sim}{r} \cdot \underset{\sim}{s}$.
7. **WE14** If the two vectors $\underset{\sim}{a} = 2\underset{\sim}{i} + 3\underset{\sim}{j} - \underset{\sim}{k}$ and $\underset{\sim}{b} = 4\underset{\sim}{i} - 3\underset{\sim}{j} + z\underset{\sim}{k}$ are orthogonal, determine the value of z.
8. The vectors $\underset{\sim}{a} = x\underset{\sim}{i} - \underset{\sim}{j} + x\underset{\sim}{k}$ and $\underset{\sim}{b} = x\underset{\sim}{i} + 5\underset{\sim}{j} - 4\underset{\sim}{k}$ are orthogonal. Determine the value(s) of x.
9. Given the vectors $\underset{\sim}{p} = 2\underset{\sim}{i} - \underset{\sim}{j} + z\underset{\sim}{k}$ and $\underset{\sim}{q} = -6\underset{\sim}{i} + 3\underset{\sim}{j} + 5\underset{\sim}{k}$, determine the value of z if:
   a. $\underset{\sim}{p}$ is parallel to the vector $\underset{\sim}{q}$
   b. $\underset{\sim}{p}$ is perpendicular to the vector $\underset{\sim}{q}$
   c. the length of the vector $\underset{\sim}{p}$ is 5.
10. Given the vectors $\underset{\sim}{r} = 6\underset{\sim}{i} + y\underset{\sim}{j} - 9\underset{\sim}{k}$ and $\underset{\sim}{s} = -4\underset{\sim}{i} + 2\underset{\sim}{j} + 6\underset{\sim}{k}$, determine the value of y if:
    a. $\underset{\sim}{r}$ is parallel to the vector $\underset{\sim}{s}$
    b. $\underset{\sim}{r}$ is perpendicular to the vector $\underset{\sim}{s}$
    c. the length of the vector $\underset{\sim}{r}$ is twice the length the vector $\underset{\sim}{s}$.
11. Given the vectors $\underset{\sim}{a} = x\underset{\sim}{i} + 2\underset{\sim}{j} - 3\underset{\sim}{k}$, $\underset{\sim}{b} = \underset{\sim}{i} - 2\underset{\sim}{j} - \underset{\sim}{k}$ and $\underset{\sim}{c} = 2\underset{\sim}{i} + 4\underset{\sim}{j} - 5\underset{\sim}{k}$, determine the value of x if:
    a. $2\underset{\sim}{a} - 3\underset{\sim}{b}$ is parallel to the yz plane
    b. $2\underset{\sim}{a} - 3\underset{\sim}{b}$ is perpendicular to the vector $\underset{\sim}{c}$
    c. the length of the vector $2\underset{\sim}{a} - 3\underset{\sim}{b}$ is 11.
12. **WE15** Given the vectors $\underset{\sim}{a} = 4\underset{\sim}{i} - 5\underset{\sim}{j} - 3\underset{\sim}{k}$ and $\underset{\sim}{b} = 2\underset{\sim}{i} - \underset{\sim}{j} + \underset{\sim}{k}$, determine the angle between the vectors $\underset{\sim}{a}$ and $\underset{\sim}{b}$, giving your answer rounded to 2 decimal places.
13. The angle between the vectors $\underset{\sim}{u} = x\underset{\sim}{i} + \underset{\sim}{j} + \underset{\sim}{k}$ and $\underset{\sim}{v} = \underset{\sim}{i} - \underset{\sim}{j} + \underset{\sim}{k}$ is 120°. Determine the value of x.
14. **WE16** If $|\underset{\sim}{a}| = 7$, $|\underset{\sim}{b}| = 5$ and $\underset{\sim}{a} \cdot \underset{\sim}{b} = 26$, evaluate $|\underset{\sim}{a} - \underset{\sim}{b}|$.
15. Given that $|\underset{\sim}{r}| = 2$, $|\underset{\sim}{s}| = 4$ and $\underset{\sim}{r} \cdot \underset{\sim}{s} = 8$, evaluate $|2\underset{\sim}{s} - \underset{\sim}{r}|$.
16. Consider the vectors $\underset{\sim}{a} = 3\underset{\sim}{i} - 2\underset{\sim}{j} + 4\underset{\sim}{k}$, $\underset{\sim}{b} = \underset{\sim}{i} + \underset{\sim}{j} - 3\underset{\sim}{k}$ and $\underset{\sim}{c} = 4\underset{\sim}{i} - 3\underset{\sim}{j} + 5\underset{\sim}{k}$.
    a. Show that $\underset{\sim}{a} \cdot (\underset{\sim}{b} + \underset{\sim}{c}) = \underset{\sim}{a} \cdot \underset{\sim}{b} + \underset{\sim}{a} \cdot \underset{\sim}{c}$.
    b. Show that $\underset{\sim}{a} \cdot (\underset{\sim}{b} - \underset{\sim}{c}) = \underset{\sim}{a} \cdot \underset{\sim}{b} - \underset{\sim}{a} \cdot \underset{\sim}{c}$.
    c. State what property **a** and **b** illustrates.
    d. State what meaning can be given to $\underset{\sim}{a} \cdot \underset{\sim}{b} \cdot \underset{\sim}{c}$.
17. **WE17** Given the vectors $\underset{\sim}{u} = 5\underset{\sim}{i} - 3\underset{\sim}{j} - 2\underset{\sim}{k}$ and $\underset{\sim}{v} = \underset{\sim}{i} - 2\underset{\sim}{j} - 2\underset{\sim}{k}$, calculate:
    a. the scalar resolute of $\underset{\sim}{u}$ in the direction of $\underset{\sim}{v}$
    b. the vector resolute of $\underset{\sim}{u}$ in the direction of $\underset{\sim}{v}$
    c. the vector resolute of $\underset{\sim}{u}$ perpendicular to $\underset{\sim}{v}$.
18. Given the vectors $\underset{\sim}{r} = 2\underset{\sim}{i} + 4\underset{\sim}{k}$ and $\underset{\sim}{s} = \underset{\sim}{i} - 4\underset{\sim}{j} - 2\underset{\sim}{k}$, calculate:
    a. the vector component of $\underset{\sim}{r}$ parallel to $\underset{\sim}{s}$
    b. the vector component of $\underset{\sim}{r}$ perpendicular to $\underset{\sim}{s}$.
19. Given the points $A(2, 3, 2)$, $B(4, p, 0)$, $C(-1, -1, 0)$ and $D(-2, 2, 1)$, determine the value of p if:
    a. $\overrightarrow{AB}$ is parallel to $\overrightarrow{DC}$
    b. $\overrightarrow{AB}$ is perpendicular to $\overrightarrow{DC}$
    c. the length of the vector $\overrightarrow{AB}$ is equal to the length of the vector $\overrightarrow{DC}$
    d. the scalar resolute of $\overrightarrow{AB}$ parallel to $\overrightarrow{DC}$ is equal to $\dfrac{4}{\sqrt{11}}$.
20. Determine the value(s) of p if the points $P(4, p, -3)$, $Q(-1, -4, -6)$ and $R(1, 6, -1)$ form a right-angled triangle at P.

21. a. The angle between the vectors $\underset{\sim}{a} = x\underset{\sim}{i} - 3\underset{\sim}{j} - 4\underset{\sim}{k}$ and $\underset{\sim}{b} = -\underset{\sim}{i} + 2\underset{\sim}{j} + \underset{\sim}{k}$ is 150°. Determine the value(s) of x.

b. The angle between the vectors $\underset{\sim}{p} = 6\underset{\sim}{i} + 2\underset{\sim}{j} + 3\underset{\sim}{k}$ and $\underset{\sim}{q} = \underset{\sim}{i} + y\underset{\sim}{j} - 2\underset{\sim}{k}$ is equal to $\cos^{-1}\left(\frac{4}{21}\right)$. Determine the value of y.

c. Determine the value of z if the cosine of the angle between the vectors $\underset{\sim}{u} = 2\underset{\sim}{i} - 2\underset{\sim}{j} + \underset{\sim}{k}$ and $\underset{\sim}{v} = 4\underset{\sim}{j} + z\underset{\sim}{k}$ is equal to $-\frac{1}{3}$.

22. If $|\underset{\sim}{u}| = 3$, $|\underset{\sim}{v}| = 4$ and $\underset{\sim}{u} . \underset{\sim}{v} = 6$, evaluate:

a. $|\underset{\sim}{u} + \underset{\sim}{v}|$
b. $|\underset{\sim}{u} - \underset{\sim}{v}|$
c. $|3\underset{\sim}{u} - 2\underset{\sim}{v}|$.

23. If $|\underset{\sim}{r}| = 4\sqrt{2}$, $|\underset{\sim}{s}| = 5\sqrt{3}$ and $\underset{\sim}{r} \cdot \underset{\sim}{s} = -6$, evaluate:

a. $|\underset{\sim}{r} + \underset{\sim}{s}|$
b. $|\underset{\sim}{r} - \underset{\sim}{s}|$
c. $|4\underset{\sim}{r} + 3\underset{\sim}{s}|$.

24. Given the vectors $\underset{\sim}{a} = 2\underset{\sim}{i} - 4\underset{\sim}{j} + \underset{\sim}{k}$ and $\underset{\sim}{b} = 3\underset{\sim}{i} - \underset{\sim}{j} - 4\underset{\sim}{k}$, determine:

a. a unit vector parallel to $\underset{\sim}{b}$
b. the scalar resolute of $\underset{\sim}{a}$ in the direction of $\underset{\sim}{b}$
c. the vector resolute of $\underset{\sim}{a}$ in the direction of $\underset{\sim}{b}$
d. the vector resolute of $\underset{\sim}{a}$ perpendicular to $\underset{\sim}{b}$
e. the angle between the vectors $\underset{\sim}{a}$ and $\underset{\sim}{b}$, giving your answer rounded to 2 decimal places.

25. Given the vectors $\underset{\sim}{p} = 3\underset{\sim}{i} + 2\underset{\sim}{j} - 5\underset{\sim}{k}$ and $\underset{\sim}{q} = 2\underset{\sim}{i} + \underset{\sim}{j} - 2\underset{\sim}{k}$, determine:

a. a unit vector parallel to $\underset{\sim}{q}$
b. the scalar resolute of $\underset{\sim}{p}$ in the direction of $\underset{\sim}{q}$
c. the component of $\underset{\sim}{p}$ in the direction of $\underset{\sim}{q}$
d. the component of $\underset{\sim}{p}$ perpendicular to $\underset{\sim}{q}$
e. the angle between the vectors $\underset{\sim}{p}$ and $\underset{\sim}{q}$, giving your answer rounded to 2 decimal places.

26. Given the vectors $\underset{\sim}{r} = 3\underset{\sim}{i} - 4\underset{\sim}{j} + \underset{\sim}{k}$ and $\underset{\sim}{s} = \underset{\sim}{i} - 2\underset{\sim}{j} + 3\underset{\sim}{k}$:

a. resolve the vector $\underset{\sim}{r}$ into two components, one parallel to $\underset{\sim}{s}$ and one perpendicular to $\underset{\sim}{s}$
b. calculate the angle between the vectors $\underset{\sim}{r}$ and $\underset{\sim}{s}$, giving your answer rounded to 2 decimal places.

27. Given the points $A(3, -2, 5)$, $B(-1, 0, 4)$ and $C(2, -1, 3)$, determine:

a. a unit vector parallel to $\overrightarrow{BC}$
b. the vector resolute of $\overrightarrow{AB}$ onto $\overrightarrow{BC}$
c. the vector resolute of $\overrightarrow{AB}$ perpendicular to $\overrightarrow{BC}$
d. the angle between $\overrightarrow{AB}$ and $\overrightarrow{BC}$, correct to 1 decimal place.

28. If $\underset{\sim}{a} = \cos(\alpha)\underset{\sim}{i} + \sin(\alpha)\underset{\sim}{j}$ and $\underset{\sim}{b} = \cos(\beta)\underset{\sim}{i} + \sin(\beta)\underset{\sim}{j}$, determine the angle between the vectors $\underset{\sim}{a}$ and $\underset{\sim}{b}$, and hence show that $\cos(\beta - \alpha) = \cos(\alpha)\cos(\beta) + \sin(\alpha)\sin(\beta)$.

29. When a force $\underset{\sim}{F}$ moves a point from A to B producing a displacement $\underset{\sim}{s} = \overrightarrow{AB}$, the work done is given by $W = \underset{\sim}{F} \cdot \underset{\sim}{s}$. Determine the work done when the force $\underset{\sim}{F} = 3\underset{\sim}{i} + 2\underset{\sim}{j} + 4\underset{\sim}{k}$ moves a point from $A(1, -2, 2)$ to the point $B(2, 1, -4)$.

30. a. Given the vectors $|\underset{\sim}{a}| = 3$, $|\underset{\sim}{b}| = 4$ and $|\underset{\sim}{c}| = 5$ such that $\underset{\sim}{a} + \underset{\sim}{b} + \underset{\sim}{c} = 0$, determine the value of $\underset{\sim}{a} \cdot \underset{\sim}{b}$.
b. Given the vectors $|\underset{\sim}{p}| = 5$, $|\underset{\sim}{q}| = 12$ and $|\underset{\sim}{r}| = 13$ and $\underset{\sim}{p} \cdot \underset{\sim}{q} = 0$, evaluate $\underset{\sim}{p} + \underset{\sim}{q} + \underset{\sim}{r}$.
c. If $\underset{\sim}{a} \cdot \underset{\sim}{b} = \underset{\sim}{b} \cdot \underset{\sim}{c}$, state what can be deduced about the vectors $\underset{\sim}{a}$, $\underset{\sim}{b}$ and $\underset{\sim}{c}$.

31. a. Show that $\underset{\sim}{u} \cdot \underset{\sim}{v} = \frac{1}{4}|\underset{\sim}{u} + \underset{\sim}{v}|^2 - \frac{1}{4}|\underset{\sim}{u} - \underset{\sim}{v}|^2$ and $|\underset{\sim}{u} + \underset{\sim}{v}|^2 + |\underset{\sim}{u} - \underset{\sim}{v}|^2 = 2(|\underset{\sim}{u}|^2 + |\underset{\sim}{v}|^2)$.

b. If $|\underset{\sim}{u} + \underset{\sim}{v}| = \sqrt{17}$ and $|\underset{\sim}{u} - \underset{\sim}{v}| = \sqrt{13}$ and the angle between the vectors $\underset{\sim}{u}$ and $\underset{\sim}{v}$ is $\cos^{-1}\left(\frac{1}{\sqrt{50}}\right)$, calculate $|\underset{\sim}{u}|$ and $|\underset{\sim}{v}|$.

Technology active

32. $OABP$ is a pyramid, where O is the origin. The coordinates of the points are $A(4, -1, -3)$, $B(3, -4, 1)$ and $P(x, y, z)$. The height of the pyramid is the length of GP, where G is a point on the base OAB such that GP is perpendicular to the base.

   a. Show using vectors that OAB is an equilateral triangle.

   b. Let M be the midpoint of AB. Given that the point G is such that $\overrightarrow{OG} = \frac{2}{3}\overrightarrow{OM}$, determine the vector $\overrightarrow{OG}$.

   c. Calculate the vector $\overrightarrow{GP}$ and, using the fact that $\overrightarrow{GP}$ is perpendicular to $\overrightarrow{OG}$, show that $7x - 5y - 2z = 26$.

   d. The faces of the pyramid, OAP, ABP and OBP, are all similar isosceles triangles, with distance $OP = AP = BP = 5\sqrt{14}$. Write another set of equations expressing this relationship in terms of x, y and z.

   e. Determine the coordinates of P.

   f. Determine the height of the pyramid.

   g. Determine the angle, in degrees correct to 2 decimal places, that the sloping edge makes with the base.

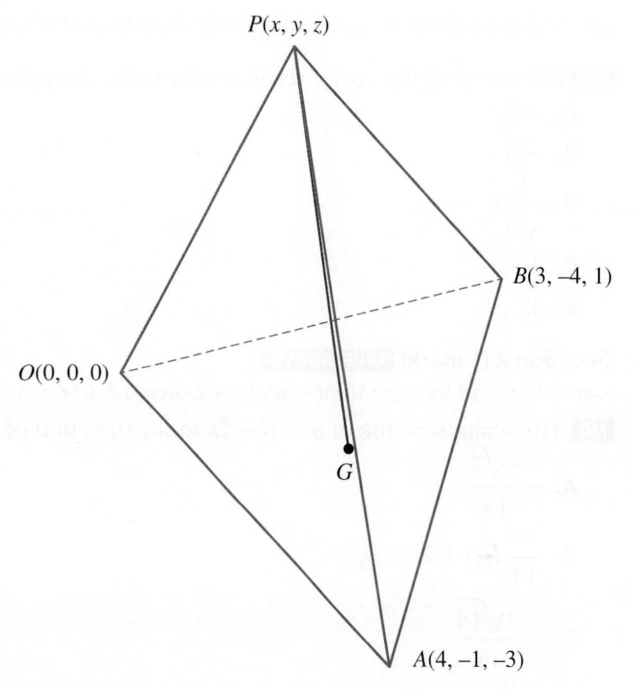

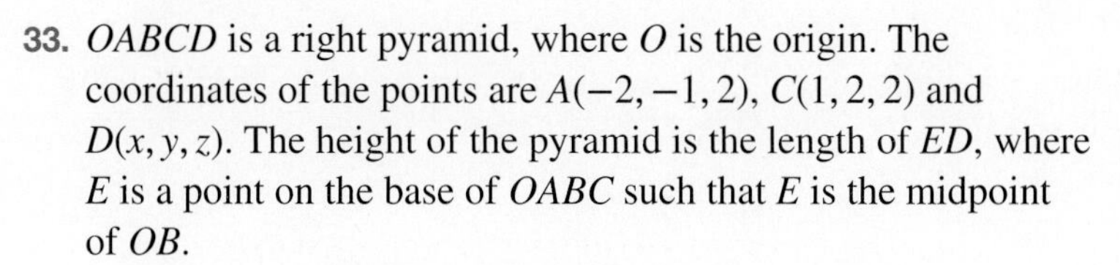

33. $OABCD$ is a right pyramid, where O is the origin. The coordinates of the points are $A(-2, -1, 2)$, $C(1, 2, 2)$ and $D(x, y, z)$. The height of the pyramid is the length of ED, where E is a point on the base of $OABC$ such that E is the midpoint of OB.

   a. Show that $OABC$ is a square, and hence show that the coordinates of B are $(-1, 1, 4)$.

   b. Determine the coordinates of the point E.

   c. If the vector $\overrightarrow{ED}$ is perpendicular to $\overrightarrow{OE}$, show that $-x + y + 4z = 9$.

   d. The faces of the pyramid, OAD, ABD, BCD and OCD, are all similar isosceles triangles, with sloping edges OD, AD, BD and CD all equal in length to $\frac{9\sqrt{22}}{2}$. Write another set of equations that can be used to solve for x, y and z.

   e. Determine the coordinates of D and hence calculate the height of the pyramid.

   f. Determine the angle, in degrees correct to 2 decimal places, that the sloping edge makes with the base.

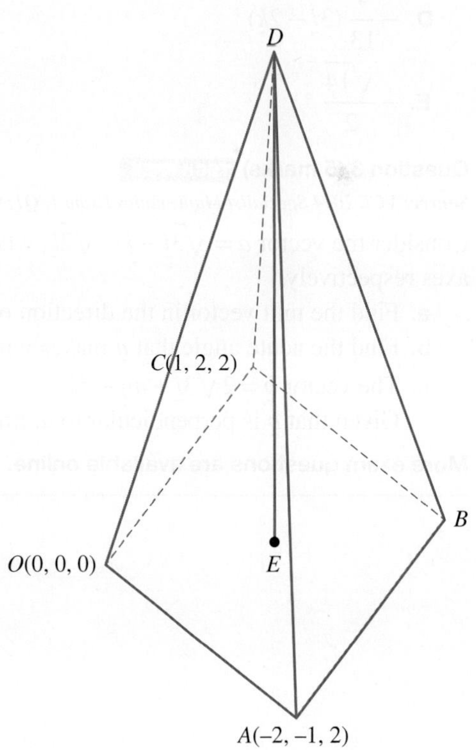

3.4 Exam questions

Question 1 (1 mark) TECH-ACTIVE

Source: *VCE 2021 Specialist Mathematics Exam 2, Section A, Q13; © VCAA.*

The scalar resolute of vector of $\underset{\sim}{a}$ in the direction of vector $\underset{\sim}{b}$ is -4.

MC If $\underset{\sim}{b} = -\sqrt{3}\underset{\sim}{i}$, the vector resolute of $\underset{\sim}{a}$ in the direction of $\underset{\sim}{b}$ is

A. $-4\underset{\sim}{i}$

B. $-3\underset{\sim}{i}$

C. $\frac{1}{\sqrt{3}}\underset{\sim}{i}$

D. $3\underset{\sim}{i}$

E. $4\underset{\sim}{i}$

Question 2 (1 mark) TECH-ACTIVE

Source: *VCE 2018 Specialist Mathematics Exam 2, Section A, Q14; © VCAA.*

MC The scalar resolute of $\underset{\sim}{a} = 3\underset{\sim}{i} - 2\underset{\sim}{k}$ in the direction of $\underset{\sim}{b} = -\underset{\sim}{i} + 2\underset{\sim}{j} + 3\underset{\sim}{k}$ is

A. $-\frac{9\sqrt{3}}{13}$

B. $\frac{-9}{14}(-\underset{\sim}{i} + 2\underset{\sim}{j} + 3\underset{\sim}{k})$

C. $-\frac{9\sqrt{14}}{14}$

D. $-\frac{9}{13}(3\underset{\sim}{i} - 2\underset{\sim}{k})$

E. $-\frac{\sqrt{14}}{2}$

Question 3 (5 marks) TECH-FREE

Source: *VCE 2014 Specialist Mathematics Exam 1, Q1; © VCAA.*

Consider the vector $\underset{\sim}{a} = \sqrt{3}\underset{\sim}{i} - \underset{\sim}{j} - \sqrt{2}\underset{\sim}{k}$, where $\underset{\sim}{i}, \underset{\sim}{j}$ and $\underset{\sim}{k}$ are unit vectors in the positive directions of the x, y and z axes respectively.

a. Find the unit vector in the direction of $\underset{\sim}{a}$. **(1 mark)**

b. Find the acute angle that $\underset{\sim}{a}$ makes with the positive direction of the x-axis. **(2 marks)**

c. The vector $\underset{\sim}{b} = 2\sqrt{3}\underset{\sim}{i} + m\underset{\sim}{j} - 5\underset{\sim}{k}$.
Given that $\underset{\sim}{b}$ is perpendicular to $\underset{\sim}{a}$, find the value of m. **(2 marks)**

More exam questions are available online.

3.5 Vector proofs using the scalar product

LEARNING INTENTION

At the end of this subtopic you should be able to:

- prove geometrical facts using vectors.

3.5.1 Applying the scalar product to vector proofs

Now that we have learnt about the scalar product of vectors we can demonstrate and prove more facts about geometric shapes.

Recall that the scalar product of perpendicular vectors equals 0. This can be used to show that a quadrilateral is a rectangle or a square.

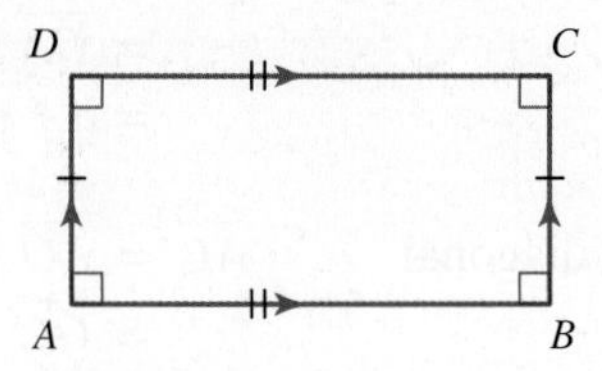

In the rectangle $ABCD$, $\overrightarrow{AB} = \overrightarrow{DC}$, $\overrightarrow{AD} = \overrightarrow{BC}$ and thus $\overrightarrow{AB} \cdot \overrightarrow{BC} = 0$, $\overrightarrow{BC} \cdot \overrightarrow{CD} = 0$, $\overrightarrow{CD} \cdot \overrightarrow{DA} = 0$ and $\overrightarrow{DA} \cdot \overrightarrow{AB} = 0$, since all these sides are perpendicular.

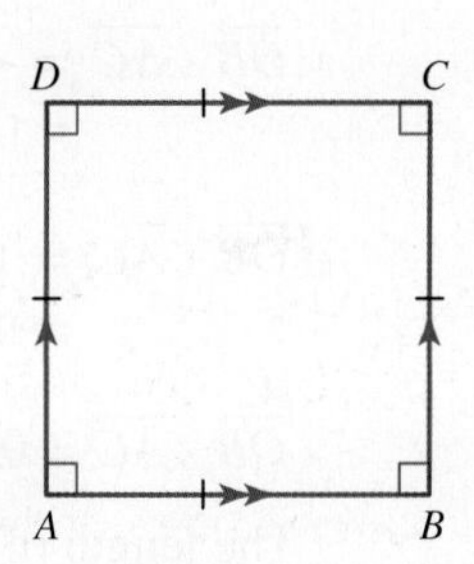

In the square $ABCD$, $\overrightarrow{AB} = \overrightarrow{DC}$, $\overrightarrow{AD} = \overrightarrow{BC}$ and thus $\overrightarrow{AB} \cdot \overrightarrow{BC} = 0$, $\overrightarrow{BC} \cdot \overrightarrow{CD} = 0$, $\overrightarrow{CD} \cdot \overrightarrow{DA} = 0$, $\overrightarrow{DA} \cdot \overrightarrow{AB} = 0$ and $\left|\overrightarrow{AB}\right| = \left|\overrightarrow{BC}\right| = \left|\overrightarrow{DC}\right| = \left|\overrightarrow{AD}\right|$.

Using vectors to prove geometrical theorems

The following statements are useful in proving geometrical theorems.

1. **If O is the origin and A and B are points, the midpoint M of the line segment AB is given by:** $\overrightarrow{OM} = \frac{1}{2}\left(\overrightarrow{OA} + \overrightarrow{OB}\right)$
2. **If two vectors $\overrightarrow{AB}$ and $\overrightarrow{CD}$ are parallel, then $\overrightarrow{AB} = \lambda\overrightarrow{CD}$ where $\lambda \in R$ is a scalar.**
3. **If two vectors $\overrightarrow{AB}$ and $\overrightarrow{CD}$ are perpendicular, then $\overrightarrow{AB} \cdot \overrightarrow{CD} = 0$.**
4. **If two vectors $\overrightarrow{AB}$ and $\overrightarrow{CD}$ are equal. Then $\overrightarrow{AB}$ is parallel to $\overrightarrow{CD}$; furthermore, these two vectors are equal in length, so that $\overrightarrow{AB} = \overrightarrow{CD} \Rightarrow \left|\overrightarrow{AB}\right| = \left|\overrightarrow{CD}\right|$.**
5. **If $\overrightarrow{AB} = \lambda\overrightarrow{BC}$, then the points A, B and C are collinear; that is, A, B and C all lie on a straight line.**

WORKED EXAMPLE 18 Geometric proof using vectors

Prove that if the diagonals of a parallelogram are perpendicular, then the parallelogram is a rhombus.

THINK	WRITE/DRAW
1. Let OABC be a parallelogram.	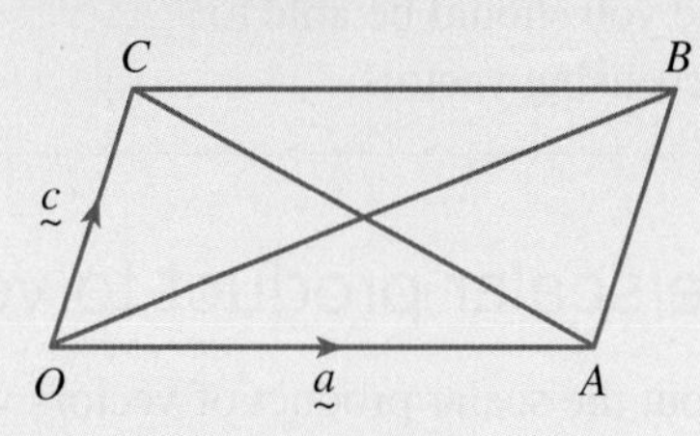 Let $\overrightarrow{OA} = \underset{\sim}{a}$ and $\overrightarrow{OC} = \underset{\sim}{c}$.
2. State the properties of the parallelogram.	Because $OABC$ is a parallelogram, $\underset{\sim}{a} = \overrightarrow{OA} = \overrightarrow{CB}$ and $\underset{\sim}{c} = \overrightarrow{OC} = \overrightarrow{AB}$.
3. Determine a vector expression for the diagonal OB in terms of $\underset{\sim}{a}$ and $\underset{\sim}{c}$.	$\overrightarrow{OB} = \overrightarrow{OA} + \overrightarrow{AB}$ $= \overrightarrow{OA} + \overrightarrow{OC}$ $= \underset{\sim}{a} + \underset{\sim}{c}$
4. Determine a vector expression for the diagonal AC in terms of $\underset{\sim}{a}$ and $\underset{\sim}{c}$.	$\overrightarrow{AC} = \overrightarrow{AO} + \overrightarrow{OC}$ $= \overrightarrow{OC} - \overrightarrow{OA}$ $= \underset{\sim}{c} - \underset{\sim}{a}$
5. The dot product of the diagonals is zero, since it was given that they are perpendicular.	$\overrightarrow{OB} \cdot \overrightarrow{AC} = (\underset{\sim}{a} + \underset{\sim}{c}) \cdot (\underset{\sim}{c} - \underset{\sim}{a})$ $= 0$
6. Expand the brackets.	$\overrightarrow{OB} \cdot \overrightarrow{AC} = -\underset{\sim}{a} \cdot \underset{\sim}{a} + \underset{\sim}{c} \cdot \underset{\sim}{a} - \underset{\sim}{a} \cdot \underset{\sim}{c} + \underset{\sim}{c} \cdot \underset{\sim}{c}$ $= 0$
7. Use the properties of the dot product: $\underset{\sim}{c} \cdot \underset{\sim}{a} = \underset{\sim}{a} \cdot \underset{\sim}{c}$ and $\underset{\sim}{a} \cdot \underset{\sim}{a} = \lvert\underset{\sim}{a}\rvert^2$.	$\overrightarrow{OB} \cdot \overrightarrow{AC} = \lvert\underset{\sim}{a}\rvert^2 - \lvert\underset{\sim}{c}\rvert^2$ $= 0$
8. State the conclusion.	$\overrightarrow{OB} \cdot \overrightarrow{AC} = 0 \Rightarrow \lvert\underset{\sim}{a}\rvert^2 = \lvert\underset{\sim}{c}\rvert^2$ so that $\left\lvert\overrightarrow{OA}\right\rvert = \left\lvert\overrightarrow{OC}\right\rvert$. The length of $\overrightarrow{OA}$ is equal to the length of $\overrightarrow{OC}$; therefore, $OABC$ is a rhombus.

3.5 Exercise

Technology free

1. WE18 Prove that the diagonals of a rhombus are perpendicular.
2. Prove that if the diagonals of a parallelogram are equal in length, then the parallelogram is a rectangle.

3. Given the points $A(8, 3, -1)$, $B(4, 5, -2)$ and $C(7, 9, -6)$, show that ABC forms a right-angled triangle at B, and hence determine the area of the triangle.

4. Given the points $A(-3, 5, 4)$, $B(2, 3, 5)$ and $C(4, 6, 1)$, show that ABC forms a right-angled triangle at B, and hence determine the area of the triangle.

5. Given the points $A(4, 7, 3)$, $B(8, 7, 1)$ and $C(6, 5, 2)$, show that ABC forms an isosceles triangle. Let M be the midpoint of AB, and show that MC is perpendicular to AB.

6. Given the points $A(3, -3, 4)$, $B(5, 3, 6)$ and $C(3, 1, 3)$, show that ABC forms an isosceles triangle. Let M be the midpoint of AB, and show that MC is perpendicular to AB.

7. Prove Pythagoras' theorem.

8. Prove that the angle inscribed in a semicircle is a right angle.

9. The diagram shows a circle of radius r with centre at the origin O on the x- and y-axes. The points A and B lie on the diameter of the circle and are on the x-axis; their coordinates are $(-r, 0)$ and $(r, 0)$ respectively. The point C has coordinates (a, b) and lies on the circle, where a, b and r are all positive real constants. Show that CA is perpendicular to CB.

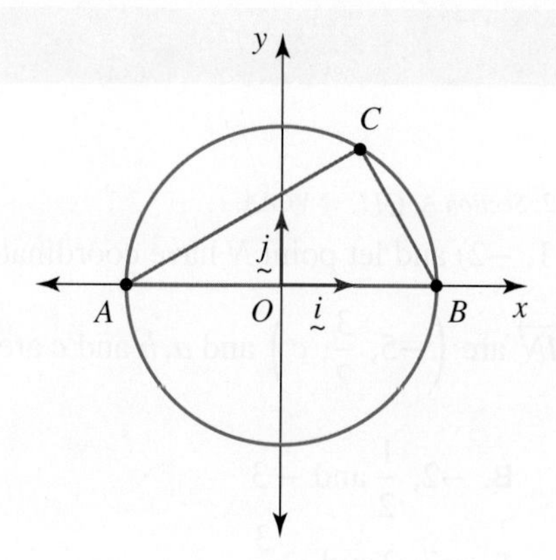

10. $OABC$ is a square. The points P, Q, R and S are the midpoints of the sides OA, AB, BC and OC respectively. Prove that $PQRS$ is a square.

11. $OABC$ is a rhombus. The points P, Q, R and S are the midpoints of the sides OA, AB, BC and OC respectively. Prove that $PQRS$ is a rectangle.

12. Prove that the line segments joining the midpoints of consecutive sides of a rectangle form a rhombus.

13. AB and CD are two diameters of a circle with centre O. Letting $\overrightarrow{OA} = \underset{\sim}{a}$ and $\overrightarrow{OC} = \underset{\sim}{c}$, prove that $ACBD$ is a rectangle.

14. OAB is a right-angled isosceles triangle with $\left|\overrightarrow{OA}\right| = \left|\overrightarrow{OB}\right|$. Let M be the midpoint of AB, and let $\overrightarrow{OA} = \underset{\sim}{a}$ and $\overrightarrow{OB} = \underset{\sim}{b}$.

   a. Express $\overrightarrow{OM}$ in terms of $\underset{\sim}{a}$ and $\underset{\sim}{b}$.
   b. Prove that $\overrightarrow{OM}$ is perpendicular to $\overrightarrow{AB}$.
   c. Show that $\left|\overrightarrow{OM}\right| = \frac{1}{2}\left|\overrightarrow{AB}\right|$.

15. Prove that the vector $\hat{\underset{\sim}{a}} + \hat{\underset{\sim}{b}}$ bisects the angle between the vectors $\underset{\sim}{a}$ and $\underset{\sim}{b}$.

16. ABC is a triangle. P, Q and R are the midpoints of the sides AC, BC and AB respectively. Perpendicular lines are drawn through the points P and Q and intersect at the point O. Let $\overrightarrow{OA} = \underset{\sim}{a}$, $\overrightarrow{OB} = \underset{\sim}{b}$ and $\overrightarrow{OC} = \underset{\sim}{c}$.

   a. Express $\overrightarrow{OP}$ and $\overrightarrow{OQ}$ in terms of $\underset{\sim}{a}$, $\underset{\sim}{b}$ and $\underset{\sim}{c}$.
   b. Show that $|\underset{\sim}{a}| = |\underset{\sim}{b}| = |\underset{\sim}{c}|$.
   c. Prove that $\overrightarrow{OR}$ is perpendicular to $\overrightarrow{AB}$.

17. The diagram shows a circle of radius r with centre at the origin O on the x- and y-axes. The three points A, B and D all lie on the circle and have coordinates $A(a, b)$, $B(a, -b)$ and $D(-r, 0)$, where a, b and r are all positive real constants.

a. Let β be the angle between the vectors $\overrightarrow{OA}$ and $\overrightarrow{OB}$. Show that $\cos(\beta) = \dfrac{a^2 - b^2}{a^2 + b^2}$.

b. Let α be the angle between the vectors $\overrightarrow{DA}$ and $\overrightarrow{DB}$. Show that $\cos(\alpha) = \dfrac{a}{\sqrt{a^2 + b^2}}$.

c. Hence show that $2\alpha = \beta$.

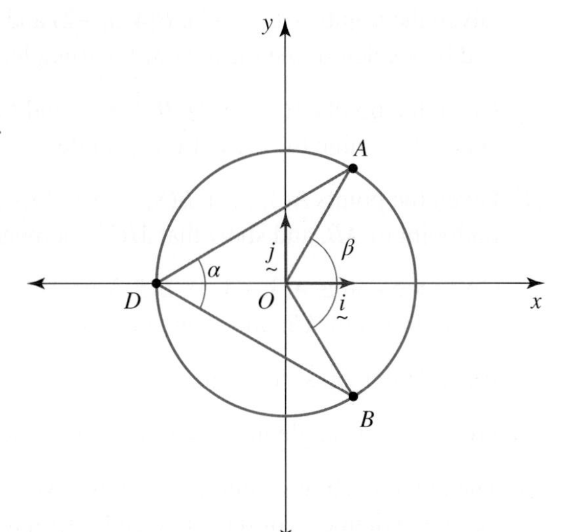

3.5 Exam questions

Question 1 (1 mark) TECH-ACTIVE

Source: *VCE 2019 Specialist Mathematics Exam 2, Section A, Q11; © VCAA.*

MC Let point M have coordinates $(a, 1, -2)$ and let point N have coordinates $(-3, b, -1)$.

If the coodinates of the midpoint of $\overrightarrow{MN}$ are $\left(-5, \dfrac{3}{2}, c\right)$ and a, b and c are real constants, then the values of a, b and c are respectively

A. -13, 2 and $-\dfrac{1}{2}$

B. -2, $\dfrac{1}{2}$ and -3

C. -7, -2 and $-\dfrac{3}{2}$

D. -2, $-\dfrac{1}{2}$ and -3

E. -7, 2 and $-\dfrac{3}{2}$

Question 2 (1 mark) TECH-ACTIVE

Source: *VCE 2018 Specialist Mathematics Exam 2, Section A, Q12; © VCAA.*

MC If $|\underset{\sim}{a} + \underset{\sim}{b}| = |\underset{\sim}{a}| + |\underset{\sim}{b}|$ and $\underset{\sim}{a}, \underset{\sim}{b} \neq \underset{\sim}{0}$, which one of the following is necessarily true?

A. $\underset{\sim}{a}$ is parallel to $\underset{\sim}{b}$

B. $|\underset{\sim}{a}| = |\underset{\sim}{b}|$

C. $\underset{\sim}{a} = \underset{\sim}{b}$

D. $\underset{\sim}{a} = -\underset{\sim}{b}$

E. $\underset{\sim}{a}$ is perpendicular to $\underset{\sim}{b}$

Question 3 (3 marks) TECH-FREE

Source: *VCE 2015 Specialist Mathematics Exam 1, Q1; © VCAA.*

Consider the rhombus $OABC$ shown below, where $\overrightarrow{OA} = a\underset{\sim}{i}$ and $\overrightarrow{OC} = \underset{\sim}{i} + \underset{\sim}{j} + \underset{\sim}{k}$, and a is a positive real constant.

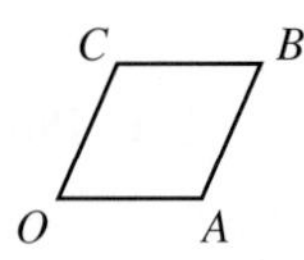

a. Find a. **(1 mark)**

b. Show that the diagonals of the rhombus $OABC$ are perpendicular. **(2 marks)**

More exam questions are available online.

3.6 Parametric equations

LEARNING INTENTION

At the end of this subtopic you should be able to:
- determine and sketch the Cartesian equation of a vector equation.

3.6.1 Parametric equations in two dimensions

A locus is a set of points traced out in the plane, satisfying some geometrical relationship. The path described by a moving particle forms a locus and can be described by a Cartesian equation. However, the Cartesian equation does not tell us where the particle is at any particular time.

The path traced out by the particle can be defined in terms of another or third variable. Here we will use the variable t as the parameter. For example, the unit circle can be described by the use of a parameter t, $x = \cos(t)$, $y = \sin(t)$. In the two-dimensional case there are two parametric equations, as both the x- and y-coordinates depend upon the parameter t.

$$x = x(t) \qquad (1)$$
$$y = y(t) \qquad (2)$$

Because a position vector is given by $\underset{\sim}{r}(t) = x\underset{\sim}{i} + y\underset{\sim}{j}$, where $\underset{\sim}{i}$ and $\underset{\sim}{j}$ are unit vectors in the x and y directions, this is also called the vector equation of the path. If we can eliminate the parameter from these two parametric equations and obtain an equation of the form $y = f(x)$, then this is called an explicit relationship and is the equation of the path.

Often we may be unable to obtain an explicit relationship but can find an implicit relationship of the form $f(x, y) = 0$. Either way, the relationship between x and y is called the Cartesian equation of the path.

WORKED EXAMPLE 19 Determining and sketching the Cartesian equation (1)

Given the vector equation $\underset{\sim}{r}(t) = (t-1)\underset{\sim}{i} + 2t^2\underset{\sim}{j}$, for $t \geq 0$, determine and sketch the Cartesian equation of the path, and state the domain and range.

THINK	WRITE/DRAW
1. Write the vector equation: $\underset{\sim}{r}(t) = x(t)\underset{\sim}{i} + y(t)\underset{\sim}{j}$.	$\underset{\sim}{r}(t) = (t-1)\underset{\sim}{i} + 2t^2\underset{\sim}{j}, t \geq 0$
2. State the parametric equations.	$x = t - 1 \quad (1)$ $y = 2t^2 \quad (2)$
3. Eliminate the parameter. Express t in terms of x.	From (1), $t = x + 1$.
4. Substitute into the second parametric equation to obtain the Cartesian equation of the path.	Substitute (1) $t = x + 1$ into (2) $y = 2t^2$: $y = 2(x+1)^2$
5. From the restriction on t, determine the domain and range.	Since $t \geq 0$, it follows from the parametric equations that $x \geq -1$ and $y \geq 0$.

6. The graph is a restricted domain function.

The graph is not the whole parabola; it is a parabola on a restricted domain, with an endpoint at (−1, 0).

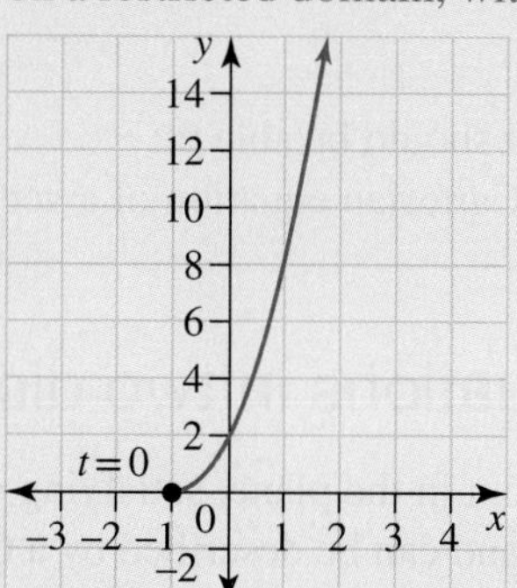

7. Although it is not required, a table of values can show the points as they are plotted and can give the direction of a particle as it moves along the curve.

t	$x = t - 1$	$y = 2t^2$
0	−1	0
1	0	2
2	1	8
3	2	18
4	3	32
5	4	50

TI | THINK

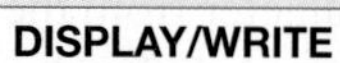

DISPLAY/WRITE

On a Graphs page, navigate to Graph Entry/Edit, Parametric and complete the entry as shown.

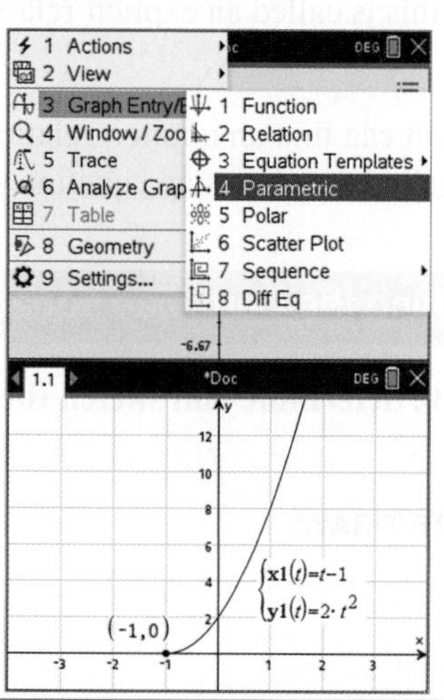

CASIO | THINK

DISPLAY/WRITE

On a Graph&Table screen, select the parametric mode and complete the entry as shown.

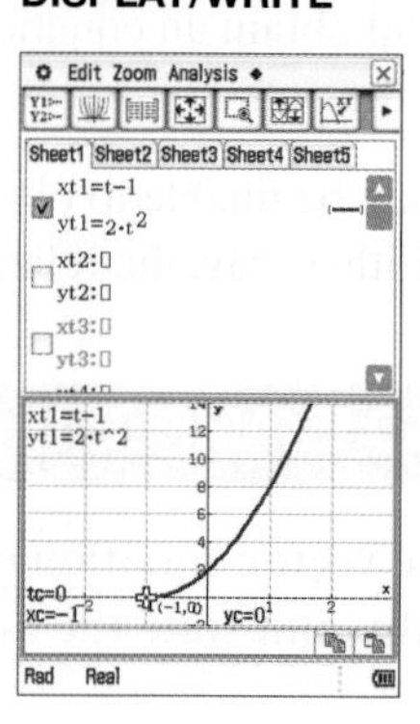

Eliminating the parameter

Eliminating the parameter is not always an easy task. Sometimes direct substitution will work; other times it is necessary to use trigonometric formulas and simple ingenuity.

WORKED EXAMPLE 20 Determining and sketching the Cartesian equation (2)

Given the vector equation $r(t) = \left(2 + 5\cos(t)\right)\underset{\sim}{i} + \left(4\sin(t) - 3\right)\underset{\sim}{j}$ for $t \geq 0$, determine and sketch the Cartesian equation of the path, and state the domain and range.

THINK

WRITE/DRAW

1. From the vector equation, $\underset{\sim}{r}(t) = x(t)\underset{\sim}{i} + y(t)\underset{\sim}{j}$.

$\underset{\sim}{r}(t) = \left(2 + 5\cos(t)\right)\underset{\sim}{i} + \left(4\sin(t) - 3\right)\underset{\sim}{j}$

2. State the parametric equations.

$x = 2 + 5\cos(t)$ (1)

$y = 4\sin(t) - 3$ (2)

3. Express the trigonometric ratios $\cos(t)$ and $\sin(t)$ in terms of x and y respectively.

$(1) \Rightarrow \cos(t) = \dfrac{x-2}{5}$

$(2) \Rightarrow \sin(t) = \dfrac{y+3}{4}$

4. Eliminate the parameter to determine the Cartesian equation of the path. In this case, the expression is given as an implicit equation.

Since $\cos^2(t) + \sin^2(t) = 1$, it follows that

$$\frac{(x-2)^2}{25} + \frac{(y+3)^2}{16} = 1$$

This is an ellipse, with centre at $(2, -3)$ and semi-major and semi-minor axes 5 and 4.

5. Determine the domain.

Since $-1 \leq \cos(t) \leq 1$, it follows from the parametric equation $x(t) = 2 + 5\cos(t)$ that the domain is $-3 \leq x \leq 7$, that is $[-3, 7]$.

6. Determine the range.

Since $-1 \leq \sin(t) \leq 1$, it follows from the parametric equation $y(t) = 4\sin(t) - 3$ that the range is $-7 \leq y \leq 1$, that is $[-7, 1]$.

7. The graph is the whole ellipse. The exact ordinates of the x- and y-intercepts are not required in this case.

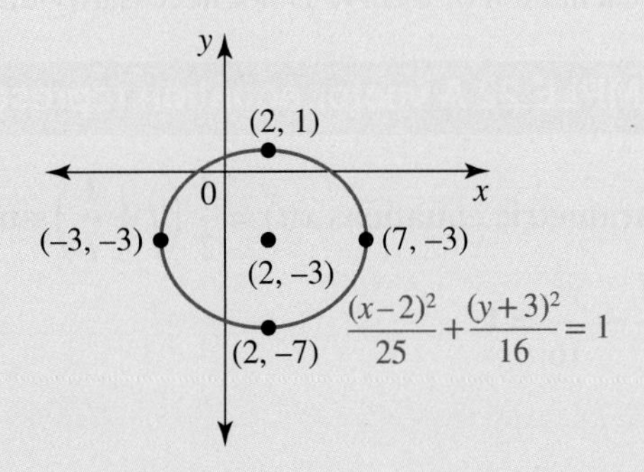

WORKED EXAMPLE 21 Determining and sketching the Cartesian equation (3)

Given the vector equation $\underset{\sim}{r}(t) = 3\sec(2t)\underset{\sim}{i} + 4\tan(2t)\underset{\sim}{j}$ for $t \geq 0$, determine and sketch the Cartesian equation of the path, and state the domain and range.

THINK

WRITE/DRAW

1. From the vector equation, $\underset{\sim}{r}(t) = x(t)\underset{\sim}{i} + y(t)\underset{\sim}{j}$.

$\underset{\sim}{r}(t) = 3\sec(2t)\underset{\sim}{i} + 4\tan(2t)\underset{\sim}{j}$

2. State the parametric equations.

$x = 3\sec(2t) \quad (1)$

$y = 4\tan(2t) \quad (2)$

3. Express the trigonometric ratios $\sec(2t)$ and $\tan(2t)$ in terms of x and y respectively.

$(1) \Rightarrow \sec(2t) = \dfrac{x}{3}$

$(2) \Rightarrow \tan(2t) = \dfrac{y}{4}$

4. Eliminate the parameter using an appropriate trigonometric identity to determine the Cartesian equation of the path. In this case, the expression is given as an implicit equation.

Since $\sec^2(2t) - \tan^2(2t) = 1$, it follows that

$$\frac{x^2}{9} - \frac{y^2}{16} = 1.$$

This is a hyperbola with centre at the origin. It has asymptotes when $\dfrac{x^2}{9} - \dfrac{y^2}{16} = 0$; that is, when $y = \pm\dfrac{4x}{3}$.

5. Determine the domain.

It follows from the parametric equation $x(t) = 3\sec(2t)$ that the domain is $(-\infty, -3] \cup [3, \infty)$.

6. Determine the range.	It follows from the parametric equation $y(t) = 4\tan(2t)$ that the range is R.
7. The graph is the whole hyperbola.	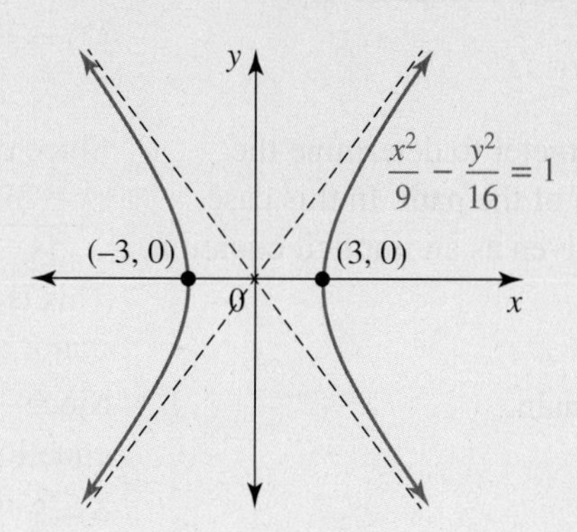

Parametric representation

The parametric representation of a curve is not necessarily unique.

WORKED EXAMPLE 22 A vector equation representing a given Cartesian equation

Show that the parametric equations $x(t) = \frac{3}{2}\left(t + \frac{1}{t}\right)$ and $y(t) = 2\left(t - \frac{1}{t}\right)$ where $t \in R\backslash\{0\}$ represent the hyperbola $\frac{x^2}{9} - \frac{y^2}{16} = 1$.

THINK	WRITE
1. State the parametric equations.	$x = \frac{3}{2}\left(t + \frac{1}{t}\right) \quad (1)$ $y = 2\left(t - \frac{1}{t}\right) \quad (2)$
2. Express the equations in a form to eliminate the parameter.	$\frac{2x}{3} = t + \frac{1}{t} \quad (1)$ $\frac{y}{2} = t - \frac{1}{t} \quad (2)$
3. Square both equations.	$\frac{4x^2}{9} = t^2 + 2 + \frac{1}{t^2} \quad (1)$ $\frac{y^2}{4} = t^2 - 2 + \frac{1}{t^2} \quad (2)$
4. Subtract the equations to eliminate the parameter.	$(1) - (2)$: $\frac{4x^2}{9} - \frac{y^2}{4} = 4$
5. Divide by 4.	$\frac{x^2}{9} - \frac{y^2}{16} = 1$ This gives the hyperbola as required.

Note: Graphing calculators and CAS calculators can draw the Cartesian equation of the path from the two parametric equations, even if the parameter cannot be eliminated.

3.6 Exercise

Students, these questions are even better in jacPLUS

Receive immediate feedback and access sample responses

Access additional questions

Track your results and progress

Find all this and MORE in jacPLUS

Technology free

1. **WE19** Given the vector equation $\underset{\sim}{r}(t) = (t+1)\underset{\sim}{i} + (t-1)^2\underset{\sim}{j}$ for $t \geq 0$, determine and sketch the Cartesian equation of the path, and state the domain and range.
2. Given the vector equation $\underset{\sim}{r}(t) = \sqrt{t}\underset{\sim}{i} + (2t+3)\underset{\sim}{j}$ for $t \geq 0$, determine and sketch the Cartesian equation of the path, and state the domain and range.
3. **WE20** Given the vector equation $\underset{\sim}{r}(t) = 3\cos(t)\underset{\sim}{i} + 4\sin(t)\underset{\sim}{j}$ for $t \geq 0$, determine and sketch the Cartesian equation of the path, and state the domain and range.
4. Given the vector equation $\underset{\sim}{r}(t) = \left(5 - 2\cos(t)\right)\underset{\sim}{i} + \left(3\sin(t) - 4\right)\underset{\sim}{j}$, for $t \geq 0$, determine and sketch the Cartesian equation of the path, and state the domain and range.
5. **WE21** Given the vector equation $\underset{\sim}{r}(t) = 5\sec(2t)\underset{\sim}{i} + 3\tan(2t)\underset{\sim}{j}$ for $t \geq 0$, determine and sketch the Cartesian equation of the path, and state the domain and range.
6. Given the vector equation $\underset{\sim}{r}(t) = 4\cot\left(\frac{t}{2}\right)\underset{\sim}{i} + 3\operatorname{cosec}\left(\frac{t}{2}\right)\underset{\sim}{j}$ for $t > 0$, determine and sketch the Cartesian equation of the path, and state the domain and range.
7. **WE22** Show that the parametric equations $x = \frac{5}{2}\left(t + \frac{1}{t}\right)$ and $y = \frac{3}{2}\left(t - \frac{1}{t}\right)$ where $t \in R\backslash\{0\}$ represent the hyperbola $\frac{x^2}{25} - \frac{y^2}{9} = 1$.
8. Show that the parametric equations $x = \frac{6t}{1+t^2}$ and $y = \frac{3(1-t^2)}{1+t^2}$ represent the circle $x^2 + y^2 = 9$.
9. Determine and sketch the Cartesian equation of the path for each of the following vector equations, and state the domain and range.
   a. $\underset{\sim}{r}(t) = 2t\underset{\sim}{i} + 4t^2\underset{\sim}{j}$ for $t \geq 0$
   b. $\underset{\sim}{r}(t) = (t-1)\underset{\sim}{i} + 3t\underset{\sim}{j}$ for $t \geq 0$
   c. $\underset{\sim}{r}(t) = 2t\underset{\sim}{i} + 8t^3\underset{\sim}{j}$ for $t \geq 0$
10. Determine and sketch the Cartesian equation of the path for each of the following vector equations, and state the domain and range.
    a. $\underset{\sim}{r}(t) = 2t\underset{\sim}{i} + \frac{1}{t}\underset{\sim}{j}$ for $t > 0$
    b. $\underset{\sim}{r}(t) = 2t\underset{\sim}{i} + (t^2 - 4t)\underset{\sim}{j}$ for $t \geq 0$
    c. $\underset{\sim}{r}(t) = \left(t + \frac{1}{t}\right)\underset{\sim}{i} + \left(t - \frac{1}{t}\right)\underset{\sim}{j}$ for $t > 0$
11. Determine and sketch the Cartesian equation of the path for each of the following vector equations, and state the domain and range.
    a. $\underset{\sim}{r}(t) = e^{-2t}\underset{\sim}{i} + e^{2t}\underset{\sim}{j}$ for $t \geq 0$
    b. $\underset{\sim}{r}(t) = e^{-t}\underset{\sim}{i} + \left(2 + e^{2t}\right)\underset{\sim}{j}$ for $t \geq 0$
    c. $\underset{\sim}{r}(t) = e^{t}\underset{\sim}{i} + \left(2 + e^{2t}\right)\underset{\sim}{j}$ for $t \geq 0$

12. Determine and sketch the Cartesian equation of the path for each of the following vector equations, and state the domain and range.

a. $\underset{\sim}{r}(t) = 3\cos(t)\underset{\sim}{i} + 3\sin(t)\underset{\sim}{j}$ for $t \geq 0$

b. $\underset{\sim}{r}(t) = 4\cos(t)\underset{\sim}{i} + 3\sin(t)\underset{\sim}{j}$ for $t \geq 0$

c. $\underset{\sim}{r}(t) = 4\sec(t)\underset{\sim}{i} + 3\tan(t)\underset{\sim}{j}$ for $t \geq 0$

13. Determine and sketch the Cartesian equation of the path for each of the following vector equations, and state the domain and range.

a. $\underset{\sim}{r}(t) = \left(1 + 3\cos(t)\right)\underset{\sim}{i} + \left(3\sin(t) - 2\right)\underset{\sim}{j}$ for $t \geq 0$

b. $\underset{\sim}{r}(t) = \left(4 + 3\cos(t)\right)\underset{\sim}{i} + \left(2\sin(t) - 3\right)\underset{\sim}{j}$ for $t \geq 0$

c. $\underset{\sim}{r}(t) = \left(2 - 3\sec(t)\right)\underset{\sim}{i} + \left(5\tan(t) - 4\right)\underset{\sim}{j}$ for $t \geq 0$

Technology active

14. Determine and sketch the Cartesian equation of the path for each of the following vector equations, and state the domain and range.

a. $\underset{\sim}{r}(t) = \cos^2(t)\underset{\sim}{i} + \sin^2(t)\underset{\sim}{j}$ for $t \geq 0$

b. $\underset{\sim}{r}(t) = \cos^3(t)\underset{\sim}{i} + \sin^3(t)\underset{\sim}{j}$ for $t \geq 0$

c. $\underset{\sim}{r}(t) = \cos^4(t)\underset{\sim}{i} + \sin^4(t)\underset{\sim}{j}$ for $t \geq 0$

15. Determine and sketch the Cartesian equation of the path for each of the following vector equations, and state the domain and range.

a. $\underset{\sim}{r}(t) = \cos^2(t)\underset{\sim}{i} + \cos(2t)\underset{\sim}{j}$ for $t \geq 0$

b. $\underset{\sim}{r}(t) = \cos(t)\underset{\sim}{i} + \cos(2t)\underset{\sim}{j}$ for $t \geq 0$

c. $\underset{\sim}{r}(t) = \sin(t)\underset{\sim}{i} + \sin(2t)\underset{\sim}{j}$ for $t \geq 0$

16. If a and b are positive real numbers, show that the following vector equations give the same Cartesian equation.

a. $\underset{\sim}{r}(t) = a\cos(t)\underset{\sim}{i} + a\sin(t)\underset{\sim}{j}$ and $\underset{\sim}{r}(t) = \left(\dfrac{2at}{1+t^2}\right)\underset{\sim}{i} + \left(\dfrac{a(1-t^2)}{1+t^2}\right)\underset{\sim}{j}$

b. $\underset{\sim}{r}(t) = a\cos(t)\underset{\sim}{i} + b\sin(t)\underset{\sim}{j}$ and $\underset{\sim}{r}(t) = \left(\dfrac{2at}{1+t^2}\right)\underset{\sim}{i} + \left(\dfrac{b(1-t^2)}{1+t^2}\right)\underset{\sim}{j}$

c. $\underset{\sim}{r}(t) = a\sec(t)\underset{\sim}{i} + b\tan(t)\underset{\sim}{j}$, $\underset{\sim}{r}(t) = \dfrac{a}{2}\left(t + \dfrac{1}{t}\right)\underset{\sim}{i} + \dfrac{b}{2}\left(t - \dfrac{1}{t}\right)\underset{\sim}{j}$ and $\underset{\sim}{r}(t) = \dfrac{a}{2}(e^{2t} + e^{-2t})\underset{\sim}{i} + \dfrac{b}{2}(e^{2t} - e^{-2t})\underset{\sim}{j}$

17. The position vector of a moving particle is given by $\underset{\sim}{r}(t) = 2\sin(t)\underset{\sim}{i} + 2\sin(t)\tan(t)\underset{\sim}{j}$, for $t > 0$. Show that the particle moves along the curve $y = \dfrac{x^2}{\sqrt{4 - x^2}}$.

18. A curve called the Witch of Agnesi is defined by the parametric equations $x = at$ and $y = \dfrac{a}{1+t^2}$. Show that Cartesian equation is given by $y = \dfrac{a^3}{a^2 + x^2}$.

19. a. Show that $\cos(3A) = 4\cos^3(A) - 3\cos(A)$.

b. A curve is defined by the parametric equations $x = 2\cos(t)$ and $y = 2\cos(3t)$. Determine the Cartesian equation of the curve.

20. a. Show that $\cos(4A) = 8\cos^4(A) - 8\cos^2(A) + 1$.

b. A curve is defined by the parametric equations $x = 2\cos^2(t)$ and $y = \cos(4t)$. Determine the Cartesian equation of the curve.

21. The position vector of a moving particle is given by $\underset{\sim}{r}(t) = 2\tan(t)\underset{\sim}{i} + 2\,\text{cosec}(2t)\underset{\sim}{j}$ for $t > 0$. Show that the particle moves along the curve $y = \dfrac{x^2 + 4}{2x}$.

22. A curve is defined by the parametric equations $x = \cos(t)\left(\sec(t) + a\cos(t)\right)$ and $y = \sin(t)\left(\sec(t) + a\cos(t)\right)$ for $t \in [0, 2\pi]$. Show that the curve satisfies the implicit equation $(x-1)(x^2+y^2) - ax^2 = 0$.

23. A curve is defined by the parametric equations $x = 4\cos(t)$ and $y = \dfrac{4\sin^2(t)}{2+\sin(t)}$ for $t \in [0, 2\pi]$. Show that the curve satisfies the implicit equation $y^2(16 - x^2) = (x^2 + 8y - 16)^2$.

24. Given the vector equations, sketch the equation of the path of the following.
 a. $\underset{\sim}{r}(t) = \cos(2t)\underset{\sim}{i} + \sin(4t)\underset{\sim}{j}$ for $t \geq 0$
 b. $\underset{\sim}{r}(t) = \cos(2t)\underset{\sim}{i} + \sin(6t)\underset{\sim}{j}$ for $t \geq 0$

25. Given the vector equations, sketch the equation of the path of the following.
 a. $\underset{\sim}{r}(t) = \cos(3t)\underset{\sim}{i} + \sin(t)\underset{\sim}{j}$ for $t \geq 0$
 b. $\underset{\sim}{r}(t) = \cos(3t)\underset{\sim}{i} + \sin(2t)\underset{\sim}{j}$ for $t \geq 0$

26. Given the vector equations, sketch the equation of the path of the following.
 a. The cycloid $\underset{\sim}{r}(t) = 2\left(t - \sin(t)\right)\underset{\sim}{i} + 2\left(1 - \cos(t)\right)\underset{\sim}{j}$ for $t \geq 0$
 b. The cardioid $\underset{\sim}{r}(t) = 2\cos(t)\left(1 + \cos(t)\right)\underset{\sim}{i} + 2\sin(t)\left(1 + \cos(t)\right)\underset{\sim}{j}$ for $t \geq 0$

27. Given the vector equations, sketch the equation of the path of the deltoid $\underset{\sim}{r}(t) = \left(2\cos(t) + \cos(2t)\right)\underset{\sim}{i} + \left(2\sin(t) - \sin(2t)\right)\underset{\sim}{j}$ for $t \geq 0$.

28. Given the vector equations, sketch the equation of the path of the hypercycloid $\underset{\sim}{r}(t) = \left(5\cos(t) + \cos(5t)\right)\underset{\sim}{i} + \left(5\sin(t) - \sin(5t)\right)\underset{\sim}{j}$.

3.6 Exam questions

Question 1 (1 mark) TECH-ACTIVE

Source: VCE 2017 Specialist Mathematics Exam 2, Section A, Q12; © VCAA.

MC Let $\underset{\sim}{r}(t) = \left(1 - \sqrt{a}\sin(t)\right)\underset{\sim}{i} + \left(1 - \dfrac{1}{b}\cos(t)\right)\underset{\sim}{j}$ for $t \geq 0$ and $a, b \in R^+$ be the path of a particle moving in the Cartesian plane.

The path of the particle will always be a circle if

A. $ab^2 = 1$
B. $a^2b = 1$
C. $ab^2 = 1$
D. $ab = 1$
E. $a^2b = 1$

Question 2 (1 mark) TECH-ACTIVE

Source: VCE 2015 Specialist Mathematics Exam 2, Section A, Q1; © VCAA.

MC The ellipse $\dfrac{(x-2)^2}{9} + \dfrac{(y-3)^2}{4} = 1$ can be expressed in parametric form as

A. $x = 2 + 3t$ and $y = 3 + 2\sqrt{1+t^2}$
B. $x = 2 + 3\sec(t)$ and $y = 3 + 2\tan(t)$
C. $x = 2 + 9\cos(t)$ and $y = 3 + 4\sin(t)$
D. $x = 3 + 2\cos(t)$ and $y = 2 + 3\sin(t)$
E. $x = 2 + 3\cos(t)$ and $y = 3 + 2\sin(t)$

Question 3 (3 marks) TECH-FREE

Source: adapted from: VCE 2014 Specialist Mathematics Exam 1, Q2; © VCAA.

The position vector of a particle at time $t \geq 0$ is given by

$$\underset{\sim}{r}(t) = (t-2)\underset{\sim}{i} + \left(t^2 - 4t + 1\right)\underset{\sim}{j}$$

a. Show that the cartesian equation of the path followed by the particle is $y = x^2 - 3$. **(1 mark)**
b. Sketch the path followed by the particle on the axes below, labelling all important features. **(2 marks)**

More exam questions are available online.

3.7 Review

3.7.1 Summary

doc-37057

Hey students! Now that it's time to revise this topic, go online to:

Access the topic summary  **Review your results** **Watch teacher-led videos**  **Practise VCAA exam questions**

Find all this and MORE in jacPLUS

3.7 Exercise

Technology free: short answer

1. Consider the points $A(4, -2, 3)$ and $B(10, 12, -3)$.
   a. Calculate the distance between the points.
   b. Determine the angle that the vector $\overrightarrow{AB}$ makes with the z-axis.
   c. Determine the coordinates of the point P which is the midpoint of AB.
   d. Determine the coordinates of the point Q which divides the line segment AB in the ratio $1:2$.
2. Given the points $A(8, -3, 5)$, $B(2, -1, 3)$ and $C(5, 4, -1)$, show that ABC forms a right angled triangle at B, and hence calculate the area of the triangle.
3. Consider the vectors $\underset{\sim}{a} = 4\underset{\sim}{i} - 3\underset{\sim}{j} - 8\underset{\sim}{k}$ and $\underset{\sim}{b} = -2\underset{\sim}{i} + y\underset{\sim}{j} + 4\underset{\sim}{k}$.
   Determine the value of the scalar y if:
   a. the length of the vector $\underset{\sim}{b}$ is 5
   b. $3\underset{\sim}{a} + 2\underset{\sim}{b}$ is parallel to the xz plane
   c. $\underset{\sim}{a}$ is parallel to the vector $\underset{\sim}{b}$
   d. $\underset{\sim}{a}$ is perpendicular to the vector $\underset{\sim}{b}$.
4. Consider the points $A(-2, 2, 1)$ and $B(1, 4, 3)$.
   a. Determine the vector component of $\overrightarrow{OB}$ perpendicular to $\overrightarrow{OA}$.
   b. Hence, determine the closest distance of the point B to the vector $\overrightarrow{OA}$.
5. P, Q and R are the three points given by $P(x, 3, -1)$, $Q(2, -1, 5)$, $R(-1, 1, 2)$.
   Determine the value(s) of x if:
   a. $\overrightarrow{PQ}$ is parallel to the yz plane
   b. $\overrightarrow{PQ}$ is parallel to $\overrightarrow{QR}$
   c. $\overrightarrow{PQ}$ is perpendicular to $\overrightarrow{QR}$
   d. the vector $\overrightarrow{PQ}$ has a length of 9
   e. the vector $\overrightarrow{PQ}$ makes an angle of $\cos^{-1}\left(\frac{-1}{\sqrt{13}}\right)$ with the x-axis.
6. Prove that the sum of the squares of the lengths of the diagonals of a parallelogram is equal to the sum of the squares of the lengths of the sides.

7. ABC is a right-angled triangle at B. Let $\underset{\sim}{a} = \overrightarrow{OA}$, $\underset{\sim}{b} = \overrightarrow{OB}$ and $\underset{\sim}{c} = \overrightarrow{OC}$. Show that:

a. $|\underset{\sim}{b}|^2 = \underset{\sim}{a} \cdot \underset{\sim}{b} + \underset{\sim}{b} \cdot \underset{\sim}{c} - \underset{\sim}{c} \cdot \underset{\sim}{a}$

b. if the angle $CAB = \theta$, use the fact that $\overrightarrow{CB} = \overrightarrow{AB} - \overrightarrow{AC}$ to prove the cosine rule.

8. A particle moves along the vector equation $\underset{\sim}{r}(t) = (2 + 3\cos(t))\underset{\sim}{i} + (4 + 2\sin(t))\underset{\sim}{j}$, for $t \geq 0$, determine and sketch the Cartesian equation of the path, stating the domain and range.

Technology active: multiple choice

9. **MC** A vector which is **not** a unit vector is

A. $\frac{1}{\sqrt{3}}(\underset{\sim}{i} + \underset{\sim}{j} + \underset{\sim}{k})$

B. $\frac{1}{3\sqrt{2}}\left(\sqrt{5}\underset{\sim}{i} + \sqrt{6}\underset{\sim}{j} + \sqrt{7}\underset{\sim}{k}\right)$

C. $\frac{1}{13}(3\underset{\sim}{i} + 4\underset{\sim}{j} + 12\underset{\sim}{k})$

D. $\frac{1}{3}\left(2\underset{\sim}{i} + \sqrt{3}\underset{\sim}{j} + 2\underset{\sim}{k}\right)$

E. $\frac{1}{\sqrt{29}}(2\underset{\sim}{i} + 3\underset{\sim}{j} + 4\underset{\sim}{k})$

10. **MC** A vector in the opposite direction to $-2\underset{\sim}{i} + \underset{\sim}{j} - 2\underset{\sim}{k}$ with magnitude 9 is

A. $9(2\underset{\sim}{i} - \underset{\sim}{j} + 2\underset{\sim}{k})$ B. $3(2\underset{\sim}{i} - \underset{\sim}{j} + 2\underset{\sim}{k})$ C. $9(-2\underset{\sim}{i} + \underset{\sim}{j} - 2\underset{\sim}{k})$ D. $3(-2\underset{\sim}{i} + \underset{\sim}{j} - 2\underset{\sim}{k})$ E. $\frac{1}{3}(2\underset{\sim}{i} - \underset{\sim}{j} + 2\underset{\sim}{k})$

11. **MC** If $\underset{\sim}{a} = \underset{\sim}{i} - m\underset{\sim}{j} - 2\underset{\sim}{k}$ and $\underset{\sim}{b} = -2\underset{\sim}{i} + \underset{\sim}{j} - n\underset{\sim}{k}$ and $\underset{\sim}{a}$ and $\underset{\sim}{b}$ are perpendicular, then it is possible that

A. $m = 2$ and $n = 2$

B. $m = -2$ and $n = -2$

C. $m = -2$ and $n = -4$

D. $m = -2$ and $n = 2$

E. $m = 2$ and $n = -2$

12. **MC** If the length of the vector $\underset{\sim}{i} - \underset{\sim}{j} + t\underset{\sim}{k}$ is 4 then the possible value(s) of t are

A. $\pm\sqrt{2}$ B. ± 2 C. 2 D. $\pm\sqrt{14}$ E. $\sqrt{2}$

13. **MC** If m and n are real constants, then the two vectors $n\underset{\sim}{i} + \sqrt{m}\underset{\sim}{j} - n\underset{\sim}{k}$ and $4\underset{\sim}{i} - 2\sqrt{m}\underset{\sim}{j} - \sqrt{m}\underset{\sim}{k}$ are parallel when

A. $n = -1$ and $m = 1$

B. $n = -2$ and $m = 4$

C. $n = -2$ and $m = 16$

D. $n = 2$ and $m = -4$

E. $n = 8$

14. **MC** If the vector $\frac{1}{2}(-\underset{\sim}{i} + \underset{\sim}{j} + z\underset{\sim}{k})$ makes an angle of 135° with the positive z-axis, then the value(s) of z are

A. $-\sqrt{2}$ B. $\pm\sqrt{2}$ C. $\sqrt{2}$ D. $-\frac{\sqrt{2}}{2}$ E. $\frac{\sqrt{2}}{2}$

15. **MC** A, B and C are points with an origin O and given the non-zero vectors $\overrightarrow{OA} = \underset{\sim}{a}$, $\overrightarrow{OB} = \underset{\sim}{b}$ and $\overrightarrow{OC} = \underset{\sim}{c}$. If $\underset{\sim}{c} - \underset{\sim}{b} = t(\underset{\sim}{a} - \underset{\sim}{b})$ where t is a real non-zero constant, then the statement which must be true is

A. $\underset{\sim}{b}$ and $\underset{\sim}{c}$ are linearly independent.

B. $\underset{\sim}{b}$ and $\underset{\sim}{c}$ are linearly dependent.

C. A, B and C are collinear and $\underset{\sim}{a}$, $\underset{\sim}{b}$ and $\underset{\sim}{c}$ are linearly dependent.

D. A, B and C are collinear and $\underset{\sim}{a}$, $\underset{\sim}{b}$ and $\underset{\sim}{c}$ are linearly independent.

E. B and C are collinear and $\underset{\sim}{b}$ and $\underset{\sim}{c}$ are linearly dependent.

16. **MC** To prove that the triangle OAB is isosceles, with $\overrightarrow{OA} = \underset{\sim}{a}$ and $\overrightarrow{OB} = \underset{\sim}{b}$ it is sufficient to prove that

A. $\underset{\sim}{a} = \underset{\sim}{b}$
B. $\underset{\sim}{a} . \underset{\sim}{b} = 0$
C. $\frac{1}{2}(\underset{\sim}{a} + \underset{\sim}{b}) \cdot (\underset{\sim}{a} + \underset{\sim}{b}) = 0$
D. $\frac{1}{2}(\underset{\sim}{b} - \underset{\sim}{a}) \cdot (\underset{\sim}{b} - \underset{\sim}{a}) = 0$
E. $\frac{1}{2}(\underset{\sim}{a} + \underset{\sim}{b}) \cdot (\underset{\sim}{b} - \underset{\sim}{a}) = 0$

17. **MC** Two vectors $\underset{\sim}{a}$ and $\underset{\sim}{b}$ are such that $\underset{\sim}{a} \cdot \underset{\sim}{b} = 0$, $\underset{\sim}{a} \cdot \underset{\sim}{a} = 1$ and $\underset{\sim}{b} \cdot \underset{\sim}{b} = 4$. The statement which is **true** is

A. $\underset{\sim}{a}$ is parallel to the vector $\underset{\sim}{b}$ and $|\underset{\sim}{a} + \underset{\sim}{b}| = \sqrt{3}$.
B. $\underset{\sim}{a}$ is parallel to the vector $\underset{\sim}{b}$ and the length of the vector $\underset{\sim}{b}$ is 2.
C. $\underset{\sim}{a}$ is a unit vector and $|\underset{\sim}{a} + \underset{\sim}{b}| = 5$.
D. $\underset{\sim}{a}$ is perpendicular to the vector $\underset{\sim}{b}$ and $|\underset{\sim}{a} + \underset{\sim}{b}| = 5$.
E. $\underset{\sim}{a}$ is perpendicular to the vector $\underset{\sim}{b}$ and $|\underset{\sim}{a} + \underset{\sim}{b}| = \sqrt{5}$.

18. **MC** If the vector equation of a moving particle is given by $\underset{\sim}{r}(t) = \left(t + \frac{1}{t}\right)\underset{\sim}{i} + \left(t^2 + \frac{1}{t^2}\right)\underset{\sim}{j}$, $t > 0$ then the particle moves as part of a

A. straight line path.
B. parabolic path.
C. circular path.
D. elliptical path.
E. hyperbolic path.

Technology active: extended response

19. The position vector of a moving particle is given by $\underset{\sim}{r}(t) = \frac{3}{2}\left(e^{2t} + e^{-2t}\right)\underset{\sim}{i} + \frac{5}{2}\left(e^{2t} - e^{-2t}\right)\underset{\sim}{j}$, for $t \geq 0$. Determine and sketch the Cartesian equation of the path, stating the domain and range.

20. A ship moves 700 metres due west, then turns and moves 4 km in a south direction, then finally turns again and moves in a direction S50°E for a further 3 km. If $\underset{\sim}{i}$, $\underset{\sim}{j}$ and $\underset{\sim}{k}$ represent unit vectors of one kilometer in the directions of east, north and vertically upwards respectively, determine the position vector of the mast of the ship which is 50 metres above the level of the ship (giving values rounded to 2 decimal places where necessary) and calculate the distance, to the nearest metre, of the mast of the ship from its initial point.

3.7 Exam questions

Question 1 (4 marks) TECH-FREE

Source: VCE 2020 Specialist Mathematics Exam 1, Q5; © VCAA.

Let $\underset{\sim}{a} = 2\underset{\sim}{i} - 3\underset{\sim}{j} + \underset{\sim}{k}$ and $\underset{\sim}{b} = \underset{\sim}{i} + m\underset{\sim}{j} - \underset{\sim}{k}$, where m is an integer.

The vector resolute of a in the direction of b is $-\frac{11}{18}(\underset{\sim}{i} + m\underset{\sim}{j} - \underset{\sim}{k})$.

a. Find the value of m. **(3 marks)**
b. Find the component of $\underset{\sim}{a}$ that is perpendicular to $\underset{\sim}{b}$. **(1 mark)**

Question 2 (1 mark) TECH-ACTIVE

Source: *VCE 2020 Specialist Mathematics Exam 2, Section A, Q16; © VCAA.*

MC Let $\underset{\sim}{a} = \underset{\sim}{i} - 2\underset{\sim}{j} + 2\underset{\sim}{k}$ and $\underset{\sim}{b} = 2\underset{\sim}{i} - 4\underset{\sim}{j} + 4\underset{\sim}{k}$, where the acute angle between these vectors is θ.

The value of $\sin(2\theta)$ is

A. $\dfrac{1}{9}$ **B.** $\dfrac{4\sqrt{5}}{9}$ **C.** $\dfrac{4\sqrt{5}}{81}$ **D.** $\dfrac{8\sqrt{5}}{81}$ **E.** $\dfrac{2\sqrt{46}}{25}$

Question 3 (11 marks) TECH-ACTIVE

Source: *VCE 2019 Specialist Mathematics Exam 2, Section B, Q4; © VCAA.*

The base of a pyramid is the parallelogram $ABCD$ with vertices at points $A(2, -1, 3)$, $B(4, -2, 1)$, $C(a, b, c)$ and $D(4, 3, -1)$. The apex (top) of the pyramid is located at $P(4, -4, 9)$.

a. Find the values of a, b and c. **(2 marks)**

b. Find the cosine of the angle between the vectors $\overrightarrow{AB}$ and $\overrightarrow{AD}$. **(2 marks)**

c. Find the area of the base of the pyramid. **(2 marks)**

d. Show that $6\underset{\sim}{i} + 2\underset{\sim}{j} + 5\underset{\sim}{k}$ is perpendicular to both $\overrightarrow{AB}$ and $\overrightarrow{AD}$, and hence find a unit vector that is perpendicular to the base of the pyramid. **(3 marks)**

e. Find the volume of the pyramid. **(2 marks)**

Question 4 (2 marks) TECH-FREE

Source: *VCE 2018 Specialist Mathematics Exam 1, Q9a; © VCAA.*

A curve is specified parametrically by $\underset{\sim}{r}(t) = \sec(t)\underset{\sim}{i} + \dfrac{\sqrt{2}}{2}\tan(t)\underset{\sim}{j},\ t \in R$.

Show that the cartesian equation of the curve is $x^2 - 2y^2 = 1$.

Question 5 (4 marks) TECH-FREE

Source: *VCE 2017 Specialist Mathematics Exam 1, Q5; © VCAA.*

Relative to a fixed origin, the points B, C and D are defined respectively by the position vectors $\underset{\sim}{b} = \underset{\sim}{i} - \underset{\sim}{j} + 2\underset{\sim}{k}$, $\underset{\sim}{c} = 2\underset{\sim}{i} - \underset{\sim}{j} + \underset{\sim}{k}$ and $\underset{\sim}{d} = a\underset{\sim}{i} - 2\underset{\sim}{j}$, where a is a real constant.

Given that the magnitude of angle BCD is $\dfrac{\pi}{3}$, find a.

More exam questions are available online.

Answers

Topic 3 Vectors

3.2 Vectors in two dimensions

3.2 Exercise

1. $3\overrightarrow{AB}$
2. $3\overrightarrow{OA}$
3. a. $3\overrightarrow{AC}$ b. $5\overrightarrow{OA}$
4. a. $6\overrightarrow{OB}$ b. $2\overrightarrow{OA}$
5. $\overrightarrow{AB} = 2\overrightarrow{BC}$
6. $\overrightarrow{PR} = 3\overrightarrow{PQ}$

7–10. Sample responses can be found in the worked solutions in the online resources.

11. a. $\underset{\sim}{a} = \overrightarrow{OA} = \overrightarrow{CB} = \overrightarrow{DE} = \overrightarrow{GF}$
    b. $\underset{\sim}{c} = \overrightarrow{OC} = \overrightarrow{AB} = \overrightarrow{EF} = \overrightarrow{DG}$
    c. $\underset{\sim}{d} = \overrightarrow{OD} = \overrightarrow{AE} = \overrightarrow{CG} = \overrightarrow{BF}$
    d. i. $\underset{\sim}{a} + \underset{\sim}{c}$ ii. $\underset{\sim}{c} - \underset{\sim}{d}$
    iii. $-(\underset{\sim}{a} + \underset{\sim}{c} + \underset{\sim}{d})$ iv. $\underset{\sim}{a} + \underset{\sim}{c} - \underset{\sim}{d}$

12–24. Sample responses can be found in the worked solutions in the online resources.

25. a–c. Sample responses can be found in the worked solutions in the online resources
    d. All are coincident at the centroid of the triangle.

26. a. $\frac{1}{4}(\underset{\sim}{a} + \underset{\sim}{b} + \underset{\sim}{c})$ b. $\frac{1}{4}(\underset{\sim}{a} + \underset{\sim}{b} + \underset{\sim}{c})$
    c. $\frac{1}{4}(\underset{\sim}{a} + \underset{\sim}{b} + \underset{\sim}{c})$ d. All are coincident.

3.2 Exam questions

Note: Mark allocations are available with the fully worked solutions online.

1. i. $\overrightarrow{AN} = \underset{\sim}{u} + \frac{1}{2}\underset{\sim}{v}$
   ii. $\overrightarrow{CM} = \frac{1}{2}(\underset{\sim}{v} - \underset{\sim}{u})$
   $\overrightarrow{BP} = -\underset{\sim}{v} - \frac{1}{2}\underset{\sim}{u}$
   iii. $\overrightarrow{AN} + \overrightarrow{CM} + \overrightarrow{BP} = \underset{\sim}{0}$
2. Sample responses can be found in the worked solutions in the online resources.
3. D

3.3 Vectors in three dimensions

3.3 Exercise

1. a. $2\underset{\sim}{i} - 2\underset{\sim}{j} - \underset{\sim}{k}$ b. $\frac{1}{3}(2\underset{\sim}{i} - 2\underset{\sim}{j} - \underset{\sim}{k})$
2. $\frac{1}{\sqrt{21}}(2\underset{\sim}{i} - 4\underset{\sim}{j} - \underset{\sim}{k})$
3. $\frac{1}{\sqrt{33}}(2\underset{\sim}{i} - 2\underset{\sim}{j} + 5\underset{\sim}{k})$
4. 13
5. 4
6. $x = -1, y = 2, z = -7$
7. a. $\pm 2\sqrt{5}$ b. $-\frac{5}{2}$
8. a. $\pm 2\sqrt{21}$ b. $-\frac{3}{2}$
9. 7
10. 7
11. a. $680.36\underset{\sim}{i} - 1615.68\underset{\sim}{j} + 150\underset{\sim}{k}$
    b. 1759.5 metres
12. a. $-2684.0\underset{\sim}{i} - 1879.4\underset{\sim}{j} + 800\underset{\sim}{k}$
    b. 3373 metres

13, 14. Sample responses can be found in the worked solutions in the online resources.

15. a. $2\underset{\sim}{i} - 4\underset{\sim}{j} + 4\underset{\sim}{k}$
    b. $6\underset{\sim}{i} - 8\underset{\sim}{j} - 24\underset{\sim}{k}$
    c. $10\underset{\sim}{i} + 6\underset{\sim}{j} - 8\underset{\sim}{k}$
16. a, b. Sample responses can be found in the worked solutions in the online resources.
    c. 4
17. a, b Sample responses can be found in the worked solutions in the online resources.
    c. $x = \frac{5}{2}, y = 6$
18. a. $4\underset{\sim}{i} + 2\underset{\sim}{j} + \underset{\sim}{k}$ b. $-3\underset{\sim}{i} + 3\underset{\sim}{j} + 2\underset{\sim}{k}$ c. $-2\underset{\sim}{i} + \underset{\sim}{j} + 3\underset{\sim}{k}$
19. a. $\underset{\sim}{c} = \frac{4}{3}\underset{\sim}{b} - \frac{1}{2}\underset{\sim}{a}$ b. 5 c. 6
20. $-2\underset{\sim}{i} - \underset{\sim}{j} + 0.5\underset{\sim}{k}$, 2.29 km
21. $-141.42\underset{\sim}{i} + 188.21\underset{\sim}{j} + 3.83\underset{\sim}{k}$, 235.45 m
22. $49.575\underset{\sim}{i} + 17.207\underset{\sim}{k}$, 52.48 km
23. 143.3°
24. $-\sqrt{3}$
25. a. $\frac{1}{\sqrt{21}}(-2\underset{\sim}{i} + 4\underset{\sim}{j} + \underset{\sim}{k})$ b. 115.9°
26. a. $\frac{1}{\sqrt{38}}(3\underset{\sim}{i} + 5\underset{\sim}{j} - 2\underset{\sim}{k})$ b. 35.8°
27. a. $\frac{1}{\sqrt{29}}(2\underset{\sim}{i} + 3\underset{\sim}{j} - 4\underset{\sim}{k})$ b. 138.0°
28. a. $\sqrt{62}$
    b. 150.5°
    c. $\frac{1}{\sqrt{11}}(-\underset{\sim}{i} + \underset{\sim}{j} + 3\underset{\sim}{k})$
29. a. -3 b. $\pm 2\sqrt{5}$
30. a. $\pm 3\sqrt{10}$ b. $-\frac{\sqrt{5}}{2}$
31. a. $\pm 2\sqrt{6}$ b. -2
32. $-1342.05\underset{\sim}{i} - 1159.03\underset{\sim}{j} + 2132.09\underset{\sim}{k}$, 2773.13 m
33. a. $\pm\sqrt{5}$ b. 60° c. 135°
34. a. $p = -4, q = -3, r = 5$
    b. $x = \pm\frac{2\sqrt{3}}{3}, y = \pm\frac{\sqrt{33}}{3}$

3.3 Exam questions

Note: Mark allocations are available with the fully worked solutions online.

1. $p \in R \backslash \{\pm\sqrt{5}\}$
2. E
3. a. $-\frac{13}{14}(\underset{\sim}{i} - 2\underset{\sim}{j} + 3\underset{\sim}{k})$ b. $d = 1$

3.4 Scalar product and applications

3.4 Exercise

1. -12
2. 15
3. a. 12 b. -30 c. -21
4. -21
5. 7
6. a. -16 b. 3 c. 0
7. -1
8. -1, 5
9. a. $-\frac{5}{3}$ b. 3 c. $\pm 2\sqrt{5}$
10. a. -3 b. 39 c. $\pm\sqrt{107}$
11. a. $\frac{3}{2}$ b. $-\frac{49}{4}$ c. $\frac{3 \pm 2\sqrt{3}}{2}$
12. 54.74°
13. $-\sqrt{6}$
14. $\sqrt{22}$
15. 6
16. a, b. Sample responses can be found in the worked solutions in the online resources.
    c. The dot product is distributive over addition and subtraction.
    d. It is meaningless; the dot product of a scalar and vector cannot be found.
17. a. 5
    b. $\frac{5}{3}(\underset{\sim}{i} - 2\underset{\sim}{j} - 2\underset{\sim}{k})$
    c. $\frac{1}{3}(10\underset{\sim}{i} + \underset{\sim}{j} + 4\underset{\sim}{k})$
18. a. $\frac{2}{7}(-\underset{\sim}{i} + 4\underset{\sim}{j} + 2\underset{\sim}{k})$
    b. $\frac{8}{7}(2\underset{\sim}{i} - \underset{\sim}{j} + 3\underset{\sim}{k})$
19. a. -3 b. $\frac{13}{3}$ c. $3 \pm \sqrt{3}$ d. 3
20. 5, -3
21. a. 5, $\frac{5}{7}$ b. 2 c. 3
22. a. $\sqrt{37}$ b. $\sqrt{13}$ c. $\sqrt{73}$
23. a. $\sqrt{95}$ b. $\sqrt{119}$ c. $\sqrt{1043}$
24. a. $\frac{1}{\sqrt{26}}(3\underset{\sim}{i} - \underset{\sim}{j} - 4\underset{\sim}{k})$ b. $\frac{6}{\sqrt{26}}$
    c. $\frac{3}{13}(3\underset{\sim}{i} - \underset{\sim}{j} - 4\underset{\sim}{k})$ d. $\frac{1}{13}(17\underset{\sim}{i} - 49\underset{\sim}{j} + 25\underset{\sim}{k})$
    e. 75.12°
25. a. $\frac{1}{3}(2\underset{\sim}{i} + \underset{\sim}{j} - 2\underset{\sim}{k})$ b. 6
    c. $2(2\underset{\sim}{i} + \underset{\sim}{j} - 2\underset{\sim}{k})$ d. $-\underset{\sim}{i} - \underset{\sim}{k}$
    e. 13.26°
26. a. $\underset{\sim}{i} - 2\underset{\sim}{j} + 3\underset{\sim}{k}, 2\underset{\sim}{i} - 2\underset{\sim}{j} - 2\underset{\sim}{k}$
    b. 42.79°
27. a. $\frac{1}{\sqrt{11}}(3\underset{\sim}{i} - \underset{\sim}{j} - \underset{\sim}{k})$ b. $-\frac{13}{11}(3\underset{\sim}{i} - \underset{\sim}{j} - \underset{\sim}{k})$
    c. $\frac{1}{11}(-5\underset{\sim}{i} + 9\underset{\sim}{j} - 24\underset{\sim}{k})$ d. 148.8°
28. Sample responses can be found in the worked solutions in the online resources.
29. 33
30. a. 0
    b. $\underset{\sim}{0}$
    c. It is possible that $\underset{\sim}{a} = \underset{\sim}{c}$. It is possible that $\underset{\sim}{b}$ is perpendicular to $\underset{\sim}{a} - \underset{\sim}{c}$.
31. a. Sample responses can be found in the worked solutions in the online resources.
    b. $|\underset{\sim}{u}| = \sqrt{5}$ and $|\underset{\sim}{v}| = \sqrt{10}$ or $|\underset{\sim}{u}| = \sqrt{10}$ and $|\underset{\sim}{v}| = \sqrt{5}$
32. a. Sample responses can be found in the worked solutions in the online resources.
    b. $\frac{1}{3}(7\underset{\sim}{i} - 5\underset{\sim}{j} - 2\underset{\sim}{k})$
    c. Sample responses can be found in the worked solutions in the online resources.
    d. $x^2 + y^2 + z^2 = 350$
    $-8x + 2y + 6z = -26$
    $-6x + 8y - 2z = -26$
    e. $(13, 9, 10)$ or $\left(-\frac{25}{3}, -\frac{37}{3}, -\frac{34}{3}\right)$
    f. $\frac{32\sqrt{3}}{3}$
    g. 80.95°
33. a. Sample responses can be found in the worked solutions in the online resources.
    b. $\left(-\frac{1}{2}, \frac{1}{2}, 2\right)$
    c. Sample responses can be found in the worked solutions in the online resources.
    d. $x^2 + y^2 + z^2 = \frac{891}{2}$
    $4x + 2y - 4z = -9$
    $2x - 2y - 8z = -18$
    $2x + 4y + 4z = 9$

e. $\left(-\frac{29}{2}, \frac{29}{2}, -5\right)$ or $\left(\frac{27}{2}, -\frac{27}{2}, 9\right)$, 21

f. 84.23°

34. a. $\frac{\pm\sqrt{6}}{18}(\underset{\sim}{i} + 7\underset{\sim}{j} + 2\underset{\sim}{k})$

b. $\frac{\pm\sqrt{19}}{57}(7\underset{\sim}{i} + \underset{\sim}{j} + 11\underset{\sim}{k})$

c. $\frac{\pm\sqrt{2}}{6}(\underset{\sim}{i} + 4\underset{\sim}{j} - \underset{\sim}{k})$

3.4 Exam questions

Note: Mark allocations are available with the fully worked solutions online.

1. E
2. C
3. a. $\hat{\underset{\sim}{a}} = \frac{1}{\sqrt{6}}(\sqrt{3}\underset{\sim}{i} - \underset{\sim}{j} - \sqrt{2}\underset{\sim}{k})$

b. 45°

c. $m = 6 + 5\sqrt{2}$

3.5 Vector proofs using the scalar product

3.5 Exercise

1, 2. Sample responses can be found in the worked solutions in the online resources.

3. $\frac{1}{2}\sqrt{861}$

4. $\frac{1}{2}\sqrt{870}$

5–13. Sample responses can be found in the worked solutions in the online resources.

14. a. $\overrightarrow{OM} = \frac{1}{2}(\underset{\sim}{a} + \underset{\sim}{b})$

b, c. Sample responses can be found in the worked solutions in the online resources.

15. Sample responses can be found in the worked solutions in the online resources.

16. a. $\overrightarrow{OP} = \frac{1}{2}(\underset{\sim}{a} + \underset{\sim}{c})$, $\overrightarrow{OQ} = \frac{1}{2}(\underset{\sim}{b} + \underset{\sim}{c})$

b, c. Sample responses can be found in the worked solutions in the online resources.

17. Sample responses can be found in the worked solutions in the online resources.

3.5 Exam questions

Note: Mark allocations are available with the fully worked solutions online.

1. E
2. A
3. a. $a = \sqrt{3}$

b. Sample responses can be found in the worked solutions in the online resources.

3.6 Parametric equations

3.6 Exercise

1. Part of parabola $y = (x - 2)^2$; domain $[1, \infty)$, range $[0, \infty)$

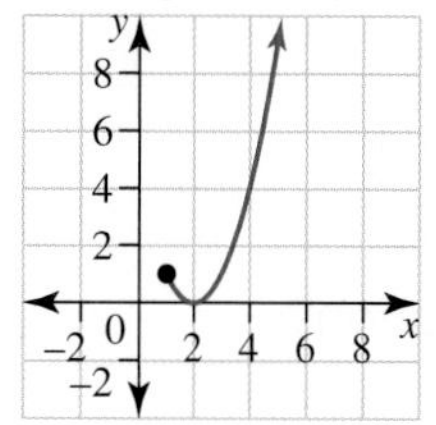

2. Part of parabola $y = 2x^2 + 3$; domain $[0, \infty)$, range $[3, \infty)$

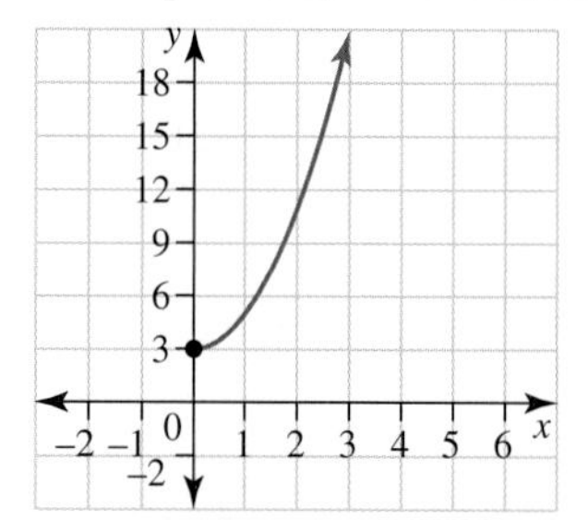

3. Ellipse $\frac{x^2}{9} + \frac{y^2}{16} = 1$; domain $[-3, 3]$, range $[-4, 4]$

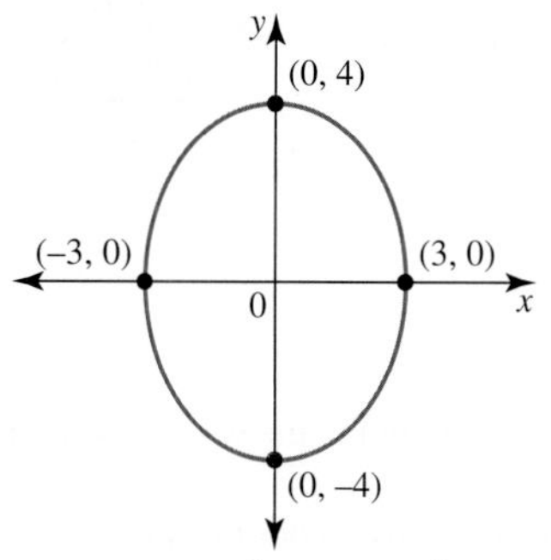

4. Ellipse $\frac{(x-5)^2}{4} + \frac{(y+4)^2}{9} = 1$; domain $[3, 7]$, range $[-7, -1]$

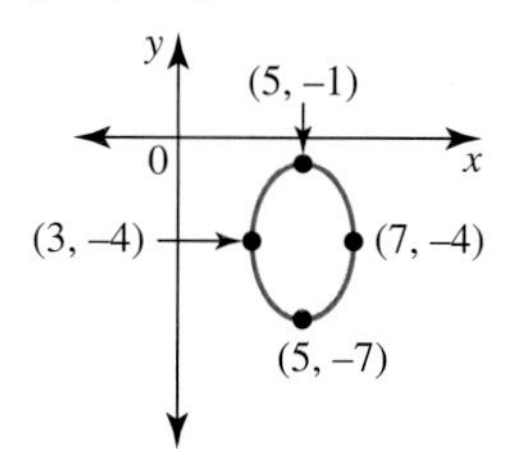

5. Hyperbola $\frac{x^2}{25} - \frac{y^2}{9} = 1$, asymptotes $y = \pm\frac{3x}{5}$; domain $(-\infty, -5] \cup [5, \infty)$, range R

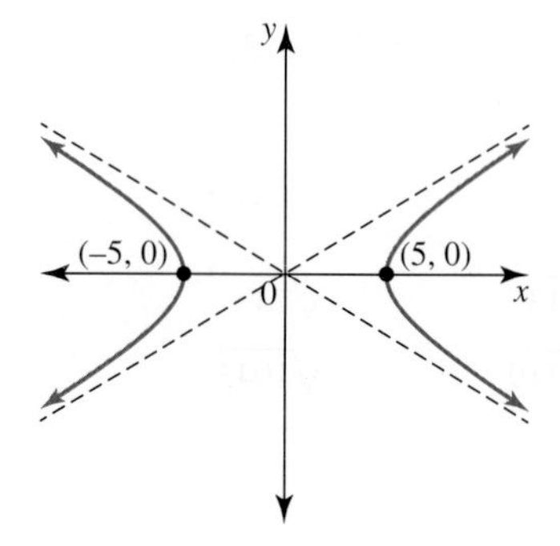

6. Hyperbola $\frac{y^2}{9} - \frac{x^2}{16} = 1$, asymptotes $y = \pm \frac{3x}{4}$; domain R, range $(-\infty, -3] \cup [3, \infty)$

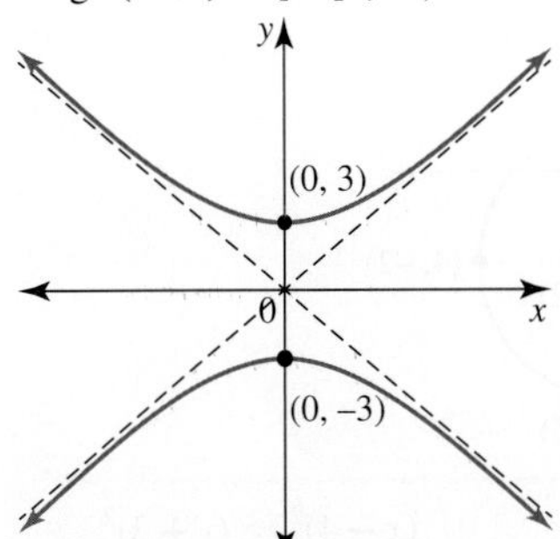

7, 8. Sample responses can be found in the worked solutions in the online resources.

9. a. Part of a parabola, $y = x^2$; domain $[0, \infty)$, range $[0, \infty)$

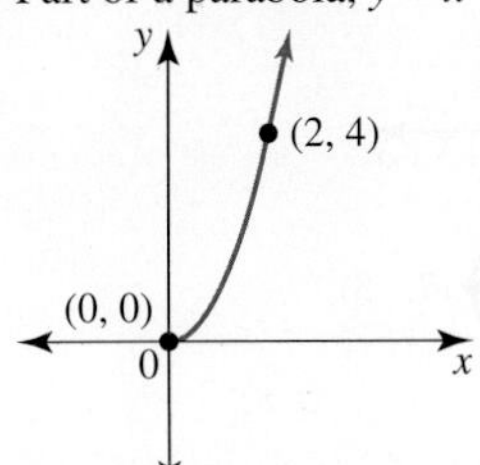

b. Part of a straight line, $y = 3x + 3$; domain $[-1, \infty)$, range $[0, \infty)$

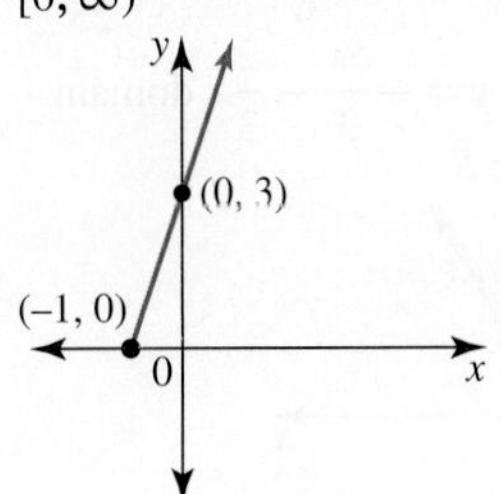

c. Part of a cubic, $y = x^3$; domain $[0, \infty)$, range $[0, \infty)$

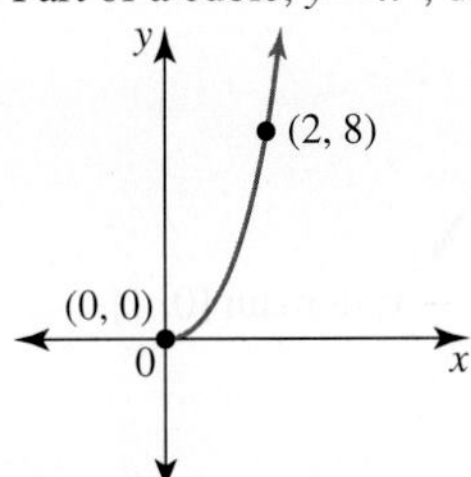

10. a. Part of hyperbola, $y = \frac{2}{x}$; domain $(0, \infty)$, range $(0, \infty)$

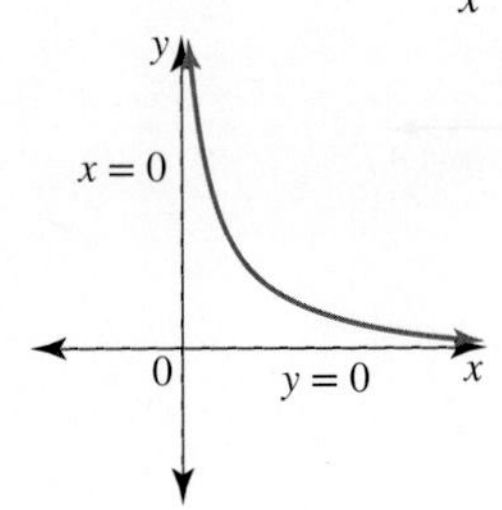

b. Part of a parabola, $y = \frac{1}{4}(x^2 - 8x)$; domain $[0, \infty)$, range $[-4, \infty)$

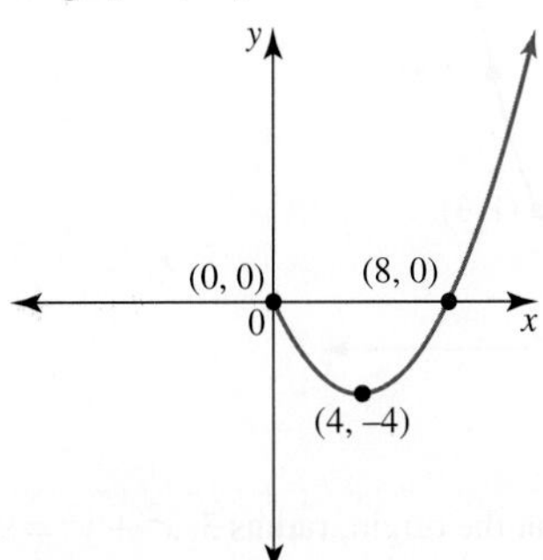

c. Part of a hyperbola, $y = \sqrt{x^2 - 4}$; domain $[2, \infty)$, range R

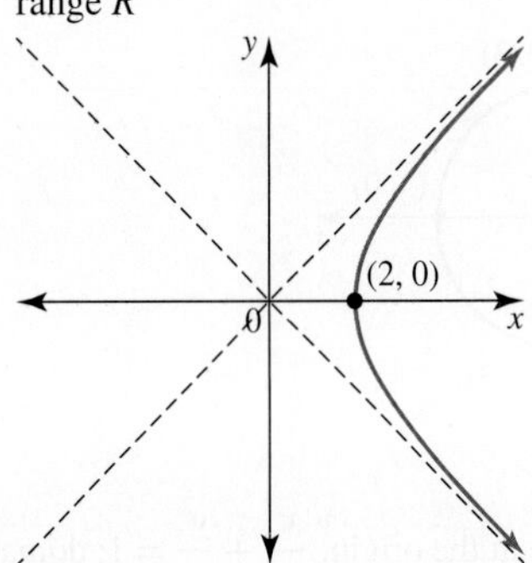

11. a. Part of a hyperbola, $y = \frac{1}{x}$; domain $(0, 1]$, range $[1, \infty)$

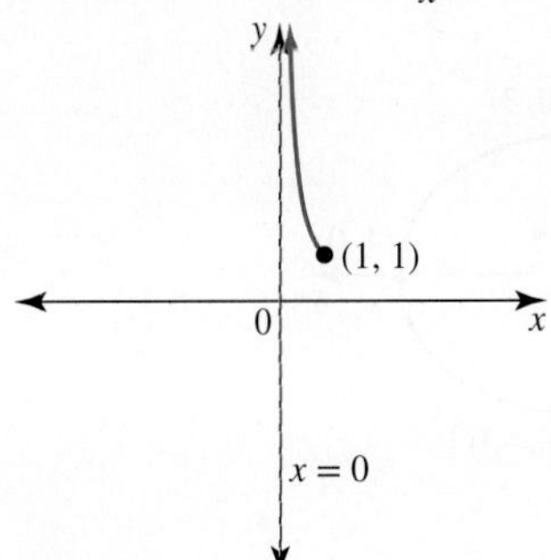

b. Part of a truncus, $y = 2 + \frac{1}{x^2}$; domain $(0, 1]$, range $[3, \infty)$

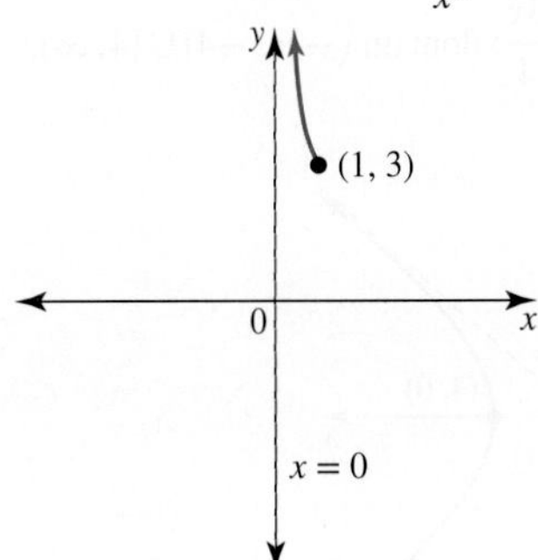

c. Part of a parabola, $y = 2 + x^2$; domain $[1, \infty)$, range $[3, \infty)$

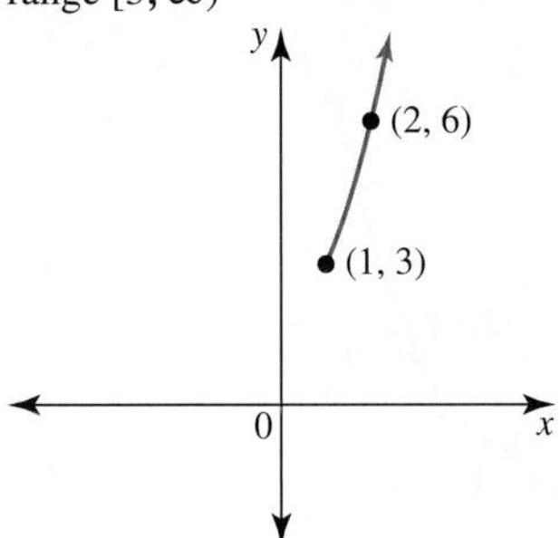

12. a. Circle with centre at the origin, radius 3, $x^2 + y^2 = 9$; domain $[-3, 3]$, range $[-3, 3]$

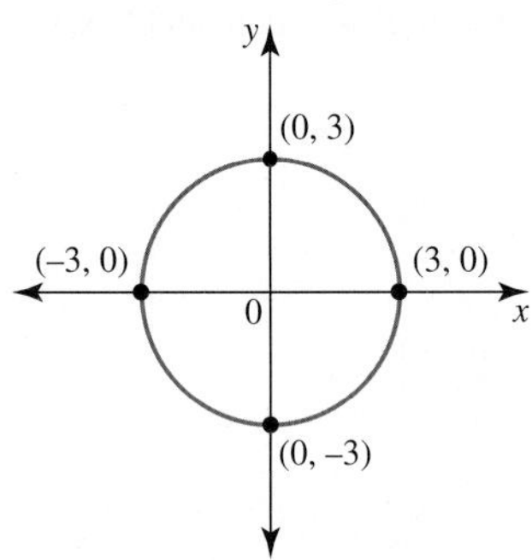

b. Ellipse with centre at the origin, $\frac{x^2}{16} + \frac{y^2}{9} = 1$; domain $[-4, 4]$, range $[-3, 3]$

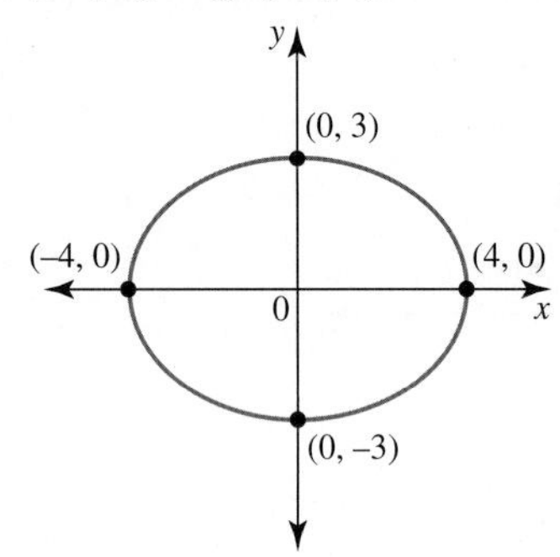

c. Hyperbola with centre at the origin, $\frac{x^2}{16} - \frac{y^2}{9} = 1$, asymptotes $y = \pm \frac{3x}{4}$; domain $(-\infty, -4] \cup [4, \infty)$, range R

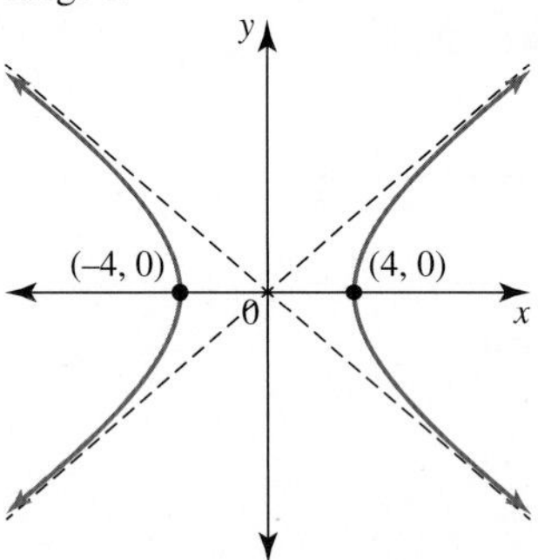

13. a. Circle with centre at $(1, -2)$, radius 3, $(x - 1)^2 + (y + 2)^2 = 9$; domain $[-2, 4]$, range $[-5, 1]$

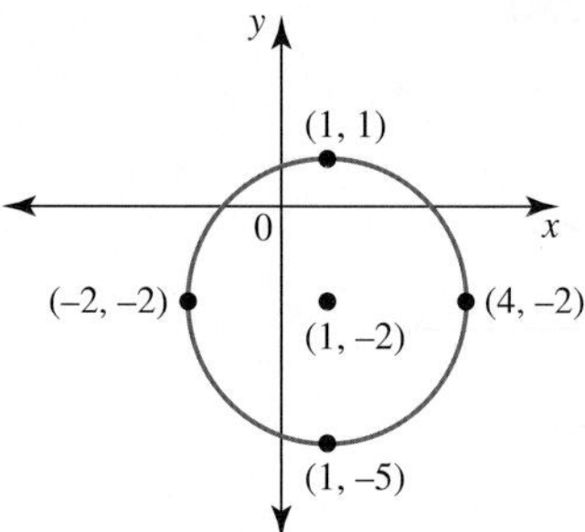

b. Ellipse with centre at $(4, -3)$, $\frac{(x-4)^2}{9} + \frac{(y+3)^2}{4} = 1$; domain $[1, 7]$, range $[-5, -1]$

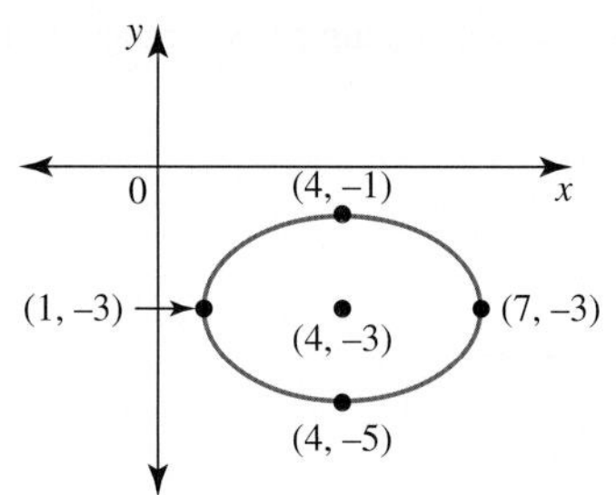

c. Hyperbola with centre $(2, -4)$, $\frac{(x-2)^2}{9} - \frac{(y+4)^2}{25} = 1$, asymptotes $y = \frac{5x}{3} - \frac{22}{3}$, $y = -\frac{5x}{3} - \frac{2}{3}$; domain $(-\infty, -1] \cup [5, \infty)$, range R

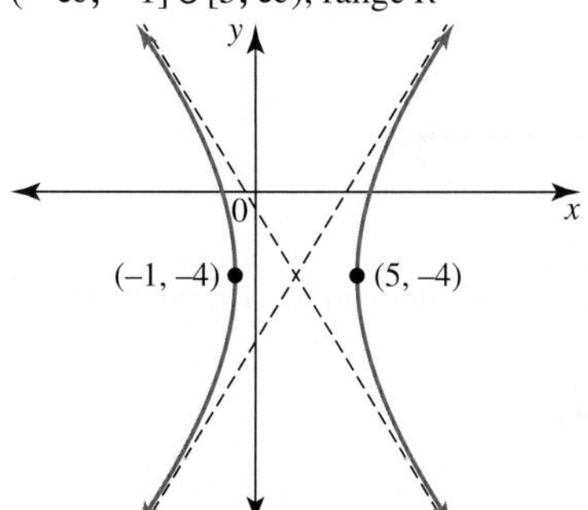

14. a. Part of a straight line, $y = 1 - x$; domain $[0, 1]$, range $[0, 1]$

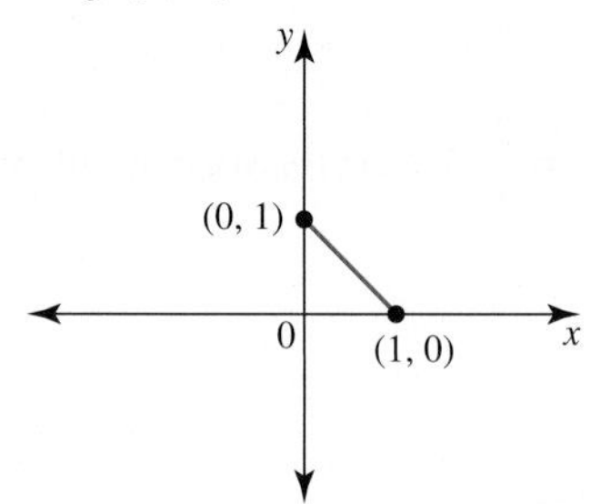

b. $x^{\frac{2}{3}} + y^{\frac{2}{3}} = 1$; domain $[-1, 1]$, range $[-1, 1]$

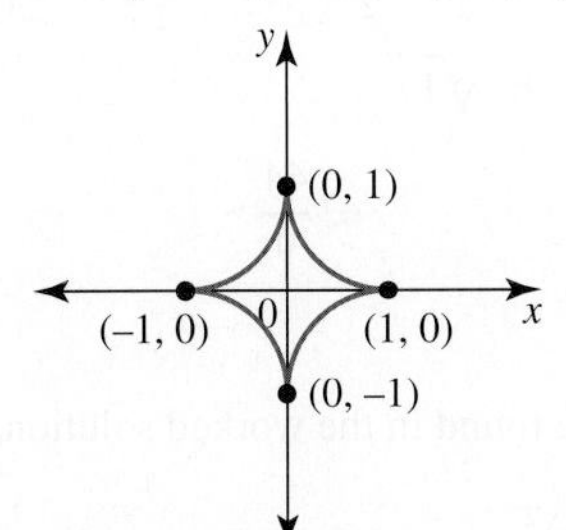

c. $\sqrt{y} + \sqrt{x} = 1$; domain $[0, 1]$, range $[0, 1]$

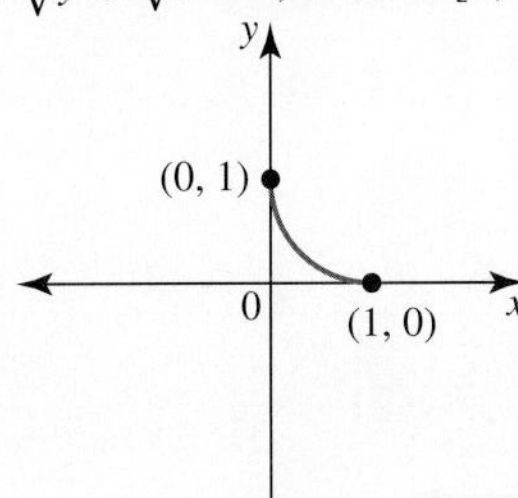

15. a. Part of a straight line, $y = 2x - 1$; domain $[0, 1]$, range $[-1, 1]$

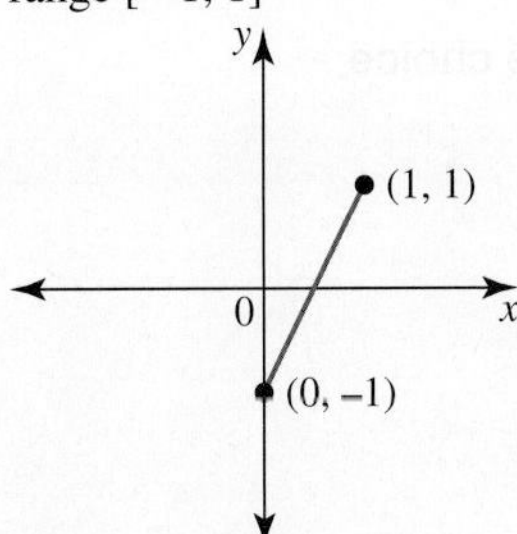

b. Part of a parabola, $y = 2x^2 - 1$; domain $[-1, 1]$, range $[-1, 1]$

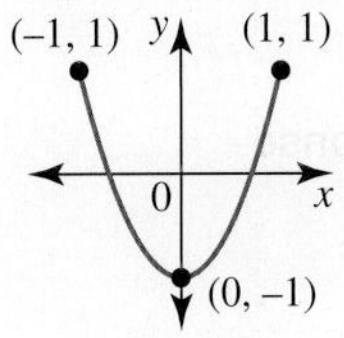

c. $y = \pm 2x\sqrt{1 - x^2}$; domain $[-1, 1]$, range $[-1, 1]$

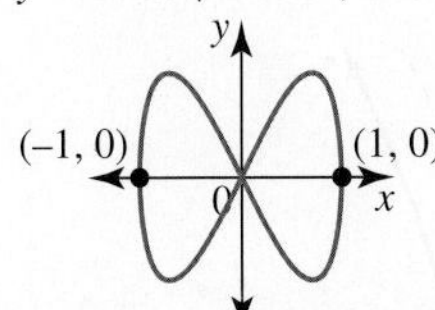

16. a. Circle, centre at the origin, radius a, $x^2 + y^2 = a^2$

b. Ellipse, centre at the origin, $\frac{x^2}{a^2} + \frac{y^2}{b^2} = 1$

c. Hyperbola, centre at the origin, $\frac{x^2}{a^2} - \frac{y^2}{b^2} = 1$

17, 18. Sample responses can be found in the worked solutions in the online resources.

19. a. Sample responses can be found in the worked solutions in the online resources.

b. $y = x^3 - 3x$

20. a. Sample responses can be found in the worked solutions in the online resources.

b. $y = 2x^2 - 4x + 1$

21–23. Sample responses can be found in the worked solutions in the online resources.

24. a.

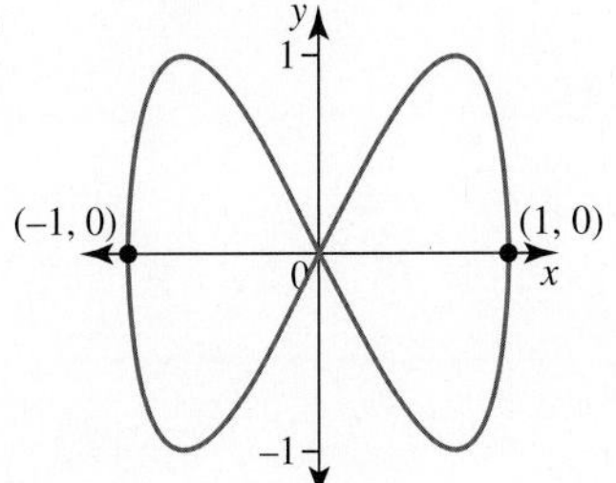

b.

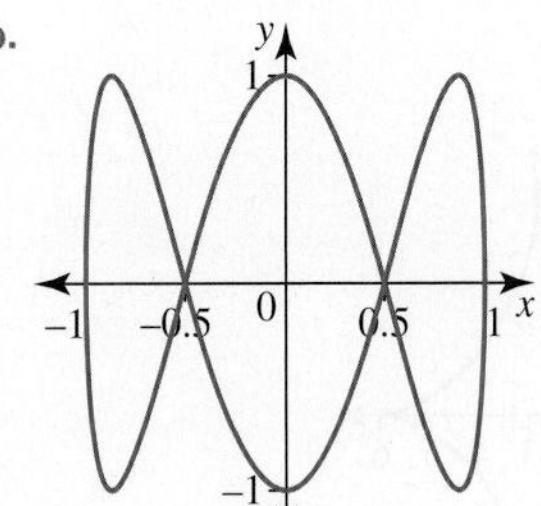

25. a.

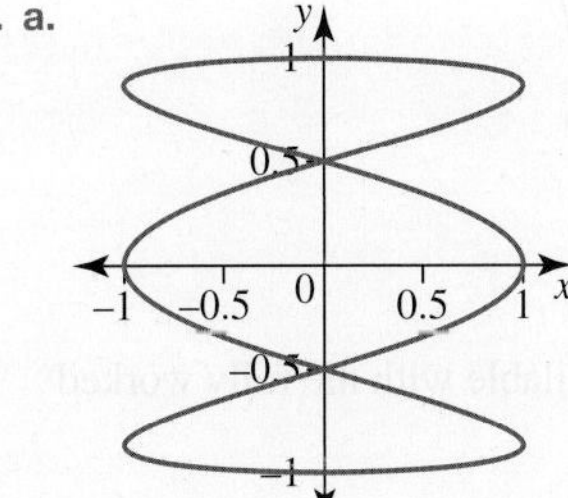

b.

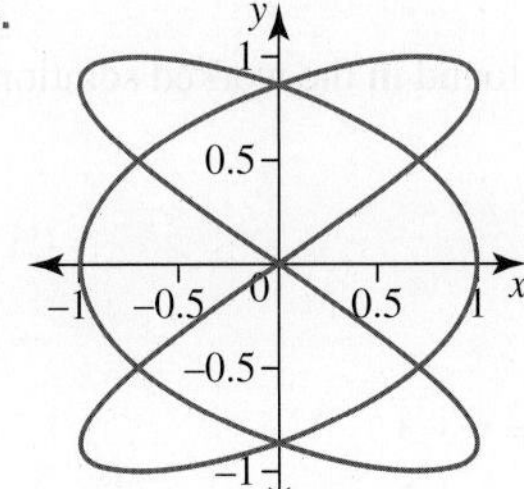

26. a.

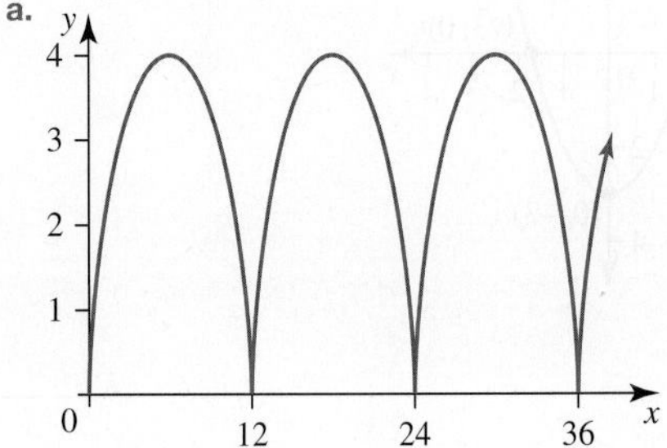

b. 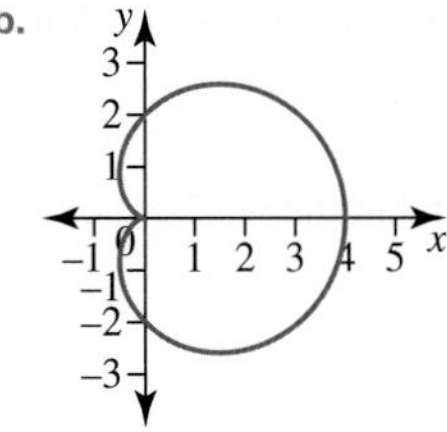

27. 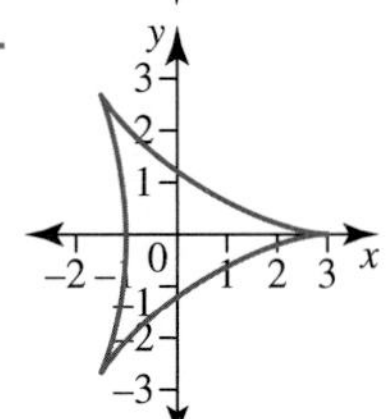

28. 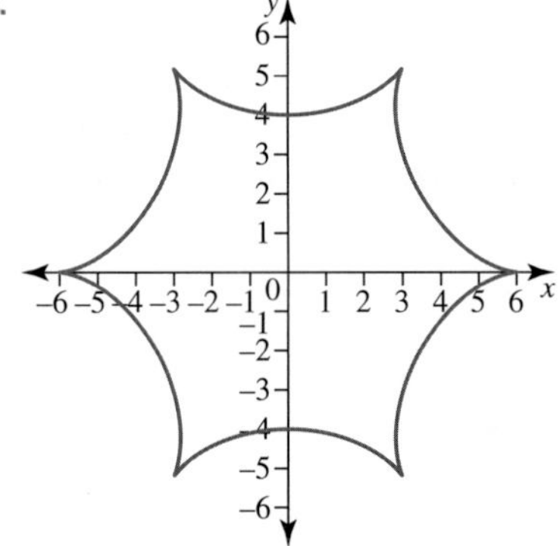

3.6 Exam questions

Note: Mark allocations are available with the fully worked solutions online.

1. A
2. E
3. a. Sample responses can be found in the worked solutions in the online resources.

b. 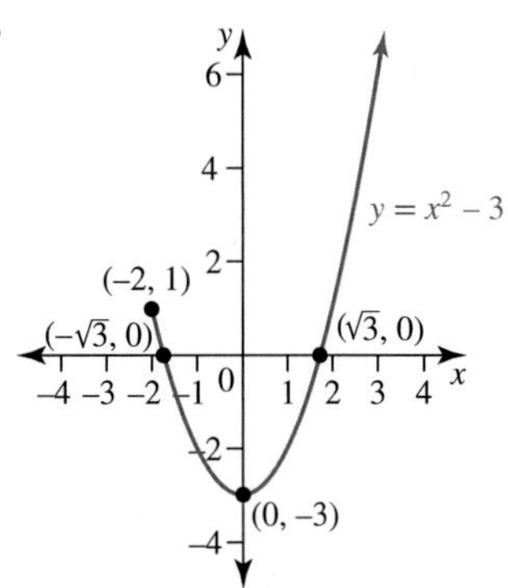

3.7 Review

3.7 Exercise

Technology free: short answer

1. a. $2\sqrt{67}$ b. $\cos^{-1}\left(\dfrac{-6}{2\sqrt{67}}\right)$
c. $(7, 5, 0)$ d. $\left(6, \dfrac{8}{3}, 1\right)$
2. $5\sqrt{22}$
3. a. $\pm\sqrt{5}$ b. $\dfrac{9}{2}$ c. $\dfrac{3}{2}$ d. $-\dfrac{40}{3}$
4. a. $3\underset{\sim}{i} + 2\underset{\sim}{j} + 2\underset{\sim}{k}$ b. $\sqrt{17}$
5. a. 2 b. -4 c. $\dfrac{44}{9}$
d. $2 \pm \sqrt{29}$ e. $-\dfrac{7}{3}$

6, 7. Sample responses can be found in the worked solutions in the online resources.

8. Ellipse $\dfrac{(x-2)^2}{9} + \dfrac{(y-4)^2}{4} = 1$, centre $(2, 4)$, domain $[-1, 5]$, range $[2, 6]$

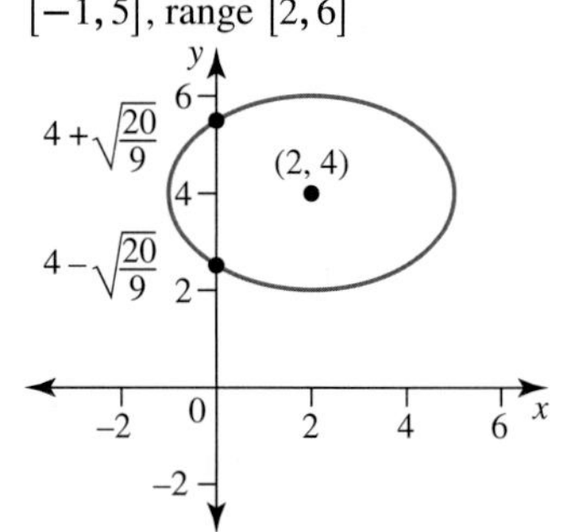

Technology active: multiple choice

9. D
10. B
11. A
12. D
13. C
14. A
15. C
16. E
17. E
18. B

Technology active: extended response

19. Hyperbola $\dfrac{x^2}{9} - \dfrac{y^2}{25} = 1$

Asymptotes $y = \pm\dfrac{5x}{3}$, domain $[3, \infty)$, range R

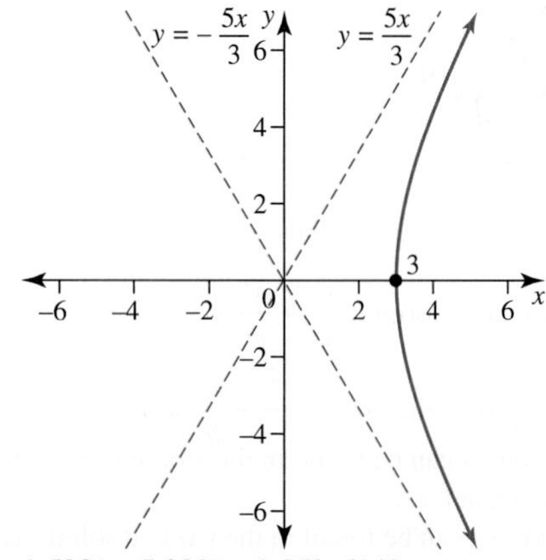

20. $1.598\underset{\sim}{i} - 5.928\underset{\sim}{j} + 0.05\underset{\sim}{k}$, 6140 m

3.7 Exam questions

Note: Mark allocations are available with the fully worked solutions online.

1. a. $m = 4$ b. $\frac{1}{18}(47\underset{\sim}{i} - 10\underset{\sim}{j} + 7\underset{\sim}{k})$
2. D
3. a. $a = 6,\ b = 2,\ c = -3$
   b. $\cos(\theta) = \frac{4}{9}$
   c. $2\sqrt{65}$
   d. Sample responses can be found in the worked solutions in the online resources.
   e. 24
4. Sample responses can be found in the worked solutions in the online resources.
5. -2

4 Vector equations of lines and planes

LEARNING SEQUENCE

Fully worked solutions for this topic are available online.

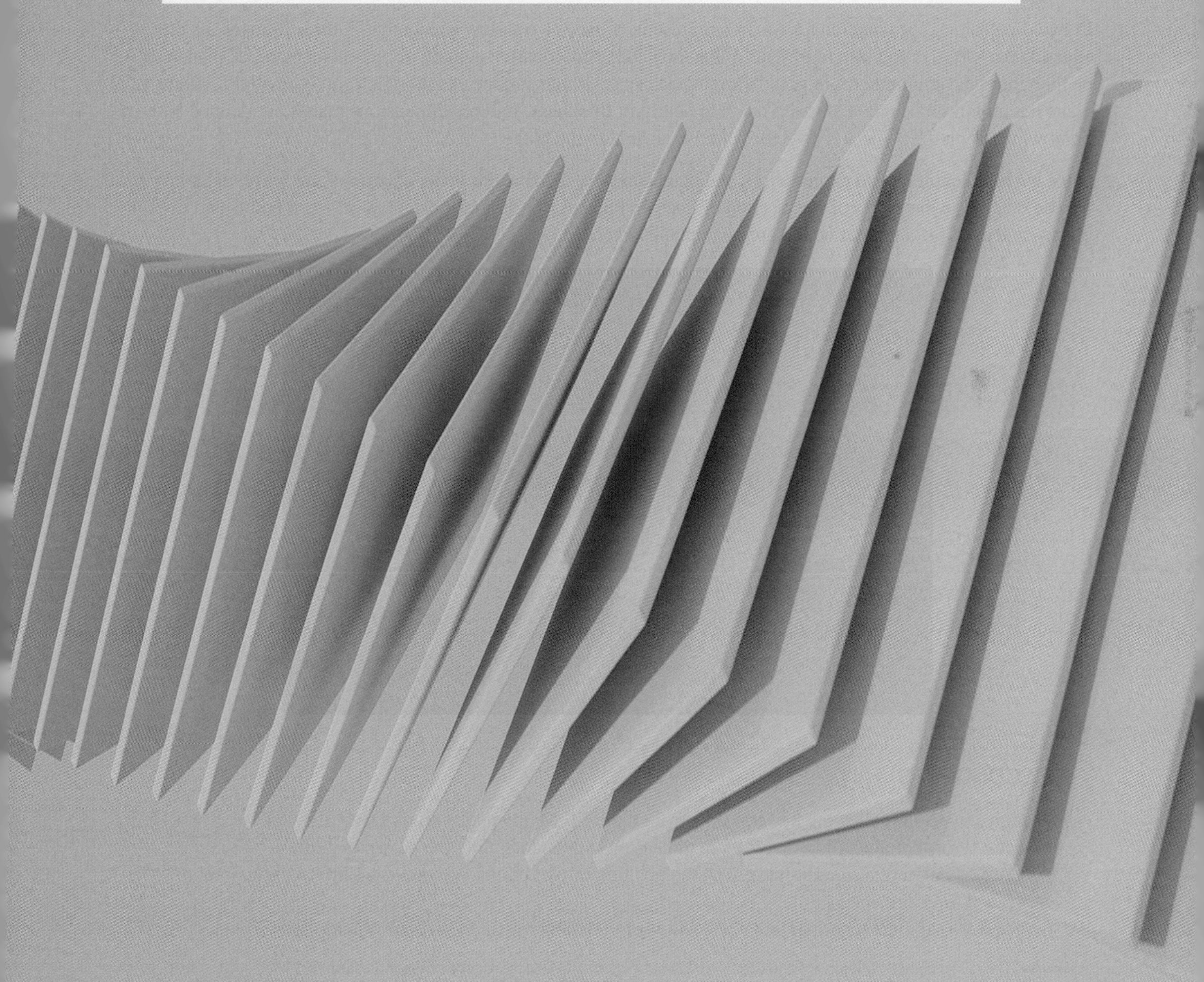

4.1 Overview

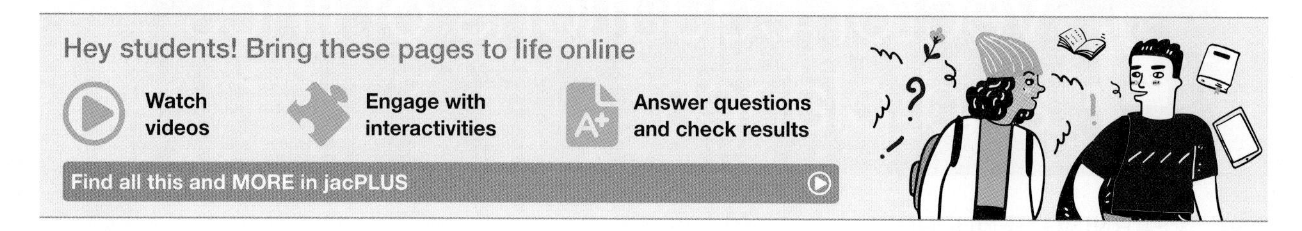

4.1.1 Introduction

In this topic we extend the concepts learned in the previous topic to cover the vector cross product, and vector, parametric and Cartesian equations of lines and planes in two and three dimensions. In subtopic 4.2 we will revise some matrix arithmetic and learn how the determinants of 3×3 matrices can be used to evaluate vector cross products.

In two dimensions, the equation $ax + by = c$ represents a line, so you may expect that in three dimensions the equation $ax + by + cz = d$ also represents a line; however, this equation actually represents a plane. A plane can be thought of as an infinite set of points forming a connected flat surface extending infinitely in all directions. A plane has an infinite length and width but zero height or thickness. In three dimensions planes are defined by a point which is on the plane and a vector which is normal to the plane.

Once we have learnt how to express lines and planes in three dimensions using equations, we will look at how we can determine the points of intersection of lines and planes, the shortest distances between points and lines or planes, and even the angles at which lines and planes intersect.

KEY CONCEPTS

This topic covers the following key concepts from the VCE Mathematics Study Design:

- vector (cross) product of two vectors in three dimensions, including the determinant form
- vector equation of a straight line, given the position of two points, or equivalent information, in both two and three dimensions
- vector cross product, normal to a plane and vector, parametric and Cartesian equations of a plane

Source: VCE Mathematics Study Design (2023–2027) extracts © VCAA; reproduced by permission.

4.2 Vector cross product

LEARNING INTENTION

At the end of this subtopic you should be able to:
- calculate the vector cross product of two vectors.

4.2.1 Vector products

We have seen that we can multiply two vectors together according to the definition of the dot or scalar product and obtain a scalar. We can also multiply two vectors together according to the vector cross product or vector product and obtain a vector.

Definition of the cross product

If $\underset{\sim}{a}$ and $\underset{\sim}{b}$ are two non-zero vectors in three dimensions and θ is the angle between $\underset{\sim}{a}$ and $\underset{\sim}{b}$ where $0 \le \theta \le \pi$ and $\hat{\underset{\sim}{n}}$ is a unit vector which is perpendicular to both $\underset{\sim}{a}$ and $\underset{\sim}{b}$, then we define the cross or vector product of $\underset{\sim}{a}$ and $\underset{\sim}{b}$ in that order as given by $\underset{\sim}{a} \times \underset{\sim}{b} = |\underset{\sim}{a}||\underset{\sim}{b}|\sin(\theta)\hat{\underset{\sim}{n}}$. This is read as $\underset{\sim}{a}$ cross $\underset{\sim}{b}$.

The vector cross product

$$\underset{\sim}{a} \times \underset{\sim}{b} = |\underset{\sim}{a}||\underset{\sim}{b}|\sin(\theta)\hat{\underset{\sim}{n}}$$

The orientation of the vector cross product is given by the right-handed coordinate system. To determine the direction of $\underset{\sim}{a} \times \underset{\sim}{b}$, use your right hand and curl your fingers from $\underset{\sim}{a}$ to $\underset{\sim}{b}$. The direction of $\underset{\sim}{a} \times \underset{\sim}{b}$ is the direction in which your thumb is pointing.

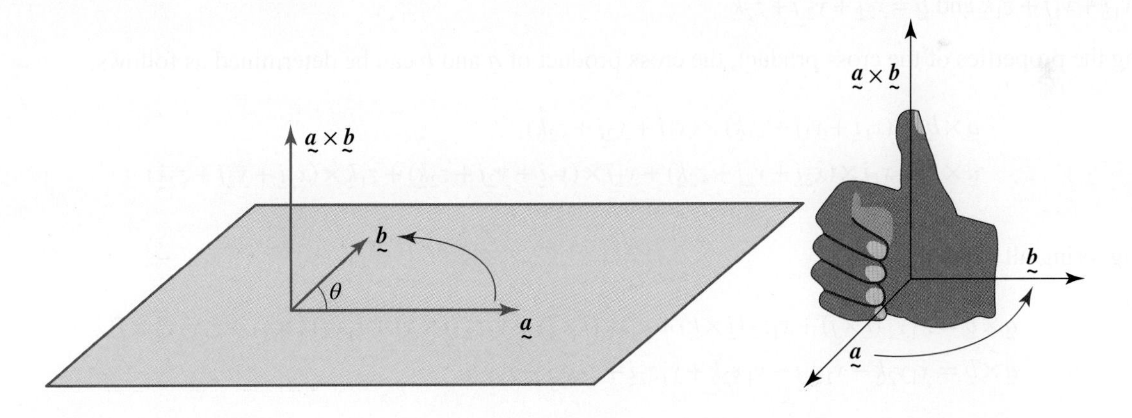

Consider taking the cross product of $\underset{\sim}{a}$ and $\underset{\sim}{b}$ in the opposite order, $\underset{\sim}{b} \times \underset{\sim}{a}$. This time you will need to curl your fingers from $\underset{\sim}{b}$ to $\underset{\sim}{a}$, giving a direction which is opposite to the direction of $\underset{\sim}{a} \times \underset{\sim}{b}$.

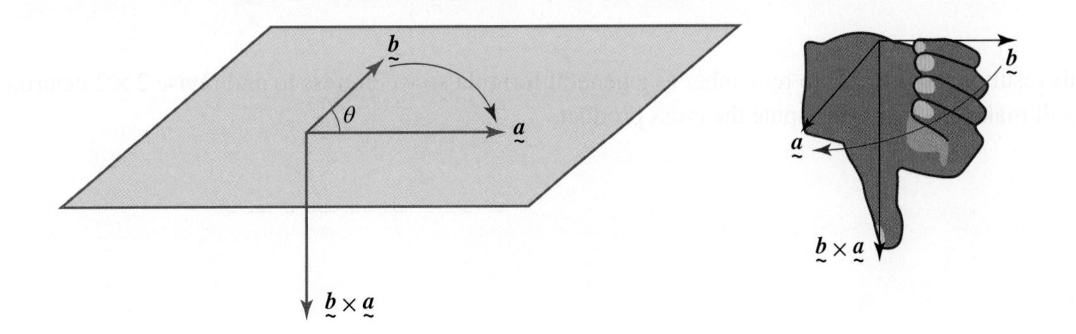

Properties of the cross product

Notice that the vector cross product is not commutative, in fact $\underset{\sim}{a} \times \underset{\sim}{b} = -\underset{\sim}{b} \times \underset{\sim}{a}$.

We must pay specific attention to the order of the factors.

If $\underset{\sim}{a} = \underset{\sim}{0}$ or $\underset{\sim}{b} = \underset{\sim}{0}$ or if $\underset{\sim}{a}$ is parallel to $\underset{\sim}{b}$ then $\theta = 0$ or π, in which case $\underset{\sim}{a} \times \underset{\sim}{b} = \underset{\sim}{0}$ (note that this is the zero vector).

These properties, and some additional properties of the cross product are summarised below.

Properties of the cross product

- $\underset{\sim}{a} \times \underset{\sim}{b} = -\underset{\sim}{b} \times \underset{\sim}{a}$
- $\underset{\sim}{a} \times \underset{\sim}{a} = \underset{\sim}{0}$
- $\underset{\sim}{a} \times \underset{\sim}{0} = \underset{\sim}{0}$
- $\underset{\sim}{a} \times (\underset{\sim}{b} + \underset{\sim}{c}) = \underset{\sim}{a} \times \underset{\sim}{b} + \underset{\sim}{a} \times \underset{\sim}{c}$
- $(\lambda \underset{\sim}{a}) \times \underset{\sim}{b} = \lambda (\underset{\sim}{a} \times \underset{\sim}{b}) = \underset{\sim}{a} \times (\lambda \underset{\sim}{b})$

Component forms

Applying these properties to the unit vectors $\underset{\sim}{i}$, $\underset{\sim}{j}$ and $\underset{\sim}{k}$:

$\underset{\sim}{i} \times \underset{\sim}{j} = \underset{\sim}{k}$, $\underset{\sim}{j} \times \underset{\sim}{k} = \underset{\sim}{i}$, $\underset{\sim}{k} \times \underset{\sim}{i} = \underset{\sim}{j}$ (by the right-hand rule)

And since $\underset{\sim}{a} \times \underset{\sim}{b} = -\underset{\sim}{b} \times \underset{\sim}{a}$, it follows that $\underset{\sim}{j} \times \underset{\sim}{i} = -\underset{\sim}{k}$, $\underset{\sim}{k} \times \underset{\sim}{j} = -\underset{\sim}{i}$, $\underset{\sim}{i} \times \underset{\sim}{k} = -\underset{\sim}{j}$.

Also recall that $\underset{\sim}{i} \times \underset{\sim}{i} = \underset{\sim}{0}$, $\underset{\sim}{j} \times \underset{\sim}{j} = \underset{\sim}{0}$ and $\underset{\sim}{k} \times \underset{\sim}{k} = \underset{\sim}{0}$.

Consider the following vectors, $\underset{\sim}{a}$ and $\underset{\sim}{b}$ which are in component form.

$\underset{\sim}{a} = x_1 \underset{\sim}{i} + y_1 \underset{\sim}{j} + z_1 \underset{\sim}{k}$ and $\underset{\sim}{b} = x_2 \underset{\sim}{i} + y_2 \underset{\sim}{j} + z_2 \underset{\sim}{k}$

Using the properties of the cross product, the cross product of $\underset{\sim}{a}$ and $\underset{\sim}{b}$ can be determined as follows,

$$\underset{\sim}{a} \times \underset{\sim}{b} = (x_1 \underset{\sim}{i} + y_1 \underset{\sim}{j} + z_1 \underset{\sim}{k}) \times (x_2 \underset{\sim}{i} + y_2 \underset{\sim}{j} + z_2 \underset{\sim}{k})$$
$$\underset{\sim}{a} \times \underset{\sim}{b} = x_1 \underset{\sim}{i} \times (x_2 \underset{\sim}{i} + y_2 \underset{\sim}{j} + z_2 \underset{\sim}{k}) + y_1 \underset{\sim}{j} \times (x_2 \underset{\sim}{i} + y_2 \underset{\sim}{j} + z_2 \underset{\sim}{k}) + z_1 \underset{\sim}{k} \times (x_2 \underset{\sim}{i} + y_2 \underset{\sim}{j} + z_2 \underset{\sim}{k})$$

and ignoring all the zero vectors,

$$\underset{\sim}{a} \times \underset{\sim}{b} = x_1 y_2 (\underset{\sim}{i} \times \underset{\sim}{j}) + x_1 z_2 (\underset{\sim}{i} \times \underset{\sim}{k}) + y_1 x_2 (\underset{\sim}{j} \times \underset{\sim}{i}) + y_1 z_2 (\underset{\sim}{j} \times \underset{\sim}{k}) + z_1 x_2 (\underset{\sim}{k} \times \underset{\sim}{i}) + z_1 y_2 (\underset{\sim}{k} \times \underset{\sim}{j})$$
$$\underset{\sim}{a} \times \underset{\sim}{b} = x_1 y_2 \underset{\sim}{k} - x_1 z_2 \underset{\sim}{j} - y_1 x_2 \underset{\sim}{k} + y_1 z_2 \underset{\sim}{i} + z_1 x_2 \underset{\sim}{j} - z_1 y_2 \underset{\sim}{i}$$

factorising the unit vectors, gives

$$\underset{\sim}{a} \times \underset{\sim}{b} = (y_1 z_2 - z_1 y_2) \underset{\sim}{i} + (x_2 z_1 - x_1 z_2) \underset{\sim}{j} + (x_1 y_2 - x_2 y_1) \underset{\sim}{k}$$

Now this result is too difficult to remember as a general formula so we digress to and revise 2×2 determinants which will make it easier to compute the cross product.

4.2.2 Determinants

Determinants of 2×2 matrices

When studying matrices, we have seen that the determinant is a number associated with a square matrix. For a 2×2 matrix, $A = \begin{bmatrix} a & b \\ c & d \end{bmatrix}$, the determinant is denoted by the symbol Δ and is written with straight lines, rather than square brackets as for the matrix, $\Delta = \det(A) = \begin{vmatrix} a & b \\ c & d \end{vmatrix}$.

To calculate the value of the determinant, multiply the elements in the leading diagonal and subtract the product of the other two elements, $\Delta = \begin{vmatrix} a & b \\ c & d \end{vmatrix} = ad - bc$.

2×2 determinants

$$\Delta = \begin{vmatrix} a & b \\ c & d \end{vmatrix} = ad - bc$$

WORKED EXAMPLE 1 Calculating the determinant of a 2×2 matrix

Evaluate $\begin{vmatrix} -5 & 4 \\ 2 & 3 \end{vmatrix}$

THINK	WRITE
1. Multiply the elements in the leading diagonal and subtract the product of the other two elements	$\begin{vmatrix} -5 & 4 \\ 2 & 3 \end{vmatrix} = -5 \times 3 - 4 \times 2$
2. State the final result, as the value of the determinant.	$= -15 - 8$ $= -23$

Determinants of 3×3 matrices

Consider a 3×3 matrix $A = \begin{bmatrix} x_1 & y_1 & z_1 \\ x_2 & y_2 & z_2 \\ x_3 & y_3 & z_3 \end{bmatrix}$, then a 3×3 determinant is of the form $\det(A) = \Delta = \begin{vmatrix} x_1 & y_1 & z_1 \\ x_2 & y_2 & z_2 \\ x_3 & y_3 & z_3 \end{vmatrix}$.

There are a couple of methods which can be used to evaluate a 3×3 determinant. The one that we will use is to proceed across the first row multiplying that element by the 2×2 determinant that remains when we delete the row and column in which that element appears. Then multiply out each 2 by 2 determinant also following the sign pattern $\begin{pmatrix} + & - & + \\ - & + & - \\ + & - & + \end{pmatrix}$ on the coefficients. So that $\Delta = \begin{vmatrix} x_1 & y_1 & z_1 \\ x_2 & y_2 & z_2 \\ x_3 & y_3 & z_3 \end{vmatrix} = x_1 \begin{vmatrix} y_2 & z_2 \\ y_3 & z_3 \end{vmatrix} - y_1 \begin{vmatrix} x_2 & z_2 \\ x_3 & z_3 \end{vmatrix} + z_1 \begin{vmatrix} x_2 & y_2 \\ x_3 & y_3 \end{vmatrix}$.

Note that there is a negative sign in front of the term y_1, since its position in the 3×3 matrix is negative in the sign pattern matrix.

3×3 determinants

$$\Delta = \begin{vmatrix} x_1 & y_1 & z_1 \\ x_2 & y_2 & z_2 \\ x_3 & y_3 & z_3 \end{vmatrix} = x_1 \begin{vmatrix} y_2 & z_2 \\ y_3 & z_3 \end{vmatrix} - y_1 \begin{vmatrix} x_2 & z_2 \\ x_3 & z_3 \end{vmatrix} + z_1 \begin{vmatrix} x_2 & y_2 \\ x_3 & y_3 \end{vmatrix}$$

WORKED EXAMPLE 2 Calculating the determinant of a 3 × 3 matrix

Evaluate $\begin{vmatrix} -1 & 2 & 1 \\ -5 & 3 & 4 \\ 2 & -1 & 3 \end{vmatrix}$

THINK	WRITE
1. Expand across the first row, multiplying each element by the 2×2 determinant which remains, and following the sign pattern.	$\Delta = \begin{vmatrix} -1 & 2 & 1 \\ -5 & 3 & 4 \\ 2 & -1 & 3 \end{vmatrix}$ $= -1\begin{vmatrix} 3 & 4 \\ -1 & 3 \end{vmatrix} - 2\begin{vmatrix} -5 & 4 \\ 2 & 3 \end{vmatrix} + 1\begin{vmatrix} -5 & 3 \\ 2 & -1 \end{vmatrix}$
2. Multiply and expand each 2×2 determinant.	$= -1(3 \times 3 - 4 \times -1) - 2(-5 \times 3 - 4 \times 2) + 1(-5 \times -1 - 3 \times 2)$ $= -1(9 + 4) - 2(-15 - 8) + 1(5 - 6)$ $= -13 + 46 - 1$ $= 32$
3. State the final result, as the value of the determinant.	$\Delta = 32$

TI | THINK

Use the command det() and insert the matrix in the brackets.

DISPLAY/WRITE

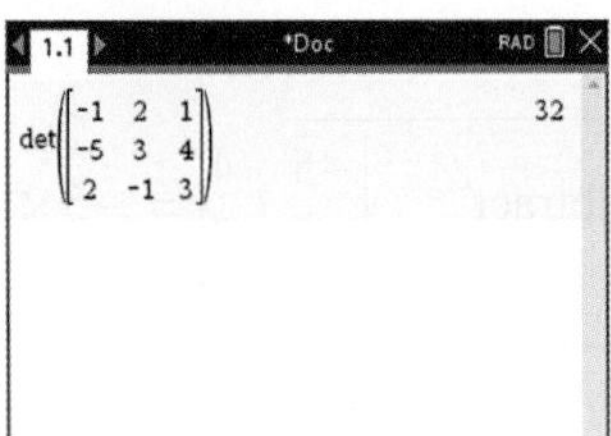

CASIO | THINK

Use the command det() and insert the matrix in the brackets.

DISPLAY/WRITE

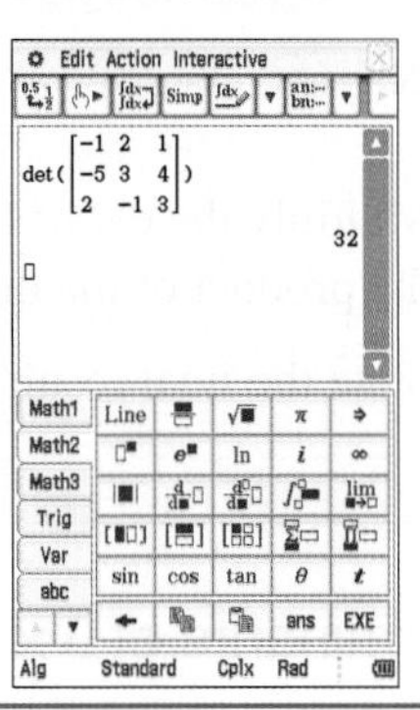

4.2.3 Using determinants to calculate the vector cross product

The vector cross product of two vectors in component form can be calculated using determinants as follows.

Vector cross product using determinants

Given $\underset{\sim}{a} = x_1\underset{\sim}{i} + y_1\underset{\sim}{j} + z_1\underset{\sim}{k}$ and $\underset{\sim}{b} = x_2\underset{\sim}{i} + y_2\underset{\sim}{j} + z_2\underset{\sim}{k}$:

$$\underset{\sim}{a} \times \underset{\sim}{b} = \begin{vmatrix} \underset{\sim}{i} & \underset{\sim}{j} & \underset{\sim}{k} \\ x_1 & y_1 & z_1 \\ x_2 & y_2 & z_2 \end{vmatrix} = \underset{\sim}{i}\begin{vmatrix} y_1 & z_1 \\ y_2 & z_2 \end{vmatrix} - \underset{\sim}{j}\begin{vmatrix} x_1 & z_1 \\ x_2 & z_2 \end{vmatrix} + \underset{\sim}{k}\begin{vmatrix} x_1 & y_1 \\ x_2 & y_2 \end{vmatrix}$$

$$\underset{\sim}{a} \times \underset{\sim}{b} = (y_1z_2 - z_1y_2)\underset{\sim}{i} + (x_2z_1 - x_1z_2)\underset{\sim}{j} + (x_1y_2 - x_2y_1)\underset{\sim}{k}$$

Now, since $\underset{\sim}{a} \times \underset{\sim}{b}$ is a vector perpendicular to both $\underset{\sim}{a}$ and $\underset{\sim}{b}$, we can determine a unit vector perpendicular to both $\underset{\sim}{a}$ and $\underset{\sim}{b}$ as $\hat{\underset{\sim}{n}} = \pm \dfrac{\underset{\sim}{a} \times \underset{\sim}{b}}{|\underset{\sim}{a} \times \underset{\sim}{b}|}$. Note that there are two possibilities, one up and one down.

WORKED EXAMPLE 3 Calculating the vector cross product

Consider the vectors $\underset{\sim}{a} = -\underset{\sim}{i} + 2\underset{\sim}{j} + \underset{\sim}{k}$ and $\underset{\sim}{b} = -5\underset{\sim}{i} + 3\underset{\sim}{j} + 4\underset{\sim}{k}$.

a. **Evaluate $\underset{\sim}{a} \times \underset{\sim}{b}$.**

b. **Determine a unit vector perpendicular to both $\underset{\sim}{a}$ and $\underset{\sim}{b}$.**

THINK	WRITE
a. 1. Use the results to express the cross product in terms of the determinant.	a. $\underset{\sim}{a} \times \underset{\sim}{b} = \begin{vmatrix} \underset{\sim}{i} & \underset{\sim}{j} & \underset{\sim}{k} \\ -1 & 2 & 1 \\ -5 & 3 & 4 \end{vmatrix}$
2. Expand across the first row, multiplying the unit vectors by the 2×2 determinant that remains, following the sign pattern.	$= \underset{\sim}{i} \begin{vmatrix} 2 & 1 \\ 3 & 4 \end{vmatrix} - \underset{\sim}{j} \begin{vmatrix} -1 & 1 \\ -5 & 4 \end{vmatrix} + \underset{\sim}{k} \begin{vmatrix} -1 & 2 \\ -5 & 3 \end{vmatrix}$
3. Evaluate the 2×2 determinants.	$= (8-3)\,\underset{\sim}{i} - (-4+5)\underset{\sim}{j} + (-3+10)\,\underset{\sim}{k}$ $= 5\underset{\sim}{i} - \underset{\sim}{j} + 7\underset{\sim}{k}$
4. State the result.	$\underset{\sim}{a} \times \underset{\sim}{b} = 5\underset{\sim}{i} - \underset{\sim}{j} + 7\underset{\sim}{k}$
b. 1. Calculate the magnitude of the cross product vector.	b. $\lvert \underset{\sim}{a} \times \underset{\sim}{b} \rvert = \sqrt{5^2 + (-1)^2 + 7^2}$ $= \sqrt{25 + 1 + 49}$ $= \sqrt{75}$ $= 5\sqrt{3}$
2. Divide the vector $\underset{\sim}{a} \times \underset{\sim}{b}$ by its magnitude. Note that there are two vectors, in either direction, so give the $\pm$ of the vectors.	$\underset{\sim}{\hat{n}} = \dfrac{\underset{\sim}{a} \times \underset{\sim}{b}}{\lvert \underset{\sim}{a} \times \underset{\sim}{b} \rvert} = \pm \dfrac{1}{5\sqrt{3}} (5\underset{\sim}{i} - \underset{\sim}{j} + 7\underset{\sim}{k})$

TI | THINK

Define the vectors $\underset{\sim}{a}$ and $\underset{\sim}{b}$.

Use the command crossP to determine the vector $\underset{\sim}{a} \times \underset{\sim}{b}$.

Use the command unitV to determine a unit vector perpendicular to $\underset{\sim}{a}$ and $\underset{\sim}{b}$ (in the direction of $\underset{\sim}{a} \times \underset{\sim}{b}$).

DISPLAY/WRITE

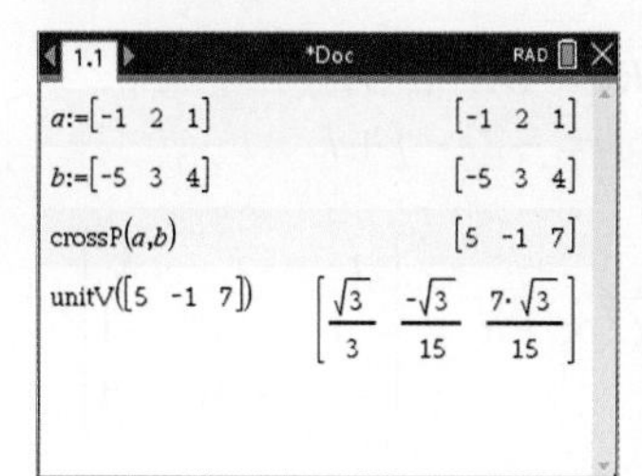

CASIO | THINK

Use the command crossP to determine the vector $\underset{\sim}{a} \times \underset{\sim}{b}$. (*Note:* Vectors can be defined as column or row matrices.)

Use the command unitV to determine a unit vector perpendicular to $\underset{\sim}{a}$ and $\underset{\sim}{b}$ (in the direction of $\underset{\sim}{a} \times \underset{\sim}{b}$).

DISPLAY/WRITE

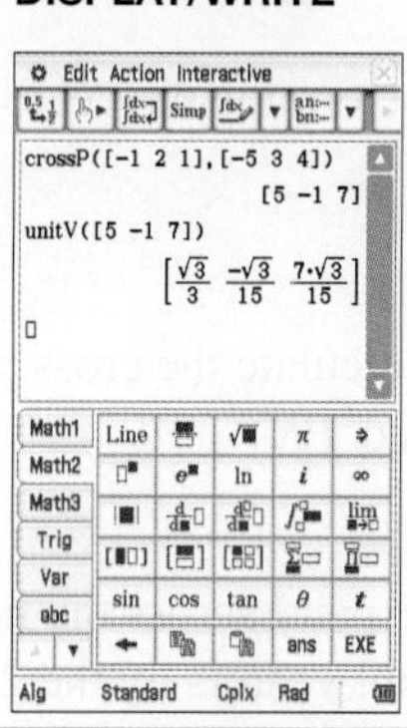

4.2.4 Applications of the cross product

Area of a triangle

Consider the triangle PQR, in three dimensional space.

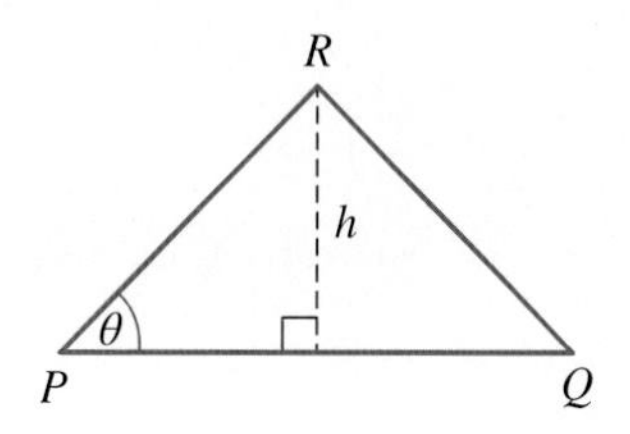

The area of the triangle is $A = \frac{1}{2} \times \text{base} \times \text{height} = \frac{1}{2}\left|\overrightarrow{PQ}\right| h$, but $\sin(\theta) = \frac{h}{\left|\overrightarrow{PR}\right|}$. Rearranging this to $h = \left|\overrightarrow{PR}\right| \sin(\theta)$ and substituting it into the area formula gives $A = \frac{1}{2}\left|\overrightarrow{PQ}\right|\left|\overrightarrow{PR}\right| \sin(\theta)$. Using the definition of the cross product and taking the magnitudes of the components: $\left|\overrightarrow{PQ} \times \overrightarrow{PR}\right| = \left|\overrightarrow{PQ}\right|\left|\overrightarrow{PR}\right| \sin(\theta)\, \hat{\underset{\sim}{n}}$. Since $\left|\hat{\underset{\sim}{n}}\right| = 1$ it follows that $\left|\overrightarrow{PQ} \times \overrightarrow{PR}\right| = \left|\overrightarrow{PQ}\right|\left|\overrightarrow{PR}\right| \sin(\theta)$.

The area of a triangle PQR is therefore equal to half of the magnitude of $\overrightarrow{PQ} \times \overrightarrow{PR}$.

Area of a triangle

The area of a triangle PQR is given by the following formula.

$$A = \frac{1}{2}\left|\overrightarrow{PQ} \times \overrightarrow{PR}\right|$$

WORKED EXAMPLE 4 Calculating the area of a triangle

Calculate the area of the triangle with vertices $P(1,3,2)$, $Q(2,-1,1)$ and $R(-1,2,3)$.

THINK	WRITE
1. Give the position vectors of the points P, Q and R.	$\overrightarrow{OP} = \underset{\sim}{i} + 3\underset{\sim}{j} + 2\underset{\sim}{k}$ $\overrightarrow{OQ} = 2\underset{\sim}{i} - \underset{\sim}{j} + \underset{\sim}{k}$ $\overrightarrow{OR} = -\underset{\sim}{i} + 2\underset{\sim}{j} + 3\underset{\sim}{k}$
2. Determine the vector $\overrightarrow{PQ}$.	$\overrightarrow{PQ} = \overrightarrow{OQ} - \overrightarrow{OP}$ $= \underset{\sim}{i} - 4\underset{\sim}{j} - \underset{\sim}{k}$
3. Determine the vector $\overrightarrow{PR}$.	$\overrightarrow{PR} = \overrightarrow{OR} - \overrightarrow{OP}$ $= -2\underset{\sim}{i} - \underset{\sim}{j} + \underset{\sim}{k}$
4. Calculate the cross product of the vectors $\overrightarrow{PQ}$ and $\overrightarrow{PR}$.	$\overrightarrow{PQ} \times \overrightarrow{PR} = \begin{vmatrix} \underset{\sim}{i} & \underset{\sim}{j} & \underset{\sim}{k} \\ 1 & -4 & -1 \\ -2 & -1 & 1 \end{vmatrix}$
5. Expand across the first row, multiplying the unit vectors by the 2×2 determinant that remains, following the sign pattern.	$= \underset{\sim}{i}\begin{vmatrix} -4 & -1 \\ -1 & 1 \end{vmatrix} - \underset{\sim}{j}\begin{vmatrix} 1 & -1 \\ -2 & 1 \end{vmatrix} + \underset{\sim}{k}\begin{vmatrix} 1 & -4 \\ -2 & -1 \end{vmatrix}$
6. Expand each 2×2 determinant.	$= \underset{\sim}{i}(-4-1) - \underset{\sim}{j}(1-2) + \underset{\sim}{k}(-1-8)$ $= -5\underset{\sim}{i} + \underset{\sim}{j} - 9\underset{\sim}{k}$
7. State the vector which represents the cross product.	$\overrightarrow{PQ} \times \overrightarrow{PR} = -5\underset{\sim}{i} + \underset{\sim}{j} - 9\underset{\sim}{k}$

8. Calculate the magnitude of this cross product vector.

$$\left|\overrightarrow{PQ} \times \overrightarrow{PR}\right| = \sqrt{(-5)^2 + 1^2 + (-9)^2} = \sqrt{25 + 1 + 81} = \sqrt{107}$$

9. The area of the triangle is half the magnitude of the cross product vector. State the result.

The area of the triangle is $\frac{1}{2}\sqrt{107}$.

4.2 Exercise

Students, these questions are even better in jacPLUS

Receive immediate feedback and access sample responses

Access additional questions

Track your results and progress

Find all this and MORE in jacPLUS

Technology free

1. **WE1** Evaluate each of the following:

a. $\begin{vmatrix} -3 & 4 \\ 2 & -1 \end{vmatrix}$ b. $\begin{vmatrix} 4 & -1 \\ -3 & 2 \end{vmatrix}$

2. **WE2** Evaluate each of the following:

a. $\begin{vmatrix} 5 & 2 & 1 \\ 3 & 4 & -2 \\ 1 & 2 & 3 \end{vmatrix}$ b. $\begin{vmatrix} 1 & -2 & 3 \\ 2 & 4 & 2 \\ 5 & 3 & 1 \end{vmatrix}$

3. a. Given the vectors $\underset{\sim}{u} = 2\underset{\sim}{i} - \underset{\sim}{j} + 4\underset{\sim}{k}$ and $\underset{\sim}{v} = -\underset{\sim}{i} + 2\underset{\sim}{j} - 3\underset{\sim}{k}$ calculate $\underset{\sim}{u} \times \underset{\sim}{v}$.
b. Evaluate $(\underset{\sim}{i} + \underset{\sim}{j} - 3\underset{\sim}{k}) \times (2\underset{\sim}{i} + \underset{\sim}{j})$.
c. Given the vectors $\underset{\sim}{r} = 2\underset{\sim}{i} - 3\underset{\sim}{j}$ and $\underset{\sim}{s} = 3\underset{\sim}{i} + 2\underset{\sim}{j} + 4\underset{\sim}{k}$ calculate $\underset{\sim}{r} \times \underset{\sim}{s}$.

4. **WE3** Determine a unit vector perpendicular to the following vectors.
a. $3\underset{\sim}{i} - \underset{\sim}{j} + 2\underset{\sim}{k}$ and $4\underset{\sim}{i} - 2\underset{\sim}{j} + 5\underset{\sim}{k}$
b. $5\underset{\sim}{i} - 2\underset{\sim}{j} - 3\underset{\sim}{k}$ and $-2\underset{\sim}{i} + 3\underset{\sim}{j} + \underset{\sim}{k}$
c. $-5\underset{\sim}{i} + 2\underset{\sim}{j} + 3\underset{\sim}{k}$ and $2\underset{\sim}{i} - \underset{\sim}{j} - 2\underset{\sim}{k}$

5. Answer the following questions given the vectors $\underset{\sim}{a} = 3\underset{\sim}{i} - 2\underset{\sim}{j} + 4\underset{\sim}{k}$, $\underset{\sim}{b} = \underset{\sim}{i} + \underset{\sim}{j} - 3\underset{\sim}{k}$ and $\underset{\sim}{c} = 4\underset{\sim}{i} - 3\underset{\sim}{j} + 5\underset{\sim}{k}$.
a. Show that $\underset{\sim}{a} \times (\underset{\sim}{b} + \underset{\sim}{c}) = \underset{\sim}{a} \times \underset{\sim}{b} + \underset{\sim}{a} \times \underset{\sim}{c}$.
b. Show that $\underset{\sim}{a} \times (\underset{\sim}{b} - \underset{\sim}{c}) = \underset{\sim}{a} \times \underset{\sim}{b} - \underset{\sim}{a} \times \underset{\sim}{c}$.
c. Describe what property parts **a** and **b** illustrate.
d. Evaluate $(\underset{\sim}{a} + \underset{\sim}{b}) \times (\underset{\sim}{a} - \underset{\sim}{b})$ and $\underset{\sim}{b} \times \underset{\sim}{a} - \underset{\sim}{a} \times \underset{\sim}{b}$.
e. Calculate $\underset{\sim}{a} \times \underset{\sim}{c}$ and hence without finding further cross products, deduce the value of $(\underset{\sim}{a} + \underset{\sim}{c}) \times (\underset{\sim}{a} - \underset{\sim}{c})$.

6. **WE4** Calculate the area of a triangle with the following vertices.
a. $A(2, -1, 3)$, $B(1, -2, 4)$ and $C(4, 3, -1)$.
b. $P(3, 1, -1)$, $Q(1, 0, 2)$ and $R(2, 2, -3)$.
c. $A(-4, 2, -1)$, $B(-2, 1, -3)$ and $C(-3, 2, 1)$.

7. If $\underset{\sim}{a}=5\underset{\sim}{i}-\underset{\sim}{j}+7\underset{\sim}{k}$ and $\underset{\sim}{b}=5\underset{\sim}{i}+3\underset{\sim}{j}+4\underset{\sim}{k}$ answer the following questions.
   a. Determine the length of the vector $\underset{\sim}{a}$.
   b. Determine the length of the vector $\underset{\sim}{b}$.
   c. Evaluate $\underset{\sim}{a}\cdot\underset{\sim}{b}$.
   d. Evaluate the angle between the vectors $\underset{\sim}{a}$ and $\underset{\sim}{b}$.
   e. Evaluate $\underset{\sim}{a}\times\underset{\sim}{b}$.
   f. Determine a unit vector $\hat{\underset{\sim}{n}}$ perpendicular to both $\underset{\sim}{a}$ and $\underset{\sim}{b}$.
   g. Verify that $\underset{\sim}{a}\times\underset{\sim}{b}=|\underset{\sim}{a}||\underset{\sim}{b}|\sin(\theta)\hat{\underset{\sim}{n}}$.
   h. Show that $|\underset{\sim}{a}\times\underset{\sim}{b}|^2=|\underset{\sim}{a}|^2\,|\underset{\sim}{b}|^2-(\underset{\sim}{a}\cdot\underset{\sim}{b})^2$.

8. If $\underset{\sim}{a}=5\underset{\sim}{i}-2\underset{\sim}{j}+4\underset{\sim}{k}$, $\underset{\sim}{b}=2\underset{\sim}{i}+3\underset{\sim}{j}-\underset{\sim}{k}$ and $\underset{\sim}{c}=\underset{\sim}{a}\times\underset{\sim}{b}$ answer the following questions.
   a. Determine the vector $\underset{\sim}{c}$.
   b. Calculate $\underset{\sim}{a}\cdot\underset{\sim}{b}$.
   c. Calculate $\underset{\sim}{a}\cdot\underset{\sim}{c}$.
   d. Calculate $\underset{\sim}{b}\cdot\underset{\sim}{c}$.
   e. Explain your answers to **c** and **d**.
   f. Determine whether the vectors $\underset{\sim}{a}$, $\underset{\sim}{b}$ and $\underset{\sim}{c}$ are linearly dependant or independent. Explain your answer.

9. Given the vectors $\underset{\sim}{p}=x\underset{\sim}{i}+y\underset{\sim}{j}-\underset{\sim}{k}$ and $\underset{\sim}{q}=4\underset{\sim}{i}-2\underset{\sim}{j}+2\underset{\sim}{k}$ determine the values of x and y if:
   a. $\underset{\sim}{p}\times\underset{\sim}{q}=\underset{\sim}{0}$
   b. $\underset{\sim}{p}\cdot\underset{\sim}{q}=0$ and $|\underset{\sim}{p}|=\sqrt{14}$.

10. a. Calculate the value of y if $\underset{\sim}{p}=3\underset{\sim}{i}+y\underset{\sim}{j}-\underset{\sim}{k}$ and $\underset{\sim}{q}=4\underset{\sim}{i}+y\underset{\sim}{j}-2\underset{\sim}{k}$ and $\underset{\sim}{p}\times\underset{\sim}{q}=3\underset{\sim}{i}+2\underset{\sim}{j}+3\underset{\sim}{k}$.
    b. If $\underset{\sim}{r}=x\underset{\sim}{i}+2\underset{\sim}{j}-\underset{\sim}{k}$ and $\underset{\sim}{s}=4\underset{\sim}{i}-4\underset{\sim}{j}+z\underset{\sim}{k}$ and $\underset{\sim}{r}\times\underset{\sim}{s}=\underset{\sim}{0}$, determine the values of x and z.
    c. Given the vectors $\underset{\sim}{a}=t\underset{\sim}{i}-\underset{\sim}{j}+2t\underset{\sim}{k}$ and $\underset{\sim}{b}=2t\underset{\sim}{i}+3\underset{\sim}{j}-t\underset{\sim}{k}$, determine the value of t if $|\underset{\sim}{a}\times\underset{\sim}{b}|=10\sqrt{6}$.

11. a. By taking the cross product of the vectors $\underset{\sim}{a}=\cos(\alpha)\underset{\sim}{i}+\sin(\alpha)\underset{\sim}{j}$ and $\underset{\sim}{b}=\cos(\beta)\underset{\sim}{i}+\sin(\beta)\underset{\sim}{j}$ and interpreting geometrically deduce that $\sin(\beta-\alpha)=\sin(\beta)\cos(\alpha)-\cos(\beta)\sin(\alpha)$.
    b. When a force $\underset{\sim}{F}$ moves a point from A to B producing a displacement $\underset{\sim}{s}=\overrightarrow{AB}$, the torque produced is $\underset{\sim}{\tau}=\underset{\sim}{F}\times\underset{\sim}{s}$. Calculate the magnitude of the torque when the force $\underset{\sim}{F}=3\underset{\sim}{i}+2\underset{\sim}{j}-\underset{\sim}{k}$ moves the point $A(1,\,-1,\,2)$ to the point $B(2,\,-1,\,4)$.

12. a. The expression $\underset{\sim}{a}\times\underset{\sim}{b}\cdot\underset{\sim}{c}$ is called the scalar triple product. Explain why the expression $(\underset{\sim}{a}\times\underset{\sim}{b})\cdot\underset{\sim}{c}=\underset{\sim}{a}\times\underset{\sim}{b}\cdot\underset{\sim}{c}$ does not require the brackets.
    For the vectors $\underset{\sim}{a}=2\underset{\sim}{i}+\underset{\sim}{j}-4\underset{\sim}{k}$, $\underset{\sim}{b}=3\underset{\sim}{i}-2\underset{\sim}{j}-\underset{\sim}{k}$ and $\underset{\sim}{c}=\underset{\sim}{i}+2\underset{\sim}{j}-3\underset{\sim}{k}$ evaluate:
    b. $\underset{\sim}{a}\times\underset{\sim}{b}\cdot\underset{\sim}{c}$
    c. $\underset{\sim}{a}\cdot\underset{\sim}{b}\times\underset{\sim}{c}$
    d. $\underset{\sim}{c}\times\underset{\sim}{a}\cdot\underset{\sim}{b}$
    e. $\begin{vmatrix} 2 & 1 & -4 \\ 3 & -2 & -1 \\ 1 & 2 & -3 \end{vmatrix}$

13. a. Expressions such as $(\underset{\sim}{a}\times\underset{\sim}{b})\times\underset{\sim}{c}$ or $\underset{\sim}{a}\times(\underset{\sim}{b}\times\underset{\sim}{c})$ are called vector triple products.
    State some meaning to the expression $\underset{\sim}{a}\times\underset{\sim}{b}\times\underset{\sim}{c}$.
    b. For the vectors $\underset{\sim}{a}=\underset{\sim}{i}-2\underset{\sim}{j}+3\underset{\sim}{k}$, $\underset{\sim}{b}=2\underset{\sim}{i}-\underset{\sim}{j}-\underset{\sim}{k}$ and $\underset{\sim}{c}=3\underset{\sim}{i}+4\underset{\sim}{j}-2\underset{\sim}{k}$ answer the following questions.
       i. Evaluate $(\underset{\sim}{a}\times\underset{\sim}{b})\times\underset{\sim}{c}$.
       ii. Evaluate $\underset{\sim}{a}\times(\underset{\sim}{b}\times\underset{\sim}{c})$.
       iii. Determine if $(\underset{\sim}{a}\times\underset{\sim}{b})\times\underset{\sim}{c}=\underset{\sim}{a}\times(\underset{\sim}{b}\times\underset{\sim}{c})$.
       iv. Show that $\underset{\sim}{a}\times(\underset{\sim}{b}\times\underset{\sim}{c})=(\underset{\sim}{a}\cdot\underset{\sim}{c})\,\underset{\sim}{b}-(\underset{\sim}{a}\cdot\underset{\sim}{b})\,\underset{\sim}{c}$.
       v. Show that $(\underset{\sim}{a}\times\underset{\sim}{b})\times\underset{\sim}{c}=(\underset{\sim}{a}\cdot\underset{\sim}{c})\,\underset{\sim}{b}-(\underset{\sim}{b}\cdot\underset{\sim}{c})\,\underset{\sim}{a}$.

Technology active

14. For each of the following, given the vector $\underset{\sim}{a} = x\underset{\sim}{i} + y\underset{\sim}{j} + z\underset{\sim}{k}$, determine the values of x, y and z if:

a. $\underset{\sim}{b} = 3\underset{\sim}{i} - \underset{\sim}{j} + 2\underset{\sim}{k}$, $\underset{\sim}{a} \times \underset{\sim}{b} = -2\underset{\sim}{i} + 8\underset{\sim}{j} + 7\underset{\sim}{k}$ and $\underset{\sim}{a} \cdot \underset{\sim}{b} = 17$

b. $\underset{\sim}{b} = 2\underset{\sim}{i} - 3\underset{\sim}{j} + 2\underset{\sim}{k}$, $\underset{\sim}{a} \times \underset{\sim}{b} = 5\underset{\sim}{i} + 18\underset{\sim}{j} + 22\underset{\sim}{k}$ and $\underset{\sim}{a}$ and $\underset{\sim}{b}$ are perpendicular

c. $\underset{\sim}{b} = 2\underset{\sim}{i} + 3\underset{\sim}{j} - 5\underset{\sim}{k}$, $\underset{\sim}{a} \times \underset{\sim}{b} = 7\underset{\sim}{i} + 12\underset{\sim}{j} + 10\underset{\sim}{k}$ and the length of the vector $\underset{\sim}{a}$ is 3

d. $\underset{\sim}{b} = 3\underset{\sim}{i} - 2\underset{\sim}{j} + 4\underset{\sim}{k}$, $\underset{\sim}{a} \times \underset{\sim}{b} = 26\underset{\sim}{i} + 23\underset{\sim}{j} - 8\underset{\sim}{k}$ and the length of the vector $\underset{\sim}{a}$ is $3\sqrt{5}$.

4.2 Exam questions

Question 1 (1 mark) TECH-ACTIVE

MC Which of the following statements imply that $\underset{\sim}{a} \times \underset{\sim}{b} = \underset{\sim}{0}$?

A. $\underset{\sim}{a}$ and $\underset{\sim}{b}$ are perpendicular
B. $\underset{\sim}{a}$ and $\underset{\sim}{b}$ are parallel
C. $|\underset{\sim}{a}| = |\underset{\sim}{b}|$
D. Neither $\underset{\sim}{a}$ or $\underset{\sim}{b}$ are the zero vector
E. $|\underset{\sim}{a}| < |\underset{\sim}{b}| < 1$

Question 2 (3 marks) TECH-FREE

Consider the points $P(3, 0, 0)$, $Q(0, -2, 0)$ and $R(0, 0, 1)$.

Determine the area of the triangle PQR using a vector method.

Question 3 (2 marks) TECH-FREE

For any two vectors $\underset{\sim}{a}$ and $\underset{\sim}{b}$, show that $|\underset{\sim}{a} \times \underset{\sim}{b}|^2 = |\underset{\sim}{a}|^2\,|\underset{\sim}{b}|^2 - (\underset{\sim}{a} \cdot \underset{\sim}{b})^2$.

More exam questions are available online.

4.3 Lines in three dimensions

LEARNING INTENTION

At the end of this subtopic you should be able to:

- determine the vector equation of a straight line, given the position of two points, or equivalent information, in both two and three dimensions.

4.3.1 The equation of a line

To uniquely define a line in three dimensions, we need to specify a point on the line and the direction of the line. Consider a line with direction $\underset{\sim}{v} = a\underset{\sim}{i} + b\underset{\sim}{j} + c\underset{\sim}{k}$ passing through a given fixed point $P_0(x_0, y_0, z_0)$. Let $P(x, y, z)$ be any general point on the line. If O is the origin, then the vector from the origin to the fixed point P_0 is given by $\underset{\sim}{r}_0 = \overrightarrow{OP_0} = x_0\underset{\sim}{i} + y_0\underset{\sim}{j} + z_0\underset{\sim}{k}$. Now consider the vector from the origin to the general point P, that is $\overrightarrow{OP}$. This vector is denoted by the position vector $\underset{\sim}{r}(t) = \overrightarrow{OP} = x\underset{\sim}{i} + y\underset{\sim}{j} + z\underset{\sim}{k}$, where t is a scalar.

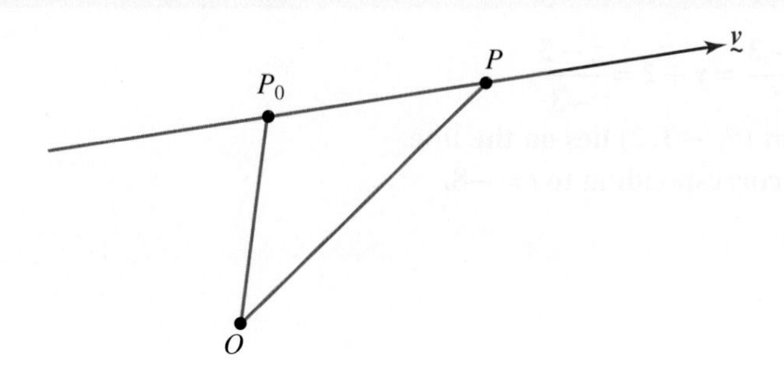

Now $\overrightarrow{OP} = \overrightarrow{OP_0} + \overrightarrow{P_0P}$, and since $\overrightarrow{P_0P}$ is a vector parallel to the direction of the line $\underset{\sim}{v}$, then it can be written as a multiple of the that vector, so $\overrightarrow{P_0P} = t\underset{\sim}{v}$ where $t \in R$.

The vector equation of the line can be written as $\underset{\sim}{r}(t) = \underset{\sim}{r}_0 + t\underset{\sim}{v}$.

Expanding the vector form into components gives

$$\begin{aligned} x\underset{\sim}{i} + y\underset{\sim}{j} + z\underset{\sim}{k} &= x_0\underset{\sim}{i} + y_0\underset{\sim}{j} + z_0\underset{\sim}{k} + t(a\underset{\sim}{i} + b\underset{\sim}{j} + c\underset{\sim}{k}) \\ &= (x_0 + ta)\underset{\sim}{i} + (y_0 + tb)\underset{\sim}{j} + (z_0 + tc)\underset{\sim}{k} \end{aligned}$$

or by equating the components of $\underset{\sim}{i}$, $\underset{\sim}{j}$ and $\underset{\sim}{k}$ gives the **parametric equations** of the line

$$x = x_0 + ta$$
$$y = y_0 + tb$$
$$z = z_0 + tc$$

where t is the parameter.

We can eliminate t from these three equations and write these equations in the **Cartesian equation** of the line: $\frac{x - x_0}{a} = \frac{y - y_0}{b} = \frac{z - z_0}{c}$, assuming that $a \neq 0$, $b \neq 0$ and $c \neq 0$.

The value of t can be interpreted as stepping along the line, as t changes, we move to different points along the line. Note that when $t = 0$, the point on the line is $P_0(x_0, y_0, z_0)$. The value of t can be positive or negative, and is usually in an integer, but does not have to be an integer, so t can be any real number, so $t \in R$.

Equation of a line

The line with direction $\underset{\sim}{v} = a\underset{\sim}{i} + b\underset{\sim}{j} + c\underset{\sim}{k}$, passing through the point $P_0(x_0, y_0, z_0)$ has a vector equation $\underset{\sim}{r}(t) = \underset{\sim}{r}_0 + t\underset{\sim}{v}$.

Expressed in parametric form:

$$x = x_0 + ta$$
$$y = y_0 + tb$$
$$z = z_0 + tc$$

Expressed in Cartesian form:

$$\frac{x - x_0}{a} = \frac{y - y_0}{b} = \frac{z - z_0}{c}$$

WORKED EXAMPLE 5 Investigating lines in 3 dimensions using vectors

Consider the line $t = \frac{x - 3}{2} = y + 2 = \frac{z - 5}{-3}$.

a. **Determine if the point $(5, -1, 2)$ lies on the line.**

b. **Determine the point corresponding to $t = -8$.**

THINK	WRITE
a. 1. Write the line in parametric form.	**a.** $t=\frac{x-3}{2}=\frac{y+2}{1}=\frac{z-5}{-3}$ Therefore: $x=3+2t,\ y=-2+t,\ z=5-3t$
2. Determine the values of x, y and z at the point $(5,-1,2)$.	Given the point $(5,-1,2)$, $x=5,\ y=-1,\ z=2$
3. Check the x-value.	$x=3+2t$ $5=3+2t$ $t=1$
4. Check the y-value.	$y=-2+t$ $-1=-2+t$ $t=1$
5. Check the z-value.	$z=5-3t$ $2=5-3t$ $t=1$
6. State the conclusion.	When $t=1, x=5,\ y=-1,\ z=2$. The point $(5,-1,2)$ therefore does lie on the line.
b. 1. Substitute $t=-8$ into the parametric equation of the line.	**b.** $x=3+2(-8)=-13$ $y=-2+(-8)=-10$ $z=5-3\times(-8)=29$
2. State the result.	When $t=-8$, the point is $(-13,-10,29)$.

Lines parallel to the xy, xz or yz plane

If say $a=0,\ b\neq 0$ and $c\neq 0$ then these equations are written in the form $x=x_0;\ \frac{y-y_0}{b}=\frac{z-z_0}{c}$, in this case the line is parallel to the yz plane.

Similarly, if $b=0$ or $c=0$, then the line is parallel to the xz or xy plane and the equation is of the form $y=y_0;\ \frac{x-x_0}{a}=\frac{z-z_0}{c}$ or $z=z_0;\ \frac{x-x_0}{a}=\frac{y-y_0}{b}$ respectively.

WORKED EXAMPLE 6 Determining the equation of a line

Determine the equation of the line in Cartesian form, passing through the point $P_0(3,-2,4)$ parallel to the vector $2\underset{\sim}{i}-\underset{\sim}{j}$.

THINK	WRITE
1. Determine $\underset{\sim}{r}_0$ using the coordinates of the point $P_0(3,-2,4)$.	$\underset{\sim}{r}_0=\overrightarrow{OP_0}=3\underset{\sim}{i}-2\underset{\sim}{j}+4\underset{\sim}{k}$
2. State the direction of the line.	$\underset{\sim}{v}=2\underset{\sim}{i}-\underset{\sim}{j}$
3. State the vector equation of the line.	$\underset{\sim}{r}(t)=\underset{\sim}{r}_0+t\underset{\sim}{v}$ $=(3\underset{\sim}{i}-2\underset{\sim}{j}+4\underset{\sim}{k})+t(2\underset{\sim}{i}-\underset{\sim}{j})$ $=(3+2t)\underset{\sim}{i}+(-2-t)\underset{\sim}{j}+4\underset{\sim}{k}$

4. State the line in parametric form.

$x = 3 + 2t$
$y = -2 - t$
$z = 4$

5. Rearrange each of the parametric equations to isolate t.

$t = \frac{x-3}{2}$
$t = \frac{y+2}{-1}$
$z = 4$

6. Eliminate t and write the line in Cartesian form. In this case the line is parallel to the xy plane.

$t = \frac{x-3}{2} = \frac{y+2}{-1}$
$\frac{x-3}{2} = \frac{y+2}{-1};\ z = 4$

Determining the equation of a line given two points

A straight line can also be specified by two distinct points on the line.

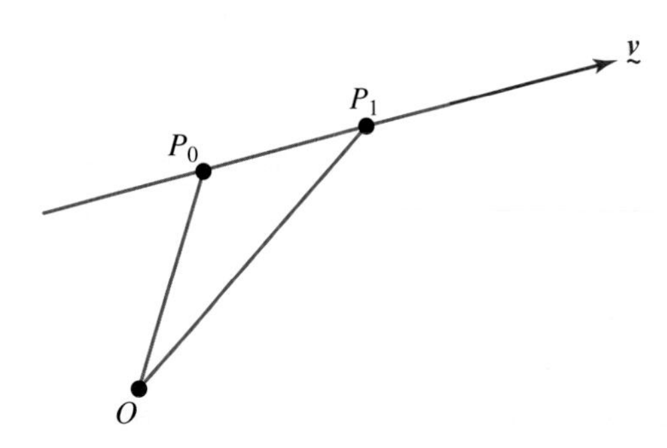

If we are given two points on the line, $P_0(x_0, y_0, z_0)$ and $P_1(x_1, y_1, z_1)$ then the vector $\overrightarrow{P_0P_1}$ is a vector parallel to the direction of the line, so that $\underset{\sim}{v} = \overrightarrow{P_0P_1} = (x_1 - x_0)\underset{\sim}{i} + (y_1 - y_0)\underset{\sim}{j} + (z_1 - z_0)\underset{\sim}{k}$.

WORKED EXAMPLE 7 Determining the equation of a line from two points

Determine the Cartesian equation of the line passing through the points $P(2, 1, -3)$ and $Q(1, 3, -2)$.

THINK

1. Write the position vectors of the two points.

WRITE

$P(2, 1, -3)$ and $Q(1, 3, -2)$
$\overrightarrow{OP} = 2\underset{\sim}{i} + \underset{\sim}{j} - 3\underset{\sim}{k}$
$\overrightarrow{OQ} = \underset{\sim}{i} + 3\underset{\sim}{j} - 2\underset{\sim}{k}$

2. Determine the vector passing through the two given points. This is the direction of the line.

$\overrightarrow{PQ} = \overrightarrow{OQ} - \overrightarrow{OP}$
$= -\underset{\sim}{i} + 2\underset{\sim}{j} + \underset{\sim}{k}$
$\underset{\sim}{v} = -\underset{\sim}{i} + 2\underset{\sim}{j} + \underset{\sim}{k}$

3. State the vector equation of the line. You can use either point P or Q.	$\underset{\sim}{r} = \overrightarrow{OP} + t\underset{\sim}{v}$ $= (2\underset{\sim}{i} + \underset{\sim}{j} - 3\underset{\sim}{k}) + t(-\underset{\sim}{i} + 2\underset{\sim}{j} + \underset{\sim}{k})$ $= (2 - t)\underset{\sim}{i} + (1 + 2t)\underset{\sim}{j} + (-3 + t)\underset{\sim}{k}$
4. State the equation of the line in parametric form.	$x = 2 - t$ $y = 1 + 2t$ $z = -3 + t$
5. Rearrange each of the parametric equations to isolate t.	$t = \frac{x - 2}{-1}$ $t = \frac{y - 1}{2}$ $t = \frac{z + 3}{1}$
6. Eliminate the parameter t. This is the Cartesian equation of the line.	$\frac{x - 2}{-1} = \frac{y - 1}{2} = \frac{z + 3}{1}$

Angle between two lines

If we are given two intersecting lines there are a couple of possible angles between the lines. Given line one L_1 specified by the direction of the line $\underset{\sim}{v}_1$ and line two L_2 specified by the direction of the line $\underset{\sim}{v}_2$, the angle between the lines could be acute or obtuse. We define the angle between the lines to be the acute angle between the directions of the lines.

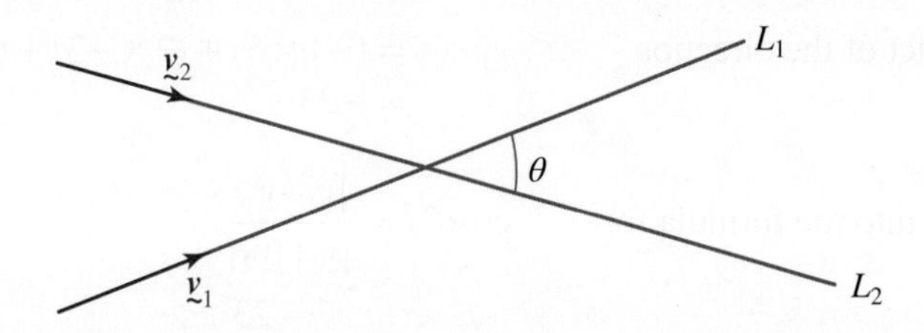

The angle, θ, between two lines is the angle between their direction vectors $\underset{\sim}{v}_1$ and $\underset{\sim}{v}_2$, such that
$\cos(\theta) = \frac{|\underset{\sim}{v}_1 \cdot \underset{\sim}{v}_2|}{|\underset{\sim}{v}_1|\,|\underset{\sim}{v}_2|} = |\hat{\underset{\sim}{v}}_1 \cdot \hat{\underset{\sim}{v}}_2|$, and $0 \le \theta \le 90°$.

Angle between two lines

$$\theta = \cos^{-1}\left(\frac{|\underset{\sim}{v}_1 \cdot \underset{\sim}{v}_2|}{|\underset{\sim}{v}_1|\,|\underset{\sim}{v}_2|}\right) = \cos^{-1}\left(|\hat{\underset{\sim}{v}}_1 \cdot \hat{\underset{\sim}{v}}_2|\right)$$

Note: Angles may be expressed as exact values in terms of the inverse cosine ratio or approximately using a calculator. In some cases in this topic you will be allowed the use of a scientific calculator to evaluate inverse cosine ratios and determine approximate values of angles.

WORKED EXAMPLE 8 Determining the angle between two lines

Given the two lines $\frac{x-1}{-1} = \frac{y-3}{2} = \frac{z+2}{3}$ and $x = -3 + 5t$, $y = 4 - 3t$, $z = 1 - 4t$, determine the angle between the lines. Use a scientific calculator to state the angle in degrees, rounded to 1 decimal place.

THINK	WRITE
1. State the direction of the first line by first converting it into parametric form.	$\frac{x-1}{-1} = \frac{y-3}{2} = \frac{z+2}{3} = t$ $x = -t + 1$ $y = 2t + 3$ $z = 3t - 2$ $\underset{\sim}{v}_1 = -\underset{\sim}{i} + 2\underset{\sim}{j} + 3\underset{\sim}{k}$
2. State the direction of the second line.	$x = -3 + 5t$ $y = 4 - 3t$ $z = 1 - 4t$ $\underset{\sim}{v}_2 = 5\underset{\sim}{i} - 3\underset{\sim}{j} - 4\underset{\sim}{k}$
3. Calculate the magnitude of $\underset{\sim}{v}_1$.	$\lvert\underset{\sim}{v}_1\rvert = \sqrt{1^2 + 2^2 + 3^2}$ $= \sqrt{14}$
4. Calculate the magnitude of $\underset{\sim}{v}_2$.	$\lvert\underset{\sim}{v}_2\rvert = \sqrt{5^2 + (-3)^2 + (-4)^2}$ $= \sqrt{50}$ $= 5\sqrt{2}$
5. Calculate the dot product of the direction vectors.	$\underset{\sim}{v}_1 \cdot \underset{\sim}{v}_2 = (-1 \times 5) + (2 \times -3) + (3 \times -4)$ $= -23$
6. Substitute these values into the formula for the angle.	$\cos(\theta) = \frac{\lvert\underset{\sim}{v}_1 \cdot \underset{\sim}{v}_2\rvert}{\lvert\underset{\sim}{v}_1\rvert\,\lvert\underset{\sim}{v}_2\rvert}$ $= \frac{23}{\sqrt{14} \times 5\sqrt{2}}$ $= \frac{23}{5\sqrt{28}} = \frac{23\sqrt{7}}{70}$
7. Determine the angle.	$\theta = \cos^{-1}\left(\frac{23\sqrt{7}}{70}\right)$
8. Use a scientific calculator to evaluate the inverse cosine in degrees.	$= 29.6°$
9. State the result.	The angle between the lines is 29.6°.

Intersection between two lines

Two lines are parallel if the vectors defining their directions are parallel, that is if $\underset{\sim}{v}_1 = \lambda\, \underset{\sim}{v}_2$ where $\lambda \in R$. In two dimensions if two lines are not parallel then they intersect in a unique point. This is not necessarily so in three dimensions. Lines that are not parallel which do not intersect are called skew lines. Any two lines which lie in parallel planes will be skew.

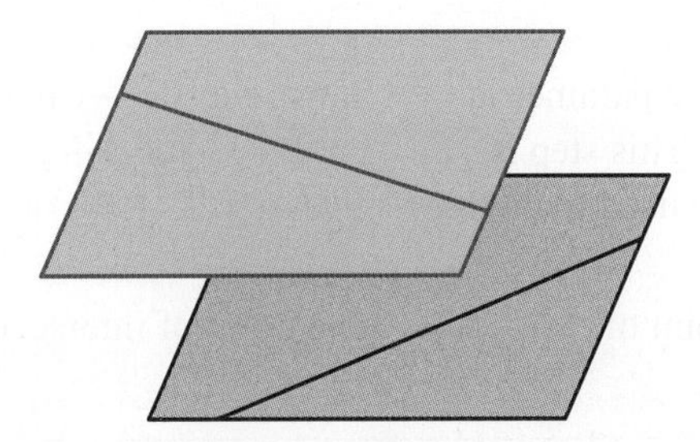

To determine the intersection of two lines, we cannot use the same parameter for each line, so we use t for one line and s for the other line. To determine the point of intersection, if it exists, we must equate the three parametric equations for x, y and z from both lines and solve. Note that we must check all three equations to check the lines do in fact intersect.

WORKED EXAMPLE 9 Determining whether two lines intersect

Given the two lines $\frac{x+3}{5}=\frac{y-4}{-3}=\frac{z-1}{-4}$ and $\frac{x-1}{-1}=\frac{y-3}{2}=\frac{z+2}{1}$, show that the lines intersect and determine the point of intersection.

THINK	WRITE
1. Write the first line in parametric form.	Let $t=\frac{x+3}{5}=\frac{y-4}{-3}=\frac{z-1}{-4}$ $x=-3+5t,\ y=4-3t,\ z=1-4t$
2. Write the second line in parametric form.	Let $s=\frac{x-1}{-1}=\frac{y-3}{2}=\frac{z+2}{1}$ $x=1-s,\ y=3+2s,\ z=-2+s$
3. Equate the x, y and z components.	(1) $-3+5t=1-s$ (2) $4-3t=3+2s$ (3) $1-4t=-2+s$
4. Solve the first equation for s in terms of t.	(1) $\Rightarrow s=4-5t$
5. Substitute for s into the second equation and solve for t.	$4-3t=3+2(4-5t)$ $4-3t=3+8-10t$ $7t=7$ $t=1$
6. Determine the corresponding value of s from the first equation.	(1) $s=4-5t$ $=4-5$ $=-1$
7. Since we have not yet used the third equation, we must check it is valid.	When $t=1$ and $s=-1$ $1-4t=-2+s$ $1-4(1)=-2+(-1)$ $-3=-3$ The third equation is consistent, so the lines intersect.
8. Substitute the value of t into the parametric equations for the first line.	Substitute $t=1$ into the first line $x=-3+5t\ y=4-3t\ z=1-4t$; this gives $x=2,\ y=1$ and $z=-3$.

9. Substitute the value of s into the parametric equations for the second line. (This step is not really necessary but can be used to double check that the lines intersect.)	Substitute $s = -1$ into the second line $x = 1 - s,\ y = 3 + 2s,\ z = -2 + s$; this also gives $x = 2,\ y = 1$ and $z = -3$.
10. State the coordinates of the point of intersection between the lines.	The point of intersection is $(2, 1, -3)$.

If the equations do not check out and we have a contradiction, or inconsistency, then the lines do not intersect, and there is no solution for the intersection between the two lines.

WORKED EXAMPLE 10 Showing that two lines don't intersect

Show that the two lines $\frac{x-2}{-5} = \frac{y-1}{3} = \frac{z+3}{4}$ and $x = -4 - t,\ y = 6 + 2t,\ z = 3 + t$ do not intersect.

THINK	WRITE
1. Write the first line in parametric form.	Let $s = \frac{x-2}{-5} = \frac{y-1}{3} = \frac{z+3}{4}$ $x = 2 - 5s,\ y = 1 + 3s,\ z = -3 + 4s$
2. Equate the x, y and z components.	(1) $2 - 5s = -4 - t$ (2) $1 + 3s = 6 + 2t$ (3) $-3 + 4s = 3 + t$
3. Solve the first equation for t in terms of s.	(1) $\Rightarrow\ t = -6 + 5s$
4. Substitute for t into the second equation and solve for s.	$1 + 3s = 6 + 2(-6 + 5s)$ $1 + 3s = 6 - 12 + 10s$ $7s = 7$ $s = 1$
5. Determine the corresponding value of t from the first equation.	(1) $t = -6 + 5s$ $= -6 + 5$ $= -1$
6. Since we have not yet used the third equation, we must check it is valid.	Substitute $t = -1$ and $s = 1$ into (3). $-3 + 4s = 3 + t$ $-3 + 4(1) = 3 + (-1)$ $1 = 2$ This is a contradiction, the third equation is inconsistent.
7. State the conclusion.	The lines do not intersect.

Distance of a point to a line

Consider calculating the distance of a point to a line. By this distance we understand it to be the closest perpendicular distance. Let the line L be specified by a point P_0 on the line and the direction of the line $\underset{\sim}{v}$. Consider calculating the distance d from this line to a given point A.

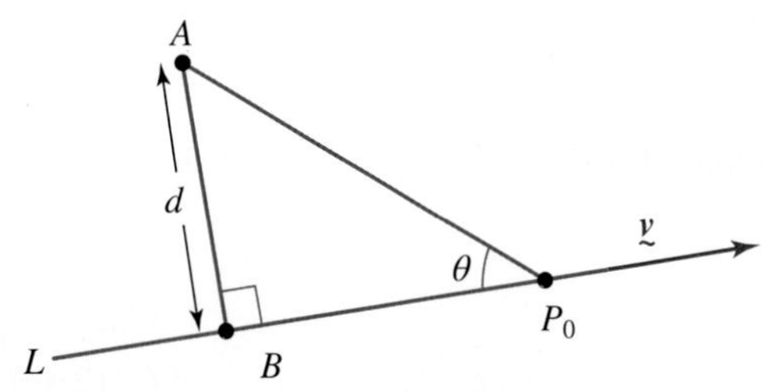

Let B be the point on the line such that AB is perpendicular to the line. Now in the triangle $\sin(\theta) = \dfrac{d}{\left|\overrightarrow{AP_0}\right|}$,

so that $d = \left|\overrightarrow{AP_0}\right| \sin(\theta)$. However, from the definition of the cross-product, we can write $\overrightarrow{AP_0} \times \underset{\sim}{v} = \left|\overrightarrow{AP_0}\right| |\underset{\sim}{v}| \sin(\theta)\, \hat{\underset{\sim}{n}}$ where $\hat{\underset{\sim}{n}}$ is a unit vector perpendicular to both $\overrightarrow{AP_0}$ and $\underset{\sim}{v}$. Rearranging and taking the magnitudes gives $\sin(\theta) = \dfrac{\left|\overrightarrow{AP_0} \times \underset{\sim}{v}\right|}{\left|\overrightarrow{AP_0}\right| |\underset{\sim}{v}|}$ since $|\hat{\underset{\sim}{n}}| = 1$. Substituting this back into $d = \left|\overrightarrow{AP_0}\right| \sin(\theta)$ gives

$$d = \left|\overrightarrow{AP_0}\right| \times \frac{\left|\overrightarrow{AP_0} \times \underset{\sim}{v}\right|}{\left|\overrightarrow{AP_0}\right| |\underset{\sim}{v}|} = \frac{\left|\overrightarrow{AP_0} \times \underset{\sim}{v}\right|}{|\underset{\sim}{v}|}.$$

Distance from a line to a point

The distance d between a point A and a line with direction $\underset{\sim}{v}$ is:

$$d = \frac{\left|\overrightarrow{AP_0} \times \underset{\sim}{v}\right|}{|\underset{\sim}{v}|} = \left|\overrightarrow{AP_0} \times \hat{\underset{\sim}{v}}\right|$$

WORKED EXAMPLE 11 Calculating the distance of a point from a line

Calculate the distance of the point (1, 3, −2) to the line $\dfrac{x-3}{2} = \dfrac{y+1}{-2} = \dfrac{z-5}{3}$.

THINK	WRITE
1. Determine the direction of the line by first converting the line into parametric form.	$\dfrac{x-3}{2} = \dfrac{y+1}{-2} = \dfrac{z-5}{3}$ $x = 3 + 2t, y = -1 - 2t, z = 5 + 3t$ $\underset{\sim}{v} = 2\underset{\sim}{i} - 2\underset{\sim}{j} + 3\underset{\sim}{k}$
2. Calculate a point on the line by determining the coordinates when $t = 0$.	When $t = 0$: $x = 3,\ y = -1,\ z = 5$ $P_0(3, -1, 5)$ $\overrightarrow{OP_0} = 3\underset{\sim}{i} - \underset{\sim}{j} + 5\underset{\sim}{k}$
3. Determine the vector $\overrightarrow{AP_0}$.	The point is $A(1, 3, -2)$. $\overrightarrow{OA} = \underset{\sim}{i} + 3\underset{\sim}{j} - 2\underset{\sim}{k}$ $\overrightarrow{AP_0} = \overrightarrow{OP_0} - \overrightarrow{OA}$ $= (3\underset{\sim}{i} - \underset{\sim}{j} + 5\underset{\sim}{k}) - (\underset{\sim}{i} + 3\underset{\sim}{j} - 2\underset{\sim}{k})$ $= 2\underset{\sim}{i} - 4\underset{\sim}{j} + 7\underset{\sim}{k}$

4. Calculate the vector cross product of $\overrightarrow{AP_0}$ with the direction of the line.

$$\overrightarrow{AP_0} \times \underset{\sim}{v} = \begin{vmatrix} \underset{\sim}{i} & \underset{\sim}{j} & \underset{\sim}{k} \\ 2 & -4 & 7 \\ 2 & -2 & 3 \end{vmatrix}$$
$$= \underset{\sim}{i}\begin{vmatrix} -4 & 7 \\ -2 & 3 \end{vmatrix} - \underset{\sim}{j}\begin{vmatrix} 2 & 7 \\ 2 & 3 \end{vmatrix} + \underset{\sim}{k}\begin{vmatrix} 2 & -4 \\ 2 & -2 \end{vmatrix}$$
$$= (-12+14)\underset{\sim}{i} - (6-14)\underset{\sim}{j} + (-4+8)\underset{\sim}{k}$$
$$= 2\underset{\sim}{i} + 8\underset{\sim}{j} + 4\underset{\sim}{k}$$

5. Calculate the magnitude of the vector $\overrightarrow{AP_0} \times \underset{\sim}{v}$.

$$\left|\overrightarrow{AP_0} \times \underset{\sim}{v}\right| = \sqrt{2^2+8^2+4^2}$$
$$= \sqrt{4+64+16}$$
$$= \sqrt{84}$$

6. Calculate the magnitude of the direction of the line.

$$|\underset{\sim}{v}| = \sqrt{2^2+(-2)^2+3^2}$$
$$= \sqrt{4+4+9}$$
$$= \sqrt{17}$$

7. Determine the distance between the point and the line and rationalise the denominator.

$$d = \frac{\left|\overrightarrow{AP_0} \times \underset{\sim}{v}\right|}{|\underset{\sim}{v}|}$$
$$= \frac{\sqrt{84}}{\sqrt{17}} \times \frac{\sqrt{17}}{\sqrt{17}}$$
$$= \frac{2\sqrt{21} \times \sqrt{17}}{17}$$

8. State the answer.

$$d = \frac{2\sqrt{357}}{17}$$

4.3 Exercise

Students, these questions are even better in jacPLUS

Receive immediate feedback and access sample responses

Access additional questions

Track your results and progress

Find all this and MORE in jacPLUS

Technology free

1. **WE5** Consider the line $t = \frac{x-5}{-3} = \frac{y+1}{-6} = \frac{z+2}{2}$.
   a. Determine whether the point $(8, 5, 4)$ lies on the line.
   b. Determine the point corresponding to $t = 6$.

2. a. Determine if the point $(1, 2, -4)$ lies on the line $\dfrac{x-4}{-1} = \dfrac{y+4}{2} = \dfrac{z-5}{-3}$.

b. Given the line, $t = \dfrac{x-6}{4} = \dfrac{y+3}{2} = \dfrac{z-5}{-3}$, determine the coordinates of the points on the line, corresponding to $t = 10, t = 5$ and $t = 0$.

3. **WE6** Determine the Cartesian equation of the following lines in Cartesian form.

a. Passing through the point $(2, -3)$ parallel to the vector $\underset{\sim}{i} + 2\underset{\sim}{j}$

b. Passing through the point $(2, -3, 4)$ parallel to the vector $\underset{\sim}{i} + 2\underset{\sim}{j} - 3\underset{\sim}{k}$

c. Passing through the point $(3, 5, 1)$ parallel to the vector $2\underset{\sim}{i} + \underset{\sim}{k}$

4. **WE7** Determine the equation of the line in Cartesian form passing through the following points.

a. $(2, -3, 4)$ and $(5, 2, -2)$

b. $(1, 0, -3)$ and $(-3, 2, 5)$

c. $(3, 4, -2)$ and $(0, 6, -2)$

5. **WE8** Given the two lines, calculate the angle in between the following pairs of intersecting lines. Use a scientific calculator to state the angle in degrees, rounded to 1 decimal place.

a. $\dfrac{x-14}{-3} = y+5 = \dfrac{z-11}{-2}$ and $x = -6 - 4t, y = -7 - 3t, z = 7 + 2t$

b. $\dfrac{x-13}{3} = \dfrac{y-4}{2} = \dfrac{z+7}{-2}$ and $x = -1 + t, y = 8 - 2t, z = 14 - 3t$

c. $\dfrac{x-7}{-3} = \dfrac{y+10}{4} = \dfrac{z+1}{2}$ and $x = -2 + 2t, y = -4 - 2t, z = 2 - t$

6. **WE9** Given the two lines, show that the lines intersect and determine the point of intersection.

a. $\dfrac{x-14}{-3} = y+5 = \dfrac{z-11}{-2}$ and $x = -6 - 4t, y = -7 - 3t, z = 7 + 2t$

b. $\dfrac{x-13}{3} = \dfrac{y-4}{2} = \dfrac{z+7}{-2}$ and $x = -1 + t, y = 8 - 2t, z = 14 - 3t$

c. $\dfrac{x-7}{-3} = \dfrac{y+10}{4} = \dfrac{z+1}{2}$ and $x = -2 + 2t, y = -4 - 2t, z = 2 - t$

7. **WE10** Show that the following lines do not intersect.

a. $\dfrac{x-2}{-5} = \dfrac{y-1}{3} = \dfrac{z+3}{4}$ and $x = -4 - t, y = 6 + 2t, z = 3 + t$

b. $\dfrac{x+7}{2} = \dfrac{y+1}{3} = z - 3$ and $x = -3 + t, y = 4 + 2t, z = -2 + t$

c. $\dfrac{x-2}{3} = \dfrac{y+3}{-4} = \dfrac{z-5}{2}$ and $x = -8 - 4t, y = 1 + 3t, z = 7 + 2t$

8. Determine the distance from the origin to each of the following lines.

a. $\dfrac{x-2}{3} = \dfrac{y+4}{-2} = z - 5$ b. $\dfrac{x+2}{5} = \dfrac{y+3}{2} = \dfrac{z-4}{-2}$ c. $\dfrac{x+2}{-3} = \dfrac{y-2}{4}; z = 1$

9. **WE11** Calculate the distance from the given point to the line.

a. $(-1, -2, 6)$ and $x = 3 + 2t, y = 2 - 2t, z = 2$

b. $(1, 0, -3)$ and $x = 2;\ \dfrac{y-2}{-3} = \dfrac{z-1}{4}$

c. $(2, -3, 4)$ and $\dfrac{x+3}{-2} = \dfrac{y-4}{5} = \dfrac{z+2}{3}$

10. a. One line passes through the two points $(1, -2, 3)$ and $(3, -4, 2)$. A second line is given by $x = 3 - 2s, y = -1 + s, z = -1 + 2s$. Determine the point of intersection and angle between the two lines. Use a scientific calculator to state the angle in degrees, rounded to 1 decimal place.

    b. One line passes through the two points $(4, 5, -2)$ and $(8, 11, -4)$. A second line is given by $x = -6 + s, y = 4 - 2s, z = -3 + s$. Determine the point of intersection and angle between the two lines. Use a scientific calculator to state the angle in degrees, rounded to 1 decimal place.

11. a. The line through the two points $(2, -5, 4)$ and $(x_0, y_0, 8)$ is parallel to the line $\dfrac{x-1}{3} = \dfrac{y+2}{2} = \dfrac{z-3}{-2}$. Determine the values of x_0 and y_0.

    b. The line in two dimensions in the xy plane, can be written as $\dfrac{x-x_0}{a} = \dfrac{y-y_0}{b}$, and in the form $y = mx + c$. Show that $m = \dfrac{b}{a}$ and $c = y_0 - \dfrac{bx_0}{a}$.

12. a. Show that the equation of the line passing through the points (x_0, y_0, z_0) and (x_1, y_1, z_1) can be written as $t = \dfrac{x-x_0}{x_1-x_0} = \dfrac{y-y_0}{y_1-y_0} = \dfrac{z-z_0}{z_1-z_0}$.

    b. Given that the lines $\dfrac{x+2}{a} = \dfrac{y-y_0}{-3} = \dfrac{z-2}{1}$ and $x = -4 + 4t, y = -6 + 2t$, $z = 9 - 2t$ intersect and are perpendicular, determine the values of a and y_0.

13. Two swords are defined by two lines. The swords intersect at the point $(1, y_0, -2)$ and the angle between the swords is 30°. One sword has the equation $x = x_0 - 2t$, $y = -1 + bt, z = -6 + t$ and the other sword has the equation $\dfrac{x+5}{a} = \dfrac{y-11}{4} = \dfrac{z-8}{c}$. It is known that a, b, c are integers, determine the values of a, b, c, x_0 and y_0.

4.3 Exam questions

Question 1 (1 mark) TECH-ACTIVE

MC Which of the following is a vector equation for the line passing through $A(-3, 0, 5)$ and $B(2, 5, -7)$?

A. $\underset{\sim}{r}(t) = (-3\underset{\sim}{i} + 5\underset{\sim}{k}) + t(5\underset{\sim}{i} + 5\underset{\sim}{j} - 12\underset{\sim}{k})$
B. $\underset{\sim}{r}(t) = (-3\underset{\sim}{i} + 5\underset{\sim}{k}) + t(-5\underset{\sim}{i} - 7\underset{\sim}{j} + 12\underset{\sim}{k})$
C. $\underset{\sim}{r}(t) = (-8\underset{\sim}{i} + 5\underset{\sim}{j} - 7\underset{\sim}{k}) + t(5\underset{\sim}{i} - 7\underset{\sim}{j} + 12\underset{\sim}{k})$
D. $\underset{\sim}{r}(t) = (2\underset{\sim}{i} - 7\underset{\sim}{j} + 17\underset{\sim}{k}) + t(-5\underset{\sim}{i} + 5\underset{\sim}{j} - 12\underset{\sim}{k})$
E. $\underset{\sim}{r}(t) = (2\underset{\sim}{i} + 5\underset{\sim}{j} + 7\underset{\sim}{k}) + t(-5\underset{\sim}{i} - 5\underset{\sim}{j} - 12\underset{\sim}{k})$

Question 2 (3 marks) TECH-FREE

Determine the vector equation of the line given by the Cartesian equation $\dfrac{x-2}{a} = \dfrac{y+1}{-2} = \dfrac{z-7}{-4}$ and which passes through the point $(8, -5, -1)$.

Question 3 (5 marks) TECH-FREE

One line passes through the two points $(5, 4, -1)$ and $(9, 14, -7)$. A second line passes through the two points $(5, 3, 10)$ and $(6, 5, 14)$. Show that the two lines intersect and calculate the point of intersection and angle between the two lines.

More exam questions are available online.

4.4 Planes

LEARNING INTENTION

At the end of this subtopic you should be able to:
- determine the normal to a plane and the Cartesian equation of a plane.

4.4.1 Equation of a plane

A plane is an infinite flat sheet. To uniquely define a plane in three dimensions, we need to specify a point on the plane and a vector perpendicular to the plane. Consider a plane containing a given fixed point with coordinates $P_0(x_0, y_0, z_0)$ and a normal vector $\underset{\sim}{n} = a\underset{\sim}{i} + b\underset{\sim}{j} + c\underset{\sim}{k}$.

Let $P(x, y, z)$ be a general point on the plane. If O is the origin, then the vector from the origin to the fixed point P_0 is given by $\overrightarrow{OP_0} = x_0\underset{\sim}{i} + y_0\underset{\sim}{j} + z_0\underset{\sim}{k}$ and the vector $\overrightarrow{OP} = x\underset{\sim}{i} + y\underset{\sim}{j} + z\underset{\sim}{k}$. Now the vector $\overrightarrow{P_0P} = \overrightarrow{OP} - \overrightarrow{OP_0} = (x - x_0)\underset{\sim}{i} + (y - y_0)\underset{\sim}{j} + (z - z_0)\underset{\sim}{k}$ is perpendicular to $\underset{\sim}{n}$ so that $\underset{\sim}{n} \cdot \overrightarrow{P_0P} = 0$.

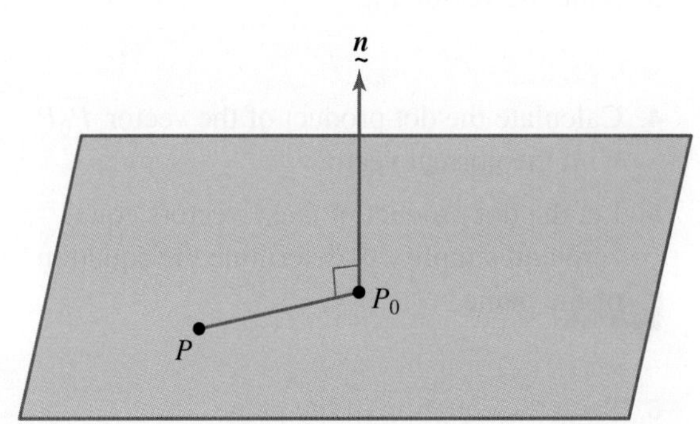

That is $(a\underset{\sim}{i} + b\underset{\sim}{j} + c\underset{\sim}{k}) \cdot \left((x - x_0)\underset{\sim}{i} + (y - y_0)\underset{\sim}{j} + (z - z_0)\underset{\sim}{k} \right) = 0$, using the definition of the dot product, the equation of the plane can be written as $a(x - x_0) + b(y - y_0) + c(z - z_0) = 0$.

We can write the equation of a plane, which passes through the fixed point with coordinates $P_0(x_0, y_0, z_0)$ and has a normal vector $\underset{\sim}{n} = a\underset{\sim}{i} + b\underset{\sim}{j} + c\underset{\sim}{k}$ in the form $ax + by + cz = d$ where $d = ax_0 + by_0 + cz_0$ and $d \geq 0$.

Equation of a plane

Given a fixed point $P_0(x_0, y_0, z_0)$ on a plane and a normal vector to the plane $\underset{\sim}{n} = a\underset{\sim}{i} + b\underset{\sim}{j} + c\underset{\sim}{k}$, the equation of the plane is:

$$ax + by + cz = d$$

where $d = ax_0 + by_0 + cz_0$ and $d \geq 0$.

WORKED EXAMPLE 12 Determining whether a point lies on a plane

Determine whether the point $(2, 1, -3)$ lies on the plane $-5x + y - 7z = 12$.

THINK	WRITE
1. Substitute the values from the coordinates of the point into the equation of the plane.	Given the point $(2, 1, -3)$, $x = 2, y = 1, z = -3$. $\text{LHS} = -5x + y - 7z$ $= -5 \times 2 + 1 - 7 \times (-3)$ $= -10 + 1 + 21$ $= 12$ $= \text{RHS}$
2. State the result.	Yes, the point $(2, 1, -3)$ does lie on the plane $-5x + y - 7z = 12$.

Consider determining the equation of the plane, given the normal to the plane and a point on the plane.

WORKED EXAMPLE 13 Determining the equation of a plane

Determine the equation of the plane perpendicular to the vector $5\underset{\sim}{i} - \underset{\sim}{j} + 7\underset{\sim}{k}$ which passes through the point $(2, 1, -3)$.

THINK	WRITE
1. State the normal vector.	$\underset{\sim}{n} = 5\underset{\sim}{i} - \underset{\sim}{j} + 7\underset{\sim}{k}$
2. State the point on the plane and the vector from the origin to this point.	$P_0(2, 1, -3)$ $\overrightarrow{OP_0} = 2\underset{\sim}{i} + \underset{\sim}{j} - 3\underset{\sim}{k}$
3. State the vector $\overrightarrow{P_0P}$.	$\overrightarrow{P_0P} = \overrightarrow{OP} - \overrightarrow{OP_0}$ $= (x-2)\underset{\sim}{i} + (y-1)\underset{\sim}{j} + (z+3)\underset{\sim}{k}$
4. Calculate the dot product of the vector $\overrightarrow{P_0P}$ with the normal vector.	$\underset{\sim}{n} \cdot \overrightarrow{P_0P} = 5(x-2) - 1(y-1) + 7(z+3)$
5. Let the dot product of these vectors equal zero and simplify to determine the equation of the plane.	$5(x-2) - 1(y-1) + 7(z+3) = 0$ $5x - y + 7z + 12 = 0$ $5x - y + 7z = -12$ $-5x + y - 7z = 12$
6. State the equation of the plane.	The equation of the plane is $-5x + y - 7z = 12$.

Determining the equation of a plane from three points on the plane

A plane can also be defined by three distinct points on the plane, provided that these three points are not collinear.

Given the three points, P, Q and R which lie on the plane, the vector cross product $\overrightarrow{PQ} \times \overrightarrow{PR}$ is a normal vector to the plane. We can then use any of the three points, to determine the equation of the plane.

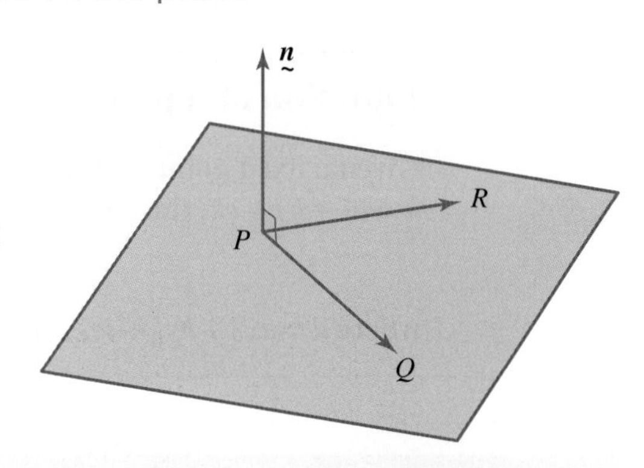

WORKED EXAMPLE 14 Determining the equation of a plane from 3 points

Determine the equation of the plane passing through the three points $P(1, 3, 2)$, $Q(2, -1, 1)$ and $R(-1, 2, 3)$.

THINK	WRITE
1. State the position vectors of the three points.	$\overrightarrow{OP} = \underset{\sim}{i} + 3\underset{\sim}{j} + 2\underset{\sim}{k}$ $\overrightarrow{OQ} = 2\underset{\sim}{i} - \underset{\sim}{j} + \underset{\sim}{k}$ $\overrightarrow{OR} = -\underset{\sim}{i} + 2\underset{\sim}{j} + 3\underset{\sim}{k}$

2. Determine two vectors which lie in the plane.

$$\overrightarrow{PQ} = \overrightarrow{OQ} - \overrightarrow{OP}$$
$$= \underset{\sim}{i} - 4\underset{\sim}{j} - \underset{\sim}{k}$$
$$\overrightarrow{PR} = \overrightarrow{OR} - \overrightarrow{OP}$$
$$= -2\underset{\sim}{i} - \underset{\sim}{j} + \underset{\sim}{k}$$

3. A vector perpendicular to both of these is a normal to the plane. Calculate the cross product of these two vectors.

$$\overrightarrow{PQ} \times \overrightarrow{PR} = \begin{vmatrix} \underset{\sim}{i} & \underset{\sim}{j} & \underset{\sim}{k} \\ 1 & -4 & -1 \\ -2 & -1 & 1 \end{vmatrix}$$
$$= \underset{\sim}{i}\begin{vmatrix} -4 & -1 \\ -1 & 1 \end{vmatrix} - \underset{\sim}{j}\begin{vmatrix} 1 & -1 \\ -2 & 1 \end{vmatrix} + \underset{\sim}{k}\begin{vmatrix} 1 & -4 \\ -2 & -1 \end{vmatrix}$$
$$= (-4-1)\underset{\sim}{i} - (1-2)\underset{\sim}{j} + (-1-8)\underset{\sim}{k}$$

4. State the normal vector to the plane.

$$\underset{\sim}{n} = -5\underset{\sim}{i} + \underset{\sim}{j} - 9\underset{\sim}{k}$$

5. Determine the equation of the plane using the normal vector and any point on the plane. In this example we are using the point P.

$$-5(x - x_0) + 1(y - y_0) - 9(z - z_0) = 0$$
$$-5(x-1) + (y-3) - 9(z-2) = 0$$
$$-5x + y - 9z + 20 = 0$$
$$5x - y + 9z = 20$$

6. State the equation of the plane.

$$5x - y + 9z = 20$$

4.4.2 Distances between points and planes

Distance of a point to a plane

Here we consider calculating the distance of a point to a plane. By this distance we understand it to be the closest perpendicular distance. Let the plane be specified $ax + by + cz = d$ and a point P_0 on the plane, the normal vector to the plane is given by $\underset{\sim}{n} = a\underset{\sim}{i} + b\underset{\sim}{j} + c\underset{\sim}{k}$. Consider calculating the distance D from this plane to a given point A.

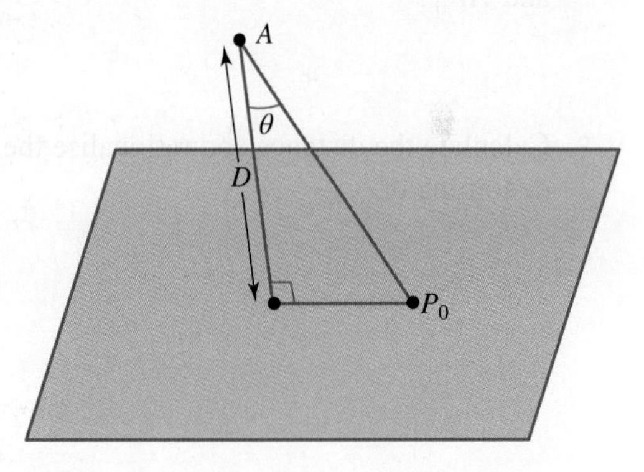

In the triangle, $\cos(\theta) = \dfrac{D}{\left|\overrightarrow{AP_0}\right|}$, so $D = \left|\overrightarrow{AP_0}\right|\cos(\theta)$. From the definition of the dot product, $\cos(\theta) = \dfrac{\hat{\underset{\sim}{n}} \cdot \overrightarrow{AP_0}}{|\underset{\sim}{n}|\left|\overrightarrow{AP_0}\right|}$. Substituting this back in gives $D = \left|\overrightarrow{AP_0}\right| \times \dfrac{\hat{\underset{\sim}{n}} \cdot \overrightarrow{AP_0}}{|\underset{\sim}{n}|\left|\overrightarrow{AP_0}\right|} = \dfrac{\hat{\underset{\sim}{n}} \cdot \overrightarrow{AP_0}}{|\underset{\sim}{n}|} = \hat{\underset{\sim}{n}} \cdot \overrightarrow{AP_0}$. However, the scalar product could give a negative number, while the distance must be positive, so we take the modulus, $D = \left|\hat{\underset{\sim}{n}} \cdot \overrightarrow{AP_0}\right|$.

Distance of a point to a plane

The distance of a point A from the plane with normal $\underset{\sim}{n}$ and containing the point P_0 is given by:

$$D = \frac{\left|\underset{\sim}{n} \cdot \overrightarrow{AP_0}\right|}{|\underset{\sim}{n}|} = \left|\hat{\underset{\sim}{n}} \cdot \overrightarrow{AP_0}\right|$$

WORKED EXAMPLE 15 Determining the distance of a point to a plane

Determine the distance of the point (1, 3, −2) to the plane $2x - y - 3z = 6$.

THINK	WRITE
1. State the position vector of the given point.	$A(1, 3, -2)$ $\overrightarrow{OA} = \underset{\sim}{i} + 3\underset{\sim}{j} - 2\underset{\sim}{k}$
2. State the normal vector to the plane.	$\underset{\sim}{n} = 2\underset{\sim}{i} - \underset{\sim}{j} - 3\underset{\sim}{k}$
3. Calculate the magnitude of the normal vector.	$\lvert\underset{\sim}{n}\rvert = \sqrt{2^2 + (-1)^2 + (-3)^2}$ $= \sqrt{14}$
4. We need a point on the plane. Any point will do here.	Let $x = 3$, then $y = 0$ and $z = 0$. $2x - y - 3z = 6$ $2(3) - 0 - 3(0) = 6$ $6 = 6$ The point $P_0(3, 0, 0)$ lies on the plane.
5. Write the position vector $\overrightarrow{OP_0}$.	$\overrightarrow{OP_0} = 3\underset{\sim}{i}$
6. Determine the vector $\overrightarrow{AP_0}$.	$\overrightarrow{AP_0} = \overrightarrow{OP_0} - \overrightarrow{OA}$ $= 3\underset{\sim}{i} - (\underset{\sim}{i} + 3\underset{\sim}{j} - 2\underset{\sim}{k})$ $= 2\underset{\sim}{i} - 3\underset{\sim}{j} + 2\underset{\sim}{k}$
7. Calculate the dot product of the normal vector and $\overrightarrow{AP_0}$.	$\underset{\sim}{n} \cdot \overrightarrow{AP_0} = (2\underset{\sim}{i} - \underset{\sim}{j} - 3\underset{\sim}{k}) \cdot (2\underset{\sim}{i} - 3\underset{\sim}{j} + 2\underset{\sim}{k})$ $= 4 + 3 - 6$ $= 1$
8. Calculate the distance and rationalise the denominator.	$D = \dfrac{\hat{\underset{\sim}{n}} \cdot \overrightarrow{AP_0}}{\lvert\underset{\sim}{n}\rvert}$ $= \dfrac{1}{\sqrt{14}} \times \dfrac{\sqrt{14}}{\sqrt{14}}$ $= \dfrac{\sqrt{14}}{14}$
9. State the answer.	$D = \dfrac{\sqrt{14}}{14}$

Distance between a plane and the origin

Consider the special case of calculating the distance of a plane to the origin. By this distance we understand it to be the closest perpendicular distance. Let the plane be specified by $ax + by + cz = d$. Consider calculating the distance D from this plane to the origin O. Let P_0 be a point on the plane. Without loss of generality we can choose $x = \dfrac{d}{a}$, assuming $a \neq 0$ and $y = z = 0$, so that $P_0\left(\dfrac{d}{a}, 0, 0\right)$. Using the results obtained earlier and substituting O for A gives, $D = \dfrac{\underset{\sim}{n} \cdot \overrightarrow{OP_0}}{\lvert\underset{\sim}{n}\rvert} = \hat{\underset{\sim}{n}} \cdot \overrightarrow{OP_0}$. Now $\overrightarrow{OP_0} = \dfrac{d}{a}\underset{\sim}{i}$ and the normal vector to the plane is

$\underset{\sim}{n} = a\underset{\sim}{i} + b\underset{\sim}{j} + c\underset{\sim}{k}$, so that $\underset{\sim}{n} \cdot \overrightarrow{OP_0} = d$. Since $|\underset{\sim}{n}| = \sqrt{a^2 + b^2 + c^2}$, it follows that the distance from the plane to the origin is given by $\dfrac{d}{\sqrt{a^2 + b^2 + c^2}}$ and since we assumed that $d \geq 0$ this distance is non-negative.

Distance between a plane and the origin

The distance between the origin and the plane defined by the equation $ax + by + cz = d$ is:

$$D = \frac{d}{\sqrt{a^2 + b^2 + c^2}}$$

Parallel planes

Two planes are parallel if their respective normals are parallel. Consider two planes with normals $\underset{\sim}{n}_1$ and $\underset{\sim}{n}_2$. The planes are parallel if $\underset{\sim}{n}_1 = \lambda \underset{\sim}{n}_2$, where $\lambda \in R$.

Two parallel planes do not intersect and we can therefore calculate the non-zero distance between the planes.

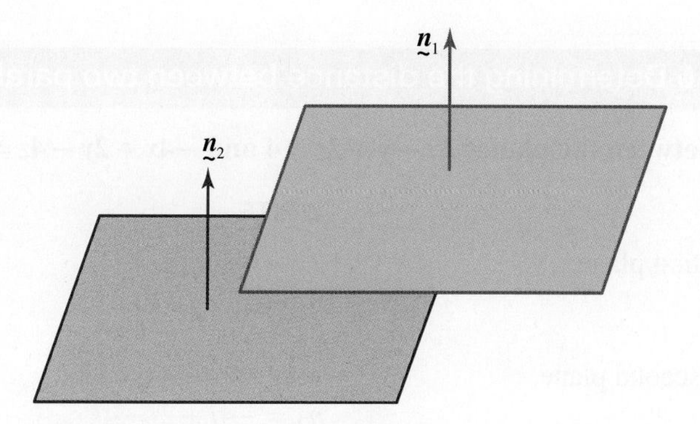

If two parallel planes are on opposite sides of the origin, their respective normals are anti-parallel and point away from each other, so that $\lambda < 0$.

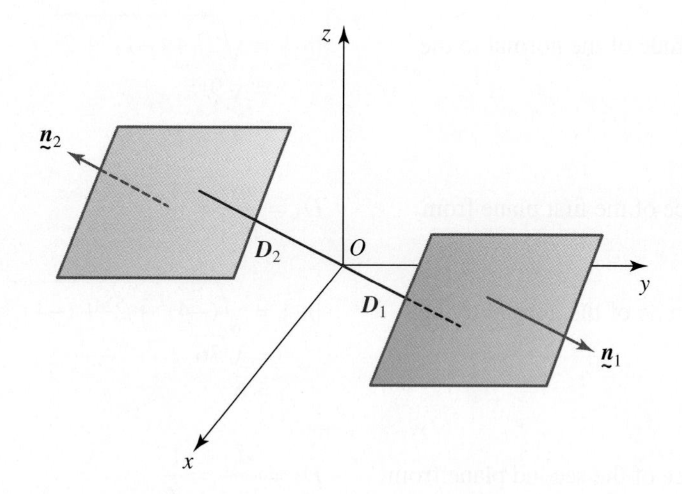

If two parallel planes are on the same side of the origin, their respective normals are pointing in the same direction, so that $\lambda > 0$.

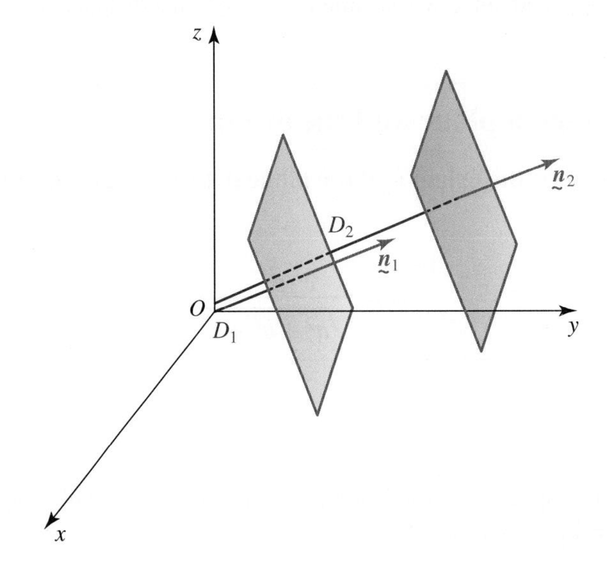

To calculate the distance between two parallel planes, we determine their respective distances from the origin and either subtract or add these distances if the planes are on the same or opposite sides of the origin.

WORKED EXAMPLE 16 Determining the distance between two parallel planes

Determine the distance between the planes $2x - y + 2z = 4$ and $-4x + 2y - 4z = 13$.

THINK	WRITE
1. State the normal to the first plane.	$2x - y + 2z = 4$ $\underset{\sim}{n}_1 = 2\underset{\sim}{i} - \underset{\sim}{j} + 2\underset{\sim}{k}$
2. State the normal to the second plane.	$-4x + 2y - 4z = 13$ $\underset{\sim}{n}_2 = -4\underset{\sim}{i} + 2\underset{\sim}{j} - 4\underset{\sim}{k}$
3. The two planes are parallel and are on opposite sides of the origin.	$-2\underset{\sim}{n}_1 = \underset{\sim}{n}_2$
4. Calculate the magnitude of the normal to the first plane.	$\lvert\underset{\sim}{n}_1\rvert = \sqrt{2^2 + (-1)^2 + 2^2}$ $= \sqrt{9}$ $= 3$
5. Calculate the distance of the first plane from the origin.	$D_1 = \dfrac{d_1}{\lvert\underset{\sim}{n}_1\rvert} = \dfrac{4}{3}$
6. Calculate the magnitude of the normal to the second plane.	$\lvert\underset{\sim}{n}_2\rvert = \sqrt{(-4)^2 + 2^2 + (-4)^2}$ $= \sqrt{36}$ $= 6$
7. Calculate the distance of the second plane from the origin.	$D_2 = \dfrac{d_2}{\lvert\underset{\sim}{n}_2\rvert} = \dfrac{13}{6}$

8. The distance between the planes is the sum of their distances from the origin. Evaluate this sum and simplify the result.	Since $\lambda < 0$ $D = D_1 + D_2$ $= \frac{4}{3} + \frac{13}{6}$ $= \frac{21}{6}$ $= \frac{7}{2}$
9. State the final result.	The distance between the planes is $\frac{7}{2}$.

4.4.3 Intersections between lines and planes

Angle between a line and a plane

If we are given a line and a plane which intersect, then there are several possible angles between them. Let α be the angle between the line with direction $\underset{\sim}{v}$ and the normal to the plane $\underset{\sim}{n}$.

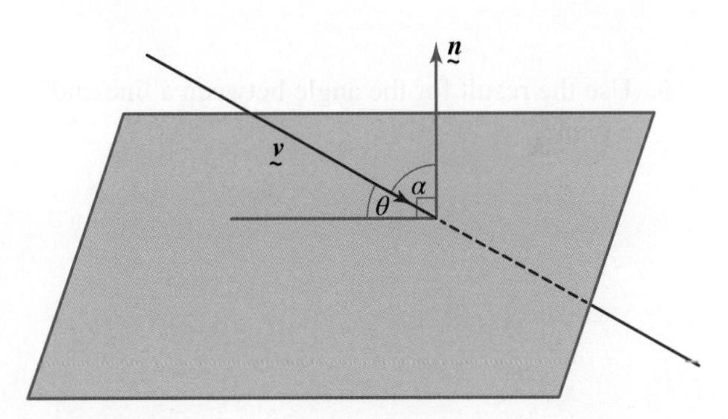

Then $\cos(\alpha) = \frac{|\underset{\sim}{v} \cdot \underset{\sim}{n}|}{|\underset{\sim}{v}|\,|\underset{\sim}{n}|}$. We define the angle between a line and the plane to be the acute angle θ. From the diagram $\theta + \alpha = 90°$, so $\cos(\alpha) = \cos(90° - \theta) = \sin(\theta)$.

Therefore $\sin(\theta) = \frac{|\underset{\sim}{v} \cdot \underset{\sim}{n}|}{|\underset{\sim}{v}|\,|\underset{\sim}{n}|}$.

Angle between a line and a plane

The acute angle θ between the line with direction $\underset{\sim}{v}$ and the plane with normal $\underset{\sim}{n}$ is given by:

$$\sin(\theta) = \frac{|\underset{\sim}{v} \cdot \underset{\sim}{n}|}{|\underset{\sim}{v}||\underset{\sim}{n}|}$$

or

$$\theta = \sin^{-1}\left(\frac{|\underset{\sim}{v} \cdot \underset{\sim}{n}|}{|\underset{\sim}{v}|\,|\underset{\sim}{n}|}\right) = \sin^{-1}\left(|\hat{\underset{\sim}{v}} \cdot \hat{\underset{\sim}{n}}|\right)$$

WORKED EXAMPLE 17 Determining the angle between a line and a plane

Determine the angle between the line $\frac{x+8}{5} = \frac{y+5}{3} = \frac{z+11}{4}$ and the plane $-5x + y - 7z = 12$. Use a scientific calculator to state the angle in degrees, rounded to 1 decimal place.

THINK	WRITE
1. State the direction of the given line.	$\frac{x+8}{5} = \frac{y+5}{3} = \frac{z+11}{4}$ $\underset{\sim}{v} = 5\underset{\sim}{i} + 3\underset{\sim}{j} + 4\underset{\sim}{k}$

2. Calculate the magnitude of the direction of the line.	$\lvert \underset{\sim}{v} \rvert = \sqrt{5^2+3^2+4^2}$ $= \sqrt{50}$ $= 5\sqrt{2}$
3. State the normal to the plane.	$-5x+y-7z=12$ $\underset{\sim}{n} = -5\underset{\sim}{i}+\underset{\sim}{j}-7\underset{\sim}{k}$
4. Calculate the magnitude of the normal to the plane.	$\lvert \underset{\sim}{n} \rvert = \sqrt{(-5)^2+1^2+(-7)^2}$ $= \sqrt{75}$ $= 5\sqrt{3}$
5. Calculate the dot product of the vector parallel to the line and the normal vector to the plane.	$\underset{\sim}{v}\cdot\underset{\sim}{n} = (5\underset{\sim}{i}+3\underset{\sim}{j}+4\underset{\sim}{k})\cdot(-5\underset{\sim}{i}+\underset{\sim}{j}-7\underset{\sim}{k})$ $= -25+3-28$ $= -50$
6. Use the result for the angle between a line and a plane.	$\sin(\theta) = \dfrac{\lvert \underset{\sim}{v}\cdot\underset{\sim}{n} \rvert}{\lvert \underset{\sim}{v} \rvert \lvert \underset{\sim}{n} \rvert}$ $= \dfrac{50}{5\sqrt{2}\times5\sqrt{3}}$ $= \dfrac{50}{25\sqrt{6}}$ $= \dfrac{2}{\sqrt{6}} = \dfrac{\sqrt{6}}{3}$
7. Determine the angle, using a scientific calculator to express it in degrees rounded to 1 decimal place.	$\theta = \sin^{-1}\left(\dfrac{\sqrt{6}}{3}\right) = 54.7°$
8. State the final result.	The angle between the line and plane is 54.7°.

Intersection between a line and a plane

Planes and lines may intersect. The intersection of a line and a plane could be a point, if the line passes through the plane.

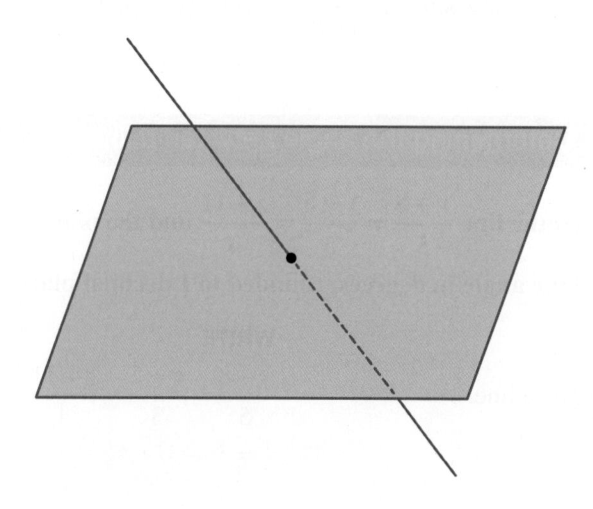

WORKED EXAMPLE 18 Determining the point of intersection between a line and a plane

Determine the intersection between the line $\frac{x+8}{5} = \frac{y+5}{3} = \frac{z+11}{4}$ and the plane $-5x + y - 7z = 12$.

THINK	WRITE
1. State the parametric equations of the line.	Let $t = \frac{x+8}{5} = \frac{y+5}{3} = \frac{z+11}{4}$, then $x = -8 + 5t, y = -5 + 3t, z = -11 + 4t$
2. Substitute the parametric equations of the line into the equation for the plane.	$-5x + y - 7z = 12$ $-5(-8 + 5t) + 1(-5 + 3t) - 7(-11 + 4t) = 12$
3. Expand and solve the equation for t.	$40 - 25t - 5 + 3t + 77 - 28t = 12$ $-50t + 112 = 12$ $50t = 100$ $t = 2$
4. Substitute the given value of t into the parametric equations of the line, to determine the values of x, y and z.	Substitute $t = 2$ $x = -8 + 5 \times 2 = 2$ $y = -5 + 3 \times 2 = 1$ $z = -11 + 4 \times 2 = -3$ The point $(2, 1, -3)$ lies on the plane.
5. State the result.	The line and the plane intersect at the point $(2, 1, -3)$.

The intersection of a line and a plane could be the whole of the line, if the line lies entirely on the plane. In this case the equations are consistent and there is an infinite number of solutions, given as the infinite number of points which lie on the line.

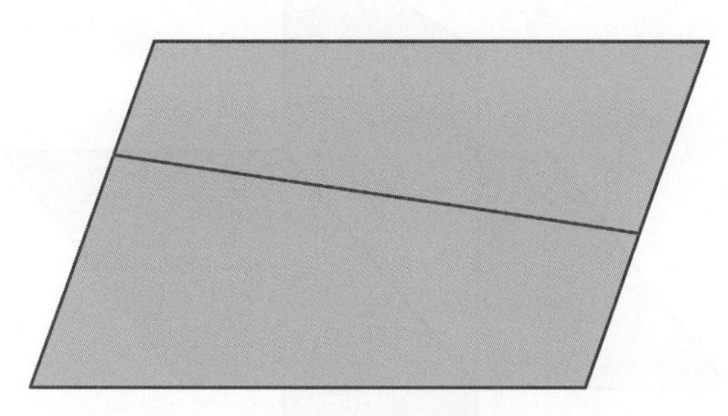

A line and a plane may not intersect at all, that is there is no solution. In this case the line lies in another plane, which is parallel to the other plane. The equation will lead to a contradiction or inconsistency.

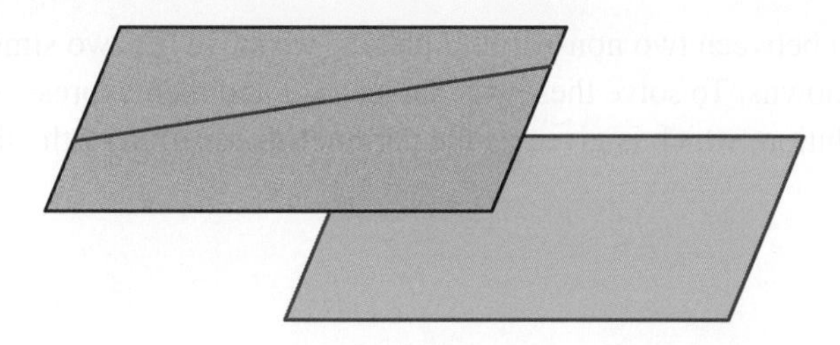

To determine which case a particular problem falls in, solve the equations simultaneously. If there is a contradiction then the line does not intersect the plane. If there is a solution to the equation calculate the dot product of the direction vector and the normal vector. If this dot product equals zero, the line is perpendicular to the normal vector and must therefore lie on the plane.

WORKED EXAMPLE 19 Determining whether a line and plane intersect

Determine whether or not the line $\frac{x+5}{4}=\frac{y+6}{2}=\frac{z-7}{5}$ and the plane $2x+y-2z=3$ intersect.

THINK	WRITE
1. State the parametric equations of the line.	Let $t=\frac{x+5}{4}=\frac{y+6}{2}=\frac{z-7}{5}$, so that $x=-5+4t, y=-6+2t, z=7+5t$
2. Substitute the parametric equations of the line into the equation for the plane.	$2x+y-2z=3$ $2(-5+4t)+(-6+2t)-2(7+5t)=3$
3. Expand and simplify the equation. Notice that this introduces a contradiction.	$-10+8t-6+2t-14-10t=3$ $-30=3$
4. State the result.	There are no solutions to the equation, therefore the line and the plane do not intersect.

4.4.4 Intersection between planes

If two planes are not parallel then they intersect in a line.

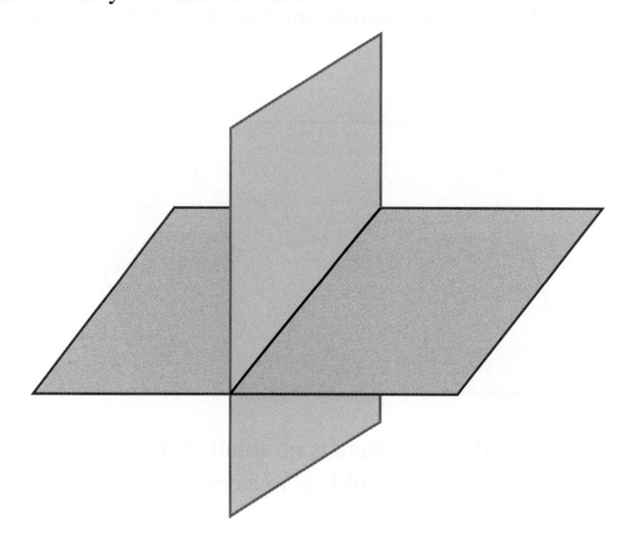

To determine the intersection between two non-parallel planes, we solve the two simultaneous linear equations; however, there are three unknowns. To solve these we can let $z=t$ and then express both x and y in terms of t. This will then give us the solution, which is given as the parametric equation of the line in terms of t.

WORKED EXAMPLE 20 Determining the intersection of two planes

Determine the line of intersection between the following pairs of planes $2x+y-z=3$ and $x-y+3z=8$.

THINK	WRITE
1. Write the two equations as equation (1) and (2). The planes are not parallel.	(1) $2x+y-z=3$ (2) $x-y+3z=8$
2. To solve these equations, let $z=t$, and rewrite the equations.	Substitute $z=t$ (1) $2x+y=3+t$ (2) $x-y=8-3t$
3. To eliminate y add these two equations and solve for x.	$3x=11-2t$ $x=\frac{1}{3}(11-2t)$
4. To solve for y use (2).	$y=x-(8-3t)$ Substitute for x $y=\frac{1}{3}(11-2t)-(8-3t)$ $y=\frac{1}{3}(7t-13)$
5. State the solution, which is a line in parametric form.	The intersection of the planes is the line $x=\frac{1}{3}(11-2t),\ y=\frac{1}{3}(7t-13),\ z=t.$

Angle between two planes

If we are given two planes which intersect in a line, then there are several possible angles between them. We define the angle θ between the two planes to be the acute angle between their respective normals.

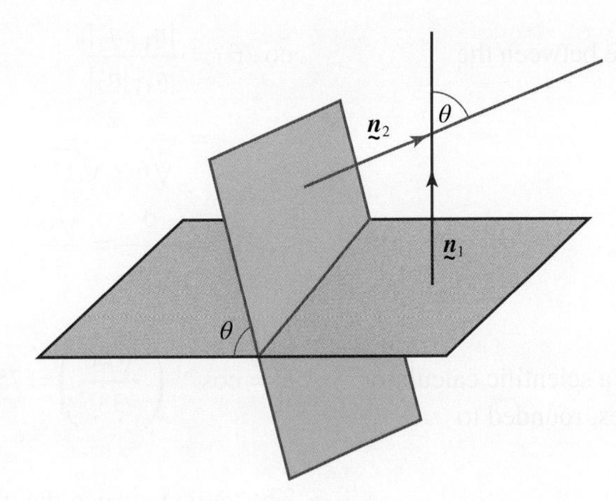

The acute angle θ between the two planes can be calculated by rearranging the formula for the dot product to isolate θ.

$$|\underset{\sim}{n}_1 \cdot \underset{\sim}{n}_2| = |\underset{\sim}{n}_1||\underset{\sim}{n}_2|\cos(\theta)$$

$$\cos(\theta) = \frac{|\underset{\sim}{n}_1 \cdot \underset{\sim}{n}_2|}{|\underset{\sim}{n}_1||\underset{\sim}{n}_2|}$$

$$\theta = \cos^{-1}\left(\frac{|\underset{\sim}{n}_1 \cdot \underset{\sim}{n}_2|}{|\underset{\sim}{n}_1||\underset{\sim}{n}_2|}\right)$$

Angle between two planes

The angle θ between two planes with normals $\underset{\sim}{n}_1$ and $\underset{\sim}{n}_2$ is given by:

$$\theta = \cos^{-1}\left(\frac{|\underset{\sim}{n}_1 \cdot \underset{\sim}{n}_2|}{|\underset{\sim}{n}_1|\ |\underset{\sim}{n}_2|}\right) = \cos^{-1}\left(|\hat{\underset{\sim}{n}}_1 \cdot \hat{\underset{\sim}{n}}_2|\right)$$

WORKED EXAMPLE 21 Determining the angle between two planes

Determine the angle between the planes $2x + y - z = 3$ and $x - y + 3z = 8$. Use a scientific calculator to state the angle in degrees, rounded to 1 decimal place.

THINK	WRITE
1. State the normal to the first plane.	$2x + y - z = 3$ $\underset{\sim}{n}_1 = 2\underset{\sim}{i} + \underset{\sim}{j} - \underset{\sim}{k}$
2. Calculate the magnitude of the normal to the first plane.	$\lvert\underset{\sim}{n}_1\rvert = \sqrt{2^2 + 1^2 + (-1)^2}$ $= \sqrt{6}$
3. State the normal to the second plane.	$x - y + 3z = 8$ $\underset{\sim}{n}_2 = \underset{\sim}{i} - \underset{\sim}{j} + 3\underset{\sim}{k}$
4. Calculate the magnitude of the normal to the second plane.	$\lvert\underset{\sim}{n}_2\rvert = \sqrt{1^2 + (-1)^2 + 3^2}$ $= \sqrt{11}$
5. Calculate the dot product of the two normal vectors.	$\underset{\sim}{n}_1 \cdot \underset{\sim}{n}_2 = (2\underset{\sim}{i} + \underset{\sim}{j} - \underset{\sim}{k}) \cdot (\underset{\sim}{i} - \underset{\sim}{j} + 3\underset{\sim}{k})$ $= 2 - 1 - 3$ $= -2$
6. Use the result for the angle between the two planes.	$\cos(\theta) = \frac{\lvert\underset{\sim}{n}_1 \cdot \underset{\sim}{n}_2\rvert}{\lvert\underset{\sim}{n}_1\rvert\,\lvert\underset{\sim}{n}_2\rvert}$ $= \frac{2}{\sqrt{6} \times \sqrt{11}}$ $= \frac{2}{\sqrt{66}} = \frac{\sqrt{66}}{33}$
7. Calculate the angle, using a scientific calculator to state the angle in degrees, rounded to 1 decimal place.	$\theta = \cos^{-1}\left(\frac{\sqrt{66}}{33}\right) = 75.7°$
8. State the answer.	The angle between the planes is 75.7°.

4.4 Exercise

Students, these questions are even better in jacPLUS

Receive immediate feedback and access sample responses

Access additional questions

Track your results and progress

Find all this and MORE in jacPLUS

Technology free

1. WE12 Determine whether the point $(5, -2, 3)$ lies on the plane $2x + 3y + z = 1$.

2. a. The point $(2, -1, -3)$ lies on the plane $x - 2y - 3z = d$, determine the value of d.
   b. The point $(x_0, 2, -4)$ lies on the plane $2x + y - 3z = 10$, determine the value of x_0.

3. WE13 Determine the equation of the plane perpendicular to the vector $\underset{\sim}{n}$ which passes through the point P_0, given the following information.
   a. $P_0(2, -3, 5)$ and $\underset{\sim}{n} = \underset{\sim}{i} + 3\underset{\sim}{j} - 2\underset{\sim}{k}$
   b. $P_0(1, 0, 4)$ and $\underset{\sim}{n} = -2\underset{\sim}{j} + 3\underset{\sim}{k}$
   c. $P_0(3, -2, -1)$ and $\underset{\sim}{n} = -\underset{\sim}{i} + \underset{\sim}{j} + 4\underset{\sim}{k}$

4. WE14 Determine the equation of the plane passing through the three points.
   a. $A(1, -2, 3)$, $B(-2, 0, 2)$ and $C(3, -4, 2)$
   b. $A(-2, 3, 1)$, $B(2, -1, 3)$ and $C(-1, -1, 0)$
   c. $A(3, -2, 5)$, $B(1, 2, 3)$ and $C(2, -1, 3)$

5. Calculate the distance from the origin to each of the following planes.
   a. $2x - y + 2z = 6$
   b. $3x + 5y - 4z = 10$
   c. $-4x - 5y + 2z = 15$

6. WE15 Calculate the distance from the point to the plane.
   a. $A(1, -2, 4)$, $2x - 2y + 3z = 6$
   b. $A(3, 2, -1)$, $-3x + 6y + 2z = 2$
   c. $A(-2, 1, -3)$, $-2x + y - 2z = 5$

7. WE16 Calculate the distance between the following pairs of parallel planes.
   a. $-x + y + 3z = 1$ and $2x - 2y - 6z = 3$
   b. $3x - 4y + 12z = 17$ and $6x - 8y + 24z = 8$
   c. $3x - 6y - 2z = 2$ and $-6x + 12y + 4z = 9$

8. WE17 Calculate the angle between the line and the plane. Use a scientific calculator to state the angle in degrees, rounded to 1 decimal place.
   a. $\dfrac{x + 16}{3} = \dfrac{y - 13}{-2} = \dfrac{z + 13}{2}$, $4x + 2y + 5z = 13$
   b. $\dfrac{x - 13}{3} = \dfrac{y - 4}{2} = \dfrac{z + 7}{-2}$, $2x + 3y - 2z = 4$
   c. $\dfrac{x - 9}{-2} = \dfrac{y + 5}{3} = \dfrac{z - 10}{-4}$, $4x - y + 3z = 2$

9. **WE18** Determine the point of intersection between the line and the plane.

a. $\dfrac{x+16}{3}=\dfrac{y-13}{-2}=\dfrac{z+13}{2}$, $4x+2y+5z=5$

b. $\dfrac{x-13}{3}=\dfrac{y-4}{2}=\dfrac{z+7}{-2}$, $2x+3y-2z=4$

c. $\dfrac{x-9}{-2}=\dfrac{y+5}{3}=\dfrac{z-10}{-4}$, $4x-y+3z=2$

10. **WE19** Determine whether or not the line and the plane intersect.

a. $\dfrac{x+3}{3}=\dfrac{y-2}{2}=\dfrac{z+1}{4}$, $2x+3y-3z=5$

b. $\dfrac{x-13}{3}=\dfrac{y-4}{2}=\dfrac{z+7}{-2}$, $2x-4y-z=17$

c. $\dfrac{x-5}{3}=\dfrac{y-3}{-2}=\dfrac{z-1}{2}$, $2x-2y-5z=2$

11. **WE20** Determine the line of intersection between the following pairs of planes.

a. $x+2y+3z=1$ and $2x-y+4z=3$
b. $-x+y+3z=1$ and $2x-2y-6z=3$
c. $2x-3y+z=11$ and $3x+y+2z=7$

12. **WE21** Calculate the angle between the following pairs of planes. Where necessary, use a scientific calculator to state the angle in degrees, rounded to 1 decimal place.

a. $x+2y+3z=1$ and $2x-y+4z=3$
b. $-x+y+3z=1$ and $2x-2y-6z=3$
c. $2x-3y+z=11$ and $3x+y+2z=7$

13. *Note:* For this question you may use a scientific calculator to state the angles in degrees rounded to 1 decimal place.

a. The blade of a knife passes through the two points $(1,-2,-3)$ and $(2,1,4)$ and intersects a chopping board with the equation $-3x-y+2z=1$. Calculate the angle and point of intersection between the knife and the chopping board.

b. A plane passes through the three points $(2,1,5)$, $(3,2,6)$ and $(1,-1,1)$. Calculate the angle and point of intersection of the plane with the line $\dfrac{x+2}{2}=\dfrac{y-1}{-1}=\dfrac{z-1}{-3}$.

c. A plane passes through the three points $(1,0,-4)$, $(2,8,1)$ and $(3,2,-1)$. A line passes through the two points $(-1,6,8)$ and $(-3,9,13)$. Calculate the angle and point of intersection between the plane and the line.

14. Consider the line $\dfrac{x-4}{2}=\dfrac{y-3}{-1}=\dfrac{z-1}{c}$ and the plane $3x-2y-2z=d$. Determine the values of c and d for which the line and the plane, have:

a. a unique point of intersection
b. an infinite number of solutions
c. no solution.

15. Consider the line $\dfrac{x-1}{2}=\dfrac{y-3}{b}=\dfrac{z-2}{-4}$ and the plane $4x-2y+3z=d$. Determine the values of b and d for which the line and the plane, have:

a. a unique point of intersection
b. an infinite number of solutions
c. no solution.

16. a. Show that the equation of the plane passing through the three points (x_0, y_0, z_0), (x_1, y_1, z_1) and (x_2, y_2, z_2) can be written as

$$\begin{vmatrix} x - x_0 & y - y_0 & z - z_0 \\ x_2 - x_1 & y_2 - y_1 & z_2 - z_1 \\ x_1 - x_0 & y_1 - y_0 & z_1 - z_0 \end{vmatrix} = 0$$

b. In two dimensions $ax + by = d$ represents a line, the closest distance of the line to the origin is given by $\frac{d}{\sqrt{a^2 + b^2}}$. Verify this result using both calculus and a geometrical consideration.

17. a. The two planes $2x - y + 2z = 6$ and $ax + by + cz = 45$ are parallel and are a distance of 3 units apart. If $a > 0$, determine the values of a, b and c.

b. The line $\frac{x-4}{a} = \frac{y+7}{b} = \frac{z+7}{c}$ and the plane $2x - y - z = d$ intersect at the point $(1, -2, -3)$ and the angle between the line and the plane is $60°$. If $a^2 + b^2 + c^2 = 50$, determine the values of a, b, c and d.

4.4 Exam questions

Question 1 (3 marks) TECH-FREE

Determine the equation of the plane which contains the points $A(1, 3, -4)$, $B(2, 0, -1)$ and $C(0, -3, 4)$.

Question 2 (2 marks) TECH-FREE

Determine the equation of the plane which is perpendicular to $\underset{\sim}{n} = 2\underset{\sim}{i} + 3\underset{\sim}{j} - \underset{\sim}{k}$ and which contains the point $P_0(1, 3, -1)$.

Question 3 (2 marks) TECH-FREE

The parametric equation $x = t + 3$, $y = 5 - t$ and $z = 2t + 1$ defines a line. The line intersects the plane $3x - 2y + z = 7$ at a point. Determine the coordinates of this point of intersection.

More exam questions are available online.

4.5 Review

4.5.1 Summary

doc-37058

Hey students! Now that it's time to revise this topic, go online to:

Access the topic summary

Review your results

Watch teacher-led videos

Practise exam questions

Find all this and MORE in jacPLUS

4.5 Exercise

Technology free: short answer

1. a. i. Determine a unit vector perpendicular to both $\underset{\sim}{a} = 3\underset{\sim}{i} - \underset{\sim}{j} + 2\underset{\sim}{k}$ and $\underset{\sim}{b} = 4\underset{\sim}{i} - 2\underset{\sim}{j} + 5\underset{\sim}{k}$.
   ii. Determine a unit vector perpendicular to both $\underset{\sim}{r} = 5\underset{\sim}{i} - 2\underset{\sim}{j} - 3\underset{\sim}{k}$ and $\underset{\sim}{s} = -2\underset{\sim}{i} + 3\underset{\sim}{j} + \underset{\sim}{k}$.
   b. Consider the three points $A(1, -2, -3)$, $B(2, 1, 4)$ and $C(3, -1, -4)$.
   i. Calculate the area of the triangle ABC.
   ii. Determine the equation of the plane passing through the three points A, B and C.
   c. i. Determine the equation of the line L passing through the points $P(-1, 3, 5)$ and $Q(5, -1, 2)$.
   ii. Determine the equation of the plane passing through the points $P(-1, 3, 5)$, $Q(5, -1, 2)$ and $R(2, 1, -1)$. Show that the intersection of the line L with this plane has an infinite number of solutions and is the line.
   iii. Calculate the area of the triangle PQR.

2. *Note:* For this question you may use a scientific calculator to state the angles in degrees rounded to 1 decimal place.
   a. Given the two lines $x = 2 - t,\ y = 3 + 2t,\ z = -4 - 2t$ and $\frac{x-1}{2} = y = \frac{z+5}{3}$, show that the lines intersect and determine the point of intersection and the angle between the lines.
   b. Given the line $\frac{x+5}{4} = \frac{y+6}{2} = \frac{z-7}{-3}$ and the plane $2x + y - z = 3$, show that the line and plane intersect and determine the point of intersection and the angle between the line and the plane.
   c. Given the two planes $2x + y - z = 3$ and $3x - 2y + 2z = 1$, determine the line of intersection and the angle between the planes.

3. A line L passes through the points $(1, -2, -3)$ and $(2, 1, 4)$. For this question you may use a scientific calculator to state the angles in degrees rounded to 1 decimal place.
   a. Show that the line L intersects the line $\frac{x+7}{3} = \frac{y-13}{-4} = \frac{z+2}{2}$ and determine the point of intersection and the angle between the lines.
   b. Determine the point of intersection and the angle between the line L and the plane $-3x - y + 2z = 1$.

4. a. Calculate the distance from the origin to the following.
   i. The line $\frac{x+1}{-2} = \frac{y-3}{4} = \frac{z+2}{5}$
   ii. The plane $-4x + 3y + 12z = 13$
   b. Calculate the distance from the point $(2, -1, -3)$ to the following.
   i. The line $\frac{x-3}{2} = \frac{y+2}{-1} = \frac{z+4}{-3}$
   ii. The plane $3x - 4z = 8$

5. a. Given the two planes $3x - 2y - 2z = 6$ and $2x + 3y = 7$, determine the line of intersection and the angle between the planes.
b. Given the two planes $3x - 2y - 2z = 6$ and $-6x + 4y + 4z = 10$, calculate the distance between the planes.
c. Calculate the distance from the plane $3x - 2y - 2z = 6$ to the point $(5, -2, 4)$.

6. Determine the intersection of the plane $5x - 2y + 4z = 12$ with each of the following lines.
a. $\frac{x+4}{2} = \frac{y-2}{-1} = \frac{z-9}{-3}$
b. $\frac{x+8}{2} = \frac{y-2}{-1} = \frac{z-11}{-3}$
c. $\frac{x-6}{-2} = \frac{y-1}{2} = \frac{z-9}{-3}$

7. Determine the intersection of the line $\frac{x-4}{5} = y + 2 = \frac{z-3}{3}$ with each of the following planes.
a. $2x - y - 3z = 1$
b. $x - 2y - z = 4$
c. $x - 3y + z = 3$

Technology active: multiple choice

8. **MC** The line $x = 2;\ y = 3;\ z = 2 + 3t$ is
A. parallel to the xy plane.
B. parallel to the yz plane.
C. parallel to the xz plane.
D. parallel to the z-axis.
E. perpendicular to the z-axis.

9. **MC** Given the points $A(0, 2, -3)$, $B(0, -6, 9)$ and the line $x = 0;\ \frac{y}{-2} = \frac{z}{3}$, then
A. the origin and both points A and B lie on the line.
B. the origin and point A lies on the line; point B does not lie on the line.
C. the origin and point B lies on the line; point A does not lie on the line.
D. neither of the points A and B lie on the line; however, the line does pass through the origin.
E. both points A and B lie on the line; however, the line does not pass through the origin.

10. **MC** Given the two lines $\frac{x-1}{2} = \frac{y-3}{-4} = \frac{z+2}{1}$ and $\frac{x-3}{-2} = \frac{y+1}{4} = \frac{z+1}{-1}$, select which of the following is true.
A. The two lines are in fact the same line.
B. The two lines do not intersect and are parallel.
C. The two lines do not intersect, they are not parallel but skew lines.
D. The two lines intersect in a unique point but are not perpendicular.
E. The two lines intersect in a unique point and are perpendicular.

11. **MC** The two lines $\frac{x-3}{3} = \frac{y+2}{2} = \frac{z-2}{2}$ and $x = 1 + 2t, y = -2t, z = 3 - t$
A. do not intersect and are parallel.
B. do not intersect and are skew.
C. intersect in a unique point and are not perpendicular.
D. intersect in a unique point and are perpendicular.
E. intersect in an infinite number of points, since they are in fact the same line.

12. **MC** The distance of the line $\frac{x}{3} = \frac{y-1}{4};\ z = 0$, from the origin is
A. $\frac{1}{5}$
B. $\frac{3}{5}$
C. $\frac{4}{5}$
D. 1
E. 5

13. **MC** The distance of the plane $-x+2y+2z=3$, from the origin is

A. 0 B. 1 C. 2 D. 3 E. 4

14. **MC** Consider the two planes $2x+y-z=0$ and $-4x-2y+2z=2$. These two planes

A. are not parallel and intersect each other at the origin.
B. are parallel and intersect each other at the origin.
C. are parallel, do not intersect, and one plane passes through the origin.
D. are perpendicular and intersect in a line.
E. are not parallel but intersect in a line.

15. **MC** Consider the point $(1,-2,3)$, the line $\frac{x-1}{1}=\frac{y+2}{-2}=\frac{z-3}{3}$ and the plane $x-y-z=0$. Then

A. the point lies on the line but not on the plane.
B. the point lies on the plane but not on the line.
C. the point does not lie on the line and does not lie on the plane; furthermore, the line and the plane do not intersect.
D. the point does not lie on the line and does not lie on the plane; however, the line and the plane do intersect in another unique point.
E. the point lies on the line and the line lies in the plane.

16. **MC** The line $x=3+t,\ y=2+6t,\ z=2-t$ and the plane $2x-y-4z=5$

A. intersect in infinitely many points as the line lies in the plane.
B. are orthogonal.
C. are parallel.
D. intersect in a unique point.
E. do not intersect.

17. **MC** Given the two points $P(1,-2,4)$ and $Q(2,-1,3)$ then which of the following is **false**?

A. The points P and Q both lie on the line $x=t-1,\ y=t-4,\ z=6-t$.
B. The points P and Q both lie on the line $x=4+t,\ y=t+1,\ z=1-t$.
C. The point Q lies on the plane $x-y+2z=10$.
D. The points P and Q both lie on the plane $-2x+5y+3z=0$.
E. The point P lies on the plane $x-y+2z=13$.

Technology active: extended response

18. a. The line $x=5-t,\ y=4t-8,\ z=6-2t$, passes through the point A in the xy plane, the point B in the yz plane and the point C in the xz plane. Determine the coordinates of the points A, B and C.
b. A plane passes through the points $P(0,2,-3),\ Q(1,0,-2)$ and $R(4,-1,0)$. Determine the equation of the plane PQR and the area of the triangle PQR.

19. a. Let $x=t+2,\ y=2t-3,\ z=3-t$ and $x=2t-1,\ y=t-3,\ z=3t-4$ be two lines which intersect. Determine the equation of the plane which contains both lines.
b. Let $x=2t+1,\ y=t+3,\ z=4-t$ and $x=2t-1,\ y=t-2,\ z=6-t$ be two lines which do not intersect. Determine the equation of the plane which contains both lines.

20. The first line $x=2t+6,\ y=t+1,\ z=-2t-1$ and the plane $4x-y-z=6$ intersect at the point A.

a. Determine the coordinates of the point A.
b. Calculate the angle between the line and the plane.
c. Another line $\frac{x-x_0}{a}=\frac{y-3}{b}=z-1$ is perpendicular to the first line and also intersects the plane at the point A. Determine the values of a, b and x_0.

4.5 Exam questions

Question 1 (1 mark) TECH-ACTIVE

MC Consider the line $\frac{x-x_1}{a_1}=\frac{y-y_1}{b_1}=\frac{z-z_1}{c_1}$ and the plane $a_2x+b_2y+c_2z=d$, the acute angle between the angle and the plane is given by

A. $\sin^{-1}\left(\frac{a_1a_2+b_1b_2+c_1c_2}{\sqrt{a_1^2+b_1^2+c_1^2}\sqrt{a_2^2+b_2^2+c_2^2}}\right)$

B. $\cos^{-1}\left(\frac{a_1a_2+b_1b_2+c_1c_2}{\sqrt{a_1^2+b_1^2+c_1^2}\sqrt{a_2^2+b_2^2+c_2^2}}\right)$

C. $\sin^{-1}\left(\frac{|a_1a_2+b_1b_2+c_1c_2|}{\sqrt{a_1^2+b_1^2+c_1^2}\sqrt{a_2^2+b_2^2+c_2^2}}\right)$

D. $\cos^{-1}\left(\frac{|a_1a_2+b_1b_2+c_1c_2|}{\sqrt{a_1^2+b_1^2+c_1^2}\sqrt{a_2^2+b_2^2+c_2^2}}\right)$

E. $\sin^{-1}\left(\frac{|a_1a_2|+|b_1b_2|+|c_1c_2|}{\sqrt{a_1^2+b_1^2+c_1^2}\sqrt{a_2^2+b_2^2+c_2^2}}\right)$

Question 2 (1 mark) TECH-ACTIVE

MC Consider the two lines $\frac{x-x_1}{a_1}=\frac{y-y_1}{b_1}=\frac{z-z_1}{c_1}$ and $\frac{x-x_2}{a_2}=\frac{y-y_2}{b_2}=\frac{z-z_2}{c_2}$, the acute angle between the lines is given by

A. $\sin^{-1}\left(\frac{a_1a_2+b_1b_2+c_1c_2}{\sqrt{a_1^2+b_1^2+c_1^2}\sqrt{a_2^2+b_2^2+c_2^2}}\right)$

B. $\cos^{-1}\left(\frac{a_1a_2+b_1b_2+c_1c_2}{\sqrt{a_1^2+b_1^2+c_1^2}\sqrt{a_2^2+b_2^2+c_2^2}}\right)$

C. $\sin^{-1}\left(\frac{|a_1a_2+b_1b_2+c_1c_2|}{\sqrt{a_1^2+b_1^2+c_1^2}\sqrt{a_2^2+b_2^2+c_2^2}}\right)$

D. $\cos^{-1}\left(\frac{|a_1a_2+b_1b_2+c_1c_2|}{\sqrt{a_1^2+b_1^2+c_1^2}\sqrt{a_2^2+b_2^2+c_2^2}}\right)$

E. $\cos^{-1}\left(\frac{|a_1a_2|+|b_1b_2|+|c_1c_2|}{\sqrt{a_1^2+b_1^2+c_1^2}\sqrt{a_2^2+b_2^2+c_2^2}}\right)$

Question 3 (1 mark) TECH-ACTIVE

MC In three dimensions the equation $y = x$ represents

A. a line parallel to the xy plane.
B. a line parallel to the z-axis.
C. a plane parallel to the xy plane.
D. a plane parallel to the z-axis.
E. a plane passing through the origin and perpendicular to the z-axis.

Question 4 (1 mark) TECH-ACTIVE

MC Consider the line $\frac{x}{2} = -y = z$ and the plane $x + y - z = 0$ then the line and the plane

A. are parallel and do not intersect.
B. are skew and do not intersect.
C. are perpendicular and intersect only at the origin.
D. intersect at two points $(2, -1, 1)$ and the origin.
E. intersect at an infinite number of points as the line lies in the plane.

Question 5 (3 marks) TECH-FREE

A triangle is formed by the points $A(-3, 5, 6)$, $B(-2, 7, 9)$ and $C(2, 1, 7)$. Classify the triangle as either scalene, isosceles or equilateral and prove the triangle is right-angled.

More exam questions are available online.

Answers

Topic 4 Vector equations of lines and planes

4.2 Vector cross product

4.2 Exercise

1. a. -5 b. 5
2. a. 60 b. -60
3. a. $-5\underset{\sim}{i} + 2\underset{\sim}{j} + 3\underset{\sim}{k}$
   b. $3\underset{\sim}{i} - 6\underset{\sim}{j} - \underset{\sim}{k}$
   c. $-12\underset{\sim}{i} - 8\underset{\sim}{j} + 13\underset{\sim}{k}$
4. a. $\frac{\pm 1}{3\sqrt{6}}(\underset{\sim}{i} + 7\underset{\sim}{j} + 2\underset{\sim}{k})$
   b. $\frac{\pm 1}{3\sqrt{19}}(7\underset{\sim}{i} + \underset{\sim}{j} + 11\underset{\sim}{k})$
   c. $\frac{\pm 1}{3\sqrt{2}}(\underset{\sim}{i} + 4\underset{\sim}{j} - \underset{\sim}{k})$
5. a. Sample responses can be found in the worked solutions in the online resources.
   b. Sample responses can be found in the worked solutions in the online resources.
   c. The cross product is distributive over addition and subtraction.
   d. $-4\underset{\sim}{i} - 26\underset{\sim}{j} - 10\underset{\sim}{k}$
   e. $2\underset{\sim}{i} + \underset{\sim}{j} - \underset{\sim}{k},\ -4\underset{\sim}{i} - 2\underset{\sim}{j} + 2\underset{\sim}{k}$
6. a. $\sqrt{2}$ b. $\frac{1}{2}\sqrt{59}$ c. $\frac{1}{2}\sqrt{41}$
7. a. $5\sqrt{3}$
   b. $5\sqrt{2}$
   c. 50
   d. $\cos^{-1}\left(\frac{\sqrt{6}}{3}\right)$
   e. $-25\underset{\sim}{i} + 15\underset{\sim}{j} + 20\underset{\sim}{k}$
   f. $\frac{1}{5\sqrt{2}}(-5\underset{\sim}{i} + 3\underset{\sim}{j} + 4\underset{\sim}{k})$
   g. Sample responses can be found in the worked solutions in the online resources.
   h. Sample responses can be found in the worked solutions in the online resources.
8. a. $-10\underset{\sim}{i} + 13\underset{\sim}{j} + 19\underset{\sim}{k}$
   b. 0
   c. 0
   d. 0
   e. $\underset{\sim}{c}$ is perpendicular to both $\underset{\sim}{a}$ and $\underset{\sim}{b}$.
   f. $\underset{\sim}{a}$, $\underset{\sim}{b}$ and $\underset{\sim}{c}$ are all mutually perpendicular, so they are linearly independent.
9. a. $x = -2, y = 1$
   b. $x = 2, y = 3$ or $x = -\frac{6}{5}, y = -\frac{17}{5}$
10. a. -3 b. $x = -2, z = 2$ c. ± 2
11. a. Sample responses can be found in the worked solutions in the online resources.
    b. $\sqrt{69}$
12. a. $\underset{\sim}{a} \times \underset{\sim}{b}$ is a vector, must then be dotted with $\underset{\sim}{c}$ to produce a scalar.
    b. -8
    c. -8
    d. -8
    e. -8
13. a. $\underset{\sim}{a} \times \underset{\sim}{b} \times \underset{\sim}{c}$ is ambiguous, could be $(\underset{\sim}{a} \times \underset{\sim}{b}) \times \underset{\sim}{c}$ or $\underset{\sim}{a} \times (\underset{\sim}{b} \times \underset{\sim}{c})$
    b. i. $-26\underset{\sim}{i} + 19\underset{\sim}{j} - \underset{\sim}{k}$
       ii. $-25\underset{\sim}{i} + 7\underset{\sim}{j} + 13\underset{\sim}{k}$
       iii. No
       iv. Sample responses can be found in the worked solutions in the online resources.
       v. Sample responses can be found in the worked solutions in the online resources.
14. a. $x = 2, y = -3, z = 4$
    b. $x = -6, y = -2, z = 3$
    c. $x = 2, y = -2, z = 1$ or $x = \frac{52}{19}, y = -\frac{17}{19}, z = -\frac{16}{19}$
    d. $x = -2, y = 4, z = 5$ or $x = -\frac{94}{29}, y = \frac{140}{29}, z = \frac{97}{29}$

4.2 Exam questions

Note: Mark allocations are available with the fully worked solutions online.

1. B
2. $\frac{7}{2}$
3. Sample responses can be found in the worked solutions in the online resources.

4.3 Lines in three dimensions

4.3 Exercise

1. a. No
   b. $(-13, -37, 10)$
2. a. Yes
   b. $(46, 17, -25), (26, 7, -10), (6, -3, 5)$
3. a. $x - 2 = \frac{y+3}{2}$
   b. $x - 2 = \frac{y+3}{2} = \frac{z-4}{-3}$
   c. $\frac{x-3}{2} = z - 1; y = 5$
4. a. $\frac{x-2}{3} = \frac{y+3}{5} = \frac{z-4}{-6}$
   b. $\frac{x-1}{-4} = \frac{y}{2} = \frac{z+3}{8}$
   c. $\frac{x-3}{-3} = \frac{y-4}{2}; z = -2$
5. a. $75.6°$ b. $71.1°$ c. $8.0°$
6. a. $(2, -1, 3)$ b. $(4, -2, -1)$ c. $(16, -22, -7)$

7. Sample responses can be found in the worked solutions in the online resources.

8. a. $\frac{\sqrt{3766}}{14}$ b. $\frac{\sqrt{1397}}{11}$ c. $\frac{\sqrt{29}}{5}$

9. a. $4\sqrt{3}$ b. $\sqrt{17}$ c. $\frac{\sqrt{131\,138}}{38}$

10. a. $(-3, 2, 5)$, 27.3°
 b. $(-2, -4, 1)$, 56.9°

11. a. $x_0 = -4$, $y_0 = -9$
 b. Sample responses can be found in the worked solutions in the online resources.

12. a. Sample responses can be found in the worked solutions in the online resources.
 b. $y_0 = 7, a = 2$

13. $a = -3, b = 1, c = 5, x_0 = 9, y_0 = 3$

4.3 Exam questions

Note: Mark allocations are available with the fully worked solutions online.

1. A
2. $\underset{\sim}{r} = (2\underset{\sim}{i} - \underset{\sim}{j} + 7\underset{\sim}{k}) + t(3\underset{\sim}{i} - 2\underset{\sim}{j} - 4\underset{\sim}{k})$
3. $(3, -1, 2)$, 90°

4.4 Planes

4.4 Exercise

1. Yes

2. a. 13
 b. -2

3. a. $-x - 3y + 2z = 17$
 b. $-2y + 3z = 12$
 c. $x - y - 4z = 9$

4. a. $-4x - 5y + 2z = 12$
 b. $-2x - y + 2z = 3$
 c. $3x + y - z = 2$

5. a. 2 b. $\sqrt{2}$ c. $\sqrt{5}$

6. a. $\frac{12\sqrt{17}}{17}$ b. $\frac{1}{7}$ c. 2

7. a. $\frac{5\sqrt{11}}{22}$ b. 1 c. $\frac{13}{14}$

8. a. 40.6° b. 70.3° c. 56.9°

9. a. $(2, 1, -1)$ b. $(4, -2, -1)$ c. $(3, 4, -2)$

10. a. The line and plane do not intersect.
 b. Line $\frac{x-13}{3} = \frac{y-4}{2} = \frac{z+7}{-2}$.
 c. The line and plane do not intersect.

11. a. $x = \frac{1}{5}(7 - 11t), y = -\frac{1}{5}(1 + 2t), z = t$
 b. Planes are parallel, no solution.
 c. $x = \frac{1}{11}(32 - 7t), y = -\frac{1}{11}(19 + t), z = t$

12. a. 45.6° b. 0° or 180° c. 69.1°

13. a. $(2, 1, 4)$, 16.2°
 b. $(4, -2, -8)$, 16.6°
 c. $(3, 0, -2)$, 36.5°

14. a. $c \neq 4, d \in R$ b. $c = 4, d = 4$ c. $c = 4, d \neq 4$

15. a. $b \neq -2, d \in R$
 b. $b = -2, d = -4$
 c. $b = -2, d \neq -4$

16. Sample responses can be found in the worked solutions in the online resources.

17. a. $a = 6, b = -3, c = 6$
 b. $a = 3, b = -5, c = -4, d = 7$

4.4 Exam questions

Note: Mark allocations are available with the fully worked solutions online.

1. $6x + 11y + 9z = 3$
2. $2x - 3y + z = 12$
3. $(4, 4, 3)$

4.5 Review

4.5 Exercise

Technology free: short answer

1. a. i. $\frac{\pm 1}{3\sqrt{6}}\left(\underset{\sim}{i} + 7\underset{\sim}{j} + 2\underset{\sim}{k}\right)$
 ii. $\frac{\pm 1}{3\sqrt{19}}(7\underset{\sim}{i} + \underset{\sim}{j} + 11\underset{\sim}{k})$
 b. i. $\frac{5\sqrt{14}}{2}$
 ii. $2x - 3y + z = 5$
 c. i. $x = -1 + 6t, y = 3 - 4t, z = 5 - 3t$
 ii. $2x + 3y = 7$
 iii. $\frac{9\sqrt{13}}{2}$

2. a. $(3, 1, -2)$, 57.7°
 b. $(3, -2, 1)$, 80.2°
 c. $x = 1; y = 1 + t, z = t$, 78.6°

3. a. $(2, 1, 4)$, 83.1° b. $(2, 1, 4)$, 16.2°

4. a. i. $\frac{\sqrt{3070}}{3}$ ii. 1
 b. i. $\frac{\sqrt{21}}{7}$ ii. 2

5. a. $x = \frac{1}{13}(32 + 6t)$, $y = \frac{1}{13}(9 - 4t)$, $z = t$, 90°
 b. $\frac{11\sqrt{17}}{17}$
 c. $\frac{5\sqrt{17}}{17}$

6. a. Line $\frac{x+4}{2} = \frac{y-2}{-1} = \frac{z-9}{-3}$.

   b. The line and plane do not intersect.

   c. Point $(2, 5, 3)$.

7. a. Line $\frac{x-4}{5} = y+2 = \frac{z-3}{3}$.

   b. The line and plane do not intersect.

   c. Point $(-6, -4, -3)$.

Technology active: multiple choice

8. D
9. A
10. A
11. D
12. B
13. B
14. C
15. E
16. E
17. C

Technology active: extended response

18. a. $A(2, 4, 0)$, $B(0, 12, -4)$, $C(3, 0, 2)$

    b. $3x - y - 5z = 13$, $\frac{1}{2}\sqrt{34}$

19. a. $7x - 5y - 3z = 20$

    b. $3x + 2y + 8z = 41$

20. a. $(2, -1, 3)$

    b. $45°$

    c. $a = 2$, $b = -2$, $x_0 = -2$

4.5 Exam questions

Note: Mark allocations are available with the fully worked solutions online.

1. C
2. D
3. D
4. E
5. Triangle ABC is a right-angled scalene triangle. Sample responses can be found in the worked solutions in the online resources.

5 Differential calculus

LEARNING SEQUENCE

Fully worked solutions for this topic are available online.

5.1 Overview

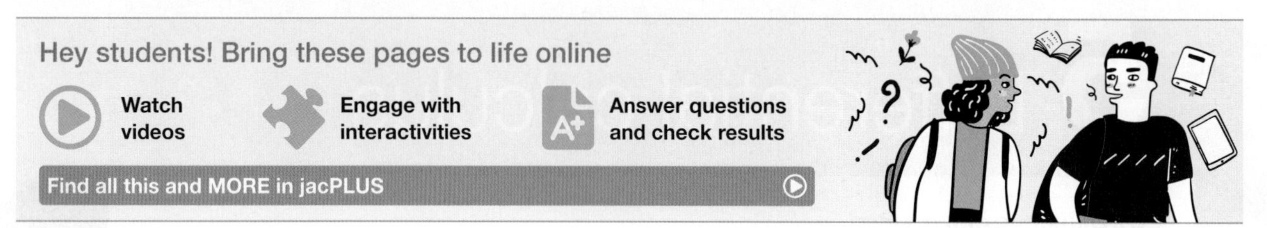

5.1.1 Introduction

Differential calculus is one branch of calculus that studies the rates at which quantities change.

Differential calculus is concerned with finding the derivative or gradient or the rate of change of a function. The process of finding the derivative is called differentiation. Geometrically, the derivative at a point is the slope of the tangent line to the graph of a function at that point, provided that the derivative exists and is defined at that point. The origin of calculus was a controversial topic for some time as two famous mathematicians, Isaac Newton (1642–1727) and Gottfried Wilhelm Leibniz (1646–1716), both claimed to have invented calculus at around the same time. Although Leibniz published his work first, Newton accused Leibniz of plagiarizing his unpublished work. These days it is agreed that both men developed their ideas independently, and we use a combination of Newton's and Leibniz's notation.

Differentiation has applications in nearly all quantitative disciplines. The derivative of the displacement of a moving body with respect to time is the velocity of the body, and the derivative of the velocity with respect to time is the acceleration.

KEY CONCEPTS

This topic covers the following points from the VCE Mathematics Study Design:

- derivatives of inverse circular functions
- second derivatives, use of notations $f''(x)$ and $\frac{d^2y}{dx^2}$, and their application to the analysis of graphs of functions, including points of inflection and concavity
- applications of chain rule to related rates of change and implicit differentiation, for example, implicit differentiation of the relations $x^2+y^2=9$, $3xy^2=x+y$ and $x\sin(y)+x^2\cos(y)=1$.

Note: *Concepts that are greyed out are covered in other topics.*

Source: VCE Mathematics Study Design (2023–2027) extracts © VCAA; reproduced by permission.

5.2 Review of differentiation techniques

LEARNING INTENTION

At the end of this subtopic you should be able to:
- differentiate composite, product and quotient functions using the chain, product and quotient rules.

5.2.1 Derivatives of basic functions

In the Mathematical Methods course, the **derivatives** of functions and the standard rules for differentiation are covered. If $y = f(x)$ is the equation of a curve, then the gradient function is given by $\frac{dy}{dx} = f'(x) = \lim_{h \to 0} \frac{f(x+h) - f(x)}{h}$. Using this formula to obtain the gradient function or the first derivative is called using the method of first principles.

Usually, the standard rules for differentiation are used to obtain the gradient function. In this section, these fundamental techniques of differentiation are revised and extended. The table below shows the basic functions for $y = f(x)$ and the corresponding gradient functions for $\frac{dy}{dx} = f'(x)$, where k and n are constants and $x \in R$.

Function: $y = f(x)$	**Gradient function: $\frac{dy}{dx} = f'(x)$**
x^n	nx^{n-1}
$\sin(kx)$	$k\cos(kx)$
$\cos(kx)$	$-k\sin(kx)$
e^{kx}	ke^{kx}
$\log_e(x)$	$\frac{1}{x}$

5.2.2 The chain rule

The chain rule is used to differentiate functions of functions. It states that if $y = f(x) = g(h(x)) = g(u)$, where $u = h(x)$, then $\frac{dy}{dx} = \frac{dy}{du}\frac{du}{dx} = g'(u)h'(x)$.

A quick proof is as follows. Let δx be the increment in x, and δy and δu be the corresponding increments in y and u. Provided that $\delta u \neq 0$ and $\delta x \neq 0$, $\frac{\delta y}{\delta x} = \frac{\delta y}{\delta u} \cdot \frac{\delta u}{\delta x}$

Now in the limit as $\delta x \to 0, \delta u \to 0$, $\frac{\delta y}{\delta x} \to \frac{dy}{dx}$, $\frac{\delta y}{\delta u} \to \frac{dy}{du}$ and $\frac{\delta u}{\delta x} \to \frac{du}{dx}$, so that $\frac{dy}{dx} = \frac{dy}{du}\frac{du}{dx}$.

The chain rule

$$\frac{dy}{dx} = \frac{dy}{du}\frac{du}{dx}$$

WORKED EXAMPLE 1 Differentiation using the chain rule (1)

a. **If $f(x) = \sqrt{4x^2 + 9}$, determine the value of $f'(-1)$.**
b. **Differentiate $2\cos^4(3x)$ with respect to x.**

THINK	WRITE
a. 1. Write the equation in index form.	**a.** $f(x) = \sqrt{4x^2+9}$ $= (4x^2+9)^{\frac{1}{2}}$
2. Express y in terms of u and u in terms of x.	$y = u^{\frac{1}{2}}$ where $u = 4x^2 + 9$
3. Differentiate y with respect to u and u with respect to x.	$\frac{dy}{du} = \frac{1}{2}u^{-\frac{1}{2}}$ and $\frac{du}{dx} = 8x$ $= \frac{1}{2\sqrt{u}}$
4. Determine $f'(x)$ using the chain rule.	$f'(x) = \frac{dy}{du}\frac{du}{dx} = \frac{1}{2\sqrt{u}} \times 8x$
5. Substitute back for u and cancel factors.	$f'(x) = \frac{4x}{\sqrt{4x^2+9}}$
6. Evaluate at the indicated point.	$f'(-1) = \frac{-4}{\sqrt{(4+9)}}$
7. In this case we need to rationalise the denominator.	$f'(-1) = -\frac{4}{\sqrt{13}} \times \frac{\sqrt{13}}{\sqrt{13}}$
8. State the final result. This represents the gradient of the curve at the indicated point.	$f'(-1) = -\frac{4\sqrt{13}}{13}$
b 1. Write the equation, expressing it in index notation.	**b.** $y = 2\cos^4(3x)$ $= 2(\cos(3x))^4$
2. Express y in terms of u and u in terms of x.	$y = 2u^4$ where $u = \cos(3x)$
3. Differentiate y with respect to u and u with respect to x.	$\frac{dy}{du} = 8u^3$ and $\frac{du}{dx} = -3\sin(3x)$
4. Determine $\frac{dy}{dx}$ using the chain rule.	$\frac{dy}{dx} = \frac{dy}{du}\frac{du}{dx} = 8u^3 \times (-3\sin(3x))$
5. Substitute back for u and state the final result.	$\frac{dy}{dx} = -24\cos^3(3x)\sin(3x)$

The chain rule can also be used in conjunction with other mixed types of functions.

WORKED EXAMPLE 2 Differentiation using the chain rule (2)

a. If $f(x) = 4\sin\left(\frac{2}{x}\right)$, evaluate $f'\left(\frac{6}{\pi}\right)$.

b. Differentiate $e^{\cos(4x)}$ with respect to x.

THINK	WRITE
a. 1. Write the equation.	**a.** $f(x) = 4\sin\left(\frac{2}{x}\right)$
2. Express y in terms of u and u in terms of x.	Let $y = 4\sin(u)$ where $u = \frac{2}{x} = 2x^{-1}$.
3. Differentiate y with respect to u and u with respect to x.	$\frac{dy}{du} = 4\cos(u)$ and $\frac{du}{dx} = -2x^{-2} = -\frac{2}{x^2}$
4. Determine $f'(x)$ using the chain rule.	$f'(x) = \frac{dy}{du}\frac{du}{dx} = 4\cos(u) \times \left(-\frac{2}{x^2}\right)$ $= -\frac{8}{x^2}\cos(u)$
5. Substitute back for u.	$f'(x) = -\frac{8}{x^2}\cos\left(\frac{2}{x}\right)$
6. Evaluate at the indicated point and simplify.	$f'\left(\frac{6}{\pi}\right) = -\frac{8}{\left(\frac{6}{\pi}\right)^2}\cos\left(\frac{2}{\frac{6}{\pi}}\right)$ $= -\frac{8\pi^2}{36}\cos\left(\frac{2\pi}{6}\right)$
7. Substitute for the trigonometric values.	$f'\left(\frac{6}{\pi}\right) = -\frac{2\pi^2}{9} \times \frac{1}{2}$
8. State the final result.	$f'\left(\frac{6}{\pi}\right) = -\frac{\pi^2}{9}$
b. 1. Write the equation.	**b.** Let $y = e^{\cos(4x)}$.
2. Express y in terms of u and u in terms of x.	$y = e^u$ where $u = \cos(4x)$
3. Differentiate y with respect to u and u with respect to x.	$\frac{dy}{du} = e^u$ and $\frac{du}{dx} = -4\sin(4x)$
4. Determine $\frac{dy}{dx}$ using the chain rule.	$\frac{dy}{dx} = \frac{dy}{du}\frac{du}{dx} = e^u \times (-4\sin(4x))$
5. Substitute back for u and state the final result.	$\frac{dy}{dx} = -4\sin(4x)e^{\cos(4x)}$

Derivatives involving logarithms

If $y = \log_e(x) = \ln(x)$ where $x > 0$, then $x = e^y$ so that $\frac{dx}{dy} = e^y$.

Since $\frac{dx}{dy} = 1/\frac{dy}{dx}$, $\frac{dy}{dx} = \frac{1}{e^y} = \frac{1}{x}$.

By applying the chain rule, we can quite easily determine the derivative of any logarithmic function of the form $y = \log_e(f(x))$. These functions are composite functions, so it follows that $\frac{dy}{dx} = \frac{1}{f(x)} \times f'(x) = \frac{f'(x)}{f(x)}$.

Derivatives of logarithmic functions

If $y = \log_e(f(x))$, then:

$$\frac{dy}{dx} = \frac{f'(x)}{f(x)}$$

WORKED EXAMPLE 3 Differentiation of logarithmic functions

a. **Differentiate $\log_e(\cos(3x))$ with respect to x.**

b. **Determine $\frac{d}{dx}\left[\log_e\left(\frac{3x+5}{3x-5}\right)\right]$.**

THINK	WRITE
a. 1. Write the equation.	a. $y = \log_e(\cos(3x))$
2. Use the general result.	$\frac{dy}{dx} = \frac{\frac{d}{dx}(\cos(3x))}{\cos(3x)}$
3. State the derivative in the numerator.	$\frac{dy}{dx} = \frac{-3\sin(3x)}{\cos(3x)}$
4. Simplify and state the final result.	$\frac{dy}{dx} = -3\tan(3x)$
b. 1. Write the equation.	b. $y = \log_e\left(\frac{3x+5}{3x-5}\right)$
2. Simplify using log laws.	$y = \log_e(3x+5) - \log_e(3x-5)$
3. Differentiate each term.	$\frac{dy}{dx} = \frac{3}{3x+5} - \frac{3}{3x-5}$
4. Form a common denominator.	$\frac{dy}{dx} = \frac{3(3x-5) - 3(3x+5)}{(3x+5)(3x-5)}$
5. Expand the numerator and denominator and simplify.	$\frac{dy}{dx} = \frac{9x - 15 - (9x+15)}{9x^2 - 25}$
6. State the final answer in simplest form.	$\frac{dy}{dx} = \frac{-30}{9x^2 - 25}$

5.2.3 The product rule

The product rule states that if $u = u(x)$ and $v = v(x)$ are two differentiable functions of x, and $y = u \cdot v$, then $\frac{dy}{dx} = u\frac{dv}{dx} + v\frac{du}{dx}$. A quick proof is as follows.

Let δx be the increment in x and δu, δv and δy be the corresponding increments in u, v and y. Then

$$y + \delta y = (u + \delta u)(v + \delta v)$$

Expanding, $$y + \delta y = uv + u\delta v + v\delta u + \delta u\delta v$$

Since $y = u \cdot v$, $$\delta y = u\delta v + v\delta u + \delta u\delta v$$

Divide each term by δx: $\frac{\delta y}{\delta x} = u\frac{\delta v}{\delta x} + v\frac{\delta u}{\delta x} + \frac{\delta u \cdot \delta v}{\delta x}$

Now in the limit as $\delta x \to 0, \delta u \to 0, \delta v \to 0, \frac{\delta y}{\delta x} \to \frac{dy}{dx}, \frac{\delta v}{\delta x} \to \frac{dv}{dx}, \frac{\delta u}{\delta x} \to \frac{du}{dx}$ and $\frac{\delta u \cdot \delta v}{\delta x} \to 0$.

Hence, $\frac{dy}{dx} = u\frac{dv}{dx} + v\frac{du}{dx}$.

The product rule

If $y = uv$, where u and v are both differentiable functions of x, then:

$$\frac{dy}{dx} = u\frac{dv}{dx} + v\frac{du}{dx}$$

WORKED EXAMPLE 4 Differentiation using the product rule

a. If $f(x) = x^4\sin(2x)$, determine $f'(x)$.

b. Determine $\frac{d}{dx}\left[x^5e^{-2x}\right]$.

THINK	WRITE
a. 1. Write the equation.	**a.** $y = f(x) = x^4\sin(2x) = u \cdot v$
2. State the functions u and v.	$u = x^4$ and $v = \sin(2x)$
3. Differentiate u and v with respect to x.	$\frac{du}{dx} = 4x^3$ and $\frac{dv}{dx} = 2\cos(2x)$
4. Determine $f'(x)$ using the product rule. Substitute for u, $\frac{dv}{dx}$, v and $\frac{du}{dx}$.	$f'(x) = u\frac{dv}{dx} + v\frac{du}{dx}$ $= 2x^4\cos(2x) + 4x^3\sin(2x)$
5. Simplify by taking out the common factors.	$f'(x) = 2x^3\left(x\cos(2x) + 2\sin(2x)\right)$
b. 1. Write the equation.	**b.** Let $y = x^5e^{-2x} = u \cdot v$.
2. State the functions u and v.	$u = x^5$ and $v = e^{-2x}$
3. Differentiate u and v with respect to x.	$\frac{du}{dx} = 5x^4$ and $\frac{dv}{dx} = -2e^{-2x}$
4. Determine $\frac{dy}{dx}$ using the product rule. Substitute for u, $\frac{dv}{dx}$, v and $\frac{du}{dx}$.	$\frac{dy}{dx} = u\frac{dv}{dx} + v\frac{du}{dx}$ $\frac{dy}{dx} = x^5 \times \left(-2e^{-2x}\right) + e^{-2x} \times 5x^4$
5. Simplify by taking out the common factors.	$\frac{dy}{dx} = x^4e^{-2x}(5 - 2x)$

Alternative notation

The product rule can be used without explicitly writing the expressions u and v. The setting out is similar, but an alternative notation is used.

1. Write the expression. This is called 'operation notation'. $\frac{d}{dx}\left[x^5 e^{-2x}\right]$
2. Use the product rule. $=x^5\frac{d}{dx}\left(e^{-2x}\right)+e^{-2x}\frac{d}{dx}\left(x^5\right)$
3. Calculate the derivatives. $=x^5\times\left(-2e^{-2x}\right)+e^{-2x}\times 5x^4$
4. Simplify by taking out the common factors and state the final result. $\frac{d}{dx}\left[x^5 e^{-2x}\right]=x^4e^{-2x}(5-2x)$

5.2.4 The quotient rule

The quotient rule states that if $u=u(x)$ and $v=v(x)$ are two differentiable functions of x and $y=\frac{u}{v}$, then $\frac{dy}{dx}=\frac{v\frac{du}{dx}-u\frac{dv}{dx}}{v^2}$. A quick proof is as follows.

Let $y=\frac{u}{v}=u\cdot v^{-1}$.

Now by the product rule, $\frac{dy}{dx}=u\frac{d}{dx}\left(v^{-1}\right)+v^{-1}\frac{du}{dx}$.

Using the chain rule,

$$\begin{aligned}\frac{dy}{dx}&=u\frac{d}{dv}\left(v^{-1}\right)\frac{dv}{dx}+v^{-1}\frac{du}{dx}\\&=-uv^{-2}\frac{dv}{dx}+v^{-1}\frac{du}{dx}\\&=\frac{v\frac{du}{dx}-u\frac{dv}{dx}}{v^2}\end{aligned}$$

The quotient rule

If $y=\frac{u}{v}$, where u and v are both differentiable functions of x, then:

$$\frac{dy}{dx}=\frac{v\frac{du}{dx}-u\frac{dv}{dx}}{v^2}$$

WORKED EXAMPLE 5 Differentiation using the quotient rule

a. **Differentiate $\dfrac{3\sin(2x)}{2x^4}$ with respect to x.**

b. **If $f(x) = \dfrac{3x}{\sqrt{4x^2+9}}$, determine $f'(x)$.**

THINK	WRITE
a. 1. Write the equation.	**a.** Let $y = \dfrac{3\sin(2x)}{2x^4} = \dfrac{u}{v}$.
2. State the functions u and v.	$u = 3\sin(2x)$ and $v = 2x^4$
3. Differentiate u and v with respect to x.	$\dfrac{du}{dx} = 6\cos(2x)$ and $\dfrac{dv}{dx} = 8x^3$
4. Determine $\dfrac{dy}{dx}$ using the quotient rule. Substitute for u, $\dfrac{dv}{dx}$, v and $\dfrac{du}{dx}$.	$\dfrac{dy}{dx} = \dfrac{v\dfrac{du}{dx} - u\dfrac{dv}{dx}}{v^2}$ $\dfrac{dy}{dx} = \dfrac{2x^4 \times 6\cos(2x) - 3\sin(2x) \times 8x^3}{\left(2x^4\right)^2}$
5. Expand and simplify by taking out the common factors.	$\dfrac{dy}{dx} = \dfrac{12x^3\left(x\cos(2x) - 2\sin(2x)\right)}{4x^8}$
6. Cancel the common factors and state the final result in simplest form.	$\dfrac{dy}{dx} = \dfrac{3}{x^5}\left(x\cos(2x) - 2\sin(2x)\right)$
b. 1. Write the equation.	**b.** Let $f(x) = \dfrac{3x}{\sqrt{4x^2+9}} = \dfrac{u}{v}$.
2. State the functions u and v.	$u = 3x$ and $v = \sqrt{4x^2+9}$
3. Differentiate u and v with respect to x, using the chain rule to determine $\dfrac{dv}{dx}$. See Worked example 1a for $\dfrac{dv}{dx}$.	$\dfrac{du}{dx} = 3$ and $\dfrac{dv}{dx} = \dfrac{4x}{\sqrt{4x^2+9}}$
4. Determine $f'(x)$ using the quotient rule. Substitute for u, $\dfrac{dv}{dx}$, v and $\dfrac{du}{dx}$.	$f'(x) = \dfrac{v\dfrac{du}{dx} - u\dfrac{dv}{dx}}{v^2}$ $f'(x) = \dfrac{\sqrt{4x^2+9} \times 3 - 3x \times \frac{4x}{\sqrt{4x^2+9}}}{\left(\sqrt{4x^2+9}\right)^2}$
5. Simplify by forming a common denominator in the numerator.	$f'(x) = \dfrac{\dfrac{3\left(4x^2+9\right) - 12x^2}{\sqrt{4x^2+9}}}{4x^2+9}$
6. Expand and simplify the terms in the numerator.	$f'(x) = \dfrac{\dfrac{12x^2 + 27 - 12x^2}{\sqrt{4x^2+9}}}{4x^2+9}$

7. Simplify using $\frac{\left(\frac{a}{b}\right)}{c} = \frac{a}{b} \times \frac{1}{c}$.

$$f'(x) = \frac{27}{\sqrt{4x^2+9}} \times \frac{1}{4x^2+9}$$
$$= \frac{27}{\left(4x^2+9\right)^{\frac{3}{2}}}$$

8. Use index laws to simplify and state the final result in simplest form.

$$f'(x) = \frac{27}{\sqrt{\left(4x^2+9\right)^3}}$$

Derivative of $\tan(kx)$

The quotient rule can be used to determine the derivative of $\tan(kx)$.

Let $y = \tan(kx) = \frac{\sin(kx)}{\cos(kx)} = \frac{u}{v}$, where $u = \sin(kx)$ and $v = \cos(kx)$.

Then $\frac{du}{dx} = k\cos(kx)$ and $\frac{dv}{dx} = -k\sin(kx)$.

Using the quotient rule, $\frac{dy}{dx} = \frac{v\frac{du}{dx} - u\frac{dv}{dx}}{v^2}$

$$\frac{dy}{dx} = \frac{\cos(kx) \times k\cos(kx) - \sin(kx) \times -k\sin(kx)}{(\cos(kx))^2}$$
$$= \frac{k\left(\cos^2(kx) + \sin^2(kx)\right)}{\cos^2(kx)} \text{ since } \cos^2(kx) + \sin^2(kx) = 1$$
$$= \frac{k}{\cos^2(kx)}$$
$$= k\sec^2(kx)$$

Hence, $\frac{d}{dx}(\tan(kx)) = k\sec^2(kx) = \frac{k}{\cos^2(kx)}$.

Derivative of tan(kx)

$$\frac{d}{dx}\left(\tan(kx)\right) = k\sec^2(kx) = \frac{k}{\cos^2(kx)}$$

WORKED EXAMPLE 6 Differentiation of tangent functions

a. Differentiate $2\tan^5\left(\frac{x}{2}\right)$ with respect to x.

b. If $f(x) = x^2\tan\left(\frac{2x}{3}\right)$, evaluate $f'\left(\frac{\pi}{2}\right)$.

THINK	WRITE
a. 1. Write the equation.	a. $y = 2\tan^5\left(\frac{x}{2}\right)$
2. Express y in terms of u and u in terms of x.	$y = 2u^5$ where $u = \tan\left(\frac{x}{2}\right)$

3. Differentiate y with respect to u and u with respect to x using $\frac{d}{dx}(\tan(kx)) = k\sec^2(kx)$ with $k = \frac{1}{2}$.

$$\frac{dy}{du} = 10u^4 \text{ and } \frac{du}{dx} = \frac{1}{2}\sec^2\left(\frac{x}{2}\right)$$

4. Determine $\frac{dy}{dx}$ using the chain rule.

$$\frac{dy}{dx} = \frac{dy}{du}\frac{du}{dx} = 10u^4 \times \frac{1}{2}\sec^2\left(\frac{x}{2}\right)$$

5. Substitute back for u and state the final result.

$$\frac{dy}{dx} = 5\tan^4\left(\frac{x}{2}\right)\sec^2\left(\frac{x}{2}\right)$$

b. 1. Write the equation.

b. $f(x) = x^2\tan\left(\frac{2x}{3}\right) = u \cdot v$

2. State the functions u and v.

$$u = x^2 \text{ and } v = \tan\left(\frac{2x}{3}\right)$$

3. Differentiate u and v with respect to x.

$$\frac{du}{dx} = 2x \text{ and } \frac{dv}{dx} = \frac{2}{3}\sec^2\left(\frac{2x}{3}\right)$$

4. Determine $f'(x)$ using the product rule. Substitute for u, $\frac{dv}{dx}$, v and $\frac{du}{dx}$.

$$f'(x) = u\frac{dv}{dx} + v\frac{du}{dx}$$

$$f'(x) = x^2 \times \frac{2}{3}\sec^2\left(\frac{2x}{3}\right) + \tan\left(\frac{2x}{3}\right) \times 2x$$

5. Substitute $x = \frac{\pi}{2}$.

$$f'\left(\frac{\pi}{2}\right) = \frac{2}{3}\left(\frac{\pi}{2}\right)^2 \times \frac{1}{\cos^2\left(\frac{\pi}{3}\right)} + \tan\left(\frac{\pi}{3}\right) \times 2 \times \frac{\pi}{2}$$

6. Substitute for the trigonometric values and simplify.

$$f'\left(\frac{\pi}{2}\right) = \frac{\pi^2}{6 \times \left(\frac{1}{2}\right)^2} + \sqrt{3}\pi$$

$$= \frac{2\pi^2}{3} + \sqrt{3}\pi$$

7. Express the final result in simplest form by taking a common denominator and the common factor out in the numerator.

$$f'\left(\frac{\pi}{2}\right) = \frac{\pi\left(2\pi + 3\sqrt{3}\right)}{3}$$

5.2 Exercise

Students, these questions are even better in jacPLUS

Receive immediate feedback and access sample responses

Access additional questions

Track your results and progress

Find all this and MORE in jacPLUS

Technology free

1. **WE1** a. If $f(x)=\dfrac{2}{3x^2+5}$, determine the value of $f'(-1)$.

   b. Differentiate $5\sin^3(2x)$ with respect to x.

2. a. Determine $\dfrac{d}{dx}\left[\dfrac{1}{2x+5}\right]$.

   b. If $f(x)=6\sqrt{\cos(4x)}$, determine the value of $f'\left(\dfrac{\pi}{12}\right)$.

3. Differentiate each of the following with respect to x.

   a. $2\sin^4(3x)$

   b. $5\cos^3(4x)$

4. Differentiate each of the following with respect to x.

   a. $x\sqrt{2x^2+9}$

   b. $\dfrac{x}{\sqrt{3x^2+5}}$

5. Determine $f'(x)$ for each of the following.

   a. $f(x)=e^{-\frac{1}{2}x^2}$

   b. $f(x)=e^{\cos(2x)}$

6. Determine $g'(x)$ for each of the following.

   a. $g(x)=\cos\left(e^{2x}\right)$

   b. $g(x)=e^{\sqrt{x}}$

7. **WE2** a. If $f(x)=6\cos\left(\dfrac{3}{x}\right)$, evaluate $f'\left(\dfrac{18}{\pi}\right)$.

   b. Differentiate $e^{\sin(2x)}$ with respect to x.

8. a. Differentiate $\sin\left(\sqrt{x}\right)$ with respect to x.

   b. If $h(x)=e^{\cos(2x)}$, evaluate $h'\left(\dfrac{\pi}{6}\right)$.

9. **WE3** a. Differentiate $\log_e\left(\sin\left(\dfrac{x}{2}\right)\right)$ with respect to x.

   b. Determine $\dfrac{d}{dx}\left[\log_e\left(\dfrac{4x^2+9}{4x^2-9}\right)\right]$.

10. **WE4** a. If $f(x)=x^3\cos(4x)$, determine $f'(x)$.

    b. Calculate $\dfrac{d}{dx}\left(x^4e^{-3x}\right)$.

11. a. Differentiate $e^{-3x}\cos(2x)$ with respect to x.

    b. If $f(x)=x^2e^{-x^2}$, evaluate $f'(2)$.

12. Determine $\frac{dy}{dx}$ for each of the following.

a. $y = x^3 \sin(5x)$

b. $y = x^4 \cos(4x)$

13. Determine each of the following.

a. $\frac{d}{dx}\left(x^3 e^{-4x}\right)$

b. $\frac{d}{dx}\left(e^{-3x}\sin(2x)\right)$

14. **WE5** a. Differentiate $\frac{3\cos(3x)}{2x^3}$ with respect to x.

b. If $f(x) = \frac{x}{\sqrt{4x+9}}$, determine $f'(x)$.

15. a. If $f(x) = \frac{1}{3xe^{2x}}$, determine $f'(x)$.

b. Determine $\frac{d}{dx}\left[\left(\frac{3x^2+5}{3x^2-5}\right)^2\right]$.

16. Calculate $\frac{dy}{dx}$ for each of the following.

a. $y = \frac{3\sin(3x)}{2x^3}$

b. $y = \frac{4\cos(2x)}{3x^4}$

17. Determine each of the following.

a. $\frac{d}{dx}\left[\frac{e^{3x}}{x^2}\right]$

b. $\frac{d}{dx}\left[\frac{1}{x^3 e^{2x}}\right]$

18. **WE6** a. Differentiate $6\tan^4\left(\frac{x}{3}\right)$ with respect to x.

b. If $f(x) = x^4 \tan\left(\frac{x}{4}\right)$, evaluate $f'(\pi)$.

19. a. If $g(x) = \frac{\tan(3x)}{x}$, evaluate $g'\left(\frac{\pi}{9}\right)$.

b. If $g(x) = 5\tan\left(\frac{2}{x}\right)$, evaluate $g'\left(\frac{12}{\pi}\right)$.

20. a. If $f(x) = \log_e\left(\sqrt{4x^2+9}\right)$, evaluate $f'(-1)$.

b. Differentiate $\cos\left(\log_e\left(\frac{x}{2}\right)\right)$ with respect to x.

21. Determine $\frac{dy}{dx}$ for $y = x^2 \log_e(5x+4)$.

22. Determine $\frac{dy}{dx}$ for each of the following.

a. $y = \log_e\left(\frac{4x-9}{4x+9}\right)$

b. $y = \log_e\left(\frac{3x^2+5}{3x^2-5}\right)$

23. a. If $g(x) = \log_e(\sin(3x))$, determine the exact value of $g'\left(\frac{\pi}{12}\right)$.

b. If $h(x) = \log_e(\tan(2x))$, determine the exact value of $h'\left(\frac{\pi}{12}\right)$.

24. a. If $f(x) = 4\cos\left(\frac{2}{x}\right)$, evaluate $f'\left(\frac{3}{2\pi}\right)$.

b. If $f(x) = 2\tan\left(\frac{3}{x}\right)$, evaluate $f'\left(\frac{18}{\pi}\right)$.

25. a. Use the chain rule to show that $\frac{d}{dx}(\sec(kx)) = k\sec(kx)\tan(kx)$.

b. Use the chain rule to show that $\frac{d}{dx}(\operatorname{cosec}(kx)) = -k\operatorname{cosec}(kx)\cot(kx)$.

c. Use the quotient rule to show that $\frac{d}{dx}(\cot(kx)) = -k\operatorname{cosec}^2(kx)$.

26. If $y = \log_e\left(\cot\left(\frac{x}{2}\right) + \operatorname{cosec}\left(\frac{x}{2}\right)\right)$, show that $\frac{dy}{dx} = -\frac{1}{2}\operatorname{cosec}\left(\frac{x}{2}\right)$.

27. If n, k, b and α are all constants, verify each of the following.

a. $\frac{d}{dx}[\sin(nx + \alpha)] = n\cos(nx + \alpha)$

b. $\frac{d}{dx}[\cos(nx + \alpha)] = -n\sin(nx + \alpha)$

28. If n, k, b and α are all constants, verify each of the following.

a. $\frac{d}{dx}\left[\sin^n(kx)\right] = nk\sin^{n-1}(kx)\cos(kx)$

b. $\frac{d}{dx}[\cos^n(kx)] = -nk\cos^{n-1}(kx)\sin(kx)$

29. If n, k, b and α are all constants, verify each of the following.

a. $\frac{d}{dx}\left[e^{kx}\sin(bx)\right] = e^{kx}(b\cos(bx) + k\sin(bx))$

b. $\frac{d}{dx}\left[e^{kx}\cos(bx)\right] = e^{kx}(k\cos(bx) - b\sin(bx))$

30. If n and k are both constants, verify the following.

$\frac{d}{dx}[x^n\sin(kx)] = x^{n-1}(n\sin(kx) + kx\cos(kx))$

31. If n and k are both constants, verify the following.

a. $\frac{d}{dx}\left[\frac{\sin(kx)}{x^n}\right] = \frac{1}{x^{n+1}}(kx\cos(kx) - n\sin(kx))$

b. $\frac{d}{dx}\left[\frac{\cos(kx)}{x^n}\right] = \frac{-1}{x^{n+1}}(kx\sin(kx) + n\cos(kx))$

32. If n and k are both constants, verify the following.

a. $\frac{d}{dx}\left[x^n e^{kx}\right] = x^{n-1}e^{kx}(n + kx)$

b. $\frac{d}{dx}\left[\frac{e^{kx}}{x^n}\right] = \frac{e^{kx}(kx - n)}{x^{n+1}}$

33. If a, b, c, d and n are all constants, verify each of the following.

a. $\frac{d}{dx}\left[\log_e\left(\sqrt{ax^2 + b}\right)\right] = \frac{ax}{ax^2 + b}$

b. $\frac{d}{dx}\left[\log_e\left(\frac{ax + b}{cx + d}\right)\right] = \frac{ad - bc}{(ax + b)(cx + d)}$

34. If a, b, c, d and n are all constants, verify each of the following.

a. $\frac{d}{dx}\left[\log_e\left(\frac{ax^2 + b}{cx^2 + d}\right)\right] = \frac{2(ad - bc)x}{\left(ax^2 + b\right)\left(cx^2 + d\right)}$

b. $\frac{d}{dx}\left[\log_e\left(\sin^n(bx)\right)\right] = \frac{nb}{\tan(bx)}$

35. If a, b, c, d and n are all constants, verify each of the following.

a. $\frac{d}{dx}\left[\log_e(\cos^n(bx))\right] = -nb\tan(bx)$

b. $\frac{d}{dx}\left[\log_e(\tan^n(bx))\right] = \frac{2nb}{\sin(2bx)}$

36. a. If $u(x)$, $v(x)$ and $w(x)$ are all functions of x and $y = u(x)v(x)w(x)$, use the product rule to show that $\frac{dy}{dx} = v(x)w(x)\frac{du}{dx} + u(x)w(x)\frac{dv}{dx} + u(x)v(x)\frac{dw}{dx}$.

b. Hence, determine $\frac{dy}{dx}$ for $y = x^3e^{-4x}\cos(2x)$.

37. Using the fundamental limit $\lim_{\theta \to 0} \frac{\sin(\theta)}{\theta} = 1$, verify each of the following, using the method of first principles where k is a constant and $x \in R$.

a. $\frac{d}{dx}(\sin(kx)) = k\cos(kx)$ b. $\frac{d}{dx}(\cos(kx)) = -k\sin(kx)$ c. $\frac{d}{dx}(\tan(kx)) = k\sec^2(kx)$

38. Using the fundamental limit $\lim_{\theta \to 0} \frac{\sin(\theta)}{\theta} = 1$, verify each of the following, using the method of first principles where k is a constant and $x \in R$.

a. $\frac{d}{dx}(\sec(kx)) = k\tan(kx)\sec(kx)$ b. $\frac{d}{dx}(\text{cosec}(kx)) = -k\cot(kx)\,\text{cosec}(kx)$

c. $\frac{d}{dx}(\cot(kx)) = -k\,\text{cosec}^2(kx)$

5.2 Exam questions

Question 1 (2 marks) TECH-FREE

Calculate the derivative of $y = \log_e\left(\sqrt{2x^2+9}\right)$.

Question 2 (2 marks) TECH-FREE

If $y = \log_e\left(\tan(3x) + \sec(3x)\right)$, show that $\frac{dy}{dx} = 3\sec(3x)$.

Question 3 (2 marks) TECH-FREE

If n and k are both constants, verify the following.

$$\frac{d}{dx}[x^n\cos(kx)] = x^{n-1}\left(n\cos(kx) - kx\sin(kx)\right)$$

More exam questions are available online.

5.3 Applications of differentiation

LEARNING INTENTION

At the end of this subtopic you should be able to:
- determine and sketch the tangent and normal to a curve at a point
- determine rates of change of one variable with respect to another.

There are many mathematical applications of differential calculus, including:
- determining tangents and normals to curves
- rates of change
- maxima and minima problems
- curve sketching
- related rate problems
- kinematics.

5.3.1 Tangents and normal to curves

Tangents to curves

As you will recall from Mathematical Methods, a tangent to a curve is a straight line that touches a curve at the point of contact. Furthermore, the gradient of the tangent is equal to the gradient of the curve at the point of contact.

WORKED EXAMPLE 7 Determining and sketching the tangent to a curve at a point

Determine the equation of the tangent to the curve $y = -x^2 + 3x + 4$ at the point where $x = 3$. Sketch the curve and the tangent.

THINK	WRITE/DRAW
1. Determine the y-coordinate corresponding to the given x-value.	$y = -x^2 + 3x + 4$ When $x = 3$, $y = -9 + 9 + 4$ $= 4$ The point is $(3, 4)$.
2. Determine the gradient of the curve at the given x-value. This will be denoted by m_T.	$\frac{dy}{dx} = -2x + 3$ When $x = 3$, $m_T = \left.\frac{dy}{dx}\right\rvert_{x=3}$ $= -3$
3. Determine the equation of the tangent, that is, the straight line passing through the given point with the given gradient. Use $y - y_1 = m_T(x - x_1)$.	$x_1 = 3, y_1 = 4$ and $m_T = -3$ $y - 4 = -3(x - 3)$ $y = -3x + 9 + 4$
4. State the equation of the tangent and sketch the graph.	$y = -3x + 13$

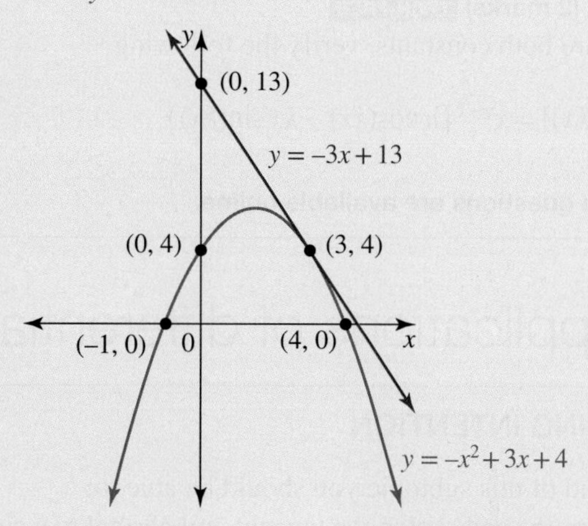

TI | THINK	DISPLAY/WRITE	CASIO | THINK	DISPLAY/WRITE
On a Calculator page select Menu: 4 Calculus 9 Tangent Line	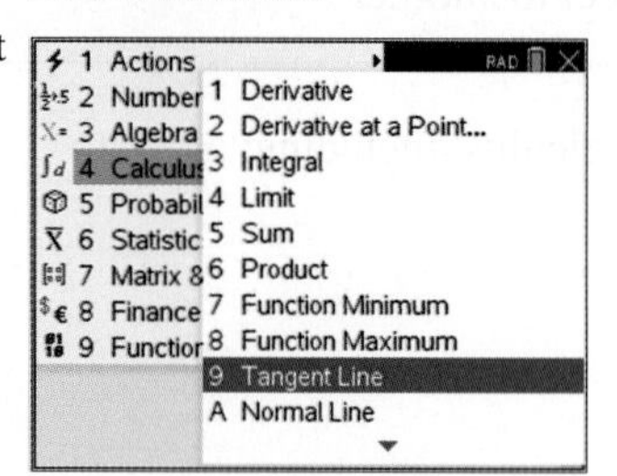	On a Main screen select Action: Calculation: line: tanLine. In the brackets type the expression, the variable and the x-value, all separated by commas as shown.	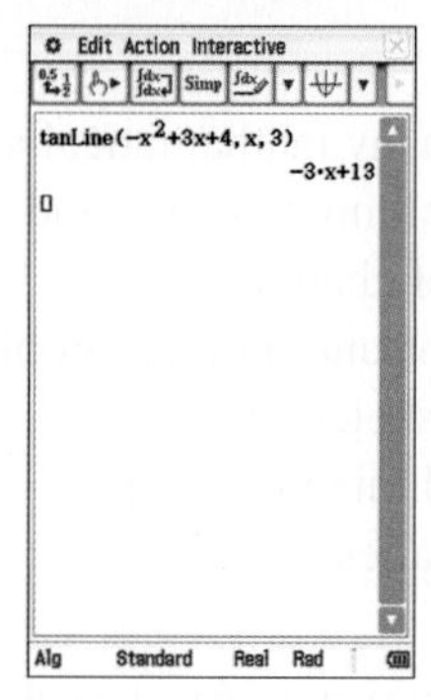

Type the expression and then the x-value into the brackets, separated by a comma, as shown.

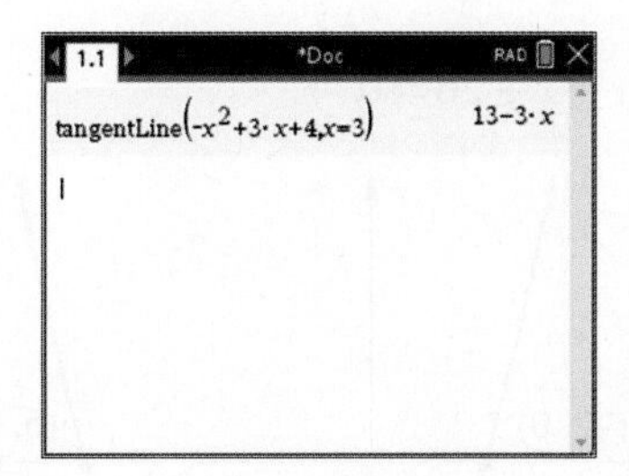

Normals to curves

The normal to a curve is a straight line perpendicular to the tangent to the curve. If two lines are perpendicular, the product of their gradients is -1. If m_T is the gradient of the tangent and m_N is the gradient of the normal, then $m_T m_N = -1$.

WORKED EXAMPLE 8 Determining and sketching the normal to a curve at a point

Determine the equation of the normal to the curve $y = x^2 - 3x - 10$ at the point where $x = 4$. Sketch the curve and the normal.

THINK	WRITE/DRAW
1. Determine the y-coordinate corresponding to the given x-value.	$y = x^2 - 3x - 10$ When $x = 4$, $y = 16 - 12 - 10$ $= -6$ The point is $(4, -6)$.
2. Determine the gradient of the curve at the given x-value. This is denoted by m_T.	$\frac{dy}{dx} = 2x - 3$ When $x = 4$, $m_T = \left.\frac{dy}{dx}\right\vert_{x=4}$ $= 5$
3. Determine the gradient of the normal, which will be denoted by m_N. Since the normal line is perpendicular to the tangent, the product of their gradients is -1.	$m_N m_T = -1$ $m_T = 5$ $\Rightarrow m_N = -\frac{1}{5}$ The normal has a gradient $m_N = -\frac{1}{5}$.
4. Determine the equation of the normal, that is, the straight line passing through the given point with the given gradient. Use $y - y_1 = m_N(x - x_1)$.	$x_1 = 4, y_1 = -6$ and $m_N = -\frac{1}{5}$ $y + 6 = -\frac{1}{5}(x - 4)$ $5y + 30 = -x + 4$

5. State the equation of the normal. To avoid fractions, give the result in the form $ax + by + k = 0$.
Sketch the graph.

$5y + x + 26 = 0$

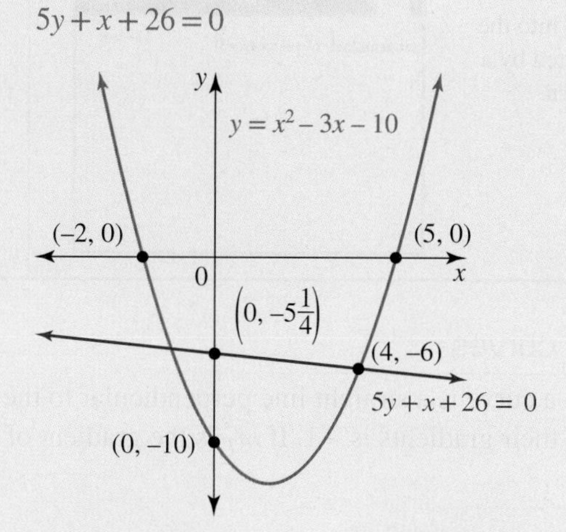

TI | THINK

On a Calculator page select Menu:
4 Calculus
A Normal Line

DISPLAY/WRITE

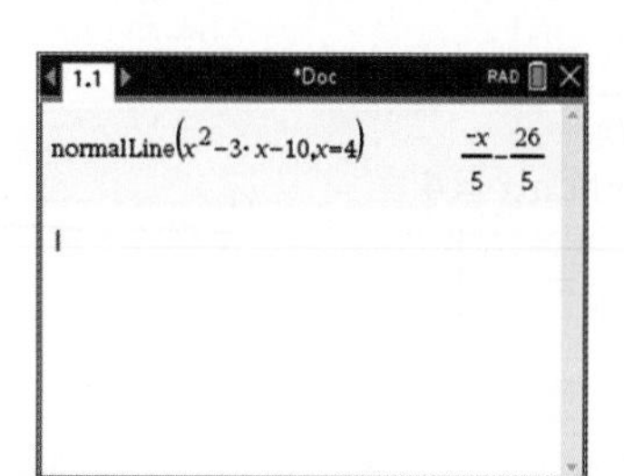

Type the expression and then the x-value into the brackets, separated by a comma, as shown.

CASIO | THINK

On a Main screen select Action: Calculation: line: normal.
In the brackets type the expression, the variable and the x-value, all separated by commas as shown.

DISPLAY/WRITE

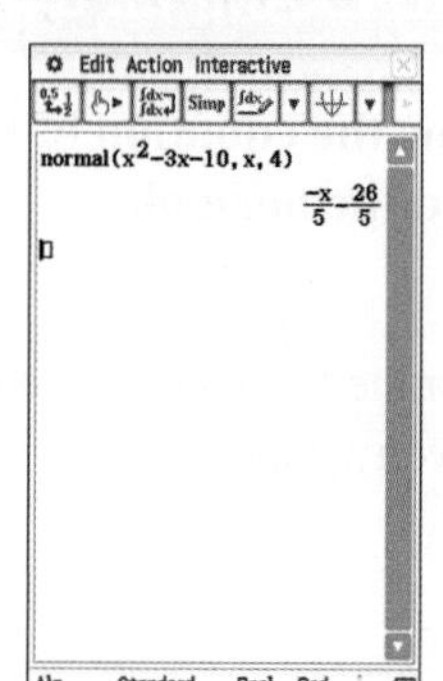

General results for determining tangents and normals to curves

In general, to determine the equation of the tangent to the curve $y = f(x)$ at the point where $x = a$, the y-value is $y = f(a)$, so the coordinates of the point are $(a, f(a))$. The gradient of the curve at this point is $m_T = f'(a)$, so the equation of the tangent is given by $y - f(a) = f'(a)(x - a)$. The normal has a gradient of $m_N = -\frac{1}{f'(a)}$, so the equation of the normal to the curve is given by $y - f(a) = -\frac{1}{f'(a)}(x - a)$.

WORKED EXAMPLE 9 Using the tangent at a point to determine the equation of a curve

The tangent to the curve $y = \sqrt{x}$ at a point is given by $6y - x + c = 0$. Determine the value of c.

THINK

1. Determine the gradient of the curve.

WRITE

$y = \sqrt{x} = x^{\frac{1}{2}}$

Then $\frac{dy}{dx} = \frac{1}{2}x^{-\frac{1}{2}} = \frac{1}{2\sqrt{x}}$

2. Determine the gradient of the tangent line.	$6y - x + c = 0$, Rearrange to make y the subject: $6y = x - c$ $y = \frac{x}{6} - \frac{c}{6}$
3. State the gradient of the tangent line.	$m_T = \frac{1}{6}$
4. At the point of contact these gradients are equal.	$\frac{1}{2\sqrt{x}} = \frac{1}{6}$
5. Solve the equation to calculate the x-value at the point of contact.	$\sqrt{x} = 3$ $x = 9$
6. Determine the coordinate at the point of contact. Substitute the x-value into the curve.	When $x = 9$, $y = \sqrt{x} = \sqrt{9} = 3$. The point is $(9, 3)$.
7. This point also lies on the tangent.	$6y - x + c = 0$ $c = x - 6y$ $= 9 - 18$
8. State the answer.	$c = -9$

Determining tangents and normals to other functions

When determining the equation of the tangent and normal to trigonometric functions or other types of functions, we must use exact values and give exact answers. That is, give answers in terms of π or surds such as $\sqrt{3}$, and do not give answers involving decimals. This principle applies across the entire course of VCE Specialist Mathematics, always give answers as exact values unless otherwise specified.

WORKED EXAMPLE 10 Determining the tangent and normal to a curve at a point

Determine the equation of the tangent and normal to the curve $y = 4\cos(2x)$ at the point where $x = \frac{\pi}{8}$. Sketch the tangent and the normal.

THINK	WRITE/DRAW
1. Determine the y-coordinate corresponding to the given x-value.	When $x = \frac{\pi}{8}$, $y = 4\cos\left(\frac{\pi}{4}\right)$ $= 2\sqrt{2}$ The point is $\left(\frac{\pi}{8}, 2\sqrt{2}\right)$.

2. Determine the gradient of the curve at the given x-value.

$$\frac{dy}{dx} = -8\sin(2x)$$

When $x = \frac{\pi}{8}$,

$$m_T = \left.\frac{dy}{dx}\right|_{x=\frac{\pi}{8}}$$

$$= -8\sin\left(\frac{\pi}{4}\right)$$

$$= -4\sqrt{2}$$

3. Determine the equation of the tangent, that is, the straight line passing through the given point with the given gradient. Use $y - y_1 = m_T(x - x_1)$

$x_1 = \frac{\pi}{8}$, $y_1 = 2\sqrt{2}$ and $m_T = -4\sqrt{2}$

$$y - 2\sqrt{2} = -4\sqrt{2}\left(x - \frac{\pi}{8}\right)$$

4. State the equation of the tangent.

$$y = -4\sqrt{2}x + \frac{\sqrt{2}\pi}{2} + 2\sqrt{2}$$

5. Determine the gradient of the normal, m_N. Since the normal line is perpendicular to the tangent, the product of their gradients is -1.

$$m_N m_T = -1$$

$$m_T = -4\sqrt{2}$$

$$\Rightarrow m_N = \frac{1}{4\sqrt{2}} \times \frac{\sqrt{2}}{\sqrt{2}}$$

$$= \frac{\sqrt{2}}{8}$$

The normal has a gradient $m_N = \frac{\sqrt{2}}{8}$.

6. Determine the equation of the normal, that is, the straight line passing through the given point with the given gradient. Use $y - y_1 = m_N(x - x_1)$.

$x_1 = \frac{\pi}{8}$, $y_1 = 2\sqrt{2}$ and $m_N = \frac{\sqrt{2}}{8}$

$$y - 2\sqrt{2} = \frac{\sqrt{2}}{8}\left(x - \frac{\pi}{8}\right)$$

7. State the equation of the normal and sketch the graph, showing the tangent and normal.

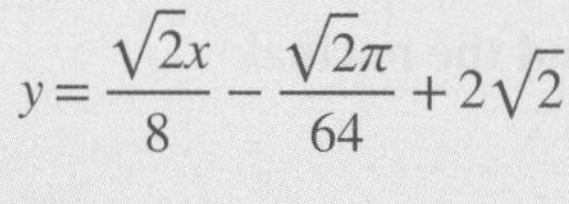

$$y = \frac{\sqrt{2}x}{8} - \frac{\sqrt{2}\pi}{64} + 2\sqrt{2}$$

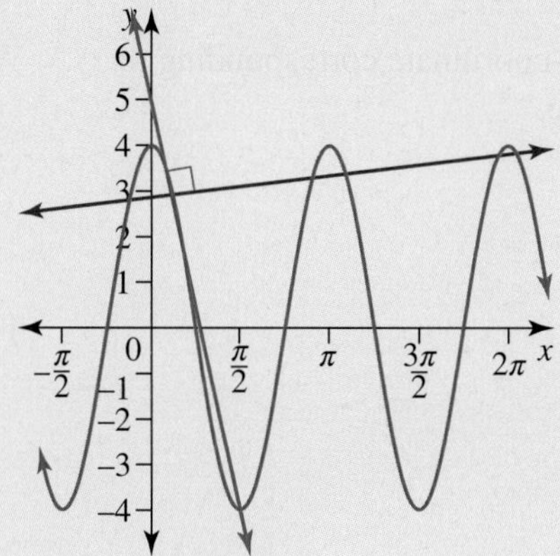

5.3.2 Rates of change

The first derivative or gradient function, $\frac{dy}{dx}$, is also a measure of the instantaneous rate of change of y with respect to x.

WORKED EXAMPLE 11 Calculating rates of change (1)

Determine the rate of change of the area of a circle with respect to the radius.

THINK	WRITE
1. Define the variables and state the area of a circle.	If the radius of the circle is r and the area is A, then $A = \pi r^2$.
2. We require the rate of change of the area of a circle with respect to the radius.	$\frac{dA}{dr}$
3. State the required result.	$\frac{dA}{dr} = 2\pi r$

Average rates of change

The average rate of change is not to be confused with the instantaneous rate of change. The average rate of change of a function $y = f(x)$ over $x \in [a, b]$ is the gradient of the line segment joining the points.

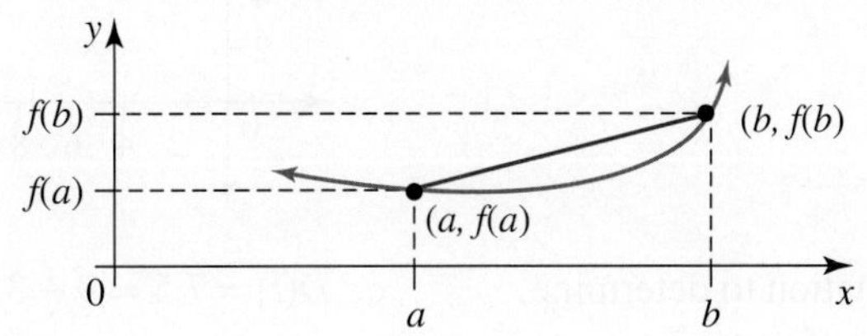

Looking at the above graph, the gradient of the line segment joining $(a, f(a))$ and $(b, f(b))$ is $\frac{f(b) - f(a)}{b - a}$.

Therefore, the average rate of change of a function $f(x)$ over $x \in [a, b]$ is $\frac{f(b) - f(a)}{b - a}$.

WORKED EXAMPLE 12 Calculating rates of change (2)

The tides in a certain bay can be modelled by

$$D(t) = 9 + 3\sin\left(\frac{\pi t}{6}\right)$$

where D is the depth of water in metres and t is the time in hours after midnight on a particular day.

a. **Calculate the depth of water in the bay at 2 am.**
b. **Sketch the graph of $D(t)$ against t for $0 \le t \le 24$.**
c. **Determine when the depth of water is below 7.5 metres.**
d. **Determine the rate of change of the depth at 2 am.**
e. **Over the first 4 hours of each day, determine the average rate of change of the depth.**

THINK	WRITE/DRAW
a. 2 am corresponds to $t = 2$. Evaluate $D(2)$, expressing your answer in simplest form.	**a.** $D(2) = 9 + 3\sin\left(\dfrac{2\pi}{6}\right)$ $= 9 + 3 \times \dfrac{\sqrt{3}}{2}$ $= \dfrac{18 + 3\sqrt{3}}{2}$ metres
b. 1. Determine the period and amplitude of the graph. State the maximum and minimum values of the depth.	**b.** $n = \dfrac{\pi}{6}$ The period is $\dfrac{2\pi}{n}$ or $\dfrac{2\pi}{\frac{\pi}{6}} = 12$. Over $0 \le t \le 24$, there are two cycles. The maximum depth is $9 + 3 = 12$ metres and the minimum depth is $9 - 3 = 6$ metres.
2. Sketch the graph on the restricted domain.	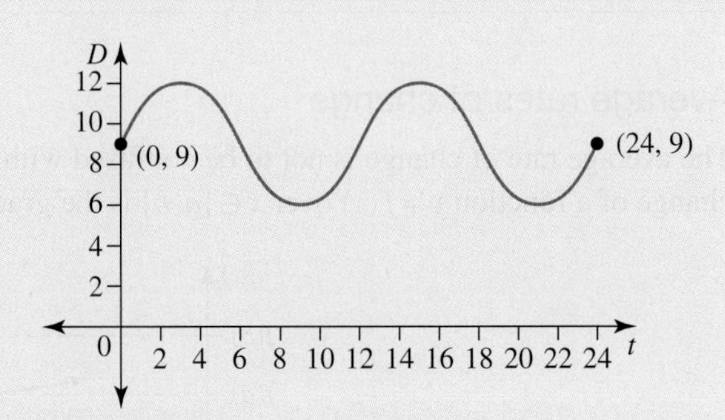
c. 1. Solve an appropriate equation to determine the times when the depth of water is 7.5 metres.	**c.** $D(t) = 7.5 = 9 + 3\sin\left(\dfrac{\pi t}{6}\right)$ $\sin\left(\dfrac{\pi t}{6}\right) = -\dfrac{1}{2}$ $\dfrac{\pi t}{6} = \dfrac{7\pi}{6}, \dfrac{11\pi}{6}, \dfrac{7\pi}{6} + 2\pi, \dfrac{11\pi}{6} + 2\pi$ $t = 7, 11, 19, 23$
2. State when the depth is below 7.5 metres.	The depth is below 7.5 metres between 7 am and 11 am and between 7 pm and 11 pm.
d. 1. Determine the rate of change of depth with respect to time.	**d.** $\dfrac{dD}{dt} = \dfrac{\pi}{2}\cos\left(\dfrac{\pi t}{6}\right)$
2. Evaluate this rate at 2 am, giving the correct units.	$\left.\dfrac{dD}{dt}\right\|_{t=2} = \dfrac{\pi}{2}\cos\left(\dfrac{\pi}{3}\right)$ $= \dfrac{\pi}{4}$ m/h

e. 1. Calculate the average rate of change over the first 4 hours.	**e.** $t \in [0, 4]$ $D(0) = 9 + 3\sin(0) = 9$ $D(4) = 9 + 3\sin\left(\frac{4\pi}{6}\right) = 9 + \frac{3\sqrt{3}}{2}$ $\frac{D(4) - D(0)}{4 - 0} = \frac{9 + \frac{3\sqrt{3}}{2} - 9}{4 - 0}$
2. State the required average rate, giving the correct units.	The average rate is $\frac{3\sqrt{3}}{8}$ m/h.

5.3 Exercise

Students, these questions are even better in jacPLUS

Receive immediate feedback and access sample responses

Access additional questions

Track your results and progress

Find all this and MORE in jacPLUS

Technology free

1. **WE7** Determine the equation of the tangent to the curve $y = x^2 - 2x - 8$ at the point where $x = 3$.
2. Determine the equation of the tangent to the curve $y = \sqrt{4x + 5}$ at the point where $x = 1$.
3. Determine the equation of the tangent to each of the following curves at the point indicated.
   a. $y = 9 - x^2$ at $x = 2$
   b. $y = 5\sin(2x)$ at $x = \frac{\pi}{3}$
4. Determine the equation of the tangent to each of the following curves at the point indicated.
   a. $y = \sqrt{2x + 1}$ at $x = 4$
   b. $y = 4\cos(3x)$ at $x = \frac{\pi}{12}$
5. **WE8** Determine the equation of the normal to the curve $y = -x^2 + 2x + 15$ at the point where $x = 2$.
6. Determine the equation of the normal to the curve $y = \frac{4}{3x - 2}$ at the point where $x = 1$.
7. **WE9** The tangent to the curve $y = x^2 - 6x + 5$ at a point is given by $y = 2x + c$. Determine the value of c.
8. The normal to the curve $y = x^2 + 4x + 12$ at a point is given by $2y - x + c = 0$. Determine the value of c.
9. **a.** Determine the value of c if $y = -12x + c$ is a tangent to the curve $y = x^2 - 8x - 9$.
   b. Determine the value of c if $y = 8x + c$ is a tangent to the curve $y = \frac{2}{(2x - 3)^2}$.
10. **a.** Determine the value of c if $3y - 4x + c = 0$ is a normal to the curve $y = \frac{3}{x^2}$.
    b. Determine the value of c if $2y + x + c = 0$ is a normal to the curve $y = \sqrt{4x - 3}$.

11. **WE10** Determine the equation of the tangent and normal to the curve $y = 4\sin\left(\frac{x}{2}\right)$ at the point where $x = \frac{\pi}{3}$.

12. Determine the equation of the tangent and normal to the curve $y = -3e^{-2x} + 4$ at the point where it crosses the y-axis.

13. Determine the equation of the normal to each of the following curves at the point indicated.
    a. $y = 16 - x^2$ at $x = 3$
    b. $y = \tan(3x)$ at $x = \frac{\pi}{9}$

14. Determine the equation of the normal to each of the following curves at the point indicated.
    a. $y = \log_e(4x - 3)$ at $x = 1$
    b. $y = \frac{3}{(2x-3)^2}$ at $x = 3$

15. **WE11** For a sphere, determine the rate of change of volume with respect to the radius.

16. For a cone, determine the rate of change of volume with respect to:
    a. the radius, assuming the height remains constant
    b. the height, assuming the radius remain constant.

17. Given the function $f: R \to R, f(x) = 5 - 3e^{-2x}$:
    a. show that the function is always increasing
    b. sketch the graph of $y = f(x)$, stating any axis intercepts and the equations of any asymptotes
    c. determine the equation of the tangent to the curve at $x = \frac{1}{2}$.

18. Given the function $f: R \to R, f(x) = 4 - 3e^{2x}$:
    a. show that the function is always decreasing
    b. sketch the graph of $y = f(x)$, stating any axis intercepts and the equations of any asymptotes
    c. determine the equation of the normal to the curve at $x = \frac{1}{2}$.

19. a. Sketch the graph of $y = 3\log_e(4x - 5)$, stating the axis intercepts, the equations of any asymptotes, and the domain and range.
    b. Show that there are no stationary points.
    c. Determine the equation of the tangent to the curve $y = 3\log_e(4x - 5)$ at the point where $x = 2$.

20. a. Sketch the graph of $f(x) = 2\log_e(5 - 2x)$, stating the axis intercepts, the equations of any asymptotes, and the domain and range.
    b. Show that the function is a one-to-one decreasing function.
    c. Determine the equation of the normal to the curve $y = 2\log_e(5 - 2x)$ at the point where $x = 1$.

21. **WE12** The tides in a certain bay can be modelled by

$$D(t) = 6 + 4\cos\left(\frac{\pi t}{12}\right)$$

where D is the depth of water in metres and t is the time in hours after midnight on a particular day.
    a. Calculate the depth of water in the bay at 6 am.
    b. Sketch the graph of $D(t)$ against t for $0 \le t \le 24$.
    c. Determine when the depth of water is below 8 metres.
    d. Determine the rate of change of the depth at 3 am.
    e. Over the first 6 hours, determine the average rate of change of the depth.

Technology active

22. The population number, $N(t)$, of a certain city can be modelled by the equation $N(t) = N_0 e^{kt}$, where N_0 and k are constants and t is the time in years after the year 2010. In the year 2010, the population number was 500 000 and in 2020 the population had grown to 750 000.
 a. Determine the values of N_0 and k.
 b. Calculate the predicted population in 2025.
 c. Determine the rate of change of population in 2025, correct to 1 decimal place.
 d. Calculate the average rate of growth over the first 30 years, correct to 1 decimal place.
 e. Sketch the graph of $N(t)$.

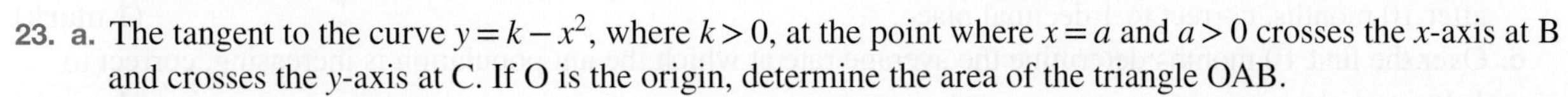

23. a. The tangent to the curve $y = k - x^2$, where $k > 0$, at the point where $x = a$ and $a > 0$ crosses the x-axis at B and crosses the y-axis at C. If O is the origin, determine the area of the triangle OAB.
 b. The normal to the curve $y = k - x^2$ at the point where $x = a$ and $a > 0$ passes through the origin. Show that $k = a^2 + \frac{1}{2}$.

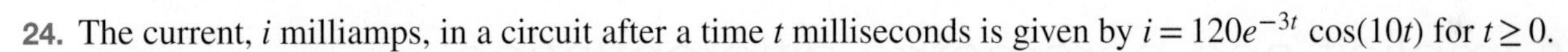

24. The current, i milliamps, in a circuit after a time t milliseconds is given by $i = 120e^{-3t} \cos(10t)$ for $t \geq 0$.
 a. Determine the rate of change of current with respect to time and evaluate it after 0.01 milliseconds, giving your answer correct to 3 decimal places.
 b. Over the time interval from $t = 0$ to $t = 0.02$, calculate the average rate of change of current, giving your answer correct to 3 decimal places.

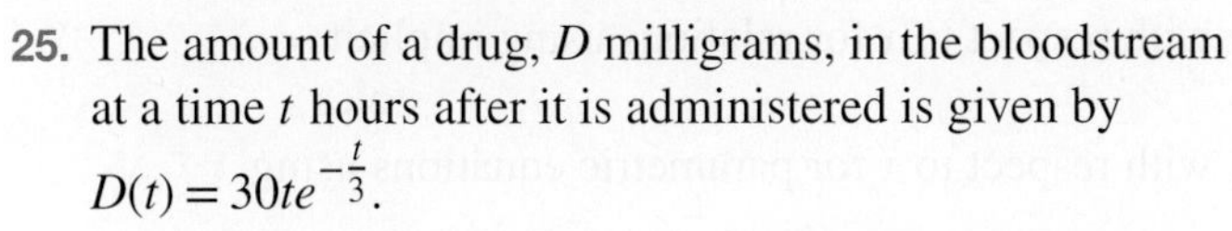

25. The amount of a drug, D milligrams, in the bloodstream at a time t hours after it is administered is given by $D(t) = 30te^{-\frac{t}{3}}$.
 a. Determine the average amount of the drug present in the bloodstream over the time from $t = 1$ to $t = 2$ hours after it is administered, correct to 2 decimal places.
 b. Calculate the instantaneous rate of change of the amount of the drug after 1.5 hours, correct to 1 decimal place.
 c. Determine the time when the amount of drug is a maximum and calculate the maximum amount of the drug in the body, correct to 2 decimal places.
 d. Determine how long the amount of the drug in the body is more than 10 milligrams, write your answer in hours correct to 2 decimal places.

5.3 Exam questions

Question 1 (4 marks) TECH-FREE

The tangent to the curve $y = \frac{1}{x}$ at the point where $x = a$ and $a > 0$ crosses the x-axis at B and crosses the y-axis at C. If O is the origin, determine the area of the triangle OBC.

Question 2 (4 marks) TECH-FREE

Given the function $f: R \to R, f(x) = x^3 e^{-2x}$:

a. determine the equations of the tangents to the curve $y = x^3 e^{-2x}$ at the points where $x = 1$ and $x = 0$ **(2 marks)**

b. show that the tangents above are the only two tangents to the curve that pass through the origin. **(2 marks)**

Question 3 (3 marks) TECH-ACTIVE

The population number, $P(t)$, of ants in a certain area is given by

$$P(t) = \frac{520}{0.3 + e^{-0.15t}}$$

where $t \geq 0$ is the time in months.

a. Calculate the initial population of the ants. **(1 mark)**

b. Determine the rate at which the ant population is increasing with respect to time and evaluate this rate after 10 months, correct to 1 decimal place. **(1 mark)**

c. Over the first 10 months determine the average rate at which the ant population is increasing, correct to 1 decimal place. **(1 mark)**

More exam questions are available online.

5.4 Implicit and parametric differentiation

LEARNING INTENTION

At the end of this subtopic you should be able to:

- determine an expression for the derivative of y with respect to x for relations using implicit differentiation
- determine an expression for the derivative of y with respect to x for parametric equations using parametric differentiation.

5.4.1 Introduction to implicit differentiation

Up until now the relationship between x and y has always been explicit: y is the dependent variable and x the independent variable, with y in terms of x written as $y = f(x)$. It is said that y depends on x and $\frac{dy}{dx}$ can be determined directly. For example:

$$\text{If } y = x^2 + 4x + 13, \text{ then } \frac{dy}{dx} = 2x + 4.$$

$$\text{If } y = 3\sin(2x) + 4e^{-2x}, \text{ then } \frac{dy}{dx} = 6\cos(2x) - 8e^{-2x}.$$

$$\text{If } y = \sqrt{16 - x^2}, \text{ then } \frac{dy}{dx} = \frac{-x}{\sqrt{16 - x^2}}.$$

$$\text{If } y = \log_e(5x + 3), \text{ then } \frac{dy}{dx} = \frac{5}{5x + 3}.$$

There are times, however, when y is not expressed explicitly in terms of x. In these cases there is a functional dependence or a so-called implicit relationship between x and y of the form $f(x, y) = c$, where c is a constant, or $g(x, y) = 0$. For example:

$$x^2 + y^2 = 16$$
$$4x^2 + 3xy - 2y^2 + 5x - 7y + 8 = 0$$
$$e^{-2xy} + 3\sin(2x - 3y) = c$$

These represent curves as an implicit relation and are not necessarily graphs of functions.

In some cases it may be possible to rearrange to make y the subject, but in most cases this is simply not possible. In this implicit form an expression for $\frac{dy}{dx}$ can still be obtained, but it may be in terms of both x and y. This can be obtained by differentiating each term in turn, with respect to x (remembering to use the chain rule as shown below).

For example, $\frac{d}{dx}\left(x^2\right) = 2x$, and in general, if n is a constant, then $\frac{d}{dx}\left(x^n\right) = nx^{n-1}$.

Recall that $\frac{d}{dx}(\text{constant}) = 0$.

Consider $\frac{d}{dx}\left(y^2\right)$. To find this, use the chain rule and let $u = y^2$. Then $\frac{d}{dx}\left(y^2\right) = \frac{du}{dx} = \frac{du}{dy} \times \frac{dy}{dx} = 2y \times \frac{dy}{dx}$; furthermore, in general $\frac{d}{dx}\left(y^n\right) = \frac{d}{dy}\left(y^n\right)\frac{dy}{dx} = ny^{n-1}\frac{dy}{dx}$.

This last result is known as **implicit differentiation**; it is just an application of the chain rule. When it is necessary to differentiate a function of y with respect to x, differentiate with respect to y and multiply by $\frac{dy}{dx}$.

WORKED EXAMPLE 13 Implicit differentiation

Given $x^2 + y^2 = 16$, determine an expression for $\frac{dy}{dx}$ in terms of both x and y.

THINK	WRITE/DRAW
1. Write the equation.	$x^2 + y^2 = 16$
2. Take $\frac{d}{dx}(\,)$ of each term in turn.	$\frac{d}{dx}\left(x^2\right) + \frac{d}{dx}\left(y^2\right) = \frac{d}{dx}(16)$
3. Use the results above together with $\frac{d}{dx}(c) = 0$, since the derivative of a constant is zero.	$2x + 2y\frac{dy}{dx} = 0$
4. Transpose the equation to make $\frac{dy}{dx}$ the subject.	$2x = -2y\frac{dy}{dx}$
5. Cancel the common factor and divide by x. State the final result, giving $\frac{dy}{dx}$ in terms of both x and y.	$\frac{dy}{dx} = -\frac{x}{y}$
Alternatively, notice that in this particular example it is possible to rearrange the equation to make y the subject	

1. Write the equation. $x^2 + y^2 = 16$

2. Rearrange to make y the subject. $y^2 = 16 - x^2$

3. There are two branches to the relation, which is a circle with centre at the origin and radius 4. Consider the branch or top half of the circle, which by itself is a function. $y = \pm\sqrt{16 - x^2}$

y
(0, 4)
$y = \sqrt{16 - x^2}$
(–4, 0)
(4, 0)
0
x

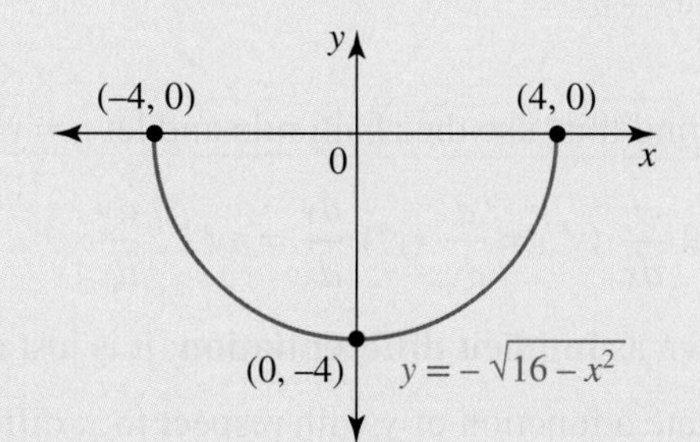

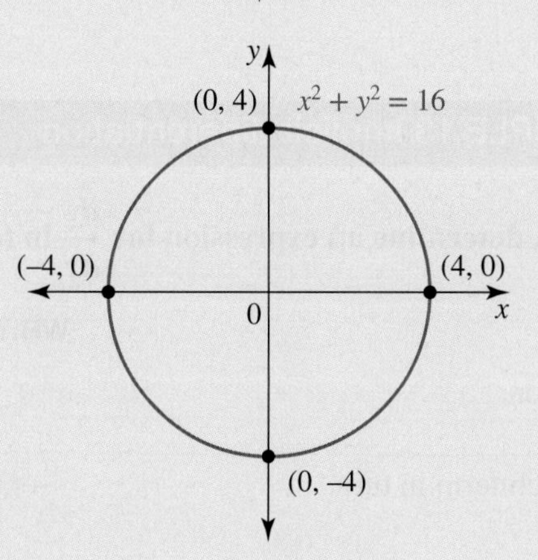

4. Rearrange to write $\frac{dy}{dx}$ in terms of x, differentiating using the chain rule.

Consider $y = \sqrt{16 - x^2}$.

$$\frac{dy}{dx} = \frac{-x}{\sqrt{16 - x^2}}$$

5. However, since $y = \sqrt{16 - x^2}$, express $\frac{dy}{dx}$ in terms of both x and y as before.

$$\frac{dy}{dx} = -\frac{x}{y}$$

Further examples involving implicit differentiation

Sometimes it may be necessary to use the product rule to obtain the required derivatives.

WORKED EXAMPLE 14 Implicit differentiation using the product rule

a. Determine $\frac{d}{dx}(3xy)$.

b. Given $4x^2 + 3xy - 2y^2 + 5x - 7y + 8 = 0$, write down an expression for $\frac{dy}{dx}$ in terms of both x and y.

THINK	WRITE/DRAW
a. 1. Write the expression.	**a.** $\frac{d}{dx}(3xy)$
2. Use the product rule.	$\frac{d}{dx}(3x \times y) = \frac{d}{dx}(u \times v)$ Let $u = 3x$ and $v = y$.
3. Calculate the derivatives.	$\frac{du}{dx} = 3$ and $\frac{dv}{dx} = \frac{d}{dy}(y)\frac{dy}{dx} = 1 \times \frac{dy}{dx} = \frac{dy}{dx}$
4. Use the product rule.	$\frac{d}{dx}(u \times v) = u\frac{dv}{dx} + v\frac{du}{dx}$
5. State the final result.	$\frac{d}{dx}(3xy) = 3x\frac{dy}{dx} + 3y$
b. 1. Write the equation.	**b.** $4x^2 + 3xy - 2y^2 + 5x - 7y + 8 = 0$
2. Take $\frac{d}{dx}(\,)$ of each term in turn.	$\frac{d}{dx}\left(4x^2\right) + \frac{d}{dx}(3xy) - \frac{d}{dx}\left(2y^2\right) + \frac{d}{dx}(5x) - \frac{d}{dx}(7y) + \frac{d}{dx}(8) = 0$
3. Substitute for the result from the second term from part **a** and use implicit differentiation on each term.	$8x + 3x\frac{dy}{dx} + 3y - 4y\frac{dy}{dx} + 5 - 7\frac{dy}{dx} + 0 = 0$
4. Transpose the equation to get all terms involving $\frac{dy}{dx}$ on one side of the equation.	$8x + 3y + 5 = 7\frac{dy}{dx} + 4y\frac{dy}{dx} - 3x\frac{dy}{dx}$
5. Factor the terms involving $\frac{dy}{dx}$ on the right-hand side of the equation.	$8x + 3y + 5 = \frac{dy}{dx}(7 + 4y - 3x)$
6. Divide and state the final result, giving $\frac{dy}{dx}$ in terms of both x and y.	$\frac{dy}{dx} = \frac{8x + 3y + 5}{7 + 4y - 3x}$

Implicit differentiation with exponential or trigonometric functions

Sometimes it may be necessary to use the derivatives of exponential or trigonometric functions together with implicit differentiation techniques to obtain the required derivative.

WORKED EXAMPLE 15 Implicit differentiation of exponential and trigonometric functions

Given $e^{-2xy} + 3\sin(2x - 3y) = c$, where c is a constant, determine an expression for $\frac{dy}{dx}$ in terms of both x and y.

THINK	WRITE
1. Write the equation.	$e^{-2xy} + 3\sin(2x-3y) = c$
2. Take $\frac{d}{dx}(\ )$ of each term in turn.	$\frac{d}{dx}\left(e^{-2xy}\right) + \frac{d}{dx}(3\sin(2x-3y)) = \frac{d}{dx}(c)$
3. Consider just the first term and use implicit differentiation.	$\frac{d}{dx}\left(e^{-2xy}\right) = \frac{d}{dx}\left(e^{-u}\right)$ where $u = 2xy$
4. Use the chain and product rule.	$\frac{d}{du}\left(e^{-u}\right)\frac{du}{dx} = -e^{-u}\frac{d}{dx}(2xy) = -e^{-u}\left(2y + 2x\frac{dy}{dx}\right)$
5. Substitute back for u.	$\frac{d}{dx}\left(e^{-2xy}\right) = -\left(2y + 2x\frac{dy}{dx}\right)e^{-2xy}$
6. Consider just the second term and use implicit differentiation.	$\frac{d}{dx}(3\sin(2x-3y)) = \frac{d}{dv}(3\sin(v))\frac{dv}{dx}$ where $v = 2x - 3y$ so that $\frac{dv}{dx} = 2 - 3\frac{dy}{dx}$. $\frac{d}{dx}(3\sin(2x-3y)) = 3\cos(v)\left(2 - 3\frac{dy}{dx}\right)$
7. Substitute back for v.	$\frac{d}{dx}(3\sin(2x-3y)) = 3\cos(2x-3y)\left(2 - 3\frac{dy}{dx}\right)$
8. Substitute for the first and second terms.	$-\left(2y + 2x\frac{dy}{dx}\right)e^{-2xy} + 3\cos(2x-3y)\left(2 - 3\frac{dy}{dx}\right) = 0$
9. Expand the brackets.	$-2ye^{-2xy} - 2xe^{-2xy}\frac{dy}{dx} + 6\cos(2x-3y) - 9\frac{dy}{dx}\cos(2x-3y) = 0$
10. Transpose the equation to get all terms involving $\frac{dy}{dx}$ on one side of the equation.	$-2ye^{-2xy} + 6\cos(2x-3y) = 2xe^{-2xy}\frac{dy}{dx} + 9\cos(2x-3y)\frac{dy}{dx}$
11. Factor the terms involving $\frac{dy}{dx}$ on the right-hand side of the equation.	$-2ye^{-2xy} + 6\cos(2x-3y) = \left(2xe^{-2xy} + 9\cos(2x-3y)\right)\frac{dy}{dx}$
12. Divide and state the final result, giving $\frac{dy}{dx}$ in terms of both x and y.	$\frac{dy}{dx} = \frac{6\cos(2x-3y) - 2ye^{-2xy}}{2xe^{-2xy} + 9\cos(2x-3y)}$

5.4.2 Parametric differentiation

In the previous section, we saw that whether the variables x and y are given in the form $y = f(x)$ as an explicit relationship or the form $f(x, y) = c$ as an implicit relationship, an expression for $\frac{dy}{dx}$ can be determined.

In this section, the two variables x and y are connected or related in terms of another variable, called a linking variable or a parameter. Often t or θ are used as parameters.

A parameter is a variable that changes from case to case but in each particular case or instant it remains the same. For example, an expression can be found for $\frac{dy}{dx}$; however, it will be in terms of the parameter by using the chain rule, since $\frac{dy}{dx} = \frac{dy}{dt} \times \frac{dt}{dx}$, and noting that $\frac{dt}{dx} = \frac{1}{\frac{dx}{dt}}$. Alternatively, it may also be possible to eliminate the parameter from the two parametric equations and obtain an implicit relationship between the two variables x and y. The following examples will illustrate these concepts.

WORKED EXAMPLE 16 Parametric differentiation

a. If $x = 3\cos(t)$ and $y = 4\sin(t)$, determine the gradient $\frac{dy}{dx}$ in terms of t.

b. For the parametric equations $x = 3\cos(t)$ and $y = 4\sin(t)$, determine an implicit relationship between x and y, and write an expression for the gradient. Hence, verify your answer to part a.

THINK	WRITE
a. 1. Differentiate x with respect to t. The dot notation is used for the derivative with respect to t.	a. $x = 3\cos(t)$ $\frac{dx}{dt} = \dot{x} = -3\sin(t)$
2. Differentiate y with respect to t.	$y = 4\sin(t)$ $\frac{dy}{dt} = \dot{y} = 4\cos(t)$
3. Use the chain rule to determine $\frac{dy}{dx}$.	$\frac{dy}{dx} = \frac{dy}{dt} \times \frac{dt}{dx} = \frac{dy}{dt} \times \frac{1}{\frac{dx}{dt}} = \dot{y} \times \frac{1}{\dot{x}} = \frac{\dot{y}}{\dot{x}}$
4. Substitute for the derivatives.	$\frac{dy}{dx} = \frac{4\cos(t)}{-3\sin(t)}$
5. State the gradient in simplest form.	$\frac{dy}{dx} = -\frac{4}{3}\cot(t)$
b. 1. Write the parametric equations.	b. $x = 3\cos(t)$ [1] $y = 4\sin(t)$ [2]
2. Express the trigonometric ratios on their own.	$\cos(t) = \frac{x}{3}$ [1] $\sin(t) = \frac{y}{4}$ [2]

3. Eliminate the parameter to determine the implicit relationship.

Since $\cos^2(t) + \sin^2(t) = 1$,

$$\frac{x^2}{9} + \frac{y^2}{16} = 1$$

4. Use implicit differentiation on the implicit form.

$$\frac{1}{9}\frac{d}{dx}\left(x^2\right) + \frac{1}{16}\frac{d}{dx}\left(y^2\right) = \frac{d}{dx}(1)$$

5. Perform the implicit differentiation.

$$\frac{2x}{9} + \frac{2y}{16}\frac{dy}{dx} = 0$$

6. Transpose the equation.

$$\frac{2x}{9} = -\frac{y}{8}\frac{dy}{dx}$$

7. Make $\frac{dy}{dx}$ the subject.

$$\frac{dy}{dx} = -\frac{16x}{9y}$$

8. From the implicit differentiation, substitute for the parametric equations.

$$\frac{dy}{dx} = -\frac{16 \times 3\cos(t)}{9 \times 4\sin(t)}$$

9. Simplify to verify the given result, as above.

$$\frac{dy}{dx} = -\frac{4}{3}\cot(t)$$

5.4 Exercise

Students, these questions are even better in jacPLUS

Receive immediate feedback and access sample responses

Access additional questions

Track your results and progress

Find all this and MORE in jacPLUS

Technology free

1. WE13 Given $x^3 + y^3 = 27$, determine an expression for $\frac{dy}{dx}$ in terms of both x and y.

2. Given $\sqrt{x} + \sqrt{y} = 4$, determine an expression for $\frac{dy}{dx}$ in terms of both x and y.

3. For each of the following implicitly defined relations, determine an expression for $\frac{dy}{dx}$ in terms of x and y.

a. $y^2 - 2x = 3$

b. $\frac{x^2}{4} + \frac{y^2}{9} = 1$

c. $\frac{x^2}{16} - \frac{y^2}{9} = 1$

d. $4x - 2y - 3x^2 + y^2 = 10$

4. WE14 a. Determine $\frac{d}{dx}\left(4x^2y^2\right)$.

b. Given $9x^3 + 4x^2y^2 - 3y^3 + 2x - 5y + 12 = 0$, determine an expression for $\frac{dy}{dx}$ in terms of both x and y.

5. Calculate the gradient of the normal to the curve $x^2 - 4xy + 2y^2 - 3x + 5y - 7 = 0$ at the point where $x = 2$ in the first quadrant.

6. For each of the following implicitly defined relations, determine an expression for $\frac{dy}{dx}$ in terms of x and y.

a. $2x^2 + 3xy - 3y^2 + 8 = 0$
b. $x^2 + x^2y^2 - 6x + 5 = 0$
c. $x^3 - 3x^2y + 3xy^2 - y^3 - 27 = 0$
d. $y^3 - y^2 - 3x - x^2 + 9 = 0$

7. **WE15** Given $2xy + e^{-(x^2+y^2)} = c$, where c is a constant, determine an expression for $\frac{dy}{dx}$ in terms of both x and y.

8. If $\sin(3x + 2y) + x^2 + y^2 = c$, where c is a constant, determine $\frac{dy}{dx}$ in terms of both x and y.

9. For each of the following implicitly defined relations, determine an expression for $\frac{dy}{dx}$ in terms of x and y.

a. $\frac{y^2 - 2x}{3x^2 + 4y} = 6x$
b. $\frac{x^3 + 8y^3}{x^2 - 2xy + 4y^2} = x^2$
c. $e^{x+y} + \cos(y) - y^2 = 0$
d. $e^{xy} + \cos(xy) + x^2 = 0$

10. For each of the following implicitly defined relations, determine an expression for $\frac{dy}{dx}$ in terms of x and y.

a. $\log_e(2x + 3y) + 4x - 5y = 10$
b. $\log_e(3xy) + x^2 + y^2 - 9 = 0$
c. $\frac{1}{x} + 2xy + \frac{1}{y} - 6 = 0$
d. $2y^3 - \frac{\sin(3x)}{\sec(2y)} + x^2 = 0$

11. a. Calculate the gradient of the tangent to the curve $y^2 = x^3(2 - x)$ at the point $(1, 1)$.
b. Calculate the equation of the tangent to the curve $x^3 + 3xy + y^3 + 13 = 0$ at the point $(1, -2)$.
c. Consider the equation $x^2 + 4x + y^2 - 3y + 6xy - 4 = 0$. Calculate the gradient of the tangent at the point $(2, -1)$.
d. A certain ellipse has the equation $3x^2 + 2xy + 4y^2 + 5x - 10y - 8 = 0$. Calculate the gradient of the normal to the ellipse at the point $(1, 2)$.

12. a. Determine the equation of the tangent to the circle $x^2 + y^2 = 25$ at the point in the fourth quadrant where $x = 3$.
b. Determine the gradient of the tangent to the ellipse $\frac{x^2}{4} + \frac{y^2}{9} = 1$ at the point in the first quadrant where $x = 1$.
c. Determine the gradient of the normal to the hyperbola $\frac{x^2}{16} - \frac{y^2}{9} = 1$ at the point in the fourth quadrant where $x = 5$.
d. Determine the gradient of the normal to the curve $2\left(x^2 + y^2\right)^2 = 25\left(x^2 - y^2\right)$ at the point $(3, 1)$.

13. **WE16** a. If $x = 4\cos(t)$ and $y = 4\sin(t)$, determine the gradient $\frac{dy}{dx}$ in terms of t.
b. For the parametric equations $x = 4\cos(t)$ and $y = 4\sin(t)$, determine an implicit relationship between x and y, and write an expression for the gradient. Hence, verify your answer to part **a**.

14. a. Given $x = t^2$ and $y = 2t - t^4$, determine $\frac{dy}{dx}$ in terms of t.
b. For the parametric equations $x = t^2$ and $y = 2t - t^4$, express y in terms of x and calculate $\frac{dy}{dx}$. Hence, verify your answer to part **a**.

15. For each of the following, determine $\frac{dy}{dx}$ in terms of the parameter.

a. The parabola $x = 2t^2$ and $y = 4t$
b. The ellipse $x = 3\sin(2t)$ and $y = 4\cos(2t)$
c. The rectangular hyperbola $x = 5t$ and $y = \frac{5}{t}$
d. The hyperbola $x = 3\sec(t)$ and $y = 4\tan(t)$

16. In each of the following, a and b are constants. Determine $\frac{dy}{dx}$ in terms of the parameter.

a. $x = at^2$ and $y = 2at$
b. $x = a\cos(t)$ and $y = b\sin(t)$
c. $x = at$ and $y = \frac{a}{t}$
d. $x = a\sec(t)$ and $y = b\tan(t)$

17. Check your answers to question **16a–d** by eliminating the parameter and determining $\frac{dy}{dx}$ by another method.

18. Rene Descartes (1596–1650) was a French mathematician and philosopher who lived in the Dutch republic. He is noted for introducing the Cartesian coordinate system in two dimensions and is credited as the father of analytical geometry, the link between algebra and geometry.

a. A curve called the folium of Descartes has the equation $x^3 - 3axy + y^3 = 0$. Determine $\frac{dy}{dx}$ for this curve, given that a is a constant.
b. Show that the folium of Descartes can be represented by the parametric equations $x = \frac{3at}{1+t^3}$ and $y = \frac{3at^2}{1+t^3}$, and calculate the gradient of the curve in terms of the parameter, t.

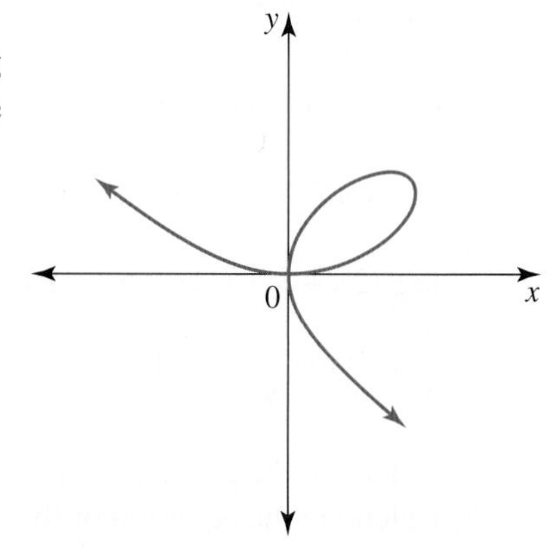

19. a. A curve called a lemniscate has the implicit equation $\left(x^2 + y^2\right)^2 = 2\left(x^2 - y^2\right)$. Determine an expression for $\frac{dy}{dx}$ in terms of x and y.

b. Show that the curve can be represented by the parametric equations

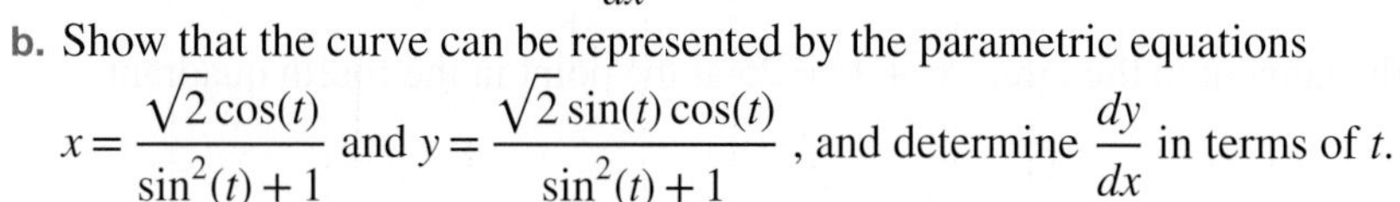

$x = \frac{\sqrt{2}\cos(t)}{\sin^2(t)+1}$ and $y = \frac{\sqrt{2}\sin(t)\cos(t)}{\sin^2(t)+1}$, and determine $\frac{dy}{dx}$ in terms of t.

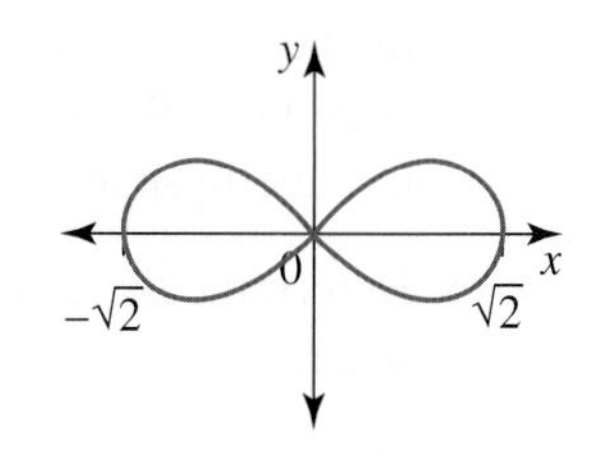

20. A curve known as the cissoid of Diocles has the equation $y^2 = \frac{x^3}{2a - x}$.

a. Determine $\frac{dy}{dx}$ for this curve, given that a is constant.
b. Show that a parametric representation of the cissoid curve is given by the equations $x = \frac{2at^2}{1+t^2}$ and $y = \frac{2at^3}{1+t^2}$, and determine $\frac{dy}{dx}$ in terms of t.
c. Show that another alternative parametric representation of the cissoid curve is given by the equations $x = 2a\sin^2(t)$ and $y = \frac{2a\sin^3(t)}{\cos(t)}$, and determine $\frac{dy}{dx}$ in terms of t.

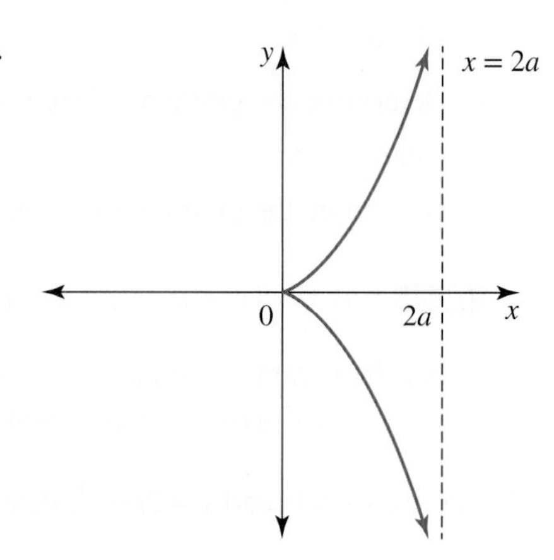

5.4 Exam questions

Question 1 (5 marks) TECH-FREE

Source: *VCE 2019 Specialist Mathematics Exam 1, Q10; © VCAA.*

Find $\frac{dy}{dx}$ at the point $\left(\frac{\sqrt{\pi}}{\sqrt{6}}, \frac{\sqrt{\pi}}{\sqrt{3}}\right)$ for the curve defined by the relation $\sin\left(x^2\right) + \cos\left(y^2\right) = \frac{3\sqrt{2}}{\pi}xy$.

Give your answer in the form $\frac{\pi - a\sqrt{b}}{\sqrt{a}\left(\pi + \sqrt{b}\right)}$, where $a, b \in Z^+$.

Question 2 (3 marks) TECH-FREE

Source: *VCE 2017 Specialist Mathematics Exam 1, Q1; © VCAA.*

Find the equation of the tangent to the curve given by $3xy^2 + 2y = x$ at the point $(1, -1)$.

Question 3 (1 mark) TECH-ACTIVE

Source: *VCE 2016 Specialist Mathematics Exam 2, Section A, Q7; © VCAA.*

MC Given that $x = \sin(t) - \cos(t)$ and $y = \frac{1}{2}\sin(2t)$, then $\frac{dy}{dx}$ in term of t is

A. $\cos(t) - \sin(t)$

B. $\cos(t) + \sin(t)$

C. $\sec(t) + \operatorname{cosec}(t)$

D. $\sec(t) - \operatorname{cosec}(t)$

E. $\frac{\cos(2t)}{\cos(t) - \sin(t)}$

More exam questions are available online.

5.5 Second derivatives

LEARNING INTENTION

At the end of this subtopic you should be able to:
- determine the second derivatives of functions.

5.5.1 Second derivatives

If $y = f(x)$ is the equation of the curve, then the first derivative is $\frac{dy}{dx} = f'(x)$ and it is the gradient of the curve. It is also the rate of change of y with respect to x, often abbreviated to 'wrt x'. This is still a function of x, so if the first derivative is differentiated again, the **second derivative** or the rate of change of the gradient is obtained. This is denoted by $\frac{d}{dx}\left(\frac{dy}{dx}\right) = \frac{d^2y}{dx^2} = f''(x)$. Notice the position of the 2s in this notation and that the two dashes after f indicate the second derivative with respect to x.

Similarly, the third derivative is denoted by $\frac{d^3y}{dx^3} = f'''(x)$.

In general, the nth derivative is denoted by $\frac{d^ny}{dx^n} = f^{(n)}(x)$.

For example, consider the general cubic equation $y = f(x) = ax^3 + bx^2 + cx + d$ where a, b, c and d are constants. The first derivative or gradient function is $\frac{dy}{dx} = f'(x) = 3ax^2 + 2bx + c$. The second derivative or rate of change of gradient function is $\frac{d^2y}{dx^2} = 6ax + 2b$. The third derivative is $\frac{d^3y}{dx^3} = 6a$, and all further derivatives are zero.

WORKED EXAMPLE 17 Calculating the value of the second derivative

If $f(x) = \dfrac{4\sqrt{x^5}}{3x^2}$, evaluate $f''(9)$.

THINK	WRITE
1. Express the function in simplified form using index laws.	$f(x) = \dfrac{4\sqrt{x^5}}{3x^2} = \dfrac{4x^{\frac{5}{2}}}{3x^2} = \dfrac{4}{3}x^{\frac{5}{2}-2} = \dfrac{4}{3}x^{\frac{1}{2}}$
2. Determine the first derivative, using the basic laws for differentiation.	$f'(x) = \dfrac{4}{3} \times \dfrac{1}{2}x^{-\frac{1}{2}} = \dfrac{2}{3}x^{-\frac{1}{2}}$
3. Determine the second derivative, using the basic laws for differentiation, by differentiating the first derivative again.	$f''(x) = \dfrac{2}{3} \times \left(\dfrac{-1}{2}\right) x^{-\frac{3}{2}} = \dfrac{-1}{3\sqrt{x^3}}$
4. Substitute in the value for x.	$f''(9) = \dfrac{-1}{3\sqrt{9^3}} = \dfrac{-1}{3 \times 27}$
5. State the final result.	$f''(9) = -\dfrac{1}{81}$

Using product and quotient rules

Often when we differentiate we may need to use rules such as the product and quotient rules.

WORKED EXAMPLE 18 Using the product and quotient rules

Determine $\dfrac{d^2y}{dx^2}$ for $y = x^2 \log_e(3x + 5)$.

THINK	WRITE
1. Write the equation.	Let $y = x^2 \log_e(3x + 5)$.
2. Use the product rule.	$\dfrac{dy}{dx} = x^2 \dfrac{d}{dx}\left(\log_e(3x+5)\right) + \dfrac{d}{dx}\left(x^2\right)\log_e(3x+5)$

THINK	WRITE
3. State the result for $\frac{dy}{dx}$.	$\frac{dy}{dx}=\frac{3x^2}{3x+5}+2x\log_e(3x+5)$
4. Differentiate with respect to x again.	$\frac{d^2y}{dx^2}=\frac{d}{dx}\left(\frac{3x^2}{3x+5}\right)+\frac{d}{dx}\left(2x\log_e(3x+5)\right)$
5. Use the quotient rule on the first term and the product rule again on the second term.	$\frac{d^2y}{dx^2}=\frac{\frac{d}{dx}\left(3x^2\right)(3x+5)-\frac{d}{dx}(3x+5)\left(3x^2\right)}{(3x+5)^2}$ $+\frac{d}{dx}(2x)\left(\log_e(3x+5)\right)+2x\frac{d}{dx}\left(\log_e(3x+5)\right)$
6. Perform the required derivatives.	$\frac{d^2y}{dx^2}=\frac{6x(3x+5)-3\left(3x^2\right)}{(3x+5)^2}+2\log_e(3x+5)+\frac{6x}{3x+5}$
7. Simplify the numerator in the first expression and state the final result.	$\frac{d^2y}{dx^2}=\frac{9x^2+30x}{(3x+5)^2}+2\log_e(3x+5)+\frac{6x}{3x+5}$
8. Add the first and last terms by forming a common denominator.	$\frac{d^2y}{dx^2}=2\log_e(3x+5)+\frac{9x^2+30x+6x(3x+5)}{(3x+5)^2}$ $\frac{d^2y}{dx^2}=2\log_e(3x+5)+\frac{9x^2+30x+18x^2+30x}{(3x+5)^2}$
9. State the final result in simplest form.	$\frac{d^2y}{dx^2}=2\log_e(3x+5)+\frac{3x(9x+20)}{(3x+5)^2}$

Determining second derivatives using implicit differentiation

Implicit differentiation techniques can be used to determine relationships between the first and second derivatives. Equations involving the function y and its first and second derivatives are called differential equations. They are explored in greater depth in later topics.

WORKED EXAMPLE 19 Using implicit differentiation

If a and b are constants and $y=x\sqrt{a+bx^2}$, show that $y\frac{d^2y}{dx^2}+\left(\frac{dy}{dx}\right)^2=a+6bx^2$.

THINK	WRITE
1. Write the equation and square both sides.	$y=x\sqrt{a+bx^2}$ $y^2=x^2\left(a+bx^2\right)$
2. Expand the brackets.	$y^2=ax^2+bx^4$
3. Take $\frac{d}{dx}()$ of each term in turn.	$\frac{d}{dx}\left(y^2\right)=\frac{d}{dx}\left(ax^2\right)+\frac{d}{dx}\left(bx^4\right)$
4. Use implicit differentiation to determine the first derivative.	$2y\frac{dy}{dx}=2ax+4bx^3$

5. Take $\frac{d}{dx}()$ of each term in turn again. $\frac{d}{dx}\left(2y\frac{dy}{dx}\right) = \frac{d}{dx}(2ax) + \frac{d}{dx}\left(4bx^3\right)$

6. Use the product rule on the first term. $2y\frac{d}{dx}\left(\frac{dy}{dx}\right) + \frac{dy}{dx}\frac{d}{dx}(2y) = 2a + 12bx^2$

7. Use implicit differentiation. $2y\frac{d^2y}{dx^2} + 2\left(\frac{dy}{dx}\right)^2 = 2a + 12bx^2$

8. Divide each term by 2 and state the final result. $y\frac{d^2y}{dx^2} + \left(\frac{dy}{dx}\right)^2 = a + 6bx^2$

5.5.2 Using parametric differentiation to determine second derivatives

When determining $\frac{d^2y}{dx^2}$ with parametric differentiation, often our first thought might be to use the chain rule as $\frac{d^2y}{dx^2} = \frac{d^2y}{dt^2} \cdot \frac{d^2t}{dx^2}$. However, this rule is *incorrect* and cannot be used, as dt^2 and d^2t are not equivalent; furthermore, $\frac{d^2y}{dx^2} \neq 1/\frac{d^2x}{dy^2}$. To correctly obtain $\frac{d^2y}{dx^2}$, we must use implicit differentiation in conjunction with parametric differentiation. Because $\frac{dy}{dx}$ is itself a function of t, we obtain the second derivative using

Parametric second derivatives

$$\frac{d^2y}{dx^2} = \frac{d}{dx}\left(\frac{dy}{dx}\right) = \frac{d}{dt}\left(\frac{dy}{dx}\right) \times \frac{dt}{dx}.$$

WORKED EXAMPLE 20 The second derivative using parametric differentiation

If $x = 3\cos(t)$ and $y = 4\sin(t)$, determine $\frac{d^2y}{dx^2}$ in terms of t.

THINK	WRITE
1. Differentiate x with respect to t. The dot notation is used for the derivative with respect to t.	$x = 3\cos(t)$ $\frac{dx}{dt} = \dot{x} = -3\sin(t)$
2. Differentiate y with respect to t.	$y = 4\sin(t)$ $\frac{dy}{dt} = \dot{y} = 4\cos(t)$
3. Use the chain rule to determine $\frac{dy}{dx}$.	$\frac{dy}{dx} = \frac{dy}{dt}\frac{dt}{dx} = \frac{\dot{y}}{\dot{x}}$
4. Substitute for the derivatives.	$\frac{dy}{dx} = \frac{4\cos(t)}{-3\sin(t)}$

5. State the gradient in simplest form.

$$\frac{dy}{dx} = -\frac{4}{3}\cot(t)$$

6. Take $\frac{d}{dx}()$ of each term.

$$\frac{d^2y}{dx^2} = \frac{d}{dx}\left(\frac{dy}{dx}\right) = \frac{d}{dx}\left(-\frac{4}{3}\cot(t)\right)$$

7. Use implicit differentiation and $\cot(t) = \frac{\cos(t)}{\sin(t)}$.

$$\frac{d^2y}{dx^2} = \frac{d}{dt}\left(-\frac{4\cos(t)}{3\sin(t)}\right)\frac{dt}{dx}$$

8. Use the quotient rule.

$$\frac{d^2y}{dx^2} = -\left(\frac{\frac{d}{dt}(4\cos(t))(3\sin(t)) - \frac{d}{dt}(3\sin(t))(4\cos(t))}{(3\sin(t))^2}\right)\frac{dt}{dx}$$

9. Perform the derivatives in the numerator.

$$\frac{d^2y}{dx^2} = -\left(\frac{-12\sin^2(t) - 12\cos^2(t)}{(3\sin(t))^2}\right)\frac{dt}{dx}$$

10. Simplify using $\sin^2(t) + \cos^2(t) = 1$.

$$\frac{d^2y}{dx^2} = \frac{12}{9\sin^2(t)}\frac{dt}{dx}$$

11. Substitute for $\frac{dt}{dx} = 1/\frac{dx}{dt}$.

$$\frac{d^2y}{dx^2} = \frac{4}{3\sin^2(t)} \times \frac{-1}{3\sin(t)}$$

12. State the final result.

$$\frac{d^2y}{dx^2} = \frac{-4}{9\sin^3(t)}$$
$$= -\frac{4}{9}\operatorname{cosec}^3(t)$$

5.5.3 Higher derivatives and mathematical induction

We can use mathematical induction to prove general results about nth derivatives of functions.

WORKED EXAMPLE 21 Using induction to determine nth derivatives of functions

Use mathematical induction to show that $\frac{d^n}{dx^n}(x^2e^x) = (x^2 + 2nx + n(n-1))e^x, \forall n \in N$.

THINK	WRITE
1 Verify that the formula is true when $n = 1$. Consider the LHS.	Proof: LHS $= \frac{d}{dx}(x^2e^x)$ using the product rule $e^x\frac{d}{dx}(x^2) + x^2\frac{d}{dx}(e^x)$ $= 2xe^x + x^2e^x$ $= (x^2 + 2x)e^x$
2 Consider the RHS.	RHS $= (x^2 + 2x + 1 \times 0)e^x = (x^2 + 2x)e^x$ So it is true when $n = 1$.

3 Write down the formula for $n = k$.	Assume that it is true for $n = k$. $\frac{d^k}{dx^k}\left(x^2 e^x\right) = \left(x^2 + 2kx + k(k-1)\right) e^x$
4 Consider the $(k+1)$th derivative, which is the derivative of the kth derivative.	$\frac{d^{k+1}}{dx^{k+1}}\left(x^2 e^x\right) = \frac{d}{dx}\left(\frac{d^k}{dx^k}\left(x^2 e^x\right)\right)$
5 Use the fact that it is true when $n = k$.	$\frac{d^{k+1}}{dx^{k+1}}\left(x^2 e^x\right) = \frac{d}{dx}\left(\left(x^2 + 2kx + k(k-1)\right) e^x\right)$
6 Use the product rule on the right-hand side.	$\frac{d^{k+1}}{dx^{k+1}}\left(x^2 e^x\right)$ $= e^x \frac{d}{dx}\left(\left(x^2 + 2kx + k(k-1)\right)\right) + \left(x^2 + 2kx + k(k-1)\right) \frac{d}{dx}\left(e^x\right)$ $= e^x (2x + 2kx) + \left(x^2 + 2kx + k(k-1)\right) e^x$
7 Take out the common factor of e^x and simplify.	$= \left(x^2 + 2kx + k(k-1) + 2x + 2k\right) e^x$ $= \left(x^2 + (2k+2)x + k^2 - k + 2k\right) e^x$ $= \left(x^2 + (2k+2)x + k^2 + k\right) e^x$ $= \left(x^2 + (2(k+1))x + (k+1)k\right) e^x$ which is of the form we want to show.
8 Write a statement to explain what you have demonstrated and concluded.	If the statement is true when $n = k$, it is also true when $n = k + 1$. The statement is true for $n = 1$; therefore, by the principle of mathematical induction, it is true $\forall n \in N$.

5.5 Exercise

Students, these questions are even better in jacPLUS

Receive immediate feedback and access sample responses

Access additional questions

Track your results and progress

Find all this and MORE in jacPLUS

Technology free

1. **WE17** If $f(x) = \frac{8\sqrt{x^3}}{3x}$, evaluate $f''(4)$.

2. If $f(x) = 8\cos\left(\frac{x}{2}\right)$, evaluate $f''\left(\frac{\pi}{3}\right)$.

3. If $f(x) = \frac{4x^2}{3\sqrt{x}}$, evaluate $f''(4)$.

4. If $f(x) = \frac{2}{3x-5}$, evaluate $f''(1)$.

5. **WE18** Determine $\frac{d^2y}{dx^2}$ for $y = x^3 \log_e\left(2x^2 + 5\right)$.

6. Determine $\frac{d^2y}{dx^2}$ for $y = \frac{x^4}{e^{3x}}$.

7. If $f(x) = 4\log_e(2x - 3)$, evaluate $f''(3)$.

8. If $f(x) = e^{x^2}$, evaluate $f''(1)$.

9. Determine $\frac{d^2y}{dx^2}$ if:

a. $y = \log_e\left(x^2 + 4x + 13\right)$ b. $y = e^{3x}\cos(4x)$ c. $y = x^3 e^{-2x}$

10. Determine $\frac{d^2y}{dx^2}$ if:

a. $y = x^2\cos(3x)$ b. $y = x\log_e(6x + 7)$ c. $y = \log_e\left(x + \sqrt{x^2 + 16}\right)$.

11. **WE19** If a and b are constants and $y^2 = a + bx^2$, show that $y\frac{d^2y}{dx^2} + \left(\frac{dy}{dx}\right)^2 = b$.

12. If a and b are constants and $y^3 = x\left(a + bx^3\right)$, show that $y^2\frac{d^2y}{dx^2} + 2y\left(\frac{dy}{dx}\right)^2 = 4bx^2$.

13. If a and b are constants, determine $\frac{d^2y}{dx^2}$ for each of the following.

a. $y^2 = 4ax$ b. $\frac{x^2}{a^2} - \frac{y^2}{b^2} = 1$ c. $xy = a^2$

14. **WE20** If $x = 4\cos(t)$ and $y = 4\sin(t)$, determine $\frac{d^2y}{dx^2}$ in terms of t.

15. Given $x = t^2$ and $y = 2t - t^4$, determine the rate of change of gradient and evaluate when $t = 2$.

16. In each of the following, a and b are constants. Determine $\frac{d^2y}{dx^2}$ in terms of the parameter.

a. $x = at^2$ and $y = 2at$ b. $x = at$ and $y = \frac{a}{t}$ c. $x = a\cos(t)$ and $y = b\sin(t)$

17. Let a and b be constants.

a. If $y^2 = a + bx$, prove that $y\frac{d^2y}{dx^2} + \left(\frac{dy}{dx}\right)^2 = 0$.

b. Given that $y = x\sqrt{a + bx}$, verify that $y\frac{d^2y}{dx^2} + \left(\frac{dy}{dx}\right)^2 = a + 3bx$.

18. If a, b and n are constants, show that:

a. $\frac{d^2}{dx^2}\left[(ax + b)^n\right] = a^2 n(n - 1)(ax + b)^{n-2}$

b. $\frac{d^2}{dx^2}\left[\log_e(ax + b)^n\right] = \frac{-a^2 n}{(ax + b)^2}$

c. $\frac{d^2}{dx^2}\left[\log_e\left(ax^2 + b\right)^n\right] = \frac{-2an\left(ax^2 - b\right)}{\left(ax^2 + b\right)^2}$

19. A circle of radius a with centre at the origin has the equation $x^2 + y^2 = a^2$.

a. Show that $\frac{d^2y}{dx^2} = -\frac{a^2}{y^3}$.

b. The radius of curvature ρ of a plane curve is given by

$$\rho = \frac{\left[1 + \left(\frac{dy}{dx}\right)^2\right]^{\frac{3}{2}}}{\frac{d^2y}{dx^2}}$$

Show that the radius of curvature of a circle has a magnitude of a.

20. An involute of a circle has the parametric equations $x = \cos(\theta) + \theta\sin(\theta)$ and $y = \sin(\theta) - \theta\cos(\theta)$. Show that the radius of curvature is θ.

21. A cycloid is the curve obtained by a point on a circle of radius a rolling along the x-axis. Its graph is shown below.

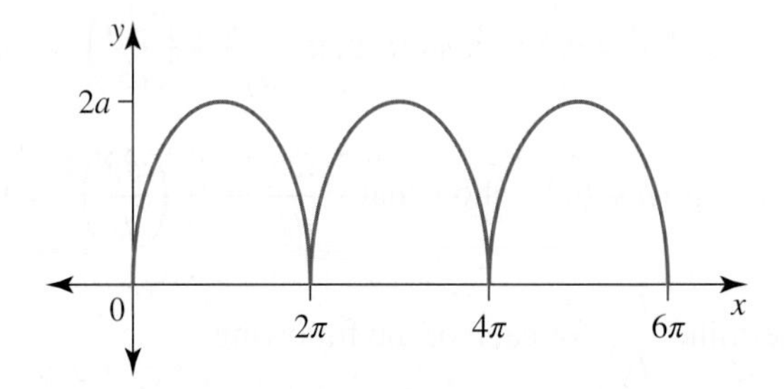

It has the parametric equations $x = a(\theta - \sin(\theta))$ and $y = a(1 - \cos(\theta))$. Show that:

a. $\frac{dy}{dx} = \cot\left(\frac{\theta}{2}\right)$

b. $\frac{d^2y}{dx^2} = \frac{-1}{4a\sin^4\left(\frac{\theta}{2}\right)}$

c. the radius of curvature of the cycloid is $-4a\sin\left(\frac{\theta}{2}\right)$.

22. a. Given the implicit equation for an astroid $x^{\frac{2}{3}} + y^{\frac{2}{3}} = a^{\frac{2}{3}}$, where a is a constant, express $\frac{dy}{dx}$ in terms of both x and y.

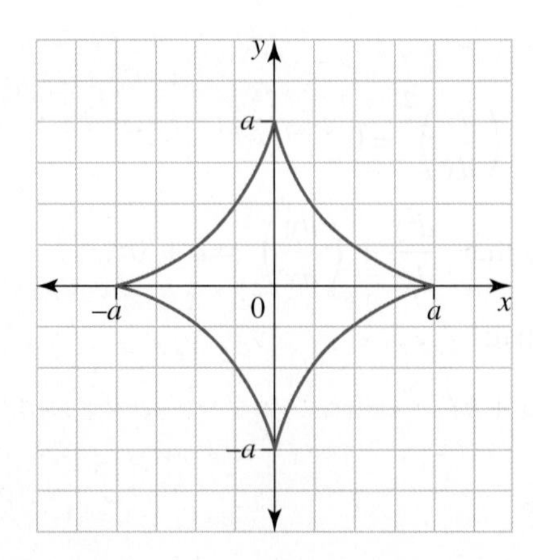

b. Show that the astroid can be expressed in parametric form as $x = a\cos^3(\theta)$ and $y = a\sin^3(\theta)$, and determine $\frac{dy}{dx}$ and $\frac{d^2y}{dx^2}$ in terms of θ.

23. a. If $y = e^{-2x}\sin(3x)$, determine $\frac{d^2y}{dx^2}$.

b. If $y = u(x)v(x)$ where $u(x)$ and $v(x)$ are differentiable functions, show that
$$\frac{d^2y}{dx^2} = u(x)\frac{d^2v}{dx^2} + 2\frac{du}{dx}\frac{dv}{dx} + v(x)\frac{d^2u}{dx^2}.$$
Use this result to verify $\frac{d^2y}{dx^2}$ for $y = e^{-2x}\sin(3x)$.

24. a. If x and y are both expressed in terms of a parameter t, that is, $x = f(t)$ and $y = g(t)$, show that
$$\frac{d^2y}{dx^2} = \frac{\dot{x}\ddot{y} - \dot{y}\ddot{x}}{\dot{x}^3}.$$

b. Show that $\frac{d^2y}{dx^2} = -\left(\frac{dy}{dx}\right)^3 \frac{d^2x}{dy^2}$.

25. Show that the following statements are true.

a. $\frac{d^3}{dx^3}\left(x^3\right) = 3!$

b. $\frac{d^4}{dx^4}\left(x^4\right) = 4!$

26. **WE21** Use mathematical induction to show that $\frac{d^n}{dx^n}(x^n) = n!, \forall n \in N$.

27. Use mathematical induction to show that $\frac{d^n}{dx^n}(xe^x) = (x+n)e^x, \forall n \in N$.

28. Use mathematical induction to show that $\frac{d^n}{dx^n}(e^{mx}) = m^n e^{mx}, \forall n \in N$.

29. Use mathematical induction to show that $\frac{d^n}{dx^n}\left(\log_e(x)\right) = \frac{(-1)^{n-1}(n-1)!}{x^n}, \forall n \in N$.

30. a. Use mathematical induction to show that $\frac{d^n}{dx^n}(\sin(mx)) = m^n \sin\left(mx + \frac{n\pi}{2}\right), \forall n \in N$.

b. Use mathematical induction to show that $\frac{d^n}{dx^n}(\cos(mx)) = m^n \cos\left(mx + \frac{n\pi}{2}\right), \forall n \in N$.

5.5 Exam questions

Question 1 (1 mark)

Source: *VCE 2017 Specialist Mathematics Exam 2, Section A, Q6; © VCAA.*

MC Given that $\frac{dy}{dx} = e^x \arctan(y)$, the value of $\frac{d^2y}{dx^2}$ at the point $(0, 1)$ is

A. $\frac{1}{2}$ **B.** $\frac{3\pi}{8}$ **C.** $-\frac{1}{2}$ **D.** $\frac{\pi}{4}$ **E.** $-\frac{8\pi}{8}$

Question 2 (4 marks) TECH-FREE

Use mathematical induction to show that $\frac{d}{dx}(x^n) = nx^{n-1}, \forall n \in N$.

Question 3 (4 marks)

If a, b and n are constants, show that $\frac{d^2}{dx^2}\left[\left(ax^2+b\right)^n\right] = 2an\left(ax^2+b\right)^{n-2}\left(a(2n-1)x^2+b\right)$.

More exam questions are available online.

5.6 Derivatives of inverse trigonometric functions

LEARNING INTENTION

At the end of this subtopic you should be able to:
- determine the derivatives of inverse trigonometric functions.

5.6.1 Derivatives of inverse trigonometric functions

In this section, the derivatives of the inverse trigonometric functions are determined. These functions have already been studied in earlier topics; recall the definitions and alternative notations.

The inverse function $y = \sin^{-1}(x)$ or $y = \arcsin(x)$ has a domain of $[-1, 1]$ and a range of $\left[-\frac{\pi}{2}, \frac{\pi}{2}\right]$. It is equivalent to $x = \sin(y)$ and $\sin\left(\sin^{-1}(x)\right) = x$ if $x \in [-1, 1]$, and $\sin^{-1}(\sin(x)) = x$ if $x \in \left[-\frac{\pi}{2}, \frac{\pi}{2}\right]$.

The inverse function $y = \cos^{-1}(x)$ or $y = \arccos(x)$ has a domain of $[-1, 1]$ and a range of $[0, \pi]$. It is equivalent to $x = \cos(y)$ and $\cos\left(\cos^{-1}(x)\right) = x$ if $x \in [-1, 1]$, and $\cos^{-1}(\cos(x)) = x$ if $x \in [0, \pi]$.

The inverse function $y = \tan^{-1}(x)$ or $y = \arctan(x)$ has a domain of R and a range of $\left(-\frac{\pi}{2}, \frac{\pi}{2}\right)$. It is equivalent to $x = \tan(y)$ and $\tan\left(\tan^{-1}(x)\right)$ if $x \in R$, and $\tan^{-1}(\tan(x)) = x$ if $x \in \left(-\frac{\pi}{2}, \frac{\pi}{2}\right)$.

The derivative of $\sin^{-1}(x)$

First we must determine the derivative of $\sin^{-1}(x)$.

Let $y = \sin^{-1}(x)$. From the definition of the inverse function, $x = \sin(y)$, so

$\frac{dx}{dy} = \cos(y)$ and $\frac{dy}{dx} = \frac{1}{\cos(y)}$.

However, we need to express the result back in terms of x. Using $\sin(y) = \frac{x}{1}$, draw a right-angled triangle. Label the angle y with the opposite side length being x, so the hypotenuse has a length of 1 unit. From Pythagoras' theorem, the adjacent side length is $\sqrt{1 - x^2}$.

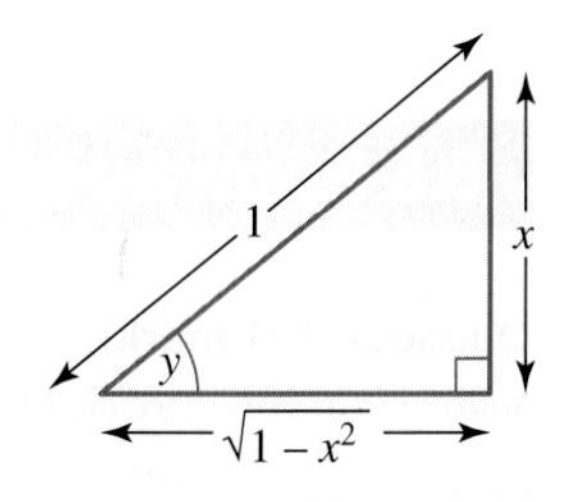

Thus, $\cos(y) = \sqrt{1 - x^2}$. It follows that $\frac{d}{dx}\left(\sin^{-1}(x)\right) = \frac{1}{\sqrt{1 - x^2}}$.

A better result is if $y = \sin^{-1}\left(\frac{x}{a}\right)$, where a is a positive real constant; then $\frac{dy}{dx} = \frac{1}{\sqrt{a^2 - x^2}}$. However, note that the maximal domain of $y = \sin^{-1}\left(\frac{x}{a}\right)$ is $|x| \le a$, whereas the domain of the derivative is $|x| < a$. Although the function is defined at the endpoints, the gradient is not defined at the endpoints; for this reason the domain of the derivative is required.

WORKED EXAMPLE 22 Determining the derivatives of inverse sine functions

Determine $\frac{dy}{dx}$ for each of the following, stating the maximal domain for which the derivative is defined.

a. $y = \sin^{-1}\left(\frac{x}{2}\right)$

b. $y = \sin^{-1}(2x)$

THINK	WRITE
a. 1. State the function and the maximal domain.	**a.** $y = \sin^{-1}\left(\frac{x}{2}\right)$ is defined for $\left\lvert\frac{x}{2}\right\rvert \le 1$; that is, $x \in [-2, 2]$.
2. Use the result $\frac{d}{dx}\left(\sin^{-1}\left(\frac{x}{a}\right)\right) = \frac{1}{\sqrt{a^2 - x^2}}$ with $a = 2$.	$\frac{dy}{dx} = \frac{1}{\sqrt{4 - x^2}}$ for $x \in (-2, 2)$
b. 1. State the function and the maximal domain.	**b.** $y = \sin^{-1}(2x)$ is defined for $\lvert 2x\rvert \le 1$; that is, $x \in \left[-\frac{1}{2}, \frac{1}{2}\right]$.
2. Express y in terms of u.	$y = \sin^{-1}(u)$ where $u = 2x$
3. Differentiate y with respect to u and u with respect to x.	$\frac{dy}{du} = \frac{1}{\sqrt{1 - u^2}}$ and $\frac{du}{dx} = 2$
4. Determine $\frac{dy}{dx}$ using the chain rule.	$\frac{dy}{dx} = \frac{dy}{du}\frac{du}{dx} = \frac{2}{\sqrt{1 - u^2}}$
5. Substitute back for u and state the final result.	$\frac{dy}{dx} = \frac{2}{\sqrt{1 - 4x^2}}$ for $x \in \left(-\frac{1}{2}, \frac{1}{2}\right)$

The derivative of $\cos^{-1}\left(\frac{x}{a}\right)$

The derivative of $\cos^{-1}\left(\frac{x}{a}\right)$ where a is a real positive constant is required.

Let $y = \cos^{-1}\left(\frac{x}{a}\right)$. From the definition of the inverse function, $\frac{x}{a} = \cos(y)$, so $x = a\cos(y)$ and $\frac{dx}{dy} = -a\sin(y)$, and $\frac{dy}{dx} = \frac{-1}{a\sin(y)}$.

We need to express the result back in terms of x. Using $\cos(y) = \frac{x}{a}$, draw a right-angled triangle. Label the angle y, with the adjacent side length being x and the length of the hypotenuse being a. From Pythagoras' theorem, the opposite side length is $\sqrt{a^2 - x^2}$.

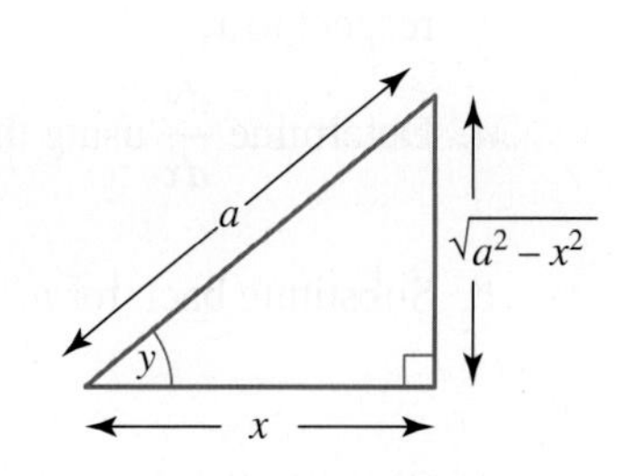

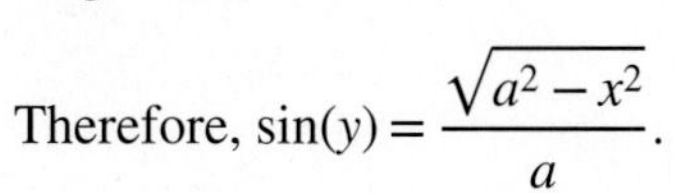

Therefore, $\sin(y) = \frac{\sqrt{a^2 - x^2}}{a}$.

It follows that $\frac{d}{dx}\left(\cos^{-1}\left(\frac{x}{a}\right)\right) = \frac{-1}{\sqrt{a^2 - x^2}}$,

and if $a = 1$, we obtain $\frac{d}{dx}\left(\cos^{-1}(x)\right) = \frac{-1}{\sqrt{1 - x^2}}$.

WORKED EXAMPLE 23 Determining the derivatives of inverse cosine functions

Determine $\frac{dy}{dx}$ for each of the following, stating the maximal domain for which the derivative is defined.

a. $y = \cos^{-1}\left(\frac{2x}{3}\right)$

b. $y = \cos^{-1}\left(\frac{5x-2}{6}\right)$

THINK	WRITE
a. 1. State the function and the maximal domain.	**a.** $y = \cos^{-1}\left(\frac{2x}{3}\right)$ is defined for $\left\lvert\frac{2x}{3}\right\rvert \le 1$; that is, $-1 \le \frac{2x}{3} \le 1$ or $x \in \left[-\frac{3}{2}, \frac{3}{2}\right]$.
2. Express y in terms of u.	$y = \cos^{-1}\left(\frac{u}{3}\right)$ where $u = 2x$
3. Differentiate y with respect to u using $\frac{d}{du}\left(\cos^{-1}\left(\frac{u}{a}\right)\right) = \frac{-1}{\sqrt{a^2-u^2}}$ with $a = 3$, and differentiate u with respect to x.	$\frac{dy}{du} = \frac{-1}{\sqrt{9-u^2}}$ and $\frac{du}{dx} = 2$
4. Determine $\frac{dy}{dx}$ using the chain rule.	$\frac{dy}{dx} = \frac{dy}{du}\frac{du}{dx} = \frac{-2}{\sqrt{9-u^2}}$
5. Substitute back for u and state the final result.	$\frac{dy}{dx} = \frac{-2}{\sqrt{9-4x^2}}$ for $x \in \left(-\frac{3}{2}, \frac{3}{2}\right)$.
b. 1. State the function and solve the inequality to determine the maximal domain of the function.	**b.** $y = \cos^{-1}\left(\frac{5x-2}{6}\right)$ is defined for $\left\lvert\frac{5x-2}{6}\right\rvert \le 1$. $-1 \le \frac{5x-2}{6} \le 1$ $-6 \le 5x-2 \le 6$ $-4 \le 5x \le 8$ $-\frac{4}{5} \le x \le \frac{8}{5}$ or $x \in \left[-\frac{4}{5}, \frac{8}{5}\right]$.
2. Express y in terms of u.	$y = \cos^{-1}\left(\frac{u}{6}\right)$ where $u = 5x-2$
3. Differentiate y with respect to u and u with respect to x.	$\frac{dy}{du} = \frac{-1}{\sqrt{36-u^2}}$ and $\frac{du}{dx} = 5$
4. Determine $\frac{dy}{dx}$ using the chain rule.	$\frac{dy}{dx} = \frac{dy}{du}\frac{du}{dx} = \frac{-5}{\sqrt{36-u^2}}$
5. Substitute back for u.	$\frac{dy}{dx} = \frac{-5}{\sqrt{36-(5x-2)^2}}$
6. Simplify the denominator using the difference of two squares.	$\frac{dy}{dx} = \frac{-5}{\sqrt{(6+(5x-2))(6-(5x-2))}}$
7. State the final result.	$\frac{dy}{dx} = \frac{-5}{\sqrt{(4+5x)(8-5x)}}$ for $x \in \left(-\frac{4}{5}, \frac{8}{5}\right)$

The derivative of $\tan^{-1}\left(\frac{x}{a}\right)$

We also need to determine the derivative of $\tan^{-1}\left(\frac{x}{a}\right)$, where a is a real positive constant.

Let $y=\tan^{-1}\left(\frac{x}{a}\right)$. From the definition of the inverse function, $\frac{x}{a}=\tan(y)$, so $x=a\tan(y)$, $\frac{dx}{dy}=a\sec^2(y)$, and $\frac{dy}{dx}=\frac{1}{a\sec^2(y)}$.

However, we need to express the result in terms of x. Using $\tan(y)=\frac{x}{a}$, draw a right-angled triangle. Label the angle y, with the opposite side length being x and the adjacent side length being a. From Pythagoras' theorem, the length of the hypotenuse is $\sqrt{a^2+x^2}$.

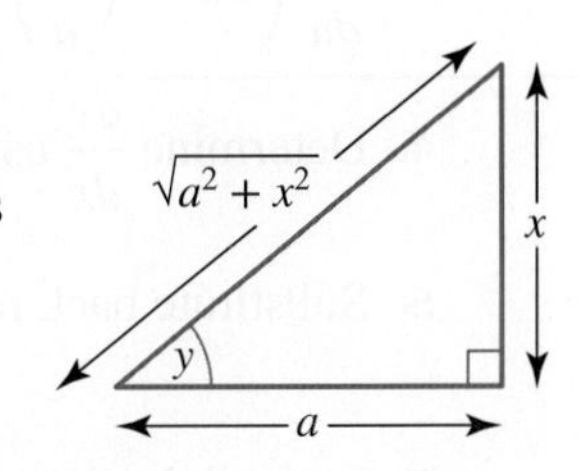

We know that $\sec^2(y)=\frac{1}{\cos^2(y)}$, and from above, $\cos(y)=\frac{a}{\sqrt{a^2+x^2}}$.

It follows that $\frac{dy}{dx}=\frac{\cos^2(y)}{a}=\frac{1}{a}\left(\frac{a}{\sqrt{a^2+x^2}}\right)^2$.

Thus, $\frac{d}{dx}\left(\tan^{-1}\left(\frac{x}{a}\right)\right)=\frac{a}{a^2+x^2}$, and if $a=1$, then $\frac{d}{dx}\left(\tan^{-1}(x)\right)=\frac{1}{1+x^2}$.

The domain is defined for all $x\in R$.

WORKED EXAMPLE 24 Determining the derivatives of inverse tangent functions

Determine $\frac{dy}{dx}$ for each of the following.

a. $y=\tan^{-1}\left(\frac{3x}{4}\right)$

b. $y=\tan^{-1}\left(\frac{5x+4}{7}\right)$

THINK	WRITE
a. 1. State the function.	**a.** $y=\tan^{-1}\left(\frac{3x}{4}\right)$
2. Express y in terms of u.	$y=\tan^{-1}\left(\frac{u}{4}\right)$ where $u=3x$
3. Differentiate y with respect to u and u with respect to x, using $\frac{d}{du}\left(\tan^{-1}\left(\frac{u}{a}\right)\right)=\frac{a}{a^2+u^2}$ with $a=4$.	$\frac{dy}{du}=\frac{4}{16+u^2}$ and $\frac{du}{dx}=3$
4. Determine $\frac{dy}{dx}$ using the chain rule.	$\frac{dy}{dx}=\frac{dy}{du}\frac{du}{dx}=\frac{12}{16+u^2}$
5. Substitute back for u and state the final result.	$\frac{dy}{dx}=\frac{12}{16+9x^2}$

b. 1. State the function.

b. $y = \tan^{-1}\left(\dfrac{5x+4}{7}\right)$

2. Express y in terms of u.

$y = \tan^{-1}\left(\dfrac{u}{7}\right)$ where $u = 5x+4$

3. Differentiate y with respect to u and u with respect to x, using $\dfrac{d}{du}\left(\tan^{-1}\left(\dfrac{u}{a}\right)\right) = \dfrac{a}{a^2+u^2}$ with $a = 7$.

$\dfrac{dy}{du} = \dfrac{7}{49+u^2}$ and $\dfrac{du}{dx} = 5$

4. Determine $\dfrac{dy}{dx}$ using the chain rule.

$\dfrac{dy}{dx} = \dfrac{dy}{du}\dfrac{du}{dx} = \dfrac{35}{49+u^2}$

5. Substitute back for u.

$\dfrac{dy}{dx} = \dfrac{35}{49+(5x+4)^2}$

6. Expand the denominator, and simplify and take out common factors.

$$\begin{aligned}\frac{dy}{dx} &= \frac{35}{49+25x^2+40x+16} \\ &= \frac{35}{25x^2+40x+65} \\ &= \frac{35}{5\left(5x^2+8x+13\right)}\end{aligned}$$

7. State the final result.

$\dfrac{dy}{dx} = \dfrac{7}{5x^2+8x+13}$

5.6.2 Applications of inverse trigonometric derivatives

Determining second derivatives

Recall that the second derivative $\dfrac{d^2y}{dx^2} = \dfrac{d}{dx}\left(\dfrac{dy}{dx}\right)$ is also the rate of change of the gradient function.

WORKED EXAMPLE 25 Second derivatives of inverse trigonometric functions

Determine $\dfrac{d^2y}{dx^2}$ if $y = \sin^{-1}\left(\dfrac{3x}{5}\right)$.

THINK

WRITE

1. State the function and the maximal domain.

$y = \sin^{-1}\left(\dfrac{3x}{5}\right)$ is defined for $\left|\dfrac{3x}{5}\right| \le 1$; that is $-1 \le \dfrac{3x}{5} \le 1$ or $x \in \left[-\dfrac{5}{3}, \dfrac{5}{3}\right]$.

2. Express y in terms of u.

$y = \sin^{-1}\left(\dfrac{u}{5}\right)$ where $u = 3x$

3. Differentiate y with respect to u and u with respect to u.

$\dfrac{dy}{du} = \dfrac{1}{\sqrt{25-u^2}}$ and $\dfrac{du}{dx} = 3$

4. Determine $\dfrac{dy}{dx}$ using the chain rule.

$\dfrac{dy}{dx} = \dfrac{dy}{du}\dfrac{du}{dx} = \dfrac{3}{\sqrt{25-u^2}}$

5 Substitute back for u. $\frac{dy}{dx}=\frac{3}{\sqrt{25-9x^2}}$

6. Write in index notation. $\frac{dy}{dx}=3\left(25-9x^2\right)^{-\frac{1}{2}}$

7. Differentiate again using the chain rule. $\frac{d^2y}{dx^2}=3\times\left(-\frac{1}{2}\right)\times(-18x)\left(25-9x^2\right)^{-\frac{3}{2}}$

8. Simplify the terms. $\frac{d^2y}{dx^2}=\frac{27x}{\left(25-9x^2\right)^{\frac{3}{2}}}$

9. State the final result in simplest form and the domain. $\frac{d^2y}{dx^2}=\frac{27x}{\sqrt{\left(25-9x^2\right)^3}}$ for $x\in\left(-\frac{5}{3},\frac{5}{3}\right)$

Further examples

Because $\sec(x)=\frac{1}{\cos(x)}$, $\operatorname{cosec}(x)=\frac{1}{\sin(x)}$ and $\cot(x)=\frac{1}{\tan(x)}$, it follows that $\sec^{-1}(x)=\cos^{-1}\left(\frac{1}{x}\right)$, $\operatorname{cosec}^{-1}(x)=\sin^{-1}\left(\frac{1}{x}\right)$ and $\cot^{-1}(x)=\tan^{-1}\left(\frac{1}{x}\right)$.

WORKED EXAMPLE 26 Further derivatives of inverse trigonometric functions

Determine $\frac{dy}{dx}$ if $y=\cos^{-1}\left(\frac{3}{x}\right)$, stating the maximal domain for which the derivative is defined.

THINK / **WRITE**

1. State the function and the maximal domain. $y=\cos^{-1}\left(\frac{3}{x}\right)$ for $\left|\frac{3}{x}\right|\le 1$

2. Solve the inequality to determine the maximal domain of the function. $-1\le\frac{3}{x}\le 1$ is equivalent to

$\frac{3}{x}\le 1 \Rightarrow \frac{x}{3}\ge 1 \Rightarrow x\ge 3$

$\frac{3}{x}\ge -1 \Rightarrow \frac{x}{3}\le -1 \Rightarrow x\le -3$

3. Express y in terms of u. $y=\cos^{-1}(u)$ where $u=\frac{3}{x}=3x^{-1}$

4. Differentiate y with respect to u and u with respect to x. $\frac{dy}{du}=\frac{-1}{\sqrt{1-u^2}}$ and $\frac{du}{dx}=-3x^{-2}=-\frac{3}{x^2}$

5. Determine $\frac{dy}{dx}$ using the chain rule. $\frac{dy}{dx}=\frac{3}{x^2\sqrt{1-u^2}}$

6. Substitute back for u. $\frac{dy}{dx}=\frac{3}{x^2\sqrt{1-\frac{9}{x^2}}}$

7. Simplify by taking a common denominator in the square root in the denominator. $\frac{dy}{dx}=\frac{3}{x^2\sqrt{\frac{x^2-9}{x^2}}}$

8. Simplify, noting that $\sqrt{x^2} = \|x\|$.	$\dfrac{dy}{dx} = \dfrac{3\|x\|}{x^2\sqrt{x^2-9}}$
9. State the final result and the domain.	$\dfrac{dy}{dx} = \dfrac{3}{\|x\|\sqrt{x^2-9}}$ for $x < -3$ or $x > 3$.

5.6 Exercise

Students, these questions are even better in jacPLUS

Receive immediate feedback and access sample responses

Access additional questions

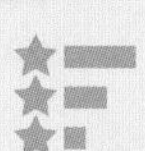
Track your results and progress

Find all this and MORE in jacPLUS

Technology free

1. WE22 Determine $\dfrac{dy}{dx}$ for each of the following, stating the maximal domain for which the derivative is defined.

   a. $y = \sin^{-1}\left(\dfrac{x}{5}\right)$

   b. $y = \sin^{-1}(5x)$

2. Given $f(x) = \sin^{-1}(4x)$, evaluate $f'\left(\frac{1}{8}\right)$.

3. Determine the derivative of each of the following, stating the maximal domain.

   a. $\sin^{-1}\left(\dfrac{x}{3}\right)$

   b. $\sin^{-1}(3x)$

4. Determine the derivative of each of the following, stating the maximal domain.

   a. $\sin^{-1}\left(\dfrac{4x}{3}\right)$

   b. $\sin^{-1}\left(\dfrac{4x+3}{5}\right)$

5. WE23 Determine $\dfrac{dy}{dx}$ for each of the following, stating the maximal domain for which the derivative is defined.

   a. $y = \cos^{-1}\left(\dfrac{3x}{4}\right)$

   b. $y = \cos^{-1}\left(\dfrac{2x-3}{5}\right)$

6. Determine the gradient of the curve $y = \cos^{-1}\left(\dfrac{3x-4}{5}\right)$ at the point where $x = \dfrac{4}{3}$.

7. Determine the derivative of each of the following, stating the maximal domain.

   a. $\cos^{-1}\left(\dfrac{x}{4}\right)$

   b. $\cos^{-1}(4x)$

8. Determine the derivative of each of the following, stating the maximal domain.

   a. $\cos^{-1}\left(\dfrac{3x}{4}\right)$

   b. $\cos^{-1}\left(\dfrac{3x+5}{7}\right)$

9. **WE24** Determine $\frac{dy}{dx}$ for each of the following.

a. $y = \tan^{-1}(4x)$

b. $y = \tan^{-1}\left(\frac{2x-3}{5}\right)$

10. Determine the gradient of the curve $y = \tan^{-1}\left(\frac{3x-5}{7}\right)$ at the point where $x = 4$.

11. Determine the derivative of each of the following.

a. $\tan^{-1}\left(\frac{x}{6}\right)$

b. $\tan^{-1}(6x)$

12. Determine the derivative of each of the following.

a. $\tan^{-1}\left(\frac{5x}{6}\right)$

b. $\tan^{-1}\left(\frac{6x+5}{4}\right)$

13. **WE25** Determine $\frac{d^2y}{dx^2}$ if $y = \cos^{-1}\left(\frac{2x}{3}\right)$.

14. Determine $\frac{d^2y}{dx^2}$ if $y = \tan^{-1}\left(\frac{4x}{3}\right)$.

15. **WE26** Determine $\frac{dy}{dx}$ if $y = \sin^{-1}\left(\frac{2}{x}\right)$, stating the maximal domain for which the derivative is defined.

16. Determine $\frac{dy}{dx}$ if $y = \tan^{-1}\left(\frac{4}{\sqrt{x}}\right)$, stating the maximal domain for which the derivative is defined.

17. Determine the derivative of each of the following, stating the maximal domain.

a. $\sin^{-1}\left(\frac{\sqrt{x}}{3}\right)$

b. $\sin^{-1}\left(\frac{3}{4x}\right)$

c. $\cos^{-1}\left(\frac{e^{2x}}{4}\right)$

18. Determine the derivative of each of the following, stating the maximal domain.

a. $\cos^{-1}\left(\frac{4}{3x}\right)$

b. $\tan^{-1}\left(\frac{x^2}{3}\right)$

c. $\tan^{-1}\left(\frac{6}{5x}\right)$

19. Determine $\frac{d^2y}{dx^2}$ if:

a. $y = \sin^{-1}\left(\frac{5x}{4}\right)$

b. $y = \cos^{-1}\left(\frac{6x}{5}\right)$

c. $y = \tan^{-1}\left(\frac{7x}{6}\right)$.

5.6 Exam questions

Question 1 (1 mark)

Source: *VCE 2020 Specialist Mathematics Exam 1, Q6a; © VCAA.*

Let $f(x) = \arctan(3x - 6) + \pi$.

Show that $f'(x) = \frac{3}{9x^2 - 36x + 37}$.

Question 2 (1 mark)

Source: *VCE 2017 Specialist Mathematics Exam 1, Q10a; © VCAA.*

Show that $\frac{d}{dx}\left(x \arccos\left(\frac{x}{a}\right)\right) = \arccos\left(\frac{x}{a}\right) - \frac{x}{\sqrt{a^2 - x^2}}$, where $a > 0$.

Question 3 (1 mark) TECH-ACTIVE

Source: *VCE 2015 Specialist Mathematics Exam 2, Section 1, Q2; © VCAA.*

The range of the function with rule $f(x) = (2 - x)\arcsin\left(\frac{x}{2} - 1\right)$ is

A. $[-\pi, 0]$

B. $\left[-\frac{\pi}{2}, \frac{\pi}{2}\right]$

C. $\left[-\frac{(2-x)\pi}{2}, \frac{(2-x)\pi}{2}\right]$

D. $[0, 4]$

E. $[0, \pi]$

More exam questions are available online.

5.7 Related rates

LEARNING INTENTION

At the end of this subtopic you should be able to:

- determine the rate of change of a variable with respect to another, indirectly related variable using the chain rule.

5.7.1 Introduction

When two or more quantities vary with time and are related by some condition, their rates of change are also related. The steps involved in solving these related rate problems are listed below.

1. Define the variables.
2. Write down the rate that is provided in the question.
3. Establish the relationship between the variables.
4. Write down the rate that needs to be determined.
5. Use a chain rule or implicit differentiation to relate the variables.

WORKED EXAMPLE 27 Related rates (1)

A circular metal plate is being heated and its radius is increasing at a rate of 2 mm/s. Calculate the rate at which the area of the plate is increasing when the radius is 30 millimetres.

THINK	WRITE
1. Define the variables.	Let r mm be the radius of the metal at time t seconds. Let A mm^2 be the area of the plate at time t seconds.
2. The radius is increasing at a rate of 2 mm/s. Note that the units also help to determine the given rate.	$\frac{dr}{dt} = 2$ mm/s.
3. The plate is circular. Write the formula for the area of a circle.	$A = \pi r^2$.

4. Determine the rate at which the area of the plate is increasing.	$\frac{dA}{dt} = ?$ and evaluate this rate when $r = 30$.
5. Form a chain rule for the required rate in terms of the given variables.	$\frac{dA}{dt} = \frac{dA}{dr} \times \frac{dr}{dt}$
6. Since one rate, $\frac{dr}{dt}$, is known, we need to determine $\frac{dA}{dr}$.	$A = \pi r^2 \Rightarrow \frac{dA}{dr} = 2\pi r$
7. Substitute the given rates into the required equations.	$\frac{dA}{dt} = \frac{dA}{dr} \times \frac{dr}{dt}$ $= 2\pi r \times 2$
8. Evaluate the required rate when $r = 30$.	$\frac{dA}{dt} = 4\pi r$ when $r = 30$
9. State the final result with the required units, leaving the answer in terms of π.	$\left.\frac{dA}{dr}\right\vert_{r=30} = 120\pi \text{ mm}^2/\text{s}$ The area is increasing at 120π mm^2/s.

Decreasing rates

If a quantity has a rate decreasing with respect to time, then the required rate is given as a negative quantity.

WORKED EXAMPLE 28 Related rates (2)

A spherical balloon has a hole in it, and the balloon's volume is decreasing at a rate of 2 cm^3/s. Determine at what rate the radius is changing when the radius of the balloon is 4 cm.

THINK	WRITE
1. Define the variables.	Let t be the time since the balloon was punctured, in seconds. Let r cm be the radius of the balloon. Let V cm^3 be the volume of the balloon.
2. The volume is decreasing at a rate of 2 cm^3/s. Notice that the units also help determine the given rate.	$\frac{dV}{dt} = -2 \text{ cm}^3/\text{s}$
3. The balloon is spherical. Write the formula for the volume of a sphere.	$V = \frac{4}{3}\pi r^3$
4. We need to determine the rate at which the radius changing.	$\frac{dr}{dt} = ?$ and evaluate this rate when $r = 4$.
5. Form a chain rule for the required rate in terms of the given variables.	$\frac{dr}{dt} = \frac{dr}{dV} \times \frac{dV}{dt}$

6. Since one rate, $\frac{dV}{dt}$, is known, we need to determine $\frac{dr}{dV}$.

Since $V = \frac{4}{3}\pi r^3 \Rightarrow \frac{dV}{dr} = 4\pi r^2$ and

$\frac{dr}{dV} = 1/\frac{dV}{dr} = \frac{1}{4\pi r^2}$

7. Substitute the given rates into the required equations.

$$\begin{aligned}\frac{dr}{dt} &= \frac{dr}{dV} \times \frac{dV}{dt} \\ &= \frac{1}{4\pi r^2} \times (-2) \\ &= -\frac{1}{2\pi r^2}\end{aligned}$$

8. Evaluate the required rate when $r = 4$.

When $r = 4$, $\left.\frac{dr}{dt}\right|_{r=4} = -\frac{1}{32\pi}$.

9. State the final result in the required units, leaving the answer in terms of π.

The radius is decreasing at a rate of $\frac{1}{32\pi}$ cm/s.

Relating the variables

It is often necessary to express a required expression in terms of only one variable instead of two. This can be achieved by determining relationships between the variables, for example, by using similar triangles.

WORKED EXAMPLE 29 Related rates (3)

A conical funnel has a height of 25 centimetres and a radius of 20 centimetres. It is positioned so that its axis is vertical and its vertex is downwards. Oil leaks out through an opening in the vertex at a rate of 4 cubic centimetres per second. Calculate the rate at which the oil level is falling when the height of the oil in the funnel is 5 centimetres. (Ignore the cylindrical section of the funnel.)

THINK	WRITE/DRAW
1. Define the variables.	Let r cm be the radius of the oil in the funnel. Let h cm the height of the oil in the funnel.
2. Write the rate given in the question.	$\frac{dV}{dt} = -4 \text{ cm}^3/\text{s}$
3. Write which rate is required to be determined.	$\frac{dh}{dt} = ?$ when $h = 5$
4. Make up a chain rule for the required rate, in terms of the given variables.	$\frac{dh}{dt} = \frac{dh}{dV} \times \frac{dV}{dt}$

5. Determine the relationship between the variables. The variables r and h change; however, the height and radius of the funnel are constant, and this can be used to determine a relationship between h and r.

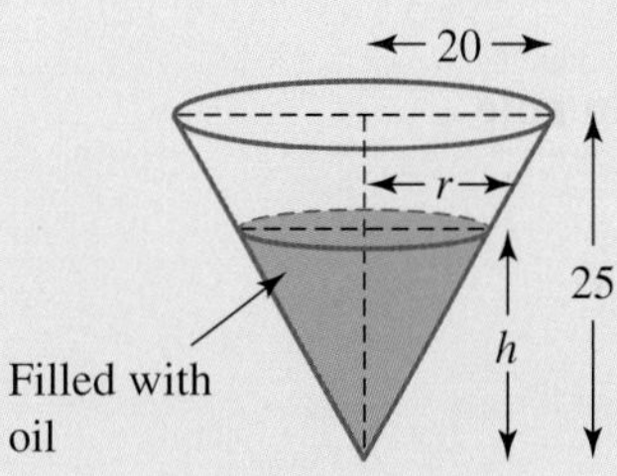

$$V = \frac{1}{3}\pi r^2 h$$

Using similar triangles,

$$\frac{20}{25} = \frac{r}{h}$$

$$r = \frac{4h}{5}$$

6. We need to express the volume in terms of h only.

Substitute into $V = \frac{1}{3}\pi r^2 h$:

$$V = \frac{1}{3}\pi\left(\frac{4h}{5}\right)^2 h$$

$$= \frac{16\pi h^3}{75}$$

7. Determine the rate to substitute into the chain rule.

$$\frac{dV}{dh} = \frac{16\pi h^2}{25}$$

$$\frac{dh}{dV} = \frac{25}{16\pi h^2}$$

8. Substitute for the required rates.

$$\frac{dh}{dt} = \frac{dh}{dV} \times \frac{dV}{dt} = \frac{25}{16\pi h^2} \times -4$$

9. Evaluate the required rate when $h = 5$.

$$\left.\frac{dh}{dt}\right|_{h=5} = \frac{25}{16\pi(5)^2} \times -4 = -\frac{1}{4\pi}$$

10. State the final result with the required units, leaving the answer in terms of π.

The height is decreasing at a rate of $\frac{1}{4\pi}$ cm/s.

Determining the required variables

An alternative method to solving related rate problems is to use implicit differentiation.

WORKED EXAMPLE 30 Related rates (4)

A ladder 3 metres long has its top end resting against a vertical wall and its lower end on horizontal ground. The top end of the ladder is slipping down at a constant speed of 0.1 metres per second. Determine the rate at which the lower end is moving away from the wall when the lower end is 1 metre from the wall.

THINK | **WRITE/DRAW**

Method 1: Using the chain rule

1. Define the variables.

Let x metres be the distance of the base of the ladder from the wall, and let y metres be the distance of the top of the ladder from the ground.

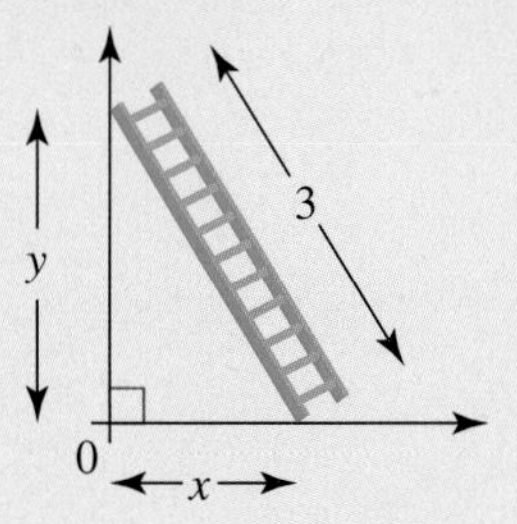

2. The top end of the ladder is slipping down at a constant speed of 0.1 m/s.

$\frac{dy}{dt} = -0.1 \text{ m/s}$

3. Apply Pythagoras' theorem.

$x^2 + y^2 = 3^2 = 9$

4. We need to determine the rate at which the lower end is moving away from the wall when the lower end is 1 m from the wall.

$\frac{dx}{dt} = ?$ when $x = 1$.

5. Construct a chain rule for the required rate in terms of the given variables.

$\frac{dx}{dt} = \frac{dx}{dy} \times \frac{dy}{dt}$

6. Express y in terms of x.

$$\begin{aligned} x^2 + y^2 &= 9 \\ y^2 &= 9 - x^2 \\ y &= \pm\sqrt{9 - x^2} \\ y &= \sqrt{9 - x^2} \text{ since } y > 0 \end{aligned}$$

7. Since one rate, $\frac{dy}{dt}$, is known, we need to determine $\frac{dx}{dy}$.

$$\frac{dy}{dx} = -\frac{x}{\sqrt{9 - x^2}}$$

$$\frac{dx}{dy} = -\frac{\sqrt{9 - x^2}}{x}$$

8. Substitute the given rates into the required equation.

$$\begin{aligned} \frac{dx}{dt} &= \frac{dx}{dy} \times \frac{dy}{dt} \\ &= -\frac{\sqrt{9 - x^2}}{x} \times (-0.1) \end{aligned}$$

9. Evaluate the required rate when $x = 1$.

$$\left.\frac{dx}{dt}\right|_{x=1} = \frac{\sqrt{8}}{10} = \frac{2\sqrt{2}}{10}$$

10. State the final result with the required units, giving an exact answer.	$\left.\frac{dx}{dt}\right\|_{x=1} = \frac{\sqrt{2}}{5}$ m/s The lower end is moving away from the wall at a rate of $\frac{\sqrt{2}}{5}$ m/s.

Method 2: Using implicit differentiation

The first 4 steps are identical to method 1 above. From this point, we use implicit differentiation to determine the required rate.

5. Take $\frac{d}{dt}(\ )$ of each term in turn.	$\frac{d}{dt}(x^2) + \frac{d}{dt}(y^2) = \frac{d}{dt}(9)$
6. Use implicit differentiation.	$2x\frac{dx}{dt} + 2y\frac{dy}{dt} = 0$
7. Rearrange to make the required rate the subject.	$\frac{dx}{dt} = -\frac{y}{x}\frac{dy}{dt}$
8. Determine the appropriate values.	When $x = 1$, $y = \sqrt{8}$ and $\frac{dy}{dt} = -0.1 = -\frac{1}{10}$
9. Substitute in the appropriate values of x and y.	$\left.\frac{dx}{dt}\right\|_{x=1} = \frac{\sqrt{8}}{10} = \frac{2\sqrt{2}}{10}$
10. State the final result as before.	$\left.\frac{dx}{dt}\right\|_{x=1} = \frac{\sqrt{2}}{5}$ m/s The lower end is moving away from the wall at a rate of $\frac{\sqrt{2}}{5}$ m/s.

5.7 Exercise

Students, these questions are even better in jacPLUS

Receive immediate feedback and access sample responses

Access additional questions

Track your results and progress

Find all this and MORE in jacPLUS

Technology free

1. WE27 A circular oil slick is expanding so that its radius increases at a rate of 0.5 m/s. Determine the rate at which the area of the oil slick is increasing when the radius of the slick is 20 metres.

2. A circular disc is expanding so that its area increases at a rate of 40π cm^2/s. Determine the rate at which the radius of the disc is increasing when the radius is 10 cm.

3. a. A square has its sides increasing at a rate of 2 centimetres per second. Calculate the rate at which the area is increasing when the sides are 4 cm long.
   b. A stone is dropped into a lake, sending out concentric circular ripples. The area of the disturbed water region increases at a rate of 2 m^2/s. Calculate the rate at which the radius of the outermost ripple is increasing when its radius is 4 metres.

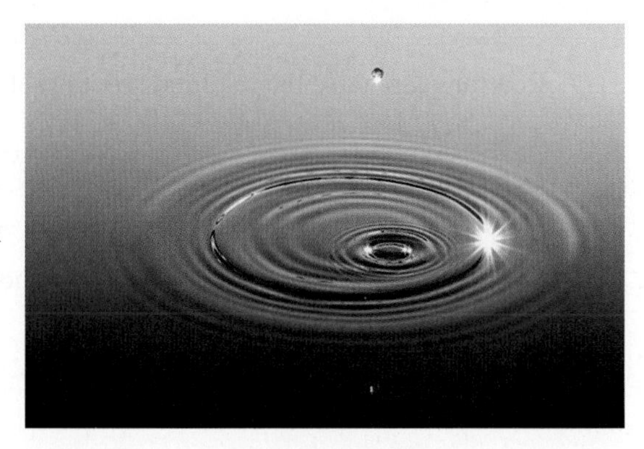

4. A spherical bubble is blown so that its radius is increasing at a constant rate of 2 millimetres per second. When its radius is 10 millimetres, calculate the rate at which:
   a. its volume is increasing
   b. its surface area is increasing.

5. WE28 A spherical basketball has a hole in it and its volume is decreasing at a rate of 6 cm^3/s. Calculate the rate at which the radius is changing when the radius of the basketball is 6 cm.

6. A metal ball is dissolving in an acid bath. Its radius is decreasing at a rate of 3 cm/s. Calculate the rate at which the ball's surface area is changing when the radius of the ball is 2 cm.

7. A mothball has its radius decreasing at a constant rate of 0.2 millimetres, per week . Assume it remains spherical.
   a. Show that the volume is decreasing at a rate that is proportional to its surface area.
   b. If its initial radius is 30 millimetres, calculate how long it takes to disappear.

8. WE29 A conical vase has a height of 40 cm and a radius of 8 cm. The axis of the vase is vertical and its vertex is downwards. Initially it is filled with water which leaks out through a small hole in the vertex at a rate of 6 cm^3/s. Calculate the rate at which the water level is falling when the height of the water is 16 cm.

9. A cone is such that its radius is always equal to half its height. If the radius is decreasing at a rate of 2 cm/s, calculate the rate at which the volume of the cone is decreasing when the radius is 4 cm.

10. The sides of an equilateral triangle are increasing at a rate of 2 centimetres per second. When the sides are $2\sqrt{3}$ centimetres, calculate the rate at which:
    a. its area is increasing
    b. its height is increasing.

Technology active

11. a. Show that the formula for the volume, V, of a right circular cone with height h is given by $V = \frac{1}{3}\pi h^3 \tan^2(\alpha)$, where α is the semi-vertex angle.

b. Falling sand forms a heap in the shape of a right circular cone whose semi-vertex angle is 71.57°. If its height is increasing at 2 centimetres per second when the heap is 5 centimetres high, calculate the rate at which its volume is increasing.

12. **WE30** A ladder 5 metres long has its top end resting against a vertical wall and its lower end on horizontal ground. The bottom end of the ladder is pushed closer to the wall at a speed of 0.3 metres per second. Calculate the rate at which the top end of the ladder is moving up the wall when the lower end is 3 metres from the wall.

13. A kite is 30 metres above the ground and is moving horizontally away at a speed of 2 metres per second from the boy who is flying it. When the length of the string is 50 metres, calculate the rate at which the string is being released.

14. a. The volume, V cm^3, of water in a hemispherical bowl of radius r cm when the depth of the water is h cm is given by $V = \frac{1}{3}\pi h^2(3r - h)$.

A hemispherical bowl of radius 10 centimetres is being filled with water at a constant rate of 3 cubic centimetres per second. Calculate the rate at which the depth of the water is increasing when the depth is 5 cm.

b. A drinking glass is in the shape of a truncated right circular cone. When the glass is filled to a depth of h cm, the volume of liquid in the glass, V cm^3, is given by $V = \frac{\pi}{432}\left(h^3 + 108h^2 + 388h\right)$. Lemonade is leaking out from the glass at a rate of 7 cm^3/s. Calculate the rate at which the depth of the lemonade is falling when the depth is 6 cm.

15. a. A rubber flotation device is being pulled into a wharf by a rope at a speed of 26 metres per minute. The rope is attached to a point on the wharf 1 metre vertically above the flotation device. Determine the rate that the flotation device is approaching the wharf when it is 10 metres from the wharf.

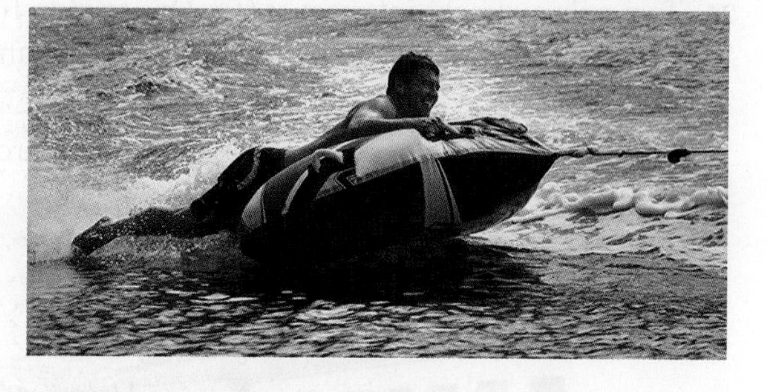

b. A car approaches the ground level of a 30-metre-tall building at a speed of 54 kilometres per hour. Calculate the rate of change of the distance from the car to the top of the building when it is 40 metres from the foot of the building.

16. The distance, q cm, between the image of an object and a certain lens in terms of p cm, the distance of the object from the lens, is given by $q = \frac{10p}{p - 10}$.

a. Show that the rate of change of distance that an image is from the lens with respect to the distance of the object from the lens is given by $\frac{dq}{dp} = \frac{-100}{(p-10)^2}$.

b. If the object distance is increasing at a rate 0.2 cm/s, calculate how fast the image distance is changing, when the distance from the object is 12 cm.

17. a. When a gas expands without a change of temperature, the pressure P and volume V are given by the relationship $PV^{1.4} = C$, where C is a constant. At a certain instant, the pressure is 1.01×10^5 pascals and the volume is 22.4×10^{-3} cubic metres. The volume is increasing at a rate of 0.005 cubic metres per second. Calculate the rate at which the pressure is changing at this instant, correct to 1 decimal place.

b. The pressure P and volume V of a certain fixed mass of gas during an adiabatic expansion are connected by the law $PV^n = C$, where n and C are constants.

Show that the time rate of change of volume satisfies $\frac{dV}{dt} = -\frac{V}{nP}\frac{dP}{dt}$.

18. A jet aircraft is flying horizontally at a speed of 300 km/h at an altitude of 1 km. It passes directly over a radar tracking station located at ground level.
Calculate the rate in degrees per second, correct to 3 decimal places, at which the radar beam to the aircraft is turning when the jet is at a horizontal distance of 30 kilometres from the station.

19. A helicopter is flying horizontally at a constant height of 300 metres. It passes directly over a light source located at ground level. The light source is always directed at the helicopter.
If the helicopter is flying at 108 kilometres per hour, calculate the rate in degrees per second, correct to 1 decimal place, at which the light source to the helicopter is turning when the helicopter is at a horizontal distance of 0.4 kilometres from the light source.

20. A man 2 metres tall is walking at 1.5 m/s. He passes under a light source 6 metres above the ground. Determine:

a. the rate at which his shadow is lengthening
b. the speed at which the end point of his shadow is increasing
c. the rate at which his head is receding from the light source when he is 8 metres from the light.

21. Two railway tracks intersect at 60°. One train is 100 km from the junction and moves towards it at 80 km/h, while another train is 120 km from the junction and moves towards the junction at 90 km/h.

a. Show that the trains do not collide.
b. Determine the rate at which the trains are approaching each other after 1 hour.

5.7 Exam questions

Question 1 (1 mark) TECH-ACTIVE

Source: *VCE 2019 Specialist Mathematics Exam 2, Section A, Q10; © VCAA.*

MC Sand falls from a chute to from a pile in the shape of a right circular cone with semi-vertex angle 60°. Sand is added to the pile at a rate of 1.5 m^3 per minute.

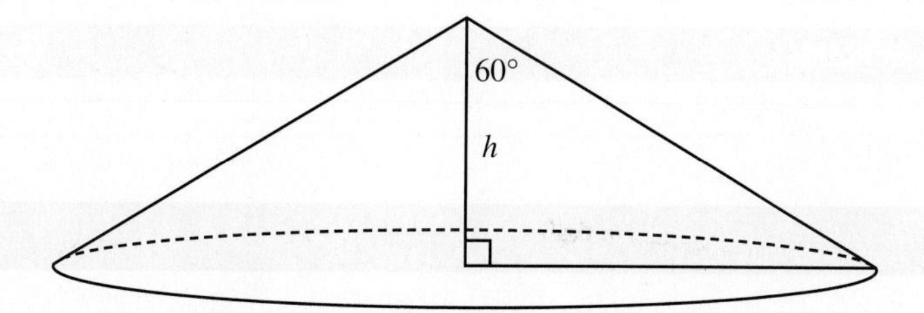

The rate at which the height h metres of the pile is increasing, in metres per minute, when the height of the pile is 0.5 m, correct to two decimal places, is

A. 0.21 **B.** 0.31 **C.** 0.64 **D.** 3.82 **E.** 3.53

Question 2 (4 marks) TECH-FREE

Source: *VCE 2016 Specialist Mathematics Exam 1, Q4; © VCAA.*

Chemicals are added to a container so that a particular crystal will grow in the shape of cube. The side length of the crystal, x millimetres, t days after the chemicals were added to the container, is given by $x = \arctan(t)$.

Find the rate at which the surface area, A square millimetres, of the crystal is growing one day after the chemicals were added. Give your answer in square millimetres per day.

Question 3 (5 marks) TECH-ACTIVE

Source: *VCE 2014 Specialist Mathematics Exam 2, Section 2, Q4 a,b; © VCAA.*

At a water fun park, a conical tank of radius 0.5 m and height 1 m is filling with water. At the same time, some water flows out from the vertex, wetting those underneath. When the tank eventually fills, it tips over and the water falls out, drenching all those underneath. The tank then returns to its original position and begins to refill.

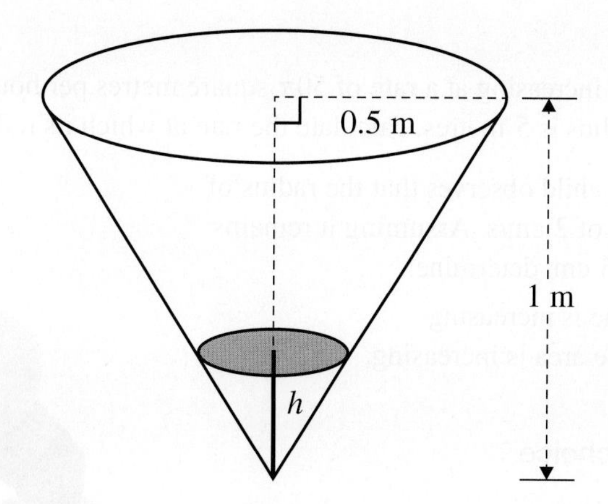

Water flows in at a constant rate of $0.02\,\pi\text{m}^3/\text{min}$ and flows out at a variable rate of $0.01\,\pi\sqrt{h}\,\text{m}^3/\text{min}$, where h metres is the depth of the water at any instant.

a. Show that the volume, V cubic metres, of water in the cone when it is filled to a depth of h metres is given by $V = \dfrac{\pi}{12}h^3$. **(1 mark)**

b. Find the rate, in m/min, at which the depth of the water in the tank is increasing when the depth is 0.25 m **(4 marks)**

More exam questions are available online.

5.8 Review

5.8.1 Summary

doc-37059

Hey students! Now that it's time to revise this topic, go online to:

 Access the topic summary  Review your results  Watch teacher-led videos  Practise VCAA exam questions

Find all this and MORE in jacPLUS

5.8 Exercise

Technology free: short answer

1. Differentiate each of the following with respect to x.

a. x^5e^{-3x}

b. $\dfrac{\cos(2x)}{x^3}$

c. $\dfrac{x}{\sqrt{3x^2+5}}$

2. Determine $\dfrac{d^2y}{dx^2}$ for each of the following.

a. $y=\sin^{-1}\left(\dfrac{4x}{5}\right)$

b. $y=e^{-3x}\sin(2x)$

c. $y=x^3\cos(2x)$

3. Determine the equation of the normal to each of the following curves at the point indicated.

a. $y=6\sin\left(\dfrac{x}{3}\right)$ at $x=\dfrac{\pi}{2}$

b. $y=\tan\left(\dfrac{2x}{3}\right)$ at $x=\dfrac{\pi}{2}$

c. $y=\sin^{-1}\left(\dfrac{\sqrt{x}}{2}\right)$ at $x=1$

4. a. Determine an expression for the gradient in terms of both x and y for the curve $2x^3-6x^2y^2+3y^3-2=0$.
   b. Determine the gradient in terms of x and y if $4x^4-7xy-3y^4-8=0$
   c. Determine the equation of the normal to the curve $x^3-4xy+y^3-15=0$ at the point $(2,-1)$.
   d. For the ellipse $4x^2+4xy+9y^2-6y-2x-23=0$, determine the equation of the tangent at the point where $x=2$ in the first quadrant.

5. An oil slick at sea has its area increasing at a rate of 50π square metres per hour. Assuming the oil slick remains circular, when the radius is 5 metres, calculate the rate at which its radius is increasing.

6. While blowing up a balloon a child observes that the radius of the balloon increases at a rate of 2 cm/s. Assuming it remains spherical, when the radius is 6 cm, determine:
   a. the rate at which the volume is increasing
   b. that rate at which its surface area is increasing.

Technology active: multiple choice

7. **MC** $\dfrac{d}{dx}\left[\sqrt{9x^2+16}\right]$ is equal to

A. $\dfrac{18x}{\sqrt{9x^2+16}}$

B. $\dfrac{9x}{\sqrt{9x^2+16}}$

C. $\dfrac{9}{\sqrt{9x^2+16}}$

D. $\dfrac{18}{\sqrt{9x^2+16}}$

E. 3

8. **MC** The equation of the normal to the curve $y = \cos\left(\frac{x}{2}\right)$ at the point $x = \frac{\pi}{3}$ is equal to

A. $y = -\frac{x}{2} + \frac{\pi}{6} + \frac{\sqrt{3}}{2}$
B. $y = -\frac{x}{4} + \frac{\pi}{12} + \frac{\sqrt{3}}{2}$
C. $y = -4x + \frac{4\pi}{3} + \frac{\sqrt{3}}{2}$
D. $y = 4x - \frac{4\pi}{3} + \frac{\sqrt{3}}{2}$
E. $y = 2x - \frac{2\pi}{3} + \frac{\sqrt{3}}{2}$

9. **MC** The gradient of the tangent to the curve $y = \sin^{-1}\left(\frac{2x}{3}\right)$ at the point where $x = \frac{3}{4}$ is equal to

A. $-\frac{4\sqrt{3}}{9}$
B. $\frac{4\sqrt{3}}{9}$
C. $-\frac{3\sqrt{3}}{4}$
D. $\frac{3\sqrt{3}}{4}$
E. $\frac{\pi}{6}$

10. **MC** If $y = x^2 \cos(2x)$, then the rate of change of y with respect to x, when $x = \frac{\pi}{6}$ is equal to

A. $\frac{\pi}{12}$
B. $\frac{\pi}{6}$
C. $\frac{\sqrt{3}\pi}{6}$
D. $\frac{\pi\left(6\sqrt{3} - \pi\right)}{36}$
E. $\frac{\pi\left(6 - \sqrt{3}\pi\right)}{36}$

11. **MC** The gradient of the curve $2x^2 + 3y^2 - 4xy - 9 = 0$ at the point $(-3, -1)$ is equal to

A. 8
B. $-\frac{4}{3}$
C. $\frac{4}{3}$
D. 2
E. $-\frac{4}{9}$

12. **MC** The average rate of change of the function with the rule $f(x) = \sqrt{2x + 1}$ between $x = 0$ and $x = 4$ is equal to

A. $\frac{1}{2}$
B. $\frac{1}{3}$
C. $\frac{3}{4}$
D. $\frac{13}{6}$
E. $\frac{\sqrt{5}}{5}$

13. **MC** If $y = \cos^{-1}(1 - 2x)$ then $\frac{dy}{dx}$ is equal to

A. $\frac{-1}{2\sqrt{x(1 - x)}}$
B. $\frac{1}{2\sqrt{x(1 - x)}}$
C. $\frac{1}{\sqrt{x(1 - x)}}$
D. $\frac{-2}{\sqrt{x(1 - x)}}$
E. $\frac{-2}{\sqrt{1 - 4x + 4x^2}}$

14. **MC** If $x = e^{-t}$ and $y = e^{2t}$ then $\frac{d^2y}{dx^2}$ is equal to

A. $6e^{4t}$
B. $4e^{3t}$
C. $-6e^{2t}$
D. $6e^{2t}$
E. $4e^{2t}$

15. **MC** The gradient of the normal to the circle $x^2 + y^2 = 169$ at the point where $x = -5$ in the third quadrant is equal to

A. $-\frac{5}{12}$
B. $\frac{5}{12}$
C. $-\frac{12}{5}$
D. $\frac{12}{5}$
E. $-\frac{5}{13}$

16. **MC** An ice block has the shape of a cube and has its volume decreasing at a rate of $2\text{ cm}^3/\text{min}$. The rate at which each of the sides are decreasing in cm/min, when the sides are 2 cm, is equal to:

A. 24
B. 16
C. 6
D. $\frac{1}{4}$
E. $\frac{1}{6}$

Technology active: extended response

17. a. Show that the equation of the tangent to the hyperbola $xy = c^2$ at the point $\left(ct, \frac{c}{t}\right)$ is given by $x + t^2y = 2ct$.

b. Show that the equation of the normal to the hyperbola $xy = c^2$ at the point $\left(ct, \frac{c}{t}\right)$ is given by $y = xt^2 - ct^3 + \frac{c}{t}$.

c. Verify that the hyperbola $xy = c^2$, can be represented by the parametric equations $x = ct$ and $y = \frac{c}{t}$.

18. a. Show that the equation of the tangent to the ellipse $\frac{x^2}{a^2} + \frac{y^2}{b^2} = 1$ at the point $(a\cos(\theta), b\sin(\theta))$ is given by $\frac{x\cos(\theta)}{a} + \frac{y\sin(\theta)}{b} = 1$.

b. i. Show that the equation of the tangent to the hyperbola $\frac{x^2}{a^2} - \frac{y^2}{b^2} = 1$ at the point (x_1, y_1) is given by $\frac{xx_1}{a} - \frac{yy_1}{b} = 1$.

ii. Show that the hyperbola $\frac{x^2}{a^2} - \frac{y^2}{b^2} = 1$ can be represented by the parametric equations $x = a\sec(\theta)$ and $y = b\tan(\theta)$.

19. If a and b are constants, show that:

a. $\frac{d^2}{dx^2}\left[x\log_e(ax+b)\right] = \frac{a(ax+2b)}{(ax+b)^2}$

b. $\frac{d^2}{dx^2}\left[x\log_e\left(ax^2+b\right)\right] = \frac{2ax\left(ax^2+3b\right)}{\left(ax^2+b\right)^2}$

c. $\frac{d^2}{dx^2}\left[x^2\log_e(ax+b)\right] = 2\log_e(ax+b) + \frac{a(3ax+4b)}{(ax+b)^2}$

20. A curve called a witch of Agnesi has an explicit equation $y = \dfrac{8a^3}{x^2 + 4a^2}$.

a. Show that $\dfrac{dy}{dx} = \dfrac{-16a^3x}{\left(x^2 + 4a^2\right)^2}$.

b. Show that the witch of Agnesi can be represented by the parametric equations $x = 2a\cot(t)$ and $y = a\left(1 - \cos(2t)\right)$.

c. Show that $\dfrac{dy}{dx} = -2\sin^3(t)\cos(t)$ and check your answer to part **a**.

21. A curve has the parametric equations $x = 2\sin(t)$ and $y = 2\sin(t)\tan(t)$.

a. Verify that $\dfrac{dy}{dx} = \dfrac{\sin(t)\left(\cos^2(t) + 1\right)}{\cos^2(t)}$.

b. Show that the curve has the implicit equation $x^4 + x^2y^2 - 4y^2 = 0$ and using implicit differentiation verify that $\dfrac{dy}{dx} = \dfrac{x\left(2x^2 + y^2\right)}{y\left(4 - x^2\right)}$.

c. Show that the implicit equation $x^4 + x^2y^2 - 4y^2 = 0$ defines $y = \dfrac{x^2}{\sqrt{4 - x^2}}$ and that $\dfrac{dy}{dx} = \dfrac{8x - x^3}{\sqrt{(4 - x^2)^3}}$.

22. A spiral curve has the parametric equations $x = t\cos(t)$ and $y = t\sin(t)$.

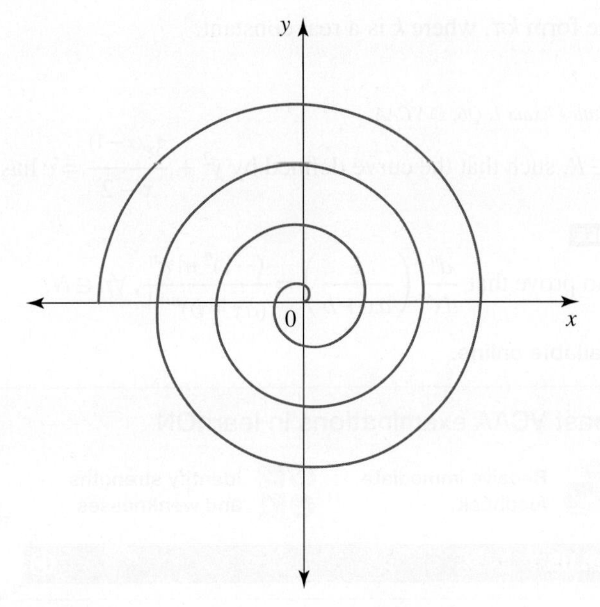

a. Show that $\dfrac{dy}{dx} = \dfrac{\sin(t) + t\cos(t)}{\cos(t) - t\sin(t)}$.

b. Show that the curve has the implicit equation $x^2 + y^2 = \left(\tan^{-1}\left(\dfrac{y}{x}\right)\right)^2$.

c. Using implicit differentiation on $x^2 + y^2 = \left(\tan^{-1}\left(\dfrac{y}{x}\right)\right)^2$ verify that $\dfrac{dy}{dx} = \dfrac{x^3 + xy^2 + y\tan^{-1}\left(\dfrac{y}{x}\right)}{x\tan^{-1}\left(\dfrac{y}{x}\right) - yx^2 - y^3}$.

5.8 Exam questions

Question 1 (3 marks)

Source: *VCE 2021 Specialist Mathematics Exam 1, Q5; © VCAA.*

Find the gradient of the curve with equation $e^x e^{2y} + e^{4y^2} = 2e^4$ at the point (2, 1).

Question 2 (4 marks)

Source: *VCE 2016 Specialist Mathematics Exam 1, Q3; © VCAA.*

Find the equation of the line perpendicular to the graph of $\cos(y) + y\,\sin(x) = x^2$ at $\left(0, -\frac{\pi}{2}\right)$.

Question 3 (6 marks)

Source: *VCE 2015 Specialist Mathematics Exam 1, Q9; © VCAA.*

Consider the curve represented by $x^2 - xy + \frac{3}{2}y^2 = 9$.

a. Find the gradient of the curve at any point (x, y). **(2 marks)**

b. Find the equation of the tangent to the curve at the point $(3, 0)$ and find the equation of the tangent to the curve at the point $\left(0, \sqrt{6}\right)$.

Write each equation in the form $y = ax + b$. **(2 marks)**

c. Find the acute angle between the tangent to the curve at the point $(3, 0)$ and the tangent to the curve at the point $\left(0, \sqrt{6}\right)$.

Give your answer in the form $k\pi$, where k is a real constant. **(2 marks)**

Question 4 (4 marks)

Source: *VCE 2013 Specialist Mathematics Exam 1, Q6; © VCAA.*

Find the value of c, where $c \in R$, such that the curve defined by $y^2 + \frac{3e^{(x-1)}}{x-2} = c$ has a gradient of 2 where $x = 1$.

Question 5 (5 marks) TECH-FREE

Use mathematical induction to prove that $\frac{d^n}{dx^n}\left(\frac{1}{ax+b}\right) = \frac{(-1)^n\, n!\, a^n}{(ax+b)^{n+1}}, \forall n \in N$.

More exam questions are available online.

Answers

Topic 5 Differential calculus

5.2 Review of differentiation techniques

5.2 Exercise

1. a. $\frac{3}{16}$ b. $30\sin^2(2x)\cos(2x)$
2. a. $\frac{-2}{(2x+5)^2}$ b. $-6\sqrt{6}$
3. a. $24\sin^3(3x)\cos(3x)$
   b. $-60\cos^2(4x)\sin(4x)$
4. a. $\frac{4x^2+9}{\sqrt{2x^2+9}}$ b. $\frac{5}{\sqrt{(3x^2+5)^3}}$
5. a. $-xe^{-\frac{1}{2}x^2}$ b. $-2\sin(2x)e^{\cos(2x)}$
6. a. $-2e^{2x}\sin\left(e^{2x}\right)$ b. $\frac{e^{\sqrt{x}}}{2\sqrt{x}}$
7. a. $\frac{\pi^2}{36}$ b. $2\cos(2x)e^{\sin(2x)}$
8. a. $\frac{\cos\left(\sqrt{x}\right)}{2\sqrt{x}}$ b. $-\sqrt{3e}$
9. a. $\frac{1}{2}\cot\left(\frac{x}{2}\right)$ b. $\frac{-144x}{16x^4-81}$
10. a. $x^2(3\cos(4x)-4x\sin(4x))$
    b. $x^3e^{-3x}(4-3x)$
11. a. $-e^{-3x}(2\sin(2x)+3\cos(2x))$
    b. $-12e^{-4}$
12. a. $x^2(5x\cos(5x)+3\sin(5x))$
    b. $4x^3(\cos(4x)-x\sin(4x))$
13. a. $x^2e^{-4x}(3-4x)$
    b. $e^{-3x}(2\cos(2x)-3\sin(2x))$
14. a. $-\frac{9}{2x^4}(x\sin(3x)+\cos(3x))$
    b. $\frac{2x+9}{\sqrt{(4x+9)^3}}$
15. a. $\frac{-(1+2x)e^{-2x}}{3x^2}$ b. $\frac{-120x\left(3x^2+5\right)}{(3x^2-5)^3}$
16. a. $\frac{9}{2x^4}(x\cos(3x)-\sin(3x))$
    b. $-\frac{8}{3x^5}(2\cos(2x)+x\sin(2x))$
17. a. $\frac{e^{3x}(3x-2)}{x^3}$ b. $\frac{-(2x+3)}{x^4e^{2x}}$
18. a. $8\tan^3\left(\frac{x}{3}\right)\sec^2\left(\frac{x}{3}\right)$ b. $\frac{\pi^3(\pi+8)}{2}$
19. a. $\frac{27\left(4\pi-3\sqrt{3}\right)}{\pi^2}$ b. $-\frac{5\pi^2}{54}$
20. a. $-\frac{4}{13}$ b. $-\frac{1}{x}\sin\left(\log_e\left(\frac{x}{2}\right)\right)$
21. $2x\log_e(5x+4)+\frac{5x^2}{5x+4}$
22. a. $\frac{72}{16x^2-81}$ b. $\frac{60x}{25-9x^4}$
23. a. 3 b. $\frac{8\sqrt{3}}{3}$
24. a. $-\frac{16\pi^2\sqrt{3}}{9}$ b. $-\frac{2\pi^2}{81}$

25–35. Sample responses can be found in the worked solutions in the online resources.

36. a. Sample responses can be found in the worked solutions in the online resources.
    b. $e^{-4x}\left[\left(3x^2-4x^3\right)\cos(2x)-2x^3\sin(2x)\right]$

37, 38. Sample responses can be found in the worked solutions in the online resources.

5.2 Exam questions

Note: Mark allocations are available with the fully worked solutions online.

1. $\frac{dy}{dx}=\frac{2x}{2x^2+9}$
2. Sample responses can be found in the worked solutions in the online resources
3. Sample responses can be found in the worked solutions in the online resources.

5.3 Applications of differentiation

5.3 Exercise

1. $y=4x-17$
2. $y=\frac{2x}{3}+\frac{7}{3}$
3. a. $y=-4x+13$
   b. $y=-5x+\frac{5\pi}{3}+\frac{5\sqrt{3}}{2}$
4. a. $3y-x-5=0$
   b. $y=-6\sqrt{2}x+\frac{\sqrt{2}\pi}{2}+2\sqrt{2}$
5. $2y-x-28=0$
6. $12y-x-47=0$
7. -11
8. -21
9. a. -13 b. -6
10. a. $\frac{23}{4}$ b. -3
11. $y_T=\sqrt{3}x-\frac{\sqrt{3}\pi}{3}+2, y_N=-\frac{\sqrt{3}x}{3}-\frac{\sqrt{3}\pi}{9}+2$
12. $y=6x+1, 6y+x-6=0$
13. a. $6y-x-39=0$ b. $12y+x-12\sqrt{3}-\frac{\pi}{9}=0$
14. a. $4y+x-1=0$
    b. $12y-27x+77=0$

15. $4\pi r^2$

16. a. $\frac{2}{3}\pi rh$ b. $\frac{1}{3}\pi r^2$

17. a. Sample responses can be found in the worked solutions in the online resources.

b. $(0, 2)$, $\left(\frac{1}{2}\log_e\left(\frac{3}{5}\right), 0\right)$; $y = 5$ is a horizontal asymptote.

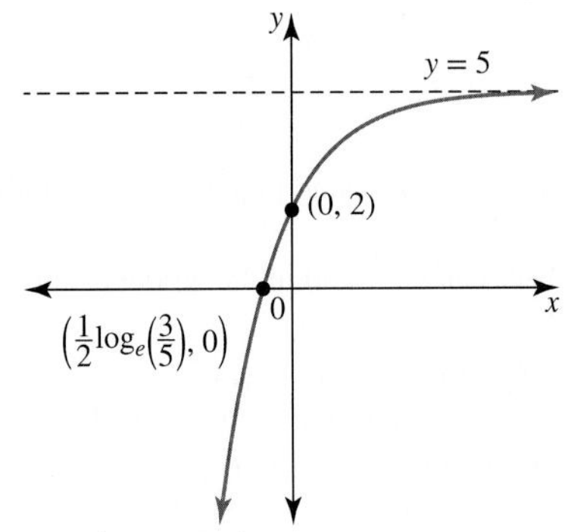

c. $y = \frac{6x}{e} + 5 - \frac{6}{e}$

18. a. Sample responses can be found in the worked solutions in the online resources.

b. $(0, 1)$, $\left(\frac{1}{2}\log_e\left(\frac{4}{3}\right), 0\right)$; $y = 4$ is a horizontal asymptote.

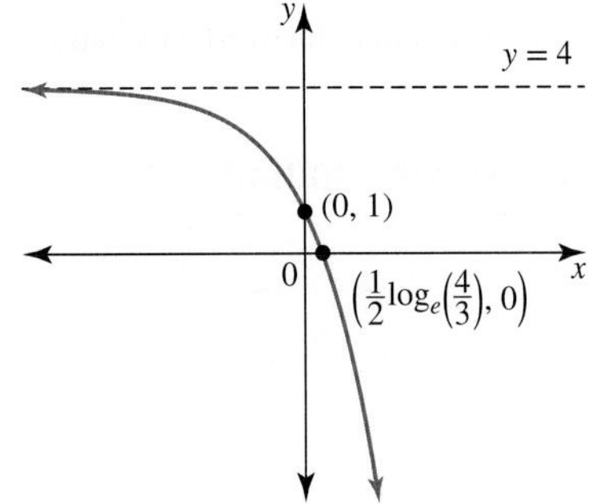

c. $y = \frac{x}{6e} - \frac{1}{12e} - 3e + 4$

19. a. Domain $\left(\frac{5}{4}, \infty\right)$, range R, $\left(\frac{3}{2}, 0\right)$; doesn't cross the y-axis;

$x = \frac{5}{4}$ is a vertical asymptote.

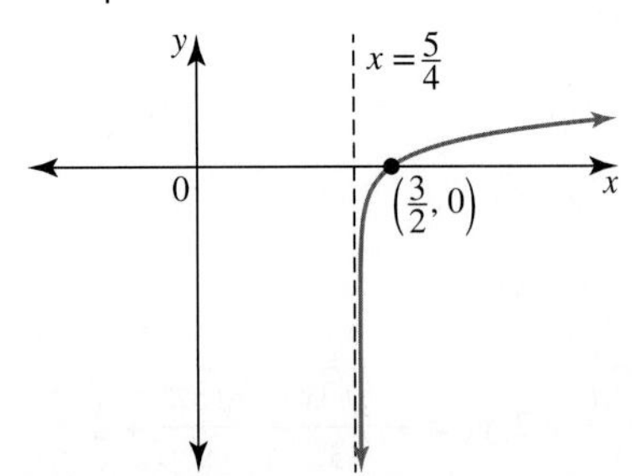

b. Sample responses can be found in the worked solutions in the online resources.

c. $y = 4x - 8 + \log_e(27)$

20. a. Domain $\left(-\infty, \frac{5}{2}\right)$, range R; intercepts $(2, 0)$, $\left(0, \log_e(25)\right)$;

$x = \frac{5}{2}$ is a vertical asymptote.

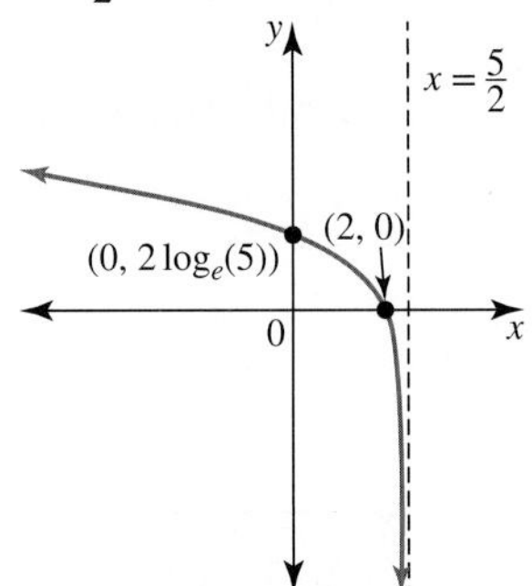

b. Sample responses can be found in the worked solutions in the online resources.

c. $y = \frac{3x}{4} - \frac{3}{4} + \log_e(9)$

21. a. 6 m

b.

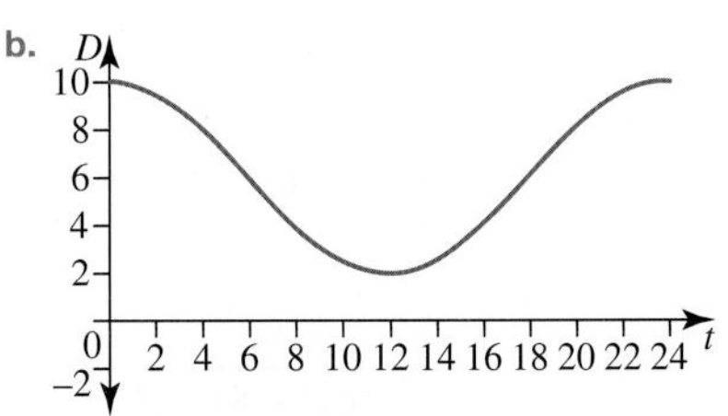

c. Between 4 am and 8 am

d. $-\frac{\sqrt{2}\pi}{6}$ m/s

e. $-\frac{2}{3}$ m/s

22. a. $N_0 = 500\,000$, $k = \frac{1}{10}\log_e\left(\frac{3}{2}\right)$

b. 918 559

c. 37 244.3

d. 39 583.3

e.

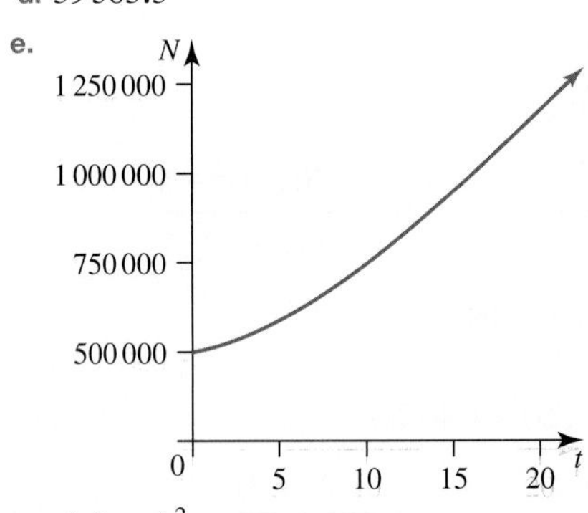

23. a. $\frac{\left(a^2 + k\right)^2}{4a}$

b. Sample responses can be found in the worked solutions in the online resources.

24. a. −463.875 mA/s

b. −462.048 mA/s

25. a. 9.31 mg
b. 9.1 mg/hour
c. 3 hours, 33.11 mg
d. 9.91 hours

5.3 Exam questions

Note: Mark allocations are available with the fully worked solutions online.

1. 2
2. a. $y = xe^{-2}$ and $y = 0$
   b. Sample responses can be found in the worked solutions in the online resources.
3. a. 400
   b. 63.6 ants per month
   c. 59.4 ants per month

5.4 Implicit and parametric differentiation

5.4 Exercise

1. $-\dfrac{x^2}{y^2}$
2. $-\sqrt{\dfrac{y}{x}}$
3. a. $\dfrac{1}{y}$
   b. $-\dfrac{9x}{4y}$
   c. $\dfrac{9x}{16y}$
   d. $\dfrac{3x-2}{y-1}$
4. a. $8x^2y\dfrac{dy}{dx} + 8xy^2$
   b. $\dfrac{27x^2 + 8xy^2 + 2}{5 + 9y^2 - 8x^2y}$
5. $-\dfrac{9}{11}$
6. a. $\dfrac{4x+3y}{6y-3x}$
   b. $\dfrac{3 - x - xy^2}{x^2y}$
   c. 1
   d. $\dfrac{2x+3}{y(3y-2)}$
7. $\dfrac{y - xe^{-\left(x^2+y^2\right)}}{ye^{-\left(x^2+y^2\right)} - x}$
8. $-\dfrac{3\cos\left(3x+2y\right) + 2x}{2\left(y + \cos\left(3x+2y\right)\right)}$
9. a. $\dfrac{27x^2 + 12y + 1}{y - 12x}$
   b. $\dfrac{4x^3 - 6x^2y - 3x^2 + 8xy^2}{24y^2 + 2x^3 - 8x^2y}$
   c. $\dfrac{e^{x+y}}{2y + \sin(y) - e^{x+y}}$
   d. $\dfrac{y\sin(xy) - ye^{xy} - 2x}{xe^{xy} - x\sin(xy)}$
10. a. $\dfrac{2 + 8x + 12y}{15y + 10x - 3}$
    b. $-\dfrac{y\left(2x^2+1\right)}{x\left(2y^2+1\right)}$
    c. $\dfrac{y^2\left(2x^2y - 1\right)}{x^2\left(1 - 2xy^2\right)}$
    d. $\dfrac{3\cos(3x)\cos(2y) - 2x}{6y^2 + 2\sin(3x)\sin(2y)}$
11. a. 1
    b. $x - 5y - 11 = 0$
    c. $-\dfrac{2}{7}$
    d. $\dfrac{8}{15}$
12. a. $3x - 4y - 25 = 0$
    b. $-\dfrac{\sqrt{3}}{2}$
    c. $\dfrac{4}{5}$
    d. $\dfrac{13}{9}$
13. a. $-\cot(t)$
    b. Sample responses can be found in the worked solutions in the online resources.
14. a. $\dfrac{1 - 2t^3}{t}$
    b. Sample responses can be found in the worked solutions in the online resources.
15. a. $\dfrac{1}{t}$
    b. $-\dfrac{4}{3}\tan(2t)$
    c. $-\dfrac{1}{t^2}$
    d. $\dfrac{4}{3}\text{cosec}(t)$
16. a. $\dfrac{1}{t}$
    b. $-\dfrac{b}{a}\cot(t)$
    c. $-\dfrac{1}{t^2}$
    d. $\dfrac{b}{a}\text{cosec}(t)$
17. a. $\dfrac{1}{t}$
    b. $-\dfrac{b}{a}\cot(t)$
    c. $-\dfrac{1}{t^2}$
    d. $\dfrac{b}{a}\text{cosec}(t)$
18. a. $\dfrac{x^2 - ay}{ax - y^2}$
    b. $\dfrac{t\left(t^3 - 2\right)}{2t^3 - 1}$
19. a. $-\dfrac{x\left(x^2 + y^2 - 1\right)}{y\left(x^2 + y^2 + 1\right)}$
    b. $\dfrac{2 - 3\cos^2(t)}{\sin(t)\left(2 + \cos^2(t)\right)}$
20. a. $\dfrac{y^2 + 3x^2}{2y(2a - x)}$
    b. $\dfrac{t\left(t^2 + 3\right)}{2}$
    c. $\dfrac{\sin(t)\left(2\cos^2(t) + 1\right)}{2\cos^3(t)}$

5.4 Exam questions

Note: Mark allocations are available with the fully worked solutions online.

1. $\dfrac{dy}{dx} = \dfrac{\pi - 2\sqrt{3}}{\sqrt{2}\left(\pi + \sqrt{3}\right)}$
2. $2y - x + 3 = 0$
3. E

5.5 Second derivatives

5.5 Exercise

1. $-\dfrac{1}{12}$
2. $-\sqrt{3}$
3. $\dfrac{1}{2}$

4. $-\frac{9}{2}$

5. $6x\log_e(2x^2+5)+\frac{20x^3(2x^2+7)}{(2x^2+5)^2}$

6. $3x^2(3x^2-8x+4)e^{-3x}$

7. $-\frac{16}{9}$

8. $6e$

9. a. $\frac{-2(x^2+4x-5)}{(x^2+4x+13)^2}$

b. $-e^{3x}(7\cos(4x)+24\sin(4x))$

c. $2xe^{-2x}(2x^2-6x+3)$

10. a. $(2-9x^2)\cos(3x)-12x\sin(3x)$

b. $\frac{12(3x+7)}{(6x+7)^2}$

c. $\frac{-x}{\sqrt{(x^2+16)^3}}$

11–12. Sample responses can be found in the worked solutions in the online resources.

13. a. $-\frac{4a^2}{y^3}$ b. $-\frac{b^4}{a^2y^3}$ c. $\frac{2a^2}{x^3}$

14. $\frac{-1}{4\sin^3(t)}$

15. $\frac{-(1+4t^3)}{2t^3}, -\frac{33}{16}$

16. a. $-\frac{1}{2at^3}$ b. $\frac{2}{at^3}$ c. $\frac{-b}{a^2\sin^3(t)}$

17–21. Sample responses can be found in the worked solutions in the online resources.

22. a. $-\sqrt[3]{\frac{y}{x}}$

b. $\frac{dy}{dx}=-\tan(\theta)$ $\frac{d^2y}{dx^2}=\frac{1}{3a\cos^4(\theta)\sin(\theta)}$

23. a. $-e^{-2x}(5\sin(3x)+12\cos(3x))$

b. Sample responses can be found in the worked solutions in the online resources.

24–30. Sample responses can be found in the worked solutions in the online resources.

5.5 Exam questions

Note: Mark allocations are available with the fully worked solutions online.

1. B
2. Sample responses can be found in the worked solutions in the online resources.
3. Sample responses can be found in the worked solutions in the online resources.

5.6 Derivatives of inverse trigonometric functions

5.6 Exercise

1. a. $\frac{1}{\sqrt{25-x^2}}, x\in(-5,5)$

b. $\frac{5}{\sqrt{1-25x^2}}, x\in\left(-\frac{1}{5},\frac{1}{5}\right)$

2. $\frac{8\sqrt{3}}{3}$

3. a. $\frac{1}{\sqrt{9-x^2}}, |x|<3$

b. $\frac{3}{\sqrt{1-9x^2}}, |x|<\frac{1}{3}$

4. a. $\frac{4}{\sqrt{9-16x^2}}, |x|<\frac{3}{4}$

b. $\frac{\sqrt{2}}{\sqrt{(x+2)(1-2x)}}, -2<x<\frac{1}{2}$

5. a. $\frac{-3}{\sqrt{16-9x^2}}, x\in\left(-\frac{4}{3},\frac{4}{3}\right)$

b. $\frac{-1}{\sqrt{(x+1)(4-x)}}, x\in(-1,4)$

6. $-\frac{3}{5}$

7. a. $\frac{-1}{\sqrt{16-x^2}}, |x|<4$

b. $\frac{-4}{\sqrt{1-16x^2}}, |x|<\frac{1}{4}$

8. a. $\frac{-3}{\sqrt{16-9x^2}}, |x|<\frac{4}{3}$

b. $\frac{-\sqrt{3}}{\sqrt{(x+4)(2-3x)}}, -4<x<\frac{2}{3}$

9. a. $\frac{4}{1+16x^2}$

b. $\frac{5}{2x^2-6x+17}$

10. $\frac{3}{14}$

11. a. $\frac{6}{x^2+36}$

b. $\frac{6}{36x^2+1}$

12. a. $\frac{30}{25x^2+36}$

b. $\frac{24}{36x^2+60x+41}$

13. $\dfrac{-8x}{\sqrt{(9-4x^2)^3}}$

14. $\dfrac{-384x}{(9+16x^2)^2}$

15. $\dfrac{-2}{|x|\sqrt{x^2-4}}, |x|>2$

16. $\dfrac{-2}{\sqrt{x}(x+16)}, x>0$

17. a. $\dfrac{1}{2\sqrt{x(9-x)}}, 0<x<9$

b. $\dfrac{-3}{|x|\sqrt{16x^2-9}}, |x|>\dfrac{3}{4}$

c. $\dfrac{-2e^{2x}}{\sqrt{16-e^{4x}}}, x<\log_e(2)$

18. a. $\dfrac{4}{|x|\sqrt{9x^2-16}}, |x|>\dfrac{4}{3}$

b. $\dfrac{6x}{x^4+9}$

c. $\dfrac{-30}{25x^2+36}, x \neq 0$

19. a. $\dfrac{125x}{(16-25x^2)^{\frac{3}{2}}}, |x|<\dfrac{4}{5}$

b. $\dfrac{-216x}{(25-36x^2)^{\frac{3}{2}}}, |x|<\dfrac{5}{6}$

c. $\dfrac{-4116x}{(49x^2+36)^2}$

5.6 Exam questions

Note: Mark allocations are available with the fully worked solutions online.

1. Sample responses can be found in the worked solutions in the online resources.
2. Sample responses can be found in the worked solutions in the online resources.
3. A

5.7 Related rates

5.7 Exercise

1. 20π m^2/s
2. 2 cm/s
3. a. 16 cm^2/s   b. $\dfrac{1}{4\pi}$ m/s
4. a. 800π mm^3/s   b. 160π mm^2/s
5. Decreasing at $\dfrac{1}{24\pi}$ cm/s
6. Decreasing at 48π cm^2/s
7. a. Sample responses can be found in the worked solutions in the online resources.
   b. 150 weeks
8. Decreasing at $\dfrac{75}{128\pi}$ cm/s
9. Decreasing at 64π cm^3/s
10. a. 6 cm^2/s   b. $\sqrt{3}$ cm/s
11. a. Sample responses can be found in the worked solutions in the online resources.
    b. 450π cm^3/s
12. 0.225 m/s
13. $\dfrac{8}{5}$ m/s
14. a. $\dfrac{1}{25\pi}$ cm/s   b. $\dfrac{27}{16\pi}$ cm/s
15. a. 26.13 m/min   b. -12 m/s
16. a. Sample responses can be found in the worked solutions in the online resources.
    b. Decreases by 5 cm/s
17. a. Decreases by 31 562.5 Pa/s
    b. Sample responses can be found in the worked solutions in the online resources.
18. 0.005°/s
19. -2.1°/s
20. a. 0.75 m/s   b. 2.25 m/s   c. 1.3 m/s
21. a. Sample responses can be found in the worked solutions in the online resources.
    b. $\dfrac{220\sqrt{7}}{7}$ km/h

5.7 Exam questions

Note: Mark allocations are available with the fully worked solutions online.

1. C
2. $\dfrac{3\pi}{2}$ mm^2/day
3. a. Sample responses can be found in the worked solutions in the online resources.
   b. 0.96 m/min

5.8 Review

5.8 Exercise

Technology free: short answer

1. a. $x^4e^{-3x}(5-3x)$
   b. $-\dfrac{1}{x^4}(2x\sin(2x)+3\cos(2x))$
   c. $\dfrac{5}{\sqrt{(3x^2+5)^3}}$
2. a. $\dfrac{64x}{\sqrt{(25-16x^2)^3}}$
   b. $e^{-3x}(5\sin(2x)-12\cos(2x))$
   c. $\left(6x-4x^3\right)\cos(2x)-12x^2\sin(2x)$

3. a. $y=-\frac{3\sqrt{3}x}{3}+\frac{\sqrt{3}\pi+18}{6}$

b. $y=-\frac{3x}{8}+\frac{3\pi}{16}+\sqrt{3}$

c. $y=-2\sqrt{3}x+2\sqrt{3}+\frac{\pi}{6}$

4. a. $\frac{2x\left(x-2y^2\right)}{y\left(4x^2-3y\right)}$

b. $\frac{16x^3-7y}{7x+12y^3}$

c. $16y+5x+6=0$

d. $10y+9x-28=0$

5. 5 m/h

6. a. 288π cm^3/s b. 96π cm^2/s

Technology active: multiple choice

7. B
8. D
9. B
10. E
11. C
12. A
13. C
14. A
15. D
16. E

Technology active: extended response

17–22. Sample responses can be found in the worked solutions in the online resources.

5.8 Exam questions

Note: Mark allocations are available with the fully worked solutions online.

1. $-\frac{1}{10}$

2. $y=-\frac{2x}{\pi}-\frac{\pi}{2}$

3. a. $\frac{dy}{dx}=\frac{2x-y}{x-3y}$

b. $T_A: y=2x-6,\ T_B: y=\frac{x}{3}+\sqrt{6}$

c. $\theta=\frac{\pi}{4}$

4. $c=-\frac{3}{4}$

5. Sample responses can be found in the worked solutions in the online resources.

6 Functions and graphs

LEARNING SEQUENCE

Fully worked solutions for this topic are available online.

6.1 Overview

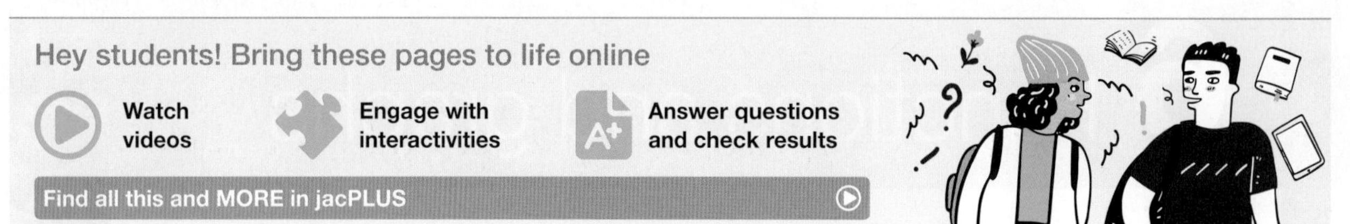

6.1.1 Introduction

Functions define special relationships between two variables — specifically, relationships where for every input value (x) there is one output value (y). Functions were first defined in the late 17th century by Leibniz as any quantity varying along a curve that is described by an equation. In 1734, Euler introduced the familiar $f(x)$ notation.

Today functions are used to model relationships between variables across a variety of disciplines, from forensic science and computer programming to climate science. For example, climate scientists use data about sea level, temperature, carbon dioxide and Arctic ice extent to define functions that allow them to create the complex models required to predict the extent of global climate change.

If we understand the relationship between two variables, it is possible to model their relationship. Knowing what a function looks like helps us understand why functions behave the way that they do. In this section, you will explore the graphing of modulus functions, reciprocal functions and rational functions.

KEY CONCEPTS

This topic covers the following points from the VCE Mathematics Study Design:

- rational functions and the expression of rational functions of low degree as sums of partial fractions
- graphs of rational functions of low degree, their asymptotic behaviour, and the nature and location of stationary points and points of inflection
- graphs of simple quotient functions, their asymptotic behaviour, and the nature and location of stationary points and points of inflection.

Source: VCE Mathematics Study Design (2023–2027) extracts © VCAA; reproduced by permission.

6.2 Sketching graphs of cubics and quartics

LEARNING INTENTION

At the end of this subtopic you should be able to:

- sketch graphs of cubic and quartic functions using calculus to determine stationary points and points of inflection.

6.2.1 Sketching graphs using key points

Throughout this topic we will be sketching graphs of all kinds of functions, the shapes of which may not be intuitively obvious to you such as they have been in the past when sketching basic functions. To sketch more complex graphs, we need a systematic method to apply to any function, $y = f(x)$, we come across.

In general, the first step to sketching a graph is to determine any points of interest such as axis intercepts and critical points. We determine these points of interest by investigating the function, its first derivative and its second derivative.

Axis intercepts

As you will recall from previous years, the y-intercept occurs when $x = 0$ and the x-intercepts occur when $y = 0$. Given an equation, substituting $x = 0$ and solving for y will determine the y-intercept, and substituting $y = 0$ and solving for x will determine the x-intercepts.

Stationary points

A **stationary point** on a curve is defined as a point where the gradient is zero; that is, $\frac{dy}{dx} = f'(x) = 0$. There are three types of stationary points: **maximum turning points, minimum turning points** and **stationary points of inflection.**

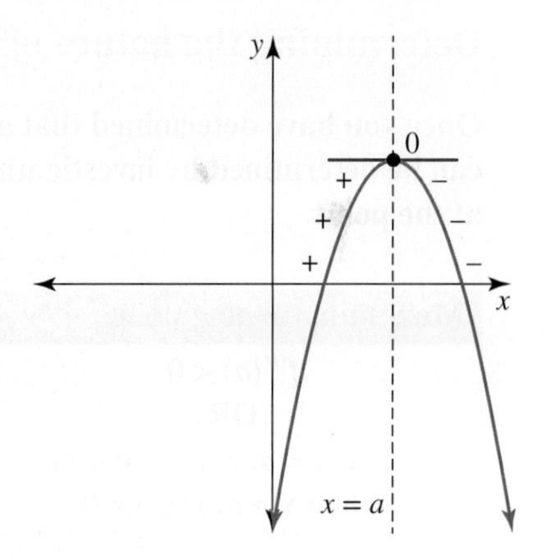

A maximum turning point is a point on the curve at which the y-coordinate has its highest value within a certain interval. Visually, a maximum turning point is like the top of a hill. At a maximum turning point the gradient of the curve changes from positive to zero to negative as x increases. The rate of change of the gradient (second derivative) is therefore negative. If there are higher y-values outside the immediate neighbourhood of this maximum, it is called a local maximum. If it is the highest y-value on the whole domain, it is called an absolute maximum. Absolute maximums can also be non-stationary points, in which case they are endpoints, so long as the graph is continuous.

A minimum turning point is a point on the curve at which the y-coordinate has its lowest value within a certain interval. Visually, a minimum turning point is like the bottom of a valley. At a minimum turning point the gradient of the curve changes from negative to zero to positive as x increases. The rate of change of the gradient (second derivative) is therefore positive. If there are lower y-values outside the immediate neighbourhood of this minimum, it is called a local minimum. If it is the lowest y-value on the whole domain, it is called an absolute minimum. Absolute minimums can also be non-stationary points, in which case they are endpoints, so long as the graph is continuous.

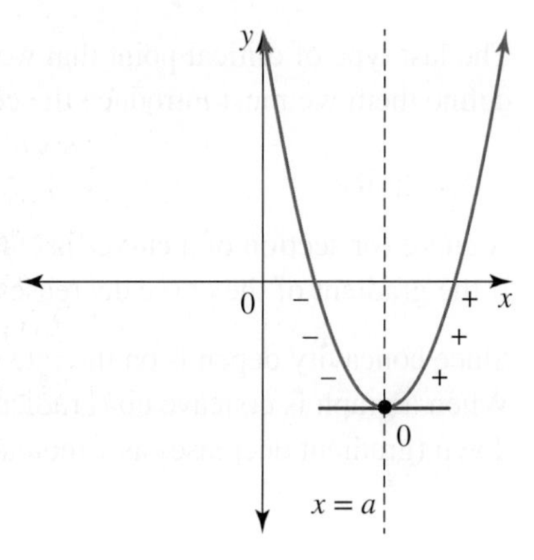

Stationary points of inflection are points on a curve where the gradient is zero, but the gradient is the same sign on both sides of this point. This type of stationary point is not a turning point, as the graph continues to increase (or decrease) on both sides of the stationary point. At a stationary point of inflection, the gradient function is also stationary, which means that the second derivative is equal to zero.

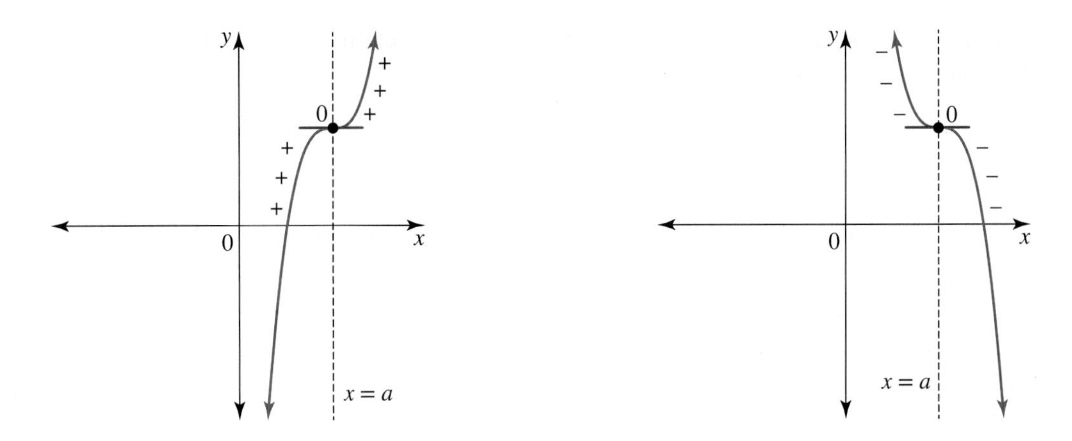

To determine the nature of a stationary point you can evaluate the gradient on either side of the stationary point, or determine the value of the second derivative at the stationary point.

If the gradient changes sign either side of the stationary point, then it will be a maximum or minimum turning point. If the sign of the gradient is the same either side of the stationary point, then it will be a stationary point of inflection.

Determining the nature of stationary points

Once you have determined that a stationary point occurs at $x = a$, the nature of the stationary point can be determined by investigating the gradient either side of it, or the value of the second derivative at the point.

Maximum turning point	Minimum turning point	Stationary point of inflection
$f''(a) < 0$ **OR** **when** $x < a, f'(x) > 0$ **and when** $x > a, f'(x) < 0$	$f''(a) > 0$ **OR** **when** $x < a, f'(x) < 0$ **and when** $x > a, f'(x) > 0$	$f''(a) = 0$ **AND** **when** $x < a$ **and** $x > a, f'(x) < 0$ **or** $f'(x) > 0$

6.2.2 Concavity and points of inflection

The last type of critical point that we will be using to sketch graphs are points of inflection, however, before we define them we must introduce the concept of concavity.

Concavity

A curve (or section of a curve) is classified as **concave up** if the gradient of the curve increases as x increases. If the gradient of the curve decreases as x increases, the curve (or section) is classified as **concave down**.

Since concavity depends on the rate of change of the gradient, it is determined by the second derivative $f''(x)$. When a graph is concave up (gradient increases as x increases), $f''(x)$ is positive, and when a graph is concave down (gradient decreases as x increases), $f''(x)$ is negative.

To test whether a curve is concave up or concave down at a point $x = a$, draw a tangent to the curve at that point and look at whether it sits above the curve, or below the curve. If the tangent sits below the curve the graph is concave up at that point, and if the tangent sits above the curve the graph is concave down at that point.

The graphs of $y = x^2$ and $y = -x^2$ are concave up and concave down respectively. Notice that the tangent at any point on $y = x^2$ sits below the graph, and the tangent at any point on $y = -x^2$ sits above the graph.

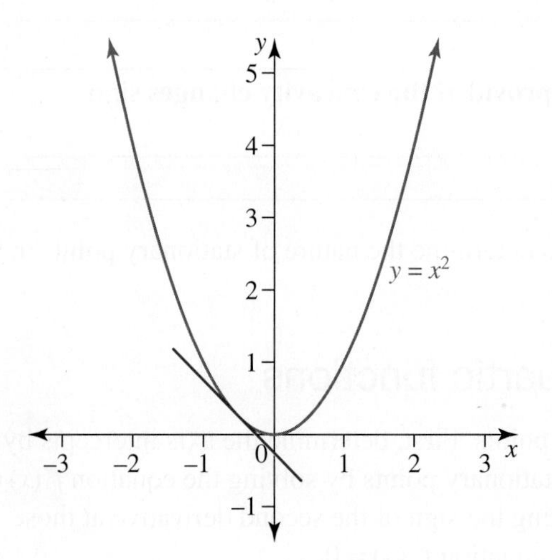

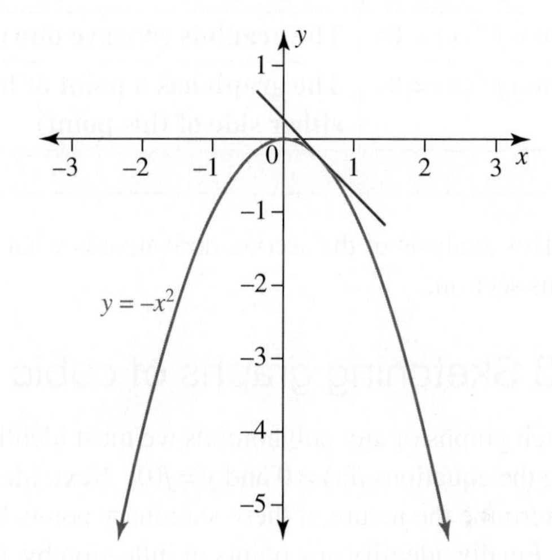

Points of inflection

Some curves are made up of sections which are concave up and sections which are concave down. The point at which the curve changes from concave up to concave down (or vice versa) is known as a **point of inflection.** Points of inflection occur when the second derivative is equal to zero and changes sign on either side of the point. At points of inflection the curve is neither concave up or concave down and the tangent to the graph sits above the curve on one side of the point and below the curve on the other side.

For example, the graph of $y = \frac{x^3}{3} - x^2 + 2$ has a point of inflection at $x = 1$. The graph is concave down when $x < 1$ and concave up when $x > 1$.

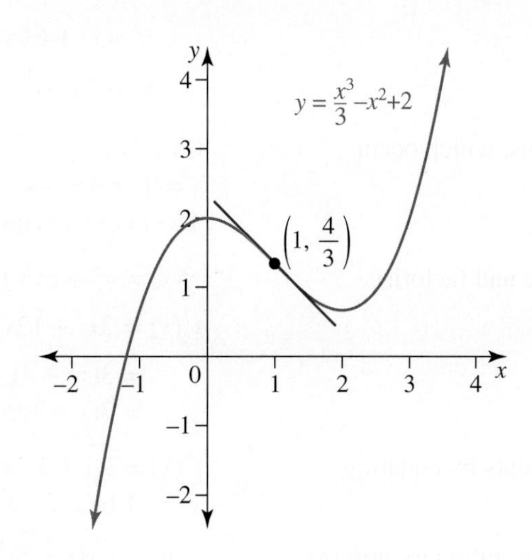

In summary, concavity and points of inflection depend on the second derivative $f''(x)$.

Concavity and points of inflection

Case	Result
When $f''(x) > 0$:	**The graph is concave up**
When $f''(x) < 0$:	**The graph is concave down**
When $f''(x) = 0$:	**The graph has a point of inflection (provided the concavity changes sign either side of this point)**

Note: This analysis of the second derivative is what we used to determine the nature of stationary points in the previous section.

6.2.3 Sketching graphs of cubic and quartic functions

To sketch graphs of any polynomials we must identify all key points. First, determine the axis intercepts by solving the equations $f(x) = 0$ and $y = f(0)$. Next, identify the stationary points by solving the equation $f'(x) = 0$ and determine the nature of these stationary points by calculating the sign of the second derivative at those points. Finally, identify any points of inflection by solving the equation $f''(x) = 0$.

Cubic functions are polynomials of degree 3 and are of the form $y = ax^3 + bx^2 + cx + d,\ a \neq 0$. They can have up to three x-intercepts, two stationary points and one point of inflection.

WORKED EXAMPLE 1 Sketching graphs of cubic functions

Sketch the graph of the function $f: R \to R, f(x) = x^3 + 6x^2 + 9x$ by determining the coordinates of all axis intercepts and stationary points and establishing their nature. Also, calculate the coordinates of the point of inflection, and determine and draw the tangent to the curve at the point of inflection.

THINK	WRITE/DRAW
1. Factorise the function.	$f(x) = x^3 + 6x^2 + 9x$ $= x(x^2 + 6x + 9)$ $= x(x+3)^2$
2. Determine the axis intercepts, which occur when $y = 0$ or $x = 0$.	$0 = x(x+3)^2$ $x = 0$ or $x = -3$ The intercepts are $(0, 0)$ and $(-3, 0)$.
3. Calculate the first derivative and factorise.	$f(x) = x^3 + 6x^3 + 9x$ $f'(x) = 3x^2 + 12x + 9$ $= 3(x^2 + 4x + 3)$ $= 3(x+3)(x+1)$
4. Determine the stationary points by equating the first derivative to zero.	$f'(x) = 3(x+3)(x+1) = 0$ $x = -1$ or $x = -3$
5. Calculate the y-values of the stationary points.	$f(x) = x(x+3)^2$ $f(-1) = -1(-1+3)^2 = -4$ $f(-3) = 0$ The stationary points are $(-3, 0)$ and $(-1, -4)$.

6. Calculate the second derivative.

$f'(x) = 3x^2 + 12x + 9$
$f''(x) = 6x + 12$
$= 6(x + 2)$

7. Test the two stationary points by using the second derivative to determine their nature.

When $x = -3, f''(-3) = -6 < 0$.
The point $(-3, 0)$ is a local maximum.
When $x = -1, f''(-1) = 6 > 0$.
The point $(-1, 4)$ is a local minimum.

8. Determine the point of inflection by equating the second derivative to zero and calculate the y-value.

$f''(x) = 6(x + 2) = 0$ when $x = -2$.
$f(-2) = -2(-2 + 3)^2 = -2$
The point $(-2, -2)$ is the point of inflection.

9. Determine the gradient at the point of inflection.

$f'(-2) = 3(-2)^2 + 12 \times -2 + 9$
$= -3$

10. Determine the equation of the tangent at the point of inflection.

Use $y - y_1 = m(x - x_1)$:
$P(-2, -2), m = -3$
$y + 2 = -3(x + 2)$
$y + 2 = -3x - 6$
$y = -3x - 8$

11. Sketch the graph and the tangent using an appropriate scale. We see that the tangent crosses the curve at the point of inflection.

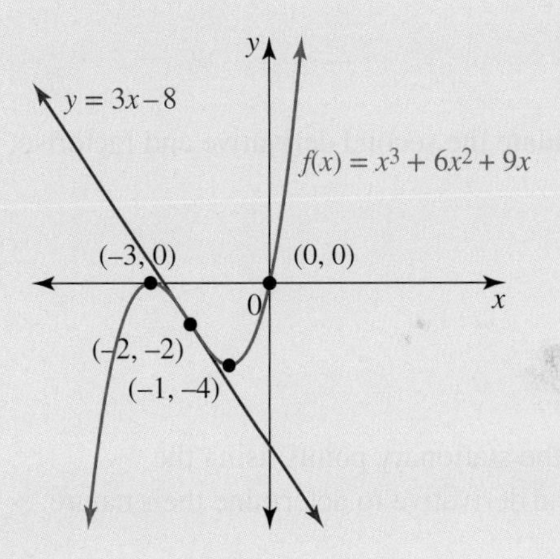

Quartic functions are polynomials of degree 4 and are of the form $y = ax^4 + bx^3 + cx^2 + dx + e,\ a \neq 0$. They can have up to four x-intercepts, three stationary points and two points of inflection.

WORKED EXAMPLE 2 Sketching graphs of quartic functions

Sketch the graph of $y = 6x^2 - x^4$ by determining the coordinates of all axis intercepts, stationary points and points of inflection, and establishing their nature.

THINK | **WRITE/DRAW**

1. Factorise.

$y = 6x^2 - x^4$
$= x^2(6 - x^2)$
$= x^2\left(\sqrt{6} - x\right)\left(\sqrt{6} + x\right)$

2. Determine the axis intercepts.

The graph crosses the x-axis when $y = 0$, at $x = 0$ or $x = \pm\sqrt{6}$.

The x-intercepts are $(0, 0)$, $\left(\sqrt{6}, 0\right)$, $\left(-\sqrt{6}, 0\right)$.

$\sqrt{6} \approx 2.45$

3. Calculate the first derivative and factorise.

$$y = 6x^2 - x^4$$
$$\begin{aligned}\frac{dy}{dx} &= 12x - 4x^3 \\ &= 4x(3 - x^2) \\ &= 4x\left(\sqrt{3} - x\right)\left(\sqrt{3} + x\right)\end{aligned}$$

4. Determine the stationary points by equating the first derivative to zero.

$$\frac{dy}{dx} = 4x\left(\sqrt{3} - x\right)\left(\sqrt{3} + x\right) = 0$$
$x = 0$ or $x = \pm\sqrt{3}$.

5. Calculate the y-values at the stationary points.

When $x = \sqrt{3}$, $y = 6\left(\sqrt{3}\right)^2 - \left(\sqrt{3}\right)^4 = 9$.

When $x = -\sqrt{3}$, $y = 6\left(-\sqrt{3}\right)^2 - \left(-\sqrt{3}\right)^4 = 9$.

The stationary points are $(0, 0)$, $\left(\sqrt{3}, 9\right)$ and $\left(-\sqrt{3}, 9\right)$.

6. Calculate the second derivative and factorise.

$$\frac{dy}{dx} = 12x - 4x^3$$
$$\begin{aligned}\frac{d^2y}{dx^2} &= 12 - 12x^2 \\ &= 12(1 - x^2) \\ &= 12(1 - x)(1 + x)\end{aligned}$$

7. Test the stationary points using the second derivative to determine their nature.

When $x = 0$, $\dfrac{d^2y}{dx^2} = 12 > 0$.

The point $(0, 0)$ is a local minimum. There are y-values lower than zero in the range of the function.

When $x = \sqrt{3}$, $\dfrac{d^2y}{dx^2} = -24 < 0$.

The point $\left(\sqrt{3}, 9\right)$ is an absolute maximum.

When $x = -\sqrt{3}$, $\dfrac{d^2y}{dx^2} = -24 < 0$.

The point $\left(\sqrt{3}, 9\right)$ is an absolute maximum.

8. Determine the points of inflection by equating the second derivative to zero and calculate the y-value.

$$\frac{d^2y}{dx^2} = 12(1 - x)(1 + x) = 0$$
$x = \pm 1$

When $x = \pm 1$, $y = 5$.

The points $(-1, 5)$ and $(1, 5)$ are both points of inflection.

9. List any other features about the graph.

Let $f(x) = 6x^2 - x^4$, then $f(-x) = f(x)$, so the graph is an even function and the graph is symmetrical about the y-axis.

10. Sketch the graph, showing all the critical points using an appropriate scale.

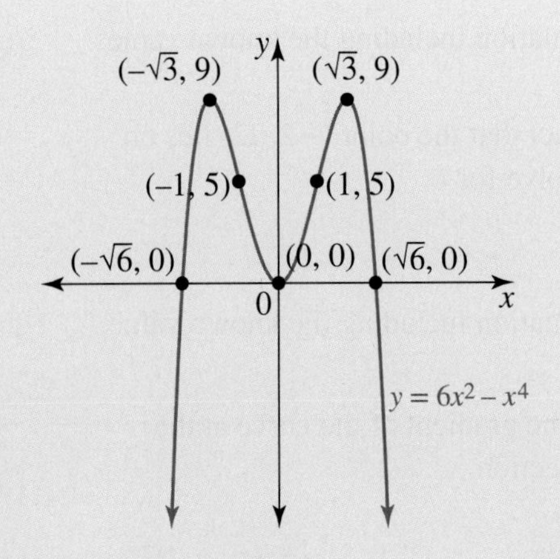

TI | THINK	DISPLAY/WRITE	CASIO | THINK	DISPLAY/WRITE
On a Graphs page, type in the equation of the function and click enter. Select Menu: 6 Analyse Graph: 1 Zero to find the x-intercepts. Select Menu: 6 Analyse Graph: 3 Maximum to find the maximum turning points. Select Menu: 6 Analyse Graph: 5 Inflection to find the points of inflection.	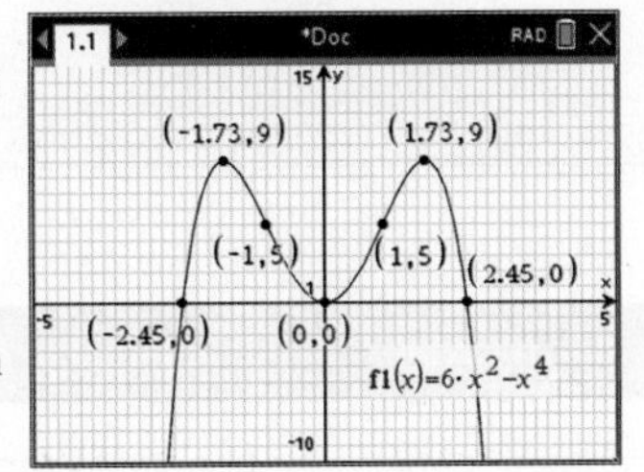	On a Graph Editor screen, type in the equation of the function. Select the graph screen to draw the image. Resize if needed. Select Analysis>G-Solve>Root to find the left-most x-intercept. Cursor right to find the others. Select Analysis>G-Solve>Max to find the left-most maximum turning point. Cursor right to find the others. Select Analysis>G-Solve> Inflection to find the leftmost point of inflection. Cursor right to find the others.	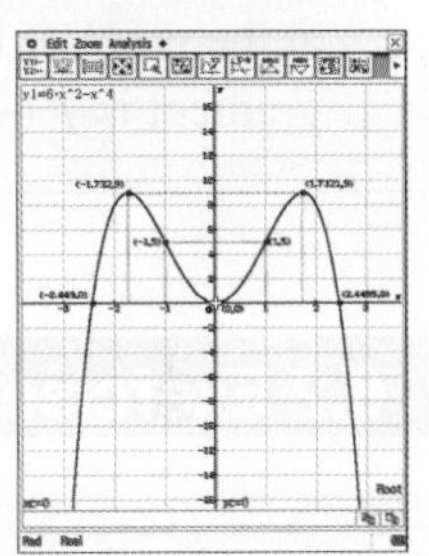

WORKED EXAMPLE 3 The equation of the tangent to a curve at the point of inflection

A cubic polynomial $y = x^3 + bx^2 + cx + d$ crosses the y-axis at $y = -10$ and has a point of inflection at $(-2, 22)$. Determine the equation of the tangent at the point of inflection.

THINK	WRITE
1. Let $y = f(x) = x^3 + bx^2 + cx + d$. The graph crosses the y-axis when $x = 0$, solve for d.	$f(0) = -10$ $d = -10$
2. Calculate the first derivative	$\frac{dy}{dx} = f'(x) = 3x^2 + 2bx + c$
3. Calculate the second derivative	$\frac{d^2y}{dx^2} = f''(x) = 6x + 2b$
4. At the point of inflection the second derivative is zero. Solve for b.	$f''(-2) = 0$ $-12 + 2b = 0$ $b = 6$

5. State the equation including the known values of b and d.	$f(x) = x^3 + 6x^2 + cx - 10$
6. Using the fact that the point $(-2, 22)$ lies on the curve, solve for c.	$f(-2) = 22$ $-8 + 24 - 2c - 10 = 22$ $2c = -16$ $c = -8$
7. State the equation including the known value of c.	$y = f(x) = x^3 + 6x^2 - 8x - 10$
8. Determine the gradient of the curve at the point of inflection.	$f'(x) = 3x^2 + 12x - 8$ $f'(-2) = 12 - 24 - 8$ $= -20$
9. Determine the equation of the tangent to curve at the point of inflection $(-2, 22)$, where the gradient is $f'(-2) = -20$.	$y - 22 = -20(x + 2)$ $y = -20x - 18$

6.2 Exercise

Technology free

1. WE1 Sketch the graph of $y = x^3 - 4x^2 + 4x$ by determining the coordinates of all axis intercepts and stationary points, and establishing their nature. Calculate the coordinates of the point of inflection. Determine and draw the tangent to the curve at the point of inflection.
2. Sketch the graphs of each of the following by determining the coordinates of all axis intercepts and any stationary points, and establishing their nature. Also calculate the coordinates of the point(s) of inflection.
   a. $y = x^3 - 27x$
   b. $y = 9x - x^3$
3. Sketch the graphs of each of the following by determining the coordinates of all axis intercepts and any stationary points, and establishing their nature. Also calculate the coordinates of the point(s) of inflection.
   a. $y = x^3 + 12x^2 + 36x$
   b. $y = -x^3 + 10x^2 - 25x$
4. Sketch the graphs of each of the following by determining the coordinates of all axis intercepts and any stationary points, and establishing their nature. Also calculate the coordinates of the point(s) of inflection.
   a. $y = x^3 - 3x^2 - 9x - 5$
   b. $y = -x^3 + 9x^2 - 15x - 25$
5. Sketch the graphs of each of the following by determining the coordinates of all axis intercepts and any stationary points, and establishing their nature. Also calculate the coordinates of the point(s) of inflection.
   a. $y = x^3 - x^2 - 16x + 16$
   b. $y = -x^3 - 5x^2 + 8x + 12$
6. WE2 Sketch the graph of $y = x^4 - 24x^2 + 80$ by determining the coordinates of all axis intercepts and stationary points, and establishing their nature. Calculate the coordinates of the point of inflection.

7. Sketch the graphs of each of the following by determining the coordinates of all axis intercepts and any stationary points, and establishing their nature. Also calculate the coordinates of the point(s) of inflection.
   a. $y = x^4 - 4x^3$
   b. $y = 4x^2 - x^4$

8. Sketch the graphs of each of the following by determining the coordinates of all axis intercepts and any stationary points, and establishing their nature. Also calculate the coordinates of the point(s) of inflection.
   a. $y = x^4 + 4x^3 - 16x - 16 = (x-2)(x+2)^3$
   b. $y = x^4 - 6x^2 + 8x - 3 = (x-1)^3(x+3)$

9. **WE3** A cubic polynomial $y = x^3 + bx^2 + cx + d$ crosses the y-axis at $y = 5$ and has a point of inflection at $(1, -21)$. Determine the equation of the tangent at the point of inflection.

10. The function $f(x) = x^3 + bx^2 + cx + d$ has a stationary point of inflection at $(1, -2)$. Determine the values of b, c and d.

11. The function $f(x) = x^3 + bx^2 + cx + d$ crosses the x-axis at $x = 3$ and has a point of inflection at $(2, -4)$. Determine the values of b, c and d.

12. A cubic polynomial $y = ax^3 + bx^2 + cx$ has a point of inflection at $x = -2$. The tangent at the point of inflection has the equation $y = 21x + 8$. Determine the values of a, b and c.

13. The graph of $y = ax^4 + bx^2 + c$ has one stationary point at $(0, -8)$ and points of inflection at $\left(\pm\sqrt{2}, 12\right)$. Determine the values of a, b and c. Find and determine the nature of any other stationary points.

14. Show that the graph of $y = x^3 - 2ax^2 + a^2x$, where $a \in R\backslash\{0\}$, crosses the x-axis at $(a, 0)$ and $(0, 0)$, has turning points at $(a, 0)$ and $\left(\frac{a}{3}, \frac{4a^3}{27}\right)$, and has an inflection point at $\left(\frac{2a}{3}, \frac{2a^3}{27}\right)$. Show that the equation of the tangent to the curve at the point of inflection is given by $y = \frac{8a^3}{27} - \frac{a^2x}{3}$.

15. a. Show that the graph of $y = x^3 - a^2x$, where $a \in R\backslash\{0\}$, crosses the x-axis at $(\pm a, 0)$ and $(0, 0)$ and has turning points at $\left(\frac{\sqrt{3}a}{3}, -\frac{2a^3\sqrt{3}}{9}\right)$ and $\left(-\frac{\sqrt{3}a}{3}, \frac{2a^3\sqrt{3}}{9}\right)$. Show that $(0, 0)$ is also an inflection point.
    b. Show that the graph of $y = (x-a)^2(x-b)$, where $a, b \in R\backslash\{0\}$, has turning points at $(a, 0)$ and $\left(\frac{a+2b}{3}, \frac{4(a-b)^3}{27}\right)$ and an inflection point at $\left(\frac{2a+b}{3}, \frac{2(a-b)^3}{27}\right)$.

16. Show that the graph of $y = (x-a)^3(x-b)$, where $a, b \in R\backslash\{0\}$, has a stationary point of inflection at $(a, 0)$, a turning point at $\left(\frac{a+3b}{4}, \frac{-27(a-b)^4}{256}\right)$ and an inflection point at $\left(\frac{a+b}{2}, \frac{-(a-b)^4}{16}\right)$.

6.2 Exam questions

Question 1 (1 mark) TECH-ACTIVE

Source: VCE 2021 Specialist Mathematics Exam 2, Section A, Q9; © VCAA.

MC Which one of the following derivatives corresponds to a graph of f that has no points of inflection?

A. $f'(x) = 2(x-2)^2 + 5$
B. $f'(x) = 2(x-3)^3 + 5$
C. $f'(x) = \frac{5}{2}(x-3)^2$
D. $f'(x) = \frac{1}{2}(x-3)^3 - 5$
E. $f'(x) = (x-3)^3 - 12x$

Question 2 (1 mark) TECH-ACTIVE

Source: *VCE 2017 Specialist Mathematics Exam 2, Section A, Q10; © VCAA.*

MC A function f, its derivative f' and its second derivative f'' are defined for $x \in R$ with the following properties.

$$f(a) = 1, f(-a) = -1$$
$$f(b) = -1, f(-b) = 1$$
$$\text{and } f''(x) = \frac{(x+a)^2(x-b)}{g(x)}, \text{ where } g(x) < 0$$

The coordinates of any points of inflection of $|f(x)|$ are

A. $(-a, 1)$ and $(b, 1)$
B. $(b, -1)$
C. $(-a, -1)$ and $(b, -1)$
D. $(-a, 1)$
E. $(b, 1)$

Question 3 (1 mark) TECH-ACTIVE

Source: *VCE 2017 Specialist Mathematics Exam 2, Section A, Q8; © VCAA.*

MC Let $f(x) = x^3 - mx^2 + 4$, where $m, x \in R$.

The **gradient** of f will always be strictly increasing for

A. $x \geq 0$
B. $x \geq \frac{m}{3}$
C. $x \leq \frac{m}{3}$
D. $x \geq \frac{2m}{3}$
E. $x \leq \frac{2m}{3}$

More exam questions are available online.

6.3 Sketching graphs of rational functions

LEARNING INTENTION

At the end of this subtopic you should be able to:

- sketch graphs of rational functions showing asymptotes and axis intercepts.

6.3.1 Sketching graphs of rational functions

A rational function is defined as $f(x) = \frac{P(x)}{Q(x)}$ where both $P(x)$ and $Q(x)$ are polynomials. In this section, we will sketch the graphs of simple rational functions where $P(x)$ is a constant, linear, quadratic or cubic function and $Q(x)$ is a simple linear or quadratic function. To sketch the graphs of these types of rational functions, we consider the following main points.

Axis intercepts of rational functions

As with all functions, the y-intercept of a rational function will occur when $x = 0$. That is, the y-intercept of $f(x) = \frac{P(x)}{Q(x)}$ is $f(0) = \frac{P(0)}{Q(0)}$.

Similarly, the x-intercepts of a rational function will occur when $y = 0$. That is, the x-intercepts occur when $P(x) = 0$.

Asymptotic behaviour of rational functions

A function is not defined when the denominator is zero. We thus have a vertical asymptote (or a point of discontinuity, which will be covered in section 6.3.4) for each value of x where $Q(x) = 0$. A vertical asymptote is never crossed.

To obtain the equations of other asymptotes, if the degree of $P(x) \geq$ the degree of $Q(x)$, we divide the denominator into the numerator to obtain an expression of the form $S(x) + \frac{R(x)}{Q(x)}$, so that $R(x)$ is of a lower degree than $Q(x)$.

Now, as $x \to \infty$, $y \to S(x)$, so $y = S(x)$ is the equation of an asymptote.

$S(x)$ may be of the form $y = c$ (a constant), in which case the non-vertical asymptote is a horizontal line, or it may be of the form $y = ax + b$, in which case we have an oblique asymptote. It is also possible that $S(x) = ax^2$, so we can even get quadratics as asymptotes.

Stationary points of rational functions

It is easiest to determine the gradient function from the divided form of these types of functions. Equate the gradient function to zero and solve for x.

Note that these stationary points may be local minima or local maxima. We are not concerned with finding the inflection points of these types of graphs. However, we can find the second derivative and use it to determine the nature of the stationary points. Using this information we can sketch the curve. Sometimes graphs of these types may not cross the x- or y-axes, or they may not have any turning points.

WORKED EXAMPLE 4 Sketching the graph of a simple rational function

Sketch the graph of $y = \frac{x^3 - 54}{9x}$. State the equations of any asymptotes. Determine the coordinates of any axis intercepts and any stationary points (rounding to two decimal places where required), and establish their nature.

THINK	WRITE/DRAW
1. Determine axis intercepts. First, determine the x-intercepts.	The graph crosses the x-axis when the numerator is zero. Solve $x^3 - 54 = 0$ $x^3 = 54$ $x = \sqrt[3]{54} \approx 3.78$ The graph crosses the x-axis at $\left(\sqrt[3]{54}, 0\right)$ or $(3.78, 0)$.
2. Vertical asymptotes occur when the denominator is zero.	$y = \frac{x^3 - 54}{9x}$ The line $x = 0$ or the y-axis is a vertical asymptote.
3. Simplify the expression by dividing the denominator into the numerator.	$y = \frac{x^3 - 54}{9x}$ $= \frac{x^3}{9x} - \frac{54}{9x}$ $= \frac{x^2}{9} - \frac{6}{x}$
4. Determine the equations of any other asymptotes.	As $x \to \infty$, $y \to \frac{x^2}{9}$ from below. As $x \to -\infty$, $y \to \frac{x^2}{9}$ from above. The quadratic $y = \frac{x^2}{9}$ is an asymptote.

5. Calculate the first derivative by differentiating the divided form.

$$y = \frac{x^2}{9} - 6x^{-1}$$

$$\frac{dy}{dx} = \frac{2x}{9} + 6x^{-2}$$

$$= \frac{2x}{9} + \frac{6}{x^2}$$

6. Stationary points occur when the gradient is zero. Equate the gradient function to zero and solve for x.

$$\frac{dy}{dx} = \frac{2x}{9} + \frac{6}{x^2} = 0$$

$$\Rightarrow \frac{2x}{9} = -\frac{6}{x^2}$$

$$x^3 = -27$$

$$x = \sqrt[3]{-27} = -3$$

7. Determine the y-value of the turning point.

When $x = -3$, $y = \dfrac{-27 - 54}{-27} = 3$.

8. Calculate the second derivative.

$$\frac{dy}{dx} = \frac{2x}{9} + 6x^{-2}$$

$$\frac{d^2y}{dx^2} = \frac{2}{9} - 12x^{-3}$$

$$= \frac{2}{9} - \frac{12}{x^3}$$

9. Determine the sign of the second derivative to determine the nature of the turning point.

When $x = -3$, $\dfrac{d^2y}{dx^2} = \dfrac{2}{9} + \dfrac{12}{27} = \dfrac{2}{3} > 0$.

The point $(-3, 3)$ is a local minimum turning point.

10. Using all of the above information, we can sketch the graph using an appropriate scale. Draw the asymptotes as dotted lines, and label the graph with all the important features.

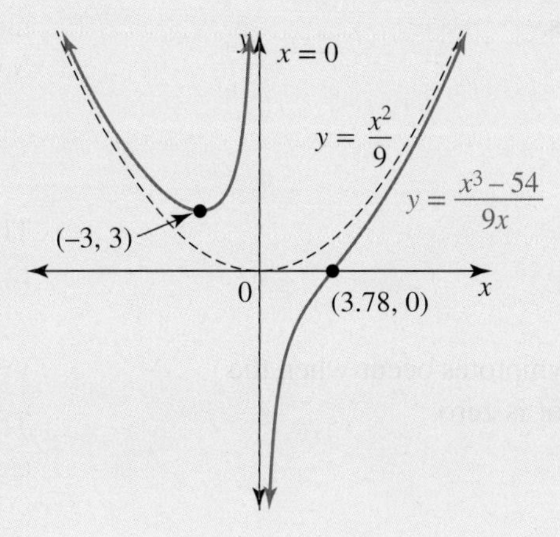

6.3.2 Sketching graphs of reciprocal functions

Reciprocal functions are the subset of rational functions which have a constant as the numerator. That is, functions of the form $f(x) = \dfrac{c}{Q(x)}$ where c is a real number and $Q(x)$ is a polynomial.

Since a constant is a polynomial of degree 0, the degree of the denominator $Q(x)$ is greater than the degree of the numerator. The function can therefore be written in the form $S(x) + \dfrac{R(x)}{Q(x)}$ with $S(x) = 0$ and $R(x) = c$ which means there will be a horizontal asymptote at $y = 0$.

As with all rational functions, reciprocal functions will have vertical asymptotes when the denominator is equal to zero.

If $Q(x)$ has a local maximum at (p, q), then $f(x)$ will have a local minimum at $\left(p, \frac{c}{q}\right)$.

If $Q(x)$ has a local minimum at (p, q), then $f(x)$ will have a local maximum at $\left(p, \frac{c}{q}\right)$.

WORKED EXAMPLE 5 Sketching the graph of a simple reciprocal function

Sketch the graph of $y = \dfrac{24}{x^2 + 2x - 24}$. State the equations of any asymptotes. Determine the coordinates of any axis intercepts and any turning points. State the maximal domain and range.

THINK	WRITE/DRAW
1. Factorise the denominator.	$y = f(x) = \dfrac{24}{x^2 + 2x - 24}$ $= \dfrac{24}{(x+6)(x-4)}$
2. Vertical asymptotes occur when the denominator is zero.	The lines $x = -6$ and $x = 4$ are both vertical asymptotes.
3. Determine axis intercepts. First determine the x-intercepts.	The graph does not cross the x-axis, as the numerator is never zero.
4. Calculate the y-intercept.	The graph crosses the y-axis when $x = 0$, $f(0) = -1$ at $(0, -1)$
5. Determine the equations of any other asymptotes.	As $x \to \pm\infty$, $y \to 0^+$. The plus indicates that the graph approaches from above the asymptote. The line $y = 0$ or the x-axis is a horizontal asymptote.
6. Use the chain rule to differentiate the function.	$f(x) = \dfrac{24}{x^2 + 2x - 24}$ $= 24(x^2 + 2x - 24)^{-1}$ $f'(x) = -24(2x + 2)(x^2 + 2x - 24)^{-2}$ $= -\dfrac{24(2x+2)}{(x^2 + 2x - 24)^2}$
7. Stationary points occur when the gradient is zero. Equate the gradient function to zero and solve for x.	$= -\dfrac{24(2x+2)}{(x^2 + 2x - 24)^2} = 0$ $\Rightarrow 2x + 2 = 0$ $x = -1$
8. Determine the y-value of the turning point. The second derivative will be complicated; however, we can determine the nature of the turning point.	Substitute $x = -1$: $f(-1) = \dfrac{24}{1 - 2 - 24} = -\dfrac{24}{25}$ Since the graph of $y = x^2 + 2x - 24$ has a local minimum at $x = -1$, the graph of $y = \dfrac{24}{x^2 + 2x - 24}$ has $\left(-1, -\frac{24}{25}\right)$ as a local maximum.

9. Using all of the above information, we can sketch the graph using an appropriate scale. Draw the asymptotes as dotted lines and label the graph with all the important features.

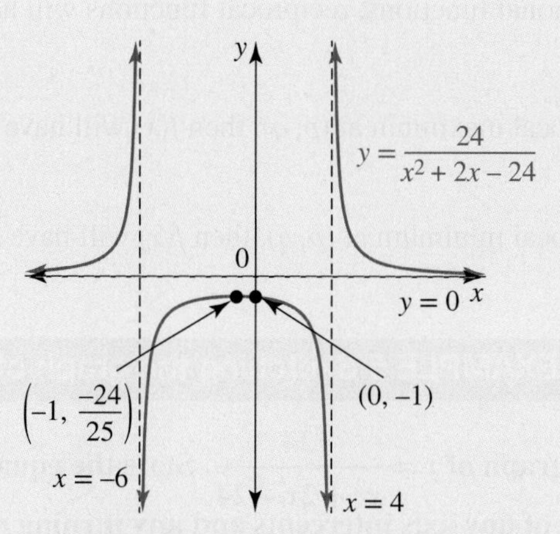

10. From the graph we can state the domain and range.

The domain is $R\backslash\{-6, 4\}$ and the range is $\left(-\infty, -\frac{24}{25}\right] \cup (0, \infty)$.

WORKED EXAMPLE 6 Determining the equations of asymptotes

The curve $y = \dfrac{A}{x^2 + bx + 8}$ has a local maximum at (−3, −2). Determine the values of A and b. State the equations of all asymptotes and the domain and range.

THINK

1. The curve will have a local maximum when the derivative of the quadratic in the denominator is zero.

WRITE

$$\text{Let } f(x) = y = \frac{A}{x^2 + bx + 8}$$
$$\text{Let } g(x) = x^2 + bx + 8$$
$$g'(x) = 2x + b$$
$$g'(-3) = -6 + b = 0$$
$$b = 6$$

2. Substitute for b and the point $(-3, -2)$ lies on the curve, solve for A.

$$f(x) = y = \frac{A}{x^2 + 6x + 8}$$
$$f(-3)\text{: } -2 = \frac{A}{9 - 18 + 8}$$
$$= -A$$
$$A = 2$$

3. The curve has vertical asymptotes when the denominator is zero.

$$g(x) = x^2 + 6x + 8$$
$$= (x + 4)(x + 2)$$

There are vertical asymptotes when $x = -4$ and $x = -2$ and a horizontal asymptote at $y = 0$.

4. Since the point $(-3, -2)$ is a local maximum, and from the features of the graph, we can state the range.

The range is $(-\infty, -2] \cup (0, \infty)$.

WORKED EXAMPLE 7 Determining values of a parameter from the number of asymptotes

State conditions on k such that the graph of $y = \dfrac{1}{kx - x^2 - 4}$ has:

a. **two vertical asymptotes**
b. **only one vertical asymptote**
c. **no vertical asymptotes.**

THINK	WRITE
The discriminant of the quadratic in the denominator is the key to answering this question.	$kx - x^2 - 4 = -x^2 + kx - 4$ $a = -1, \; b = k, \; c = -4$ $\Delta = b^2 - 4ac$ $\Delta = k^2 - 16$
a. For two vertical asymptotes, the discriminant is positive. Solve for k.	a. $\Delta = k^2 - 16 > 0$ $\|k\| > 4$ $(-\infty, -4) \cup (4, \infty)$
b. For only one vertical asymptote the discriminant is zero. Solve for k.	b. $\Delta = k^2 - 16 = 0$ $k = \pm 4$
c. For no vertical asymptotes, the discriminant is negative. Solve for k.	c. $\Delta = k^2 - 16 < 0$ $\|k\| < 4$ $(-4, 4)$

6.3.3 Using CAS to help determine the critical points

Some rational functions are difficult to differentiate (or differentiate twice) by hand. For such functions we must use technology to help differentiate and determine the coordinates of critical points. You will be able to differentiate functions and solve equations using your CAS calculator for questions which allow the use of technology.

WORKED EXAMPLE 8 Sketching more rational functions

Sketch the graph of each of the following, stating the equations of any asymptotes and the coordinates of any axial intercepts, stationary points and points of inflection. State the domain and range.

a. $y = \dfrac{x+2}{x^2 - 9}$

b. $y = \dfrac{x^2 - 9}{x + 2}$

THINK	WRITE/DRAW
a. 1. Factorise the denominator.	a. $y = f(x) = \dfrac{x+2}{(x+3)(x-3)}$
2. Vertical asymptotes occur when the denominator is zero.	The lines $x = -3$ and $x = 3$ are both vertical asymptotes.
3. Determine axial intercepts. The graph crosses the x-axis, when the numerator is zero.	$x + 2 = 0$ $x = -2$ $(-2, 0)$

4. Calculate the y-intercept.

The graph crosses the y-axis when $x = 0$.

$f(0) = -\frac{2}{9}$ at $\left(0, -\frac{2}{9}\right)$

5. Determine the equations of any other asymptotes.

As $x \to \infty, y \to 0^+$ and as $x \to -\infty, y \to 0^-$

The line $y = 0$ or the x-axis is a horizontal asymptote.

Note that the graph crosses the horizontal asymptote.

6. Use the quotient rule to differentiate the function.

$$f'(x) = \frac{1\left(x^2 - 9\right) - 2x(x + 2)}{\left(x^2 - 9\right)^2}$$

$$f'(x) = \frac{-\left(x^2 + 4x + 9\right)}{\left(x^2 - 9\right)^2}$$

7. Stationary points occur when the gradient is zero. Equate the gradient function to zero and solve for x.

$$f'(x) = \frac{-\left(x^2 + 4x + 9\right)}{\left(x^2 - 9\right)^2} = 0$$

But $x^2 + 4x + 9 = (x + 2)^2 + 5 > 0$

So there are no stationary points.

8. Use CAS to calculate the second derivative and equate to zero for inflection points.

$$f''(x) = \frac{2\left(x^3 + 6x^2 + 27x + 18\right)}{\left(x^2 - 9\right)^3} = 0$$

Solving $x^3 + 6x^2 + 27x + 18 = 0$ gives one solution as $x = -0.786$ and $f(-0.786) = -0.145$, so $(-0.786, -0.145)$ is the inflection point.

9. Using all of the above information we can sketch the graph using an appropriate scale. Draw the asymptotes as dotted lines, labelling the graph with all important features.

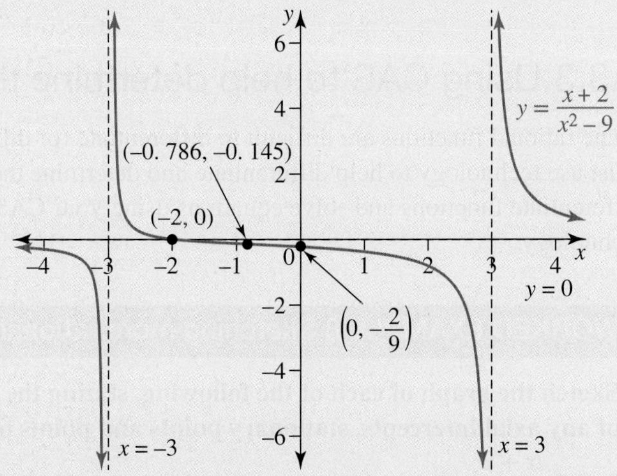

10. From the graph we can state the domain and range.

The domain is $R\backslash\{\pm 3\}$ and the range is R.

b. 1. Vertical asymptotes occur when the denominator is zero.

b. $y = \frac{x^2 - 9}{x + 2}$, the line $x = -2$ is a vertical asymptote.

2. Express a proper fraction.

$$y = \frac{x^2 - 9}{x + 2} = \frac{x(x + 2) - 2(x + 2) - 5}{x + 2}$$

$$y = x - 2 - \frac{5}{x + 2}$$

The line $y = x - 2$ is an oblique asymptote.

3. Determine axial intercepts. The graph crosses the x-axis, when the numerator is zero.

$x^2 - 9 = 0$
$x = \pm 3$
$(-3, 0)$, $(3, 0)$

4. Calculate the y-intercept, which is the point corresponding to $x = 0$.

$f(0) = -\dfrac{9}{2}$
The y-intercept is $\left(0, -\dfrac{9}{2}\right)$.

5. Use the chain rule to differentiate the proper function expression.

$y = x - 2 - 5(x+2)^{-1}$
$\dfrac{dy}{dx} = 1 + \dfrac{5}{(x+2)^2}$

6. Stationary points occur when the gradient is zero.

But $\dfrac{dy}{dx} = 1 + \dfrac{5}{(x+2)^2} > 0$
So there are no stationary points.

7. Calculate the second derivative.

$\dfrac{d^2y}{dx^2} = \dfrac{-10}{(x+2)^3}$

8. Inflection points occur when the second derivative is zero.

But $\dfrac{d^2y}{dx^2} = \dfrac{-10}{(x+2)^3} < 0$
So there are no inflection points.

9. Using all of the above information we can sketch the graph using an appropriate scale. Draw the asymptotes as dotted lines, labelling the graph with all important features.

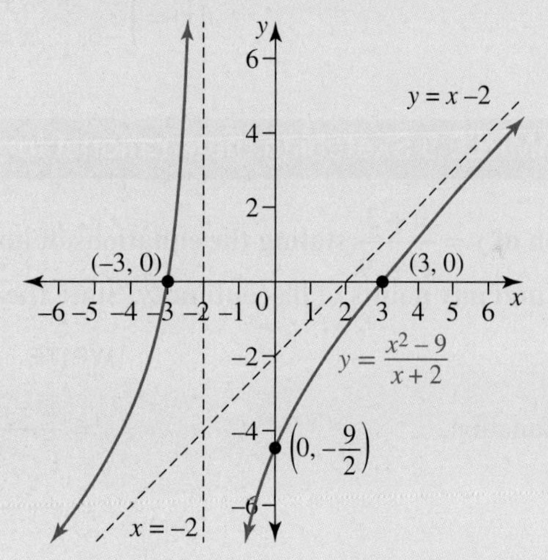

10. From the graph we can state the domain and range.

The domain is $R\backslash\{-2\}$ and the range is R.

6.3.4 Points of removable discontinuity

We have stated that we get vertical asymptotes when the denominator is zero, but this statement is not exactly correct.

A **removable discontinuity** or commonly called a **point of discontinuity** occurs when the graph of a function has a hole. A hole appears when we cancel a factor.

For example, consider the function $f(x) = \dfrac{x^2 - 9}{x+3}$. By factorising $f(x) = \dfrac{(x+3)(x-3)}{x+3}$ the function is $y = x - 3, \quad x \neq -3$ as the set of factors $(x+3)$ has cancelled out.

Note that the limit exists, $\lim_{x \to -3} \left(\frac{x^2 - 9}{x + 3} \right) = \lim_{x \to -3} (x - 3) = -6$, so the graph is the straight line $y = x - 3$ but with a hole in it at the point $(-3, -6)$.

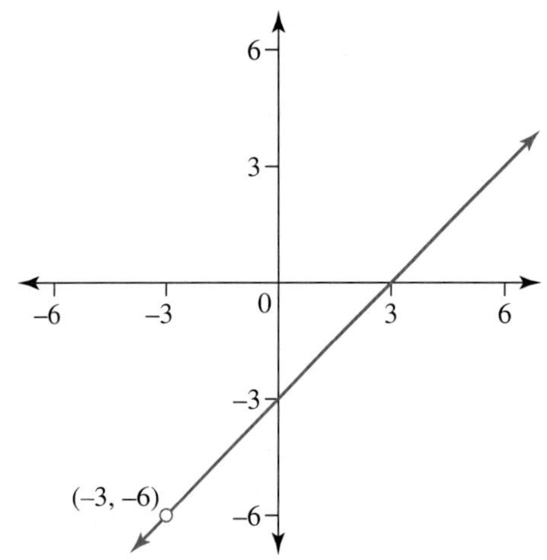

Removable discontinuities can be "filled in" if you make the function a piecewise function and define a part of the function at the point where the hole is. In the example above, to make the function continuous we could redefine the function as

$$f(x) = \begin{cases} x - 3, & x \neq -3 \\ -6, & x = -3 \end{cases}$$

WORKED EXAMPLE 9 Sketching more rational functions with points of discontinuity

Sketch the graph of $y = \frac{x + 3}{x^2 - 9}$ stating the equations of any asymptotes and the coordinates of any axial intercepts and any points of discontinuity. State the domain and range.

THINK	WRITE
1. Simplify the function.	$y = \frac{x+3}{x^2-9} = \frac{x+3}{(x+3)(x-3)}$ $y = \frac{1}{x-3}, \quad x \neq -3$
2. Vertical asymptotes occur when the denominator is zero.	The line $x = 3$ is a vertical asymptote, however the graph has a removable point of discontinuity at $x = -3$ or a hole in the graph at $x = -3$.
3. Determine the equations of any other asymptotes.	As $x \to \infty$ $y \to 0^+$ and as $x \to -\infty$ $y \to 0^-$. The line $y = 0$ or the x-axis is a horizontal asymptote.
4. Determine axial intercepts.	The numerator of $y = \frac{1}{x-3}$, $x \neq -3$ is never zero, the graph does not cross the x-axis. The graph crosses the y-axis when $x = 0$ at $y = -\frac{1}{3}$, $\left(0, -\frac{1}{3}\right)$
5. The graph is a hyperbola but has a point of discontinuity at $x = -3$. This is indicated by an open circle on the graph.	$\lim_{x \to -3} \left(\frac{x+3}{x^2-9} \right) = \lim_{x \to -3} \left(\frac{1}{x-3} \right) = -\frac{1}{6}$ Draw an open circle at the point $\left(-3, -\frac{1}{6}\right)$

6. Using calculus to determine stationary or inflection points.

$$\frac{dy}{dx} = \frac{-1}{(x-3)^2},\ x \neq -3$$
$$\frac{d^2y}{dx^2} = \frac{2}{(x-3)^3},\ x \neq -3$$

Both the first and second derivatives are never equal to zero, the graph has no stationary points or points of inflection.

7. Using all of the above information we can sketch the graph using an appropriate scale. Draw the asymptotes as dotted lines, labelling the graph with all important features.

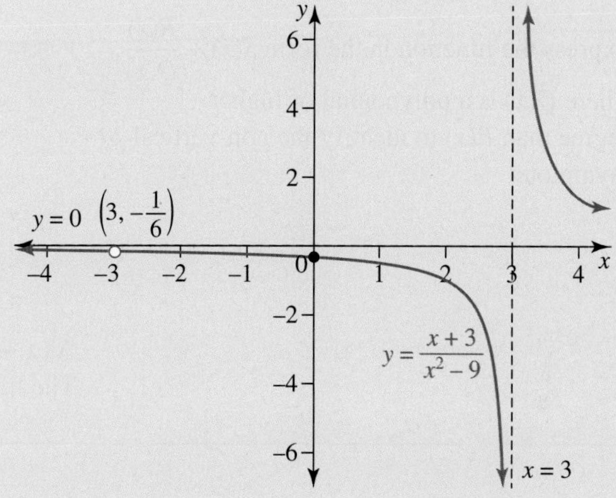

8. From the graph we can state the domain and range.

The domain is $R\backslash\{\pm 3\}$ and the range is $R\backslash\{-\frac{1}{6}, 0\}$.

WORKED EXAMPLE 10 Determining vertical asymptotes and points of discontinuity

Consider the function $f(x) = \frac{x^2 + 2x - 15}{x^2 - 2x - 3}$. State the equations of all asymptotes and/or points of discontinuity.

THINK

WRITE

1. Simplify the function.

$$f(x) = \frac{x^2 + 2x - 15}{x^2 - 2x - 3}$$
$$= \frac{(x+5)(x-3)}{(x+1)(x-3)}$$
$$= \frac{x+5}{x+1},\ x \neq 3$$

2. Vertical asymptotes occur when the denominator of the simplified function equals zero.

The line $x = -1$ is a vertical asymptote.

3. The factor that cancels out is $(x-3)$. Therefore there is a point of discontinuity at $x=3$.

$$\lim_{x \to 3} = \frac{x^2+2x-15}{x^2-2x-3}$$
$$\lim_{x \to 3} = \frac{x+5}{x+1}$$
$$= \frac{3+5}{3+1}$$
$$= 2$$

There is a point of discontinuity at $(3, 2)$.

4. Express the function in the form $S(x) + \dfrac{R(x)}{Q(x)}$, where $Q(x)$ is a polynomial of higher degree than $R(x)$ to identify the non-vertical asymptote.

$$f(x) = \frac{x^2+2x-15}{x^2-2x-3}$$
$$= \frac{x+5}{x+1}, \ x \neq 3$$
$$= \frac{x+1+4}{x+1}, \ x \neq 3$$
$$= 1 + \frac{4}{x+1}, \ x \neq 3$$

As $x \to \infty$, $y \to 1^+$ and as $x \to -\infty$, $y \to 1^-$.
The line $y=1$ is a horizontal asymptote.

6.3 Exercise

Technology free

1. WE4 Sketch the graph of $y = \dfrac{16-x^3}{4x}$. State the equations of any asymptotes. Determine the coordinates of any axis intercepts and any stationary points, and establish their nature.

2. Sketch the graph of $y = \dfrac{x^2+9}{2x}$. State the equations of any asymptotes. Determine the coordinates of any axis intercepts and any stationary points, and establish their nature.

3. Sketch the graphs of each of the following. State the equations of any asymptotes. Determine the coordinates of any axis intercepts and any stationary points, and establish their nature.

   a. $y = \dfrac{x^2+4}{2x}$

   b. $y = \dfrac{x^3+16}{2x}$

4. WE5 Sketch the graph of $y = \dfrac{16}{16-x^2}$. State the equations of any asymptotes. Determine the coordinates of any axis intercepts and any turning points. State the maximal domain and range.

5. Sketch the graph of $y = \dfrac{12}{x^2 - 4x - 12}$ by calculating the equations of all straight-line asymptotes. Determine the coordinates of any axis intercepts and turning points. State the maximal domain and range.

6. Sketch the graphs of each of the following. State the equations of any asymptotes. Determine the coordinates of any axis intercepts and any turning points. State the domain and range.

   a. $y = \dfrac{18}{x^2 - 9}$

   b. $y = \dfrac{18}{8 + 2x - x^2}$

7. Sketch the graphs of each of the following. State the equations of any asymptotes. Determine the coordinates of any axis intercepts and any stationary points, and establish their nature.

   a. $y = \dfrac{x^3 - 32}{2x^2}$

   b. $y = \dfrac{-(x^3 + 4)}{x^2}$

8. Sketch the graphs of each of the following. State the equations of any asymptotes. Determine the coordinates of any axis intercepts and any stationary points, and establish their nature.

   a. $y = \dfrac{x^4 - 81}{2x^2}$

   b. $y = \dfrac{-(x^4 + 16)}{2x^2}$

9. **WE6** The curve $y = \dfrac{A}{x^2 + bx + 7}$ has a local maximum at $\left(-4, -\dfrac{4}{3}\right)$. Determine the values of A and b. State the equations of all straight-line asymptotes and the domain and range.

10. The curve $y = \dfrac{A}{bx + c - x^2}$ has a local minimum at $(3, 2)$ and a vertical asymptote at $x = 8$. Determine the values of A, b and c. State the equations of all straight-line asymptotes and the domain and range.

11. **WE7** State conditions on k such that the graph of $y = \dfrac{1}{kx - 4x^2 - 25}$ has:

    a. two vertical asymptotes

    b. only one vertical asymptote

    c. no vertical asymptotes.

12. Sketch the graph of $y = \dfrac{18}{x^2 + 9}$ by calculating the coordinates of any stationary points and establishing their nature. Also determine the coordinates of the points of inflection, and determine the equation of the tangent to the curve at the point of inflection where $x > 0$.

13. Consider the function $f(x) = \dfrac{ax^3 + b}{x^2}$ where $a, b \in R\backslash\{0\}$. State the equations of all the asymptotes and show that the graph has a turning point at $\left(\sqrt[3]{\dfrac{2b}{a}}, \dfrac{3\sqrt[3]{2a^2b}}{2}\right)$ and an x-intercept at $x = \sqrt[3]{-\dfrac{b}{a}}$.

14. Consider the function $f(x) = \dfrac{ax^3 + b}{x}$ where $a, b \in R\backslash\{0\}$. State the equations of all the asymptotes and show that the graph has a turning point at $\left(\sqrt[3]{\dfrac{b}{2a}}, \dfrac{3\sqrt[3]{2ab^2}}{2}\right)$ and an x-intercept at $x = \sqrt[3]{-\dfrac{b}{a}}$.

15. Consider the function $f(x) = \dfrac{ax^4 + b}{x^2}$ where $a, b \in R\backslash\{0\}$. State the equations of all the asymptotes and show that if $ab > 0$, the graph has two turning points at $\left(\pm\sqrt[4]{\dfrac{b}{a}}, 2\sqrt{ab}\right)$ and does not cross the x-axis. However, if $ab < 0$, then there are no turning points, but the graph crosses the x-axis at $x = \pm\sqrt[4]{-\dfrac{b}{a}}$.

16. Consider the function $f(x)=\frac{ax^2+b}{x}$ where $a, b \in R\backslash\{0\}$. State the equations of all the asymptotes and show that if $ab>0$, then the graph has two turning points at $\left(\sqrt{\frac{b}{a}}, 2\sqrt{ab}\right)$ and $\left(-\sqrt{\frac{b}{a}}, -2\sqrt{ab}\right)$ and does not cross the x-axis. However, if $ab<0$, then there are no turning points, but the graph crosses the x-axis at $x=\pm\sqrt{-\frac{b}{a}}$.

17. Sketch the graphs of each of the following, give the equations of any asymptotes and the coordinates of any intercepts and any turning points.

a. $y=\frac{x^2+5x+4}{x}$ b. $y=\frac{2x^2+x-6}{x}$ c. $y=\frac{5x-6-x^2}{x}$

18. For the function $f(x)=\frac{ax^2+bx+c}{x}$ state the equations of all the asymptotes and show that if $ac>0$ then the function has two turning points at $\left(\sqrt{\frac{c}{a}}, b+2\sqrt{ac}\right)$ and $\left(-\sqrt{\frac{c}{a}}, b-2\sqrt{ac}\right)$. Additionally, show that if $ac<0$ then there are no turning points and if $c\neq 0$ then there are no inflection points.

Technology active

19. **WE8** Sketch the graph of each of the following, stating the equations of any asymptotes and the coordinates of any axial intercepts and any stationary points. State the domain and range.

a. $y=\frac{x+2}{x^2-16}$ b. $y=\frac{x^2-16}{x+2}$

20. **WE9** Sketch the graph of $y=\frac{x+4}{x^2-16}$ stating the equations of any asymptotes and the coordinates of any axial intercepts and any points of discontinuity. State the domain and range.

21. **WE10** Consider the function $f(x)=\frac{x^2-2x-8}{x^2-x-6}$, state the equations of all asymptotes and/or points of discontinuity.

22. Consider the function $f(x)=\frac{x^3-6x^2+9x}{x^3-9x}$, state the equations of all asymptotes and/or points of discontinuity.

23. Sketch the graphs of each of the following, stating the coordinates of all axial intercepts, stationary points and their nature, inflection points (if any) and the equations of any asymptotes.

a. $y=\frac{x^2-9}{x^2-4}$ b. $y=\frac{x^2+9}{x^2-4}$ c. $y=\frac{x^2-9}{x^2+4}$

24. Sketch the graphs of each of the following, stating the coordinates of all axial intercepts, stationary points and their nature, inflection points and the equations of any asymptotes.

a. $y=\frac{x^2-4}{x^2-9}$ b. $y=\frac{x^2+4}{x^2-9}$ c. $y=\frac{x^2+4}{x^2+9}$

25. Sketch the graphs of each of the following, stating the coordinates of all axial intercepts, stationary points and their nature, inflection points and the equations of any asymptotes.

a. $y=\frac{x^2+4}{x^2-5x+4}$ b. $y=\frac{2x^2+2x+3}{2x^2-2x+5}$

26. Consider the function $f(x) = \dfrac{x^2 + bx + c}{x^2 + px + q}$, where $b \neq p$ and $c \neq q$.

a. State the values of b and c for which the graph of the function crosses the x-axis at two distinct points.
b. State the values of b and c for which the graph of the function does not crosses the x-axis.
c. State the values of p and q for which the graph of the function has two vertical asymptotes.
d. State the values of p and q for which the graph of the function has no vertical asymptotes.
e. State the values of b, p and q for which the graph of the function has at least two inflection points.
f. Determine the value of x where it crosses the horizontal asymptote.

6.3 Exam questions

Question 1 (4 marks) TECH-FREE

Source: VCE 2018 Specialist Mathematics Exam 1, Q5; © VCAA.

Sketch the graph of $f(x) = \dfrac{x+1}{x^2 - 4}$ on the axes provided below, labelling any asymptotes with their equations and any intercepts with their coordinates.

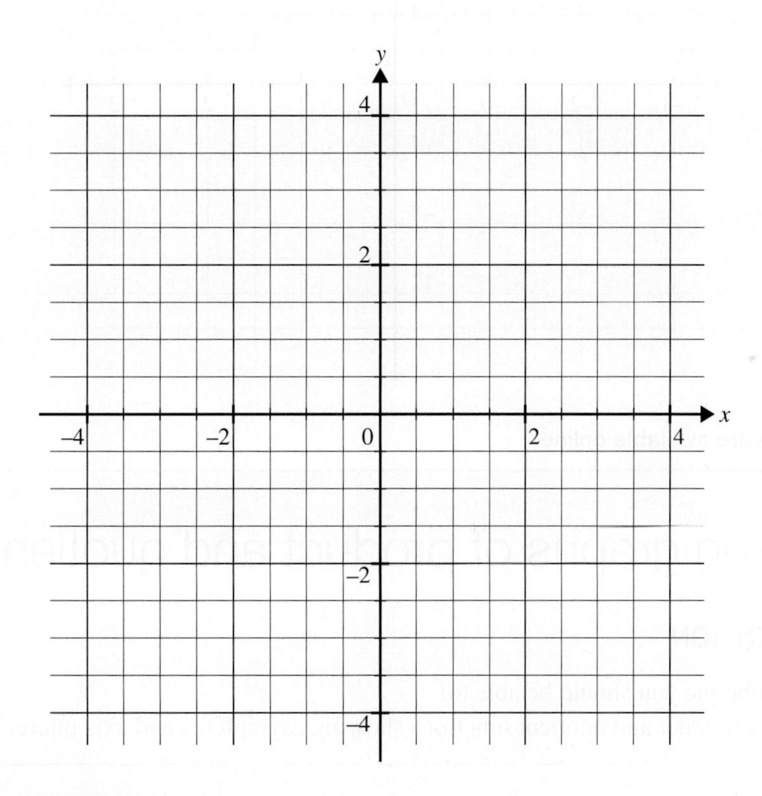

Question 2 (1 mark) TECH-ACTIVE

Source: VCE 2016 Specialist Mathematics Exam 2, Section A, Q3; © VCAA.

MC The straight-line asymptote(s) of the graph of the function with rule $f(x) = \dfrac{x^3 - ax}{x^2}$, where a is a non-zero real constant, is given by

A. $x = 0$ only.
B. $x = 0$ and $y = 0$ only.
C. $x = 0$ and $y = x$ only.
D. $x = 0, x = \sqrt{a}$ and $x = -\sqrt{a}$ only.
E. $x = 0$ and $y = a$ only.

Question 3 (6 marks) TECH-ACTIVE

Source: *VCE 2016 Specialist Mathematics Exam 2, Section B, Q1 a, b, c; © VCAA.*

a. Find the stationary point of the graph of $f(x) = \dfrac{4 + x^2 + x^3}{x}$, $x \in R \backslash \{0\}$. Express your answer in coordinate form, giving values correct to two decimal places. **(1 mark)**

b. Find the point of inflection of the graph given in part **a**. Express your answer in coordinate form, giving values correct to two decimal places. **(2 marks)**

c. Sketch the graph of $f(x) = \dfrac{4 + x^2 + x^3}{x}$ for $x \in [-3, 3]$ on the axes below, labelling the turning point and the point of inflection with their coordinates, correct to two decimal places. **(3 marks)**

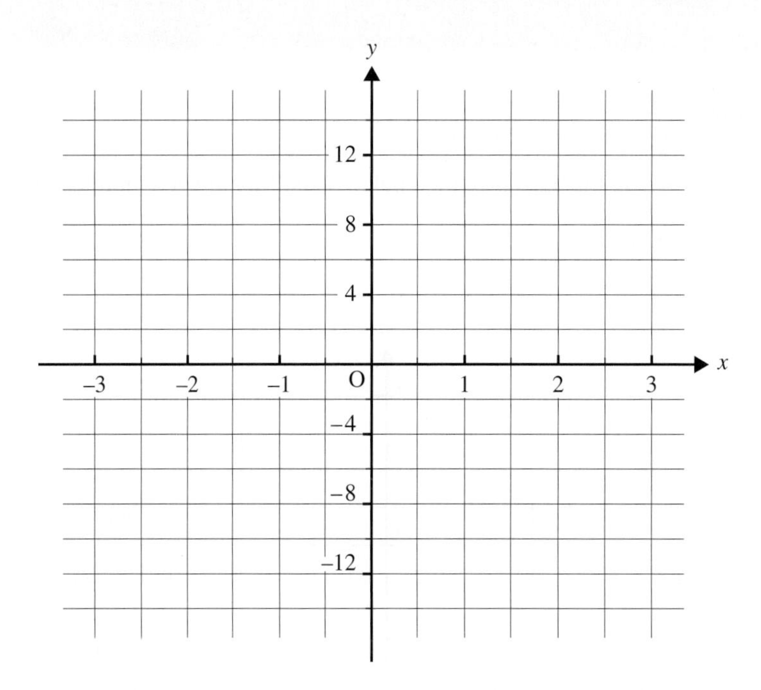

More exam questions are available online.

6.4 Sketching graphs of product and quotient functions

LEARNING INTENTION

At the end of this subtopic you should be able to:

- sketch graphs of product and quotient functions showing asymptotes and axis intercepts.

A rational function was defined as a ratio of two polynomial functions, we can also use calculus and sketch the graphs of quotients and products of other types of functions, for examples logarithmic, exponential, trigonometric, inverse trigonometric or radical functions.

First we revise the properties of the inverse trigonometric functions.

6.4.1 Inverse trigonometric functions

All circular functions are periodic and are many-to-one functions therefore their inverses will be one-to-many relations, not functions. However, by restricting the domain so that the circular functions are one-to-one functions, the inverses are functions.

The inverse sine function

The sine function, $y = \sin(x)$ is a many-to-one function.

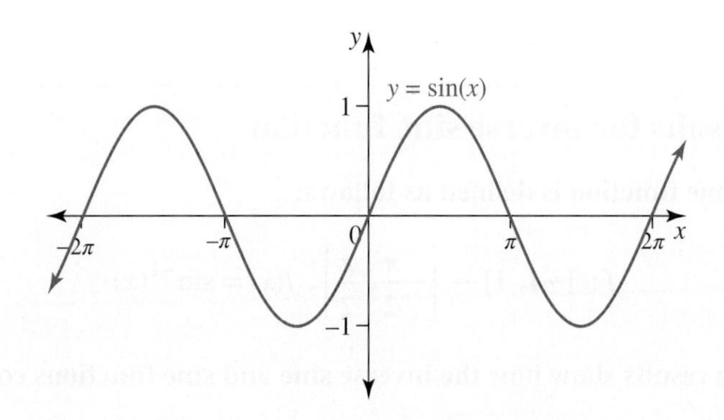

Therefore its inverse does not exist as a function. However, there are many restrictions of the domain $\left[-\frac{3\pi}{2}, -\frac{\pi}{2}\right]$ or $\left[-\frac{\pi}{2}, \frac{\pi}{2}\right]$ or $\left[\frac{\pi}{2}, \frac{3\pi}{2}\right]$ so that it is a one-to-one function. Convention states that we use the domain $\left[-\frac{\pi}{2}, \frac{\pi}{2}\right]$ for the restricted sine function.

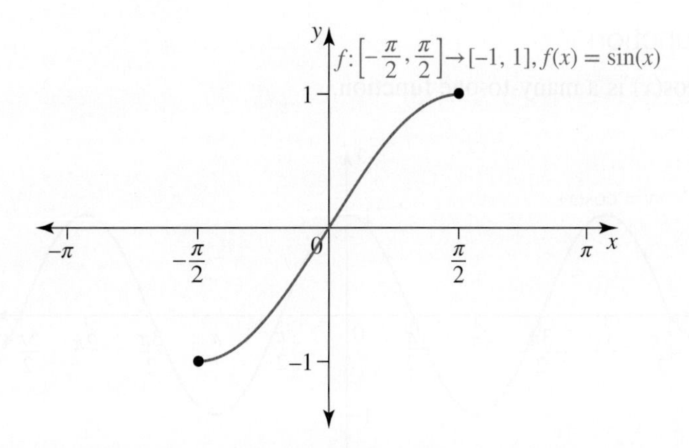

Therefore, it is a one-to-one function and its inverse exists.

The inverse of this function is denoted by $\sin^{-1}$, or alternatively arcsin.

The graph of $y = \sin^{-1}(x)$ is obtained from the graph of $y = \sin(x)$ by reflecting in the line $y = x$.

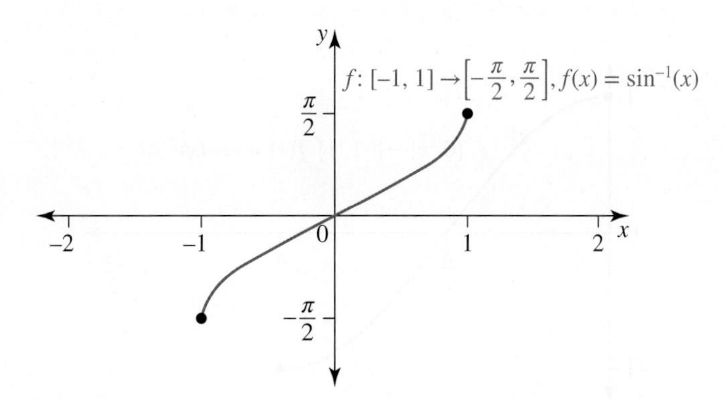

If we were to solve the equation $\sin(x) = \frac{1}{2}$ we might recognise that there is an infinite number of solutions, such as $\frac{\pi}{6}$, $2\pi + \frac{\pi}{6}$, $4\pi + \frac{\pi}{6}$ since we can add any multiple of 2π to a solution. However, if we try to solve the equation $x = \sin^{-1}\left(\frac{1}{2}\right)$, there is only one solution, $x = \frac{\pi}{6}$, since the range of the inverse sine function is $x \in \left[-\frac{\pi}{2}, \frac{\pi}{2}\right]$.

General results for inverse sine function

The inverse sine function is defined as follows:

$$f: [-1, 1] \rightarrow \left[-\frac{\pi}{2}, \frac{\pi}{2}\right], f(x) = \sin^{-1}(x)$$

The following results show how the inverse sine and sine functions combine:

$$\sin\left(\sin^{-1}(x)\right) = x \text{ if } x \in [-1, 1]$$

and

$$\sin^{-1}\left(\sin(x)\right) = x \text{ if } x \in \left[-\frac{\pi}{2}, \frac{\pi}{2}\right]$$

The inverse cosine function

The cosine function, $y = \cos(x)$ is a many-to-one function.

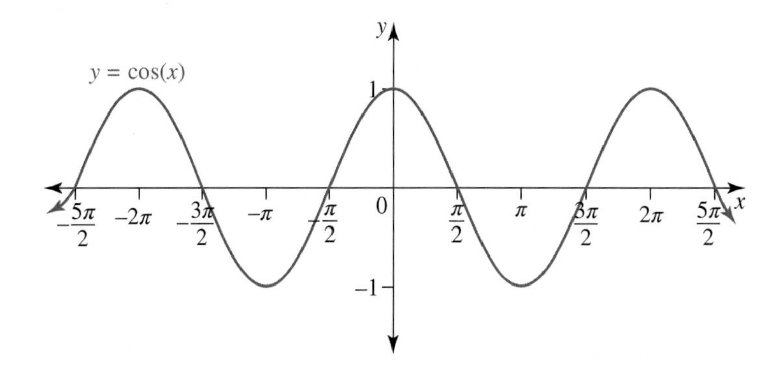

Therefore its inverse does not exist as a function. However, there are many restrictions of the domain $[-\pi, 0]$ or $[0, \pi]$ or $[\pi, 2\pi]$ so that it is a one-to-one function. Convention states that we use the domain $[0, \pi]$ for the restricted cosine function.

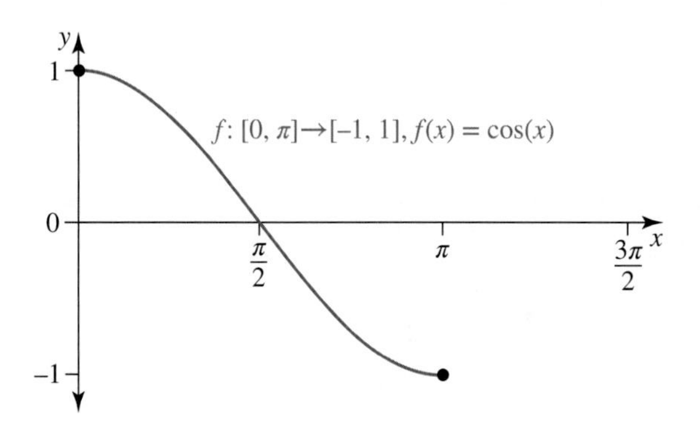

Therefore, it is a one-to-one function and its inverse exists.

The inverse of this function is denoted by $\cos^{-1}$, or alternatively arccos.

The graph of $y = \cos^{-1}(x)$ is obtained from the graph of $y = \cos(x)$ by reflecting in the line $y = x$.

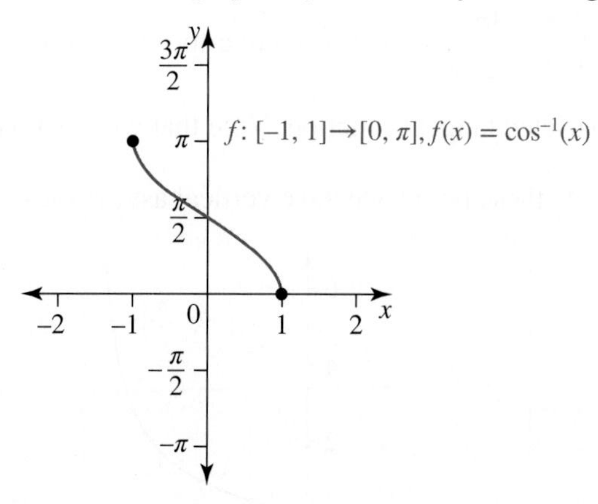

If we were to solve the equation $\cos(x) = \frac{\sqrt{2}}{2}$ we might recognise that there is an infinite number of solutions, such as $\frac{\pi}{4}$, $2\pi + \frac{\pi}{4}$, $4\pi + \frac{\pi}{4}$ ... $2\pi - \frac{\pi}{4}$, $4\pi - \frac{\pi}{4}$ since we can add any multiple of 2π to a solution. However, if we try to solve the equation $x = \cos^{-1}\left(\frac{\sqrt{2}}{2}\right)$, there is only one solution $x = \frac{\pi}{4}$, since the range of the inverse cosine function is $x \in [0, \pi]$.

General results for inverse cosine function

The inverse cosine function is defined as follows:

$$f: [-1, 1] \to [0, \pi], f(x) = \cos^{-1}(x)$$

The following results show how the inverse cosine and cosine functions combine:

$$\cos\left(\cos^{-1}(x)\right) = x \text{ if } x \in [-1, 1]$$

and

$$\cos^{-1}\left(\cos(x)\right) = x \text{ if } [0, \pi]$$

The inverse tangent function

The tangent function, $y = \tan(x)$ is a many-to-one function.

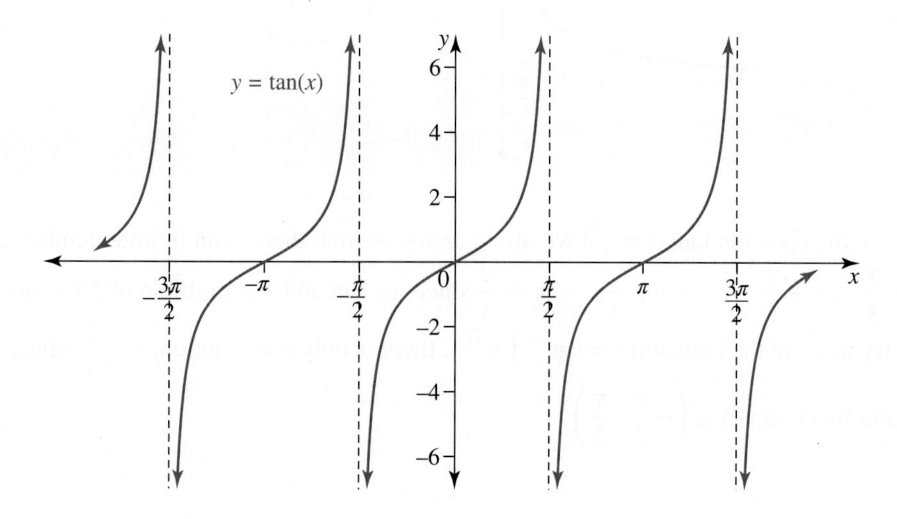

Therefore, its inverse does not exist as a function. However, there are many restrictions of the domain $\left(-\frac{3\pi}{2}, -\frac{\pi}{2}\right)$ or $\left(-\frac{\pi}{2}, \frac{\pi}{2}\right)$ or $\left(\frac{\pi}{2}, \frac{3\pi}{2}\right)$ so that it is a one-to-one function. Convention states that we use the domain $\left(-\frac{\pi}{2}, \frac{\pi}{2}\right)$ for the restricted tangent function. Note that we must have an open interval, since the function is not defined at $x = \pm\frac{\pi}{2}$. At these points we have vertical asymptotes.

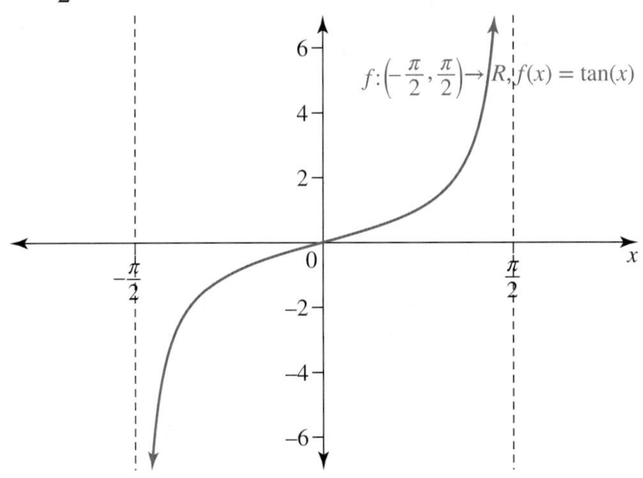

Therefore, it is a one-to-one function and its inverse exists.

The inverse of this function is denoted by $\tan^{-1}$, or alternatively arctan.

The graph of $y = \tan^{-1}(x)$ is obtained from the graph of $y = \tan(x)$ by reflecting in the line $y = x$.

Notice that there are horizontal asymptotes at $y = \pm\frac{\pi}{2}$.

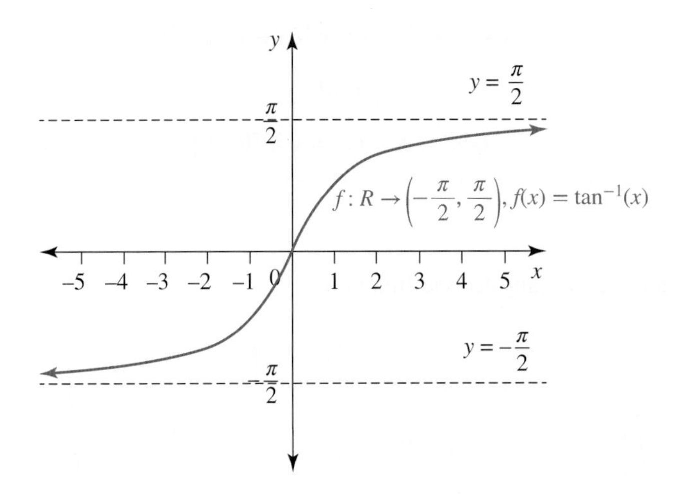

If we were to solve the equation $\tan(x) = \sqrt{3}$ we might recognise that there is an infinite number of solutions, such as $\frac{\pi}{3}$, $\pi + \frac{\pi}{3}$, $2\pi + \frac{\pi}{3}$ $\ldots$ $-\pi + \frac{\pi}{3}$, $-2\pi + \frac{\pi}{3}$ since we can add any multiple of 2π to any angle. However, if we try to solve the equation $x = \tan^{-1}\left(\sqrt{3}\right)$, there is only one solution, $x = \frac{\pi}{3}$, since the range of the inverse tangent function is $x \in \left(-\frac{\pi}{2}, \frac{\pi}{2}\right)$.

General results for the inverse tan function

The inverse tangent function is defined as follows:

$$f: R \to \left(-\frac{\pi}{2}, \frac{\pi}{2}\right),\ f(x) = \tan^{-1}(x)$$

The following results show how the inverse tangent and tangent functions combine:

$$\tan\left(\tan^{-1}(x)\right) = x \text{ if } x \in R$$

and

$$\tan^{-1}\left(\tan(x)\right) = x \text{ if } x \in \left(-\frac{\pi}{2}, \frac{\pi}{2}\right)$$

WORKED EXAMPLE 11 Sketching an inverse trigonometric function

Sketch the graph of the function $f(x) = \frac{6}{\pi}\sin^{-1}\left(\frac{2x-5}{3}\right)$ stating the coordinates of all axial intercepts and endpoints. Determine if the graph has any stationary points or points of inflection.

THINK	WRITE
1. Determine the maximal domain.	Since the maximal domain of $\sin^{-1}(x)$ is $[-1, 1]$ the maximal domain is: $\left\lvert\frac{2x-5}{3}\right\rvert \le 1$ $-3 \le 2x-5 \le 3$ $2 \le 2x \le 8$ $1 \le x \le 4$ or $[1, 4]$
2. Calculate the coordinates of the endpoints.	The endpoints are $f(1) = \frac{6}{\pi}\sin^{-1}(-1) = \frac{6}{\pi} \times \frac{-\pi}{2} = -3,\ (1, -3)$ $f(4) = \frac{6}{\pi}\sin^{-1}(1) = \frac{6}{\pi} \times \frac{\pi}{2} = 3,\ (4, 3)$
3. Determine the equations of any asymptotes.	The graph has no asymptotes.
4. Calculate where the graph crosses the coordinate axes.	The graph does not cross the y-axis, but crosses the x-axis when $y = 0$, that is when $2x - 5 = 0$ $x = \frac{5}{2}$ at $\left(\frac{5}{2}, 0\right)$.

5. Calculate the first derivative.

We can differentiate using the chain rule.

$y = \frac{6}{\pi}\sin^{-1}\left(\frac{u}{3}\right)$ where $u = 2x - 5$

$\frac{dy}{du} = \frac{6}{\pi\sqrt{9-u^2}} \qquad \frac{du}{dx} = 2$

$\frac{dy}{dx} = \frac{12}{\pi\sqrt{9-(2x-5)^2}}$

$\frac{dy}{dx} = \frac{12}{\pi\sqrt{9-(4x^2-20x+25)}}$

$\frac{dy}{dx} = \frac{12}{\pi\sqrt{-4(x^2-5x+4)}}$

$\frac{dy}{dx} = \frac{6}{\pi\sqrt{-4+5x-x^2}}$

Since $\frac{dy}{dx} > 0$ for $x \in (1, 4)$ there are no stationary points, the function is an increasing function.

6. State the range of the function.

The range of the function is $[-3, 3]$.

7. Calculate the second derivative.

$\frac{dy}{dx} = \frac{6}{\pi}\left(-x^2+5x+4\right)^{-\frac{1}{2}}$

$\frac{d^2y}{dx^2} = -\frac{1}{2} \times \frac{6}{\pi} \times (-2x+5)\left(-x^2+5x+4\right)^{-\frac{3}{2}}$

For inflection points,

$\frac{d^2y}{dx^2} = 0 \qquad -2x+5=0, \quad x = \frac{5}{2}$

$\left.\frac{d^2y}{dx^2}\right|_{x=2} = -\frac{3\sqrt{2}}{4\pi} < 0$ and $\left.\frac{d^2y}{dx^2}\right|_{x=3} = \frac{3\sqrt{2}}{4\pi} > 0$

The point $\left(\frac{5}{2}, 0\right)$ is a point of inflection point as the second derivative changes sign.

8. Sketch the graph, showing the critical points using an appropriate scale.

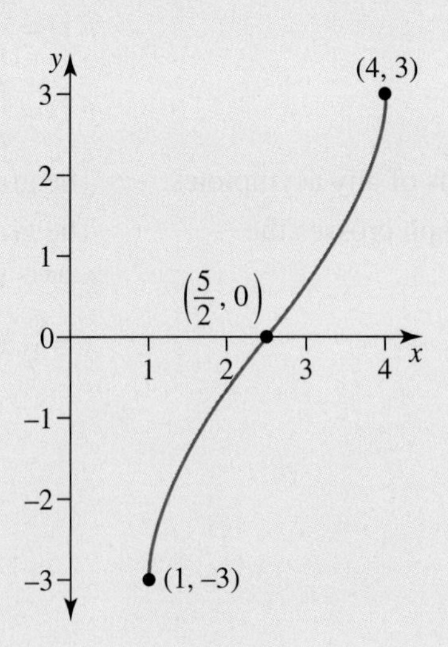

6.4.2 Sketching graphs of quotient and product functions

Quotient functions are functions of the form $f(x) = \frac{g(x)}{h(x)}$ and product functions are functions of the form $f(x) = g(x)h(x)$. $g(x)$ and $h(x)$ can be any type of function. When sketching quotient and product functions we use the same method as we used for rational functions. First, determine the key features such as axis intercepts, stationary points, inflection points and equations of any asymptotes and then put all of this information together to create a sketch of the graph.

You may need to use CAS to determine some derivatives or solve some equations.

WORKED EXAMPLE 12 Sketching a quotient function involving a logarithmic function

Sketch the graph of the function $f(x) = \frac{\log_e(x)}{x}$, stating the domain and range, equations of any asymptotes, the coordinates of any axial intercepts, the stationary points and points of inflection, giving all answers correct to two decimal places.

THINK	WRITE
1. Determine the maximal domain.	Since the maximal domain of $\log_e(x)$ is $x > 0$ this is the domain of the function.
2. State the equations of the asymptote.	The denominator is undefined when $x = 0$, so this (the y-axis) is a vertical asymptote.
3. Determine the equations of any other asymptotes.	Since $\lim\limits_{x\to\infty}\left(\frac{\log_e(x)}{x}\right) = 0$, so $y = 0$ (the x-axis) is a horizontal asymptote.
4. Calculate where the graph crosses the coordinate axes.	The graph does not cross the y-axis, but crosses the x-axis when $y = 0$, that is when $\log_e(x) = 0$ $x = 1$ at $(1, 0)$. So the graph crosses the horizontal asymptote, note that if $x > 1$ then $y > 0$ and if $0 < x < 1$ then $y < 0$.
5. Determine the coordinates of the stationary points.	We can differentiate using the quotient rule $y = \frac{u}{v}$ where $u = \log_e(x)$ and $v = x$. $\frac{du}{dx} = \frac{1}{x}$ and $\frac{dv}{dx} = 1$ $\frac{dy}{dx} = f'(x) = \frac{\frac{1}{x} \times x - \log_e(x)}{x^2} = \frac{1 - \log_e(x)}{x^2} = 0$ for stationary points.
6. Establish the nature of the stationary point.	When $\log_e(x) = 1$ that is $x = e$, $y = f(e) = \frac{1}{e} = e^{-1}$. To establish the nature of this stationary point, use a sign test, if $x < e$, $\frac{dy}{dx} > 0$ and when $x > e$, $\frac{dy}{dx} < 0$, the point $\left(e, e^{-1}\right) \approx (2.72, 0.37)$ is an absolute maximum turning point.
7. State the range of the function.	The range of the function is $\left\{y : y \le e^{-1}\right\} = \left(-\infty, e^{-1}\right]$.

8. Use CAS to calculate the second derivative.

$\frac{d^2y}{dx^2}=f''(x)=\frac{2\log_e(x)-3}{x^3}=0$ for inflection points,

$x=\sqrt{e^3}\approx 4.48$,

$y=f\left(\sqrt{e^3}\right)=\frac{3}{2\sqrt{e^3}}\approx 0.33$

The point (4.48, 0.33) is the inflection point.

9. Sketch the graph, showing the critical points using an appropriate scale.

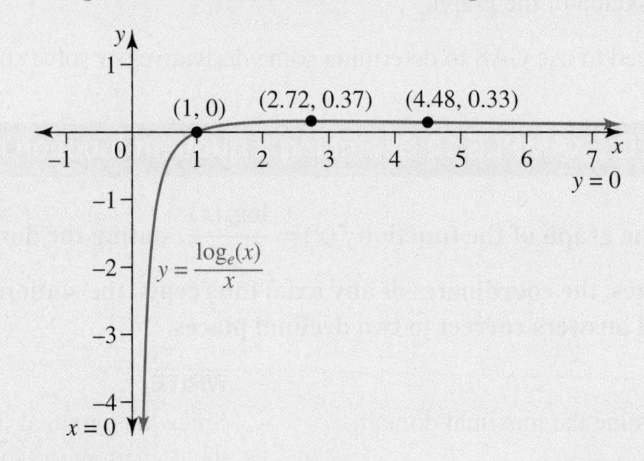

TI | THINK	DISPLAY/WRITE	CASIO | THINK	DISPLAY/WRITE
On a Graphs page, type in the equation of the function and click enter. Select Menu: 6 Analyse Graph: 1 Zero to find the x-intercept. Select Menu: 6 Analyse Graph: 3 Maximum to find the maximum turning point. Select Menu: 6 Analyse Graph: 5 Inflection to find the point of inflection.	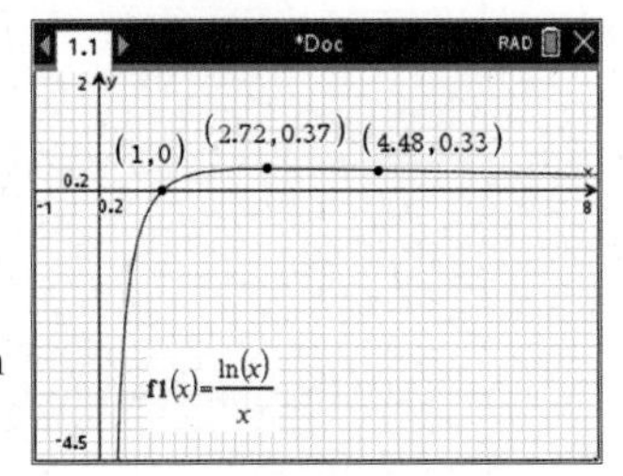	On a Graph Editor screen, type in the equation of the function. Select the graph screen to draw the image. Resize if needed. Select Analysis>G-Solve>Root to find the x-intercept. Select Analysis>G-Solve>Max to find the maximum turning point. Select Analysis>G-Solve >Inflection to find the point of inflection.	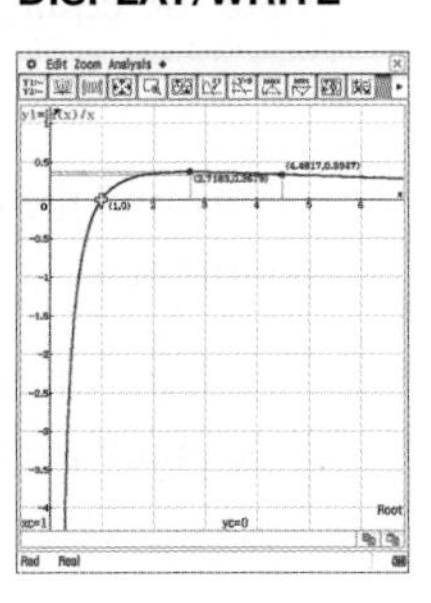

Radical functions

A radical function is a function of the form $\sqrt{f(x)}$ where $f(x)$ is a polynomial. The key thing to consider when sketching functions involving radical functions is that the function inside the square root is not defined when it is negative. This will restrict the domain of the function.

WORKED EXAMPLE 13 Sketching a quotient function involving a radical function

Sketch the graph of the function $f(x)=\frac{\sqrt{x+4}}{x+2}$ stating the maximal domain and range, the equations of all asymptotes, any axial intercepts and the coordinates of any stationary points or inflection points.

THINK	WRITE
1. Determine the maximal domain. The maximal domain of $\sqrt{x+4}$ is $x\geq -4$, but $x+2\neq 0$ so $x\neq -2$.	The maximal domain of $f(x)$ is $[-4,-2)\cup(-2,\infty)$.
2. State the equation of the vertical asymptote, which occurs when the denominator is equal to zero.	There is a vertical asymptote at $x=-2$.

3. Determine the equations of any other asymptotes.

$\lim\limits_{x \to \infty}\left(\frac{\sqrt{x+4}}{x+2}\right) \to 0$ so the y-axis is a horizontal asymptote.

4. Calculate where the graph crosses the coordinate axes.

When $y = 0$, $x = -4$, so the x-intercept is $(-4, 0)$.

When $x = 0$, $f(0) = \frac{\sqrt{4}}{2} = 1$, so the y-intercept is $(0, 1)$.

5. Calculate the first derivative and determine the coordinates of the stationary points.

We can differentiate using CAS.

$$f'(x) = \frac{-x-6}{(2x^2+8x+8)\sqrt{x+4}}$$

Since $x \in [-4, -2) \cup (-2, \infty)$, $\frac{dy}{dx} \neq 0$ so there are no turning points.

6. Use CAS to calculate the second derivative and determine the coordinates of any inflection points.

$$f''(x) = \frac{3x^2+36x+92}{(4x^3+24x^2+48x+32)\sqrt{(x+4)^3}}$$

For inflection points, $f''(x) = 0$: $3x^2 + 36x + 92 = 0$.

Solving for x in the domain, gives $x = -3.69$, $f(-3.69) = -0.33$.

$(-3.69, -0.33)$ is the point of inflection.

7. State the range of the function.

The range is R.

8. Sketch the graph, showing the critical points using an appropriate scale.

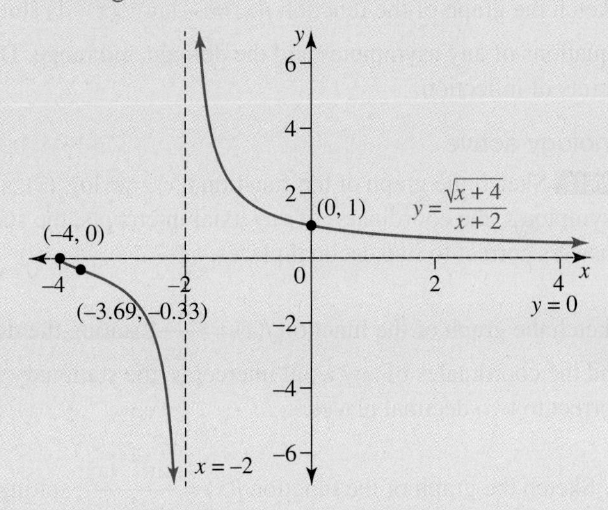

6.4 Exercise

Students, these questions are even better in jacPLUS

Receive immediate feedback and access sample responses

Access additional questions

Track your results and progress

Find all this and MORE in jacPLUS

Technology free

1. Sketch the graph of the function $f(x) = \dfrac{x^4}{e^{2x}}$ by calculating the coordinates of the absolute minimum turning point, the local maximum turning point and the points of inflection.

2. WE11 Sketch the graph of the function $f(x) = \dfrac{3}{\pi}\sin^{-1}\left(\dfrac{3x-2}{4}\right)$ stating the coordinates of all axial intercepts and endpoints. Determine if the graph has any stationary points or points of inflection.

3. Consider the function $f(x) = \dfrac{2}{\pi}\cos^{-1}\left(\dfrac{5-2x}{4}\right)$, sketch the graph stating the coordinates of all axial intercepts and endpoints. Determine if the graph has any stationary points or points of inflection.

4. Sketch the graph of the function $f(x) = \dfrac{4}{\pi}\tan^{-1}(x-1)$ stating the coordinates of all axial intercepts, the equations of any asymptotes and the domain and range. Determine if the graph has any stationary points or points of inflection.

Technology active

5. WE12 Sketch the graph of the function $f(x) = x^2\log_e(x)$, stating the domain and range, equations of any asymptotes, the coordinates of any axial intercepts, the stationary points and points of inflection, giving all answers correct to two decimal places.

6. Sketch the graph of the function $f(x) = \dfrac{e^{-2x}}{x}$, stating the domain and range, equations of any asymptotes and the coordinates of any axial intercepts, the stationary points and points of inflection, giving all answers correct to two decimal places.

7. a. Sketch the graph of the function $f(x) = \dfrac{\sin^{-1}(x)}{x}$, stating the domain and range, and coordinates of the endpoints. Determine if the graph has any stationary points, inflection points or points of discontinuity.

   b. Sketch the graph of the function $f(x) = \dfrac{1}{\sin^{-1}\left(\frac{x}{2}\right)}$, stating the domain and range, equations of any asymptotes and the coordinates of the endpoints. Determine if the graph has any stationary points or inflection points.

8. Sketch the graph of the function $f(x) = \dfrac{\tan^{-1}(x)}{x}$, stating the domain and range. Determine if the graph has any stationary points or inflection or points of discontinuity.

9. a. Sketch the graph of the function $f(x) = \sin\left(\dfrac{1}{x}\right)$, stating the domain and range and the equations of any asymptotes. Classify the point at the origin.

   b. Sketch the graph of the function $f(x) = \dfrac{\sin(x)}{x}$, stating the coordinates of the absolute minimum and the first local maximum turning points and the equations of any asymptotes. Classify the point at the origin.

10. **WE13** Sketch the graph of the function $f(x) = \dfrac{\sqrt{x+9}}{x+3}$ stating the maximal domain and range, the equations of all asymptotes, any axial intercepts and the coordinates of any stationary points or inflection points.

11. Sketch the graph of the function $f(x) = \dfrac{\sqrt{x+4}}{x-1}$, stating the maximal domain and range, the equations of all asymptotes, any axial intercepts and the coordinates of any stationary points or inflection points.

12. Sketch the graph of the function $f(x) = \dfrac{\sqrt{x^2+4}}{x^2-4}$, stating the maximal domain and range, the equations of all asymptotes, any axial intercepts and the coordinates of any stationary points or inflection points.

13. Consider the function $f(x) = \dfrac{\sqrt{x^2+4x+4}}{(x+2)^2}$.

    a. Sketch the graph of $y = f(x)$ stating the domain and range, classifying all critical points.

    b. Sketch the graph of $y = \dfrac{1}{f(x)}$ stating the domain and range, classifying all critical points.

14. Sketch the graph of the function $f(x) = \dfrac{(x+2)^2}{\sqrt{x^2-4}}$ stating the maximal domain and range, the equations of all asymptotes, any axial intercepts and the coordinates of any stationary points, points of inflection or points of discontinuity.

15. Consider the graph of the function $f(x) = x^n \log_e(x+1)$, where $n \in Z^+$. State the maximal domain, the equations of all asymptotes and classify the point at the origin for values of n.

16. Consider the function $f: R \to R,\ f(x) = x^n e^{-x^2}$ where $n \in Z^+$. State the equations of all asymptotes and determine how many points of inflection and turning points the function has for various values of n.

17. a. Consider the function $f: [-\pi, \pi] \to R,\ f(x) = x^n \sin(x)$ where $n \in Z^+$.
    Determine how many points of inflection and turning points the function has for various values of n.

    b. Consider the function $f: [-\pi, \pi] \to R,\ f(x) = x^n \cos(x)$ where $n \in Z^+$.
    For various values of n determine how many points of inflection and turning points the function has.

18. Consider the function $f(x) = \dfrac{40}{\pi}\sin^{-1}\left(\dfrac{x^2-16}{16}\right)$.

    a. Sketch the graph of $y = f(x)$ stating the coordinates of all axial intercepts and endpoints.

    b. Define the function $f'(x)$ and sketch its graph stating all critical features.

19. Consider the function $f(x) = \dfrac{30}{\pi}\cos^{-1}\left(\dfrac{x^2-49}{49}\right)$.

    a. Sketch the graph of $y = f(x)$ stating the coordinates of all axial intercepts and endpoints.

    b. Define the function $f'(x)$ and sketch its graph stating all critical features.

20. a. The amount of a certain drug, A milligrams, in the bloodstream at a time t hours after it is administered is given by

    $A(t) = 30te^{-\frac{t}{3}}$, for $0 \le t \le 12$.

    i. Determine the time when the amount of the drug, A, in the body is a maximum, and calculate the maximum amount.

    ii. Determine the point of inflection on the graph of A versus t.

    b. The amount of a drug, B milligrams, in the bloodstream at a time t hours after it is administered is given by $B(t) = 15t^2e^{-\frac{t}{2}}$, for $0 \le t \le 12$.

i. Determine the time when the amount of the drug, B, in the body is a maximum, and calculate the maximum amount.

ii. Calculate the points of inflection correct to 2 decimal places on the graph of B versus t.

c. Sketch the graphs of $A(t)$ and $B(t)$ on one set of axes and determine the percentage of the time when the amount of drug B is greater than drug A in the bloodstream.

6.4 Exam questions

Question 1 (1 mark) TECH-ACTIVE

Source: *VCE 2021 Specialist Mathematics Exam 2, Section A, Q3; © VCAA.*

MC The coordinates of the local maxima of the graph of $y = \dfrac{1}{(\cos(ax)+1)^2+3}$, where $a \in R\backslash\{0\}$, are

A. $\left(\dfrac{2\pi k}{a}, \dfrac{1}{7}\right), k \in Z$

B. $\left(\dfrac{2\pi k}{a}, \dfrac{1}{3}\right), k \in Z$

C. $\left(\dfrac{(1+2k)\pi}{2a}, \dfrac{1}{4}\right), k \in Z$

D. $\left(\dfrac{\pi(1+2k)}{a}, \dfrac{1}{4}\right), k \in Z$

E. $\left(\dfrac{\pi(1+2k)}{a}, \dfrac{1}{3}\right), k \in Z$

Question 2 (10 marks) TECH-ACTIVE

Source: *VCE 2020 Specialist Mathematics Exam 2, Section B, Q3; © VCAA.*

Let $f(x) = x^2e^{-x}$.

a. Find an expression for $f'(x)$ and state the coordinates of the stationary points of $f(x)$. **(2 marks)**

b. State the equation(s) of any asymptotes of $f(x)$. **(1 mark)**

c. Sketch the graph of $y = f(x)$ on the axes provided below, labelling the local maximum stationary point and all points of inflection with their coordinates, correct to two decimal places. **(3 marks)**

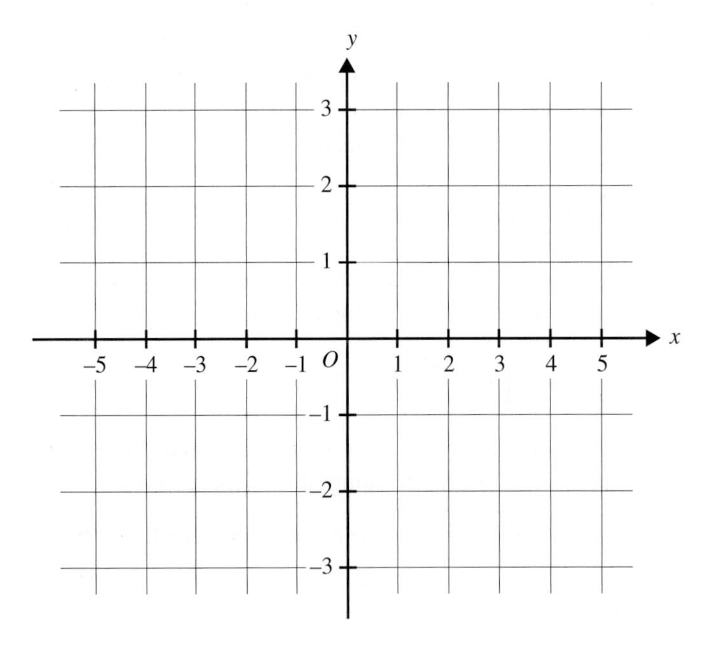

Let $g(x) = x^n e^{-x}$, where $n \in Z$.

d. Write down an expression for $g''(x)$. **(1 mark)**

e. i. Find the non-zero values of x for which $g''(x) = 0$. **(1 mark)**

ii. Complete the following table by stating the values(s) of n for which the graph of $g(x)$ has the given number of points of inflection. **(2 marks)**

Number of points of inflection	Value(s) of n (where $n \in Z$)
0	
1	
2	
3	

Question 3 (1 mark) TECH-ACTIVE

Source: VCE 2019 Specialist Mathematics Exam 2, Section A, Q1; © VCAA.

MC The graph of $f(x) = \dfrac{e^x}{x-1}$ does **not** have a

A. horizontal asymptote.
B. vertical asymptote.
C. local minimum.
D. vertical axis intercept.
E. point of inflection.

More exam questions are available online.

6.5 Review

6.5.1 Summary

doc-37060

Hey students! Now that it's time to revise this topic, go online to:

Access the topic summary

Review your results

Watch teacher-led videos

Practise exam questions

Find all this and MORE in jacPLUS

6.5 Exercise

Technology free: short answer

1. Sketch the graphs of each of the following by finding the coordinates of all axial intercepts, stationary points and points of inflection.
   a. $y = x^3 - 12x$
   b. $y = x^4 - 12x^3 + 48x^2 - 64x$

2. a. Determine for what values of x the graph of $y = x^4 - 8x^3$ is concave up.
   b. For the graph of $y = x^4 + 2bx^3 + 3cx^2$ where $c > 0$, express b in terms of c if the graph has no inflection points.

3. The graph of $y = x^3 - 6x^2 + px + q$ has a stationary point at $(1, 9)$. Determine the values of p and q and calculate the coordinates of all other stationary points, and the inflection point.

4. Sketch the graphs of each of the following. State the equations of any asymptotes and determine the coordinates of any turning points. State the domain and range.
   a. $y = \dfrac{12}{x^2 + 4x}$
   b. $y = \dfrac{8}{7 - 6x - x^2}$

5. The curve $y = \dfrac{A}{x^2 + bx + 12}$ has a local maximum at $(4, -2)$. Determine the values of A and b. State the equations of all asymptotes and the domain and range.

6. Sketch the graph of $y = \dfrac{x - 5}{x^2 - 25}$ stating the equations of any asymptotes and the coordinates of any axial intercepts and any points of discontinuity. State the domain and range.

7. a. Sketch the graph of the function $f(x) = \dfrac{\sqrt{x^2 - 6x + 9}}{(x - 3)^2}$ classifying all critical points.
   b. Sketch the graph of the function $f(x) = \dfrac{(x - 3)^2}{\sqrt{x^2 - 6x + 9}}$ classifying all critical points.

8. Show that the graph of $y = \dfrac{x^2 - a^2}{x - b}$ where $a \neq b$ has stationary points at $x = b \pm \sqrt{b^2 - a^2}$ provided that $|b| > |a|$ and does not have any inflection points.

Technology active: multiple choice

9. **MC** Let $f(x) = x^3 + 3kx^2$, where $x, k \in R$. The gradient of f is decreasing for values of x

A. $[-k, \infty)$ B. $(-\infty, -k]$ C. $[-2k, \infty)$ D. $(-\infty, -2k]$ E. $[0, \infty)$

10. **MC** The graph of the function $y = \dfrac{x^2 + 4x - 12}{x - 2}$

A. has a point of discontinuity at $x = 2$.
B. has only one horizontal asymptote at $y = x$.
C. has only one vertical asymptote at $x = 2$.
D. has a horizontal asymptote at $y = x$ and a vertical asymptote at $x = 2$.
E. crosses the x-axis at $x = -6$ and $x = 2$.

11. **MC** For the graph of the function $y = \dfrac{x - 2}{x^2 - 2x - 8}$ which of the following is false?

A. The graph has a point of discontinuity at $x = 2$.
B. The graph has two vertical asymptotes at $x = -2$ and $x = 4$.
C. The graph has a horizontal asymptote at $y = 0$ and crosses the asymptote at the point $(2, 0)$.
D. The graph crosses the x-axis at $x = 2$.
E. The graph crosses the y-axis at $y = \dfrac{1}{4}$.

12. **MC** The graph of the $y = \dfrac{\sqrt{x}}{x + 2}$ defined on its maximal domain, does not have

A. a horizontal asymptote.
B. a vertical asymptote.
C. an absolute maximum turning point.
D. an axial intercept.
E. a point of inflection.

13. **MC** The graph of the function $y = \dfrac{4}{x^2 + 4}$ has

A. no asymptotes.
B. one horizontal asymptote at $y = 1$.
C. one horizontal asymptote at $y = 0$.
D. two vertical asymptotes at $x = \pm 2$.
E. two vertical asymptotes at $x = \pm 2$ and one horizontal asymptote at $y = 0$.

14. **MC** The graph of $y = \dfrac{1}{ax^2 + b(1 - a)x - b^2}$ where a and b are non-zero real constants, has asymptotes at

A. $x = b,\ x = -\dfrac{b}{a}$ and $y = 1$.
B. $x = b,\ x = -\dfrac{b}{a}$ and $y = 0$.
C. $x = -b,\ x = \dfrac{b}{a}$ and $y = 0$.
D. $x = -b$ and $x = \dfrac{b}{a}$ only.
E. $x = b$ and $x = -\dfrac{b}{a}$ only.

15. **MC** For the graph of $y = \dfrac{x^2 + a^2}{x^2 - a^2}$ where a is a non-zero real constant, then

A. the maximal domain is $R\backslash\{\pm a\}$ and the range is $R\backslash\{0\}$.
B. the graph has vertical asymptotes at $x = \pm a$ and a horizontal asymptote at $y = 0$.
C. the graph crosses the y-axis at $y = -1$ and this point $(0, -1)$ is a minimum turning point.
D. the graph has vertical asymptotes at $x = \pm a$ and a horizontal asymptote at $y = -1$.
E. the graph does not cross the x-axis, the line $y = 1$ is a horizontal asymptote and the range is $(-\infty, -1] \cup (1, \infty)$.

16. **MC** The graph shown, could be that of a function f whose rule is

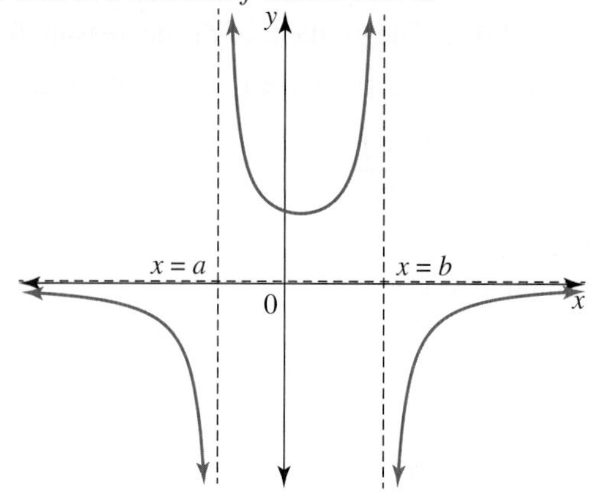

A. $f(x) = \dfrac{1}{x^2 + (a - b)x - ab}$

B. $f(x) = \dfrac{1}{x^2 - (a + b)x + ab}$

C. $f(x) = \dfrac{1}{ab - (a - b)x - x^2}$

D. $f(x) = \dfrac{1}{ab - (a + b)x - x^2}$

E. $f(x) = \dfrac{1}{(a + b)x - ab - x^2}$

17. **MC** Consider the graph of the function with the rule $f(x) = \dfrac{1}{c - 4x - x^2}$ over its maximal domain, where c is a non-zero real constant. Which one of the following statements is false?

A. The x-axis is a horizontal asymptote.

B. The graph crosses the y-axis at $y = \dfrac{1}{c}$.

C. The graph has a minimum turning point at $x = -2$.

D. If $c < -4$, then the graph has two vertical asymptotes.

E. If $c = -4$, then the graph has only one vertical asymptote.

18. **MC** Consider the graph of the function with the rule $f(x) = \dfrac{1}{x^2 + 2bx + 9}$ over its maximal domain, where b is a real number. Which one of the following statements is false?

A. If $|b| < 3$ then the graph has no vertical asymptotes.

B. If $|b| > 3$ then the graph has two vertical asymptotes.

C. The x-axis is a horizontal asymptote.

D. The graph has a minimum turning point at $\left(-b, \dfrac{1}{9 - b^2}\right)$.

E. The graph crosses the y-axis at $y = \dfrac{1}{9}$.

Technology active: extended response

19. a. Sketch the graph of $f(x) = \dfrac{\sqrt{x + 9}}{x - 3}$ stating the maximal domain and range, the equations of all asymptotes, any axial intercepts and the coordinates of any stationary points or inflection points.

b. Consider the graph of $f(x) = \dfrac{\sqrt{x + a}}{x + b}$ where a and b are real.

i. Determine a relationship between a and b such that the graph does not cross the x-axis.

ii. Determine a relationship between a and b such that the graph does not have a vertical asymptote.

20. Sketch the graph of the function $f: [-2\pi, 2\pi] \to R,\ f(x) = \dfrac{x}{\cos(x)}$ stating the equations of all asymptotes, the coordinates of the endpoints, stationary points and inflection points.

6.5 Exam questions

Question 1 (10 marks) TECH-ACTIVE

Source: *VCE 2021 Specialist Mathematics Exam 2, Section B, Q1; © VCAA.*

Let $f(x) = \dfrac{(2x-3)(x+5)}{(x-1)(x+2)}$.

a. Express $f(x)$ in the form $A + \dfrac{Bx + C}{(x-1)(x+2)}$, where A, B and C are real constants. **(1 mark)**

b. State the equations of the asymptotes of the graph of f. **(2 marks)**

c. Sketch the graph of f on the set of axes below. Label the asymptotes with their equations, and label the maximum turning point and the point of inflection with their coordinates, correct to two decimal places. Label the intercepts with the coordinate axes. **(3 marks)**

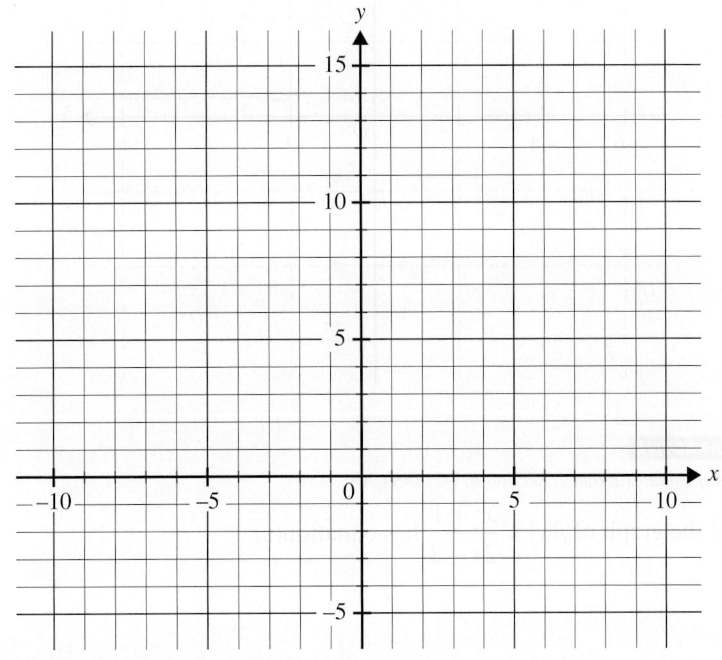

d. Let, $g_k(x) = \dfrac{(2x-3)(x+5)}{(x-k)(x+2)}$ where k is a real constant.

i. For what values of k will the graph of g_k have two asymptotes? **(2 marks)**

ii. Given that the graph of g_k has more than two asymptotes, for what values of k will the graph of g_k have no stationary points? **(2 marks)**

Question 2 (5 marks) TECH-FREE

Source: *VCE 2020 Specialist Mathematics Exam 1, Q6; © VCAA.*

Let $f(x) = \arctan(3x - 6) + \pi$.

a. Show that $f'(x) = \dfrac{3}{9x^2 - 36x + 37}$. **(1 mark)**

b. Hence, show that the graph of f has a point of inflection at $x = 2$. **(2 marks)**

c. Sketch the graph of $y = f(x)$ on the axes provided below. Label any asymptotes with their equations and the point of inflection with its coordinates. **(2 marks)**

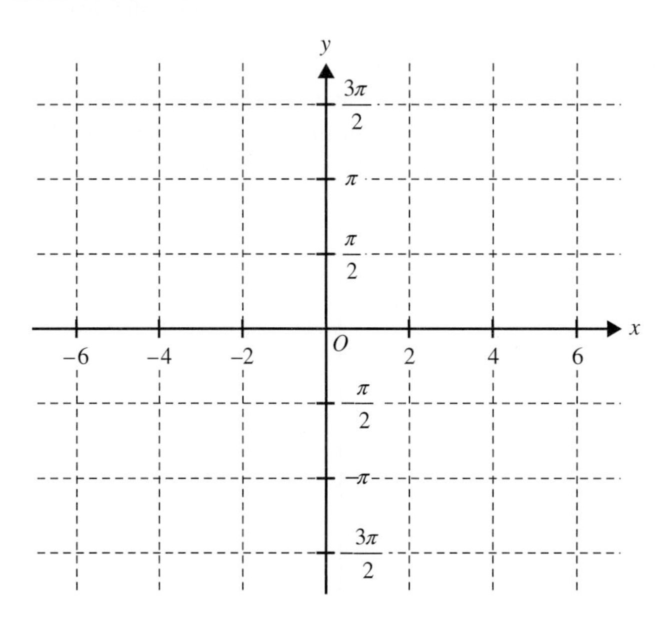

Question 3 (1 mark) TECH-ACTIVE

Source: *VCE 2019 Specialist Mathematics Exam 2, Section A, Q2; © VCAA.*

MC The asymptote(s) of the graph of $f(x) = \dfrac{x^2 + 1}{2x - 8}$ has equation(s)

A. $x = 4$

B. $x = 4$ and $y = \dfrac{x}{2}$

C. $x = 4$ and $y = \dfrac{x}{2} + 2$

D. $x = 8$ and $y = \dfrac{x}{2}$

E. $x = 8$ and $y = 2x + 2$

Question 4 (8 marks) TECH-ACTIVE

Source: *VCE 2017 Specialist Mathematics Exam 2, Section B, Q1 a,b; © VCAA.*

Let $f: D \to R, f(x) = \dfrac{x}{1 + x^3}$, where D is the maximal domain of f.

a. i. Find the equations of any asymptotes of the graph of f. **(1 mark)**

ii. Find $f'(x)$ and state the coordinates of any stationary points of the graph of f, correct to two decimal places. **(2 marks)**

iii. Find the coordinates of any point of inflection of the graph of f, correct to two decimal places. **(2 marks)**

b. Sketch the graph of $f(x) = \dfrac{x}{1 + x^3}$ form $x = -3$ to $x = 3$ on the axes provided below, marking all stationary points, points of inflection and intercepts with axes, labelling them with their coordinates. Show any asymptotes and label them with their equations. **(3 marks)**

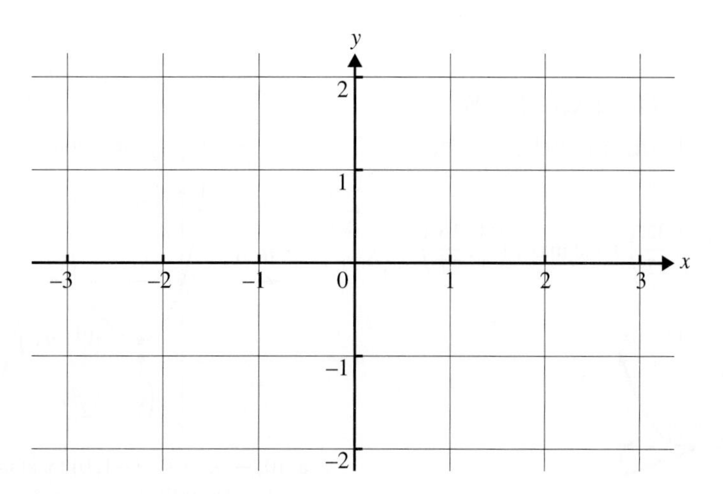

Question 5 (1 mark) TECH-ACTIVE

***Source:** VCE 2014 Specialist Mathematics Exam 2, Section A, Q3; © VCAA.*

MC The features of the graph of the function with rule $f(x) = \dfrac{x^2 - 4x + 3}{x^2 - x - 6}$ include

A. asymptotes at $x = 1$ and $x = -2$
B. asymptotes at $x = 3$ and $x = -2$
C. an asymptote at $x = 1$ and a point of discontinuity at $x = 3$
D. an asymptote at $x = -2$ and a point of discontinuity at $x = 3$
E. an asymptote at $x = 3$ and a point of discontinuity at $x = -2$

More exam questions are available online.

Answers

Topic 6 Functions and graphs

6.2 Sketching graphs of cubics and quartics

6.2 Exercise

1. $(0,0)$, $(2,0)$ local min., $\left(\frac{2}{3},\frac{32}{27}\right)$ local max., $\left(\frac{4}{3},\frac{16}{27}\right)$ inflection

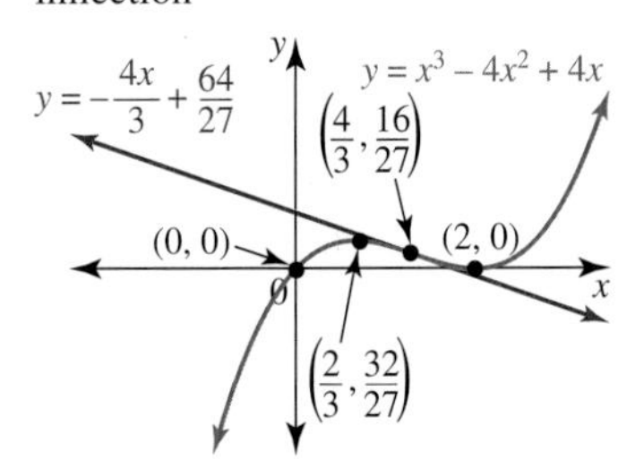

2. a. $(0,0)$, $\left(\pm 3\sqrt{3},0\right)$, $(3,-54)$ local min., $(-3,54)$ local max., $(0,0)$ inflection

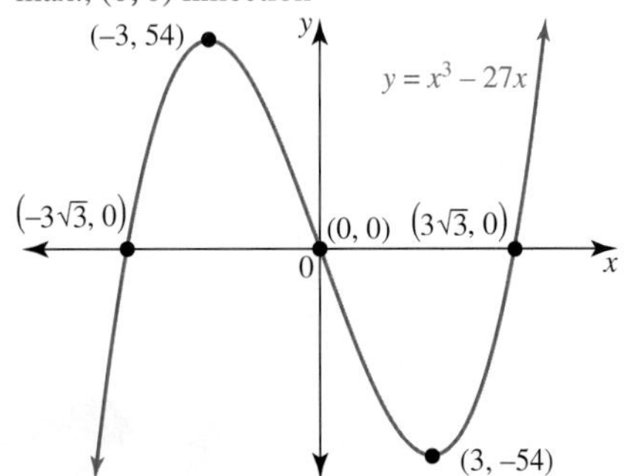

b. $(0,0)$, $(\pm 3,0)$, $\left(-\sqrt{3},-6\sqrt{3}\right)$ local min., $\left(\sqrt{3},6\sqrt{3}\right)$ local max., $(0,0)$ inflection

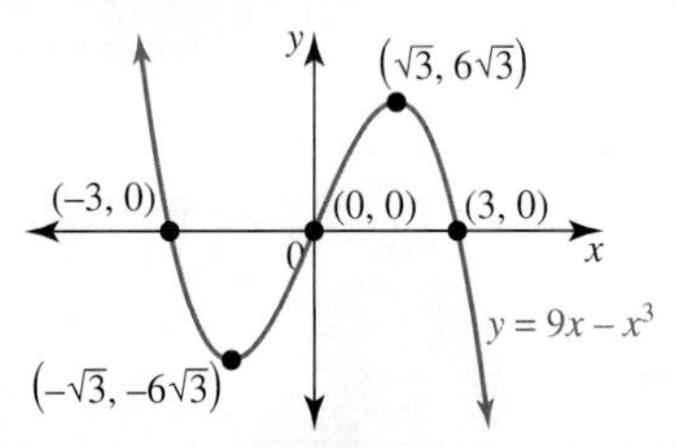

3. a. $(0,0)$, $(-6,0)$ local max., $(-2,-32)$ local min., $(-4,-16)$ inflection

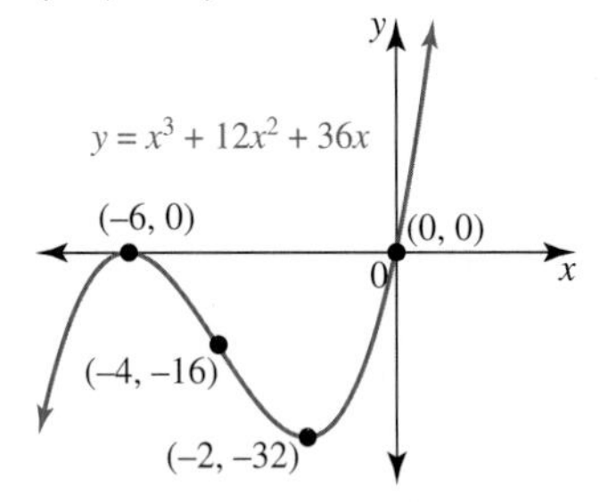

b. $(0,0)$, $(5,0)$ local max., $\left(\frac{5}{3},-18\frac{14}{27}\right)$ local min., $\left(\frac{10}{3},-9\frac{7}{27}\right)$ inflection

4. a. $(0,-5)$, $(5,0)$, $(-1,0)$ local max., $(3,-32)$ local min., $(1,-16)$ inflection

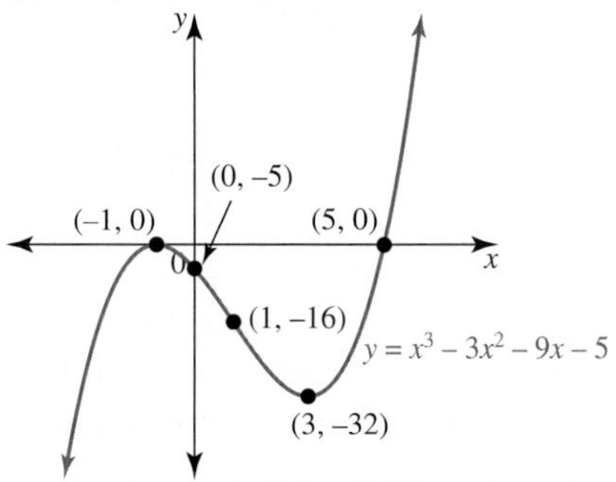

b. $(0,-25)$, $(-1,0)$, $(5,0)$ local max., $(1,-32)$ local min., $(3,-16)$ inflection

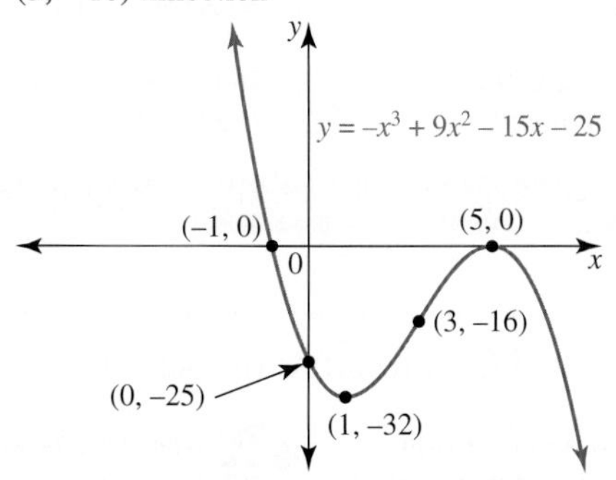

5. a. $(0,16)$, $(\pm 4,0)$, $(1,0)$, $(-2,36)$ local max., $\left(\frac{8}{3},-14\frac{22}{27}\right)$ local min., $\left(\frac{1}{3},10\frac{16}{27}\right)$ inflection

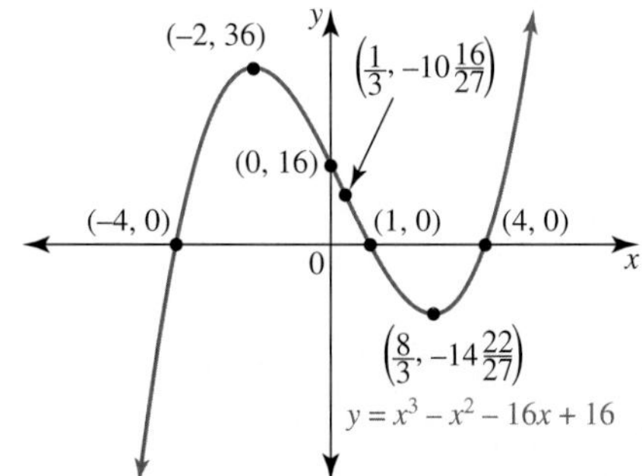

b. $(0,12)$, $(-6,0)$, $(-1,0)$, $(2,0)$, $\left(\frac{2}{3},14\frac{22}{27}\right)$ local max., $(-4,-36)$ local min., $\left(-\frac{5}{3},-10\frac{16}{27}\right)$ inflection

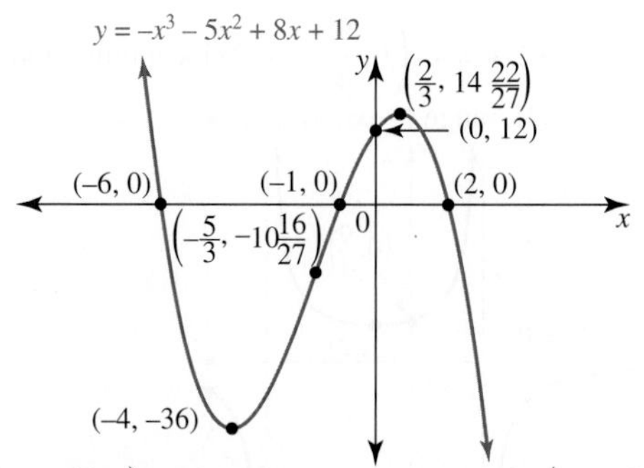

6. $\left(\pm 2\sqrt{5}, 0\right), (\pm 2, 0), (0, 80)$ local max., $\left(\pm 2\sqrt{3}, -64\right)$ min., $(\pm 2, 0)$ inflection

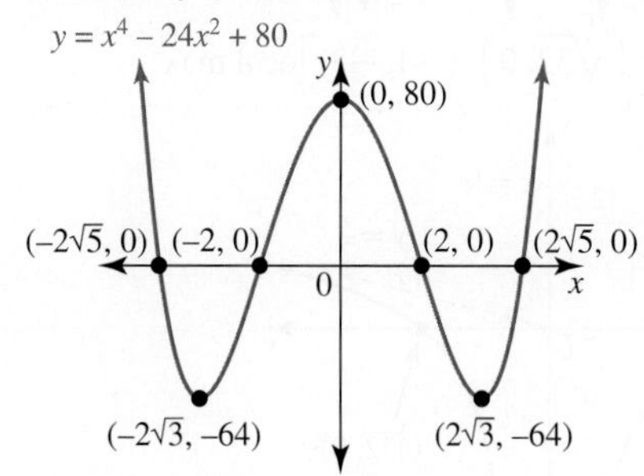

7. a. $(0, 0)$, $(4, 0)$ $(3, -27)$ min., $(2, -16)$ inflection, $(0, 0)$ stationary point of inflection

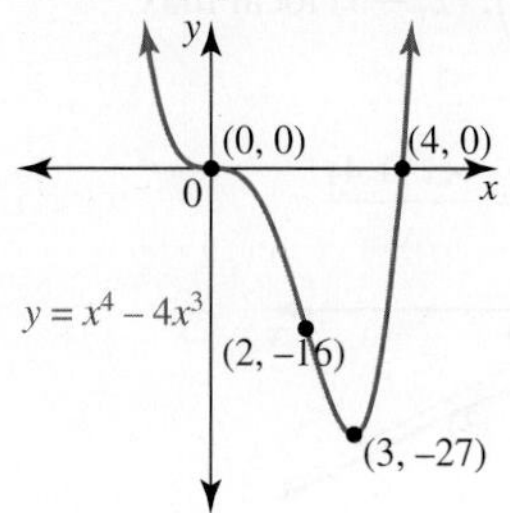

b. $(\pm 2, 0), (0, 0)$ local min., $\left(\pm \sqrt{2}, 4\right)$ max., $\left(\pm \frac{\sqrt{6}}{3}, 2\frac{2}{9}\right)$ inflection

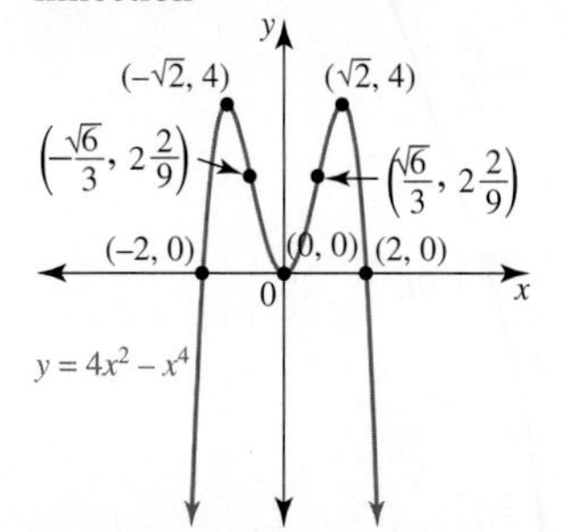

8. a. $(0, -16)$, $(2, 0)$, $(1, -27)$ min., $(-2, 0)$ stationary point of inflection

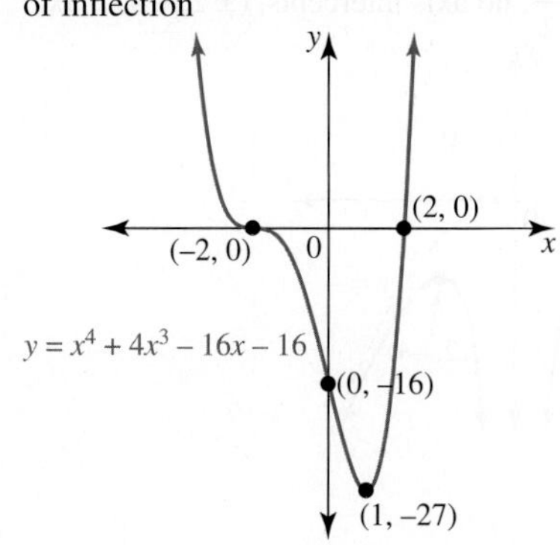

b. $(0, -3)$, $(-3, 0)$, $(-2, -27)$ min., $(-1, -16)$ inflection, $(1, 0)$ stationary point of inflection

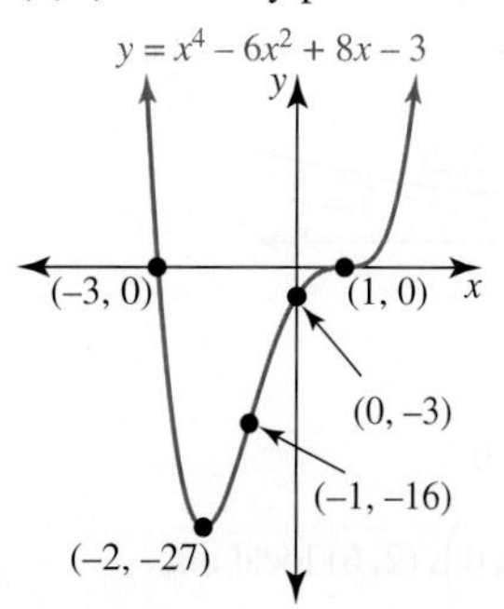

9. $y = 6 - 27x$
10. $b = -3, c = 3, d = -3$
11. $b = -6, c = 15, d = -18$
12. $a = -1, b = -6, c = 9$
13. $a = -1,\ b = 12,\ c = -8;\ \left(\pm \sqrt{6}, 28\right)$ max.
14. Sample responses can be found in the worked solutions in the online resources.
15. Sample responses can be found in the worked solutions in the online resources.
16. Sample responses can be found in the worked solutions in the online resources.

6.2 Exam questions

Note: Mark allocations are available with the fully worked solutions online.

1. B
2. E
3. B

6.3 Sketching graphs of rational functions

6.3 Exercise

1. $x = 0, y = -\frac{x^2}{4}; \left(\sqrt[3]{16}, 0\right), (-2, -3)$ max.

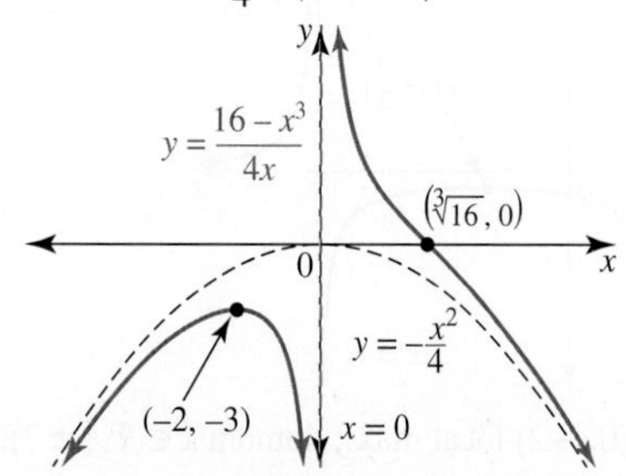

2. $x = 0, y = \frac{x}{2}; (3, 3)$ min., $(-3, -3)$ max.

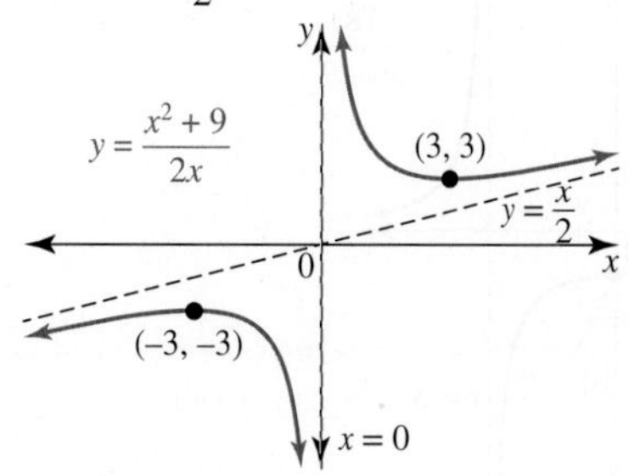

3. a. $x=0, y=\frac{x}{2}$; $(2,2)$ local min., $(-2,-2)$ local max.

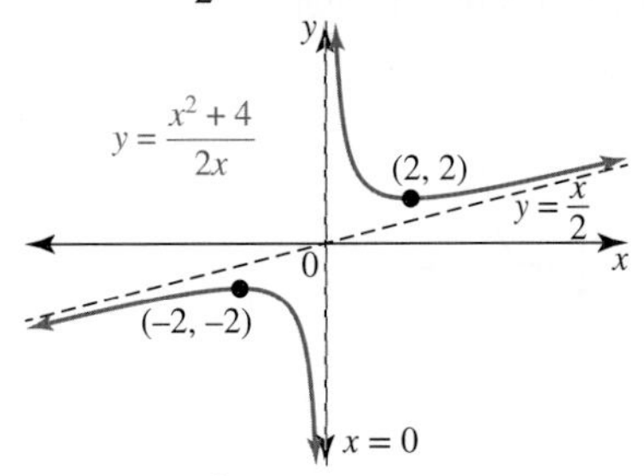

b. $x=0, y=\frac{x^2}{2}$; $\left(\sqrt[3]{-16}, 0\right)$, $(2,6)$ local min.

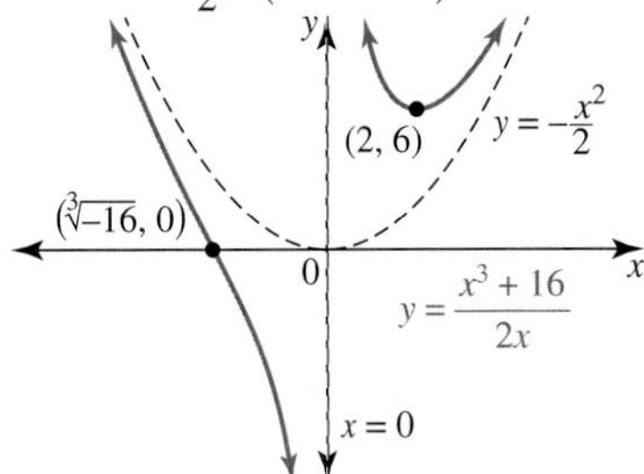

4. $y=0, x=\pm 4$; $(0,-1)$ local min.; domain $x \in R\backslash\{\pm 4\}$, range $(-\infty, 0) \cup [1, \infty)$

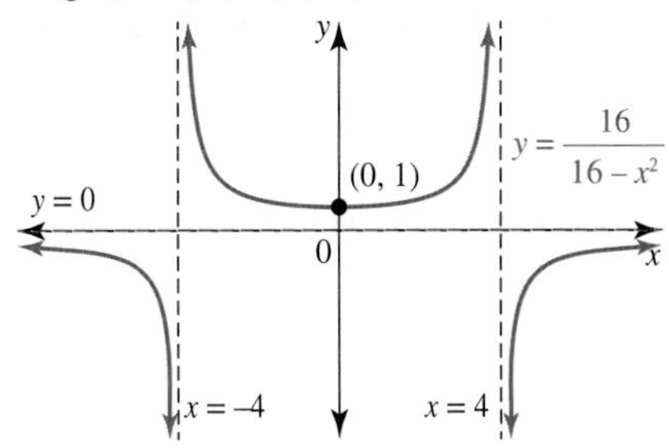

5. $y=0, x=-2, x=6$; $\left(2, -\frac{3}{4}\right)$ local max.; domain $x \in R\backslash\{-2, 6\}$, range $\left(-\infty, -\frac{3}{4}\right] \cup (0, \infty)$

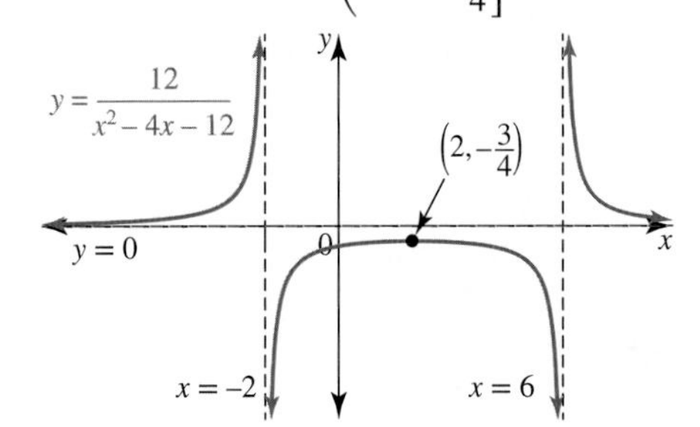

6. a. $y=0, x=\pm 3$; $(0,-2)$ local max.; domain $x \in R\backslash\{\pm 3\}$, range$(-\infty, -2] \cup (0, \infty)$

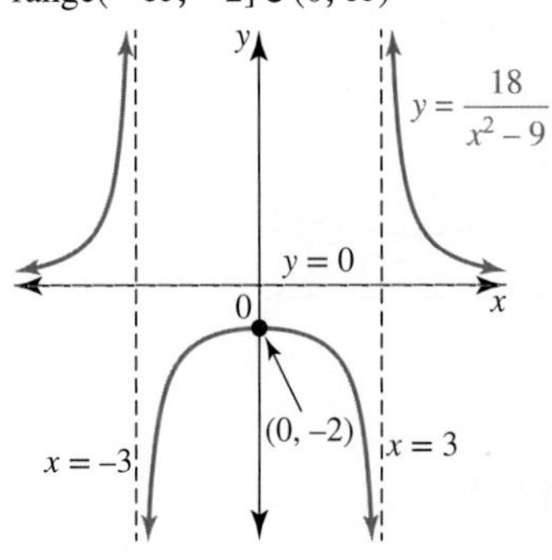

b. $y=0, x=-2, x=4$; $\left(0, \frac{9}{4}\right)$, $(1,2)$ local min.; domain $x \in R\backslash\{-2, 4\}$, range $(-\infty, 0) \cup [2, \infty)$

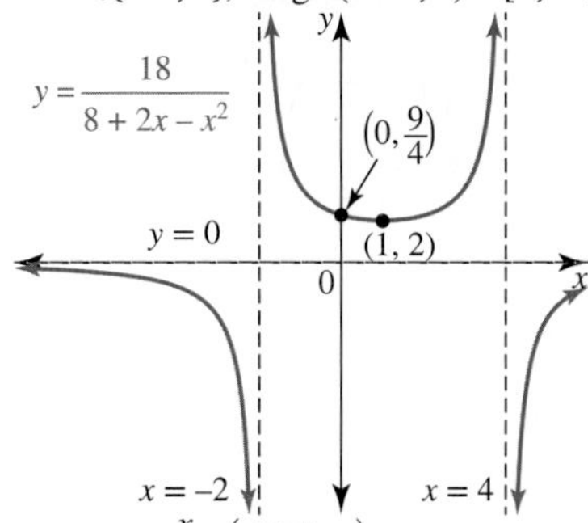

7. a. $x=0, y=\frac{x}{2}$; $\left(\sqrt[3]{32}, 0\right)$, $(-4,-3)$ local max.

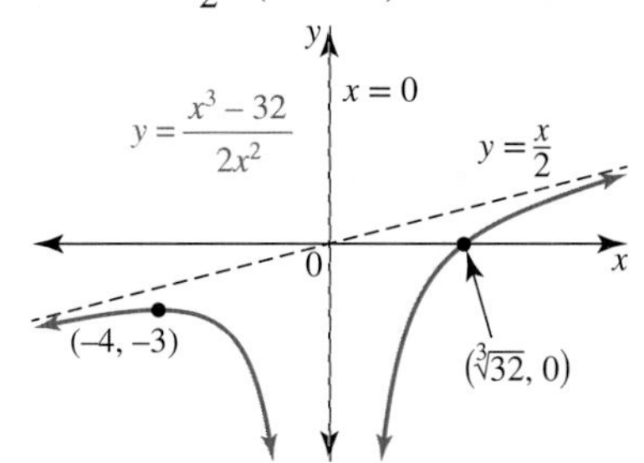

b. $x=0, y=-x$; $\left(\sqrt[3]{-4}, 0\right)$, $(2,-3)$ local max.

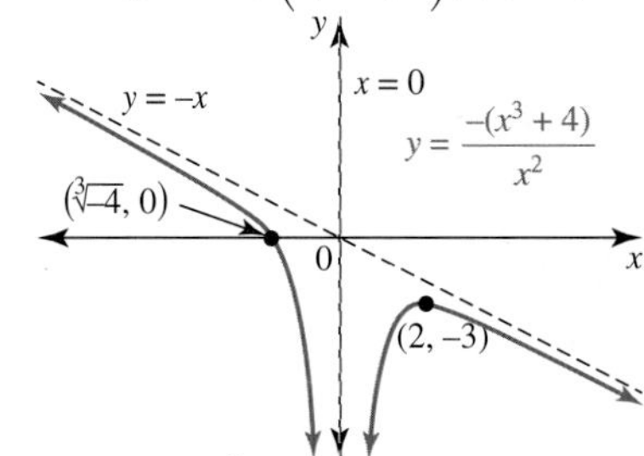

8. a. $x=0, y=\frac{x^2}{2}$; $(\pm 3, 0)$, no turning points

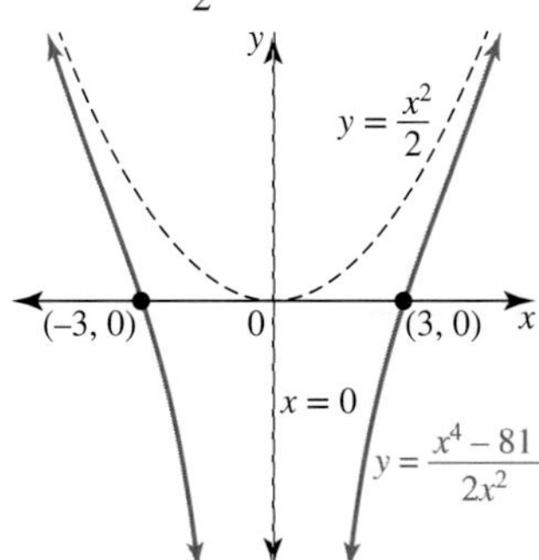

b. $x=0, y=-\frac{x^2}{2}$, no axis intercepts, $(\pm 2, -4)$ max.

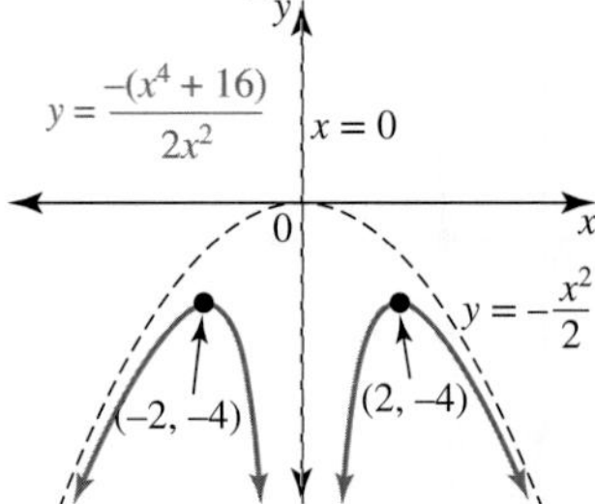

9. $A = 12, b = 8, y = 0, x = -7, x = -1$; domain $x \in R\backslash\{-7, -1\}$, range $\left(-\infty, -\frac{4}{3}\right] \cup (0, \infty)$

10. $A = 50, b = 6, c = 16, y = 0, x = -2, x = 8$; domain $x \in R\backslash\{-2, 8\}$, range $(-\infty, 0) \cup [2, \infty)$

11. a. $(-\infty, -20) \cup (20, \infty)$
    b. ± 20
    c. $(-20, 20)$

12. $(0, 2)$ max., $\left(\pm\sqrt{3}, \frac{3}{2}\right)$ inflection, $y = -\frac{\sqrt{3}x}{4} + \frac{9}{4}$

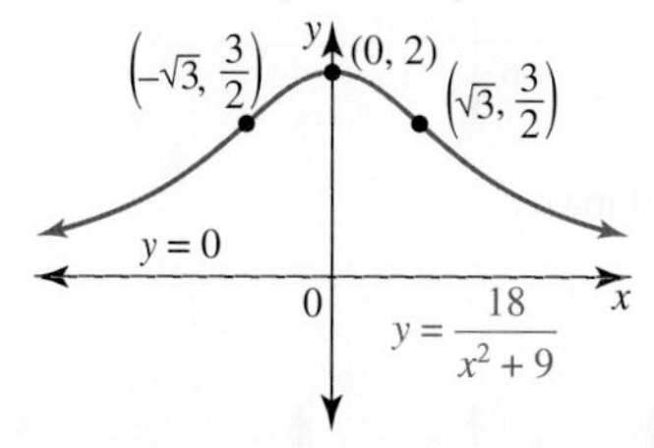

13. $x = 0, y = ax$

14. $x = 0, y = ax^2$

15. $x = 0, y = ax^2$

16. $x = 0, y = ax$

17. a. $y = \frac{x^2 + 5x + 4}{x} = x + 5 + \frac{4}{x}$
$(-1, 0)\ (-4, 0)$
Does not cross the y-axis
$y = x + 5$ oblique asymptote
$x = 0$ vertical asymptote
$(2, 9)$ local min
$(-2, 1)$ local max

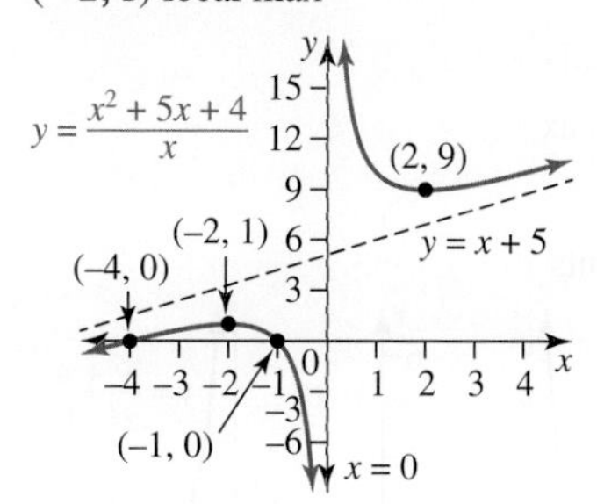

b. $y = \frac{2x^2 + x - 6}{x} = 2x + 1 - \frac{6}{x}$
$(-2, 0)\ \left(\frac{3}{2}, 0\right)$
Does not cross the y-axis
$y = 2x + 1$ oblique asymptote
$x = 0$ vertical asymptote
No turning points

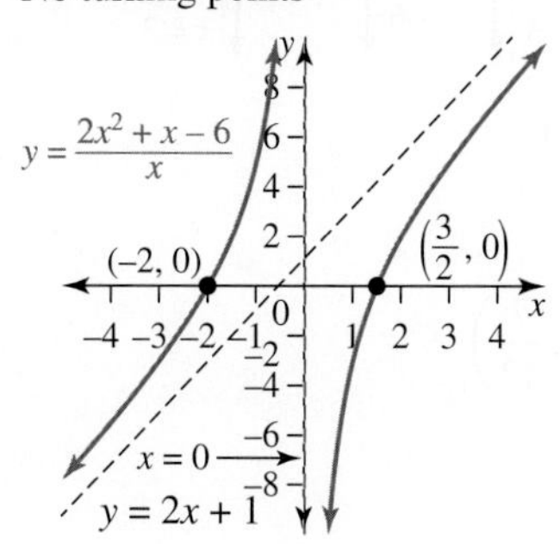

c. $y = \frac{5x - 6 - x^2}{x} = 5 - x - \frac{6}{x}$
Crosses the x-axis $(2, 0)\ (3, 0)$
Does not cross the y-axis
$y = 5 - x$ oblique asymptote
$x = 0$ vertical asymptote
$(2.45, 0.101)$ local max
$(-2.45, 9.9)$ local min

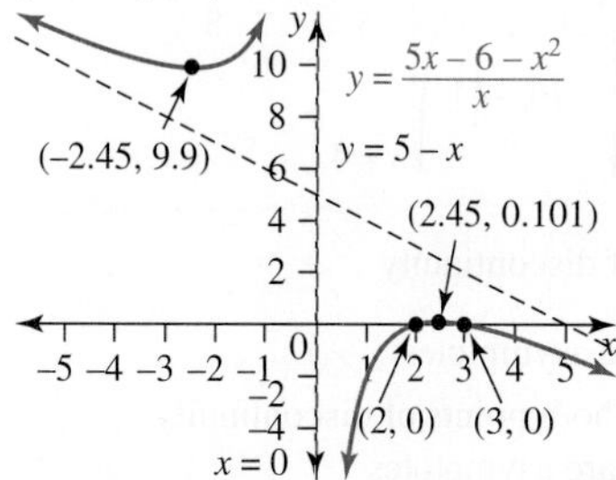

18. $x = 0, y = ax + b$

19. a. $y = \frac{x + 2}{x^2 - 16}$
Crosses the x-axis $(-2, 0)$
Cross the y-axis $\left(0, -\frac{1}{8}\right)$
$y = 0$ horizontal asymptote
$x = \pm 4$ vertical asymptote
No turning points
$(-0.724, -0.082)$ point of inflection

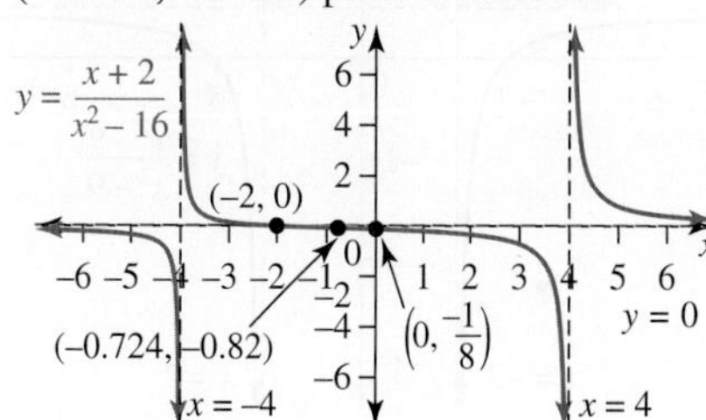

b. $y = \frac{x^2 - 16}{x + 2} = x - 2 - \frac{12}{x + 2}$
Crosses the x-axis $(\pm 4, 0)$
Crosses the y-axis $(0, -8)$
$y = x - 2$ oblique asymptote
$x = -2$ vertical asymptote
No turning points
No points of inflection

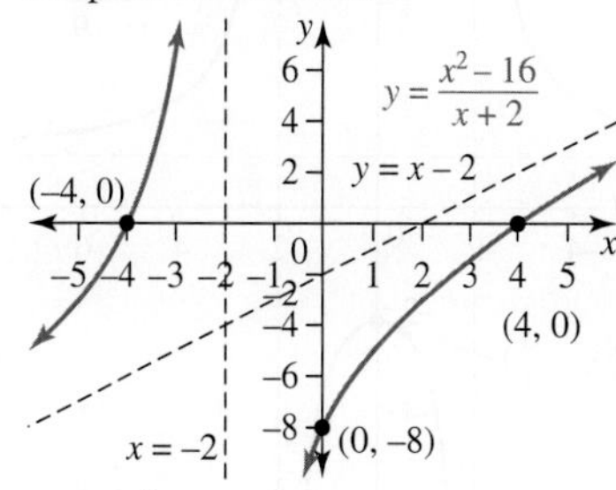

20. $y = \frac{x + 4}{x^2 - 16} = \frac{1}{x - 4}, x \neq -4$
Does not cross the x-axis
Crosses the y-axis $\left(0, -\frac{1}{4}\right)$
Point of discontinuity at $\left(-4, -\frac{1}{8}\right)$
$y = 0$ horizontal asymptote

$x = -4$ vertical asymptote
No turning points
No points of inflection

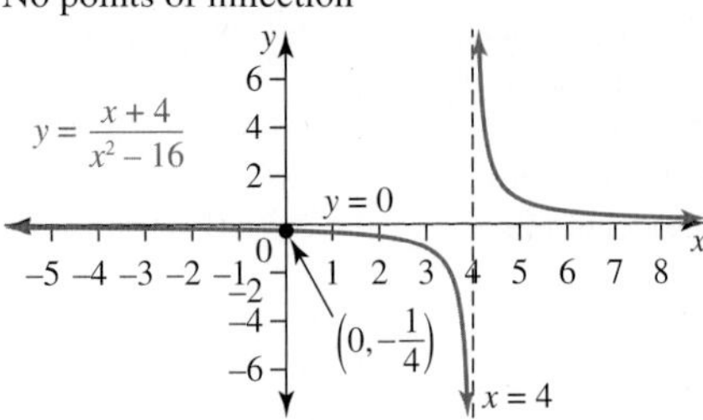

21. $\left(-2, \frac{6}{5}\right)$ point of discontinuity
$x = 3$ and $y = 1$ are asymptotes

22. $(0, -1), (3, 0)$ are both points of discontinuity
$x = -3$ and $y = 1$ are asymptotes

23. a. $(\pm 3, 0)$ $\left(0, \frac{9}{4}\right)$ local min
$x = \pm 2, y = 1$
No inflection points

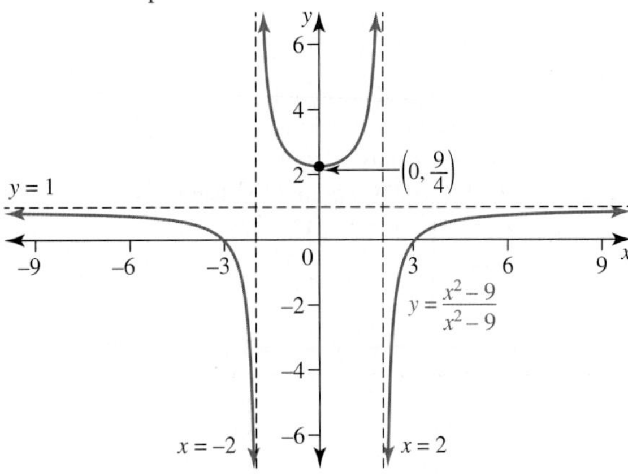

b. Doesn't cross x-axis
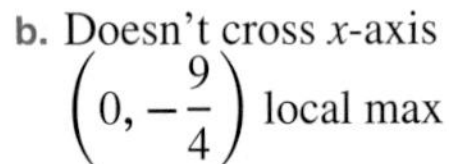
$\left(0, -\frac{9}{4}\right)$ local max
$x = \pm 2, y = 1$
No inflection points

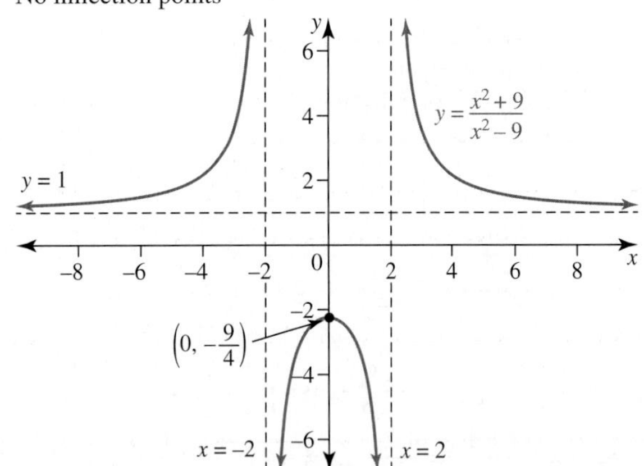

c. $(\pm 3, 0)$ $\left(0, -\frac{9}{4}\right)$ local min
$y = 1$
Inflection $\left(\pm\frac{2\sqrt{3}}{3}, -\frac{23}{16}\right)$

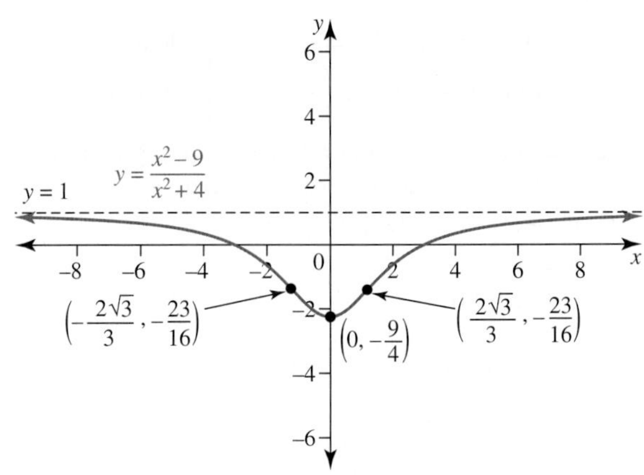

24. a. $(\pm 2, 0)$ $\left(0, \frac{4}{9}\right)$ local max
$x = \pm 3, y = 1$
No inflection points

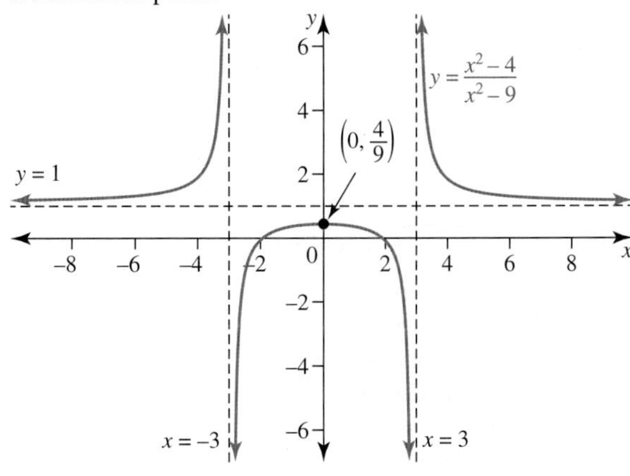

b. Doesn't cross x-axis
$\left(0, -\frac{4}{9}\right)$ local max
$x = \pm 3, y = 1$
No inflection points

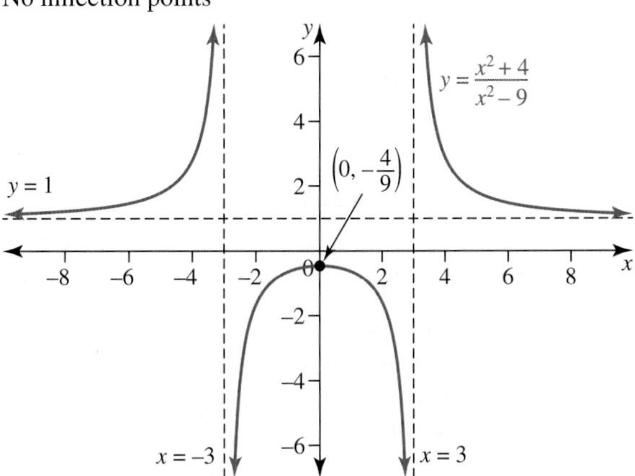

c. Doesn't cross x-axis
$\left(0, \frac{4}{9}\right)$ local min
$y = 1$
Inflection $\left(\pm\sqrt{3}, \frac{7}{12}\right)$

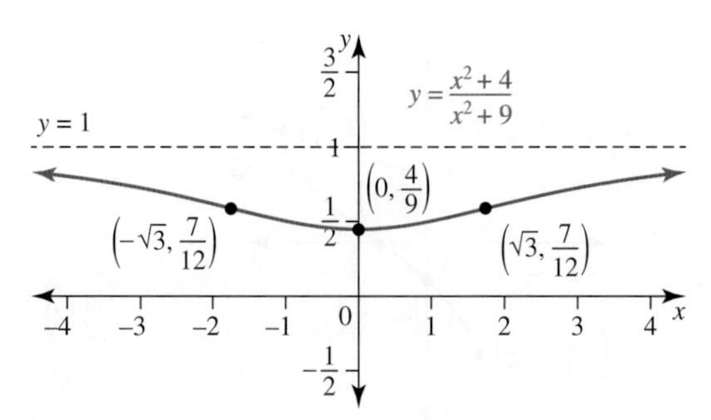

25. a. Doesn't cross x-axis

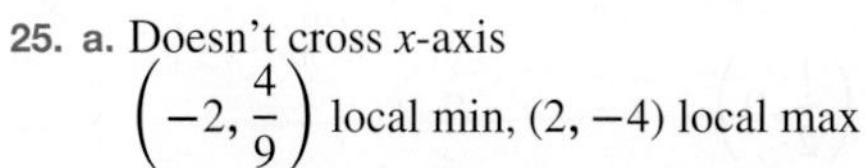

$\left(-2, \frac{4}{9}\right)$ local min, $(2, -4)$ local max

$x = 1,\ x = 4,\ y = 1$ crosses horizontal asymptote at $x = 0$

Inflection $(-4.11, 0.5)$

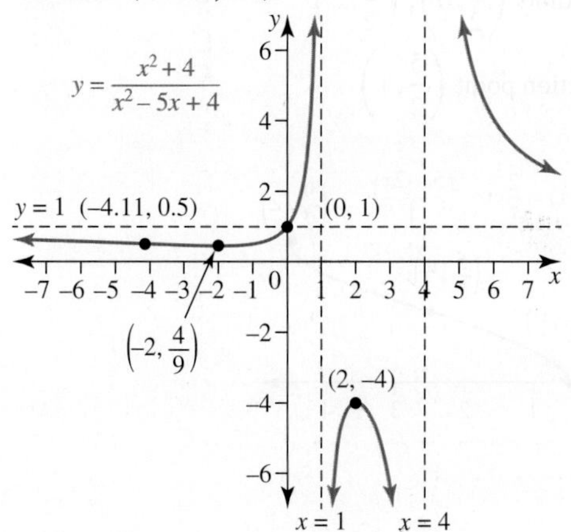

b. Doesn't cross x-axis $\left(0, \frac{3}{5}\right)$

$\left(-1, \frac{1}{3}\right)$ min, $\left(2, \frac{5}{3}\right)$ max

$y = 1$ crosses asymptote at $x = \frac{1}{2}$

Inflection $(-2.1, 0.43)$, $\left(\frac{1}{2}, 1\right)$, $(3.1, 1.58)$

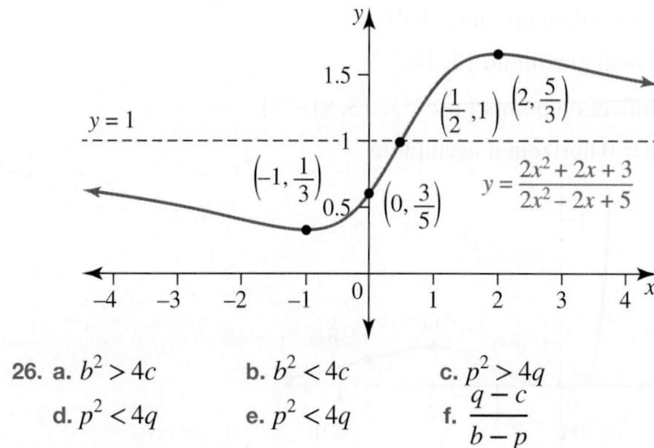

26. a. $b^2 > 4c$ **b.** $b^2 < 4c$ **c.** $p^2 > 4q$

d. $p^2 < 4q$ **e.** $p^2 < 4q$ **f.** $\frac{q-c}{b-p}$

6.3 Exam questions

Note: Mark allocations are available with the fully worked solutions online.

1. See graph at the bottom of the page*

2. C

***1.**

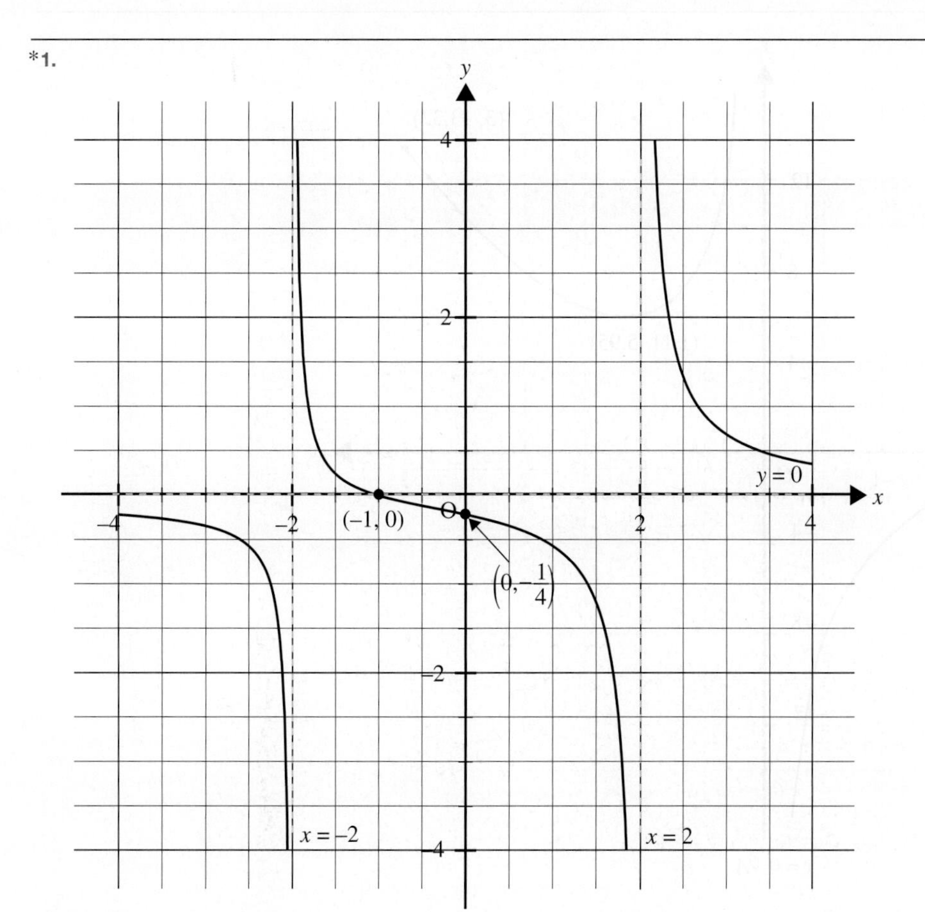

3. a. $(1.11,\ 5.95)$

b. $(-1.59,\ -1.59)$

c. See graph at the bottom of the page*

6.4 Sketching graphs of product and quotient functions

6.4 Exercise

1. Absolute minimum $(0, 0)$

Local maximum $\left(2, 16e^{-4}\right)$

Inflection points $\left(1, e^{-2}\right)$, $\left(3, 81e^{-6}\right)$

$y = 0$ horizontal asymptote

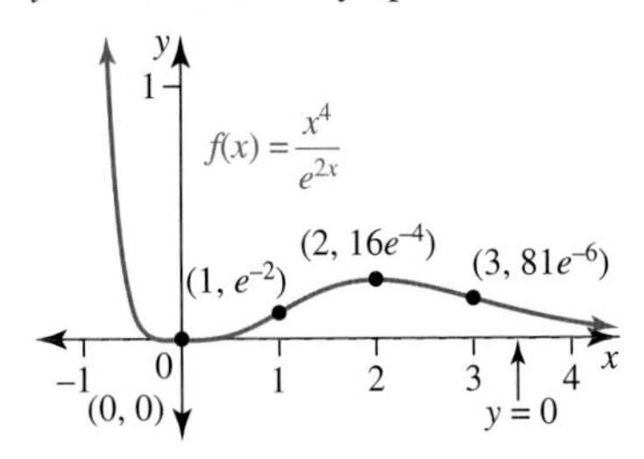

2. Axial intercepts $\left(\frac{2}{3}, 0\right)$, $\left(0, -\frac{1}{2}\right)$

Endpoints $\left(2, \frac{3}{2}\right)$, $\left(-\frac{2}{3}, -\frac{3}{2}\right)$

Inflection point $\left(\frac{2}{3}, 0\right)$

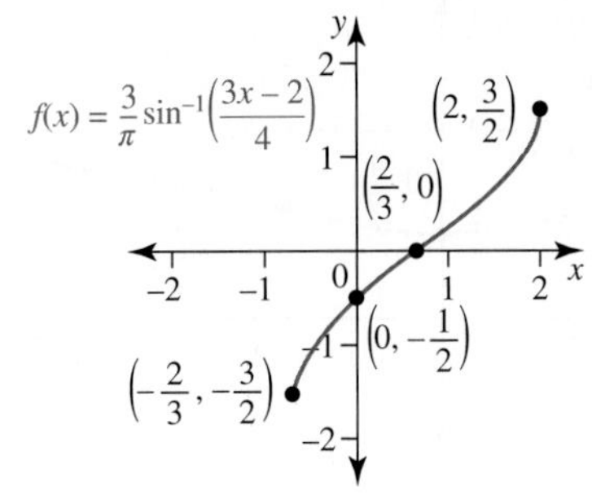

3. Crosses x-axis $\left(\frac{1}{2}, 0\right)$

Doesn't cross y-axis

Endpoints $\left(\frac{1}{2}, 0\right)$, $\left(\frac{9}{2}, 2\right)$

Inflection point $\left(\frac{5}{2}, 1\right)$

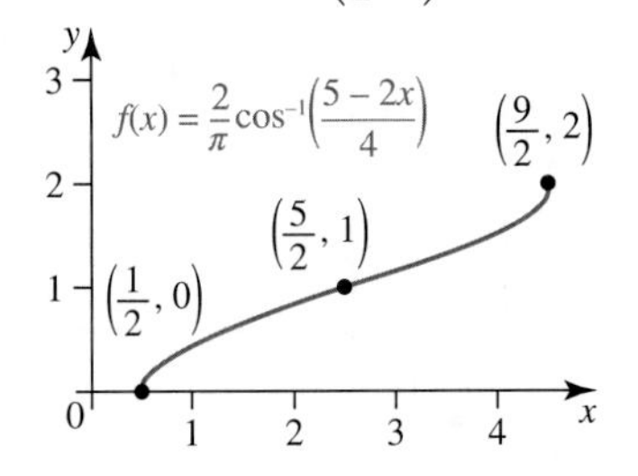

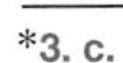
*3. c.

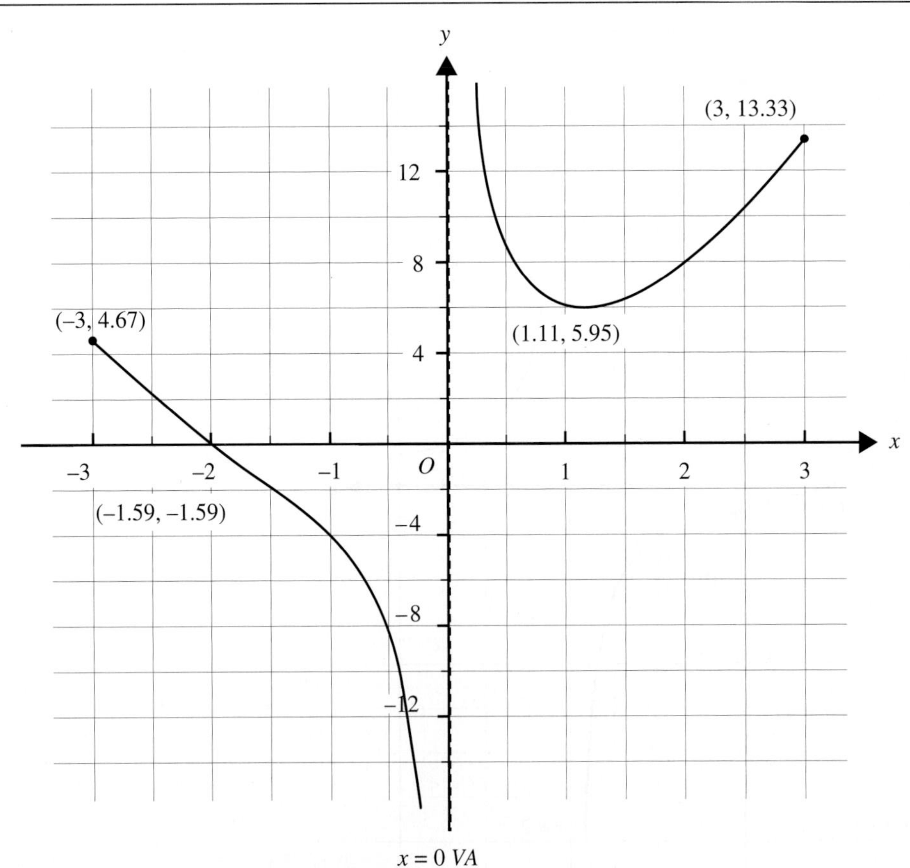

4. Axial intercepts $(1, 0), (0, -1)$

Inflection point $(1, 0)$

$y = \pm 2$ horizontal asymptotes

Domain R, range $(-2, 2)$

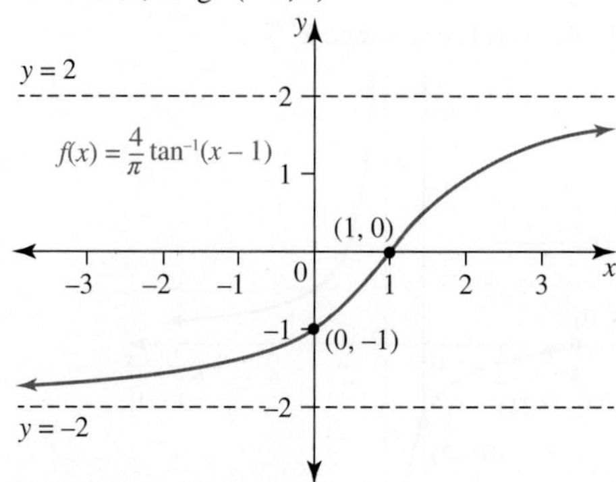

5. Crosses the x-axis $(1, 0)$

Does not cross the y-axis

$y = x^2$ quadratic asymptote

Domain $x > 0$

Range $\left[-\frac{1}{2e}, \infty\right)$

Local min $(0.61, -0.18)$

Inflection point $(0.22, -0.07)$

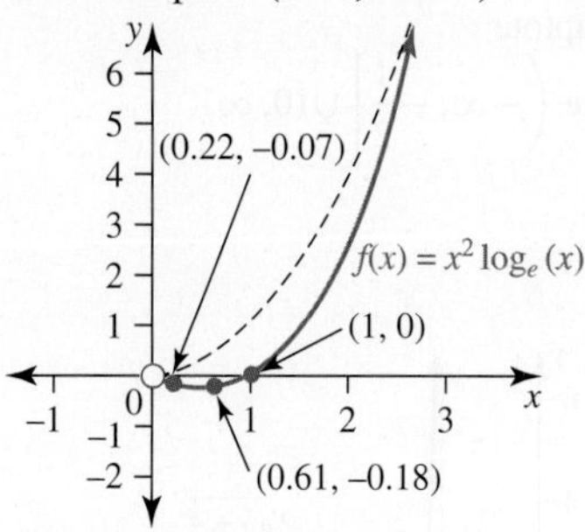

6. Does not cross the x- or y-axis

$x = 0$ vertical asymptote

$y = 0$ horizontal asymptote

Domain $R\backslash\{0\}$

Range $(-\infty, -2e] \cup (0, \infty)$

Local max $\left(-\frac{1}{2}, -2e\right)$

No inflection points

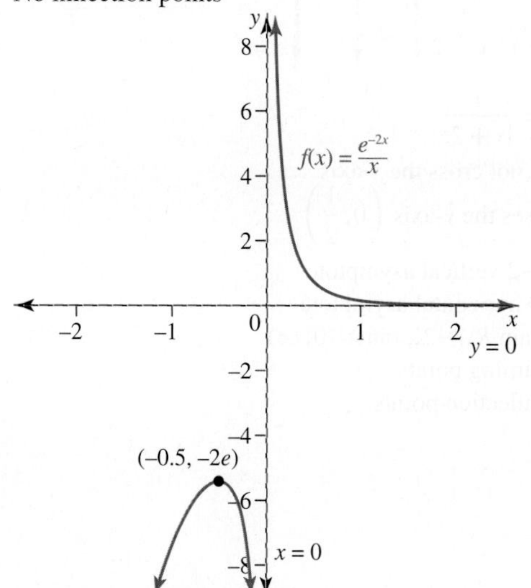

7. a. Endpoints $\left(-1, -\frac{\pi}{2}\right), \left(1, -\frac{\pi}{2}\right)$

Does not cross the x- or y-axis

$x = 0$ is a point of discontinuity

Domain $[-1, 0) \cup (0.1]$

Range $\left(1, \frac{\pi}{2}\right]$

No turning points

No inflection points

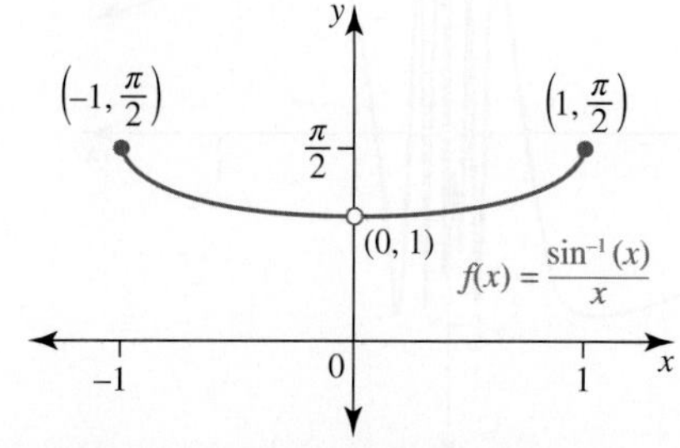

b. Endpoints $\left(-2, -\frac{2}{\pi}\right), \left(2, \frac{2}{\pi}\right)$

Does not cross the x- or y-axis

$x = 0$ vertical asymptote

Domain $[-2, 0) \cup (0, 2]$

Range $\left(-\infty, -\frac{2}{\pi}\right] \cup \left[\frac{2}{\pi}, \infty\right)$

No turning points

No inflection points

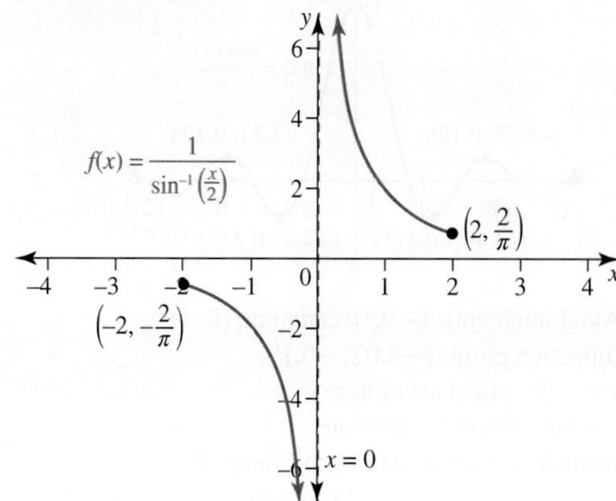

8. Does not cross the x- or y-axis

$x = 0$ is a point of discontinuity

$y = 0$ horizontal asymptote

Domain $R\backslash\{0\}$

Range $(0, 1)$

No turning points

$(\pm 0.824, 0.836)$ inflection points

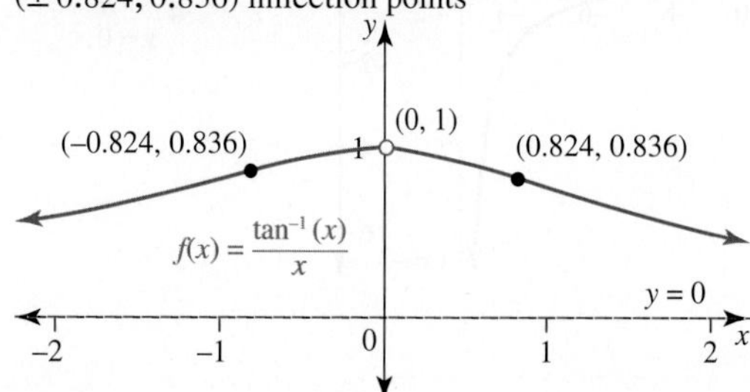

9. a. $x = 0$ is a point of discontinuity
Domain $R\backslash\{0\}$
Range $[-1, 1]$
Infinitely many turning points
Infinitely many inflection points

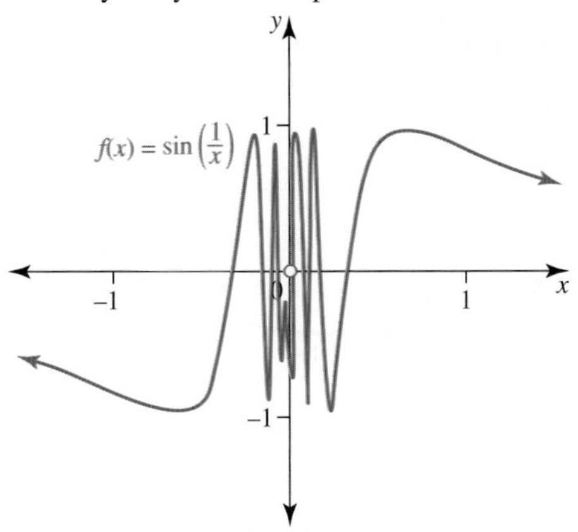

b. $x = 0$ is a point of discontinuity
Domain $R\backslash\{0\}$
Range $[-0.217, 1)$
$(\pm 4.49, -0.217)$ absolute min
Infinitely many turning points
Infinitely many inflection points

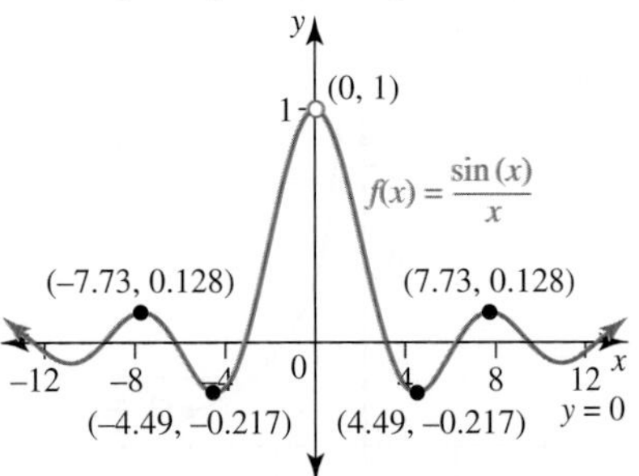

10. Axial intercepts: $(-9, 0)$ endpoint, $(0, 1)$
Inflection point: $(-8.07, -0.19)$
$x = -3$ vertical asymptote
$y = 0$ horizontal asymptote
Domain $[-9, -3) \cup (-3, \infty)$, range R

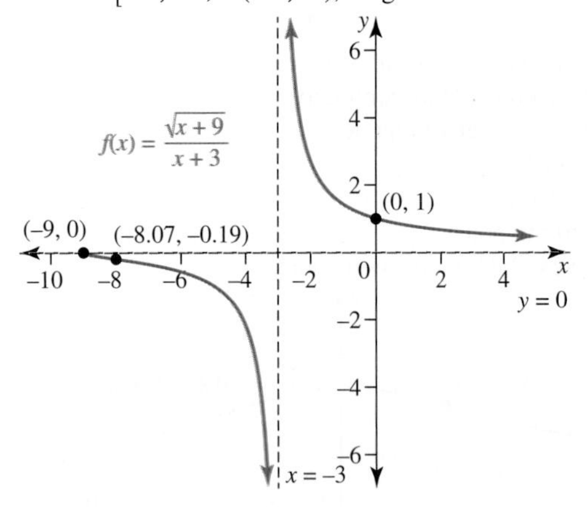

11. Axial intercepts: $(-4, 0)$ endpoint, $(0, -2)$
Inflection point: $(-3.23, -0.21)$
$x = 1$ vertical asymptote
$y = 0$ horizontal asymptote
Domain $[-4, 1) \cup (1, \infty)$, range R

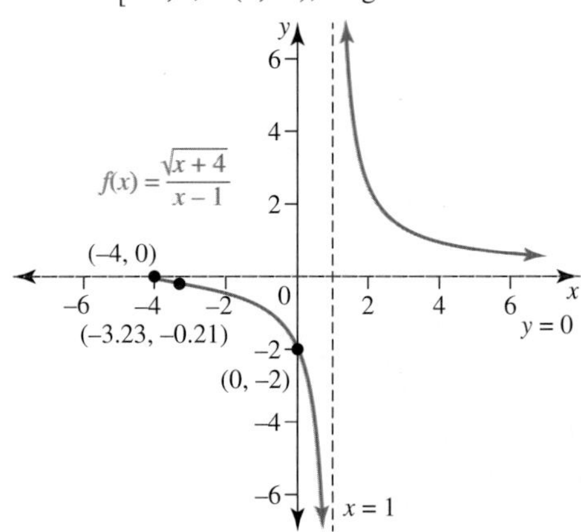

12. Does not cross the x-axis
Crosses the y-axis $\left(0, -\frac{1}{2}\right)$
$x = \pm 2$ vertical asymptote
$y = 0$ horizontal asymptote
Domain $R\backslash\{\pm 2\}$, range $\left(-\infty, -\frac{1}{2}\right] \cup (0, \infty)$
Local max $\left(0, -\frac{1}{2}\right)$
No inflection points

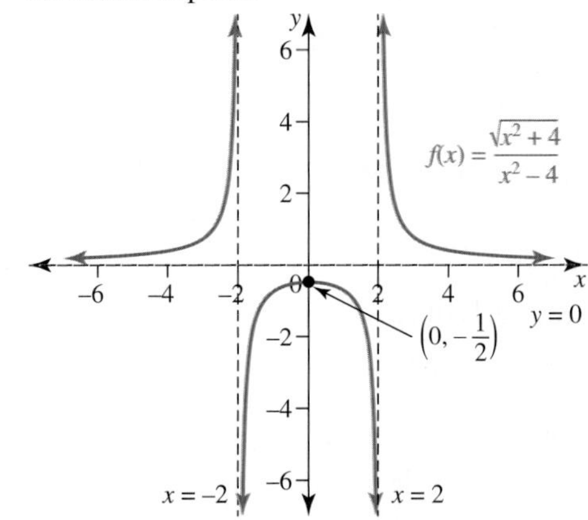

13. a. $f(x) = \frac{1}{|x+2|}$
Does not cross the x-axis
Crosses the y-axis $\left(0, \frac{1}{2}\right)$
$x = -2$ vertical asymptote
$y = 0$ horizontal asymptote
Domain $R\backslash\{-2\}$, range $(0, \infty)$
No turning points
No inflection points

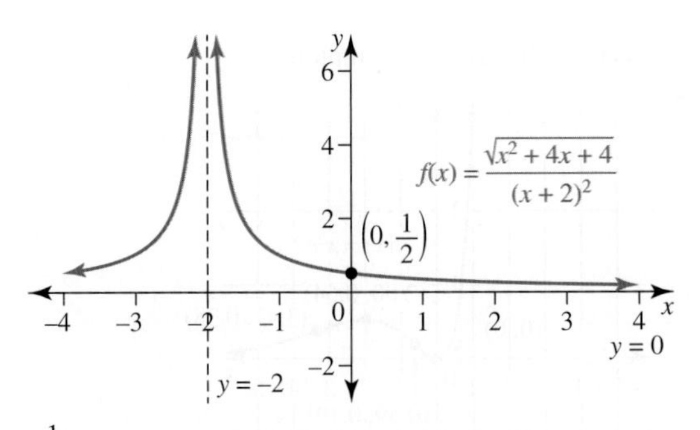

b. $\frac{1}{f(x)} = |x + 2|, \quad x \neq -2$

$x = -2$ is a point of discontinuity
Does not cross the x-axis
Crosses the y-axis $(0, 2)$
Domain $R \backslash \{-2\}$, range $(0, \infty)$
No turning points
No inflection points

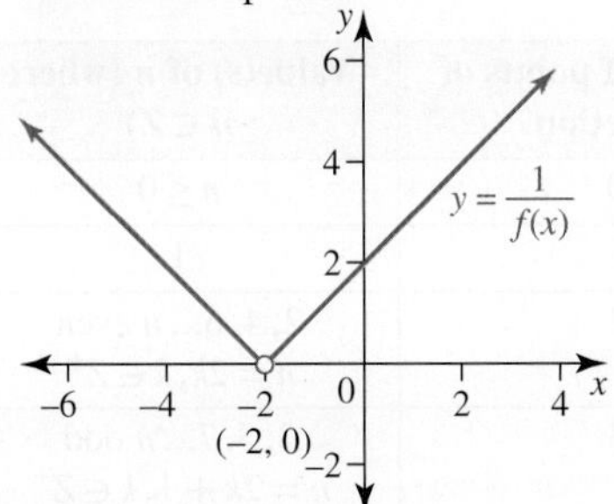

14. $x = -2$ is a point of discontinuity
Does not cross the x- or y-axis
Domain $(-\infty, -2) \cup (2, \infty)$, range $(0, \infty)$
Local min $\left(4, 6\sqrt{3}\right)$
No inflection points

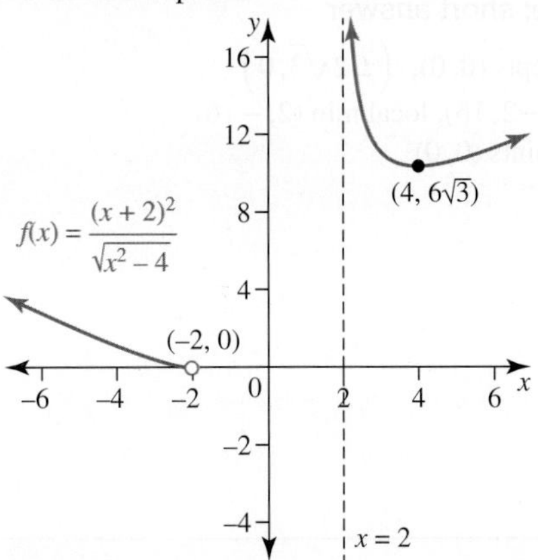

15. $x = -1$ vertical asymptote
Domain $(-1, \infty)$
$y = x^n$ non-linear asymptote
Crosses $y = x^n$ at $x = e - 1$
n even $(0, 0)$ is an inflection point.
n odd $(0, 0)$ is an absolute minimum turning point.

16. $y = 0$ horizontal asymptote
n odd $(0, 0)$ is an inflection point,
four other inflection points, one absolute maximum and one absolute minimum turning points.
n even $(0, 0)$ is an absolute minimum turning point, four other inflection points, two absolute maximum turning points.

17. a. n odd $(0, 0)$ is an absolute minimum turning point, two other inflection points, two absolute maximum turning points.
n even $(0, 0)$ is an inflection point, two other inflection points, one absolute maximum and minimum turning points.

b. n odd $(0, 0)$ is an inflection point, two inflection points, one local maximum and one local minimum turning points.
n even $(0, 0)$ is a minimum turning point, two other inflection points, two absolute maximum turning points.

18. a. Axial intercepts $(\pm 4, 0)$, $(0, -20)$
Endpoints $\left(\pm 4\sqrt{2}, 20\right)$

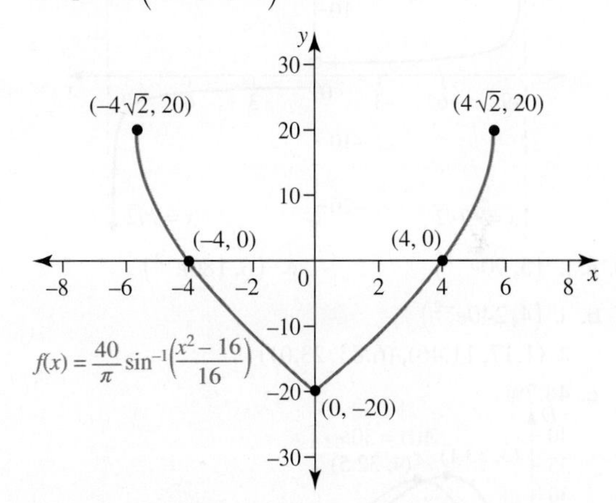

b. $f'(x) = \begin{cases} \frac{80}{\pi\sqrt{32 - x^2}}, & 0 < x < 4\sqrt{2} \\ \frac{-80}{\pi\sqrt{32 - x^2}}, & -4\sqrt{2} < x < 0 \end{cases}$

Points of discontinuity $\left(0, \pm \frac{10\sqrt{2}}{\pi}\right)$
$x = \pm 4\sqrt{2}$ vertical asymptotes

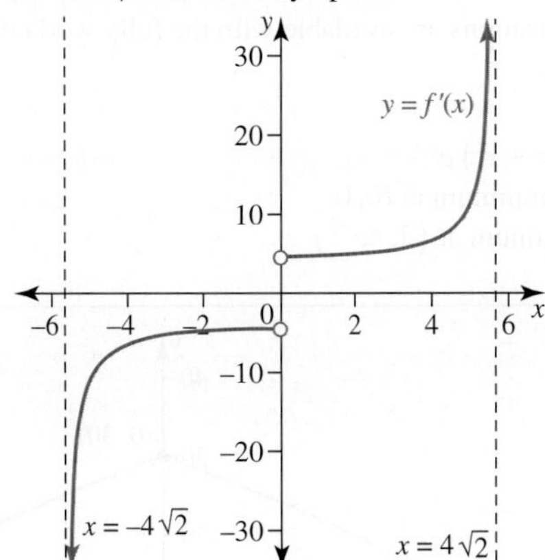

19. a. Axial intercepts $\left(\pm 7\sqrt{2}, 0\right)$, $(0, 30)$

Endpoints $\left(\pm 7\sqrt{2}, 0\right)$

See graph at the bottom of the page*

b. $f'(x) = \begin{cases} \dfrac{-60}{\pi\sqrt{98 - x^2}}, & 0 < x < 7\sqrt{2} \\ \dfrac{60}{\pi\sqrt{98 - x^2}}, & -7\sqrt{2} < x < 0 \end{cases}$

Points of discontinuity $\left(0, \pm\dfrac{30\sqrt{2}}{7\pi}\right)$

$x = \pm 7\sqrt{2}$ vertical asymptotes

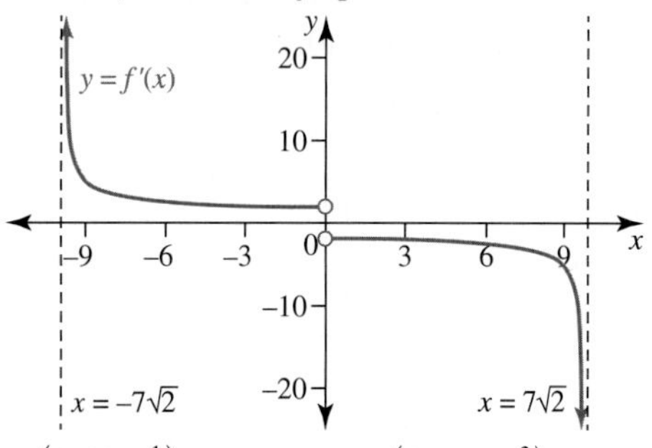

20. a. i. $\left(3, 90e^{-1}\right)$ **ii.** $\left(6, 180e^{-2}\right)$

b. i. $\left(4, 240e^{-2}\right)$

ii. $(1.17, 11.46)$, $(6.83, 23.01)$

c. 44.7%

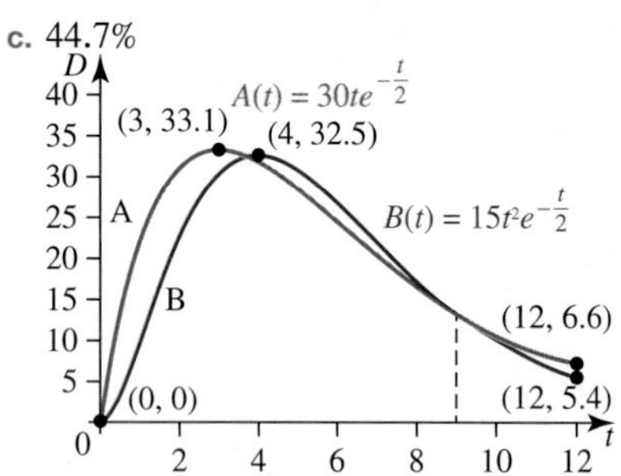

6.4 Exam questions

Note: Mark allocations are available with the fully worked solutions online.

1. E

2. a. $f'(x) = \left(2x - x^2\right)e^{-x}$

Absolute minimum at $(0, 0)$

Local maximum at $\left(2, 4e^{-2}\right)$

b. $y = 0$ is a horizontal asymptote

c.

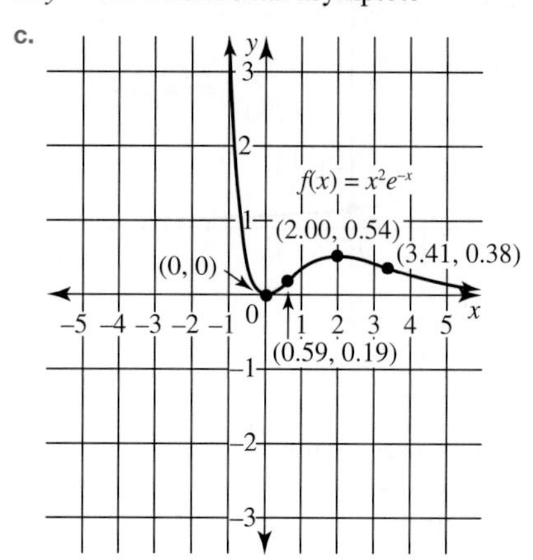

d. $g''(x) = x^{x-2}\left(x^2 - 2nx + n(n-1)\right)e^{-x}$

e. i. $x = n \pm \sqrt{n}$

ii.

Number of points of inflection	Value(s) of n (where $n \in Z$)
0	$n \leq 0$
1	1
2	$2, 4, 6\ldots n$ even $n = 2k, k \in Z^+$
3	$3, 5, 7\ldots n$ odd $n = 2k + 1, k \in Z^+$

3. E

6.5 Review

6.5 Exercise

Technology free: short answer

1. a. Axial intercepts $(0, 0)$, $\left(\pm 2\sqrt{3}, 0\right)$

Local max $(-2, 16)$, local min $(2, -16)$

Inflection points $(0, 0)$

*19. a.

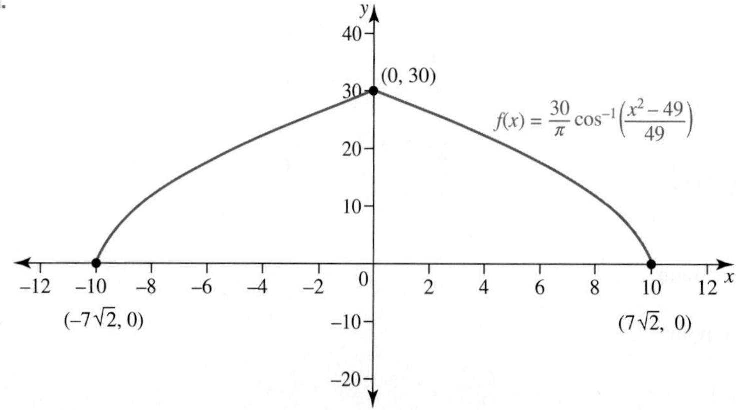

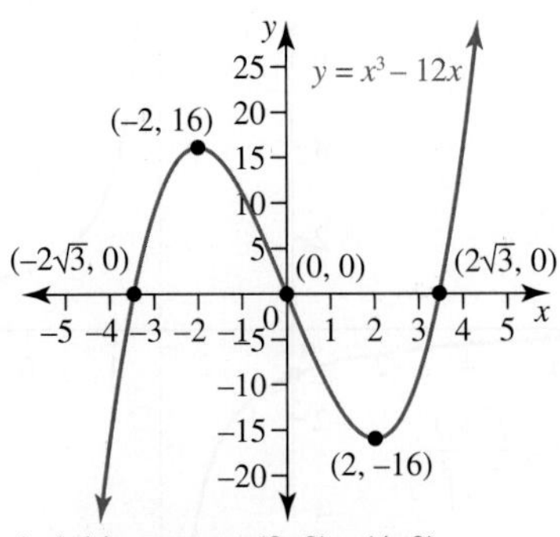

b. Axial intercepts $(0, 0)$, $(4, 0)$
Absolute min $(1, -27)$
Inflection points $(2, -16)$, $(4, 0)$

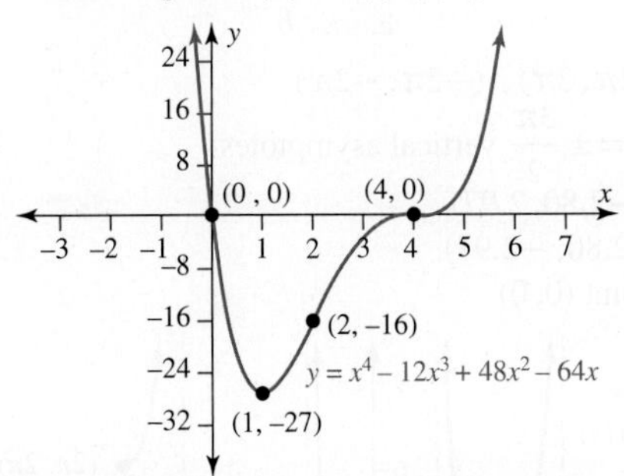

2. a. $(-\infty, 0) \cup (4, \infty)$
b. $\left(-\sqrt{2c}, \sqrt{2c}\right)$

3. $p = 9$, $q = 5$ $(3, 5)$ turning point
$(2, 7)$ inflection point

4. a. $y = 0, x = -4, x = 0$; $(-2, -3)$ local max.; domain $x \in R \setminus \{-4, 0\}$, range $(-\infty, -3] \cup (0, \infty)$

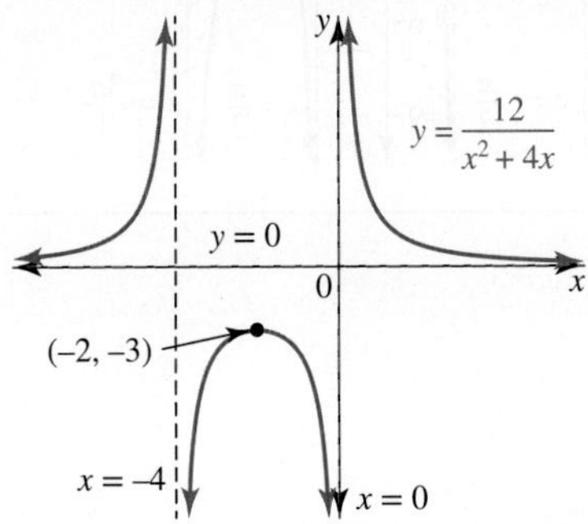

b. $y = 0, x = -7, x = 1$; $\left(0, \frac{8}{7}\right), \left(-3, \frac{1}{2}\right)$ local min.;

domain $x \in R \setminus \{-7, 1\}$, range $(-\infty, 0) \cup \left[\frac{1}{2}, \infty\right)$

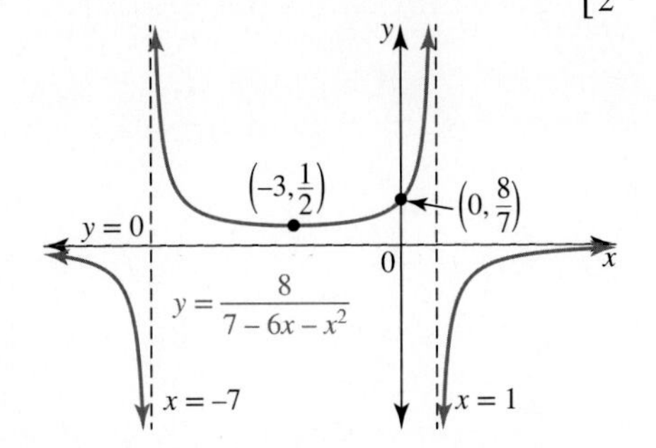

5. $b = -8$, $A = 8$
$y = 0$ horizontal asymptote
$x = 2$, $x = 6$ vertical asymptote
Domain $R \setminus \{2, 6\}$
Range $(-\infty, -2] \cup (0, \infty)$

6. $y = \frac{x - 5}{x^2 - 25} = \frac{1}{x + 5}, x \neq 5$
Does not cross the x-axis
Crosses the y-axis $\left(0, \frac{1}{5}\right)$
Point of discontinuity at $\left(5, \frac{1}{10}\right)$
$y = 0$ horizontal asymptote
$x = -5$ vertical asymptote
No turning points
No points of inflection
Domain $R \setminus \{\pm 5\}$
Range $R \setminus \left\{0, \frac{1}{10}\right\}$

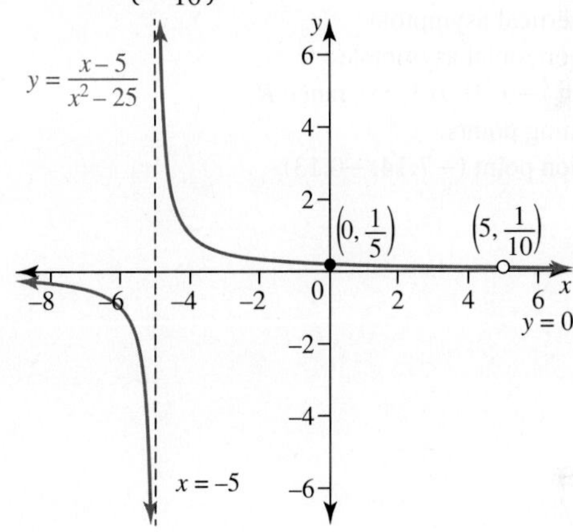

7. a. See graph at the bottom of the page*

b. See graph at the bottom of the page**

8. Sample responses can be found in the worked solutions in the online resources.

Technology active: multiple choice

9. B
10. A
11. A
12. B
13. C
14. B
15. E
16. E
17. D
18. D

Technology active: extended response

19. a. Crosses x axis $(-9, 0)$
Crosses y-axis $(0, -1)$
$x = 3$ vertical asymptote
$y = 0$ horizontal asymptote
Domain $[-9, 3) \cup (3, \infty)$, range R
No turning points
Inflection point $(-7.14, -0.13)$

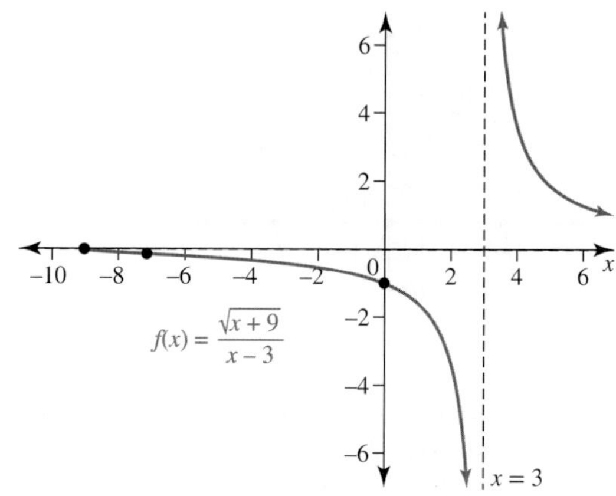

b. i. $a = b$ ii. $a < b$

20. Endpoints $(2\pi, 2\pi)$, $(-2\pi, -2\pi)$
$x = \pm\frac{\pi}{2}$, $x = \pm\frac{3\pi}{2}$ vertical asymptotes
Local min $(-2.80, 2.97)$
Local max $(2.80, -2.97)$
Inflection point $(0, 0)$

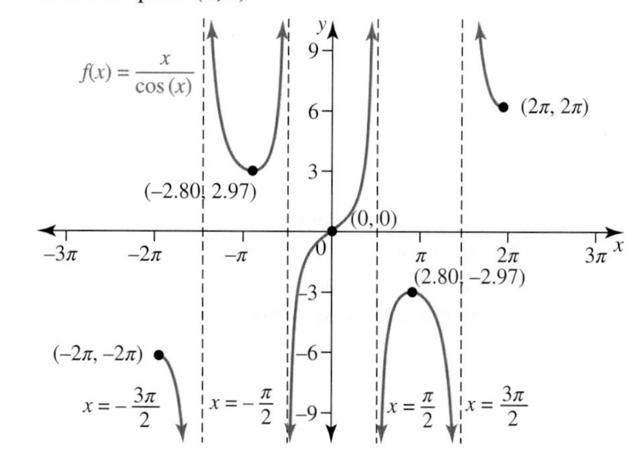

*7. a.

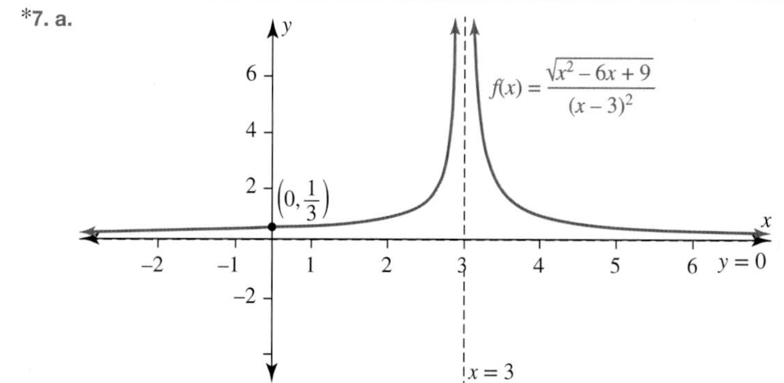

**7. b.

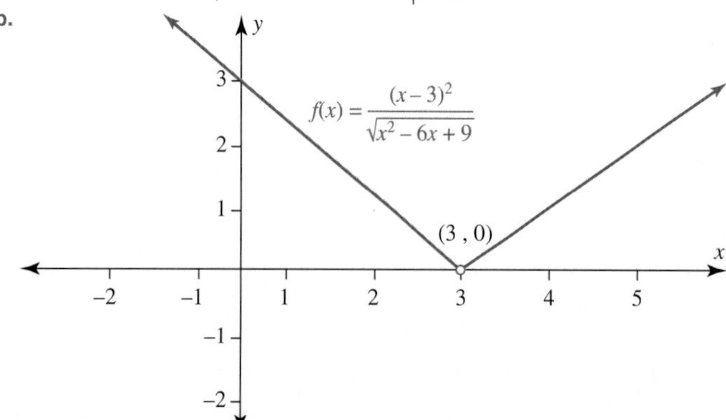

6.5 Exam questions

Note: Mark allocations are available with the fully worked solutions online.

1. a. $f(x) = 2 + \dfrac{5x - 11}{(x - 1)(x + 2)}$

   b. Vertical asymptotes $x = 1,\ x = -2$
   Horizontal asymptote $y = 2$

   c. See graph at the bottom of the page*

   d. i. $k = -5,\quad g_k(x) = \dfrac{2x - 3}{x + 2}\quad k \in R$

   $k = \dfrac{3}{2}\quad g_k(x) = \dfrac{2(x + 5)}{x + 2}$

   $k = -2\quad g_k(x) = \dfrac{(2x - 3)(x + 5)}{(x + 2)^2}$

   ii. $k < -5$ or $k > \dfrac{3}{2}$

2. a. Sample responses can be found in the worked solutions in the online resources.

   b. Sample responses can be found in the worked solutions in the online resources.

   c.

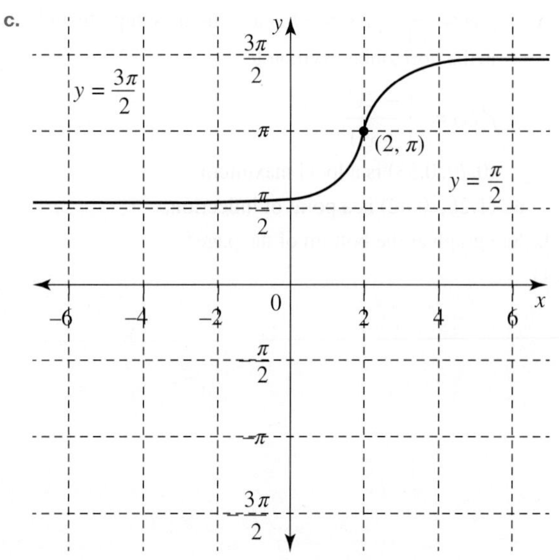

3. C

*1. c.

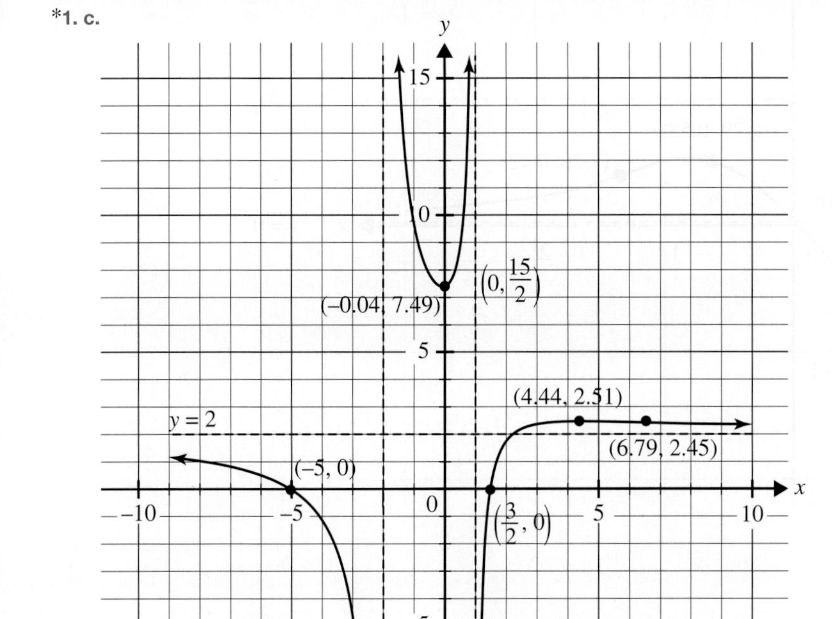

4. a. i. $f(x) = \dfrac{x}{1+x^3}$, $x = -1$ is a vertical asymptote and $y = 0$ is the horizontal asymptote.

ii. $f'(x) = \dfrac{1-2x^3}{(1+x^3)^2}$

$(0.79, 0.53)$ is a local maximum.

iii. $(1.26,\ 0.42)$ is a point of inflection.

b. See graph at the bottom of the page*

5. D

*4. b.

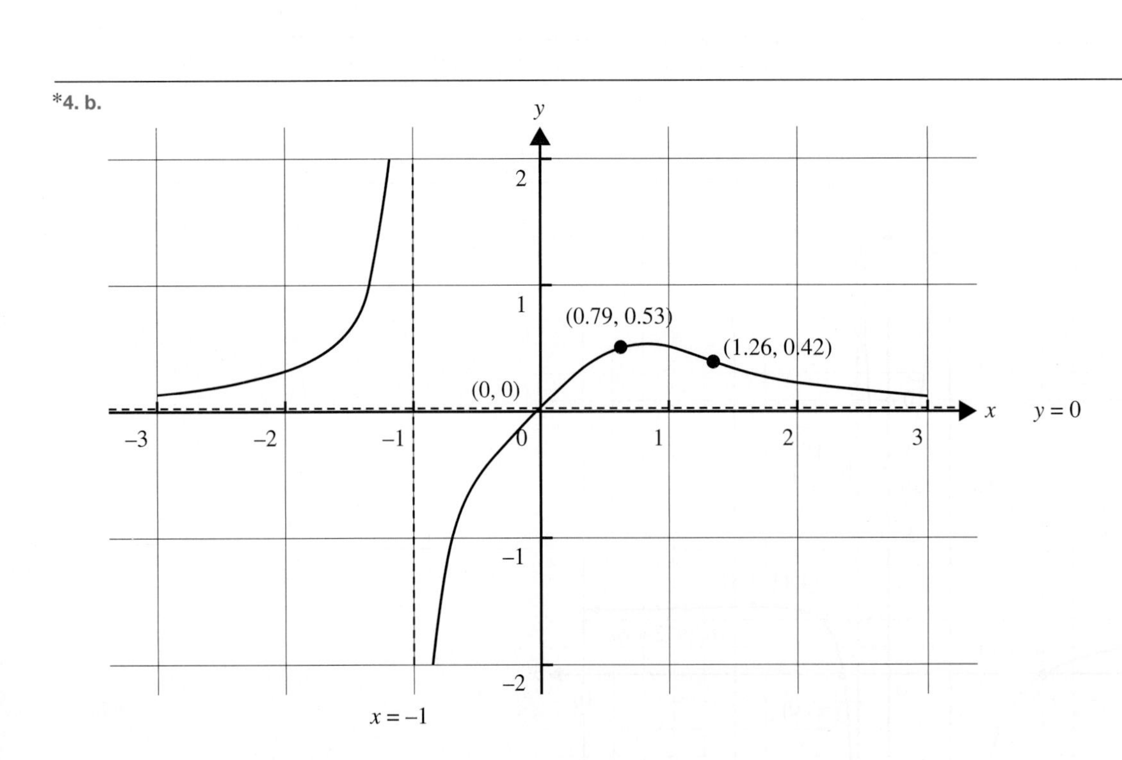

7 Integral calculus

LEARNING SEQUENCE

Fully worked solutions for this topic are available online.

7.1 Overview

Hey students! Bring these pages to life online

 **Watch videos**

 **Engage with interactivities**

 **Answer questions and check results**

Find all this and MORE in jacPLUS

7.1.1 Introduction

Calculus is concerned with both differentiation and integration.

The process of finding integrals is called integration.

How do you find the area under a curve? Integral calculus and evaluating definite integrals give us the means to answer this question and many more applications of integral calculus in the next couple of topics. When first discovered, again by Newton and Leibnitz independently, it was surprising that the integral calculus was the opposite procedure to that of differential calculus.

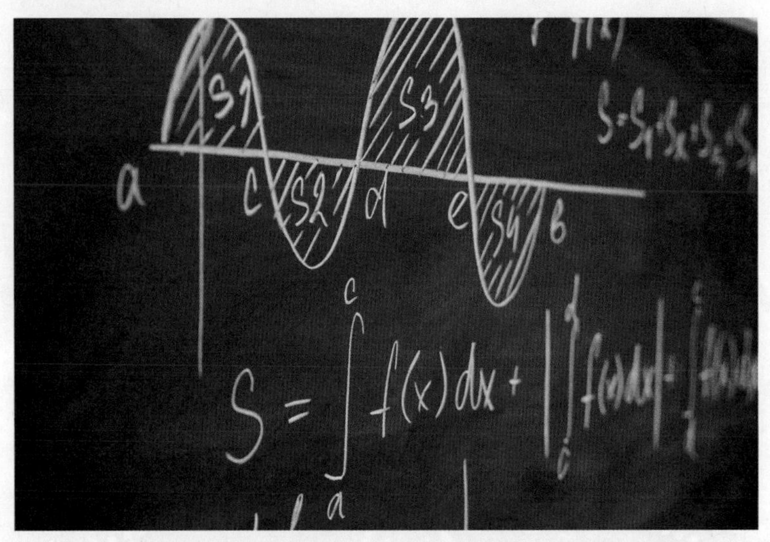

KEY CONCEPTS

This topic covers the following points from the VCE Mathematics Study Design:

- techniques of anti-differentiation and for the evaluation of definite integrals:
  - anti-differentiation of $\frac{1}{x}$ to obtain $\log_e |x|$
- techniques of anti-differentiation and for the evaluation of definite integrals:
  - anti-differentiation of $\frac{1}{\sqrt{a^2 - x^2}}$ and $\frac{a}{a^2 + x^2}$ for $a > 0$ by recognition that they are derivatives of corresponding inverse circular functions
  - use of the trigonometric identities $\sin^2(ax) = \frac{1}{2}\left(1 - \cos(2ax)\right)$ and $\cos^2(ax) = \frac{1}{2}\left(1 + \cos(2ax)\right)$ in anti-differentiation techniques
  - use of the substitution $u = g(x)$ to anti-differentiate expressions
  - anti-differentiation using partial fractions of rational functions.

7.2 Areas under and between curves

LEARNING INTENTION

At the end of this subtopic you should be able to:
- calculate the area under and between curves.

7.2.1 Area between a curve and the x-axis

Basic integration techniques and evaluating areas bounded by curves and the x-axis, have been covered in the Mathematical Methods course. The examples and theory presented here are a review of this material.

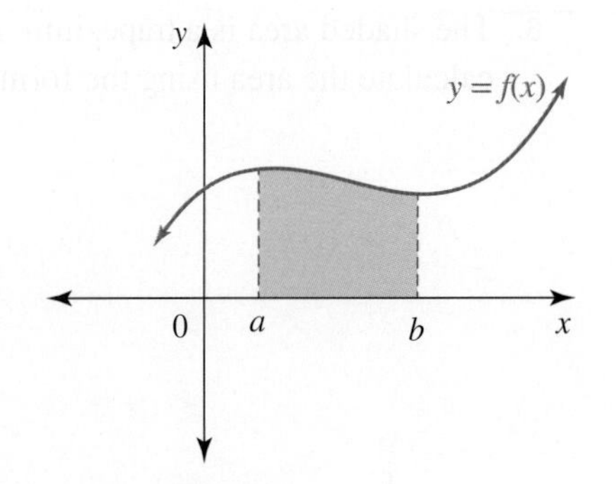

Recall that the definite integral, $A = \int_a^b f(x)dx$, gives a measure of the area A bounded by the curve $y = f(x)$, the x-axis and the lines $x = a$ and $x = b$. This result is known as the Fundamental Theorem of Calculus.

Area under a curve

The area under the curve of $f(x)$, between $x = a$ and $x = b$ is:

$$\int_a^b f(x)\,dx = F(b) - F(a)$$

WORKED EXAMPLE 1 Calculating the area under a linear graph

Determine the area bounded by the line $y = 2x + 3$, the x-axis and the lines $x = 2$ and $x = 6$.

THINK

1. Draw a diagram to identify the required area and shade this area.

WRITE

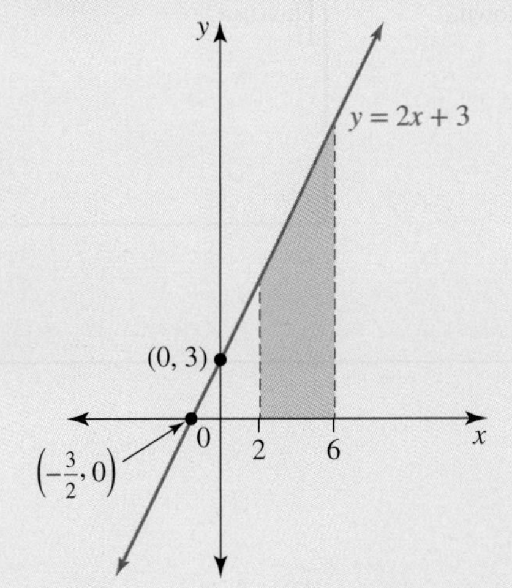

2. The required area is given by a definite integral.

$$A = \int_2^6 (2x + 3)dx$$

3. Perform the integration using square bracket notation.

$$A = [x^2 + 3x]_2^6$$

4. Evaluate.

$$A = (6^2 + 3 \times 6) - (2^2 + 3 \times 2)$$
$$= (36 + 18) - (4 + 6)$$
$$= 44$$

5. State the value of the required area region.

The area is 44 square units.

6. The shaded area is a trapezium. As a check on the result, calculate the area using the formula for a trapezium.

The width of the trapezium is $h = 6 - 2 = 4$, and since $y = 2x + 3$:

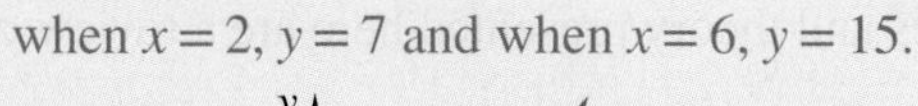

when $x = 2$, $y = 7$ and when $x = 6$, $y = 15$.

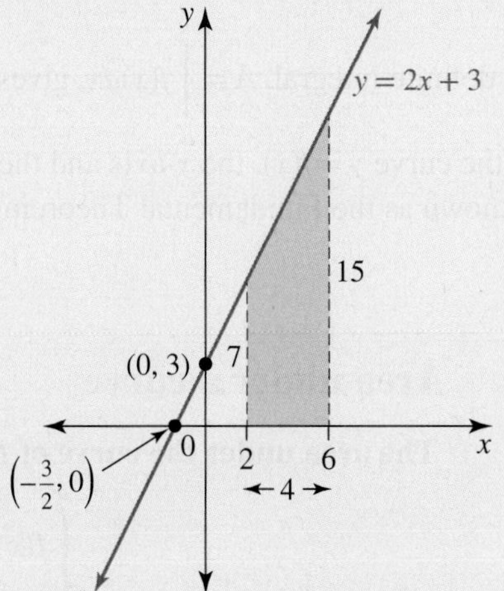

The area of a trapezium is $\frac{h}{2}(a + b)$.

The area is $\frac{4}{2}(7 + 15) = 44$ square units.

TI | THINK	DISPLAY/WRITE	CASIO | THINK	DISPLAY/WRITE
On a Calculator page, complete the definite integral as shown.	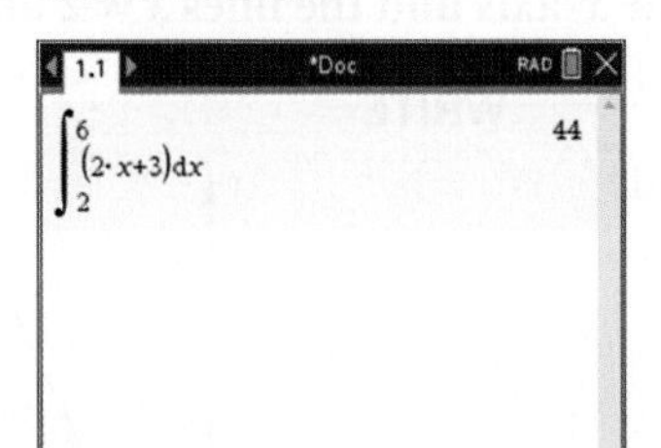	On a Main screen, complete the definite integral as shown.	

Using symmetry

Sometimes symmetry can be used to simplify the area calculation.

WORKED EXAMPLE 2 Calculating the area under a curve using symmetry

Calculate the area bounded by the curve $y = 16 - x^2$, the x-axis and the lines $x = \pm 3$.

THINK	WRITE
1. Factorise the quadratic to determine the x-intercepts.	$y = 16 - x^2$ $y = (4 - x)(4 + x)$ The graph crosses the x-axis at $x = \pm 4$ and crosses the y-axis at $y = 16$.
2. Draw a diagram to identify the required area and shade this area.	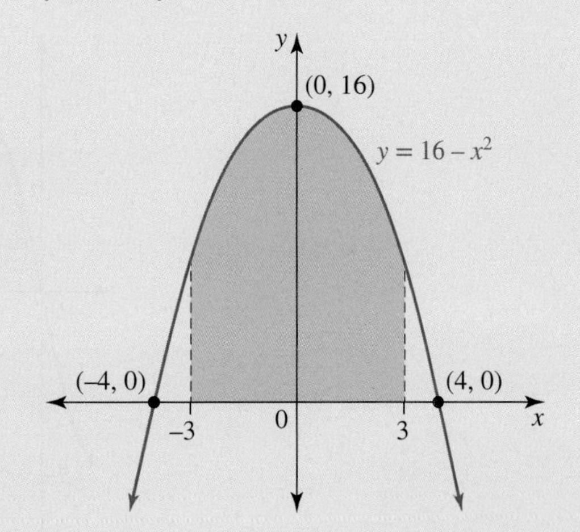
3. The required area is given by a definite integral; however, we can use symmetry.	$A = \int_{-3}^{3} (16 - x^2)dx$ $= \int_{-3}^{0} (16 - x^2)dx + \int_{0}^{3} (16 - x^2)dx$ However, $\int_{-3}^{0} (16 - x^2)dx = \int_{0}^{3} (16 - x^2)dx$ $A = 2\int_{0}^{3} (16 - x^2)dx$
4. Perform the integration using square bracket notation.	$= 2\left[16x - \frac{1}{3}x^3\right]_0^3$
5. Evaluate.	$= 2\left[\left(16 \times 3 - \frac{1}{3}(3)^3\right) - 0\right]$ $= 2(48 - 9)$ $= 78$
6. State the value of the required area.	The area is 78 square units.

Areas involving basic trigonometric functions

For integrals and area calculations involving the basic trigonometric functions, we use the results $\int \cos(kx)dx = \frac{1}{k}\sin(kx) + c$ and $\int \sin(kx)dx = -\frac{1}{k}\cos(kx) + c$, where $x \in R$, $k \neq 0$, and k and c are constants.

WORKED EXAMPLE 3 Calculating the area under a trigonometric curve

Calculate the area under one arch of the sine curve $y = 5\sin(3x)$.

THINK	WRITE
1. Draw a diagram to identify the required area and shade this area.	$y = 5\sin(3x)$ has an amplitude of 5 and a period of $\frac{2\pi}{3}$. 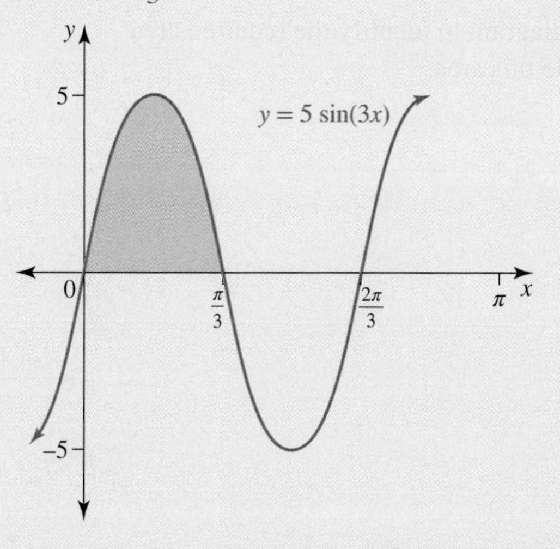
2. One arch is defined to be the area under one half-cycle of the sine wave.	The graph crosses the x-axis at $\sin(3x) = 0$, when $x = 0$, $\frac{\pi}{3}$ and $\frac{2\pi}{3}$. The required area is $A = \int_0^{\frac{\pi}{3}} 5\sin(3x)dx$
3. Perform the integration using square bracket notation.	$= 5\left[-\frac{1}{3}\cos(3x)\right]_0^{\frac{\pi}{3}}$
4. Evaluate, taking the constant factors outside the brackets.	$= -\frac{5}{3}[\cos(\pi) - \cos(0)]$ $= -\frac{5}{3}[-1 - 1]$ $= \frac{10}{3}$
5. State the value of the required area.	The area is $\frac{10}{3}$ square units.

Areas involving basic exponential functions

For integrals and area calculations involving the basic exponential functions, we use the result $\int e^{kx}dx = \frac{1}{k}e^{kx} + c$, where $x \in R$, $k \neq 0$, and k and c are constants.

WORKED EXAMPLE 4 Calculating the area under an exponential curve

Calculate the area bounded by the coordinate axes, the graph of $y = 4e^{-2x}$ and the line $x = 1$.

THINK	WRITE
1. Draw a diagram to identify the required area and shade this area.	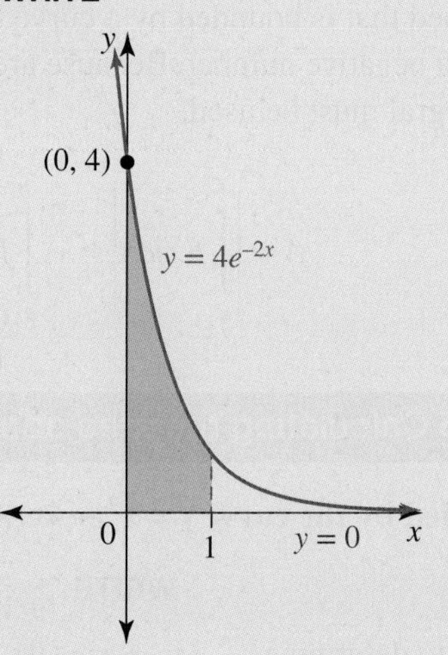
2. The required area is given by a definite integral.	$A = \int_0^1 4e^{-2x}dx$
3. Perform the integration using square bracket notation.	$= 4\left[-\frac{1}{2}e^{-2x}\right]_0^1$ $= -2\left[e^{-2x}\right]_0^1$
4. Evaluate.	$= -2[e^{-2} - e^0]$ $= -2(e^{-2} - 1)$
5. State the value of the required area.	The exact area is $2(1 - e^{-2})$ square units.

7.2.2 Areas involving signed areas

When evaluating a definite integral, the result is a number; this number can be positive or negative. A definite integral that represents an area is a signed area; that is, it may also be positive or negative. However, areas cannot be negative.

Areas above the x-axis

When a function is such that $f(x) \geq 0$ for $a \leq x \leq b$, where $b > a$, that is, the function lies above the x-axis, then the definite integral that represents the area A is positive:

$$A = \int_a^b f(x)dx > 0$$

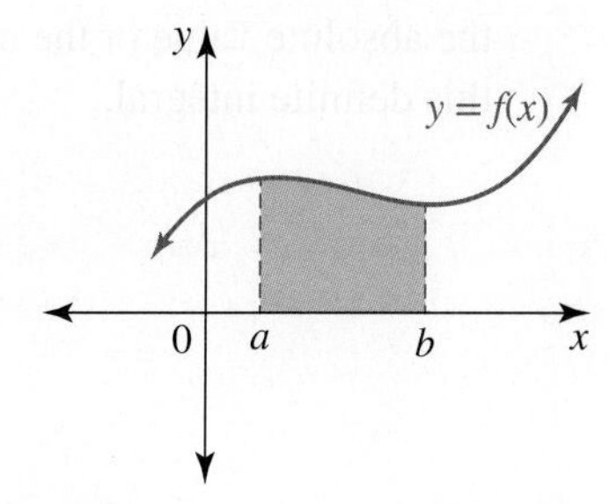

Areas below the x-axis

When a function is such that $f(x) \leq 0$ for $a \leq x \leq b$, where $b > a$, that is, the function lies below the x-axis, then the definite integral that represents the area A is negative:

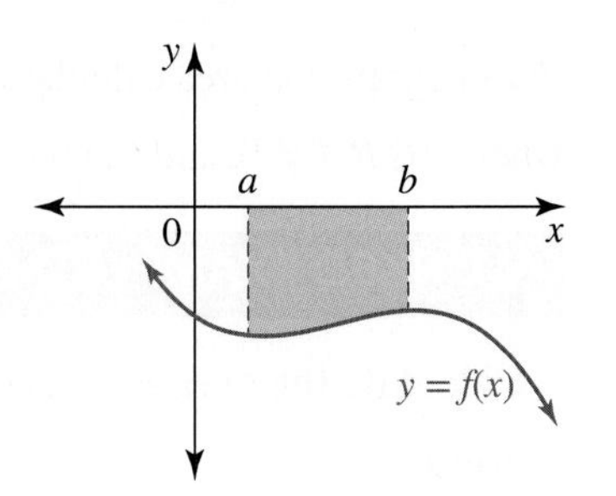

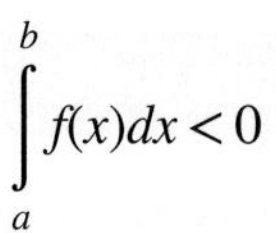

$$\int_a^b f(x)dx < 0$$

So, when an area is determined that is bounded by a curve that is entirely below the x-axis, the result will be a negative number. Because areas cannot be negative, the absolute value of the integral must be used.

$$A = \left|\int_a^b f(x)dx\right| = -\int_a^b f(x)dx = \int_b^a f(x)dx$$

WORKED EXAMPLE 5 Calculating the area above a curve and below the x-axis

Calculate the area bounded by the curve $y = x^2 - 4x + 3$ and the x-axis.

THINK	WRITE
1. Factorise the quadratic to determine the x-intercepts.	$y = x^2 - 4x + 3$ $= (x-3)(x-1)$ The graph crosses the x-axis at $x = 1$ and $x = 3$ and crosses the y-axis at $y = 3$.
2. Sketch the graph, shading the required area.	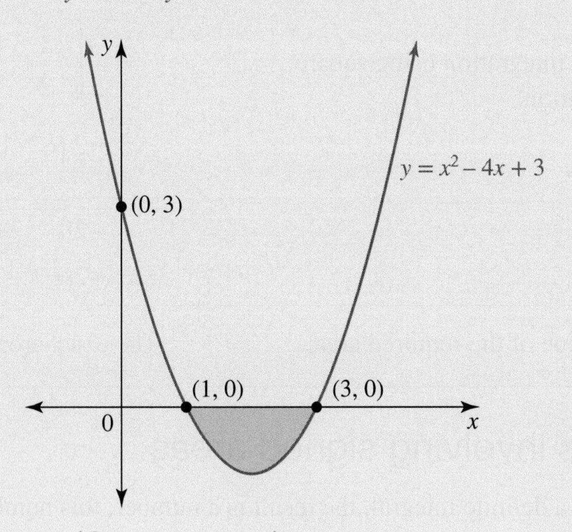
3. The required area is below the x-axis and will evaluate to a negative number. The area must be given by the absolute value or the negative of this definite integral.	$A = \left\|\int_1^3 (x^2 - 4x + 3)dx\right\|$ $= -\int_1^3 (x^2 - 4x + 3)dx$

4. Perform the integration using square bracket notation.	$= -\left[\frac{1}{3}x^3 - 2x^2 + 3x\right]_1^3$
5. Evaluate the definite integral.	$= -\left[\left(\frac{1}{3} \times 3^3 - 2 \times 3^2 + 3 \times 3\right) - \left(\frac{1}{3} \times 1^3 - 2 \times 1^2 + 3 \times 1\right)\right]$ $= \frac{4}{3}$
6. State the value of the required area.	The area is $\frac{4}{3}$ square units.

Areas both above and below the x-axis

When dealing with areas that are both above and below the x-axis, each area must be evaluated separately.

Since $A_1 = \int_a^b f(x)dx > 0$ and

$A_2 = \int_b^c f(x)dx < 0$, the required area is

$$A = A_1 + |A_2|$$
$$= \int_a^b f(x)dx + \left|\int_b^c f(x)dx\right|$$
$$= \int_a^b f(x)dx + \int_c^b f(x)dx$$

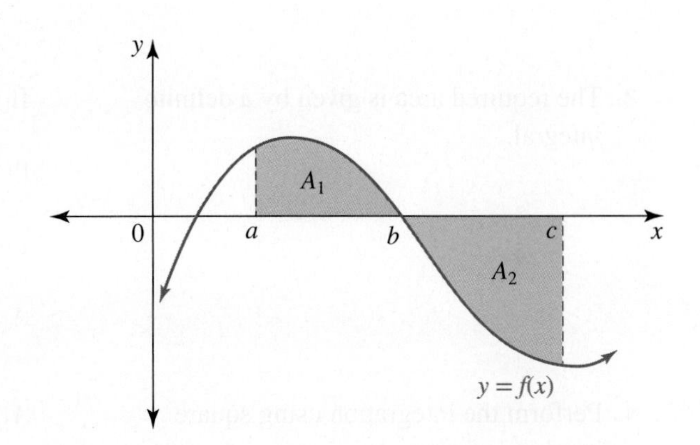

Areas above and below the x-axis

Areas above the x-axis are positive and areas below the x-axis are negative.

Remember to add the magnitude of negative areas when calculating the total area.

WORKED EXAMPLE 6 Calculating the total area bounded by a curve and the x-axis

Calculate the area bounded by the curve $y = x^3 - 9x$ and the x-axis.

THINK	WRITE
1. Factorise the cubic to determine the x-intercepts.	$y = x^3 - 9x$ $= x(x^2 - 9)$ $= x(x + 3)(x - 3)$ The graph crosses the x-axis at $x = 0$ and $x = \pm 3$.

2. Sketch the graph, shading the required area.

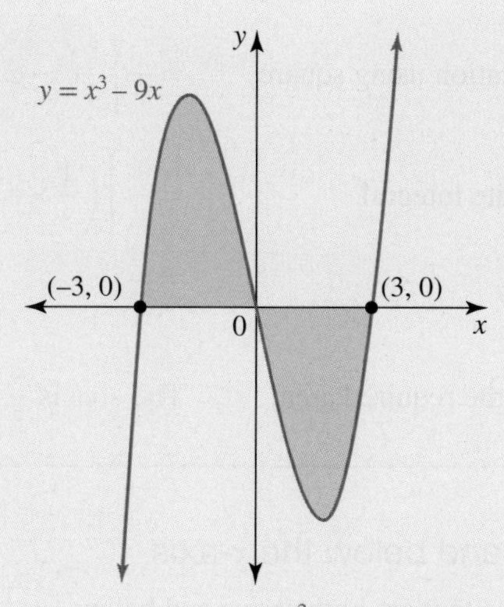

3. The required area is given by a definite integral.

If we work out $A = \int_{-3}^{3} (x^3 - 9x)dx$, it comes to zero, as the positive and negative area have cancelled out.

Let $A_1 = \int_{-3}^{0} (x^3 - 9x)dx$ and $A_2 = \int_{0}^{3} (x^3 - 9x)dx$, so that $A_1 > 0$ and $A_2 < 0$, but $A_1 = |A_2|$ by symmetry.

4. Perform the integration using square bracket notation.

$$A_1 = \int_{-3}^{0} (x^3 - 9x)dx$$

$$= \left[\frac{1}{4}x^4 - \frac{9}{2}x^2\right]_{-3}^{0}$$

5. Evaluate the definite integral.

$$= \left[\left(0 - \frac{1}{4} \times (-3)^4 + \frac{9}{2} \times (-3)^2\right)\right]$$

$$= \frac{81}{4}$$

$$= 20\frac{1}{4}$$

$$A_2 = \int_{0}^{3} (x^3 - 9x)dx = -\frac{81}{4} = -20\frac{1}{4}$$

$$A_1 + |A_2| = \frac{81}{4} + \frac{81}{4} = \frac{81}{2} = 40\frac{1}{2}$$

6. State the value of the required area.

The area is $40\frac{1}{2}$ square units.

7.2.3 Area between curves

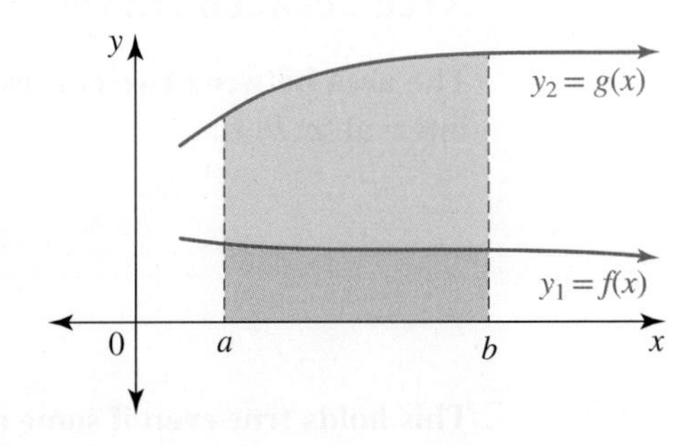

If $y_1 = f(x)$ and $y_2 = g(x)$ are two continuous curves that do not intersect between $x = a$ and $x = b$, then the area between the curves is obtained by simple subtraction.

The area A_1 is the entire shaded area bounded by the curve $y_2 = g(x)$, the x-axis and the lines $x = a$ and $x = b$, so $A_1 = \int_a^b g(x)dx$. The pink area, A_2, is the area bounded by the curve $y_1 = f(x)$, the x-axis and the lines $x = a$ and $x = b$, so $A_2 = \int_a^b f(x)dx$.

The required area is the blue area, which is the area between the curves.

$A = A_1 - A_2 = \int_a^b g(x)dx - \int_a^b f(x)dx$, and by the properties of definite integrals

$$A = A_1 - A_2 = \int_a^b \big(g(x) - f(x)\big)dx = \int_a^b (y_2 - y_1)dx.$$

Note that when calculating areas between curves, it does not matter if some of the area is above or below the x-axis.

We can translate both curves up k units parallel to the y-axis so that the area between the curves lies entirely above the x-axis as shown below right.

$$\begin{aligned} A &= \int_a^b \big(g(x) + k\big)dx - \int_a^b \big(f(x) + k\big)dx \\ &= \int_a^b \big(g(x) - f(x)\big)dx \\ &= \int_a^b (y_2 - y_1)dx \end{aligned}$$

Provided that $y_2 \geq y_1$ for $a < x < b$, it does not matter if some or all of the area is above or below the x-axis, as the required area between the curves will be a positive number. Note that only one definite integral is required, that is $y_2 - y_1 = g(x) - f(x)$. Evaluate this as one definite integral.

Area between curves

The area between two curves $f(x)$ and $g(x)$ where $f(x) \geq g(x)$ over the entire interval (a, b) is:

$$\int_a^b f(x) - g(x)\, dx$$

This holds true even if some of the required region is below the x-axis.

WORKED EXAMPLE 7 Calculating the area between two curves

Calculate the area between the parabola $y = x^2 - 2x - 15$ and the straight line $y = 2x - 3$.

THINK	WRITE
1. Factorise the quadratic to determine the x-intercepts.	$y = x^2 - 2x - 15$ $y = (x-5)(x+3)$ The parabola crosses the x-axis at $x = 5$ and $x = -3$ and crosses the y-axis at $y = -15$. The straight line crosses the x-axis at $x = \frac{3}{2}$ and crosses the y-axis at $y = -3$.
2. Determine the x-values of the points of intersection between the parabola and the straight line.	Let $y_1 = x^2 - 2x - 15$ and $y_2 = 2x - 3$. To determine the points of intersection, solve $y_1 = y_2$. $x^2 - 2x - 15 = 2x - 3$ $x^2 - 4x - 12 = 0$ $(x-6)(x+2) = 0$ $x = 6, -2$
3. Sketch the graph of the parabola and the straight line on one set of axes, shading the required area.	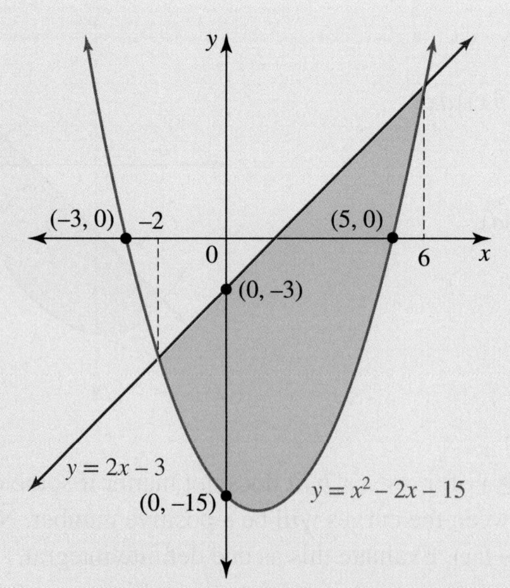

4. The required area is given by a definite integral.	$A = \int_a^b (y_2 - y_1)dx$ with $a = -2, b = 6$, $y_1 = x^2 - 2x - 15$ and $y_2 = 2x - 3$. $y_2 - y_1 = -x^2 + 4x + 12$ $A = \int_{-2}^{6} (-x^2 + 4x + 12)dx$
5. Perform the integration using square bracket notation.	$= \left[-\frac{1}{3}x^3 + 2x^2 + 12x\right]_{-2}^{6}$
6. Evaluate the definite integral.	$= \left[\left(-\frac{1}{3} \times 6^3 + 2 \times 6^2 + 12 \times 6\right) - \left(-\frac{1}{3} \times (-2)^3 + 2 \times (-2)^2 + 12 \times (-2)\right)\right]$ $= 85\frac{1}{3}$
7. State the value of the required area between the parabola and the straight line.	The area between the straight line and the parabola is $85\frac{1}{3}$ square units.

7.2 Exercise

Students, these questions are even better in jacPLUS

Receive immediate feedback and access sample responses

Access additional questions

Track your results and progress

Find all this and MORE in jacPLUS

Technology free

1. **WE1** Calculate the area bounded by the line $y = 4x + 5$, the x-axis and the lines $x = 1$ and $x = 3$. Check your answer algebraically.

2. Calculate the area bounded by the line $y = 4 - \frac{3x}{2}$ and the coordinate axes. Check your answer algebraically.

3. a. Calculate the area between the line $y = 6 - 2x$ and the coordinate axis. Check your answer algebraically.
   b. Calculate the area between the line $y = 3x + 5$, the x-axis, $x = 1$ and $x = 4$. Check your answer algebraically.

4. **WE2** Calculate the area bounded by the curve $y = 9 - x^2$, the x-axis and the lines $x = \pm 2$.

5. The area bounded by the curve $y = b - 3x^2$, the x-axis and the lines $x = \pm 1$ is equal to 16. Given that $b > 3$, determine the value of b.

6. Calculate the area bounded by:
   a. the curve $y = 12 - 3x^2$ and the x-axis
   b. the curve $y = 12 - 3x^2$, the x-axis and the lines $x = \pm 1$.

7. Determine the area bounded by:

   a. the graph of $y = x^2 - 25$ and the x-axis
   b. the graph of $y = x^2 - 25$, the x-axis and the lines $x = \pm 3$.

8. WE3 Calculate the area under one arch of the sine curve $y = 4\sin(2x)$.

9. Calculate the area under one arch of the curve $y = 3\cos\left(\frac{x}{2}\right)$.

10. a. Calculate the area under one arch of the sine curve $y = 6\sin\left(\frac{\pi x}{3}\right)$.

    b. Calculate the area under one arch of the curve $y = 4\cos\left(\frac{\pi x}{2}\right)$.

    c. Calculate the area under one arch of the sine curve $y = a\sin(nx)$, where a and $n \in R^+$.

11. WE4 Calculate the area bounded by the coordinate axes, the graph of $y = 6e^{3x}$ and the line $x = 2$.

12. Calculate the area bounded by the graph of $y = 6(e^{-2x} + e^{-2x})$, the x-axis and $x = \pm 1$.

13. WE5 Calculate the area bounded by the curve $y = x^2 - 5x + 6$ and the x-axis.

14. Calculate the area bounded by the curve $y = 3x^2 - 10x - 8$ and the x-axis.

15. Calculate the area bounded by the curve $y = x^2 - 2x - 15$ and the x-axis.

16. Calculate the area under the graph of $y = \frac{1}{x}$ between the x-axis and:

    a. $x = 1$ and $x = 4$
    b. $x = 1$ and $x = e$
    c. $x = 1$ and $x = a$, where $a > e$.

17. Calculate the area under the graph of $y = \frac{1}{x^2}$ between the x-axis and:

    a. $x = 1$ and $x = 3$
    b. $x = 1$ and $x = a$, where $a > 1$.

18. WE6 Calculate the area bounded by the curve $y = x^3 - 4x$ and the x-axis.

19. Calculate the area bounded by the curve $y = 16x - x^3$, the x-axis, $x = -2$ and $x = 4$.

20. Calculate the area bounded by the curve $y = x^2 + 3x - 18$, the x-axis and the lines $x = -3$ and $x = 6$.

21. Calculate the area bounded by:

    a. the curve $y = x^3 - 36x$ and the x-axis
    b. the curve $y = x^3 - 36x$, the x-axis and the lines $x = -3$ and $x = 6$.

22. WE7 Calculate the area between the parabola $y = x^2 - 3x - 18$ and the straight line $y = 4x - 10$.

23. Calculate the area corresponding to the region $\{y \geq x^2 - 2x - 8\} \cap \{y \leq 1 - 2x\}$.

24. If a is a positive constant, calculate the area bounded by the curve $y = x^2 - a^2$ and the x-axis.

25. Calculate the area between the curves:

    a. $y = x^2$ and $y = x$
    b. $y = x^3$ and $y = x$
    c. $y = x^4$ and $y = x$
    d. $y = x^5$ and $y = x$.

26. a. Calculate the area between the parabola $y = x^2 - 2x - 35$ and the x-axis.
    b. Calculate the area between the parabola $y = x^2 - 2x - 35$ and the straight line $y = 4x - 8$.

27. a. Calculate the area between the parabola $y = x^2 + 5x - 14$ and the x-axis.
    b. Calculate the area corresponding to the region

$$\{y \geq x^2 + 5x - 14\} \cap \{y \leq 2x + 4\}.$$

28. a. Calculate the area between the line $2y + x - 5 = 0$ and the hyperbola $y = \frac{2}{x}$.

b. Calculate the area between the line $9y + 3x - 10 = 0$ and the hyperbola $y = \frac{1}{3x}$.

29. a. Prove using calculus methods that the area of a right-angled triangle of base length a and height b is given by $\frac{1}{2}ab$.

b. Prove using calculus that the area of a trapezium of side lengths a and b, and width h is equal to $\frac{h}{2}(a + b)$.

Technology active

30. Consider the graphs of $y = \frac{x^2}{3}$ and $y = 4\sin\left(\frac{x}{2}\right)$.

a. Determine the coordinates of the point of intersection between the graphs that sits in the first quadrant, giving your answer rounded to 4 decimal places.

b. Calculate the area between the graphs, the origin and this point of intersection, giving your answer correct to 4 decimal places.

31. Consider the graphs of $y = 5e^{-\frac{x}{4}}$ and $y = \frac{x}{2}$.

a. Determine the coordinates of the point of intersection between the graphs, giving your answer rounded to 4 decimal places.

b. Calculate the area between the curves, the y-axis and this point of intersection, giving your answer correct to 4 decimal places.

32. Consider the graphs of $y = 23e^{\frac{x}{2}}$ and $y = 45\sin\left(\frac{2x}{3}\right) + 42$ for $x \geq 0$.

a. Determine the coordinates of the point of intersection between the graphs, giving your answer rounded to 4 decimal places.

b. Calculate the area between the curves, the y-axis and this point of intersection, giving your answer correct to 4 decimal places.

7.2 Exam questions

Question 1 (1 mark) TECH-FREE

Calculate the area bounded by the curve $y = x^2 - 2x - 24$, the x-axis and the lines $x = 2$ and $x = 8$.

Question 2 (2 marks) TECH-FREE

If a is a positive constant, calculate the area bounded by the curve $y = x^3 - a^2x$ and the x-axis.

Question 3 (2 marks) TECH-ACTIVE

Consider the graphs of $y = \frac{190}{x^2} - 5$ and $y = -32\cos\left(\frac{x}{5}\right)$ for $x \geq 0$.

a. Determine the coordinates of the first two points of intersection, giving your answer rounded to 4 decimal places. **(1 mark)**

b. Calculate the area between the curves and these first two points of intersection, giving your answer correct to 4 decimal places. **(1 mark)**

More exam questions are available online.

7.3 Linear substitutions

LEARNING INTENTION

At the end of this subtopic you should be able to:
- evaluate integrals by using linear substitutions.

7.3.1 Calculating integrals of the form $\int (ax+b)^n dx$ where $n \in Z$

Integrals of the form $\int (ax+b)^n dx$, where a and b are non-zero real numbers and n is a positive integer, can be performed using a linear substitution with $u = ax + b$. The derivative $\frac{du}{dx} = a$ is a constant, and this constant factor can be taken outside the integral sign by the properties of indefinite and definite integrals. The integration process can then be completed in terms of u. Note that since u has been introduced in this solution process, the final answer must be given back in terms of the original variable, x.

WORKED EXAMPLE 8 Integration of a linear function to an integer power

Calculate $\int (2x-5)^4 dx$.

THINK	WRITE
1. Although we could expand and integrate term by term, it is preferable and easier to use a linear substitution.	Let $u = 2x - 5$. $\int (2x-5)^4 dx = \int u^4 dx$
2. Differentiate u with respect to x.	$u = 2x - 5$ $\frac{du}{dx} = 2$
3. Express dx in terms of du by inverting both sides.	$\frac{dx}{du} = \frac{1}{2}$ $dx = \frac{1}{2} du$
4. Substitute for dx.	$\int u^4 dx = \int u^4 \frac{1}{2} du$
5. Use the properties of indefinite integrals to transfer the constant factor outside the front of the integral sign.	$= \frac{1}{2} \int u^4 du$
6. Perform the integration using $\int u^n du = \frac{1}{n+1} u^{n+1}$ with $n = 4$ so that $n + 1 = 5$, and add in the constant $+c$.	$\frac{1}{2} \times \frac{1}{5} u^5 + c$ $= \frac{1}{10} u^5 + c$
7. Substitute back for $u = 2x - 5$ and express the final answer in terms of x only and an arbitrary constant $+c$.	$\int (2x-5)^4 dx = \frac{1}{10}(2x-5)^5 + c$

Calculating particular integrals of the form $\int (ax+b)^n dx$ where $n \in Q$

Integrals of the form $\int (ax+b)^n dx$, where a and b are non-zero real numbers and n is a rational number, can also be performed using a linear substitution with $u = ax + b$. First express the integrand (the function being integrated) as a power using the index laws.

WORKED EXAMPLE 9 Integration of a linear function to a rational power

The gradient of a curve is given by $\frac{1}{\sqrt{4x+9}}$. Determine the particular curve that passes through the origin.

THINK	WRITE
1. Recognise that the gradient of a curve is given by $\frac{dy}{dx}$.	$\frac{dy}{dx} = \frac{1}{\sqrt{4x+9}}$
2. Integrate both sides to give an expression for y.	$y = \int \frac{1}{\sqrt{4x+9}} dx$
3. Use index laws to express the integrand as a function to a power and use a linear substitution.	Let $u = 4x + 9$. $y = \int (4x+9)^{-\frac{1}{2}} dx$ $y = \int u^{-\frac{1}{2}} dx$
4. The integral cannot be done in this form, so differentiate.	$u = 4x + 9$ $\frac{du}{dx} = 4$
5. Express dx in terms of du by inverting both sides.	$\frac{dx}{du} = \frac{1}{4}$ $dx = \frac{1}{4} du$
6. Substitute for dx.	$y = \int u^{-\frac{1}{2}} \frac{1}{4} du$
7. Use the properties of indefinite integrals to transfer the constant factor outside the front of the integral sign.	$y = \frac{1}{4} \int u^{-\frac{1}{2}} du$
8. Perform the integration process using $\int u^n du = \frac{1}{n+1} u^{n+1}$ with $n = -\frac{1}{2}$ so that $n + 1 = \frac{1}{2}$, and add in the constant $+c$.	$y = \frac{1}{4} \times \frac{1}{\frac{1}{2}} u^{\frac{1}{2}} + c$ $y = \frac{2}{4} u^{\frac{1}{2}} + c$ $y = \frac{1}{2} \sqrt{u} + c$
9. Simplify and substitute for $u = 4x + 9$ to express the answer in terms of x and an arbitrary constant $+c$.	$y = \frac{1}{2} \sqrt{4x+9} + c$

10. The arbitrary constant $+c$ in this particular case can be evaluated using the given condition that the curve passes through the origin.	Substitute $y=0$ and $x=0$ to evaluate c: $0=\frac{1}{2}\sqrt{0+9}+c$ $c=-\frac{3}{2}$
11. Substitute back for $+c$.	$y=\frac{1}{2}\sqrt{4x+9}-\frac{3}{2}$
12. State the equation of the particular curve in a factorised form.	$y=\frac{1}{2}\left(\sqrt{4x+9}-3\right)$

Calculating integrals of the form $\int (ax+b)^n dx$ when $n=-1$

Integrals of the form $\int (ax+b)^n dx$, when $n=-1$, $a \neq 0$ and $b \in R$, involve the logarithm function, since $\int \frac{1}{x} dx = \log_e(|x|)+c$.

WORKED EXAMPLE 10 Integration of a linear function to the power of –1

Anti-differentiate $\frac{1}{5x+4}$.

THINK	WRITE
1. Write the required integral.	$\int \frac{1}{5x+4} dx$
2. Use a linear substitution.	Let $u=5x+4$. $\int \frac{1}{5x+4} dx = \int \frac{1}{u} dx$
3. The integral cannot be done in this form, so differentiate.	$u=5x+4$ $\frac{du}{dx}=5$
4. Express dx in terms of du by inverting both sides.	$\frac{dx}{du}=\frac{1}{5}$ $dx=\frac{1}{5}du$
5. Substitute for dx.	$\int \frac{1}{5x+4} dx = \int \frac{1}{u} \times \frac{1}{5} du$
6. Use the properties of indefinite integrals to transfer the constant factor outside the front of the integral sign.	$=\frac{1}{5}\int \frac{1}{u} du$
7. Perform the integration process using $\int \frac{1}{u} du = \log_e \|u\| + c$.	$\int \frac{1}{5x+4} dx = \frac{1}{5}\log_e(\|u\|)+c$
8. Simplify and substitute for $u=5x+4$ to express the final answer in terms of x only and an arbitrary constant $+c$, in simplest form.	$\int \frac{1}{5x+4} dx = \frac{1}{5}\log_e(\|5x+4\|)+c$

We can generalise the results from the last three examples as follows.

Integration of powers of linear functions

$$\int (ax+b)^n dx = \begin{cases} \dfrac{1}{a(n+1)}(ax+b)^{n+1} + c & n \neq -1 \\ \dfrac{1}{a}\log_e(|ax+b|) + c & n = -1 \end{cases}$$

7.3.2 Evaluating definite integrals using a linear substitution

When we evaluate a definite integral, the result is a number. This number is also independent of the original variable used. When using a substitution, change the terminals to the new variable and evaluate this definite integral in terms of the new variable with new terminals. The following worked example clarifies this process.

WORKED EXAMPLE 11 Evaluating a definite integral using a linear substitution

Evaluate $\displaystyle\int_0^1 \frac{4}{(3x+2)^2}dx.$

THINK	WRITE
1. Write the integrand as a power using index laws and transfer the constant factor outside the front of the integral sign.	$\displaystyle\int_0^1 \frac{4}{(3x+2)^2}dx = 4\int_0^1 (3x+2)^{-2}dx$
2. Use a linear substitution. Note that the terminals in the definite integral refer to x-values.	Let $u = 3x+2$. $\displaystyle 4\int_0^1 (3x+2)^{-2}dx = 4\int_{x=0}^{x=1} u^{-2}dx$
3. The integral cannot be done in this form, so differentiate. Express dx in terms of du by inverting both sides.	$u = 3x+2$ $\dfrac{du}{dx} = 3$ $\dfrac{dx}{du} = \dfrac{1}{3}$ $dx = \dfrac{1}{3}du$
4. Change the terminals to the new variable.	When $x = 0 \Rightarrow u = 2$ and when $x = 1 \Rightarrow u = 5$.
5. Substitute for dx and the new terminals.	$\displaystyle 4\int_{u=2}^{u=5} u^{-2}\frac{1}{3}du$
6. Transfer the constant factor outside the front of the integral sign.	$\displaystyle = \frac{4}{3}\int_2^5 u^{-2}du$

7. The value of this definite integral has the same value as the original definite integral. There is no need to substitute back for x, and there is no need for the arbitrary constant when evaluating a definite integral.	$=\frac{4}{3}\left[-\frac{1}{u}\right]_2^5$
8. Evaluate this definite integral.	$=\frac{4}{3}\left[-\frac{1}{5}-\left(-\frac{1}{2}\right)\right]$ $=\frac{4}{3}\left(\frac{1}{2}-\frac{1}{5}\right)$ $=\frac{4}{3}\left(\frac{5-2}{10}\right)$ $=\frac{2}{5}$
9. State the final result.	$\int_0^1 \frac{4}{(3x+2)^2}dx=\frac{2}{5}$

7.3.3 Evaluating integrals using a back substitution

Integrals of the form $\int x(ax+b)^n dx$ can also be performed using a linear substitution with $u = ax + b$. Since the derivative $\frac{du}{dx} = a$ is a constant, this constant can be taken outside the integral sign. However, we must express the integrand in terms of u only before integrating. We can do this by expressing x in terms of u; that is, $x = \frac{1}{a}(u - b)$. However, the final result for an indefinite integral must be given in terms of the original variable, x.

WORKED EXAMPLE 12 Integration using back substitution (1)

Determine:

a. $\int x(2x-5)^4 dx$

b. $\int \frac{6x-5}{4x^2-12x+9}dx.$

THINK	WRITE
a. 1. Use a linear substitution.	a. Let $u = 2x - 5$. $\int x(2x-5)^4 dx = \int xu^4 dx$
2. The integral cannot be done in this form, so differentiate and express dx in terms of du by inverting both sides.	$u = 2x - 5$ $\frac{du}{dx} = 2$ $\frac{dx}{du} = \frac{1}{2}$ $dx = \frac{1}{2}du$
3. Express x in terms of u.	$2x = u + 5$ $x = \frac{1}{2}(u+5)$

4. Substitute for x and dx.

$$\int x(2x-5)^4 dx = \int \frac{1}{2}(u+5)u^4 \frac{1}{2} du$$

5. Use the properties of indefinite integrals to transfer the constant factors outside the front of the integral sign and expand the integrand.

$$= \frac{1}{4}\int (u^5 + 5u^4) du$$

6. Perform the integration, integrating term by term and adding in the constant.

$$= \frac{1}{4} \times \left(\frac{1}{6}u^6 + u^5\right) + c$$

7. Simplify the result by expanding.

$$= \frac{1}{24}u^6 + \frac{1}{4}u^5 + c$$

8. Substitute $u = 2x - 5$ and express the final answer in terms of x only and an arbitrary constant $+c$.

$$\int x(2x-5)^4 dx = \frac{1}{24}(2x-5)^6 + \frac{1}{4}(2x-5)^5 + c$$

9. Alternatively, the result can be expressed in a simplified form by taking out the common factors.

$$\frac{1}{24}u^6 + \frac{1}{4}u^5 + c = \frac{u^5}{24}(u+6) + c$$

10. Substitute back for $u = 2x - 5$ and simplify.

$$= \frac{(2x-5)^5}{24}(2x-5+6) + c$$

11. Express the final answer in terms of x only and an arbitrary constant $+c$.

$$\int x(2x-5)^4 dx = \frac{1}{24}(2x-5)^5(2x+1) + c$$

b. 1. Factorise the denominator as a perfect square.

b. $$\int \frac{6x-5}{4x^2-12x+9} dx = \int \frac{6x-5}{(2x-3)^2} dx$$

2. Use a linear substitution. Differentiate and express dx in terms of du by inverting both sides.

Let $u = 2x - 3$.

$$\frac{du}{dx} = 2$$

$$\frac{dx}{du} = \frac{1}{2}$$

$$dx = \frac{1}{2}du$$

3. Express the numerator $6x - 5$ in terms of u.

$$2x = u + 3$$
$$6x = 3(u+3)$$
$$6x = 3u + 9$$
$$6x - 5 = 3u + 4$$

4. Substitute for $6x - 5$, u and dx.

$$\int \frac{6x-5}{(2x-3)^2} dx = \int \frac{3u+4}{u^2} \times \frac{1}{2} du$$

5. Use the properties of indefinite integrals to transfer the constant factor outside the front of the integral sign.

$$= \frac{1}{2}\int \left(\frac{3u+4}{u^2}\right) du$$

6. Simplify the integrand.

$$= \frac{1}{2}\int \left(\frac{3}{u} + \frac{4}{u^2}\right) du$$

7. Write in index form. $= \frac{1}{2}\int\left(\frac{3}{u}+4u^{-2}\right)du$

8. Perform the integration, adding in the constant. The first term is a log, but in the second term, we use $\int u^n du = \frac{1}{n+1}u^{n+1}$ with $n=-2$, so that $n+1=-1$.

$$= \frac{1}{2}\left(3\log_e(|u|)-4u^{-1}\right)+c$$
$$= \frac{1}{2}\left(3\log_e(|u|)-\frac{4}{u}\right)+c$$

9. Substitute $u=2x-3$ and express the final answer in terms of x only and an arbitrary constant $+c$, as before.

$$\int \frac{6x-5}{4x^2-12x+9}dx$$
$$= \frac{3}{2}\log_e\left(|2x-3|\right)-\frac{2}{2x-3}+c$$

WORKED EXAMPLE 13 Integration using back substitution (2)

Calculate $\int \frac{2x}{4x-3}dx$.

THINK / **WRITE**

Method 1

1. Use a linear substitution. Differentiate and express dx in terms of du by inverting both sides.

$$u = 4x-3$$
$$\frac{du}{dx} = 4$$
$$\frac{dx}{du} = \frac{1}{4}$$
$$dx = \frac{1}{4}du$$

2. Express the numerator $2x$ in terms of u.

$$4x = u+3$$
$$2x = \frac{1}{2}(u+3)$$

3. Substitute for $2x$, u and dx.

$$\int \frac{2x}{4x-3}dx = \int \frac{\frac{1}{2}(u+3)}{u}\times\frac{1}{4}du$$

4. Use the properties of indefinite integrals to transfer the constant factors outside the front of the integral sign.

$$= \frac{1}{8}\int\left(\frac{u+3}{u}\right)du$$

5. Simplify the integrand.

$$= \frac{1}{8}\int\left(1+\frac{3}{u}\right)du$$

6. Perform the integration, adding in the constant.

$$= \frac{1}{8}\left(u+3\log_e(|u|)\right)+c$$
$$= \frac{u}{8}+\frac{3}{8}\log_e(|u|)+c$$

7. Substitute $u=4x-3$ and express the final answer in terms of x only and an arbitrary constant $+c$.

$$\int \frac{2x}{4x-3}dx = \frac{4x-3}{8}+\frac{3}{8}\log_e(|4x-3|)+c$$

Method 2

1. Express the numerator as a multiple of the denominator (in effect, use long division to divide the denominator into the numerator).

$$\int \frac{2x}{4x-3}dx = \frac{1}{2}\int \frac{4x}{4x-3}dx$$
$$= \frac{1}{2}\int \frac{(4x-3)+3}{4x-3}dx$$

2. Simplify the integrand.

$$= \frac{1}{2}\int \left(1+\frac{3}{4x-3}\right)dx$$

3. Perform the integration, adding in the constant.

$$= \frac{1}{2}\left(x+\frac{3}{4}\log_e(|4x-3|)\right)+c$$

4. State the final answer.

$$\int \frac{2x}{4x-3}dx = \frac{x}{2}+\frac{3}{8}\log_e(|4x-3|)+c$$

5. Although the two answers do not appear to be the same, the log terms are identical. However, since $\frac{4x-3}{8}=\frac{x}{2}-\frac{3}{8}$, the two answers are equivalent in x and differ in the constant only, $c_1=-\frac{3}{8}+c$.

$$= \frac{4x-3}{8}+\frac{3}{8}\log_e(|4x-3|)+c$$
$$= \frac{x}{2}+\frac{3}{8}\log_e(|4x-3|)+c_1$$

TI | THINK

On a Calculator page, complete the indefinite integral as shown.
If you wish, expand the result to see the expression in expanded form.
Remember to add an arbitrary constant $(+c)$ to your answer.

DISPLAY/WRITE

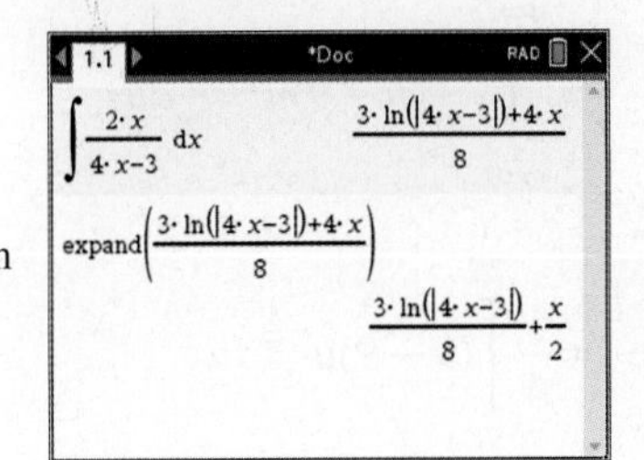

CASIO | THINK

On a Main screen, complete the indefinite integral as shown.
Remember to add an arbitrary constant $(+c)$ to your answer.

DISPLAY/WRITE

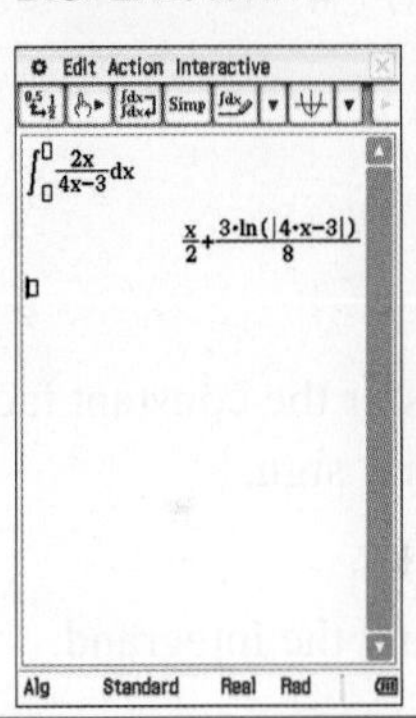

The situation above can often happen when evaluating indefinite integrals. Answers may not appear to be identical, but after some algebraic or trigonometric simplification, they are revealed to be equivalent and may differ by a constant only.

Definite integrals using a back substitution

When evaluating definite integrals, recall that the result is simply a number, so we can change the terminals to fit the variable used in the substitution and evaluate the integral without having to substitute back for the original variable.

WORKED EXAMPLE 14 Evaluating definite integrals using back substitution

Evaluate $\int_0^8 \frac{x}{\sqrt{2x+9}}dx$.

THINK

1. Write the integrand as a power using index laws.

WRITE

$$\int_0^8 \frac{x}{\sqrt{2x+9}}dx = \int_0^8 x(2x+9)^{-\frac{1}{2}}dx$$

2. Use a linear substitution.	Let $u = 2x + 9$. $\int_0^8 x(2x+9)^{-\frac{1}{2}}dx = \int_{x=0}^{x=8} xu^{-\frac{1}{2}}dx$
3. The integral cannot be done in this form, so differentiate. Express dx in terms of du by inverting both sides.	$\frac{du}{dx} = 2$ $\frac{dx}{du} = \frac{1}{2}$ $dx = \frac{1}{2}du$
4. Express x back in terms of u.	$u = 2x + 9$ $2x = u - 9$ $x = \frac{1}{2}(u - 9)$
5. Change the terminals to the new variable.	When $x = 0$, $u = 9$, and when $x = 8$, $u = 25$.
6. Substitute for dx, x and the new terminals.	$\int_{x=0}^{x=8} xu^{-\frac{1}{2}}dx$ $= \int_{u=9}^{u=25} \frac{1}{2}(u-9)u^{-\frac{1}{2}}\frac{1}{2}du$
7. Transfer the constant factors outside the front of the integral sign.	$= \frac{1}{4}\int_9^{25} (u-9)u^{-\frac{1}{2}}du$
8. Expand the integrand.	$= \frac{1}{4}\int_9^{25} \left(u^{\frac{1}{2}} - 9u^{-\frac{1}{2}}\right) du$
9. Perform the integration.	$= \frac{1}{4}\left[\frac{2}{3}u^{\frac{3}{2}} - 18u^{\frac{1}{2}}\right]_9^{25}$
10. Evaluate the definite integral.	$= \frac{1}{4}\left[\left(\frac{2}{3}(25)^{\frac{3}{2}} - 18(25)^{\frac{1}{2}}\right) - \left(\frac{2}{3}(9)^{\frac{3}{2}} - 18(9)^{\frac{1}{2}}\right)\right]$ $= \frac{1}{4}\left[\left(\frac{2}{3} \times 125 - 18 \times 5\right) - \left(\frac{2}{3} \times 27 - 18 \times 3\right)\right]$ $= \frac{1}{4}\left[\left(-\frac{20}{3}\right) - \left(-\frac{108}{3}\right)\right]$ $= \frac{1}{4}\left(\frac{88}{3}\right)$
11. State the final result.	$\int_0^8 \frac{x}{\sqrt{2x+9}}dx = \frac{22}{3}$

7.3 Exercise

Students, these questions are even better in jacPLUS

Receive immediate feedback and access sample responses

Access additional questions

Track your results and progress

Find all this and MORE in jacPLUS

Technology free

1. **WE8** Calculate $\int (5x-9)^6 dx$.

2. Calculate $\int (3x+4)^7 dx$.

3. Integrate each of the following with respect to x.

   a. $(3x+5)^6$
   b. $\dfrac{1}{(3x+5)^2}$
   c. $\dfrac{1}{(3x+5)^3}$
   d. $\dfrac{1}{\sqrt[3]{3x+5}}$

4. **WE9** A particular curve has a gradient equal to $\dfrac{1}{\sqrt{16x+25}}$. Determine the particular curve that passes through the origin.

5. Given that $f'(x)=\dfrac{1}{(3x-7)^2}$ and $f(2)=3$, determine the value of $f(1)$.

6. **WE10** Anti-differentiate $\dfrac{1}{3x-5}$.

7. Determine an anti-derivative of $\dfrac{1}{7-2x}$.

8. Calculate each of the following.

   a. $\int (6x+7)^8 dx$
   b. $\int \dfrac{1}{\sqrt{6x+7}} dx$
   c. $\int \dfrac{1}{6x+7} dx$
   d. $\int \dfrac{1}{(6x+7)^2} dx$

9. **WE11** Evaluate $\displaystyle\int_{-1}^{0} \frac{9}{(2x+3)^3} dx$.

10. Determine the area bounded by the graph of $y=\dfrac{6}{\sqrt{3x+4}}$, the coordinate axes and $x=4$.

11. **WE12** Calculate each of the following.

    a. $\int x(5x-9)^5 dx$
    b. $\int \dfrac{2x-1}{9x^2-24x+16} dx$.

12. Calculate each of the following.

    a. $\int \dfrac{x}{(2x+7)^3} dx$
    b. $\int \dfrac{x}{\sqrt{6x+5}} dx$.

13. Integrate each of the following with respect to x.

a. $x(3x+5)^6$ b. $\dfrac{x}{(3x+5)^2}$ c. $\dfrac{x}{(3x+5)^3}$ d. $\dfrac{x}{\sqrt[3]{3x+5}}$

14. Calculate each of the following.

a. $\displaystyle\int x(6x+7)^8\,dx$ b. $\displaystyle\int \frac{x}{\sqrt{6x+7}}\,dx$ c. $\displaystyle\int \frac{x}{6x+7}\,dx$ d. $\displaystyle\int \frac{x}{(6x+7)^2}\,dx$

15. **WE13** Calculate $\displaystyle\int \frac{6x}{3x+4}\,dx$.

16. Evaluate $\displaystyle\int_0^1 \frac{4x}{2x-5}\,dx$.

17. **WE14** Evaluate $\displaystyle\int_0^5 \frac{x}{\sqrt{3x+1}}\,dx$.

18. Evaluate $\displaystyle\int_0^1 \frac{15x}{(3x+2)^2}\,dx$.

19. a. Given that $\dfrac{dx}{dt} = \dfrac{1}{(2-5t)^2}$ and $x(0)=0$, express x in terms of t.

b. A certain curve has its gradient given by $\dfrac{1}{\sqrt{3-2x}}$ for $x < \dfrac{3}{2}$. If the point $\left(-\dfrac{1}{2}, -2\right)$ lies on the curve, determine the equation of the curve.

20. Given that $f'(x) = \dfrac{3}{3-2x}$ and that $f(0)=0$, evaluate $f(1)$.

21. A certain curve has a gradient of $\dfrac{x}{\sqrt{2x+9}}$. Determine the particular curve that passes through the origin.

22. Evaluate each of the following.

a. $\displaystyle\int_1^2 (3x-4)^5\,dx$ b. $\displaystyle\int_1^2 x(3x-4)^5\,dx$ c. $\displaystyle\int_0^{13} \frac{1}{\sqrt[3]{2t+1}}\,dt$ d. $\displaystyle\int_0^{13} \frac{t}{\sqrt[3]{2t+1}}\,dt$

23. a. Sketch the graph of $y=\sqrt{2x+1}$. Determine the area between the curve, the coordinates axes and the line $x=4$.

b. Sketch the graph of $y=\dfrac{1}{3x+5}$. Determine the area bounded by the curve, the coordinate axes and $x=3$.

24. a. Sketch the curve $y=\dfrac{1}{(2x+3)^2}$. Determine the area bounded by this curve and the x-axis between $x=1$ and $x=2$.

b. Determine the area bounded by the curve $y=\dfrac{x}{\sqrt{16-3x}}$, the coordinate axes and $x=5$.

25. a. Determine the area of the region enclosed by the curves with the equations:
$y=4\sqrt{x-1}$ and $y=4\sqrt{3-x}$

b. Determine the area between the curve $y^2=4-x$ and the y-axis.

26. Given that a and b are non-zero real constants, calculate each of the following.

a. $\int \sqrt{ax+b}\,dx$ b. $\int x\sqrt{ax+b}\,dx$ c. $\int \frac{1}{ax+b}dx$ d. $\int \frac{x}{ax+b}dx$

27. Given that a and b are non-zero real constants, calculate each of the following.

a. $\int \frac{1}{\sqrt{ax+b}}dx$ b. $\int \frac{x}{\sqrt{ax+b}}dx$ c. $\int \frac{1}{(ax+b)^2}dx$ d. $\int \frac{x}{(ax+b)^2}dx$

28. Given that a, b, c and d are non-zero real constants, calculate each of the following.

a. $\int \frac{cx+d}{ax+b}dx$ b. $\int \frac{cx+d}{(ax+b)^2}dx$ c. $\int \frac{cx^2+d}{(ax+b)^2}dx$ d. $\int \frac{cx^2+d}{ax+b}dx$

29. Given that a and b are non-zero real constants, calculate each of the following.

a. $\int \frac{x^2}{ax+b}dx$ b. $\int \frac{x^2}{(ax+b)^2}dx$ c. $\int \frac{x^2}{(ax+b)^3}dx$ d. $\int \frac{x^2}{\sqrt{ax+b}}dx$

Technology active

30. a. Determine the area of the loop with the equation $y^2 = x^2(4-x)$.
b. Determine the area between the curve $y^2 = a - x$ where $a > 0$ and the y-axis.
c. Determine the area of the loop with the equation $y^2 = x^2(a-x)$, where $a > 0$.

7.3 Exam questions

Question 1 (4 marks) TECH-FREE

Source: VCE 2020, Specialist Mathematics Exam 1, Q2; © VCAA.

Evaluate $\int_{-1}^{0} \frac{1+x}{\sqrt{1-x}}dx$. Given your answer in the form $a\sqrt{b}+c$, where $a, b, c \in R$.

Question 2 (1 mark) TECH-ACTIVE

Source: VCE 2019, Specialist Mathematics Exam 2, Section A, Q8; © VCAA.

MC With a suitable substitution, $\int_{1}^{5} (2x-1)\sqrt{2x+1}\,dx$ can be expressed as

A. $\frac{1}{2}\int_{1}^{5}\left(u^{\frac{3}{2}}+u^{\frac{1}{2}}\right)du$

B. $2\int_{3}^{11}\left(u^{\frac{3}{2}}+u^{\frac{1}{2}}\right)du$

C. $2\int_{1}^{5}\left(u^{\frac{3}{2}}+2u^{\frac{1}{2}}\right)du$

D. $2\int_{3}^{11}\left(u^{\frac{3}{2}}+2u^{\frac{1}{2}}\right)du$

E. $\frac{1}{2}\int_{3}^{11}\left(u^{\frac{3}{2}}-2u^{\frac{1}{2}}\right)du$

Question 3 (1 mark) TECH-ACTIVE

Source: *VCE 2017, Specialist Mathematics Exam 2, Section A, Q7; © VCAA.*

MC With a suitable substitution, $\int_1^2 x^2\sqrt{2-x}\,dx$ can be expressed as

A. $-\int_1^2 \left(4u^{\frac{1}{2}} - 4u^{\frac{3}{2}} + u^{\frac{5}{2}}\right) du$

B. $\int_1^2 \left(4u^{\frac{1}{2}} - 4u^{\frac{3}{2}} + u^{\frac{5}{2}}\right) du$

C. $\int_0^1 \left(-4u^{\frac{1}{2}} + 4u^{\frac{3}{2}} - u^{\frac{5}{2}}\right) du$

D. $-\int_1^0 \left(4u^{\frac{1}{2}} - 4u^{\frac{3}{2}} + u^{\frac{5}{2}}\right) du$

E. $\int_1^0 \left(4u^{\frac{1}{2}} - 4u^{\frac{3}{2}} - u^{\frac{5}{2}}\right) du$

More exam questions are available online.

7.4 Non-linear substitutions

LEARNING INTENTION

At the end of this subtopic you should be able to:
- integrate functions using non-linear substitutions.

7.4.1 Non-linear substitutions

The basic idea of a non-linear substitution is to reduce the integrand to one of the standard u forms shown in the first column of the table below. Remember that after making a substitution, x or the original variable should be eliminated. The integral must be entirely in terms of the new variable u.

$f(u)$	$\int f(u)du$
$u^n, n \neq -1$	$\frac{u^{n+1}}{n+1}$
$\frac{1}{u}$	$\log_e(\lvert u \rvert)$
e^u	e^u
$\cos(u)$	$\sin(u)$
$\sin(u)$	$-\cos(u)$
$\sec^2(u)$	$\tan(u)$

WORKED EXAMPLE 15 Integration using non-linear substitution (1)

Calculate $\int \frac{3x}{(x^2+9)^2}dx$.

THINK	WRITE
1. Write the integrand as a power using index laws.	$\int 3x(x^2+9)^{-2}dx$

2. Use a non-linear substitution.

Let $u = x^2 + 9$.

$$\int 3x(x^2+9)^{-2}dx = \int 3xu^{-2}dx$$

3. The integral cannot be done in this form, so differentiate. Express dx in terms of du by inverting both sides.

$$\frac{du}{dx} = 2x$$

$$\frac{dx}{du} = \frac{1}{2x}$$

$$dx = \frac{1}{2x}du$$

4. Substitute for dx, noting that the terms involving x will cancel.

$$\int 3xu^{-2}dx = \int 3xu^{-2} \times \frac{1}{2x}du$$

5. Transfer the constant factors outside the front of the integral sign.

$$= \frac{3}{2}\int u^{-2}du$$

6. The integral can now be done. Anti-differentiate using $\int u^n du = \frac{u^{n+1}}{n+1}$ with $n = -2$, so that $n + 1 = -1$.

$$= -\frac{3}{2}u^{-1} + c$$

7. Write the expression with positive indices.

$$= -\frac{3}{2u} + c$$

8. Substitute back for x, and state the final result.

$$\int \frac{3x}{(x^2+9)^2}dx = -\frac{3}{2(x^2+9)} + c$$

Integrals involving the logarithm function

The result $\int u^n du = \frac{u^{n+1}}{n+1}$ is true provided that $n \neq -1$. When $n = -1$ we have the special case

$\int \frac{1}{u}du = \log_e(|u|) + c$.

WORKED EXAMPLE 16 Integration using non-linear substitution (2)

Calculate $\int \frac{x-3}{x^2-6x+13}dx$.

THINK

WRITE

Method 1

1. Use a non-linear substitution.

Let $u = x^2 - 6x + 13$.

$$\frac{du}{dx} = 2x - 6$$

$$= 2(x-3)$$

$$\frac{dx}{du} = \frac{1}{2(x-3)}$$

$$dx = \frac{1}{2(x-3)}du$$

2. Substitute for dx and u, noting that the terms involving x cancel.

$$\int \frac{x-3}{x^2-6x+13}dx = \int \frac{x-3}{u} \times \frac{1}{2(x-3)}du$$

3. Transfer the constant outside the front of the integral sign.

$$= \frac{1}{2}\int \frac{1}{u} du$$

4. The integration can now be done. Anti-differentiate using $\int \frac{1}{u} du = \log_e(|u|)$.

$$= \frac{1}{2}\log_e(|u|) + c$$

5. In this case, since $x^2 - 6x + 13 = (x-3)^2 + 4 > 0$, for all values of x, the modulus is not needed. Substitute back for x and state the final result.

Note that since $\frac{d}{dx}\left[\log_e(f(x))\right] = \frac{f'(x)}{f(x)}$, it follows that $\int \frac{f'(x)}{f(x)} dx = \log_e(|f(x|) + c$.

$$\int \frac{x-3}{x^2-6x+13} dx = \frac{1}{2}\log_e(x^2 - 6x + 13) + c$$

Method 2

1. To make the numerator the derivative of the denominator, multiply the numerator by 2, and take the constant factor outside the front of the integral sign to retain the equality.

$$\int \frac{x-3}{x^2-6x+13} dx = \frac{1}{2}\int \frac{2(x-3)}{(x^2-6x+13)} dx$$
$$= \frac{1}{2}\int \frac{2x-6}{x^2-6x+13} dx$$

2. Use the result $\int \frac{f'(x)}{f(x)} dx = \log_e(|f(x)|) + c$, with $f(x) = x^2 - 6x + 13$.

$$\int \frac{x-3}{x^2-6x+13} dx = \frac{1}{2}\log_e(x^2 - 6x + 13) + c$$

Examples involving trigonometric functions

For trigonometric functions we use the results $\int \cos(u) du = \sin(u) + c$ and $\int \sin(u) du = -\cos(u) + c$.

WORKED EXAMPLE 17 Integration using non-linear substitution (3)

Calculate $\int \frac{1}{x^2}\cos\left(\frac{1}{x}\right) dx$.

THINK

1. Use a non-linear substitution. Let $u = \frac{1}{x}$. We choose this as the derivative of $\frac{1}{x}$ is $-\frac{1}{x^2}$, which is present in the integrand.

WRITE

$$u = \frac{1}{x}$$
$$= x^{-1}$$
$$\frac{du}{dx} = -x^{-2}$$
$$= -\frac{1}{x^2}$$
$$\frac{dx}{du} = -x^2$$
$$dx = \left(-x^2\right) dx$$

2. Substitute for u and dx, noting that the x^2 terms cancel.

$$\int \frac{1}{x^2}\cos\left(\frac{1}{x}\right)dx = \int \frac{1}{x^2}\cos(u)\times\left(-x^2\right)du$$

3. Transfer the negative sign outside the integral sign.

$$= -\int \cos(u)du$$

4. Anti-differentiate, using $\int \cos(u)du = \sin(u) + c$.

$$= -\sin(u) + c$$

5. Substitute back for x and state the answer.

$$\int \frac{1}{x^2}\cos\left(\frac{1}{x}\right)dx = -\sin\left(\frac{1}{x}\right) + c$$

Examples involving exponential functions

For exponential functions we use the result $\int e^u du = e^u + c$.

WORKED EXAMPLE 18 Integration using non-linear substitution (4)

Calculate:

a. $\int \sin(2x)\, e^{\cos(2x)}dx$

b. $\int \frac{\sin(\sqrt{x})}{\sqrt{x}}dx.$

THINK / **WRITE**

a. 1. Use a non-linear substitution. Let $u = \cos(2x)$. Choose this as the derivative of $\cos(2x)$ is $-2\sin(2x)$, which is present in the integrand. Note that the substitution $u = \sin(2x)$ will not work.

a. $u = \cos(2x)$

$$\frac{du}{dx} = -2\sin(2x)$$

$$\frac{dx}{du} = \frac{-1}{2\sin(2x)}$$

$$dx = \frac{-1}{2\sin(2x)}du$$

2. Substitute for u and dx.

$$\int \sin(2x)e^{\cos(2x)}dx$$

$$= \int \sin(2x)e^u \frac{-1}{2\sin(2x)}du$$

3. Transfer the constant factor outside the integral sign.

$$= -\frac{1}{2}\int e^u du$$

4. Anti-differentiate using $\int e^u du = e^u + c$.

$$= -\frac{1}{2}e^u + c$$

5. Substitute $u = \cos(2x)$ and state the answer.

$$\int \sin(2x)e^{\cos(2x)}dx = -\frac{1}{2}e^{\cos(2x)} + c$$

b. 1. Use a non-linear substitution. Let $u = \sqrt{x}$, but only replace u in the numerator.
Express dx in terms of du by inverting both sides.

b. $u = \sqrt{x} = x^{\frac{1}{2}}$

$$\frac{du}{dx} = \frac{1}{2}x^{-\frac{1}{2}} = \frac{1}{2\sqrt{x}}$$

$$\frac{dx}{du} = 2\sqrt{x}$$

$$dx = 2\sqrt{x}\,du$$

2. Substitute for u and dx, noting that the $\sqrt{x}$ terms cancel.

$$\int \frac{\sin(\sqrt{x})}{(\sqrt{x})}dx = \int \sin(\sqrt{x}) \times \frac{1}{\sqrt{x}}dx$$

$$= \int \sin(u) \times \frac{1}{\sqrt{x}} \times 2\sqrt{x}\,du$$

3. Transfer the constant factor outside the integral sign.

$$= 2\int \sin(u)du$$

4. Anti-differentiate using $\int \sin(u)du = -\cos(u) + c$.

$$= -2\cos(u) + c$$

5. Substitute $u = \sqrt{x}$ and state the answer.

$$\int \frac{\sin(\sqrt{x})}{\sqrt{x}}dx = -2\cos(\sqrt{x}) + c$$

7.4.2 Definite integrals involving non-linear substitutions

When evaluating a definite integral, recall that the result is a number independent of the original or dummy variable. In these cases, instead of substituting the original variable back into the integral, change the terminals and work with the new definite integral obtained.

WORKED EXAMPLE 19 Evaluating definite integrals using non-linear substitution (1)

Evaluate $\int_0^{\sqrt{5}} \frac{t}{\sqrt{t^2+4}}dt$**.**

THINK

WRITE

1. Write the integrand as a power using index laws.

$$\int_0^{\sqrt{5}} \frac{t}{\sqrt{t^2+4}}dt = \int_0^{\sqrt{5}} t(t^2+4)^{-\frac{1}{2}}dt$$

2. Use a non-linear substitution.

Let $u = t^2 + 4$.

$$\int_0^{\sqrt{5}} t(t^2+4)^{-\frac{1}{2}}dt = \int_{t=0}^{t=\sqrt{5}} tu^{-\frac{1}{2}}dt$$

3. The integral cannot be done in this form, so differentiate. Express dt in terms of du by inverting both sides.

$$\frac{du}{dt} = 2t$$

$$\frac{dt}{du} = \frac{1}{2t}$$

$$dt = \frac{1}{2t}du$$

4. Change the terminals to the new variable.

When $t = 0$, $u = 4$, and when $t = \sqrt{5}$, $u = 9$.

5. Substitute for dt and the new terminals, noting that the dummy variable t cancels.

$$\int_{t=0}^{t=\sqrt{5}} tu^{-\frac{1}{2}}dt = \int_{u=4}^{u=9} tu^{-\frac{1}{2}}\frac{1}{2t}du$$

6. Transfer the multiplying constant outside the front of the integral sign.

$$= \frac{1}{2}\int_{4}^{9} u^{-\frac{1}{2}}du$$

7. Perform the integration using $\int u^n du = \dfrac{u^{n+1}}{n+1}$

with $n = -\dfrac{1}{2}$, so that $n + 1 = \dfrac{1}{2}$.

$$= \frac{1}{2}\left[2u^{\frac{1}{2}}\right]_4^9$$

$$= \left[u^{\frac{1}{2}}\right]_4^9$$

8. Evaluate the definite integral.

$$= [\sqrt{9} - \sqrt{4}]$$
$$= 3 - 2$$
$$= 1$$

9. State the answer.

$$\int_0^{\sqrt{5}} \frac{t}{\sqrt{t^2+4}}dt = 1$$

Definite integrals involving inverse trigonometric functions

Recall that $\dfrac{d}{dx}\left(\sin^{-1}\left(\dfrac{x}{a}\right)\right) = \dfrac{1}{\sqrt{a^2-x^2}}$, $\dfrac{d}{dx}\left(\cos^{-1}\left(\dfrac{x}{a}\right)\right) = \dfrac{-1}{\sqrt{a^2-x^2}}$ for $a > 0$ and $|x| < a$ and $\dfrac{d}{dx}\left(\tan^{-1}\left(\dfrac{x}{a}\right)\right) = \dfrac{a}{a^2+x^2}$ for $x \in R$.

WORKED EXAMPLE 20 Evaluating definite integrals using non-linear substitution (2)

Evaluate $\displaystyle\int_0^{\frac{1}{2}} \frac{\cos^{-1}(2x)}{\sqrt{1-4x^2}}dx.$

THINK

1. Use a non-linear substitution.

WRITE

Let $u = \cos^{-1}(2x)$.

$$\int_0^{\frac{1}{2}} \frac{\cos^{-1}(2x)}{\sqrt{1-4x^2}}dx = \int_{x=0}^{x=\frac{1}{2}} \frac{u}{\sqrt{1-4x^2}}dx$$

2. The integral cannot be done in this form, so differentiate. Express dx in terms of du by inverting both sides.

$$\frac{du}{dx} = \frac{-2}{\sqrt{1-4x^2}}$$

$$\frac{dx}{du} = -\frac{1}{2}\sqrt{1-4x^2}$$

$$dx = -\frac{1}{2}\sqrt{1-4x^2}\,du$$

3. Change the terminals to the new variable.

When $x = \frac{1}{2}$, $u = \cos^{-1}(1) = 0$, and when $x = 0$, $u = \cos^{-1}(0) = \frac{\pi}{2}$.

4. Substitute for dx and the new terminals, noting that the x terms cancel. Transfer the constant multiple outside the front of the integral sign.

$$\int_{x=0}^{x=\frac{1}{2}} \frac{u}{\sqrt{1-4x^2}}\,dx = -\frac{1}{2}\int_{u=\frac{\pi}{2}}^{u=0} \frac{1}{\sqrt{1-4x^2}}u\sqrt{1-4x^2}\,du$$

$$= -\frac{1}{2}\int_{\frac{\pi}{2}}^{0} u\,du$$

5. Using the properties of the definite integral, swap the terminals to change the sign.

$$\int_a^b f(t)dt = -\int_b^a f(t)dt$$

$$= \frac{1}{2}\int_0^{\frac{\pi}{2}} u\,du$$

6. Perform the integration.

$$= \frac{1}{2}\left[\frac{1}{2}u^2\right]_0^{\frac{\pi}{2}}$$

$$= \frac{1}{4}\left[u^2\right]_0^{\frac{\pi}{2}}$$

7. Evaluate the definite integral.

$$= \frac{1}{4}\left[\left(\frac{\pi}{2}\right)^2 - 0^2\right]$$

8. State the answer.

$$\int_0^{\frac{1}{2}} \frac{\cos^{-1}(2x)}{\sqrt{1-4x^2}}\,dx = \frac{\pi^2}{16}$$

7.4 Exercise

Students, these questions are even better in jacPLUS

Receive immediate feedback and access sample responses

Access additional questions

Track your results and progress

Find all this and MORE in jacPLUS

Technology free

1. **WE15** Calculate $\int \frac{8x}{(x^2+16)^3}\,dx$.

2. Calculate $\int \frac{5x}{\sqrt{2x^2+3}}\,dx$.

3. Calculate each of the indefinite integrals shown.

   a. $\int x(x^2+4)^5\,dx$ **b.** $\int \frac{x}{x^2+4}\,dx$ **c.** $\int \frac{x}{(x^2+9)^2}\,dx$ **d.** $\int \frac{x}{\sqrt{x^2+9}}\,dx$

4. **WE16** Calculate $\int \frac{x+2}{x^2+4x+29}\,dx$.

5. Calculate $\int \frac{x^2}{x^3+9}\,dx$.

6. Calculate each of the indefinite integrals shown.

   a. $\int \frac{x^2}{(x^3+27)^3}\,dx$ **b.** $\int \frac{x^2}{\sqrt{x^3+27}}\,dx$ **c.** $\int \frac{x^2}{x^3+8}\,dx$ **d.** $\int x^2(x^3+8)^3\,dx$

7. Calculate each of the indefinite integrals shown.

   a. $\int (x-2)(x^2-4x+13)^3\,dx$ **b.** $\int \frac{x-2}{(x^2-4x+13)^2}\,dx$

   c. $\int \frac{4-x}{\sqrt{x^2-8x+25}}\,dx$ **d.** $\int \frac{4-x}{x^2-8x+25}\,dx$

8. **WE17** Calculate $\int \frac{1}{x^2}\sin\left(\frac{1}{x}\right)dx$.

9. Calculate $\int x\cos(x^2)dx$.

10. **WE18** Calculate each of the following.

    a. $\int \cos(3x)e^{\sin(3x)}dx$ **b.** $\int \frac{\cos(\sqrt{x})}{\sqrt{x}}\,dx$.

11. **a.** Calculate $\int \sec^2(2x)e^{\tan(2x)}dx$. **b.** Calculate $\int \frac{\log_e(3x)}{4x}\,dx$.

12. Calculate each of the indefinite integrals shown.

a. $\int \frac{e^{2x}}{4e^{2x}+5}\,dx$ b. $\int \frac{e^{-3x}}{(2e^{-3x}-5)^2}\,dx$ c. $\int \frac{e^{-2x}}{(3e^{-2x}+4)^3}\,dx$ d. $\int \frac{2e^{2x}+1}{(e^{2x}+x)^2}\,dx$

13. **WE19** Evaluate $\int_0^{2\sqrt{2}} \frac{s}{\sqrt{2s^2+9}}\,ds$.

14. Evaluate $\int_0^1 \frac{p}{(3p^2+5)^2}\,dp$.

15. Evaluate each of the following definite integrals.

a. $\int_3^5 \frac{3t}{\sqrt{t^2-9}}\,dt$ b. $\int_0^2 \frac{x}{4x^2+9}\,dx$ c. $\int_1^2 \frac{x}{(2x^2+1)^2}\,dx$ d. $\int_{-1}^1 \frac{s}{\sqrt{2s^2+3}}\,ds$

16. Evaluate each of the following definite integrals.

a. $\int_4^5 \frac{3-x}{x^2-6x+34}\,dx$ b. $\int_2^3 \frac{2-x}{(x^2-4x+5)^2}\,dx$ c. $\int_1^4 \frac{e^{\sqrt{p}}}{\sqrt{p}}\,dp$ d. $\int_0^{\frac{\pi}{3}} \sec^2(2\theta)e^{\tan(2\theta)}\,d\theta$

17. Evaluate each of the following definite integrals.

a. $\int_0^{\frac{\pi}{4}} \sin(2\theta)e^{\cos(2\theta)}\,d\theta$ b. $\int_{\frac{1}{2}}^{\frac{e}{2}} \frac{\log_e(2t)}{3t}\,dt$ c. $\int_{\frac{6}{\pi}}^{\frac{3}{\pi}} \frac{1}{x^2}\cos\left(\frac{1}{x}\right)dx$ d. $\int_1^4 \frac{\sin\sqrt{x}}{\sqrt{x}}\,dx$

18. **WE20** Evaluate $\int_0^{\frac{1}{3}} \frac{\sin^{-1}(3x)}{\sqrt{1-9x^2}}\,dx$.

19. Evaluate $\int_0^4 \frac{\tan^{-1}\left(\frac{x}{4}\right)}{16+x^2}\,dx$.

20. Calculate each of the indefinite integrals shown.

a. $\int \frac{1}{x}\sin\left(\log_e(4x)\right)dx$ b. $\int \frac{1}{x}\cos\left(\log_e(3x)\right)dx$

c. $\int \frac{\sin^{-1}\left(\frac{x}{2}\right)}{\sqrt{4-x^2}}\,dx$ d. $\int \frac{\tan^{-1}(2x)}{1+4x^2}\,dx$

21. a. A certain curve has a gradient given by $x\sin(x^2)$. Determine the equation of the particular curve that passes through the origin.

b. If $\frac{dy}{dx}=\frac{1}{x^2}\sec^2\left(\frac{1}{x}\right)$ and when $x=\frac{4}{\pi}$, $y=0$, evaluate y when $x=\frac{3}{\pi}$.

c. Given that $f'(x)=\frac{5-x}{x^2-10x+29}$ and $f(0)=0$, evaluate $f(1)$.

d. If $\frac{dy}{dx}=\sin(2x)e^{\cos(2x)}$ and when $x=\frac{\pi}{4}$, $y=0$, evaluate y when $x=0$.

22. a. The graph of $y=\dfrac{x}{x^2-c}$ has vertical asymptotes at $x=\pm 2$. Determine the value of c and the area bounded by the curve, the x-axis and the lines $x=3$ and $x=5$.

b. Determine the area bounded by the curve $y=x\cos(x^2)$, the x-axis and the lines $x=0$ and $x=\dfrac{\sqrt{\pi}}{2}$.

23. a. Determine the area bounded by the curve $y=xe^{-x^2}$, the x-axis and the lines $x=0$ and $x=2$.

b. Determine the area bounded by the curve $y=\dfrac{x}{\sqrt{x^2+4}}$, the x-axis and the lines $x=0$ and $x=2\sqrt{3}$.

24. If $a, b \in R\backslash\{0\}$, then calculate each of the following.

a. $\displaystyle\int \frac{x}{\sqrt{ax^2+b}}\,dx$

b. $\displaystyle\int \frac{x}{(ax^2+b)^2}\,dx$

c. $\displaystyle\int x(ax^2+b)^n\,dx \quad n\neq -1$

d. $\displaystyle\int \frac{x}{ax^2+b}\,dx$

25. Deduce the following indefinite integrals, where $f(x)$ is any function of x.

a. $\displaystyle\int \frac{f'(x)}{f(x)}\,dx$

b. $\displaystyle\int \frac{f'(x)}{(f(x))^2}\,dx$

c. $\displaystyle\int \frac{f'(x)}{\sqrt{f(x)}}\,dx$

d. $\displaystyle\int f'(x)e^{f(x)}\,dx$

7.4 Exam questions

Question 1 (1 mark) TECH-ACTIVE

Source: VCE 2019, Specialist Mathematics Exam 2, Section A, Q8; © VCAA.

Using a suitable substitution $\displaystyle\int_0^{\frac{\pi}{6}} \tan^2(x)\sec^2(x)dx$ can be expressed as

A. $\displaystyle\int_0^{\frac{1}{\sqrt{3}}} \left(u^4+u^2\right) du$

B. $\displaystyle\int_0^{\frac{2}{\sqrt{3}}} \left(u^4+u^2\right) du$

C. $\displaystyle\int_0^{\frac{1}{\sqrt{3}}} u\,du$

D. $\displaystyle\int_0^{\frac{\pi}{6}} u^2\,du$

E. $\displaystyle\int_0^{\frac{1}{\sqrt{3}}} u^2\,du$

Question 2 (1 mark) TECH-ACTIVE

Source: VCE 2016, Specialist Mathematics Exam 2, Section A, Q8; © VCAA.

Using a suitable substitution $\displaystyle\int_a^b \left(x^3e^{2x^4}\right) dx$, where a and b are real constants, can be written as

A. $\displaystyle\int_a^b \left(e^{2u}\right) du$

B. $\displaystyle\int_{a^4}^{b^4} \left(e^{2u}\right) du$

C. $\displaystyle\frac{1}{8}\int_a^b \left(e^{u}\right) du$

D. $\displaystyle\frac{1}{4}\int_{a^4}^{b^4} \left(e^{2u}\right) du$

E. $\displaystyle\frac{1}{8}\int_{8a^3}^{8b^3} \left(e^{u}\right) du$

Question 3 (1 mark) TECH-ACTIVE

Source: *VCE 2013 Specialist Mathematics Exam 2, Section A, Q9; © VCAA.*

MC The definite integral $\int_{e^3}^{e^4} \frac{1}{x\log_e(x)}dx$ can be written in the form $\int_a^b \frac{1}{u}du$ where

A. $u=\log_e(x), a=\log_e(3), b=\log_e(4)$

B. $u=\log_e(x), a=3, b=4$

C. $u=\log_e(x), a=e^3, b=e^4$

D. $u=\frac{1}{x}, a=e^{-3}, b=e^{-4}$

E. $u=\frac{1}{x}, a=e^3, b=e^4$

More exam questions are available online.

7.5 Integrals of powers of trigonometric functions

LEARNING INTENTION

At the end of this subtopic you should be able to:

- integrate functions involving powers of trigonometric functions.

7.5.1 Integrals involving powers of sine and cosine

In this section, we examine indefinite and definite integrals involving powers of trigonometric functions of the form $\int \sin^m(kx)\cos^n(kx)dx$ where $m, n \in N$.

Integrals involving $\sin^2(kx)$ and $\cos^2(kx)$

The trigonometric double-angle formulas

$$(1)\quad \sin(2A) = 2\sin(A)\cos(A)$$

and

$$\begin{aligned}(2)\quad \cos(2A) &= \cos^2(A) - \sin^2(A)\\ &= 2\cos^2(A) - 1\\ &= 1 - 2\sin^2(A)\end{aligned}$$

are useful in integrating certain powers of trigonometric functions. Rearranging (2) gives

$$(3)\quad \sin^2(A) = \frac{1}{2}\left(1-\cos(2A)\right) \text{ and}$$

$$(4)\quad \cos^2(A) = \frac{1}{2}\left(1+\cos(2A)\right)$$

To integrate $\sin^2(kx)$ use (3); to integrate $\cos^2(kx)$ use (4).

WORKED EXAMPLE 21 Integration involving powers of trigonometric functions

Calculate:

a. $\int \cos^2(3x)dx$

b. $\int \sin^2(3x)\cos^2(3x)dx$.

THINK	WRITE
a. 1. Use the double-angle formula $\cos^2(A) = \frac{1}{2}\left(1+\cos(2A)\right)$ with $A = 3x$. Transfer the constant factor outside the front of the integral sign.	**a.** $\int \cos^2(3x)dx = \frac{1}{2}\int\left(1+\cos(6x)\right)dx$
2. Integrate term by term, using $\int \cos(kx) = \frac{1}{k}\sin(kx)+c$ with $k = 6$.	$= \frac{1}{2}\left[x+\frac{1}{6}\sin(6x)\right]+c$
3. Expand and state the final result.	$\int \cos^2(3x)dx = \frac{x}{2}+\frac{1}{12}\sin(6x)+c$
b. 1. Use the double-angle formula $2\sin(A)\cos(A) = \sin(2A)$ with $A = 3x$, so that $\sin^2(A)\cos^2(A) = \frac{1}{4}\sin^2(2A)$.	**b.** $\int \sin^2(3x)\cos^2(3x)dx = \frac{1}{4}\int\left(2\sin(3x)\cos(3x)\right)^2dx$ $= \frac{1}{4}\int \sin^2(6x)dx$
2. Use the double-angle formula $\sin^2(A) = \frac{1}{2}\left(1-\cos(2A)\right)$ with $A = 6x$.	$= \frac{1}{4}\int \frac{1}{2}\left(1-\cos(12x)\right)dx$ $= \frac{1}{8}\int\left(1-\cos(12x)\right)dx$
3. Integrate term by term, using $\int \cos(kx) = \frac{1}{k}\sin(kx)+c$ with $k = 12$.	$= \frac{1}{8}\left[x-\frac{1}{12}\sin(12x)\right]+c$
4. Expand and state the final result.	$\int \sin^2(3x)\cos^2(3x)dx = \frac{x}{8}-\frac{1}{96}\sin(12x)+c$

Note that as well as the double-angle formulas, there are many other relationships between trigonometric functions, for example, $\sin^2(A)+\cos^2(A) = 1$.

Often answers to integrals involving trigonometric functions can be expressed in several different ways, for example, as powers or multiple angles. Answers derived from CAS calculators may appear different, but often they are actually identical and differ in the constant term only.

Integrals involving $\sin(kx)\cos^m(kx)$ and $\cos(kx)\sin^m(kx)$ where $m > 1$

Integrals of the forms $\sin(kx)\cos^m(kx)$ and $\cos(kx)\sin^m(kx)$ where $m > 1$ can be performed using non-linear substitution, as described in the previous section.

WORKED EXAMPLE 22 Integration of the product of trigonometric functions

Calculate $\int \sin(3x)\cos^4(3x)dx$.

THINK

1. Use a non-linear substitution. Let $u = \cos(3x)$. We choose this as the derivative of $\cos(3x)$ is $-3\sin(3x)$, which is present in the integrand.

WRITE

$$u = \cos(3x)$$
$$\frac{du}{dx} = -3\sin(3x)$$
$$\frac{dx}{du} = \frac{-1}{3\sin(3x)}$$
$$dx = \frac{-1}{3\sin(3x)}du$$

2. Substitute for u and dx, noting that the x terms cancel.

$$\int \sin(3x)\cos^4(3x)dx = \int \sin(3x)u^4 \times \frac{-1}{3\sin(3x)}du$$

3. Transfer the constant factor outside the integral sign.

$$= -\frac{1}{3}\int u^4 du$$

4. Anti-differentiate.

$$= -\frac{1}{3} \times \frac{u^5}{5} + c$$
$$= -\frac{1}{15}u^5 + c$$

5. Substitute back for x and state the final result.

$$\int \sin(3x)\cos^4(3x)dx = -\frac{1}{15}\cos^5(3x) + c$$

Integrals involving odd powers

Anti-differentiation of $\sin^m(kx)\cos^n(kx)$ when at least one of m or n is an odd power can be performed using a non-linear substitution and the formula $\sin^2(A) + \cos^2(A) = 1$.

WORKED EXAMPLE 23 Integration of odd powers of trigonometric functions

Determine an anti-derivative of $\sin^3(2x)\cos^4(2x)$.

THINK

1. Write the required anti-derivative.

WRITE

$$\int \sin^3(2x)\cos^4(2x)\,dx$$

2. Break the odd power. $\sin^3(2x) = \sin(2x)\sin^2(2x)$, and $\sin^2(A) = 1 - \cos^2(A)$.

$$\int \sin(2x)\sin^2(2x)\cos^4(2x)dx$$
$$= \int \sin(2x)(1 - \cos^2(2x))\cos^4(2x)dx$$

3. Use a non-linear substitution. Let $u = \cos(2x)$. We choose this as the derivative of $\cos(2x)$ is $-2\sin(2x)$, which is present in the integrand and will cancel.

$$u = \cos(2x)$$
$$\frac{du}{dx} = -2\sin(2x)$$
$$\frac{dx}{du} = \frac{-1}{2\sin(2x)}$$
$$dx = \frac{-1}{2\sin(2x)}du$$

4. Substitute for u and dx, noting that the sin (2x) terms cancel.

$$= \int \sin(2x)(1-u^2)u^4 \times \frac{-1}{2\sin(2x)}\,du$$

5. Transfer the constant factor outside the integral sign and expand.

$$= -\frac{1}{2}\int (1-u^2)u^4\,du$$

$$= -\frac{1}{2}\int (u^4-u^6)\,du$$

6. Anti-differentiate.

$$= -\frac{1}{2} \times \left(\frac{1}{5}u^5 - \frac{1}{7}u^7\right) + c$$

$$= \frac{1}{14}u^7 - \frac{1}{10}u^5 + c$$

7. Substitute back for x and state the final result.

$$\int \sin^3(2x)\cos^4(2x)\,dx$$

$$= \frac{1}{14}\cos^7(2x) - \frac{1}{10}\cos^5(2x) + c$$

Integrals involving even powers

Anti-differentiation of $\sin^m(kx)\cos^n(kx)$ when both m and n are even powers must be performed using the double angle formulas.

WORKED EXAMPLE 24 Integration of even powers of trigonometric functions

Determine an anti-derivative of $\cos^4(2x)$.

THINK | **WRITE**

1. Write the required anti-derivative.

$$\int \cos^4(2x)\,dx = \int \left(\cos^2(2x)\right)^2 dx$$

2. Since there is no odd power, we must use the double angle formula $\cos^2(A) = \frac{1}{2}\left(1 + \cos(2A)\right)$ with $A = 2x$.

$$= \int \left(\frac{1}{2}\left(1+\cos(4x)\right)\right)^2 dx$$

3. Expand the integrand.

$$= \frac{1}{4}\int \left(1 + 2\cos(4x) + \cos^2(4x)\right) dx$$

4. Replace $\cos^2(A) = \frac{1}{2}\left(1+\cos(2A)\right)$ with $A = 4x$ and expand. The integrand is now in a form that we can integrate term by term.

$$= \frac{1}{4}\int \left(1 + 2\cos(4x) + \frac{1}{2}\left(1+\cos(8x)\right)\right) dx$$

$$= \frac{1}{4}\int \left(\frac{3}{2} + 2\cos(4x) + \frac{1}{2}\cos(8x)\right) dx$$

5. Anti-differentiate term by term.

$$= \frac{1}{4}\left[\frac{3x}{2} + \frac{2}{4}\sin(4x) + \frac{1}{16}\sin(8x)\right] + c$$

6. State the final result.

$$\int \cos^4(2x)\,dx = \frac{3x}{8} + \frac{1}{8}\sin(4x) + \frac{1}{64}\sin(8x) + c$$

7.5.2 Integrals involving powers of $\tan(kx)$

In this section, we examine indefinite and definite integrals involving powers of the tangent function, that is, integrals of the form $\int \tan^n(kx)dx$ where $n \in Z$.

The result $\tan(A) = \dfrac{\sin(A)}{\cos(A)}$ is used to integrate $\tan(A)$.

WORKED EXAMPLE 25 Integration of tangent functions

Calculate $\int \tan(2x)dx$.

THINK	WRITE
1. Use $\tan(A) = \dfrac{\sin(A)}{\cos(A)}$ with $A = 2x$.	$\int \tan(2x)dx = \int \dfrac{\sin(2x)}{\cos(2x)}dx$
2. Use a non-linear substitution. Let $u = \cos(2x)$. We choose this as the derivative of $\cos(2x)$ is $-2\sin(2x)$, which is present in the numerator integrand and will cancel.	$u = \cos(2x)$ $\dfrac{du}{dx} = -2\sin(2x)$ $\dfrac{dx}{du} = \dfrac{-1}{2\sin(2x)}$ $dx = \dfrac{-1}{2\sin(2x)}du$
3. Substitute for u and dx, noting that the x terms cancel.	$\int \tan(2x)dx = \int \dfrac{\sin(2x)}{u} \times \dfrac{-1}{2\sin(2x)}du$
4. Transfer the constant factor outside the integral sign.	$= -\dfrac{1}{2}\int \dfrac{1}{u}du$
5. Anti-differentiate.	$= -\dfrac{1}{2}\log_e(\lvert u \rvert) + c$
6. Substitute back for x and state the final result. Again note that there are different answers using log laws and trigonometric identities.	$\int \tan(2x)dx = -\dfrac{1}{2}\log_e(\lvert\cos(2x)\rvert) + c$ $= \dfrac{1}{2}\log_e(\lvert\cos(2x)\rvert)^{-1} + c$ $= \dfrac{1}{2}\log_e(\lvert\sec(2x)\rvert) + c$

Integrals involving $\tan^2(kx)$

To calculate $\int \tan^2(kx)dx$, we use the results $1 + \tan^2(A) = \sec^2(A)$ and $\dfrac{d}{dx}(\tan(kx)) = k\sec^2(kx)$, so that $\int \sec^2(kx) = \dfrac{1}{k}\tan(kx) + c$.

WORKED EXAMPLE 26 Integration of squared tangent functions

Calculate $\int \tan^2(2x)dx$.

THINK	WRITE
1. Use $1+\tan^2(A)=\sec^2(A)$ with $A=2x$.	$\int \tan^2(2x)dx = \int \left(\sec^2(2x)-1\right)dx$
2. Use $\int \sec^2(kx) = \frac{1}{k}\tan(kx)+c$ and integrate term by term. State the final result.	$\int \tan^2(2x)dx = \frac{1}{2}\tan(2x)-x+c$

7.5 Exercise

Technology free

1. **WE21** Calculate each of the following.

   a. $\int \sin^2\left(\frac{x}{4}\right)dx$ b. $\int \sin\left(\frac{x}{4}\right)\cos\left(\frac{x}{4}\right)dx.$

2. Evaluate each of the following.

   a. $\int_0^{\frac{\pi}{6}} 4\sin^2(2x)dx$ b. $\int_0^{\frac{\pi}{3}} \sin^2(2x)\cos^2(2x)dx.$

3. **WE22** a. Calculate $\int \cos(4x)\sin^5(4x)dx.$ b. Evaluate $\int_0^{\frac{\pi}{4}} \sin(2x)\cos^3(2x)dx.$

4. Calculate each of the following.

   a. $\int \cos(2x)\sin(2x)dx$ b. $\int \left(\cos^2(2x)+\sin^2(2x)\right)dx$

5. Calculate each of the following.

   a. $\int \cos^3(2x)\sin(2x)dx$ b. $\int \cos(2x)\sin^3(2x)dx$

6. **WE23** a. Determine an anti-derivative of $\cos^5(4x)\sin^2(4x)$.

   b. Evaluate $\int_0^{\frac{\pi}{12}} \sin^3(3x)dx.$

7. Evaluate each of the following.

a. $\int_0^{\frac{\pi}{4}} \cos(2x)\sin^4(2x)dx$

b. $\int_0^{\frac{\pi}{4}} \cos^2(2x)\sin^3(2x)dx$

8. WE24 a. Determine an anti-derivative of $\sin^4(2x)$.

b. Evaluate $\int_0^{\frac{\pi}{12}} \cos^4(3x)dx$.

9. Evaluate each of the following.

a. $\int_0^{\frac{\pi}{4}} \cos^2(2x)\sin^2(2x)dx$

b. $\int_0^{\frac{\pi}{4}} \cos^3(2x)\sin^2(2x)dx$

10. Calculate each of the following.

a. $\int \cos^2(4x)\sin^2(4x)dx$

b. $\int \cos^2(4x)\sin^3(4x)dx$

11. Calculate each of the following.

a. $\int \cos^3(4x)\sin^2(4x)dx$

b. $\int \cos^3(4x)\sin^4(4x)dx$

12. Evaluate each of the following.

a. $\int_0^{\frac{\pi}{6}} \sin^2(3x)dx$

b. $\int_0^{\frac{\pi}{6}} \cos^3(3x)dx$

13. Evaluate each of the following.

a. $\int_0^{\frac{\pi}{6}} \sin^4(3x)dx$

b. $\int_0^{\frac{\pi}{6}} \cos^5(3x)dx$

14. a. Calculate $\int (\cos(2x) + \sin(2x))^2 dx$.

b. Calculate $\int \cos^3(2x) + \sin^3(2x)dx$.

15. Consider $\int \sin^3(2x)\cos^3(2x)dx$. Show that this integration can be done using:

a. a double angle formula

b. the substitution $u = \cos(2x)$

c. the substitution $u = \sin(2x)$.

16. WE25 a. Calculate $\int \tan\left(\frac{x}{2}\right) dx$.

b. Evaluate $\int_0^{\frac{\pi}{12}} \tan(4x)dx$.

17. WE26 a. Calculate $\int \tan^2\left(\frac{x}{3}\right) dx$.

b. Evaluate $\int_0^{\frac{\pi}{16}} \tan^2(4x)dx$.

18. Calculate each of the following.

a. $\int \tan(3x)dx$

b. $\int \cot(3x)dx$

19. Calculate each of the following.

a. $\int \tan(3x)\sec^2(3x)dx$

b. $\int \tan^2(3x)\sec^2(3x)dx$

20. Evaluate each of the following.

a. $\int_0^{\frac{\pi}{20}} \tan(5x)dx$

b. $\int_0^{\frac{\pi}{20}} \tan^2(5x)dx$

21. Evaluate each of the following.

a. $\int_0^{\frac{\pi}{20}} \tan^3(5x)\sec^2(5x)dx$

b. $\int_0^{\frac{\pi}{20}} \tan^2(5x)\sec^4(5x)dx$

22. Determine an anti-derivative of each of the following.

a. $\text{cosec}^2(2x)\cos(2x)$

b. $\sec^2(2x)\sin(2x)$

c. $\dfrac{\sin(2x)}{\cos^3(2x)}$

d. $\dfrac{\cos(2x)}{\sin^3(2x)}$

23. Given that $n \neq -1$ and $a \in R$, calculate each of the following.

a. $\int \sin(ax)\cos^n(ax)dx$

b. $\int \cos(ax)\sin^n(ax)dx$

c. $\int \tan^n(ax)\sec^2(ax)dx$

24. Integrate each of the following where $a \in R\backslash\{0\}$.

a. $\tan(ax)$

b. $\tan^2(ax)$

c. $\tan^3(ax)$

d. $\tan^4(ax)$

25. Integrate each of the following where $a \in R\backslash\{0\}$.

a. $\sin^2(ax)$

b. $\sin^3(ax)$

c. $\sin^4(ax)$

d. $\sin^5(ax)$

26. Integrate each of the following where $a \in R\backslash\{0\}$.

a. $\cos^2(ax)$

b. $\cos^3(ax)$

c. $\cos^4(ax)$

d. $\cos^5(ax)$

7.5 Exam questions

Question 1 (1 mark) TECH-ACTIVE

MC The exact value of $\int_0^{\frac{\pi}{3}} \cos^2\left(\frac{3x}{2}\right) dx$ is

A. $\frac{\pi}{3}$

B. $\frac{\pi}{6}$

C. $\frac{\pi}{12}$

D. $\frac{2\pi}{3}$

E. 0

Question 2 (1 mark) TECH-ACTIVE

MC State which of the following is an anti-derivative of $\sec^2(3x)\tan^2(3x)$.

A. $3\tan^3(3x) + c$

B. $3\sec^3(3x) + c$

C. $\frac{1}{3}\tan^3(3x) + c$

D. $\frac{1}{9}\tan^3(3x) + c$

E. $\frac{1}{3}\sec^3(3x) + c$

Question 3 (2 marks) TECH-FREE

Evaluate $\int_0^p \left(\sin^2(5x) - \cos^2(5x)\right) dx$ in terms of p.

More exam questions are available online.

7.6 Integrals involving inverse trigonometric functions

LEARNING INTENTION

At the end of this subtopic you should be able to:
- integrate functions which are derivatives of inverse trigonometric functions.

7.6.1 Integrals involving the inverse sine function

Since $\frac{d}{dx}\left(\sin^{-1}\left(\frac{x}{a}\right)\right) = \frac{1}{\sqrt{a^2 - x^2}}$, it follows that $\int \frac{1}{\sqrt{a^2 - x^2}}\,dx = \sin^{-1}\left(\frac{x}{a}\right) + c$ for $a > 0$ and $|x| < a$.

WORKED EXAMPLE 27 Integration using inverse sine functions

Calculate:

a. $\int \frac{12}{\sqrt{36 - x^2}}\,dx$

b. $\int \frac{4}{\sqrt{49 - 36x^2}}\,dx.$

THINK	WRITE
a. Use $\int \frac{1}{\sqrt{a^2 - x^2}}\,dx = \sin^{-1}\left(\frac{x}{a}\right) + c$ with $a = 6$.	**a.** $\int \frac{12}{\sqrt{36 - x^2}}\,dx = 12\int \frac{1}{\sqrt{36 - x^2}}\,dx$ $= 12\sin^{-1}\left(\frac{x}{6}\right) + c$
b. 1. Use a linear substitution with $u = 6x$ and express dx in terms of du by inverting both sides.	**b.** $u = 6x$ $u^2 = 36x^2$ $\frac{du}{dx} = 6$ $\frac{dx}{du} = \frac{1}{6}$ $dx = \frac{1}{6}\,du$
2. Substitute for dx and u.	$\int \frac{4}{\sqrt{49 - 36x^2}}\,dx = \int \frac{4}{\sqrt{49 - u^2}} \times \frac{1}{6}\,du$
3. Use the properties of indefinite integrals to transfer the constant factor outside the front of the integral sign.	$= \frac{2}{3}\int \frac{1}{\sqrt{49 - u^2}}\,du$
4. Use $\int \frac{1}{\sqrt{a^2 - u^2}}\,du = \sin^{-1}\left(\frac{u}{a}\right) + c$ with $a = 7$.	$= \frac{2}{3}\sin^{-1}\left(\frac{u}{7}\right) + c$
5. Substitute back for u and express the final answer in terms of x only and an arbitrary constant $+c$.	$\int \frac{4}{\sqrt{49 - 36x^2}}\,dx = \frac{2}{3}\sin^{-1}\left(\frac{6x}{7}\right) + c$

7.6.2 Integrals involving the inverse cosine function

Since $\frac{d}{dx}\left(\cos^{-1}\left(\frac{x}{a}\right)\right)=\frac{-1}{\sqrt{a^2-x^2}}$, it follows that $\int \frac{1}{\sqrt{a^2-x^2}}\,dx=\cos^{-1}\left(\frac{x}{a}\right)+c$ where $a>0$ and $|x|<a$.

WORKED EXAMPLE 28 Integration using inverse cosine functions

On a certain curve the gradient is given by $\frac{-4}{\sqrt{81-25x^2}}$. Determine the equation of the curve that passes through the origin.

THINK	WRITE
1. Recognise that the gradient of a curve is given by $\frac{dy}{dx}$.	$\frac{dy}{dx}=\frac{-4}{\sqrt{81-25x^2}}$
2. Integrating both sides gives an expression for y.	$y=\int \frac{-4}{\sqrt{81-25x^2}}\,dx$
3. Use a linear substitution with $u=5x$ and express dx in terms of du by inverting both sides.	$u=5x$ $u^2=25x^2$ $\frac{du}{dx}=5$ $\frac{dx}{du}=\frac{1}{5}$ $dx=\frac{1}{5}\,du$
4. Substitute for dx and u.	$y=\int \frac{-4}{\sqrt{81-u^2}}\times\frac{1}{5}\,du$
5. Use the properties of indefinite integrals to transfer the constant factor outside the front of the integral sign.	$=\frac{4}{5}\int \frac{-1}{\sqrt{81-u^2}}\,du$
6. Use $\int \frac{-1}{\sqrt{a^2-u^2}}\,du=\cos^{-1}\left(\frac{u}{a}\right)+c$ with $a=9$.	$=\frac{4}{5}\cos^{-1}\left(\frac{u}{9}\right)+c$
7. Substitute back for u and express the final answer in terms of x only and an arbitrary constant $+c$.	$=\frac{4}{5}\cos^{-1}\left(\frac{5x}{9}\right)+c$
8. Since the curve passes through the origin, we can let $y=0$ when $x=0$ to determine c.	$0=\frac{4}{5}\cos^{-1}(0)+c$ $0=\frac{4}{5}\times\frac{\pi}{2}+c$ $c=-\frac{2\pi}{5}$
9. Substitute back for c.	$y=\frac{4}{5}\cos^{-1}\left(\frac{5x}{9}\right)-\frac{2\pi}{5}$

10. State the equation of the particular curve in a factored form.	$= \frac{4}{5}\left(\cos^{-1}\left(\frac{5x}{9}\right) - \frac{\pi}{2}\right)$
11. Note that an alternative answer is possible, since $\sin^{-1}(x) + \cos^{-1}(x) = \frac{\pi}{2}$.	$= -\frac{4}{5}\sin^{-1}\left(\frac{5x}{9}\right)$

7.6.3 Integrals involving the inverse tangent function

Since $\frac{d}{dx}\left(\tan^{-1}\left(\frac{x}{a}\right)\right) = \frac{a}{a^2+x^2}$, it follows that $\int \frac{1}{a^2+x^2} = \frac{1}{a}\tan^{-1}\left(\frac{x}{a}\right) + c$ for $x \in R$.

WORKED EXAMPLE 29 Integration using inverse tangent functions

Anti-differentiate each of the following with respect to x.

a. $\frac{12}{36+x^2}$

b. $\frac{4}{49+36x^2}$

THINK	WRITE
a. 1. Write the required anti-derivative.	a. $\int \frac{12}{36+x^2}\,dx$
2. Use $\int \frac{1}{a^2+x^2} = \frac{1}{a}\tan^{-1}\left(\frac{x}{a}\right) + c$ with $a = 6$.	$12\int \frac{1}{36+x^2}\,dx$ $= 12 \times \frac{1}{6}\tan^{-1}\left(\frac{x}{6}\right) + c$ $= 2\tan^{-1}\left(\frac{x}{6}\right) + c$
b. 1. Write the required anti-derivative.	b. $\int \frac{4}{49+36x^2}\,dx$
2. Use a linear substitution with $u = 6x$ and express dx in terms of du by inverting both sides.	$u = 6x$ $u^2 = 36x^2$ $\frac{du}{dx} = 6$ $dx = \frac{1}{6}\,du$
3. Substitute for dx and u.	$\int \frac{4}{49+36x^2}\,dx = \int \frac{4}{49+u^2} \times \frac{1}{6}\,du$
4. Transfer the constant factor outside the front of the integral sign.	$= \frac{2}{3}\int \frac{1}{49+u^2}\,du$
5. Use $\int \frac{1}{a^2+x^2} = \frac{1}{a}\tan^{-1}\left(\frac{x}{a}\right) + c$ with $a = 7$.	$= \frac{2}{3} \times \frac{1}{7}\tan^{-1}\left(\frac{u}{7}\right) + c$
6. Substitute back for u and express the final answer in terms of x only and an arbitrary constant $+c$.	$\int \frac{4}{49+36x^2}\,dx = \frac{2}{21}\tan^{-1}\left(\frac{6x}{7}\right) + c$

7.6.4 Definite integrals involving inverse trigonometric functions

WORKED EXAMPLE 30 Definite integrals involving inverse trigonometric functions

Evaluate:

a. $\displaystyle\int_0^{\frac{2}{9}} \frac{-2}{\sqrt{16-81x^2}}dx$ **b.** $\displaystyle\int_0^{\frac{4}{9}} \frac{2}{16+81x^2}dx.$

THINK / **WRITE**

a. 1. Use a linear substitution with $u = 9x$ and express dx in terms of du by inverting both sides.

a.
$$u = 9x$$
$$u^2 = 81x^2$$
$$\frac{du}{dx} = 9$$
$$\frac{dx}{du} = \frac{1}{9}$$
$$dx = \frac{1}{9}du$$

2. Change the terminals to the new variable.

When $x = \frac{2}{9}$, $u = 2$, and when $x = 0$, $u = 0$.

3. Substitute for dx, u and the new terminals.

$$\int_0^{\frac{2}{9}} \frac{-2}{\sqrt{16-81x^2}}dx$$
$$= \int_0^{2} \frac{-2}{\sqrt{16-u^2}} \times \frac{1}{9}\,du$$

4. Transfer the constant factor outside the front of the integral sign.

$$= \frac{2}{9}\int_0^{2} \frac{-1}{\sqrt{16-u^2}}\,du$$

5. Perform the integration using $\displaystyle\int \frac{-1}{\sqrt{a^2-u^2}}\,du = \cos^{-1}\left(\frac{u}{a}\right) + c$ with $a = 4$.

$$= \frac{2}{9}\left[\cos^{-1}\left(\frac{u}{4}\right)\right]_0^2$$

6. Evaluate the definite integral.

$$= \frac{2}{9}\left[\cos^{-1}\left(\frac{1}{2}\right) - \cos^{-1}(0)\right]$$
$$= \frac{2}{9}\left[\frac{\pi}{3} - \frac{\pi}{2}\right]$$
$$= \frac{2}{9}\left(\frac{\pi(2-3)}{6}\right)$$

7. State the final result. Note that since this is just evaluating a definite integral and not determining an area, the answer stays as a negative number.

$$\int_0^{\frac{2}{9}} \frac{-2}{\sqrt{81-16x^2}}\,dx = -\frac{\pi}{27}$$

THINK	WRITE
b. 1. Use a linear substitution with $u = 9x$ and express dx in terms of du by inverting both sides.	**b.** $u = 9x$ $u^2 = 81x^2$ $\frac{du}{dx} = 9$ $\frac{dx}{du} = \frac{1}{9}$ $dx = \frac{1}{9}\,du$
2. Change the terminals to the new variable.	When $x = \frac{4}{9}$, $u = 4$, and when $x = 0$, $u = 0$.
3. Substitute for dx, u and the new terminals.	$\int_0^{\frac{4}{9}} \frac{2}{16+81x^2}\,dx$ $= \int_0^4 \frac{2}{16+u^2} \times \frac{1}{9}\,du$
4. Transfer the constant factor outside the front of the integral sign.	$= \frac{2}{9}\int_0^4 \frac{1}{16+u^2}\,du$
5. Perform the integration using $\int \frac{1}{a^2+x^2} = \frac{1}{a}\tan^{-1}\left(\frac{x}{a}\right) + c$ with $a = 4$.	$= \frac{2}{9}\left[\frac{1}{4}\tan^{-1}\left(\frac{u}{4}\right)\right]_0^4$ $= \frac{1}{18}\left[\tan^{-1}\left(\frac{u}{4}\right)\right]_0^4$
6. Evaluate this definite integral.	$= \frac{1}{18}\left[\tan^{-1}(1) - \tan^{-1}(0)\right]$ $= \frac{1}{18}\left[\frac{\pi}{4} - 0\right]$
7. State the final result.	$\int_0^{\frac{4}{9}} \frac{2}{16+81x^2}\,dx = \frac{\pi}{72}$

Integrals involving completing the square

A quadratic expression in the form $ax^2 + bx + c$ can be expressed in the form $a(x+h)^2 + k$ (the completing the square form). This can be used to integrate expressions of the form $\frac{1}{ax^2+bx+c}$ with $a > 0$ and $\Delta = b^2 - 4ac < 0$, which will then involve the inverse tangent function.

Integrating expressions of the form $\frac{1}{\sqrt{ax^2+bx+c}}$ with $a < 0$ and $\Delta = b^2 - 4ac > 0$ will involve the inverse sine or cosine functions.

WORKED EXAMPLE 31 Integrals involving completing the square

Calculate each of the following with respect to x.

a. $\displaystyle\int \frac{1}{9x^2+12x+29}\,dx$

b. $\displaystyle\int \frac{1}{\sqrt{21-12x-9x^2}}\,dx$

THINK	WRITE
a. 1. Express the quadratic factor in the denominator in the completing the square form by making it into a perfect square.	a. $9x^2+12x+29 = \left(9x^2+12x+4\right)+25$ $= (3x+2)^2+25$
2. Write the denominator as the sum of two squares.	$\displaystyle\int \frac{1}{9x^2+12x+29}\,dx = \int \frac{1}{(3x+2)^2+25}\,dx$
3. Use a linear substitution with $u=3x+2$ and express dx in terms of du by inverting both sides.	$u=3x+2$ $\dfrac{du}{dx}=3$ $\dfrac{dx}{du}=\dfrac{1}{3}$ $dx=\dfrac{1}{3}\,du$
4. Substitute for dx and u.	$\displaystyle\int \frac{1}{(3x+2)^2+25}\,dx = \int \frac{1}{u^2+25}\times\frac{1}{3}\,du$
5. Transfer the constant factor outside the front of the integral sign.	$\displaystyle = \frac{1}{3}\int \frac{1}{u^2+25}\,du$
6. Use $\displaystyle\int \frac{1}{a^2+u^2} = \frac{1}{a}\tan^{-1}\left(\frac{u}{a}\right)+c$ with $a=5$.	$\displaystyle = \frac{1}{3}\times\frac{1}{5}\tan^{-1}\left(\frac{u}{5}\right)+c$ $\displaystyle = \frac{1}{15}\tan^{-1}\left(\frac{u}{5}\right)+c$
7. Substitute back for u and express the final answer in terms of x only and an arbitrary constant $+c$.	$\displaystyle\int \frac{1}{9x^2+12x+29}\,dx = \frac{1}{15}\tan^{-1}\left(\frac{3x+2}{5}\right)+c$
b. 1. Express the quadratic factor in the denominator in the completing the square form by making it into the difference of two squares.	b. $21-12x-9x^2 = 21-\left(9x^2+12x\right)$ $= 21-\left(9x^2+12x+4\right)+4$ $= 25-(3x+2)^2$
2. Write the denominator as the difference of two squares under the square root.	$\displaystyle\int \frac{1}{\sqrt{21-12x-9x^2}}\,dx = \int \frac{1}{\sqrt{25-(3x+2)^2}}\,dx$
3. Use a linear substitution with $u=3x+2$ and express dx in terms of du by inverting both sides.	$u=3x+2$ $\dfrac{du}{dx}=3$ $\dfrac{dx}{du}=\dfrac{1}{3}$ $dx=\dfrac{1}{3}\,du$

4. Substitute for dx and u.

$$\int \frac{1}{\sqrt{25-(3x+2)^2}}\,dx = \int \frac{1}{\sqrt{25-u^2}} \times \frac{1}{3}\,du$$

5. Transfer the constant factor outside the front of the integral sign.

$$= \frac{1}{3}\int \frac{1}{\sqrt{25-u^2}}\,du$$

6. Use $\int \frac{1}{\sqrt{a^2-u^2}} = \sin^{-1}\left(\frac{u}{a}\right) + c$ with $a = 5$.

$$= \frac{1}{3}\sin^{-1}\left(\frac{u}{5}\right) + c$$

7. Substitute back for u and express the final answer in terms of x only and an arbitrary constant $+c$.

$$\int \frac{1}{\sqrt{21-12x-9x^2}}\,dx = \frac{1}{3}\sin^{-1}\left(\frac{3x+2}{5}\right) + c$$

Integrals involving substitutions and inverse trigonometric functions

We can break up complicated integrals into two manageable integrals using the following property of indefinite integrals: $\int (f(x) \pm g(x))dx = \int f(x)dx \pm \int g(x)dx$.

WORKED EXAMPLE 32 Integrals involving substitution

Calculate $\int \frac{4x+5}{\sqrt{25-16x^2}}dx$**.**

THINK

1. If the x was not present in the numerator, the integral would involve an inverse sine function. If the 5 was not present in the numerator, the integral would involve a non-linear substitution. Break the integral into two distinct problems: one involving a non-linear substitution and one involving the inverse trigonometric function.
2. Write the first integral as a power using index laws.

WRITE

1.
$$\int \frac{4x+5}{\sqrt{25-16x^2}}\,dx$$
$$= \int \frac{4x}{\sqrt{25-16x^2}}\,dx + \int \frac{5}{\sqrt{25-16x^2}}\,dx$$

2.
$$\int \frac{4x}{\sqrt{25-16x^2}}\,dx = \int 4x\left(25-16x^2\right)^{-\frac{1}{2}}dx$$

3. Use a non-linear substitution with $u = 25 - 16x^2$ and express dx in terms of du by inverting both sides.

$$u = 25 - 16x^2$$
$$\frac{du}{dx} = -32x$$
$$\frac{dx}{du} = -\frac{1}{32x}$$
$$dx = -\frac{1}{32x}\,du$$

4. Substitute for dx and u, noting that the x terms cancel.

$$\int 4x(25 - 16x^2)^{-\frac{1}{2}}dx = \int 4xu^{-\frac{1}{2}} \times -\frac{1}{32x}\,du$$

5. Transfer the constant factor outside the integral sign.

$$= -\frac{1}{8}\int u^{-\frac{1}{2}}du$$

6. Perform the integration using $\int u^n du = \frac{1}{n+1}\,u^{n+1}$ with $n = -\frac{1}{2}$, so that $n + 1 = \frac{1}{2}$.

$$= -\frac{1}{8} \times \frac{1}{\frac{1}{2}}u^{\frac{1}{2}}$$
$$= -\frac{1}{4}\sqrt{u}$$

7. Substitute back for u.

$$\int \frac{4x}{\sqrt{25 - 16x^2}}\,dx = -\frac{1}{4}\sqrt{25 - 16x^2}$$

8. Consider the second integral. Use a linear substitution with $v = 4x$ and express dx in terms of dv by inverting both sides.

$$\int \frac{5}{\sqrt{25 - 16x^2}}\,dx$$
$$v = 4x$$
$$\frac{dv}{dx} = 4$$
$$\frac{dx}{dv} = \frac{1}{4}$$
$$dx = \frac{1}{4}\,dv$$

9. Substitute for dx and v.

$$\int \frac{5}{\sqrt{25 - 15x^2}}\,dx = \int \frac{5}{\sqrt{25 - v^2}} \times \frac{1}{4}\,dv$$

10. Use the properties of indefinite integrals to transfer the constant factor outside the front of the integral sign.

$$= \frac{5}{4}\int \frac{1}{\sqrt{25 - v^2}}\,dv$$

11. Use $\int \frac{1}{\sqrt{a^2 - v^2}}\,dv = \sin^{-1}\left(\frac{v}{a}\right) + c$ with $a = 5$.

$$= \frac{5}{4}\,\sin^{-1}\left(\frac{v}{5}\right)$$

12. Substitute back for v.

$$\int \frac{5}{\sqrt{25 - 4x^2}}\,dx = \frac{5}{4}\,\sin^{-1}\left(\frac{4x}{5}\right)$$

13. Express the original integral as the sum as the two integrals, adding in only one arbitrary constant c.

$$\int \frac{4x + 5}{\sqrt{25 - 16x^2}}\,dx = \frac{5}{4}\,\sin^{-1}\left(\frac{4x}{5}\right) - \frac{1}{4}\sqrt{25 - 16x^2} + c$$

7.6 Exercise

Students, these questions are even better in jacPLUS

Receive immediate feedback and access sample responses

Access additional questions

Track your results and progress

Find all this and MORE in jacPLUS

Technology free

1. WE27 Calculate each of the following.

   a. $\displaystyle\int \frac{1}{\sqrt{100-x^2}}dx$

   b. $\displaystyle\int \frac{12}{\sqrt{64-9x^2}}dx.$

2. Integrate each of the following with respect to x.

   a. $\dfrac{1}{\sqrt{36-25x^2}}$

   b. $\dfrac{x}{\sqrt{36-25x^2}}$

3. Integrate each of the following with respect to x.

   a. $\dfrac{1}{\sqrt{16-x^2}}$

   b. $\dfrac{1}{\sqrt{1-16x^2}}$

4. Integrate each of the following with respect to x.

   a. $\dfrac{10}{\sqrt{49-25x^2}}$

   b. $\dfrac{10x}{\sqrt{49-25x^2}}$

5. a. WE28 On a certain curve the gradient is given by $\dfrac{-2}{\sqrt{25-16x^2}}$. Determine the equation of the curve that passes through the origin.

   b. On a certain curve the gradient is given by $\dfrac{-2}{\sqrt{4-x^2}}$. Determine the equation of the curve that passes through the point $(2, 3)$.

6. Calculate each of the following.

   a. $\displaystyle\int \frac{-1}{\sqrt{4-x^2}}dx$

   b. $\displaystyle\int \frac{-2}{\sqrt{1-4x^2}}dx$

7. Calculate each of the following.

   a. $\displaystyle\int \frac{-3x}{\sqrt{36-49x^2}}dx$

   b. $\displaystyle\int \frac{-3}{\sqrt{36-49x^2}}dx$

8. WE29 Anti-differentiate each of the following with respect to x.

   a. $\dfrac{1}{100+x^2}$

   b. $\dfrac{12}{64+9x^2}$

9. Integrate each of the following with respect to x.

   a. $\dfrac{1}{36+25x^2}$

   b. $\dfrac{5x}{64+25x^2}$

10. WE30 Evaluate each of the following.

a. $\int_0^{\frac{9}{8}} \frac{1}{\sqrt{81-16x^2}}dx$

b. $\int_0^{\frac{9}{4}} \frac{1}{81+16x^2}dx.$

11. Determine a set of anti-derivatives for each of the following.

a. $\frac{21}{49+36x^2}$

b. $\frac{8}{1+16x^2}$

12. Evaluate each of the following.

a. $\int_0^{\frac{1}{\sqrt{2}}} \frac{1}{3+2x^2}dx$

b. $\int_0^{\frac{1}{\sqrt{2}}} \frac{1}{\sqrt{2-3x^2}}dx.$

13. Evaluate each of the following.

a. $\int_0^{3} \frac{x}{\sqrt{9-x^2}}dx$

b. $\int_0^{3} \frac{1}{\sqrt{9-x^2}}dx$

14. Evaluate each of the following.

a. $\int_0^{3} \frac{1}{9+x^2}dx$

b. $\int_0^{3} \frac{x}{9+x^2}dx$

15. Evaluate each of the following.

a. $\int_0^{\frac{5}{6}} \frac{1}{\sqrt{25-9x^2}}dx$

b. $\int_0^{\frac{5}{3}} \frac{x}{25+9x^2}dx$

16. Evaluate each of the following.

a. $\int_0^{\frac{5}{3}} \frac{x}{\sqrt{25-9x^2}}dx$

b. $\int_0^{\frac{5}{3}} \frac{1}{25+9x^2}dx$

17. a. On a certain curve the gradient is given by $\frac{1}{\sqrt{4-x^2}}$. Determine the equation of the curve that passes through the point $(\sqrt{3}, \pi)$.

b. On a certain curve the gradient is given by $\frac{1}{1+4x^2}$. Determine the equation of the curve that passes through the point $\left(\frac{1}{2}, \pi\right)$.

18. a. If $\frac{dy}{dx}+\frac{1}{3+x^2}=0$, and when $x=1$, $y=0$, evaluate y when $x=0$.

b. If $\frac{dy}{dx}+\frac{1}{\sqrt{6-x^2}}=0$, and when $x=\sqrt{3}$, $y=0$, evaluate y when $x=0$.

19. Evaluate each of the following.

a. $\int_0^1 \frac{1}{\sqrt{4-3x^2}}dx$

b. $\int_0^1 \frac{1}{1+3x^2}dx$

20. Evaluate each of the following.

a. $\int_0^1 \frac{x}{1+3x^2}dx$

b. $\int_0^1 \frac{x}{\sqrt{4-3x^2}}dx$

21 **WE31** Calculate each of the following.

a. $\int \frac{1}{4x^2-12x+25}dx$

b. $\int \frac{1}{\sqrt{7+12x-4x^2}}dx$

22. Calculate each of the following.

a. $\int \frac{1}{25x^2-20x+13}dx$

b. $\int \frac{1}{\sqrt{12+20x-25x^2}}dx$

23. Calculate each of the following.

a. $\int \frac{2}{\sqrt{5-4x-x^2}}dx$

b. $\int \frac{2}{x^2+4x+13}dx$

24. Calculate each of the following.

a. $\int \frac{6}{\sqrt{24-30x-9x^2}}dx$

b. $\int \frac{6}{74+30x+9x^2}dx$

25 a. **WE32** Calculate $\int \frac{5-3x}{\sqrt{25-9x^2}}dx.$

b. Calculate $\int \frac{3x+5}{9x^2+25}dx.$

26. Calculate each of the following.

a. $\int \frac{3x-4}{\sqrt{9-16x^2}}dx$

b. $\int \frac{3+4x}{9+16x^2}dx$

27. Calculate each of the following.

a. $\int \frac{5-2x}{\sqrt{5-2x^2}}dx$

b. $\int \frac{5-2x}{25+2x^2}dx$

28. If a, b, p and q are positive real constants, calculate each of the following.

a. $\int \frac{1}{\sqrt{p^2-q^2x^2}}dx$

b. $\int \frac{1}{p^2+q^2x^2}dx$

c. $\int \frac{ax+b}{\sqrt{p^2-q^2x^2}}dx$

d. $\int \frac{ax+b}{p^2+q^2x^2}dx$

29. a. i. Use the substitution $x=4\sin(\theta)$ to calculate $\int \sqrt{16-x^2}dx.$

ii. Evaluate $\int_0^4 \sqrt{16-x^2}dx$. Describe the area that this represents.

b. i. Use the substitution $x=\frac{6}{5}\cos(\theta)$ to calculate $\int \sqrt{36-25x^2}dx.$

ii. Evaluate $\int_0^{\frac{6}{5}} \sqrt{36-25x^2}dx$. Describe the area that this represents.

30. a. Prove that the total area inside the circle $x^2 + y^2 = r^2$ is given by πr^2.

b. Prove that the total area inside the ellipse $\frac{x^2}{a^2} + \frac{y^2}{b^2} = 1$ is given by πab.

31. Evaluate each of the following.

a. $\int_3^5 \frac{x-4}{x^2 - 6x + 13}\,dx$

b. $\int_{\frac{3}{2}}^{\frac{7}{2}} \frac{2x+1}{4x^2-12x+25}\,dx.$

32. Evaluate each of the following.

a. $\int_1^4 \frac{3x+2}{\sqrt{8+2x-x^2}}\,dx$

b. $\int_{-2}^{1} \frac{2x-3}{\sqrt{12-8x-4x^2}}\,dx$

7.6 Exam questions

Question 1 (3 marks) TECH-FREE

Source: *VCE 2021, Specialist Mathematics Exam 1, Q2; © VCAA.*

Evaluate $\int_0^1 \frac{2x+1}{x^2+1}\,dx.$

Question 2 (1 mark) TECH-ACTIVE

MC $\int \frac{dx}{2+5x^2}$ is equal to

A. $\tan^{-1}\left(\frac{\sqrt{5}x}{\sqrt{2}}\right) + c$

B. $\frac{\sqrt{2}}{2}\tan^{-1}\left(\frac{\sqrt{10}x}{2}\right) + c$

C. $\frac{\sqrt{5}}{5}\tan^{-1}\left(\frac{\sqrt{10}x}{2}\right) + c$

D. $\frac{\sqrt{10}}{10}\tan^{-1}\left(\frac{\sqrt{10}x}{2}\right) + c$

E. $\frac{1}{10x}\log_e\left(2+5x^2\right) + c$

Question 3 (2 marks) TECH-FREE

Determine an anti-derivative of $\frac{\frac{a}{b}}{\sqrt{(b^2 - a^2x^2)}}$.

More exam questions are available online.

7.7 Integrals involving partial fractions

LEARNING INTENTION

At the end of this subtopic you should be able to:
- integrate rational functions by breaking the function into partial fractions.

7.7.1 Integration of rational functions

A rational function is a ratio of two functions, both of which are polynomials. For example, $\dfrac{mx+k}{ax^2+bx+c}$ is a rational function; it has a linear function in the numerator and a quadratic function in the denominator.

In the preceding section, we integrated certain expressions of this form when $a>0$ and $\Delta=b^2-4ac<0$, meaning that the quadratic function was expressed as the sum of two squares. In this section, we examine cases when $a\neq 0$ and $\Delta=b^2-4ac>0$.

This means that the quadratic function in the denominator can now be factorised into linear factors. Integrating expressions of this kind does not involve a new integration technique, just an algebraic method of expressing the integrand into its partial fractions decomposition.

Converting expressions into equivalent forms is useful for integration. We have seen this when expressing a quadratic in the completing the square form or converting trigonometric powers into multiple angles.

However, if the derivative of the denominator is equal to the numerator or a constant multiple of it, that is, if $\dfrac{d}{dx}\left(ax^2+bx+c\right)=2ax+b$, then $\displaystyle\int\frac{2ax+b}{ax^2+bx+c}\,dx=\log_e(|ax^2+bx+c|)$.

The method of partial fractions involves factorising the denominator, for example, $\dfrac{mx+k}{ax^2+bx+c}=\dfrac{mx+k}{(ax+\alpha)(x+\beta)}$, and then expressing as $\dfrac{A}{(ax+\alpha)}+\dfrac{B}{(x+\beta)}$, where A and B are constants to be found. We can evaluate A and B by equating coefficients and then solving simultaneous equations, or by substituting in specific values of x. See the following worked examples.

Equating coefficients means, for example, that if $Ax+B=3x-4$, then $A=3$ from equating the coefficient of x, since $Ax=3x$, and $B=-4$ from the term independent of x (the constant term).

WORKED EXAMPLE 33 Integrals involving partial fractions (1)

Calculate each of the following.

a. $\displaystyle\int\frac{12}{36-x^2}\,dx$

b. $\displaystyle\int\frac{4}{49-36x^2}\,dx.$

THINK	WRITE
a. 1. Factorise the denominator into linear factors using the difference of two squares.	a. $\dfrac{12}{36-x^2}=\dfrac{12}{(6+x)(6-x)}$
2. Write the integrand in its partial fractions decomposition, where A and B are constants to be found.	$=\dfrac{A}{6+x}+\dfrac{B}{6-x}$
3. Add the fractions by forming a common denominator.	$=\dfrac{A(6-x)+B(6+x)}{(6+x)(6-x)}$

4. Expand the numerator, factor in x and expand the denominator.

$$= \frac{6A - 6Ax + 6B + 6Bx}{(6+x)(6-x)}$$
$$= \frac{6x(B-A) + 6(A+B)}{36 - x^2}$$

5. Because the denominators are equal, the numerators are also equal. Equate the coefficients of x and the term independent of x. This gives two simultaneous equations for the two unknowns, A and B.

$6x(B-A) + 6(A+B) = 12$
(1) $B - A = 0$
(2) $6(A+B) = 12$

6. Solve the simultaneous equations.

(1) $\Rightarrow A = B$; substitute into
(2) $12A = 4 \Rightarrow A = B = 1$

7. An alternative method can be used to evaluate the unknowns A and B. Equate the numerators from the working above.

$12 = A(6-x) + B(x+6)$

8. Substitute an appropriate value of x.

Substitute $x = 6$:
$12 = 12B$
$B = 1$

9. Substitute an appropriate value of x.

Substitute $x = -6$:
$12 = 12A$
$A = 1$

10. Express the integrand as its partial fractions decomposition.

$$\frac{12}{36-x^2} = \frac{1}{6+x} + \frac{1}{6-x}$$

11. Instead of integrating the original expression, we integrate the partial fractions expression, since these expressions are equal.

$$\int \frac{12}{36-x^2}\,dx = \int \left(\frac{1}{6+x} + \frac{1}{6-x}\right) dx$$

12. Integrate term by term, using the result $\int \frac{1}{ax+b}\,dx = \frac{1}{a}\log_e(|ax+b|)$.

$$= \log_e(|6+x|) - \log_e(|6-x|) + c$$

13. Use the log laws to express the final answer as a single log term.

$$\int \frac{12}{36-x^2}\,dx = \log_e\left(\left|\frac{6+x}{6-x}\right|\right) + c$$

b. 1. Factorise the denominator into linear factors using the difference of two squares.

b. $$\frac{4}{49-36x^2} = \frac{4}{(7-6x)(7+6x)}$$

2. Write the integrand in its partial fractions decomposition, where A and B are constants to be found.

$$= \frac{A}{7-6x} + \frac{B}{7+6x}$$

3. Add the fractions by forming a common denominator.

$$= \frac{A(7+6x) + B(7-6x)}{(7-6x)(7+6x)}$$

4. Expand the numerator, factor in x and expand the denominator.

$$= \frac{7A + 6Ax + 7B - 6Bx}{(7-6x)(7+6x)}$$
$$= \frac{6x(A-B) + 7(A+B)}{49 - 36x^2}$$

5. Because the denominators are equal, the numerators are also equal. Equate coefficients of x and the term independent of x. This gives two simultaneous equations for the two unknowns, A and B.

$6x(A-B)+7(A-B)=4$
$(1)\ A-B=0$
$(2)\ 7(A+B)=4$

6. Solve the simultaneous equations

$(1) \Rightarrow A=B$: substitute into
$(2)\ 14A=4 \Rightarrow A=B=\frac{2}{7}$

7. Express the integrand as its partial fractions decomposition.

$$\frac{4}{49-36x^2}=\frac{2}{7(7-6x)}+\frac{2}{7(7+6x)}$$

8. Instead of integrating the original expression, we integrate the partial fractions expression, since these expressions are equal. Transfer the constant factors outside the integral sign.

$$\int \frac{4}{49-36x^2}\,dx$$
$$=\int\left(\frac{2}{7(7-6x)}+\frac{2}{7(7+6x)}\right)dx$$
$$=\frac{2}{7}\int\left(\frac{1}{7-6x}+\frac{1}{7+6x}\right)dx$$

9. Integrate term by term using the result $\int \frac{1}{ax+b}\,dx=\frac{1}{a}\log_e(|ax+b|)$.

$$=\frac{2}{7}\left[-\frac{1}{6}\log_e(|7-6x|)+\frac{1}{6}\log_e(|7+6x|)\right]+c$$

10. Use the log laws to express the final answer as a single log term.

$$\int \frac{4}{49-36x^2}\,dx=\frac{1}{21}\log_e\left(\left|\frac{7+6x}{7-6x}\right|\right)+c$$

WORKED EXAMPLE 34 Integrals involving partial fractions (2)

Calculate $\int \frac{x+10}{x^2-x-12}\,dx$.

THINK

1. Factorise the denominator into linear factors.

2. Write the integrand in its partial fractions decomposition, where A and B are constants to be found.

3. Add the fractions by forming a common denominator.

4. Expand the numerator, factor in x and expand the denominator.

WRITE

$$\frac{x+10}{x^2-x-12}=\frac{x+10}{(x-4)(x+3)}$$
$$=\frac{A}{x-4}+\frac{B}{x+3}$$
$$=\frac{A(x+3)+B(x-4)}{(x-4)(x+3)}$$
$$=\frac{Ax+3A+Bx-4B}{(x-4)(x+3)}$$
$$=\frac{x(A+B)+3A-4B}{x^2-x-12}$$

5. Because the denominators are equal, the numerators are also equal. Equate the coefficients of x and the term independent of x to give two simultaneous equations for the two unknowns, A and B.

$x(A+B)+3A-4B=x+10$
$(1)\ A+B=1$
$(2)\ 3A-4B=10$

6. Solve the simultaneous equations by elimination. Add the two equations to eliminate B and solve for A.

$4\times(1)\ 4A+4B=4$
$(2)\ 3A-4B=10$
$4\times(1)+(2)\ 7A=14$
$A=2$

7. Substitute back into (1) to determine B.

$2+B=1$
$B=-1$

8. An alternative method can be used to evaluate the unknowns A and B. Equate the numerators from the working above.

$A(x+3)+B(x-4)=x+10$

9. Let $x=4$.

$7A=14$
$A=2$

10. Let $x=-3$.

$-7B=7$
$B=-1$

11. Express the integrand as its partial fractions decomposition.

$$\int \frac{x+10}{x^2-x-12}\,dx=\int\left(\frac{2}{x-4}-\frac{1}{x+3}\right)dx$$

12. Integrate term by term using the result $\int \frac{1}{ax+b}\,dx=\frac{1}{a}\log_e(|ax+b|)$, and state the final simplified answer using log laws.

$$=2\log_e(|x-4|)-\log_e(|x+3|)+c$$
$$=\log_e\left(\frac{(x-4)^2}{|x+3|}\right)$$

Perfect squares

If an expression is of the form $\frac{mx+k}{ax^2+bx+c}$ when $a\neq 0$ and $\Delta=b^2-4ac=0$, the quadratic function in the denominator is now a perfect square.

Integrating by the method of partial fractions in this case involves writing $\frac{mx+k}{ax^2+bx+c}=\frac{mc+k}{(px+\alpha)^2}$ and then expressing it as $\frac{A}{(px+\alpha)}+\frac{B}{(px+\alpha)^2}$, where A and B are constants to be found. We can calculate A and B by equating coefficients and then solving the simultaneous equations, as before.

WORKED EXAMPLE 35 Integrals involving partial fractions and perfect squares

Calculate $\int \frac{6x-5}{4x^2-12x+9}\,dx$.

THINK | **WRITE**

1. Factorise the denominator as a perfect square.

$$\frac{6x-5}{4x^2-12x+9}=\frac{6x-5}{(2x-3)^2}$$

2. Write the integrand in its partial fractions decomposition, where A and B are constants to be found.

$$= \frac{A}{2x-3} + \frac{B}{(2x-3)^2}$$

3. Add the fractions by forming the lowest common denominator and expanding the numerator.

$$= \frac{A(2x-3)+B}{(2x-3)^2}$$

$$= \frac{2Ax+B-3A}{(2x-3)^2}$$

4. Because the denominators are equal, the numerators are also equal. Equate coefficients of x and the term independent of x to give two simultaneous equations for the two unknowns, A and B.

$2Ax + B - 3A = 6x - 5$
(1) $2A = 6$
(2) $B - 3A = -5$

5. Solve the simultaneous equations.

(1) $\Rightarrow A = 3$: substitute into (2):
$B - 9 = -5 \Rightarrow B = 4$

6. Express the integrand as its partial fractions decomposition.

$$\int \frac{6x-5}{4x^2-12x+9}dx = \int \left(\frac{3}{2x-3} + \frac{4}{(2x-3)^2} \right) dx$$

7. Integrate term by term, using the result $\int \frac{1}{ax+b}dx = \frac{1}{a}\log_e(|ax+b|)$ in the first term and $\int \frac{1}{(ax+b)^2}dx = -\frac{1}{a(ax+b)}$ for the second term.

$$= \frac{3}{2}\log_e(|2x-3|) - \frac{2}{2x-3} + c$$

Note that this example can also be done using a linear substitution with a back substitution. See Worked example 12b.

Rational functions involving ratios of two quadratic functions

When the degree of the polynomial in the numerator is greater than or equal to the degree of the polynomial in the denominator, the rational function is said to be a non-proper rational function. In this case we have to divide the denominator into the numerator to obtain a proper rational function.

For example, when we have quadratic functions in both the numerator and denominator, that is, the form $\frac{rx^2+sx+t}{ax^2+bx+c}$ where $r \neq 0$ and $a \neq 0$, we can use long division to divide the denominator into the numerator to express the function as $\frac{rx^2+sx+t}{ax^2+bx+c} = q + \frac{mx+k}{ax^2+bx+c}$ where $q = \frac{r}{a}$.

WORKED EXAMPLE 36 Integrals involving improper partial fractions

Calculate $\displaystyle\int \frac{2x^2 + 5x + 3}{x^2 + 3x - 4} dx$.

THINK

1. Rewrite the numerator as a multiple of the denominator and simplify.
   Note: $11 - x$ is the remainder, if you were to perform long division.
2. Factorise the denominator into linear factors.
3. Write the integrand in its partial fractions decomposition, where A and B are constants to be found.
4. Add the fractions by forming a common denominator and expanding the numerator.
5. Because the denominators are equal, the numerators are also equal. Equate coefficients of x and the term independent of x to give two simultaneous equations for the two unknowns, A and B.
6. Solve the simultaneous equations by elimination. Add the two equations to eliminate B.
7. Express the integrand as its partial fractions decomposition.
8. Integrate term by term using the result $\displaystyle\int \frac{1}{ax+b} dx = \frac{1}{a} \log_e(|ax+b|)$ and state the final answer.

WRITE

1.
$$\frac{2x^2 + 5x + 3}{x^2 + 3x - 4} = \frac{2(x^2 + 3x - 4) + 11 - x}{x^2 + 3x - 4}$$
$$= 2 + \frac{11 - x}{x^2 + 3x - 4}$$

2.
$$= 2 + \frac{11 - x}{(x-1)(x+4)}$$

3.
$$= 2 + \frac{A}{x-1} + \frac{B}{x+4}$$

4.
$$= 2 + \frac{A(x+4) + B(x-1)}{(x-1)(x+4)}$$
$$= 2 + \frac{x(A+B) + 4A - B}{x^2 + 3x - 4}$$

5.
$x(A+B) + 4A - B = 11 - x$
(1) $A + B = -1$
(2) $4A - B = 11$

6.
$5A = 10$
$A = 2 \Rightarrow B = -3$

7.
$$\int \frac{2x^2 + 5x + 3}{x^2 + 3x - 4} dx$$
$$= \int \left(2 + \frac{2}{x-1} - \frac{3}{x+4}\right) dx$$

8.
$$= 2x + 2\log_3(|x-1|) - 3\log_e(|x+4|) + c$$
$$= 2x + \log_e\left(\frac{(x-1)^2}{|x+4|^3}\right) + c$$

Rational functions involving non-linear factors

If the denominator does not factorise into linear factors, then we proceed with $\displaystyle\frac{px^2 + qx + r}{(x+a)(x^2+b^2)} = \frac{A}{x+a} + \frac{Bx + C}{x^2 + b^2}$.

Note that we will need three simultaneous equations to solve for the three unknowns: A, B and C.

WORKED EXAMPLE 37 Integrals involving partial fractions with cubic denominators

Calculate $\displaystyle\int \frac{x^2+7x+2}{x^3+2x^2+4x+8}dx$.

THINK	WRITE
1. First try to factorise the cubic in the denominator using the factor theorem.	$f(x) = x^3 + 2x^2 + 4x + 8$ $f(1) = 1 + 2 + 4 + 8 \neq 0$ $f(2) = 8 + 8 + 8 + 8 \neq 0$ $f(-2) = -8 + 8 - 8 + 8 = 0$ $(x + 2)$ is a factor.
2. Factorise the denominator into factors.	$x^3 + 2x^2 + 4x + 8 = (x + 2)(x^2 + 4)$
3. Write the integrand in its partial fractions decomposition, where A, B and C are constants to be found.	$\dfrac{x^2+7x+2}{x^3+2x^2+4x+8} = \dfrac{A}{x+2} + \dfrac{Bx+C}{x^2+4}$
4. Add the fractions by forming a common denominator and expanding the numerator.	$= \dfrac{A(x^2+4)+(x+2)(Bx+C)}{(x+2)(x^2+4)}$ $= \dfrac{Ax^2+4A+Bx^2+2Bx+Cx+2C}{(x+2)(x^2+4)}$ $= \dfrac{x^2(A+B)+x(2B+C)+4A+2C}{x^3+2x^2+4x+8}$
5. Because the denominators are equal, the numerators are also equal. Equate the coefficients of x^2 and x and the term independent of x to give three simultaneous equations for the three unknowns: A, B and C.	(1) $A + B = 1$ (2) $2B + C = 7$ (3) $4A + 2C = 2$
6. Solve the simultaneous equations by elimination.	$2 \times (1)$ $2A + 2B = 2$ (2) $2B + C = 7$ Subtracting gives (4) $C - 2A = 5$ $2 \times (4)$ $2C - 4A = 10$ (3) $4A + 2C = 2$ Adding gives $4C = 12$ $C = 2$
7. Back substitute to determine the remaining unknowns.	Substitute into (2): $2B + 3 = 7 \Rightarrow B = 2$ Substitute into (1): $\Rightarrow A = -1$
8. Express the integrand as its partial fractions decomposition.	$\dfrac{x^2+7x+2}{x^3+2x^2+4x+8} = \dfrac{-1}{x+2} + \dfrac{2x+3}{x^2+4}$
9. Separate the last term into two expressions.	$\displaystyle\int \frac{x^2+7x+2}{x^3+2x^2+4x+8}dx$ $= \displaystyle\int \left(\frac{-1}{x+2} + \frac{2x+3}{x^2+4}\right) dx$ $= \displaystyle\int \left(\frac{2x}{x^2+4} - \frac{1}{x+2} + \frac{3}{x^2+4}\right) dx$

10. Integrate term by term, using the results $\int \frac{f'(x)}{f(x)}dx = \log_e(|f(x)|)$, $\int \frac{1}{ax+b}dx = \frac{1}{a}\log_e(|ax+b|)$ and $\int \frac{1}{x^2+a^2}dx = \frac{1}{a}\tan^{-1}\left(\frac{x}{a}\right)$, and state the final answer.

$$= \log_e(x^2+4) - \log_e(|x+2|) + \frac{3}{2}\tan^{-1}\left(\frac{x}{2}\right) + c$$

$$= \log_e\left(\frac{x^2+4}{|x+2|}\right) + \frac{3}{2}\tan^{-1}\left(\frac{x}{2}\right) + c$$

7.7 Exercise

Students, these questions are even better in jacPLUS

Receive immediate feedback and access sample responses

Access additional questions

Track your results and progress

Find all this and MORE in jacPLUS

Technology free

1. **WE33** Calculate each of the following.

a. $\int \frac{1}{100-x^2}dx$

b. $\int \frac{12}{64-9x^2}dx.$

2. Calculate each of the following.

a. $\int \frac{1}{36-25x^2}dx$

b. $\int \frac{x}{36-25x^2}dx.$

3. Determine an anti-derivative of each of the following.

a. $\frac{1}{x^2-4}$

b. $\frac{2}{16+x^2}$

4. Determine an anti-derivative of each of the following.

a. $\frac{x}{x^2-25}$

b. $\frac{2x-3}{x^2-36}$

5. **WE34** Calculate each of the following.

a. $\int \frac{x+13}{x^2+2x-15}dx$

b. $\int \frac{x-11}{x^2+3x-4}dx$

6. Integrate each of the following with respect to x.

a. $\frac{x+11}{x^2+x-12}$

b. $\frac{5x-9}{x^2-2x-15}$

7. Integrate each of the following with respect to x.

a. $\frac{2x-19}{x^2+x-6}$

b. $\frac{11}{x^2-3x-28}$

8. WE35 Calculate each of the following.

a. $\int \frac{2x+1}{x^2+6x+9}\,dx$

b. $\int \frac{2x-1}{9x^2-24x+16}\,dx$

9. Integrate each of the following with respect to x.

a. $\frac{2x+3}{x^2-6x+9}$

b. $\frac{2x-5}{x^2+4x+4}$

10. Integrate each of the following with respect to x.

a. $\frac{4x}{4x^2+12x+9}$

b. $\frac{6x-19}{9x^2-30x+25}$

11. Determine $\int \frac{1}{x^2+kx+25}\,dx$ for the cases when:

a. $k=0$

b. $k=26$

c. $k=-10$.

12. WE36 Calculate each of the following.

a. $\int \frac{3x^2+10x-4}{x^2+3x-10}\,dx$

b. $\int \frac{-2x^2-x+20}{x^2+x-6}\,dx$

13. Calculate an anti-derivative of each of the following.

a. $\frac{x^2-4x-11}{x^2+x-12}$

b. $\frac{-3x^2-4x-5}{x^2+2x-3}$

14. Calculate an anti-derivative of each of the following.

a. $\frac{4x^2-17x-26}{x^2-4x-12}$

b. $\frac{-2x^3+12x^2-17x}{x^2-6x+8}$

15. Evaluate each of the following.

a. $\int_{1}^{2} \frac{1}{x^2+4x}\,dx$

b. $\int_{5}^{6} \frac{1}{x^2-16}\,dx$

16. Evaluate each of the following.

a. $\int_{-1}^{1} \frac{3x+8}{x^2+6x+8}\,dx$

b. $\int_{3}^{4} \frac{x-6}{x^2-4x+4}\,dx$

17. a. Determine the area bounded by the curve $y=\frac{5}{x^2+x-6}$, the coordinate axes and $x=-1$.

b. Determine the area bounded by the curve $y=\frac{2x-3}{4+3x-x^2}$, the x-axis and the lines $x=2$ and $x=3$.

18. a. Determine the area bounded by the curve $y=\frac{21}{40-11x-2x^2}$, the coordinate axes and $x=-5$.

b. Determine the area bounded by the curve $y=\frac{x^3-9x+9}{x^2-9}$, the x-axis and the lines $x=4$ and $x=6$.

19. a. Determine the value of a if $\int_{1}^{2} \frac{2}{4x - x^2} dx = \log_e (a)$.

b. Determine the value of b if $\int_{1}^{2} \frac{2}{\sqrt{4x - x^2}} dx = \pi b$.

c. Determine the value of c if $\int_{3}^{4} \frac{3x}{x^2 - x - 2} dx = \log_e (c)$.

d. Determine the value of d if $\int_{\frac{1}{2}}^{2} \frac{3}{x^2 - x + 1} dx = \pi d$.

20. If $y = \frac{16x^2 + 25}{16x^2 - 25}$, calculate:

a. $\int y \, dx$

b. $\int \frac{1}{y} dx$.

21. Evaluate each of the following.

a. $\int_{0}^{1} \frac{27x^2}{16 - 9x^2} dx$

b. $\int_{0}^{2} \frac{20x^2}{4x^2 + 4x + 1} dx$

22. a. WE37 Calculate $\int \frac{19 - 3x}{x^3 - 2x^2 + 9x - 18} dx$.

b. Calculate $\int \frac{25}{x^3 + 3x^2 + 16x + 48} dx$.

23. Evaluate each of the following.

a. $\int_{\sqrt{3}}^{3} \frac{x^2 - 2x + 9}{x^3 + 9x} dx$

b. $\int_{2}^{3} \frac{4x^2 - 16x + 19}{(2x - 3)^3} dx$

24. If a and b are non-zero real constants, calculate each of the following.

a. $\int \frac{1}{b^2x^2 - a^2} dx$

b. $\int \frac{x}{b^2x^2 - a^2} dx$

25. If a, b, p and q are all non-zero real constants, calculate each of the following.

a. $\int \frac{x}{(ax - b)^2} dx$

b. $\int \frac{1}{(px + a)(qx + b)} dx$

26. Show that $\int_{0}^{1} \frac{x^4 (1 - x)^4}{1 + x^2} dx = \frac{22}{7} - \pi$.

7.7 Exam questions

Question 1 (1 mark) TECH-ACTIVE

Source: *VCE 2018 Specialist Mathematics Exam 2, Section A, Q3; © VCAA.*

Which one of the following, where A, B, C and D are non-zero real numbers, is the partial fraction form for the expression $\dfrac{2x^2+3x+1}{(2x+1)^3\left(x^2-1\right)}$?

A. $\dfrac{A}{2x+1}+\dfrac{B}{x-1}+\dfrac{C}{x+1}$

B. $\dfrac{A}{2x+1}+\dfrac{B}{(2x+1)^2}+\dfrac{C}{(2x+1)^3}+\dfrac{Dx}{x^2-1}$

C. $\dfrac{A}{2x+1}+\dfrac{Bx+C}{x^2-1}$

D. $\dfrac{A}{2x+1}+\dfrac{B}{(2x+1)^2}+\dfrac{C}{x-1}$

E. $\dfrac{A}{2x+1}+\dfrac{Bx+C}{(2x+1)^2}+\dfrac{D}{x-1}$

Question 2 (4 marks) TECH-FREE

Source: *VCE 2017 Specialist Mathematics Exam 1, Q2; © VCAA.*

Find $\displaystyle\int_1^{\sqrt{3}} \frac{1}{x\left(1+x^2\right)}dx$ expressing your answer in the form $\log_e\left(\sqrt{\dfrac{a}{b}}\right)$ where a and b are positive integers.

Question 3 (1 mark) TECH-ACTIVE

MC If a is a positive real constant, then $\displaystyle\int \frac{a}{x(x-a)}dx$ for $x > a$ is equal to

A. $\log_e\left(\dfrac{a-x}{x}\right)$

B. $\log_e\left(\dfrac{x-a}{x}\right)$

C. $-\log_e\left(\dfrac{x-a}{x}\right)$

D. $-\log_e\left(\dfrac{a-x}{x}\right)$

E. $-\log_e(x(x-a))$

More exam questions are available online.

7.8 Review

7.8.1 Summary

doc-37061

Hey students! Now that it's time to revise this topic, go online to:

 **Access the topic summary**

 **Review your results**

 **Watch teacher-led videos**

 **Practise exam questions**

Find all this and MORE in jacPLUS

7.8 Exercise

Technology free: short answer

1. Calculate the following.

a. $\displaystyle\int \frac{1}{\sqrt{5-2x}}\,dx$

b. $\displaystyle\int \frac{1}{\sqrt{25-4x^2}}\,dx$

c. $\displaystyle\int \frac{x}{\sqrt{25-4x^2}}\,dx$

d. $\displaystyle\int \frac{x}{25-4x^2}\,dx$

2. Calculate the following.

a. $\displaystyle\int \frac{1}{25-4x^2}\,dx$

b. $\displaystyle\int \frac{1}{25+4x^2}\,dx$

c. $\displaystyle\int \frac{1}{5-2x}\,dx$

d. $\displaystyle\int \frac{x^2}{25-4x^2}\,dx$

3. Calculate the following.

a. $\displaystyle\int \frac{x}{\left(25-4x^2\right)^2}\,dx$

b. $\displaystyle\int \frac{x}{5-2x}\,dx$

c. $\displaystyle\int \frac{x}{(5-2x)^2}\,dx$

d. $\displaystyle\int \frac{x}{\sqrt{5-2x}}\,dx$

4. Evaluate the following.

a. $\displaystyle\int_0^1 \frac{1}{4-3x}\,dx$

b. $\displaystyle\int_0^1 \frac{1}{\sqrt{4-3x}}\,dx$

c. $\displaystyle\int_0^1 \frac{1}{(4-3x)^2}\,dx$

d. $\displaystyle\int_0^1 \frac{x}{(4-3x)^2}\,dx$

5. Evaluate the following.

a. $\displaystyle\int_0^1 \frac{x}{4-3x}\,dx$

b. $\displaystyle\int_0^1 \frac{x}{\sqrt{4-3x}}\,dx$

c. $\displaystyle\int_0^{\frac{2}{3}} \frac{1}{\sqrt{16-9x^2}}\,dx$

d. $\displaystyle\int_0^{\frac{4}{3}} \frac{x}{\sqrt{16-9x^2}}\,dx$

6. Evaluate the following.

a. $\displaystyle\int_0^{\frac{4}{3}} \frac{1}{16+9x^2}\,dx$

b. $\displaystyle\int_0^{\frac{4}{3}} \frac{x}{16+9x^2}\,dx$

c. $\displaystyle\int_0^{\frac{2}{3}} \frac{1}{16-9x^2}\,dx$

d. $\displaystyle\int_0^{\frac{2}{3}} \frac{x}{16-9x^2}\,dx$

Technology active: multiple choice

7. **MC** Let $f: R\backslash\left\{\frac{5}{2}\right\} \to R,\ f(x)=\dfrac{2}{5-2x}$. Using a substitution, the area bounded by the graph of f, the x-axis and the lines $x=3$ and $x=4$ can be found by evaluating

A. $\displaystyle\int_3^4 \frac{2}{u}\,du$

B. $\displaystyle\int_3^4 \frac{1}{u}\,du$

C. $\displaystyle\int_{-1}^{-3} \frac{1}{u}\,du$

D. $\displaystyle\int_{-3}^{-1} \frac{2}{u}\,du$

E. $\displaystyle\int_{-1}^{-3} \frac{2}{u}\,du$

8. **MC** The area between the parabola $y=x^2-4x$ and the line $y=2x$ in square units is equal to

A. 0

B. $10\frac{2}{3}$

C. $26\frac{2}{3}$

D. 36

E. $37\frac{1}{3}$

9. **MC** Using a suitable substitution, $\displaystyle\int_1^2 \frac{1}{x\sqrt{2x-1}}\,dx$ can be expressed in terms of u as

A. $\displaystyle\int_1^3 \frac{1}{(u+1)\sqrt{u}}\,du$

B. $\displaystyle 4\int_1^3 \frac{1}{(u+1)\sqrt{u}}\,du$

C. $\displaystyle \frac{1}{4}\int_1^3 \frac{1}{(u+1)\sqrt{u}}\,du$

D. $\displaystyle \frac{1}{2}\int_1^3 \frac{1}{(u+1)\sqrt{u}}\,du$

E. $\displaystyle \frac{1}{2}\int_1^2 \frac{1}{(u+1)\sqrt{u}}\,du$

10. **MC** Using a suitable substitution, $\displaystyle\int_0^2 \frac{x^3}{\sqrt{1+2x^2}}\,dx$ can be expressed in terms of u as

A. $\displaystyle \frac{1}{4}\int_0^2 \frac{u-1}{\sqrt{u}}\,du$

B. $\displaystyle \frac{1}{4}\int_1^9 \frac{u-1}{\sqrt{u}}\,du$

C. $\displaystyle \frac{1}{8}\int_1^9 \frac{u-1}{\sqrt{u}}\,du$

D. $\displaystyle\int_0^4 \frac{u}{\sqrt{1+2u}}\,du$

E. $\displaystyle \frac{1}{2}\int_0^2 \frac{u}{\sqrt{1+2u}}\,du$

11. **MC** With a suitable substitution $\int_1^4 \frac{e^{\sqrt{x}}}{\sqrt{x}}\,dx$ can be expressed as

A. $\int_1^4 \frac{e^{\sqrt{u}}}{u}\,du$
B. $\int_1^2 \frac{e^{\sqrt{u}}}{u}\,du$
C. $2\int_1^2 \frac{e^{\sqrt{u}}}{u}\,du$
D. $\frac{1}{2}\int_1^2 e^{u}\,du$
E. $2\int_1^2 e^{u}\,du$

12. **MC** Using a suitable substitution, $\int_0^{\frac{\pi}{6}} \cos^3(x)\sin^4(x)dx$ can be expressed in terms of u as

A. $\int_0^{\frac{\pi}{6}} \left(u^4 - u^6\right)\,du$
B. $\int_0^{\frac{1}{2}} \left(u^4 - u^6\right)\,du$
C. $\int_0^{\frac{1}{2}} \left(u^6 - u^4\right)\,du$
D. $\int_0^{\frac{\sqrt{3}}{2}} \left(u^4 - u^6\right)\,du$
E. $\int_0^{\frac{\sqrt{3}}{2}} \left(u^6 - u^4\right)\,du$

13. **MC** $\int \sin^2(2x)\,dx$ is equal to

A. $\frac{x}{2} + \frac{1}{8}\sin(4x) + c$
B. $\frac{x}{2} - \frac{1}{8}\sin(4x) + c$
C. $4\cos(2x) + c$
D. $\frac{1}{3}\sin^3(2x) + c$
E. $\frac{1}{6}\sin^3(2x) + c$

14. **MC** $\int \frac{1}{25 + 16x^2}\,dx$ is equal to

A. $\frac{1}{5}\sin^{-1}\left(\frac{4x}{5}\right) + c$
B. $\frac{1}{32x}\log_e\left(25 + 16x^2\right) + c$
C. $\frac{1}{32}\log_e\left(25 + 16x^2\right) + c$
D. $\frac{1}{4}\tan^{-1}\left(\frac{4x}{5}\right) + c$
E. $\frac{1}{20}\tan^{-1}\left(\frac{4x}{5}\right) + c$

15. **MC** $\int \frac{1}{\sqrt{25-16x^2}}\,dx$ is equal to

A. $\frac{1}{4}\sin^{-1}\left(\frac{4x}{5}\right)+c$

B. $\frac{1}{5}\sin^{-1}\left(\frac{4x}{5}\right)+c$

C. $\frac{1}{20}\tan^{-1}\left(\frac{4x}{5}\right)+c$

D. $\log_e\left(\sqrt{25-16x^2}\right)+c$

E. $-\frac{1}{16x}\log_e\left(\sqrt{25-16x^2}\right)+c$

16. **MC** $\frac{4x+1}{(x+4)^2\left(x^2+4\right)}$ can be expressed in partial fractions as

A. $\frac{A}{(x+4)^2}+\frac{B}{x^2+4}$

B. $\frac{A}{x+4}+\frac{B}{x+2}+\frac{C}{x-2}$

C. $\frac{A}{x+4}+\frac{B}{(x+4)^2}+\frac{C}{x^2+4}$

D. $\frac{A}{x+4}+\frac{B}{(x+4)^2}+\frac{Cx+D}{x^2+4}$

E. $\frac{A}{(x+4)^2}+\frac{Bx+C}{x^2+4}$

Technology active: extended response

17. Calculate each of the following, showing all steps of working.

a. $\int \sin^2(4x)\,dx$

b. $\int \cos(4x)\sin^2(4x)\,dx$

c. $\int \cos^2(4x)\sin^2(4x)\,dx$

d. $\int \cos^3(4x)\sin^2(4x)\,dx$

18. Evaluate each of the following, showing all steps of working.

a. $\int_0^{\frac{4}{3}} \frac{x^3}{\sqrt{16-9x^2}}\,dx$

b. $\int_0^{\frac{2}{3}} \frac{x^2}{16-9x^2}\,dx$

c. $\int_0^{\frac{4}{3}} \sqrt{16-9x^2}\,dx$. *Hint:* Let $x=\frac{4}{3}\sin(\theta)$.

d. $\int_0^{\frac{4}{3}} \frac{x^2}{\sqrt{16-9x^2}}\,dx$. *Hint:* Let $x=\frac{4}{3}\sin(\theta)$.

19. Evaluate each of the following, showing all steps of working.

a. $\int_0^{\frac{\pi}{12}} \cos(4x)\sin^3(4x)\,dx$

b. $\int_0^{\frac{\pi}{12}} \cos^2(4x)\sin^3(4x)\,dx$

c. $\int_0^{\frac{\pi}{12}} \cos^3(4x)\sin^3(4x)\,dx$

d. $\int_0^{\frac{\pi}{12}} \cos^4(4x)\sin^3(4x)\,dx$

20. Calculate each of the following, showing all steps of working.

a. $\int \frac{x^3}{\sqrt{25-4x^2}}\,dx$

b. $\int x^3\sqrt{25-4x^2}\,dx$

c. $\int \sqrt{25-4x^2}\,dx$. *Hint:* Let $x = \frac{5}{2}\sin(\theta)$.

d. $\int x^2\sqrt{25-4x^2}\,dx$. *Hint:* Let $x = \frac{5}{2}\sin(\theta)$.

7.8 Exam questions

Question 1 (1 mark) TECH-ACTIVE

Source: VCE 2020, Specialist Mathematics Exam 2, Section A, Q11; © VCAA.

MC With a suitable substitution $\int_{\frac{\pi}{4}}^{\frac{\pi}{3}} \frac{\sec^2(x)}{\sec^2(x)-3\tan(x)+1}\,dx$ can be expressed as

A. $\int_1^{\frac{1}{\sqrt{3}}} \left(\frac{1}{u-1}-\frac{1}{u-2}\right) du$

B. $\int_1^{\sqrt{3}} \left(\frac{1}{3(u-3)}-\frac{1}{3u}\right) du$

C. $\int_1^{\sqrt{3}} \left(\frac{1}{u-2}-\frac{1}{u-1}\right) du$

D. $\int_1^{\sqrt{3}} \left(\frac{1}{u-1}-\frac{1}{u-2}\right) du$

E. $\int_{\frac{\pi}{4}}^{\frac{\pi}{3}} \left(\frac{1}{3(u-1)}-\frac{1}{3(u+2)}\right) du$

Question 2 (2 marks) TECH-ACTIVE

Source: *VCE 2015 Specialist Mathematics Exam 2, Section A, Q1a; © VCAA.*

Show that $\int \tan(2x)dx = \frac{1}{2}\log_e|\sec(2x)| + c$.

Question 3 (5 marks) TECH-FREE

Source: *VCE 2014 Specialist Mathematics Exam 1, Q7; © VCAA.*

Consider $f(x) = 3x \arctan(2x)$.

a. Write down the range of f. **(1 mark)**

b. Show that $f'(x) = 3\arctan(2x) + \frac{6x}{1+4x^2}$. **(1 mark)**

c. Hence evaluate the area enclosed by the graph of $g(x) = \arctan(2x)$, the x-axis and the lines $x = \frac{1}{2}$ and $x = \frac{\sqrt{3}}{2}$. **(3 marks)**

Question 4 (1 mark) TECH-ACTIVE

MC If a and b are positive real constants, then $\int \frac{a^2x^2+b^2}{a^2x^2-b^2}dx$ is equal to

A. $x + \frac{2b}{a}\tan^{-1}\left(\frac{bx}{a}\right)$

B. $x - b\log_e\left(\left|\frac{ax+b}{ax-b}\right|\right)$

C. $x + \frac{b}{a}\log_e\left(\left|\frac{ax-b}{ax+b}\right|\right)$

D. $x + \frac{b}{a}\log_e\left(\left|\frac{ax+b}{ax-b}\right|\right)$

E. $x + \frac{b}{a}\log_e\left(\left|a^2x^2 - b^2\right|\right)$

Question 5 (1 mark) TECH-FREE

Evaluate $\int_a^b \left(\text{cosec}^2(3x)e^{3\cot(3x)}\right) dx$ in terms of a and b.

More exam questions are available online.

Answers

Topic 7 Integral calculus

7.2 Areas under and between curves

7.2 Exercise

1. 26
2. $\frac{16}{3}$
3. a. 9
   b. $37\frac{1}{2}$
4. $\frac{92}{3}$
5. 9
6. a. 32
   b. 22
7. a. $166\frac{2}{3}$
   b. 132
8. 4
9. 12
10. a. $\frac{36}{\pi}$
    b. $\frac{16}{\pi}$
    c. $\frac{2a}{n}$
11. $2\left(e^6 - 1\right)$
12. $6\left(e^2 - e^{-2}\right)$
13. $\frac{1}{6}$
14. $\frac{1372}{27}$
15. $85\frac{1}{3}$
16. a. $\log_e(4)$
    b. 1
    c. $\log_e(a)$
17. a. $\frac{2}{3}$
    b. $1 - \frac{1}{a}$
18. 8
19. 128
20. $139\frac{1}{2}$
21. a. 648
    b. $465\frac{3}{4}$
22. $121\frac{1}{2}$
23. 36
24. $\frac{4a^3}{3}$
25. a. $\frac{1}{6}$
    b. $\frac{1}{2}$
    c. $\frac{3}{10}$
    d. $\frac{2}{3}$
26. a. 288
    b. 288
27. a. $121\frac{1}{2}$
    b. $121\frac{1}{2}$
28. a. $\frac{15}{4} - 2\log_e(4)$
    b. $\frac{40}{27} - \frac{1}{3}\log_e(9)$
29. Sample responses can be found in the worked solutions in the online resources.
30. a. (3.4443, 3.9543)
    b. 4.6662
31. a. (3.8343, 1.9172)
    b. 8.6558
32. a. (2.6420, 86.1855)
    b. 64.8779

7.2 Exam questions

Note: Mark allocations are available with the fully worked solutions online.

1. $81\frac{1}{3}$
2. $\frac{1}{2}a^4$
3. a. (7.5882, −1.7003), (24.2955, −4.6781)
   b. 384.3732

7.3 Linear substitutions

7.3 Exercise

1. $\frac{1}{35}(5x-9)^7 + c$
2. $\frac{1}{24}(3x+4)^8 + c$
3. a. $\frac{1}{21}(3x+5)^7 + c$
   b. $\frac{-1}{3(3x+5)} + c$
   c. $\frac{-1}{6(3x+5)^2} + c$
   d. $\frac{1}{2}\sqrt[3]{(3x+5)^2} + c$

4. $y = \frac{1}{8}\left(\sqrt{16x + 25} - 5\right)$

5. $\frac{11}{4}$

6. $\frac{1}{3}\log_e(|3x - 5|) + c$

7. $-\frac{1}{2}\log_e(|7 - 2x|) + c$

8. a. $\frac{1}{54}(6x + 7)^9 + c$

b. $\frac{1}{3}\sqrt{6x + 7} + c$

c. $\frac{1}{6}\log_e(|6x + 7|) + c$

d. $\frac{-1}{6(6x + 7)} + c$

9. 2

10. 8

11. a. $\frac{1}{175}(5x - 9)^7 + \frac{3}{50}(5x - 9)^6 + c$

$= \frac{1}{350}(10x + 3)(5x - 9)^6 + c$

b. $\frac{2}{9}\log_e(|3x - 4|) - \frac{5}{9(3x - 4)} + c$

12. a. $\frac{-(4x + 7)}{8(2x + 7)^2} + c$

b. $\frac{1}{27}(3x - 5)\sqrt{6x + 5} + c$

13. a. $\frac{1}{72}(3x + 5)^8 - \frac{5}{63}(3x + 5)^7 + c$

$= \frac{1}{504}(21x - 5)(3x + 5)^7 + c$

b. $\frac{1}{9}\log_e(|3x + 5|) + \frac{5}{9(3x + 5)} + c$

c. $\frac{-(6x + 5)}{18(3x + 5)^2} + c$

d. $\frac{1}{10}(2x - 5)(3x + 5)^{\frac{2}{3}} + c$

14. a. $\frac{1}{360}(6x + 7)^{10} - \frac{7}{324}(6x + 7)^9 + c$

$= \frac{1}{3240}(54x - 7)(6x + 7)^9 + c$

b. $\frac{1}{27}(3x - 7)\sqrt{6x + 7} + c$

c. $\frac{x}{6} - \frac{7}{36}\log_e(|6x + 7|) + c$

d. $\frac{7}{36(6x + 7)} + \frac{1}{36}\log_e(|6x + 7|) + c$

15. $2x - \frac{8}{3}\log_e(|3x + 4|) + c$

16. $2 + 5\log_e(\frac{3}{5})$

17. 4

18. $\frac{5}{3}\log_e(\frac{5}{2}) - 1$

19. a. $\frac{t}{2(2 - 5t)}$

b. $y = -\sqrt{3 - 2x}$

20. $\frac{3}{2}\log_e(3)$

21. $y = \frac{1}{3}(x - 9)\sqrt{2x + 9} + 9$

22. a. $\frac{7}{2}$

b. $\frac{47}{7}$

c. 6

d. $\frac{333}{10}$

23. a. $\frac{26}{3}$

b. $\frac{1}{3}\log_e\left(\frac{14}{5}\right)$

24. a. $\frac{1}{35}$

b. 6

25. a. $\frac{16}{3}$

b. $\frac{32}{3}$

26. a. $\frac{2}{3a}(ax + b)^{\frac{3}{2}} + c$

b. $\frac{2}{15a^2}(3ax - 2b)(ax + b)^{\frac{3}{2}} + c$

c. $\frac{1}{a}\log_e(|ax + bx|) + c$

d. $\frac{x}{a} - \frac{b}{a^2}\log_e(|ax + b|) + c$

27. a. $\frac{2}{a}\sqrt{ax + b} + c$

b. $\frac{2}{3a^2}(ax - 2b)\sqrt{ax + b} + c$

c. $\frac{-1}{a(ax + b)} + c$

d. $\frac{b}{a^2(ax + b)} + \frac{1}{a^2}\log_e(|ax + b|) + c$

28. a. $\frac{ad–bc}{a^2}\log_e(|ax + b|) + \frac{cx}{a} + k$

b. $\frac{c}{a^2}\log_e(|ax + b|) - \frac{ad - bc}{a^2(ax + b)} + k$

c. $\frac{cx}{a^2} - \frac{2bc}{a^3}\log_e(|ax + b|) - \frac{a^2d + cb^2}{a^3(ax + b)} + k$

d. $\frac{a^2d + cb^2}{a^3}\log_e(|ax + b|) + \frac{cx^2}{2a} - \frac{bcx}{a^2} + k$

29. a. $\frac{1}{2a^3}(ax–3b)(ax + b) + \frac{b^2}{a^3}\log_e(|ax + b|) + c$

b. $\dfrac{x(xa+2b)}{a^2(ax+b)} - \dfrac{2b}{a^3}\log_e(|ax+b|)+c$

c. $\dfrac{1}{a^3}\log_e(|ax+b|) + \dfrac{b(4ax+3b)}{2a^3(ax+b)^2}+c$

d. $\dfrac{2}{15a^3}\left(3a^2x^2-4abx+8b^2\right)\sqrt{ax+b}+c$

30. a. $\dfrac{256}{15}$

b. $\dfrac{4}{3}\sqrt{a^3}$

c. $\dfrac{8}{15}\sqrt{a^5}$

7.3 Exam questions

Note: Mark allocations are available with the fully worked solutions online.

1. $\dfrac{8\sqrt{2}}{3}-\dfrac{10}{3}$
2. E
3. D

7.4 Non-linear substitutions

7.4 Exercise

1. $\dfrac{-2}{(x^2+16)^2}+c$

2. $\dfrac{5}{2}\sqrt{2x^2+3}+c$

3. a. $\dfrac{1}{12}(x^2+4)^6+c$

b. $\dfrac{1}{2}\log_e(x^2+4)+c$

c. $\dfrac{-1}{2(x^2+9)}+c$

d. $\sqrt{x^2+9}+c$

4. $\dfrac{1}{2}\log_e(x^2+4x+29)+c$

5. $\dfrac{1}{3}\log_e(|x^3+9|)+c$

6. a. $\dfrac{-1}{6(x^3+27)^2}+c$

b. $\dfrac{2}{3}\sqrt{x^3+27}+c$

c. $\dfrac{1}{3}\log_e(|x^3+8|)+c$

d. $\dfrac{1}{12}(x^3+8)^4+c$

7. a. $\dfrac{1}{8}(x^2-4x+13)^4+c$

b. $\dfrac{-1}{2(x^2-4x+13)}+c$

c. $-\sqrt{x^2-8x+25}+c$

d. $-\dfrac{1}{2}\log_e(x^2-8x+25)+c$

8. $\cos\left(\dfrac{1}{x}\right)+c$

9. $\dfrac{1}{2}\sin(x^2)+c$

10. a. $\dfrac{1}{3}e^{\sin(3x)}+c$

b. $2\sin\left(\sqrt{x}\right)+c$

11. a. $\dfrac{1}{2}e^{\tan(2x)}+c$

b. $\dfrac{1}{8}\left(\log_e(3x)\right)^2+c$

12. a. $\dfrac{1}{8}\log_e(4e^{2x}+5)+c$

b. $\dfrac{1}{6(2e^{-3x}-5)}+c$

c. $\dfrac{1}{12(3e^{-2x}+4)^2}+c$

d. $\dfrac{-1}{e^{2x}+x}+c$

13. 1

14. $\dfrac{1}{80}$

15. a. 12

b. $\dfrac{1}{4}\log_e\left(\dfrac{5}{3}\right)$

c. $\dfrac{1}{18}$

d. 0

16. a. $\dfrac{1}{2}\log_e\left(\dfrac{26}{29}\right)$

b. $-\dfrac{1}{4}$

c. $2e(e-1)$

d. $\dfrac{1}{2}(e-1)$

17. a. $\dfrac{1}{2}(e-1)$

b. $\dfrac{1}{6}$

c. $\dfrac{1}{2}(1-\sqrt{3})$

d. $2(\cos(1)-\cos(2))$

18. $\dfrac{\pi^2}{24}$

19. $\dfrac{\pi^2}{128}$

20. a. $-\cos\left(\log_e(4x)\right)+c$

b. $\sin\left(\log_e(3x)\right)+c$

c. $\dfrac{1}{2}\left(\sin^{-1}\left(\dfrac{x}{2}\right)\right)^2+c$

d. $\dfrac{1}{4}\left(\tan^{-1}(2x)\right)^2+c$

21. a. $y = \frac{1}{2}\left(1 - \cos(x^2)\right)$

b. $1 - \sqrt{3}$

c. $\frac{1}{2}\log_e\left(\frac{29}{20}\right)$

d. $\frac{1}{2}(1 - e)$

22. a. $c = 4, \frac{1}{2}\log_e\left(\frac{21}{5}\right)$

b. $\frac{\sqrt{2}}{4}$

23. a. $\frac{1}{2}(1 - e^{-4})$

b. 2

24. a. $\frac{1}{a}\sqrt{ax^2 + b} + c$

b. $\frac{-1}{2a\,(ax^2 + b)} + c$

c. $\frac{1}{2a\,(n+1)}\left(ax^2 + b\right)^{n+1} + c$

d. $\frac{1}{2a}\log_e(|ax^2 + b|) + c$

25. a. $\log_e(|f(x)|) + c$

b. $-\frac{1}{f(x)} + c$

c. $2\sqrt{f(x)} + c$

d. $e^{f(x)} + c$

7.4 Exam questions

Note: Mark allocations are available with the fully worked solutions online.

1. E
2. D
3. B

7.5 Integrals of powers of trigonometric functions

7.5 Exercise

1. a. $\frac{x}{2} - \sin\left(\frac{x}{2}\right) + c$

b. $-\cos\left(\frac{x}{2}\right) + c$

2. a. $\frac{4\pi - 3\sqrt{3}}{12}$

b. $\frac{\pi}{24} - \frac{\sqrt{3}}{128}$

3. a. $\frac{1}{24}\sin^6(4x) + c$

b. $\frac{1}{8}$

4. a. $-\frac{1}{8}\cos(4x) + c$ b. $x + c$

5. a. $-\frac{1}{8}\cos^4(2x) + c$

b. $\frac{1}{8}\sin^4(2x) + c$

6. a. $\frac{1}{12}\sin^3(4x) - \frac{1}{10}\sin^5(4x) + \frac{1}{28}\sin^7(4x)$

b. $\frac{8 - 5\sqrt{2}}{36}$

7. a. $\frac{1}{10}$

b. $\frac{1}{15}$

8. a. $\frac{3x}{8} - \frac{1}{8}\sin(4x) + \frac{1}{64}\sin(8x) + c$

b. $\frac{\pi}{32} + \frac{1}{12}$

9. a. $\frac{\pi}{32}$

b. $\frac{1}{15}$

10. a. $\frac{x}{8} - \frac{1}{128}\sin(16x) + c$

b. $\frac{1}{20}\cos^5(4x) - \frac{1}{12}\cos^3(4x) + c$

11. a. $\frac{1}{12}\sin^3(4x) - \frac{1}{20}\sin^5(4x) + c$

b. $\frac{1}{20}\sin^5(4x) - \frac{1}{28}\sin^7(4x) + c$

12. a. $\frac{\pi}{12}$

b. $\frac{2}{9}$

13. a. $\frac{\pi}{16}$

b. $\frac{8}{45}$

14. a. $x - \frac{1}{4}\cos(4x) + c$

b. $\frac{1}{2}(\sin(2x) - \cos(2x)) + \frac{1}{6}\left(\cos^3(2x) - \sin^3(2x)\right) + c$

15. a. $\frac{1}{96}\cos^3(4x) - \frac{1}{32}\cos(4x) + c_1$

b. $\frac{1}{12}\cos^6(2x) - \frac{1}{8}\cos^4(2x) + c_2$

c. $\frac{1}{8}\sin^4(2x) - \frac{1}{12}\sin^6(2x) + c_3$

16. a. $2\log_e\left(\left|\sec\left(\frac{x}{2}\right)\right|\right) + c$

b. $\frac{1}{4}\log_e(2)$

17. a. $3\tan\left(\frac{x}{3}\right) - x + c$

b. $\frac{4 - \pi}{16}$

18. a. $\frac{1}{3}\log_e(|\sec(3x)|)+c$

b. $\frac{1}{3}\log_e(|\sin(3x)|)+c$

19. a. $\frac{1}{6}\tan^2(3x)+c$

b. $\frac{1}{9}\tan^3(3x)+c$

20. a. $\frac{1}{10}\log_e(2)$

b. $\frac{4-\pi}{20}$

21. a. $\frac{1}{20}$

b. $\frac{8}{75}$

22. a. $-\frac{1}{2}\operatorname{cosec}(2x)+c$

b. $\frac{1}{2}\sec(2x)+c$

c. $\frac{1}{4}\sec^2(2x)+c$

d. $-\frac{1}{4}\operatorname{cosec}^2(2x)+c$

23. a. $\frac{-1}{a(n+1)}\cos^{n+1}(ax)+c$

b. $\frac{1}{a(n+1)}\sin^{n+1}(ax)+c$

c. $\frac{1}{a(n+1)}\tan^{n+1}(ax)+c$

24. a. $-\frac{1}{a}\log_e(|\cos(ax)|)+c$

b. $\frac{1}{a}\tan(ax)-x+c$

c. $\frac{1}{2a}\tan^2(ax)+\frac{1}{a}\log_e(|\cos(ax)|)+c$

d. $\frac{1}{3a}\tan^3(ax)-\frac{1}{a}\tan(ax)+x+c$

25. a. $\frac{x}{2}-\frac{1}{4a}\sin(2ax)+c$

b. $\frac{1}{3a}\cos^3(ax)-\frac{1}{a}\cos(ax)+c$

c. $\frac{3x}{8}-\frac{1}{4a}\sin(2ax)+\frac{1}{32a}\sin(4ax)+c$

d. $-\frac{1}{a}\cos(ax)+\frac{2}{3a}\cos^3(ax)-\frac{1}{5a}\cos^5(ax)+c$

26. a. $\frac{x}{2}+\frac{1}{4a}\sin(2ax)+c$

b. $\frac{1}{a}\sin(ax)-\frac{1}{3a}\sin^3(ax)+c$

c. $\frac{3x}{8}+\frac{1}{4a}\sin(2ax)+\frac{1}{32a}\sin(4ax)+c$

d. $\frac{1}{a}\sin(ax)-\frac{2}{3a}\sin^3(ax)+\frac{1}{5a}\sin^5(ax)+c$

7.5 Exam questions

Note: Mark allocations are available with the fully worked solutions online.

1. B
2. D
3. $-\frac{1}{10}\sin(10p)$

7.6 Integrals involving inverse trigonometric functions

7.6 Exercise

1. a. $\sin^{-1}\left(\frac{x}{10}\right)+c$

b. $4\sin^{-1}\left(\frac{3x}{8}\right)+c$

2. a. $\frac{1}{5}\sin^{-1}\left(\frac{5x}{6}\right)+c$

b. $-\frac{1}{25}\sqrt{36-25x^2}+c$

3. a. $\sin^{-1}\left(\frac{x}{4}\right)+c$

b. $\frac{1}{4}\sin^{-1}(4x)+c$

4. a. $2\sin^{-1}\left(\frac{5x}{7}\right)+c$

b. $-\frac{2}{5}\sqrt{49-25x^2}+c$

5. a. $-\frac{1}{2}\sin^{-1}\left(\frac{4x}{5}\right)=\frac{1}{4}\left(2\cos^{-1}\left(\frac{4x}{5}\right)-\pi\right)$

b. $2\cos^{-1}\left(\frac{x}{2}\right)+3$

6. a. $\cos^{-1}\left(\frac{x}{2}\right)+c$

b. $\cos^{-1}(2x)+c$

7. a. $\frac{3}{49}\sqrt{36-49x^2}+c$

b. $\frac{3}{7}\cos^{-1}\left(\frac{7x}{6}\right)+c$

8. a. $\frac{1}{10}\tan^{-1}\left(\frac{x}{10}\right)+c$

b. $\frac{1}{2}\tan^{-1}\left(\frac{3x}{8}\right)+c$

9. a. $\frac{1}{30}\tan^{-1}\left(\frac{5x}{6}\right)+c$

b. $\frac{1}{10}\log_e(64+25x^2)+c$

10. a. $\frac{\pi}{24}$

b. $\frac{\pi}{144}$

11. a. $\frac{1}{2}\tan^{-1}\left(\frac{6x}{7}\right)+c$

b. $2\tan^{-1}(4x)+c$

12. a. $\frac{\pi\sqrt{6}}{36}$

b. $\frac{\pi\sqrt{3}}{9}$

13. a. 3

b. $\frac{\pi}{2}$

14. a. $\frac{\pi}{12}$

b. $\frac{1}{2}\log_e(2)$

15. a. $\frac{\pi}{18}$

b. $\frac{1}{18}\log_e(2)$

16. a. $\frac{5}{9}$

b. $\frac{\pi}{60}$

17. a. $y = \sin^{-1}\left(\frac{x}{2}\right) + \frac{2\pi}{3}$

b. $y = \frac{1}{2}\tan^{-1}(2x) + \frac{7\pi}{8}$

18. a. $\frac{\pi\sqrt{3}}{18}$

b. $\frac{\pi}{4}$

19. a. $\frac{\sqrt{3}\pi}{9}$

b. $\frac{\sqrt{3}\pi}{9}$

20. a. $\frac{1}{3}\log_e(2)$

b. $\frac{1}{3}$

21. a. $\frac{1}{8}\tan^{-1}\left(\frac{2x-3}{4}\right) + c$

b. $\frac{1}{2}\sin^{-1}\left(\frac{2x-3}{4}\right) + c$

22. a. $\frac{1}{15}\tan^{-1}\left(\frac{5x-2}{3}\right) + c$

b. $\frac{1}{5}\sin^{-1}\left(\frac{5x-2}{3}\right) + c$

23. a. $2\sin^{-1}\left(\frac{x+2}{3}\right) + c$

b. $\frac{2}{3}\tan^{-1}\left(\frac{x+2}{3}\right) + c$

24. a. $2\sin^{-1}\left(\frac{3x+5}{7}\right) + c$

b. $\frac{2}{7}\tan^{-1}\left(\frac{3x+5}{7}\right) + c$

25. a. $\frac{5}{3}\sin^{-1}\left(\frac{3x}{5}\right) + \frac{1}{3}\sqrt{25-9x^2} + c$

b. $\frac{1}{6}\log_e\left(9x^2+25\right) + \frac{1}{3}\tan^{-1}\left(\frac{3x}{5}\right) + c$

26. a. $\cos^{-1}\left(\frac{4x}{3}\right) - \frac{3}{16}\sqrt{9-16x^2} + c$

b. $\frac{1}{4}\tan^{-1}\left(\frac{4x}{3}\right) + \frac{1}{8}\log_e(16x^2+9) + c$

27. a. $\sqrt{5-2x^2} + \frac{5\sqrt{2}}{2}\sin^{-1}\left(\frac{\sqrt{10}x}{5}\right) + c$

b. $\frac{\sqrt{2}}{2}\tan^{-1}\left(\frac{\sqrt{2}x}{5}\right) - \frac{1}{2}\log_e(2x^2+25) + c$

28. a. $\frac{1}{q}\sin^{-1}\left(\frac{qx}{p}\right) + c$

b. $\frac{1}{pq}\tan^{-1}\left(\frac{qx}{p}\right) + c$

c. $\frac{b}{q}\sin^{-1}\left(\frac{qx}{p}\right) - \frac{a}{q^2}\sqrt{p^2-q^2x^2} + c$

d. $\frac{b}{pq}\tan^{-1}\left(\frac{qx}{p}\right) + \frac{a}{2q^2}\log_e(q^2x^2+p^2) + c$

29. a. i. $8\sin^{-1}\left(\frac{x}{4}\right) + \frac{x}{2}\sqrt{16-x^2} + c$

ii. 4π; one-quarter of the area of a circle of radius 4

b. i. $\frac{x}{2}\sqrt{36-25x^2} - \frac{18}{5}\cos^{-1}\left(\frac{5x}{6}\right) + c$

ii. $\frac{9\pi}{5}$; one-quarter of the area of an ellipse with semi-minor and major axes of $\frac{6}{5}$ and 6

30. Sample responses can be found in the worked solutions in the online resources.

31. a. $\frac{1}{2}\log_e(2) - \frac{\pi}{8}$

b. $\frac{1}{4}\log_e(2) + \frac{\pi}{8}$

32. a. $9 + \frac{5\pi}{2}$

b. $\sqrt{3} - \frac{5\pi}{3}$

7.6 Exam questions

Note: Mark allocations are available with the fully worked solutions online.

1. $\log_e(2) + \frac{\pi}{4}$

2. D

3. $\frac{1}{b}\sin^{-1}\left(\frac{ax}{b}\right)$

7.7 Integrals involving partial fractions

7.7 Exercise

1. a. $\frac{1}{20}\log_e\left(\left|\frac{x+10}{x-10}\right|\right) + c$

b. $\frac{1}{4}\log_e\left(\left|\frac{3x+8}{3x-8}\right|\right) + c$

2. a. $\frac{1}{60}\log_e\left(\left|\frac{5x+6}{5x-6}\right|\right)+c$

b. $-\frac{1}{50}\log_e(|25x^2-36|)+c$

3. a. $\frac{1}{4}\log_e\left(\left|\frac{x-2}{x+2}\right|\right)+c$

b. $\frac{1}{2}\tan^{-1}\left(\frac{x}{4}\right)+c$

4. a. $\frac{1}{2}\log_e(|x^2-25|)+c$

b. $\frac{1}{4}\log_e(|(x-6)^3(x+6)^5|)+c$

5. a. $\log_e\left(\frac{(x-3)^2}{|x+5|}\right)+c$

b. $\log_e\left(\frac{|x+4|^3}{(x-1)^2}\right)+c$

6. a. $\log_e\left(\frac{(x-3)^2}{|x+4|}\right)+c$

b. $\log_e((x-5)^2|x+3|^3)+c$

7. a. $\log_e\left(\frac{|x+3|^5}{|x-2|^3}\right)+c$

b. $\log_e\left(\left|\frac{x-7}{x+4}\right|\right)+c$

8. a. $\frac{5}{x+3}+2\log_e(|x+3|)+c$

b. $\frac{2}{9}\log_e(|3x-4|)-\frac{5}{9(3x-4)}+c$

9. a. $2\log_e(|x-3|)-\frac{9}{x-3}+c$

b. $2\log_e(|x+2|)+\frac{9}{x+2}+c$

10. a. $\log_e(|2x+3|)+\frac{3}{2x+3}+c$

b. $\frac{2}{3}\log_e(|3x-5|)+\frac{3}{3x-5}+c$

11. a. $\frac{1}{5}\tan^{-1}\left(\frac{x}{5}\right)+c$

b. $\frac{1}{24}\log_e\left(\left|\frac{x+1}{x+25}\right|\right)+c$

c. $-\frac{1}{x-5}+c$

12. a. $3x+\log_e\left(\frac{(x-2)^4}{|x+5|^3}\right)+c$

b. $\log_e\left(\frac{(x-2)^2}{|x+3|}\right)-2x+c$

13. a. $x-\log_e(|(x-3)^2(x+4)^3|)+c$

b. $-3x+\log_e\left(\left|\frac{(x+3)^5}{(x-1)^3}\right|\right)+c$

14. a. $4x+\log_e\left(\frac{(x-6)^2}{|x+2|^3}\right)+c$

b. $-x^2+\log_e\left(\frac{|x-2|}{(x-4)^2}\right)+c$

15. a. $\frac{1}{4}\log_e\left(\frac{5}{3}\right)$

b. $\frac{1}{8}\log_e\left(\frac{9}{5}\right)$

16. a. $\log_e\left(\frac{25}{3}\right)$

b. $\log_e(2)-2$

17. a. $\log_e\left(\frac{9}{4}\right)$

b. $\log_e\left(\frac{3}{2}\right)$

18. a. $\log_e(8)$

b. $10+\frac{3}{2}\log_e\left(\frac{7}{3}\right)$

19. a. $\sqrt{3}$ b. $\frac{1}{3}$ c. 5 d. $\frac{2\sqrt{3}}{3}$

20. a. $x+\frac{5}{4}\log_e\left(\left|\frac{4x-5}{4x+5}\right|\right)+c$

b. $x-\frac{5}{2}\tan^{-1}\left(\frac{4x}{5}\right)+c$

21. a. $2\log_e(7)-3$

b. $12-5\log_e(5)$

22. a. $\log_e\left(\frac{|x-2|}{\sqrt{x^2+9}}\right)-\frac{5}{3}\tan^{-1}\left(\frac{x}{3}\right)+c$

b. $\log_e\left(\frac{|x+3|}{\sqrt{x^2+16}}\right)+\frac{3}{4}\tan^{-1}\left(\frac{x}{4}\right)+c$

23. a. $\frac{1}{2}\log_e(3)-\frac{\pi}{18}$

b. $\frac{2}{9}+\frac{1}{2}\log_e(3)$

24. a. $\frac{1}{2ab}\log_e\left(\frac{|bx-a|}{|bx+a|}\right)+c$

b. $\frac{1}{2b^2}\log_e(|b^2x^2-a^2|)+c$

25. a. $\frac{1}{a^2}\log_e(|ax-b|)-\frac{b}{a^2(ax-b)}+c$

b. $\frac{1}{aq-bp}\log_e\left(\frac{|qx+b|}{|px+a|}\right)+c$

26. Sample responses can be found in the worked solutions in the online resources.

7.7 Exam questions

Note: Mark allocations are available with the fully worked solutions online.

1. D

2. $\log_e\left(\sqrt{\frac{3}{2}}\right)$

3. B

7.8 Review

7.8 Exercise

Technology free: short answer

1. a. $-\sqrt{5-2x}+c$

   b. $\frac{1}{2}\sin^{-1}\left(\frac{2x}{5}\right)+c$

   c. $-\frac{1}{4}\sqrt{25-4x^2}+c$

   d. $-\frac{1}{8}\log_e(|25-4x^2|)+c$

2. a. $\frac{1}{20}\log_e\left(\left|\frac{5+2x}{5-2x}\right|\right)+c$

   b. $\frac{1}{10}\tan^{-1}\left(\frac{2x}{5}\right)+c$

   c. $-\frac{1}{2}\log_e(|5-2x|)+c$

   d. $\frac{5}{16}\log_e\left(\left|\frac{5+2x}{5-2x}\right|\right)-\frac{x}{4}+c$

3. a. $\frac{1}{8(25-4x^2)}+c$

   b. $-\frac{5}{4}\log_e(|5-2x|)-\frac{x}{2}+c$

   c. $\frac{1}{4}\log_e(|5-2x|)+\frac{5}{4(5-2x)}+c$

   d. $-\frac{1}{3}(x+5)\sqrt{5-2x}+c$

4. a. $\frac{2}{3}\log_e(2)$

   b. $\frac{2}{3}$

   c. $\frac{1}{4}$

   d. $\frac{1}{9}\left(3-\log_e(4)\right)$

5. a. $\frac{1}{9}\left(8\log_e(2)-3\right)$

   b. $\frac{10}{27}$

   c. $\frac{\pi}{18}$

   d. $\frac{4}{9}$

6. a. $\frac{\pi}{48}$

   b. $\frac{1}{18}\log_e(2)$

   c. $\frac{1}{24}\log_e(3)$

   d. $\frac{1}{18}\log_e\left(\frac{4}{3}\right)$

Technology active: multiple choice

7. C
8. D
9. A
10. C
11. E
12. B
13. B
14. E
15. A
16. D

Technology active: extended response

17. a. $\frac{x}{2}-\frac{1}{16}\sin(8x)+c$

    b. $\frac{1}{12}\sin^3(4x)+c$

    c. $\frac{x}{8}-\frac{1}{128}\sin(16x)+c$

    d. $\frac{1}{12}\sin^3(4x)-\frac{1}{20}\sin^5(4x)+c$

18. a. $\frac{128}{243}$

    b. $\frac{2}{27}\left(\log_e(3)-1\right)$

    c. $\frac{4\pi}{3}$

    d. $\frac{4\pi}{27}$

19. a. $\frac{9}{256}$

    b. $\frac{47}{1920}$

    c. $\frac{9}{512}$

    d. $\frac{233}{17\,920}$

20. a. $-\frac{1}{24}\left(2x^2+25\right)\sqrt{25-4x^2}+c$

    b. $-\frac{1}{120}\left(6x^2+25\right)\left(25-4x^2\right)^{\frac{3}{2}}+c$

    c. $\frac{25}{4}\sin^{-1}\left(\frac{2x}{5}\right)+\frac{x}{2}\sqrt{25-4x^2}+c$

    d. $\frac{625}{64}\sin^{-1}\left(\frac{2x}{5}\right)+\frac{x}{32}\left(8x^2-25\right)\sqrt{25-4x^2}+c$

7.8 Exam questions

Note: Mark allocations are available with the fully worked solutions online.

1. C
2. Sample responses can be found in the worked solutions in the online resources.
3. a. $[0, \infty)$
   b. Sample responses can be found in the worked solutions in the online resources.
   c. $\frac{\left(4\sqrt{3}-3\right)\pi}{24} - \frac{1}{4}\log_e(2)$
4. C
5. $-\frac{1}{9}\left(e^{3\cot(3b)} - e^{3\cot(3a)}\right)$

8 Differential equations

LEARNING SEQUENCE

Fully worked solutions for this topic are available online.

8.1 Overview

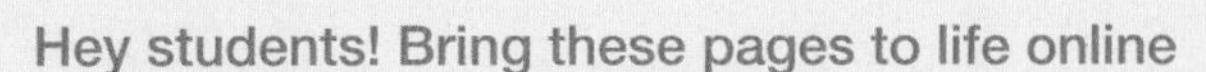

Find all this and MORE in jacPLUS

8.1.1 Introduction

The rate of change of one quantity with respect to another quantity, that is, the derivative, is a measure of how the quantities are related or how a change in one quantity will effect a change in the other. As you have already learned, rates of change form the basis of the study of calculus. Differential equations build on the concepts already covered in differential and integral calculus. Modelling the world around us using complex differential equations and sophisticated computers to solve these equations has led to an ever-increasing understanding of our world and the universe we inhabit.

One example of the use of differential equations is in the science of meteorology. Supercomputers can analyse enormous amounts of weather data, but it is the use of differential equations to model long- and short-term patterns that has enabled meteorologists to determine how changes in temperature and moisture in the atmosphere may lead to changes in the weather. This can mean the difference between life and death as communities now have advance warning of extreme weather conditions.

KEY CONCEPTS

This topic covers the following point from the VCE Mathematics Study Design:

- solution of simple differential equations of the form $\frac{dy}{dx} = f(x)$, $\frac{dy}{dx} = g(y)$ and in general differential equations of the form $\frac{dy}{dx} = f(x)g(y)$ using separation of variables and differential equations of the form $\frac{d^2y}{dx^2} = f(x)$.

Source: VCE Mathematics Study Design (2023–2027) extracts © VCAA; reproduced by permission.

8.2 Verifying solutions to differential equations

LEARNING INTENTION

At the end of this subtopic you should be able to:

- verify solutions to differential equations
- determine the values of a constant in an equation that is a solution to a differential equation.

8.2.1 Classification of differential equations

A differential equation is an equation involving derivatives. It is of the form $g\left(x, y, \frac{dy}{dx}, \frac{d^2y}{dx^2}, \ldots\right)=0$.

A differential equation contains the function $y=f(x)$ with y as the dependent variable, x as the independent variable, and various derivatives. In this topic, only differential equations that contain functions of one variable, $y=f(x)$, are considered.

Differential equations can be classified according to their order and degree. The **order** of a differential equation is the order of the highest derivative present. The **degree** of a differential equation is the degree of the highest power of the highest derivative.

A linear differential equation is one in which all variables including the derivatives are raised to the power of 1.

Some examples of differential equations are:

(a) $\frac{dy}{dx}=ky$

(b) $a\frac{d^2y}{dx^2}+b\frac{dy}{dx}+cy=0$

(c) $\ddot{x}-t\dot{x}+2x=t$

(d) $\frac{d^2y}{dx^2}+n^2y=0$

(e) $x\left(\frac{dy}{dx}\right)^3+3\frac{dy}{dx}+5y=0$

(f) $D_t^3x=\sqrt{x^2+1}$

Note that (a) and (e) are first order; (b), (c) and (d) are second order; and (f) is third order. Equation (e) has a degree of 3, whereas all the others have a degree of 1. Equations (a), (b), (c) and (d) are linear; (e) and (f) are non-linear. Note also that there are many different notations for derivatives; for example, second-order derivatives can be expressed as $\ddot{x}=\frac{d^2x}{dt^2}$ and $D_t^2x=\frac{d^2x}{dt^2}$.

8.2.2 Verifying solutions to differential equations

To check that a given solution satisfies the differential equation, use the process of differentiation and substitution. Generally only first- or second-order differential equations will be considered in this topic.

When setting out a proof, it is necessary to show that the left-hand side (LHS) of the equation is equal to the right-hand side (RHS), while working on each side independently of the other side.

WORKED EXAMPLE 1 Verifying a solution to a differential equation (1)

Verify that $y = x^3$ is a solution of the differential equation $x^3\frac{d^2y}{dx^2} - \left(\frac{dy}{dx}\right)^2 + 3xy = 0$.

THINK	WRITE
1. Use basic differentiation to determine the first derivative.	$y = x^3$ $\frac{dy}{dx} = 3x^2$
2. Calculate the second derivative.	$\frac{d^2y}{dx^2} = 6x$
3. Substitute the expressions for y, the first derivative and the second derivative into the LHS of the differential equation.	$\text{LHS} = x^3\frac{d^2y}{dx^2} - \left(\frac{dy}{dx}\right)^2 + 3xy$ $= x^3 \times (6x) - \left(3x^2\right)^2 + 3x \times \left(x^3\right)$
4. Simplify and expand so that LHS = RHS = 0, thus proving the given solution does satisfy the differential equation.	$= 6x^4 - 9x^4 + 3x^4$ $= 0$ $= \text{RHS}$

Differential equations involving unknowns

When we verify a given solution to a differential equation involving algebraic, trigonometric or exponential functions, there may also be an unknown value that must be determined for which the given solution satisfies the differential equation.

WORKED EXAMPLE 2 Determining a constant value when given a solution

Given that $y = e^{kx}$ is a solution of the differential equation $\frac{d^2y}{dx^2} - 2\frac{dy}{dx} - 8y = 0$, determine the values of the real constant k.

THINK	WRITE
1. Use the rule for differentiation of exponential functions, $\frac{d}{dx}\left(e^{kx}\right) = ke^{kx}$, to determine the first derivative.	$y = e^{kx}$ $\frac{dy}{dx} = ke^{kx}$
2. Differentiate again to calculate the second derivative.	$\frac{d^2y}{dx^2} = k^2e^{kx}$
3. Substitute the expressions for y, the first derivative and the second derivative into the given differential equation.	$\frac{d^2y}{dx^2} - 2\frac{dy}{dx} - 8y = 0$ $k^2e^{kx} - 2ke^{kx} - 8e^{kx} = 0$
4. Take out the common factor.	$e^{kx}\left(k^2 - 2k - 8\right) = 0$
5. Factorise the quadratic equation involving the unknown.	$e^{kx} \neq 0 \Rightarrow k^2 - 2k - 8 = 0$ $(k-4)(k+2) = 0$

6. Solve the resulting equation for the unknown and state the answer.

When $k = 4$ or $k = -2$, $y = e^{kx}$ is a solution of $\frac{d^2y}{dx^2} - 2\frac{dy}{dx} - 8y = 0$.

TI | THINK

On a Calculator page, complete the entry as shown.

DISPLAY/WRITE

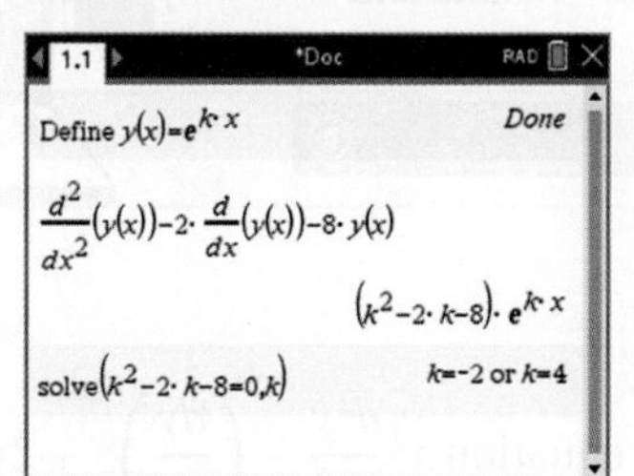

CASIO | THINK

On a Main screen, complete the entry as shown.

DISPLAY/WRITE

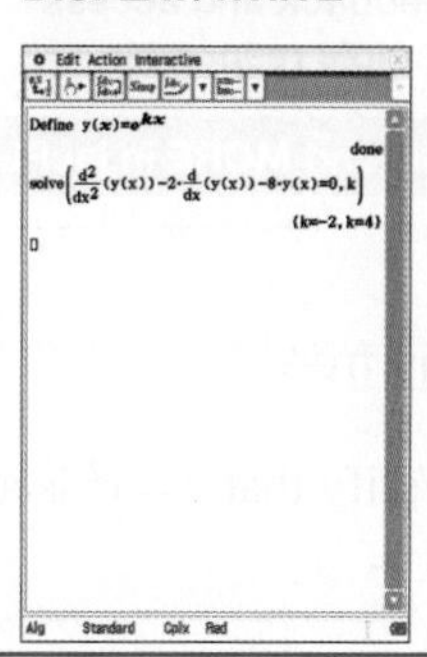

Solutions to a differential equation involving products

When verifying solutions to a differential equation involving a mixture of algebraic, trigonometric or exponential functions, it may be necessary to use the product or quotient rules for differentiation.

WORKED EXAMPLE 3 Verifying a solution to a differential equation (2)

Verify that $y = xe^{-2x}$ is a solution of the differential equation $\frac{d^2y}{dx^2} + 4\frac{dy}{dx} + 4y = 0$.

THINK / **WRITE**

1. Use the product rule for differentiation to determine the first derivative.

$$y = xe^{-2x}$$
$$\frac{dy}{dx} = x\frac{d}{dx}\left(e^{-2x}\right) + e^{-2x}\frac{d}{dx}(x)$$
$$= -2xe^{-2x} + e^{-2x}$$

2. Simplify the first derivative by taking out the common factor.

$$= e^{-2x}(1 - 2x)$$

3. Calculate the second derivative, using the product rule again.

$$\frac{d^2y}{dx^2} = e^{-2x}\frac{d}{dx}(1 - 2x) + (1 - 2x)\frac{d}{dx}\left(e^{-2x}\right)$$
$$= -2e^{-2x} - 2(1 - 2x)e^{-2x}$$

4. Simplify the second derivative by taking out the common factor.

$$= e^{-2x}(-2 - 2(1 - 2x))$$
$$= e^{-2x}(4x - 4)$$

5. Substitute the expressions for y, the first derivative and the second derivative into the LHS of the differential equation.

$$\text{LHS} = \frac{d^2y}{dx^2} + 4\frac{dy}{dx} + 4y$$
$$= e^{-2x}(4x - 4) + 4e^{-2x}(1 - 2x) + 4xe^{-2x}$$

6. Take out the common factor and simplify so that LHS = RHS = 0, thus proving the given solution does satisfy the differential equation.

$$= e^{-2x}[(4x - 4) + 4(1 - 2x) + 4x]$$
$$= e^{-2x}[4x - 4 + 4 - 8x + 4x]$$
$$= 0$$
$$= \text{RHS}$$

8.2 Exercise

Students, these questions are even better in jacPLUS

Receive immediate feedback and access sample responses

Access additional questions

Track your results and progress

Find all this and MORE in jacPLUS

Technology free

1. **WE1** Verify that $y = x^2$ is a solution of the differential equation $x^2 \frac{d^2y}{dx^2} - \left(\frac{dy}{dx}\right)^2 + 2y = 0$.

2. For the differential equation $x^4 \frac{d^2y}{dx^2} - \left(\frac{dy}{dx}\right)^2 + 4x^2y = 0$, show that $y = x^4$ is a solution.

3. If $y = 2x^2 - 3x + 5$, show that $\left(\frac{dy}{dx}\right)^2 - 8y + 31 = 0$.

4. a. Given the differential equation $\frac{d^2y}{dx^2} - 2x\frac{dy}{dx} + 6y + 6x^2 = 0$, show that $y = x^3 - 3x^2 - \frac{3x}{2} + 1$ is a solution.

   b. If $y = ax^3 + bx^2$ where a and b are real constants, show that $x^2 \frac{d^2y}{dx^2} - 4x\frac{dy}{dx} + 6y = 0$.

5. **WE2** Given that $y = e^{kx}$ is a solution of the differential equation $\frac{d^2y}{dx^2} + 3\frac{dy}{dx} - 10y = 0$, determine the values of the real constant k.

6. If $y = \cos(kx)$ is a solution of the differential equation $\frac{d^2y}{dx^2} + 9y = 0$, determine the values of the real constant k.

7. a. Determine the constants a, b and c if $y = a + bx + cx^2$ is a solution of the differential equation $\frac{d^2y}{dx^2} + 2\frac{dy}{dx} + 4y = 4x^2$.

   b. Determine the constants a, b, c and d if $y = ax^3 + bx^2 + cx + d$ is a solution of the differential equation $\frac{d^2y}{dx^2} + 2\frac{dy}{dx} + y = x^3$.

8. a. Show that $y = x^n$ is a solution of the differential equation $x^2y\frac{d^2y}{dx^2} - x^2\left(\frac{dy}{dx}\right)^2 + ny^2 = 0$.

   b. If the differential equation $x^2\frac{d^2y}{dx^2} - 2x\frac{dy}{dx} - 10y = 0$ has a solution $y = x^n$, determine the possible values of n.

9. a. Given that $x = e^{3t} + e^{-4t}$ show that $\frac{d^2x}{dt^2} + \frac{dx}{dt} - 12x = 0$.

   b. If $y = Ae^{3x} + Be^{-3x}$ where A and B are real constants, show that $\frac{d^2y}{dx^2} - 9y = 0$.

10. Determine the values of the real constant k such that $y = e^{kx}$ satisfies the differential equation $\frac{d^2y}{dx^2} + 5\frac{dy}{dx} - 6y = 0$.

11. a. If $y = 3\sin(2x) + 4\cos(2x)$, show that $\dfrac{d^2y}{dx^2} + 4y = 0$.

b. Show that $y = A\sin(3x) + B\cos(3x)$, where A and B are constants, is a solution of the differential equation $\dfrac{d^2y}{dx^2} + 9y = 0$.

12. a. If $y = a\sin(nx) + b\cos(nx)$, where n is a positive real number, show that $\dfrac{d^2y}{dx^2} + n^2y = 0$.

b. Given that $x = a\sin(pt)$ satisfies $\dfrac{d^2x}{dt^2} + 9x = 0$, determine the value(s) of the real constant p.

13. **WE3** Verify that $y = xe^{3x}$ is a solution of the differential equation $\dfrac{d^2y}{dx^2} - 6\dfrac{dy}{dx} + 9y = 0$.

14. Given that $y = Ax\cos(2x)$ is a solution of the differential equation $\dfrac{d^2y}{dx^2} + 4y = 8\sin(2x)$, determine the value of the real constant A.

15. a. Show that $y = e^{x^2}$ satisfies the differential equation $\dfrac{d^2y}{dx^2} - 2x\dfrac{dy}{dx} - 2y = 0$.

b. Verify that $y = \cos(x^2)$ satisfies the differential equation $x\dfrac{d^2y}{dx^2} - \dfrac{dy}{dx} + 4x^3y = 0$.

16. a. If $y = ax + b\sqrt{x^2+1}$ where a and b are constants, show that $\left(x^2+1\right)\dfrac{d^2y}{dx^2} + x\dfrac{dy}{dx} - y = 0$.

b. Given that $y = \log_e\left(x + \sqrt{x^2-9}\right)$, show that $\left(x^2-9\right)\dfrac{d^2y}{dx^2} + x\dfrac{dy}{dx} = 0$.

17. a. Show that $y = \tan(ax)$, where $a \in R\backslash\{0\}$, satisfies the differential equation $\dfrac{d^2y}{dx^2} = 2a^2y\left(1+y^2\right)$.

b. Verify that $y = \tan^2(ax)$, where $a \in R\backslash\{0\}$, is a solution of the differential equation $\dfrac{d^2y}{dx^2} = 2a^2\left(3y^2 + 4y + 1\right)$.

18. Show that $y = \log_e(ax+b)$, where $a, b \in R\backslash\{0\}$, is a solution of the differential equation $\dfrac{d^2y}{dx^2} + a^2e^{-2y} = 0$.

19. a. Show that $y = \sin^{-1}(3x)$ is a solution of the differential equation $\left(1-9x^2\right)\dfrac{d^2y}{dx^2} - 9x\dfrac{dy}{dx} = 0$.

b. Verify that $y = \cos^{-1}\left(\dfrac{x}{4}\right)$ is a solution of the differential equation $\left(16-x^2\right)\dfrac{d^2y}{dx^2} - x\dfrac{dy}{dx} = 0$.

20. a. A parachutist of mass m falls from rest in the Earth's gravitational field and is subjected to air resistance. Their velocity v is given by

$$v = \frac{mg}{k}\left(1 - e^{-kt}\right)$$

at time t where g and k are constants. Show that $\dfrac{dv}{dt} + kv = mg$.

b. In a transient circuit, the current i amperes at a time t seconds is given by $i = 3e^{-2t}\sin(3t)$.

Show that $\dfrac{d^2i}{dt^2} + 4\dfrac{di}{dt} + 13i = 0$.

21. a. Verify that $y = e^{3x}\cos(2x)$ satisfies the differential equation $\frac{d^2y}{dx^2} - 6\frac{dy}{dx} + 13y = 0$.

b. Determine the real constants a and b if $x = t(a\cos(3t) + b\sin(3t))$ is a solution of the differential equation $\frac{d^2x}{dt^2} + 9x = 6\cos(3t)$.

22. a. Given that $y = xe^{-3x}$ is a solution of the differential equation $\frac{d^2y}{dx^2} + a\frac{dy}{dx} + by = 0$, determine the values of the real constants a and b.

b. Show that $y = e^{kx}(Ax + B)$, where A, B and k are all real constants, is a solution of the differential equation $\frac{d^2y}{dx^2} - 2k\frac{dy}{dx} + k^2y = 0$.

23. a. Given that $y = Ax^2e^{-3x}$ is a solution of the differential equation $\frac{d^2y}{dx^2} + 6\frac{dy}{dx} + 9y = 10e^{-3x}$, determine the value of A.

b. If $y = Ax^2e^{-kx}$ is a solution of the differential equation $\frac{d^2y}{dx^2} + 2k\frac{dy}{dx} + k^2y = Be^{-kx}$, show that $B = 2A$.

24. Show that $y = \sqrt{\frac{\pi}{x}}\sin(x)$ is a solution of Bessel's equation, $4x^2\frac{d^2y}{dx^2} + 4x\frac{dy}{dx} + \left(4x^2 - 1\right)y = 0$.

25. Adrien-Marie Legendre (1752–1833) was a famous French mathematician. He made many mathematical contributions in the areas of elliptical integrals, number theory and the calculus of variations. He is also known for the differential equation named after him. Legendre's differential equation of order n is given by $\left(1 - x^2\right)\frac{d^2y}{dx^2} - 2x\frac{dy}{dx} + n(n+1)y = 0$ for $|x| < 1$, and the solutions of the differential equation are given by the polynomials $P_n(x)$. The first few polynomials are given by

$$P_0(x) = 1$$
$$P_1(x) = x$$
$$P_2(x) = \frac{1}{2}\left(3x^2 - 1\right)$$
$$P_3(x) = \frac{1}{2}\left(5x^3 - 3x\right)$$
$$P_4(x) = \frac{1}{8}\left(35x^4 - 30x^2 + 3\right)$$

a. Verify the solution of the Legendre's differential equation for the cases when $n = 3$ and $n = 4$.

b. The Legendre polynomials also satisfy many other mathematical properties. One such relation is $P_n(x) = \frac{1}{2^n n!}\frac{d^n}{dx^n}\left[\left(x^2 - 1\right)^n\right]$. Use this result to obtain $P_2(x)$ and $P_3(x)$.

c. The Legendre polynomials also satisfy $\int_{-1}^{1} P_n(x)P_m(x)\,dx = 0$ when $m \neq n$ and $\int_{-1}^{1} (P_n(x))^2 dx = \frac{2}{2n+1}$.

Verify these results for $P_2(x)$ and $P_3(x)$.

8.2 Exam questions

Question 1 (1 mark) TECH-ACTIVE

***Source:** VCE 2015, Specialist Mathematics Exam 2, Section A, Q14; © VCAA.*

MC A differential equation that has $y = x\sin(x)$ as a solution is

A. $\frac{d^2y}{dx^2} + y = 0$ **B.** $x\frac{d^2y}{dx^2} + y = 0$ **C.** $\frac{d^2y}{dx^2} + y = -\sin(x)$

D. $\frac{d^2y}{dx^2} + y = -2\cos(x)$ **E.** $\frac{d^2y}{dx^2} + y = 2\cos(x)$

Question 2 (3 marks) TECH-FREE

Find the value of m where $m \in C$ if $y = e^{mx}$ satisfies $\frac{d^2y}{dx^2} + 4\frac{dy}{dx} + 13y = 0$.

Question 3 (4 marks) TECH-FREE

Verify that $y = \tan^{-1}(2x)$ is a solution of the differential equation $\left(1 + 4x^2\right)\frac{d^2y}{dx^2} + 8x\frac{dy}{dx} = 0$.

More exam questions are available online.

8.3 Solving Type 1 differential equations, $\frac{dy}{dx} = f(x)$

LEARNING INTENTION

At the end of this subtopic you should be able to:

- solve Type 1 differential equations of the form $\frac{dy}{dx} = f(x)$ and state the largest domain over which the solution is valid.

8.3.1 Classifying solutions to a differential equation

The solution of a differential equation is usually obtained by the process of integration. Because the integration process produces an arbitrary constant of integration, the solutions of a differential equation are classified as follows.

A **general solution** is one that contains arbitrary constants of integration and satisfies the differential equation.

A **particular solution** is one that satisfies the differential equation and some other initial value condition, also known as a boundary value, that enables the constant(s) of integration to be found.

In general, the number of arbitrary constants of integration to be found is equal to the order of the differential equation. Throughout this course we study and solve special types of first- and second-order differential equations.

8.3.2 Type 1 differential equations, $\frac{dy}{dx} = f(x)$

Direct integration

In this section we solve first-order differential equations of the form $\frac{dy}{dx} = f(x)$, $y(x_0) = y_0$. Differential equations of this form can be solved by direct integration. Hence, it is necessary to be familiar with all the integration techniques studied so far. Anti-differentiating both sides gives $y = \int f(x)dx + c$. This is the general solution, which can be thought of as a family of curves.

If we use the given condition $x = x_0$ when $y = y_0$, we can determine the value of the constant of integration c in this particular case, which thus gives us the particular solution.

WORKED EXAMPLE 4 Determining solutions to Type 1 differential equations

a. Determine the general solution to $\frac{dy}{dx} + 12x = 0$.

b. Determine the particular solution of $\frac{dy}{dx} + 6x^2 = 0, y(1) = 2$.

THINK	WRITE
a. 1. Rewrite the equation to make $\frac{dy}{dx}$ the subject.	a. $\frac{dy}{dx} + 12x = 0$ $\frac{dy}{dx} = -12x$
2. Anti-differentiate to obtain y.	$y = -\int 12x\,dx$
3. Write the general solution in terms of a constant.	$y = -6x^2 + c$
b. 1. Rewrite the equation to make $\frac{dy}{dx}$ the subject.	b. $\frac{dy}{dx} + 6x^2 = 0$ $\frac{dy}{dx} = -6x^2$
2. Anti-differentiate to obtain y.	$y = -\int 6x^2 dx$
3. Express y in terms of x with an arbitrary constant.	$y = -2x^3 + c$
4. Substitute and use the given conditions to determine the value of the constant.	$y(1) = 2$: $\Rightarrow x = 1$ when $y = 2$ $2 = -2(1)^3 + c$ $c = 4$
5. Substitute back for c and state the particular solution.	$y = 4 - 2x^3$

TI | THINK	DISPLAY/WRITE	CASIO | THINK	DISPLAY/WRITE
a. On a Calculator page, select: Menu: 4 Calculus: D Differential Equation Solver and complete the fields as shown, then click OK. Click enter to see the solution.	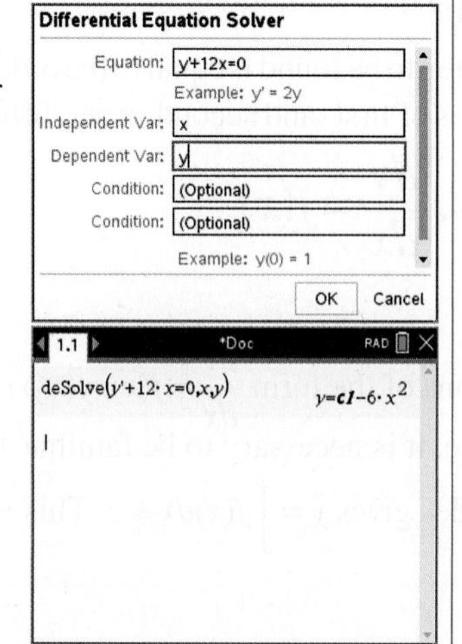	a. On a Main screen, select Action: Advanced: dSolve and complete the entry as shown.	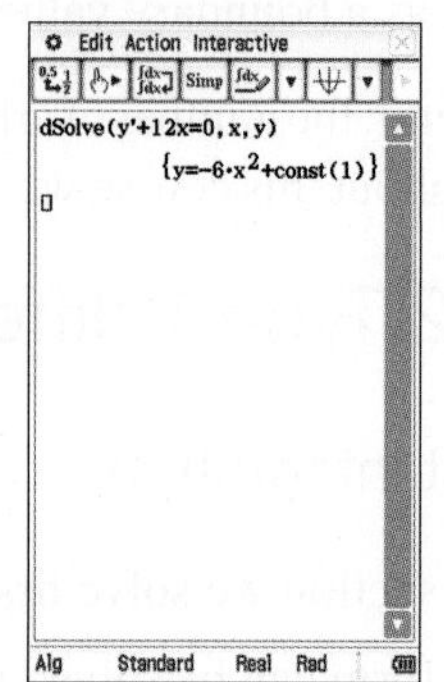

b. On a Calculator page, navigate to the Differential Equation Solver and complete the fields as shown, then click OK.
Click enter to see the solution.

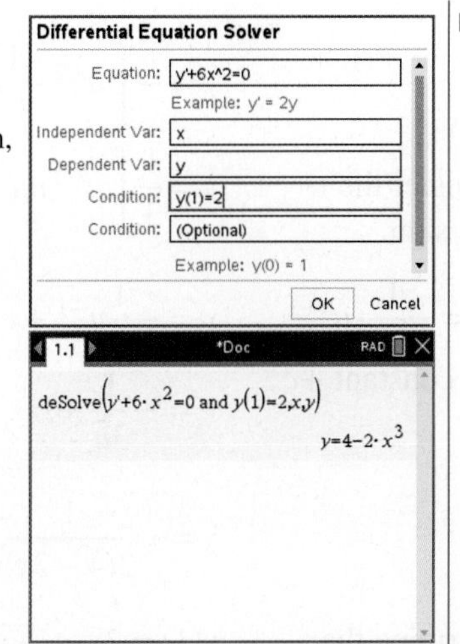

b. On a Main screen, select Action: Advanced: dSolve and complete the entry as shown.

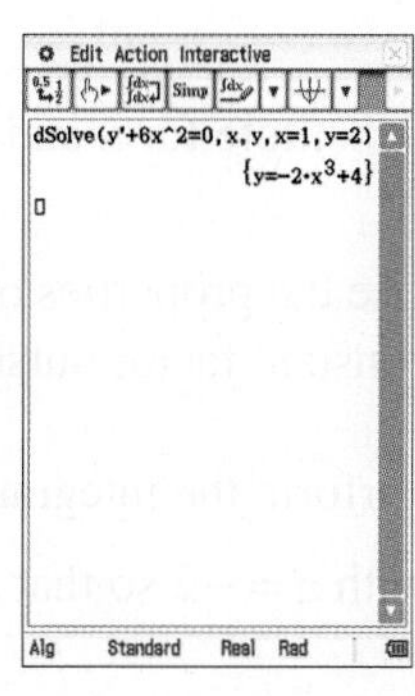

Determining particular solutions

In Topic 7, linear substitutions were used to integrate linear expressions. The example presented in Worked example 5 is a review of this process in the context of differential equations.

WORKED EXAMPLE 5 Solving Type 1 differential equations (1)

Solve the differential equation $(4-3x)^2\dfrac{dy}{dx}+1=0, y(1)=2$.

THINK	WRITE
1. Rewrite the equation to make $\dfrac{dy}{dx}$ the subject.	$(4-3x)^2\dfrac{dy}{dx}+1=0$ $(4-3x)^2\dfrac{dy}{dx}=-1$ $\dfrac{dy}{dx}=\dfrac{-1}{(4-3x)^2}$
2. Anti-differentiate to obtain y.	$y=\displaystyle\int\dfrac{-1}{(4-3x)^2}dx$
3. Use index laws to express the integrand as a function to a power.	$y=-\displaystyle\int(4-3x)^{-2}dx$
4. Use a linear substitution. Express dx in terms of du by inverting both sides.	Let $u=4-3x$. $\dfrac{du}{dx}=-3$ $\dfrac{dx}{du}=-\dfrac{1}{3}$ $dx=-\dfrac{1}{3}du$

5. Substitute for u and dx. $y = -\int u^{-2} \frac{-1}{3} du$

6. Use the properties of indefinite integrals to transfer the constant factor outside the front of the integral sign. $y = \frac{1}{3}\int u^{-2}\, du$

7. Perform the integration process, using $\int u^n du = \frac{1}{n+1}u^{n+1}$ with $n = -2$ so that $n + 1 = -1$, and add in the constant $+c$.

$$y = -\frac{1}{3}u^{-1} + c$$
$$y = -\frac{1}{3u} + c$$

8. Substitute back for x. $y = -\frac{1}{3(4-3x)} + c$

9. Substitute and use the given conditions to determine the value of the constant.

$$y(1) = 2$$
$$\Rightarrow x = 1 \text{ when } y = 2$$
$$2 = -\frac{1}{3} + c$$
$$c = 2 + \frac{1}{3}$$
$$c = \frac{7}{3}$$

10. Substitute back for c and state the particular solution. Although this is a possible answer, this result can be simplified. $y = \frac{-1}{3(4-3x)} + \frac{7}{3}$

11. Form the lowest common denominator. $y = \frac{-1 + 7(4-3x)}{3(4-3x)}$

12. Expand the brackets in the numerator; do not expand the brackets in the denominator. $y = \frac{-1 + 28 - 21x}{3(4-3x)}$

13. Simplify and take out common factors that cancel.

$$y = \frac{27 - 21x}{3(4-3x)}$$
$$= \frac{3(9-7x)}{3(4-3x)}$$

14. State the final answer in simplified form. Note the maximal domain for which the solution is valid. $y = \frac{9-7x}{4-3x}$ for $x \neq \frac{4}{3}$

15. Note that as a check, we can use the given condition to check the value of y.

Substitute $x = 1$:

$$y = \frac{9-7}{4-3} = 2$$

Stating the domain for which the solution is valid

As seen in the last example, the maximal domain for which the solution is valid is important. When solving differential equations, unless the solution is defined for all values of x, that is for $x \in R$, we are required to state the largest subset of R for which the given differential equation and solution are valid.

WORKED EXAMPLE 6 Solving Type 1 differential equations (2)

Solve the differential equation $\sqrt{3x-5}\dfrac{dy}{dx}+6=0$, $y(7)=2$, stating the largest domain for which the solution is valid.

THINK

1. Rewrite the equation to make $\dfrac{dy}{dx}$ the subject.

WRITE

$$\sqrt{3x-5}\frac{dy}{dx}+6=0$$
$$\sqrt{3x-5}\frac{dy}{dx}=-6$$
$$\frac{dy}{dx}=\frac{-6}{\sqrt{3x-5}}$$

2. Anti-differentiate to obtain y.

$$y=\int\frac{-6}{\sqrt{3x-5}}dx$$

3. Use the properties of indefinite integrals to transfer the constant factor outside the front of the integral sign.

$$y=-6\int\frac{1}{\sqrt{3x-5}}dx$$

4. Use index laws to express the integrand as a function to a power.

$$y=-6\int(3x-5)^{-\frac{1}{2}}dx$$

5. Use a linear substitution. Express dx in terms of du by inverting both sides.

Let $u=3x-5$.

$$\frac{du}{dx}=3$$
$$\frac{dx}{du}=\frac{1}{3}$$
$$dx=\frac{1}{3}du$$

6. Substitute for u and dx.

$$y=-6\int u^{-\frac{1}{2}}\frac{1}{3}du$$

7. Use the properties of indefinite integrals to transfer the constant factor outside the front of the integral sign.

$$y=-2\int u^{-\frac{1}{2}}\,du$$

8. Perform the integration process using $\int u^n du=\dfrac{1}{n+1}u^{n+1}$ with $n=-\dfrac{1}{2}$, so that $n+1=\dfrac{1}{2}$, and add in the constant c.

$$y=-4\sqrt{u}+c$$

9. Substitute back for x.

$$y=-4\sqrt{3x-5}+c$$

10. Substitute and use the given conditions to determine the value of the constant.

$y(7)=2$

$\Rightarrow x=7$ when $y=2$

$$2=-4\sqrt{16}+c$$
$$c=18$$

11. Substitute back for c and state the particular solution.

$$y=18-4\sqrt{3x-5}$$

12. Determine the domain for which the solution is valid from the differential equation.

$$\frac{dy}{dx}=\frac{-6}{\sqrt{3x-5}}\text{ for }3x-5>0$$

13. Solve the inequality for x to state the largest domain for which the solution is valid for the given differential equation. State the answer.	$3x > 5$ $x > \frac{5}{3}$ The solution $y = 18 - 4\sqrt{3x-5}$ is valid for $x > \frac{5}{3}$.

Solving first-order differential equations involving inverse trigonometric functions

The results $\int \frac{1}{\sqrt{a^2 - x^2}}\,dx = \sin^{-1}\left(\frac{x}{a}\right) + c$, $\int \frac{-1}{\sqrt{a^2 - x^2}}\,dx = \cos^{-1}\left(\frac{x}{a}\right) + c$ and $\int \frac{1}{a^2 + x^2}\,dx = \frac{1}{a}\tan^{-1}\left(\frac{x}{a}\right) + c$ are used throughout this topic.

WORKED EXAMPLE 7 Solving Type 1 differential equations (3)

Solve the differential equation $\sqrt{16 - x^2}\,\frac{dy}{dx} + 2 = 0$, $y(0) = 0$, stating the largest domain for which the solution is valid.

THINK	WRITE
1. Rewrite the equation to make $\frac{dy}{dx}$ the subject.	$\sqrt{16 - x^2}\,\frac{dy}{dx} + 2 = 0,\ y(0) = 0$ $\sqrt{16 - x^2}\,\frac{dy}{dx} = -2$ $\frac{dy}{dx} = \frac{-2}{\sqrt{16 - x^2}}$
2. Anti-differentiate to obtain y.	$y = \int \frac{-2}{\sqrt{16 - x^2}}\,dx$
3. Perform the integration process using $\int \frac{-1}{\sqrt{a^2 - x^2}}\,dx = \cos^{-1}\left(\frac{x}{a}\right) + c$.	$y = 2\cos^{-1}\left(\frac{x}{4}\right) + c$
4. Substitute and use the given conditions to determine the value of the constant.	$y(0) = 0$ $\Rightarrow x = 0$ when $y = 0$ $0 = 2\cos^{-1}(0) + c$ $c = -2\cos^{-1}(0)$ $c = -2 \times \frac{\pi}{2}$ $c = -\pi$
5. Substitute back for c and state the particular solution.	$y = 2\cos^{-1}\left(\frac{x}{4}\right) - \pi$

6. Determine the domain for which the solution is valid from the differential equation.	Since the domain of inverse cosine is $[-1,1]$: $1 \leq \frac{x}{4} \leq 1$ $-4 \leq x \leq 4$ $\lvert x \rvert < 4$
7. State the answer.	The solution $y = 2\cos^{-1}\left(\frac{x}{4}\right) - \pi$ is valid for $\lvert x \rvert \leq 4$.

8.3.3 Using numerical integration in differential equations

For many first order particular differential equations, the integral obtained cannot be evaluated using integration techniques. In these situations we must determine an expression for the definite integral and use calculators which can give a numerical approximation.

Consider the differential equation or so called initial value problem, $\frac{dy}{dx} = f(x)$ and $y(x_0) = y_0$. We want to determine the value of y, say $y = y_1$, when $x = x_1$. Integrating, $y = \int_0^x f(t)\, dt + c$, where we have arbitrarily used zero as the lower terminal and t as a dummy variable (note that x is maintained as the independent variable of the solution function). Using the given initial condition, $y = y_0$ when $x = x_0$, the equation becomes $y_0 = \int_0^{x_0} f(t)\, dt + c$.

Rearranging this equation we get an expression for the constant of integration: $c = y_0 - \int_0^{x_0} f(t)\, dt$. Substituting this expression for c back into the initial equation gives:

$$\begin{aligned} y &= \int_0^x f(t)\, dt + y_0 - \int_0^{x_0} f(t)\, dt \\ &= y_0 + \int_0^x f(t)\, dt - \int_0^{x_0} f(t)\, dt \\ &= y_0 + \int_0^x f(t)\, dt + \int_{x_0}^0 f(t)\, dt \\ &= y_0 + \int_{x_0}^x f(t)\, dt \end{aligned}$$

When $x = x_1$, this becomes $y = y_0 + \int_{x_0}^{x_1} f(t)\, dt$, which can then be evaluated by a calculator.

WORKED EXAMPLE 8 Numerical solutions to first order differential equations

Given that $\frac{dy}{dx} = e^{x^2}$, $y(1) = 2$.

a. **Determine a definite integral for y in terms of x.**

b. **Calculate the value of y, correct to 4 decimal places, when $x = 1.5$.**

THINK	WRITE
a. 1. Anti-differentiate the differential equation.	a. $\frac{dy}{dx} = e^{x^2}$ $y = \int_0^x e^{t^2}\,dt + c$
2. Use the given initial conditions to determine the value of the constant of integration.	Substitute $x = 1$ when $y = 2$. $2 = \int_0^1 e^{t^2}\,dt + c$ $c = 2 - \int_0^1 e^{t^2}\,dt$
3. Substitute back for the constant, and simplify using the properties of definite integrals.	$y = \int_0^x e^{t^2}\,dt + 2 - \int_0^1 e^{t^2}\,dt$ $y = 2 + \int_0^x e^{t^2}\,dt + \int_1^0 e^{t^2}\,dt$
4. State the solution for y as a definite integral involving x.	$y = 2 + \int_1^x e^{t^2}\,dt$
b. 1. Calculate the value of y at the required x-value.	b. Substitute $x = 1.5$. $y = 2 + \int_1^{1.5} e^{t^2}\,dt$
2. This definite integral must be evaluated using a calculator. State the final result.	$y = 4.6005$

8.3 Exercise

Students, these questions are even better in jacPLUS

Receive immediate feedback and access sample responses

Access additional questions

Track your results and progress

Find all this and MORE in jacPLUS

Technology free

1. a. WE4 Determine the general solution of $\frac{dy}{dx} + 12x^3 = 0$.

   b. Determine the particular solution of $\frac{dy}{dx} + 6x = 0$, $y(2) = 1$.

2. a. Determine the general solution of $\frac{dy}{dx} + 12\cos(2x) = 0$.

   b. Solve the differential equation $\frac{dy}{dx} + 6\sin(3x) = 0$, $y(0) = 0$, and express y in terms of x.

3. Determine the general solution to each of the following.

   a. $\frac{dy}{dx} - 4x = 3$

   b. $\frac{dy}{dx} - (3x - 5)(x + 4) = 0$

4. WE5 Solve the differential equation $(5 - 4x)^2 \frac{dy}{dx} + 1 = 0$, $y(1) = 2$.

5. Solve the differential equation $(7 - 4x)\frac{dy}{dx} + 2 = 0$, $y(2) = 3$.

6. WE6 Solve the differential equation $\sqrt{2x - 5}\frac{dy}{dx} + 1 = 0$, $y(3) = 0$, stating the largest domain for which the solution is valid.

7. Solve the differential equation $\sqrt{x}\frac{dy}{dx} + 2 = 0$, $y(4) = 3$, expressing y in terms of x, and state the largest domain for which the solution is valid.

8. WE7 Solve the differential equation $\sqrt{64 - x^2}\frac{dy}{dx} - 6 = 0$, $y(4) = 0$, stating the largest domain for which the solution is valid.

9. Solve the differential equation $(16 + x^2)\frac{dy}{dx} + 4 = 0$, $y(4) = \frac{\pi}{4}$, stating the largest domain for which the solution is valid.

For questions 10–18, solve each of the differential equations given and state the maximal domain for which the solution is valid.

10. a. $3x\frac{dy}{dx} - 2x^2 = 5$, $y(1) = 3$

    b. $\frac{dy}{dx} = 6(e^{-3x} + e^{3x})$, $y(0) = 0$

11. a. $\frac{dy}{dx} - 4\sin(2x) = 0$, $y(0) = 2$

    b. $\frac{dy}{dx} + 6\cos(3x) = 0$, $y\left(\frac{\pi}{2}\right) = 5$

12. a. $\frac{dy}{dx} - 8\sin^2(2x) = 0$, $y(0) = 0$

    b. $\frac{dy}{dx} - 12\cos^2(3x) = 0$, $y(0) = 0$

13. a. $\frac{dy}{dx} = \frac{1}{\sqrt{4x + 9}}$, $y(0) = 0$

    b. $\frac{dy}{dx} + \frac{1}{3 - 2x} = 0$, $y(2) = 1$

14. a. $\frac{dy}{dx}=\frac{1}{(3x-5)^2}, y(2)=3$

b. $\frac{dy}{dx}=\frac{8}{7-4x}, y(2)=5$

15. a. $(x^2+9)\frac{dy}{dx}-3x=0, y(0)=0$

b. $\sqrt{x^2+4}\,\frac{dy}{dx}+x=0, y(0)=0$

16. a. $(x^2+6x+13)\,\frac{dy}{dx}-x=3, y(0)=0$

b. $(x^2-4x+9)\,\frac{dy}{dx}+x=2, y(0)=0$

17. $\sec(2x)\frac{dy}{dx}+\sin^3(2x)=0, y(0)=0$

18. a. $\frac{dy}{dx}+\ln(2x)=4, y\left(\frac{1}{2}\right)=1$

b. $e^x\frac{dy}{dx}+x=5, y(0)=0$

Technology active

19. **WE8** Given that $\frac{dy}{dx}=e^{\frac{1}{x}}, y(1)=3$.

a. Determine a definite integral for y in terms of x.

b. Calculate the value of y, correct to 3 decimal places, when $x=2$.

20. Given that $\frac{dy}{dx}=\sin\left(x^2\right), y(0.1)=1$.

a. Determine a definite integral for y in terms of x.

b. Calculate the value of y, correct to 3 decimal places, when $x=0.5$.

21. Solve the following differential equations and state the maximal domain for which the solution is valid.

a. $\left(4x^2+9\right)\,\frac{dy}{dx}+2x=3, y(0)=0$

b. $\sqrt{9-4x^2}\,\frac{dy}{dx}+2x=3, y(0)=0$

22. a. If $a>0$ and $b\neq 0$, solve the following differential equations, stating the maximal domains for which the solution is valid.

i. $\sqrt{a^2-x^2}\frac{dy}{dx}+b=0, y(0)=0$

ii. $\left(a^2-x^2\right)\,\frac{dy}{dx}+b=0, y(0)=0$

iii. $(a+bx)^2\frac{dy}{dx}+1=0, y(0)=0$

b. Solve the differential equation $e^{2x}\,\frac{dy}{dx}+\cos(3x)=0, y(0)=0$.

8.3 Exam questions

Question 1 (2 marks) TECH-FREE

Determine the general solution to the differential equation $e^{2x}\frac{dy}{dx}+6=2e^{4x}$.

Question 2 (2 marks) TECH-FREE

Determine the general solution to the differential equation $\sqrt{x^2+9}\frac{dy}{dx}-x=0$.

Question 3 (3 marks) TECH-FREE

Solve the following differential equation and state the maximal domain for which the solution is valid.

$$\operatorname{cosec}(3x)\frac{dy}{dx}+9\cos^2(3x)=0,\ y(0)=0$$

More exam questions are available online.

8.4 Solving Type 2 differential equations, $\frac{dy}{dx} = f(y)$

LEARNING INTENTION

At the end of this subtopic you should be able to:

- solve Type 2 differential equations of the form $\frac{dy}{dx} = f(y)$ and state the largest domain over which the solution is valid.

8.4.1 Invert, integrate and transpose

Solving first-order differential equations of the form $\frac{dy}{dx} = f(y)$, $y(x_0) = y_0$ is studied in this section. In this situation it is not possible to integrate directly. The first step in the solution process is to invert both sides.

From $\frac{dx}{dy} = \frac{1}{\left(\frac{dy}{dx}\right)}$, we obtain $\frac{dx}{dy} = \frac{1}{f(y)}$.

Integrate both sides with respect to y to obtain

$$x = \int \frac{1}{f(y)}\, dy + c$$

This gives the general solution. The initial condition can be used to determine the value of the constant c. The resulting equation can be rearranged to express y in terms of x, which gives the particular solution.

Determining general solutions

Determining a general solution means writing the solution in terms of an arbitrary constant.

WORKED EXAMPLE 9 Determining the general solutions to a Type 2 differential equation

Determine the general solution to the differential equation $\frac{dy}{dx} - 4\sqrt{y} = 0$.

THINK	WRITE
1. Rewrite the equation to make $\frac{dy}{dx}$ the subject.	$\frac{dy}{dx} - 4\sqrt{y} = 0$ $\frac{dy}{dx} = 4\sqrt{y}$
2. Invert both sides.	$\frac{dx}{dy} = \frac{1}{4\sqrt{y}}$
3. Anti-differentiate to obtain x in terms of y.	$x = \int \frac{1}{4\sqrt{y}}\,dy$
4. Use the properties of indefinite integrals to transfer the constant factor outside the front of the integral sign.	$x = \frac{1}{4}\int \frac{1}{\sqrt{y}}\,dy$

5. Use index laws to express the integrand as a power. $x = \frac{1}{4}\int y^{-\frac{1}{2}}dy$

6. Perform the integration process using $\int u^n du = \frac{1}{n+1}u^{n+1}$ with $n = -\frac{1}{2}$, so that $n+1 = \frac{1}{2}$, and add in the constant of integration. $x = \frac{1}{4} \times \frac{2}{1}y^{\frac{1}{2}} + c$

7. Simplify. $x = \frac{1}{2}y^{\frac{1}{2}} + c$

$x = \frac{1}{2}\sqrt{y} + c$

8. Transpose to make y the subject. $\frac{1}{2}\sqrt{y} = x - c$

$\sqrt{y} = 2x - 2c$

9. Since c is a constant, $2c$ is also a constant. Let $A = 2c$.

$\sqrt{y} = 2x - A$

10. Square both sides and state the answer in terms of an arbitrary constant A. $y = (2x - A)^2$

Determining particular solutions

Determining particular solutions involves solving the differential equation and expressing y in terms of x, and evaluating the constant of integration. Note that this is not always possible, and so it is valid to express a solution as x in terms of y, particularly when x is in terms of a polynomial in y which is difficult/impossible to rearrange to make y the subject

WORKED EXAMPLE 10 Solving Type 2 differential equations (1)

Solve the differential equation $\frac{dy}{dx} + (4-3y)^2 = 0, y(2) = 1$.

THINK	WRITE
1. Rewrite the equation to make $\frac{dy}{dx}$ the subject.	$\frac{dy}{dx} + (4-3y)^2 = 0$ $\frac{dy}{dx} = -(4-3y)^2$
2. Invert both sides.	$\frac{dx}{dy} = -\frac{1}{(4-3y)^2}$
3. Express x as an integral of y.	$x = \int \frac{-1}{(4-3y)^2}dy$
4. Use index laws to express the integrand as a function to a power.	$x = -\int (4-3y)^{-2}dy$

5. Use a linear substitution. Express dy in terms of du by inverting both sides.

Let $u = 4 - 3y$.

$$\frac{du}{dy} = -3$$

$$\frac{dy}{du} = -\frac{1}{3}$$

$$dy = -\frac{1}{3}\,du$$

6. Substitute for u and dy.

$$x = -\int u^{-2}\frac{-1}{3}\,du$$

7. Use the properties of indefinite integrals to transfer the constant factor outside the front of the integral sign.

$$x = \frac{1}{3}\int u^{-2}\,du$$

8. Perform the integration process using $\int u^n du = \frac{1}{n+1}u^{n+1}$ with $n = -2$, so that $n + 1 = -1$, and add in the constant $+c$.

$$x = -\frac{1}{3}u^{-1} + c$$

$$x = -\frac{1}{3u} + c$$

9. Substitute back for y.

$$x = -\frac{1}{3(4 - 3y)} + c$$

10. Substitute and use the given conditions to determine the value of the constant.

$$y(2) = 1$$

$$\Rightarrow x = 2 \text{ when } y = 1$$

$$2 = -\frac{1}{3} + c$$

$$c = 2 + \frac{1}{3}$$

$$c = \frac{7}{3}$$

11. Substitute back for c.

$$x = -\frac{1}{3(4 - 3y)} + \frac{7}{3}$$

12. To begin making y the subject, transpose the equation.

$$\frac{1}{3(4 - 3y)} = \frac{7}{3} - x$$

13. Form a common denominator on the right-hand side.

$$\frac{1}{3(4 - 3y)} = \frac{7 - 3x}{3}$$

14. Cancel the common factor and invert both sides.

$$4 - 3y = \frac{1}{7 - 3x}$$

15. Rearrange to make y the subject.

$$3y = 4 - \frac{1}{7 - 3x}$$

16. Express the right-hand side of the equation with a common denominator.

$$3y = \frac{4(7 - 3x) - 1}{7 - 3x}$$

17. Expand the brackets in the numerator.

$$3y=\frac{28-12x-1}{4-3x}$$

18. Simplify and take out the common factor.

$$3y=\frac{27-12x}{7-3x}$$

$$3y=\frac{3(9-4x)}{7-3x}$$

19. State the final answer in a simplified form and state the maximal domain.

$$y=\frac{9-4x}{7-3x}\text{ for }x\neq\frac{7}{3}$$

TI | THINK

On a Calculator page, select: Menu: 4 Calculus: D Differential Equation Solver and complete the fields as shown, then click OK.
Click enter to see the solution.

DISPLAY/WRITE

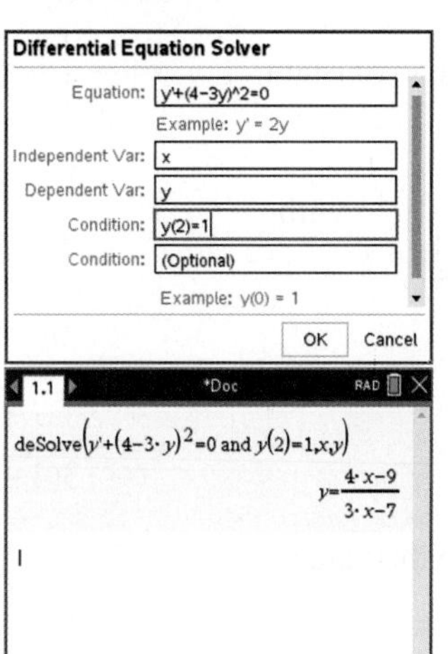

CASIO | THINK

On a Main screen, select Action: Advanced: dSolve and complete the entry as shown.

DISPLAY/WRITE

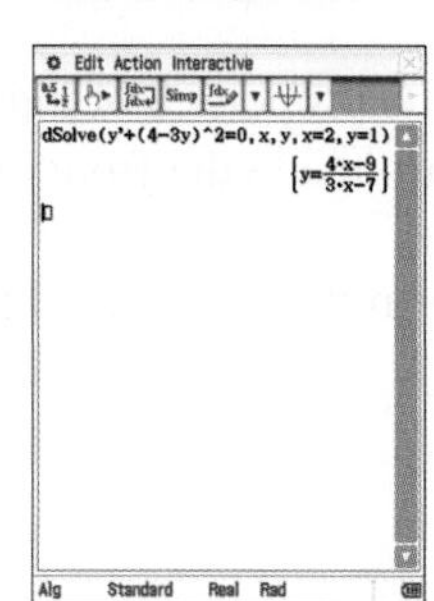

Evaluating c or rearrange to make y the subject?

When solving these types of differential equations, it is necessary to evaluate the constant of integration and also rearrange to make y the subject. Sometimes the order in which we do these operations can make the processes simpler.

WORKED EXAMPLE 11 Solving Type 2 differential equations (2)

Solve the differential equation $\frac{dy}{dx}+4y=0, y(0)=3$.

THINK

1. Rewrite the equation to make $\frac{dy}{dx}$ the subject.

2. Invert both sides.

3. Integrate both sides.

4. Take the constant factor outside the front of the integral sign.

WRITE

$$\frac{dy}{dx}+4y=0$$

$$\frac{dy}{dx}=-4y$$

$$\frac{dx}{dy}=-\frac{1}{4y}$$

$$x=-\int\frac{1}{4y}dy$$

$$x=-\frac{1}{4}\int\frac{1}{y}dy$$

5. Use $\int \frac{1}{u} du = \log_e(|u|) + c$ to express x in terms of y and the constant of integration c.

$$x = -\frac{1}{4}\log_e(|y|) + c$$

From this point forward, we have two processes to complete: evaluating c, and transposing the equation to make y the subject.

Method 1: Evaluate c first, then transpose to make y the subject.

6. Substitute and use the given conditions to determine the value of the constant.

$$y(0) = 3$$
$$\Rightarrow x = 0 \text{ when } y = 3$$
$$0 = -\frac{1}{4}\log_e(|3|) + c$$
$$c = \frac{1}{4}\log_e(3)$$

7. Substitute back for c and take out the common factor.

$$x = -\frac{1}{4}\log_e(|y|) + \frac{1}{4}\log_e(3)$$
$$x = \frac{1}{4}\left[\log_e(3) - \log_e(|y|)\right]$$

8. Use the logarithm laws to simplify the expression.

$$x = \frac{1}{4}\log_e\left(\frac{3}{|y|}\right)$$
$$4x = \log_e\left(\frac{3}{|y|}\right)$$

9. Use the definition of the logarithm.

$$e^{4x} = \frac{3}{|y|}$$

10. Invert both sides again in attempting to make y the subject.

$$\frac{|y|}{3} = \frac{1}{e^{4x}}$$
$$|y| = 3e^{-4x}$$

11. Since the initial condition is that $y(0) = 3$, the modulus is removed, leaving the particular solution as shown.

$$y = 3e^{-4x}$$

Method 2: Make y the subject and then evaluate the constant c.

12. Rearrange to make y the subject.

$$x = -\frac{1}{4}\log_e(|y|) + c$$
$$\frac{1}{4}\log_e(|y|) = c - x$$
$$\log_e(|y|) = 4c - 4x$$

13. Since c is a constant, $4c$ is also a constant.

Let $B = 4c$.
$$\log_e(|y|) = B - 4x$$

14. Use the definition of the logarithm.

$$|y| = e^{B - 4x}$$
$$= e^B e^{-4x}$$

15. Since B is a constant, e^B is also a constant.

Let $A = e^B$.
$$|y| = Ae^{-4x}$$

16. Substitute and use the given condition to determine the value of the constant.

$$y(0) = 3$$
$$\Rightarrow x = 0 \text{ when } y = 3$$
$$3 = Ae^{-0}$$
$$3 = A$$

17. Since the initial condition is that $y(0) = 3$, the modulus is removed, leaving the particular solution as shown.

$$y = 3e^{-4x}$$

Stating the domain for which the solution is valid

As discussed in the previous section, the solution to a differential equation should include the largest domain for which the solution is valid.

WORKED EXAMPLE 12 Solving Type 2 differential equations (3)

Solve the differential equation $2\frac{dy}{dx} + \sqrt{16 - y^2} = 0$, $y(0) = 0$, stating the largest domain for which the solution is valid.

THINK / **WRITE**

1. Rewrite the equation to make $\frac{dy}{dx}$ the subject.

$$2\frac{dy}{dx} + \sqrt{16 - y^2} = 0$$
$$2\frac{dy}{dx} = -\sqrt{16 - y^2}$$
$$\frac{dy}{dx} = \frac{-\sqrt{16 - y^2}}{2}$$

2. Invert both sides.

$$\frac{dx}{dy} = \frac{-2}{\sqrt{16 - y^2}}$$

3. Integrate with respect to y.

$$x = \int \frac{-2}{\sqrt{16 - y^2}} dy$$

4. Perform the integration process using $\int \frac{-1}{\sqrt{a^2 - x^2}} dx = \cos^{-1}\left(\frac{x}{a}\right) + c$.

$$x = 2\cos^{-1}\left(\frac{y}{4}\right) + c$$

5. Substitute and use the given conditions to determine the value of the constant.

$$y(0) = 0$$
$$\Rightarrow x = 0 \text{ when } y = 0$$
$$0 = 2\cos^{-1}(0) + c$$
$$c = -2\cos^{-1}(0)$$
$$c = -2 \times \frac{\pi}{2}$$
$$c = -\pi$$

6. Substitute back for c.

$$x = 2\cos^{-1}\left(\frac{y}{4}\right) - \pi$$

7. Rewrite the equation.

$$2\cos^{-1}\left(\frac{y}{4}\right) = x + \pi$$
$$\cos^{-1}\left(\frac{y}{4}\right) = \frac{x + \pi}{2}$$

8. Take the cosine of both sides to make y the subject.	$\frac{y}{4} = \cos\left(\frac{x}{2} + \frac{\pi}{2}\right)$ $y = 4\cos\left(\frac{x}{2} + \frac{\pi}{2}\right)$
9. Expand using trigonometric compound-angle formulas.	$y = 4\left(\cos\left(\frac{x}{2}\right)\cos\left(\frac{\pi}{2}\right) - \sin\left(\frac{x}{2}\right)\sin\left(\frac{\pi}{2}\right)\right)$ $= 4\left(\cos\left(\frac{x}{2}\right) \times 0 - \sin\left(\frac{x}{2}\right) \times 1\right)$
10. State the particular solution.	$y = -4\sin\left(\frac{x}{2}\right)$
11. Determine the domain for which the solution is valid.	$\cos^{-1}\left(\frac{y}{4}\right) = \frac{x+\pi}{2}$ The range of $y = \cos^{-1}(x)$ is $[0, \pi]$, but $\lvert y \rvert < 4$, so $0 < \frac{x+\pi}{2} < \pi$
12. Solve the inequality for x to state the largest domain for which the solution is valid. State the answer.	$0 < x + \pi < 2\pi$ $-\pi < x < \pi$ The solution $y = -4\sin\left(\frac{x}{2}\right)$ is valid for $-\pi < x < \pi$.

8.4 Exercise

Students, these questions are even better in jacPLUS

Receive immediate feedback and access sample responses

Access additional questions

Track your results and progress

Find all this and MORE in jacPLUS

Technology free

1. **WE9** Determine the general solution to the differential equation $\sqrt{y}\frac{dy}{dx} + 4 = 0$.

2. Determine the general solution to the differential equation $\frac{dy}{dx} - \tan(2y) = 0$.

3. Determine the general solution to each of the following.

   a. $\frac{dy}{dx} = \frac{y^2}{4}$

   b. $\frac{dy}{dx} = y + 4$

4. Determine the general solution to each of the following.

   a. $\frac{dy}{dx} = \frac{y}{4}$

   b. $\frac{dy}{dx} = \frac{4}{y^2}$

5. **WE10** Solve the differential equation $\frac{dy}{dx} + (5 - 4y)^2 = 0$, $y(1) = 2$.

6. Solve the differential equation $\frac{dy}{dx}+4y-7=0$, $y(0)=3$.

7. **WE11** Solve the differential equation $\frac{dy}{dx}+3y=0$, $y(0)=5$.

8. Given the differential equation $\frac{dy}{dx}-5y=0$, $y(0)=3$, express y in terms of x.

9. Solve the following differential equations, expressing y in terms of x.

a. $\frac{dy}{dx}+5y=0$, $y(0)=4$

b. $\frac{dy}{dx}-3y=0$, $y(1)=2$

10. Solve the following differential equations, expressing y in terms of x.

a. $\frac{dy}{dx}+2y=5$, $y(0)=3$

b. $\frac{dy}{dx}-3y+4=0$, $y(0)=2$

11. **WE12** Solve the differential equation $\sqrt{64-y^2}-6\frac{dy}{dx}=0$, $y(0)=8$, stating the largest domain for which the solution is valid.

12. Solve the differential equation $16+y^2-4\frac{dy}{dx}=0$, $y(0)=0$, stating the largest domain for which the solution is valid.

For questions 13–18, solve each of the differential equations given, and where appropriate state the largest domain for which the solution is valid.

13. a. $\frac{dy}{dx}=\sqrt{y}$, $y(1)=4$

b. $\frac{dy}{dx}=y^2$, $y(1)=3$

14. a. $\frac{dy}{dx}=4e^{2y}$, $y(2)=0$

b. $\frac{dy}{dx}+6e^{3y}=0$, $y(1)=0$

15. a. $\frac{dy}{dx}=(5-2y)^2$, $y(1)=3$

b. $\frac{dy}{dx}+(7-3y)^2=0$, $y(3)=2$

16. a. $\frac{dy}{dx}-\sqrt{4y+9}=0$, $y(0)=0$

b. $\frac{dy}{dx}-4y^2=9$, $y(0)=0$

17. a. $\frac{dy}{dx}+4y=y^2$, $y(0)=3$

b. $\frac{dy}{dx}-3y=y^2$, $y(0)=6$

18. a. $\frac{dy}{dx}+7y=y^2+12$, $y(0)=0$

b. $\frac{dy}{dx}-6y-y^2=8$, $y(0)=0$

Technology active

19. If k and y_0 are constants, solve the differential equation $\frac{dy}{dx}+ky=0$, $y(0)=y_0$.

20. Given that a, b and c are non-zero real constants, solve the differential equation $\frac{dy}{dx}+ay=b$, $y(0)=c$.

21. Given that a and b are non-zero real constants, solve the differential equations:

a. $\frac{dy}{dx}=(ay+b)^2$, $y(0)=0$

b. $\frac{dy}{dx}=b^2y^2+a^2$, $y(0)=0$.

22. If a and b are constants with $a > b > 0$:

a. solve the differential equation $\frac{dy}{dx} = (y + a)(y + b)$, $y(0) = 0$

b. determine $\lim\limits_{x \to \infty} y(x)$.

8.4 Exam questions

Question 1 (1 mark) TECH-ACTIVE

Source: *VCE 2015, Specialist Mathematics Exam 2, Section 1, Q12; © VCAA.*

MC Given $\frac{dy}{dx} = 1 - \frac{y}{3}$ and $y = 4$ when $x = 2$, then

A. $y = e^{\frac{-(x-2)}{3}} - 3$

B. $y = e^{\frac{-(x-2)}{3}} + 3$

C. $y = 4e^{\frac{-(x-2)}{3}}$

D. $y = e^{\frac{4(y-x-2)}{3}}$

E. $y = e^{\frac{(x-2)}{3}} + 3$

Question 2 (4 marks) TECH-FREE

Solve the following differential equation and state the maximal domain for which the solution is valid.

$$\frac{dy}{dx} + 6\text{cosec}\left(\frac{y}{2}\right) = 0,\ y\left(\frac{1}{3}\right) = 0$$

Question 3 (5 marks) TECH-FREE

Given that a, b and c are non-zero real constants, solve the differential equation $\frac{dy}{dx} + ay = by^2$, $y(0) = c$.

More exam questions are available online.

8.5 Solving Type 3 differential equations, $\frac{dy}{dx} = f(x)g(y)$

LEARNING INTENTION

At the end of this subtopic you should be able to:

- solve Type 3 differential equations of the form $\frac{dy}{dx} = f(x)g(y)$ and state the largest domain over which the solution is valid.

8.5.1 Separation of variables

Differential equations of the form $\frac{dy}{dx} = f(x)g(y)$, $y(x_0) = y_0$ are called separable equations, as it is possible to separate all the x terms onto one side of the equation and all the y terms onto the other side of the equation.

For $\frac{dy}{dx} = f(x)g(y)$, divide both sides by $g(y)$, since $g(y) \neq 0$. This gives

$$\frac{1}{g(y)}\frac{dy}{dx} = f(x)$$

Integrate both sides of the equation with respect to x.

$$\int \frac{1}{g(y)} \frac{dy}{dx} dx = \int f(x)\, dx$$

$$\text{Thus, } \int \frac{1}{g(y)}\, dy + c_1 = \int f(x)\, dx + c_2.$$

$$\int \frac{1}{g(y)} dy = \int f(x)\, dx + c, \text{ since } c = c_2 - c_1.$$

After performing the integration, an implicit relationship between x and y is obtained. However, in specific cases it may be possible to rearrange to make y the subject.

WORKED EXAMPLE 13 Solving Type 3 differential equations (1)

Determine the general solution to the differential equation $\frac{dy}{dx} = \frac{x+4}{y^2+4}$.

THINK	WRITE
1. Write the differential equation.	$\frac{dy}{dx} = \frac{x+4}{y^2+4}$
2. Separate the variables and integrate both sides.	$\int \left(y^2+4\right) dy = \int (x+4)\, dx$
3. Perform the integration and add the constant on one side only.	$\frac{1}{3}y^3 + 4y = \frac{1}{2}x^2 + 4x + c$
4. The general solution is given as an implicit equation, as in this case it is impossible to solve this equation explicitly for y.	$\frac{1}{3}y^3 + 4y - \frac{1}{2}x^2 - 4x = c$

Determining particular solutions

Determining particular solutions involves solving the differential equation, expressing y in terms of x where possible, and evaluating the constant of integration.

WORKED EXAMPLE 14 Solving Type 3 differential equations (2)

Solve the differential equation $\frac{dy}{dx} + y = 6x^2y,\ y(0) = 1$.

THINK	WRITE
1. Rewrite the equation to make $\frac{dy}{dx}$ the subject.	$\frac{dy}{dx} + y = 6x^2y$
	$\frac{dy}{dx} = 6x^2y - y$
2. Factor the RHS.	$\frac{dy}{dx} = y\left(6x^2 - 1\right)$
3. Separate the variables and integrate both sides.	$\int \frac{1}{y} dy = \int \left(6x^2 - 1\right) dx$

4. Perform the integration and add in the constant on one side only.	$\log_e(\lvert y\rvert) = 2x^3 - x + c$
5. Substitute and use the given conditions to determine the value of the constant.	$y(0) = 1 \Rightarrow x = 0$ when $y = 1$ $\log_e(\lvert 1\rvert) = 0 + c$ $c = 0$
6. Substitute back for c and use the definition of a logarithm to state the solution explicitly as y in terms of x. Note that the initial condition $y(0) = 1$ controls the removal of the modulus. Initially you'd get $y = e^{(2x^3 - x)}$ and then given that $y(0) = 1 > 0$ the result is the positive equation.	$\log_e(\lvert y\rvert) = 2x^3 - x$ $y = e^{2x^3 - x}$

TI | THINK	DISPLAY/WRITE	CASIO | THINK	DISPLAY/WRITE
On a Calculator page, select: Menu: 4 Calculus: D Differential Equation Solver and complete the fields as shown, then click OK. Click enter to see the solution.	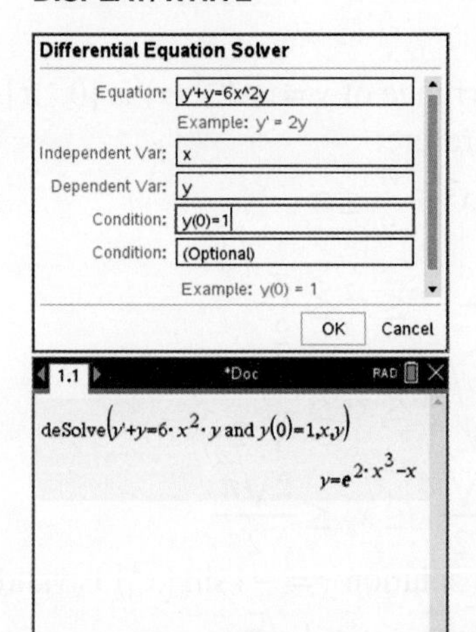	On a Main screen, select Action: Advanced: dSolve and complete the entry as shown.	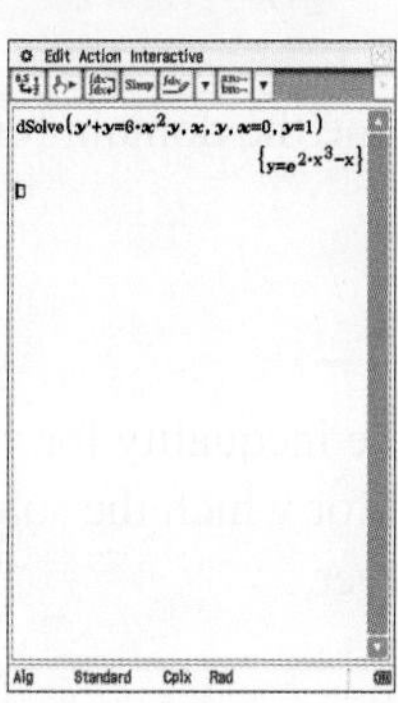

Stating the domain for which the solution is valid

As previously stated, when solving differential equations it is necessary to state the largest domain for which the solution is valid.

WORKED EXAMPLE 15 Solving Type 3 differential equations (3)

Solve the differential equation $\frac{dy}{dx} + 2x\sqrt{16 - y^2} = 0$, $y(0) = 0$, stating the largest domain for which the solution is valid.

THINK	WRITE
1. Rewrite the equation to make $\frac{dy}{dx}$ the subject.	$\frac{dy}{dx} + 2x\sqrt{16 - y^2} = 0,\ y(0) = 0$ $\frac{dy}{dx} = -2x\sqrt{16 - y^2}$
2. Separate the variables and integrate both sides.	$\int \frac{-1}{\sqrt{16 - y^2}}\,dy = \int 2x\,dx$
3. Perform the integration and add the constant on one side only.	$\cos^{-1}\left(\frac{y}{4}\right) = x^2 + c$

4. Substitute and use the given conditions to determine the value of the constant.

$y(0) = 0$
$\Rightarrow x = 0$ when $y = 0$
$\cos^{-1}(0) = c$
$c = \frac{\pi}{2}$

5. Substitute back for c.

$\cos^{-1}\left(\frac{y}{4}\right) = x^2 + \frac{\pi}{2}$

6. Take the cosine of both sides to make y the subject and use the identity $\cos\left(\theta + \frac{\pi}{2}\right) = -\sin(\theta)$ to rewrite it in terms of sine.

$\frac{y}{4} = \cos\left(x^2 + \frac{\pi}{2}\right)$
$y = -4\sin\left(x^2\right)$

7. Determine the domain for which the solution is valid.

The range of $y = \cos^{-1}(x)$ is $[0, \pi]$.
Therefore:
$0 \le x^2 + \frac{\pi}{2} \le \pi$

8. Solve the inequality for x to state the largest domain for which the solution is valid. State the answer.

$-\frac{\pi}{2} \le x^2 \le \frac{\pi}{2}$
$-\sqrt{\frac{\pi}{2}} \le x \le \sqrt{\frac{\pi}{2}}$
$-\frac{2\sqrt{\pi}}{2} \le x \le \frac{2\sqrt{\pi}}{2}$
The solution $y = -4\sin\left(x^2\right)$ is valid for
$-\frac{2\sqrt{\pi}}{2} \le x \le \frac{2\sqrt{\pi}}{2}$.

8.5 Exercise

Technology free

1. WE13 Determine the general solution to the differential equation $\frac{dy}{dx} = \frac{x+2}{y^3+8}$.

2. Obtain an implicit relationship of the form $f(x, y) = c$ for $\frac{dy}{dx} = \frac{y^2+4}{x^2y^2}$.

3. Obtain an implicit relationship of the form $f(x, y) = c$ for each of the following differential equations.

a. $\dfrac{dy}{dx} = \dfrac{x^2+4}{y^2+4}$

b. $\dfrac{dy}{dx} = \dfrac{xy}{y^2+4}$

4. Obtain an implicit relationship of the form $f(x, y) = c$ for each of the following differential equations.

a. $\dfrac{dy}{dx} = \dfrac{x^2y^2}{y^2+4}$

b. $\dfrac{dy}{dx} = \dfrac{xy^2e^{x^2}}{y^3+8}$

5. **WE14** Solve the differential equation $\dfrac{dy}{dx} - y = 3x^2y$, $y(0) = 1$.

6. Given the differential equation $\dfrac{dy}{dx} + y^2 = 2xy^2$, $y(2) = 1$, express y in terms of x.

7. **WE15** Solve the differential equation $\dfrac{dy}{dx} - 2x\sqrt{64-y^2} = 0$, $y(0) = 0$, stating the largest domain for which the solution is valid.

8. Solve the differential equation $2\dfrac{dy}{dx} - x(16+y^2)$, $y(0) = 0$, stating the largest domain for which the solution is valid.

For questions 9–17, solve each of the given differential equations and express y in terms of x.

9. a. $\dfrac{dy}{dx} - \dfrac{y^2}{x} = 0$, $y(1) = 1$

b. $\dfrac{dy}{dx} + 12y^2\sin(4x) = 0$, $y(\pi) = 1$

10. a. $\dfrac{dy}{dx} + \dfrac{x}{y} = 0$, $y(1) = 2$

b. $\dfrac{dy}{dx} + 6y^2x^2 = 0$, $y(1) = 3$

11. a. $\dfrac{dy}{dx} + 18x^3y^2 = 0$, $y(-1) = 2$

b. $\dfrac{dy}{dx} - \dfrac{y^2}{x^2} = 0$, $y(1) = 2$

12. a. $\dfrac{dy}{dx} = y^2e^{2x}$, $y(0) = 1$

b. $\dfrac{dy}{dx} + 12x^5y^2 = 0$, $y(1) = 2$

13. a. $\dfrac{dy}{dx} + y = 3x^2y$, $y(0) = 1$

b. $\dfrac{dy}{dx} + 6x^2y^2 = y^2$, $y(-1) = 2$

14. a. $\dfrac{dy}{dx} + 2xy^2 = y^2$, $y(1) = 2$

b. $\dfrac{dy}{dx} + 8x^3y^4 = y^4$, $y(0) = 1$

15. a. $x\dfrac{dy}{dx} + 2y = y^2$, $y(1) = 1$

b. $x\dfrac{dy}{dx} - 4y = y^2$, $y(1) = 1$

16. a. $(4+x^2)\dfrac{dy}{dx} - 2xy = 0$, $y(0) = 1$

b. $\dfrac{y^2+4}{x^2+9} - \dfrac{y}{x}\dfrac{dy}{dx} = 0$, $y(0) = 2$

17. a. $\dfrac{dy}{dx} - x\left(25+y^2\right) = 0$, $y(0) = 0$

b. $\dfrac{dy}{dx} + 4x\sqrt{25-y^2} = 0$, $y(0) = 5$

Technology active

18. Use the substitution $v=\frac{y}{x}$ to show that $\frac{dy}{dx}=v+x\frac{dv}{dx}$, and hence reduce to a separable differential equation and determine the solution to the following.

$x\frac{dy}{dx}+3y=4x,\ y(2)=1$

19. Use the substitution $v=\frac{y}{x}$ to show that $\frac{dy}{dx}=v+x\frac{dv}{dx}$, and hence reduce to a separable differential equation and determine the solution to the following.

$x\frac{dy}{dx}-y=4x,\ y(1)=2$

20. Use the substitution $v=\frac{y}{x}$ to show that $\frac{dy}{dx}=v+x\frac{dv}{dx}$. Hence, reduce the differential equation $x\frac{dy}{dx}+ay=bx$ to a separable differential equation and determine its general solution for the cases when:

a. $a=-1$

b. $a\neq-1$.

8.5 Exam questions

Question 1 (4 marks) TECH-FREE

Source: *VCE 2019 Specialist Mathematics Exam 1, Q1; © VCAA.*

Solve the differential equation $\frac{dy}{dx}=\frac{2ye^{2x}}{1+e^{2x}}$ given that $y(0)=\pi$.

Question 2 (1 mark) TECH-ACTIVE

Source: *VCE 2018 Specialist Mathematics Exam 2, Section A, Q9; © VCAA.*

MC A solution to the differential equation $\frac{dy}{dx}=\frac{2}{\sin(x+y)-\sin(x-y)}$ can be obtained from

A. $\int 1\,dx=\int 2\sin(y)dy$

B. $\int \cos(y)\,dy=\int \text{cosec}(x)dx$

C. $\int \cos(x)\,dx=\int \text{cosec}(y)dy$

D. $\int \sec(x)\,dx=\int \sin(y)dy$

E. $\int \sec(x)\,dx=\int \text{cosec}(y)dy$

Question 3 (2 marks) TECH-FREE

Source: *VCE 2017 Specialist Mathematics Exam 1, 8b; © VCAA.*

Solve the differential equation $\frac{dy}{dx}=\frac{-x}{1+y^2}$ with the condition $y(-1)=1$. Express your answer in the form $ay^3+by+cx^2+d=0$, where a, b, c and d are integers.

More exam questions are available online.

8.6 Solving Type 4 differential equations, $\frac{d^2y}{dx^2} = f(x)$

LEARNING INTENTION

At the end of this subtopic you should be able to:

- solve Type 4 differential equations of the form $\frac{d^2y}{dx^2} = f(x)$, giving the particular equation when the required information is specified in the question
- apply Type 4 differential equations to calculate the deflection of a beam under weight which is supported either at both ends or at one end only.

8.6.1 Integrate twice

In this section, solutions of second-order differential equations of the form $\frac{d^2y}{dx^2} = f(x)$ are required. This type of differential equation can be solved by direct integration, since $\frac{d^2y}{dx^2} = \frac{d}{dx}\left(\frac{dy}{dx}\right)$. Integrating both sides with respect to x gives $\frac{dy}{dx} = \int f(x)\,dx + c_1$. This is now in the Type 1 form and can be solved by direct integration.

Determining a general solution involves giving the solution in terms of two arbitrary constants, which we usually denote as c_1 and c_2.

WORKED EXAMPLE 16 Solving Type 4 differential equations (1)

Determine the general solution to the differential equation $\frac{d^2y}{dx^2} + 36x^2 = 0$.

THINK	WRITE
1. Rewrite the equation to make $\frac{d^2y}{dx^2}$ the subject.	$\frac{d^2y}{dx^2} + 36x^2 = 0$ $\frac{d^2y}{dx^2} = -36x^2$
2. Integrate both sides with respect to x.	$\frac{dy}{dx} = \int -36x^2 dx$
3. Perform the integration.	$\frac{dy}{dx} = -12x^3 + c_1$
4. Integrate both sides again with respect to x.	$y = \int \left(-12x^3 + c_1\right) dx$
5. Perform the integration and state the general solution in terms of two arbitrary constants.	$y = -3x^4 + c_1x + c_2$

Determining particular solutions

To solve $\frac{d^2y}{dx^2} = f(x)$ and obtain a particular solution, we need two sets of initial conditions to evaluate the two constants of integration. These are usually of the form $y(x_0) = y_0$ and $y'(x_1) = y_1$.

WORKED EXAMPLE 17 Solving Type 4 differential equations (2)

Solve the differential equation $\frac{d^2y}{dx^2} + 36x = 0, y(1) = 3, y'(1) = 2.$

THINK	WRITE
1. Rewrite the equation to make $\frac{d^2y}{dx^2}$ the subject.	$\frac{d^2y}{dx^2} + 36x = 0$ $\frac{d^2y}{dx^2} = -36x$
2. Integrate both sides with respect to x.	$\frac{dy}{dx} = \int -36x\,dx$ $= -18x^2 + c_1$
3. Substitute and use the given condition to determine the value of the first constant of integration.	$y'(1) = 2$ $\Rightarrow \frac{dy}{dx} = 2$ when $x = 1$ $2 = -18 + c_1$ $c_1 = 20$
4. Substitute back for c_1.	$\frac{dy}{dx} = -18x^2 + 20$
5. Integrate both sides again with respect to x.	$y = \int (-18x^2 + 20)\,dx$
6. Perform the integration.	$y = -6x^3 + 20x + c_2$
7. Substitute and use the given condition to determine the value of the second constant of integration.	$y(1) = 3$ $\Rightarrow y = 3$ when $x = 1$ $3 = -6 + 20 + c_2$ $c_2 = -11$
8. Substitute back for c_2 and state the particular solution.	$y = -6x^3 + 20x - 11$

TI | THINK	DISPLAY/WRITE
On a Calculator page, select: Menu: 4 Calculus: D Differential Equation Solver and complete the fields as shown, then click OK. Click enter to see the solution.	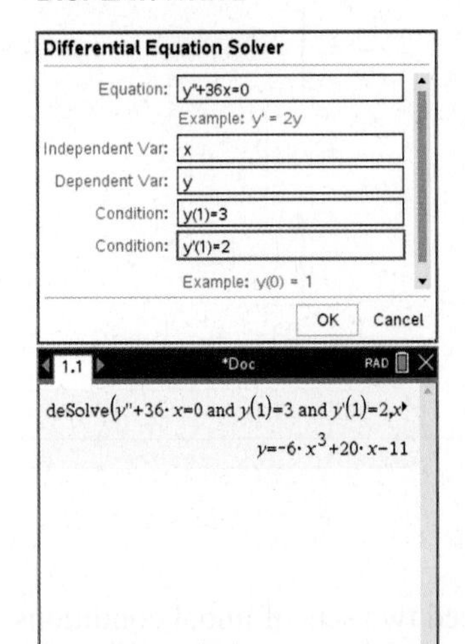

CASIO | THINK	DISPLAY/WRITE
On a Main screen, select Action: Advanced: dSolve and complete the entry as shown.	

Simplifying the answer

We have seen earlier that answers can often be given in a simplified form.

WORKED EXAMPLE 18 Solving Type 4 differential equations (3)

Solve the differential equation $\dfrac{d^2y}{dx^2}+\dfrac{2}{(2x+9)^3}=0, y(0)=0, y'(0)=0.$

THINK	WRITE
1. Rewrite the equation to make $\frac{d^2y}{dx^2}$ the subject.	$\frac{d^2y}{dx^2}+\frac{2}{(2x+9)^3}=0$ $\frac{d^2y}{dx^2}=\frac{-2}{(2x+9)^3}$
2. Integrate both sides with respect to x.	$\frac{dy}{dx}=\int\frac{-2}{(2x+9)^3}dx$
3. Transfer the constant factor outside the front of the integral and use index laws to express the integrand as a function to a power.	$\frac{dy}{dx}=-2\int(2x+9)^{-3}\,dx$
4. Use a linear substitution. Express dx in terms of du by inverting both sides.	Let $u=2x+9$. $\frac{du}{dx}=2$ $\frac{dx}{du}=\frac{1}{2}$ $dx=\frac{1}{2}du$
5. Substitute for u and dx, and simplify.	$\frac{dy}{dx}=-2\int u^{-3}\frac{1}{2}\,du$ $=-\int u^{-3}du$
6. Perform the integration, adding in the first constant of integration and substitute back for x.	$=\frac{1}{2}u^{-2}+c_1$ $=\frac{1}{2(2x+9)^2}+c_1$
7. Use the given condition to determine the value of the first constant of integration.	$y'(0)=0$ $\Rightarrow$ when $x=0, \frac{dy}{dx}=0$ $0=\frac{1}{162}+c_1$ $c_1=-\frac{1}{162}$

8. Substitute back for c_1.

$$\frac{dy}{dx}=\frac{1}{2(2x+9)^2}-\frac{1}{162}$$

9. Integrate both sides again with respect to x.

$$y=\int\left(\frac{1}{2(2x+9)^2}-\frac{1}{162}\right)dx$$

10. Simplify the integrand.

$$=\int\left(\frac{1}{2(2x+9)^2}\right)dx-\frac{x}{162}$$

11. Use the substitution $u=2x+9$ again.

$$=\int\left(\frac{1}{2}u^{-2}\right)\frac{1}{2}\,du-\frac{x}{162}$$

$$=\frac{1}{4}\int u^{-2}du-\frac{x}{162}$$

12. Perform the integration and add in the second constant of integration.

$$=-\frac{1}{4}u^{-1}-\frac{x}{162}+c_2$$

13. Substitute back for x.

$$=-\frac{1}{4(2x+9)}-\frac{x}{162}+c_2$$

14. Substitute and use the given condition to determine the value of the second constant of integration.

$$y(0)=0$$

$$\Rightarrow y=0 \text{ when } x=0$$

$$0=-\frac{1}{36}+c_2$$

$$c_2=\frac{1}{36}$$

15. Substitute back for c_2 and state the particular solution. Although this is a possible answer, this result can be simplified.

$$y=-\frac{1}{4(2x+9)}-\frac{x}{162}+\frac{1}{36}$$

16. Form the lowest common denominator.

$$=\frac{-81-2x(2x+9)+9(2x+9)}{324\,(2x+9)}$$

17. Expand and simplify the numerator.

$$=\frac{-81-4x^2-18x+18x+81}{324(2x+9)}$$

$$=\frac{-4x^2}{324(2x+9)}$$

18. State the particular solution in simplest form.

$$y=\frac{-x^2}{81(2x+9)}$$

8.6.2 Applications of Type 4 differential equations — beam deflections

One application of Type 4 differential equations, $\frac{d^2y}{dx^2}=f(x)$, is beam deflections. A cantilever or a beam can be fixed at one end and have a weight at the other end. The weight at the unfixed end causes the beam to bend so that the downwards deflection, y, at a distance, x, measured along the beam from the fixed point satisfies a differential equation of this type. In this situation the maximum deflection occurs at the end of the beam.

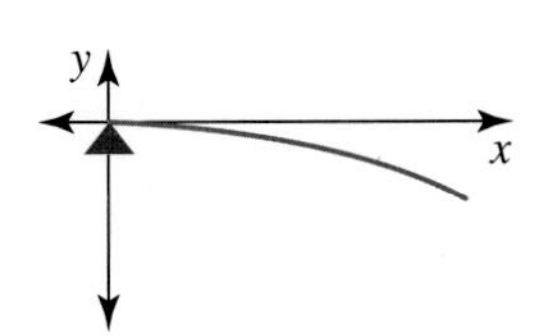

Another type of beam deflection is the case of a beam fixed at both ends. The weight of the beam causes the beam to bend so that the downwards deflection, y, at a distance, x, measured along the beam from the fixed point satisfies a differential equation of this type. In this situation we can show that the maximum deflection occurs in the middle of the beam.

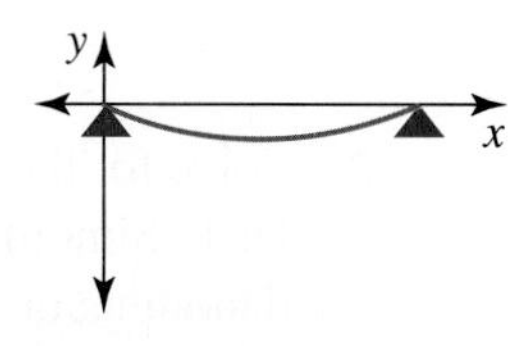

WORKED EXAMPLE 19 Application of Type 4 differential equations — beam deflections

A beam of length $2L$ rests with its end on two supports at the same horizontal level. The downward deflection, y, from the horizontal satisfies the differential equation $\frac{d^2y}{dx^2} = kx(x - 2L)$ for $0 \le x \le 2L$, where x is the horizontal distance from one end of the beam and k is a constant related to the stiffness and bending moment of the beam.

a. Determine the deflection, y, in terms of x and show that the maximum deflection occurs in the middle of the beam.

b. Calculate the maximum deflection of the beam.

THINK	WRITE
a. 1. Expand.	a. $\frac{d^2y}{dx^2} = kx(x-2L)$ $= k\left(x^2 - 2Lx\right)$
2. Integrate both sides with respect to x.	$\frac{dy}{dx} = k\int \left(x^2 - 2Lx\right)\, dx$
3. Perform the integration.	$\frac{dy}{dx} = k\left(\frac{x^3}{3} - Lx^2 + c_1\right)$
4. Since the beam is fixed at both ends, $x = 0$ when $y = 0$ and $y = 0$ when $x = 2L$. We cannot determine the first constant of integration at this stage. Integrate both sides with respect to x again.	$y = k\int \left(\frac{x^3}{3} - Lx^2 + c_1\right) dx$
5. Perform the integration.	$y = k\left(\frac{x^4}{12} - \frac{Lx^3}{3} + c_1x + c_2\right)$
6. To evaluate the second constant of integration, c_2, use $x = 0$ when $y = 0$.	Substitute $x = 0$ when $y = 0$: $c_2 = 0$
7. To evaluate the first constant of integration, c_1, use $y = 0$ when $x = 2L$ and simplify.	Substitute $y = 0$ when $x = 2L$: $0 = k\left(\frac{(2L)^4}{12} - \frac{L(2L)^3}{3} + 2Lc_1\right)$ $0 = k\left(\frac{16L^4}{12} - \frac{8L^4}{3} + 2Lc_1\right)$

8. Solve for the first constant and substitute back. Simplify the result by taking a common denominator. This gives the deflection, y, in terms of x.

$$c_1 = \frac{2L^3}{3}$$
$$y = k\left(\frac{x^4}{12} - \frac{Lx^3}{3} + \frac{2L^3x}{3}\right)$$
$$= \frac{k}{12}\left(x^4 - 4Lx^3 + 8L^2x\right)$$

9. Determine the first derivative.

$$\frac{dy}{dx} = \frac{k}{12}(4x^3 - 12Lx^2 + 8L^2)$$
$$= \frac{k}{3}(x^3 - 3Lx^2 + 2L^3)$$

10. To show that the maximum deflection occurs in the middle of the beam, show that $\frac{dy}{dx} = 0$ when $x = L$.

Substitute $x = L$:
$$\frac{dy}{dx} = \frac{k}{3}\left(L^3 - 3L^3 + 2L^3\right)$$
$$= 0$$
So the maximum deflection occurs in the middle of the beam.

b. 1. To calculate the maximum deflection, substitute $x = L$ into the result for y.

b. $$y_{\max} = y(L)$$
$$= \frac{k}{12}\left(L^3 - 4L^3 + 8L^3\right)$$
$$= \frac{5L^3k}{12}$$

2. State the maximum deflection of the beam.

The maximum deflection occurs in the middle of the beam and is $\frac{5L^3k}{12}$.

8.6 Exercise

Technology free

1. WE16 Determine the general solution to the differential equation $\frac{d^2y}{dx^2} + 30x^4 = 0$.

2. Determine the general solution to the differential equation $\frac{d^2y}{dx^2} + 36\sin(3x) = 0$.

3. Determine the general solution to each of the following.

a. $x^3\frac{d^2y}{dx^2} + 4 = 0$

b. $\frac{d^2y}{dx^2} + (x+4)(2x-5) = 0$

4. Determine the general solution to each of the following.

a. $x^3 \frac{d^2y}{dx^2} + 2x - 5 = 0$

b. $e^{3x} \frac{d^2y}{dx^2} + 5 = 2e^{2x}$

5. WE17 Solve the differential equation $\frac{d^2y}{dx^2} + 24x^2 = 0,\ y(-1) = 2,\ y'(-1) = 3.$

6. Solve the differential equation $\frac{d^2y}{dx^2} + 12\sin(2x) = 0,\ y\left(\frac{\pi}{4}\right) = 4,\ y'\left(\frac{\pi}{4}\right) = 6.$

7. WE18 Solve the differential equation $\frac{d^2y}{dx^2} + \frac{12}{(3x+16)^3} = 0,\ y(0) = 0,\ y'(0) = 0.$

8. Solve the differential equation $\frac{d^2y}{dx^2} + \frac{12}{\sqrt{(2x+9)^3}} = 0,\ y(0) = 0,\ y'(0) = 1.$

For questions 9–13, solve each of the given differential equations.

9. $\frac{d^2y}{dx^2} + 6x = 0,\ y(1) = 2,\ y(2) = 3$

10. $\frac{d^2y}{dx^2} + 8\left(e^{2x} + e^{-2x}\right) = 0,\ x = 0,\ \frac{dy}{dx} = 0,\ y = 0$

11. a. $\frac{d^2y}{dx^2} + 64\sin(4x) = 0,\ y(0) = 4,\ y'(0) = 8$

b. $\frac{d^2y}{dx^2} + 27\cos(3x) = 0,\ y\left(\frac{\pi}{6}\right) = 3,\ y'\left(\frac{\pi}{6}\right) = 9$

12. a. $\frac{d^2y}{dx^2} + 32\sin^2(2x) = 0,\ y(0) = 0,\ y'(0) = 0$

b. $\frac{d^2y}{dx^2} + 16\cos^2(4x) = 0,\ y(0) = 0,\ y'(0) = 0$

13. a. $\frac{d^2y}{dx^2} = \frac{1}{(3x+2)^3},\ y(0) = 0,\ y'(0) = 0$

b. $\frac{d^2y}{dx^2} + \frac{1}{\sqrt{(2x+9)^3}} = 0,\ y(0) = 0,\ y'(0) = 0$

14. a. At all points on a certain curve, the rate of change of gradient is constant. Show that the family of curves with this property are parabolas.

b. At all points on a certain curve, the rate of change of the gradient is -12. If the curve has a turning point at $(-2, 4)$, determine the equation of the particular curve.

15. a. At all points on a certain curve, the rate of change of the gradient is proportional to the x-coordinate. Show that this family of curves are cubics.

b. At all points (x, y) on a certain curve, the rate of change of the gradient is $18x$. If the curve has a turning point at $(-2, 0)$, determine the equation of the particular curve.

16. a. Solve $\frac{d^2y}{dx^2} + \frac{20}{\sqrt{4x+9}} = 0,\ y(0) = 0$ and $y'(0) = 0.$

b. Solve $\frac{d^2y}{dx^2} + \frac{16}{(4x+9)^2} = 0,\ y(0) = 0$ and $y'(0) = 0.$

17. WE19 A beam of length L has both ends simply supported at the same horizontal level and the downward deflection, y, satisfies the differential equation $\frac{d^2y}{dx^2} = k\left(x^2 - Lx\right)$ for $0 \le x \le L$, where k is a constant.

a. Determine the deflection, y, in terms of x and show that the maximum deflection occurs in the middle of the beam.

b. Calculate the maximum deflection of the beam.

Technology active

18. A cantilever of length L is rigidly fixed at one end and is horizontal in the unstrained position. If a load is added at the free end of the beam, the downward deflection, y, at a distance, x, along the beam satisfies the differential equation $\frac{d^2y}{dx^2} = k(L - x)$ for $0 \le x \le L$, where k is a constant. Determine the deflection, y, in terms of x and hence calculate the maximum deflection of the beam.

19. a. A diving board of length L is rigidly fixed at one end and has a girl of weight W standing at the free end. The downward deflection, y, measured at a distance, x, along the beam satisfies the differential equation

$$EI\frac{d^2y}{dx^2} = \frac{W}{2}(L - x)^2 \text{ for } 0 \le x \le L.$$

The deflection and inclination to the horizontal are both zero at the fixed end, and the product EI is a constant related to the stiffness of the beam. Determine the formula for y in terms of x and calculate the maximum deflection of the beam.

b. A uniform beam of length L carries a load of W per unit length and has both ends clamped horizontally at the same horizontal level. The downward deflection, y, measured at any distance, x, from one end along the beam satisfies the differential equation

$$EI\frac{d^2y}{dx^2} = \frac{W}{2}\left(x^2 - Lx + \frac{L^2}{6}\right) \text{ for } 0 \le x \le L$$

where W, E and I are constants. Prove that the maximum deflection occurs in the middle of the beam, and determine the maximum deflection of the beam.

20. If a and b are positive real constants, determine the particular solution to the following differential equation.

$$\frac{d^2y}{dx^2} + \frac{1}{(ax+b)^3} = 0,\ y(0) = 0 \text{ and } y'(0) = 0$$

21. If a and b are positive real constants, determine the particular solution to the following differential equation.

$$\frac{d^2y}{dx^2} + \frac{1}{(ax+b)^2} = 0,\ y(0) = 0 \text{ and } y'(0) = 0$$

22. a. Show that $\frac{d}{dx}\left[\frac{x}{\sqrt{9+4x^2}}\right] = \frac{9}{\sqrt{(9+4x^2)^3}}$.

b. Hence, find the general solution to $\frac{d^2y}{dx^2} + \frac{9}{\sqrt{(9+4x^2)^3}} = 0$.

8.6 Exam questions

Question 1 (1 mark) TECH-FREE

Solve the differential equation $\frac{d^2y}{dx^2} + 24x^2 = 0,\ y(1) = 2,\ y(2) = 3.$

Question 2 (2 marks) TECH-FREE

Solve the differential equation $e^x \frac{d^2y}{dx^2} + 4e^{-2x} = 5,\ x = 0,\ \frac{dy}{dx} = 0,\ y = 0.$

Question 3 (2 marks) TECH-FREE

a. If a and b are positive real constants, show that $\frac{d}{dx}\left[\frac{x}{\sqrt{a + bx^2}}\right] = \frac{a}{\sqrt{(a + bx^2)^3}}$. **(1 mark)**

b. Hence, find the general solution to $\frac{d^2y}{dx^2} + \frac{1}{\sqrt{(a + bx^2)^3}} = 0.$ **(1 mark)**

More exam questions are available online.

8.7 Review

8.7.1 Summary

doc-37062

Hey students! Now that it's time to revise this topic, go online to:

Review your results

Watch teacher-led videos

Practise exam questions

Find all this and MORE in jacPLUS

8.7 Exercise

Technology free: short answer

1. Complete the following.
   a. Show that $y = \sin^2(3x)$ satisfies the differential equation $\frac{d^2y}{dx^2} + 36y - 18 = 0$.
   b. Verify that $y = \sin^{-1}\left(\frac{x}{3}\right)$ satisfies the differential equation $\left(9 - x^2\right)\frac{d^2y}{dx^2} - x\frac{dy}{dx} = 0$.
2. Complete the following.
   a. Verify that $y = \log_e\left(x + \sqrt{x^2 + 9}\right)$ satisfies the differential equation $\left(x^2 + 9\right)\frac{d^2y}{dx^2} + x\frac{dy}{dx} = 0$.
   b. Determine the value of the constant A if $y = A\,x\sin(3x)$ is a solution of the differential equation $\frac{d^2y}{dx^2} + 9y = 18\cos(3x)$.

For questions 3–6, determine the particular solution to each of the following differential equations, stating the maximal domain, where appropriate.

3. a. $\frac{dy}{dx} - x^2 = 4,\ y(1) = 2$
   b. $x^2\frac{dy}{dx} + 4 = 2x,\ y(1) = 2$
4. a. $\left(x^2 + 9\right)\frac{dy}{dx} + 3 = 0,\ y(0) = 2$
   b. $\sqrt{9 + x^2}\frac{dy}{dx} + 3x = 0,\ y(0) = 2$
   c. $\left(9 - x^2\right)\frac{dy}{dx} + 3 = 0,\ y(0) = 2$
5. a. $\frac{dy}{dx} + 9y^2 = 0,\ y(0) = 1$
   b. $\frac{dy}{dx} + \frac{9}{y} = 0,\ y(0) = 1$
   c. $\frac{dy}{dx} + 9y^3 = 0,\ y(0) = 1$
6. a. $\frac{dy}{dx} + 4y = 7,\ y(0) = 2$
   b. $\frac{dy}{dx} + (5 - 2y)^2 = 0,\ y(1) = 2$
   c. $\frac{dy}{dx} + \sqrt{9 - y^2} = 0,\ y(0) = 3$

Technology active: multiple choice

7. **MC** If $y = x^n$ satisfies $x^2\frac{d^2y}{dx^2} + x\frac{dy}{dx} - y = 0$, then
   A. $n = 0$ only
   B. $n = 1$ only
   C. $n = -1$ only
   D. $n = -1$ or $n = 1$
   E. $n \in R\backslash\{-1, 1\}$
8. **MC** The value(s) of k for which $y = e^{-kx}$ satisfies $\frac{d^2y}{dx^2} - \frac{dy}{dx} - 6y = 0$ are:
   A. $k = -3$ or $k = 2$
   B. $k = -2$ or $k = 3$
   C. $k = -3$ only
   D. $k = 3$ only
   E. $k \in R\backslash\{-3, 2\}$

9. **MC** $y = x\cos(3x)$ satisfies the differential equation:

A. $\frac{d^2y}{dx^2} + 9y = 0$
B. $\frac{d^2y}{dx^2} + 9x = 0$
C. $\frac{d^2y}{dx^2} + 9y + 6\sin(3x) = 0$
D. $\frac{d^2y}{dx^2} + 9y + 6\cos(3x) = 0$
E. $x\frac{d^2y}{dx^2} + 9y = 0$

10. **MC** $y = 2\sin\left(\frac{x}{2}\right)$ satisfies the differential equation

A. $\frac{d^2y}{dx^2} + y = 0$
B. $\frac{d^2y}{dx^2} + 4y = 0$
C. $\frac{d^2y}{dx^2} + 4x = 0$
D. $4\frac{d^2y}{dx^2} + y = 0$
E. $4\frac{d^2y}{dx^2} + x = 0$

11. **MC** The general solution of the differential equation $\frac{dy}{dx} + 36x = 0$ is

A. $y = 18x^2 + c$
B. $y = c - 18x^2$
C. $y = c\sin(6x)$
D. $y = c\cos(6x)$
E. $y = ce^{-6x}$

12. **MC** The particular solution of the differential equation $\frac{dy}{dx} = \frac{1}{(7x-6)^2}$, $y(0) = 0$ is

A. $y = \frac{1}{7(7x-6)}$
B. $y = \frac{-1}{7(7x-6)}$
C. $y = \frac{x}{6(6-7x)}$
D. $y = \frac{1}{7(7x-6)} + \frac{1}{42}$
E. $y = \log_e\left(\frac{(7x-6)^2}{36}\right)$

13. **MC** The general solution of the differential equation $3\frac{dy}{dx} + y = 0$, where A is a constant, is

A. $y = Ae^{-\frac{x}{3}}$
B. $y = Ae^{-3x}$
C. $y = Ae^{\frac{x}{3}}$
D. $y = Ae^{3x}$
E. $y = -\frac{x}{3} + A$

14. **MC** The general solution of the differential equation $\frac{dy}{dx} + 3x^2y^2 = 0$, where A and c are real constants, is

A. $y = Ae^{-x^3}$
B. $y = Ae^{x^3}$
C. $y = \frac{1}{x^3} + c$
D. $y = \frac{1}{c - x^3}$
E. $y = \frac{1}{c + x^3}$

15. **MC** The particular solution of the differential equation $\frac{dy}{dx} = y^2 + 4$, $y(0) = 0$ is given by

A. $y = \sqrt{2e^{2x} - 4}$
B. $y = e^{2x} - 1$
C. $y = \tan^{-1}\left(\frac{x}{2}\right)$
D. $y = 2\tan(x)$
E. $y = 2\tan(2x)$

16. **MC** The general solution of the differential equation $\frac{d^2y}{dx^2} + 36x = 0$, where c_1 and c_2 are constants, is given by

A. $y = 6x^3 + c_1x + c_2$
B. $y = -6x^3 + c_1x + c_2$
C. $y = e^{-6x}(c_1\sin(6x) + c_2\cos(6x))$
D. $y = c_1\sin(6x) + c_2\cos(6x)$
E. $y = c_1e^{-6x} + c_2e^{6x}$

Technology active: extended response

17. Solve the following differential equations, expressing y in terms of x, showing all working.

a. $\frac{dy}{dx} - y^2 = 4$, $y(0) = 0$
b. $\frac{dy}{dx} - y^2 = 4y$, $y(0) = 2$
c. $16\frac{dy}{dx} + e^{4y} = 0$, $y(0) = 0$

18. Solve the following differential equations, expressing y in terms of x, showing all working.

a. $\frac{dy}{dx} - 6x^2y^2 = 0$, $y(1) = 1$
b. $\frac{dy}{dx} = 4x^3e^{-y}$, $y(0) = 0$
c. $\frac{dy}{dx} - y^2 = 4xy^2$, $y(1) = 1$

19. Solve the following differential equations, expressing y in terms of x, showing all working.

a. $x^2\frac{d^2y}{dx^2}-4=2\sqrt{x}, y(1)=3\ ,\ y'(1)=2$

b. $\frac{d^2y}{dx^2}+\frac{49}{(4x+7)^3}=0, y(0)=0\ ,\ y'(0)=0$

c. $\frac{d^2y}{dx^2}+\frac{42}{\sqrt{7x+4}}=0, y(0)=0\ ,\ y'(0)=0$

20. a. Let $y=\cos^{-1}\left(\frac{2x}{3}\right)$. Determine the value of a given that $\frac{d^2y}{dx^2}-ax\left(\frac{dy}{dx}\right)^3=0$.

b. Let $y=bx^2e^{-3x}$. Determine the value of b given that $\frac{d^2y}{dx^2}+6\frac{dy}{dx}+9y=8e^{-3x}$.

8.7 Exam questions

Question 1 (3 marks) TECH-FREE

Show that $y=\sin\left(x^2\right)$ satisfies the differential equation $x\frac{d^2y}{dx^2}-\frac{dy}{dx}+4x^3y=0$.

Question 2 (5 marks) TECH-FREE

Given that $y=4e^{-2x}\sin(3x)$ is a solution of the differential equation $\frac{d^2y}{dx^2}+a\frac{dy}{dx}+by=0$, determine the values of a and b.

Question 3 (3 marks) TECH-FREE

Find the particular solution to the following differential equation, stating the maximal domain over which the solution is valid. $(5x+3)^2\frac{dy}{dx}+4=0, y(-1)=2$

Question 4 (1 mark) TECH-ACTIVE

MC $y=\tan^{-1}\left(\frac{x}{3}\right)$ satisfies which of the following differential equations?

A. $\left(x^2+9\right)\frac{dy}{dx}-3=0$

B. $\left(x^2+9\right)\frac{dy}{dx}+3=0$

C. $\left(x^2-3\right)\frac{dy}{dx}-9=0$

D. $\left(9-x^2\right)\frac{dy}{dx}+3=0$

E. $\left(9-x^2\right)\frac{dy}{dx}+3x^2=0$

Question 5 (3 marks) TECH-FREE

Let $y=ct\sin(2t)$. Find the value of c given that $\frac{d^2y}{dt^2}+4y=8\cos(2t)$.

More exam questions are available online.

Answers

Topic 8 Differential equations

8.2 Verifying solutions to differential equations

8.2 Exercise

1-4. Sample responses can be found in the worked solutions in the online resources.

5. $-5, 2$
6. ± 3
7. a. $a = 0, b = -1, c = 1$
   b. $a = 1, b = -6, c = 18, d = -24$
8. a. Sample responses can be found in the worked solutions in the online resources.
   b. $-2, 5$
9. Sample responses can be found in the worked solutions in the online resources.
10. $-6, 1$
11. Sample responses can be found in the worked solutions in the online resources.
12. a. Sample responses can be found in the worked solutions in the online resources
    b. ± 3
13. Sample responses can be found in the worked solutions in the online resources.
14. -2

15–20. Sample responses can be found in the worked solutions in the online resources.

21. a. Sample responses can be found in the worked solutions in the online resources.
    b. $a = 0, \; b = 1$
22. a. $a = 6, \; b = 9$
    b. Sample responses can be found in the worked solutions in the online resources.
23. a. 5
    b. Sample responses can be found in the worked solutions in the online resources.

24, 25. Sample responses can be found in the worked solutions in the online resources.

8.2 Exam questions

Note: Mark allocations are available with the fully worked solutions online.

1. E
2. $-2 \pm 3i$
3. Sample responses can be found in the worked solutions in the online resources.

8.3 Solving Type 1 differential equations, $\frac{dy}{dx} = f(x)$

8.3 Exercise

1. a. $y = c - 3x^4$
   b. $y = 13 - 3x^2$
2. a. $y = c - 6\sin(2x)$
   b. $y = 2(\cos(3x) - 1)$
3. a. $y = 2x^2 + 3x + c$
   b. $y = x^3 + \frac{7}{2}x^2 - 20x + c$
4. $y = \frac{9x - 11}{4x - 5}, x \in R\backslash\left\{\frac{5}{4}\right\}$
5. $y = 3 + \frac{1}{2}\log_e(|7 - 4x|), x \in R\backslash\left\{\frac{7}{4}\right\}$
6. $y = 1 - \sqrt{2x - 5}, x > \frac{5}{2}$
7. $y = 11 - 4\sqrt{x}, x > 0$
8. $y = 6\sin^{-1}\left(\frac{x}{8}\right) - \pi, x \in (-8, 8)$
9. $y = \frac{\pi}{2} - \tan^{-1}\left(\frac{x}{4}\right), x \in R$
10. a. $y = \frac{1}{3}[5\log_e(|x|) + x^2 + 8], x \in R\backslash\{0\}$
    b. $y = 2(e^{3x} - e^{-3x}), x \in R$
11. a. $y = 4 - 2\cos(2x), x \in R$
    b. $y = 3 - 2\sin(3x), x \in R$
12. a. $y = 4x - \sin(4x), x \in R$
    b. $y = 6x + \sin(6x), x \in R$
13. a. $y = \frac{1}{2}\left(\sqrt{4x + 9} - 3\right), x > -\frac{9}{4}$
    b. $y = 1 + \frac{1}{2}\log_e(|2x - 3|), x \in R\backslash\left\{\frac{3}{2}\right\}$
14. a. $y = \frac{10x - 17}{3x - 5}, x \in R\backslash\left\{\frac{5}{3}\right\}$
    b. $y = 5 - 2\log_e(|7 - 4x|), x \in R\backslash\left\{\frac{7}{4}\right\}$
15. a. $y = \frac{3}{2}\log_e\left(\frac{x^2 + 9}{9}\right), x \in R$
    b. $y = 2 - \sqrt{x^2 + 4}, x \in R$
16. a. $y = \frac{1}{2}\log_e\left(\frac{x^2 + 6x + 13}{13}\right), x \in R$
    b. $y = \log_e\left(\frac{3}{\sqrt{x^2 - 4x + 9}}\right), x \in R$
17. $y = -\frac{1}{8}\sin^4(2x), x \in R$
18. a. $y = 5x - x\log_e(2x) - \frac{3}{2}, x > 0$
    b. $y = (x - 4)e^{-x} + 4, x \in R$
19. a. $\int_1^x e^{\frac{1}{t}}\,dt + 3$ b. 5.020
20. a. $\int_{0.1}^x \sin^{-1}\left(u^2\right)\,du + 1$ b. 1.042
21. a. $y = \frac{1}{2}\tan^{-1}\left(\frac{2x}{3}\right) + \frac{1}{4}\log_e\left(\frac{9}{4x^2 + 9}\right)$

b. $y = \frac{3}{2}\left(\sin^{-1}\left(\frac{2x}{3}\right) - 1\right) + \frac{\sqrt{9-4x^2}}{2}, |x| < \frac{2}{3}$

22. a. i. $y = -b\sin^{-1}\left(\frac{x}{a}\right), |x| < a$

ii. $y = \frac{b}{2a}\log_e\left(\frac{|a-x|}{|a+x|}\right), |x| < a$

iii. $y = \frac{-x}{a(a+bx)}, x \neq -\frac{a}{b}$

b. $y = \frac{e^{-2x}}{13}\left(2\cos(3x) - 3\sin(3x)\right) - \frac{2}{13}$

8.3 Exam questions

Note: Mark allocations are available with the fully worked solutions online.

1. $y = e^{2x} + 3e^{-2x} + c$
2. $y = \sqrt{x^2+9} + c$
3. $y = \cos^3(3x) - 1,\ x \in R\backslash \frac{n\pi}{3},\ n \in Z$

8.4 Solving Type 2 differential equations, $\frac{dy}{dx} = f(y)$

8.4 Exercise

1. $y = \sqrt[3]{(B-6x)^2}$
2. $y = \frac{1}{2}\sin^{-1}(Be^{2x})$
3. a. $y = \frac{4}{c-x}$

   b. $y = Ae^x - 4$
4. a. $y = Ae^{\frac{x}{4}}$

   b. $y = \sqrt[3]{12x+A}$
5. $y = \frac{15x-13}{12x-11}, x \neq \frac{11}{12}$
6. $y = \frac{1}{4}(7+5e^{-4x})$
7. $y = 5e^{-3x}$
8. $y = 3e^{5x}$
9. a. $y = 4e^{-5x}$

   b. $y = 2e^{3x-3}$
10. a. $y = \frac{1}{2}(5+e^{-2x})$

    b. $y = \frac{2}{3}(2+e^{3x})$
11. $y = 8\cos\left(\frac{x}{6}\right), -6\pi < x < 0$
12. $y = 4\tan(x), -\frac{\pi}{2} < x < \frac{\pi}{2}$
13. a. $y = \frac{1}{4}(x+3)^2,\ x \in R$

    b. $y = \frac{3}{4-3x}, x \neq \frac{4}{3}$
14. a. $y = -\frac{1}{2}\log_e(17-8x), x < \frac{17}{8}$

    b. $y = -\frac{1}{3}\log_e(18x-17), x > \frac{17}{18}$
15. a. $y = \frac{5x-8}{2x-3}, x \neq \frac{3}{2}$

    b. $y = \frac{7x-23}{3x-10}, x \neq \frac{10}{3}$
16. a. $y = x^2 + 3x$

    b. $y = \frac{2}{3}\tan(6x), -\frac{\pi}{12} < x < \frac{\pi}{12}$
17. a. $y = \frac{12}{3+e^{4x}}$

    b. $y = \frac{6e^{3x}}{3-2e^{3x}}$
18. a. $y = \frac{12(e^x-1)}{4e^x-3}$

    b. $y = \frac{4(1-e^{2x})}{e^{2x}-2}, x \neq \log_e\left(\sqrt{2}\right)$
19. $y = y_0e^{-kx}$
20. $y = \left(c - \frac{b}{a}\right)e^{-ax} + \frac{b}{a}$
21. a. $y = \frac{b^2x}{1-abx}, x \neq \frac{1}{ab}$

    b. $y = \frac{a}{b}\tan(abx)$
22. a. $y = \frac{ab(1-e^{-(a-b)x})}{ae^{-(a-b)x}-b}$

    b. $-a$

8.4 Exam questions

Note: Mark allocations are available with the fully worked solutions online.

1. B
2. $y = 2\cos^{-1}(3x), |x| \leq \frac{1}{3}$
3. $y = \frac{ac}{(a-bc)e^{ax}+bc}$

8.5 Solving Type 3 differential equations, $\frac{dy}{dx} = f(x)g(y)$

8.5 Exercise

1. $\frac{y^4}{4} + 8y - \frac{x^2}{2} - 2x = c$
2. $y - 2\tan^{-1}\left(\frac{y}{2}\right) + \frac{1}{x} = c$
3. a. $\frac{1}{3}y^3 + 4y - \frac{1}{3}x^3 - 4x = c$

   b. $\frac{1}{2}y^2 + 4\log_e(|y|) - \frac{1}{2}x^2 = c$
4. a. $y - \frac{4}{y} - \frac{1}{3}x^3 = c$

   b. $\frac{1}{2}y^2 - \frac{8}{y} - \frac{1}{2}e^{x^2} = c$

5. $y = e^{x^3 + x}$

6. $y = \frac{1}{3 + x - x^2}, x \neq \frac{1 \pm \sqrt{13}}{2}$

7. $y = 8\sin(x^2), |x| < \frac{\sqrt{2\pi}}{2}$

8. $y = 4\tan(x^2), |x| < \frac{\sqrt{2\pi}}{2}$

9. a. $y = \frac{1}{1 - \log_e(|x|)}, x \neq 0$

b. $y = \frac{1}{4 - 3\cos(4x)}$

10. a. $y = \sqrt{5 - x^2}, |x| \leq \sqrt{5}$

b. $y = \frac{3}{6x^3 - 5}, x \neq \sqrt[3]{\frac{5}{6}}$

11. a. $y = \frac{2}{9x^4 - 8}, x \neq \pm\sqrt[4]{\frac{8}{9}}$

b. $y = \frac{2x}{2 - x}, x \neq 2$

12. a. $y = \frac{2}{3 - e^{2x}}, x \neq \log_e\left(\sqrt{3}\right)$

b. $y = \frac{2}{4x^6 - 3}$

13. a. $y = e^{x^3 - x}$

b. $y = \frac{2}{4x^3 - 2x + 3}$

14. a. $y = \frac{2}{2x^2 - 2x + 1}$

b. $y = \frac{1}{\sqrt[3]{6x^4 - 3x + 1}}$

15. a. $y = \frac{2}{1 + x^2}$

b. $y = \frac{4x^4}{5 - x^4}, x \neq \pm\sqrt[4]{5}$

16. a. $y = \frac{1}{4}\left(x^2 + 4\right)$

b. $y = \frac{2}{3}\sqrt{2x^2 + 9}$

17. a. $y = 5\tan\left(\frac{5x^2}{2}\right), |x| < \frac{\sqrt{5\pi}}{5}$

b. $y = 5\cos\left(2x^2\right), |x| \leq \frac{\sqrt{2\pi}}{2}$

18. $y = \frac{x^4 - 8}{x^3}, x \neq 0$

19. $y = 2x(1 + 2\log_e(|x|)), x \neq 0$

20. a. $y = x(c + b\log_e(|x|)), x \neq 0$

b. $y = \frac{bx}{a + 1} + \frac{c}{x^a}$

8.5 Exam questions

Note: Mark allocations are available with the fully worked solutions online.

1. $y = \frac{\pi}{2}\left(1 + e^{2x}\right)$

2. D

3. $2y^3 + 6y + 3x^2 - 11 = 0$

8.6 Solving Type 4 differential equations, $\frac{d^2y}{dx^2} = f(x)$

8.6 Exercise

1. $y = c_2 + c_1x - x^6$

2. $y = c_2 + c_1x + 4\sin(3x)$

3. a. $y = c_2 + c_1x - \frac{2}{x}, x \neq 0$

b. $y = c_2 + c_1x + 10x^2 - \frac{x^3}{2} - \frac{x^4}{6}$

4. a. $y = c_2 + c_1x + 2\log_e(|x|) + \frac{5}{2x}, x \neq 0$

b. $y = c_2 + c_1x + 2e^{-x} - \frac{5}{9}e^{-3x}$

5. $y = -2x^4 - 5x - 1$

6. $y = 3\sin(2x) + 6x + 1 - \frac{3\pi}{2}$

7. $y = \frac{-3x^2}{128(3x + 16)}, x \neq -\frac{16}{3}$

8. $y = 12\sqrt{2x + 9} - 3x - 36, x > -\frac{9}{2}$

9. $y = -x^3 + 8x - 5$

10. $y = 4 - 2e^{2x} - 2e^{-2x}$

11. a. $y = 4\sin(4x) - 8x + 4$

b. $y = 3\cos(3x) + 18x - 3\pi + 3$

12. a. $y = 1 - \cos(4x) - 8x^2$

b. $y = \frac{1}{8}\cos(8x) - 4x^2 - \frac{1}{8}$

13. a. $y = \frac{x^2}{8(3x + 2)}, x \neq -\frac{2}{3}$

b. $y = \sqrt{2x + 9} - \frac{x}{3} - 3, x > -\frac{9}{2}$

14. a. Sample responses can be found in the worked solutions in the online resources.

b. $y = -6x^2 - 24x - 20$

15. a. Sample responses can be found in the worked solutions in the online resources.

b. $y = 3x^3 - 36x - 48$

16. a. $y = 30x + 45 - \frac{5}{3}\sqrt{(4x + 9)^3}, x > -\frac{9}{4}$

b. $y = \log_e\left(\frac{|4x + 9|}{9}\right) - \frac{4x}{9}, x \neq -\frac{9}{4}$

17. a. $y = \frac{k}{12}\left(x^4 - 2Lx^3 + L^3x\right)$

b. $\frac{5kL^4}{192}$

18. $y = \frac{k}{6}\left(3Lx^2 - x^3\right), \frac{kL^3}{3}$

19. a. $y = \frac{W}{24EI}(6L^2x^2 - 4Lx^3 + x^4), \frac{WL^4}{8EI}$

b. $y = \frac{Wx^2}{24EI}(x - L)^2, \frac{WL^4}{384EI}$

20. $y = \frac{-x^2}{2b^2(ax+b)}, x \neq -\frac{b}{a}$

21. $y = \frac{1}{a^2}\log_e\left(\frac{|ax+b|}{b}\right) - \frac{x}{ab}, x \neq -\frac{b}{a}$

22. a. Sample responses can be found in the worked solutions in the online resources.

b. $y = c_2 + c_1x - \frac{1}{4}\sqrt{9 + 4x^2}$

8.6 Exam questions

Note: Mark allocations are available with the fully worked solutions online.

1. $y = -2x^4 + 31x - 27$

2. $y = 5e^{-x} - \frac{4}{9}e^{-3x} + \frac{11x}{3} - \frac{41}{9}$

3. a. Sample responses can be found in the worked solutions in the online resources.

b. $y = c_2 + c_1x - \frac{1}{ab}\sqrt{a + bx^2}$

8.7 Review

8.7 Exercise

Technology active: short answer

1. Sample responses can be found in the worked solutions in the online resources.

2. a. Sample responses can be found in the worked solutions in the online resources.

b. 3

3. a. $y = 4x + \frac{1}{3}\left(x^3 - 7\right)$

b. $y = 2\log_e(|x|) - 2 + \frac{4}{x} \quad x \neq 0$

4. a. $y = 2 - \tan^{-1}\left(\frac{x}{3}\right)$

b. $y = 11 - 3\sqrt{x^2 + 9}$

c. $y = 2 + \frac{1}{2}\log_e\left(\left|\frac{x-3}{x+3}\right|\right), x \neq \pm 3$

5. a. $y = \frac{1}{9x+1}, x \neq -\frac{1}{9}$

b. $y = \sqrt{1 - 18x}, x < \frac{1}{18}$

c. $y = \frac{1}{\sqrt{18x+1}}, x > -\frac{1}{18}$

6. a. $y = \frac{1}{4}\left(7 + e^{-4x}\right)$

b. $y = \frac{5x-7}{2x-3}, x \neq \frac{3}{2}$

c. $y = 3\cos(x), 0 \leq x \leq \pi$

Technology active: multiple choice

7. D
8. B
9. C
10. D
11. B
12. C
13. A
14. E
15. E
16. B

Technology active: extended response

17. a. $y = 2\tan(2x), -\frac{\pi}{4} < x < \frac{\pi}{4}$

b. $y = \frac{4e^{4x}}{3 - e^{4x}}, x \neq \frac{1}{4}\log_e(3)$

c. $y = \frac{1}{4}\log_e\left(\frac{4}{x+4}\right), x > -4$

18. a. $y = \frac{1}{3 - 2x^3}$

b. $y = \log_e\left(x^4 + 1\right)$

c. $y = \frac{1}{4 - x - 2x^2}$

19. a. $y = 10x + 1 - 8\sqrt{x} - 4\log_e(x), x > 0$

b. $y = \frac{-x^2}{2(4x+7)} \quad x \neq -\frac{7}{4}$

c. $y = \frac{8}{7}\left(21x + 8 - \sqrt{(7x+4)^3}\right) \quad x > -\frac{4}{7}$

20. a. 1

b. 4

8.7 Exam questions

Note: Mark allocations are available with the fully worked solutions online.

1. Sample responses can be found in the worked solutions in the online resources.

2. $a = 4$
$b = 13$

3. $y = \frac{12x+8}{5x+3}, x \neq \frac{-3}{5}$

4. A

5. $c = 2$

9 Further integration techniques and applications

LEARNING SEQUENCE

Fully worked solutions for this topic are available online.

9.1 Overview

Hey students! Bring these pages to life online

Watch videos

Engage with interactivities

Answer questions and check results

Find all this and MORE in jacPLUS

9.1.1 Introduction

We have already learnt how to manipulate and evaluate many integrals. In this topic we will cover a few more advanced integration techniques and applications. Specifically, we will learn how integration can be extended into 3 dimensions to calculate the surface area and volume of shapes formed by rotating a curve $f(x)$ around the x- or y-axes. Some examples of 3-dimensional shapes that can be formed by such rotations are cylinders, cones, spheres, bowls, bottles and any other shape which is symmetrical about an axis of rotation.

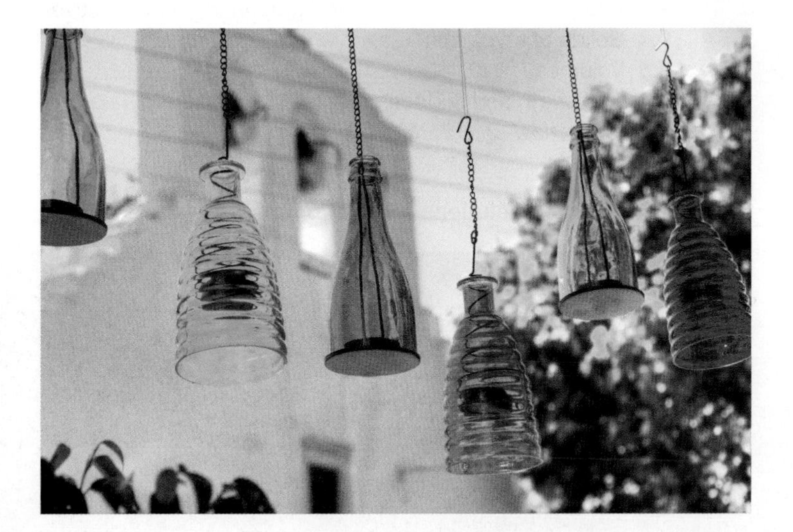

KEY CONCEPTS

This topic covers the following points from the VCE Mathematics Study Design:

- the relationship between the graph of a function and the graphs of its anti-derivative functions
- techniques of anti-differentiation and for the evaluation of definite integrals:
  - anti-differentiation of $\frac{1}{x}$ to obtain $\log_e |x|$
  - anti-differentiation of $\frac{1}{\sqrt{a^2 - x^2}}$ and $\frac{a}{a^2 + x^2}$ for $a > 0$ by recognition that they are derivatives of corresponding inverse circular functions
  - use of the substitution $u = g(x)$ to anti-differentiate expressions
  - use of the trigonometric identities $\sin^2(ax) = \frac{1}{2}(1 - \cos(2ax))$ and $\cos^2(ax) = \frac{1}{2}(1 + \cos(2ax))$ in anti-differentiation techniques
  - anti-differentiation using partial fractions of rational functions
  - integration by parts
- numerical and symbolic integration using technology
- application of integration, areas of regions bounded by curves, arc lengths for parametrically determined curves, surface area of solids of revolution, volumes of solids of revolution of a region about either coordinate axis.

Note: *Concepts shown in grey are covered in other topics.*

Source: VCE Mathematics Study Design (2023-2027) extracts © VCAA; reproduced by permission.

9.2 Integration by parts

LEARNING INTENTION

At the end of this subtopic you should be able to:
- integrate products of functions using integration by parts.

9.2.1 The anti-product rule

Integration by parts is a technique used to integrate products of two different types of functions, just as the product rule has been used to differentiate products of two different types of functions. Consider the product rule $\frac{d}{dx}(uv) = u\frac{dv}{dx} + v\frac{du}{dx}$, if we transpose this equation we get $u\frac{dv}{dx} = \frac{d}{dx}(uv) - v\frac{du}{dx}$. Now integrate both sides with respect to x to obtain $\int u\frac{dv}{dx}dx = uv - \int v\frac{du}{dx}dx$.

Integration by parts

$$\int u\frac{dv}{dx}dx = uv - \int v\frac{du}{dx}dx$$

To use this result we choose one part of the integral on the left-hand side as u which is easily differentiated to obtain $\frac{du}{dx}$ and the other part is chosen as $\frac{dv}{dx}$ which is easily integrated to obtain v. Then we substitute these into the result on the right-hand side. The resulting integral on the right hand side should then be easier than the original integral on the left-hand side. Finally for indefinite integrals, just add in the arbitrary constant $+c$ at the end, do not include an arbitrary constant $+c$ in the term for v.

WORKED EXAMPLE 1 Using integration by parts to integrate products of functions

Evaluate $\int x\sin(2x)\,dx$.

THINK	WRITE/DRAW
1. Use integration by parts, choose one term of the product as u and the other as $\frac{dv}{dx}$.	$\int x\sin(2x)\,dx = \int u\frac{dv}{dx}dx$ Let $u = x$ and $\frac{dv}{dx} = \sin(2x)$
2. Differentiate u and integrate $\frac{dv}{dx}$ to determine v.	$\frac{du}{dx} = 1$ and $v = \int \sin(2x)dx$ $v = -\frac{1}{2}\cos(2x)$
3. Substitute into the right-hand side of the integration by parts formula.	$\int u\frac{dv}{dx}dx = uv - \int v\frac{du}{dx}dx$ $\int x\sin(2x)\,dx = -\frac{1}{2}x\cos(2x) - \int -\frac{1}{2}\cos(2x)dx$ $= -\frac{1}{2}x\cos(2x) + \int \frac{1}{2}\cos(2x)dx$

4. Perform the integral on the right-hand side, add in the constant, and state the final result.

$$\int x\sin(2x)\,dx = -\frac{1}{2}x\cos(2x) + \frac{1}{4}\sin(2x) + c$$

Notice that if we had made the other incorrect choice, that is, choosing $u = \sin(2x)$ and $\frac{dv}{dx} = x$, then $\frac{du}{dx} = 2\cos(2x)$ and $v = \frac{1}{2}x^2$, so that $\int x\sin(2x)\,dx = \frac{1}{2}x^2\sin(2x) - \int x^2\cos(2x)dx$, which although this equation is correct, the integral on the right-hand side, is even more complicated than the original integral that we started with.

Applying integration by parts to functions which are not products

Sometimes it can be useful to use the integration by parts formula for functions which are not a product of two terms. We can write $\int f(x)\,dx = \int 1 \cdot f(x)\,dx$ and choose $u = f(x)$ and $\frac{dv}{dx} = 1$.

WORKED EXAMPLE 2 Using integration by parts to integrate a function

Evaluate $\int \cos^{-1}(2x)\,dx$.

THINK	WRITE/DRAW
1. Rewrite the integral as the product of the original function with 1.	$\int \cos^{-1}(2x)\,dx = \int 1 \cdot \cos^{-1}(2x)\,dx$ $= \int u\frac{dv}{dx}dx$ where $u = \cos^{-1}(2x)$ and $\frac{dv}{dx} = 1$
2. Differentiate u and integrate $\frac{dv}{dx}$ to determine v.	$\frac{du}{dx} = \frac{-2}{\sqrt{1-4x^2}}$ and $v = \int 1\,dx$ $v = x$
3. Substitute into the integration by parts formula.	$\int u\frac{dv}{dx}dx = uv - \int v\frac{du}{dx}dx$ $\int \cos^{-1}(2x)\,dx = x\cos^{-1}(2x) - \int \frac{-2x}{\sqrt{1-4x^2}}dx$ $= x\cos^{-1}(2x) + \int \frac{2x}{\sqrt{1-4x^2}}dx$
4. Consider the integral on the right-hand side which still needs to be evaluated. Write the integrand as a power, using index laws.	$\int \frac{2x}{\sqrt{1-4x^2}}dx = \int 2x(1-4x^2)^{-\frac{1}{2}}dx$

5. Use a non-linear substitution. Take the derivate and rearrange the expression to write dx in terms of dt.	Let $t = 1 - 4x^2$ $\frac{dt}{dx} = -8x$ $\frac{dx}{dt} = -\frac{1}{8x}$ $dx = -\frac{1}{8x}\,dt$
6. Substitute t and dx into the equation.	$\int 2x(1-4x^2)^{-\frac{1}{2}}dx = \int 2xt^{-\frac{1}{2}}dx$ $= \int 2xt^{-\frac{1}{2}} \times \frac{-1}{8x}\,dt$ $= -\frac{1}{4}\int t^{-\frac{1}{2}}dt$
7. Perform the integration.	$= -\frac{1}{2}t^{\frac{1}{2}}$
8. Express the result, in square root form, and substitute back for x.	$= -\frac{1}{2}\sqrt{t}$ $= -\frac{1}{2}\sqrt{1-4x^2}$
9. State the final answer to the original integral, adding in the constant.	$\int \cos^{-1}(2x)\,dx = x\cos^{-1}(2x) - \frac{1}{2}\sqrt{1-4x^2} + c$

TI | THINK	DISPLAY/WRITE	CASIO | THINK	DISPLAY/WRITE
On a Calculator page, complete the entry as shown.	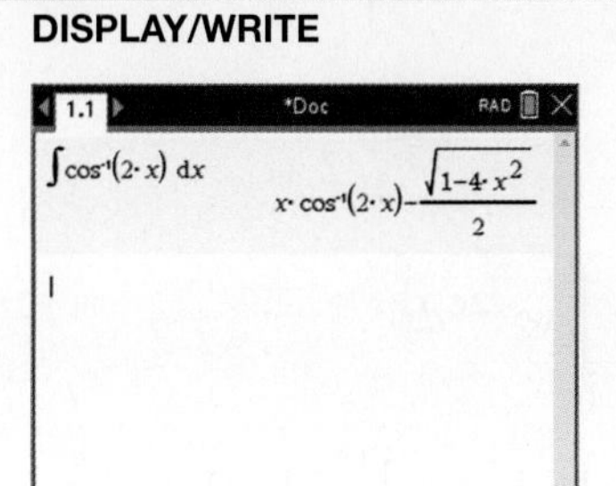	On a Main screen, complete the entry as shown.	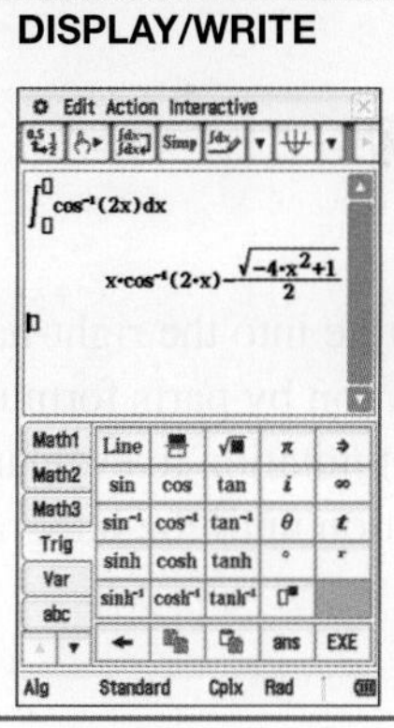

9.2.2 Evaluating definite integrals

When determining the value of a definite integral or evaluating an area, recall that there is no need for the constant of integration. Also, it is usually easier to leave the final evaluation to the last stage, so proceed as follows.

WORKED EXAMPLE 3 Evaluating definite integrals using integration by parts

Determine the area bounded by the function $f: R \to R$, $f(x) = xe^{-2x}$, the x-axis, the origin and $x = 1$.

THINK

1. Draw a diagram to identify the required area and shade this area.

WRITE/DRAW

The graph passes through the origin, and as $x \to \infty$, $y \to 0$, so that the positive x-axis is an asymptote.

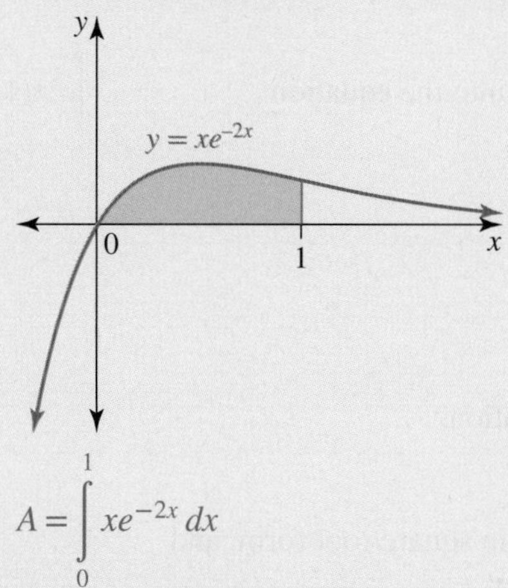

2. The required area is given by a definite integral.

$$A = \int_0^1 xe^{-2x}\,dx$$

3. To evaluate the definite integral, use integration by parts. Choose one term of the product as u and the other as $\frac{dv}{dx}$.

Let $u = x$ and $\frac{dv}{dx} = e^{-2x}$

4. Differentiate u and integrate $\frac{dv}{dx}$ to determine v.

$\frac{du}{dx} = 1$ and $v = \int e^{-2x}dx$

$$v = -\frac{1}{2}e^{-2x}$$

5. Substitute into the right-hand side of the integration by parts formula, using square bracket notation to evaluate, but leave the evaluation until the integration is completed.

$$A = \int_0^1 xe^{-2x}\,dx$$

$$= \left[-\frac{1}{2}xe^{-2x}\right]_0^1 - \int_0^1 -\frac{1}{2}e^{-2x}\,dx$$

$$= \left[-\frac{1}{2}xe^{-2x}\right]_0^1 + \int_0^1 \frac{1}{2}e^{-2x}\,dx$$

6. Perform the integration.

$$= \left[-\frac{1}{2}xe^{-2x}\right]_0^1 - \left[\frac{1}{4}e^{-2x}\right]_0^1$$

7. Simplify the expression to be evaluated, by taking out the common factor.

$$= \left[-\frac{1}{4}e^{-2x}(2x+1)\right]_0^1$$

8. Now evaluate, by substituting in the terminals.

$$= \left(-\frac{1}{4}e^{-2}(2+1)\right) - \left(-\frac{1}{4}e^{-0}(0+1)\right)$$

9. State the final answer, as an exact value.

$$A = \frac{1}{4}(1 - 3e^{-2})\text{units}^2$$

TI | THINK	DISPLAY/WRITE	CASIO | THINK	DISPLAY/WRITE
On a Calculator page, complete the entry as shown.	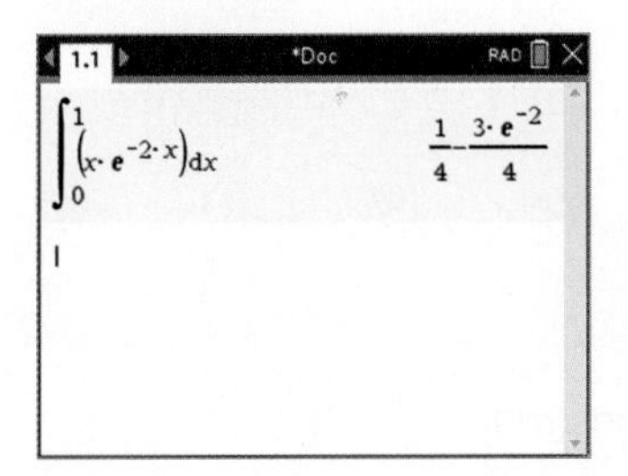	On a Main screen, complete the entry as shown.	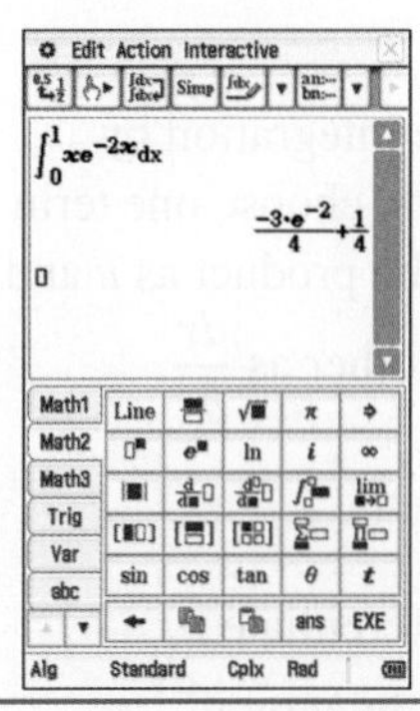

Determining which part becomes *u*

How do we know which function to choose as u and which to choose as $\frac{dv}{dx}$?

In general, we have looked at five different types of functions.

L logarithmic

I inverse trigonometric

A algebraic

T trigonometric

E exponential

The first two logarithmic and inverse trigonometric are chosen as u as they are easily differentiated, the final two trigonometric and exponential are chosen as $\frac{dv}{dx}$ as when they are integrated, they remain trigonometric or exponential. The middle of the road algebraic function can be chosen as either depending if it is combined with either of the other types.

A quick and easy way to remember which functions should become u and which should become $\frac{dv}{dx}$ is the acronym LIATE. When deciding between two functions, let the function which comes first in LIATE become u.

Reduction formulae

Sometimes, the operation of integration by parts, may have to be applied more than once to reduce an integral to a suitable form. This process can lead to a general statement called a reduction or recurrence formula which can be used to simplify integrals of products.

WORKED EXAMPLE 4 Developing and using a reduction formula

a. If m is a positive integer, and a is a positive real number, use integration by parts to show that

$$\int x^m \cos(ax)\,dx = \frac{1}{a}x^m \sin(ax) + \frac{m}{a^2}x^{m-1}\cos(ax) - \frac{m\,(m-1)}{a^2}\int x^{m-2}\cos(ax)\,dx.$$

b. Hence, determine $\int x^4 \cos(3x)\,dx$.

THINK	WRITE/DRAW
a. 1. Use integration by parts, choose one term of the product as u and the other as $\frac{dv}{dx}$.	**a.** $\int x^m \cos(ax)\,dx$ $= \int u\frac{dv}{dx}dx$ Let $u = x^m$ and $\frac{dv}{dx} = \cos(ax)$.
2. Differentiate u, and integrate $\frac{dv}{dx}$ to determine v.	$\frac{du}{dx} = mx^{m-1}$ and $v = \int \cos(ax)\,dx$ $= \frac{1}{a}\sin(ax)$
3. Substitute into the right-hand side of the integration by parts formula, and transfer the constant factor outside the front of the integral.	$\int u\frac{dv}{dx}dx = uv - \int v\frac{du}{dx}dx$ $\int x^m \cos(ax)\,dx = \frac{1}{a}x^m \sin(ax) - \int \frac{m}{a}x^{m-1}\sin(ax)dx$ $= \frac{1}{a}x^m \sin(ax) - \frac{m}{a}\int x^{m-1}\sin(ax)dx \quad (1)$
4. Use integration by parts again, choose one term of the product as u and the other as $\frac{dv}{dx}$, these are different u's and v's from the steps above.	Consider now, just the integral on the right-hand side. $\int x^{m-1}\sin(ax)\,dx = \int u\frac{dv}{dx}dx$ Let $u = x^{m-1}$ and $\frac{dv}{dx} = \sin(ax)$.
5. Differentiate u, and integrate $\frac{dv}{dx}$ to determine v.	$\frac{du}{dx} = (m-1)x^{m-2}$ and $v = \int \sin(ax)dx = -\frac{1}{a}\cos(ax)$
6. Substitute into the right-hand side of the integration by parts formula.	$\int u\frac{dv}{dx}dx = uv - \int v\frac{du}{dx}dx$ $\int x^{m-1}\sin(ax)dx = -\frac{1}{a}x^{m-1}\cos(ax) - \int -\frac{m-1}{a}x^{m-2}\cos(ax)dx$ $= -\frac{1}{a}x^{m-1}\cos(ax) + \frac{m-1}{a}\int x^{m-2}\cos(ax)dx \quad (2)$
7. Substitute for the integral from (2), back into the result obtained in (1) above.	$\int x^m \cos(ax)\,dx$ $= \frac{1}{a}x^m \sin(ax) - \frac{m}{a}\left[-\frac{1}{a}x^{m-1}\cos(ax) + \frac{m-1}{a}\int x^{m-2}\cos(ax)dx\right]$
8. Expand the brackets and simplify, to obtain the required result.	$\int x^m \cos(ax)\,dx$ $= \frac{1}{a}x^m \sin(ax) + \frac{m}{a^2}x^{m-1}\cos(ax) - \frac{m(m-1)}{a^2}\int x^{m-2}\cos(ax)\,dx$

b. 1. Use the appropriate values and substitute these into the reduction formulae.

b. To determine $\int x^4 \cos(3x)\,dx$ substitute $m = 4$ and $a = 3$

$$\int x^4 \cos(3x)\,dx$$
$$= \frac{1}{3}x^4 \sin(3x) + \frac{4}{9}x^3 \cos(3x) - \frac{12}{9}\int x^2 \cos(3x(\,dx$$
$$= \frac{1}{3}x^4 \sin(3x) + \frac{4}{9}x^3 \cos(3x) - \frac{4}{3}\int x^2 \cos(3x)\,dx \qquad (1)$$

2. Use the appropriate values and substitute these into the reduction formulae again.

To determine $\int x^2 \cos(3x)\,dx$ let $m = 2$ and $a = 3$

$$\int x^2 \cos(3x)\,dx$$
$$= \frac{1}{3}x^2 \sin(3x) + \frac{2}{9}x \cos(3x) - \frac{2}{9}\int x^0 \cos(3x)\,dx$$
$$= \frac{1}{3}x^2 \sin(3x) + \frac{2}{9}x \cos(3x) - \frac{2}{9}\int \cos(3x)\,dx$$

3. Perform the integral on the right.

$$\int x^2 \cos(3x)\,dx$$
$$= \frac{1}{3}x^2 \sin(3x) + \frac{2}{9}x \cos(3x) - \frac{2}{27}\sin(3x)$$

4. Substitute this expression back into the expression in (1).

$$\int x^4 \cos(3x)\,dx$$
$$= \frac{1}{3}x^4 \sin(3x) + \frac{4}{9}x^3 \cos(3x)$$
$$-\frac{4}{3}\left[\frac{1}{3}x^2 \sin(3x) + \frac{2}{9}x \cos(3x) - \frac{2}{27}\sin(3x)\right]$$

5. Simplify and state the final result.

$$\int x^4 \cos(3x)\,dx$$
$$= \frac{1}{3}x^4 \sin(3x) + \frac{4}{9}x^3 \cos(3x) - \frac{4}{9}x^2 \sin(3x) - \frac{8}{27}x \cos(3x) + \frac{8}{81}\sin(3x)$$
$$= \frac{4x}{27}(3x^2 - 2)\cos(3x) + \frac{1}{81}(27x^4 - 36x^2 + 8)\sin(3x)$$

TI | THINK	DISPLAY/WRITE	CASIO | THINK	DISPLAY/WRITE
On a Calculator page, complete the entry as shown.	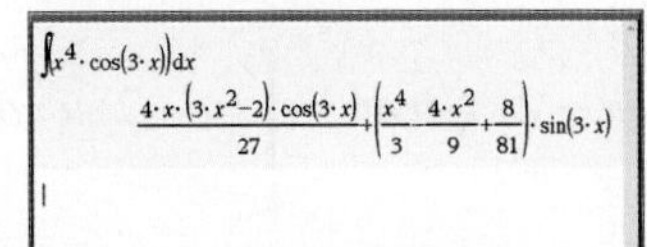	On a Main screen, complete the entry as shown.	

9.2.3 Induction with integration by parts

Using integration by parts and recurrence relations, we can solve problems involving definite integrals together with mathematical induction.

WORKED EXAMPLE 5 Induction with recurrence relations

Let $I_n = \int_0^{-2} x^n \sqrt{x+2}\,dx$ for integers $n \geq 0$.

a. **Show that $I_0 = \dfrac{-4\sqrt{2}}{3}$ and $I_1 = \dfrac{16\sqrt{2}}{15}$.**

b. **Let $I_{k+1} = \int_0^{-2} x^{k+1} \sqrt{x+2}\,dx$. Using integration by parts show that $I_{k+1} = -\dfrac{8(k+1)}{2k+5} I_k$.**

c. **Hence, prove by induction that $I_n = \dfrac{-4n}{2n+3} I_{n-1}$ for $n = 1, 2, 3...$**

THINK	WRITE
a. 1. Substitute $n = 0$ and evaluate the definite integral.	a. $I_0 = \int_0^{-2} \sqrt{x+2}\,dx = \int_0^{-2} (x+2)^{\frac{1}{2}}\,dx$ $I_0 = \left[\frac{2}{3}(x+2)^{\frac{3}{2}}\right]_0^{-2}$ $I_0 = \frac{2}{3}\left[(0) - 2^{\frac{3}{2}}\right]$ $I_0 = -\frac{4\sqrt{2}}{3}$
2. Substitute $n = 1$. Use a substitution and change the terminals. Write the definite integral in terms of the new variable.	$I_1 = \int_0^{-2} x\sqrt{x+2}\,dx$ Let $u = x+2,\ x = u-2,\ \frac{du}{dx} = 1$ $x = -2,\ u = 0,\ x = 0,\ u = 2$ $I_1 = \int_2^0 (u-2)\sqrt{u}\,du = \int_2^0 (u-2)u^{\frac{1}{2}}\,du$
3. Expand and perform the integration.	$I_1 = \int_2^0 \left(u^{\frac{3}{2}} - 2u^{\frac{1}{2}}\right) du$ $I_1 = \left[\frac{2}{5}u^{\frac{5}{2}} - \frac{4}{3}u^{\frac{3}{2}}\right]_2^0$

4. Evaluate the definite integral.

$$I_1 = \left[0 - \left(\frac{2}{5}\left(\sqrt{2}\right)^5 - \frac{4}{3}\left(\sqrt{2}\right)^3\right)\right]$$

$$I_1 = -\frac{8\sqrt{2}}{5} + \frac{8\sqrt{2}}{3} = \sqrt{2}\left(\frac{-24+40}{15}\right)$$

$$I_1 = \frac{16\sqrt{2}}{15}$$

b. 1. Use integration by parts.

b. $$I_{k+1} = \int_0^{-2} x^{k+1}\sqrt{x+2}\,dx$$

$$u = x^{k+1} \qquad \frac{dv}{dx} = \sqrt{x+2} = (x+2)^{\frac{1}{2}}$$

$$\frac{du}{dx} = (k+1)x^k \qquad v = \frac{2}{3}(x+2)^{\frac{3}{2}}$$

2. Substitute back using the integration by parts formulae.
Note that the first term to be evaluated is zero.

$$I_{k+1} = \left[\frac{2x^{k+1}}{3}(x+2)^{\frac{3}{2}}\right]_0^{-2} - \frac{2(k+1)}{3}\int_0^{-2} x^k (x+2)^{\frac{3}{2}}\,dx$$

$$I_{k+1} = -\frac{2(k+1)}{3}\int_0^{-2} x^k (x+2)^{\frac{3}{2}}\,dx$$

3. Write $(x+2)^{\frac{3}{2}} = (x+2)\sqrt{x+2}$ and expand.

$$I_{k+1} = -\frac{2(k+1)}{3}\int_0^{-2} \left(x^k\right)(x+2)\sqrt{x+2}\,dx$$

$$I_{k+1} = -\frac{2(k+1)}{3}\int_0^{-2} \left(x^{k+1} + 2x^k\right)\sqrt{x+2}\,dx$$

4. Use the properties of definite integrals to express the integrand as the sum of two integrals, and identify that $I_{k+1} = \int_0^{-2} x^{k+1}\sqrt{x+2}\,dx$.

$$I_{k+1} = -\frac{2(k+1)}{3}\left[\int_0^{-2} x^{k+1}\sqrt{x+2}\,dx + 2\int_0^{-2} x^k\sqrt{x+2}\,dx\right]$$

$$I_{k+1} = -\frac{2(k+1)}{3}\left[I_{k+1} + 2I_k\right]$$

5. Solve for I_{k+1} to obtain the result as required.

$$I_{k+1}\left(1 + \frac{2(k+1)}{3}\right) = -\frac{4(k+1)}{3}I_k$$

$$\left(\frac{2k+5}{3}\right)I_{k+1} = -\frac{4(k+1)}{3}I_k$$

$$I_{k+1} = \frac{-4(k+1)}{2k+5}I_k$$

c. 1. Identify the propositional statement that we need to show using mathematical induction.

c. $$I_n = \int_0^{-2} x^n\sqrt{x+2}\,dx \quad \text{then} \quad I_n = \frac{-4n}{2n+3}I_{n-1}$$

2. We have shown the base case in **a** when $n = 1$. Verify the RHS for this case.	When $n = 1$ $\text{LHS} = I_1 = \frac{16\sqrt{2}}{15}$ $\text{RHS} = \frac{-4}{3+2} I_0 = -\frac{4}{5} \times \frac{-4\sqrt{2}}{3} = \frac{16\sqrt{2}}{15}$ So it is true when $n = 1$.
3. Consider when $n = k + 1$.	$I_{k+1} = \int_0^{-2} x^{k+1}\sqrt{x+2}\,dx$
4. Use the result we have shown in **b**.	$I_{k+1} = -\frac{4(k+1)}{2k+5} I_k$ $I_{k+1} = -\frac{4(k+1)}{2(k+1)+3} I_k$
5. State the principle of mathematical induction.	So it is true for $n = k + 1$; by the principle of induction it is true for $n = 1, 2, 3...$

9.2 Exercise

Technology free

For questions 1 to 5, calculate each of the following integrals.

1. WE1 **a.** $\int xe^{-2x}\,dx$ **b.** $\int xe^{\frac{x}{3}}\,dx$

2. a. $\int x\sin(3x)\,dx$ **b.** $\int x\cos\left(\frac{x}{2}\right)\,dx$

3. a. $\int \log_e(3x)dx$ **b.** $\int x\log_e\left(\frac{x}{2}\right)\,dx$

4. WE2 **a.** $\int \sin^{-1}\left(\frac{x}{3}\right)\,dx$ **b.** $\int \cos^{-1}(4x)\,dx$

5. a. $\int \tan^{-1}\left(\frac{x}{5}\right)\,dx$ **b.** $\int x\tan^{-1}(2x)\,dx$

For questions 6 and 7, evaluate each of the following.

6. a. $\displaystyle\int_{\frac{1}{3}}^{1} \frac{\log_e(3x)}{x^3}dx$

b. $\displaystyle\int_{0}^{\frac{\pi}{8}} x\sin(2x)\cos(2x)\,dx$

7. a. $\displaystyle\int_{0}^{\frac{\pi}{8}} x\cos^2(2x)\,dx$

b. $\displaystyle\int_{0}^{\frac{\pi}{9}} x\sin^2(3x)\,dx$

8. **WE3** Determine the area bounded by the function $f{:}\left\{x : |x| \le \frac{1}{2}\right\} \to R,\ f(x) = \cos^{-1}(2x)$ the x-axis, $x = 0$ and $x = \frac{1}{2}$.

9. Determine the area bounded by the curve $y = \sin^{-1}(3x)$, the x-axis, the origin and the line $x = \frac{1}{3}$.

10. Calculate the area bounded by the graph of $y = \tan^{-1}\left(\frac{x}{4}\right)$, the x-axis, the origin and the line $x = 4$.

For questions 11 to 14, a is a positive real number. Calculate each of the following.

11. a. $\displaystyle\int xe^{ax}\,dx$

b. $\displaystyle\int x\cos(ax)\,dx$

12. a. $\displaystyle\int x\sin(ax)\,dx$

b. $\displaystyle\int x^n\log_e(ax)\,dx$. What happens if $n = -1$?

13. a. $\displaystyle\int \sin^{-1}\left(\frac{x}{a}\right)dx$

b. $\displaystyle\int \cos^{-1}\left(\frac{x}{a}\right)dx$

14. a. $\displaystyle\int \tan^{-1}\left(\frac{x}{a}\right)dx$

b. $\displaystyle\int x\tan^{-1}\left(\frac{x}{a}\right)dx$

15. **WE4** **a.** If m is a positive integer, and a is a positive real number, then:
use integration by parts to show that

$$\int x^m\sin(ax)\,dx = -\frac{1}{a}x^m\cos(ax) + \frac{m}{a^2}x^{m-1}\sin(ax) - \frac{m(m-1)}{a^2}\int x^{m-2}\sin(ax)\,dx.$$

b. Hence, determine $\displaystyle\int x^4\sin(2x)\,dx$.

16. a. If n is a positive integer, and $I_n(x) = \displaystyle\int \sin^n(x)dx$ use integration by parts to show that

$$I_n(x) = -\frac{1}{n}\sin^{n-1}(x)\cos(x) + \frac{n-1}{n}I_{n-2}(x).$$

b. Hence, evaluate $\displaystyle\int_{0}^{\frac{\pi}{2}} \sin^5(x)dx$.

17. a. If n is a positive integer, and $C_n(x) = \displaystyle\int \cos^n(x)dx$ use integration by parts to show that

$$C_n(x) = \frac{1}{n}\cos^{n-1}(x)\sin(x) + \frac{n-1}{n}C_{n-2}(x).$$

b. Hence, evaluate $\displaystyle\int_{0}^{\frac{\pi}{2}} \cos^5(x)dx$.

18. Let $SL(x)=\int \sin(\log_e(x))dx$ and $CL(x)=\int \cos(\log_e(x))dx$.

a. Use integration by parts on $SL(x)$ to show that $SL(x)=x\sin(\log_e(x))-CL(x)$.

b. Use integration by parts on $CL(x)$ to show that $CL(x)=x\cos(\log_e(x))+SL(x)$.

c. Hence show that $SL(x)=\frac{x}{2}\left(\sin(\log_e(x))-\cos(\log_e(x))\right)$ and $CL(x)=\frac{x}{2}\left(\cos(\log_e(x))+\sin\left(\log_e(x)\right)\right)$.

19. Let $S(x)=\int e^{ax}\sin(bx)dx$ and $C(x)=\int e^{ax}\cos(bx)dx$, where a and b are real numbers.

a. Use integration by parts on $S(x)$ to show that $S(x)=-\frac{1}{b}e^{ax}\cos(bx)+\frac{a}{b}C(x)$.

b. Use integration by parts on $C(x)$ to show that $C(x)=\frac{1}{b}e^{ax}\sin(bx)-\frac{a}{b}S(x)$.

c. Hence show that $C(x)=\frac{e^{ax}}{a^2+b^2}(a\cos(bx)+b\sin(bx))$ and $S(x)=\frac{e^{ax}}{a^2+b^2}(a\sin(bx)-b\cos(bx))$.

d. Use the results of part **c** to determine the area enclosed between the curve $y=e^{-2x}\sin(3x)$, the x-axis the origin and the first intercept it makes with the positive x-axis.

20. a. Use the product rule and mathematical induction to show that $\frac{d}{dx}(x^n)=nx^{n-1}$ for $n\in Z$.

b. Use integration by parts and mathematical induction to show that $\int x^n=\frac{1}{n+1}x^{n+1}$ for $n\in Z$.

21. For a positive integer, n let $I_n(x)=\int_0^x t^n e^{-t}\,dt$ for $n\in N$ $n\geq 0$.

Prove using mathematical induction $I_n(x)=n!\left[1-e^{-x}\left(1+x+\frac{x^2}{2!}+\frac{x^3}{3!}+\ldots+\frac{x^n}{n!}\right)\right]$.

22. **WE5** Let $I_n=\int_0^{-4} x^n\sqrt{x+4}\,dx$ for integers $n\geq 0$.

a. Show that $I_0=\frac{-16}{3}$ and $I_1=\frac{128}{15}$.

b. Let $I_{k+1}=\int_0^{-4} x^{k+1}\sqrt{x+4}\,dx$. Using integration by parts show that $I_{k+1}=-\frac{8(k+1)}{2k+5}I_k$.

c. Hence, prove by induction that $I_n=\frac{-8n}{2n+3}I_{n-1}$ for $n=1,2,3\ldots$

23. Let $I_n=\int_1^{e^2}\left(\log_e(x)\right)^n dx$ for integers $n\geq 0$.

a. Show that $I_0=e^2-1$ and $I_1=e^2+1$.

b. Show that $I_{k+1}=2^{k+1}e^2-(k+1)I_k$ and hence calculate I_4.

c. Hence, prove by induction that $I_n=2^ne^2-nI_{n-1}$ for $n=1,2,3\ldots$

24. Let $I_n = \int_0^1 x^{2n+1} e^{x^2} \, dx$ for integers $n \geq 0$.

a. Show that $I_0 = \frac{1}{2}(e-1)$.

b. Show that $I_n = \frac{e}{2} - nI_{n-1}$ and hence find I_4.

9.2 Exam questions

Question 1 (2 marks) TECH-FREE

Find $\int x^3 \log_e(2x) dx$.

Question 2 (4 marks) TECH-FREE

In determining the Fourier series to represent the function $f(x) = x$ in the interval $-\pi \leq x \leq \pi$, the Fourier coefficients are given by $a_n = \frac{1}{\pi} \int_{-\pi}^{\pi} x \cos(nx) \, dx$ and $b_n = \frac{1}{\pi} \int_{-\pi}^{\pi} x \sin(nx) \, dx$, where n is a positive integer.

Show that $a_n = 0$ and $b_n = -\frac{2(-1)^n}{n}$.

Question 3 (3 marks) TECH-FREE

Let $J_n(x) = \int x^n e^{kx} dx$.

a. Use integration by parts to show that $J_n(x) = \frac{1}{k} x^n e^{kx} - \frac{n}{k} J_{n-1}(x)$. **(1 mark)**

b. Hence, evaluate $\int_0^1 x^3 e^{2x} \, dx$. **(2 marks)**

More exam questions are available online.

9.3 Integration by recognition and graphs of anti-derivatives

LEARNING INTENTION

At the end of this subtopic you should be able to:

- integrate functions which are scalar multiples of a known derivative function using integration by recognition
- sketch the graph of the anti-derivative from the graph of a function.

9.3.1 Integration by recognition

Deducing an anti-derivative

Since differentiation and integration are inverse processes, if we differentiate a function $f(x)$ with respect to x and obtain another function $g(x)$, then it follows that an anti-derivative of the function $g(x)$ with respect to x, is the function $f(x)$, we need to add in an arbitrary constant $+c$. In mathematical notation, if $\frac{d}{dx}(f(x)) = g(x)$

then $\int g(x)\,dx = f(x) + c$. However functions may not be exactly the derivatives of other functions, but may differ only by a constant multiple. Let k be a non-zero constant, then if $\frac{d}{dx}(f(x)) = k\,g(x)$ it follows that $\int kg(x)dx = k\int g(x)\,dx = f(x)$, since constant multiples can be taken outside the front of integral signs. Dividing both sides by the constant multiple k we obtain $\int g(x)\,dx = \frac{1}{k}f(x) + c$, since we can add in the arbitrary constant at the end. While this is not a new integration technique and problems must be worded in a way to give both functions, we can often deduce an anti-derivative using this technique, it is also called integration by recognition.

WORKED EXAMPLE 6 Integration by recognition

Differentiate $\cos^{-1}\left(\frac{\sqrt{x}}{4}\right)$ and hence evaluate $\int_0^8 \frac{1}{\sqrt{16x - x^2}}\,dx$.

THINK	WRITE/DRAW
1. Express y in terms of u and u in terms of x.	Let $y = \cos^{-1}\left(\frac{\sqrt{x}}{4}\right)$ $= \cos^{-1}\left(\frac{u}{4}\right)$ where $u = \sqrt{x} = x^{\frac{1}{2}}$
2. Differentiate y with respect to u and u with respect to x.	$\frac{dy}{du} = \frac{-1}{\sqrt{16 - u^2}}$ and $\frac{du}{dx} = \frac{1}{2}x^{-\frac{1}{2}} = \frac{1}{2\sqrt{x}}$
3. Determine $\frac{dy}{dx}$ using the chain rule.	$\frac{dy}{dx} = \frac{dy}{du}\frac{du}{dx} = \frac{1}{2\sqrt{x}} \times \frac{-1}{\sqrt{16 - u^2}}$
4. Substitute back for u.	$\frac{dy}{dx} = \frac{-1}{2\sqrt{x}\sqrt{16 - x}}$ since $0 < x < 16$ $\frac{dy}{dx} = \frac{-1}{2\sqrt{16x - x^2}}$
5. Write the result as a derivative of one function.	$\frac{d}{dx}\left[\cos^{-1}\left(\frac{\sqrt{x}}{4}\right)\right] = \frac{-1}{2\sqrt{16x - x^2}}$
6. Use integration by recognition.	$\int \frac{-1}{2\sqrt{16x - x^2}}\,dx = \cos^{-1}\left(\frac{\sqrt{x}}{2}\right)$
7. Take the constant factor outside the integral sign.	$-\frac{1}{2}\int \frac{1}{\sqrt{16x - x^2}}\,dx = \cos^{-1}\left(\frac{\sqrt{x}}{4}\right)$
8. Multiply by the constant factor, add in the arbitrary constant $+c$.	$\int \frac{1}{\sqrt{16x - x^2}}\,dx = -2\cos^{-1}\left(\frac{\sqrt{x}}{4}\right) + c$
9. However in this we are required to evaluate a definite integral.	$\int_0^8 \frac{1}{\sqrt{16x - x^2}}\,dx = \left[-2\cos^{-1}\left(\frac{\sqrt{x}}{4}\right)\right]_0^8$

10. Substitute in the upper and lower terminals and simplify.	$\left(-2\cos^{-1}\left(\frac{\sqrt{8}}{4}\right)\right) - (-2\cos^{-1}(0))$ $= \left(-2\cos^{-1}\left(\frac{\sqrt{2}}{2}\right)\right) - (-2\cos^{-1}(0))$
11. Evaluate.	$= -2 \times \frac{\pi}{4} + 2 \times \frac{\pi}{2}$
12. State the final value of the definite integral.	$\int_0^8 \frac{1}{\sqrt{16x - x^2}}\, dx = \frac{\pi}{2}$

9.3.2 Graphs of anti-derivatives of functions

Given a function $f(x)$, we can sketch the graph of the anti-derivative $F(x) = \int f(x)\, dx$, by noting key features and $F'(x) = f(x)$. The table below shows the relationships between the graphs. Note that the graph of the anti-derivative cannot be completely determined as it includes a constant of integration, which is just a vertical translation of the graph of $F(x)$ parallel to the y-axis.

Graph of function f	Graph of anti-derivative F
$f(x) < 0$ for $x \in (a, b)$	$F(x)$ has a negative gradient, or is decreasing for $x \in (a, b)$
$f(x) > 0$ for $x \in (a, b)$	$F(x)$ has a positive gradient, or is increasing for $x \in (a, b)$
$f(x)$ cuts the x-axis at $x = a$ from negative to positive	$F(x)$ has a local minimum at $x = a$
$f(x)$ cuts the x-axis at $x = a$ from positive to negative	$F(x)$ has a local maximum at $x = a$
$f(x)$ touches the x-axis at $x = a$	$F(x)$ has stationary point of inflection at $x = a$
$f(x)$ has a turning point at $x = a$	$F(x)$ has a point of inflexion at $x = a$ (non-stationary unless $f(x) = 0$)

Recall that the derivative of a polynomial is a polynomial which is one degree lower. Therefore, if $f(x)$ is a polynomial of degree n, then the anti-derivative $F(x)$ will be a polynomial of degree $n + 1$. For example, when $f(x)$ is a linear function, $F(x)$ will be a quadratic function.

WORKED EXAMPLE 7 Sketching graphs of anti-derivatives

a. Given the graph below, sketch a possible graph of the anti-derivative.

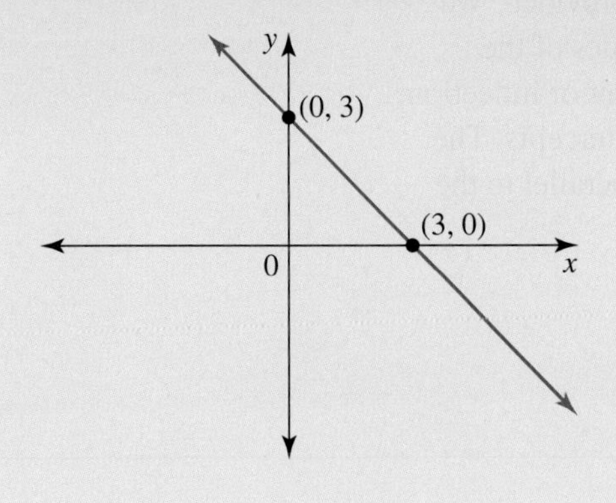

b. The graph of the gradient function is shown below. Sketch a possible graph of the function.

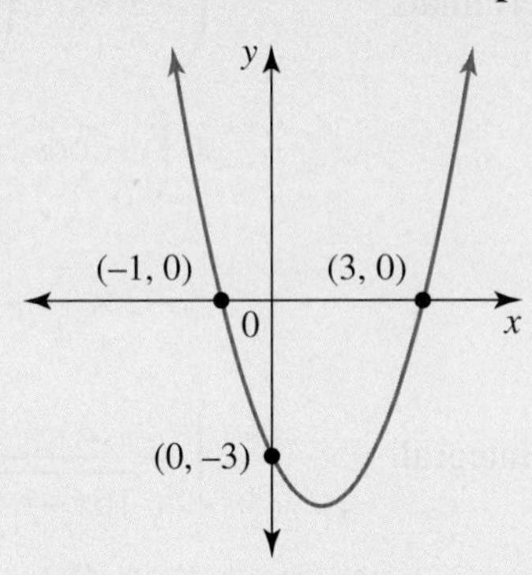

THINK

a. 1. The given graph crosses the x-axis at $x = 3$, so the graph of the anti-derivative has a stationary point at $x = 3$.

2. At $x = 3$ the given graph changes from positive to negative as x increases, so the stationary point is a maximum turning point.

3. No further information is provided, so we cannot determine the y-value of the turning point or any values of the axis intercepts. The graph of the anti-derivative could be translated parallel to the y-axis.

b. 1. The given graph crosses the x-axis at $x = -1$ and $x = 3$, so the graph of the anti-derivative has stationary points at these values.

2. At $x = -1$ the gradient changes from positive to negative as x increases, so the stationary point is a maximum turning point. At $x = 3$ the gradient changes from negative to positive as x increases, so the stationary point is a minimum turning point.

3. The gradient function has a turning point at $x = 1$ so the graph of the anti-derivative has a point of inflection at $x = 1$.

4. No further information is provided. We cannot determine the y-values of the stationary points or the point of inflection, or any values of the axis intercepts. The graph could be translated parallel to the y-axis.

WRITE/DRAW

a.

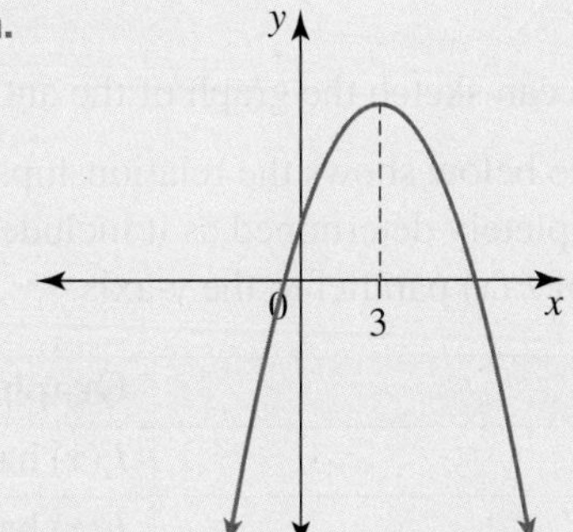

b.

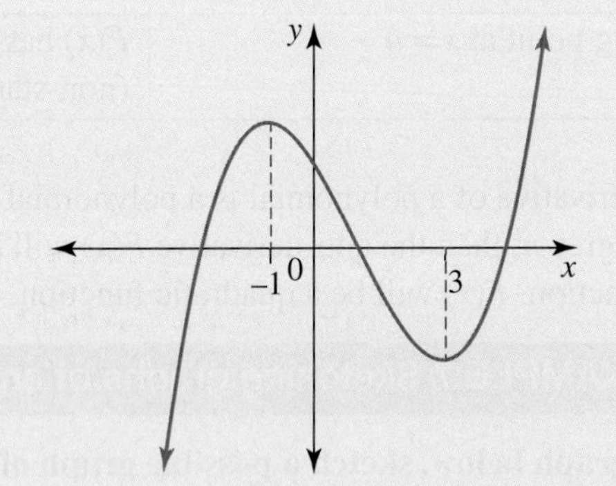

9.3 Exercise

Students, these questions are even better in jacPLUS

Receive immediate feedback and access sample responses

Access additional questions

Track your results and progress

Find all this and MORE in jacPLUS

Technology free

1. **WE6** Differentiate $\sin^{-1}\left(\frac{\sqrt{x}}{2}\right)$ and hence evaluate $\int_0^4 \frac{1}{\sqrt{4x-x^2}}\,dx$.

2. Differentiate $\arcsin\left(\frac{2}{x}\right)$ and hence evaluate $\int_{\frac{4\sqrt{3}}{3}}^{4} \frac{1}{x\sqrt{x^2-4}}\,dx$.

3. If $f(x)=\sin^{-1}\left(\frac{x^3}{8}\right)$ determine $f'(x)$ and hence evaluate $\int_2^{\sqrt[3]{4}} \frac{x^2}{\sqrt{64-x^6}}\,dx$.

4. If $f(x)=\cos^{-1}\left(\frac{4}{\sqrt{x}}\right)$ determine $f'(x)$ and hence evaluate $\int_{\frac{64}{3}}^{64} \frac{1}{x\sqrt{x-16}}\,dx$.

5. Determine $\frac{d}{dx}\left[\arccos\left(\frac{6}{x}\right)\right]$ and hence evaluate $\int_{4\sqrt{3}}^{12} \frac{1}{x\sqrt{x^2-36}}\,dx$.

6. If $f(x)=\arctan\left(\frac{4}{\sqrt{x}}\right)$ determine $f'(x)$ and hence evaluate $\int_{\frac{16}{3}}^{16} \frac{1}{\sqrt{x}(x+16)}\,dx$.

7. Determine $\frac{d}{dx}\left[\tan^{-1}\left(\frac{\sqrt{x}}{3}\right)\right]$ and hence evaluate $\int_9^{27} \frac{1}{\sqrt{x}(9+x)}\,dx$.

8. Determine $\frac{d}{dx}\left[\tan^{-1}\left(\frac{x^2}{4}\right)\right]$ and hence evaluate $\int_0^2 \frac{x}{x^4+16}\,dx$.

9. Differentiate $\log_e(\tan(x))$ and hence determine the area bounded by the curve $y=\text{cosec}(2x)$ the x-axis and the lines $x=\frac{\pi}{6}$ and $x=\frac{\pi}{3}$.

10. If $y = \log_e\left(\sqrt{\dfrac{\sin(2x)+\cos(2x)}{\sin(2x)-\cos(2x)}}\right)$ show that $\dfrac{dy}{dx} = 2\sec(4x)$. Hence determine the area enclosed between the curve $y = \sec(4x)$, the x-axis and the lines $x = \dfrac{\pi}{6}$ and $x = \dfrac{\pi}{4}$.

11. a. WE7 Given the graph below, sketch a possible graph of the anti-derivative.

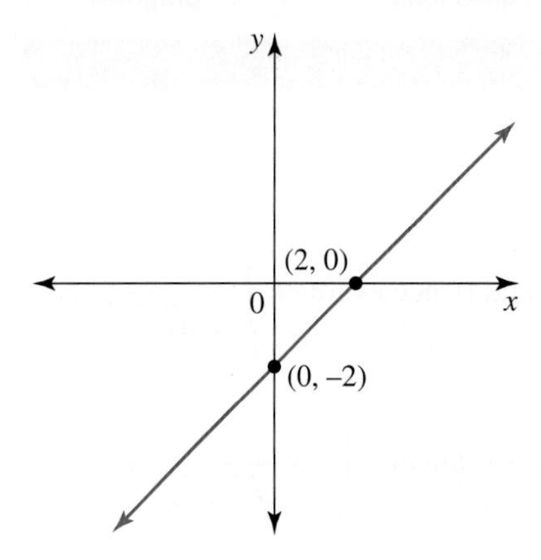

b. The graph of the gradient function is shown below. Sketch a possible graph of the function.

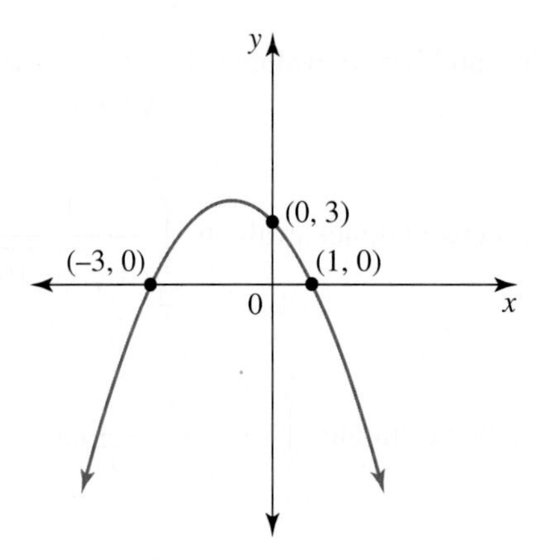

12. The graph of $y = f'(x)$ is shown. For the graph of $y = f(x)$, state the x-values of the stationary points and their nature.

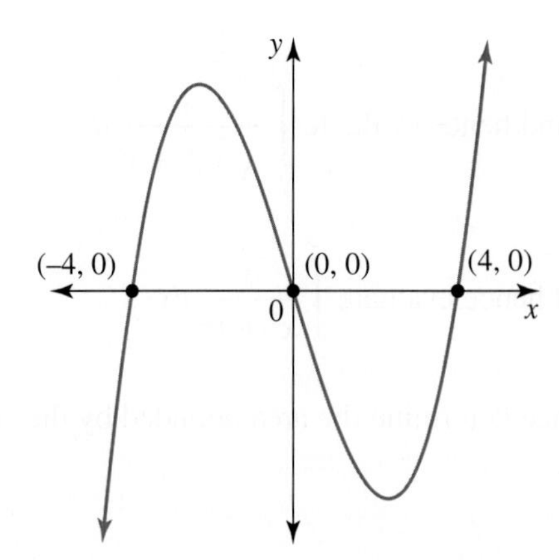

13. The following graphs are of gradient functions. In each case sketch a possible graph of the original function.

a.

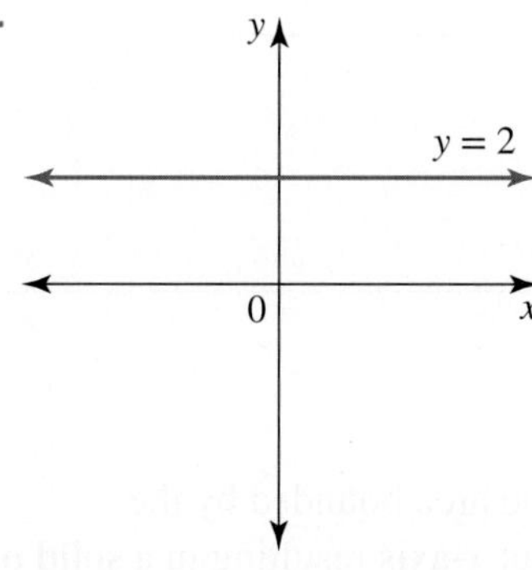

b.

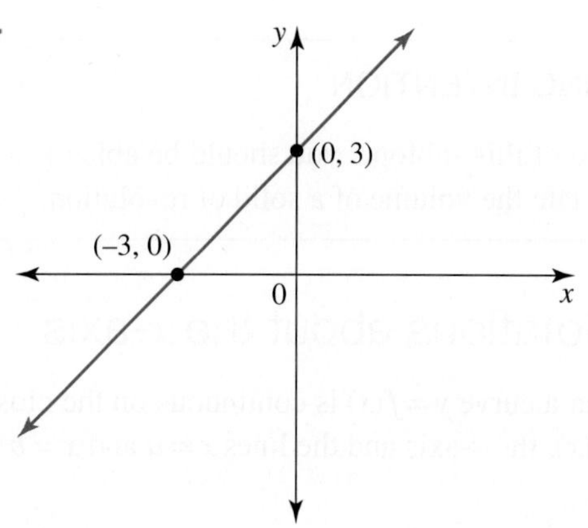

14. Given the differential equation $\frac{dy}{dx} = \tan\left(\frac{1}{x}\right), y\left(\frac{\pi}{8}\right) = \frac{1}{2}$.

Determine a definite integral for y in terms of x.

9.3 Exam questions

Question 1 (4 marks) TECH-FREE

Determine $\frac{d}{dx}\left[\cos^{-1}\left(\frac{4}{x^2}\right)\right]$ and hence evaluate $\int_2^{2\sqrt{2}} \frac{1}{x\sqrt{x^4 - 16}}dx$.

Question 2 (4 marks) TECH-FREE

Find $\frac{d}{dx}\left[\log_e(\tan(x) + \sec(x))\right]$ and hence determine the area enclosed between the curve $y = \sec(x)$, the coordinate axes and the line $x = \frac{\pi}{4}$.

Question 3 (2 marks) TECH-FREE

Consider the following graph of a gradient function, $f'(x)$.

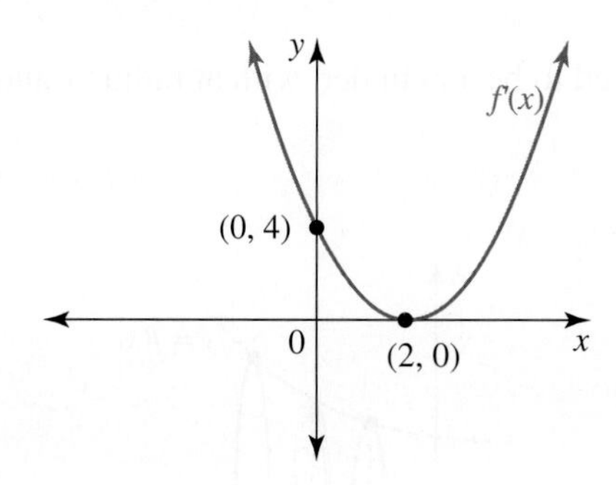

Sketch the graph of the original function, $f(x)$.

More exam questions are available online.

9.4 Solids of revolution

LEARNING INTENTION

At the end of this subtopic you should be able to:
- calculate the volume of a solid of revolution.

9.4.1 Rotations about the x-axis

Suppose that a curve $y = f(x)$ is continuous on the closed interval $a \le x \le b$. The area bounded by the curve $y = f(x)$, the x-axis and the lines $x = a$ and $x = b$ can be rotated 360° about x-axis resulting in a **solid of revolution**.

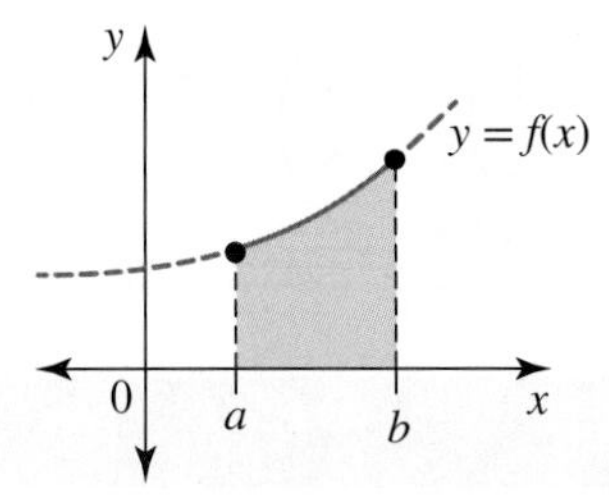

The solid generated is symmetrical about the x-axis and any vertical cross-section is a circle with a radius equal to the value of y at that point. For example, the radius of the circle at $x = a$ is $f(a)$.

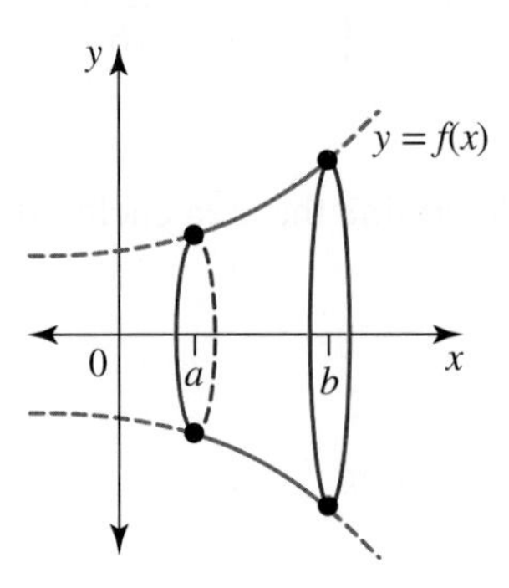

A thin vertical slice could be considered to be a cylinder, with at radius y and a height δx. The volume of the cylinder is $\pi y^2 \delta x$.

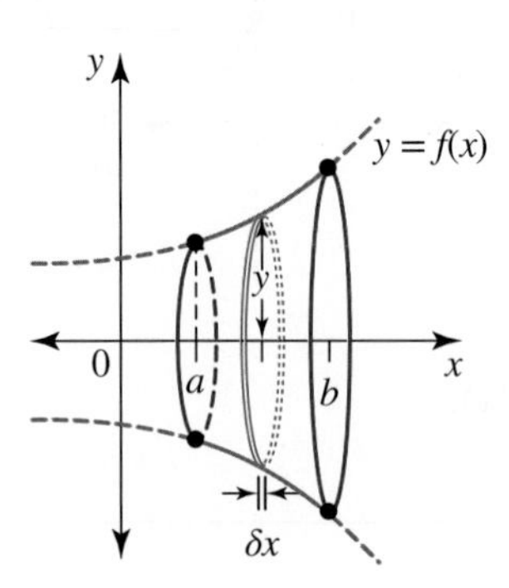

To determine the volume of the solid of revolution, the volumes of all of the cylinders between $x = a$ and $x = b$ are added together and the limit as $\delta x \to 0$ determined.

$$V = \lim_{\delta x \to 0} \sum_{x=a}^{x=b} \pi y^2 \delta x$$

$$= \int_a^b \pi y^2 dx$$

$$= \pi \int_a^b [f(x)]^2 dx$$

WORKED EXAMPLE 8 Calculating volumes of solids of revolution rotated about the x-axis

a. Sketch the graph of $y = 2x$ and identify the region bounded by the graph, the x-axis and the line $x = 2$.

b. Calculate the volume of the solid of revolution when the region is rotated about the x-axis.

THINK	WRITE
a. Sketch the lines $y = 2x$ and $x = 2$ and shade the area bounded by $y = 2x$, $x = 2$ and the x-axis.	a. 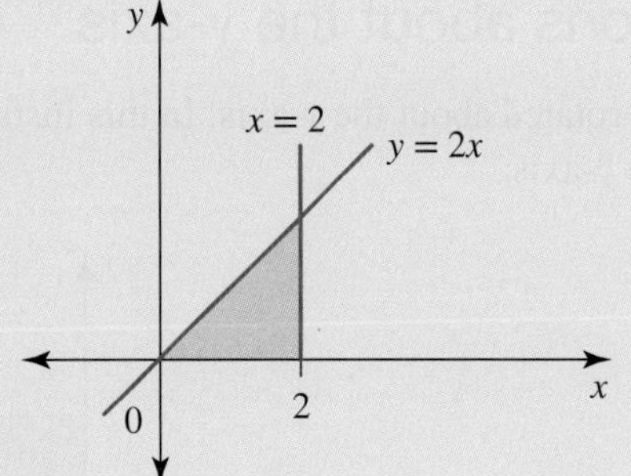
b. 1. The volume generated is bounded by $x = 0$ and $x = 2$. The formula for volume of a solid formed by rotation about the x-axis is $\pi \int_b^a [f(x)]^2 dx$.	b. $V = \pi \int_0^2 [f(x)]^2 dx$
2. The integrand needs to be expressed in terms of x. Substitute $f(x) = 2x$.	$= \pi \int_0^2 (2x)^2 dx$
3. Simplify the integrand.	$= 4\pi \int_0^2 x^2 dx$
4. Determine the anti-derivative.	$= 4\pi \left[\frac{1}{3}x^3\right]_0^2$ $= 4\pi \left[\frac{1}{3}(2)^3 - \frac{1}{3}(0)^3\right]$

5. Evaluate the integral. $= 4\pi\left[\frac{8}{3} - 0\right]$

$= \frac{32\pi}{3}$

6. State the volume. The volume is $\frac{32\pi}{3}$ units^3.

In the worked example above, the solid generated is a cone. The height of the cone is 2 units and the radius of the base is 4 units. Using the formula for volume of a cone:

$$\begin{aligned}V &= \frac{1}{3}\pi r^2 h \\ &= \frac{1}{3}\pi(4)^2(2) \\ &= \frac{32\pi}{3} \text{ units}^3\end{aligned}$$

This is the same result as determined in the worked example.

9.4.2 Rotations about the y-axis

Curves can also be rotated about the y-axis. In this instance, the region is bounded by the curve $x = f(y)$, the lines $y = a$, $y = b$ and the y-axis.

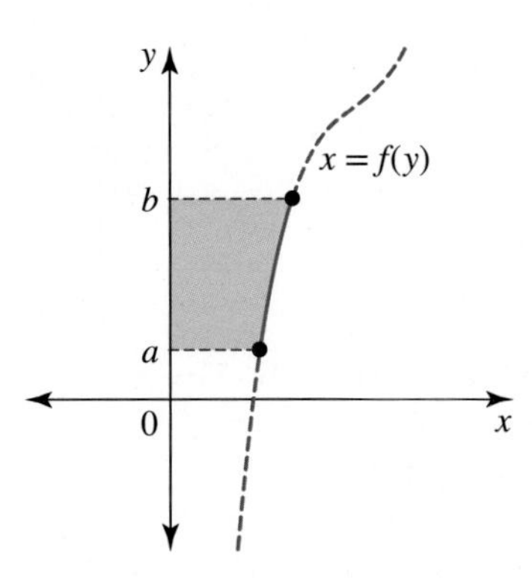

The solid generated is symmetrical about the y-axis and any horizontal cross-section is a circle with a radius equal to the value of x at that point. For example, the radius of the circle at $y = a$ is $f(a)$.

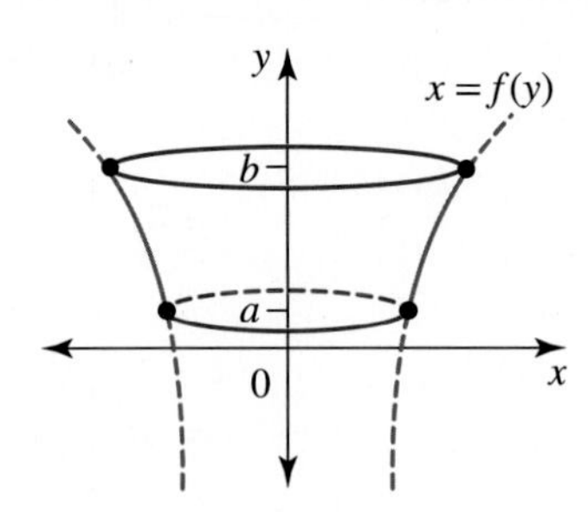

A thin horizontal slice could be considered to be a cylinder, with at radius x and a height δy. The volume of the cylinder is $\pi x^2 \delta y$.

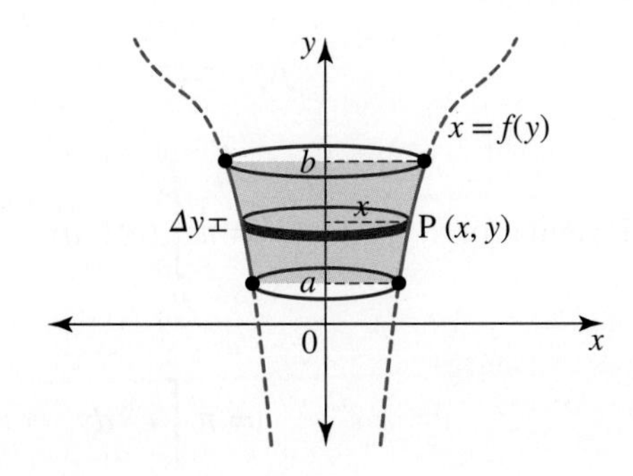

To determine the volume of the solid of revolution, the volumes of all of the cylinders between $y = a$ and $y = b$ are added together and the limit as $\delta y \to 0$ determined.

$$
\begin{aligned}
V &= \lim_{\delta y \to 0} \sum_{y=a}^{y=b} \pi x^2 \delta y \\
&= \int_a^b \pi x^2 dy \\
&= \pi \int_a^b [f(y)]^2 \, dy
\end{aligned}
$$

WORKED EXAMPLE 9 Calculating volumes of solids of revolution rotated about the y-axis

a. Sketch the region bounded by the curve $y = \log_e(x)$, the x-axis, the y-axis and the line $y = 2$.

b. Calculate the volume of the solid of revolution formed when the region is rotated about the y-axis.

THINK	WRITE/DRAW
a. Sketch the curve $y = \log_e(x)$ and the line $y = 2$. Shade the area bounded by $y = \log_e(x)$, the x-axis, the y-axis and the line $y = 2$.	a. 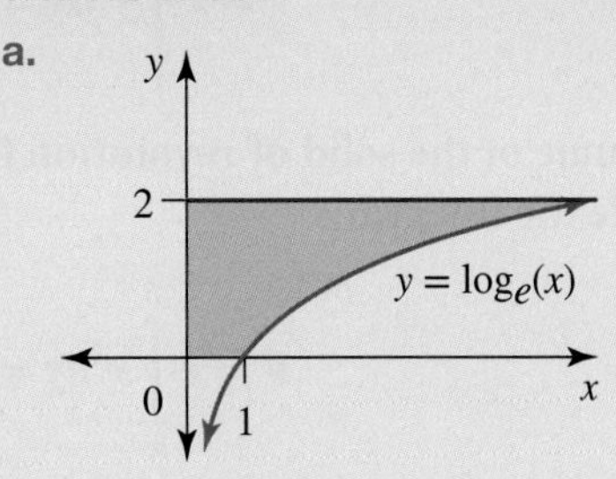
b. 1. The volume generated is bounded by $y = 0$ and $y = 2$. The formula for volume of a solid formed by rotation about the y-axis is $\pi \int_b^a [f(y)]^2 \, dy$.	b. 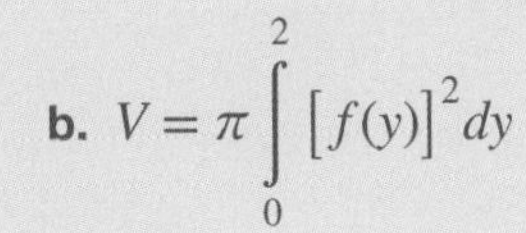 $V = \pi \int_0^2 [f(y)]^2 dy$

2. The integrand needs to be expressed in terms of y. Rearrange $y = \log_e(x)$ to make x the subject.

$$y = \log_e(x)$$
$$x = e^y$$
$$f(y) = e^y$$

3. Substitute $f(y) = e^y$ into the formula for volume.

$$V = \pi \int_0^2 (e^y)^2 \, dy$$

4. Simplify the integrand.

$$= \pi \int_0^2 e^{2y} dy$$

5. Determine the anti-derivative.

$$= \pi \left[\frac{1}{2}e^{2y}\right]_0^2$$

6. Evaluate the integral.

$$= \pi \left[\frac{1}{2}e^4 - \frac{1}{2}e^0\right]$$
$$= \frac{\pi}{2}\left(e^4 - 1\right)$$

7. State the volume.

The volume is $\frac{\pi}{2}\left(e^4 - 1\right) \approx 84.19$ units^3.

Volumes of solids of revolution

To determine the volume of the solid of revolution formed by rotating $y = f(x)$ about the x-axis from $x = a$ to $x = b$, evaluate the integral:

$$V = \pi \int_a^b y^2 dx = \pi \int_a^b [f(x)]^2 \, dx$$

To determine the volume of the solid of revolution formed by rotating $x = f(y)$ about the y-axis from $y = a$ to $y = b$, evaluate the integral:

$$V = \pi \int_a^b x^2 dy = \pi \int_a^b [f(y)]^2 \, dy$$

9.4.3 Applications

The volume of some common geometric shapes can now be found using calculus, by rotation lines or curves about the x- or y-axis and using the techniques described.

WORKED EXAMPLE 10 Calculating volumes of everyday objects using integration

A drinking glass has a base diameter of 5 cm, a top diameter of 7 cm and a height of 11 cm. Determine the capacity of the glass to the nearest mL.

THINK

WRITE/DRAW

1. Sketch the graph and identify the area to rotate. Write the coordinates of the points A and B.

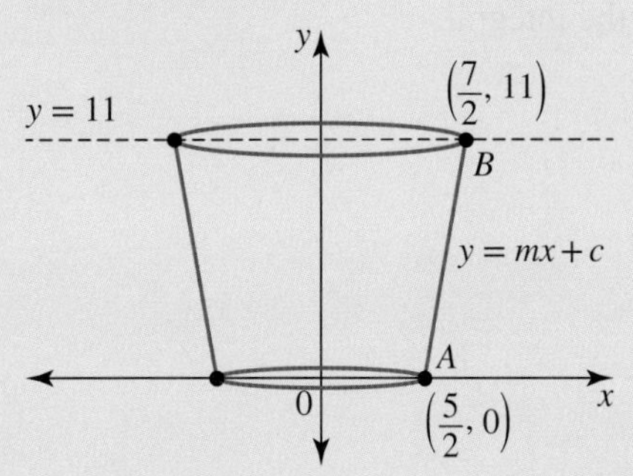

2. Determine the gradient of the line segment AB joining the points $A\left(\frac{5}{2},0\right)$ and $B\left(\frac{7}{2},11\right)$.

$$m=\frac{11-0}{\frac{7}{2}-\frac{5}{2}}=11$$

3. Use $y-y_1=m\left(x-x_1\right)$ to determine the equation of the line segment AB.

$$y-0=11\left(x-\frac{5}{2}\right)$$

$$y=11x-\frac{55}{2}$$

4. The glass is formed by rotating AB about the y-axis. The formula for rotation about the y-axis is $V=\pi\int_a^b x^2dy$. We are interested in the region between $y=0$ and $y=11$.

$$V=\pi\int_a^b\left[f(y)\right]^2dy$$

$$=\pi\int_0^{11}x^2dy$$

5. The integrand needs to be expressed in terms of y. Rearrange $y=11x-\frac{55}{2}$ to make x the subject.

$$y=11x-\frac{55}{2}$$

$$y+\frac{55}{2}=11x$$

$$x=\frac{1}{11}y+\frac{5}{2}$$

$$f(y)=\frac{1}{11}y+\frac{5}{2}$$

6. Substitute the equation for x into the formula for volume so that the integrand is expressed in terms of y only.

$$V = \pi \int_0^{11} [f(y)]^2 \, dy$$

$$= \pi \int_0^{11} \left(\frac{1}{11}y + \frac{5}{2}\right)^2 dy$$

7. Determine the anti-derivative.

$$= \pi \left[\frac{1}{3}\left(\frac{1}{11}y + \frac{5}{2}\right)^3 \times 11\right]_0^{11}$$

$$= \frac{11\pi}{3}\left[\left(\frac{1}{11}y + \frac{5}{2}\right)^3\right]_0^{11}$$

8. Evaluate the integral.

$$= \frac{11\pi}{3}\left[\left(1 + \frac{5}{2}\right)^3 - \left(0 + \frac{5}{2}\right)^3\right]$$

$$= \frac{11\pi}{3}\left(\left(\frac{7}{2}\right)^3 - \left(\frac{5}{2}\right)^3\right)$$

$$= \frac{11\pi}{3} \times \frac{109}{4}$$

$$= \frac{1199\pi}{12}$$

$$\approx 314$$

9. State the volume in cm^3. Use $1\,cm^3 = 1\,mL$ to state the volume of the glass to the nearest mL.

The volume is $314\,cm^3$. This means that the glass can hold 314 mL.

9.4 Exercise

Students, these questions are even better in jacPLUS

Receive immediate feedback and access sample responses

Access additional questions

Track your results and progress

Find all this and MORE in jacPLUS

Technology free

1. a. WE8 Sketch the graph of $y = \sqrt{x}$ and identify the region bounded by the graph, the x-axis and the line $x = 4$.
   b. Calculate the volume of the solid of revolution when the region is rotated about the x-axis.

2. The area bounded by the curve $y = 4 - x^2$ and the coordinate axis is rotated about the x-axis to form a solid of revolution. Determine the volume of the solid formed.

3. a. WE9 Sketch the region bounded by the curve $y = \sqrt{x}$, the y-axis, the origin and the line $y = 2$.
   b. Calculate the volume of the solid of revolution formed when the region is rotated about the y-axis.

4. The area bounded by the curve $y = 4 - x^2$ and the coordinate axes is rotated about the y-axis to form a solid of revolution. Calculate the volume of the solid.

5. Consider the line $y = 3x$.
   a. Determine the volume of the solid of revolution formed by rotating the section of the line between the origin and $x = 5$ about the x-axis.
   b. Determine the volume of the solid of revolution formed by rotating the section of the line between the origin and $y = 5$ about the y-axis.

6. A cone is formed by rotating the line segment of $2x + 3y = 6$ cut off by the coordinate axes about:
   a. the x-axis
   b. the y-axis.

Determine the volume of the solid formed in each case.

7. If the region bounded by the curve $y = \sec(2x)$, the coordinate axes and $x = \frac{\pi}{8}$ is rotated 360° about the x-axis, determine the volume of the solid formed.

8. If the region bounded by the curve $y = 2e^{\frac{x}{2}}$, the coordinate axes and $x = 2$ is rotated 360° about the x-axis, determine the volume of the solid formed.

9. **WE10** A plastic bucket has a base diameter of 20 cm, a top diameter of 26 cm and a height of 24 cm. Determine the capacity of the bucket to the nearest litre.

10. A soup bowl has a base radius of 6 cm, a top radius of 8 cm, and a height of 7 cm. The edge of the bowl is a parabola of the form $y = ax^2 + c$. Determine the capacity of the soup bowl to the nearest mL.

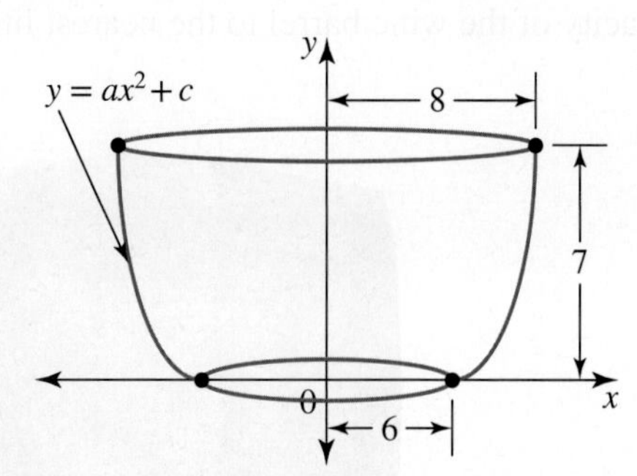

11. a. If the region bounded by the curve $y = 3\sin(2x)$, the origin, the x-axis and the first intercept the curve makes with the x-axis is rotated 360° about the x-axis, determine the volume of the solid formed.
b. If the region bounded by the curve $y = 4\cos(3x)$, the origin, the x-axis and the first intercept the curve makes with the x-axis is rotated 360° about the x-axis, determine the volume of the solid formed.

12. Consider the curve $y = 3x^2 + 4$.
a. Determine the volume of the solid formed by rotating the area between the curve, the coordinate axes and the line $x = 2$ about the x-axis.
b. Determine the volume of the solid formed by rotating the area between the curve, $y = 4$ and $y = 10$ about the y-axis.

13. Consider the area between the curve $y = x^2 - 9$ and the x-axis.
a. If the area described is rotated 360° about the x-axis, determine the volume of the solid formed.
b. If the area described is rotated 360° about the y-axis, determine the volume of the solid formed.

14. a. Determine the volume of the solid formed by rotating the ellipse $\frac{x^2}{25} + \frac{y^2}{16} = 1$ about the x-axis.
b. Determine the volume of the solid formed by rotating the ellipse $\frac{x^2}{25} + \frac{y^2}{16} = 1$ about the y-axis.
c. Determine the volume of the solid formed by rotating the ellipse $\frac{x^2}{a^2} + \frac{y^2}{b^2} = 1$ about the x-axis.
d. Determine the volume of the solid formed by rotating the ellipse $\frac{x^2}{a^2} + \frac{y^2}{b^2} = 1$ about the y-axis.
e. An egg can be regarded as an ellipsoid (formed by rotating an ellipse). If the egg has a total length of 57 mm and its diameter at the centre is 44 mm, determine an equation in the form $\frac{x^2}{a^2} + \frac{y^2}{b^2} = 1$ that can be rotated to form the required ellipsoid.
f. Use the equation from **e** to determine the volume of the egg correct to the nearest cubic millimetre.

15. The diagram below shows a wine barrel. The barrel has a total length of 60 cm, a total height of 60 cm at the middle and a total height of 40 cm at the ends.
a. If the upper arc can be represented by a parabola, demonstrate that its equation is given by $y = -\frac{x^2}{90} + 30$ for $-30 \le x \le 30$.
b. If the arc is rotated about the x-axis, determine the capacity of the wine barrel to the nearest litre.

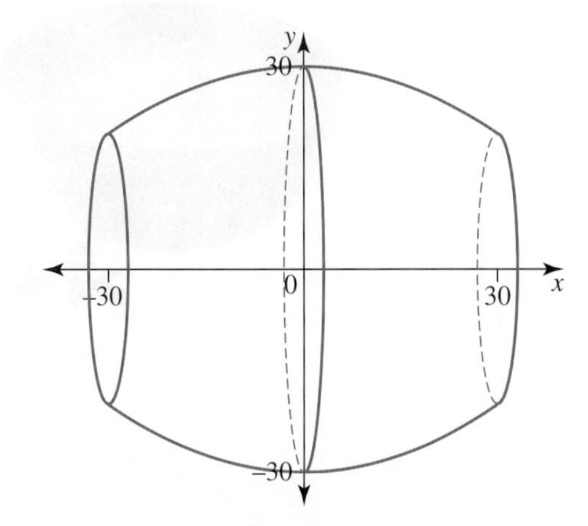

16. A vase has a base radius of 6 cm, its top radius is 8 cm and its height is 20 cm. Determine the volume of the vase in cubic centimetres, correct to 2 decimal places, if the side is modelled by:

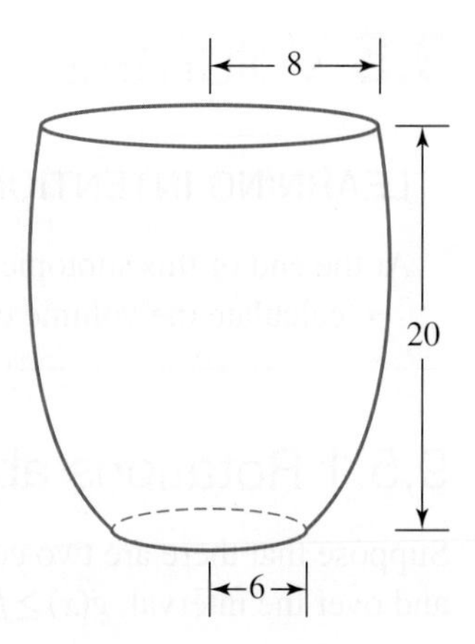

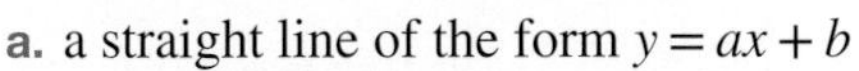

   a. a straight line of the form $y = ax + b$
   b. a parabola of the form $y = ax^2 + c$
   c. a cubic $y = ax^3 + c$
   d. a quartic of the form $y = ax^4 + c$.

17. a. The region bounded by the curve $\frac{1}{x} + \frac{1}{y} = \frac{1}{4}$, the x-axis, $x = 0$ and $x = 2$ is rotated about the x-axis. Determine the volume of the solid formed.
   b. The region bounded by the curve $\sqrt{x} + \sqrt{y} = 2$ and the coordinate axes is rotated about the x-axis. Determine the volume of the solid formed.

18. A fish bowl consists of a portion of a sphere of radius 20 cm. The bowl is filled with water so that the radius of the water at the top is 16 cm and the base of the bowl has a radius of 12 cm.
If the total height of the water in the bowl is 28 cm, determine the volume of water in the bowl.

9.4 Exam questions

Question 1 (5 marks) TECH-FREE

Source: *VCE 2020, Specialist Mathematics Exam 1, Q8; © VCAA.*

Find the volume, V of the solid of revolution formed when the graph of $y = 2\sqrt{\dfrac{x^2 + x + 1}{(x+1)\left(x^2+1\right)}}$ is rotated about the x-axis over the interval $[0, \sqrt{3}]$.

Give your answer in the form $V = 2\pi\left(\log_e(a) + b\right)$, where $a, b, \in R$.

Question 2 (4 marks) TECH-FREE

Source: *VCE 2019, Specialist Mathematics Exam 1, Q8; © VCAA.*

Find the volume of the solid of revolution formed when the graph of $y = \sqrt{\dfrac{1+2x}{1+x^2}}$ is rotated about the x-axis over the interval $[0, 1]$.

Question 3 (3 marks) TECH-FREE

Source: *VCE 2015, Specialist Mathematics Exam 1, Q5; © VCAA.*

Find the volume generated when the region bounded by the graph of $y = 2x^2 - 3$, the line $y = 5$ and the y-axis is rotated about the y-axis.

More exam questions are available online.

9.5 Volumes

LEARNING INTENTION

At the end of this subtopic you should be able to:
- calculate the volume of complex solids of revolution.

9.5.1 Rotations about the x-axis

Suppose that there are two curves $y_1 = f(x)$ and $y_2 = g(x)$ that are continuous over the closed interval $a \leq x \leq b$ and over the interval, $g(x) \geq f(x)$.

If the area bounded by the functions and the lines $x = a$ and $x = b$ is rotated about the x-axis, the resulting object will be hollow in the middle.

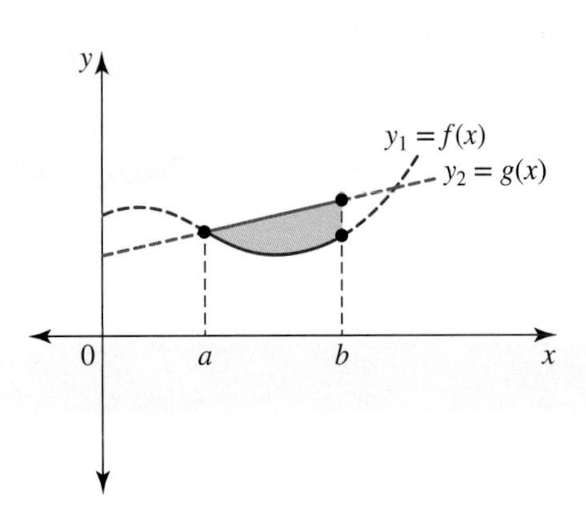

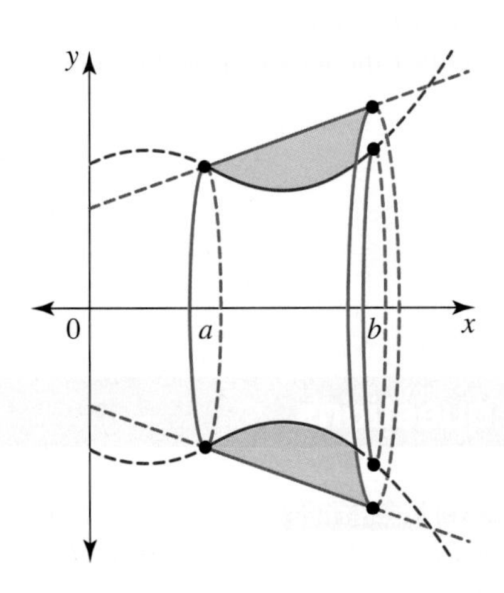

A cross-section of the shape would look like a circular washer. The inner radius, r_1, would be equal to $y_1 = f(x)$ and the outer radius, r_2, would be equal to $y_2 = g(x)$.

The area of this cross-section can be found be determining $\pi(r_2^2 - r_1^2)$.

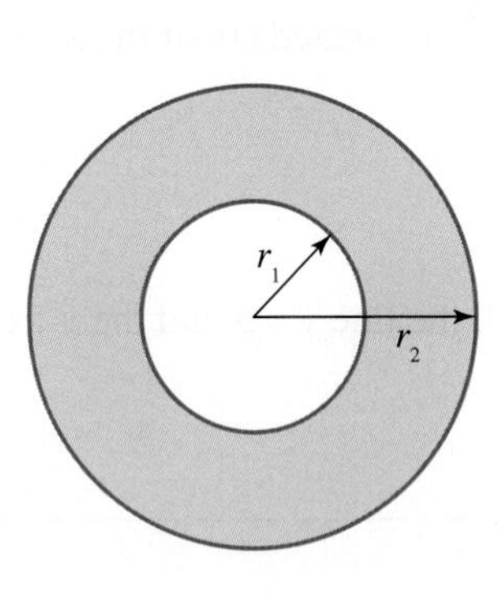

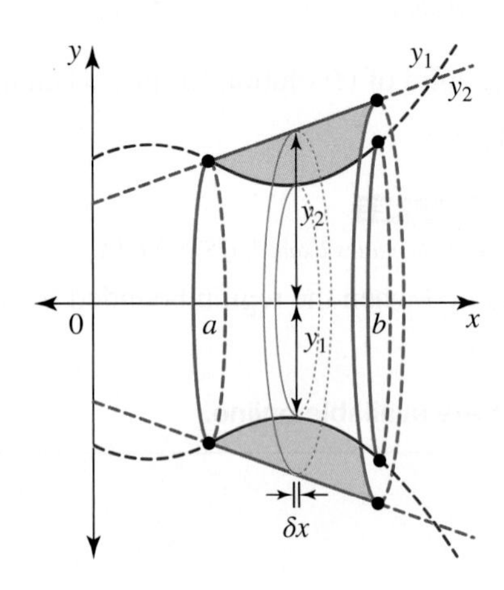

The volume of the thin vertical slice of height δx would have a volume of $\pi(y_2^2 - y_1^2)\delta x$.

The volume of the solid of revolution can be found by adding the volume of the thin circular washers between $x = a$ and $x = b$ and finding the limit as $\delta x \to 0$.

$$V = \sum_{x=a}^{x=b} \pi \left(y_2^2 - y_1^2\right) \delta x$$

$$= \int_a^b \pi \left(y_2^2 - y_1^2\right) dx$$

$$= \pi \int_a^b \left([g(x)]^2 - [f(x)]^2\right) dx$$

WORKED EXAMPLE 11 Calculating volumes of revolution (1)

Determine the volume formed when the area bounded by the curve $y = 4 - x^2$ and the line $y = 3$ is rotated about:

a. **the x-axis**

b. **the y-axis.**

THINK	WRITE
a. 1. Determine the point of intersection between the curve and the line.	a. Let $y_1 = 4 - x^2$ and $y_2 = 3$ $y_1 = y_2$. $4 - x^2 = 3$ $x^2 = 1$ $x = \pm 1$
2. Sketch the graph and identify the area to rotate about the x-axis.	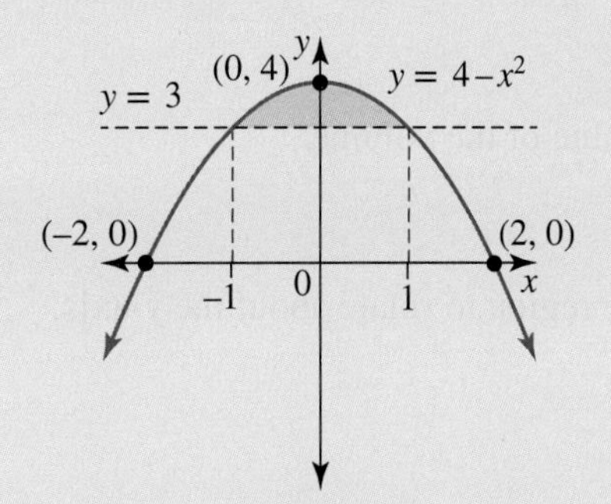 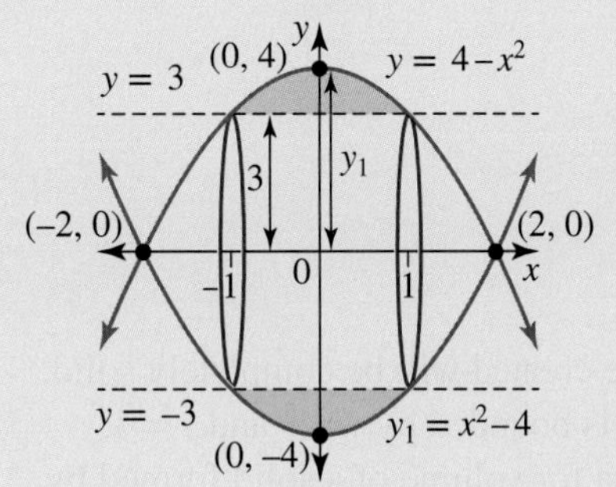

3. The volume can be calculated using

$$V = \pi \int_a^b \left([g(x)]^2 - [f(x)]^2 \right) dx.$$

$a = -1$ and $b = 1$.

$$V = \pi \int_{-1}^{1} \left([g(x)]^2 - [f(x)]^2 \right) dx$$

4. The integrand needs to be expressed in terms of x. In the shaded area, $4 - x^2 \geq 3$, therefore $y_2 = 4 - x^2$ and $y_1 = 3$.

$$= \pi \int_{-1}^{1} \left((4 - x^2)^2 - 3^2 \right) dx$$

5. Simplify the integrand so that it can be determined.

$$= \pi \int_{-1}^{1} \left(16 - 8x^2 + x^4 - 9 \right) dx$$

$$= \pi \int_{-1}^{1} \left(7 - 8x^2 + x^4 \right) dx$$

6. Determine the anti-derivative.

$$= \pi \left[7x - \frac{8}{3}x^3 + \frac{1}{5}x^5 \right]_{-1}^{1}$$

7. Evaluate the integral.

$$= \pi \left[\left(7 - \frac{8}{3} + \frac{1}{5} \right) - \left(-7 + \frac{8}{3} - \frac{1}{5} \right) \right]$$

$$= 2\pi \left(7 - \frac{40}{15} + \frac{3}{15} \right)$$

$$= \frac{136\pi}{15}$$

8. State the value of the volume.

The volume is $\frac{136\pi}{15}$ units^3.

b. 1. Identify the region to rotate about the y-axis.

b.

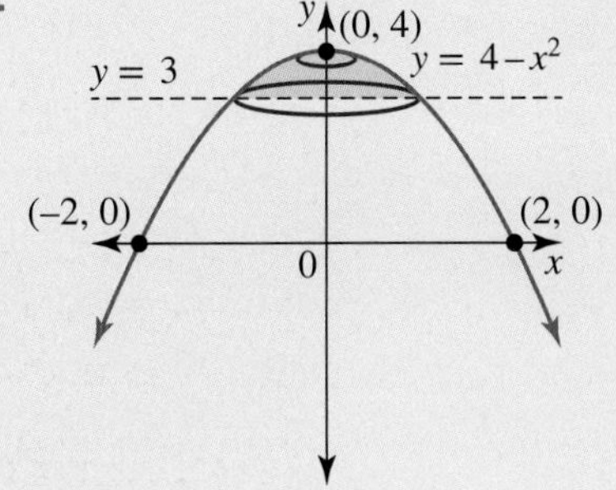

2. The volume created will be completely solid. The region is bounded by $y = 3$ and $y = 4$. The formula for volume of a solid formed by rotation about the y-axis is $\pi \int_a^b [f(y)]^2 \, dy$.

$$V = \pi \int_3^4 [f(y)]^2 \, dy$$

3. The integrand needs to be expressed in terms of y. Rearrange $y = 4 - x^2$ to make x^2 the subject.

$$y = 4 - x^2$$
$$x^2 = 4 - y$$
$$[f(y)]^2 = 4 - y$$

4. Substitute $[f(y)]^2 = 4 - y$ into the volume formula.

$$V = \pi \int_3^4 (4 - y)\, dy$$

5. Determine the anti-derivative.

$$= \pi \left[4y - \frac{1}{2}y^2\right]_3^4$$

6. Evaluate the integral.

$$= \pi \left[\left(4 \times 4 - \frac{1}{2} \times 4^2\right) - \left(4 \times 3 - \frac{1}{2} \times 3^2\right)\right]$$
$$= \pi \left(16 - 8 - 12 + \frac{9}{2}\right)$$
$$= \frac{\pi}{2}$$

7. State the volume.

The volume is $\frac{\pi}{2}$ units^3.

9.5.2 Rotations about the y-axis

It is also possible to rotate two curves about the y-axis. If $x_1 = f(y)$ and $x_2 = g(y)$ and $g(y) \geq f(y)$ over the interval $a \leq y \leq b$.

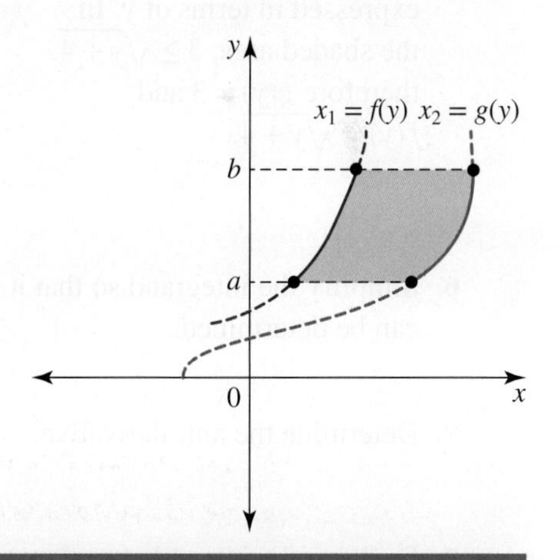

The solid formed will be hollow in the middle. The volume of the solid formed will be

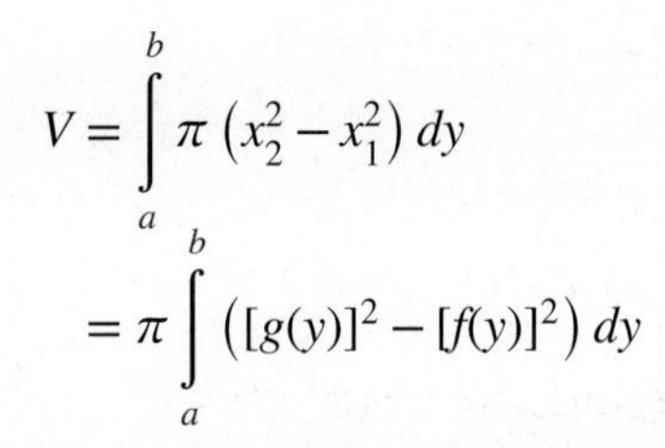

$$V = \int_a^b \pi \left(x_2^2 - x_1^2\right) dy$$
$$= \pi \int_a^b \left([g(y)]^2 - [f(y)]^2\right) dy$$

WORKED EXAMPLE 12 Calculating volumes of revolution (2)

Determine the volume formed when the area bounded by the curve $y = x^2 - 4$, the x-axis and the line $x = 3$ is rotated about:

a. the y-axis

b. the x-axis.

THINK

a. 1. Determine the point of intersection between the curve and the line.

WRITE

a. $y = x^2 - 4$ and $x = 3$.

$$y = (3)^2 - 4$$
$$= 9 - 4$$
$$= 5$$

2. Sketch the graph and identify the area to rotate about the y-axis.

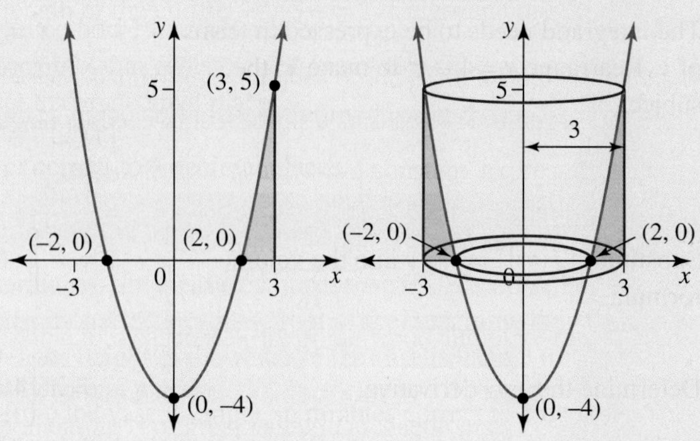

3. The volume can be calculated using $V = \pi \int_a^b \left([g(y)]^2 - [f(y)]^2 \right) dy$. $a = 0$ and $b = 5$.

$$V = \pi \int_0^5 \left([g(y)]^2 - [f(y)]^2 \right) dy$$

4. Rearrange $y = x^2 - 4$ to make x^2 the subject.

$$y = x^2 - 4$$
$$x^2 = y + 4$$

5. The integrand needs to be expressed in terms of y. In the shaded area, $3 \geq \sqrt{y+4}$, therefore $g(y) = 3$ and $f(y) = \sqrt{y+4}$.

$$V = \pi \int_0^5 (3^2 - (y+4))\, dy$$
$$= \pi \int_0^5 (9 - (y+4))\, dy$$

6. Simplify the integrand so that it can be determined.

$$= \pi \int_0^5 (5 - y)\, dy$$

7. Determine the anti-derivative.

$$= \pi \left[5y - \frac{1}{2}y^2 \right]_0^5$$

8. Evaluate the integral.

$$= \pi \left[\left(25 - \frac{1}{2} \times 5^2 \right) - 0 \right]$$
$$= \frac{25\pi}{2}$$

9. State the volume.

The volume is $\frac{25\pi}{2}$ units^3.

b. 1. Identify the region to rotate about the x-axis.

b.

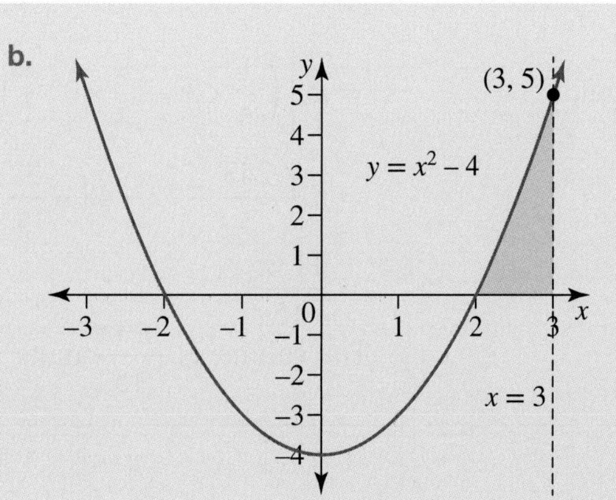

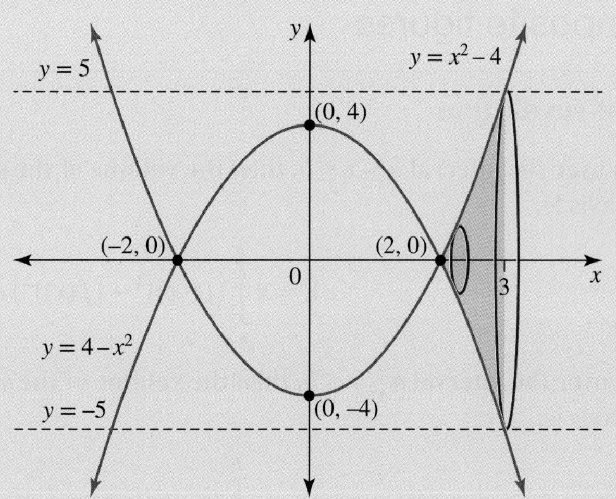

2. The volume created will be completely solid. The region is bounded by $x = 2$ and $x = 3$. The formula for volume of a solid formed by rotation about the x-axis is $\int_a^b \pi y^2 dx$.

$$V = \pi \int_2^3 \left[f(x)\right]^2 dx$$

3. Substitute $f(x) = x^2 - 4$ into the integrand so that it is expressed entirely in terms of x.

$$= \pi \int_2^3 \left(x^2 - 4\right)^2 dx$$

4. Simplify the integrand so that it can be determined.

$$= \pi \int_2^3 \left(x^4 - 8x^2 + 16\right) dx$$

5. Determine the anti-derivative.

$$= \pi \left[\frac{1}{5}x^5 - \frac{8}{3}x^3 + 16x\right]_2^3$$

6. Evaluate the integral.

$$= \pi\left[\left(\frac{1}{5}\times 3^5 - \frac{8}{3}\times 3^3 + 16\times 3\right) - \left(\frac{1}{5}\times 2^5 - \frac{8}{3}\times 2^3 + 16\times 2\right)\right]$$

$$= \pi\left(\frac{243}{5} - 72 + 48 - \frac{32}{5} + \frac{64}{3} - 32\right)$$

$$= \frac{113\pi}{15}$$

7. State the volume.

The volume is $\frac{113\pi}{15}$ units^3.

9.5.3 Composite figures

Volumes of revolution

If $g(x) \geq f(x)$ over the interval $a \leq x \leq b$, then the volume of the shape formed by rotating the region about the x-axis is:

$$V = \pi\int_a^b \left([g(x)]^2 - [f(x)]^2\right) dx$$

If $g(y) \geq f(y)$ over the interval $a \leq y \leq b$, then the volume of the shape formed by rotating the region about the y-axis is:

$$V = \pi\int_a^b \left([g(y)]^2 - [f(y)]^2\right) dy$$

It may be necessary to identify a number of smaller regions so that within each region the functions do not intersect.

WORKED EXAMPLE 13 Calculating volumes of revolution (3)

Determine the volume formed when the area between the curves $y = x^2$ and $y = 8 - x^2$ is rotated about:

a. the x-axis **b. the y-axis.**

THINK

a. 1. Determine the point of intersection between the two curves.

WRITE

a. Let $y_1 = x^2$ and $y_2 = 8 - x^2$.

$$\begin{aligned} y_1 &= y_2 \\ x^2 &= 8 - x^2 \\ 2x^2 &= 8 \\ x^2 &= 4 \\ x &= \pm 2 \end{aligned}$$

When $x = \pm 2$, $y = 4$.

2. Identify the region to be rotated about the x-axis.

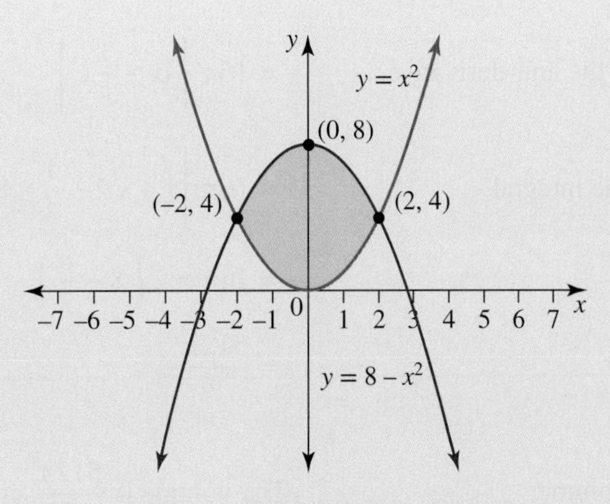

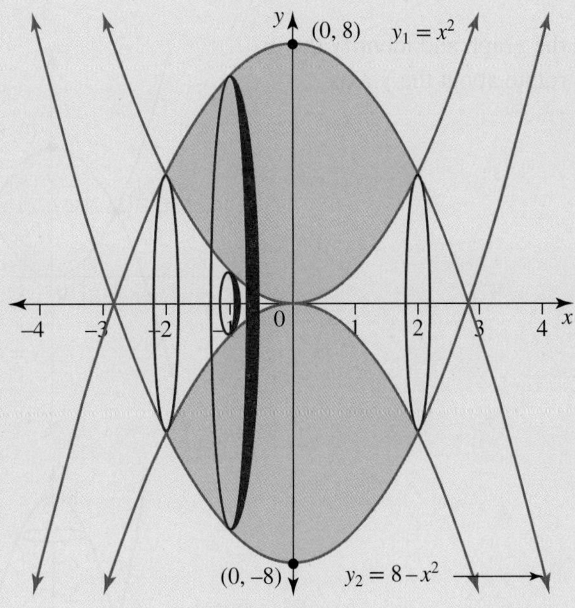

3. Over the region $-2 \le x \le 2$, $8 - x^2 \ge x^2$, therefore only one integral is required. $a = -2$, $b = 2$ and $V = \pi \int_a^b \left([g(x)]^2 - [f(x)]^2 \right) dx$.

$$V = \pi \int_{-2}^{2} \left([g(x)]^2 - [f(x)]^2 \right) dx$$

4. As $8 - x^2 \ge x^2$, $g(x) = 8 - x^2$ and $f(x) = x^2$. Substitute in the integrand.

$$= \pi \int_{-2}^{2} \left((8 - x^2)^2 - (x^2)^2 \right) dx$$

5. Simplify the integrand so that it can be determined.

$$= \pi \int_{-2}^{2} (64 - 16x^2 + x^4 - x^4)\, dx$$

$$= 16\pi \int_{-2}^{2} (4 - x^2)\, dx$$

6. Determine the anti-derivative. $= 16\pi\left[4x - \frac{1}{3}x^3\right]_{-2}^{2}$

7. Evaluate the integral. $= 16\pi\left[\left(4\times2 - \frac{1}{3}\times8\right) - \left(4\times-2 - \frac{1}{3}\times-8\right)\right]$

$= 16\pi\times2\left(8 - \frac{8}{3}\right)$

$= \frac{512\pi}{3}$

8. State the volume. The volume is $\frac{512\pi}{3}$ units^3.

b. 1. Sketch the graph and identify the area to rotate about the y-axis.

b.

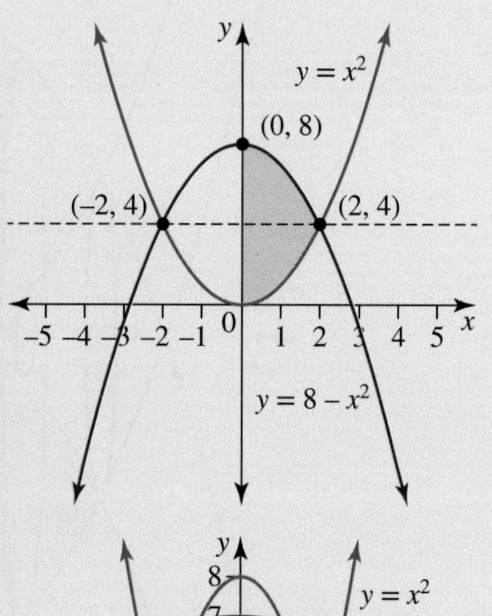

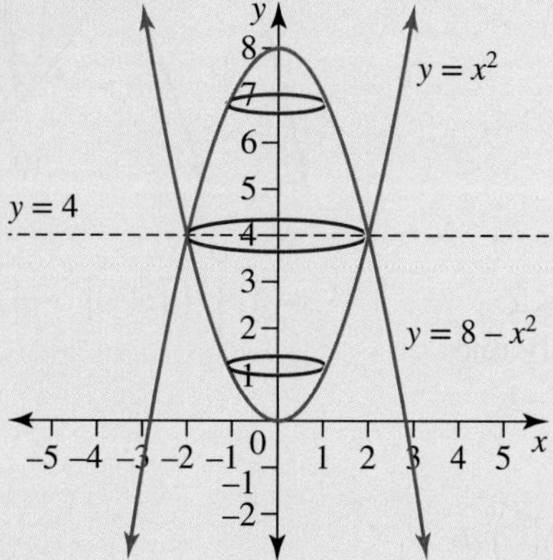

2. For $0 \le y \le 4$ the region is bounded by the curve $y_1 = x^2$. For $4 \le y \le 8$, the region is bounded by the curve $y_2 = 8 - x^2$. In both cases, the volume can be determined using $\pi\int_b^a [f(y)]^2\,dy$.

$$V = \pi\int_0^4 [f(y)]^2\,dy + \pi\int_4^8 [g(y)]^2\,dy$$

3. The integrand needs to be expressed in terms of y. Rearrange y_1 and y_2 for x^2 so they can be easily substituted into the formula.
$y_1 = x^2 \leftrightarrow x^2 = y_1$
$y_2 = 8 - x^2 \Rightarrow x^2 = 8 - y_2$

$$= \pi \int_0^4 y\,dy + \pi \int_4^8 (8-y)\,dy$$

4. Determine the anti-derivative.

$$= \pi \left[\frac{1}{2}y^2\right]_0^4 + \pi \left[8y - \frac{1}{2}y^2\right]_4^8$$

5. Evaluate the integral.

$$= \pi\left(\frac{1}{2}\times 4^2 - 0\right) + \pi\left[\left(8\times 8 - \frac{1}{2}\times 8^2\right) - \left(8\times 4 - \frac{1}{2}\times 4^2\right)\right]$$
$$= 8\pi + \pi(64 - 32 - 32 + 8)$$
$$= 16\pi$$

6. State the value of the volume. The volume is 16π units^3.

9.5 Exercise

Students, these questions are even better in jacPLUS

Receive immediate feedback and access sample responses

Access additional questions

Track your results and progress

Find all this and MORE in jacPLUS

Technology free

1. **WE11** Determine the volume formed when the area bounded by the curve $y = 9 - x^2$ and the line $y = 5$ is rotated about:
   a. the x-axis
   b. the y-axis.
2. Determine the volume formed when the area bounded by the curve $y = \sqrt{x}$ and the line $y = 2$ is rotated about:
   a. the x-axis
   b. the y-axis.
3. **WE12** Determine the volume formed when the area bounded by the curve $y = x^2 - 9$, the x-axis and the line $x = 4$ is rotated about:
   a. the y-axis
   b. the x-axis.
4. Determine the volume formed when the area bounded by the curve $y = \sqrt{x}$, the x-axis and the line $x = 4$ is rotated about:
   a. the y-axis
   b. the x-axis.
5. Consider the area bounded by the curve $y = 16 - x^2$ and the line $y = 12$.
   a. If the area is rotated about the x-axis, determine the volume of the object created.
   b. If the area is rotated about the y-axis, determine the volume of the object created.

6. Consider the area bounded by the curve $y = 16 - x^2$, the x-axis and the line $x = 5$.
   a. If the area is rotated about the y-axis, determine the volume of the object created.
   b. If the area is rotated about the x-axis, determine the volume of the object created.
7. Let R be the region bounded by the curve $y = x^3$, the x-axis and the line $x = 3$.
   a. If the region R is rotated about the y-axis, determine the volume of the object formed.
   b. If the region R is rotated about the x-axis, determine the volume of the object formed.
8. The area between the curve $y = \sqrt{2x - 8}$, the x-axis and the line $x = 6$ is rotated about:
   a. the y-axis; determine the volume of the resulting solid of revolution
   b. the x-axis; determine the volume of the resulting solid of revolution.
9. a. Determine the volume formed when the area bounded by the curve $y = 4\cos^{-1}\left(\frac{x}{2}\right)$ and the lines $x = 2$ and $y = 2\pi$ is rotated about the y-axis.
   b. Determine the volume formed when the area bounded by the curve $y = 2\sin^{-1}\left(\frac{x}{3}\right)$, the x-axis and the line $x = 3$ is rotated about the y-axis.
10. Determine the volume formed when the area between the curves $y = 2x$ and $y = x^2$ is rotated about:
    a. the x-axis
    b. the y-axis.
11. Determine the volume formed when the area between the curves $y = x^2$ and $y = x^3$ is rotated about:
    a. the x-axis
    b. the y-axis.
12. a. If the area between two curves $y = 2\sqrt{x}$ and $y = \frac{x^2}{4}$ is rotated about the x-axis, determine the volume formed.
    b. If the area between two curves $y = \sqrt{ax}$ and $y = \frac{x^2}{a}$ is rotated about the x-axis, determine the volume formed.
13. **WE13** Determine the volume formed when the area between the curves $y = x^2$ and $y = 18 - x^2$ is rotated about:
    a. the x-axis
    b. the y-axis.
14. Determine the volume obtained when the area between the curves $y = x^2 - 4x + 4$ and $y = 4 + 4x - x^2$ is rotated about the x-axis.

Technology active

15. Determine the volume obtained when the area between the curves $y = 8\sin\left(\frac{\pi x}{4}\right)$ and $y = 2x^2$ is rotated about the x-axis.
16. a. A cylindrical hole of radius 2 cm is cut vertically through the centre of a solid sphere of cheese, with radius 4 cm. Determine the volume of cheese remaining.
    b. A cylindrical hole of radius $\frac{a}{2}$ is cut vertically through the centre of a solid sphere of radius a. Determine the volume remaining.

17. a. Determine the volume of a torus (a doughnut-shaped figure) formed when the area bounded by the circle $x^2 + (y - 8)^2 = 16$ is rotated about the x-axis.
    b. Determine the volume of a torus (a doughnut-shaped figure) formed when the area bounded by the circle $(x - a)^2 + y^2 = r^2$, $a > r > 0$, is rotated about the y-axis.

18. Determine the volume formed when the area bounded by the curve $y = a^2 - x^2$ and the line $y = \frac{a^2}{4}$, where $a > 0$, is rotated about

a. the x-axis

b. the y-axis.

9.5 Exam questions

Question 1 (3 marks) TECH-FREE

Source: VCE 2018, Specialist Mathematics Exam 1, Q9bc; © VCAA.

a. Find the x-coordinates of the points of intersection of the curve $x^2 - 2y^2 = 1$ and the line $y = x - 1$. **(1 mark)**

b. Find the volume of the solid of revolution formed when the region bounded by the curve and the line is rotated about the x-axis. **(2 marks)**

Question 2 (1 mark) TECH-FREE

Source: VCE 2013, Specialist Mathematics Exam 2, Section 1, Q10; © VCAA.

MC The region bounded by the lines $x = 0$, $y = 3$ and the graph of $y = x^{\frac{4}{3}}$ where $x \geq 0$ is rotated about the y-axis form a solid of revolution. The volume of this solid is

A. $\frac{81\pi\, 3^{\frac{2}{3}}}{11}$

B. $\frac{12\pi\, 3^{\frac{1}{4}}}{7}$

C. $\frac{27\pi\, 3^{\frac{1}{3}}}{7}$

D. $\frac{18\pi\, 3^{\frac{1}{2}}}{5}$

E. $\frac{6\pi\, 3^{\frac{1}{2}}}{5}$

Question 3 (4 marks) TECH-FREE

Source: VCE 2013, Specialist Mathematics Exam 1, Q9; © VCAA.

The shaded region below is enclosed by the graph of $y = \sin(x)$ and the lines $y = 3x$ and $x = \frac{\pi}{3}$.

This region is rotated about the x-axis.

Find the volume of the resulting solid of revolution.

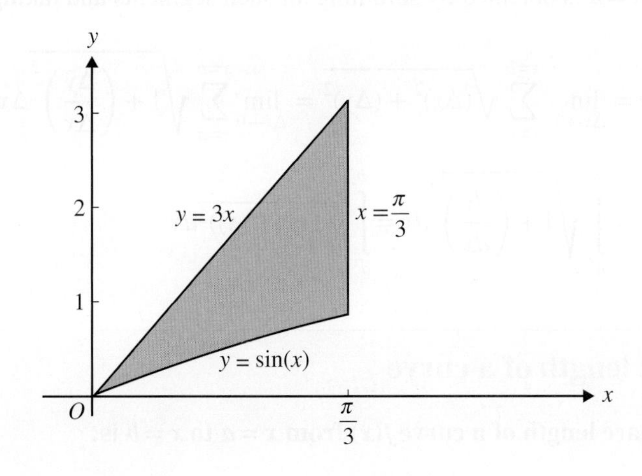

More exam questions are available online.

9.6 Arc length and surface area

LEARNING INTENTION

At the end of this subtopic you should be able to:
- calculate the arc length of a section of a continuous function
- calculate the surface area of a surface of revolution.

9.6.1 Length of a curve

In this section integration will be used to determine the arc length, s, of a plane curve. Suppose that the curve $y = f(x)$ is a continuous curve on the closed interval $a \leq x \leq b$. The curve can be thought of as being made up of infinitely many short line segments as shown.

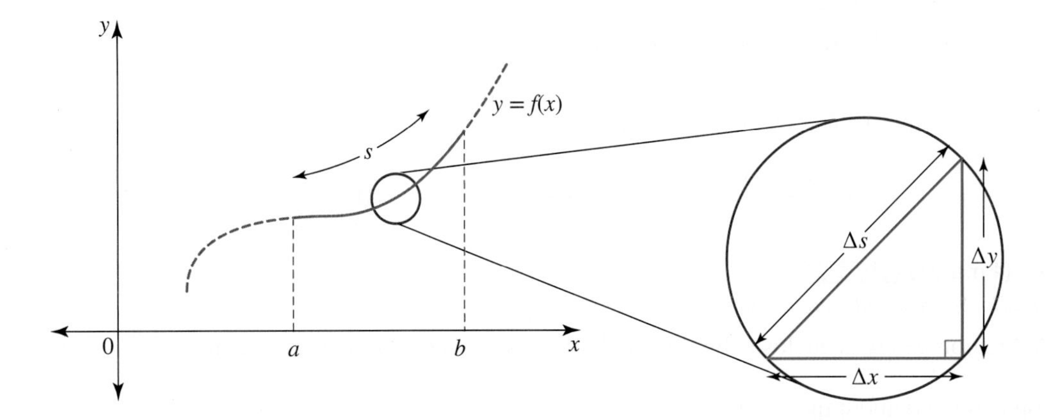

By Pythagoras' theorem, the length of a typical small segment Δs is equal to $\sqrt{(\Delta x)^2 + (\Delta y)^2}$. The total length of the curve s from $x = a$ to $x = b$ is obtained by summing all such segments and taking the limit as $\Delta x \to 0$.

$$s = \lim_{\Delta x \to 0} \sum_{x=a}^{x=b} \sqrt{(\Delta x)^2 + (\Delta y)^2} = \lim_{\Delta x \to 0} \sum_{x=a}^{x=b} \sqrt{1 + \left(\frac{\Delta y}{\Delta x}\right)^2} \Delta x$$

$$s = \int_a^b \sqrt{1 + \left(\frac{dy}{dx}\right)^2} dx = \int_a^b \sqrt{1 + (f'(x))^2} dx$$

Arc length of a curve

The arc length of a curve $f(x)$ from $x = a$ to $x = b$ is:

$$s = \int_a^b \sqrt{1 + \left(\frac{dy}{dx}\right)^2} dx = \int_a^b \sqrt{1 + (f'(x))^2} dx$$

WORKED EXAMPLE 14 Calculating arc length

Calculate the length of the curve $y = \frac{x^3}{2} + \frac{1}{6x}$ from $x = 1$ to $x = 5$.

THINK	WRITE/DRAW
1. Determine the gradient function $\frac{dy}{dx}$ by differentiating and express back with positive indices.	$y = \frac{x^3}{2} + \frac{1}{6x}$ $= \frac{x^3}{2} + \frac{1}{6}x^{-1}$ $\frac{dy}{dx} = \frac{3x^2}{2} - \frac{1}{6}x^{-2}$ $= \frac{3x^2}{2} - \frac{1}{6x^2}$
2. Substitute into the formula for arc length with $a = 1$ and $b = 5$.	$s = \int_a^b \sqrt{1 + \left(\frac{dy}{dx}\right)^2}\, dx$ $= \int_1^5 \sqrt{1 + \left(\frac{3x^2}{2} - \frac{1}{6x^2}\right)^2}\, dx$
3. Expand using $(a - b)^2 = a^2 - 2ab + b^2$ and then simplify.	$= \int_1^5 \sqrt{1 + \left(\left(\frac{3x^2}{2}\right)^2 - 2 \times \frac{3x^2}{2} \times \frac{1}{6x^2} + \left(\frac{1}{6x^2}\right)^2\right)}\, dx$ $= \int_1^5 \sqrt{1 + \left(\frac{9x^4}{4} - \frac{1}{2} + \frac{1}{36x^4}\right)}\, dx$ $= \int_1^5 \sqrt{\left(\frac{9x^4}{4} + \frac{1}{2} + \frac{1}{36x^4}\right)}\, dx$
4. Recognise that the integrand is a perfect square.	$= \int_1^5 \sqrt{\left(\frac{3x^2}{2} + \frac{1}{6x^2}\right)^2}\, dx$ $= \int_1^5 \left(\frac{3x^2}{2} + \frac{1}{6x^2}\right) dx$ $= \int_1^5 \left(\frac{3x^2}{2} + \frac{1}{6}x^{-2}\right) dx$
5. Perform the integration and evaluate.	$= \left[\frac{x^3}{2} - \frac{1}{6}x^{-1}\right]_1^5$ $= \left[\frac{x^3}{2} - \frac{1}{6x}\right]_1^5$ $= \left(\frac{5^3}{2} - \frac{1}{6 \times 5}\right) - \left(\frac{1^3}{2} - \frac{1}{6 \times 1}\right)$ $= \frac{932}{15}$ units

TI | THINK	DISPLAY/WRITE	CASIO | THINK	DISPLAY/WRITE
On a Calculator page, complete the entry as shown.	arcLen$\left(\frac{x^3}{2}+\frac{1}{6\cdot x},x,1,5\right)$ $\frac{932}{15}$ $\int_1^5\left(\frac{3\cdot x^2}{2}+\frac{1}{6\cdot x^2}\right)dx$ $\frac{932}{15}$	On a Main screen, complete the entry as shown.	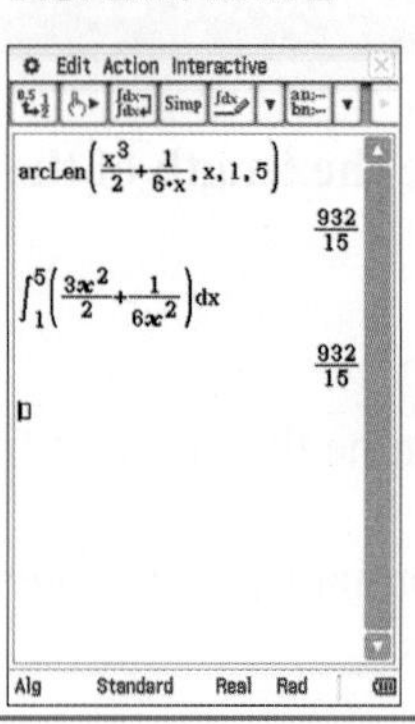

9.6.2 Surface area of surfaces of revolution

Suppose that the curve $y = f(x)$ is continuous on the closed interval $a \le x \le b$.

When the curve is rotated 360° about the x-axis, it forms a surface of revolution.

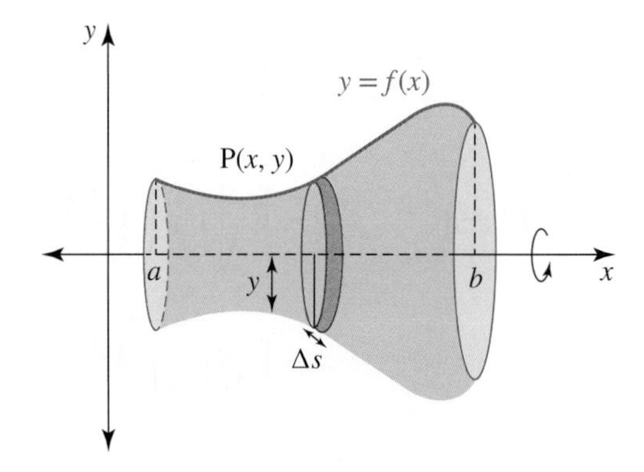

Consider a typical point with coordinates $P(x, y)$, when rotated about the x-axis this forms a circular disc with a radius of y. Consider a typical disc with a width of Δx, the circumference of a typical small segment is given by $2\pi y$ and its surface area is $S(y) = 2\pi y\Delta s$. Where $\Delta s = \sqrt{(\Delta x)^2 + (\Delta y)^2}$ is the arc length of this disc. The total surface area S of a surface of revolution is found by summing over all such discs between $x = a$ and $x = b$ and taking the limit as $\Delta x \to 0$ and is given by

$$
\begin{aligned}
S &= \lim_{\Delta x\to 0} \sum_{x=a}^{x=b} 2\pi y\Delta s \\
&= \lim_{\Delta x\to 0} \sum_{x=a}^{x=b} 2\pi y\sqrt{(\Delta x)^2 + (\Delta y)^2} \\
&= \lim_{\Delta x\to 0} \sum_{x=a}^{x=b} 2\pi y\sqrt{1 + \left(\frac{\Delta y}{\Delta x}\right)^2}\,\Delta x \\
&= 2\pi \int_a^b y\sqrt{1 + \left(\frac{dy}{dx}\right)^2}\,dx = 2\pi \int_a^b f(x)\sqrt{1 + (f'(x))^2}\,dx
\end{aligned}
$$

Surface area of a surface of revolution

The surface area of the surface of revolution formed by rotating the section of the curve $y = f(x)$ from $x = a$ to $x = b$ around the x-axis is:

$$S = 2\pi \int_a^b y\sqrt{1+\left(\frac{dy}{dx}\right)^2}\,dx = 2\pi \int_a^b f(x)\sqrt{1+(f'(x))^2}\,dx$$

WORKED EXAMPLE 15 Calculating the surface area of a surface of revolution

Calculate the surface area obtained by rotating the curve $y = \sqrt{3x+4}$ from $x = 0$ to $x = 2$ about the x-axis.

THINK

1. Determine the gradient function $\frac{dy}{dx}$.

WRITE/DRAW

$$y = \sqrt{3x+4}$$
$$= (3x+4)^{\frac{1}{2}}$$
$$\frac{dy}{dx} = 3 \times \frac{1}{2} \times (3x+4)^{-\frac{1}{2}}$$
$$= \frac{3}{2\sqrt{3x+4}}$$

2. Substitute into the formula for surface area with $a = 0$ and $b = 2$ and simplify the integral until it is in a form which can be integrated easily.

$$S = 2\pi \int_a^b y\sqrt{1+\left(\frac{dy}{dx}\right)^2}\,dx$$
$$= 2\pi \int_0^2 \sqrt{3x+4}\sqrt{1+\left(\frac{3}{2\sqrt{3x+4}}\right)^2}\,dx$$
$$= 2\pi \int_0^2 \sqrt{3x+4}\sqrt{1+\frac{9}{4(3x+4)}}\,dx$$
$$= 2\pi \int_0^2 \sqrt{3x+4}\sqrt{\frac{4(3x+4)+9}{4\sqrt{3x+4}}}\,dx$$
$$= 2\pi \int_0^2 \sqrt{\frac{4(3x+4)+9}{4}}\,dx$$
$$= 2\pi \int_0^2 \frac{1}{2}\sqrt{12x+25}\,dx$$
$$= \pi \int_0^2 (12x+25)^{\frac{1}{2}}\,dx$$

3. Perform the integration.	$S = \pi \left[\frac{1}{12} \times \frac{2}{3}(12x+25)^{\frac{3}{2}}\right]_0^2$
	$= \frac{\pi}{18}\left[(12x+25)^{\frac{3}{2}}\right]_0^2$
4. Evaluate the definite integral.	$= \frac{\pi}{18}\left(49^{\frac{3}{2}} - 25^{\frac{3}{2}}\right)$
	$= \frac{\pi}{18}(343 - 125)$
	$= \frac{\pi}{18}(218)$
	$= \frac{109\pi}{9}$ units^2

9.6.3 Numerical integration

For many simple curves, the arc length or surface area formulae results in definite integrals which cannot be evaluated by techniques of integration. In these situations we must determine an expression for the definite integral and use calculators which can give a numerical approximation.

WORKED EXAMPLE 16 Calculating arc length and surface area numerically

Consider the function $f: [0, 2] \to R,\ f(x) = x^4$.

a. Express the arc length of the curve $y = x^4$ from $x = 0$ to $x = 2$ as a definite integral and hence calculate the arc length, giving your answer correct to 4 decimal places.

b. When the graph of $y = x^4$ from $x = 0$ to $x = 2$ is rotated about the x-axis, it forms a surface of revolution.
Express the surface area formed as a definite integral and hence calculate the surface area, giving your answer correct to 4 decimal places.

THINK	WRITE/DRAW
a. 1. Determine the gradient function $\frac{dy}{dx}$ by differentiating.	a. $y = x^4$ $\frac{dy}{dx} = 4x^3$
2. Substitute into the formula for arc length with $a = 0$ and $b = 2$.	$s = \int_a^b \sqrt{1 + \left(\frac{dy}{dx}\right)^2}\, dx$
	$= \int_0^2 \sqrt{1 + (4x^3)^2}\, dx$
	$= \int_0^2 \sqrt{1 + 16x^6}\, dx$

3. This definite integral cannot be evaluated using integration techniques and must be evaluated using a calculator. Type the integral into a CAS calculator and state the final result.

$s = 16.6469$ units

b. 1. Determine the gradient function $\frac{dy}{dx}$ by differentiating.

b. $y = x^4$

$\frac{dy}{dx} = 4x^3$

2. Substitute into the formula for surface area with $a = 0$ and $b = 2$.

$$S = 2\pi \int_a^b y\sqrt{1 + \left(\frac{dy}{dx}\right)^2}\, dx$$

$$= 2\pi \int_0^2 y\sqrt{1 + (4x^3)^2}\, dx$$

$$= 2\pi \int_0^2 x^4\sqrt{1 + 16x^6}\, dx$$

3. This definite integral cannot be evaluated using integration techniques and must be evaluated using a calculator. Type the integral into a CAS calculator and state the final result.

$S = 805.7179$ units^2

9.6 Exercise

Technology free

1. WE14 Calculate the length of the curve $y = \frac{2}{3}x^{\frac{3}{2}}$ from $x = 0$ to $x = 3$.

2. Calculate the length of the curve $y = \sqrt{x^3}$ from $x = 0$ to $x = 5$.

3. WE15 Calculate the surface area obtained by rotating the curve $y = \sqrt{4x + 5}$ from $x = 0$ to $x = 4$, about the x-axis.

4. a. Calculate the length of line $y = 3x + 5$ from $x = 1$ to $x = 6$.
   b. Calculate the surface area formed when the line $y = 3x + 5$ between $x = 1$ to $x = 6$, is rotated about the x-axis.

5. Determine the surface area obtained by rotating the curve $y = x^3$ from $x = 0$ to $x = 1$, about the x-axis.

6. a. Calculate the length of the curve $y=\frac{x^3}{6}+\frac{1}{2x}$ from $x=1$ to $x=2$.

b. Calculate the surface area formed when the curve $y=\frac{x^3}{6}+\frac{1}{2x}$ from $x=1$ to $x=2$, is rotated about the x-axis.

7. a. Determine the length of the curve $y=\frac{x^6+2}{8x^2}$ from $x=1$ to $x=2$.

b. Determine the surface area formed when the curve $y=\frac{x^6+2}{8x^2}$ from $x=1$ to $x=2$, is rotated about the x-axis.

8. a. Calculate the arc length of $y=\frac{1}{2}(e^x+e^{-x})$ from $x=0$ to $x=1$.

b. Calculate the surface area formed when the curve $y=\frac{1}{2}(e^x+e^{-x})$ from $x=0$ to $x=1$, is rotated about the x-axis.

9. Determine the length of the curve $y=\frac{4\sqrt{2}}{3}x^{\frac{3}{2}}-1$ from $x=0$ to $x=1$.

10. For the curve $27y^2=4(x-2)^3$ determine the arc length from $x=3$ to $x=8$.

11. Calculate the length of the curve $y=\frac{2}{3}\sqrt{(x-1)^3}$ from $x=1$ to $x=9$.

12. Calculate the length of the curve $y=\frac{2}{3}\sqrt{(2x-3)^3}$ from $x=\frac{5}{2}$ to $x=\frac{9}{2}$.

13. a. Determine the arc length of the function $f: [1,4]\to R, f(x)=\frac{\sqrt{x}(4x-3)}{6}$.

b. Determine the surface area formed when the function $f: [1,4]\to R, f(x)=\frac{\sqrt{x}(4x-3)}{6}$ is rotated about the x-axis.

14. For the line $y=mx+c$, verify that the arc length formulae, gives the distance along the line between the points $x=a$ and $x=b$.

15. a. Prove that the total surface area of a cylinder of height h and radius r is $2\pi r$.

b. Prove that the total surface area of a cone of height h and radius r is $\pi r\sqrt{h^2+r^2}$.

16. a. For the curve $y=\sqrt{9-x^2}$ determine the length from $x=0$ to $x=3$ and state what this length represents.

b. Calculate the surface area formed when the curve $y=\sqrt{9-x^2}$ from $x=0$ to $x=3$, is rotated about the x-axis and state what this area represents.

17. a. Show that $\frac{d}{dx}\left[\frac{1}{2}\left(\log_e\left(x+\sqrt{x^2+1}\right)+x\sqrt{x^2+1}\right)\right]=\sqrt{x^2+1}$.

b. Hence determine the length of the curve $y=x^2$ from $x=0$ to $x=1$.

18. Calculate the surface area formed when the curve $y=\sin(x)$ is rotated about the x-axis from $x=0$ to $x=\frac{\pi}{2}$.

19. Calculate the surface area formed when the curve $y=\cos(x)$ is rotated about the x-axis from $x=0$ to $x=\pi$.

20. a. Show that $\frac{d}{dx}\left[\log_e(\sec(x)+\tan(x))\right]=\sec(x)$.

b. Hence, determine the length of the curve $y=\log_e(\sec(x))$ from $x=0$ to $x=\frac{\pi}{3}$.

Technology active

21. Consider the function $f(x)=4\sin^{-1}\left(\frac{x}{3}\right)$.

a. Sketch the graph of $y=4\sin^{-1}\left(\frac{x}{3}\right)$ stating the coordinates of the endpoints.

b. Calculate the area bounded by the graph of $y=4\sin^{-1}\left(\frac{x}{3}\right)$ and the x-axis.

c. Calculate the length of the curve $y=4\sin^{-1}\left(\frac{x}{3}\right)$, giving your answer correct to 4 decimal places,

d. When the graph of $y=4\sin^{-1}\left(\frac{x}{3}\right)$ is rotated about the x-axis, it forms a volume of revolution. Determine this volume.

e. When the graph of $y=4\sin^{-1}\left(\frac{x}{3}\right)$ is rotated about the x-axis, it forms a surface of revolution. Determine this surface area giving your answer correct to 4 decimal places.

f. When the graph of $y=4\sin^{-1}\left(\frac{x}{3}\right)$ is rotated about the y-axis, it forms a volume of revolution. Determine this volume.

g. When the graph of $y=4\sin^{-1}\left(\frac{x}{3}\right)$ is rotated about the y-axis, it forms a surface of revolution. Determine this surface area. When rotating about the y-axis the surface area is given by

$$S=2\pi\int_a^b x\sqrt{1+\left(\frac{dx}{dy}\right)^2}\,dy.$$

22. a. **WE16** Set up a definite integral for the length of the $y=3e^{-2x}$ from $x=0$ to $x=1$, and hence determine the arc length giving your answer correct to 4 decimal places.

b. When the function $f:[0,1]\to R, f(x)=3e^{-2x}$ is rotated about the x-axis, it forms a surface of revolution. Set up a definite integral for the surface area, and hence determine the surface area giving your answer correct to 4 decimal places.

23. a. Set up a definite integral for the length of $y=\log_e(2x+1)$ from $x=0$ to $x=3$, and hence determine the arc length giving your answer correct to 4 decimal places.

b. When the function $f:[0,3]\to R, f(x)=\log_e(2x+1)$ is rotated about the x-axis, it forms a surface of revolution. Set up a definite integral for the surface area, and hence determine the surface area giving your answer correct to 4 decimal places.

24. a. Set up a definite integral for the length of $y=3\cos^{-1}\left(\frac{x}{2}\right)$, and hence determine the arc length giving your answer correct to 4 decimal places.

b. When the function $f:[-2,2]\to R, f(x)=3\cos^{-1}\left(\frac{x}{2}\right)$ is rotated about the x-axis, it forms a surface of revolution. Set up a definite integral for the surface area, and hence determine the surface area giving your answer correct to 4 decimal places.

25. a. Set up a definite integral for the total length of the ellipse $\frac{x^2}{9}+\frac{y^2}{4}=1$. Determine this length, giving your answer correct to 4 decimal places.

b. When the ellipse $\frac{x^2}{9}+\frac{y^2}{4}=1$ is rotated about the x-axis it forms a surface of revolution. Determine the area of this surface, giving your answer correct to 4 decimal places.

c. When the ellipse $\frac{x^2}{9}+\frac{y^2}{4}=1$ is rotated about the y-axis it forms a surface of revolution. When rotating about the y-axis the surface area is given by $s=2\pi\int_a^b x\sqrt{1+\left(\frac{dx}{dy}\right)^2}\,dy$. Determine the area of this surface, giving your answer correct to 4 decimal places.

26. The two lampshades shown have base radii of 20 cm, top radii of 10 cm, and heights of 20 cm. The edges of the lampshades can be modelled by the functions.

a. f: $[10, 20] \to R, f(x)=mx+c$

b. f: $[10, 20] \to R, f(x)=\frac{a}{x}+k$

The lowest point on the edges of the lampshades is (20, 10) and the upper point is (10, 30). Determine the values of m, c, a and k and calculate the surface area of each of the lampshades, giving your answers correct to 2 decimal places in cubic centimetres.

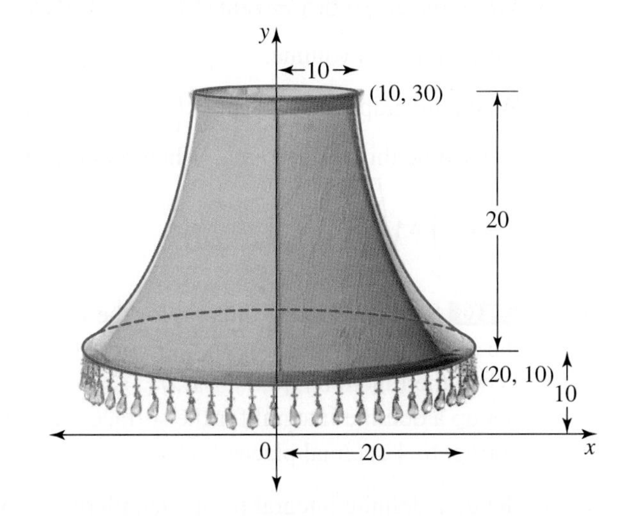

9.6 Exam questions

Question 1 (4 marks) TECH-FREE

Source: VCE 2016, Specialist Mathematics Exam 1, Q7; © VCAA.

Find the arc length of the curve $y=\frac{1}{3}(x^2+2)^{\frac{3}{2}}$ from $x=0$ to $x=2$.

Question 2 (3 marks) TECH-FREE

Find the surface area obtained by rotating the curve $y=\sqrt{x}$ from $x=0$ to $x=1$, about the x-axis.

Question 3 (8 marks) TECH-FREE

Find the surface area obtained by rotating the curve $y=\sqrt{x}$ from $x=0$ to $x=1$, about the x-axis.

a. Prove that the circumference of a circle of radius r is $2\pi r$. **(4 marks)**

b. Prove that the total surface area of a sphere of radius r is $4\pi r^2$. **(4 marks)**

More exam questions are available online.

9.7 Water flow

LEARNING INTENTION

At the end of this subtopic you should be able to:
- calculate volumes of solids and determine how long it will take for a leaking vessel of water to empty.

9.7.1 Torricelli's theorem

Evangelista Torricelli (1608–1647) was an Italian scientist interested in mathematics and physics. He invented the barometer to measure atmospheric pressure and was also one of the first to correctly describe what causes the wind. He also designed telescopes and microscopes. Modern weather forecasting owes much to the work of Torricelli. His main achievement is the theorem named after him, Torricelli's theorem, which describes the relationship between fluid leaving a container through a small hole and the height of the fluid in the container. Basically, the theorem states that the rate at which the volume of fluid leaves the container is proportional to the square root of the height of the fluid in the tank. This theorem applies for all types of containers.

Problem solving

In solving problems involving fluid flow, we use the techniques of calculating volumes and use related rate problems to set up and solve differential equations. Sometimes we use numerical methods to evaluate a definite integral.

WORKED EXAMPLE 17 Determining how long it will take a vessel of water to empty (1)

A vase has a circular base and top with radii of 4 cm and 9 cm respectively, and a height of 16 cm. The origin, O, is at the centre of the base. The vase is formed when the curve $y = a\sqrt{x} + b$ is rotated about the y-axis. Initially the vase is filled with water, but the water leaks out at a rate equal to $4\sqrt{h}$ cm^3/min, where h cm is the height of the water remaining in the vase after t minutes. Set up the differential equation for h and t, and determine how much time it takes for the vase to become empty.

THINK	WRITE/DRAW
1. Set up simultaneous equations which can be solved for a and b. The height of the vase is 16 and the height of the water in the vase is h, so $0 \le h \le 16$. Note all dimensions are in centimetres.	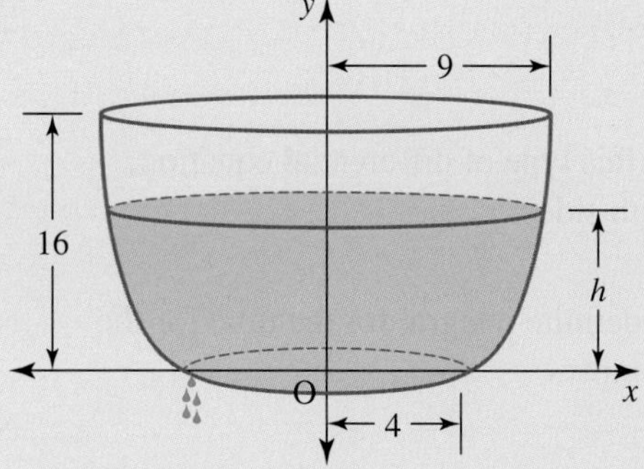 The curve $y = a\sqrt{x} + b$ passes through the points (4, 0) and (9, 16). Substituting: $(4, 0) \Rightarrow 0 = 2a + b \qquad (1)$ $(9, 16) \Rightarrow 16 = 3a + b \qquad (2)$

2. Determine the values of a and b.	$(2)-(1) \Rightarrow a=16$, so $b=-32$. The vase is formed when $y=16\sqrt{x}-32$ for $4 \le x \le 9$ is rotated about the y-axis.
3. Determine the volume of the vase.	When a curve is rotated about the y-axis, the volume is $V=\pi\int_0^h x^2 dy$.
4. Rearrange the equation to make x the subject.	$y=16\sqrt{x}-32$ $16\sqrt{x}=y+32$ $\sqrt{x}=\frac{1}{16}(y+32)$ $x=\frac{1}{256}(y+32)^2$ $x^2=\frac{(y+32)^4}{65\,536}$
5. Determine a definite integral for the volume of water when the vase is filled to a height of h cm.	$V=\pi\int_0^h \frac{(y+32)^4}{65\,536}dy$
6. Determine the given rates in terms of time, t. The rate is negative as it is a decreasing rate.	Since the water leaks out at a rate proportional to the square root of the remaining height of the water, $\frac{dV}{dt}=-4\sqrt{h}$.
7. Note the result used is from the numerical techniques described earlier.	$V=\pi\int_0^h \frac{(y+32)^4}{65\,536}dy$ $\frac{dV}{dh}=\frac{\pi(h+32)^4}{65\,536}$
8. Use related rates and a chain rule.	$\frac{dh}{dt}=\frac{dh}{dV}\frac{dV}{dt}$
9. Set up the differential equation for the height h at time t.	Substitute for the rates and use $\frac{dh}{dV}=1/\frac{dV}{dh}$: $\frac{dh}{dt}=\frac{-65\,536\times 4\sqrt{h}}{\pi(h+32)^4}$
10. To solve this type of differential equation, invert both sides.	$\frac{dt}{dh}=\frac{-\pi(h+32)^4}{262\,144\sqrt{h}}$
11. Set up a definite integral for the time for the vase to empty.	$t=\int_{13}^{0}\frac{-\pi(h+32)^4}{262\,144\sqrt{h}}dh$ Note the order of the terminals from $h=16$ to $h=0$.
12. Use a calculator to numerically evaluate the definite integral.	$t=205.59$; note that the time is positive.
13. State the result.	The tank is empty after a total time of 205.59 minutes.

WORKED EXAMPLE 18 Determining how long it will take a vessel of water to empty (2)

Another vase is formed when part of the curve $\frac{x^2}{16} - \frac{y^2}{500} = 1$ for $4 \le x \le 6, y \ge 0$, is rotated about the y-axis to form a volume of revolution. The x- and y-coordinates are measured in centimetres. The vase has a small crack in the base, and the water leaks out at a rate proportional to the square root of the remaining height of the water. Initially the vase was full, and after 10 minutes the height of the water in the vase is 16 cm. Calculate how much longer it will be before the vase is empty.

THINK	WRITE/DRAW
1. The vase is formed when the given curve is rotated about the y-axis.	$V = \pi \int_a^b x^2 dy$
2. Determine the height of the vase.	Determine the value of y when $x = 6$. $\frac{y^2}{500} = \frac{6^2}{16} - 1 = \frac{5}{4}$; $y^2 = 625$; $y = 25$ as $y > 0$
3. Sketch the region of the hyperbola which forms the vase. The height of the vase is 25 cm and the height of the water in the vase is h, so $0 \le h \le 25$.	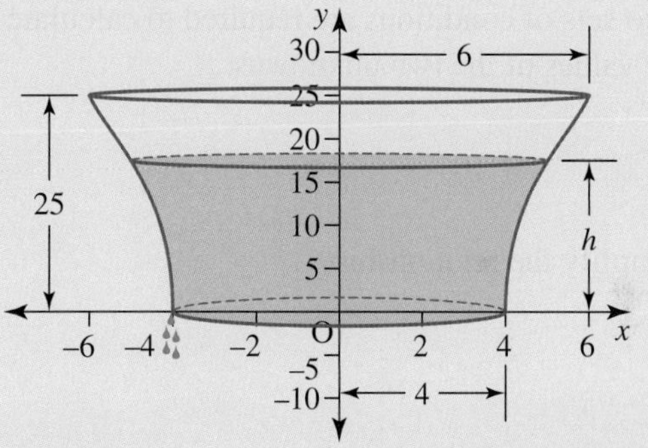
4. Transpose the equation to make x^2 the subject.	$\frac{x^2}{16} - \frac{y^2}{500} = 1$; $\frac{x^2}{16} = 1 + \frac{y^2}{500}$; $x^2 = \frac{16\left(500 + y^2\right)}{500}$; $x^2 = \frac{4\left(500 + y^2\right)}{125}$
5. Determine a definite integral for the volume of water when the vase is filled to a height of h cm, where $0 \le h \le 25$.	$V = \frac{4\pi}{125} \int_0^h (500 + y^2) dy$
6. Note the result used is from the numerical techniques described earlier.	$\frac{dV}{dh} = \frac{4\pi\left(500 + h^2\right)}{125}$

7. Determine the given rates in terms of time t in minutes.	Since the water leaks out at a rate proportional to the square root of the remaining height of the water, $\frac{dV}{dt} = -k\sqrt{h}$, where k is a positive constant.
8. Use related rates and a chain rule.	$\frac{dh}{dt} = \frac{dh}{dV}\frac{dV}{dt}$
9. Set up the differential equation for the height h at time t.	Substitute for the related rates, using $\frac{dh}{dV} = 1/\frac{dV}{dh}$ $\frac{dh}{dt} = \frac{-125k\sqrt{h}}{4\pi\left(h^2+500\right)}$
10. Incorporate the constants into one constant. To solve this type 2 differential equation, invert both sides.	$\frac{dt}{dh} = \frac{-A\left(h^2+500\right)}{\sqrt{h}}$ where $A = \frac{4\pi}{125k}$ $\frac{dt}{dh} = -A\left(h^{\frac{3}{2}} + 500h^{-\frac{1}{2}}\right)$
11. Integrate with respect to h.	$t = -A\int\left(h^{\frac{3}{2}} + 500h^{-\frac{1}{2}}\right)dh$
12. Perform the integration.	$t = -A\left[\frac{2}{5}h^{\frac{5}{2}} + 1000h^{\frac{1}{2}}\right] + c$, where c is the constant of integration.
13. Two sets of conditions are required to calculate the values of the two unknowns.	Initially, when $t = 0$, $h = 25$, since the vase was full. Substitute $t = 0$ and $h = 25$: $0 = -A\left[\frac{2}{5}(25)^{\frac{5}{2}} + 1000 \times 25^{\frac{1}{2}}\right] + c$
14. Simplify the relationship.	$0 = -A\left[\frac{2}{5} \times 3125 + 1000 \times 5\right] + c$ $c - 6250A = 0$ $A = \frac{c}{6250}$
15. Determine another relationship between the unknowns.	Since after 10 minutes the height of the water in the vase is 16 cm, substitute $t = 10$ and $h = 16$: $10 = -A\left[\frac{2}{5}(16)^{\frac{5}{2}} + 1000 \times 16^{\frac{1}{2}}\right] + c$
16. Simplify the relationship and solve for the constant of integration.	Substitute for A: $10 = -\frac{c}{6250}\left[\frac{2}{5} \times 1024 + 1000 \times 4\right] + c$ $10 = c\left(1 - \frac{1}{6250}\left[\frac{2}{5} \times 1024 + 4000\right]\right)$ $c = \frac{156\,250}{4601}$ $c = 33.96$

17. Determine when the vase will be empty.

Since $t = -A\left[\frac{2}{5}h^{\frac{5}{2}} + 1000h^{\frac{1}{2}}\right] + c$, the vase is empty when $h = 0$, that is at time when $t = c$.

18. State the final result.

The vase is empty after a total time of 33.96 minutes, so it takes another 23.96 minutes to empty.

9.7 Exercise

Students, these questions are even better in jacPLUS

Receive immediate feedback and access sample responses

Access additional questions

Track your results and progress

Find all this and MORE in jacPLUS

Technology active

1. **WE17** A vase has a circular base and top with radii of 4 cm and 9 cm respectively, and a height of 16 cm. The origin, O, is at the centre of the base of the vase. The vase is formed when the line $y = ax + b$ is rotated about the y-axis. Initially the vase is filled with water, but the water leaks out at a rate equal to $2\sqrt{h}\ \text{cm}^3/\text{min}$, where h cm is the height of the water remaining in the vase after t minutes.
Set up the differential equation for h and t, and determine how long it will be before the vase is empty, in minutes correct to 1 decimal place.

2. A vase has a circular base and top with radii of 4 cm and 9 cm respectively, and a height of 16 cm. The origin, O, is at the centre of the base of the vase. The vase is formed when the curve $y = ax^2 + b$ is rotated about the y-axis. Initially the vase is filled with water, but the water leaks out at a rate equal to $2\sqrt{h}\ \text{cm}^3/\text{min}$, where h cm is the height of the water remaining in the vase after t minutes.
Set up the differential equation for h and t, and determine how long before the vase is empty, in minutes correct to 1 decimal place.

3. **WE18** A vase is formed when part of the curve $\frac{x^2}{16} - \frac{65y^2}{4096} = 1$ for $4 \le x \le 9,\ y \ge 0$, is rotated about the y-axis to form a volume of revolution. The x- and y-coordinates are measured in centimetres. The vase has a small crack in the base, and the water leaks out at a rate proportional to the square root of the remaining height of the water. Initially the vase was full, and after 10 minutes the height of the water in the vase is 9 cm.
Determine how much time it will take for the vase to become empty, in minutes correct to 1 decimal place.

4. A vase has a base radius of 4 cm, a top radius is 9 cm, and a height of 16 cm. The origin, O, is at the centre of the base of the vase. Initially the vase is filled with water, but the water leaks out at a rate proportional to the square root of the remaining height of the water. The side of the vase is modelled by a quadratic $y = ax^2 + b$. Initially the vase was full, and after 10 minutes the height of the water in the vase is 9 cm.
Determine how much time it will take for the vase to become empty, in minutes correct to 1 decimal place.

5. a. A cylindrical coffee pot has a base radius of 10 cm, a height of 49 cm and is initially filled with hot coffee. Coffee is removed from the pot at a rate equal to $2\sqrt{h}$ cm^3/sec, where h cm is the height of the coffee remaining in the coffee pot after t seconds. Set up the differential equation for h and t, and determine how long it will be before the coffee pot is empty, in minutes correct to 2 decimal place.

   b. A small cylindrical teapot with a base radius of 5 cm and a height of 16 cm is initially filled with hot water. The hot water is removed from the teapot at a rate proportional to $\sqrt{h}$ cm^3/ min, where h cm is the height of the hot water remaining in the teapot after t minutes. Set up the differential equation for h and t. If after 10 minutes the height of the hot water is 9 cm, determine how much time elapses before the teapot is empty.

6. a. A rectangular bathtub has a length of 1.5 metres and is 0.6 metres wide. It is filled with water to a height of 1 metre. When the plug is pulled, the water flows out of the bath at a rate equal to $2\sqrt{h}$ m^3/min, where h is the height of the water in metres in the bathtub at a time t minutes after the plug is pulled. Determine how long it will take for the bathtub to empty, to the nearest second.

   b. A cylindrical hot water tank with a capacity of 160 litres is 169 cm tall and is filled with hot water. Hot water starts leaking out through a crack in the bottom of the tank at a rate equal to $k\sqrt{h}$ cm^3/ min, where h cm is the depth of water remaining in the tank after t minutes. If the tank is empty after 90 minutes, determine the value of k, correct to 1 decimal place.

7. a. The volume of a hemispherical bowl is given by $\frac{\pi h^2}{3}(30-h)$ cm^3, where h cm is the depth of the water in the bowl. Initially the bowl has water to a depth of 9 cm. The water starts leaking out through a small hole in the bowl at a rate equal to $k\sqrt{h}$ cm^3/min. After 1314 minutes the bowl is empty. Determine the exact value of k.

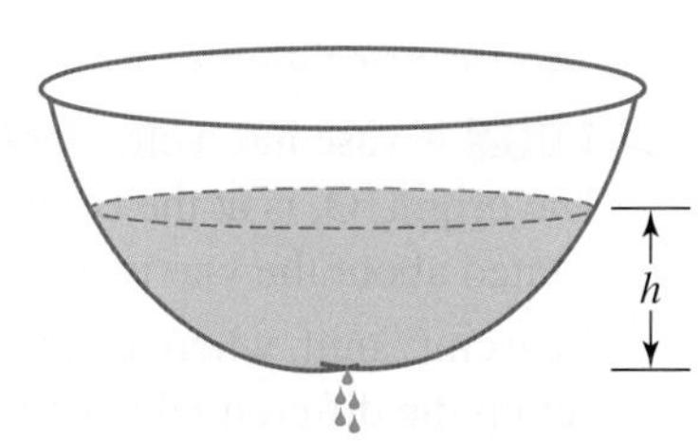

   b. A drinking trough has a length of 2 m. Its cross-sectional face is in the shape of a trapezium with a height of 25 cm and with lengths 40 cm and 90 cm. Both sloping edges are at an angle of 45° to the vertical, as shown in the diagram. The trough contains water to a height of h cm. The water leaks out through a crack in the base of the trough at a rate proportional to $\sqrt{h}$ cm^3/min. Initially the trough is full, and after 20 minutes the height of the water in the trough is 16 cm. Determine how long it will be before the trough is empty, in minutes correct to 2 decimal places.

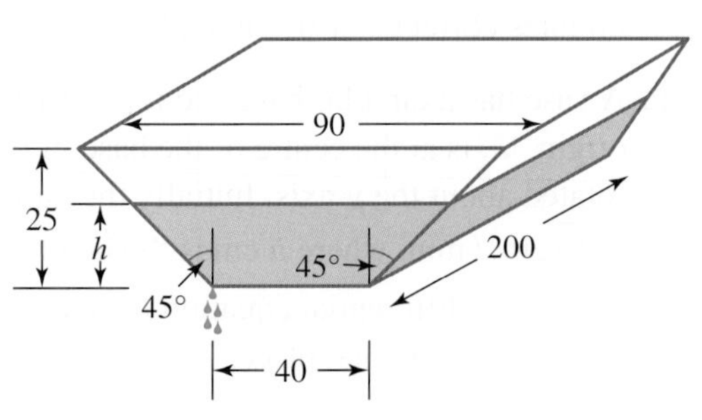

8. a. A plastic coffee cup has a base diameter of 5 cm, a top diameter of 8 cm and a height of 9 cm. Initially the cup is filled with coffee. However, the coffee leaks at a rate equal to $k\sqrt{h}$ cm^3/ min, where h cm is the height of the coffee in the cup. If the cup is empty after 3 minutes, determine the exact value of k.

   b. A plastic bucket has a base diameter of 20 cm, a top diameter of 26 cm and a height of 24 cm. The side of the bucket is straight. Initially the bucket is filled with water to a height of 16 cm. However, there is a small hole in the bucket, and the water leaks out at a rate proportional to $\sqrt{h}$ cm^3/ min, where h cm is the height of the water in the bucket. When the height of the water in the bucket is 16 cm, the height is decreasing at a rate of 0.1 cm/ min. Determine how long it will be before the bucket is empty, in minutes correct to 2 decimal places.

9. a. A conical vessel with its vertex downwards has a height of 20 cm and a radius of 10 cm. Initially it contains water to a depth of 16 cm. Water starts flowing out through a hole in the vertex at a rate proportional to the square root of the remaining depth of the water in the vessel. If after 10 minutes the depth is 9 cm, determine how much longer it will be before the vessel is empty, in minutes correct to 2 decimal places.

b. A conical funnel with its vertex downwards has a height of 25 cm and a radius of 20 cm. Initially the funnel is filled with oil. The oil flows out through a hole in the vertex at a rate equal to $k\sqrt{h}$ cm^3/s, where h cm is the height of the oil in the funnel. If the funnel is empty after 40 seconds, determine the exact value of k.

10. When filled to a depth of h metres, a fountain contains V litres of water, where $V = 500\left(h^2 - \frac{h^4}{4}\right)$.

Initially the fountain is empty. Water is pumped into the fountain at a rate of 300 litres per hour and leaks out at a rate equal to $2\sqrt{h}$ litres per hour.

a. Determine the rate in metres per hour at which the water level is rising when the depth is 0.5 metres, correct to 2 decimal places.

b. The fountain is considered full when the height of the water in the fountain is 1 metre. Determine how long it takes to fill the fountain, in hours correct to 2 decimal places.

c. When the fountain is full, water is no longer pumped into the fountain, but water still leaks out at the same rate. Determine how long it will be before the fountain is empty again, in hours correct to 1 decimal place.

11. a. A wine glass is formed when the arc OB with the equation $y = x^{\frac{3}{2}}$ is rotated about the y-axis. The dimensions of the glass are given in centimetres.

The wine leaks out through a crack in the base of the glass at a rate proportional to $\sqrt{h}$ cm^3/min. Initially the glass is full, and after 3 minutes the height of the wine in the glass is 1 cm. Determine what further time elapses before the wine glass is empty, in minutes correct to 2 decimal places.

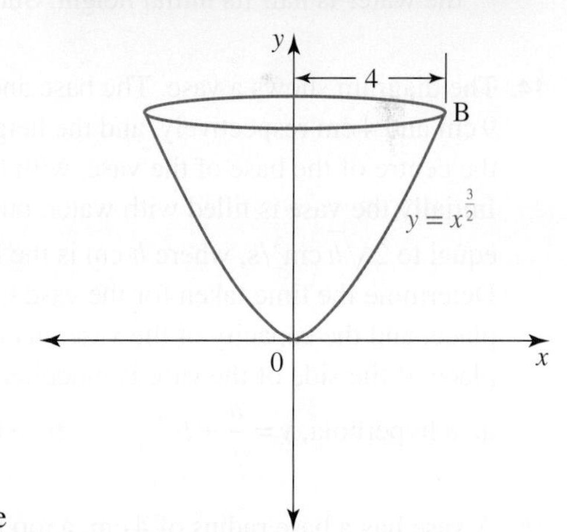

b. A large beer glass is formed when a portion of the curve with the equation $y = x^{\frac{4}{3}}$ is rotated about the y-axis between the origin and $y = 16$. The dimensions of this glass are given in centimetres. Beer leaks out through a crack in the base at a rate proportional to $\sqrt{h}$ cm^3/min. Initially the glass is full, and after 3 minutes the height of the beer in this glass is 12 cm. Calculate how much longer it will be before the glass is empty, in minutes correct to 2 decimal places.

12. a. An ornamental vase has a circular top and base, both with radii of 4 centimetres, and a height of 16 cm. The origin, O, is at the centre of the base of the vase, and the vase is formed when the hyperbola

$$\frac{25x^2}{144} - \frac{(y-8)^2}{36} = 1$$

is rotated about the y-axis, with dimensions in centimetres. Initially the vase is filled with water, but the water leaks out at a rate equal to $2\sqrt{h}\,\text{cm}^3/\text{s}$, where h cm is the height of the water in the vase. Determine the time taken for the vase to empty, in seconds, and the capacity of the vase, in cubic centimetres. Express both answers correct to 1 decimal place.

b. A different ornamental vase has a circular top and base, both with radii of 3.6 cm, and a height of 16 cm. The origin, O, is at the centre of the base of the vase, and the vase is formed when the ellipse

$$\frac{x^2}{36} + \frac{(y-8)^2}{100} = 1$$

is rotated about the y-axis, with dimensions in centimetres. Initially this vase is filled with water, but the water leaks out at a rate equal to $2\sqrt{h}\,\text{cm}^3/\text{s}$, where h cm is the height of the water in the vase. Determine the time taken for this vase to empty, in seconds, and the capacity of the vase, in cubic centimetres. Express both answers correct to 1 decimal place.

13. a. A cylindrical vessel is initially full of water. Water starts flowing out through a hole in the bottom of the vessel at a rate proportional to the square root of the remaining depth of the water. After a time of T, the depth of the water is half its initial height. Show that the vessel is empty after a time of $\dfrac{2T}{2-\sqrt{2}}$.

b. A conical tank is initially full of water. Water starts flowing out through a hole in the bottom of the tank at a rate proportional to the square root of the remaining depth of the water. After a time of T, the depth of the water is half its initial height. Show that the tank is empty after a time of $\dfrac{T}{1-\sqrt{2^{-5}}}$.

14. The diagram shows a vase. The base and the top are circular with radii of 9 cm and 4 cm respectively, and the height is 16 cm. The origin, O, is at the centre of the base of the vase, with the coordinate axes as shown. Initially the vase is filled with water, but the water leaks out at a rate equal to $2\sqrt{h}\,\text{cm}^3/\text{s}$, where h cm is the height of the water in the vase. Determine the time taken for the vase to empty, in seconds to 1 decimal place, and the capacity of the vase, in cubic centimetres to 1 decimal place, if the side of the vase is modelled by:

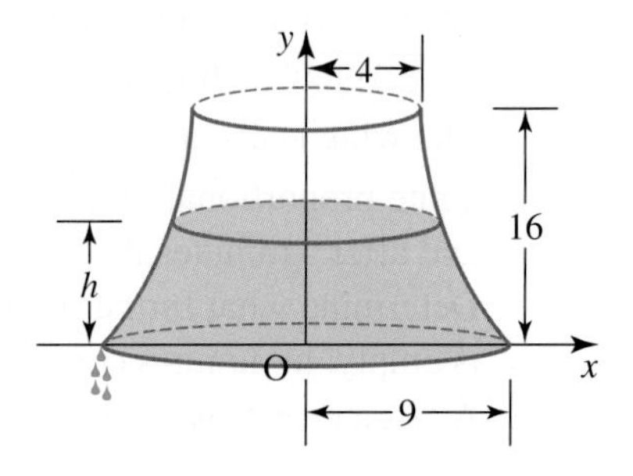

a. a hyperbola, $y = \dfrac{a}{x} + b$ b. a truncus, $y = \dfrac{a}{x^2} + b$.

15. A vase has a base radius of 4 cm, a top radius of 9 cm and a height of 16 cm. The origin, O, is at the centre of the base of the vase, and the vase can be represented by a curve rotated about the y-axis. Initially the vase is filled with water, but the water leaks out at a rate equal to $2\sqrt{h}\,\text{cm}^3/\text{s}$, where h cm is the height of the water in the vase.
Determine the time taken for the vase to empty, in seconds to 1 decimal place, and the capacity of the vase, in cubic centimetres to 1 decimal place, if the side of the vase is modelled by:

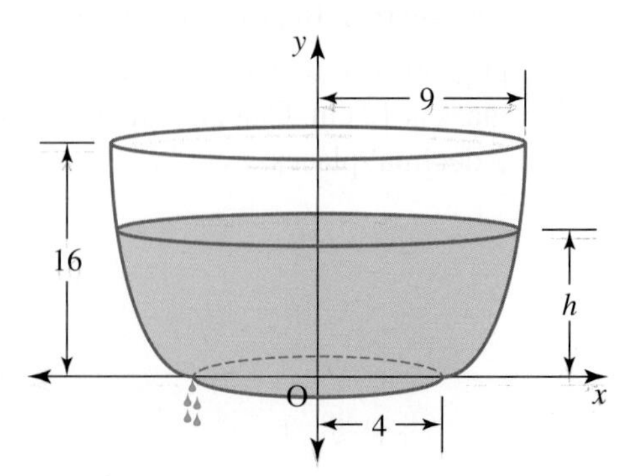

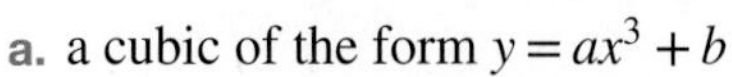
a. a cubic of the form $y = ax^3 + b$
b. a quartic of the form $y = ax^4 + b$.

16. A hot water tank has a capacity of 160 litres and initially contains 100 litres of water. Water flows into the tank at a rate of $12\sqrt{t}\sin^2\left(\frac{\pi t}{4}\right)$ litres per hour over the time $0 \le t \le T$ hours, where $4 < T < 8$. During the time interval $0 \le t \le 4$, water flows out of the tank at a rate of $5\sqrt{t}$ litres per hour.

a. After 3 hours, determine if the water level in the tank is increasing or decreasing.
b. After 4 hours, determine if the water level in the tank is increasing or decreasing.
c. Determine the time at which the inflow and outflow rates are equal, stating the answer in hours correct to 2 decimal places.
d. After 4 hours, determine the volume of water in the tank, in litres correct to 2 decimal places.
e. After 4 hours, no water flows out of the tank. However, the inflow continues at the same rate until the tank is full. Determine the time, T, required to fill the tank, in hours correct to 2 decimal places.

9.7 Exam questions

Question 1 (10 marks) TECH-ACTIVE

Source: VCE 2021, Specialist Mathematics Exam 2, Section B, Q3; © VCAA.

A thin-walled vessel is produced by rotating the graph of $y = x^3 - 8$ about the y-axis for $0 \le y \le H$. All lengths are measured in centimetres.

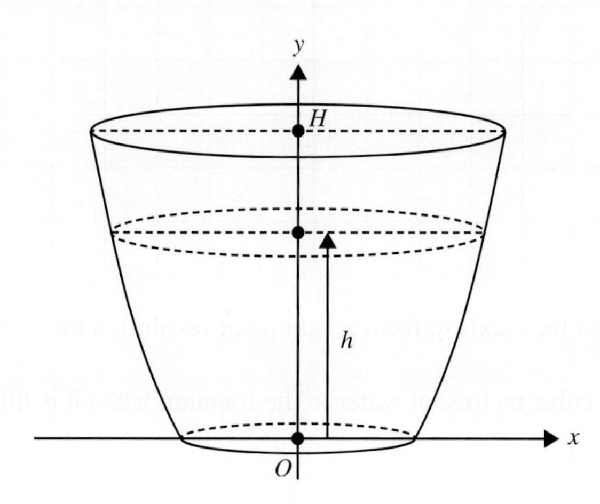

a. i. Write down a definite integral in terms of y and H for the volume of the vessel in cubic centimetres. **(1 mark)**

ii. Hence, find an expression for the value of the vessel in terms of H. **(1 mark)**

Water is poured into the vessel. However, due to a crack in the base, water leaks out a rate proportional to the square root of the depth h of water in the vessel, that is $\frac{dv}{dt} = -4\sqrt{h}$, where V is the volume of water remaining in the vessel, in cubic centimetres, after t minutes.

b. i. Show that $\frac{dh}{dt} = \frac{-4\sqrt{h}}{\pi(h+8)^{\frac{2}{3}}}$ **(2 marks)**

ii. Find the maximum rate, in centimetres per minute, at which the depth of water in the vessel decreases, correct to two decimal places, and find the corresponding depth in centimetres. **(2 marks)**

iii. Let $H = 50$ for a particular vessel. The vessel is initially fully and water continues to leak out at a rate of $4\sqrt{h}\,\text{cm}^3\ \text{min}^{-1}$. Find the maximum rate at which can be added, in cubic centimetres per minute, without the vesel overflowing. **(1 mark)**

c. The vessel is initially full where $H = 50$ and water leaks out at a rate of $4\sqrt{h}\,\text{cm}^3\ \text{min}^{-1}$. When the depth of the water drops to 25 cm, extra water is poured in at a rate of $40\sqrt{2}\,\text{cm}^3\ \text{min}^{-1}$.
Find how long it takes for the vessel to refill completely from a depth of 25 cm. Give your answer in minutes, correct to one decimal place. **(3 marks)**

Question 2 (11 marks) TECH-ACTIVE

Source: *VCE 2018, Specialist Mathematics Exam 2, Section B, Q3a-d,f; © VCAA.*

Part of the graph of $y = \frac{1}{2}\sqrt{4x^2 - 1}$ is shown below.

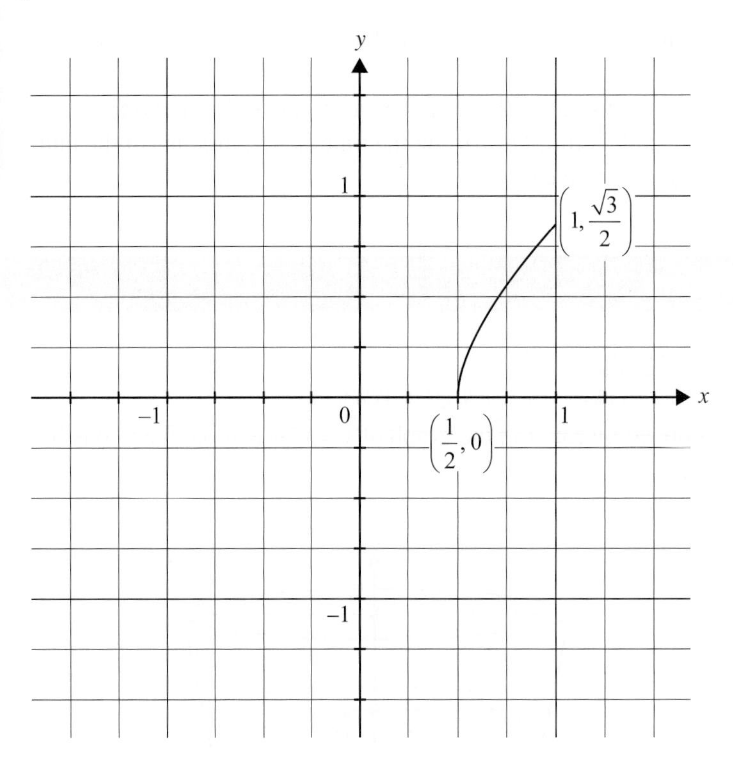

The curve shown is rotated about the y-axis to form a volume of revolution that is to model a fountain, where length units are in metres.

a. Show that the volume, V cubic metres, of water in the fountain when it is filled to a depth of h metres is given by $V = \frac{\pi}{4}\left(\frac{4}{3}h^3 + h\right)$. **(2 marks)**

b. Find the depth h when the fountain is filled to half its volume. Give your answer in metres, correct to two decimal places. **(2 marks)**

The fountain is initially empty. A vertical jet of water in the centre fills the fountain at a rate of 0.04 cubic metres per second and, at the same time, water flows out from the bottom of the fountain at a rate of $0.05\sqrt{h}$ cubic metres per second when the depth is h metres.

c. **i.** Show that $\frac{dh}{dt} = \frac{4 - 5\sqrt{h}}{25\pi\left(4h^2 + 1\right)}$. **(2 marks)**

ii. Find the rate, in metres per second, correct to four decimal places, at which the depth is increasing when the depth is 0.25 m. **(1 mark)**

d. Express the time taken for the depth to reach 0.25 m as a definite integral and evaluate this integral correct to the nearest tenth of a second. **(2 marks)**

e. How far from the top of the fountain does the water level ultimately stabilise? Give your answer in metres, correct to two decimal places. **(2 marks)**

Question 3 (12 marks) TECH-ACTIVE

***Source:** VCE 2014, Specialist Mathematics Exam 2, Section 2, Q4; © VCAA.*

At a water fun park, a conical tank of radius 0.5 m and height 1 m is filling with water. At the same time, some water flows out from the vertex wetting those underneath. When the tank eventually fills, it tips over and the water falls out, drenching all those underneath. The tank then returns to its original position and begins to refill.

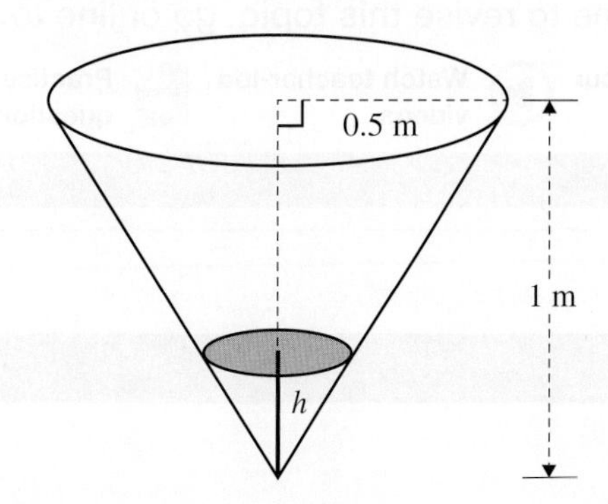

Water flows in at a constant rate of $0.02\pi \text{ m}^3/\text{min}$ and flows out at a variable rate of $0.01\pi\sqrt{h} \text{ m}^3/\text{min}$, where h metres is the depth of the
water at any instant.

a. Show that the volume, V cubic metres, of water in the cone when it is filled to a depth of h metres is given by $V = \frac{\pi}{12}h^3$. **(1 mark)**

b. Find the rate, in m/ min at which the depth of the water in the tank is increasing when the depth is 0.25 m. **(4 marks)**

The tank is empty at time $t = 0$ minutes.

c. By using an appropriate definite integral, find the time it takes for the tank to fill. Give your answer in minutes, correct to one decimal place. **(2 marks)**

Another water tank, shown below, has the shape of a large bucket (part of a cone) with the dimensions given. Water fills the tank at a rate of $0.05\pi \text{ m}^3/\text{min}$, but no water leaks out.

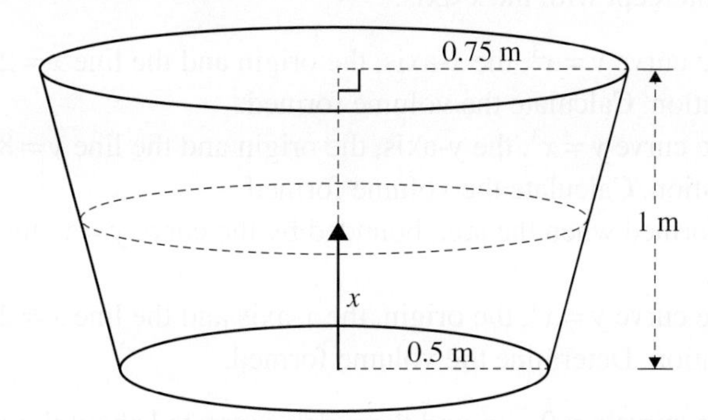

When filled to a depth of x metres, the volume of water, V cubic metres, in the tank is given by

$$V = \frac{\pi}{48}\left(x^3 + 6x^2 + 12x\right)$$

d. Given that the tank is initially empty, find the depth, x metres, as a function of time t. **(5 marks)**

More exam questions are available online.

9.8 Review

9.8.1 Summary

doc-37063

Hey students! Now that it's time to revise this topic, go online to:

 Access the topic summary  Review your results  Watch teacher-led videos  Practise exam questions

Find all this and MORE in jacPLUS

9.8 Exercise

Technology free: short answer

1. Determine each of the following.

   a. $\int x\sin(4x)dx$

   b. $\int \cos^{-1}(5x)dx$

   c. $\int \sin^{-1}\left(\frac{2x}{3}\right)dx$

   d. $\int x^2\log_e(3x)dx$

2. a. Calculate the area bounded by the graph of $y = xe^{-\frac{x}{3}}$ the x-axis the origin and the line $x = 3$.

   b. Calculate the area bounded by the graph of $y = \tan^{-1}\left(\frac{x}{5}\right)$ the x-axis the origin and the line $x = 5$.

   c. Given the function f: $[0, \infty] \to R, f(x) = x\cos(2x)$ determine the area bounded by the curve, the x-axis, the origin and the first intercept with the x-axis.

   d. Given the function f: $(0, \infty) \to R, f(x) = x\log_e\left(\frac{x}{3}\right)$ determine the area bounded by the curve, the x-axis, the origin and the first intercept with the x-axis.

3. a. The area bounded by the curve $y = x^3$, the x-axis, the origin and the line $x = 2$, is rotated about the x-axis to form a solid of revolution. Calculate the volume formed.

   b. The area bounded by the curve $y = x^3$, the y-axis, the origin and the line $y = 8$, is rotated about the y-axis to form a solid of revolution. Calculate the volume formed.

   c. Determine the volume formed when the area bounded by the curve $y = x^3$ the y-axis, and the line $y = 8$ is rotated about the x-axis.

   d. The area bounded by the curve $y = x^3$, the origin, the x-axis and the line $x = 2$, is rotated about the y-axis to form a solid of revolution. Determine the volume formed.

4. a. The area bounded by the curve $y = 9 - x^2$ and the x-axis is rotated about the x-axis to form a solid of revolution. Calculate the volume formed.

   b. The area bounded by the curve $y = 9 - x^2$ and the x-axis is rotated about the y-axis to form a solid of revolution. Calculate the volume formed.

5. a. The area bounded by the curve $y = x^4$, the x-axis, the origin and the line $x = 2$, is rotated about the x-axis to form a solid of revolution. Determine the volume formed.

   b. The area bounded by the curve $y = x^4$, the x-axis, the origin and the line $x = 2$, is rotated about the y-axis to form a solid of revolution. Determine the volume formed.

6. A cylindrical hole of radius r is drilled through a solid sphere of radius R. Show that volume remaining is given by $\frac{4\pi R^3}{3}\left(1 - \frac{r^2}{R^2}\right)^{\frac{3}{2}}$.

7. a. Calculate the length of the curve $y = \frac{x^3}{3} + \frac{1}{4x}$ from $x = 1$ to $x = 3$.

b. Calculate the length of the curve $y = \frac{2x^4 + 6}{12x}$ from $x = 1$ to $x = 5$.

8. A unidentified flying object, UFO, is formed when the area between the curves $y = \frac{1}{2}(1 - x^2)$ and $y = \frac{1}{4}(x^4 - 1)$ is rotated about the y-axis. Calculate the volume of the UFO.

Technology active: multiple choice

9. **MC** $\int x^2 \cos\left(\frac{x}{2}\right) dx$ is equal to

A. $2x^2 \sin\left(\frac{x}{2}\right) - 4\int x \sin\left(\frac{x}{2}\right) dx$

B. $2x^2 \sin\left(\frac{x}{2}\right) - \int x \sin\left(\frac{x}{2}\right) dx$

C. $x^2 \sin\left(\frac{x}{2}\right) + \int x \sin\left(\frac{x}{2}\right) dx$

D. $-2x^2 \sin\left(\frac{x}{2}\right) + \int x \sin\left(\frac{x}{2}\right) dx$

E. $2x \cos\left(\frac{x}{2}\right) - 4\int x \cos\left(\frac{x}{2}\right) dx$

10. **MC** A cylindrical tank has a base radius of 5 cm and a height of 9 cm and is initially filled with water. The water flows out through a hole in the bottom of the tank at a rate of $5\sqrt{h}$ cm^3/min, where h cm is the height of the water in the tank at time t. The time taken in minutes for the tank to empty is given by

A. $\frac{1}{5\pi}\int_0^9 \frac{1}{\sqrt{h}} dh$ B. $5\pi \int_9^0 \sqrt{h}\, dh$ C. $5\pi \int_0^9 \sqrt{h}\, dh$ D. $5\pi \int_0^9 \frac{1}{\sqrt{h}} dh$ E. $5\pi \int_9^0 \frac{1}{\sqrt{h}} dh$

11. **MC** The area bounded by the curve $y = \sqrt{x - 2}$, the x-axis and the line $x = 6$ is rotated about the x-axis, the volume formed is equal to

A. 6π B. 8π C. $\frac{16\pi}{3}$ D. $\frac{5744\pi}{5}$ E. $\frac{6464\pi}{5}$

12. **MC** The area bounded by the curve $y = \log_e(x + 1)$, the y-axis and the line $y = \log_e(3)$ is rotated about the y-axis, the volume formed is closest to

A. 2.83 B. 3.23 C. 3.45 D. 12.57 E. 50.33

13. **MC** The area bounded by the curve $y = 2\sin(2x)$, the y-axis and the line $y = 2$ is rotated about the x-axis, the volume formed is equal to

A. $4\pi \int_0^{\frac{\pi}{4}} \cos^2(2x)dx$

B. $4\pi \int_0^{\frac{\pi}{4}} \sin^2(2x)dx$

C. $\frac{\pi}{4}\int_0^{\frac{\pi}{4}} \left(\sin^{-1}\left(\frac{x}{2}\right)\right)^2 dx$

D. $\pi \int_0^{\frac{\pi}{4}} \left(\frac{\pi^2}{16} - \frac{1}{4}\left(\sin^{-1}\left(\frac{x}{2}\right)\right)^2\right) dx$

E. $\pi \int_0^{\frac{\pi}{4}} \left(\frac{\pi^2}{16} - 4\sin^2(2x)\right) dx$

14. **MC** The length of the curve $y = \log_e(3x)$ from $x = 1$ to $x = 2$ is given by

A. $\int_1^2 \frac{\sqrt{9x^2 + 1}}{3x}\,dx$

B. $\int_1^2 \frac{\sqrt{x^2 + 1}}{x}\,dx$

C. $\int_1^2 \frac{\sqrt{x^2 + 9}}{x}\,dx$

D. $\frac{1}{3}\int_1^2 \sqrt{9 + e^{2x}}\,dx$

E. $\frac{1}{3}\int_1^2 \sqrt{9 + x^2}\,dx$

15. **MC** The length of the curve $y = \tan(2x)$ from $x = 0$ to $x = \frac{\pi}{8}$ is closest to

A. 0.64 B. 0.88 C. 1.08 D. 1.62 E. 2.04

16. **MC** When the area bounded by the curve $y = \tan(2x)$ the x-axis $x = -\frac{\pi}{8}$ to $x = \frac{\pi}{8}$ is rotated about the x-axis, the surface area formed is given by

A. $2\pi \int_0^{\frac{\pi}{8}} \tan(2x)\sqrt{1 + 4\sec^4(2x)}\,dx$

B. $2\pi \int_0^{\frac{\pi}{8}} \tan(2x)\sqrt{1 + 4\sec^4(2x)}\,dx$

C. $2\pi \int_0^{\frac{\pi}{8}} \tan(2x)\sqrt{1 + \sec^2(2x)}\,dx$

D. $4\pi \int_0^{\frac{\pi}{8}} \tan(2x)\sqrt{1 + \sec^2(2x)}\,dx$

E. $2\pi \int_0^{\frac{\pi}{8}} \sqrt{1 + 4\sec^2(2x)}\,dx$

17. **MC** When the area bounded by the curve $y = \tan^{-1}(2x)$ the x-axis $x = 0$ to $x = \pi$ is rotated about the x-axis, the surface area formed is closest to

A. 3291.55 B. 81.15 C. 3.71 D. 3.81 E. 23.89

18. **MC** If $\frac{dy}{dx} = \sqrt{2x^6 + 1}$ and when $y = 5$ when $x = 1$, then the value of y when $x = 4$ is given by

A. $\int_1^4 \left(\sqrt{2t^6 + 1} + 5\right)\,dt$

B. $\int_1^4 \sqrt{2t^6 + 1}\,dt$

C. $\int_1^4 \left(\sqrt{2t^6 + 1} - 5\right)\,dt$

D. $\int_1^4 \sqrt{2t^6 + 1}\,dt - 5$

E. $\int_1^4 \sqrt{2t^6 + 1}\,dt + 5$

Technology active: extended response

19. Consider the function $f: [2, 11] \to R,\ f(x) = \sqrt{11 - x}$.

a. Determine the area bounded by the graph of the function and the x-axis.

b. When the graph of the function is rotated about the x-axis it forms a solid of revolution. Calculate the volume formed.

c. Show that the length of the function can be expressed as the definite integral $\frac{1}{2}\int_2^{11} \sqrt{\frac{45 - 4x}{11 - x}}\,dx$.

d. When the graph of the function is rotated about the x-axis it forms a solid of revolution. Determine the surface area of this solid.

20. The region R is bounded by the line $x = 2$, the curve $y = 2\log_e(3x)$ and the x-axis. Give all answers correct to 4 decimal places.

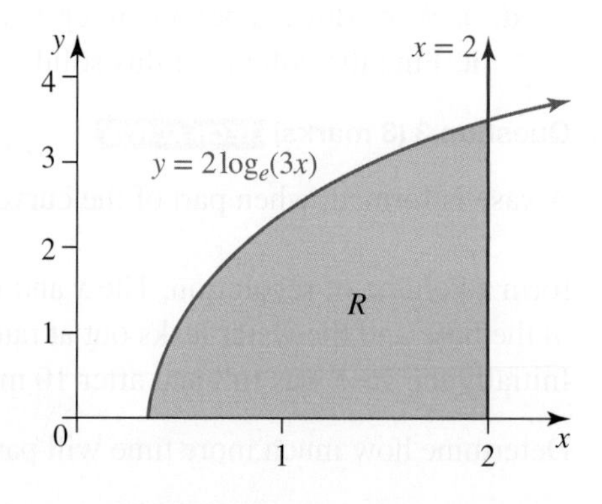

a. Calculate the total perimeter of the region R.
b. Calculate the area of the region R.
c. When the region R is rotated about the x-axis it forms a solid of revolution. Determine the volume of this solid of revolution.
d. When the region R is rotated about the x-axis it forms a solid of revolution. Determine the surface area of this solid of revolution.
e. The line $x = k$ divides the area of the region R in half. Determine the value of k.
f. When the region R is rotated about the x-axis it forms a solid of revolution. The line $x = k$ divides the volume of the of the region R in half. Determine the value of k.

9.8 Exam questions

Question 1 (2 marks) TECH-ACTIVE

Source: *VCE 2019, Specialist Mathematics Exam 2, Section B, Q1e; © VCAA.*

The portion of the curve given by $y = \sqrt{x^2 - 2x}$ for $x \in [2, 4]$ is rotated about the y-axis to form a solid of revolution.

Write down, but do not evaluate, a definite integral in terms of t that gives the volume of the solid formed.

Question 2 (11 marks) TECH-ACTIVE

Source: *VCE 2014, Specialist Mathematics Exam 2, Section 2, Q1; © VCAA.*

Consider the function f with rule $f(x) = \dfrac{9}{(x+2)(x-4)}$ over its maximal domain.

a. Find the coordinates of the stationary point(s). **(3 marks)**
b. State the equation of all asymptotes of the graph of f. **(2 marks)**
c. Sketch the graph of f for $x \in [-6, 6]$ on the axes below, showing asymptotes, the values of the coordinates of any intercepts with the axes, and the stationary point(s). **(3 marks)**

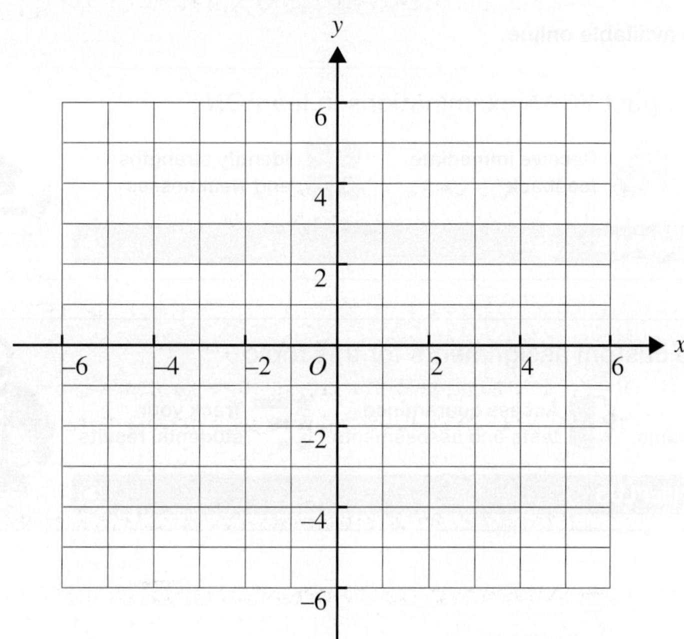

The region bounded by the coordinate axes, the graph of f and the line $x = 3$, is rotated about the x-axis to form a solid of revolution.

d. i. Write down a definite integral in terms of x that gives the volume of this solid of revolution. **(2 marks)**

ii. Find the volume of this solid, correct to two decimal places. **(1 mark)**

Question 3 (3 marks) TECH-ACTIVE

A vase is formed, when part of the curve $\frac{x^2}{16} - \frac{65y^2}{4096} = 1$ for $4 \le x \le 9$, $y \ge 0$, is rotated about the y-axis to form a volume of revolution. The x and y coordinates are measured in centimetres. The vase has a small crack in the base and the water leaks out at rate proportional to the square root of the remaining height of the water. Initially the vase was full and after 10 minutes the height of the water in the vase is 9 cm.

Determine how much more time will pass before the vase is empty, in minutes correct to 1 decimal place.

Question 4 (6 marks) TECH-FREE

If n is a positive integer, and a is a positive real number then:
show that

a. $\int \cos^n(ax)\,dx = \frac{1}{an}\sin(ax)\cos^{n-1}(ax) + \frac{n-1}{n}\int \cos^{n-2}(ax)\,dx$ **(3 marks)**

b. Hence, determine $\int \cos^5(3x)\,dx$. **(3 marks)**

Question 5 (4 marks) TECH-ACTIVE

A soup bowl has a circular base and top, with radii of 4 and 9 cm respectively, its height is 16 cm. The origin O is at the centre of the base of the bowl. The bowl is formed when the curve $y = ax^2 + b$ is rotated about the y-axis. Initially the bowl is filled with soup, but the soup leaks out at a rate equal to $2\sqrt{h}\,\text{cm}^3/\text{min}$, where h cm is the height of the soup remaining in the bowl after t minutes.

Set up the differential equation for h and t and determine how long before the bowl is empty, in minutes correct to one decimal place.

More exam questions are available online.

Answers

Topic 9 Further integration techniques and applications

9.2 Integration by parts

9.2 Exercise

1. a. $-\frac{1}{4}e^{-2x}(1+2x)+c$

   b. $3e^{\frac{x}{3}}(x-3)+c$

2. i. $\frac{1}{9}(\sin(3x)-3x\cos(3x))+c$

   ii. $2\left(x\sin\left(\frac{x}{2}\right)+2\cos\left(\frac{x}{2}\right)\right)+c$

3. i. $x\left(\log_e(3x)-1\right)+c$

   ii. $\frac{1}{2}x^2\left(\log_e\left(\frac{x}{2}\right)-2\right)+c$

4. a. $x\sin^{-1}\left(\frac{x}{3}\right)+\sqrt{9-x^2}+c$

   b. $x\cos^{-1}(4x)-\frac{1}{4}\sqrt{1-16x^2}+c$

5. a. $x\tan^{-1}\left(\frac{x}{5}\right)-\frac{5}{2}\log_e\left(x^2+25\right)+c$

   b. $\frac{1}{8}\left(1+4x^2\right)\tan^{-1}(2x)-\frac{x}{4}+c$

6. a. $2-\frac{1}{2}\log_e(3)$ b. $\frac{1}{32}$

7. a. $\frac{1}{256}\left(\pi^2+4\pi-8\right)$

   b. $\frac{1}{1296}\left(4\pi^2-6\pi\sqrt{3}+27\right)$

8. $\frac{1}{2}$ units^2

9. $\frac{1}{6}(\pi-2)$ units^2

10. $\pi-\log_e(4)$ units^2

11. a. $\frac{1}{a^2}e^{ax}(ax-1)+c$

    b. $\frac{1}{a^2}(ax\sin(ax)+\cos(ax))+c$

12. a. $\frac{1}{a^2}(\sin(ax)-ax\cos(ax))+c$

    b. $\begin{cases}\frac{x^{n+1}}{(n+1)^2}\left((n+1)\log_e(ax)-1\right)+c, & n\neq-1\\ \frac{1}{2}\left(\log_e(ax)\right)^2+c, & n=-1\end{cases}$

13. a. $x\sin^{-1}\left(\frac{x}{a}\right)+\sqrt{a^2-x^2}+c$

    b. $x\cos^{-1}\left(\frac{x}{a}\right)-\sqrt{a^2-x^2}+c$

14. a. $x\tan^{-1}\left(\frac{x}{a}\right)-\frac{a}{2}\log_e\left(x^2+a^2\right)+c$

    b. $\frac{1}{2}\left(x^2+a^2\right)\tan^{-1}\left(\frac{x}{a}\right)-\frac{ax}{2}+c$

15. a. Sample responses can be found in the worked solutions in the online resources.

    b. $-\frac{1}{2}x^4\cos(2x)+x^3\sin(2x)+\frac{3}{2}x^2\cos(2x)-\frac{3}{2}x\sin(2x)-\frac{3}{4}\cos(2x)+c$

16. a. Sample responses can be found in the worked solutions in the online resources.

    b. $\frac{8}{15}$

17. a. Sample responses can be found in the worked solutions in the online resources.

    b. $\frac{8}{15}$

18. Sample responses can be found in the worked solutions in the online resources.

19. a. Sample responses can be found in the worked solutions in the online resources.

    b. Sample responses can be found in the worked solutions in the online resources.

    c. Sample responses can be found in the worked solutions in the online resources.

    d. $\frac{3}{13}\left(1+e^{-\frac{2\pi}{3}}\right)$ units^2

20. Sample responses can be found in the worked solutions in the online resources.

21, 22. Sample responses can be found in the worked solutions in the online resources.

23. a. Sample responses can be found in the worked solutions in the online resources.

    b. $8e^2-24$

24. a. Sample responses can be found in the worked solutions in the online resources.

    b. $\frac{9e}{2}-12$

9.2 Exam questions

Note: Mark allocations are available with the fully worked solutions online.

1. $\frac{1}{16}x^4\left(4\log_e(2x)-1\right)+c$

2. Sample responses can be found in the worked solutions in the online resources.

3. a. Sample responses can be found in the worked solutions in the online resources.

   b. $\frac{1}{8}\left(e^2+3\right)$

9.3 Integration by recognition and graphs of anti-derivatives

9.3 Exercise

1. $\frac{2\pi}{3}$
2. $\frac{\pi}{12}$
3. $-\frac{\pi}{9}$
4. $\frac{\pi}{12}$
5. $\frac{\pi}{36}$
6. $\frac{\pi}{24}$
7. $\frac{\pi}{18}$
8. $\frac{\pi}{32}$
9. $\log_e\left(\sqrt{3}\right)$
10. $\frac{1}{4}\log_e\left(\sqrt{3}+2\right)$
11. a. $x = 2$ is a minimum turning point.

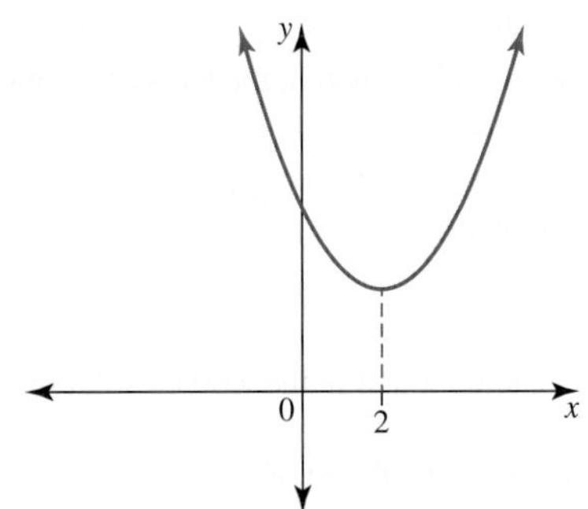

b. $x = -3$ is a minimum turning point, $x = 1$ is a maximum turning point, $x = -1$ is an inflection point.

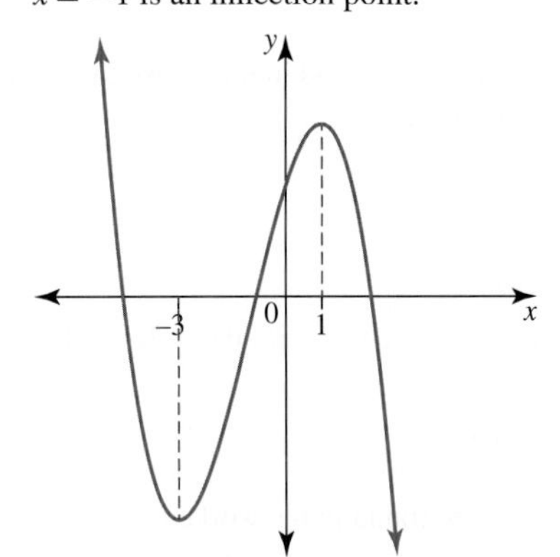

12. $x = \pm 4$ are minimum turning points, $x = 0$ is a maximum turning point.
13. a. A straight line with a gradient of 2.

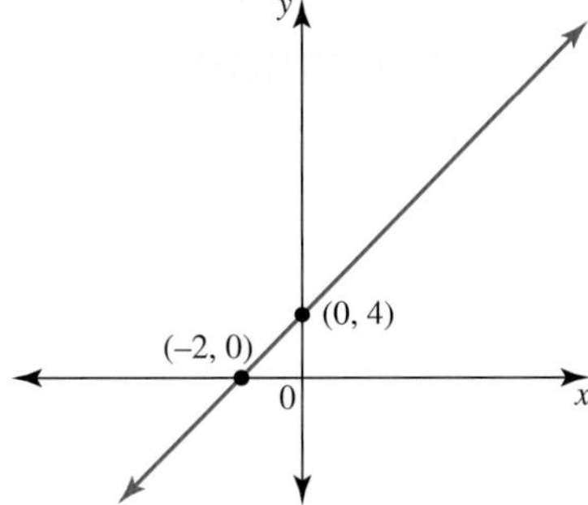

b. $x = -3$ is a minimum turning point.

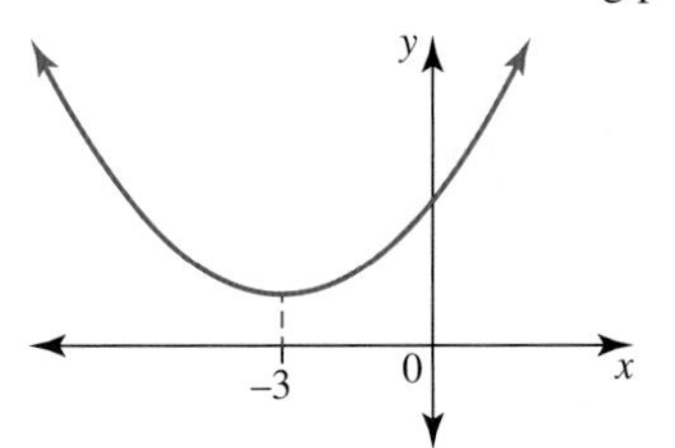

14. $\int_{\frac{\pi}{8}}^{x} \tan\left(\frac{1}{t}\right) dt + \frac{1}{2}$

9.3 Exam questions

Note: Mark allocations are available with the fully worked solutions online.

1. $\frac{\pi}{24}$
2. $\log_e\left(1+\sqrt{2}\right)$
3. A stationary point of inflection at $x = 2$.

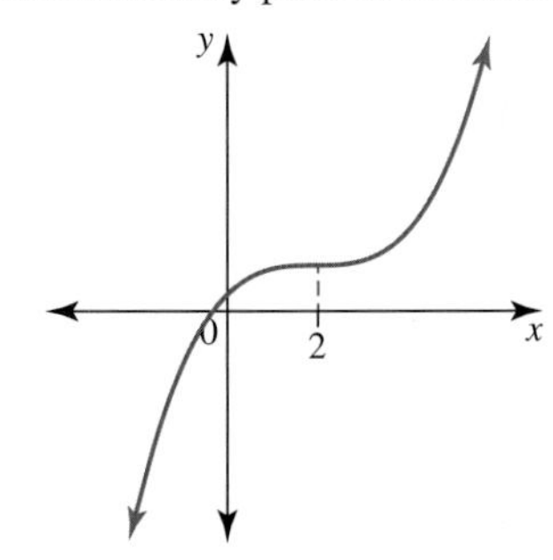

9.4 Solids of revolution

9.4 Exercise

1. a.

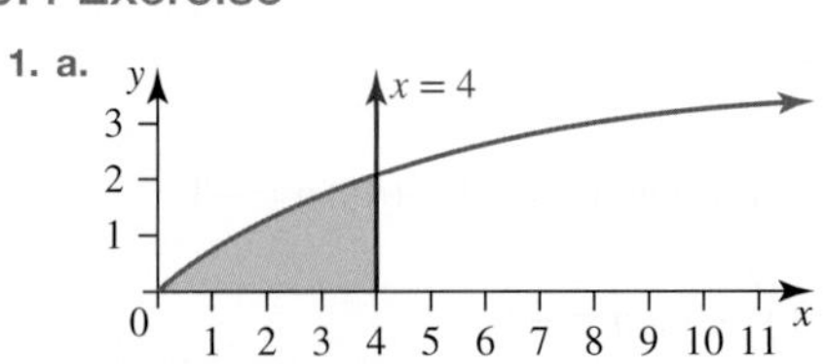

b. 8π units^3

2. $\frac{512\pi}{15}$ units^3

3. a. 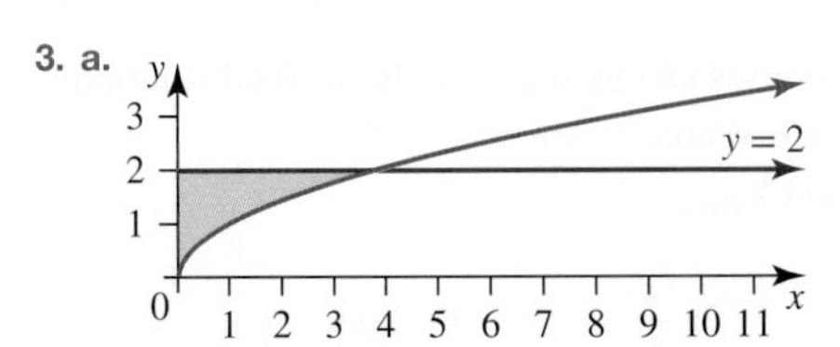

b. $\frac{32\pi}{5}$ units^3

4. 8π units^3

5. a. 375π units^3 b. $\frac{125\pi}{27}$ units^3

6. a. 4π units^3 b. 6π units^3

7. $\frac{\pi}{2}$ units^3

8. $4\pi\left(e^2 - 1\right)$ units^3

9. 10 L

10. 1100 mL

11. a. $\frac{9\pi^2}{4}$ units^3 b. $\frac{4\pi^2}{3}$ units^3

12. a. $\frac{768\pi}{5}$ units^3 b. 6π units^3

13. a. $\frac{1296\pi}{5}$ units^3 b. $\frac{81\pi}{2}$ units^3

14. a. $\frac{320\pi}{3}$ units^3

b. $\frac{400\pi}{3}$ units^3

c. $\frac{4}{3}\pi ab^2$ units^3

d. $\frac{4}{3}\pi a^2 b$ units^3

e. If rotating about the x-axis: $\frac{4x^2}{57^2} + \frac{y^2}{22^2} = 1$

If rotating about the y-axis: $\frac{x^2}{22^2} + \frac{4y^2}{57^2} = 1$

f. 57 780 mm^3

15. a. Sample responses can be found in the worked solutions in the online resources.

b. 136 L

16. a. 3099.70 cm^3 b. 3141.59 cm^3

c. 3183.03 cm^3 d. 3223.69 cm^3

17. a. $32\pi\left(3 - 4\log_e(2)\right)$ units^3

b. $\frac{64\pi}{15}$ units^3

18. $\frac{27\,776\pi}{3}$ cm^3

9.4 Exam questions

Note: Mark allocations are available with the fully worked solutions online.

1 $V = 2\pi\left(\log_e\left(2 + 2\sqrt{3}\right) + \frac{\pi}{3}\right)$ units^3

2 $V = \pi\left(\frac{\pi}{4} + \log_e(2)\right)$ units^3

3 16π units^3

9.5 Volumes

9.5 Exercise

1. a. $\frac{704\pi}{5}$ units^3 b. 8π units^3

2. a. 8π units^3 b. $\frac{32\pi}{5}$ units^3

3. a. $\frac{49\pi}{2}$ units^3 b. $\frac{76\pi}{5}$ units^3

4. a. $\frac{128\pi}{5}$ units^3 b. 8π units^3

5. a. $\frac{4352\pi}{15}$ units^3 b. 8π units^3

6. a. $\frac{81\pi}{2}$ units^3 b. $\frac{383\pi}{15}$ units^3

7. a. $\frac{486\pi}{5}$ units^3 b. $\frac{2187\pi}{7}$ units^3

8. a. $\frac{416\pi}{15}$ units^3 b. 4π units^3

9. a. $4\pi^2$ units^3 b. $\frac{9\pi^2}{2}$ units^3

10. a. $\frac{64\pi}{15}$ units^3 b. $\frac{8\pi}{3}$ units^3

11. a. $\frac{2\pi}{35}$ units^3 b. $\frac{\pi}{10}$ units^3

12. a. $\frac{96\pi}{5}$ units^3 b. $\frac{3\pi a^3}{10}$ units^3

13. a. 1296π units^3 b. 81π units^3

14. $\frac{512\pi}{3}$ units^3

15. $\frac{192\pi}{5}$ units^3

16. a. $32\sqrt{3}\pi$ units^3 b. $\frac{a^3\pi\sqrt{3}}{2}$ units^3

17. a. 256π units^3 b. $2\pi^2 ar^2$ units^3

18. a. $\frac{11\sqrt{3}\pi a^5}{40}$ units^3 b. $\frac{9\pi a^4}{32}$ units^3

9.5 Exam questions

Note: Mark allocations are available with the fully worked solutions online.

1. a. $x = 1,\ x = 3$ b. $V = \dfrac{2\pi}{3}$ unit^3

2. D

3. $V = \dfrac{\pi^4}{9} - \dfrac{\pi^2}{6} - \dfrac{\sqrt{3}\pi}{8}$

9.6 Arc length and surface area

9.6 Exercise

1. $\dfrac{14}{3}$
2. $\dfrac{335}{27}$
3. $\dfrac{98\pi}{3}$
4. a. $5\sqrt{10}$ b. $155\pi\sqrt{10}$
5. $\dfrac{\pi}{27}\left(10\sqrt{10} - 1\right)$
6. a. $\dfrac{17}{12}$ b. $\dfrac{47\pi}{16}$
7. a. $\dfrac{33}{16}$ b. $\dfrac{1179\pi}{256}$
8. a. $\dfrac{1}{2}\left(e - \dfrac{1}{e}\right)$ b. $\dfrac{\pi}{4}\left(e^2 + 4 - \dfrac{1}{e^2}\right)$
9. $\dfrac{13}{6}$
10. $\dfrac{38\sqrt{3}}{9}$
11. $\dfrac{52}{3}$
12. $\dfrac{49}{6}$
13. a. $\dfrac{31}{6}$ b. $\dfrac{89\pi}{4}$
14. Sample responses can be found in the worked solutions in the online resources.
15. Sample responses can be found in the worked solutions in the online resources.
16. a. $\dfrac{3\pi}{2}$; this represents $\dfrac{1}{4}$ of the circumference of a circle of radius 3.
    b. 18π; this represents $\dfrac{1}{2}$ of the surface area of a sphere of radius 3.
17. a. Sample responses can be found in the worked solutions in the online resources.
    b. $\dfrac{1}{4}\log_e\left(2 + \sqrt{5}\right) + \dfrac{\sqrt{5}}{2}$ units
18. $\pi\left(\log_e\left(1 + \sqrt{2}\right) + \sqrt{2}\right)$ units^2
19. $\pi\left(\log_e\left(1 + \sqrt{2}\right) + \sqrt{2}\right)$ units^2
20. a. Sample responses can be found in the worked solutions in the online resources.
    b. $\log_e\left(2 + \sqrt{3}\right)$ units
21. a.

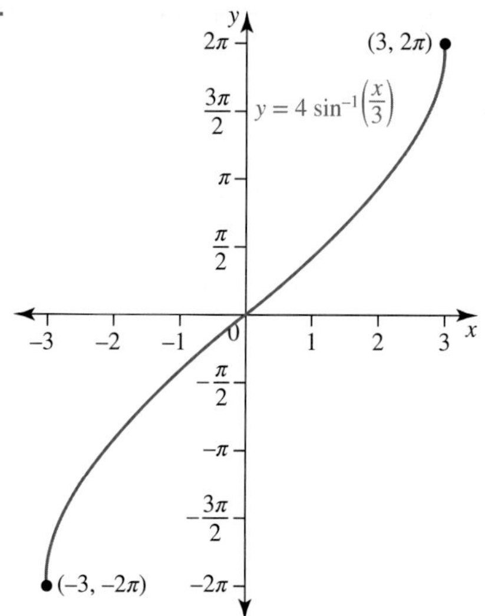

    b. $12(\pi - 2)$ units^2
    c. 14.1808 units
    d. $24\pi\left(\pi^2 - 8\right)$ units^3
    e. 267.3693 units^2
    f. $18\pi^2$units^3
    g. $2\pi\left(15 + 16\log_e(2)\right)$ units^2
22. a. 2.8323 units b. 29.2171 units^2
23. a. 3.6837 units b. 27.1426 units^2
24. a. 10.3978 units b. 307.8661 units^2
25. a. 15.8654 units b. 67.6729 units^2 c. 89.0007 units^2
26. a. $m = -2,\ c = 50,\ S = 2107.44\text{ cm}^3$
    b. $a = 400,\ k = -10,\ S = 2006.57\text{ cm}^3$

9.6 Exam questions

Note: Mark allocations are available with the fully worked solutions online.

1. $s = \dfrac{14}{3}$ units
2. $\dfrac{\pi}{6}\left(5\sqrt{5} - 1\right)$ units^2
3. Sample responses can be found in the worked solutions in the online resources.

9.7 Water flow

9.7 Exercise

1. 431.4 min
2. 473.3 min
3. 10.8 min
4. 12.8 min
5. a. 36.65 min b. 20 min
6. a. 54 s b. 273.5
7. a. $\dfrac{\pi}{5}$ b. 50.25 min

8. a. $\frac{92\pi}{5}$ b. 253.63 min

9. a. 3.11 min b. 20π

10. a. 0.68 m/h b. 1.26 h c. 261.9 h

11. a. 0.07 min b. 3.86 min

12. a. 132.4 s, 461.1 cm^3
 b. 317.3 s, 1423.5 cm^3

13. Sample responses can be found in the worked solutions in the online resources.

14. a. 609.1 s, 1809.6 cm^3
 b. 560.7 s, 1625.5 cm^3

15. a. 515.3 s, 2631.6 cm^3
 b. 555.6 s, 2802.8 cm^3

16. a. Increasing b. Decreasing c. 0.89, 3.11 h
 d. 106.65 L e. 7.04 h

9.7 Exam questions

Note: Mark allocations are available with the fully worked solutions online.

1. a. i. $V = \pi \int_0^H (y+8)^{\frac{2}{3}}\, dy$

 ii. $V(H) = \frac{3\pi}{5}\left[(H+8)^{\frac{5}{3}} - 32\right]$

 b. i. Sample answers can be found in the worked solutions in the online resources.

 ii. $h = 24$, decreases at 0.62 cm/min

 iii. $20\sqrt{2}\,\text{cm}^3/\text{min}$

 c. 3.1 min

2. a. Sample responses can be found in the worked solutions in the online resources.

 b. $h = 0.59$ m

 c. i. Sample responses can be found in the worked solutions in the online resources.

 ii. 0.0153 m/s

 d. $t = \int_0^{0.25} \left(\frac{25\pi\left(4h^2+1\right)}{4-5\sqrt{h}}\right) dh$
 $= 9.8$ s

 e. 0.23 m

3. a. Sample responses can be found in the worked solutions in the online resources.

 b. 0.96 m/ min

 c. 7.4 minutes

 d. $x(t) = 2\sqrt[3]{(0.3t+1)} - 2$ for $0 \le t \le \frac{95}{12}$

9.8 Review

9.8 Exercise

Technology free: short answer

1. a. $\frac{1}{16}(\sin(4x) - 4x\cos(4x)) + c$

 b. $x\cos^{-1}(5x) - \frac{1}{5}\sqrt{1-25x^2} + c$

 c. $x\sin^{-1}\left(\frac{2x}{3}\right) + \frac{1}{2}\sqrt{9-4x^2} + c$

 d. $\frac{1}{9}x^3\left(3\log_e(3x) - 1\right) + c$

2. a. $9 - \frac{18}{e}$ units^2

 b. $\frac{5}{4}\left(\pi - 2\log_e(2)\right)$ units^2

 c. $\frac{1}{8}(\pi - 2)$ units^2

 d. $\frac{9}{4}$ units^2

3. a. $\frac{128\pi}{7}$ units^3 b. $\frac{96\pi}{5}$ units^3
 c. $\frac{768\pi}{7}$ units^3 d. $\frac{64\pi}{3}$ units^3

4. a. $\frac{1296\pi}{5}$ units^3 b. $\frac{81\pi}{2}$ units^3

5. a. $\frac{512\pi}{9}$ units^3 b. $\frac{64\pi}{3}$ units^3

6. Sample responses can be found in the worked solutions in the online resources.

7. a. $\frac{53}{6}$ units b. $\frac{316}{15}$ units

8. $\frac{5\pi}{12}$ units^3

Technology active: multiple choice

9. A
10. D
11. B
12. C
13. A
14. B
15. C
16. B
17. E
18. E

Technology active: extended response

19. a. 18 units^2

 b. $\frac{81\pi}{2}$ units^3

 c. Sample responses can be found in the worked solutions in the online resources.

 d. $\frac{\pi}{3}\left(37\sqrt{37} - 1\right)$ units^2

20. a. 9.2718 units b. 3.8337 units^2
 c. 32.5105 units^3 d. 47.6287 units^2
 e. 1.4111 f. 1.5344

9.8 Exam questions

Note: Mark allocations are available with the fully worked solutions online.

1. $V = \pi \int_0^{\tan^{-1}\left(2\sqrt{2}\right)} (1 + \sec(t))^2 \sec^2(t)\, dt$

2. a. $(1, -1)$

   b. Vertical asymptotes at $x = 4$ and $x = -2$, horizontal asymptote at $y = 0$

   c. 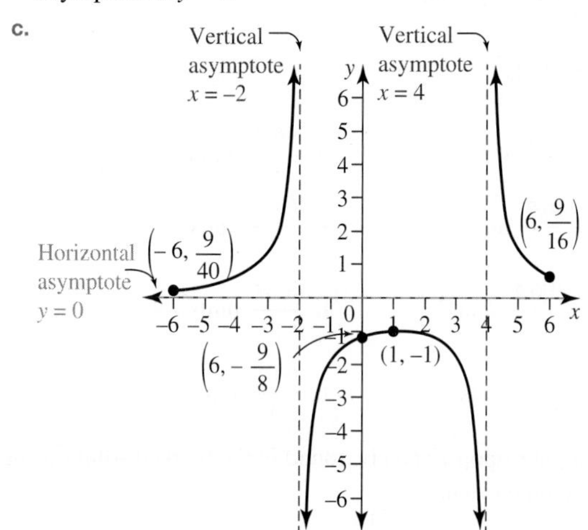

   d. i. $V = \pi \int_0^3 \frac{81}{(x+2)^2(x-4)^2} dx$

   ii. $V = 12.85$

3. 10.8 minutes

4. a. Sample responses can be found in the worked solutions in the online resources.

   b. $\frac{1}{45}\sin(3x)\left(3\cos^4(3x) + 4\cos^2(3x) + 8\right) + c$

5. 473.3 minutes

10 Applications of first-order differential equations

LEARNING SEQUENCE

Fully worked solutions for this topic are available online.

10.1 Overview

Hey students! Bring these pages to life online

Watch videos

Engage with interactivities

Answer questions and check results

Find all this and MORE in jacPLUS

10.1.1 Introduction

Differential equations first came into existence through the study and discovery of the concepts of calculus by Isaac Newton (1643–1727) and Gottfried Leibniz (1646–1716). A differential equation is a mathematical equation that relates a physical quantity with its rate of change. Because such relations are extremely common, differential equations play a major role in engineering, physics, economics and nature. In mathematics, differential equations are studied for several purposes, such as determining their solution set and their ability to model many practical applications. In 1822 Jean-Baptiste Fourier (1768–1830) published his work on heat flow in *Théorie analytique de la chaleur* (The Analytic Theory of Heat), in which he based his reasoning on Newton's law of cooling, namely, that the flow of heat between two adjacent molecules is proportional to the small difference of their temperatures. This partial differential equation is now taught to most secondary school and university students of physics. Differential equations play an integral role in modelling virtually every physical, technical or biological process, from celestial motion, to bridge design, to interactions between neurons. Many fundamental laws of physics can be formulated as differential equations. There are many different types of differential equations, however we only consider a limited variety of them in Specialist Mathematics.

KEY CONCEPTS

This topic covers the following point from the VCE Mathematics Study Design:

- formulation of differential equations from contexts in, for example, chemistry, biology and economics, in situations where rates are involved (including some differential equations whose analytic solutions are not required, but can be solved numerically using technology)
- the logistic differential equation
- verification of solutions of differential equations and their representation using direction (slope) fields
- numerical solution by Euler's method (first order approximation).

10.2 Growth and decay

LEARNING INTENTION

At the end of this subtopic you should be able to:

- set up and solve differential equations modelling natural population growth and radioactive decay.

10.2.1 Introduction

In Topic 8, methods of solving first-order differential equations were discussed. In this topic, some applications of first-order differential equations are explored. There are many applications of differential equations in business and science; in fact, they are useful whenever a rate of change needs to be considered.

10.2.2 The law of natural growth

Consider the differential equation $\frac{dy}{dx} = ky$, where k is a constant.

To solve this equation:

Invert both sides of the equation: $\frac{dx}{dy} = \frac{1}{ky}$

Integrate both sides with respect to y: $\int \frac{dx}{dy} dy = \int \frac{1}{ky} dy$

$$x = \int \frac{1}{ky} dy$$

$$x = \frac{1}{k} \int \frac{1}{y} dy$$

Integrate: $x = \frac{1}{k} \log_e |y| + c$, where c is a constant of integration and $y \neq 0$.

Rearrange: $\log_e |y| = kx + A$, where $A = -kc$ is another constant

Set both sides to base e: $|y| = e^{kx+A}$

$$y = \pm e^A \times e^{kx}$$

$$y = Be^{kx}, \text{ where } B = \pm e^A$$

Assume that $y(0) = y_0$, then $B = y_0$.

Therefore, the particular solution of the differential equation $\frac{dy}{dx} = ky$, $y(0) = y_0$ is given by $y = y_0 e^{kx}$. This can easily be verified using differentiation.

If $y = y_0 e^{kx}$, then

$$\begin{aligned} \frac{dy}{dx} &= k y_0 e^{kx} \\ &= k(y_0 e^{kx}) \\ &= ky \end{aligned}$$

Thus, there is a function, $y = y_0 e^{kx}$, whose derivative is proportional to itself. This result gives rise to several applications.

10.2.3 Population growth

The law of natural growth states that the rate of increase of population is proportional to the current population at that time. Let $N = N(t)$ represent the population number of a certain quantity at a time t, where $t \geq 0$. This leads to the equation

$$\frac{dN}{dt} \propto N$$
$$\Rightarrow \frac{dN}{dt} = kN, \text{ where } k \text{ is a positive constant.}$$

Assuming that the initial population number is $N_0 = N(0)$, the solution of this differential equation is given by $N = N_0 e^{kt}$. This is obtained by simply substituting N for y and t for x.

This equation has been found to model population growth, although it applies over only a limited time frame. Also, it does not include factors for population-changing events such as war or famine.

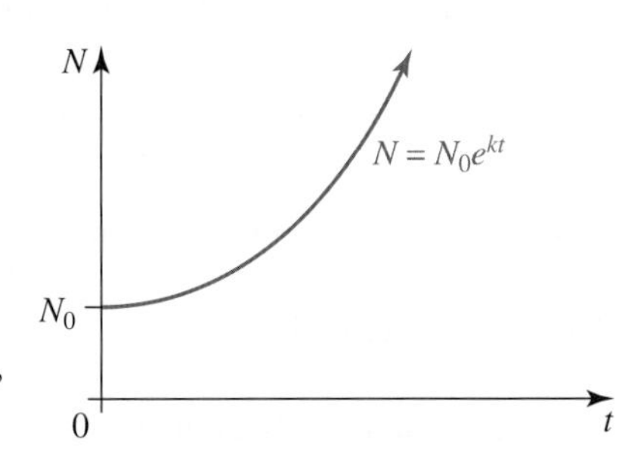

The equation $N = N_0 e^{kt}$ is called the law of natural growth or exponential growth.

The constant k can be interpreted as the excess birth rate over the death rate, and is called the annual growth rate.

Note that mathematically, N is treated as a continuous variable. However, when these equations are used to model populations, N is really a discrete quantity, the number of individuals in the population. In these situations the final answer may need to be rounded to the nearest whole number.

WORKED EXAMPLE 1 Applying the law of natural growth

The population of a certain town increases at a rate proportional to the current population. At the start of 2009 the population was 250 000, and at the start of 2017 the population was 750 000.

a. **Express the population number in millions, N, in terms of t, the time in years after the start of 2009.**

b. **Determine the predicted population at the start of 2025.**

c. **Calculate which year the population reaches 5 million.**

THINK	WRITE
a. 1. Let $N = N(t)$ represent the population number (in millions) of the town after 2009.	a. $\frac{dN}{dt} \propto N$ $\therefore \frac{dN}{dt} = kN$ $\Rightarrow N = N_0 e^{kt}$ $\frac{1}{4} = N_0 e^0$ $N_0 = \frac{1}{4}$
2. In the year 2009, that is at $t = 0$, the population was 250 000, or one-quarter of a million. Use this to determine the value of N_0.	Substitute $N_0 = \frac{1}{4}$: $N(t) = \frac{1}{4} e^{kt}$

3. In the year 2017 ($t = 8$), the population was 750 000, which is three-quarters of a million. Use this to determine the value of k.	Substitute $t = 8$ and $N = \frac{3}{4}$: $\frac{3}{4} = \frac{1}{4}e^{8k}$
4. Solve the equation for k by taking the natural logarithms of both sides.	$e^{8k} = 3$ $8k = \log_e(3)$ $k = \frac{1}{8}\log_e(3)$
5. Express the population number in millions at a time t years after 2009.	$N(t) = \frac{1}{4}e^{\frac{t}{8}\log_e(3)}$
b. 1. Determine the population in 2025, that is, when $t = 16$.	**b.** Substitute $t = 16$: $N(10) = \frac{1}{4}e^{\frac{16}{8}\log_e(3)}$ $N(10) = \frac{1}{4}e^{2\log_e(3)}$ $= \frac{1}{4}e^{\log_e(9)}$ $= \frac{1}{4} \times 9$ $= 2.25$
2. State the result.	The population in 2025 is predicted to be 2.25 million.
c. 1. Determine the year by calculating the value of t when $N = 5$.	**c.** $5 = \frac{1}{4}e^{\frac{t}{8}\log_e(3)}$
2. Solve the equation for t by taking the natural logarithms of both sides.	$20 = e^{\frac{t}{8}\log_e(3)}$ $\frac{t}{8}\log_e(3) = \log_e(20)$ $t = \frac{8\log_e(20)}{\log_e(3)}$ ≈ 21.815
3. State the result.	The population reaches 5 million late in the year 2030.

10.2.4 Radioactive decay

Chemical experiments have shown that when a radioactive substance decays, it changes into another element or an isotope of the same element. Scientists have determined that the rate of decay is proportional to the mass of the radioactive substance present at that time.

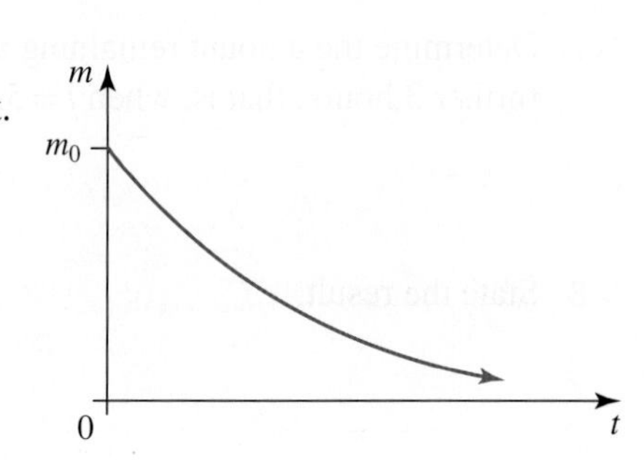

Let $m = m(t)$ represent the mass of the radioactive substance at a time, t, where $t \geq 0$.

Then

$$\frac{dm}{dt} \propto -m$$
$$\Rightarrow \quad \frac{dm}{dt} = -km, \text{ where } k > 0 \text{ and is called the decay constant.}$$

The solution of this differential equation is given by $m = m_0 e^{-kt}$, where $m_0 = m(0)$.

This equation is called the law of exponential decay.

WORKED EXAMPLE 2 Radioactive decay

The rate of decay of a radioactive substance is proportional to the amount of the substance present at that time. Initially 30 milligrams of a radioactive substance is present. After 2 hours it is observed that 20% has disintegrated.
Determine the amount remaining after a further 3 hours, correct to 3 decimal places.

THINK	WRITE
1. Let $m = m(t)$ represent the amount of the radioactive substance present in milligrams (mg) after a time t hours.	$\frac{dm}{dt} \propto -m$ $\therefore \frac{dm}{dt} = -km$ $\Rightarrow m = m_0 e^{-kt}$ [1]
2. Initially, when $t = 0$, the mass is 30 mg. Use this to determine the value of m_0.	$m_0 = m(0) = 30$
3. After 2 hours, 20% has disintegrated, so 80% remains of the initial 30 mg. Use this to determine the amount left after 2 hours.	When $t = 2$, $m = 0.8 \times m_0$: $m(2) = 0.8 \times 30$ $= 24$
4. Substitute $m_0 = 30$, $t = 2$ and $m = 24$ into the law of exponential decay, $m = m_0 e^{-kt}$, where $m_0 = m(0)$.	Substitute $m_0 = 30$, $t = 2$ and $m = 24$ into [1]: $24 = 30e^{-2k}$
5. Solve the equation for k by taking the natural logarithms of both sides.	$e^{-2k} = \frac{24}{30}$ $-2k = \log_e\left(\frac{4}{5}\right)$ $k = -\frac{1}{2}\log_e\left(\frac{4}{5}\right)$
6. Express the mass remaining in milligrams after t hours.	$m(t) = 30e^{\frac{t}{2}\log_e\left(\frac{4}{5}\right)}$
7. Determine the amount remaining after a further 3 hours, that is, when $t = 5$.	Substitute $t = 5$: $m(t) = 30e^{\frac{5}{2}\log_e\left(\frac{4}{5}\right)}$ ≈ 17.173
8. State the result.	17.173 mg remains after 5 hours.

10.2.5 Half-lives

For $m = m_0 e^{-kt}$ where $k > 0$, as $t \to \infty$, $m \to 0$.

An infinite time is required for all of the radioactive material to disintegrate. For this reason the rate of disintegration is often measured in terms of the **half-life** of the radioactive material. The half-life, T, is the time it takes for half of the original mass to disintegrate.

To determine the half-life, the value of T when $m = \frac{1}{2}m_0$ is required.

Substitute $m = \frac{1}{2}m_0$ into the equation $m = m_0 e^{-kt}$.

This gives: $\frac{1}{2}m_0 = m_0 e^{-kT}$

$$\frac{1}{2} = e^{-kT}$$

$$e^{kT} = 2$$

Solving for T gives: $T = \frac{1}{k}\log_e(2)$

Notice that this half-life formula does not depend upon the initial mass m_0 and is thus independent of the time when observations began. Half-lives for radioactive substances can range from milliseconds to billions of years.

WORKED EXAMPLE 3 Calculating the half-life of a radioactive substance

It takes 2 years for 30% of a particular radioactive substance to disintegrate. Determine the half-life of the substance, correct to 3 decimal places.

THINK	WRITE
1. Let $m = m(t)$ represent the amount of the radioactive substance present after time t in years.	$m = m_0 e^{-kt}$
2. Since 30% has disintegrated, 70% remains.	When $t = 2$, $m = 0.7m_0$.
3. Substitute $m = 0.7m_0$ and $t = 2$.	$0.7m_0 = m_0 e^{-2k}$ $0.7 = e^{-2k}$
4. Solve the equation for k by taking the natural logarithms of both sides.	$-2k = \log_e(0.7)$ $k = -\frac{1}{2}\log_e(0.7)$
5. Determine the half-life.	$T = \frac{1}{k}\log_e(2)$ $T = \frac{1}{-\frac{1}{2}\log_e(0.7)}\log_e(2)$ $T \approx 3.887$
6. State the final result.	The half-life of the substance is 3.887 years.

10.2 Exercise

Students, these questions are even better in jacPLUS

Receive immediate feedback and access sample responses

Access additional questions

Track your results and progress

Find all this and MORE in jacPLUS

Technology active

1. **WE1** The population of a city increases at a rate proportional to the current population. In 2015 the population was 1.79 million, and in 2019 the population was 2 million.
   a. Express the population number, N, in terms of t, the time in years after 2015.
   b. Determine the predicted population at the end of 2030, correct to the nearest ten thousand.
   c. Determine when the population reaches 5 million.

2. The number of frogs in a colony increases at a rate proportional to the current number. After 15 months there are 297 frogs present, and this number grows to 523 after 26 months. Determine the initial number of frogs present in the colony.

3. a. The rate of growth of a population of insects is proportional to its present size. A colony of insects initially contains 600 individuals and is found to contain 1300 after 2 weeks. Determine the number of insects after a further 3 weeks.
   b. Initially there are 10 000 fish in a reservoir. The number of fish in the reservoir grows continuously at a rate of 4% per year. Determine the number of fish in the reservoir after 3 years.

4. The number of possums in a certain area grows at a rate proportional to the current number. In 4 months the number of possums has increased from 521 to 678. Determine the number of possums in the area after a further 5 months.

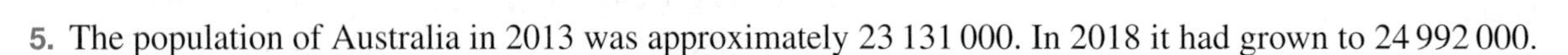

5. The population of Australia in 2013 was approximately 23 131 000. In 2018 it had grown to 24 992 000.
   a. Determine the predicted population in 2025.
   b. Calculate what year the population will first exceed 30 million.

6. In 2011 the world population reached 7 billion. In 2018 the world population was approximately 7 632 819 325.
   a. Determine the estimated world population in 2024.
   b. Determine the year in which the world population will reach 10 billion.

7. **WE2** The rate of decay of a radioactive substance is proportional to the amount of the substance present at that time. Initially 80 milligrams of a radioactive substance is present, and after 1 hour it is observed that 10% has disintegrated. Determine the amount remaining after a further 2 hours, in milligrams correct to 2 decimal places.

8. The rate of decay of a radioactive substance is proportional to the amount of the substance present at that time. After 2 hours, 64 milligrams of a radioactive substance is present, and after 4 hours, 36 milligrams is present. Determine the initial amount of the substance present, in milligrams correct to 2 decimal places.

9. WE3 For a radioactive substance, it takes 3 years for 15% to disintegrate. Determine the half-life of the substance, correct to the nearest tenth of a year.

10. An isotope of radium has a half-life of 1601 years. Determine the percentage remaining after 1000 years, correct to the nearest whole percent.

11. a. The rate of decay of a radioactive substance is proportional to the amount of the substance present at any time. If 1 gram of radioactive material decomposes to half a gram in 22 652 years, determine how much remains after 40 000 years, in grams correct to 2 decimal places.
    b. If 15% of a radioactive element disintegrates in 6 years, determine the half-life of this element, in years correct to 2 decimal places.

12. Iodine-131 is present in radioactive waste and decays exponentially to form a substance that is not radioactive. The half-life of iodine-131 is 8 days. If 120 milligrams is considered a safe level, determine how long will it take 2 grams of iodine-131 to decay to a safe level, in days correct to 2 decimal places.

13. Strontium-90 is an unpleasant radioactive isotope that is a byproduct of a nuclear explosion. Strontium-90 has a half-life of 28.9 years. Calculate how many years elapse before 60% of the strontium-90 in a sample has decayed, in years correct to 1 decimal place.

14. Curium-243 has a half-life of 29.1 years. It decays into plutonium-239 through alpha decay. Determine the percentage of curium-243 that remains in a sample after 10 years, correct to 1 decimal place.

15. Uranium-238 has a half-life of 4.468 billion years. Estimate the percentage of the original amount of uranium-238 that remains in the universe, assuming we began with that original amount 13.8 billion years ago, correct to 2 decimal places.

16. A biologist is investigating two different types of bacteria. Both types grow at a rate proportional to the number present. Initially 900 type A bacteria are present, and after 10 hours 3000 are present. The number of type B bacteria doubles after 4 hours, and there are 2500 type B bacteria after 8 hours. Determine the time at which equal numbers of type A and type B bacteria are present, in hours correct to 2 decimal places.

17. The table below shows the comparative population in millions of Sydney and Melbourne, in certain years.

City	2013	2021
Melbourne	4.217	5.061
Sydney	4.386	4.992

Assume the growth rates are proportional to the current population and these growth rates continue.

  a. Show that Melbourne's population growth rate is greater than Sydney's population growth rate.
  b. Determine what year the population of Melbourne will first exceed 8 million.
  c. Determine what year the population of Sydney will first exceed 8 million.

18. In the 1950s, W F Libby and others at the University of Chicago devised a method of estimating the age of organic material based on the decay rate of carbon-14, a radioactive isotope of carbon. Carbon-14 dating can be used on objects ranging from a few hundred years old to 40 000 years old. Carbon-14 obeys the law of radioactive decay. If Q denotes the amount of carbon-14 present at time t years, then $\frac{dQ}{dt} = -kQ$.

a. If Q_0 represents the amount of carbon-14 present at time t_0 and k is a constant, express Q in terms of Q_0, t_0 and k.

b. Given that the half-life of carbon-14 is 5730 years, determine the value of k. Write your answer rounded to 6 decimal places.

In residual amounts of carbon-14 it is possible to measure $\frac{Q}{Q_0}$ for some wood and plant remains, and hence determine the elapsed time since the death of these remains, that is, the period during which decay has been taking place. This technique of radiocarbon dating is of great value to archaeologists.

c. For a particular specimen, $\frac{Q}{Q_0} = \frac{2}{7}$. Determine the age of the specimen.

10.2 Exam questions

Question 1 (9 marks) TECH-ACTIVE

Source: *VCE 2019, Specialist Mathematics Exam 2, Section B, Q3; © VCAA.*

a. The growth and decay of a quantity P with respect to time t is modelled by the differential equation

$$\frac{dP}{dt} = kP$$

where $t \geq 0$.

i. Given that $P(a) = r$ and $P(b) = s$, where P is a function of t, show that $k = \frac{1}{a-b}\log_e\left(\frac{r}{s}\right)$. **(2 marks)**

ii. Specify the condition(s) for which $k > 0$. **(2 marks)**

b. The growth of another quantity Q with respect to time t is modelled by the differential equation

$$\frac{dQ}{dt} = e^{t-Q}$$

where $t \geq 0$ and $Q = 1$ when $t = 0$.

i. Express this differential equation in the form $\int f(Q)\,dQ = \int h(t)\,dt$. **(1 mark)**

ii. Hence, show that $Q = \log_e\left(e^t + e - 1\right)$. **(2 marks)**

iii. Show that the graph of Q as a function of t does not have a point of inflection. **(2 marks)**

Question 2 (2 marks) TECH-ACTIVE

Plutonium-239 is a silvery metal that is used for the production of nuclear weapon. If 3 micrograms of plutonium-239 decomposes to 1 microgram in 38 213 years, determine the half-life of plutonium-239, correct to the nearest year.

Question 3 (1 mark) TECH-ACTIVE

MC The population of Germany in 2010 was approximately 80.827 million and by 2021 it had grown to 83.900 million. Assuming the population of Germany follows the law of natural growth, the year in which the population will reach 100 million is

A. 2062 B. 2063 C. 2066 D. 2072 E. 2073

More exam questions are available online.

10.3 Other applications of first-order differential equations

LEARNING INTENTION

At the end of this subtopic you should be able to:
- set up and solve differential equations where growth is proportional to a function of the current population
- set up and solve differential equations which relate to other real-world applications.

10.3.1 Miscellaneous types

There are many other applications of first-order differential equations that are similar to the growth and decay models. Some examples are described in the section below, however it is far from being a comprehensive list of applications.

Investments and money matters

When an amount of money, $\$P$, is invested with continuously compounding interest at $R\%$, the rate at which the accumulated amount grows is proportional to the amount invested. In fact, the proportionality constant is the interest, so $\frac{dP}{dt} = Pr$, where t is the time ($t \geq 0$) and $r = \frac{R}{100}$.

Pressure and height in the atmosphere

The rate of decrease of pressure in the atmosphere, P, with respect to the height above sea level, h, is proportional to the pressure at that height. For a proportionality constant k (such that $k > 0$), $\frac{dP}{dh} = -kP$.

Drug disappearance in the body

For many drugs, the rate of disappearance from the body is proportional to the amount, D, of the drug still present in the body at time t. If k is the proportionality constant ($k > 0$), $\frac{dD}{dt} = -kD$.

Light intensity and depth

When a beam of light passes through a medium, it loses its intensity as it penetrates more deeply. If I is the intensity of light at depth x, the rate of loss of intensity with respect to the depth is proportional to the intensity at that depth. If k is the proportionality constant ($k > 0$), then $\frac{dI}{dx} = -kI$.

WORKED EXAMPLE 4 Compound interest

When money is invested in a bank at a constant rate of $R\%$ with continuously compounding interest, the accumulated amount $\$P$, t years after the start of the investment satisfies the differential equation $\frac{dP}{dt} = Pr$.

a. **Assuming an initial investment of $\$P_0$, solve the differential equation to determine the amount $\$P$ after a time t years.**

b. **Determine the amount to which \$10 000 will grow in 6 years if invested at 4.5%.**

THINK	WRITE
a. 1. To solve the differential equation, invert both sides. The initial investment is $\$P_0$.	a. $\frac{dP}{dt} = Pr, P(0) = P_0$ $\frac{dt}{dP} = \frac{1}{Pr}$
2. Because r is a constant, it can be taken outside the integral sign.	$t = \frac{1}{r}\int \frac{1}{P}\,dP$
3. Integrate.	$t = \frac{1}{r}\log_e(\lvert P\rvert) + c$ $= \frac{1}{r}\log_e(P) + c$, as $P > 0$
4. Substitute $t = 0$, $P = P_0$.	$0 = \frac{1}{r}\log_e(P_0) + c$ $c = -\frac{1}{r}\log_e(P_0)$
5. Rewrite the equation, substituting $c = -\frac{1}{r}\log_e(P_0)$, and use logarithm laws to simplify.	$t = \frac{1}{r}\log_e(P) - \frac{1}{r}\log_e(P_0)$ $t = \frac{1}{r}\log_e\left(\frac{P}{P_0}\right)$
6. Rearrange and write as an exponential.	$rt = \log_e\left(\frac{P}{P_0}\right)$ $\frac{P}{P_0} = e^{rt}$
7. Solve for P and state the solution of the differential equation.	$P = P_0e^{rt}$
b. 1. Substitute $P_0 = 10\,000$ and $r = 4.5\% = \frac{4.5}{100}$.	b. $P(t) = 10\,000e^{\frac{4.5t}{100}}$
2. Determine the amount after 6 years.	$P(6) = 10\,000e^{\frac{4.5\times 6}{100}}$ $= 13\,099.64$
3. State the total amount.	The principal has grown to \$13 099.64 after 6 years.

10.3.2 Other population models

In the natural growth model, the growth rate of a population is proportional to the current population. However, population models can have other growth rates. For example, the growth rate may be proportional to a power of the current population other than 1. In these models, the same approaches can be used as for natural growth.

WORKED EXAMPLE 5 Increasing proportionally to the square root of the population

The population of a certain town increases at a rate proportional to the square root of the current population. In 2014 the population was 250 000, and in 2022 the population was 640 000. Express the population number, N, in terms of t, the time in years after 2014. Calculate the predicted population in 2030.

THINK	WRITE
1. Let $N = N(t)$ represent the population number of the town after 2014.	$\frac{dN}{dt} = k\sqrt{N}, N(0) = 250\,000$
2. To solve the differential equation, invert both sides and write in index notation.	$\frac{dt}{dN} = \frac{1}{k\sqrt{N}}$ $\frac{dt}{dN} = \frac{1}{k}N^{-\frac{1}{2}}$
3. Integrate both sides with respect to N and multiply through by k.	$\int 1dt = \frac{1}{k}\int N^{-\frac{1}{2}}dN$ $kt = \int N^{-\frac{1}{2}}dN$
4. Integrate.	$kt = 2N^{\frac{1}{2}} + c$ $kt = 2\sqrt{N} + c$
5. Substitute $t = 0$ and $N = 250\,000$ into the equation.	$0 = 2\sqrt{250\,000} + c$ $c = -1000$
6. Substitute $c = -1000$.	$kt = 2\sqrt{N} - 1000$
7. Substitute $t = 8$ and $N = 640\,000$ to solve for k.	$8k = 2\sqrt{640\,000} - 1000$ $8k = 600$ $k = 75$
8. Substitute $k = 75$ into the equation $kt = 2\sqrt{N} - 1000$ and rearrange to make N the subject.	$75t = 2\sqrt{N} - 1000$ $2\sqrt{N} = (75t + 1000)$ $\sqrt{N} = \frac{1}{2}(75t + 1000)$
9. State the particular solution of the differential equation.	$N(t) = \frac{1}{4}(75t + 1000)^2$
10. Substitute $t = 16$.	$N(16) = \frac{1}{4}(75 \times 16 + 1000)^2$ $= 1\,210\,000$
11. State the final result.	The predicted population in 2030 is 1 210 000.

10.3.3 Population models with regular removal

Another model that can limit but not bound the population is to remove some of the population at regular intervals. This has applications for situations such as people emigrating from a country or leaving a certain city yearly. The growth rate k still represents the excess of births over deaths and is often given as a percentage.

WORKED EXAMPLE 6 Population growth with regular removal

A farm initially has 200 sheep. The number of sheep on the farm grows at a rate of 10% per year. Each year the farmer sells 15 sheep.

a. **Write the differential equation modelling the number of sheep, N, on the farm after t years.**

b. **Solve the differential equation to express the number of sheep, N, on the farm after t years.**

c. **Hence, determine the number of sheep on the farm after 5 years.**

THINK	WRITE
a. 1. Let $N(t)$ be the number of sheep on the farm at time t years. The growth rate is 10%, and 15 are sold each year.	a. $k = 0.1$ $\frac{dN}{dt} = 0.1N - 15$ $\frac{dN}{dt} = \frac{N}{10} - 15$
2. Write the differential equation in simplest form, along with the initial condition. The initial number of sheep is 200.	$\frac{dN}{dt} = \frac{N - 150}{10}, N(0) = 200$
b. 1. To solve the differential equation, invert both sides.	b. $\frac{dt}{dN} = \frac{10}{N - 150}$
2. Integrate both sides.	$t = \int \frac{10}{N - 150}\, dN$
3. Perform the integration.	$t = 10\log_e(\lvert N - 150\rvert) + c$ $= 10\log_e(N - 150) + c$, as $N > 150$
4. Substitute $t = 0$ and $N = 200$.	$0 = 10\log_e(200 - 150) + c$ $c = -10\log_e(50)$
5. Substitute $c = -10\log_e(50)$ and use log laws to simplify.	$t = 10\log_e(N - 150) - 10\log_e(50)$ $t = 10\left[\log_e(N - 150) - \log_e(50)\right]$ $\frac{t}{10} = \log_e\left(\frac{N - 150}{50}\right)$
6. Rearrange and solve for N.	$\frac{N - 150}{50} = e^{\frac{t}{10}}$ $N - 150 = 50e^{\frac{t}{10}}$

7. State the solution of the differential equation.	$N = 150 + 50e^{\frac{t}{10}}$ $N = 50\left(3 + e^{\frac{t}{10}}\right)$
c. 1. Substitute $t = 5$.	c. $N(5) = 50\left(3 + e^{\frac{1}{2}}\right) \approx 232.44$
2. State the number of sheep present after 5 years. Note that we always round down, and express the answer as an integer.	There are 232 sheep present after 5 years.

10.3 Exercise

Technology active

1. WE4 When money is invested in a bank at a constant rate of r% with continuously compounding interest, the accumulated amount \$$P$ at a time t years after the start of the investment satisfies the differential equation $\frac{dP}{dt} = P \times r$. Determine what initial investment is required if \$10 000 is the target in 2 years' time and the interest rate is 5%.

2. The price of houses in a certain area grows at a rate proportional to their current value. In 2002 a particular house was purchased for \$315 000; in 2018 the house was sold for \$1 260 000. Calculate the value of the house in 2010.

3. a. The rate of decrease of air pressure, P, with respect to the height above sea level, h, is proportional to the pressure at that height. The constant of proportionality is k such that $k > 0$. If P_0 is the air pressure at sea level, solve the differential equation to determine the pressure at height h.
   b. If the air pressure at sea level is 76 cm of mercury, and the pressure at a height of 1 kilometre above sea level is 62.2235 cm of mercury, determine the pressure at a height of 2 kilometres above sea level, in cm correct to 2 decimal places.

4. a. For many drugs, the rate of disappearance from the body is proportional to the amount of the drug still present in the body. If $D = D(t)$ is the amount of the drug present at time t and k is the constant of proportionality where $k > 0$, solve the differential equation expressing D in terms of t given the initial dosage of D_0.
   b. The elimination time for alcohol (the time it takes for a person's body to remove alcohol from their bloodstream) is measured as $T = \frac{1}{k}$. The process varies from person to person, so the value of k varies. For one person the elimination time is 3 hours. For this person, determine how long will it take the excess level of alcohol in their bloodstream to be reduced from 0.10% to the legal adult driving level of 0.05%, in hours correct to 2 decimal places.

5. When light passes through a medium, it loses its intensity as it penetrates deeper. If I is the intensity of light at depth x, the rate of loss of intensity with respect to the depth is proportional to the intensity at that depth, so that $\frac{dI}{dx} = -kI$, where k is a positive constant.
   a. Solve the differential equation to show that $I = I_0 e^{-kx}$.
   b. If 5% of light is lost in penetrating 25% of a glass slab, determine what percentage is lost in penetrating the whole slab, correct to 1 decimal place.

6. When a rope is wrapped around a pole, the rate of change of tension, T (in newtons), in the rope with respect to the angle, θ (in radians), is proportional to the tension at that instant. The constant of proportionality is the coefficient of friction, μ.
   a. Given that $\frac{dT}{d\theta} = \mu T$ and the tension when the angle is zero is T_0, show that $T = T_0 e^{\mu\theta}$.
   b. A man holds one end of a rope with a pull of 80 newtons. The rope goes halfway around a tree, and the coefficient of friction in this case is 0.3. Determine what force the other end can sustain, in newtons correct to 2 decimal places.

7. **WE5** The population of a country increases at a rate proportional to the square root of the current population. In 2013 the population was 4 million, and in 2018 the population was 9 million. Express the population number, N, in terms of t, the time in years after 2013. Calculate the predicted population in 2028.

8. The number of insects in a colony increases at a rate inversely proportional to the current number. At first there are 20 insects present, and this number grows to 80 after 5 months. Determine the number of insects present in the colony after a further 11 months.

9. The variation of resistance, R ohms, of a copper conductor with temperature T degrees Celsius satisfies the differential equation $\frac{dR}{dT} = \alpha R$.
   a. If the resistance of copper is R_0 at 0°C, show that $R = R_0 e^{\alpha T}$.
   b. If $\alpha = 0.004$ per degree Celsius and the resistance at 60°C is 40 ohms, determine the resistance at 30°C, in ohms correct to 2 decimal places.

10. In an electrical circuit consisting of a resistance of R ohms and an inductance of L henries, the current, i amperes, after a time t seconds satisfies the differential equation $L\frac{di}{dt} + Ri = 0$. The initial current is i_0. Solve the differential equation to determine the current at any time t.

11. **WE6** A certain area initially contains 320 rabbits. The number of rabbits in the area grows at a rate of 25% per year. Each year 40 rabbits are culled to try to limit the population.
   a. Write the differential equation modelling the number of rabbits in the area, N, after t years.
   b. Solve the differential equation to express the number of rabbits in the area after t years.
   c. Hence, determine the number of rabbits in the area after 8 years.

12. An area initially contains N_0 rabbits. The number of rabbits in this area grows at a rate of 20% per year. Each year K rabbits are culled to limit the population. After 10 years, there are $6K$ rabbits present. Show that $N_0 = K\left(5 + \frac{1}{e^2}\right)$.

13. A farm initially has 200 cows. The number of cows on the farm grows at a rate of 5% per year. Each year the farm sells 5 cows.
   a. Write the differential equation modelling the number of cows, N, on the farm after t years.
   b. Solve the differential equation to express the number of cows on the farm after t years.
   c. Hence, determine the number of cows on the farm after 3 years.

14. A certain country has a 4% population growth rate, and additionally every year 10 000 immigrate to the country. If in 2019 the population of the country was half a million, estimate the population in 2024.

15. Koalas on a plantation have a population growth rate of 10%. Initially 400 koalas are present.
   a. Every year 50 koalas are sold to zoos. Determine how many years elapse before the number of koalas is reduced to 200.
   b. After the population has reached 200, calculate the maximum number of koalas the managers of the plantation can sell to zoos each year so that the number of koalas at the plantation will increase above 200.

16. Solve the differential equation $\frac{dN}{dt} = kN^{\alpha}$, $N(0) = N_0$ for when:
   a. $\alpha = \frac{1}{2}$
   b. $\alpha = -1$
   c. $\alpha = \frac{3}{2}$
   d. $\alpha = 2$.

17. a. Show that the solution of the differential equation $\frac{dN}{dt} = kN - c$, where k and c are positive constants and $N(0) = N_0$, is given by $N(t) = \left(N_0 - \frac{c}{k}\right)e^{kt} + \frac{c}{k}$.
   b. Show that:
      i. N increases if $N_0 > \frac{c}{k}$
      ii. N decreases if $N_0 < \frac{c}{k}$
      iii. if $N_0 = \frac{c}{k}$, then N remains stable.

18. The number of bacteria in a culture increases at a rate proportional to the number of bacteria present in the culture at any time. By natural increase, the culture will double in $\log_e(8)$ days. If the initial number of bacteria present is N_0 and bacteria are removed from the colony at a constant rate of Q per day, show that the number of bacteria:
   a. will increase if $N_0 > 3Q$
   b. will remain stationary if $N_0 = 3Q$
   c. will decrease if $N_0 < 3Q$.

10.3 Exam questions

Question 1 (2 marks) TECH-ACTIVE

The charge, Q units, on a plate conductor t seconds after it starts to discharge is proportional to the charge at that instant. If the charge is 500 units after a half-second and falls to 250 units after 1 second, determine:

a. the original charge **(1 mark)**

b. the time needed for the charge to fall to 125 units. **(1 mark)**

Question 2 (3 marks) TECH-ACTIVE

In an electrical circuit consisting of a resistance of R ohms and a capacitance of C farads, the charge, Q coulombs, at a time t seconds decays according to the differential equation $\frac{dQ}{dt} + \frac{Q}{RC} = 0$. Assuming the initial charge is Q_0, solve the differential equation to determine the charge at any time t after discharging.

Question 3 (3 marks) TECH-ACTIVE

When light passes through a medium it loses its intensity as it penetrates depths. The rate of loss of intensity with respect to the depth is proportional to the intensity at that depth. If 5% of light is lost in penetrating 40% of a glass slab, calculate the percentage lost penetrating the whole slab. Give your answer correct to 1 decimal place.

More exam questions are available online.

10.4 Bounded growth and Newton's law of cooling

LEARNING INTENTION

At the end of this subtopic you should be able to:

- set up and solve differential equations relating to bounded growth or decay.

10.4.1 Bounded growth models

If $N = N(t)$ is the population, then the law of exponential growth states that $\frac{dN}{dt} = kN$, where k is a positive constant. Assuming that the initial population number is $N_0 = N(0)$, the solution of this differential equation is given by $N = N_0 e^{kt}$. This model predicts that, as time increases, the population number increases without bounds; that is, as $t \to \infty$, $N \to \infty$. In reality this is not accurate as populations cannot grow infinitely. A more realistic model is one in which the growth of a quantity is not unlimited, but instead approaches some maximum value at which point the growth stops and the population reaches equilibrium. This is called bounded growth. In such cases, the growth rate is proportional to the difference between the population number and the equilibrium value.

The differential equation to model this type of behaviour is $\frac{dN}{dt} = k(P - N)$, where $k > 0$ and $P > 0$. P is called the carrying capacity or the equilibrium value.

Assuming an initial condition $N_0 = N(0)$, if $N_0 < P$, then as $t \to \infty$, $N \to P$ from below.

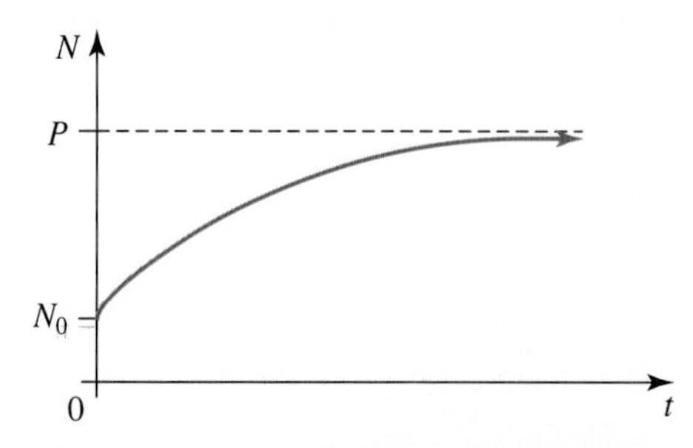

However, if $N_0 > P$, then as $t \to \infty$, $N \to P$ from above.

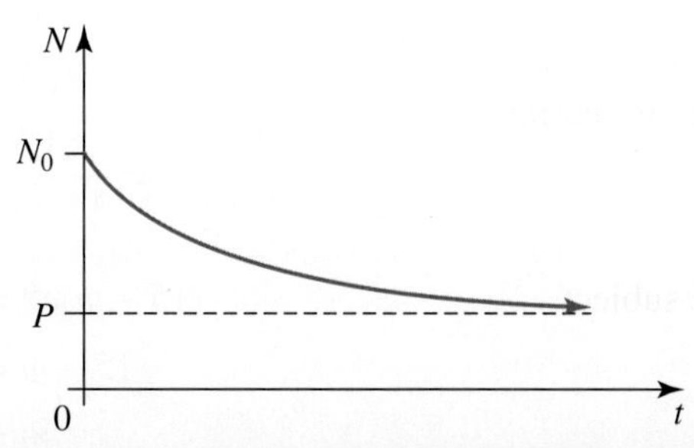

WORKED EXAMPLE 7 Bounded growth

The mass of medium-sized dogs are found to follow a bounded growth model. The initial mass of a typical newborn puppy is 0.5 kg, and after 16 weeks its mass has increased to 4.5 kg. It is known that the dog's mass will never exceed 15 kg.

a. Determine the mass of the dog after one year.

b. Sketch the graph of mass versus time for this model.

THINK	WRITE/DRAW
a. 1. Let $m = m(t)$ be the mass in kg of the dog after t weeks. The equilibrium value is $P = 15$ and the initial mass is 0.5 kg. State the differential equation to model the mass.	**a.** $\frac{dm}{dt} = k(P - m)$ $\frac{dm}{dt} = k(15 - m), m(0) = 0.5$
2. To solve this differential equation, invert both sides.	$\frac{dt}{dm} = \frac{1}{k(15 - m)}$
3. Separate the variables and multiply by the constant k.	$t = \frac{1}{k}\int \frac{1}{15 - m} dm$ $kt = \int \frac{1}{15 - m} dm$
4. Perform the integration.	$kt = -\log_e(\lvert 15 - m \rvert) + c$ $kt = -\log_e(15 - m) + c \quad \text{as } 0.5 < m < 15$
5. The initial mass of a newborn puppy is 0.5 kg. Use this to determine the value of c.	$m(0) = 0.5$: $0 = -\log_e(15 - 0.5) + c$ $c = \log_e(14.5)$
6. Substitute $c = \log_e(14.5)$ and simplify.	$kt = -\log_e(15 - m) + \log_e(14.5)$ $kt = \log_e\left(\frac{14.5}{15 - m}\right)$
7. After 16 weeks the mass is 4.5 kg. Use this to determine the value of k.	$m(16) = 4.5$: $16k = \log_e\left(\frac{14.5}{15 - 4.5}\right)$ $k = \frac{1}{16}\log_e\left(\frac{29}{21}\right)$

8. Leave the equation in terms of k and use the definition of the logarithm to rearrange.

$$kt = \log_e\left(\frac{14.5}{15 - m}\right)$$

$$e^{kt} = \frac{14.5}{15 - m}$$

9. Rearrange to make m the subject.

$$(15 - m)e^{kt} = 14.5$$

$$15 - m = 14.5e^{-kt}$$

$$m = 15 - 14.5e^{-kt}$$

10. State the particular solution of the differential equation.

$$m(t) = 15 - 14.5e^{-\frac{t}{16}\log_e\left(\frac{29}{21}\right)}$$

11. Now determine the mass of the dog after one year, that is $t = 52$ weeks.

$$m(52) = 15 - 14.5e^{-\frac{52}{16}\log_e\left(\frac{29}{21}\right)}$$

$$\approx 9.92$$

12. State the final result.

The mass of the dog after one year is 9.92 kg.

b. Sketch the graph of m versus t.

b. 

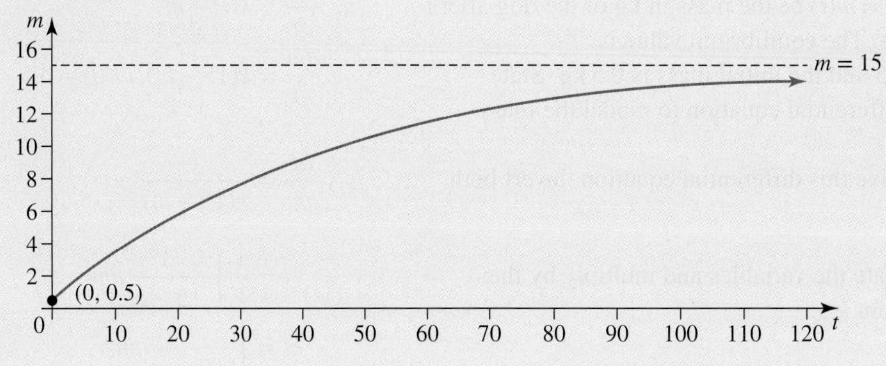

10.4.2 Newton's law of cooling

Newton's law of cooling states that the rate of change of the temperature of a body is proportional to the difference between the temperature of the body and the surrounding medium. Let T denote the temperature of a body after a time t and T_m denote the temperature of the surrounding medium, which is assumed to be constant. Newton's law of cooling can be written as $\frac{dT}{dt} \propto (T - T_m)$ or $\frac{dT}{dt} = k(T - T_m)$, where k is a constant of proportionality. If the body is cooling, then $k < 0$.

To solve this differential equation, let θ be the difference in temperature, so that $\theta = T - T_m$. Because T_m is a constant, $\frac{dT}{dt} = \frac{d\theta}{dt}$; therefore, $\frac{d\theta}{dt} = k\theta$. The particular solution of the differential equation is $\theta = \theta_0 e^{kt}$, where $\theta(0) = \theta_0 = T_0 - T_m$ and T_0 is the initial temperature of the body. Newton's law of cooling is another example of a bounded growth or decay model.

WORKED EXAMPLE 8 Newton's law of cooling

A metal ball is heated to a temperature of 200°C and is then placed in a room that is maintained at a constant temperature of 30°C. After 5 minutes the temperature of the ball has dropped to 150°C. Assuming Newton's law of cooling applies, determine:

a. **the equation for the difference between the temperature of the ball and the temperature of the room**

b. **the temperature of the ball after a further 10 minutes**

c. **how long it will take for the temperature of the ball to reach 40°C.**

THINK	WRITE
a. 1. Let T denote the temperature of the ball after a time t minutes. The constant room temperature is $T_m = 30$°C.	a. $\frac{dT}{dt} = k(T - T_m)$ $\frac{dT}{dt} = k(T - 30)$
2. Let θ be the difference in temperature between the ball and the room.	$\theta = T - 30$ $\frac{d\theta}{dt} = k\theta$
3. Write the solution of the particular differential equation.	$\theta = \theta_0 e^{kt}$
4. To determine the value of θ_0, use the given initial conditions.	Initially, $t = 0$ and $T_0 = 200$°C. $\theta_0 = T_0 - T_m$ $= 200 - 30$ $= 170$
5. To determine the value of k, use the other given condition.	When $t = 5$, $T = 150$°C, so $\theta_5 = 150 - 30$ $= 120$
6. Substitute in the given values.	$\theta_0 = 170, \theta = 120$ and $t = 5$: $120 = 170e^{5k}$
7. Rearrange and use the definition of the logarithm.	$\frac{120}{170} = e^{5k}$ $5k = \log_e\left(\frac{12}{17}\right)$
8. Solve for k.	$k = \frac{1}{5}\log_e\left(\frac{12}{17}\right)$
9. Write the solution of the differential equation.	$\theta(t) = 170e^{\frac{t}{5}\log_e\left(\frac{12}{17}\right)}$
b. 1. Substitute $t = 15$ to determine the temperature after a further 10 minutes.	b. $\theta(15) = 170e^{\frac{15}{5}\log_e\left(\frac{12}{17}\right)}$ ≈ 59.79
2. State the temperature of the ball at this time.	Since $T = \theta + 30$, the temperature of the ball is 89.79°C.
c. 1. To determine when the temperature of the ball reaches 40°C, use $T = 40$.	c. $T = \theta + 30$ $40 = \theta + 30$ $\theta = 40 - 30$ $\theta = 10$

2. Write an equation involving the unknown time.

Substitute $\theta = 10$

$$10 = 170e^{\frac{t}{5}\log_e\left(\frac{12}{17}\right)}$$

3. Solve the equation for t, using the definition of the logarithm.

$$\frac{10}{170} = e^{\frac{t}{5}\log_e\left(\frac{12}{17}\right)}$$

$$\frac{t}{5}\log_e\left(\frac{12}{17}\right) = \log_e\left(\frac{10}{170}\right)$$

$$t = \frac{5\log_e\left(\frac{10}{170}\right)}{\log_e\left(\frac{12}{17}\right)}$$

$$\approx 40.67$$

4. State the required result.

The temperature of the ball reaches 40°C after 40.67 minutes.

10.4 Exercise

Technology active

1. WE7 The mass of a bird is found to follow a bounded growth model. The initial mass of a baby bird is 30 grams, and after 10 weeks its mass is 100 grams. If it is known that the mass of the bird will never exceed 200 grams, determine the mass of the bird after 30 weeks, in grams correct to 1 decimal place.

2. The number of birds in a flock follows a bounded growth model. Initially there are 200 birds in the flock. After 5 months the number has grown to 800. The number of birds in the flock can never exceed 3000. Calculate after how many months there are 1500 birds in the flock.

3. a. The mass of cats are found to follow a bounded growth model. The initial mass of a typical newborn cat is 0.1 kg, and after 30 weeks the cat's mass has increased to 4 kg. If it is known that the cat's mass will never exceed 5 kg, determine the mass of the cat after 40 weeks, in kg correct to 1 decimal place.

   b. The mass of toy poodle dogs are found to follow a bounded growth model. The initial mass of a puppy toy poodle is 0.2 kg, and after 14 weeks its mass has increased to 1.3 kg. If it is known that the dog's mass will never exceed 4.5 kg, determine the mass of the toy poodle after one year, stating your answer in kg rounded to 2 decimal places.

4. The number of fish in a lake follows a bounded growth model. Initially there are 50 fish in the lake, and after 10 months the number of fish has increased to 500. The number of fish in the lake can never exceed 1000.

   a. Determine the number of fish in the lake after 20 months.

   b. Calculate after how many months there are 900 fish in the lake, stating your answer rounded to 1 decimal place.

5. A woman is on a diet. Her initial mass is 84 kg, and she knows that her mass will always be above 70 kg. After 10 weeks she has lost 7 kg. Assuming a bounded decay model, calculate her total mass loss after 20 weeks, in kg correct to 1 decimal place.

6. **WE8** A warm can of soft drink at a temperature of 25°C is chilled by placing it in a refrigerator at a constant temperature of 2°C. If the temperature of the drink falls to 22°C in 5 minutes, determine how long it will take for the temperature of the drink to reach 13°C, stating your answer in minutes, rounded to 2 decimal places.

7. An iron is preheated to a temperature of 250°C. After 6 minutes the temperature of the iron is 210°C. Brent is ironing in a room that is kept at a constant temperature of 20°C. Brent knows that it is best to iron a cotton T-shirt when the temperature of the iron is between 195°C and 205°C. Determine how long Brent has to iron the T-shirt, in minutes correct to 2 decimal places.

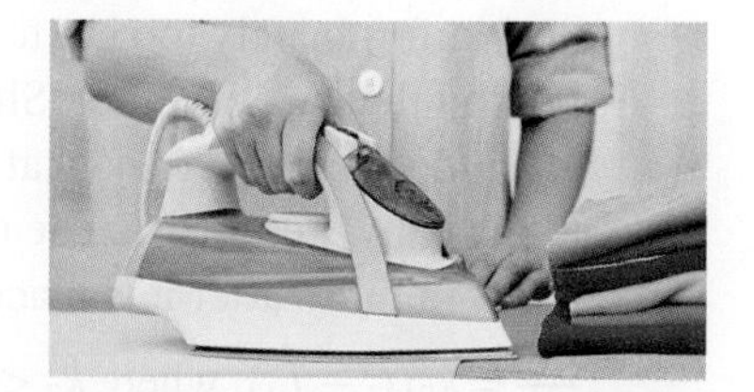

8. a. A cold can of soft drink is taken from a refrigerator at a temperature of 3°C and placed in a sunroom at a temperature of 30°C. If after 2 minutes the temperature of the can is 4°C, determine its temperature after a further 3 minutes, rounded to 2 decimal places.
   b. A body at an unknown temperature is placed in a room that has a temperature of 18°C. If after 10 minutes the body has a temperature of 22°C, and after a further 10 minutes its temperature is 20°C, determine the initial temperature of the body, correct to the nearest whole degree.

9. A mother is giving her baby a bath. The bathtub contains hot water that is initially at a temperature of 42°C in a room where the temperature is constant at 20°C. The water cools, and after 2 minutes its temperature is 40°C. The mother knows that babies like to be bathed when the temperature of the water is between 34°C and 38°C, otherwise the bath water is either too hot or too cold. Determine how long the baby should stay in the bath, in minutes correct to 2 decimal places.

10. A frozen chicken should be thawed before cooking. A freezer is maintained at a constant temperature of −18°C. A chicken is taken out of the freezer at 8:00 am and placed on a kitchen bench. At 11:00 am the temperature of the chicken is 0°C, and at 2:00 pm its temperature is 10°C. At 5:00 pm the chicken has thawed and is placed in an oven which has been preheated to a temperature of 200°C. At 6:00 pm the temperature of the chicken in the oven is 70°C. Determine the temperature of the chicken when it is removed from the oven at 7:00 pm, in degrees correct to 2 decimal places.

11. A capacitor of C farads is charged by applying a steady voltage of E volts through a resistance of R ohms. The potential difference, v volts between the plates satisfies the differential equation $RC\frac{dv}{dt} + v = E$. Assuming the initial voltage is zero, determine the voltage v at any time t.

12. Show that the particular solution of the differential equation $\frac{dN}{dt} = k(P - N)$, $N(0) = N_0 > 0$, where $P > 0$, is given by $N(t) = P + (N_0 - P)e^{-kt}$. Consider both cases when $N_0 > P$ and $N_0 < P$.

13. a. Show that the solution of the differential equation $\frac{dT}{dt} = k(T - T_m)$, where k and T_m are positive constants and $T(0) = T_0$, is given by $T(t) = T_m + (T_0 - T_m)e^{kt}$.
    b. Police discover a dead body in a hotel room at 8:00 am. At that time its temperature is 30.4°C. One hour later the temperature of the body is 29.9°C. The room was kept at a constant temperature of 17°C overnight. If normal body temperature is 37°C, determine the time of death.

14. A refrigerator is kept at a constant temperature of 3°C. A baby's bottle is taken from the fridge, but the bottle is too cold to give to the baby. The mother places the bottle in a saucepan full of boiling water (maintained at 95°C) for 90 seconds. When the bottle is removed from the saucepan, its temperature is 45°C.

a. Let $H = H(t)$ be the temperature of the bottle at a time t minutes while it is being heated. If $\frac{dH}{dt} = k_1(H - T_h)$, where $k_1 < 0$ for $0 \le t \le t_1$, state the values of t_1 and T_h.

b. The mother knows that the bottle should be at a temperature of 35°C for the baby to drink it. She realises that the bottle is now unfortunately too hot to give to the baby, so she places it back in the fridge to cool. She takes the bottle out of the fridge after 45 seconds. It is now at the correct temperature to give to the baby. Let $C = C(t)$ be the temperature of the bottle at a time t minutes after it is placed back in the fridge. If $\frac{dC}{dt} = k_2(C - T_c)$, where $k_2 < 0$ for $0 \le t \le t_2$, state the values of t_2 and T_c.

c. Determine the ratio of $\frac{k_1}{k_2}$, correct to 2 decimal places.

15. Let $\$P_0$ be the initial amount of money borrowed from a lender at an annual interest rate of r. Assume that the interest is compounded continuously and that the borrower makes equal payments of $\$m$ per year to reduce the amount borrowed. If $\$P$ is the amount of money owing at time t, the differential equation modelling the repayments is given by $\frac{dP}{dt} = rP - m$.

a. If the loan is completely paid off after a time T years, show that $\frac{m}{m - rP_0} = e^{rT}$.

b. If Jared borrows $300 000 for a house loan at 6% per annum and pays off the loan after 20 years, determine his monthly repayments and the total amount of interest paid.

c. Sharon can afford to pay $350 a month over 5 years at 8% per annum compounded continuously. Determine the maximum amount she can borrow for a car loan.

d. Ashley wants to buy a block of land and will have to borrow $120 000 to do it. He can afford to make monthly repayments of $1200 on the loan, compounded continuously at 6.3%. Calculate how long it will take Ashley to pay off the loan, in years correct to 1 decimal place.

e. Ryan pays off a loan of $8000 by paying back $150 per month over 6 years. Assuming the interest is compounded continuously, determine the annual interest rate as a percentage correct to 2 decimal places.

10.4 Exam questions

Question 1 (5 marks) TECH-FREE

Source: VCE 2013, Specialist Mathematics Exam 1, Q5; © VCAA.

A container of water is heated to boiling point (100°C) and then placed in a room that has a constant temperature of 20°C. After five minutes the temperature of the water is 80°C.

a. Use Newton's law of cooling $\frac{dT}{dt} = -k(T - 20)$, where T°C is the temperature of the water at time t minutes after the water is placed in the room, to show that $e^{-5k} = \frac{3}{4}$. **(2 marks)**

b. Find the temperature of the water 10 minutes after it is placed in the room. **(3 marks)**

Question 2 (6 marks) TECH-FREE

An RL series circuit consists of a resistance, R ohms, and an inductance, L henries, connected to a voltage source, E volts. The rise of current, i amperes, after a time t seconds satisfies the differential equation $L\frac{di}{dt}+Ri=E$.

a. Assuming the initial current is zero, determine the current at any time t. **(4 marks)**

b. Show that the time required for the current to reach half its ultimate value is given by $\frac{L}{R}\log_e\left(\frac{1}{2}\right)$. **(2 marks)**

Question 3 (2 marks) TECH-ACTIVE

The temperature of a room is 25°C. A thermometer which was in the room is taken outdoors and in five minutes it reads 15°C. Five minutes later the thermometer reads 10°C. Determine the temperature outdoors.

More exam questions are available online.

10.5 Chemical reactions and dilution problems

LEARNING INTENTION

At the end of this subtopic you should be able to:

- verify solutions to dilution problems with different inflow and outflow rates
- solve dilution problems where the inflow and outflow rates are equal
- set up and solve differential equations modelling chemical reactions.

10.5.1 Input–output mixing problems

Consider a tank that initially holds V_0 litres of a solution in which a kilograms of salt have been dissolved. Another solution containing b kilograms of salt per litre is poured into the tank at a rate of f litres per minute. The well-stirred mixture leaves the tank at a rate of g litres per minute. The problem is to determine the amount of salt in the tank at any time t.

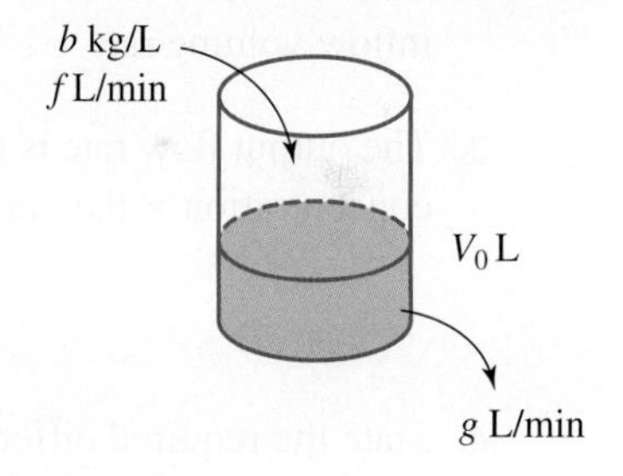

Let Q denote the amount of salt in kilograms in the tank at a time t minutes. The rate of change of Q is equal to the rate at which the salt flows into the tank minus the rate at which the salt flows out of the tank.

Differential equation for mixing problems

$$\frac{dQ}{dt} = \text{inflow rate} - \text{outflow rate}$$

We know b kilograms of salt per litre is being poured into the tank at a rate of f litres per minute; multiplying these together gives the inflow rate of the salt as bf kilograms per minute.

To determine the outflow rate, first we determine the volume $V(t)$ of the tank at time t. The initial volume is V_0, but f litres per minute flow in, while g litres per minute flow out. Therefore, the volume at time t minutes is given by $V(t)=V_0+(f-g)t$. To get the required units, we multiply the solution's outflow rate of g litres per minute by the mass and divide by the volume to get the outflow rate of the salt as $\frac{gQ}{V_0+(f-g)t}$ kilograms per minute.

Therefore, the differential equation is

$$\frac{dQ}{dt}=bf-\frac{Qg}{V_0+(f-g)t}, \text{ and } Q(0)=a.$$

(Note that concentration, c, is the mass divided by the volume, $c = \frac{\text{mass}}{\text{volume}}$.)

When $g \neq f$, this equation cannot be solved by hand using the techniques of integration studied in this course (but can be solved by computers). However, if we are given a solution, differentiation and substitution can be used to verify the solution.

WORKED EXAMPLE 9 Salt concentration problems

A tank has a capacity of 300 litres and contains 200 litres of water in which 100 kilograms of salt have been dissolved. A salt solution of concentration 2 kilograms per litre is poured into the tank at a rate of 3 litres per minute, and the well-stirred mixture flows out at a rate of 2 litres per minute.

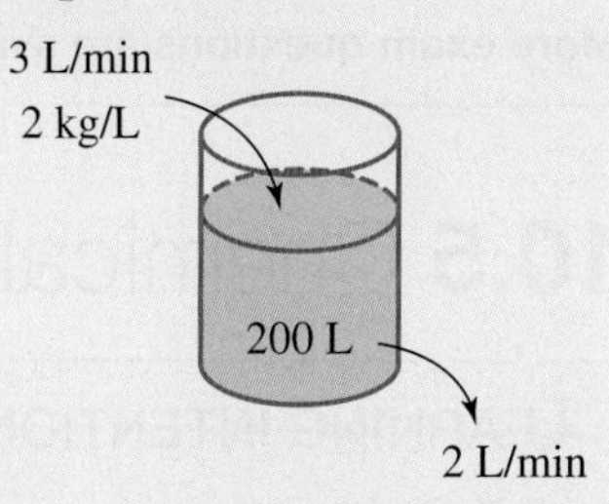

a. **State the differential equation for Q, the amount of salt in kilograms in the tank after t minutes.**

b. **Verify that $Q = 2(200 + t) + \frac{C}{(200 + t)^2}$ is a general solution of the differential equation.**

c. **Determine the value of C.**

d. **Determine the concentration of salt in kilograms per litre when the tank overflows, correct to 3 decimal places.**

THINK	WRITE
a. 1. The initial volume in the tank is 200 litres; 3 litres per minute flows in and 2 litres per minute flows out. Let $V(t)$ be the volume in litres at time t minutes.	a. $V(t) = 200 + (3 - 2)t$ $= 200 + t$
2. The inflow rate is the concentration × the inflow volume rate.	Inflow rate $= 3 \times 2$ kg/min $= 6$ kg/min
3. The output flow rate is the outflow concentration × the outflow volume rate.	Outflow rate $= \frac{2Q}{V(t)}$ kg/min $= \frac{2Q}{200 + t}$ kg/min
4. State the required differential equation along with the initial condition.	$\frac{dQ}{dt} =$ inflow rate − outflow rate $= 6 - \frac{2Q}{200 + t}$ [1] $Q(0) = 100$
b. 1. To verify the given solution, first write the solution in index form.	b. $Q = 2(200 + t) + \frac{C}{(200 + t)^2}$ [2] $= 2(200 + t) + C(200 + t)^{-2}$
2. Now differentiate the given solution as the left-hand side.	LHS $= \frac{dQ}{dt}$ $= 2 - 2C(200 + t)^{-3}$ $= 2 - \frac{2C}{(200 + t)^3}$

3. Substitute for Q into the right-hand side of [1].

$$\text{RHS} = 6 - \frac{2Q}{200+t}$$
$$= 6 - \frac{2}{200+t}\left[2(200+t) + \frac{C}{(200+t)^2}\right]$$

4. Expand the right-hand side.

$$= 6 - \frac{2\times 2(200+t)}{200+t} - \frac{2C}{(200+t)^3}$$

5. Simplify the right-hand side. Because the left-hand side is equal to the right-hand side, we have verified that the given solution does satisfy the differential equation.

$$= 6 - 4 - \frac{2C}{(200+t)^3}$$
$$= 2 - \frac{2C}{(200+t)^3}$$

c. 1. To determine the value of C, use the given initial conditions.

c. Substitute $t=0$ and $Q=100$ into [2].

2. Substitute the given values in the given solution.

$$100 = 2\times 200 + \frac{C}{200^2}$$

3. Solve for C.

$$100 = 400 + \frac{C}{200^2}$$
$$\frac{C}{200^2} = -300$$
$$C = -12\,000\,000$$

d. 1. Determine the time when the tank overflows. The capacity of the tank is 300 litres.

d. $300 = 200 + t$

The tank overflows when $t = 100$ minutes.

2. Determine the value of Q when the tank overflows.

Substitute for C:

$$Q(t) = 2(200+t) - \frac{12\,000\,000}{(200+t)^2}$$
$$Q(100) = 600 - \frac{12\,000\,000}{300^2}$$
$$= 466\frac{2}{3} \text{ kg}$$

3. Determine the concentration when the tank overflows.

$$\text{Concentration} = \frac{\text{mass}}{\text{volume}}$$
$$\frac{Q(100)}{V(100)} = \frac{466\frac{2}{3}}{300}$$
$$\approx 1.556 \text{ kg/litre}$$

4. State the answer.

When the tank overflows, the concentration of salt in the solution is 1.556 kg/litre.

10.5.2 Equal input and output flow rates

In the special case when the inflow rate is equal to the outflow rate, that is, when $f = g$ so the volume remains constant, the differential equation can be solved using the separable integration techniques learnt in topic 8. In fact, in this case the differential equation is just another application of a bounded growth or decay model.

WORKED EXAMPLE 10 Equal input and output flow rates

A tank contains 5 litres of water in which 10 grams of salt have been dissolved. A salt solution containing 4 grams per litre is poured into the tank at a rate of 2 litres per minute, and the mixture is kept uniform by stirring. The mixture leaves the tank at a rate of 2 litres per minute.

a. Set up the differential equation for the amount of salt, Q grams, in the tank at time t minutes.
b. Solve the differential equation to determine Q at any time t.
c. Determine how much salt is in the tank and the concentration after 5 minutes, correct to 3 decimal places.
d. Show that the salt solution can never exceed 20 grams.
e. Sketch the graph of Q versus t.

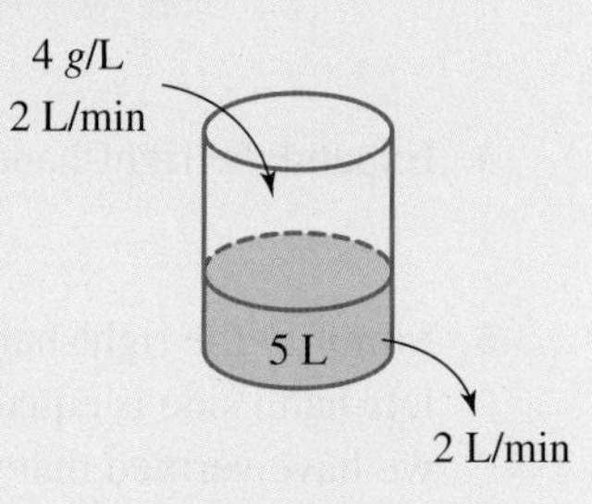

THINK / **WRITE/DRAW**

a. 1. The inflow rate is equal to the outflow rate. The volume in the tank remains constant at 5 litres.

a. $V(t) = 5$

2. The inflow rate is inflow concentration × the inflow volume rate. The output flow rate is the outflow concentration × the outflow volume rate.

$$\frac{dQ}{dt} = \text{inflow rate} - \text{output rate}$$
$$= 4 \times 2 - \frac{2Q}{V(t)}$$

3. State the differential equation.

$$\frac{dQ}{dt} = 8 - \frac{2Q}{5}$$

4. State the differential equation in simplest form, along with the initial condition.

$$\frac{dQ}{dt} = \frac{2(20-Q)}{5}, Q(0) = 10$$

b. 1. To solve the differential equation, invert both sides.

b. $$\frac{dt}{dQ} = \frac{5}{2(20-Q)}$$

2. Use the separation of variables technique, taking the constant factors to one side.

$$\int \frac{2}{5} dt = \int \frac{1}{22-Q} dQ$$

3. Perform the integration.

$$\frac{2t}{5} = -\log_e(|20-Q|) + c$$

4. Since $Q(0) = 10$, the modulus signs are not needed. Use the initial condition to determine the constant of integration.

Substitute $t = 0$ when $Q = 10$:
$$0 = -\log_e(20-10) + c$$
$$c = \log_e(10)$$

5. Substitute back for C and simplify using log laws.

$$\frac{2t}{5} = -\log_e|20-Q| + \log_e(10)$$
$$\frac{2t}{5} = \log_e\left|\frac{10}{20-Q}\right|$$

6. Use the definition of the logarithm.
Remove the absolute value sign as $e^{\frac{2t}{5}} > 0$

$$e^{\frac{2t}{5}} = \frac{10}{20-Q}$$

7. Invert both sides.

$$\frac{20-Q}{10} = e^{-\frac{2t}{5}}$$

8. Rearrange and solve for Q.

$$20 - Q = 10e^{-\frac{2t}{5}}$$

$$Q = 20 - 10e^{-\frac{2t}{5}}$$

9. State the solution of the differential equation in simplest form.

$$Q(t) = 10\left(2 - e^{-\frac{2t}{5}}\right)$$

c. Determine how much salt is in the tank and the concentration after 5 minutes.

c. Substitute $t = 5$:

$$Q(5) = 10(2 - e^{-2})$$
$$= 18.647 \text{ grams}$$

The concentration is

$$c = \frac{Q}{V}$$
$$= \frac{18.647}{5}$$
$$= 3.729 \text{ grams per litre}$$

d. Determine the limit as t approaches infinity.

d. As $t \to \infty, e^{-\frac{2t}{5}} \to 0, \therefore Q \to 20.$
Therefore, the amount of salt can never exceed 20 grams.

e. The graph starts at $Q(0) = 10$ and approaches the horizontal asymptote of $Q = 20$.

e.

Q; 20; 15; 10; 5; 0; 2; 4; 6; 8; 10; 12; t; Q = 20; (0, 10)

TI | THINK	DISPLAY/WRITE	CASIO | THINK	DISPLAY/WRITE
b. On a Calculator page, select: Menu: 4: CalculusD: Differential Equation Solver and complete the fields as shown, then click OK. Click ENTER to see the solution.	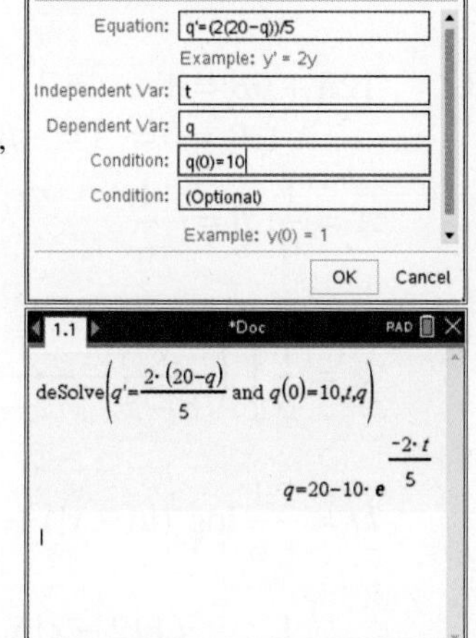	**b.** On a Main screen, select Action: Advanced: dSolve and complete the entry as shown.	

10.5.3 Chemical reaction rates

In a chemical reaction, the rate at which a new substance is formed is proportional to the unused amounts of the reacting substances. Differential equations can be set up and solved to model the amount of the new substance as it forms over time.

WORKED EXAMPLE 11 Chemical reaction problems

In a biomolecular chemical reaction between 3 grams of substance A and 6 grams of substance B, the rate of the reaction is proportional to the product of the unused amounts of A and B respectively. A and B combine in equal parts to form substance X. Initially no X is present, but after 3 minutes, 3 grams of X have formed.

a. Set up the differential equation for the amount of substance X at time *t* minutes.

b. Determine the amount of X present after a further 3 minutes.

THINK	WRITE
a. 1. Let $x = x(t)$ be the amount in grams of substance X formed at time t minutes.	**a.** $[A + B \rightarrow X]$
2. The rate of the reaction is proportional to the product of the unused amounts of A and B, so $\frac{x}{2}$ is used from both A and B.	$\frac{dx}{dt} \propto \left(3 - \frac{x}{2}\right)\left(6 - \frac{x}{2}\right)$ $\frac{dx}{dt} = k(6-x)(12-x), x(0) = 0$ where k is a constant to be found.
b. 1. To solve the differential equation, invert both sides.	**b.** $\frac{dx}{dt} = \frac{1}{k(6-x)(12-x)}$
2. Take the constant k to the other side and use separation of variables.	$kt = \int \frac{1}{(6-x)(12-x)} dx$
3. To determine the integral on the right-hand side, we need to use partial fractions.	$\frac{1}{(6-x)(12-x)} = \frac{A}{6-x} + \frac{B}{12-x}$
4. Add the partial fractions and simplify.	$\frac{1}{(6-x)(12-x)} = \frac{A(12-x) + B(6-x)}{(6-x)(12-x)}$ $= \frac{12A + 6B - x(A+B)}{(6-x)(12-x)}$
5. Equate the coefficients and solve for A and B.	$12A + 6B = 1$ [1] $A + B = 0 \Rightarrow A = -B$ [2] $A = \frac{1}{6}, B = -\frac{1}{6}$
6. Substitute back for the partial fraction decomposition.	$kt = \frac{1}{6}\int \left(\frac{1}{6-x} - \frac{1}{12-x}\right) dx$
7. Perform the integration.	$kt = -\frac{1}{6}\log_e(\lvert 6-x \rvert) + \frac{1}{6}\log_e(\lvert 12-x \rvert) + c$
8. Take out the common factors and use log laws.	$kt = \frac{1}{6}\log_e\left(\left\lvert \frac{12-x}{6-x} \right\rvert\right) + c$

9. Since $x(0) = 0$ and $0 \le x < 6$, the modulus signs are not needed. Use the initial condition to determine the constant of integration.

$$0 = \frac{1}{6}\log_e(2) + c$$
$$c = -\frac{1}{6}\log_e(2)$$

10. Substitute back for c and use log laws again.

$$kt = \frac{1}{6}\log_e\left(\frac{12-x}{6-x}\right) - \frac{1}{6}\log_e(2)$$
$$kt = \frac{1}{6}\log_e\left(\frac{12-x}{2(6-x)}\right) \qquad [3]$$

11. We can now determine the value of k by substituting $t = 3$ when $x = 3$.

Substitute $t = 3$ when $x = 3$:

$$3k = \frac{1}{6}\log_e\left(\frac{9}{6}\right)$$
$$3k = \frac{1}{6}\log_e\left(\frac{3}{2}\right)$$
$$k = \frac{1}{18}\log_e\left(\frac{3}{2}\right) \qquad [4]$$

12. Attempt to make x the subject, but leave the result in terms of k, since k has a known value.

From [3], $6kt = \log_e\left(\frac{12-x}{2(6-x)}\right)$

13. Use the definition of a logarithm.

$$\frac{12-x}{2(6-x)} = e^{6kt}$$

14. Invert both sides.

$$\frac{12-2x}{12-x} = e^{-6kt}$$

15. Remove the denominator.

$$12 - 2x = (12-x)e^{-6kt}$$

16. Expand the brackets.

$$12 - 2x = 12e^{-6kt} - xe^{-6kt}$$

17. Transfer the x to one side and factorise.

$$12 - 12e^{-6kt} = 2x - xe^{-6kt}$$
$$12\left(1 - e^{-6kt}\right) = x\left(2 - e^{-6kt}\right)$$

18. State the solution of the differential equation.

$$x = x(t) = \frac{12\left(1 - e^{-6kt}\right)}{2 - e^{-6kt}}$$
where $k = \frac{1}{18}\log_e\left(\frac{3}{2}\right)$

19. Determine the value of x after a further 3 minutes. When $t = 6$

$$x(6) = \frac{12\left(1 - e^{-36k}\right)}{2 - e^{-36k}}$$

From [4], $36k = 2\log_e\left(\frac{3}{2}\right)$

20. Use exact values.	$36k = \log_e\left(\frac{9}{4}\right)$ $e^{-36k} = \frac{4}{9}$
21. Evaluate	$x(6) = \frac{12\left(1-\frac{4}{9}\right)}{2-\frac{4}{9}}$ $= \frac{30}{7}$
22. State the final result. Note that as $t \to \infty$, $x \to 12$.	After 6 minutes, $4\frac{2}{7}$ grams of X have formed.

10.5 Exercise

Technology active

1. **WE9** A vessel initially contains 200 litres of a salt solution with a concentration of 0.1 kg/litre. The salt solution is drawn off at a rate of 3 litres per minute, and at the same time a mixture containing salt of concentration 1.5 kg/litre is added to the vessel at a rate of 2 litres per minute. The contents of the vessel are kept well stirred.
   a. Set up the differential equation for Q, the amount of salt in kilograms in the vessel after t minutes.
   b. Verify that $Q = \frac{3}{2}(200-t) + C(200-t)^3$ is a general solution of the differential equation.
   c. Determine the value of C.
   d. Determine the concentration of salt after 100 minutes.

2. A vat initially contains 50 litres of a sugar solution. More sugar solution containing b grams per litre is poured into the vat at a rate of 6 litres per minute, and simultaneously the well-stirred mixture leaves the vat at a rate of 3 litres per minute.
   a. Set up the differential equation for Q, the amount of sugar in grams in the vat after t minutes.
   b. Verify that $Q = 2(50+3t) + \frac{C}{50+3t}$ is a general solution of the differential equation, and determine the value of b.

3. A container initially contains 30 litres of water. A brine solution of concentration 4 grams per litre is added to the container at a rate of 1 litre per minute. The well-stirred mixture is drawn off at 2 litres per minute.
   a. Set up the differential equation for Q, the amount of brine in grams in the container after t minutes.
   b. Verify that $Q(t) = \frac{2t}{15}(30-t)$ is the particular solution of the differential equation.

4. A vessel initially contains 20 litres of water. A sugar solution with a concentration of 3 grams per litre is added at a rate of 2 litres per minute. The well-stirred mixture is drawn off at 1 litre per minute.
   a. Set up the differential equation for Q, the amount of sugar in grams in the vessel after t minutes.
   b. Verify that $Q(t) = \dfrac{3t(t+40)}{20+t}$ is the particular solution of the differential equation.
5. A trough initially contains 20 litres of water. A dye solution with a concentration of 4 grams per litre is added to the trough at a rate of 2 litres per minute. The well-stirred mixture is sent down a drain at a rate of 3 litres per minute.
   a. Set up the differential equation for Q, the amount of dye in grams in the trough after t minutes.
   b. Verify that $Q(t) = \dfrac{t}{100}(t-40)(t-20)$ is a particular solution of the differential equation.
6. A vessel initially contains 64 litres of a salt solution with a concentration of 2 grams per litre. The solution is drawn off at a rate of 3 litres per minute, and at the same time a mixture containing salt at a concentration of 4 grams per litre is added to the vessel at a rate of 5 litres per minute. The contents of the vessel are kept well stirred.
   a. Set up the differential equation for Q, the amount of salt in grams in the vessel after t minutes.
   b. Verify that $Q = 4(64+2t) + C(64+2t)^{-\frac{3}{2}}$ is a general solution of the differential equation.
   c. Determine the value of C.
   d. Determine the concentration when the volume of the vessel is 100 litres, in g/L correct to 2 decimal places.
7. A vat initially contains 40 litres of water. A brine solution containing b grams per litre is poured into the vat at a rate of 2 litres per minute, and simultaneously the well-stirred mixture leaves the vat at a rate of 5 litres per minute. If Q is the amount of brine in grams in the vessel after a time t minutes and $Q = 3(40-3t) + C(40-3t)^{\frac{5}{3}}$ is a general solution of the differential equation, determine the value of b.
8. **WE10** A tank initially contains 20 litres of a salt solution that has a concentration of 0.25 grams per litre. The salt solution is drawn off at a rate of 3 litres per minute, and at the same time a mixture containing salt of concentration 4 grams per litre is added to the tank at a rate of 3 litres per minute. The contents of the tank are kept well stirred.
   a. Set up the differential equation for Q, the amount of salt in grams in the tank after t minutes.
   b. Solve the differential equation to determine Q at any time t.
   c. Determine the concentration of salt after 1.6 minutes, in g/L correct to 2 decimal places.
   d. Show that the amount of salt can never exceed 80 grams.
   e. Sketch the graph of Q versus t.
9. A vat contains 15 litres of pure water. A sugar solution containing 5 grams per litre is poured into the vat at a rate of 4 litres per minute, and simultaneously the well-stirred mixture leaves the vat at the same rate.
   a. Set up the differential equation for the amount of sugar, Q grams, in the vat at time t minutes.
   b. Solve the differential equation to determine Q at any time t.
   c. Calculate after how long the sugar concentration is 4 grams per litre, in minutes correct to 2 decimal places.
10. A tank contains 600 litres of water in which 30 kilograms of salt have been dissolved. Water is poured into the tank at a rate of 5 litres per minute, and the mixture is kept uniform by stirring. The mixture leaves the tank at a rate of 5 litres per minute.
   a. Determine the amount of salt in kilograms in the tank at any time t minutes.
   b. Determine the amount of salt in kilograms after 2 hours, correct to 2 decimal places.
   c. Determine how long before there are 15 kilograms of salt in the tank, in minutes correct to 2 decimal places.

11. A sink contains 50 litres of water for washing dishes, in which 50 grams of detergent have been dissolved. A dishwashing solution containing 3 grams per litre of detergent is poured into the sink at a rate of 4 litres per minute and the mixture is kept uniform. The mixture leaves the sink down the drain at the same rate. Determine the concentration of detergent after 5 minutes, in g/L correct to 3 decimal places.

12. a. A 10 litre urn is full of boiling water. Caterers pour the entire contents of a 300 gram jar of coffee into the urn and mix them thoroughly. While the caterers are making cups of coffee, the coffee is drawn out of the urn at a rate of 0.2 litres per minute, and at the same time boiling water is added at the same rate.

    i. Lilly likes to drink her coffee when the concentration of the coffee is 25 grams per litre. Determine how long after the process starts should she wait to get her cup of coffee, in minutes correct to 2 decimal places.
    ii. When the concentration of coffee in the urn falls below 15 grams per litre, more coffee must be added to the urn. Determine how long after the process starts must more coffee be added, in minutes correct to 2 decimal places.

    b. The urn now contains hot water at a temperature of 93°C. Water is poured from the urn to make a cup of coffee. The coffee cup is placed in a room where the temperature is constant at 17°C. After 1 minute the temperature of the coffee is 88°C. Lilly likes to drink her coffee when its temperature is between 50°C and 65°C. Determine how long Lilly has to drink the coffee, in minutes correct to 1 decimal place.

13. **WE11** In a chemical reaction between 1 gram of substance A and 3 grams of substance B, the rate of the reaction is proportional to the product of the unused amounts of A and B. A and B combine in equal parts to form substance X, and initially no X is present. After 3 minutes, 1 gram of X has formed.
    a. Set up the differential equation to determine the amount of substance X at time t minutes.
    b. Solve the differential equation to determine the amount of substance X at time t minutes.
    c. Determine the amount of X present after a further 3 minutes.

14. In a chemical reaction between 4 grams of substance A and 4 grams of substance B, the rate of the reaction is proportional to the product of the unused amounts of A and B. A and B combine in equal parts to form substance X. Initially, no X is present, but after 2 minutes, 3 grams of X has formed.
    a. Set up the differential equation for the amount of substance X at time t minutes.
    b. Solve the differential equation to determine the amount of X in grams present after a time t minutes.
    c. Determine how much more time passes before 6 grams of X are present.
    d. Calculate the maximum amount of X that could eventually be formed.

15. In a chemical reaction between 2 grams of substance A and 4 grams of substance B, the rate of the reaction is proportional to the product of the unused amounts of A and B. A and B combine in equal parts to form substance X. Initially, no X is present, and after 2 minutes, 1 gram of X has formed.
    a. Set up the differential equation to determine the amount of substance X at time t minutes.
    b. Solve the differential equation to determine the amount of substance X at time t minutes.
    c. Determine the amount of X present after a further 2 minutes.
    d. Calculate the ultimate amount of substance X that can eventually be formed.

16. In a chemical reaction between 5 grams of substance A and 2 grams of substance B, the rate of the reaction is proportional to the product of the unused amounts of A and B. A and B combine in equal parts to form a substance X. Initially, no X is present, and after 3 minutes, 2 grams of X have formed.
    a. Determine how long before 3 grams of X have formed, in minutes correct to 2 decimal places.
    b. After 6 minutes, calculate how much of X has formed.
    c. Calculate the ultimate amount of substance X that can eventually be formed.

17. Many chemical reactions follow Wilhelmy's Law, which states that the rate of the reaction is proportional to the concentration of the reacting substance. In such a reaction containing a grams of a reagent, the amount of substance X formed, x grams, after a time t minutes is given by $\frac{dx}{dt} = k(a - x)$, where a and k are both positive constants.

    a. If initially there is no X present, show that $x(t) = a(1 - e^{-kt})$.
    b. If $a = 5$ and after 4 minutes 2 grams of X is present, determine the amount of X present after 10 minutes, in grams correct to 2 decimal places.

18. In a chemical reaction, equal amounts of A and B combine to form substance X. Initially there are b grams of each A and B present, and no X is present. If x grams of X have formed after t minutes, the rate of the reaction is proportional to the product of the unused amounts of A and B.

    a. Given that $\frac{dx}{dt} = k\left(b - \frac{x}{2}\right)^2$ where b and k are both positive constants, show that $x(t) = \frac{2b^2kt}{2 + bkt}$.
    b. If $b = 5$ and after 4 minutes 2 grams of X is present, determine the amount of X present after 10 minutes, in grams correct to 2 decimal places.
    c. Calculate the ultimate amount of substance X that can eventually be formed.

19. A vat initially contains V_0 litres of water. A brine solution of concentration b grams per litre is added to the vat at a rate of f litres per minute. The well-stirred mixture is drawn off at the same rate.

    a. Set up the differential equation for Q, the amount of brine in grams in the vat after t minutes.
    b. Use integration to show that $Q(t) = bV_0\left(1 - e^{-\frac{ft}{V_0}}\right)$ is a particular solution of the differential equation.

20. A container initially contains V_0 litres of a brine solution in which q_0 grams of brine have been dissolved. A brine solution of concentration b grams per litre is added to the container at a rate of f litres per minute. The well-stirred mixture leaves the container at the same rate.

    a. Set up the differential equation for Q, the amount of brine in grams in the container after t minutes.
    b. Use integration to show that $Q(t) = bV_0 + (q_0 - bV_0)e^{-\frac{ft}{V_0}}$ is a particular solution of the differential equation.

21. In a trimolecular chemical reaction, three substances, A, B and C, react to form a single substance X. The rate of the reaction is proportional to the product of the unreacted amounts of A, B and C. Substances A, B and C combine in equal parts to form substance X, and initially no X is present. Given that $\frac{dx}{dt} = k\left(a - \frac{x}{3}\right)^3$ where a is the initial amount of all substances A, B and C present and $k > 0$, show that $t = \frac{3x(6a - x)}{2a^2k(3a - x)^2}$ for $0 \le x < 3a$.

22. In a chemical reaction between a grams of substance A and b grams of substance B, the rate of the reaction is proportional to the product of the unused amounts of A and B. Substances A and B combine in equal parts to form substance X, and initially no X is present. Given that $\frac{dx}{dt} = k\left(a - \frac{x}{2}\right)\left(b - \frac{x}{2}\right)$ where $k > 0$ is a constant, x is the amount of substance X formed at a time t minutes, and $a > b > 0$, show that

$$x(t) = \frac{2ab\left(1 - e^{-\frac{(a-b)kt}{2}}\right)}{a - be^{-\frac{(a-b)kt}{2}}}$$ and deduce that the limiting amount of X present is $2b$.

23. A pond initially contains 200 litres of water with a mineral concentration of 0.01 grams per litre. A mineral solution of variable concentration of $2 + \sin\left(\frac{t}{6}\right)$ grams per litre is added to the pond at a rate of 2 litres per minute, where $t \geq 0$ is the time in minutes. The mixed water spills off at the same rate.

a. Set up the differential equation for Q, the amount of minerals in grams in the pond at a time t minutes.
b. Using CAS, solve the differential equation to express Q terms of t.
c. After 100 minutes of this process, determine the concentration of the minerals in the pond, in g/L correct to 1 decimal place.

24. A tank has a capacity of 325 litres and initially contains 25 litres of water. A chemical solution of variable concentration of $3e^{-\frac{t}{2}}$ grams per litre is added to the tank at a rate of 10 litres per minute, where $t \geq 0$ is the time in minutes. The mixed water leaves the tank at a rate of 5 litres per minute.

a. Set up the differential equation for Q, the amount of the chemical in grams in the tank at a time t minutes.
b. Using CAS, solve the differential equation to express Q in term of t.
c. Determine the time, in minutes correct to 2 decimal places, when the amount of the chemical in the tank is a maximum, and determine the maximum concentration of the chemical, in g/L to 2 decimal places.
d. Determine the concentration of the chemical in the tank when the tank overflows, in g/L to 2 decimal places.
e. From the start until the time when the tank overflows, determine the total amount of the chemical which has flowed out of the tank, in grams to 2 decimal places.

10.5 Exam questions

Question 1 (4 marks) TECH-FREE

Source: *VCE 2018 Specialist Mathematics Exam 1, Q8; © VCAA.*

A tank initially holds 16 L of water in which 0.5 kg of salt has been dissolved. Pure water then flows into the tank at a rate of 5 L per minute. The mixture is stirred continuously and flows out of the tank at a rate of 3 L per minute.

a. Show that the differential equation for Q, the number of kilograms of salt in the tank after t minutes, is given by:

$$\frac{dQ}{dt} = -\frac{3Q}{16+2t}$$

(1 mark)

b. Solve the differential equation given in part **a.** to find Q as a function of t. Express your answer in the form $Q = \frac{a}{(16+2t)^{\frac{b}{c}}}$, where a, b and c are positive integers. **(3 marks)**

Question 2 (1 mark) TECH-ACTIVE

Source: *VCE 2013 Specialist Mathematics Exam 2, Section A, Q13; © VCAA.*

MC Water containing 2 grams of salt per litre flows at the rate of 10 litres per minute into a tank that initially contained 50 litres of pure water. The concentration of salt in the tank is kept uniform by stirring and the mixture flows out of the tank at the rate of 6 litres per minute.

If Q grams is the amount of salt in the tank t minutes after the water begins to flow, the differential equation relating Q to t is

A. $\frac{dQ}{dt} = 20 - \frac{3Q}{25+2t}$

B. $\frac{dQ}{dt} = 10 - \frac{3Q}{25+2t}$

C. $\frac{dQ}{dt} = 20 - \frac{3Q}{25-2t}$

D. $\frac{dQ}{dt} = 10 - \frac{3Q}{25-2t}$

E. $\frac{dQ}{dt} = 20 - \frac{3Q}{25}$

Question 3 (3 marks) TECH-FREE

A tank initially contains 50 litres of water. A chemical solution is drawn off at a rate of 5 litres per minute, and at the same time a mixture containing the chemical at a concentration of 3 grams per litre is added to the tank at a rate of 4 litres per minute. The contents of the tank are kept well stirred.

a. Set up the differential equation for Q, the amount of the chemical in grams in the tank after t minutes. **(1 mark)**

b. Verify that $Q(t) = 3(50 - t) + C(50 - t)^5$ is a general solution of the differential equation. **(2 marks)**

More exam questions are available online.

10.6 The logistic equation

LEARNING INTENTION

At the end of this subtopic you should be able to:

- set up and solve logistic growth differential equations
- identify the point at which growth reaches a maximum in a logistic growth scenario.

10.6.1 Introduction

In previous sections we considered growth when it is directly proportional to itself; that is, the law of natural growth, $\frac{dN}{dt} = kN$; and also when the growth rate is proportional to the difference between the number and the equilibrium value; that is, bounded growth, $\frac{dN}{dt} = k(P - N)$. In this section we consider growth rates that are proportional to the product of both of these: $\frac{dN}{dt} \propto N(P - N)$.

10.6.2 Logistic growth

To turn this proportionality into an equation, we use a constant of proportionality of $\frac{k}{P}$, where k is the growth rate and P is the carrying capacity. The resulting equation is $\frac{dN}{dt} = \frac{k}{P}N(P - N) = kN\left(1 - \frac{N}{P}\right)$.
For an initial number of $N(0) = N_0$, assuming $k > 0$ and $P > N_0 > 0$, it can be shown that $N(t) = \frac{PN_0}{N_0 + (P - N_0)e^{-kt}} = \frac{P}{1 + \left(\frac{P}{N_0} - 1\right)e^{-kt}}$. (Showing this is left as an exercise for you to do yourself.) In this relationship, as $t \to \infty$, $N \to P$.

The graph of the logistic equation always has the S shape shown below.

The growth pattern modelled by this type of differential equation is called logistic growth. It was devised by the Belgian mathematician Pierre François Verhulst (1804–1849). His idea was a response to the work of the English scholar Thomas Robert Malthus, who published a paper in 1798, *An Essay on the Principle of Population*, predicting unlimited population growth. Verhulst disagreed with Malthus's model and used his own model to show the characteristics of bounded population growth.

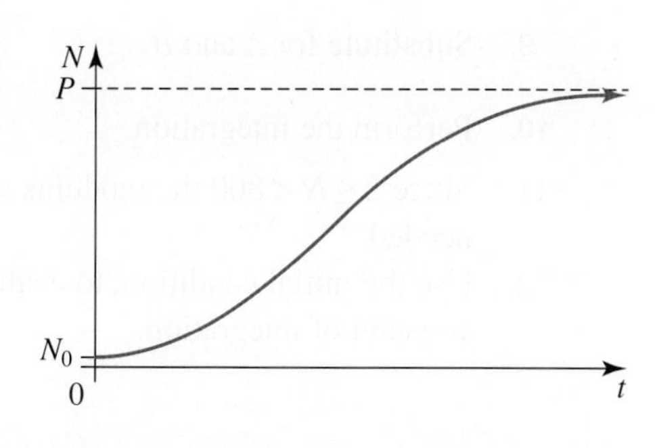

WORKED EXAMPLE 12 Logistic growth (1)

A rumour is spreading in a neighbourhood area with 800 people. The rumour spreads at a rate proportional to both the number who have heard the rumour and the number who have not as yet heard the rumour. Initially only 2 people in the area have heard the rumour and after 4 days 15% of the residence in the area have heard it.

a. **Let $N = N(t)$ be the number who have heard the rumour, after t days, write down and solve the differential equation.**

b. **Calculate the number of residence in the neighbourhood who have heard the rumour after 5 more days.**

c. **Determine after how many days 50% of the residence in the area have heard the rumour, correct to 2 decimal places.**

d. **Sketch the graph.**

THINK	WRITE/DRAW
a. 1. The growth rate is proportional to the product of N and $P - N$.	a. The maximum number is $P = 800$. $\frac{dN}{dt} \propto N\left(1 - \frac{N}{800}\right)$
2. Write down the differential equation and the initial condition.	$\frac{dN}{dt} = \frac{kN}{800}(800 - N),\ N(0) = 2$
3. To solve the differential equation, invert both sides.	$\frac{dt}{dN} = \frac{800}{kN(800 - N)}$
4. Use the separation of variables.	$kt = \int \frac{800}{N(800 - N)}\, dN$
5. Use partial fractions on the integrand on the right hand side.	$\frac{800}{N(800 - N)} = \frac{A}{N} + \frac{B}{800 - N}$
6. Add the partial fractions.	$\frac{800}{N(800 - N)} = \frac{A(800 - N) + BN}{N(800 - N)}$ $= \frac{N(B - A) + 800A}{N(800 - N)}$
7. Write down equations which can be solved to determine the values of A and B.	From the coefficient of N: (1) $B - A = 0$ From the term independent of N: (2) $800A = 800$
8. Solve the equations and state the values of A and B.	$A = B = 1$
9. Substitute for A and B.	$kt = \int \left(\frac{1}{N} + \frac{1}{800 - N}\right) dN$
10. Perform the integration.	$kt = \log_e(\lvert N \rvert) - \log_e(\lvert 800 - N \rvert) + c$
11. Since $2 \le N < 800$ the modulus are not needed.	$kt = \log_e(N) - \log_e(800 - N) + c$
12. Use the initial condition, to evaluate the constant of integration.	Substitute $t = 0$ when $N = 2$ $0 = \log_e(2) - \log_e(800 - 2) + c$

13.	Solve for c and use log laws.	$c = \log_e(798) - \log_e(2)$ $= \log_e\left(\frac{798}{2}\right)$ $= \log_e(399)$
14.	Substitute back for c and use log laws again.	$kt = \log_e(N) - \log_e(800 - N) + \log_e(399)$ $= \log_e\left(\frac{399N}{800 - N}\right)$
15.	After 4 days 15%, that is 120 have heard the rumour. Use this to evaluate k.	Substitute $t = 4$ when $N = 120$ $4k = \log_e\left(\frac{399 \times 120}{800 - 120}\right)$
16.	Solve for k.	$k = \frac{1}{4}\log_e\left(\frac{1197}{17}\right) \approx 1.06359$
17.	Use the definition of the logarithm, but leave the result in terms k, since k has a known value.	$\frac{399N}{800 - N} = e^{kt}$
18.	Rearrange to make N the subject.	$399N = (800 - N)e^{kt}$ $800 - N = 399Ne^{-kt}$ $800 = N + 399Ne^{-kt}$
19.	Take out the common factor of N.	$800 = N(1 + 399e^{-kt})$
20.	State the solution of the differential equation.	$N(t) = \frac{800}{1 + 399e^{-1.0636t}}$
b. 1.	Calculate the value of N after 5 more days.	**b.** Substitute $t = 9$ $N(9) = \frac{800}{1 + 399e^{-1.0636 \times 9}} \approx 778.37$
2.	State the required number, rounding down.	778 have heard the rumour.
c. 1.	Determine t when 50% of 800 have heard the rumour, that is when $N = 400$. It is easier to use an earlier equation.	**c.** $kt = 1.06359t$ $= \log_e\left(\frac{399N}{800 - N}\right)$ Substitute $N = 400$ $1.06359t = \log_e\left(\frac{399 \times 400}{800 - 400}\right)$
2.	Solve for t.	$t = \frac{1}{1.06359}\log_e\left(\frac{399 \times 400}{400}\right)$ $= \frac{1}{1.06359}\log_e(399) \approx 5.63$
3.	State the required time.	After 5.63 days 50% have heard the rumour.

d. Sketch the graph of $N(t) = \dfrac{800}{1 + 399e^{-1.0636t}}$.

(Note that as $t \to \infty,\ N \to 800$)

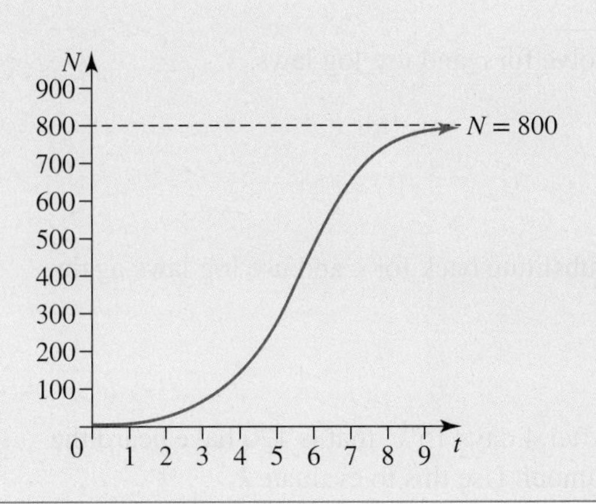

Solving for the parameters

We can see from the equation that in a logistic growth model, there are three parameters: the initial value, $N(0) = N_0$; the maximum value or ultimate value, P; and the constant of proportionality, k. If these parameters are not known, three conditions need to be given and the values substituted into the logistic growth model:

$$N(t) = \frac{PN_0}{N_0 + (P - N_0)e^{-kt}}$$

$$= \frac{P}{1 + \left(\frac{P}{N_0} - 1\right)e^{-kt}}$$

The equations can be solved using algebra or technology.

WORKED EXAMPLE 13 Solving logistic growth equations using technology

The table shows the population, in millions, of Adelaide.

Year	2001	2005	2009
Population (millions)	1.066	1.129	1.187

Assuming the population follows a logistic growth rate, determine:

a. **the maximum population of Adelaide, correct to the nearest thousand**

b. **the year in which the population of Adelaide will first exceed 1.5 million.**

THINK	WRITE
a. 1. Let $N = N(t)$ be the population of Adelaide, in millions, after 2001.	a. $N(t) = \dfrac{PN_0}{N_0 + (P - N_0)e^{-kt}}$ In 2001, $t = 0$ and $N(0) = N_0 = 1.066$.
2. In 2005, the population was 1.129 million.	Substitute $N_0 = 1.066$, $t = 4$ and $N = 1.129$: $1.129 = \dfrac{1.066P}{1.066 + (P - 1.066)e^{-4k}}$ [1]
3. In 2009, the population was 1.187 million.	Substitute $N_0 = 1.066$, $t = 8$ and $N = 1.187$: $1.187 = \dfrac{1.066P}{1.066 + (P - 1.066)e^{-8k}}$ [2]
4. Solve equations [1] and [2] using a CAS calculator.	The solution is $P = 1.57264\ldots$, $k = 0.047552\ldots$

5. The maximum or limiting population of Adelaide is the value of P. Give the answer to an appropriate number of decimal places (i.e. matching the values in the table).

The maximum population of Adelaide is 1.573 million.

b. 1. Write the solution for $N(t)$.

b. Substitute $P = 1.57264\ldots$, $k = 0.047552\ldots$ and $N_0 = 1.066$ into the formula:

$$N(t) = \frac{P}{N_0 + \left(\frac{P}{N_0} - 1\right)e^{-kt}}$$

$$N(t) = \frac{1.57264}{1 + 0.4743e^{-0.0476t}}$$

2. Determine the value of t when $N = 1.5$.

Substitute $N = 1.5$:

$$1.5 = \frac{1.57264}{1 + 0.4743e^{-0.0476t}}$$

3. Solve the equation for t.

$$1 + 0.4743e^{-0.0476t} = \frac{1.57264}{1.5}$$

$$0.4743e^{-0.0476t} = 0.048427$$

$$e^{-0.0476t} = \frac{0.048427}{0.4743}$$

$$e^{-0.0476t} = 0.10189$$

4. Use the definition of the logarithm.

$$-0.0467t = \log_e(0.10189)$$

5. Determine the value of t.

$$t = \frac{\log_e(0.10189)}{-0.0475}$$

$$= 48.1$$

6. State the answer.

The population of Adelaide will first exceed 1.5 million in the year 2049.

10.6.3 Analysis of the logistic solution

The differential equation for the logistic curve in general is given by $\frac{dN}{dt} = kN\left(1 - \frac{N}{P}\right)$. For an initial number $N(0) = N_0$, assuming $k > 0$ and $P > N_0 > 0$, the solution is given by $N(t) = \frac{PN_0}{N_0 + (P - N_0)e^{-kt}} = \frac{P}{1 + \left(\frac{P}{N_0} - 1\right)e^{-kt}}$ for $t \geq 0$.

The graph of the logistic curve always has this so-called S-shaped curve. The graph is relatively flat at the ends, so the gradient is close to zero as $t \to \infty$ and $N \to P$. The line $N = P$, is in fact a horizontal asymptote.

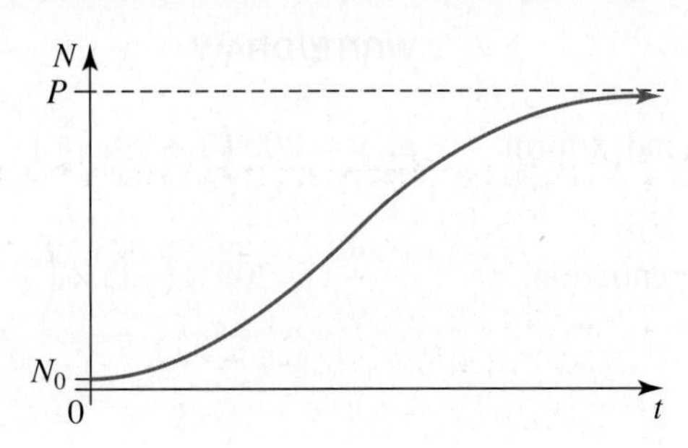

Because $\frac{dN}{dt} = kN\left(1 - \frac{N}{P}\right)$, we can write $\frac{dN}{dt} = \frac{k}{P}(PN - N^2)$. Taking the derivative again with respect to t, because k and P are constants, and using implicit differentiation or the chain rule, then

$$\begin{aligned}\frac{d^2N}{dt^2} &= \frac{d}{dt}\left(\frac{dN}{dt}\right)\\ &= \frac{k}{P}\frac{d}{dt}(PN - N^2)\\ &= \frac{k}{P}\frac{d}{dN}(PN - N^2)\frac{dN}{dt}\\ &= \frac{k}{P}(P - 2N)\frac{dN}{dt}\end{aligned}$$

There is a point on the graph at which the gradient or rate of change of growth is a maximum. Assume $k > 0$, $N_0 > 0$ and $\frac{dN}{dt} > 0$. This means that in the equation $\frac{d^2N}{dt^2} = \frac{k}{P}(P - 2N)\frac{dN}{dt}$, the expressions $\frac{k}{P}$ and $\frac{dN}{dt}$ are both positive.

To keep the two sides of the equation equal, if $\frac{d^2N}{dt^2} > 0$, then $P - 2N > 0$. Therefore, $0 < N < \frac{P}{2}$ is where the curve is concave up.

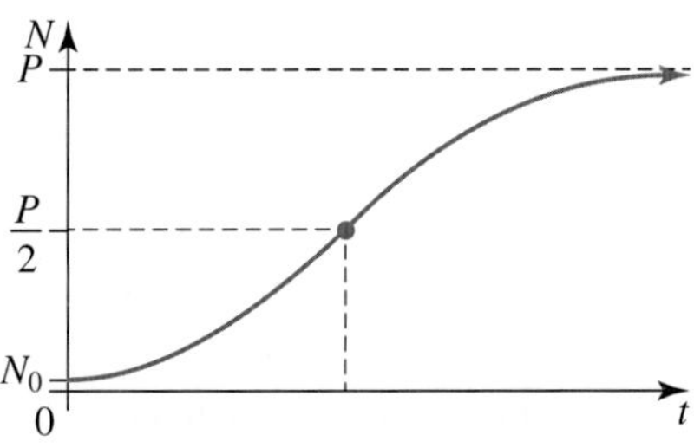

Conversely, if $\frac{d^2N}{dt^2} < 0$, then $P - 2N < 0$.

Therefore, $\frac{P}{2} < N < P$ is where the curve is concave down.

Recall that if the second derivative on a curve is zero, there is a point of inflection. For a logistic curve, the point of inflection occurs when $\frac{d^2N}{dt^2} = 0$; that is, for $N = \frac{P}{2}$, or halfway up the curve.

The point of inflection is the point of maximum growth rate, or where the value is increasing most rapidly. It can be shown in general that when $N = \frac{P}{2}$, the corresponding value for t is given by $\frac{1}{k}\log_e\left(\frac{P}{N_0} - 1\right)$.

WORKED EXAMPLE 14 Logistic growth (2)

A logistic equation has the solution $y(x) = \frac{200}{1 + 99e^{-\frac{x}{2}}}$.

a. Using differentiation, show that the solution satisfies the differential equation $\frac{dy}{dx} = \frac{y(200 - y)}{400}$.

b. Show that $\frac{d^2y}{dx^2} = \frac{y(100 - y)(200 - y)}{80\,000}$.

c. Hence determine the coordinates of the point of inflection.

d. Sketch the graph of y versus x.

THINK	WRITE/DRAW
a. 1. Write the given solution in index form.	a. $y = 200\left(1 + 99e^{-\frac{x}{2}}\right)^{-1}$
2. Use the chain rule for differentiation.	$\frac{dy}{dx} = 200 \times (-1) \times \left(-\frac{1}{2}\right) \times 99e^{-\frac{x}{2}}\left(1 + 99e^{-\frac{x}{2}}\right)^{-2}$

3. Simplify the expression for $\frac{dy}{dx}$.

$$\frac{dy}{dx} = \frac{100 \times 99e^{-\frac{x}{2}}}{\left(1 + 99e^{-\frac{x}{2}}\right)^2}$$

4. We need to express x in terms of y.

$$\frac{200}{1 + 99e^{-\frac{x}{2}}} = y \qquad [1]$$

$$\frac{1}{1 + 99e^{-\frac{x}{2}}} = \frac{y}{200} \qquad [2]$$

5. Express the exponential in terms of y.

$$1 + 99e^{-\frac{x}{2}} = \frac{200}{y}$$

$$99e^{-\frac{x}{2}} = \frac{200}{y} - 1$$

$$99e^{-\frac{x}{2}} = \frac{200 - y}{y} \qquad [3]$$

6. Rewrite the expression for $\frac{dy}{dx}$.

$$\frac{dy}{dx} = \frac{100 \times 99e^{-\frac{x}{2}}}{\left(1 + 99e^{-\frac{x}{2}}\right)^2}$$

7. Replace $99e^{-\frac{x}{2}}$ with $\frac{200 - y}{y}$.

Substitute [3] into $\frac{dy}{dx} . \frac{dy}{dx} = \frac{100 \times \left(\frac{200-y}{y}\right)}{\left(1 + \frac{200-y}{y}\right)^2}$

8. Simplify.

$$\frac{dy}{dx} = \frac{y(200 - y)}{400}$$

b. 1. To determine the second derivative, differentiate with respect to x again. However, since $\frac{dy}{dx}$ is in terms of y, use implicit differentiation.

b.
$$\frac{d^2y}{dx^2} = \frac{d}{dx}\left(\frac{dy}{dx}\right)$$

$$\frac{d^2y}{dx^2} = \frac{d}{dy}\left(\frac{dy}{dx}\right) \times \frac{dy}{dx}$$

$$\frac{d^2y}{dx^2} = \frac{d}{dy}\left(\frac{y(200 - y)}{400}\right) \times \left(\frac{y(200 - y)}{400}\right)$$

2. Expand the brackets.

$$\frac{d^2y}{dx^2} = \frac{d}{dy}\left(\frac{1}{400}(200 - y^2)\right) \times \left(\frac{y(200 - y)}{400}\right)$$

3. Take the derivative and leave the constant factor.

$$\frac{d^2y}{dx^2} = \frac{1}{400}(200 - 2y) \times \left(\frac{y(200 - y)}{400}\right)$$

4. Take out the common factor of 2 and simplify.

$$\frac{d^2y}{dx^2} = \frac{y(100 - y)(200 - y)}{80\,000}$$

c. 1. The inflection point occurs when the second derivative is zero.

c. When $\frac{d^2y}{dx^2} = 0, y = 0, 100, 200$. However $y \neq 0$, so the graph never crosses the x-axis, and $y \neq 200$. The lines and $y = 0$ are horizontal asymptotes. Therefore, $y = 100$ is the only possible result.

2. Determine the corresponding x-value.	Substitute $y = 100$: $100 = \dfrac{200}{1 + 99e^{-\frac{x}{2}}}$
3. Solve for x.	$1 + 99e^{-\frac{x}{2}} = 2$ $99e^{-\frac{x}{2}} = 1$ $e^{-\frac{x}{2}} = \dfrac{1}{99}$ $e^{\frac{x}{2}} = 99$
4. Use the definition of the logarithm.	$\dfrac{x}{2} = \log_e(99)$ $x = 2\log_e(99)$
5. State the coordinates of the inflection point.	$(2\log_e(99), 100) \approx (9.2, 100)$

d. Sketch the graph.

d. For $y = \dfrac{200}{1 + 99e^{-\frac{x}{2}}}$, $y \to 200$ as $x \to \infty$

and the inflection point is at

$(2\log_e(99), 100) \approx (9.2, 100)$.

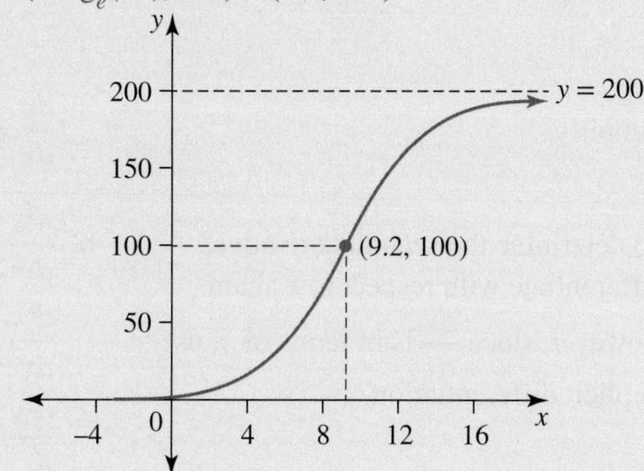

10.6 Exercise

Students, these questions are even better in jacPLUS

Receive immediate feedback and access sample responses

Access additional questions

Track your results and progress

Find all this and MORE in jacPLUS

Technology active

1. WE12 A rumour is spreading in an elderly nursing home which contains 100 pensioners. The rumour spreads at a rate proportional to both the number who have heard the rumour and the number who have not yet heard the rumour. At first, 4 pensioners in the home have heard the rumour, and after 2 weeks, 20 of the pensioners have heard it.
   a. Let $N = N(t)$ be the number who have heard the rumour after t weeks. Write and solve the differential equation.
   b. Determine the number of pensioners in the home who have heard the rumour after 3 more weeks.
   c. Calculate how long it is before 50 pensioners in the home have heard the rumour, in weeks correct to 2 decimal places.
2. An infection is spreading in another elderly nursing home which contains 80 pensioners. The infection spreads at a rate proportional to both the number who have the infection and the number who have not yet caught the infection. At first, only 2 pensioners in the home have the infection. After 3 days, 25% of the pensioners have the infection.
   a. Let $N = N(t)$ be the number who have the infection after t days. Write and solve the differential equation.
   b. Determine the number of pensioners in the home who have the infection after 4 more days.
   c. Calculate long it is before 60% of the pensioners in the home have the infection, in days correct to 2 decimal places.
3. WE13 The table shows the population of Midgar in millions of people.

Year	2011	2015	2019
Population (millions)	3.948	4.256	4.504

   Assuming the population follows a logistic growth rate:
   a. determine the maximum population of Midgar, correct to the nearest thousand
   b. determine the year in which the population of Midgar first exceeds 5 million.
4. The table shows the population of Hillwood in millions of people.

Year	2011	2013	2019
Population (millions)	1.609	1.735	2.004

   Assuming the population follows a logistic growth rate:
   a. determine the year in which the population first exceeded 2 million
   b. estimate the population of Hillwood in 2030.

5. **WE14** A logistic equation has the solution $y(x) = \dfrac{500}{1 + 9e^{-3x}}$.

a. Using differentiation, show that it satisfies the differential equation $\dfrac{dy}{dx} = \dfrac{3y(500 - y)}{500}$.

b. Show that $\dfrac{d^2y}{dx^2} = \dfrac{9y(250 - y)(500 - y)}{125\,000}$.

c. Hence, determine the coordinates of the point of inflection.

d. Sketch the graph of y versus x.

6. A logistic equation has the solution $y(x) = \dfrac{400}{1 + 199e^{-2x}}$.

a. Using differentiation, show that it satisfies the differential equation $\dfrac{dy}{dx} = \dfrac{y(400 - y)}{200}$.

b. Show that $\dfrac{d^2y}{dx^2} = \dfrac{y(200 - y)(400 - y)}{20\,000}$.

c. Hence, determine the coordinates of the point of inflection.

d. Sketch the graph of y versus x.

7. A logistic equation has the solution $y(x) = \dfrac{600}{1 + 99e^{-\frac{x}{3}}}$.

a. Using differentiation, show that it satisfies the differential equation $\dfrac{dy}{dx} = \dfrac{y(600 - y)}{1800}$.

b. Show that $\dfrac{d^2y}{dx^2} = \dfrac{y(300 - y)(600 - y)}{1\,620\,000}$.

c. Hence, determine the coordinates of the point of inflection.

d. Sketch the graph of y versus x.

8. An area of the ocean initially contains 500 fish. The area can support a maximum of 4000 fish. The number of fish in the area grows continuously at a rate of 8% per year; that is, the constant of proportionality is 8%.

a. Write the differential equation for the number of fish, $N(t)$, in the area at a time t years, assuming a logistic growth model.

b. Solve the differential equation to determine the number of fish in the area after t years.

c. Hence, determine the number of fish in the area after 5 years.

d. Determine after how many years the number of fish will reach 3000, correct to 2 decimal places.

9. In a movie, people in a town are turned into zombies by a touch from an existing zombie. The town contains 360 people, and initially there is only one zombie. After 3 days, there are 60 zombies. The number of zombies in the town grows at a rate proportional to both the number of zombies and the number of people in the town who are not yet zombies.

a. Write the differential equation for the number of zombies, N, after t days.

b. Determine after how many days 75% of the people in the town will become zombies, correct to 1 decimal place.

10. A rumour is spreading in a school with 2000 students. The rumour spreads at a rate proportional to both the number who have heard the rumour and the number who have not yet heard it. Initially, only 2 students have heard the rumour, but after 3 days, 10% of the school has heard it.

a. Write the differential equation for the number of students, N, in the school after t days who have heard the rumour, assuming a logistic growth model.

b. Determine after how many days half the students have heard the rumour, correct to 1 decimal place.

c. Determine when the rumour is spreading most rapidly, in days correct to 1 decimal place.

11. A disease is spreading through an area with a population of 10000. Initially, 4 people in the area have the disease. After 4 days, 50 people have the disease. The disease is spreading at a rate proportional to both the number of people who have the disease and the number of people who do not yet have the disease. Determine how many people in the area have the disease after 2 weeks.

12. The number of children who contract the common cold from a kindergarten is found to be proportional to those who have the cold and those yet to get the cold. Initially, only 1 child from the kindergarten has the common cold. After 2 days, 5 children have the cold, and after a further, 2 days 15 children have the cold.
   a. Calculate how many children from the kindergarten will eventually get the cold.
   b. Determine after how many days the cold is spreading most rapidly, correct to 2 decimal places.
   c. Calculate how many children from the kindergarten have the cold after 6 days.
   d. Sketch the graph of the number of children with the cold against the time in days.

10.6 Exam questions

Question 1 (2 marks) TECH-FREE

If a logistic equation has as its solution $N(t) = \dfrac{200}{1 + 99e^{-kt}}$ where $k > 0$, write an expression for $\dfrac{dN}{dt}$.

Question 2 (1 mark) TECH-FREE

A logistic equation has the solution $N(t) = \dfrac{500}{1 + 9e^{-kt}}$ where N is the population number of a city in millions at a time t years and k is a positive real constant. Determine the value of the population when the population increasing most rapidly.

Question 3 (7 marks) TECH-ACTIVE

Given the general logistic equation $\dfrac{dN}{dt} = kN\left(1 - \dfrac{N}{P}\right)$, $N(0) = N_0$, where k is a positive constant and $P > N_0 > 0$:

a. show, using integration, that $N(t) = \dfrac{PN_0}{N_0 + (P - N_0)^{e^{-kt}}}$ **(2 marks)**

b. show that $\lim\limits_{n \to \infty} N(t) = P$ **(1 mark)**

c. show that $\dfrac{d^2N}{dt^2} = k^2N\left(1 - \dfrac{2N}{P}\right)\left(1 - \dfrac{N}{P}\right)$ and hence that there is a point of inflection at $\left(\dfrac{1}{k}\log_e\left(\dfrac{P}{N_o} - 1\right), \dfrac{P}{2}\right)$ **(2 marks)**

d. verify, using differentiation, that $N(t) = \dfrac{PN_0}{N_0 + (P - N_0)^{e^{-kt}}}$ is a solution of $\dfrac{dN}{dt} = kN\left(1 - \dfrac{N}{P}\right)$, $N(0) = N_0$. **(2 marks)**

More exam questions are available online.

10.7 Euler's method

LEARNING INTENTION

At the end of this subtopic you should be able to:
- approximate solutions to differential equations using Euler's method.

10.7.1 Introduction

The Swiss mathematician Leonhard Euler (1707–1783) is considered one of the greatest mathematicians of all time. He made significant discoveries in many areas of mathematics, including calculus, graph theory, fluid dynamics, optics and astronomy. It is his notation for functions that we still use today.

10.7.2 Numerical solution of a differential equation

Many first-order differential equations can be solved by integration, obtaining the particular solution. A differential equation with a given initial condition is often called an 'initial value problem', abbreviated to IVP.

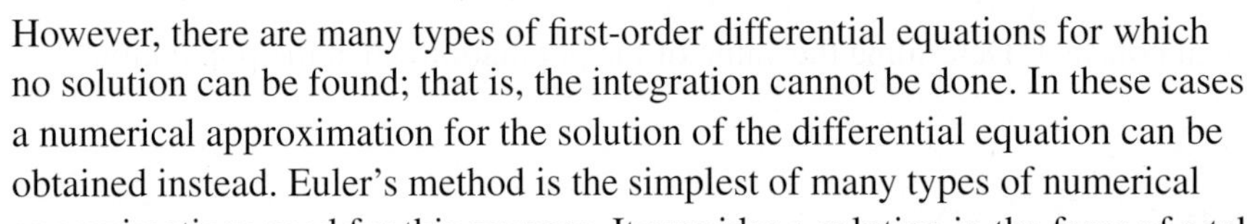

However, there are many types of first-order differential equations for which no solution can be found; that is, the integration cannot be done. In these cases a numerical approximation for the solution of the differential equation can be obtained instead. Euler's method is the simplest of many types of numerical

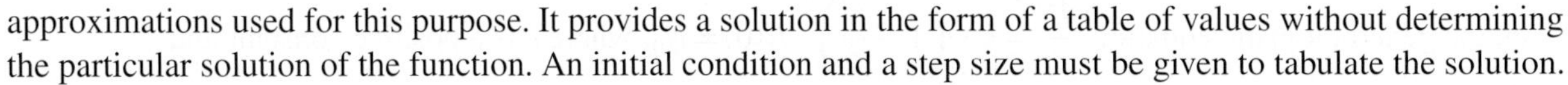

approximations used for this purpose. It provides a solution in the form of a table of values without determining the particular solution of the function. An initial condition and a step size must be given to tabulate the solution.

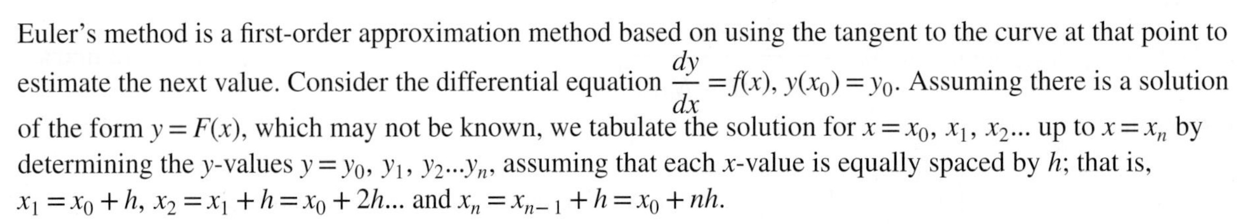

Euler's method is a first-order approximation method based on using the tangent to the curve at that point to estimate the next value. Consider the differential equation $\frac{dy}{dx} = f(x)$, $y(x_0) = y_0$. Assuming there is a solution of the form $y = F(x)$, which may not be known, we tabulate the solution for $x = x_0, x_1, x_2 ...$ up to $x = x_n$ by determining the y-values $y = y_0, y_1, y_2 ... y_n$, assuming that each x-value is equally spaced by h; that is, $x_1 = x_0 + h$, $x_2 = x_1 + h = x_0 + 2h ...$ and $x_n = x_{n-1} + h = x_0 + nh$.

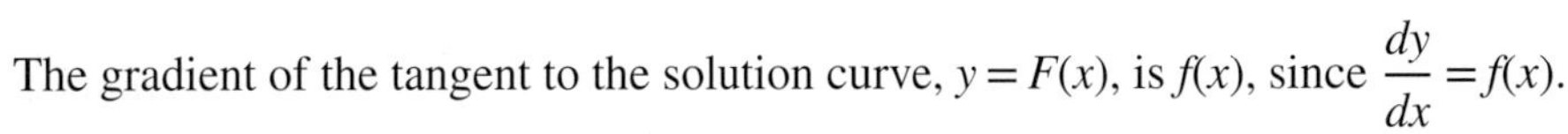

The gradient of the tangent to the solution curve, $y = F(x)$, is $f(x)$, since $\frac{dy}{dx} = f(x)$.

Let m_T be the gradient of the tangent to the solution curve at (x_0, y_0).

For $m_T = f(x_0)$, the equation of the tangent to the solution curve is given by $y - y_0 = m_T(x - x_0) = f(x_0)(x - x_0)$.

When $x = x_1$, the next value of y is given by
$y_1 = y_0 + f(x_0)(x_1 - x_0) = y_0 + hf(x_0)$.

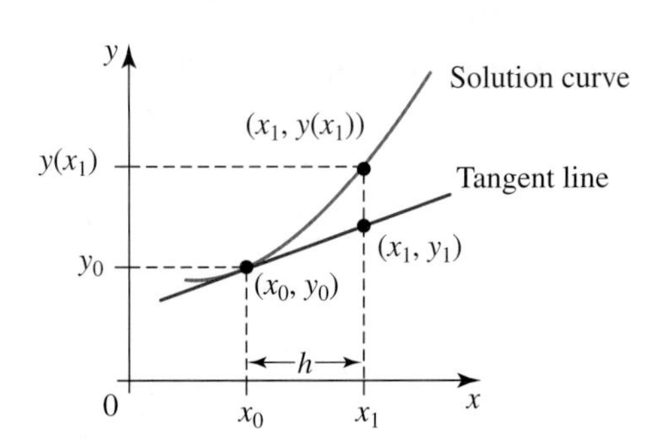

This iterative process is repeated, with each coordinate being used to determine the next coordinate.

In general, to use Euler's method, iterate and use the result $y_{n+1} = y_n + hf(x_n)$. To obtain y_n, the value of y when $x = x_n = x_0 + nh$, repeat the procedure n times.

The result obtained by Euler's method can be compared to the exact solution.

WORKED EXAMPLE 15 Using Euler's formula to approximate solutions

a. Use Euler's method to tabulate the solutions to the differential equation $\frac{dy}{dx} = \frac{1}{x^2+4}$, $y(0) = 0$, using $h = \frac{1}{2}$ to approximate $y(2)$, correct to 4 decimal places.

b. Solve the differential equation $\frac{dy}{dx} = \frac{1}{x^2+4}$, $y(0) = 0$ and determine $y(2)$. Determine the percentage error, correct to 1 decimal place, for the result obtained by Euler's method compared to the exact value.

THINK	WRITE
a. 1. State the initial values and the step size.	a. $f(x) = \frac{1}{x^2+4}$, $x_0 = 0$, $y_0 = 0$, $h = \frac{1}{2}$ When $x = 2$, we require y_4 from x_4, so the procedure must be repeated 4 times.
2. Use Euler's method to determine the value of y_1.	$y_1 = y_0 + hf(x_0)$ $y_1 = 0 + \frac{1}{2}\left(\frac{1}{0+4}\right)$ $= \frac{1}{8}$ $= 0.125$
3. Use Euler's method to determine the value of y_2.	$y_2 = y_1 + hf(x_1)$, $x_1 = x_0 + h = \frac{1}{2} = 0.5$ $y_2 = \frac{1}{8} + \frac{1}{2}\left(\frac{1}{0.5^2+4}\right)$ $= \frac{33}{136}$ $= 0.242\,65$
4. Use Euler's method to determine the value of y_3.	$y_3 = y_2 + hf(x_2)$, $x_2 = x_1 + h = 1$ $y_3 = \frac{33}{136} + \frac{1}{2}\left(\frac{1}{1^2+4}\right)$ $= \frac{233}{680}$ $= 0.3426$
5. Use Euler's method to determine the value of y_4.	$y_4 = y_3 + hf(x_3)$, $x_3 = x_2 + h = \frac{3}{2} = 1.5$ $y_4 = \frac{233}{680} + \frac{1}{2}\left(\frac{1}{1.5^2+4}\right)$ $= \frac{1437}{3400}$ $= 0.4226$
6. Summarise the results. Note that exact fractions could be given in this particular case, but the answers are given to 4 decimal places.	see table below

x	0	0.5	1.0	1.5	2
y	0	0.125	0.2426	0.3426	0.4226

b. 1. To solve the differential equation, integrate with respect to x.

b. $\frac{dy}{dx} = \frac{1}{x^2+4}$

$y = \int \frac{1}{x^2+4} dx$

2. Perform the integration.

$y = \frac{1}{2}\tan^{-1}\left(\frac{x}{2}\right) + c$

3. Determine the value of the constant of integration.

Since $y(0) = 0$, substitute $x = 0$ when $y = 0$:

$0 = \frac{1}{2}\tan^{-1}(0) + c$

$c = 0$

4. State the particular solution.

$y = \frac{1}{2}\tan^{-1}\left(\frac{x}{2}\right)$

5. Determine the required value.

Substitute $x = 2$:

$$\begin{aligned} y &= \frac{1}{2}\tan^{-1}(1) \\ &= \frac{1}{2} \times \frac{\pi}{4} \\ &= \frac{\pi}{8} \\ &\approx 0.3927 \end{aligned}$$

6. Compare the exact result with the result obtained by Euler's method.

In this case Euler's method overestimates the exact result by $\frac{0.4226 - 0.3927}{0.3927} \times 100 = 7.6\%$.

10.7.3 Using Euler's method to solve Type 2 differential equations

Type 2 first-order differential equations are of the form $\frac{dy}{dx} = f(y)$, $y(x_0) = y_0$. Euler's method can also be used to approximate solutions for these equations, but in this case the iterative result $y_{n+1} = y_n + hf(y_n)$ is used.

WORKED EXAMPLE 16 Using Euler's formula with Type 2 differential equations

a. Use Euler's method to tabulate the solutions to the differential equation $\frac{dy}{dx} - e^{-y} = 0$, $y(1) = 0$, using $h = \frac{1}{4}$ to approximate $y(2)$, correct to 4 decimal places.

b. Solve the differential equation $\frac{dy}{dx} - e^{-y} = 0$, $y(1) = 0$ and determine $y(2)$. Determine the percentage error, correct to 1 decimal place, for the result obtained by Euler's method compared to the exact value.

THINK | **WRITE**

a. 1. Write the differential equation in standard form and state the initial values and the step size.

a. $\dfrac{dy}{dx} - e^{-y}, y(1) = 0$

$f(y) = e^{-y}, x_0 = 1, y_0 = 0, h = \dfrac{1}{4}$

When $x = 2$, we require y_4 from y_3, so the procedure must be repeated 4 times.

2. Use Euler's method to determine the value of y_1.

$$y_1 = y_0 + hf(y_0)$$
$$y_1 = 0 + \frac{1}{4}e^{-0}$$
$$= \frac{1}{4}$$
$$= 0.25$$

3 Use Euler's method to determine the value of y_2.

$$y_2 = y_1 + hf(y_1)$$
$$= 0.25 + \frac{1}{4} \times e^{-0.25}$$
$$= 0.4447$$

4. Use Euler's method to determine the value of y_3.

$$y_3 = y_2 + hf(y_2)$$
$$= 0.4447 + \frac{1}{4} \times e^{-0.4447}$$
$$= 0.6050$$

5. Use Euler's method to determine the value of y_4.

$$y_4 = y_3 + hf(y_3)$$
$$= 0.6050 + \frac{1}{4} \times e^{-0.6050}$$
$$= 0.7415$$

6. Summarise the results. (Note that in the table they are given correct to 4 decimal places, but more decimal places were used in the working to avoid rounding errors.)

x	1	1.25	1.5	1.75	2
y	0	0.25	0.4447	0.6050	0.7415

b. 1. To solve the differential equation, first invert both sides, then integrate with respect to y.

b. $\dfrac{dy}{dx} = e^{-y}$

$$\frac{dy}{dx} = \frac{1}{e^{-y}}$$
$$\frac{dy}{dx} = e^{y}$$
$$x = \int e^{y}\,dy$$

2. Perform the integration.

$$x = e^{y} + c$$

3. Determine the value of the constant of integration.

Since $y(1) = 0$, substitute $x = 1$ when $y = 0$:

$$1 = e^{0} + c$$
$$c = 0$$

4. Rearrange to make y the subject and state the particular solution.

$$x = e^{y}$$
$$y = \log_e(x), x > 0$$

5. Determine the required value.	Substitute $x = 2$: $y = \log_e(2) \approx 0.6931$
6. Compare the exact result with the result obtained by Euler's method.	In this case Euler's method overestimates the exact result by $\frac{0.7415 - 0.6931}{0.6931} \times 100 = 7.0\%$.

10.7.4 Using Euler's method to solve Type 3 differential equations

Recall that first-order differential equations of the form $\frac{dy}{dx} = f(x, y),\ y(x_0) = y_0$ have been solved. Euler's method can also be applied to these equations, but in this case the iterative result $y_{n+1} = y_n + hf(x_n, y_n)$ is used.

WORKED EXAMPLE 17 Using Euler's formula with Type 3 differential equations

a. Use Euler's method to tabulate the solutions to the differential equation $\frac{dy}{dx} + 2xy^2 = 0,\ y(0) = 2$ using $h = \frac{1}{4}$ to approximate $y(1)$, correct to 4 decimal places.

b. Solve the differential equation $\frac{dy}{dx} + 2xy^2 = 0,\ y(0) = 2$ and determine $y(1)$. Determine the percentage error, correct to 1 decimal place, for the result obtained by Euler's method compared to the exact value.

THINK	WRITE
a. 1. Write the differential equation in standard form and state the initial values and the step size.	a. $\frac{dy}{dx} = -2xy^2,\ y(0) = 2$ $f(x, y) = -2xy^2,\ x_0 = 0,\ y_0 = 2,\ h = \frac{1}{4}$ When $x = 1$, we require y_4, so the procedure must be repeated 4 times.
2. Use Euler's method to determine the value of y_1.	$y_1 = y_0 + hf(x_0, y_0)$ $y_1 = 2 + \frac{1}{4} \times (-2 \times 0 \times 2^2)$ $= 2$
3. Use Euler's method to determine the value of y_2.	$y_2 = y_1 + hf(x_1, y_1),\ x_1 = x_0 + h = \frac{1}{4}$ $y_2 = 2 + \frac{1}{4} \times \left(-2 \times \frac{1}{4} \times 2^2\right)$ $= \frac{3}{2}$ $= 1.5$
4. Use Euler's method to determine the value of y_3.	$y_3 = y_2 + hf(x_2, y_2)\ x_2 = x_1 + h = \frac{1}{2},$ $y_3 = \frac{3}{2} + \frac{1}{4} \times \left(-2 \times \frac{1}{2} \times \left(\frac{3}{2}\right)^2\right)$ $= \frac{15}{16}$ $= 0.9375$

5. Use Euler's method to determine the value of y_4.

$$y_4 = y_3 + hf(x_3, y_3),\ x_3 = x_2 + h = \frac{3}{4}$$

$$y_4 = \frac{15}{16} + \frac{1}{4} \times \left(-2 \times \frac{3}{4} \times \left(\frac{15}{16}\right)^2\right)$$

$$= \frac{1245}{2048}$$

$$= 0.6079$$

6. Summarise the results. (Note that they are given correct to 4 decimal places in the table, but exact answers were used when performing the calculations.)

x	0	0.25	0.5	0.75	1.0
y	2	2	1.5	0.9375	0.6079

b. 1. To solve the differential equation, separate the variables, setting all x's on one side and all y's on the other side.

b. $$\frac{dy}{dx} = -2xy^2$$

$$\int 2x\,dx = \int \frac{-1}{y^2}dy$$

2. Perform the integration.

$$x^2 = \frac{1}{2} + c$$

3. Determine the value of the constant of integration.

Since $y(0) = 2$, substitute $x = 0$ when $y = 2$:

$$0 = \frac{1}{2} + c$$

$$c = -\frac{1}{2}$$

4. Rearrange to make y the subject and state the particular solution.

$$x^2 = \frac{1}{y} - \frac{1}{2}$$

$$\frac{1}{y} = x^2 + \frac{1}{2}$$

$$\frac{1}{y} = \frac{2x^2 + 1}{2}$$

$$y = \frac{2}{2x^2 + 1}$$

5. Determine the required value.

Substitute $x = 1$:

$$y = \frac{2}{3}$$

$$= 0.6667$$

6. Compare the exact result with the result obtained by Euler's method.

In this case Euler's method underestimates the exact result by $\frac{0.6079 - 0.6667}{0.6667} \times 100 = -8.8\%$

10.7.5 Comparing Euler's method

Euler's method can overestimate or underestimate the exact solution of a differential equation. When a curve is concave up, the tangent line lies under the curve, so in these cases Euler's method will underestimate the exact solution.

When a curve is concave down, the tangent line lies above the curve, so in these cases Euler's method will overestimate the exact solution.

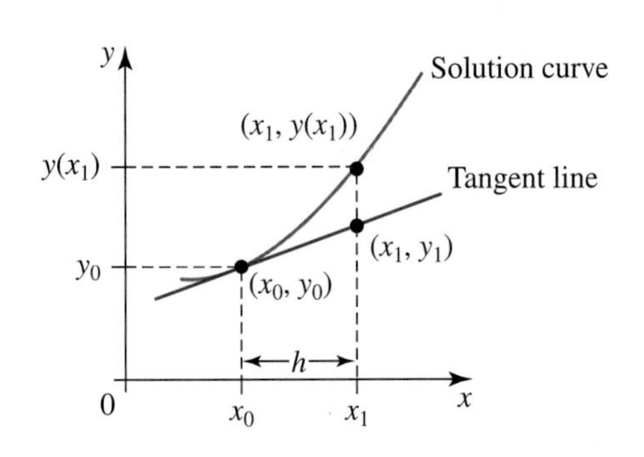

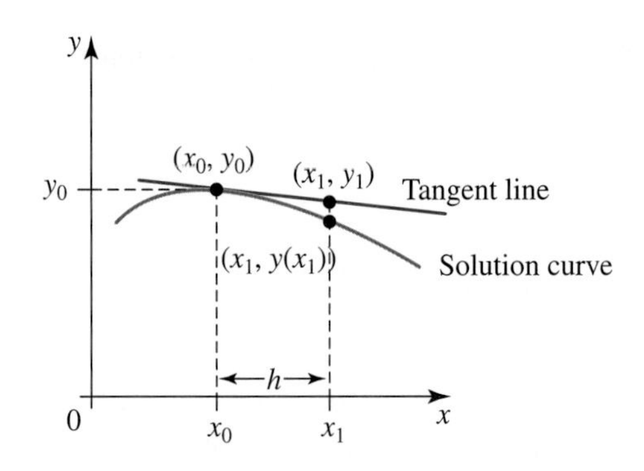

It is true that if the step size h is made smaller and the value of n larger, that is, the iterative process is repeated more times, then the percentage error can be reduced. However, there are occasions when Euler's method diverges and results are significantly wrong and incorrect.

10.7 Exercise

Students, these questions are even better in jacPLUS

Receive immediate feedback and access sample responses

Access additional questions

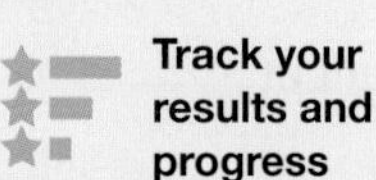
Track your results and progress

Find all this and MORE in jacPLUS

Technology active

1. a. WE15 Use Euler's method to tabulate the solutions to the differential equation $\frac{dy}{dx} = 3\sqrt{x}$, $y(4) = 1$, using $h = \frac{1}{4}$ to approximate $y(5)$, correct to 4 decimal places.

   b. Solve the differential equation $\frac{dy}{dx} = 3\sqrt{x}$, $y(4) = 1$ and determine $y(5)$, correct to 4 decimal places.

   Determine the percentage error, correct to 1 decimal place, for the result obtained by Euler's method compared to the exact value.

The following instruction relates to questions 2–4.

Use Euler's method to determine the value of y_n indicated for each of the following initial value problems, using the given value of h. Give your answers correct to 4 decimal places, and compare the approximated answer to the exact answer.

2. For $\frac{dy}{dx} = 2\cos\left(\frac{x}{2}\right)$, $y(0) = 2$ determine y_2 with $h = \frac{1}{2}$.
3. For $\frac{dy}{dx} = \sin(3x)$, $y(0) = 3$ determine y_3 with $h = \frac{1}{3}$.
4. For $\frac{dy}{dx} = 4e^{-2x}$, $y(0) = 2$ determine y_4 with $h = \frac{1}{4}$.
5. Use Euler's method to tabulate the solutions to the differential equation $\frac{dy}{dx} + 6x = 0$, $y(1) = 2$ up to $x = 2$ using the following step sizes.

   a. $h = \frac{1}{2}$
   b. $h = \frac{1}{3}$
   c. $h = \frac{1}{4}$
   d. $h = \frac{1}{5}$

6. Consider the differential equation $\frac{dy}{dx} = 6x^2$ with $x_0 = 1$ and $y_0 = k$. When Euler's method is used with a step size of $\frac{1}{3}$, $y_3 = 12$. Determine the value of k.

7. Consider the differential equation $\frac{dy}{dx} + \frac{1}{x^2} = 0$ with $x_0 = 1$ and $y_0 = k$. When Euler's method is used with a step size of $\frac{1}{3}$, $y_3 = \frac{431}{1200}$. Determine the value of k.

8. Consider the differential equation $\frac{dy}{dx} = \frac{2}{x}$ with $x_0 = 1$ and $y_0 = k$. When Euler's method is used with a step size of 0.25, $y_4 = 2$. Determine the value of k.

9. Consider the differential equation $\frac{dy}{dx} = 4x^3$, $y(0) = k$. Use Euler's method with a step size of $h = \frac{1}{4}$ and the value of $y_4 = y(1) = 1$. Determine the value of k.

10. Consider the differential equation $\frac{dy}{dx} = \log_e(3x+2)$, $y(1) = 2$. Use Euler's method with a step size of $\frac{1}{3}$. Show that when $x = 2$, $y_3 = 2 + \frac{1}{3}\log_e(210)$.

11. Consider the differential equation $\frac{dy}{dx} = \log_e(4x+1)$, $y(3) = 5$. Use Euler's method with a step size of 0.25. Show that when $x = 4$, $y_4 = 5 + \frac{1}{4}\log_e(43\,680)$.

12. a. The solution to the initial value problem $\frac{dy}{dx} = \log(2x+5)$, $y(2) = 4$ is approximated using Euler's method with a step size of 0.5. Show that when $x = 3$, $y_2 = 4 + \frac{1}{2}\log_e(90)$.

    b. Solve the differential equation $\frac{dy}{dx} = \log_e(2x+5)$, $y(2) = 4$ to determine the value of y when $x = 3$.

13. **WE16** a. Use Euler's method to tabulate the solutions to the differential equation $\frac{dy}{dx} - \frac{2}{y} = 0$, $y(2) = 3$, using $h = \frac{1}{4}$ to approximate $y(3)$, correct to 4 decimal places.

    b. Solve the differential equation $\frac{dy}{dx} - \frac{2}{y} = 0$, $y(2) = 3$ and determine $y(3)$, correct to 4 decimal places. Determine the percentage error, correct to 1 decimal place, for the result obtained by Euler's method compared to the exact value.

14. Use Euler's method with a step size of $h = 0.1$ to determine y_3 for the differential equation $\frac{dy}{dx} = \tan(y)$, $y(0.2) = 0.4$, correct to 4 decimal places.

15. For $\frac{dy}{dx} = \frac{y}{3}$, given $(x_0, y_0) = (0, 4)$:

    a. use Euler's method to determine the value of y_2 with $h = \frac{1}{2}$

    b. use Euler's method to determine the value of y_3 with $h = \frac{1}{3}$.

16. For $\frac{dy}{dx} = \frac{y}{2}(5 - y)$, given $(x_0, y_0) = (0, 1)$:

a. use Euler's method to determine the value of y_2 with $h = \frac{1}{2}$

b. use Euler's method to determine the value of y_3 with $h = \frac{1}{3}$.

17. a. **WE17** Use Euler's method to tabulate the solutions to the differential equation $\frac{dy}{dx} + 2x^3y^2 = 0$, $y(0) = 2$, using $h = \frac{1}{4}$ to approximate $y(1)$, correct to 4 decimal places.

b. Solve the differential equation $\frac{dy}{dx} + 2x^3y^2 = 0$, $y(0) = 2$ and determine $y(1)$. Determine the percentage error, correct to 1 decimal place, for the result obtained by Euler's method compared to the exact value.

18. a. Consider the differential equation $\frac{dy}{dx} - 2y\cos(x) = 0$, $y(0) = 1$. Use Euler's method with a step size of 0.1 to approximate y_3, correct to 4 decimal places.

b. Solve the differential equation $\frac{dy}{dx} - 2y\cos(x) = 0$, $y(0) = 1$ and determine $y(0.3)$. Determine the percentage error, correct to 1 decimal place, for the result obtained by Euler's method compared to the exact value.

19. Use Euler's method to tabulate the solutions to the differential equation $\frac{dy}{dx} + xy^2 = 0$, $y(1) = 2$ up to $x = 2$, giving your answers correct to 4 decimal places, with a step size of $\frac{1}{4}$.

20. Consider the differential equation $\frac{dy}{dx} = y\sqrt{x^2 + 5}$ if $(x_0, y_0) = (2, 5)$. Use Euler's method to determine (correct to 4 decimal places):

a. y_3 with $h = \frac{1}{3}$

b. y_4 with $h = \frac{1}{4}$.

21. Consider the differential equation $\frac{dy}{dx} + \frac{e^{-2x}}{y} = 0$ if $(x_0, y_0) = (1, 5)$. Use Euler's method to determine (correct to 4 decimal places):

a. y_3 with $h = \frac{1}{3}$

b. y_4 with $h = \frac{1}{4}$.

22. a. Given the initial value problem $\frac{dy}{dx} = 4x - y + 2$, $y(0) = 2$, tabulate the solutions using Euler's method up to $x = 1$ with a step size of 0.25. Give your answers correct to 4 decimal places.

b. Given that $y = ax + b + ce^{kx}$ is a solution of the differential equation $\frac{dy}{dx} = 4x - y + 2$, $y(0) = 2$, determine the values of a, b, c and k. Hence, determine the value of y when $x = 1$, correct to 4 decimal places.

23. a. Given the initial value problem $\frac{dy}{dx} = 5x + 2y + 1$, $y(0) = 1$, tabulate the solutions, using Euler's method with $h = \frac{1}{4}$ up to $x = 1$. Give your answers correct to 4 decimal places.

b. Given that $y = ax + b + ce^{kx}$ is a solution of the differential equation $\frac{dy}{dx} = 5x + 2y + 1$, $y(0) = 1$, determine the values of a, b, c and k. Hence, determine the value of y when $x = 1$, correct to 4 decimal places.

24. Given the initial value problem $\frac{dy}{dx} = x\sin\left(\frac{x}{2}\right)$, $y(0) = 2$:

a. use Euler's method with $h = \frac{1}{3}$ to show that $y_3 = 2 + \frac{1}{9}\left[\sin\left(\frac{1}{6}\right) + 2\sin\left(\frac{1}{3}\right)\right]$

b. use Euler's method with $h = \frac{1}{4}$ to show that $y_4 = 2 + \frac{1}{16}\left[\sin\left(\frac{1}{8}\right) + 2\sin\left(\frac{1}{4}\right) + 3\sin\left(\frac{3}{8}\right)\right]$

c. Solve the differential equation $\frac{dy}{dx} = x\sin\left(\frac{x}{2}\right)$, $y(0) = 2$ to determine the value of y when $x = 1$.

25. Given the initial value problem $\frac{dy}{dx} = xe^{-3x}$, $y(0) = 4$:

a. use Euler's method with $h = \frac{1}{3}$ to show that $y_3 = 4 + \frac{1}{9}\left(e^{-1} + 2e^{-2}\right)$

b. use Euler's method with $h = \frac{1}{4}$ to show that $y_4 = 4 + \frac{1}{16}\left(e^{-\frac{3}{4}} + 2e^{-\frac{3}{2}} + 3e^{-\frac{9}{4}}\right)$

c. solve the differential equation $\frac{dy}{dx} = xe^{-3x}$, $y(0) = 4$ to determine the value of y when $x = 1$.

26. Show that when the differential equation $\frac{dy}{dx} = f(x)$, $y(x_0) = y_0$ is approximated using Euler's method, for a small value of h, $y_n = y_0 + h\sum_{k=0}^{n-1} f(x_0 + kh)$.

10.7 Exam questions

Question 1 (1 mark) TECH-ACTIVE

Source: VCE 2021 Specialist Mathematics Exam 2, Section A, Q8; © VCAA.

Euler's method, with a step size of 0.1, is used to approximated the solution of the differential equation $\frac{dy}{dx} = y\sin(x)$.

Given that $y = 2$ when $x = 1$, the value of y correct to three decimal places, when $x = 1.2$ is

A. 2.168 **B.** 2.178 **C.** 2.362 **D.** 2.370 **E.** 2.381

Question 2 (1 mark) TECH-ACTIVE

Source: VCE 2020 Specialist Mathematics Exam 2, Section A, Q12; © VCAA.

If $\frac{dy}{dx} = e^{\cos(x)}$ and $y_0 = e$ when $x_0 = 0$ then, using Euler's formula with step size 0.1, y_3 is equal to

A. $e + 0.1\left(1 + e^{\cos(0.1)}\right)$

B. $e + 0.1\left(1 + e^{\cos(0.1)} + e^{\cos(0.2)}\right)$

C. $e + 0.1\left(e + e^{\cos(0.1)} + e^{\cos(0.2)}\right)$

D. $e + 0.1\left(e^{\cos(0.1)} + e^{\cos(0.2)} + e^{\cos(0.3)}\right)$

E. $e + 0.1\left(e + e^{\cos(0.1)} + e^{\cos(0.2)} + e^{\cos(0.3)}\right)$

Question 3 (1 mark) TECH-ACTIVE

Source: VCE 2017 Specialist Mathematics Exam 2, Section A, Q9; © VCAA.

Consider $\frac{dy}{dx} = 2x^2 + x + 1$ where $y(1) = y_0 = 2$.

Using Euler's method with a step size of 0.1 an approximation to $y(0.8) = y_2$ is given by

A. 0.94 **B.** 1.248 **C.** 1.6 **D.** 2.4 **E.** 2.852

More exam questions are available online.

10.8 Slope fields

LEARNING INTENTION

At the end of this subtopic you should be able to:
- sketch slope fields for differential equations
- determine the differential equation from a given slope filed
- determine the slope field from a given differential equation.

Slope fields, also known as direction fields, are a tool for graphically visualising the solutions to a first-order differential equation. A slope field is simply a graph on the Cartesian coordinate system showing short line segments that represent the slopes of the possible solutions at each point.

10.8.1 Type 1, $\frac{dy}{dx} = f(x)$

Type 1 first-order differential equations are of the form $\frac{dy}{dx} = f(x)$. In this section you will draw the slope fields for a simple differential equation of this type.

WORKED EXAMPLE 18 Sketching slope fields for Type 1 differential equations

Sketch the slope field for the differential equation $\frac{dy}{dx} = \frac{x}{2}$ for $y = -2, -1, 0, 1, 2$ at each of the values $x = -2, -1, 0, 1, 2$ on the grid shown.

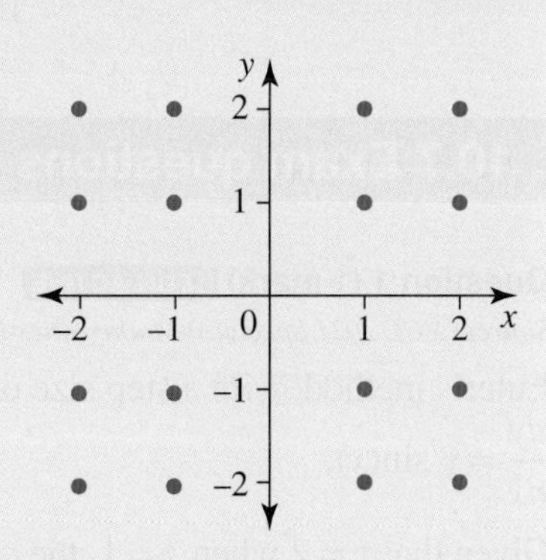

THINK	WRITE/DRAW
1. Make some observations about the slope.	When $x = 0$ (i.e. along the y-axis), the slope $\frac{dy}{dx}$ is zero. The y-values do not affect the slope.
2. Determine the slope at appropriate values of x.	When $x = 1$, the slope is $\frac{dy}{dx} = \frac{1}{2}$. A slope of $\frac{1}{2}$ makes an angle of $\tan^{-1}\left(\frac{1}{2}\right) \approx 27°$ with the positive x-axis.
3. Consider another x-value.	When $x = 2$, the slope is $\frac{dy}{dx} = 1$. A slope of 1 makes an angle of $\tan^{-1}(1) = 45°$ with the positive x-axis.

4. Consider the negative values of x.

When $x = -1$, the slope is $\dfrac{dy}{dx} = -\dfrac{1}{2}$.
A slope of $-\dfrac{1}{2}$ makes an angle of $\tan^{-1}\left(-\dfrac{1}{2}\right) \approx 153°$ with the positive x-axis.

5. Consider the final x-value.

When $x = -2$, the slope is $\dfrac{dy}{dx} = -1$.
A slope of -1 makes an angle of $\tan^{-1}(-1) = 135°$ with the positive x-axis.

6. Draw short line segments with the slopes found at each of the points.

The slope field in this case is symmetrical about the y-axis.

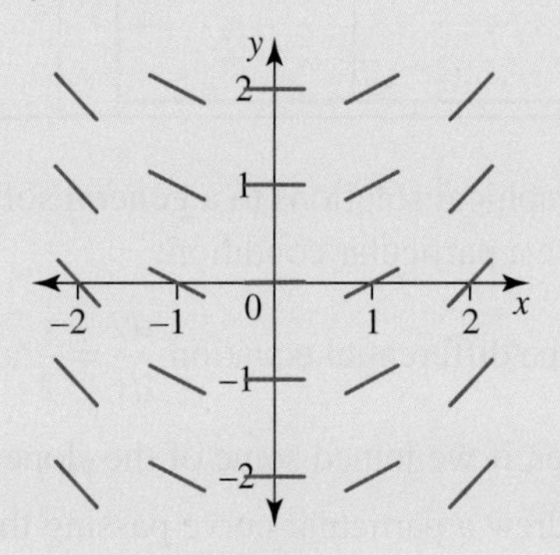

TI | THINK — **DISPLAY/WRITE**

1. Open a Graphs page.

2. Slope fields can be graphed by selecting:
MENU
3: Graph Entry/Edit
8: Diff Eq

3. Complete the entry line as $\dfrac{x}{2}$ and press ENTER.

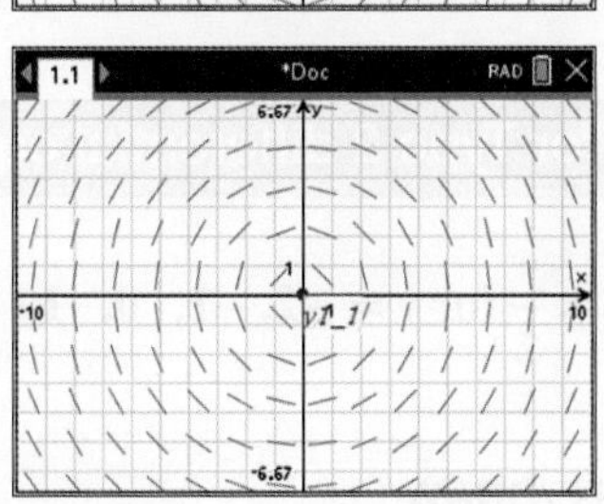

CASIO | THINK — **DISPLAY/WRITE**

1. Open a DiffEq-Graph screen.

2. Complete the entry line as $y' = \dfrac{x}{2}$.

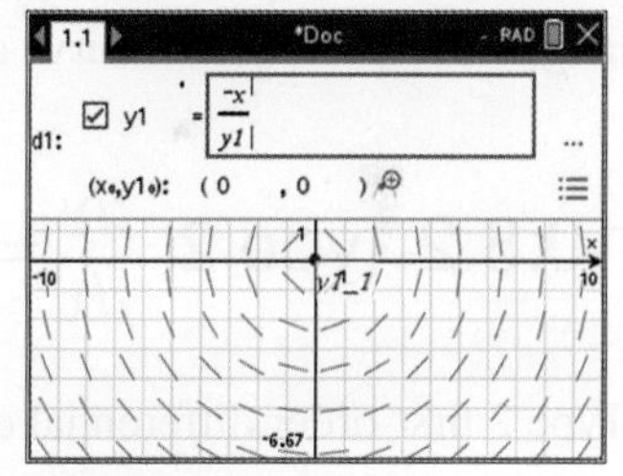

3. A particular solution can be drawn if the initial conditions are known. To input the initial conditions $(x, y) = (0, 0)$:
Select the IC tab and enter the conditions
$xi = 0$
$yi = 0$
as shown.

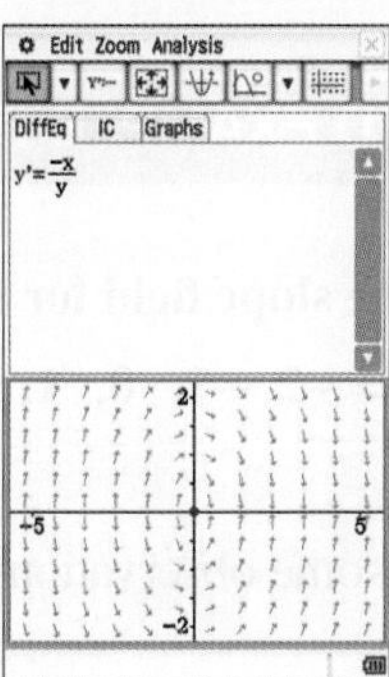

4. A particular solution can be drawn if the initial conditions are known. To input the initial conditions $(x, y) = (0, 0)$: Press TAB to view the entry line and insert 0 and 0 into the initial conditions coordinate template as shown. Press ENTER to see the particular solution.

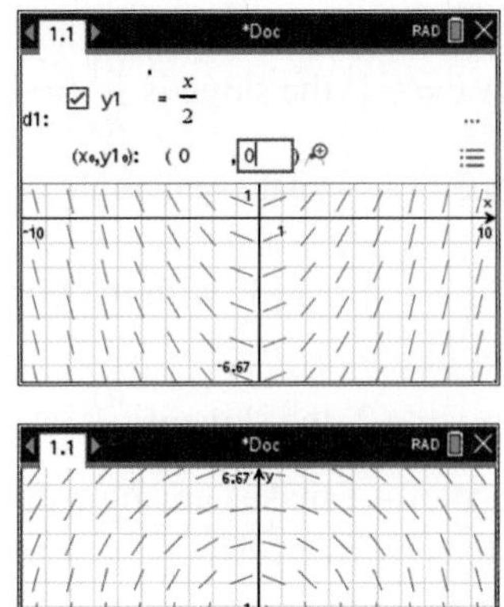

Note that we obtain these graphical solutions to a general solution of a first-order differential equation; that is, we do not necessarily require a particular condition.

In general, the solutions to the differential equation $\frac{dy}{dx} = \frac{x}{2}$ are a family of curves of the form $y = \frac{x^2}{4} + c$. However, if we joined some of the slopes shown in the answer to Worked example 18 and drew a particular curve passing through the origin, the result would represent the solution curve $y = \frac{x^2}{4}$, which is the particular solution to the differential equation $\frac{dy}{dx} = \frac{x}{2}$, $y(0) = 0$. The slope field simply represents tangents to the solution curve at the points.

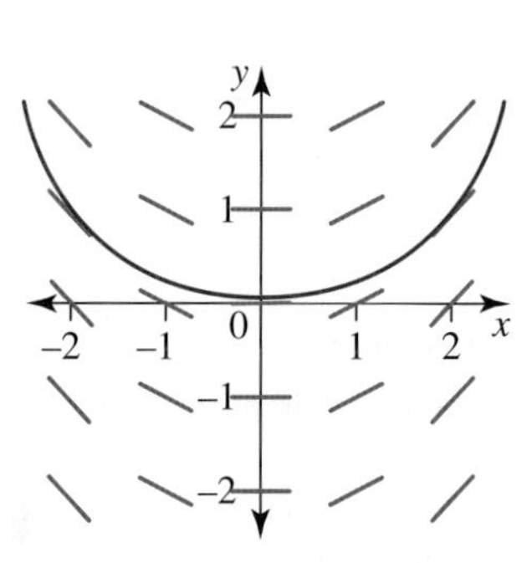

10.8.2 Type 2, $\frac{dy}{dx} = f(y)$

Type 2 first-order differential equations are of the form $\frac{dy}{dx} = f(y)$.

WORKED EXAMPLE 19 Sketching slope fields for Type 2 differential equations

Sketch the slope field for the differential equation $\frac{dy}{dx} = \frac{y}{2}$ for $y = -2, -1, 0, 1, 2$ at each of the values $x = -2, -1, 0, 1, 2$ on a grid.

THINK	WRITE/DRAW
1. Make some observations about the slope.	When $y = 0$ (i.e. along the x-axis), the slope is zero. The x-values do not affect the slope.
2. Determine the slope at appropriate values of y.	When $y = 1$, the slope is $\frac{dy}{dx} = \frac{1}{2}$. A slope of $\frac{1}{2}$ makes an angle of $\tan^{-1}\left(\frac{1}{2}\right) \approx 27°$ with the positive x-axis.

3. Consider another y-value.

When $y = 2$, the slope is $\frac{dy}{dx} = 1$.
A slope of 1 makes an angle of $\tan^{-1}(1) = 45°$ with the positive x-axis.

4. Consider the negative values of y.

When $y = -1$, the slope is $\frac{dy}{dx} = -\frac{1}{2}$.
A slope of $-\frac{1}{2}$ makes an angle of $\tan^{-1}\left(-\frac{1}{2}\right) \approx 153°$ with the positive x-axis.

5. Consider the final y-value.

When $y = -2$, the slope is $\frac{dy}{dx} = -1$.
A slope of -1 makes an angle of $\tan^{-1}(-1) = 135°$ with the positive x-axis.

6. Draw short line segments with the slopes found at each of the points.

The slope field in this case is symmetrical about the x-axis.

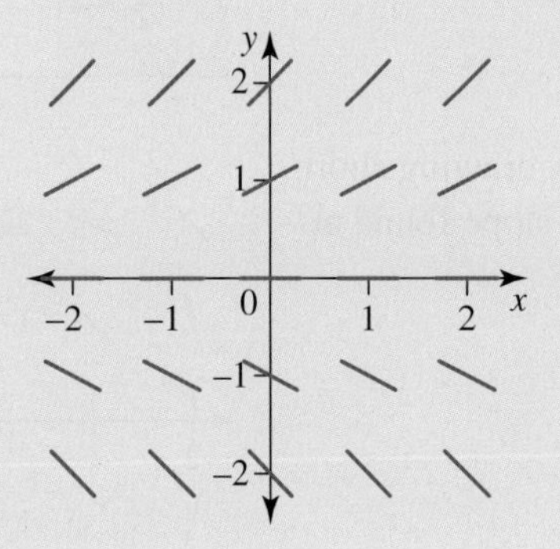

Note that, if we joined the slope field lines in Worked example 19 and drew curves, they would represent the curve $y = y_0 e^{\frac{x}{2}}$, which is the particular solution to the differential equation $\frac{dy}{dx} = \frac{y}{2}, y(0) = y_0$. The two curves shown have $y_0 > 0$ and $y_0 < 0$.

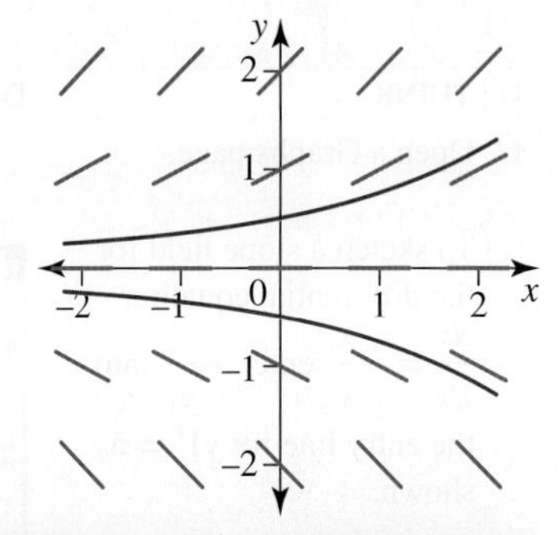

10.8.3 Type 3, $\frac{dy}{dx} = f(x, y)$

Type 3 first-order differential equations are of the form $\frac{dy}{dx} = f(x, y)$.

WORKED EXAMPLE 20 Sketching slope fields for Type 3 differential equations

Sketch the slope field for the differential equation $y\frac{dy}{dx} + x = 0$ for $y = -2, -1, 0, 1, 2$ at each of the values $x = -2, -1, 0, 1, 2$ on a grid.

THINK

WRITE/DRAW

1. Rearrange the differential equation to make $\frac{dy}{dx}$ the subject.

$$y\frac{dy}{dx} + x = 0$$
$$y\frac{dy}{dx} = -x$$
$$\frac{dy}{dx} = -\frac{x}{y}$$

2. We have to evaluate the slope at each of the 25 points on the grid. Rather than substituting point by point, draw up a table.

Substitute $x = 2$ and $y = 2$:
The slope $\frac{dy}{dx} = -\frac{2}{2} = -1$.

3. Complete the values. Note that an undefined slope is one that is parallel to the y-axis.

y \ x	**−2**	**−1**	**0**	**1**	**2**
−2	−1	$-\frac{1}{2}$	0	$\frac{1}{2}$	1
−1	−2	−1	0	1	2
0	Undef.	Undef.	Undef.	Undef.	Undef.
1	2	1	0	−1	−2
2	1	$\frac{1}{2}$	0	$-\frac{1}{2}$	−1

4. Draw the slope field by drawing short line segments with the slope found at each point.

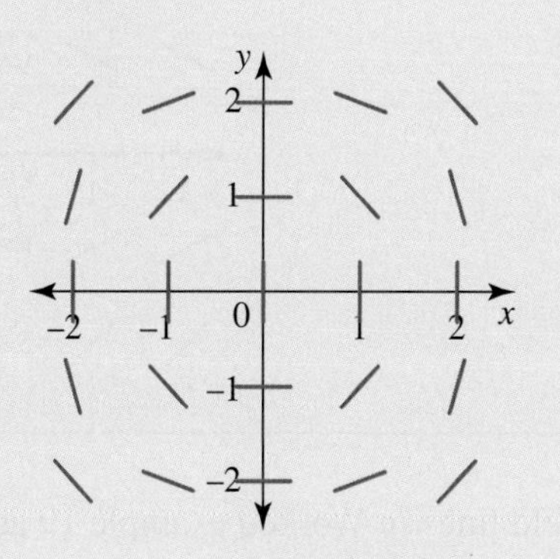

TI | THINK — **DISPLAY/WRITE**

1. Open a Graphs page.

2. To sketch a slope field for the differential equation $\frac{dy}{dx} = -\frac{x}{y}$, enter $-\frac{x}{y_1}$ into the entry line for $y1' =$ as shown.

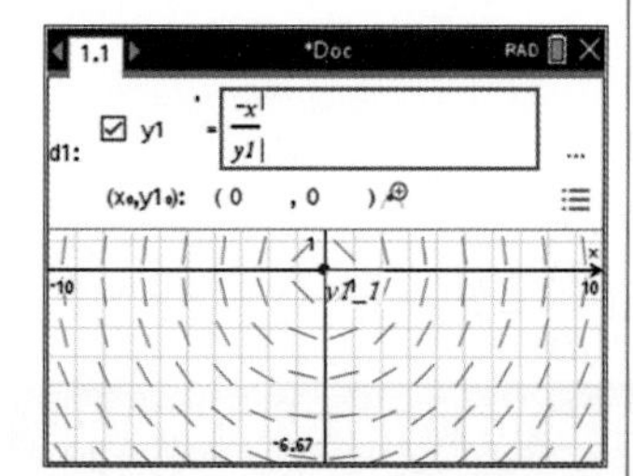

3. Press ENTER and the slope field will appear.

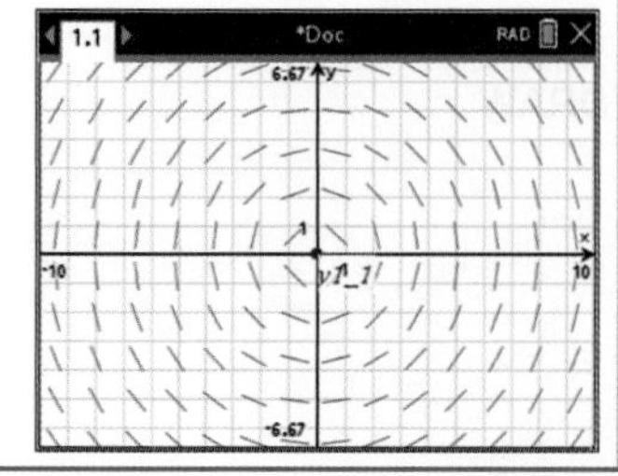

CASIO | THINK — **DISPLAY/WRITE**

1. Open a DiffEq- Graph screen.

2. Complete the entry line as $y' = -\frac{x}{y}$.

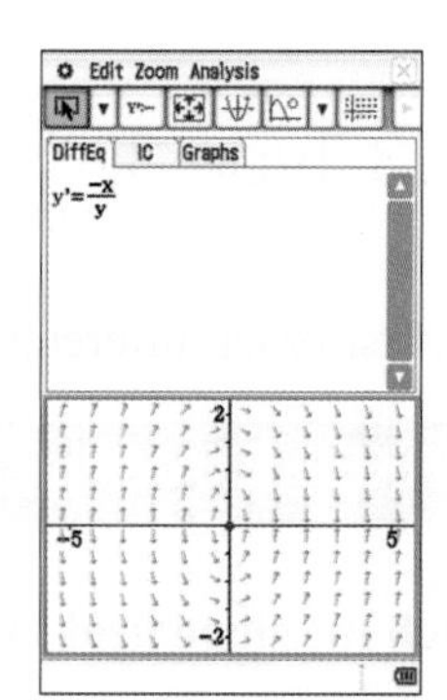

In the answer for Worked example 20, if the line segments were joined, the field would look like a series of circles of varying radii centred at the origin. In fact, if $x^2 + y^2 = r^2$, using implicit differentiation gives $2x + 2y\frac{dy}{dx} = 0$ or $x + y\frac{dy}{dx} = 0$.

10.8.4 Interpreting a slope field

As has been shown in Worked examples 18 and 20, technology can be used to draw slope fields and questions involving slope fields are sometimes designed to analyse the fields. There are four common question types:

- Given a differential equation, choose the correct slope field.
- Given a solution of a differential equation, choose the correct slope field.
- Given a slope field, choose the correct differential equation.
- Given a slope field, choose the solution of the differential equation.

When analysing a slope field, the following approaches are useful:

- Determine the values where the slope is zero; that is, the values of x and y for which $\frac{dy}{dx} = 0$.
- Determine values for which the slope is parallel to the y-axis. In this case, the slope is infinite, and if the differential equation is of the form $\frac{dy}{dx} = \frac{g(x, y)}{h(x, y)}$, then the denominator $h(x, y)$ equals 0.
- Determine the values of the slopes along the x- and y-axes.
- Determine if the slopes are independent of x and therefore depend only on the y-value. In this case, the differential equation is of the form $\frac{dy}{dx} = f(y)$.
- Determine if the slopes are independent of y and therefore depend only on the x-value. In this case, the differential equation is of the form $\frac{dy}{dx} = f(x)$. (Differential equations of only one variable of the types $\frac{dy}{dx} = f(x)$ or $\frac{dy}{dx} = f(y)$ are called autonomous differential equations.)
- Determine where the slopes are positive and where the slopes are negative.
- The symmetry of the slope field and the slopes in each of the four quadrants can give information on the solution.
- Determine if there is no slope field for certain values. This could indicate a required domain.

Note that we are not solving the differential equation when we analyse a slope field. However, if we are given a solution, we can use implicit differentiation to understand the families of curves that are represented.

WORKED EXAMPLE 21 Determining the differential equation from a slope field

The differential equation that best represents the slope field shown is:

A. $\frac{dy}{dx} = \frac{y^2}{x}$

B. $\frac{dy}{dx} = \frac{x^2}{y}$

C. $\frac{dy}{dx} = yx^2$

D. $\frac{dy}{dx} = y^2x$

E. $\frac{dy}{dx} = \frac{y^3}{x}$

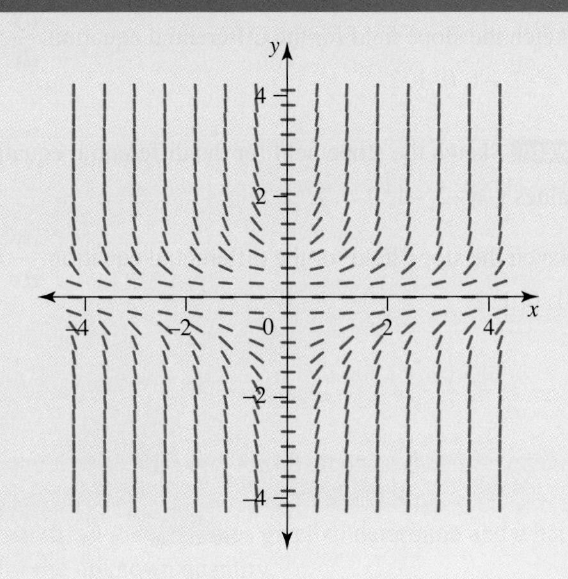

THINK	WRITE
1. Consider the y-axis.	When $x = 0$ (i.e. along the y-axis), the slopes are all zero. Options A and E are incorrect.
2. Consider the first quadrant.	When $x = 1$ and $y = 1$, the slopes are 1. This satisfies B, C and D.
3. Consider the second quadrant.	When $x = -1$ and $y = 1$, the slopes are -1. Option C is incorrect.
4. Consider the third quadrant.	When $x = -1$ and $y = -1$, the slopes are 1. Options B and D are still valid.
5. Consider the fourth quadrant.	When $x = 1$ and $y = -1$, the slopes are 1. Option B is incorrect.
6. State the result.	Option D, $\frac{dy}{dx} = y^2x$, is the only differential equation that is represented by this slope field.

10.8 Exercise

Technology free

1. WE18 Sketch the slope field for the differential equation $\frac{dy}{dx} = \frac{2}{x}$ for y = −2, −1, 0, 1, 2 at each of the values x = −2, −1, 0, 1, 2.
2. Sketch the slope field for the differential equation $\frac{dy}{dx} = \frac{1}{2\sqrt{x}}$ for y = −2, −1, 0, 1, 2 at each of the values x = −2, −1, 0, 1, 2.
3. WE19 Sketch the slope field for the differential equation $\frac{dy}{dx} = \frac{2}{y}$ for y = −2, −1, 0, 1, 2 at each of the values x = −2, −1, 0, 1, 2.
4. Sketch the slope field for the differential equation $\frac{dy}{dx} = 2\sqrt{y}$ for y = −2, −1, 0, 1, 2 at each of the values x = −2, −1, 0, 1, 2.
5. WE20 Sketch the slope field for the differential equation $y\frac{dy}{dx} - x = 0$ for y = −2, −1, 0, 1, 2 at each of the values x = −2, −1, 0, 1, 2.
6. Sketch the slope field for the differential equation $\frac{dy}{dx} = xy$ for y = −2, −1, 0, 1, 2 at each of the values x = −2, −1, 0, 1, 2.

Technology active

7. 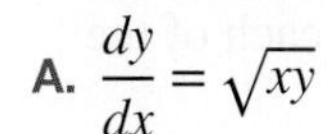 WE21 MC The differential equation that best represents the slope field shown is:

A. $\frac{dy}{dx} = \sqrt{xy}$

B. $\frac{dy}{dx} = x\sqrt{y}$

C. $\frac{dy}{dx} = y\sqrt{x}$

D. $\frac{dy}{dx} = \frac{1}{\sqrt{xy}}$

E. $\frac{dy}{dx} = x^2\sqrt{y}$

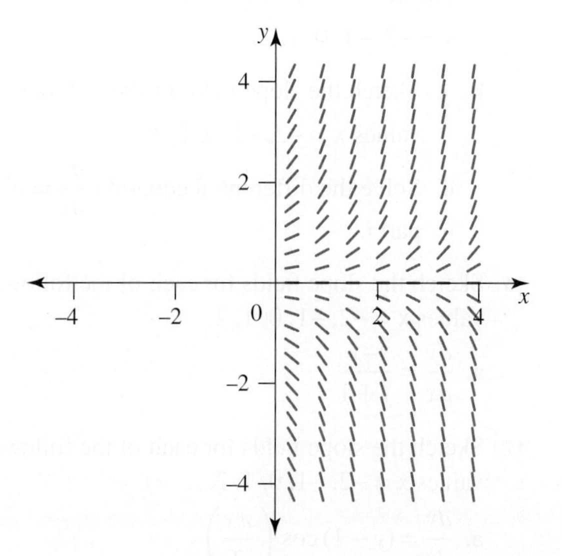

8. MC The differential equation that best represents the slope field shown is:

A. $\frac{dy}{dx} = \cos(x + y)$

B. $\frac{dy}{dx} = \cos(x - y)$

C. $\frac{dy}{dx} = \sin(x - y)$

D. $\frac{dy}{dx} = \sin(x + y)$

E. $\frac{dy}{dx} = \tan(x - y)$

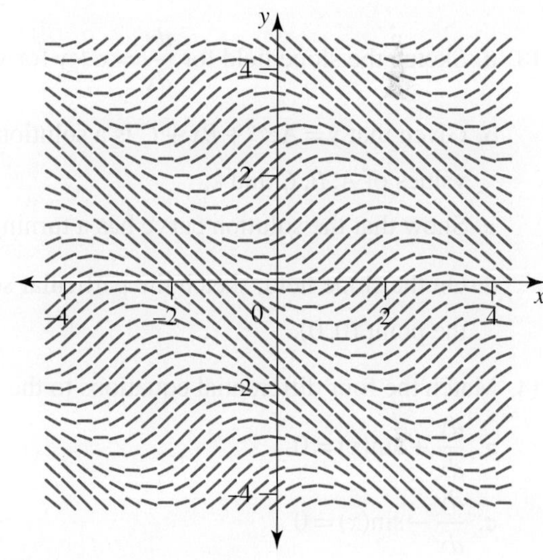

9. a. Sketch the slope field for the differential equation $\frac{dy}{dx} = \frac{2}{(x-1)^2}$ for $y = -2, -1, 0, 1, 2$ at each of the values $x = -2, -1, 0, 1, 2$.

b. i. Sketch the slope field for the differential equation $\frac{dy}{dx} = \frac{1}{x^2+1}$ for $y = -2, -1, 0, 1, 2$ at each of the values $x = -2, -1, 0, 1, 2$.

ii. Solve the differential equation $\frac{dy}{dx} = \frac{1}{x^2+1}$, $y(0) = 0$ and sketch the solution curve on the diagram for part i.

10. a. Sketch the slope field for the differential equation $\frac{dy}{dx} = y^2$ for y = −2, −1, 0, 1, 2 at each of the values x = −2, −1, 0, 1, 2.

b. i. Sketch the slope field for the differential equation $\frac{dy}{dx} = y^2 + 1$ for y = −2, −1, 0, 1, 2 at each of the values x = −2, −1, 0, 1, 2.

ii. Solve the differential equation $\frac{dy}{dx} = y^2 + 1, y(0) = 0$ and sketch the solution curve on the diagram for part i.

11. Sketch the slope fields for each of the following differential equations for y = −2, −1, 0, 1, 2 at each of the values x = −2, −1, 0, 1, 2.

a. $\frac{dy}{dx} = \frac{-x}{y+1}$

b. $\frac{dy}{dx} = \frac{x-2}{y}$

12. Sketch the slope fields for each of the following differential equations for y = −2, −1, 0, 1, 2 at each of the values x = −2, −1, 0, 1, 2.

a. $\frac{dy}{dx} = (y-1)\cos\left(\frac{\pi x}{2}\right)$

b. $\frac{dy}{dx} = y^2 \sin\left(\frac{\pi x}{2}\right)$

13. a. Sketch the slope field for $\frac{dy}{dx} = x + y$ for y = −2, −1, 0, 1, 2 at each of the values x = −2, −1, 0, 1, 2.

b. Given that $y = Ae^{kx} + Bx + C$ is a solution of the initial value problem $\frac{dy}{dx} = x + y, y(0) = 0$, determine the values of A, B, C and k.

c. Show that the solution curve has a turning point at $(0, 0)$.

d. On the slope field, sketch the particular solution to the differential equation $\frac{dy}{dx} = x + y$ that passes through the point $(0, 0)$.

14. Match the four differential equations to the four slope fields shown.

a. $\frac{dy}{dx} + \sin(x) = 0$

b. $\frac{dy}{dx} + \cos(x) = 0$

c. $\frac{dy}{dx} - \sin(x) = 0$

d. $\frac{dy}{dx} - \cos(x) = 0$

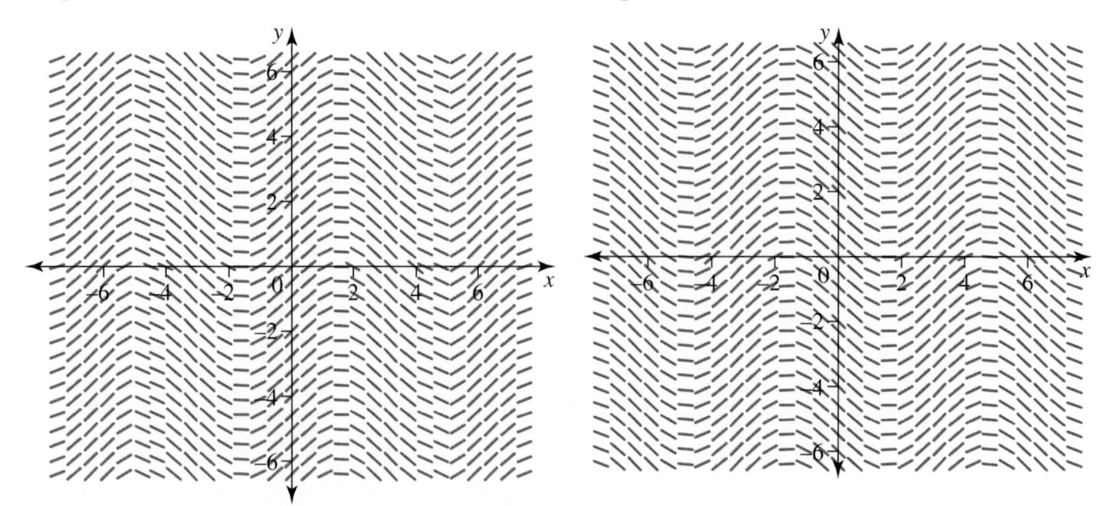

Slope field 3 Slope field 4

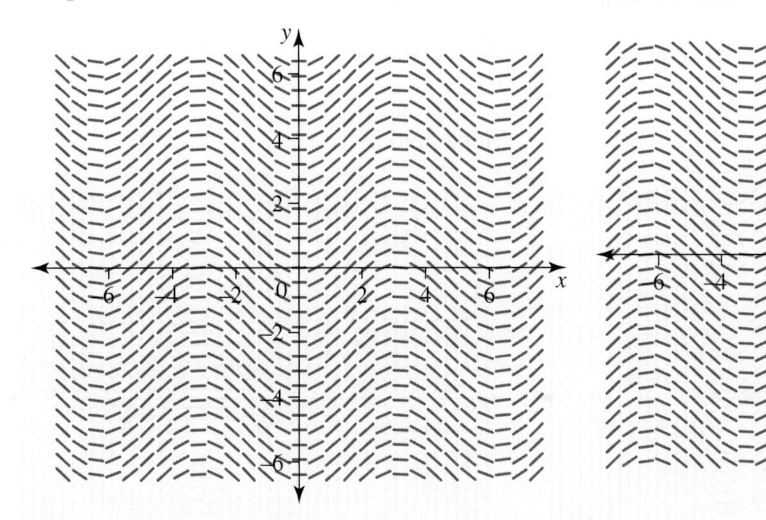

15. Match the four differential equations to the four slope fields shown.

a. $\dfrac{dy}{dx} = \sin(x^2)$ b. $\dfrac{dy}{dx} = \cos(x^2)$ c. $\dfrac{dy}{dx} = \sin(y^2)$ d. $\dfrac{dy}{dx} = \cos(y^2)$

Slope field 1 Slope field 2

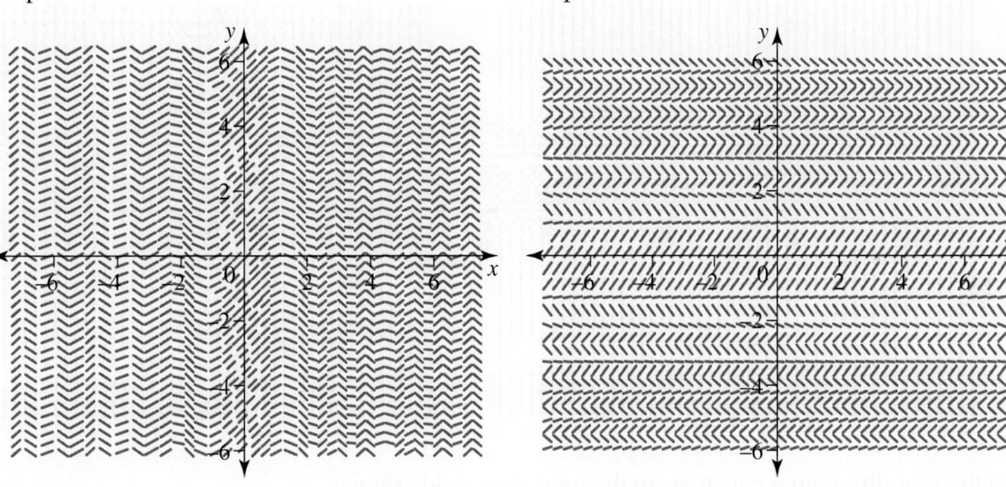

Slope field 3 Slope field 4

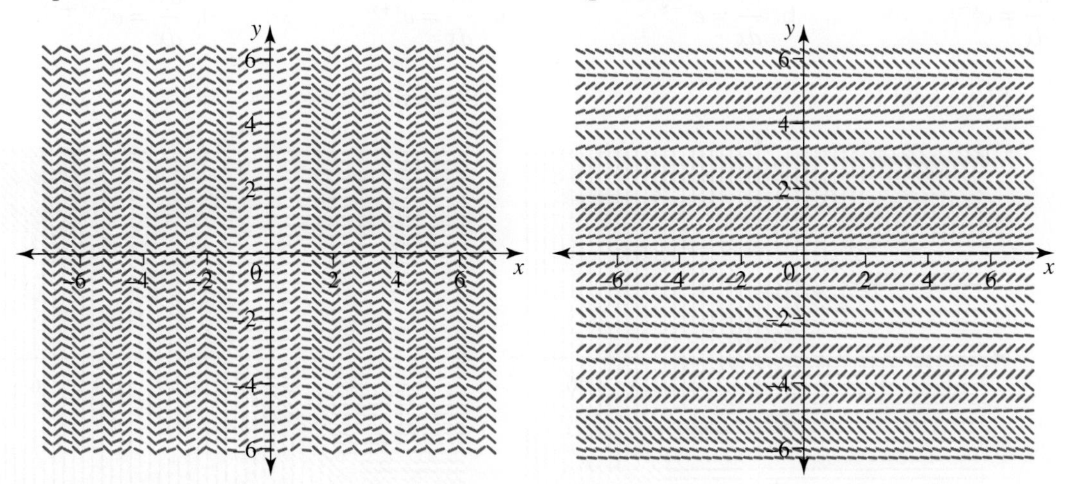

16. Match the four differential equations to the four slope fields shown.

a. $\dfrac{dy}{dx} = x(x - 2)$ b. $\dfrac{dy}{dx} = y(y - 2)$ c. $\dfrac{dy}{dx} = y(x - 2)$ d. $\dfrac{dy}{dx} = x(y - 2)$

Slope field 1

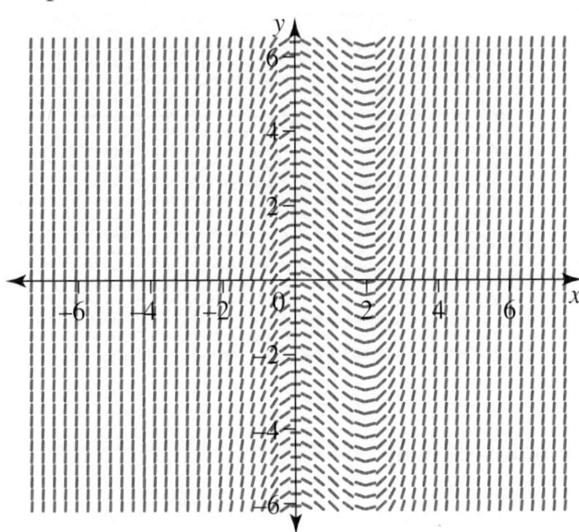

Slope field 2

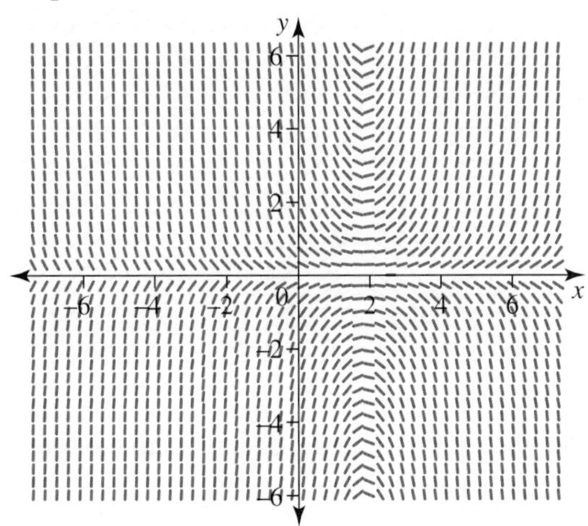

Slope field 3

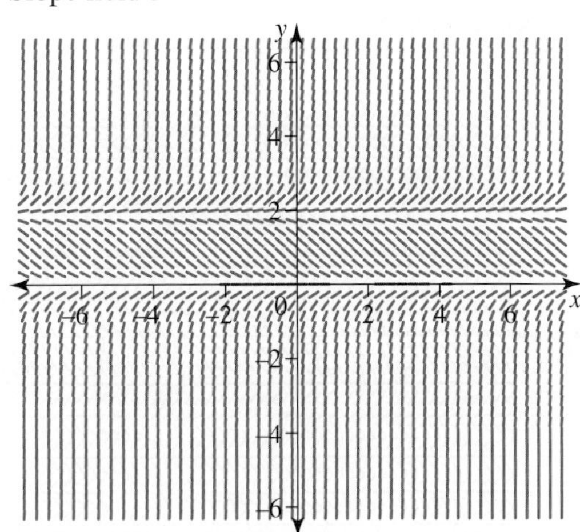

Slope field 4

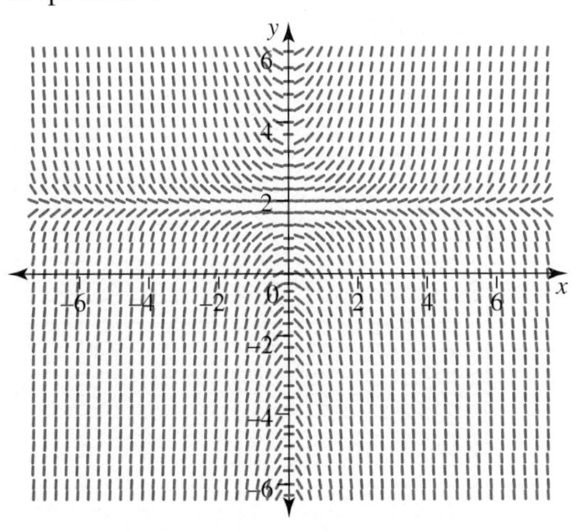

17. Match the four differential equations to the four slope fields shown.

a. $\dfrac{dy}{dx} = e^{x-y}$ b. $\dfrac{dy}{dx} = e^{y-x}$ c. $\dfrac{dy}{dx} = e^{x+y}$ d. $\dfrac{dy}{dx} = e^{-(x+y)}$

Slope field 1

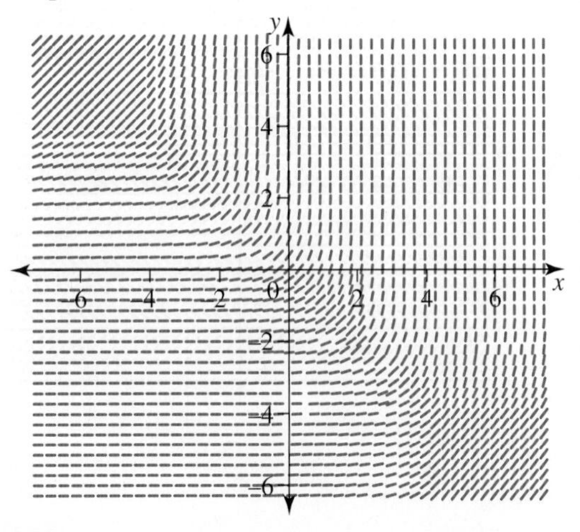

Slope field 2

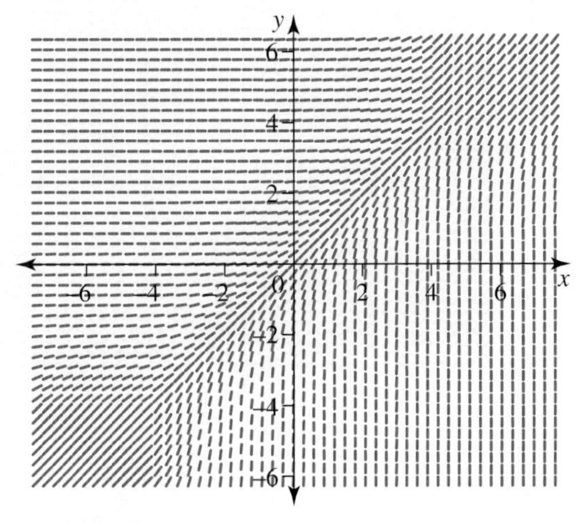

Slope field 3

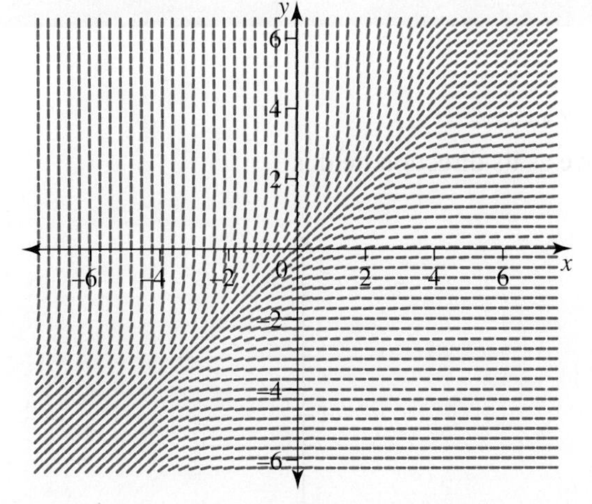

Slope field 4

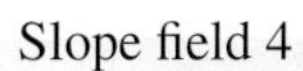

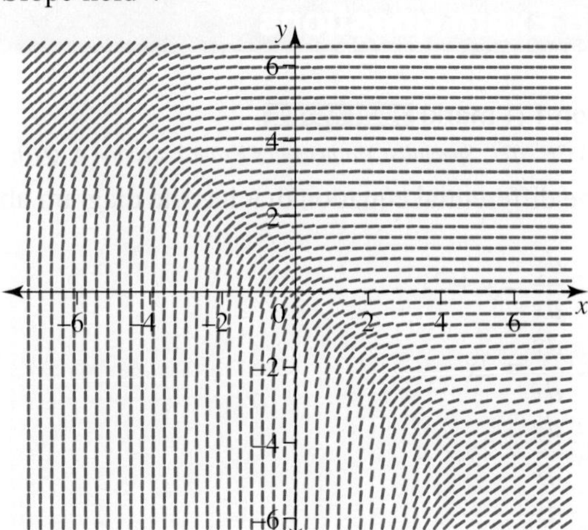

18. Match the four differential equations to the four slope fields shown.

a. $\dfrac{dy}{dx} = \sqrt{\dfrac{y}{x}}$ **b.** $\dfrac{dy}{dx} = \sqrt{\dfrac{x}{y}}$ **c.** $\dfrac{dy}{dx} = y\sqrt{x}$ **d.** $\dfrac{dy}{dx} = x\sqrt{y}$

Slope field 1

Slope field 2

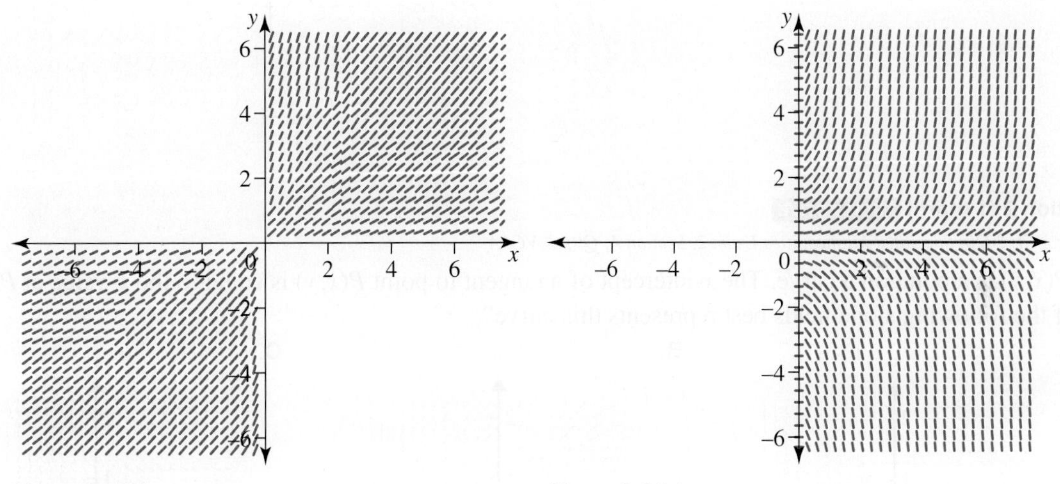

Slope field 3

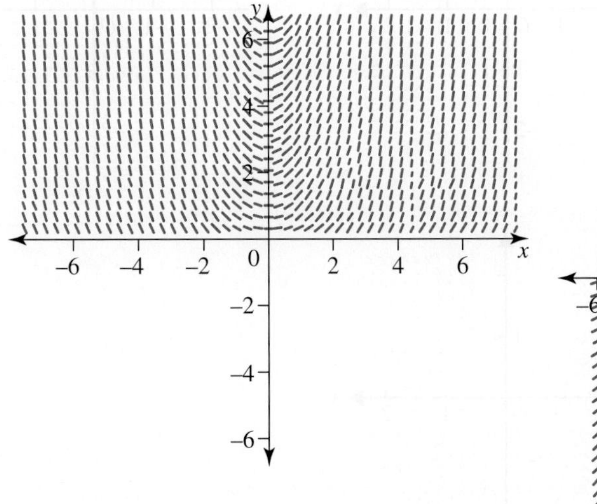

Slope field 4

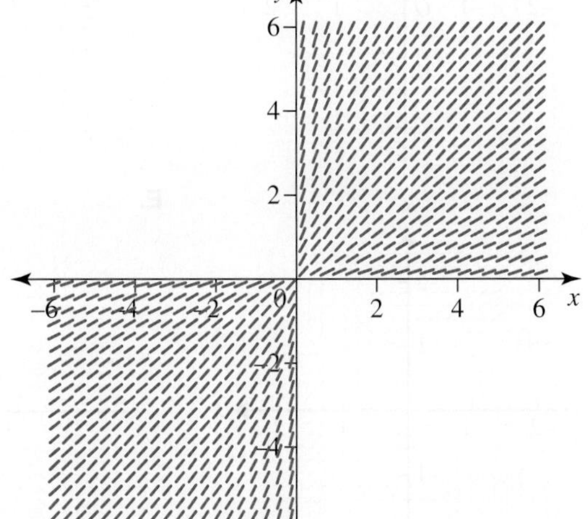

10.8 Exam questions

Question 1 (1 mark) TECH-ACTIVE

Source: *VCE 2021, Specialist Mathematics Exam 2, Section A, Q10; © VCAA.*

MC The differential equation that has the diagram above as its direction field is

A. $\frac{dy}{dx} = y + 2x$

B. $\frac{dy}{dx} = 2x - y$

C. $\frac{dy}{dx} = 2y - x$

D. $\frac{dy}{dx} = y - 2x$

E. $\frac{dy}{dx} = x + 2y$

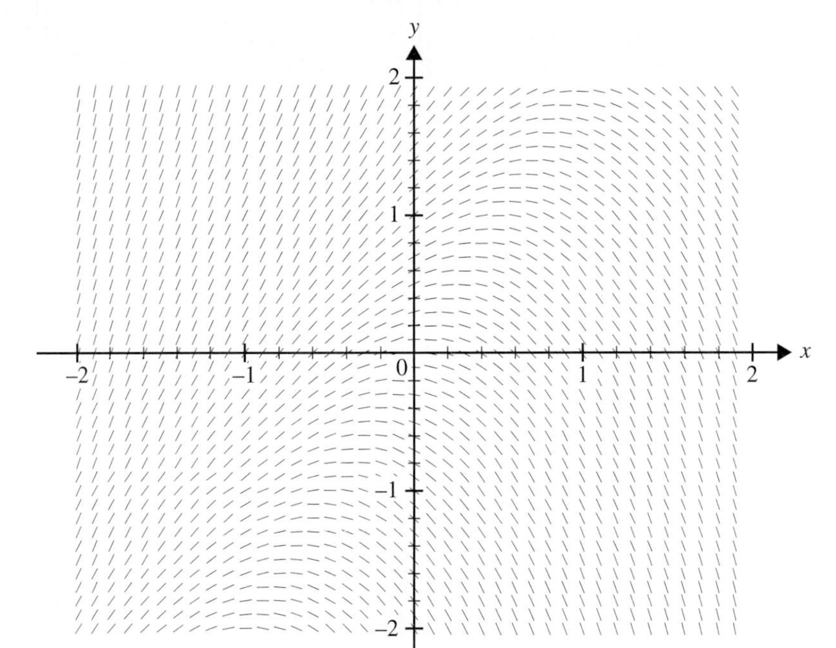

Question 2 (1 mark) TECH-ACTIVE

Source: *VCE 2020, Specialist Mathematics Exam 2, Section A, Q9; © VCAA.*

MC $P(x, y)$ is a point on a curve. The x-intercept of a tangent to point $P(x, y)$ is equal to the y-value at P. Which one of the following slope fields best represents this curve?

A.

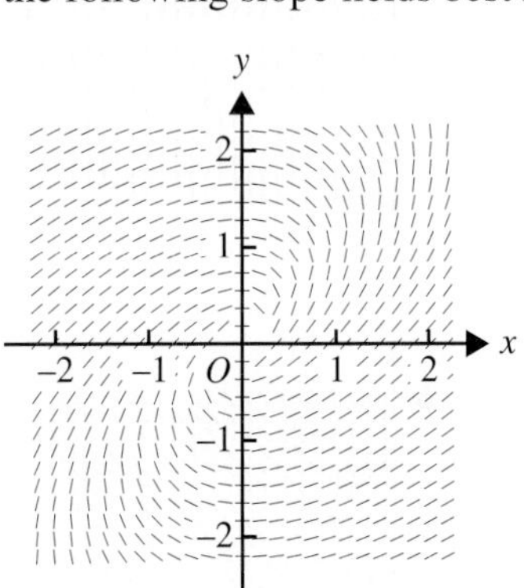

B.

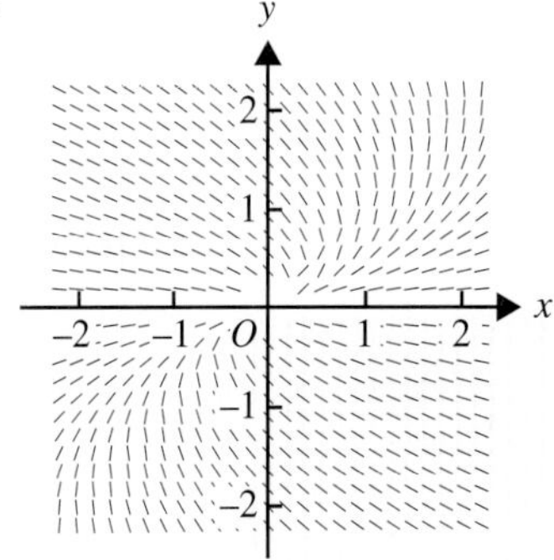

C.

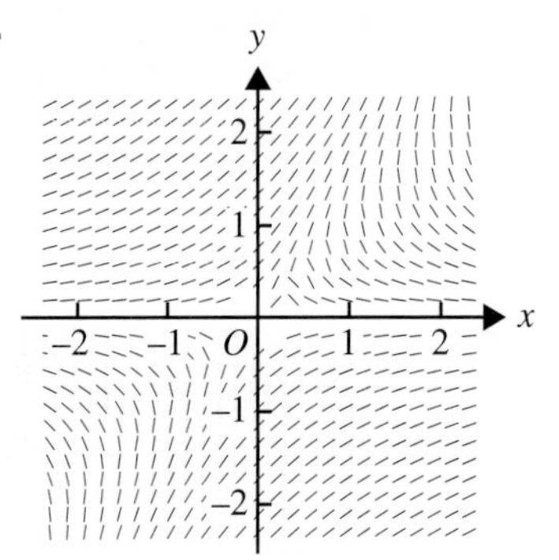

D.

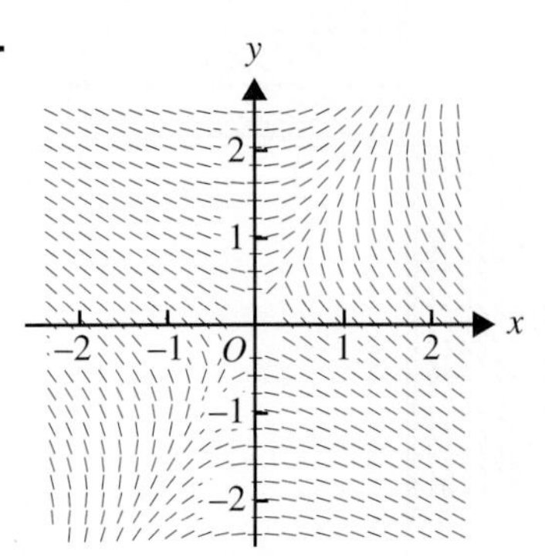

E.

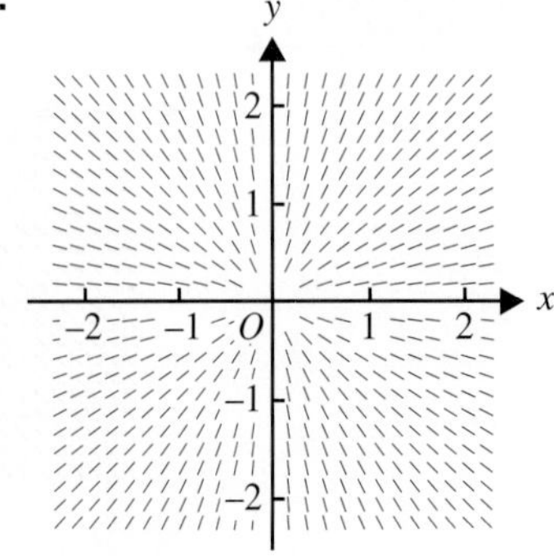

Question 3 (1 mark) TECH-ACTIVE

Source: *VCE 2019, Specialist Mathematics Exam 2, Section A, Q9; © VCAA.*

MC The differential equation that has the diagram above as its direction field is

A. $\frac{dy}{dx} = \sin(y - x)$

B. $\frac{dy}{dx} = \cos(y - x)$

C. $\frac{dy}{dx} = \sin(x - y)$

D. $\frac{dy}{dx} = \frac{1}{\cos(y - x)}$

E. $\frac{dy}{dx} = \frac{1}{\sin(y - x)}$

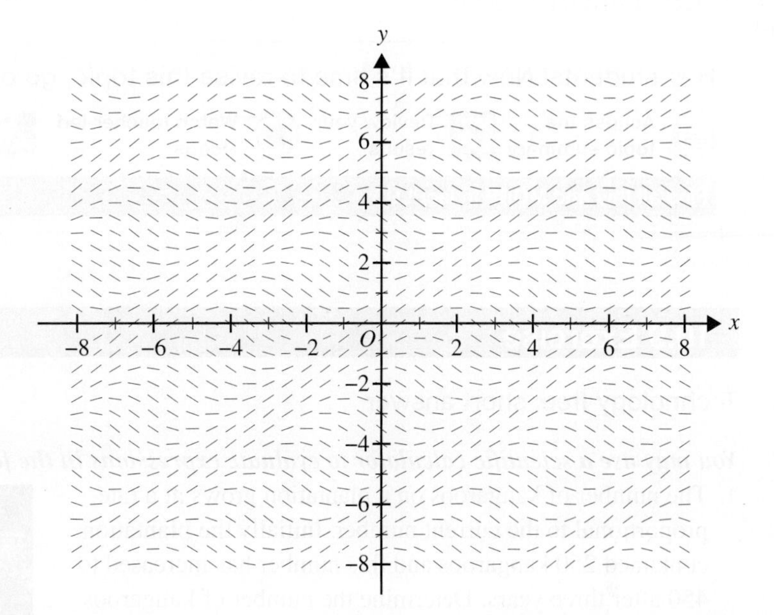

More exam questions are available online.

10.9 Review

10.9.1 Summary

doc-37064

Hey students! Now that it's time to revise this topic, go online to:

Access the topic summary  Review your results  Watch teacher-led videos Practise exam questions

Find all this and MORE in jacPLUS

10.9 Exercise

Technology free: short answer

You may use a scientific calculator to evaluate expressions in the following questions.

1. The number of kangaroos on a plantation grows at a rate proportional to the current number. Initially the plantation contained 250 kangaroos and this number has increased to 450 after three years. Determine the number of kangaroos on the plantation after a further five years.

2. A condenser of capacitance K Farads is initially charged to a potential of v_0 volts, and is discharged through a resistor of R ohms. The potential v at any time t seconds after discharging commences satisfies the differential equation $\frac{dv}{dt} + \frac{v}{KR} = 0$.
   a. Solve the equation to determine the potential v at any time t
   b. Calculate the time when the potential is half its initial value.
3. A new iPhone is being released. It is estimated that 70% of the population will purchase it and the sales growth will follow a logistic growth pattern. Sales records show that 20% of the population will purchase it within 4 weeks of its release. Determine how many weeks elapse before 50% of the population have the new iPhone. Write your answer in weeks correct to 1 decimal place.
4. Although most forms of gold are stable, there is one form of gold Au-195 that is synthetically produced which decays into a form of platinum, with a half-life of 186.1 days. Determine the percentage of gold Au-195 present after 90 days, correct to 1 decimal place.

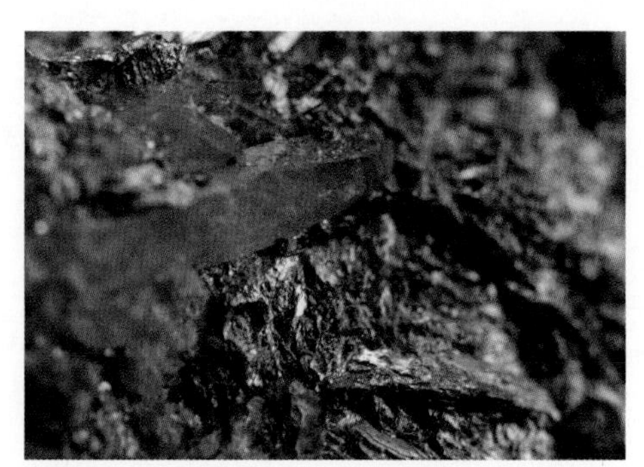

5. When light passes through a medium it loses its intensity as it penetrates depths. If I is the intensity of light at depth x, the rate of loss of intensity with respect to the depth is proportional to the intensity at that depth, so that $\frac{dI}{dx} = -kI$, where k is a positive constant. Solve the differential equation to show that $I = I_0 e^{-kx}$.

6. A gas BBQ is used for cooking and is outdoors where the surrounding temperature is constant at 15°C. When the gas is ignited the BBQ heats up at a rate 25°C per minute, but also cools according to Newton's law of cooling. When the temperature of the BBQ grill is 75°C, the temperature of the BBQ is increasing at a rate of 45°C per minute.
   a. Show that differential equation representing the temperature T°C of the grill on the BBQ, at a time t after the gas has been ignited, satisfies $\frac{dT}{dt} = \frac{60+T}{3}$.
   b. Solve the differential equation.
   c. Determine how long before the grill of the BBQ reaches a suitable cooking temperature of 185°C.

Technology active: multiple choice

7. **MC** The number of rabbits in a certain area increases at a rate proportional to the number present. If the number of rabbits doubles in four months, determine how many months elapse before the number of rabbits triples.

   A. $\frac{\log_e(81)}{\log_e(2)}$ B. $\frac{\log_e(3)}{\log_e(16)}$ C. $\log_e\left(\frac{81}{2}\right)$ D. $\frac{\log_e(3)}{\log_e(4)}$ E. $\frac{\log_e(9)}{\log_e(4)}$

8. **MC** A radioactive substance has a half-life of 5 years. After 3 years

   A. 66% has disintegrated B. 60% has disintegrated C. 40% has disintegrated
   D. 38% has disintegrated E. 34% has disintegrated

9. **MC** When light passes through a medium it loses its intensity as it penetrates depths. The rate of loss of intensity with respect to the depth is proportional to the intensity at that depth. If 5% of light is lost in penetrating 40% of a glass slab, then the percentage lost penetrating the whole slab is closest to

   A. 12% B. 12.5% C. 87.5% D. 88% E. 95%

10. **MC** A container initially holds 10 litres of water. A salt solution containing 5 kg per litre is poured into the tank at a rate of 6 litres per minute and the solution is kept uniform by stirring. If the mixture flows out at a rate of 4 litres per minute and Q kg is the amount of salt in the tank at time t minutes, then Q satisfies

    A. $\frac{dQ}{dt} = 30 - \frac{4Q}{10-2t}$
    B. $\frac{dQ}{dt} = 30 - \frac{4Q}{10+2t}$
    C. $\frac{dQ}{dt} = 12 - \frac{6Q}{10-2t}$
    D. $\frac{dQ}{dt} = 12 - \frac{6Q}{10+2t}$
    E. $\frac{dQ}{dt} = 12 + \frac{6Q}{10+2t}$

11. **MC** A cake is taken out of an oven, which has been kept at a constant temperature of 180°C, and placed on a table where the temperature is 18°C. According to Newton's law of cooling, the temperature T°C, of the cake at a time t minutes after it has been placed on the table, satisfies the differential equation.

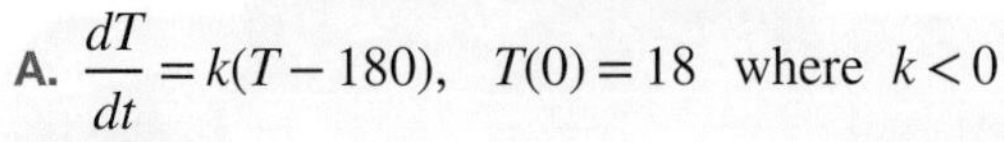

A. $\frac{dT}{dt} = k(T - 180), \quad T(0) = 18 \quad \text{where } k < 0$

B. $\frac{dT}{dt} = k(T - 180), \quad T(0) = 18 \quad \text{where } k > 0$

C. $\frac{dT}{dt} = k(T - 18), \quad T(0) = 180 \quad \text{where } k < 0$

D. $\frac{dT}{dt} = k(T - 18), \quad T(0) = 180 \quad \text{where } k > 0$

E. $\frac{dT}{dt} = k(T - 162), \quad T(0) = 18 \quad \text{where } k > 0$

12. **MC** Let $W(t)$ represent the number of wolves in a certain area at a time t months. Initially there are 200 wolves present and the number of wolves is increasing at a rate directly proportional to the product of the number of wolves and to $200 - W(t)$. The wolves are also being killed off at 20 per month. If k is a positive constant, then the differential equation for the number of wolves is best represented by

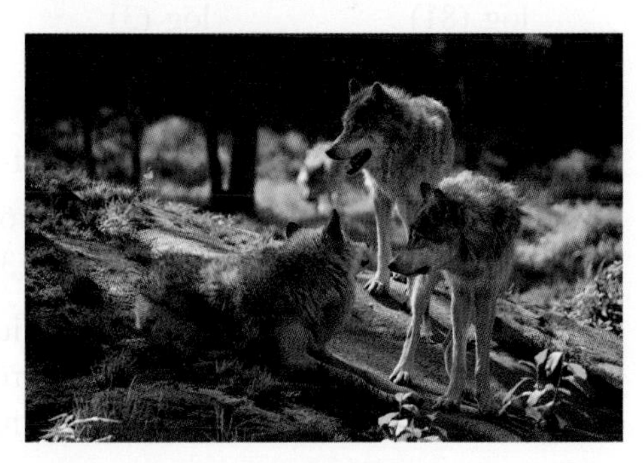

A. $\frac{dW}{dt} = k(200 - W) - 20, \quad W(0) = 200$

B. $\frac{dW}{dt} = kW(200 - W) - 20, \quad W(0) = 200$

C. $\frac{dW}{dt} = kW\,t(W - 200), \quad W(0) = 200$

D. $\frac{dW}{dt} = kW(200 - W) - 20t, \quad W(0) = 20$

E. $\frac{dW}{dt} = k(200 - W) - 20\,t, \quad W(0) = 20$

13. **MC** A logistic equation the solution $N(t) = \frac{200}{1 + 99e^{-kt}}$ where $k > 0$, then

A. $\frac{dN}{dt} = kN(200 - N), \; N(0) = 2$

B. $\frac{dN}{dt} = kN(N - 200), \; N(0) = 2$

C. $\frac{dN}{dt} = kN\left(1 - \frac{N}{200}\right), \; N(0) = 2$

D. $\frac{dN}{dt} = kN\left(\frac{N}{200} - 1\right), \; N(0) = 2$

E. $\frac{dN}{dt} = kN(200 - N), \; N(0) = 20$

14. **MC** A logistic equation the solution $N(t) = \frac{500}{1 + 9e^{-kt}}$ where N is the population number of a city in millions at a time t years and k is a positive real constant. The population is increasing most rapidly when the population is equal to

A. 10 B. 25 C. 50 D. 250 E. 500

15. **MC** Euler's method, with a step size of 0.5, is used to approximate the differential equation $\frac{dy}{dx} + \frac{1}{9^x} = 0$, with initial condition $y_0 = 1$ when $x_0 = 0$. The value obtained for y_2 is closest to

A. 1.66 B. 1.50 C. 0.60 D. 0.50 E. 0.33

16. **MC** The slope field for a certain differential equation is shown. The differential equation could be

A. $\frac{dy}{dx} = -\left(\frac{x+2}{y-2}\right)$

B. $\frac{dy}{dx} = \frac{y+2}{x-2}$

C. $\frac{dy}{dx} = \frac{x-2}{y+2}$

D. $\frac{dy}{dx} = \frac{x+2}{y-2}$

E. $\frac{dy}{dx} = \frac{y-2}{x+2}$

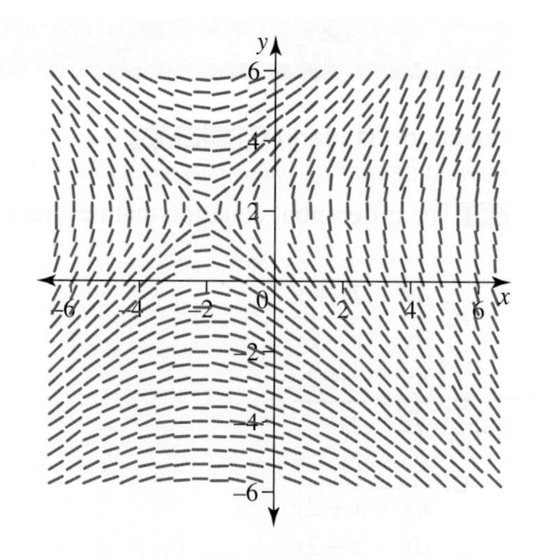

Technology active: extended response

17. A tank contains 20 litres of water. A salt solution of concentration 2 grams per litre is poured into the tank at a rate of 6 litres per minute, and the well-stirred mixture flows out at a rate of 4 litres per minute.
   a. Set up the differential equation for Q the amount of salt in grams in the tank after a time t minutes.
   b. Verify, by differentiation that $Q = \frac{4t(t^2 + 30t + 300)}{(t+10)^2}$ is solution of the differential equation.

18. A metal bar has been heated to a temperature of 400°C and is placed in a room at which the temperature is 18°C. After 10 minutes the bar has a temperature of 250°C.
   Calculate how much longer before the temperature of the bar has fallen to 120°C. Write your answer in minutes correct to 1 decimal place.

19. Ice-cream is taken out of a freezer at a temperature of −20°C and is placed in a warm room. If after one minute the temperature of the ice-cream is −16°C and after a further minute its temperature is $-12\frac{1}{3}$°C. Calculate:
   a. the room temperature, correct to the nearest degree
   b. the temperature of the ice-cream after a further three minutes, correct to 2 decimal places.

20. a. Sketch the slope field for the differential equation $\frac{dy}{dx} = \frac{y}{2}(4-y)$ for $y = 0, 1, 2, 3, 4$ at each of the values of $x = -3, -2, -1, 0, 1, 2, 3$ on a grid.
   b. Use Euler's method to approximate the solution to the differential equation $\frac{dy}{dx} = \frac{y}{2}(4-y)$, $y(0) = 1$, when $x = 1$ using a step size of $\frac{1}{3}$, correct to 4 decimal places.
   c. Solve the differential equation $\frac{dy}{dx} = \frac{y}{2}(4-y)$, $y(0) = 1$, using integral calculus.
   d. Using differential calculus verify that $y = \frac{4}{1+3e^{-2x}}$ is a solution of the initial value problem $\frac{dy}{dx} = \frac{y}{2}(4-y)$, $y(0) = 1$.
   e. Show that $\frac{d^2y}{dx^2} = \frac{y}{2}(2-y)(4-y)$ and hence determine the coordinates of the point of inflection on the graph $y = \frac{4}{1+3e^{-2x}}$.
   f. Sketch the solution curve for the differential equation $\frac{dy}{dx} = \frac{y}{2}(4-y)$, $y(0) = 1$ on the diagram in part **a**.

10.9 Exam questions

Question 1 (1 mark) TECH-ACTIVE

Source: *VCE 2018, Specialist Mathematics Exam 2, Section A, Q10; © VCAA.*

MC The differential equation that best represents the direction field shown is

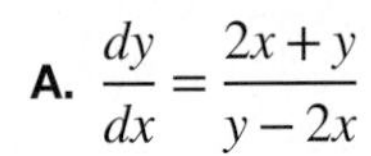

A. $\frac{dy}{dx} = \frac{2x+y}{y-2x}$

B. $\frac{dy}{dx} = \frac{x+2y}{2x-y}$

C. $\frac{dy}{dx} = \frac{2x-y}{x+2y}$

D. $\frac{dy}{dx} = \frac{x-2y}{y-2x}$

E. $\frac{dy}{dx} = \frac{2x+y}{2y-x}$

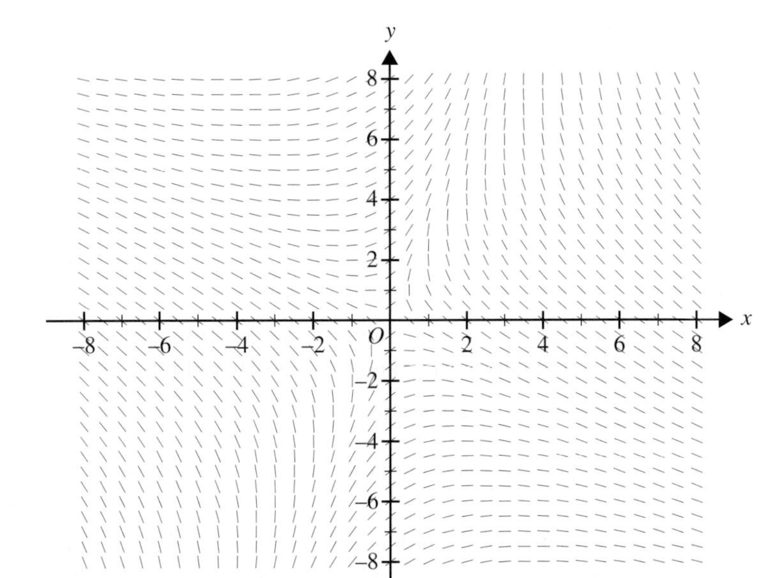

Question 2 (1 mark) TECH-ACTIVE

Source: *VCE 2016, Specialist Mathematics Exam 2, Section A, Q9; © VCAA.*

MC If $f(x) = \frac{dy}{dx} = 2x^2 - x$, where $y_0 = 0 = y(2)$, then y_3 using Euler's formula with step size 0.1 is

A. $0.1f(2)$
B. $0.6 + 0.1f(2.1)$
C. $1.272 + 0.1f(2.2)$
D. $2.02 + 0.1f(2.3)$
E. $2.02 + 0.1f(2.2)$

Question 3 (11 marks) TECH-ACTIVE

Source: *VCE 2013, Specialist Mathematics Exam 2, Section 2, Q3; © VCAA.*

The number of mobile phones, N, owned in a certain community after t years, may be modelled by $\log_e(N) = 6 - 3e^{-0.4t}$, $t \geq 0$.

a. Verify by substitution that $\log_e(N) = 6 - 3e^{-0.4t}$ satisfies the differential equation

$\frac{1}{N}\frac{dN}{dt} + 0.4\log_e(N) - 2.4 = 0.$ **(2 marks)**

b. Find the initial number of phones owned in the community. Give your answer correct to the nearest integer. **(1 mark)**

c. Using this mathematical model, find the limiting number of mobile phones that would eventually be owned in the community. Give your answer correct to the nearest integer. **(2 marks)**

The differential equation in part **a.** can also be written in the form $\frac{dN}{dt} = 0.4N(6 - \log_e(N))$.

d. i. Find $\frac{d^2N}{dt^2}$ in terms of N and $\log_e(N)$. **(2 marks)**

ii. The graph of N as a function of t has a point of inflection.
Find the values of the coordinates of this point.
Give the value of t correct to one decimal place and the value of N correct to the nearest integer. **(2 marks)**

e. Sketch the graph of N as a function of t on the axes below for $0 \le t \le 15$. **(2 marks)**

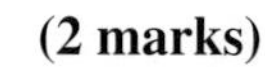

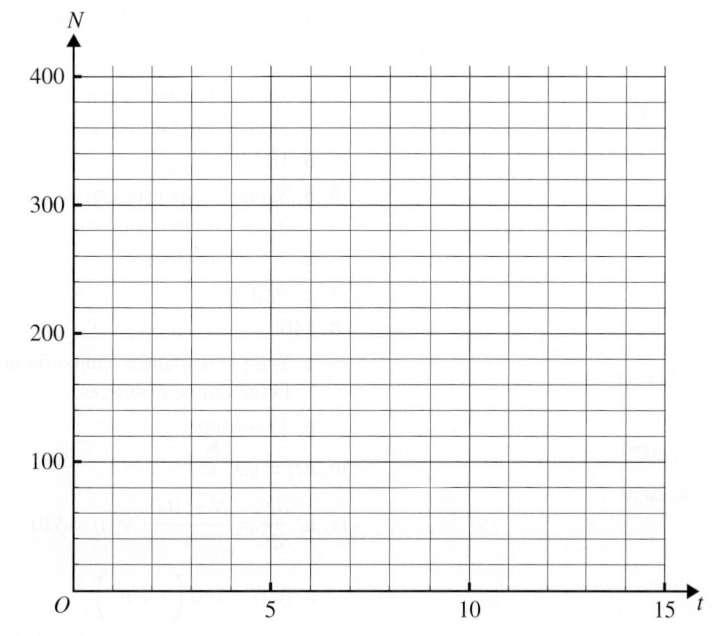

Question 4 (4 marks) TECH-ACTIVE

A lake can withstand a maximum of 50 ducks. Initially there are 4 ducks in the lake. The number of ducks in the lake grows at a rate proportional to the difference between the maximum number and the current number of ducks in the lake. After 4 months there are 16 ducks in the lake.

a. Write down and solve the differential equation for the number of ducks n in the lake after t months. **(3 marks)**

b. Calculate how many ducks are in the lake after 8 months. **(1 mark)**

Question 5 (3 marks) TECH-ACTIVE

The table shows the population in millions of Canada.

Year	2010	2015	2020
Population	34.148	36.027	37.742

Assuming the population follows a logistic growth rate:

a. determine the maximum population of Canada, correct to the nearest thousand **(2 marks)**

b. calculate the population of Canada in 2000, correct to the nearest thousand. **(1 mark)**

More exam questions are available online.

Answers

Topic 10 Applications of first-order differential equations

10.2 Growth and decay

10.2 Exercise

1. a. $N = 1.79 \times 10^6 e^{0.0277t}$
   b. 2.71 million
   c. 2052
2. 137
3. a. 4146 b. 11 275
4. 942
5. a. 27 851 607 b. 2029
6. a. 8 220 578 135 b. 2039
7. 58.31 mg
8. 113.78 mg
9. 12.8 years
10. 65%
11. a. 0.29 g b. 25.59 years
12. 32.47 days
13. 38.2 years
14. 78.8%
15. 11.76%
16. 6.89 hours
17. a. Sample responses can be found in the worked solutions in the online resources.
    b. 2041
    c. 2050
18. a. $Q(t) = Q_0 e^{-k(t - t_0)}$
    b. 0.000121
    c. 10 356 years

10.2 Exam questions

Note: Mark allocations are available with the fully worked solutions online.

1. a. i. Sample responses can be found in the worked solutions in the online resources.
   ii. $a > b$ and $r > s > 0$ or $a < b$ and $s > r > 0$.
   b. i. $\int e^Q \, dQ = \int e^t \, dt$
   ii. Sample responses can be found in the worked solutions in the online resources.
   iii. Sample responses can be found in the worked solutions in the online resources.
2. 24 109 years
3. D

10.3 Other applications of first-order differential equations

10.3 Exercise

1. \$9048.37
2. \$630 000
3. a. $P(h) = P_0 e^{kh}, k > 0$ b. 50.94 cm
4. a. $D(t) = D_0 e^{-kt}$ b. 2.08 hours
5. a. Sample responses can be found in the worked solutions in the online resources.
   b. 18.5%
6. a. Sample responses can be found in the worked solutions in the online resources.
   b. 205.31 newtons
7. 25 million
8. 140
9. a. Sample responses can be found in the worked solutions in the online resources.
   b. 35.48 ohms
10. $i(t) = i_0 e^{-\frac{Rt}{L}}$
11. a. $\frac{dN}{dt} = \frac{N - 160}{4}, N(0) = 320$
    b. $N(t) = 160\left(1 + e^{\frac{t}{4}}\right)$
    c. 1342
12. Sample responses can be found in the worked solutions in the online resources.
13. a. $\frac{dN}{dt} = \frac{N - 100}{20}, N(0) = 200$
    b. $N(t) = 100\left(1 + e^{\frac{t}{20}}\right)$
    c. 216
14. 666 052
15. a. 11 years b. 19
16. a. $N(t) = \left(\frac{1}{2}kt + \sqrt{N_0}\right)^2$
    b. $N(t) = \sqrt{2kt + N_0^2}$
    c. $N(t) = \frac{4N_0}{\left(2 - kt\sqrt{N_0}\right)^2}$
    d. $N(t) = \frac{N_0}{1 - ktN_0}$
17. Sample responses can be found in the worked solutions in the online resources.
18. Sample responses can be found in the worked solutions in the online resources.

10.3 Exam questions

Note: Mark allocations are available with the fully worked solutions online.

1. a. 1000 units b. 1.5 seconds
2. $Q(t) = Q_0 e^{-\frac{t}{RC}}$
3. 12.0%

10.4 Bounded growth and Newton's law of cooling

10.4 Exercise

1. 165.4 g
2. 13 months
3. a. 4.4 kg b. 3.06 kg
4. a. 736 b. 35.1 months
5. 12.75 kg
6. 26.34 min
7. 1.75 min
8. a. 5.43°C b. 26°C
9. 5.27 min
10. 108.37°C
11. $v(t) = E\left(1 - e^{-\frac{t}{RC}}\right)$
12. Sample responses can be found in the worked solutions in the online resources.
13. a. Sample responses can be found in the worked solutions in the online resources.
    b. 9:28 pm
14. a. $t_1 = \frac{3}{2}, T_h = 95$
    b. $t_2 = \frac{3}{4}$, $T_c = 3$
    c. 1.12
15. a. Sample responses can be found in the worked solutions in the online resources.
    b. Monthly repayments: \$2146.52, total interest: \$215 164.60
    c. \$17 308.20
    d. 11.8 years
    e. 10.56%

10.4 Exam questions

Note: Mark allocations are available with the fully worked solutions online.

1. a. Sample responses can be found in the worked solutions in the online resources.
   b. 65°C
2. a. $i(t) = \frac{E}{R}\left(1 - e^{-\frac{Rt}{L}}\right)$
   b. Sample responses can be found in the worked solutions in the online resources.
3. 5°C

10.5 Chemical reactions and dilution problems

10.5 Exercise

1. a. $\frac{dQ}{dt} = 3 - \frac{3Q}{200 - t}, Q(0) = 20$
   b. Sample responses can be found in the worked solutions in the online resources.
   c. $-\frac{7}{200\,000}$
   d. 1.15 kg/L
2. a. $\frac{dQ}{dt} = 6b - \frac{3Q}{50 + 3t}$ b. 2
3. a. $\frac{dQ}{dt} = 4 - \frac{2Q}{30 - t}, Q(0) = 0$
   b. Sample responses can be found in the worked solutions in the online resources.
4. a. $\frac{dQ}{dt} = 6 - \frac{Q}{20 + t}, Q(0) = 0$
   b. Sample responses can be found in the worked solutions in the online resources.
5. a. $\frac{dQ}{dt} = 8 - \frac{3Q}{20 - t}, Q(0) = 0$
   b. Sample responses can be found in the worked solutions in the online resources.
6. a. $\frac{dQ}{dt} = 20 - \frac{3Q}{64 + 2t}, Q(0) = 128$
   b. Sample responses can be found in the worked solutions in the online resources.
   c. -2^{16}
   d. 3.34 g/L
7. 3
8. a. $\frac{dQ}{dt} = \frac{3(80 - Q)}{20}, Q(0) = 5$
   b. $Q(t) = 5\left(16 - 15e^{-\frac{3t}{20}}\right)$
   c. 1.05 g/L
   d. Sample responses can be found in the worked solutions in the online resources.
   e. 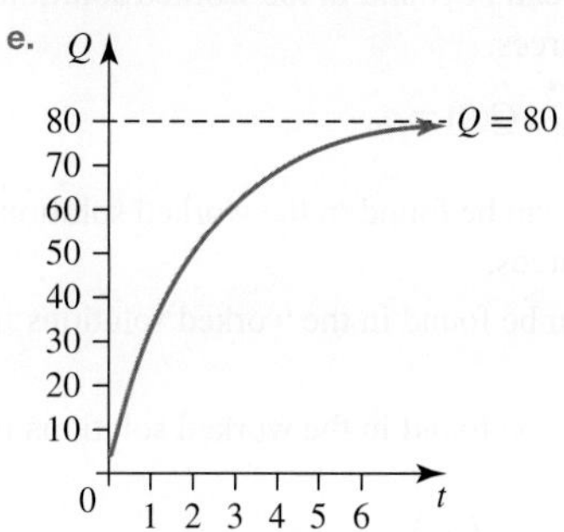

9. a. $\frac{dQ}{dt} = \frac{4(75 - Q)}{15}, Q(0) = 0$
   b. $Q(t) = 75\left(1 - e^{-\frac{4t}{15}}\right), t \to \infty, Q \to 75$
   c. 6.04 min
10. a. $30e^{-\frac{t}{120}}$ b. 11.04 kg c. 83.18 min
11. 1.659 g/L
12. a. i. 9.12 min
    ii. 34.66 min
    b. 5.5 min
13. a. $\frac{dx}{dt} = k(2 - x)(6 - x)$, $x(0) = 0$
    b. $x(t) = \frac{6(1 - e^{-4kt})}{3 - e^{-4kt}}$ where $k = \frac{1}{12}\log_e\left(\frac{5}{3}\right)$
    c. $\frac{16}{11}$g

14. a. $\frac{dx}{dt} = k(8-x)^2, x(0) = 0$

b. $x(t) = \frac{24t}{3t+10}$

c. 8 minutes (for 10 minutes total after the start of the reaction)

d. 8 g

15. a. $\frac{dx}{dt} = k(4-x)(8-x),\ x(0) = 0$

b. $x(t) = \frac{8(1-e^{-4kt})}{2-e^{-4kt}}$ where $k = \frac{1}{8}\log_e\left(\frac{7}{6}\right)$

c. $\frac{52}{31}$ g

d. 8 g

16. a. 6.57 minutes

b. $\frac{26}{9}$ g

c. 4 g

17. a. Sample responses can be found in the worked solutions in the online resources.

b. 3.61 g

18. a. Sample responses can be found in the worked solutions in the online resources.

b. 3.85 g

c. 10 g

19. a. $\frac{dQ}{dt} = \frac{f}{V_0}(bV_0 - Q), Q(0) = 0$

b. Sample responses can be found in the worked solutions in the online resources.

20. a. $\frac{dQ}{dt} = \frac{f}{V_0}(bV_0 - Q), Q(0) = q_0$

b. Sample responses can be found in the worked solutions in the online resources.

21. Sample responses can be found in the worked solutions in the online resources.

22. Sample responses can be found in the worked solutions in the online resources.

23. a. $\frac{dQ}{dt} + \frac{Q}{100} = 4 + 2\sin\left(\frac{t}{6}\right),\ Q(0) = 2$

b. $Q(t) = \frac{1}{2509}\left(1800\sin\left(\frac{t}{66}\right) - 30\,000\cos\left(\frac{t}{6}\right) - 968582e^{\frac{t}{100}}\right)$

c. 1.3 g/L

24. a. $30e^{-\frac{t}{2}} - \frac{Q}{5+t},\ Q(0) = 0$

b. $Q(t) = \frac{420}{t+5} - \frac{60(t+7)e^{-\frac{t}{2}}}{t+5}$

c. 0.82 g/L

d. 0.02 g/L

e. 53.54 g

10.5 Exam questions

Note: Mark allocations are available with the fully worked solutions online.

1. a. Sample responses can be found in the worked solutions in the online resources.

b. $Q = \frac{32}{(16+2t)^{\frac{3}{2}}}$

2. A

3. a. $\frac{dQ}{dt} = 12 - \frac{5Q}{50-t}$

b. Sample responses can be found in the worked solutions in the online resources.

10.6 The logistic equation

10.6 Exercise

1. a. $N(t) = \frac{100}{1+24e^{-0.8959t}}$

b. 78

c. 3.55 weeks

2. a. $N(t) = \frac{80}{1+39e^{-0.855t}}$

b. 72

c. 4.76 days

3. a. 5.236 million

b. 2033

4. a. 2018

b. 2.197 million

5. a, b. Sample responses can be found in the worked solutions in the online resources.

c. $\left(\frac{1}{3}\log_e(9), 250\right)$

d.

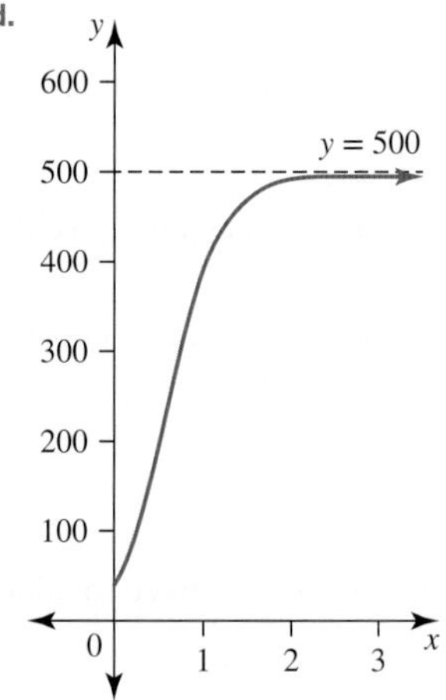

6. a, b. Sample responses can be found in the worked solutions in the online resources.

c. $\left(\frac{1}{2}\log_e(199), 200\right)$

d. 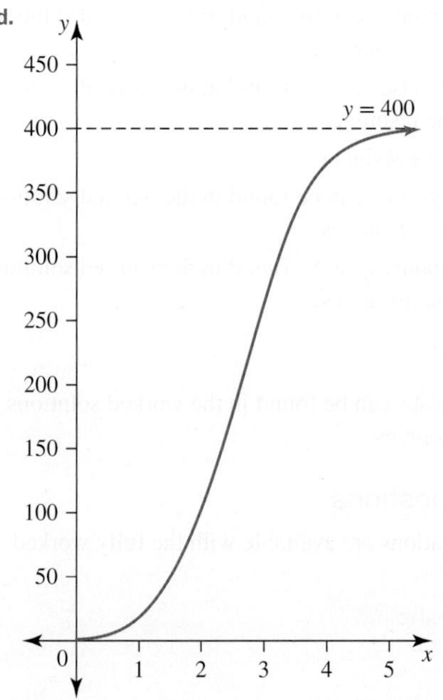

7. a, b. Sample responses can be found in the worked solutions in the online resources.

c. $(3 \log_e(99), 300)$

d. 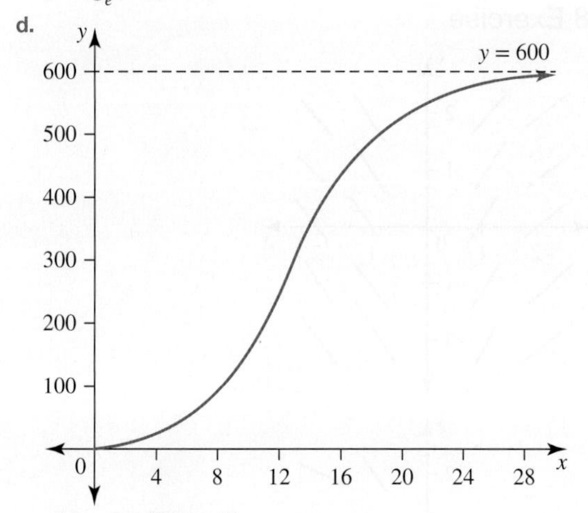

8. a. $\dfrac{dN}{dt} = \dfrac{N(4000 - N)}{50\,000}, N(0) = 500$

b. $N(t) = \dfrac{4000}{1 + 7e^{-\frac{2t}{5}}}$

c. 2054

d. 7.61 years

9. a. $\dfrac{dN}{dt} = kN(360 - N), N(0) = 1$

b. 4.9 days

10. a. $\dfrac{dN}{dt} = kN(2000 - N), N(0) = 2$

b. 4.4 days

c. 4.4 days

11. 7374

12. a. 25

b. 3.55 days

c. 22

d. 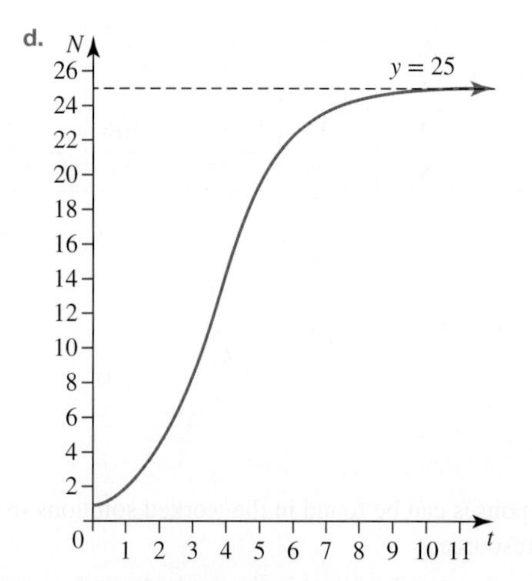

10.6 Exam questions

Note: Mark allocations are available with the fully worked solutions online.

1. $\dfrac{dN}{dt} = kN\left(1 - \dfrac{N}{200}\right), N(0) = 2$

2. 250

3. Sample responses can be found in the worked solutions in the online resources.

10.7 Euler's method

10.7 Exercise

1. a.

x	4	4.25	4.5	4.75	5
y	1	2.5	4.0462	5.6372	7.2717

b. 7.3607 underestimates by 1.2%

2. Approximated value $y_2 = 3.9689$; exact value of $y(1) = 3.9177$

3. Approximated value $y_3 = 3.5836$; exact value of $y(1) = 3.6633$

4. Approximated value $y_4 = 4.1975$; exact value of $y(1) = 3.7293$

5. a.

x	1	$\frac{3}{2}$	2
y	2	-1	$-\frac{11}{2}$

b.

x	1	$\frac{4}{3}$	$\frac{5}{3}$	2
y	2	0	$-\frac{8}{3}$	-6

c.

x	1	$\frac{5}{4}$	$\frac{3}{2}$	$\frac{7}{4}$	2
y	2	$\frac{1}{2}$	$-\frac{11}{8}$	$-\frac{29}{8}$	$-\frac{25}{4}$

d.

x	1	$\frac{6}{5}$	$\frac{7}{5}$	$\frac{8}{5}$	$\frac{9}{5}$	2
y	2	$\frac{4}{5}$	$-\frac{16}{25}$	$-\frac{58}{25}$	$-\frac{106}{25}$	$-\frac{32}{5}$

6. $\frac{8}{9}$

7. 1

8. $\frac{101}{210}$

9. $\frac{7}{16}$

10. Sample responses can be found in the worked solutions in the online resources.

11. Sample responses can be found in the worked solutions in the online resources.

12. a. Sample responses can be found in the worked solutions in the online resources.

b. $\frac{11}{2}\log_e(11) - \frac{9}{2}\log_e(9) + 3$

13. a.

x	2	2.25	2.5	2.75	3
y	3	3.1667	3.3246	3.4750	3.6188

b. 3.6056, overestimates by 0.4%

14.

x	0.2	0.3	0.4	0.5
y	0.4	0.4423	0.4896	0.5429

15. a. $\frac{49}{9}$ b. $\frac{4000}{729}$

16. a. $\frac{7}{2}$ b. $\frac{7945}{2187}$

17. a.

x	0	0.25	0.5	0.75	1
y	2	2	1.9688	1.7265	1.0977

b. 1 overestimates by 9.8%

18. a. 1.7208

b. 1.8059, underestimates by 4.7%.

19.

x	1	1.25	1.5	1.75	2
y	2	1	0.6875	0.5103	0.3963

20. a. 44.8696

b. 54.4038

21. a. 4.9840

b. 4.9851

22. a.

x	0	0.25	0.5	0.75	1
y	2	2	2.25	2.6875	3.2656

b. $a = 4, b = -2, c = 4, k = -1; y(1) = 3.4715$

23. a.

x	0	0.25	0.5	0.75	1
y	1	1.75	3.1875	5.6563	9.6719

b. $a = -\frac{5}{2}, b = -\frac{7}{4}, c = \frac{11}{4}, k = 2; y(1) = 16.0699$

24. a. Sample responses can be found in the worked solutions in the online resources.

b. Sample responses can be found in the worked solutions in the online resources.

c. $2 - 4\cos(1) + 4\sin(1)$

25. a. Sample responses can be found in the worked solutions in the online resources.

b. Sample responses can be found in the worked solutions in the online resources.

c. $\frac{1}{9}\left(37 - 4e^{-3}\right)$

26. Sample responses can be found in the worked solutions in the online resources.

10.7 Exam questions

Note: Mark allocations are available with the fully worked solutions online.

1. C
2. C
3. B

10.8 Slope fields

10.8 Exercise

1.

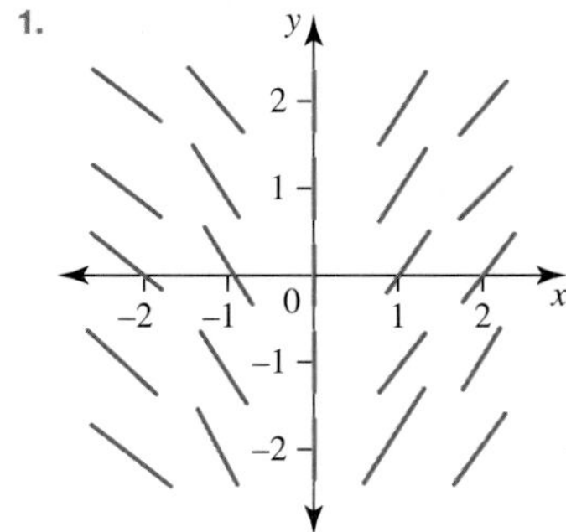

2.

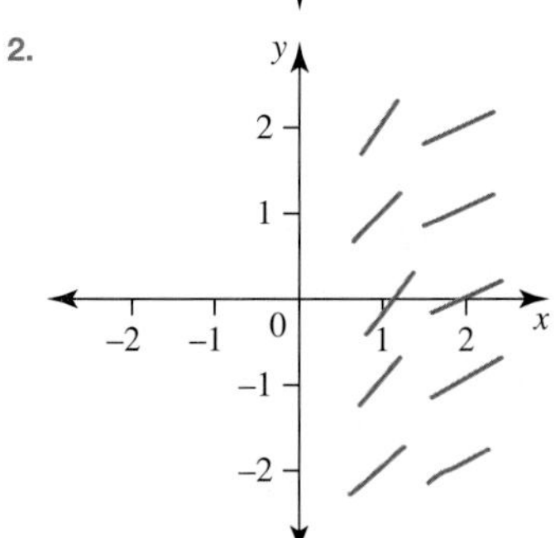

3.

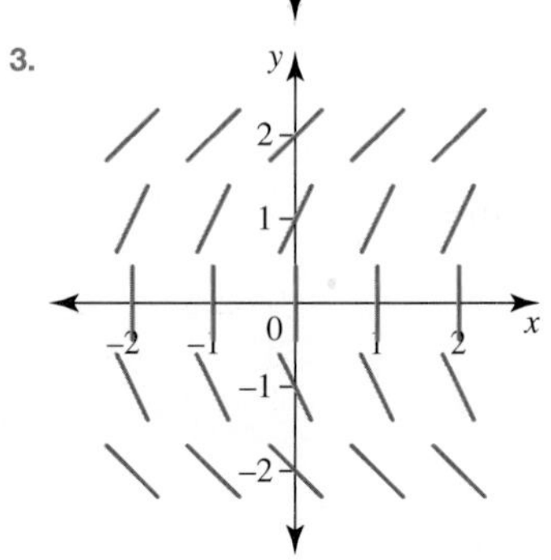

4.

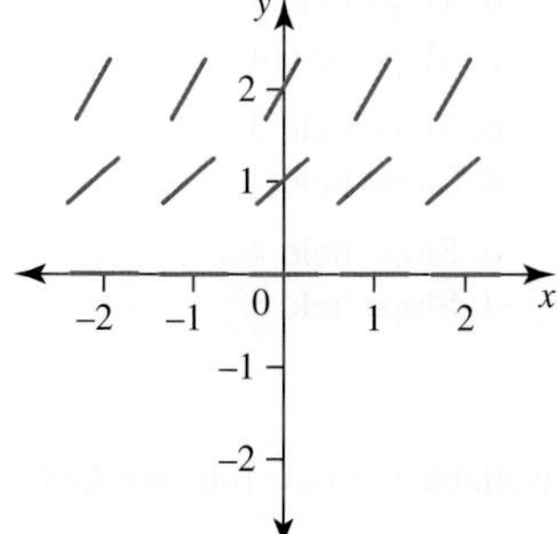

5.

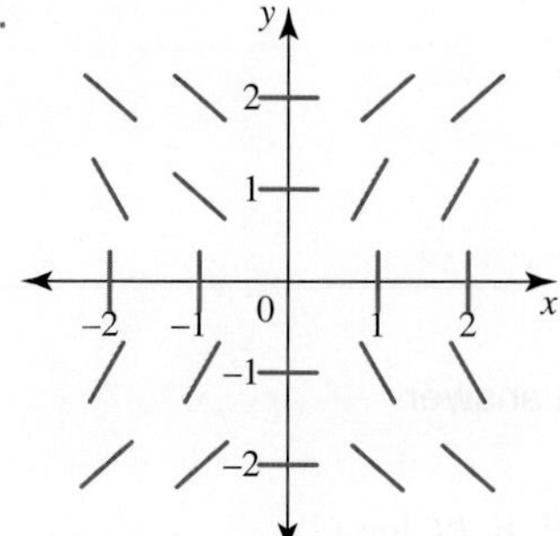

6.

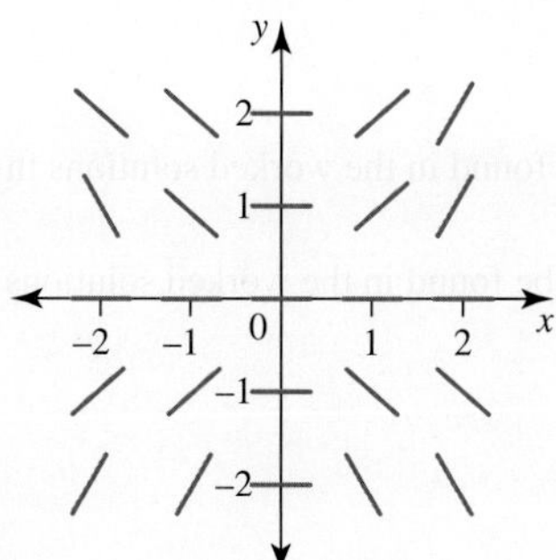

7. C

8. D

9. a.

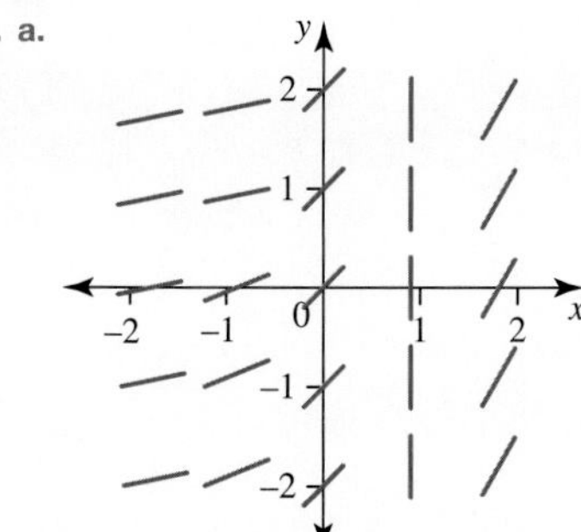

b. i.

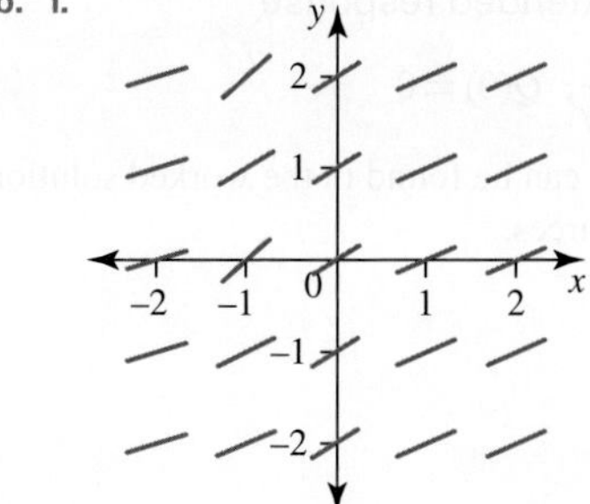

ii.

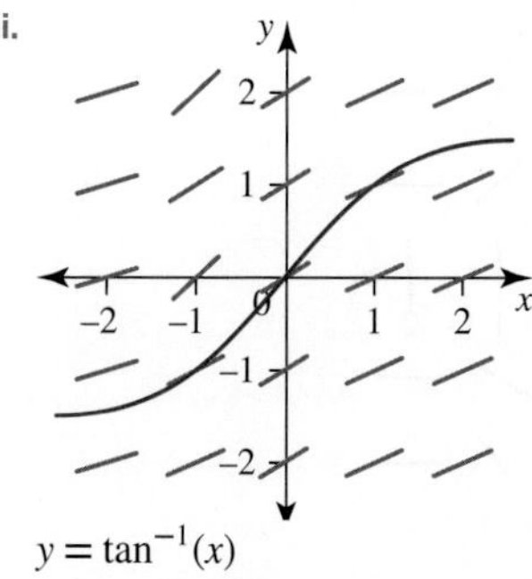

$y = \tan^{-1}(x)$

10. a.

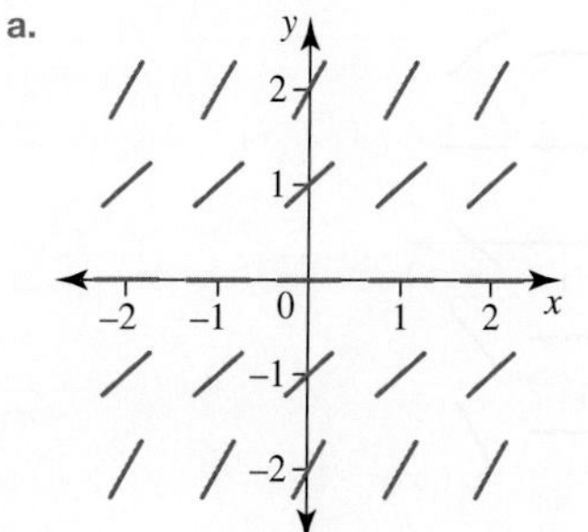

b. i.

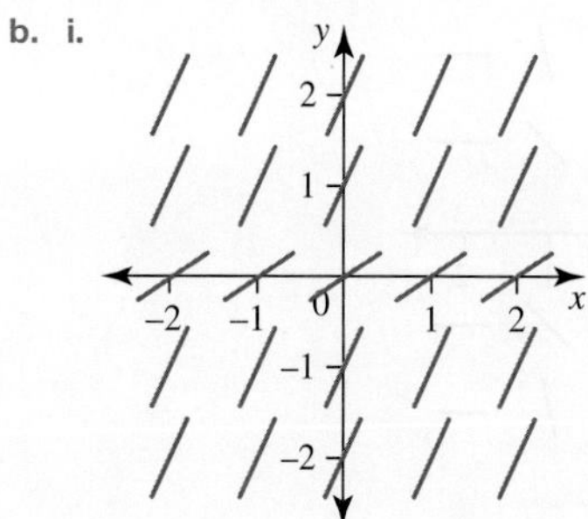

ii.

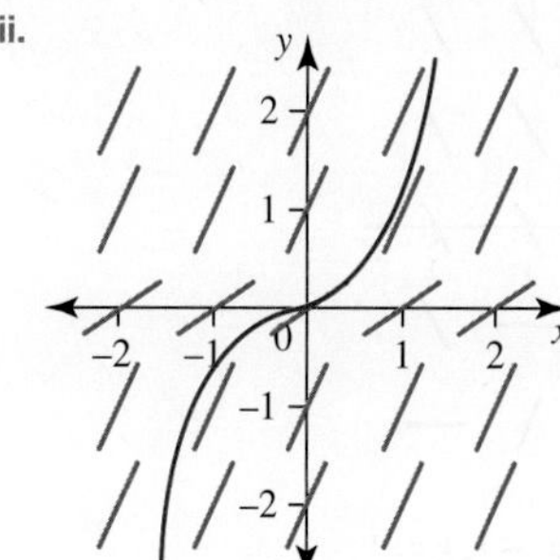

11. a.

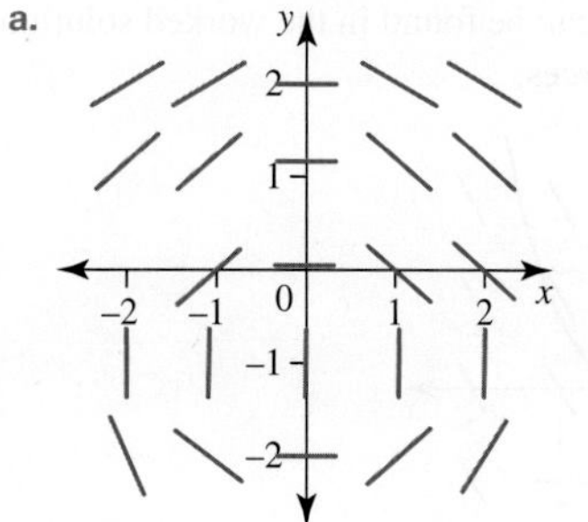

b.

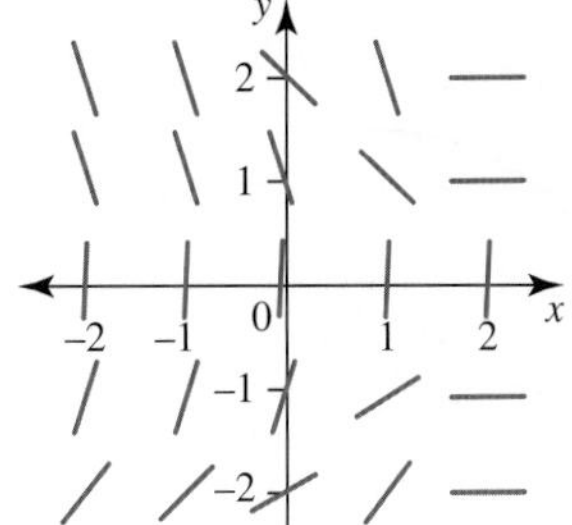

12. a.

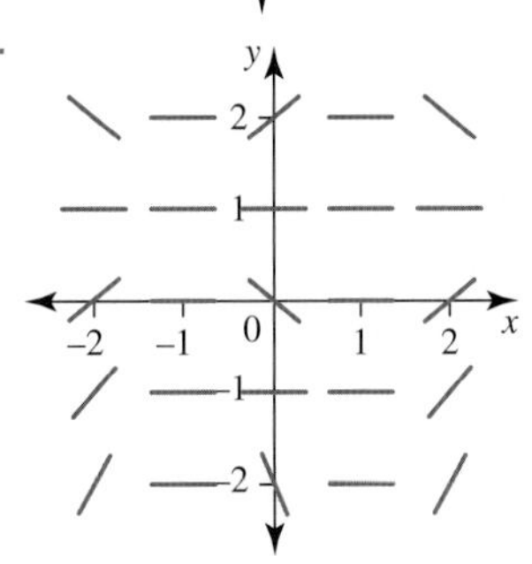

b.

13. a.

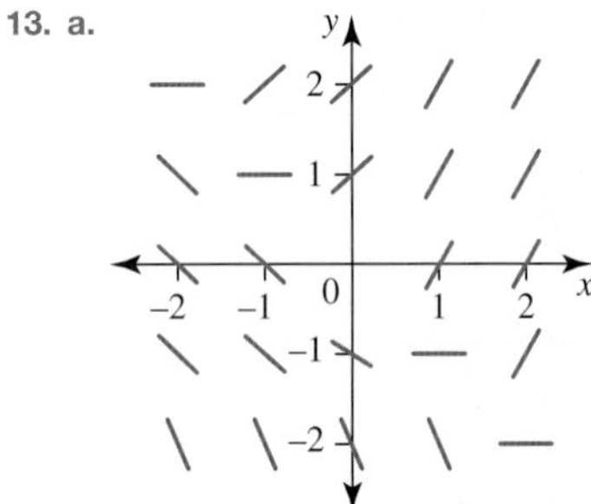

b. $A = 1, B = -1, C = -1, k = 1$

c. Sample responses can be found in the worked solutions in the online resources.

d.

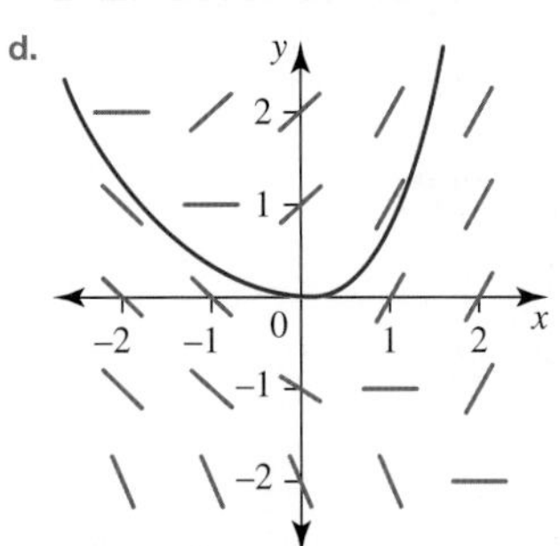

14. a. Slope field 4 b. Slope field 2
    c. Slope field 3 d. Slope field 1

15. a. Slope field 3 b. Slope field 1
    c. Slope field 4 d. Slope field 2

16. a. Slope field 1 b. Slope field 3
    c. Slope field 2 d. Slope field 4

17. a. Slope field 2 b. Slope field 3
    c. Slope field 1 d. Slope field 4

18. a. Slope field 1 b. Slope field 4
    c. Slope field 2 d. Slope field 3

10.8 Exam questions

Note: Mark allocations are available with the fully worked solutions online.

1. D
2. B
3. B

10.9 Review

10.9 Exercise

Technology active: short answer

1. 1198

2. a. $v(t) = v_0 e^{-\frac{t}{RC}}$ b. $RC \log_e(2)$

3. 14.8 weeks

4. 28.5%

5. Sample responses can be found in the worked solutions in the online resources.

6. a. Sample responses can be found in the worked solutions in the online resources.

   b. $T(t) = 15\left(5e^{\frac{t}{3}} - 4\right)$

   c. $3\log_e\left(\frac{49}{15}\right)$ minutes

Technology active: multiple choice

7. A
8. E
9. A
10. B
11. C
12. B
13. C
14. D
15. E
16. D

Technology active: extended response

17. a. $\frac{dQ}{dt} = 12 - \frac{2Q}{10+t},\ Q(0) = 0$

    b. Sample responses can be found in the worked solutions in the online resources.

18. 16.5 minutes

19. a. 28°C

    b. −3.07°C

20. a.

b.

x	0	$\frac{1}{3}$	$\frac{2}{3}$	1
y	1	1.5	2.125	2.7891

c. $y = \dfrac{4}{1 + 3e^{-2x}}$

d. Sample responses can be found in the worked solutions in the online resources.

e. Sample responses can be found in the worked solutions in the online resources.

f.

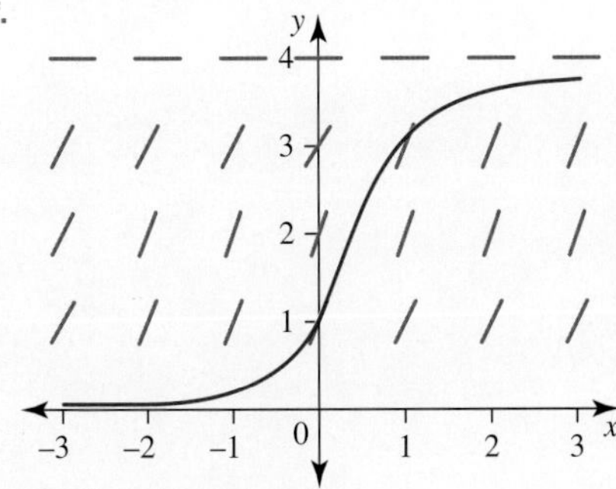

10.9 Exam questions

Note: Mark allocations are available with the fully worked solutions online.

1. A

2. C

3. a. Sample responses can be found in the worked solutions in the online resources.

b. 20

c. 403

d. i. $\dfrac{4N}{25}(6 - \log_e(N))(5 - \log_e(N))$

ii. $N = 148,\ t = 2.7$ years

e.

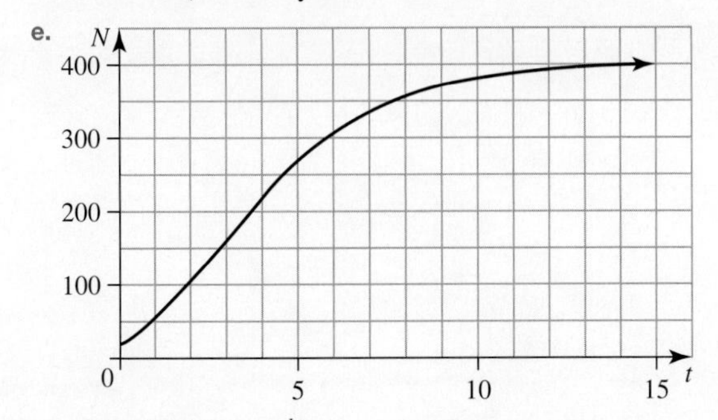

4. a. $n(t) = 2(25 - 23e^{-kt})$ **b.** 25

5. a. 48.742 million **b.** 29.962 million

11 Kinematics: rectilinear motion

LEARNING SEQUENCE

Fully worked solutions for this topic are available online.

11.1 Overview

Hey students! Bring these pages to life online

Watch videos

Engage with interactivities

Answer questions and check results

Find all this and MORE in jacPLUS

11.1.1 Introduction

The word "kinematics" comes from a Greek word "kinesis" meaning motion, and is related to other English words such as "cinema" (movies) and "kinesiology" the study of human motion. Kinematics is the study of the motion of a particle without consideration of the causes of the motion. In this chapter we will consider rectilinear motion, which is the motion of a particle moving along a straight-line path. Associated with the motion of a particle are certain quantities, these are the position, velocity, and acceleration which may depend on time.

Galileo Galilei (1564–1642) was a famous Italian scientist and mathematician. Among his many contributions to science and mathematics was his discovery that all objects fall with the same, constant acceleration due to gravity (provided air resistance is minimised or ignored). Galileo illustrated this property by dropping various sized cannon balls from the top of the Leaning Tower of Pisa.

The acceleration due to gravity depends on the distance from the centre of the Earth, it varies slightly at different places on the Earth's surface — being greater at the poles than at the equator and less at high altitudes. It is usual to state the constant of acceleration due to gravity as 9.8 m/s^2 unless stated otherwise.

KEY CONCEPTS

This topic covers the following point from the VCE Mathematics Study Design:

- use of velocity–time graphs to describe and analyse rectilinear motion
- application of differentiation, anti-differentiation and solution of differential equations to rectilinear motion of a single particle, including the different derivative forms for acceleration

$$a = \frac{d^2x}{dt^2} = \frac{dv}{dt} = v\frac{dv}{dx} = \frac{d}{dx}\left(\frac{1}{2}v^2\right).$$

11.2 Differentiating position and velocity

LEARNING INTENTION

At the end of this subtopic you should be able to:
- calculate the velocity and acceleration functions from a position function.

11.2.1 Rectilinear motion

In this topic we will consider only motion in a straight line, **rectilinear motion**, and all objects will be treated as points for mathematical convenience: that is, the objects do not rotate or change shape. First, some basic concepts are explained.

Position

The **position** of a particle moving in a straight line measures its location relative to a fixed point O called the origin. Since we are working with motion in a straight line we can represent the position in one dimension using the variable x.

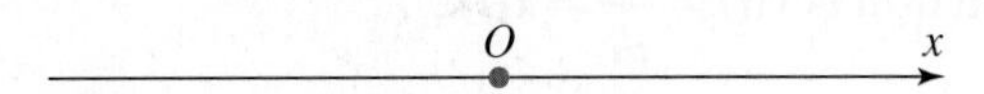

We usually take the positive direction to be to the right. Therefore:

If $x > 0$, then the particle is to the right of O.

If $x < 0$, then the particle is to the left of O.

When the position changes over time we express it as a function of time $x(t)$ where $t \geq 0$. The initial position occurs when $t = 0$ and so can be calculated by evaluating $x(0)$.

Unless otherwise specified, the units for position are metres and the units for time are seconds.

Displacement

The **displacement** of a moving particle is its change in position, Δx. It is not necessarily the distance travelled. The displacement of a particle from $t = t_1$ to $t = t_2$ is $\Delta x = x(t_2) - x(t_1)$.

Velocity

The **velocity** v is defined as the rate of change of position with respect to time. The function representing the instantaneous velocity v at a time t is therefore $v(t) = \frac{dx}{dt}$. An alternate notation of the velocity function is $\dot{x}(t)$, where the dot refers to the derivative with respect to time.

The average velocity over the time interval (t_1, t_2) is the change in position over the change in time, that is:

$$\text{average velocity} = \frac{\Delta x}{\Delta t} = \frac{x(t_2) - x(t_1)}{t_2 - t_1}.$$

The units for velocity are the units for distance per the units for time. Taking the default units for distance and time, the units for velocity are 'metres per second', abbreviated to ms^{-1} or m/s.

Note that velocity is a vector and considers both magnitude and direction, whereas **speed** is a scalar and is the magnitude of the velocity.

If $v > 0$, then the particle is moving to the right.

If $v < 0$, then the particle is moving to the left.

If $v = 0$, then the particle is at rest.

Acceleration

The **acceleration** a is defined as the rate of change of velocity with respect to time. The function representing the instantaneous acceleration a at a time t, is therefore $a(t) = \frac{dv}{dt}$.

Other notations for the acceleration are $a(t) = \ddot{x}(t) = \frac{d^2x}{dt^2}$, where the two dots refers to the second derivative with respect to time.

The average acceleration over the time interval (t_1, t_2) is the change in velocity over the change in time, that is: average acceleration $= \frac{\Delta v}{\Delta t} = \frac{v(t_2) - v(t_1)}{t_2 - t_1}$.

The units for the acceleration are 'metres per second per second', or 'metres per second squared', abbreviated to ms^{-2} or m/s^2.

Position, velocity and acceleration functions

The position function is $x(t)$.

The velocity function is $v(t) = \frac{dx}{dt} = \dot{x}(t)$.

The acceleration function is $a(t) = \frac{dv}{dt} = \frac{d^2x}{dt^2} = \ddot{x}(t)$.

WORKED EXAMPLE 1 Calculating position and velocity of a particle

The position of a particle x metres from the origin at a time t seconds is given by $x = 2t^3 - 9t^2 + 12t + 6$. Determine:

a. **its initial position**
b. **its initial velocity**
c. **when and where it is at rest**
d. **its average velocity over the first three seconds**
e. **its average speed over the first three seconds.**

THINK	WRITE
a. The initial position is the value of x when $t = 0$.	a. $x(t) = 2t^3 - 9t^2 + 12t + 6$ $x(0) = 6$ Initially the particle is 6 metres to the right of the origin O.
b. 1. Differentiate the position with respect to time, to determine the velocity.	b. $v(t) = \frac{dx}{dt} = 6t^2 - 18t + 12$.
2. The initial velocity is the value of v when $t = 0$.	$v(0) = 12$ The initial velocity of the particle is 12 m/s to the right.
c. 1. The particle is at rest when $v = 0$, solve to calculate the values of t.	c. $6t^2 - 18t + 12 = 0$ $6\left(t^2 - 3t + 2\right) = 0$ $(t-2)(t-1) = 0$ $t = 1, 2$

2. To determine where the particle is at rest, substitute these values of t into the position function.

$x(1) = 2 - 9 + 12 + 6 = 11$ m
$x(2) = 16 - 36 + 24 + 6 = 10$ m
The particle is at rest after one second at a position of 11 metres to the right of the origin and after two seconds at a position of 10 metres to the right of the origin.

d. 1. Determine the position after three seconds.

d. $x(3) = 54 - 81 + 36 + 6 = 15$
After three seconds the particle is 15 metres to the right of the origin.

2. Calculate the average velocity over the first three seconds.

$\frac{x(3) - x(0)}{3 - 0} = \frac{15 - 6}{3} = 3$ m/s

e. 1. Determine the total distance travelled over the first three seconds.

e. When $t = 0,\ x = 6$, when $t = 1,\ x = 11$, when $t = 2,\ x = 10$ and when $t = 3,\ x = 15$.
The total distance moved by the particle is $5 + 1 + 5 = 11$ m.

2. Calculate the average speed over the first three seconds, that is, the distance travelled devided by the time taken.

The average speed over the first three seconds is $\frac{11\text{ m}}{3\text{ s}} = \frac{11}{3}$ m/s.

Sometimes the position of a particle or object is representing height or distances in other directions such as North-South or East-West. The position function may be written using a different pronumeral to x but the same concepts apply. The velocity is always the derivative of the position, and the acceleration is always the derivative of the velocity.

WORKED EXAMPLE 2 Calculating position, velocity and acceleration of a ball

A tennis ball is projected vertically upwards, from a machine so that its height h metres above the ground at a time t seconds is given by $h(t) = 19.6t - 4.9t^2 + 2$.

a. Determine from what height the ball is projected.
b. Calculate the initial speed of projection.
c. Determine when the ball reaches its maximum height.
d. Calculate the maximum height the ball reaches.
e. Determine when the ball hits the ground.
f. Calculate the speed of the ball when it strikes the ground.
g. Show that its acceleration is always constant.

THINK

WRITE

a. The initial height is the value of h when $t = 0$.

a. $h(t) = 19.6t - 4.9t^2 + 2$
$h(0) = 2$
The ball was projected from a height of 2 metres above the ground.

b. 1. Differentiate the position with respect to time, to determine the velocity.

b. $h(t) = 19.6t - 4.9t^2 + 2$
$v(t) = \frac{dh}{dt} = 19.6 - 9.8t$

2. The initial velocity is the value of v when $t=0$.	$v(0)=19.6$ The ball was projected vertically upwards with a speed of 19.6 m/s.
c. The ball reaches its maximum height when the velocity is zero.	**c.** $v(t)=0$ $0=19.6-9.8t$ $9.8t=19.6$ $t=2$ The ball reaches a maximum height after 2 seconds.
d. The maximum height is the value of h when $t=2$.	**d.** $h_{max}=19.6(2)-4.9(2)^2+2$ $=39.2-19.6+2$ $=21.6\text{ m}$ The maximum height above ground level is 21.6 metres
e. The ball hits the ground when $h=0$.	**e.** $h(t)=0$ $19.6t-4.9t^2+2=0$ $-4.9t^2+19.6t+2=0$ Using the quadratic formula: $a=-4.9\ b=19.6\ c=2$ $\Delta=b^2-4ac=423.36$ $t=\dfrac{-19.6\pm\sqrt{423.36}}{-9.8}=4.1,-0.1$ since $t\geq 0$ reject $t=-0.1$. The ball hits the ground after 4.1 seconds
f. The speed at which the ball strikes the ground is the value of v when $t=4.1$	**f.** $v(4.1)=19.6-9.8(4.1)$ $=19.6-40.2$ $=-20.6\text{ m/s}$ The ball strikes the ground with a speed of 20.6 m/s.
g. 1. Determine an expression for the acceleration by differentiating the velocity function.	**g.** $a(t)=\dfrac{dv}{dt}=-9.8\text{ m s}^{-2}$
2. Note that the acceleration is constant regardless of the time. It is 9.8 m/s^2 downwards throughout the entire motion.	The acceleration is independent of time t and is constant throughout the motion.

11.2 Exercise

Students, these questions are even better in jacPLUS

Receive immediate feedback and access sample responses

Access additional questions

Track your results and progress

Find all this and MORE in jacPLUS

Technology free

1. A particle moves in a straight-line path so that its position x metres from an origin O at a time t seconds is given by $x(t) = t^2 - 4t - 12$. Determine:
   a. its initial position
   b. its initial velocity
   c. when and where it is at rest
   d. when it passes through the origin and its velocity then.

2. A body moves according to the law, $x(t) = \frac{1}{18}(2t^3 - 3t^2 - 12t + 8)$. Calculate its acceleration at the instant when it comes to rest.

3. **WE1** The position of a particle x metres from the origin at a time t seconds is given by $x(t) = t^3 - 9t^2 + 15t + 3$. Determine:
   a. its initial position
   b. its initial velocity
   c. when and where it is at rest
   d. its average velocity over the first five seconds
   e. its average speed over the first five seconds.

4. The position of a particle moving in a straight line from a fixed position, at time t seconds is given by $x = t^3 - 9t^2 + 24t + 5$. Determine:
   a. when and where the particle is at rest
   b. its average velocity over the first four seconds
   c. its average speed over the first four seconds
   d. its acceleration after two seconds.

5. The height h metres of a body projected vertically upwards from the Earth's surface t seconds after projection is given by $h = 40t - 5t^2$. Determine the:
   a. time to reach maximum height
   b. maximum height reached
   c. time taken to return to Earth
   d. times when the speed is 20 m/s
   e. total distance travelled
   f. average speed over the journey.

Technology active

6. **WE2** At a time t seconds a bullet is fired vertically upwards, its height h metres above the surface of the Earth is given by $h = 10 + 49t - 4.9t^2$.
   a. Determine from what height it was projected.
   b. Calculate its initial speed of projection.
   c. Determine how long before it reaches its maximum height.
   d. Calculate the maximum height reached.
   e. Determine how long before it reaches ground level, in seconds to 1 decimal place.
   f. Calculate its speed when it strikes the ground, in m/s to 2 decimal places.
   g. Show that its acceleration is always constant.

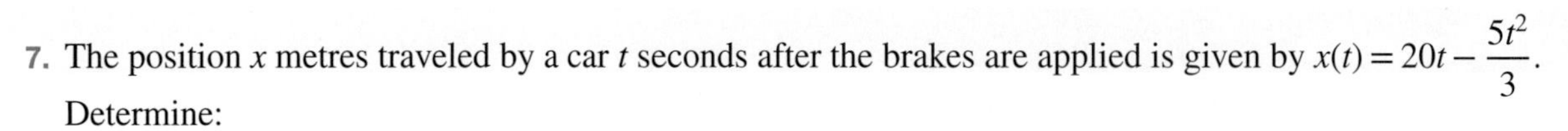

7. The position x metres traveled by a car t seconds after the brakes are applied is given by $x(t) = 20t - \frac{5t^2}{3}$. Determine:
   a. the speed of the car initially, (in km/h) that is, at the instant when the brakes are applied
   b. how long it takes for the car to stop
   c. the distance the car travels before it stops.
8. A bus moves from a point A to point B. Its position from A after a time t seconds is given by $x(t) = \frac{t^2}{6} - \frac{t^3}{1620}$ metres. Determine:
   a. its maximum speed in km/h
   b. how long in minutes does it take to reach B
   c. the distance in kilometres between A and B
   d. the average speed over the journey in km/h.

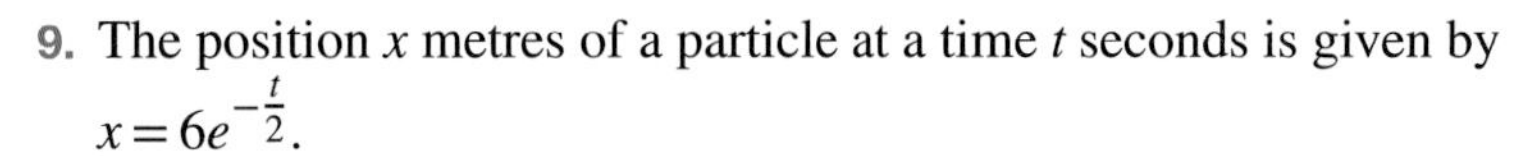

9. The position x metres of a particle at a time t seconds is given by

$x = 6e^{-\frac{t}{2}}$.
   a. Determine the initial position.
   b. Calculate the initial velocity.
   c. Determine its initial acceleration.
   d. If its acceleration is a show that $a = \frac{x}{4}$.
10. The position x metres of a particle at a time t seconds is given by $x = 9e^{-\frac{t}{3}} - 6e^{-t}$.
    a. Determine the initial position.
    b. Calculate the initial velocity.
    c. Determine the time it comes to rest.
    d. If its acceleration is a, its velocity v and its position x and $a + bv + cx = 0$, determine the values of b and c.
11. The position x metres of a particle from an origin O at a time t seconds is given by $x = 4\cos\left(\frac{t}{2}\right)$.
    a. Determine the initial position.
    b. Calculate the initial velocity.
    c. If its acceleration is a, show that $x = -4a$.
    d. Determine the times when it is 2 metres from the origin.
    e. Calculate the time when it comes to rest.
12. A particle moves back and forth along a straight-line track so that its position x metres at a time t seconds is given by $x(t) = 9\cos\left(\frac{t}{3}\right) + 18\sin\left(\frac{t}{3}\right)$.
    a. Determine its initial position.
    b. Calculate its initial velocity.
    c. Determine its initial acceleration.
    d. If its velocity is v show that $v^2 = 45 - \frac{x^2}{9}$.
    e. If its acceleration is a show that $a = -\frac{x}{9}$.

11.2 Exam questions

Question 1 (1 mark) TECH-ACTIVE

MC The position x metres of a particle from an origin O at a time t seconds is given by $x = 9\sin\left(\frac{t}{3}\right)$, if its acceleration is a then

A. $9a + x = 0$ **B.** $9a - x = 0$ **C.** $a + 9x = 0$ **D.** $a - 9x = 0$ **E.** $a - x = 0$

Question 2 (5 marks) TECH-FREE

A train moves in a straight line between two stations and its position is given by $x(t) = \frac{t^2}{4} - \frac{t^3}{720}$ metres, at a time t seconds after leaving the first station. Determine:

a. the time intervals for which its speed is increasing and decreasing **(1 mark)**

b. its maximum speed in km/h **(1 mark)**

c. how long in minutes the journey is between the stations **(1 mark)**

d. the distance in kilometres between the stations **(1 mark)**

e. the average speed over the journey in km/h. **(1 mark)**

Question 3 (4 marks) TECH-FREE

A particle has a height, $x = 49t - 390 + 490e^{-\frac{t}{10}}$ metres above the ground after a time t seconds.

a. Determine its initial height above the ground. **(1 mark)**

b. Calculate its initial velocity. **(1 mark)**

c. Determine its acceleration $a(t)$ at time t. **(1 mark)**

d. If $v(t)$ is its velocity at a time t seconds, show that $a = \frac{1}{10}(49 - v)$. **(1 mark)**

More exam questions are available online.

11.3 Constant acceleration

LEARNING INTENTION

At the end of this subtopic you should be able to:

- apply the constant acceleration formulae to determine position, initial or final velocity, acceleration or time.

11.3.1 Constant acceleration formulae

When considering motion of an object in a straight line in which the acceleration is constant, a number of formulae can be used. These will be proven later in the chapter.

Let v (m/s) be the velocity, u (m/s) the initial velocity, a the acceleration $\left(\text{m/s}^2\right)$ at t time (s) and s the displacement (m).

Constant acceleration formulae

$$s = ut + \frac{1}{2}at^2$$

$$v = u + at$$

$$v^2 = u^2 + 2as$$

$$s = \frac{1}{2}(u + v)\,t$$

In applying these rules, we must consider the following conditions:

- These formulae can only apply when acceleration is constant.
- Deceleration implies that acceleration is negative.
- The variable s is the displacement of an object. It is not necessarily the distance travelled by the object.
- In solving constant acceleration problems it is important to list the quantities given to determine and what is required and use an appropriate equation, to solve for the unknown quantity.

- Sometimes we may need to solve two simultaneous equations for two sets of unknowns.
- It is essential to have all quantities converted to compatible units such as m and m/s from km and km/h.

WORKED EXAMPLE 3 Using the constant acceleration formulae (1)

A tram accelerates uniformly from a speed of 5 m/s to 15 m/s, while travelling 300 metres. Calculate:

a. **the acceleration**
b. **the distance covered in the first 6 seconds**
c. **the time taken.**

THINK	WRITE
a. 1. List all the quantities given and what is required.	a. $u = 5$ m/s, $v = 15$ m/s, $s = 300$ m $a = ?$
2. Determine the appropriate equation that will solve for the unknown given all the quantities.	$v^2 = u^2 + 2as$
3. Substitute the quantities into the equation and solve for the unknown value of the acceleration a.	$15^2 = 5^2 + 2 \times 300a$ $225 = 25 + 600a$ $600a = 200$ $a = \frac{1}{3}$ m/s^2
b. 1. List all the quantities given and what is required.	b. $u = 5$ m/s, $a = \frac{1}{3}$m/s^2, $t = 6$ s, $s = ?$
2. Determine the appropriate equation that will solve for the unknown given all the quantities.	$s = ut + \frac{1}{2}at^2$
3. Substitute the quantities into the equation and solve for the unknown value of s.	$s = 5 \times 6 + \frac{1}{2} \times \frac{1}{3} \times 6^2$ $s = 30$ m After 6 seconds, the tram has travelled 30 metres.
c. 1. List all the quantities given and what is required.	c. $u = 5$ m/s, $v = 15$ m/s, $s = 300$ m, $t = ?$
2. Determine the appropriate equation that will solve for the unknown given all the quantities.	$s = \frac{1}{2}(u + v)\, t$

3. Substitute the quantities into the equation and solve for the unknown value of t.

$\frac{1}{2}(5+15)t=300$

$t=30$

The time taken is 30 seconds.

4. Alternatively, this shows there is often more than one formulae which can be used to solve these types of problems.

$v=u+at$

$15=5+\frac{t}{3}$

$\frac{t}{3}=10$

$t=30$

WORKED EXAMPLE 4 Using the constant acceleration formulae (2)

A driver of a car travelling at 57.6 km/h applies the brakes and the speed is reduced to 14.4 km/h after five seconds, assume that the deceleration is constant.

a. **Determine the acceleration of the train while breaking.**

b. **Calculate the distance traveled in this time.**

c. **Determine how long it will take to stop from its initial speed of 57.6 km/h, assuming the same deceleration.**

d. **Calculate how far it will travel until it comes to rest, from its initial speed of 57.6 km/h, assuming the same deceleration.**

THINK	WRITE
a. 1. List all the quantities given and change to appropriate units.	a. $57.6\text{ km/h}=\frac{57.6\times 1000}{60\times 60}=16\text{ m/s}$ $u=16$ $14.4\text{ km/h}=\frac{14.4\times 1000}{60\times 60}=4\text{ m/s}$ $v=4$ $t=5,\ a=?$
2. Determine the appropriate equation that will solve for the unknown given all the quantities.	$v=u+at$
3. Substitute the quantities into the equation and solve for the unknown value of a.	$4=16+5a$ $a=-\frac{12}{5}=-2.4\text{ m/s}^2$ The acceleration is negative as it is a deceleration. This is the constant throughout the whole motion.
b. 1. List all the quantities given and what is required.	b. $u=16\text{ m/s},\ v=4\text{ m/s},\ t=5\text{ s}$
2. Determine the appropriate equation that will solve for the unknown given all the quantities.	$s=\frac{1}{2}(u+v)\,t$
3. Substitute the quantities into the equation and solve for the unknown value of s.	$s=\frac{1}{2}(16+4)\times 5$ $s=50$

c. 1. Consider the motion, of the car from the start to the finish, when the car comes to rest. List all the quantities given and what is required.

c. $u = 16$ m/s, $v = 0$, $a = -2.4$ m/s^2
$t = ?$

2. Determine the appropriate equation that will solve for the unknown given all the quantities.

$v = u + at$

3. Substitute the quantities into the equation and solve for the unknown value of t.

$$16 = 0 - \frac{12t}{5}$$

$$t = \frac{80}{12} = 6\frac{2}{3} \text{ s}$$

The car comes to rest after $6\frac{2}{3}$ s.

d. 1. Consider the motion, of the car from the start to the finish, when the car comes to rest. List all the quantities given and what is required.

d. $u = 16$ m/s, $v = 0$, $a = -2.4$ m/s^2
$s = ?$

2. Determine the appropriate equation that will solve for the unknown given all the quantities.

$v^2 = u^2 + 2as$

Substitute the quantities into the equation and solve for the unknown value of s.

$$0 = 16^2 + 2 \times \left(-\frac{12}{4}\right) s$$

$$s = \frac{160}{3} = 53\frac{1}{3} \text{ m}$$

The car travels a total distance of $53\frac{1}{3}$ m before coming to rest.

11.3.2 Vertical motion

Vertical motion is when an object in the Earth's gravitional field, is moving vertically upwards or downwards.

In solving vertical motion problems, we take the upward direction as positive and the downwards direction as negative, the acceleration due to gravity is $g = 9.8$ m/s^2 and always acts downwards, so in vertical motion problems, the acceleration is $a = -g = -9.8$ m/s^2.

WORKED EXAMPLE 5 Vertical motion involving constant acceleration (1)

A parachutist jumps from a plane which is flying at constant altitude and falls vertically downwards for 10 seconds before their parachute opens. Calculate the speed (in km/h) that the parachutist is travelling at when their parachute opens and the distance that they have fallen in metres.

THINK	WRITE
1. List all the quantities given and what is required.	$u = 0, t = 10\text{ s}, a = -9.8\text{ m/s}^2$ $v = ?$
2. Determine the appropriate equation that will solve for v given all the quantities.	$v = u + at$
3. Substitute the quantities into the equation and solve for the unknown value of v. Convert to the required units.	$v = 0 - 9.8 \times 10 = -98\text{ m/s}$ $v = \dfrac{-98 \times 60 \times 60}{1000}$ $v = -352\text{ km/h}$ The negative indicates the parachutist is moving downwards, his speed is 352 km/h.
4. List all the quantities given and what is required.	$u = 0, t = 10\text{ s}, a = -9.8\text{ m/s}^2$ $s = ?$
5. Determine the appropriate equation that will solve for s given all the quantities.	$s = ut + \dfrac{1}{2}at^2$
6. Substitute the quantities into the equation and solve for the unknown value of s.	$s = 0 - \dfrac{1}{2} \times 9.8 \times 10^2 = -490\text{ m}$ The negative indicates that the displacement is below the initial drop point, the parachutist has fallen a distance of 490 m.

WORKED EXAMPLE 6 Vertical motion involving constant acceleration (2)

A tennis ball is projected vertically upwards by a machine with a speed of 19.6 m/s from a point 2 metres above the ground. Calculate the time taken for the tennis ball to hit the ground.

THINK	WRITE
1. List all the quantities given and what is required. Note that when the ball hits the ground, $s = -2$ as it is below the point of projection. Note that we can solve the problem in one calculation, we do not need to consider the upward motion and downward motion separately.	$u = 19.6\text{ m/s}, s = -2\text{ m}, a = -9.8\text{ m/s}^2\ t = ?$
2. Determine the appropriate equation that will solve for distance given all the quantities.	$s = ut + \dfrac{1}{2}at^2$
3. Substitute the quantities into the equation and solve for the unknown value of t. Compare this to Worked example 2.	$-2 = 19.6t - 4.9t^2$ $4.9t^2 - 19.6t - 2 = 0$ Using the quadratic formula $a = 4.9\ b = -19.6\ c = -2$ $\Delta = b^2 - 4ac = 423.36$ $t = \dfrac{19.6 \pm \sqrt{423.36}}{9.8} = 4.1, -0.1$ Since $t \geq 0$ reject $t = -0.1$ The ball reaches the ground after 4.1 seconds.

WORKED EXAMPLE 7 Vertical motion involving constant acceleration (3)

At a certain instance, a hot air balloon is rising vertically with an acceleration of 0.8 m/s^2, its upwards speed is 4 m/s and its height above the ground is h metres. A stone is dropped from the hot air balloon and hits the ground 5 seconds later. Determine the value of h.

THINK	WRITE
1. List all the quantities given and what is required. Note that the stone takes on the initial upwards speed of the balloon, but not the acceleration.	$u = 4$ m/s, $s = h = ?$, $a = -9.8$ m/s^2 $t = 5$ s
2. Determine the appropriate equation that will solve for distance given all the quantities.	$s = ut + \frac{1}{2}at^2$
3. Substitute the quantities into the equation and solve for the unknown value of s.	$s = 4 \times 5 - 4.9 \times 5^2$ $s = 36 - 122.5$ $s = -86.5$ m
4. State the required result.	The balloon was at a height of 86.5 m above the ground when the stone was dropped.

11.3 Exercise

Technology free

1. WE3 A cyclist accelerates uniformly from a speed of 3 m/s to 9 m/s in 6 seconds. Determine the:
   a. acceleration
   b. distance traveled in the first 6 seconds
   c. time taken to reach a speed of 12 m/s.

2. a. A car is traveling at 72 km/h when it sees a dangerous situation ahead. The brakes are applied and provide a constant deceleration, causing the car to stop in 20 metres. Determine the time taken to come to a stop.
   b. A car traveling at 72 km/h is stopped in 20 seconds. Calculate the distance travelled in this time.
3. a. A car travels a distance of 600 metres in 30 seconds after accelerating uniformly from rest. Determine the car's speed after 15 seconds.
   b. A car travels a distance of D metres in T seconds after accelerating uniformly from rest. Determine the car's speed after half this time.

4. a. A bus travels a distance of one kilometre in 40 seconds after accelerating uniformly from rest. Calculate its velocity after it has travelled 500 metres.
   b. A bus travels a distance of D metres in T seconds after accelerating uniformly from rest. Calculate its velocity after traveling half this distance.

5. **WE4** A train traveling at 54 km/h, applies the brakes, reducing the speed to 18 km/h in 6 seconds. Assuming that the deceleration is constant, calculate:
   a. the acceleration of the train while braking
   b. the distance the train traveled in this time
   c. how long it will take for the train to come to a complete stop, assuming the same deceleration
   d. how far the train would travel before it stops, assuming the same deceleration.

6. a. A tram moves with uniform acceleration and has a speed of 18 km/h and 36 km/h respectively when passing two points.
      i. Determine its speed halfway between these two points.
      ii. Calculate the average speed.
   b. A tram moves with uniform acceleration and has a speed of U m/s and V m/s respectively when passing two points.
      i. Determine its speed half-way between these two points.
      ii. Calculate the average speed.

7. a. A train has a speed of 10 m/s after 4 seconds and a speed of 14 m/s after a further two seconds. Determine the acceleration and initial speed of the train.
   b. A train has a speed of V_1 after a time of T_1 and a speed of V_2 after a time of T_2. Determine the acceleration and its initial speed.

8. a. A man is jogging along a straight road with constant acceleration and travels a distance of 28 metres in the first 4 seconds and a distance of 45 metres in the first 6 seconds. Calculate the acceleration and initial velocity of the man.
   b. A car traveling along a straight road with constant acceleration is observed to cover a distance of d_1 metres in T_1 seconds and a distance of d_2 metres in T_2 seconds. Calculate the acceleration of the car.

9. a. A bus moving along a straight road has a speed of $5\sqrt{2}$ m/s when it is a distance of 25 metres from the last intersection, and a speed of $5\sqrt{5}$ m/s when it is a distance of 100 metres from the last intersection. Calculate the acceleration and the initial speed of the bus as it left the last intersection.
   b. A body moving in a straight line has a speed of V_1 at a distance of d_1 from a point O and a speed of V_2 at a distance of d_2 from O. Calculate the acceleration and the initial speed at O.

10. a. A ball is dropped from the top of a cliff and hits the ground two seconds later. Determine the speed of the ball as it hits the ground and the height of the cliff.
    b. A ball is dropped from the top of a cliff and hits the ground T seconds later. Determine the speed of the ball as it hits the ground and the height of the cliff.

Technology active

11. a. **WE5** A coin is dropped from the top of a building that is 253 metres above the ground level. Calculate the speed that the ball hits the ground at (in km/h to 1 decimal place) and the time it takes to hit the ground (in seconds to 2 decimal places).
    b. A cricketer at an indoor sports stadium tries to throw a ball vertically upwards to touch the roof of the stadium, which is 38 m high. If he throws the ball from a point one metre above the ground, calculate the initial speed (in km/h to 2 decimal places) so that the ball just touches the roof, and how long it takes for the ball to touch the roof (in seconds to 2 decimal places).

12. **WE6** A bullet is fired vertically upwards from a point one metre above the ground, with an initial speed of 49 m/s. Determine:

a. the time to reach maximum height
b. the maximum height reached
c. the time taken to hit the ground, in seconds to 2 decimal places
d. the times when its speed is 20 m/s, in seconds to 2 decimal places.

13. a. **WE7** A hot air balloon is descending at 3 m/s. When it is 112.5 metres above the ground, a stone is thrown upwards from the balloon at a speed of 5 m/s. Determine how long it takes the stone to hit the ground.

b. A tennis ball is thrown vertically upwards with a speed of U m/s from a point H metres above the ground. Determine the time taken for the ball to hit the ground.

14. a. A stone is projected vertically upwards from the top of a building and reaches the ground in 3 seconds. If projected vertically downwards with the same speed of projection it reaches the ground in 2 seconds.

i. Calculate the height of the building.
ii. Determine how long it would take to reach the ground if it were simply dropped.

b. A stone is projected vertically upwards from the top of a building and reaches the ground in T_1 seconds. If projected vertically downwards with the same speed of projection it reaches the ground in T_2 seconds.

i. Calculate the height of the building.
ii. Determine how long it would take to reach the ground if it were simply dropped.

15. A stone is dropped into a well and reaches the water with a speed of 40 m/s. The sound of the splash is heard exactly 4.3196 seconds after it was dropped. From this data, calculate the speed of sound in air. Write your answer correct to the nearest integer in m/s.

16. A particle is dropped from the top of a building and is observed to pass a one metre high window in 0.03 seconds. If the window is one metre above the ground level, determine the height of the building, in metres to 2 decimal places.

17. a. A stone is projected vertically upwards with an initial speed of 5 m/s from the top of a cliff of height 10 metres. After T seconds, a second stone is dropped from the same point. If they both hit the ground at the same time, determine the value of T, in seconds to 4 decimal places.

b. A stone is projected vertically upwards with an initial speed of U m/s from the top of a cliff of height H. After T seconds, a second stone is dropped from the same point. If they both hit the ground at the same time, show that $T = \dfrac{U + \sqrt{U^2 + 2gH} - \sqrt{2gH}}{g}$.

18. a. A stone is dropped from the top of a cliff of height 10 metres. When the stone is halfway down a second stone is thrown vertically downwards from the top of the cliff, such that both stones hit the ground at exactly the same time. Calculate the speed at which the second stone is thrown at, in m/s rounded to 2 decimal places.

b. A stone is dropped from the top of a cliff of height H. When the stone is halfway down a second stone is thrown vertically downwards from the top of the cliff such that both stones hit the ground at exactly the same time. Show that the speed of the second stone is given by $\frac{1}{2}\sqrt{gH}\left(\sqrt{2}+3\right)$.

11.3 Exam questions

Question 1 (1 mark) TECH-ACTIVE

***Source:** VCE 2018, Specialist Mathematics Exam 2, Section A Q17; © VCAA.*

MC A tourist standing in the basket of a hot air balloon is ascending at 2 ms^{-1}. The tourist drops a camera over the side when the balloon is 50 m above the ground.

Neglecting air resistance, the time in seconds, correct to the nearest tenth of a second, taken for the camera to hit the ground is

A. 2.3 **B.** 2.4 **C.** 3.0 **D.** 3.2 **E.** 3.4

Question 2 (1 mark) TECH-ACTIVE

***Source:** VCE 2015, Specialist Mathematics Exam 2, Section 1, Q20; © VCAA.*

MC An object is moving in a straight line, initially at 5 ms^{-1}. Sixteen seconds later, it is moving at 11 ms^{-1} in the opposite direction to its initial velocity.

Assuming that the acceleration of the object is constant, after 16 seconds the distance, in metres, of the object from its starting point is

A. 24 **B.** 48 **C.** 73 **D.** 96 **E.** 128

Question 3 (1 mark) TECH-ACTIVE

***Source:** VCE 2013, Specialist Mathematics Exam 2, Section 1, Q19; © VCAA.*

MC A tourist in a hot air balloon, which is rising at 2 m/s, accidentally drops a camera over the side and it falls 100 m to the ground.

Neglecting the effect of air resistance on the camera, the time taken for the camera to hit the ground, correct to the nearest tenth of a second, is

A. 4.3 s **B.** 4.5 s **C.** 4.7 s **D.** 4.9 s **E.** 5.0 s

More exam questions are available online.

11.4 Velocity–time graphs

LEARNING INTENTION

At the end of this subtopic you should be able to:

- solve problems involving motion in a straight line by drawing and analysing velocity–time graphs.

Velocity–time graphs are a useful visual representation of motion of an object moving in a straight line. We can use velocity–time graphs to solve kinematic problems.

The following properties of velocity–time graphs make this possible.

Since $a = \frac{dv}{dt}$, the gradient of the velocity–time graph at time t gives the instantaneous acceleration at time t.

Since $v = \frac{dx}{dt}$, the displacement is found by evaluating the definite integral $s = \int_{t_1}^{t_2} v(t)\, dt$ $\left(\text{since } s = x(t_2) - x(t_1)\right)$. The distance is found by determining the magnitude of the signed area under the curve bounded by the graph and the t axis, $\int_{t_1}^{t_2} |v(t)| dt$. Distance travelled cannot be a negative value.

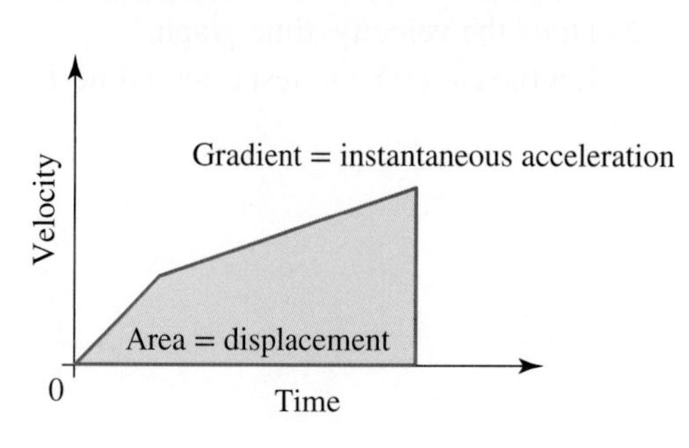

Displacement and distance

If $v(t)$ represents velocity with respect to time:

The displacement over the time interval (t_1, t_2) is $\displaystyle\int_{t_1}^{t_2} v(t)\,dt$

The distance travelled over the time interval (t_1, t_2) is $\displaystyle\int_{t_1}^{t_2} |v(t)|\,dt$.

It is useful to know the area formulae of basic shapes to assist in calculating the displacement without using calculus.

Area of some basic shapes

Area of a triangle: $A = \frac{1}{2}bh$

Area of a rectangle: $A = lw$

Area of a trapezium: $A = \frac{1}{2}(a + b)h$

WORKED EXAMPLE 8 Solving problems by drawing a velocity–time graph (1)

A driver of a car travelling at 57.6 km/h applies the brakes and the speed is reduced to 14.4 km/h after five seconds, assuming that the deceleration is constant calculate the total distance travelled and how much longer it takes for the car to come to rest.

THINK	WRITE
1. Convert the speeds into m/s.	$u = 57.6\text{ km/h}$ $= \frac{57.6 \times 1000}{60 \times 60}$ $= 16\text{ m/s}$ $v = 14.4\text{ km/h}$ $= \frac{14.4 \times 1000}{60 \times 60}$ $= 4\text{ m/s}$
2. Draw the velocity–time graph. Let the car come to rest after a time T.	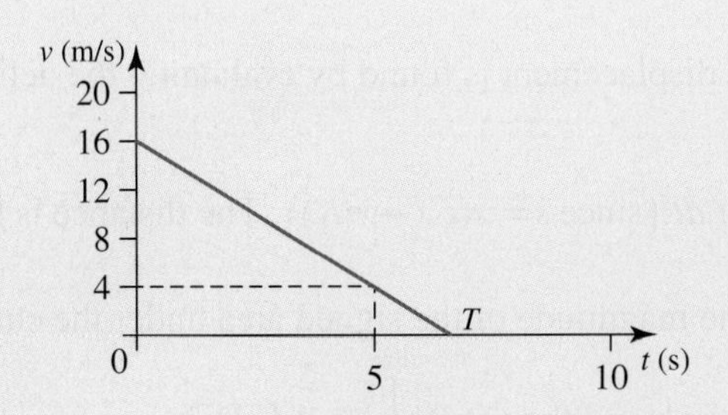

3. Use similar triangles.

$$\frac{T-5}{4} = \frac{T}{16}$$
$$16(T-5) = 4T$$
$$12T = 80$$
$$T = \frac{20}{3} = 6\frac{2}{3}\text{ s}$$

4. The total distance travelled is the area under the graph, that is the area of the triangle.

Total distance
$$\frac{1}{2} \times \frac{20}{3} \times 16 = \frac{160}{3} = 53\frac{1}{3}$$

5. State the required result. Compare this with the method used to solve this same problem in Worked example 4.

The car travels a total distance of $53\frac{1}{3}$ m before coming to rest.

WORKED EXAMPLE 9 Solving problems by drawing a velocity–time graph (2)

An aeroplane accelerates from rest down a straight runway at a uniform rate of 2 m/s^2, the plane reaches a speed of 252 km/h, when the pilot notices that something is wrong. The pilot aborts the take-off and immediately decelerates at a constant rate bringing the plane to rest. If the total distance travelled by the plane was 2.205 km, determine the length of time that the plane was decelerating for.

THINK | **WRITE**

1. Convert to the speed and distance into the required units.

$$v = 252\text{ km/h}$$
$$v = \frac{252 \times 1000}{60 \times 60}$$
$$v = 70\text{ m/s}$$
$$s = 2.205\text{ km}$$
$$s = 2205\text{ m}$$

2. Determine the time while accelerating.

$$v = u + at$$
$$u = 0$$
$$a = \frac{v}{t}$$
$$2 = \frac{70}{t}$$
$$t = 35$$

3. Draw the velocity–time graph.

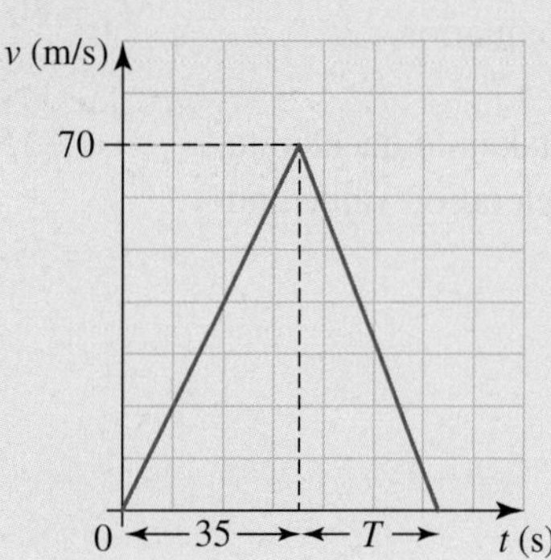

4. The total distance travelled is the area under the graph, that is the area of the two triangles.

Total distance
$s = 2.205$ km
$s = 2205 = \frac{1}{2} \times 70 \times 35 + \frac{1}{2} \times 70 \times T$
$T = \frac{2205}{35} - 35$
$T = 63 - 35$
$T = 28$

5. State the required result.

The plane was decelerating for 28 seconds.

WORKED EXAMPLE 10 Solving problems by drawing a velocity–time graph (3)

Puffing Billy runs on a straight track between Belgrave and Lakeside stations, a distance of 10 km apart.
It accelerates at 0.1 m/s² from rest at Belgrave station until it reaches its maximum speed of 8 m/s. It maintains this speed for a time, then decelerates at 0.15 m/s² and comes to rest at Lakeside station.
Determine the time taken for the Puffing Billy journey.

THINK

1. Draw a velocity–time graph.
Let the time it accelerates for be T_1, the time it travels at constant speed of 8 m/s be T_2 and the time it decelerates for be T_3.

WRITE

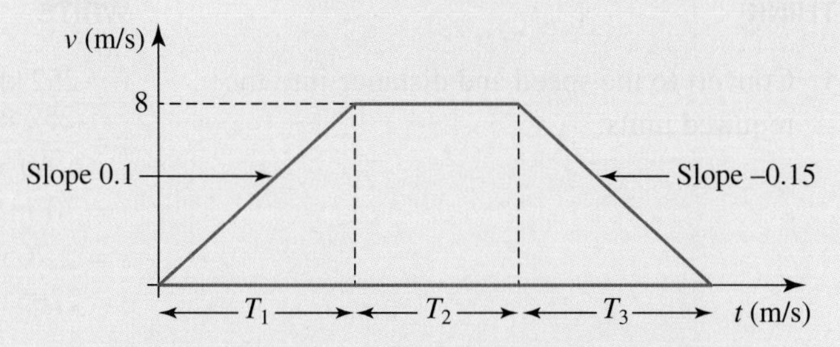

2. To determine the time taken for Puffing Billy's journey, we need to use the total distance travelled 10 km = 10 000 m and area under the graph.

Total distance $s = \frac{1}{2} \times 8T_1 + 8T_2 + \frac{1}{2} \times 8T_3 = 10\,000$

3. To determine the time taken for Puffing Billy to accelerate, T_1, we can use the acceleration and slope equation.

$a = 0.1 = \frac{8}{T_1}$
$T_1 = 80$

4. To determine the time taken to decelerate, T_3, we can use the deceleration value and slope equation.

$a = 0.15 = \frac{8}{T_3}$
$T_3 = 53\frac{1}{3}$

5. Substituting the time values for the train to accelerate and decelerate, we can determine the time taken for the train to move at a constant speed.

$$\frac{1}{2}\times 8\times 80+8T_2+\frac{1}{2}\times 8\times 53\frac{1}{3}=10000$$
$$T_2=1183\frac{1}{3}$$

6. Calculate the total time.

$$T=T_1+T_2+T_3$$
$$T=80+1183\frac{1}{3}+53\frac{1}{3}$$
$$T=1316\frac{2}{3}$$

The total time taken is $1316\frac{2}{3}$ seconds or 21 minutes and 57 seconds.

WORKED EXAMPLE 11 Solving problems using a velocity–time graph and calculus (1)

The velocity v m/s of a body at a time t s is given by

$$v(t)=\begin{cases} 5t & 0\le t\le 2 \\ 10 & 2<t\le 6 \\ 10-\frac{5}{18}(t-6)^2 & 6<t\le 14 \end{cases}$$

a. Sketch the velocity–time graph.
b. Determine the distance travelled in the first 14 seconds.
c. Calculate its displacement after 14 seconds.

THINK	WRITE
a. Draw the velocity–time diagram	a. 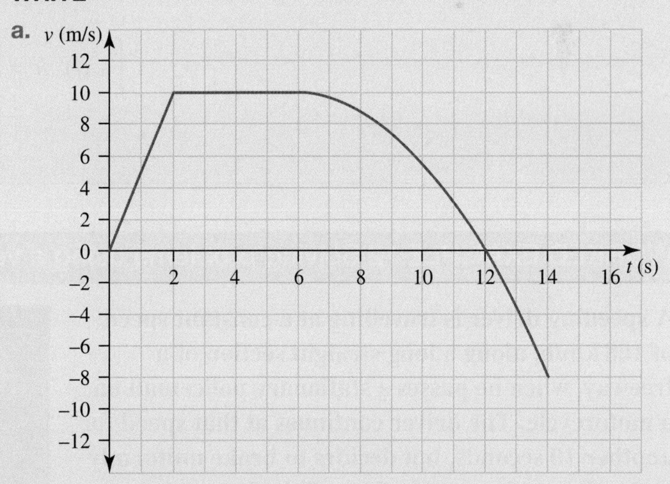
b. 1. Evaluate a definite integral.	b. $\int_6^{12} v(t)\,dt=\int_6^{12}\left(10-\frac{5}{18}(t-6)^2\right)dt$ $=\left[10t-\frac{5}{54}(t-6)^3\right]_6^{12}$ $=\left(24-\frac{5}{24}\times 6^3-60-0\right)$ $=40$

2. Evaluate a definite integral.

$$\int_{12}^{14} v(t)\,dt = \int_{12}^{14} \left(10 - \frac{5}{18}(t-6)^2\right) dt$$

$$= \left[10t - \frac{5}{54}(t-6)^3\right]_{12}^{14}$$

$$= \left(140 - \frac{5}{24} \times 8^3 - 120 + \frac{5}{24} \times 6^3\right)$$

$$= -\frac{200}{27}$$

3. The distance is the total area of the triangle and the rectangle and the signed area of the quadratic

$$d = \frac{1}{2} \times 2 \times 10 + 4 \times 10 + \int_{6}^{12} v(t)\,dt - \int_{12}^{14} v(t)\,dt$$

$$= 10 + 40 + 40 - \left(-\frac{200}{27}\right)$$

$$= \frac{2630}{27}$$

$$\int_{0}^{14} |v(t)|\,dt = 97.4074$$

c. The displacement is the area under the velocity–time graph, taking the negative area from the positive area.

c. $$s = \frac{1}{2} \times 2 \times 10 + 4 \times 10 + \int_{6}^{12} v(t)\,dt + \int_{12}^{14} v(t)\,dt$$

$$= 10 + 40 + 40 + \left(-\frac{200}{27}\right)$$

$$= \frac{2630}{27}$$

$$\int_{0}^{14} v(t)\,dt = 82.5926$$

WORKED EXAMPLE 12 Solving problems using a velocity–time graph and calculus (2)

A speeding driver is travelling at a constant speed of 108 km/h, along a long straight section of a freeway, when he passes a stationary policeman on a motorcycle. The driver continues at that speed for another 10 seconds, but decides to brake uniformly and reduces his speed to 90 km/h in 5 seconds and then continues at that speed. The policeman on the motorcycle begins to chase the speeding driver and starts 5 seconds after the speeding car was first seen. The policeman accelerates for 15 seconds, until he reaches a speed of 108 km/h which he then maintains, until he draws level with the speeding motorist. Determine how long it takes the policeman to catch up to the speeding driver.

THINK	WRITE
1. The distance travelled by the speeding driver and the policeman on the motor cycle will be the same when they draw level. Convert to the speeds into the required units.	$v = 108 \text{ km/h}$ $v = \dfrac{108 \times 1000}{60 \times 60}$ $v = 30 \text{ m/s}$ $v = 90 \text{ km/h}$ $v = \dfrac{90 \times 1000}{60 \times 60}$ $v = 25 \text{ m/s}$
2. Draw the velocity–time graphs for both the speeding driver and the policeman on the motorcycle. Let T be the time when they draw level.	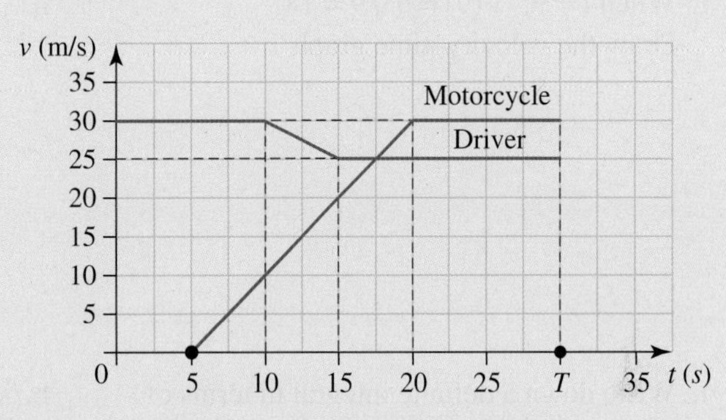
3. For the speeding driver, the distance is the area of the first rectangle, then the area of the trapezium, and finally the area of the last rectangle.	$30 \times 10 + \dfrac{1}{2} \times 5 \times (30 + 25) + 25\,(T - 15)$ $= 25T + \dfrac{125}{2}$
4. For the policeman on the motorcycle, the distance is the area of the triangle and the area of the rectangle.	$\dfrac{1}{2} \times 15 \times 30 + 30\,(T - 20) = 30T - 375$
5. Equate and solve for T.	$25T + \dfrac{125}{2} = 30T - 375$ $5T = \dfrac{875}{2}$ $T = \dfrac{175}{2} = 87.5 \text{ s}$
6. State the required result.	The policeman on the motorcycle takes 82.5 seconds, to catch the speeding driver.

WORKED EXAMPLE 13 Solving problems using a velocity–time graph and calculus (3)

A car is stopped at traffic light A. It then accelerates from rest for nine seconds. Over this stage of the journey, the velocity v m/s of the car at a time t seconds, is given by $v = 6\sqrt{t}$ for $0 \le t \le 9$. The car then travels at a constant speed for a further twenty seconds and finally decelerates until it comes to rest at another traffic light B. During deceleration its velocity v m/s is given by $v = 18\cos\left(\dfrac{\pi}{22}(t - 29)\right)$ for $29 < t \le 40$.

a. **Draw the velocity–time graph which shows the motion of the car as it travels from traffic light A to traffic light B.**
b. **Determine the distance the car travels in the first nine seconds of the motion.**
c. **Calculate the distance the car travels while it is decelerating.**
d. **Determine the total distance travelled by the car as it moves from traffic light A to traffic light B.**
e. **The speed limit on the road is 60 km/h, calculate the percentage of the time when the car was travelling between the traffic lights when it was exceeding the speed limit.**
f. **Determine the acceleration of the car after 4 seconds.**

THINK	WRITE
a. 1. When $t = 9$, $v(9) = 6\sqrt{9} = 18$ Draw the velocity–time graph	a. 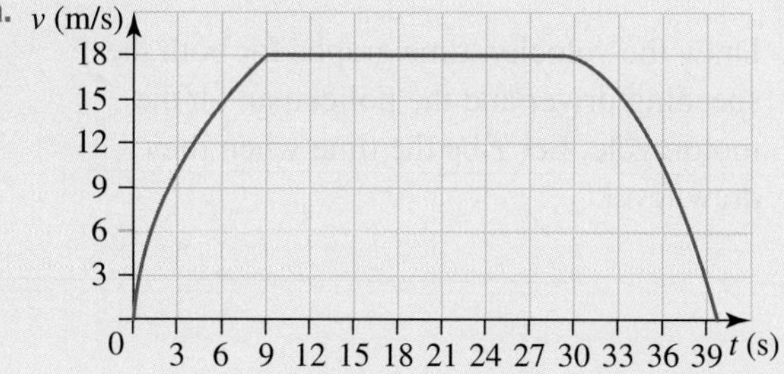
b. 1. Write down a definite integral in terms of t which gives the distance travelled while accelerating.	b. $d_1 = \int_0^9 6\sqrt{t}\,dt$
2. Evaluate the distance while accelerating.	$= \int_0^9 6t^{\frac{1}{2}}\,dt$ $= 6\left[\frac{2}{3}t^{\frac{3}{2}}\right]_0^9 = 4\left[9^{\frac{3}{2}} - 0\right]$ $= 108\text{ m}$
c. 1. Write down a definite integral in terms of t which gives the distance travelled while decelerating.	c. $d_3 = \int_{29}^{40} 18\cos\left(\frac{\pi}{22}(t-29)\right)dt$
2. Evaluate the distance while decelerating.	$= \left[18 \times \frac{22}{\pi}\sin\left(\frac{\pi}{22}(t-29)\right)\right]_{29}^{40}$ $= \frac{396}{\pi}\left[\sin\left(\frac{\pi}{2}\right) - \sin(0)\right]$ $= \frac{396}{\pi}\text{ m}$
d. Determine the total distance travelled by the car.	d. $D = d_1 + d_2 + d_3$ $= 108 + 20 \times 18 + \frac{396}{\pi}$ $= 594.05\text{ m}$

e. 1. Convert to the speeds into the required units.

e. $v = 60\,\text{km/h}$

$= \dfrac{60 \times 1000}{60 \times 60}\,\text{m/s}$

$= 16.\dot{6}\,\text{m/s}$

2. Determine the first time when the car is at the speed limit.

Solve $v = 6\sqrt{t} = 16.\dot{6}$
gives $t = 7.716$

3. Determine the next time when the car is at the speed limit. Note we must use a CAS calculator to solve this equation.

Solve $v = 18\cos\left(\dfrac{\pi}{22}(t-29)\right) = 16.\dot{6}$
with $29 \le t \le 40$
gives $t = 31.7123$

4. Determine the percentage of the time when the car was exceeding the speed limit.

$\dfrac{31.7123 - 7.7160}{40} \times 100 = 60\%$

f. Since $v = 6\sqrt{t}$ for $0 \le t \le 9$.
The acceleration is the derivative.
Evaluate the acceleration when $t = 4$.

f. $v = 6\sqrt{t} = 6t^{\frac{1}{2}}$

$a(t) = \dfrac{dv}{dt} = 3t^{-\frac{1}{2}} = \dfrac{3}{\sqrt{t}}$ for $0 < t < 9$

$a(4) = \dfrac{3}{\sqrt{4}} = 1.5\,\text{m/s}^2$

11.4 Exercise

Students, these questions are even better in jacPLUS

Receive immediate feedback and access sample responses

Access additional questions

Track your results and progress

Find all this and MORE in jacPLUS

Technology free

1. **WE8** A train traveling at 54 km/h, applies the brakes, reducing the speed to 18 km/h, in 6 seconds, assuming the same constant deceleration calculate how much longer it takes for the train to come to rest and the total distance travelled.

2. A car traveling on a straight road at 9 km/h accelerates uniformly, increasing its speed to 18 km/h in five seconds. Assuming that the acceleration is constant calculate how much longer it will take for the car to reach a speed of 72 km/h and determine the total distance travelled.

3. A car traveling at an initial speed of U m/s applies the brakes, reducing its speed to V m/s in T seconds.
   a. Show that the distance travelled in this time is $\dfrac{1}{2}(U+V)T$ metres.
   b. Assuming the same constant deceleration, calculate the total time for the car to come to rest and determine the total distance travelled.

4. **WE9** A car travelling on a straight road, accelerates at $0.5\,\text{m/s}^2$ from rest for 40 seconds, then immediately brakes uniformly coming to rest after travelling one km, calculate the time spent decelerating.

5. A car accelerates uniformly from rest reaches a top speed of 90 km/h and immediately applies the brakes uniformly coming to rest after travelling 900 metres. Determine the total time of the trip.

6. A train travels 1.5 km between two stations on a straight-line track. It accelerates uniformly from rest to reach a maximum speed of 54 km/h, then immediately brakes uniformly coming to rest at the next station. If the magnitude of the deceleration is double the magnitude of the acceleration, determine the length of time that the train was accelerating for.

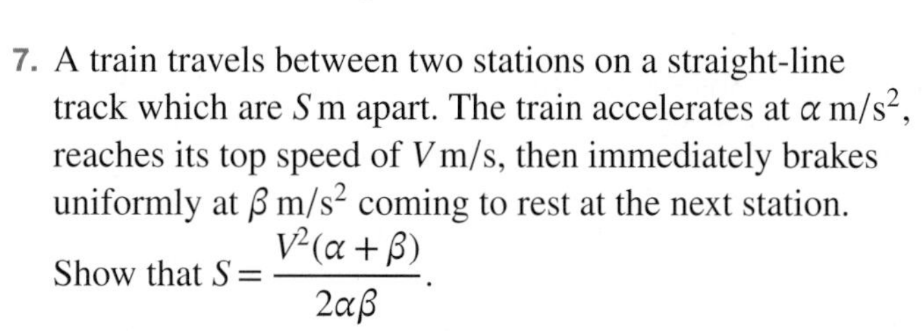

7. A train travels between two stations on a straight-line track which are S m apart. The train accelerates at α m/s^2, reaches its top speed of V m/s, then immediately brakes uniformly at β m/s^2 coming to rest at the next station. Show that $S = \dfrac{V^2(\alpha + \beta)}{2\alpha\beta}$.

8. **WE10** A bus takes 6 minutes to travel from rest to rest between two stops 4.86 km apart on a long straight road. It travels with a constant acceleration until it reaches its maximum speed, then at this maximum speed for 3 minutes, then decelerates uniformly. Determine the maximum speed.

9. A train travelling between two stations has its velocity v m/s at time t s, given by

$$v(t) = \begin{cases} \dfrac{t}{4} & 0 \le t \le 60 \\ 15 & 60 < t \le 300 \\ \dfrac{1}{2}(330 - t) & 300 < t \le 330 \end{cases}$$

Determine the distance between the two stations.

10. A girl goes for a three-minute bike ride which ends with her stopping at her friend's house. Her velocity v m/s at time t s, is given by

$$v(t) = \begin{cases} \dfrac{t}{b} & 0 \le t \le 30 \\ U & 30 \le t \le 120 \\ \dfrac{1}{12}(180 - t) & 120 \le t \le T \end{cases}$$

Determine the values of b, T and U and the total distance travelled by the girl over the three minute ride.

11. A bus starts from rest, accelerates uniformly to a maximum speed of 16 m/s in 45 seconds and then maintains this speed for two minutes, after which the brakes are applied, and the bus is brought to rest with uniform deceleration. If the total distance traveled by the bus is 2500 metres, calculate the total time for the journey.

12. A train runs on a straight-line track between two stations, a distance of 6 km apart. It accelerates at 0.2 m/s^2 from rest at the first station until it reaches its maximum speed of 15 m/s. It maintains this speed for a time, then decelerates at 0.1 m/s^2 to come to rest at the next station. Determine the time taken for the train to travel between the two stations.

13. A car accelerates at α m/s^2. The car reaches its maximum speed of V m/s which it maintains for a while, then has a constant deceleration of β m/s^2 and comes to rest after travelling a total distance of S m. Show that $S > \dfrac{V^2(\alpha+\beta)}{2\alpha\beta}$ and the total time in seconds for the journey is $\dfrac{V}{2}\left(\dfrac{1}{\alpha}+\dfrac{1}{\beta}\right)+\dfrac{S}{V}$.

14. **WE11** The velocity of the body v m/s at time t s is given by $v(t) = \begin{cases} 4t & 0 \le t \le 5 \\ 20 & 5 < t \le 10 \\ 20 - \frac{5}{4}(t-10)^2 & 10 < t \le 16 \end{cases}$

a. Sketch the velocity–time graph.
b. Calculate the distance travelled in the first 16 seconds.
c. Calculate the displacement after 16 seconds.

15. The velocity–time graph of the body v m/s at time t s is shown.

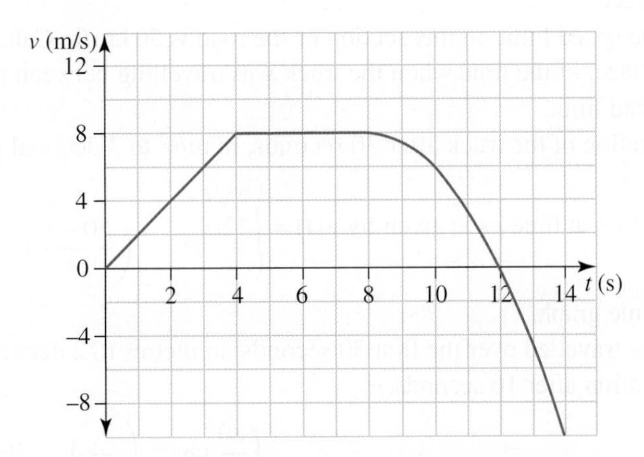

The velocity of the body is given as $v(t) = \begin{cases} at & 0 \le t \le 4 \\ b & 4 < t \le 8 \\ c - d(t-8)^2 & 8 < t \le 14 \end{cases}$

a. Determine the values of a, b and c.
b. Calculate the distance travelled in the first 14 seconds.
c. Calculate the displacement after 14 seconds.

16. **WE12** A speeding driver is travelling at a constant speed of 126 km/h, along a long straight section of a freeway, when they pass a stationary policeman on a motorcycle. The driver continues at that speed for another 4 seconds, before braking uniformly to 90 km/h in 4 seconds and then continuing at that speed. The policeman on the motorcycle begins to chase the speeding driver 2 seconds after the speeding car passed. The policeman accelerates for 16 seconds, until he reaches a speed of 108 km/h which he then maintains until he draws level with the speeding motorist. Determine how long it takes the policeman to catch up to the speeding driver.

Technology active

17. **WE13** A truck moves along a straight road from rest at a traffic light A and accelerates for one minute. Over this stage of the journey, the velocity v m/s of the truck at a time t seconds, is given by $v = \dfrac{30}{\pi}\sin^{-1}\left(\dfrac{t}{60}\right)$ for $0 \le t \le 60$.

The truck maintains its speed for 2 minutes, before the driver sees a red light at traffic light B and decelerates to a stop. The truck's velocity v m/s over this period is given by $v = 15\cos\left(\dfrac{\pi}{80}(t - 180)\right)$ for $180 < t \le 220$.

a. Draw the velocity–time graph which shows the motion of the truck as it travels from traffic light A to traffic light B.

b. Determine the distance the truck travels in the first minute of the motion, in metres to 2 decimal places.

c. Determine the distance the truck travels while it is decelerating, in metres to 2 decimal places.

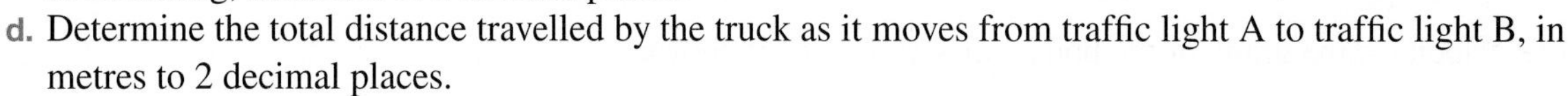

d. Determine the total distance travelled by the truck as it moves from traffic light A to traffic light B, in metres to 2 decimal places.

e. Due to road works, the speed limit on this section of the road is 50 km/h. Calculate the percentage, correct to 1 decimal place, of the time when the truck was travelling between the traffic lights when it was exceeding the speed limit.

f. Determine the acceleration of the truck after 30 seconds, in m/s^2 to 3 decimal places.

18. The velocity of a body v m/s at time t s is given by $v(t) = \begin{cases} 6\sqrt{t} & 0 \le t \le 25 \\ \dfrac{120}{\pi}\tan^{-1}\left(\dfrac{50 - t}{25}\right) & 25 < t \le 50 \end{cases}$

a. Sketch the velocity–time graph.

b. Determine the distance travelled over the first 50 seconds, in metres to 2 decimal places.

c. Determine the acceleration after 16 seconds.

19. The velocity of a body v m/s at time t s is given by $v(t) = \begin{cases} \dfrac{80}{\pi}\tan^{-1}\left(\dfrac{t}{16}\right) & 0 \le t \le 16 \\ 20 - \dfrac{5}{49}(t - 16)^2 & 16 < t \le 36 \end{cases}$

a. Sketch the velocity–time graph.

b. Determine the distance travelled over the first 36 seconds, in metres to 2 decimal places.

c. Determine its displacement after 36 seconds, in metres to 2 decimal places.

d. Determine the acceleration after 23 seconds.

20. A body accelerates for a time, then decelerates at a constant rate, then brakes and comes to rest. Its velocity v m/s at time t s is given by $v(t) = \begin{cases} \dfrac{80}{\pi}\tan^{-1}\left(\dfrac{t}{90}\right) & 0 \le t \le 90 \\ a + bt & 90 < t \le 270 \\ \dfrac{30}{\pi}\cos^{-1}\left(\dfrac{t - 270}{90}\right) & 270 < t \le 360 \end{cases}$

a. Determine the values of a and b.

b. Sketch the velocity–time graph.

c. Determine the instantaneous acceleration after 45 seconds of motion, in m/s^2 to 3 decimal places.

d. Determine the distance travelled over the first 6 minutes, in metres to 2 decimal places.

11.4 Exam questions

Question 1 (1 mark) TECH-ACTIVE

Source: VCE 2015, Specialist Mathematics Exam 2, Section 1, Q11; © VCAA.

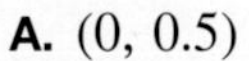

MC The velocity–time graph for a body moving along a straight line is shown.

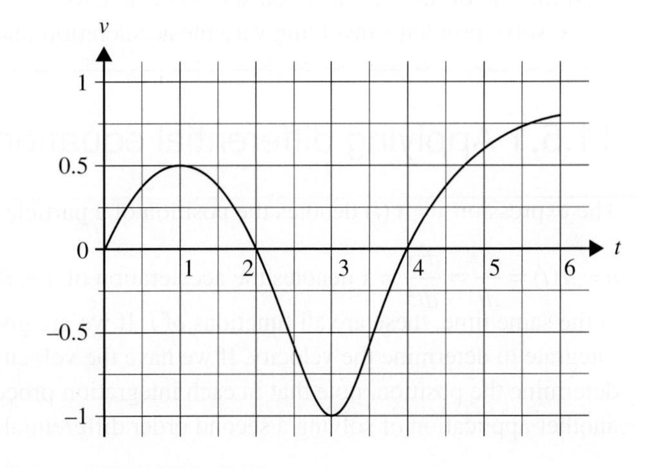

The body first returns to its initial position within the time interval

A. (0, 0.5)
B. (0.5, 1.5)
C. (1.5, 2.5)
D. (2.5, 3.5)
E. (3.5, 5)

Question 2 (1 mark) TECH-ACTIVE

Source: VCE 2014, Specialist Mathematics Exam 2, Section 1, Q22; © VCAA.

MC The velocity–time graph shows the motion of a body travelling in a straight line, where v ms^{-1} is its velocity after t seconds.

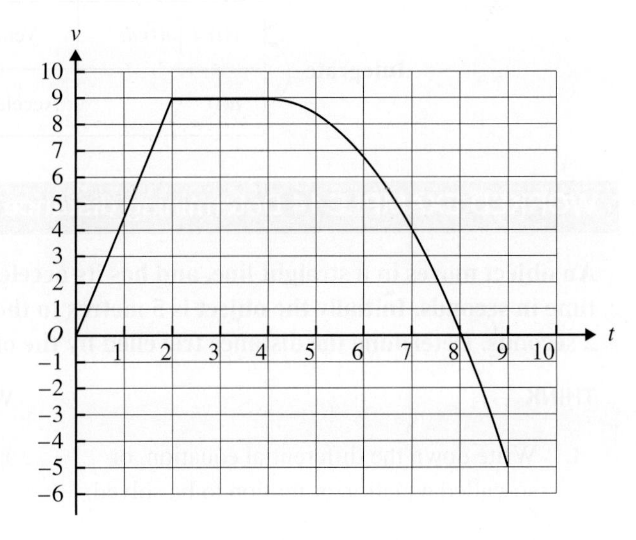

The velocity of the body over the time interval $t \in [4, 9]$ is given by $v(t) = -\frac{9}{16}(t-4)^2 + 9$.

The distance, in metres, travelled by the body over nine seconds is closest to

A. 45.6
B. 47.5
C. 48.6
D. 51.0
E. 53.4

Question 3 (4 marks) TECH-FREE

On a certain straight section of tram line the stopping points are 2.76 kilometres apart. The tram accelerates uniformly to reach its maximum speed of 16 m/s which it maintains for a while, then brakes uniformly. If the magnitude of the deceleration is double the magnitude of the acceleration and the time between the stops is four minutes, calculate the distance travelled at top speed.

More exam questions are available online

11.5 Acceleration that depends on time

LEARNING INTENTION

At the end of this subtopic you should be able to:

- solve problems involving variable acceleration that is expressed as a function of time.

11.5.1 Applying differential equations to rectilinear motion

The expression $x = x(t)$ denotes the position of a particle, $v = v(t) = \frac{dx}{dt} = \dot{x}$ denotes the velocity and $a = a(t) = \frac{dv}{dt} = \frac{d^2x}{dt^2} = \ddot{x}$ denotes the acceleration of a particle at a time t. Since a particle cannot be in two places at the same time, these are all functions of t. If we are given the acceleration as a function of t, then we can integrate to determine the velocity. If we have the velocity as a function of t, then we can integrate again to determine the position, note that in each integration process we need to add in an arbitrary constant. This is just another application of solving a second order differential equation as we have considered in earlier chapters.

Integrate	$x(t) = \int v(t)\,dt$	Position	$x(t)$	**Differentiate**
	$v(t) = \int a(t)\,dt$	Velocity	$v(t) = \frac{dx}{dt}$	
Integrate	$a(t)$	Acceleration	$a(t) = \frac{dv}{dt}$	**Differentiate**

WORKED EXAMPLE 14 Determining distance travelled given an acceleration function (1)

An object moves in a straight line, and has its acceleration given as $a(t) = 6t - 12$ m/s^2, where t is the time in seconds. Initially the object is 5 metres to the right of the origin and comes to rest after 2 seconds. Determine the distance travelled by the object in the first 2 seconds.

THINK	WRITE
1. Write down the differential equation, or so-called equation of motion to be solved.	$\ddot{x} = \frac{d^2x}{dt^2} = \frac{dv}{dt} = 6t - 12$
2. Integrate both sides with respect to t.	$v = \int (6t - 12)dt$
3. Perform the integration, using basic integration techniques, adding in the first constant of integration.	$v = 3t^2 - 12t + c_1$
4. Use the given initial condition to determine the first constant of integration.	When $t = 2$, the object is at rest, so $v = 0$. Substituting to determine c_1. $0 = 12 - 24 + c_1$ $c_1 = 12$
5. Substitute back for the constant of integration.	$v = \frac{dx}{dt} = 3t^2 - 12t + 12$
6. Integrate both sides again with respect to t.	$x = \int \left(3t^2 - 12t + 12\right) dt$
7. Perform the integration, adding in a second constant of integration.	$x = t^3 - 6t^2 + 12t + c_2$

8. Use the given initial condition to determine the second constant of integration.

Initially when $t=0, x=5$.
Substituting to determine c_2.
$5=0+c_2$

9. Substitute back for the constant of integration.

$x=x(t)=t^3-6t^2+12t+5$

10. Calculate the position at the required time.

Substitute $t=2$.
$x(2)=8-24+24+5=13$

11. State the required result.

The object moves from $t=0, x=5$ to $t=2, x=13$, the distance travelled is 8 metres.

12. An alternative method to determine the distance travelled can be used.

Since the distance travelled is the area under the velocity–time graph, this distance is given by the definite integral

$$D=\int_0^2 \left(3t^2-12t+12\right) dt$$
$$=\left[t^3-6t^2+12t\right]_0^2$$
$$=(8-24+24)-0$$
$$=8\text{ m}$$

WORKED EXAMPLE 15 Determining distance travelled given an acceleration function (2)

A particle moves back and forth along the x-axis and is subjected to an acceleration $a(t)=-4(\sin(2t)+\cos(2t))$ m/s^2 at time t seconds. If its initial velocity is 2 m/s and initially the particle starts from a point 3 metres to the right of the origin, express the position x metres in terms of t.

THINK

WRITE

1. Write down the equation of motion to be solved.

$$\ddot{x}=\frac{d^2x}{dt^2}=\frac{dv}{dt}=-4\sin(2t)-4\cos(2t)$$

2. Integrate both sides with respect to t.

$$v=\int (-4\sin(2t)-4\cos(2t))\,dt$$

3. Perform the integration using the results $\int \sin(kx)dx=-\frac{1}{k}\cos(kx)$ and $\int \cos(kt)dx=\frac{1}{k}\sin(kt)$ with $k=2$.

$$=2\cos(2t)-2\sin(2t)+c_1$$

4. Use the given initial condition to determine the first constant of integration.

Since initially when $t=0, v=2$, then substituting
$2=2\cos(0)-2\sin(0)+c_1$
$c_1=0$

5. Substitute back for the constant of integration.

$$v=\frac{dx}{dt}=2\cos(2t)-2\sin(2t)$$

6. Integrate both sides again with respect to t.

$$x=\int (2\cos(2t)-2\sin(2t))\,dt$$

7. Perform the integration, adding in a second constant of integration.

$$=\sin(2t)+\cos(2t)+c_2$$

8. Use the given initial condition to determine the second constant of integration.

Since initially when $t=0$, $x=3$, then substituting
$3=\sin(0)+\cos(0)+c_2$
$3=1+c_2$
$c_2=2$

9. Substitute back for the constant of integration. This is the required result.

$x=\sin(2t)+\cos(2t)+2$

Sometimes the brakes on a decelerating car can be applied to give a non-constant deceleration. In these cases the acceleration can be considered a function of the time.

WORKED EXAMPLE 16 Determining braking distance with non-constant deceleration

A car is moving along a straight road at a speed of 90 km/h when the driver brakes, the acceleration is given by $a(t)=-(20-8t)$ m/s^2 where t is the time in seconds after the driver applies the brakes. Determine:

a. the time in seconds after which speed of the car has been reduced to 57.6 km/h

b. the distance travelled in metres in this time.

THINK	WRITE
a. 1. Write down the equation of motion to be solved.	a. $\ddot{x}=\frac{d^2x}{dt^2}=\frac{dv}{dt}=-(20-8t)=8t-20$
2. Integrate both sides with respect to t.	$v=\int(8t-20)dt$
3. Perform the integration using the basic integration techniques, adding in the first constant of integration.	$=4t^2-20t+c_1$
4. Use the given initial conditions to determine the first constant of integration. We need to use correct units.	The initial speed is 90 km/h. Convert km/h to m/s $\frac{90\times 1000}{60\times 60}=25$ m/s Initially when $t=0$, $v=25$ $25=0+c_1$ $c_1=25$
5. Substitute back for the constant of integration.	$v=4t^2-20t+25$
6. Determine the braking time.	The final speed is 57.6 km/h. Convert km/h to m/s $\frac{57.6\times 1000}{60\times 60}=16$ m/s Determine t when $v=16$
7. The earlier time is the one required	$16=4t^2-20t+25$ $4t^2-20t+9=0$ $(2t-1)(2t-9)=0$ $t=\frac{1}{2},\frac{9}{2}$ $T=\frac{1}{2}$

8. State the required result. After 0.5 seconds the car's speed has been reduced from 90 to 57.6 km/h.

b. 1. Express the velocity in terms of time t.

b. $v = \frac{dx}{dt} = 4t^2 - 20t + 25$

2. Integrate both sides again with respect to t.

$x = \int (4t^2 - 20t + 25)\, dt$

3. Perform the integration using basic integration techniques, adding in a second constant of integration.

$= \frac{4t^3}{3} - 10t^2 + 25t + c_2$

4. Use the given initial condition to determine the second constant of integration.

Since we are after the distance travelled from first braking, assume that when $t = 0$, $x = 0$, so that $c_2 = 0$.

5. Express the position in terms of time.

$x = x(t) = \frac{4t^3}{3} - 10t^2 + 25t$

6. Calculate the distance travelled D metres while braking.

$D = x(T)$

$D = \frac{4 \times (0.5)^3}{3} - 10 \times 0.5^2 + 25 \times 0.5 = 10\frac{1}{6}$

7. State the final result.

The distance travelled while braking is $10\frac{1}{6}$ metres.

8. An alternative method to determine the distance travelled can be used.

Since the distance travelled is the area under the velocity–time graph, this distance is given by the definite integral

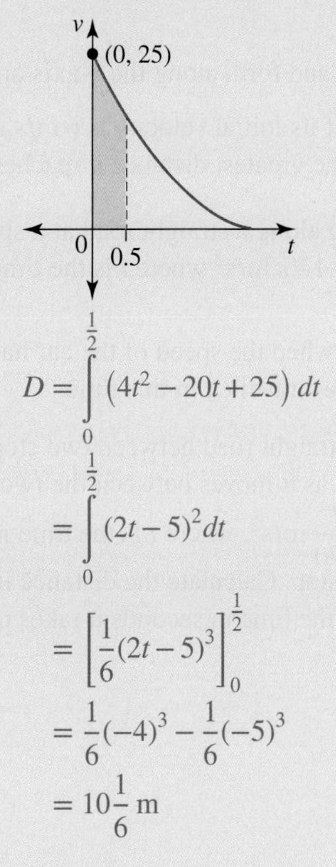

$$D = \int_0^{\frac{1}{2}} (4t^2 - 20t + 25)\, dt$$

$$= \int_0^{\frac{1}{2}} (2t - 5)^2 dt$$

$$= \left[\frac{1}{6}(2t - 5)^3\right]_0^{\frac{1}{2}}$$

$$= \frac{1}{6}(-4)^3 - \frac{1}{6}(-5)^3$$

$$= 10\frac{1}{6}\text{ m}$$

11.5 Exercise

Students, these questions are even better in jacPLUS

Receive immediate feedback and access sample responses

Access additional questions

Track your results and progress

Find all this and MORE in jacPLUS

Technology free

1. **WE14** An object moves in a straight line, and has its acceleration given as $a(t) = 3 - 6t\,\text{m/s}^2$, where t is the time in seconds. Initially the object is at rest at the origin and comes to rest after one second. Calculate the distance travelled by the object in the first second.

2. A particle is moving along the x-axis and at time $t \geq 0$ seconds has an acceleration $a(t) = 24t - 6\,\text{m/s}^2$. If after one second its position is 4 metres and its initial velocity is 1 m/s, determine its position x metres at a time t seconds.

3. **WE15** A particle moves back and forth along the x-axis and its acceleration is given by $a(t) = -2\sin\left(\frac{t}{3}\right)\,\text{m/s}^2$ at time $t \geq 0$ seconds. If its initial velocity is 6 m/s and initially the particle starts at the origin, express the position x metres in terms of t.

4. A particle moves back and forth along the x-axis and has an acceleration $a(t) = -2\cos\left(\frac{t}{2}\right)\,\text{m/s}^2$ at time $t \geq 0$ seconds. If initially the particle is at rest and starts from a point 4 metres from the origin, determine the position in metres of the particle after a time of $\frac{\pi}{2}$ seconds.

5. A particle moves back and forth along the x-axis and has an acceleration given by $a(t) = -3\sin\left(\frac{t}{2}\right)\,\text{m/s}^2$ at time $t \geq 0$ seconds. If its initial velocity is 6 m/s and initially the particle starts from a point 4 metres from the origin, determine the greatest distance it reaches from the origin.

6. **WE16** A car is moving along a straight road at a speed of 72 km/h when the driver brakes. The acceleration is given by $a(t) = 50 - 120t\,\text{m/s}^2$ where t is the time in seconds after the driver applies the brakes. Determine:
   a. the time in seconds when the speed of the car has been reduced to 36 km/h
   b. the distance in metres travelled in this time.

7. A bus moves along a straight road between two stops. The acceleration of the bus as it moves between the two stops is given by $a(t) = \frac{1}{3} - \frac{t}{270}\,\text{m/s}^2$, where t is the time in seconds after it leaves the first stop. Calculate the distance in metres between the stops and the time in seconds it takes to travel between the two stops.

8. A car moves along a straight road. When travelling at 60 km/h the driver applies the brakes, its acceleration is given by $a(t) = -\frac{100t}{3}$ m/s^2, where $t \geq 0$ is the time in seconds after the driver applies the brakes. Determine the distance travelled in metres until the car comes to rest.

9. A car is moving along a straight road at a speed of 90 km/h when the driver slams on the brakes. The deceleration is given by $\frac{200}{(t+2)^3}$ m/s^2 where t is the time in seconds after the driver applies the brakes. Determine:

   a. the time taken in seconds for the speed of the car to be reduced to 57.6 km/h
   b. the distance in metres travelled in this time.

10. A particle moves in a straight line so that at time t seconds its acceleration is given by $a(t) = \frac{-800}{(t+5)^3}$ m/s^2. If the initial velocity of the particle is 16 m/s, determine how far it has travelled in the first 5 seconds.

11. A particle is moving along the x-axis and at time t has its acceleration given by $a(t) = 8e^{2t} - 4$ m/s^2. Initially the body is at rest at the origin. Express x in terms of t.

12. A particle moves in a straight line. At time t seconds, $t \geq 0$, its position from a fixed origin O is x metres and its acceleration is a m/s^2, is given by $\frac{49}{5}e^{-\frac{t}{5}}$. If the particle starts from rest at the point where $x = 20$, express x in terms of t.

13. A body is moving in a straight line path on a horizontal table and has its acceleration given by $a(t) = be^{-kt}$ m/s^2, where b and k are positive constants. If its initial speed is U m/s, show that after a time t seconds its position is given by $x = Ut + \frac{b}{k}\left[t + \frac{1}{k}\left(e^{-kt} - 1\right)\right]$ metres.

Technology active

14. A particle moves in a straight line. At time t seconds, $t \geq 0$ its position from a fixed origin O is x metres and its acceleration is a m/s^2, is given by $a(t) = 4 - 4e^{-0.1t}$. If initially it is moving away from the origin with a velocity of 12 m/s, determine how far in metres, correct to 3 decimal places, it has travelled in the first 5 seconds.

15. A car is moving along a straight road at a speed of 90 km/h when the driver brakes, the acceleration is given $a(t) = \frac{-36900\sqrt{2}}{\sqrt{(369t+128)^3}}$ m/s^2 where t is the time in seconds after the driver applies the brakes. After a time T seconds the speed of the car has been reduced to 57.6 km/h and in this time the car has travelled a distance of D metres. Determine the values of T and D.

16. A particle moves along a straight line path with an acceleration of $a(t) = t\cos\left(t^2\right)$ m/s^2, where $t \geq 0$ is the time in seconds. Initially the particle is at rest at the origin, determine the distance travelled in metres over the first two seconds of its motion, correct to 3 decimal places.

17. A particle moving in a straight line has an acceleration given by $a(t) = -(12\cos(3t) + 5\sin(3t))e^{-2t}$ m/s^2, where $t \geq 0$ is the time in seconds. Initially the particle is at the origin moving with a speed of 3 m/s. Calculate the distance travelled in metres over the first second of its motion, correct to 3 decimal places.

18. A particle has an acceleration given by $a(t) = 2\cos(2t)e^{\sin(2t)}$ m/s^2, where $t \geq 0$ is the time in seconds. Initially the particle is at rest at the origin, determine the distance travelled in metres over the first two seconds of its motion, correct to 3 decimal places.

11.5 Exam questions

Question 1 (9 marks) TECH-ACTIVE

Source: *adapted from: VCE 2016, Specialist Mathematics Exam 2, Section B, Q5; © VCAA.*

A model rocket is launched from rest and travels vertically up, with a vertical propulsion force of $(50 - 10t)$ newtons after t seconds of flight, where $t \in [0, 5]$. Assume that the rocket is subject only to the vertical propulsion force and gravity, and that air resistance is negligible.

Let v ms^{-1} be the velocity of the rocket t seconds after it is launched.

The rate of change of the velocity with respect to time is given by $\frac{dv}{dt} = \frac{76}{5} - 5t$.

a. Find the velocity, in ms^{-1}, of the rocket after five seconds. **(2 marks)**

b. Find the height of the rocket after five seconds. Give your answer in metres, correct to two decimal places. **(2 marks)**

c. After five seconds, when the vertical propulsion force has stopped, the rocket is subject only to gravity. Find the maximum height reached by the rocket. Give your answer in metres, correct to two decimal places. **(2 marks)**

d. Having reached its maximum height, the rocket falls directly to the ground. Assumming negligible air resistance during this final stage of motion, find the time for which the rocket was in flight. Give you answer in seconds, correct to one decimal place. **(3 marks)**

Question 2 (4 marks) TECH-FREE

A particle moves back and forth along the x-axis and has an acceleration a m/s^2 at time $t \geq 0$ seconds, given by $a(t) = 36\cos(3t)$. If initially it is at rest 2 metres from the origin, determine the furthest distance it reaches from the origin.

Question 3 (4 marks) TECH-FREE

A car moves along a straight road, when its speed is U m/s the driver applies the brakes. The deceleration of the car is given by $-kt$ m/s^2, where $t \geq 0$, is the time in seconds after the driver applies the brakes and k is a positive constant, show that the car comes to rest after a time of $\sqrt{\frac{2U}{k}}$ seconds and travels a distance of $\frac{2U}{3}\sqrt{\frac{2U}{k}}$ metres.

More exam questions are available online.

11.6 Acceleration that depends on velocity

LEARNING INTENTION

At the end of this subtopic you should be able to:

- solve problems involving variable acceleration that is expressed as a function of velocity.

11.6.1 Acceleration given as a function of velocity

If the acceleration of an object depends upon the velocity v then we can write it as a function of the velocity, $a = a(v)$.

Recall that $a = \frac{dv}{dt}$. Inverting both sides gives $\frac{dt}{dv} = \frac{1}{a(v)}$. Integrating both sides with respect to v gives $t = \int \frac{1}{a(v)} dv$, with initial conditions on t and v. This gives a relationship between v and t.

Another way of working with acceleration that is a function of v is to consider using the chain rule as follows: $a = \frac{dv}{dt} = \frac{dx}{dt} \cdot \frac{dv}{dx} = v\frac{dv}{dx}$.

Dividing by v and then inverting both sides gives $\frac{dx}{dv} = \frac{v}{a(v)}$. Both sides can then be integrated with respect to v giving $x = \int \frac{v}{a(v)} dv$, with initial conditions on x and v. This gives a relationship between v and x.

Acceleration as a function of velocity

When the acceleration is written as a function of velocity, $a = a(v)$:

- **If initial conditions on v and t are given and a relationship between v and t is required use $\ddot{x} = \frac{dv}{dt} = a(v)$ and solve to give $t = \int \frac{1}{a(v)} dv$.**
- **If initial conditions on v and x are given and a relationship between v and x is required use $\ddot{x} = v\frac{dv}{dx} = a(v)$ and solve $\frac{dx}{dv} = \frac{v}{a(v)}$ to give $x = \int \frac{v}{a(v)} dv$.**

WORKED EXAMPLE 17 Determining the position function, given the acceleration as a function of velocity

A body moves in a straight line, and is subjected to an acceleration of $a(v) = -2(v + 3)$, where v is the velocity in m/s. Initially the body is moving at 3 m/s and is at the origin. Show that the position x at time t is given by $x = 3\left(1 - t - e^{-2t}\right)$.

THINK	WRITE
1. Write down the equation of motion to be solved.	$\ddot{x} = \frac{d^2x}{dt^2} = \frac{dv}{dt} = a(v) = -2(v+3)$
2. Invert both sides or separate the variables.	$\frac{dt}{dv} = -\frac{1}{2(v+3)}$ $-2\frac{dt}{dv} = \frac{1}{v+3}$
3. Integrate both sides with respect to v.	$-2t = \int \frac{1}{v+3} dv$
4. Perform the integration; place the first constant of integration on one side of the equation.	$-2t = \log_e(\lvert v+3 \rvert) + c_1$
5. Use the given initial conditions to determine the first constant of integration.	Initially when $t = 0$, $v = 3$ $0 = \log_e(6) + c_1$ $c_1 = -\log_e(6)$
6. Substitute back for the first constant of integration.	$-2t = \log_e(\lvert v+3 \rvert) - \log_e(6)$
7. Use the laws of logarithms.	$-2t = \log_e\left(\frac{\lvert v+3 \rvert}{6}\right)$
8. When removing the modulus symbol, include a ± symbol.	$\frac{\lvert v+3 \rvert}{6} = e^{-2t}$ $\lvert v+3 \rvert = 6e^{-2t}$ $\pm(v+3) = 6e^{-2t}$

9. Use the initial conditions to determine if the positive or negative should be taken. (In this case we take the positive.)	Initially when $t=0,\ v=3$ $\pm\,(3+3)=6e^{0}$
10. Solve for v.	$v+3=6e^{-2t}$ $v=6e^{-2t}-3$
11. Use $v=\dfrac{dx}{dt}$.	$\dfrac{dx}{dt}=6e^{-2t}-3$
12. Integrate both sides with respect to t.	$x=\displaystyle\int\left(6e^{-2t}-3\right)dt$
13. Perform the integration, place the second constant of integration on one side of the equation.	$x=-3e^{-2t}-3t+c_2$
14. Use the given initial conditions to determine the second constant of integration.	Initially when $t=0,\ x=0$ $0=-3+c_2$ $c_2=3$
15. Substitute back for the second constant of integration.	$x=-3e^{-2t}-3t+3$
16. Factorise and the result is shown.	$x=3\left(1-t-e^{-2t}\right)$

Sometimes the brakes on a decelerating car can create a deceleration that is a function of the car's velocity. We can use the same skills as in the previous worked example to solve these problems.

WORKED EXAMPLE 18 Deceleration as a function of velocity

A car is moving along a straight road at a speed of 90 km/h when the driver brakes. The resulting acceleration is $a(v)=-4\sqrt{v}$ m/s^2 where v is the speed in m/s after the driver applies the brakes.

a. Determine the time taken in seconds for the speed of the car to be reduced to 57.6 km/h.

b. Determine the distance travelled in metres in this time.

THINK	WRITE
a. 1. Write down the equation of motion to be solved.	**a.** $\ddot{x}=\dfrac{d^2x}{dt^2}=\dfrac{dv}{dt}=-4\sqrt{v}$
2. Invert both sides or separate the variables.	$\dfrac{dt}{dv}=-\dfrac{1}{4\sqrt{v}}$ $-4\dfrac{dt}{dv}=\dfrac{1}{\sqrt{v}}=v^{-\frac{1}{2}}$
3. Integrate both sides with respect to v.	$-4t=\displaystyle\int v^{-\frac{1}{2}}dv$
4. Perform the integration; place the first constant of integration on one side of the equation.	$-4t+c_1=2v^{\frac{1}{2}}=2\sqrt{v}$
5. Use the given initial conditions to determine the first constant of integration. We need to use correct units.	Since 90 km/h = 25 m/s Initially when $t=0,\ v=25$ $c_1=2\sqrt{25}=10$

6. Substitute back for the constant of integration.	$10 - 4t = 2\sqrt{v}$
7. Determine the braking time.	Since 57.6 km/h = 16 m/s Calculate t when $v = 16$ $10 - 4t = 2\sqrt{16} = 8$ $4t = 2$ $t = \frac{1}{2}$
8. State the required result.	After 0.5 seconds the car's speed has been reduced from 90 to 57.6 km/hr.
b. 1. Use the alternative form for the acceleration.	**b.** $\ddot{x} = v\frac{dv}{dx} = -4\sqrt{v}$ $\frac{dv}{dx} = -\frac{4\sqrt{v}}{v}$ $= -\frac{4}{\sqrt{v}}$
2. Invert both sides or separate the variables.	$-4\frac{dx}{dv} = v^{\frac{1}{2}}$ $-4\int dx = \int v^{\frac{1}{2}} dv$
3. Perform the integration; place the second constant of integration on one side of the equation.	$-4x + c_2 = \frac{2}{3}v^{\frac{3}{2}}$
4. Use the given initial condition to determine the second constant of integration.	When $v = 25$, $x = 0$ $c_2 = \frac{2}{3}\sqrt{25^3} = \frac{250}{3}$
5. Substitute back for the second constant of integration.	$\frac{250}{3} - 4x = \frac{2}{3}v^{\frac{3}{2}}$
6. Determine the distance travelled.	Solve for x when $v = 16$ $4x = \frac{250}{3} - \frac{2}{3}\sqrt{16^3}$ $4x = \frac{250 - 128}{3}$ $x = \frac{61}{6}$
7. State the distance travelled while braking.	The distance travelled while braking is $10\frac{1}{6}$ metres.

11.6.2 General cases

When an object moves, there is always a force that acts in the opposing direction (such as drag or friction) and generally it is proportional to some power of the velocity. That is the acceleration can be expressed as $a(v) = -kv^n$; typical values of n are $1, 2, 3, 4, 5, \frac{1}{2}, \frac{3}{2}$ etc. Using these and generalizing from the last example, some general expressions can be derived.

WORKED EXAMPLE 19 Deceleration that is proportional to a power of the velocity

A car moves along a straight road at a speed of U m/s before the driver brakes. The deceleration due to braking is proportional to the fourth power of the velocity v in m/s. After a time T seconds the speed of the car has been reduced to $\frac{1}{2}U$ m/s and in this time the car has travelled a distance of D metres. Show that $\frac{D}{T} = \frac{9U}{14}$.

THINK	WRITE
1. Write down the equation of motion to be solved, where λ is a positive proportionality constant to be determined.	$a(v) \alpha v^4$ Use $\ddot{x} = \frac{dv}{dt} = -\lambda v^4$
2. Invert both sides or separate the variables.	$\frac{dt}{dv} = \frac{1}{\lambda v^4}$
3. Integrate both sides with respect to v.	$\int -\lambda dt = \int v^{-4} dv$
4. Perform the integration; place the first constant of integration on one side of the equation.	$-\lambda t + c_1 = -\frac{1}{3}v^{-3}$
5. Use the given initial conditions to determine the first constant of integration. We are using correct units.	Initially when $t = 0$, $v = U$ $c_1 = -\frac{1}{3}U^{-3} = -\frac{1}{3U^3}$
6. Substitute back for the constant of integration.	$-\lambda t - \frac{1}{3U^3} = -\frac{1}{3v^3}$ $\lambda t + \frac{1}{3U^3} = \frac{1}{3v^3}$
7. Obtain a relationship between the parameters.	Now when $t = T$, $v = \frac{1}{2}U$ $\lambda T + \frac{1}{3U^3} = \frac{1}{3\left(\frac{1}{2}U\right)^3} = \frac{8}{3U^3}$
8. Simplify this relationship and express λ in terms of T and U.	$\lambda T = \frac{8}{3U^3} - \frac{1}{3U^3}$ $\lambda T = \frac{7}{3U^3}$ $\lambda = \frac{7}{3TU^3}$
9. Next obtain a relationship between v and x.	Use $x = v\frac{dv}{dx} = -\lambda v^4$ so that $\frac{dv}{dx} = -\lambda v^3$
10. Invert both sides or separate the variables.	$\frac{dx}{dv} = -\frac{1}{\lambda v^3}$
11. Integrate both sides with respect to v.	$\int -\lambda dx = \int v^{-3} dv$

12. Perform the integration, place the second constant of integration on one side of the equation.	$-\lambda x + c_2 = -\frac{1}{2}v^{-2}$
13. Use the given initial conditions to determine the second constant of integration.	Initially when $t=0$, $x=0$ and $v=U$ $c_2 = -\frac{1}{2}U^{-2} = -\frac{1}{2U^2}$
14. Substitute back for the second constant of integration.	$-\lambda x - \frac{1}{2U^2} = -\frac{1}{2v^2}$ $\lambda x + \frac{1}{2U^2} = \frac{1}{2v^2}$
15. Obtain a relationship between the parameters.	Now when $x = D$, $v = \frac{1}{2}U$ $\lambda D + \frac{1}{2U^2} = \frac{1}{2\left(\frac{1}{2}U\right)^2} = \frac{2}{U^2}$
16. Simplify this relationship and express λ in terms of D and U.	$\lambda D = \frac{2}{U^2} - \frac{1}{2U^2}$ $\lambda D = \frac{3}{2U^2}$ $\lambda = \frac{3}{2DU^2}$
17. Eliminate λ be equating the two expressions for λ.	$\lambda = \frac{7}{3TU^3} = \frac{3}{2DU^2}$
18. Simplify the resulting expression.	Hence $\frac{D}{T} = \frac{9U}{14}$ as required.

11.6.3 Vertical motion

When a body moves vertically, we must consider gravity as part of its equation of motion.

Downwards motion

Consider a body moving vertically downwards. The acceleration on the body, is made up of the component due to gravity which always acts vertically downwards, and a component upwards (resisting the downwards motion) which is proportional to some power of its velocity. Considering downwards as the positive direction, the equation of motion is given by $a = g - kv^n$. As the body falls, it reaches a so-called terminal or limiting velocity, v_T. This value can be obtained from $v_T = \lim\limits_{t\to\infty} v(t)$ or, since it is a constant speed, when the acceleration is zero, $a = g - kv_T^n = 0$.

Upwards motion

Consider a body moving vertically upwards. The acceleration on the body, has its gravity component which always acts vertically downwards, and a component downwards (resisting the upwards motion) which is proportional to some power of its velocity. Considering upwards as the positive direction, its equation of motion is given by $a = -g - kv^n$. As before typical values of n are $1, 2, 3, 4, 5, \frac{1}{2}, \frac{3}{2}$.

WORKED EXAMPLE 20 Vertical acceleration as a function of velocity

A large brick is accidentally dropped from a high-rise construction site. As it falls vertically downwards its acceleration is given by $a(v) = 9.8 - \frac{v^2}{500}$ m/s^2, where v m/s is the velocity of the brick at a time t seconds after it was dropped. It has travelled a distance of x metres in this time.

a. **Show that $x = 250 \log_e \left(\frac{4900}{4900 - v^2} \right)$.**

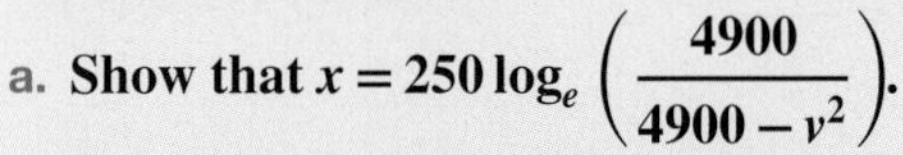

b. **Determine the speed of the brick after it has fallen a distance of 100 metres.**

c. **Show that $t = \frac{25}{7} \log_e \left(\frac{70 + v}{70 - v} \right)$.**

d. **Calculate the terminal velocity of the brick.**

e. **Determine the time taken for the brick to fall a distance of 100 metres.**

THINK	WRITE
a. 1. Formulate the equation of motion to be solved. Simplify the required result.	a. $\ddot{x} = a(v) = 9.8 - \frac{v^2}{500}$ $\ddot{x} = \frac{4900 - v^2}{500}$
2. To obtain a relationship between v and x use $\ddot{x} = v\frac{dv}{dx}$ and invert both sides.	$v\frac{dv}{dx} = \frac{4900 - v^2}{500}$ $\frac{dx}{dv} = \frac{500v}{4900 - v^2}$
3. Integrate both sides with respect to v.	$x = 500 \int \frac{v}{4900 - v^2} dv$
4. Perform the integration, place the first constant of integration on one side of the equation.	$x = -\frac{500}{2} \log_e (4900 - v^2) + c_1$
5. Use the given initial conditions to determine the first constant of integration.	Since the brick was dropped, when $t = 0, x = 0, v = 0$. $0 = -250 \log_e(4900) + c_1$ $c_1 = 250 \log_e(4900)$
6. Substitute back for the first constant of integration and take out common factors.	$x = -250 \log_e (4900 - v^2) + 250 \log_e(4900)$ $= 250 (\log_e(4900) - \log_e (4900 - v^2))$
7. Use log laws, to show the required result.	$x = 250 \log_e \left(\frac{4900}{4900 - v^2} \right)$
b. 1. Determine the speed when the brick has fallen the required distance.	b. When $x = 100, v = ?$ $100 = 250 \log_e \left(\frac{4900}{4900 - v^2} \right)$

2. Use the definition of the logarithm and transpose to make v the subject.

$$e^{0.4} = \frac{4900}{4900 - v^2}$$
$$4900 - v^2 = 4900e^{-0.4}$$
$$v^2 = 4900\left(1 - e^{-0.4}\right)$$

3. Calculate the speed when the brick has fallen this required distance.

$$v = 70\sqrt{(1 - e^{-0.4})}$$
$$v = 40.19 \text{ m/s}$$

c. 1. To obtain a relationship between v and t use $x = \frac{dv}{dt}$ and invert both sides.

c. $$\frac{dv}{dt} = \frac{4900 - v^2}{500}$$
$$\frac{dt}{dv} = \frac{500}{4900 - v^2}$$

2. Integrate both sides with respect to v.

$$t = \int \frac{500}{4900 - v^2} dv$$

3. To calculate this integral, use partial fractions. Express the integrand into its partial fractions decomposition.

$$\frac{500}{4900 - v^2} = \frac{A}{70 - v} + \frac{B}{70 + v}$$
$$= \frac{A(70 + v) + B(70 - v)}{(70 - v)(70 + v)}$$
$$= \frac{70(A + B) + v(A - B)}{4900 - v^2}$$

4. Determine the values of the constants A and B.

By equating coefficients,

$A - B = 0 \quad \Rightarrow A + B$

$70(A + B) = 500$

So that $A = B = \frac{500}{140}$

5. Express the integrand in a form where we can perform the integration.

$$t = \int \frac{500}{4900 - v^2} dv$$
$$= \frac{500}{140} \int \left(\frac{1}{70 + v} + \frac{1}{70 - v} \right) dv$$

6. Perform the integration, and place the second constant of integration on one side of the equation.

$$t = \frac{25}{7}\left(\log_e(|70 + v|) - \log_e(|70 - v|) + c_2\right)$$

7. Use the given initial condition to determine the second constant of integration.

Since when $x = 0$, $t = 0$, $v = 0$.

$$0 = \frac{25}{7}\left(\log_e(70) - \log_e(70) + c_2\right)$$
$$c_2 = 0$$

8. Substitute back for the second constant of integration and use log laws again, the required result is shown.

$$t = \frac{25}{7}\left(\log_e(|70 + v|) - \log_e(|70 - v|)\right)$$
$$t = \frac{25}{7}\log_e\left(\left|\frac{70 + v}{70 - v}\right|\right)$$

But since $0 \leq v < 70$ the modulus signs are not needed.

$$t = \frac{29}{7}\log_e\left(\frac{70 + v}{70 - v}\right)$$

d. The terminal velocity can be calculated when the acceleration is zero.	**d.** $\frac{4900 - v^2}{500} = 0$ $v^2 = 4900$ $v_T = \sqrt{4900}$ $v_T = 70$ The terminal velocity is 70 m/s.
e. 1. Determine the time taken for the brick to fall the required distance.	**e.** $t = ?$ when $v = 40.19$ $t = \frac{25}{7} \log_e \left(\frac{70 + 40.19}{70 - 40.19}\right)$ $= 4.76$
2. State the time to fall the required distance.	The time taken to fall is 4.67 seconds
3. An alternative method to calculate the time is to numerically evaluate a definite integral.	$t = \int_0^{40.19} \frac{500}{4900 - v^2} dv = 4.67$

11.6 Exercise

Students, these questions are even better in jacPLUS

Receive immediate feedback and access sample responses

Access additional questions

Track your results and progress

Find all this and MORE in jacPLUS

Technology free

1. **WE17** A body moves along a straight-line path and has an acceleration $a(v) = \frac{1}{2}(8 - v)$ m/s^2 where v is the velocity in m/s. Initially the body is at rest at the origin. Show that the position x metres at time t is given by $x = 8\left(1 + 2\left(e^{-\frac{t}{2}} - 1\right)\right)$.

2. A body moving in a straight line, has its acceleration $a(v) = -\frac{1}{3}(v - 6)$ m/s^2, where v is the velocity in m/s. Initially the body is moving at 12 m/s and is at the origin. Show that the position x metres at time t is given by $x = 6\left(t + 3\left(1 - e^{-\frac{t}{3}}\right)\right)$.

3. A body is moving in a straight line, and has an acceleration given by $a(v) = -2v$ m/s^2, where v is the velocity in m/s. Initially the body is moving at 1 m/s and is at the origin. Show that the position, x metres, at time t is given by $x = \frac{1}{2}\left(1 - e^{-2t}\right)$.

4. A body moving in a straight line, and has an acceleration given by $a(v) = -3(v + 4)$ m/s^2, where v is the velocity in m/s. Initially the body is moving at 2 m/s and is at the origin. Show that the position at time t is given by $x = 2\left(1 - e^{-3t} - 2t\right)$.

5. **WE18** A car is moving along a straight road at a speed of 90 km/h when the driver brakes. The resulting acceleration is given by $a(v) = -\frac{\sqrt{v^3}}{5}$ m/s^2, where v is the speed in m/s after the driver applies the brakes. Determine:

   a. the time taken in seconds for the speed of the car to be reduced to 57.6 km/h
   b. the distance travelled in metres in this time.

6. A sports car is moving along a level road at a speed of 57.6 km/h when the driver applies the brakes. The acceleration is given by $a(v) = -\frac{v^{\frac{3}{2}}}{10}$ m/s^2, where v m/s is the speed of the car at a time t seconds. After a time T seconds, the speed of the car is 14.4 km/h, and in this time it has travelled a distance of D metres. Determine the values of:

   a. T
   b. D

7. A boat is sailing in a straight line at a speed of 57.6 km/h when the driver disengages the engine. The acceleration is given by is $a(v) = -\frac{4\sqrt{v}}{5}$ m/s^2 where v m/s is the speed of the boat at a time t seconds. Determine:

   a. the time taken for the speed of the boat to be reduced to 14.4 km/h
   b. the distance travelled in this time.

8. A car is moving along a straight road at a speed of 90 km/h when the driver brakes. The acceleration is given by $a(v) = -\frac{369v^3}{160\,000}$ m/s^2, where v is the speed in m/s after the driver applies the brakes. After a time T seconds the speed of the car has been reduced to 57.6 km/h and in this time the car has travelled a distance of D metres. Determine the values of T and D.

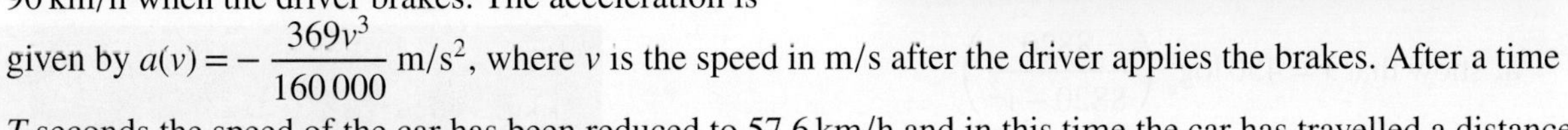

9. **WE19** A car moves along a straight road at a speed of U m/s when the driver brakes. The deceleration is proportional to the cube of the velocity v, where v is the speed in m/s after the driver applies the brakes. After a time T seconds the speed of the car has been reduced to $\frac{1}{2}U$ m/s and in this time the car has travelled a distance of D metres. Show that $\frac{D}{T} = \frac{2U}{3}$.

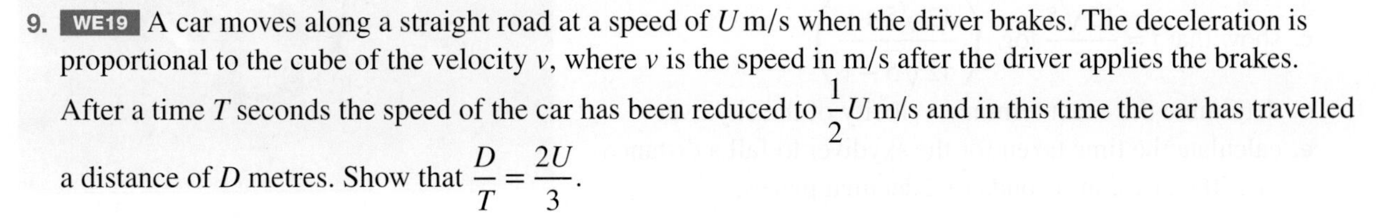

10. A car moves along a straight road at a speed of U m/s when the driver brakes. The deceleration is proportional to the square root of the cube of the velocity v, where v is the speed in m/s after the driver applies the brakes. After a time T seconds the speed of the car has been reduced to $\frac{1}{2}U$ m/s and in this time the car has travelled a distance of D metres. Show that $\frac{D}{T} = \frac{\sqrt{2}U}{2}$.

11. A body is moving in a straight line path on a horizontal surface and has an acceleration given by $a(v) = -kv$ m/s^2 where k is a positive constant. If its initial speed is U m/s, show that:

    a. its speed v m/s at a time t seconds satisfies $v = Ue^{-kt}$
    b. its position x metres after a time t seconds is given by $x = \frac{U}{k}\left(1 - e^{-kt}\right)$
    c. its speed v after moving a distance x is given by $v = U - kx$.

12. A block is moving in a straight line path on a smooth horizontal surface and has its acceleration given by $a(v) = -kv^2$ m/s^2 where k is a positive constant. If its initial speed is U m/s, show that:

a. its speed v m/s after moving a distance of x metres is given by $v = Ue^{-kx}$

b. its speed v m/s at a time t seconds satisfies $v = \dfrac{U}{1 + kUt}$.

13. A train is moving in a straight line path and has its acceleration given by $a(v) = -(p + qv)$ m/s^2 where p and q are positive constants and v m/s is its speed at any time t seconds. If its initial speed is U m/s, and it travels a distance of D metres before coming to rest in a time of T seconds, show that:

a. $T = \dfrac{1}{q}\log_e\left(1 + \dfrac{qU}{p}\right)$

b. $D = \dfrac{1}{q}(U - pT)$.

14. A motor car when travelling with a speed of v m/s along a level section of road when the brakes are applied has its acceleration given by $a(v) = -\left(p + qv^2\right)$ m/s^2, where p and q are positive constants. Show that with the engine disengaged, the brakes will bring the car from an initial speed of U m/s to rest in a time T seconds and that the car will travel a distance of D metres where

$$T = \frac{1}{\sqrt{pq}}\tan^{-1}\left(U\sqrt{\frac{q}{p}}\right) \text{ and } D = \frac{1}{2q}\log_e\left(1 + \frac{qU^2}{p}\right).$$

Technology active

15. WE20 A skydiver falls vertically from rest from a plane. While falling vertically downwards a distance of x metres, after a time t second her speed is v m/s her acceleration is a m/s^2, given that $a(v) = \dfrac{8820 - v^2}{900}$:

a. show that $x = 450\log_e\left(\dfrac{8820}{8820 - v^2}\right)$

b. determine the speed of the skydiver after she has fallen a distance of 150 metres, in m/s to 3 decimal places

c. show that $t = \dfrac{15\sqrt{5}}{7}\log_e\left(\dfrac{42\sqrt{5} + v}{42\sqrt{5} - v}\right)$

d. determine the exact terminal velocity of the skydiver

e. calculate the time taken for the skydiver to fall a distance of 150 metres, in seconds to 2 decimal places.

16. A body falls vertically from rest. While falling vertically downwards the acceleration is given by $a(v) = g - kv^2$ m/s^2, where k is a positive constant and v m/s is the speed of the body at a time t seconds. Show that the terminal speed is given by $v_T = \sqrt{\dfrac{g}{k}}$ and that $t = \dfrac{1}{2kv_T}\log_e\left(\dfrac{v_T + v}{v_T - v}\right)$.

17. A body falls from rest in the Earth's gravitational field, as it falls its acceleration is given by $a(v) = g - kv$ where k is a positive constant and v m/s is its speed at a time t seconds.

a. Show that $v = \dfrac{g}{k}\left(1 - e^{-kt}\right)$.

b. Show that the limiting (or terminal speed) is given by $\dfrac{g}{k}$.

c. Show that when the speed is half the terminal speed, the distance of the particle below the point of projection is given by $\dfrac{g}{k^2}\left(\log_e(2) - \dfrac{1}{2}\right)$ metres.

18. During a snow storm, a small block of ice falls from the sky. While falling the acceleration of the ice block is given by $a(v) = \dfrac{49 - v^2}{5}$, where v m/s is its speed at a time t seconds after falling a distance of x metres.

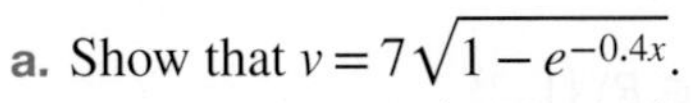

a. Show that $v = 7\sqrt{1 - e^{-0.4x}}$.

b. Show that $v = \dfrac{7\left(1 - e^{-2.8t}\right)}{1 + e^{-2.8t}}$ and hence determine the terminal speed of the ice block.

c. Show that $x = 5\log_e\left(\dfrac{e^{1.4t} + e^{-1.4t}}{2}\right)$.

19. A ball is projected vertically upwards, with an initial speed of U m/s from ground level. While travelling upwards or downwards it is subjected to a gravity and a deceleration which is equal to $-kv^2$ m/s^2, where k is a positive constant and v is its velocity in m/s.

a. Show that the ball reaches a maximum height of $\dfrac{1}{2k}\log_e\left(1 + \dfrac{kU^2}{g}\right)$.

b. Show that the time required for the ball to reach its maximum height is given by $\dfrac{1}{\sqrt{gk}}\tan^{-1}\left(U\sqrt{\dfrac{k}{g}}\right)$.

c. Show that the ball returns to its original point with a speed of $U\sqrt{\dfrac{g}{g + kU^2}}$.

20. A bullet is fired horizontally from a gun and experiences a deceleration proportional to the cube of its speed, the constant of proportionality being k. If the initial speed of the bullet is U m/s, show that after a time t seconds, if its position is x metres, then $t = \dfrac{x}{U} + \dfrac{kx^2}{2}$ assuming that the motion remains horizontal.

21. A bullet is fired into a bullet-proof vest. While moving horizontally its acceleration is given by $a(v) = -400\left(v^2 + 10\,000\right)$ m/s^2 where v is its velocity in m/s and t is the time in seconds after impact. The initial speed of the bullet was 400 m/s

a. Express the time t in terms of the velocity v and hence determine how long before the bullet comes to rest, in milliseconds rounded to 3 decimal places.

b. Determine how far the bullet penetrates the bullet-proof vest before coming to rest, stating your answer in millimetres, rounded to 1 decimal place.

22. A cyclist travelling horizontally at a speed U m/s reaches a level section of road and begins to freewheel. She observes that after travelling a distance of D metres in time T seconds along this section of road her speed has fallen to V m/s, where $0 < V < U$. If during the freewheeling her deceleration is proportional to the:

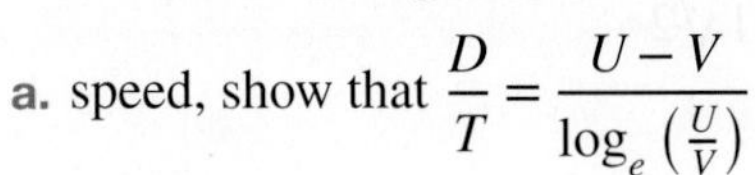

a. speed, show that $\dfrac{D}{T} = \dfrac{U - V}{\log_e\left(\frac{U}{V}\right)}$

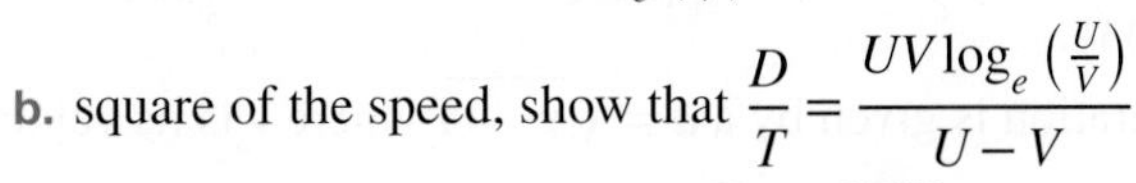

b. square of the speed, show that $\dfrac{D}{T} = \dfrac{UV\log_e\left(\frac{U}{V}\right)}{U - V}$

c. cube of the speed, show that $\dfrac{D}{T} = \dfrac{2UV}{U + V}$

d. fourth power of the speed, show that $\dfrac{D}{T} = \dfrac{3UV(U + V)}{2\left(U^2 + UV + V^2\right)}$

e. fifth power of the speed, show that $\frac{D}{T}=\frac{4UV\left(U^3-V^3\right)}{3\left(U^4-V^4\right)}$

f. square root of the speed cubed, show that $\frac{D}{T}=\sqrt{UV}$

g. nth power of the speed, show that $\frac{D}{T}=\frac{\left(U^{2-n}-V^{2-n}\right)(1-n)}{\left(U^{1-n}-V^{1-n}\right)(2-n)}$ where $n\in R\backslash\{1,2\}$

h. if we assume a constant deceleration, show that $\frac{D}{T}=\frac{U+V}{2}$.

11.6 Exam questions

Question 1 (10 marks) TECH-ACTIVE

Source: *VCE 2017, Specialist Mathematics Exam 2, Section B, Q2; © VCAA.*

A helicopter is hovering at a constant height above a fixed location. A skydiver falls from rest for two seconds from the helicopter. The skydiver is subject to gravitational acceleration and air resistance is negligible for the first two seconds. Let downward displacement be positive.

a. Find the distance, in metres, fallen the first two seconds. **(2 marks)**

b. Show that the speed of the skydiver after two seconds is $19.6\,\text{ms}^{-1}$.

After two seconds, air resistance is significant and the acceleration of the skydiver is given by $a = g - 0.01v^2$. **(1 mark)**

c. Find the limiting (terminal) velocity, in ms^{-1}, that the skydiver would reach. **(1 mark)**

d. i. Write down an expression involving a definite integral that gives the time taken for the skydiver to reach a speed of $30\,\text{ms}^{-1}$. **(2 marks)**

ii. Hence, find the time, in seconds, taken to reach a speed of $30\,\text{ms}^{-1}$, correct to the nearest tenth of a second. **(1 mark)**

e. Write down an expression involving a definite integral that gives the distance through which the skydiver falls to reach a speed of $30\,\text{ms}^{-1}$. Find this distance, giving your answer in metres, correct to the nearest metre. **(3 marks)**

Question 2 (1 mark) TECH-ACTIVE

Source: *VCE 2015, Specialist Mathematics Exam 1, Section 1, Q22; © VCAA.*

MC A ball is thrown vertically up with an initial velocity of $7\sqrt{6}\,\text{ms}^{-1}$, and is subject to gravity and air resistance.

The acceleration of the ball is given by $\ddot{x}=-\left(9.8+0.1v^2\right)$, where x metres is its vertical displacement, and $v\,\text{ms}^{-1}$ is its velocity at time t seconds.

The time taken for the ball to reach its maximum height is

A. $\frac{\pi}{3}$ **B.** $\frac{5\pi}{21\sqrt{2}}$ **C.** $\log_e(4)$ **D.** $\frac{10\pi}{21\sqrt{2}}$ **E.** $10\log_e(4)$

Question 3 (1 mark) TECH-ACTIVE

Source: *VCE 2013, Specialist Mathematics Exam 2, Section 1, Q18; © VCAA.*

MC A particle moves in a straight line such that its acceleration is given by a $a=\sqrt{v^2-1}$ where v is its velocity and x is its displacement from a fixed point.

Given that $v=\sqrt{2}$ when $x=0$, the velocity v in terms of x is

A. $v=\sqrt{2+x}$ **B.** $v=1+|x+1|$ **C.** $v=\sqrt{2+x^2}$

D. $v=\sqrt{1+(1+x)^2}$ **E.** $v=\sqrt{1+(x-1)^2}$

More exam questions are available online.

11.7 Acceleration that depends on position

LEARNING INTENTION

At the end of this subtopic you should be able to:

- solve problems involving variable acceleration that is expressed as a function of position.

11.7.1 Acceleration as a function of position

If the acceleration $a = a(x)$ is a function of the position x, then by the chain rule:
$a = \ddot{x} = \frac{d^2x}{dt^2} = \frac{dv}{dt} = \frac{dx}{dt} \cdot \frac{dv}{dx} = v\frac{dv}{dx} = \frac{d\left(\frac{1}{2}v^2\right)}{dv}\frac{dv}{dx} = \frac{d\left(\frac{1}{2}v^2\right)}{dx}$. So $\frac{d\left(\frac{1}{2}v^2\right)}{dx} = a(x)$. Integrating both sides with respect to x gives $\frac{1}{2}v^2 = \int a(x)dx$ with initial conditions on x and v. This gives a relationship between v and x.

Acceleration as a function of position

When the acceleration $\ddot{x} = a(x)$ is a function of the position use $\ddot{x} = \frac{d}{dx}\left(\frac{1}{2}v^2\right) = a(x)$ and solve to give $\frac{1}{2}v^2 = \int a(x)dx$.

WORKED EXAMPLE 21 Acceleration as a function of position

A particle moves so that at a time t seconds, its position is x metres from a fixed origin. The particle has an acceleration $a(x) = \frac{-(5+2x)}{x^3}$ m/s^2 and the particle is at rest at a distance of 5 metres from the origin. Determine where else the particle comes to rest.

THINK	WRITE
1. Write down the equation of motion to be solved.	$\ddot{x} = \frac{d\left(\frac{1}{2}v^2\right)}{dx} = \frac{-(5+2x)}{x^3}$
2. Integrate both sides with respect to x.	$\frac{1}{2}v^2 = \int \left(\frac{-(5+2x)}{x^3}\right) dx$
3. Express the integrand in index notation.	$\frac{1}{2}v^2 = \int \left(-5x^{-3} - 2x^{-2}\right) dx$
4. Perform the integration; place the constant of integration on one side of the equation.	$\frac{1}{2}v^2 = \frac{5}{2}x^{-2} - 2x^{-1} + c$
5. Write the expression with positive indices.	$\frac{1}{2}v^2 = \frac{5}{2x^2} - \frac{2}{x} + c$
6. Use the given initial conditions to determine the constant of integration.	When $v = 0$, $x = 5$ $0 = \frac{5}{2(5)^2} - \frac{2}{5} + c$ $c = \frac{2}{5} - \frac{1}{10} = \frac{3}{10}$
7. Substitute back for the constant of integration.	$\frac{1}{2}v^2 = \frac{5}{2x^2} - \frac{2}{x} + \frac{3}{10}$

8. Form a common denominator.	$\frac{1}{2}v^2 = \frac{25-20x+3x^2}{10x^2}$
9. Factorise the quadratic in the numerator.	$v^2 = \frac{(3x-5)(x-5)}{5x^2}$
10. Express the velocity in terms of x.	$v = \frac{1}{\lvert x\rvert}\sqrt{\frac{(3x-5)(x-5)}{5}}$
11. Determine the values of x when the particle comes to rest.	If $v=0$ then $(3x-5)(x-5)=0$ so that $x=5$ or $x=\frac{5}{3}$. Since we were given $x=5$ when $v=0$, the required solution is $x=\frac{5}{3}$.
12. State the final result.	The particle comes to rest $\frac{5}{3}$ metres from the origin.

11.7.2 Velocity in terms of position

So far, we have used integration techniques, to obtain relationships between time t, position x, velocity v and the acceleration a. However we can also use differentiation to obtain relations, using

$$a = \ddot{x} = \frac{d^2x}{dt^2} = \frac{dv}{dt} = \frac{dx}{dt}\cdot\frac{dv}{dx} = v\frac{dv}{dx} = \frac{d\left(\frac{1}{2}v^2\right)}{dv}\frac{dv}{dx} = \frac{d\left(\frac{1}{2}v^2\right)}{dx}.$$

Note that $(t, v(t))$ and $(t, a(t))$ define functions, since an object can not have more than one velocity or more than one acceleration at any given time, however a relation between position x and speed v need not necessarily define a function.

WORKED EXAMPLE 22 Velocity in terms of position

A body is moving in a straight line. Its velocity v m/s is given by $v = \sqrt{9-4x^2}$ when it is x m from the origin at time t seconds. Show that the acceleration $a = -4x$.

THINK	WRITE
1. Differentiate using the chain rule.	$v = \sqrt{9-4x^2} = \left(9-4x^2\right)^{\frac{1}{2}}$ $\frac{dv}{dx} = \frac{1}{2}\times(-8x)\times\left(9-4x^2\right)^{-\frac{1}{2}}$ $= \frac{-4x}{\sqrt{9-4x^2}}$
2. Use an expression for the acceleration.	$a = v\frac{dv}{dx}$ $= \sqrt{9-4x^2}\times\frac{-4x}{\sqrt{9-4x^2}}$ $= -4x$
3. Alternatively square the velocity, half it, then differentiate with respect to x.	$v^2 = 9-4x^2$ $\frac{1}{2}v^2 = \frac{9}{2}-2x^2$ $\frac{d\left(\frac{1}{2}v^2\right)}{dx} = -4x$

11.7.3 Expressing position in terms of time

In some cases it may be possible to rearrange and express the position x in terms of t, by solving for v and using $v = \frac{dx}{dt}$.

WORKED EXAMPLE 23 Position in terms of time

A body moves in a straight line and has its acceleration given by $a(x) = -4x$ m/s^2, where x is its position in metres from a fixed origin. Initially the body is 1 metre from the origin and the initial velocity of the body is 2 m/s. Express x in terms of t where t is the time in seconds.

THINK	WRITE
1. Write down the equation of motion to be solved.	$\ddot{x} = \frac{d\left(\frac{1}{2}v^2\right)}{dx} = -4x$
2. Integrate both sides with respect to x.	$\frac{1}{2}v^2 = \int -4x\,dx$
3. Perform the integration; place the constant of integration on one side of the equation.	$\frac{1}{2}v^2 = -2x^2 + c_1$
4. Use the given initial conditions to determine the first constant of integration.	Initially, when $t = 0$, $v = 2$ and $x = 1$ $2 = -2 + c_1$ $\Rightarrow c_1 = 4$
5. Substitute back for the first constant of integration.	$\frac{1}{2}v^2 = -2x^2 + 4$
6. Rearrange and solve for v.	$v^2 = 8 - 4x^2$ $v^2 = 4\left(2 - x^2\right)$ $v = \pm 2\sqrt{2 - x^2}$
7. Since initially when $t = 0$, $x = 1$ and $v = 2 > 0$, we can take the positive root only.	$v = \frac{dx}{dt} = 2\sqrt{2 - x^2}$
8. Invert both sides.	$\frac{dt}{dx} = \frac{1}{2\sqrt{2 - x^2}}$
9. Integrate both sides with respect to x.	$t = \frac{1}{2}\int \frac{1}{\sqrt{2 - x^2}}dx$
10. Perform the integration.	$t = \frac{1}{2}\sin^{-1}\left(\frac{x}{\sqrt{2}}\right) + c_2$
11. Use the given initial conditions to determine the second constant of integration.	When $t = 0$, $v = 2$ and $x = 1$ $0 = \frac{1}{2}\sin^{-1}\left(\frac{1}{\sqrt{2}}\right) + c_2$ $c_2 = -\frac{\pi}{8}$

12. Substitute back for the second constant of integration.	$t=\frac{1}{2}\sin^{-1}\left(\frac{x}{\sqrt{2}}\right)-\frac{\pi}{8}$
13. Rearrange to make x the subject.	$\sin^{-1}\left(\frac{x}{\sqrt{2}}\right)=2t+\frac{\pi}{4}$ $\frac{x}{\sqrt{2}}=\sin\left(2t+\frac{\pi}{4}\right)$
14. Expand using compound angle formulae. $\sin(A+B)$ $=\sin(A)\cos(B)+\cos(A)\sin(B)$	$\frac{x}{\sqrt{2}}=\sin(2t)\cos\left(\frac{\pi}{4}\right)+\cos(2t)\sin\left(\frac{\pi}{4}\right)$ $\frac{x}{\sqrt{2}}=\frac{1}{\sqrt{2}}\sin(2t)+\frac{1}{\sqrt{2}}\cos(2t)$
15. State the final result.	$x=\sin(2t)+\cos(2t)$

11.7.4 Using calculus to establish the constant acceleration formulae

If the acceleration a is constant, and u is the initial velocity, then

$$\frac{dv}{dt}=a$$

$$v(t)=\int a\,dt=at+c_1$$

$$v(0)=u \;\Rightarrow\; c_1=u$$

$$\boldsymbol{v=u+at}$$

$$\frac{dx}{dt}=u+at$$

$$x(t)=\int (u+at)\,dt$$

$$=ut+\frac{1}{2}at^2+c_2$$

$$x(0)=0 \;\Rightarrow\; c_2=0,\; x=s$$

$$\boldsymbol{s=ut+\frac{1}{2}at^2}$$

$$\frac{d}{dx}\left(\frac{1}{2}v^2\right)=a$$

$$\frac{1}{2}v^2=\int a\,dx$$

$$=ax+c_3$$

$$x=0,\; v=u \;\Rightarrow\; c_1=\frac{1}{2}u^2,\; x=s$$

$$\boldsymbol{v^2=u^2+2as}$$

11.7 Exercise

Students, these questions are even better in jacPLUS

Receive immediate feedback and access sample responses

Access additional questions

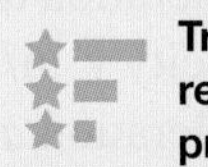
Track your results and progress

Find all this and MORE in jacPLUS

Technology free

1. **WE21** A particle moves so that at a time t seconds, its position is x metres from a fixed origin. The particle has an acceleration $a(x) = \frac{7x-6}{x^3}$ m/s^2 and the particle is at rest at a distance of 3 metres from the origin. Determine where else the particle comes to rest.

2. A body moves in a straight line so that at a time t seconds, its position is x metres from a fixed origin and its acceleration is given by $a(x) = \frac{3-x}{2x^3}$ m/s^2. The particle is at rest at a distance of one metre from the origin, determine where else the particle comes to rest.

3. A particle has its position is x metres from a fixed origin. The particle moves in a straight line and has an acceleration given by $a(x) = \frac{10x-8}{x^3}$ m/s^2 and the particle is at rest at a distance of 2 metres from the origin. Determine where else the particle comes to rest.

4. A particle moves along a straight line track, so that its position is x metres from a fixed origin. The particle has an acceleration given by $a(x) = \frac{1+bx}{x^3}$ m/s^2 and the particle is at rest when $x = 1$ and also when $x = -\frac{1}{5}$. Determine the value of the constant b.

5. A block moves along a horizontal table top and its acceleration is given by $a(x) = \frac{-(p+qx)}{x^3}$ m/s^2, where p and q are constants and its position is x metres. If initially the block is at rest when the position is p metres, show that the block next comes to rest again when the position is $\frac{-p}{2q+1}$.

6. A particle moves in a straight line and has its acceleration given by $a(x) = 4x - 6$ m/s^2 where x is its position in metres from a fixed origin. Initially the particle is 2 metres from the origin and the initial velocity of the body is 1 m/s. Express x in terms of t where t is the time in seconds.

7. A particle moves along a straight line track, its position is x metres from a fixed origin O and its acceleration given by $a(x) = 9x - 15$ m/s^2. The particle starts at the point $x = 2$ with a velocity of 1 m/s. Express x in terms of t where t is the time in seconds.

8. An object moves in a straight line and has an acceleration given by $a(x) = x - 1$ m/s^2, where x is its position in metres from a fixed origin. Initially the object is at the origin and the initial velocity of the body is 1 m/s. Express x in terms of t where t is the time in seconds.

9. A particle moves along a straight track and has a deceleration given by e^{-2x} m/s^2, where x is its position in metres from a fixed origin. Initially the particle is at the origin and the initial velocity of the particle is 1 m/s. Express x in terms of t where t is the time in seconds, and $0 \le t \le \frac{1}{2}$.

10. A body moves in a straight line and has an acceleration given by $a(x) = 2e^{4x}$ m/s^2, where x is its position in metres from a fixed origin. Initially the body is at the origin and the initial velocity of the body is 1 m/s. Express x in terms of t where t is the time in seconds.

11. WE22 a. A body is moving in a straight line. Its velocity v m/s is given by $v = x^2$ when it is x m from the origin at time t seconds. Show that the acceleration $a = 2x^3$.

b. A body is moving in a straight line. Its velocity v m/s is given by $v = \sqrt{16 - 25x^2}$ when it is x m from the origin at time t seconds. Show that the acceleration $a = -25x$.

c. A body is moving in a straight line. Its velocity v m/s is given by $v = e^{2x} + e^{-2x}$ when it is x m from the origin at time t seconds. Show that the acceleration $a = 2\left(e^{4x} - e^{-4x}\right)$.

12. a. A body is moving in a straight line. Its velocity v m/s is given by $v = x^4$ when it is x m from the origin at time t seconds. Show that the acceleration $a = 4x^7$.

b. A body is moving in a straight line. Its velocity v m/s is given by $v = \sqrt{4 + 9x^2}$ when it is x m from the origin at time t seconds. Show that the acceleration $a = 9x$.

c. A body is moving in a straight line. Its velocity v m/s at time t seconds is given by $v = 3 - 2e^{-2t}$. Show that the acceleration $a = 2(v - 3)$.

13. a. A body is moves along a straight path. Its velocity v m/s is given by $v = x^n$ where n is a constant, when it is x m from the origin at time t seconds. Show that the acceleration $a = nx^{2n-1}$.

b. A body is moving in a straight line. Its velocity v m/s is given by $v = \sqrt{b - n^2x^2}$ where n and b are constants when it is x m from the origin at time t seconds. Show that the acceleration $a = -n^2x$.

14. A body is moving in a straight line. Its velocity v m/s is given by $v = e^{nx} + e^{-nx}$ where n is a constant, when it is x m from the origin at time t seconds. Show that the acceleration $a = n\left(e^{2nx} - e^{-2nx}\right)$.

15. WE23 A body moves in a straight line and has an acceleration given by $a(x) = -9x$ m/s^2, where x is its position in metres from a fixed origin. Initially the body is 1 metre from the origin and the initial velocity of the body is 3 m/s. Express x in terms of t where t is the time in seconds.

Technology active

16. A body moves in a straight line and has an acceleration $a(x) = -\frac{x}{4}$ m/s^2, where x is its position in metres from a fixed origin. Initially the body is at rest 8 metres from the origin. Determine the times when the body is 4 metres from the origin.

17. A body moves in a straight line and has an acceleration given by $a(x) = -16x$ m/s^2, where x is its position in metres from a fixed origin. Initially the body is 3 metres from the origin and initially at rest. Determine the times when the body passes through the origin.

18. A body moves in a straight line with an acceleration of $a(x) = -\frac{x}{9}$ m/s^2, where x is its position in metres from a fixed origin. Initially the body is at the origin, moving with a speed of 6 m/s. Determine the times when the body is 9 metres from the origin.

19. A block moves back and forth along a straight line track. Its acceleration is given by $a(x) = -n^2x$ where x is the distance in metres from the origin O and n is constant. Initially the particle starts from rest when the position from O is p. Show that if its speed is v m/s at any time t seconds then $v^2 = n^2\left(p^2 - x^2\right)$ and $x = p\cos(nt)$.

20. A particle moves in a straight line. When its position from a fixed origin O is x metres its acceleration a m/s^2, is given by $a(x) = -\frac{k}{x^3}$, where k is a positive constant. Initially the particle starts from rest when its distance from O is p, where p is a positive constant. Show that its speed v m/s is given by $v = \frac{\sqrt{k\left(p^2 - x^2\right)}}{px}$ where x metres is its distance from O at a time t seconds.

21. A particle moves so that at a time t seconds, its position is x metres from a fixed origin. The particle has an acceleration $p(px+q)$ m/s^2 where p and q are non-zero real constants. Initially the particle is $\frac{q}{p}$ metres from the origin, moving with a speed of $2q$ m/s. Show that $x = \frac{q}{p}\left(2e^{pt} - 1\right)$.

22. A particle moves so that at a time t seconds, its position is x metres from a fixed origin. The particle has an acceleration $a(x) = 8x\left(x^2 - 9\right)$ m/s^2 and initially is 2 metres from the origin, moving with a speed of 10 m/s.
   a. Express x in terms of t.
   b. As t approaches infinity, the particle approaches a fixed position, determine that position.

23. A particle moves in a straight line and has its acceleration given by $a(x) = -\frac{8}{x^2}$ m/s^2, where x is its distance in metres from a fixed point O. If the particle starts from rest when it is 4 metres from O, determine the time taken correct to 2 decimal places to travel to the origin.

24. A particle moves in a straight line and has an acceleration $a(x) = -\frac{k}{x^2}$ m/s^2 where k is a positive constant and x metres is its distance from the origin$O$. The particle starts from rest when its distance from O is p metres, show that the speed v m/s at a distance x from O satisfies $v = \sqrt{\frac{2k(p-x)}{px}}$.

11.7 Exam questions

Question 1 (1 mark) TECH-ACTIVE

Source: VCE 2014, Specialist Mathematics Exam 2, Section A, Q21; © VCAA.

MC The acceleration, in ms^2, of a particle moving in a straight line is given by $-4x$, where x metres is its displacement from a fixed origin O.

If the particle is at rest where $x = 5$, the speed of the particle, in ms^{-1}, where $x = 3$ is

A. 8 **B.** $8\sqrt{2}$ **C.** 12 **D.** $4\sqrt{2}$ **E.** $2\sqrt{34}$

Question 2 (3 marks) TECH-FREE

A car is travelling at a speed of U m/s along a level road when the driver applies the brakes. The deceleration is given by $-kx$ m/s^2, where x is the distance travelled in metres after the driver applies the brakes and k is a positive constant. Show that when the car comes to rest it has travelled a distance of $U\sqrt{\frac{1}{k}}$ metres.

Question 3 (2 marks) TECH-FREE

A body is moving in a straight line. Its velocity v m/s at time t seconds is given by $v = b - ne^{-nt}$, where n is a constant. Show that the acceleration $a = -n(v - b)$.

More exam questions are available online.

11.8 Review

11.8.1 Summary

doc-37065

Hey students! Now that it's time to revise this topic, go online to:

 Access the topic summary  Review your results  Watch teacher-led videos Practise exam questions

Find all this and MORE in jacPLUS

11.8 Exercise

Technology free: short answer

1. The velocity of an object which is initially 3 metres to the left of an origin O is given by $v(t) = \frac{1}{6}\left(t^2 - t - 2\right)$ m/s, where $t \geq 0$ is the time in seconds. Determine:

   a. the position of the object at time t
   b. when and where the object comes to rest
   c. the acceleration when the object comes to rest.

2. A golf cart travels in a straight line, between two holes on a golf course 400 metres apart. It starts from rest accelerates uniformly for a time reaching a maximum speed of 5 m/s which it maintains for a while, before decelerating and coming to rest after travelling 400 metres in 100 seconds. If the time while accelerating is equal to the time while decelerating, calculate the initial acceleration.

3. A body moves in a straight line so that at time t seconds its acceleration is given by $a(t) = -\frac{24}{\sqrt{(3t+1)^3}}$ m/s^2. If the initial velocity of the body is 16 m/s, determine how far it has travelled in the first 5 seconds.

4. A particle moves back and forth along the x-axis and its acceleration is given by $a(t) = -(16\cos(2t) + 12\sin(2t))$ m/s^2 at time t seconds. If its initial velocity is 6 m/s and initially the particle starts from a point 4 metres to the right of the origin, determine its position after $\frac{\pi}{8}$ seconds.

5. A cyclist is riding a bike along a level road at a speed of 57.6 km/h when she begins to freewheel. The deceleration is $\frac{112}{15\sqrt{v}}$ m/s^2, where v m/s is her speed after she begins to freewheel. When her speed has decreased to 14.4 km/h she has travelled a distance of D metres in a time of T seconds. Determine the values of T and D.

6. A particle moves so that at a time t seconds, its position is x metres from a fixed origin. The particle is acted upon by a force resulting in an acceleration of $\dfrac{17x-12}{x^3}$ m/s^2 and the particle is at rest at a distance of 3 metres from the origin, determine where else the particle comes to rest.

Technology active: multiple choice

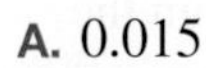

7. **MC** An athlete travels a distance of 1.5 km in 100 seconds when accelerating uniformly from rest. His acceleration in m/s^2 is equal to

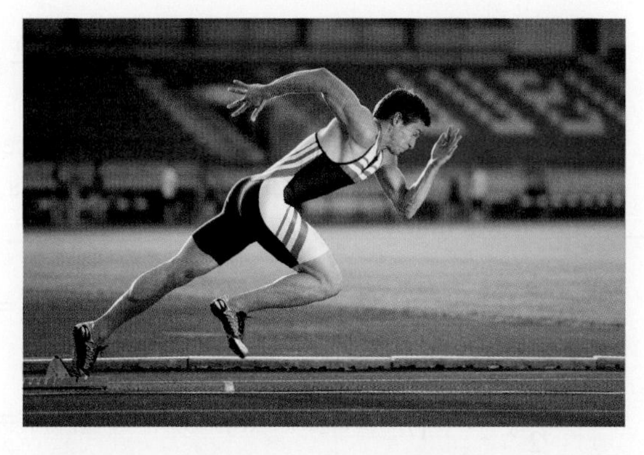

A. 0.015
B. 0.15
C. 0.75
D. 0.3
E. 0.0003

8. **MC** A triathlete on a bicycle reduces her speed from 21 m/s to 6 m/s over 200 m. Assuming the deceleration is constant, how much further, in metres, she will travel on her bicycle before she comes to rest is closest to

A. 417.78 B. 217.78 C. 17.78 D. 20.74 E. 17.46

9. **MC** A basketball is thrown vertically upwards with a velocity of 5 m/s. The maximum height, in metres, the basketball reaches above the point of projection is closest to

A. 1.25 B. 1.276 C. 1.3 D. 5.4 E. 44.6

10. **MC** A body moves so that its acceleration is equal to $2t$ where t is the time in seconds when its velocity is v m/s. If the initial velocity of the body is u m/s, then

A. $v=u+t^2$ B. $v=u+2t^2$ C. $v=u+2$ D. $v^2=u^2+4t$ E. $v^2=u^2+\dfrac{4t^4}{3}$

11. **MC** A body moves in a straight line so that its velocity v m/s at a time $t\geq 0$ seconds is given by $v(t)=4-2t$ m/s, over the first six seconds the distance, in metres, travelled by the body is

A. 4 B. 12 C. 16 D. 20 E. 48

12. **MC** A body moves so that its acceleration $a=v$ m/s^2, when its velocity is v m/s at a time t seconds. The initial velocity of the body is u m/s, then

A. $v=u+at$ B. $v=ue^t$ C. $v=ue^{-t}$ D. $v=ue^{2t}$ E. $v=ue^{-2t}$

13. **MC** A body is moving in a straight line. When it is x metres from the origin it is travelling at v m/s. Given that $v=\sin\left(\dfrac{x}{2}\right)$, then the acceleration in m/s^2 of the body is given by

A. $\dfrac{1}{2}\cos\left(\dfrac{x}{2}\right)$ B. $-2\cos\left(\dfrac{x}{2}\right)$ C. $\dfrac{1}{2}\cos^2\left(\dfrac{x}{2}\right)$ D. $\dfrac{1}{2}\sin(x)$ E. $\dfrac{1}{4}\sin(x)$

14. **MC** A body moves in a straight line, its velocity is v m/s at a time t seconds and its acceleration is $a(t)$ m/s^2. Given that $v=v_1$ when $t=t_1$ and $v=v_2$ when $t=t_2$, it follows that

A. $v_2-v_1=a(t_2)-a(t_1)$ B. $a=\dfrac{v_2^2-v_1^2}{t_1-t_2}$ C. $a=\dfrac{v_2-v_1}{t_2-t_1}$

D. $\dfrac{1}{2}\left(v_2^2-v_1^2\right)=a(t_2)-a(t_1)$ E. $v_2-v_1=\displaystyle\int_{t_1}^{t_2} a(t)dt$

15. **MC** A body moves in a straight line, its velocity is v at a time t seconds. Its acceleration is $a(v)$ m/s^2. Given that $v = v_1$ when $t = t_1$ and $v = v_2$ when $t = t_2$, it follows that

A. $t_2 - t_1 = \dfrac{1}{a(v_2) - a(v_1)}$

B. $t_2 - t_1 = \displaystyle\int_{v_1}^{v_2} a(v)dv$

C. $t_2 - t_1 = \displaystyle\int_{v_1}^{v_2} \frac{1}{a(v)}dv$

D. $t_2 - t_1 = (a(v_2) - a(v_1))$

E. $\dfrac{v_2 - v_1}{t_2 - t_1} = a(v_2) - a(v_1)$

16. **MC** A body moves in a straight line. The acceleration of the body is $a(x)$ m/s^2, when it is x m from the origin, at a time t seconds when its velocity is v m/s. Given that $x = x_1$ when $v = v_1$ and $x = x_2$ when $v = v_2$, it follows that

A. $\dfrac{1}{2}\left(v_2^2 - v_1^2\right) = \displaystyle\int_{x_1}^{x_2} a(x)dx$

B. $v_2 - v_1 = a(x_2) - a(x_1)$

C. $v_2 - v_1 = \displaystyle\int_{x_1}^{x_2} a(x)dx$

D. $\dfrac{1}{2}\left(v_2^2 - v_1^2\right) = a(x_2) - a(x_1)$

E. $t_2 - t_1 = \dfrac{1}{a(x_2) - a(x_1)}$

Technology active: extended response

17. The velocity of a body v m/s at time t s is given by $v(t) = \begin{cases} 15 & 0 \le t \le 4 \\ 15 - \dfrac{5}{12}(t-4)^2 & 4 < t \le 12 \end{cases}$

   a. Sketch the velocity–time graph
   b. Determine the distance travelled in the first 12 seconds.
   c. Calculate its position after 12 seconds.
   d. Determine the acceleration after 6 seconds.

18. A body starts from rest, accelerates, moves at a constant speed for a while, then decelerates coming to rest. The velocity of the body v m/s at time t s is modeled by

$$v(t) = \begin{cases} a\sin\left(\dfrac{\pi t}{180}\right) & 0 \le t \le 90 \\ 20 & 90 < t \le 240 \\ b\cos^{-1}\left(\dfrac{t-240}{60}\right) & 240 < t \le 300 \end{cases}$$

   a. Determine the values of a and b.
   b. Sketch the velocity–time graph.
   c. Show that the distance travelled over the first four minutes is $\dfrac{3600}{\pi} + 3000$ m.
   d. Calculate the total distance travelled over the first 5 minutes, giving your answer to the nearest metre.
   e. Determine the instantaneous acceleration after 45 seconds of motion.
   f. Determine the instantaneous deceleration after 270 seconds of motion.

19. A cyclist travelling at a speed U m/s reaches a level section of road and begins to freewheel. He observes that after travelling a distance of D metres in time T seconds along this section of road his speed has fallen to V m/s, where $U > V > 0$. If during the freewheeling the deceleration is proportional to the square root of the speed, show that $\dfrac{D}{T} = \dfrac{1}{3}\left(U + \sqrt{UV} + V\right)$.

20. A car moves along a straight road at a speed of U m/s when the driver brakes. After a time T seconds the speed of the car has been reduced to $\dfrac{1}{2}U$ m/s and in this time the car has travelled a distance of D metres.

   a. If the deceleration is proportional to the velocity, show that $\dfrac{D}{T} = \dfrac{U}{2\log_e(2)}$.

b. If the deceleration is proportional to the velocity squared, show that $\frac{D}{T} = U\log_e(2)$.

c. If the deceleration is proportional to the square root of velocity, show that $\frac{D}{T} = \frac{\left(\sqrt{2}+3\right)U}{6}$.

d. If the deceleration is constant, show that $\frac{D}{T} = \frac{3U}{4}$.

11.8 Exam questions

Question 1 (5 marks) TECH-FREE

Source: *VCE 2021, Specialist Mathematics Exam 1, Q7; © VCAA.*

The velocity of a particle satisfies the differential equation $\frac{dx}{dt} = x\sin(t)$, where x centimetres is its displacement relative to a fixed point O at time t seconds.

Initially, the displacement of the particle is 1 cm.

a. Find an expression for x in terms of t. **(3 marks)**

b. Find the maximum displacement of the particle and the times at which this occurs. **(2 marks)**

Question 2 (1 mark) TECH-ACTIVE

Source: *VCE 2019, Specialist Mathematics Exam 2, Section A, Q16; © VCAA.*

MC A variable force acts on a particle, causing it to move in a straight line. At time t seconds, where $t \geq 0$, its velocity v metres per second and position x metres from the origin are such that $v = e^x\sin(x)$.

The acceleration of the particle, in ms^{-2}, can be expressed as

A. $e^{2x}\left(\sin^2(x) + \frac{1}{2}\sin(2x)\right)$ **B.** $e^x\sin(x)(\sin(x) + \cos(x))$ **C.** $e^x(\sin(x) + \cos(x))$

D. $\frac{1}{2}e^{2x}\sin^2(x)$ **E.** $e^x\cos(x)$

Question 3 (7 marks) TECH-ACTIVE

Source: *adapted from: VCE 2015, Specialist Mathematics Exam 2, Section 2, Q5; © VCAA.*

A boat ramp at the edge of a deep lake is inclined at an angle of 10° to the horizontal. A boat trailer on the ramp is unhitched from a car and a man attempts to lower the trailer down the ramp using a rope parallel to the ramp, as shown in the diagram below.

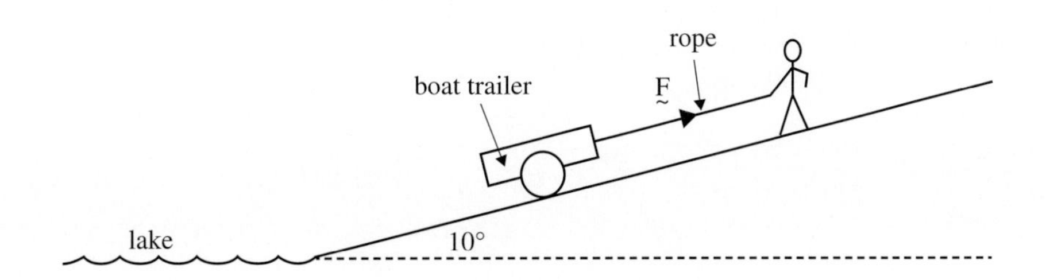

When the trailer rolls into the water, it stops, then sinks vertically from rest so that its depth x metres after t seconds is given by the differential equation

$$\frac{d^2x}{dt^2} = 1.4\left(7 - \frac{dx}{dt}\right)$$

a. Show that the above differential equation can be written as
$1.4\frac{dx}{dv} = -1 + \frac{7}{7-v}$, where $v = \frac{dx}{dt}$. **(2 marks)**

b. **Hence**, show by integration that $1.4x = -v - 7\log_e(7-v) + 7\log_e(7)$. **(1 mark)**

When the trailer has sunk to a depth of D metres, it is descending at a rate of 5 ms^{-1}.

c. Find D, correct to one decimal place. **(1 mark)**

d. Write down a definite integral for the time, in seconds, taken for the trailer to sink to the depth of D metres and evaluate this integral correct to one decimal place. **(3 marks)**

Question 4 (4 marks) TECH-FREE

Source: *VCE 2015, Specialist Mathematics Exam 1, Q6; © VCAA.*

The acceleration $a\text{ ms}^{-2}$ of a body moving in a straight line in terms of the velocity $v\text{ ms}^{-1}$ is given by $a = 4v^2$. Given that $v = e$ when $x = 1$, where x is the displacement of the body in metres, find the velocity of the body when $x = 2$.

Question 5 (3 marks) TECH-ACTIVE

Initially cyclist A is travelling at a speed of 5 m/s along a straight line path, passes cyclist B who is stationary. Three seconds later, cyclist B accelerates in the same direction as cyclist A, for 12 seconds in such a way that his velocity v m/s is given by $v = \frac{30}{\pi}\tan^{-1}\left(\frac{t-3}{24}\right)$

a. Sketch the velocity–time graph for both cyclists. **(2 marks)**

b. Determine the time when cyclist B passes cyclist A. **(1 mark)**

More exam questions are available online.

Answers

Topic 11 Kinematics: rectilinear motion

11.2 Differentiating position and velocity

11.2 Exercise

1. a. -12 m b. -4 m/s
   c. -16 m d. 6 s, 8 m/s
2. 1 m/s^2
3. a. 3 m
   b. 15 m/s
   c. 1 s and 5 s, 10 m, -22 m
   d. -5 m/s
   e. 7.8 m/s
4. a. 2 and 4 s, 25 m, 21 m
   b. 4 m/s
   c. 6 m/s
   d. -6 m/s^2
5. a. 4 s b. 80 m c. 8 s
   d. 2 s and 6 s e. 160 m f. 20 m/s
6. a. 10 m b. 49 m/s c. 5 s
   d. 132.5 m e. 10.2 s f. 50.96 m/s
   g. $a = -9.8$ m/s^2
7. a. 72 km/h b. 6 s c. 60 m
8. a. 54 km/h b. 3 min
   c. 1.8 km d. 36 km/h
9. a. 6 m
   b. -3 m/s
   c. $\frac{3}{2}$ m/s^2
   d. Sample responses can be found in the worked solutions in the online resources.
10. a. 3 m b. 3 m/s
    c. $\frac{3}{2}\log_e(2)$ d. $b = \frac{4}{3},\ c = \frac{1}{3}$
11. a. 4 m
    b. -2 m/s
    c. Sample responses can be found in the worked solutions in the online resources.
    d. $\frac{2\pi}{3}(6k \pm 1),\ k \in Z^+$
    e. $2k\pi,\ k \in Z^+$
12. a. 9 m
    b. -6 m/s
    c. -1 m/s^2
    d. Sample responses can be found in the worked solutions in the online resources.
    e. Sample responses can be found in the worked solutions in the online resources.

11.2 Exam questions

Note: Mark allocations are available with the fully worked solutions online.

1. A
2. a. $(0, 60),\quad (60, 120)$
   b. 54 km/h
   c. 2 min
   d. 1.2 km
   e. 36 km/h
3. a. 100 m
   b. 0 m/s
   c. $4.9e^{-\frac{t}{10}}$
   d. Sample responses can be found in the worked solutions in the online resources.

11.3 Constant acceleration

11.3 Exercise

1. a. 1 m/s^2 b. 36 m c. 9 s
2. a. 2 s b. 200 m
3. a. 20 m/s b. $\frac{D}{T}$
4. a. $25\sqrt{2}$ m/s b. $\frac{\sqrt{2}D}{T}$
5. a. $-\frac{5}{3}$ m/s^2 b. 60 m
   c. 9 s d. 67.5 m
6. a. i. $\frac{5\sqrt{10}}{2}$ m/s ii. 7.5 m/s
   b. i. $\sqrt{\frac{U^2 + V^2}{2}}$ ii. $\frac{U + V}{2}$
7. a. 2 m/s^2, 2 m/s
   b. $\frac{V_1T_2 - V_2T_1}{T_2 - T_1}$
8. a. 0.5 m/s^2, 6 m/s
   b. $\frac{2\left(d_2T_1 - d_1T_2\right)}{T_1T_2\left(T_2 - T_1\right)}$
9. a. 0.5 m/s^2, 5 m/s
   b. $\frac{V_2^2 - V_1^2}{2\left(d_2 - d_2\right)}, \sqrt{\frac{V_1^2d_2 - V_2^2d_1}{d_2 - d_1}}$
10. a. 19.6 m/s, 19.6 m
    b. $gT, \frac{1}{2}gT^2$
11. a. 253.5 km/h, 7.19 s
    b. 96.96 km/h, 2.75 s
12. a. 5 s b. 123.5 m
    c. 10.02 s d. 7.04, 2.96 s
13. a. 5 s b. $\frac{U + \sqrt{U^2 + 2gH}}{g}$

14. a. i. $\sqrt{6}$ s

ii. 29.4 m

b. i. $\sqrt{T_1 T_2}$

ii. $\frac{1}{2} g T_1 T_2$

15. 343 m/s
16. 58.19 m
17. a. 0.5986 s

b. Sample responses can be found in the worked solutions in the online resources.

18. a. 21.85 m/s

b. Sample responses can be found in the worked solutions in the online resources.

11.3 Exam questions

Note: Mark allocations are available with the fully worked solutions online.

1. E
2. B
3. C

11.4 Velocity–time graphs

11.4 Exercise

1. 3 s, 67.5 m
2. 30 s, 393.75 m
3. a. Sample responses can be found in the worked solutions in the online resources.

b. $\dfrac{V^2 T}{2(V-U)}$

4. 60 s
5. 72 s
6. $133\frac{1}{3}$ s
7. Sample responses can be found in the worked solutions in the online resources.
8. 18 m/s
9. 4275 m
10. $T = 180,\ b = 6,\quad U = 5, D = 675$ m
11. 192.5 s
12. 512.5 s
13. Sample responses can be found in the worked solutions in the online resources.
14. a. See graph at the bottom of the page*

b. $226\frac{2}{3}$ m

c. 180 m

15. a. $a = 2,\ b = 8,\ c = 8,\ d = \frac{1}{2}$

b. $78\frac{2}{3}$ m

c. 60 m

16. 72 s
17. a. See graph at the bottom of the page*

b. 327.04 m

c. 381.97 m

d. 2509.01 m

e. 59.2 %

f. 0.184 m/s^2

*14. a.

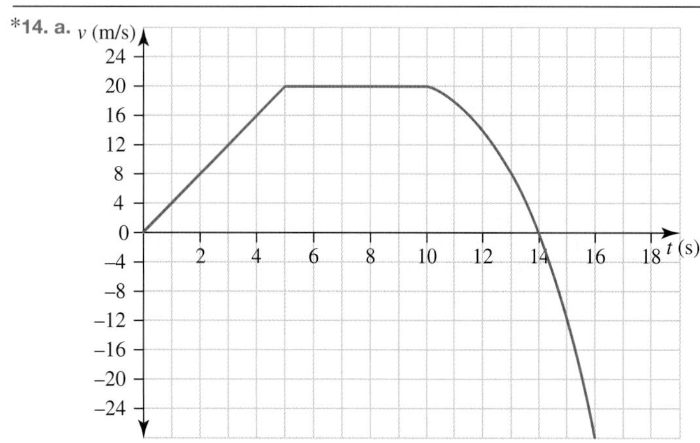

*17. a.

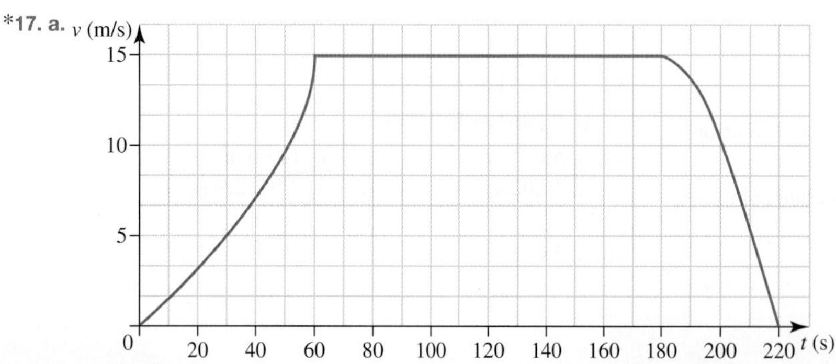

18. a. See graph at the bottom of the page*

b. 919.05 m

c. 0.75 m/s^2

19. a. See graph at the bottom of the page*

b. 424.24 m

c. 306.68 m

d. $-\dfrac{10}{7}$ m/s^2

20. a. $a = \dfrac{45}{2}$, $b = -\dfrac{1}{36}$

b. See graph at the bottom of the page*

c. 0.226 m/s^2

d. 5015.15 m

11.4 Exam questions

Note: Mark allocations are available with the fully worked solutions online.

1. D
2. E
3. 1680 m

11.5 Acceleration that depends on time

11.5 Exercise

1. $\dfrac{1}{2}$ m
2. $x = 4t^3 - 3t^2 + t + 2$
3. $x = 18\sin\left(\dfrac{t}{3}\right)$
4. $4\left(\sqrt{2} - 1\right)$ m
5. 12 m
6. a. 1 s b. 25 m
7. 1800 m, 180 s
8. $11\dfrac{1}{9}$ m

*18. a.

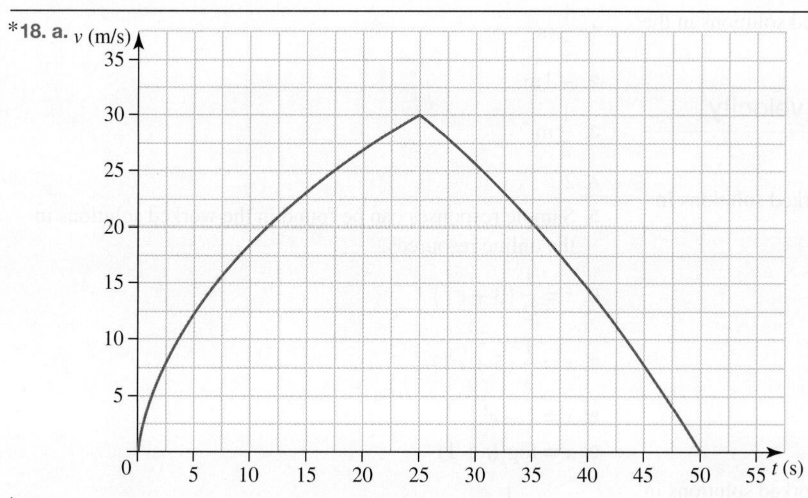

*19. a.

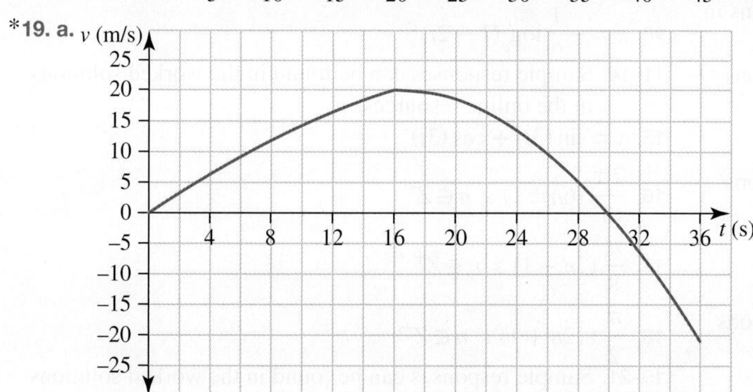

*20. b.

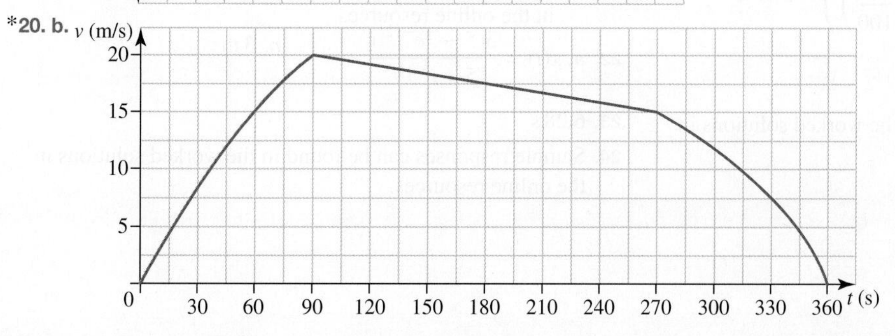

9. a. $\frac{1}{2}$ s b. 10 m

10. 40 m

11. $x = 2\left(e^{2t} - 1\right) - 2t^2 - 4t$

12. $x = 49t + 245e^{-\frac{t}{5}} - 225$

13. Sample responses can be found in the worked solutions in the online resources.

14. 67.388 m

15. $\frac{1}{2}$ s, $9\frac{31}{40}$ m

16. 0.492 m

17. 0.845 m

18. 1.669 m

11.5 Exam questions

Note: Mark allocations are available with the fully worked solutions online.

1. a. 13.5 m/s b. 85.83 m
   c. 95.13 m d. 10.8 s

2. 10 m

3. Sample responses can be found in the worked solutions in the online resources.

11.6 Acceleration that depends on velocity

11.6 Exercise

1–4. Sample responses can be found in the worked solutions in the online resources.

5. a. $\frac{1}{2}$ s b. 10 m

6. a. 5 b. 40

7. a. 5 s b. $46\frac{2}{3}$ m

8. $T = \frac{1}{2}, D = \frac{400}{41}$

9–14. Sample responses can be found in the worked solutions in the online resources.

15. a. Sample responses can be found in the worked solutions in the online resources.
    b. 50.002 s
    c. Sample responses can be found in the worked solutions in the online resources.
    d. $42\sqrt{5}$ m/s
    e. 5.69 s

16–20. Sample responses can be found in the worked solutions in the online resources.

21. a. $t = \frac{1}{40\,000}\left(\tan^{-1}(4) - \tan^{-1}\left(\frac{v}{100}\right)\right)$
    0.033 ms
    b. 3.5 mm

22. Sample responses can be found in the worked solutions in the online resources.

11.6 Exam questions

Note: Mark allocations are available with the fully worked solutions online.

1. a. 19.6 m
   b. Sample responses can be found in the worked solutions in the online resources.
   c. $14\sqrt{5}$ m/s
   d. i. $t = \int_{19.6}^{30} \frac{1}{g - 0.01v^2}\,dv + 2$
      ii. 5.8 s
   e. $x = \int_{19.6}^{30} \frac{v}{g - 0.01v^2}\,dv + 19.6$
      $= 120$ m

2. D

3. D

11.7 Acceleration that depends on position

11.7 Exercise

1. $\frac{1}{2}$ m

2. −3 m

3. $\frac{1}{2}$ m

4. 2

5. Sample responses can be found in the worked solutions in the online resources.

6. $x = \frac{1}{2}\left(3 + e^{2t}\right)$

7. $x = \frac{1}{3}\left(5 + e^{3t}\right)$

8. $x = 1 - e^t$

9. $x = \log_e(t + 1)$

10. $x = -\frac{1}{2}\log_e(1 - 2t)$

11–14. Sample responses can be found in the worked solutions in the online resources.

15. $x = \sin(3t) + \cos(3t)$

16. $\frac{2\pi}{3}(6n \pm 1)$ s $n \in Z^+$

17. $\frac{\pi}{8}(2n - 1)$ s $n \in Z^+$

18. $\frac{\pi}{2}(12n + 1)$ s $n \in Z^+$

19–21. Sample responses can be found in the worked solutions in the online resources.

22. a. $x(t) = \frac{3(5 - 3e^{-12t})}{5 + e^{-12t}}$ b. 3 m

23. 6.28 s

24. Sample responses can be found in the worked solutions in the online resources.

11.7 Exam questions

Note: Mark allocations are available with the fully worked solutions online.

1. A
2. Sample responses can be found in the worked solutions in the online resources.
3. Sample responses can be found in the worked solutions in the online resources.

11.8 Review

11.8 Exercise

Technology free: short answer

1. a. $x = \frac{1}{36}\left(2t^3 - 3t^2 - 12t - 108\right)$ m

   b. $-\frac{32}{9}$ m

   c. $\frac{1}{2}$ m/s^2

2. 0.25 m/s^2
3. 32 m
4. $\frac{7\sqrt{2}}{2}$ m
5. $T = 5,\ D = 53\frac{1}{7}$
6. 0.4 m

Technology active: multiple choice

7. D
8. C
9. B
10. A
11. D
12. B
13. E
14. E
15. C
16. A

Technology active: extended response

17. a.

   b. $131\frac{1}{9}$ m

   c. $108\frac{8}{9}$ m

   d. $-\frac{5}{3}$ m/s^2

18. a. $a = 20,\ b = \frac{40}{\pi}$

   b. See graph at the bottom of the page*

   c. Sample responses can be found in the worked solutions in the online resources.

   d. 4910 m

   e. $\frac{\pi\sqrt{2}}{18}$ m/s^2

   f. $-\frac{4\sqrt{3}}{9\pi}$ m/s^2

19, 20. Sample responses can be found in the worked solutions in the online resources.

11.8 Exam questions

Note: Mark allocations are available with the fully worked solutions online.

1. a. $x = e^{1-\cos(t)}$

   b. $x_{max} = e^2$ and occurs at $t = (2n - 1)\pi$ s, $n \in N$

2. A

*18. b.

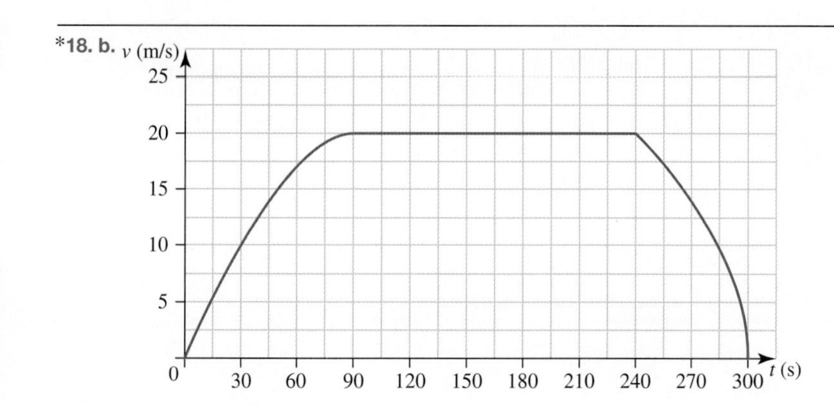

3. a. Sample responses can be found in the worked solutions in the online resources.

b. Sample responses can be found in the worked solutions in the online resources.

c. 2.7 m

d. $t = \dfrac{1}{1.4}\displaystyle\int_0^5 \dfrac{1}{7-v}\,dv$

$= 0.9$ s

4. $v = e^5$ ms^{-1}

5. a.

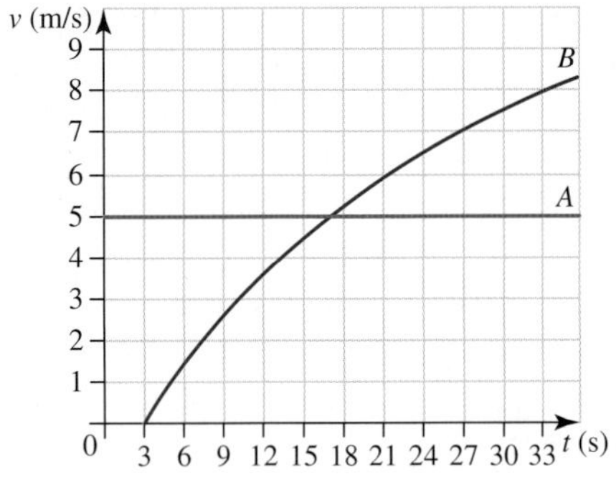

b. 37.175 s

12 Vector calculus

LEARNING SEQUENCE

Fully worked solutions are available online.

12.1 Overview

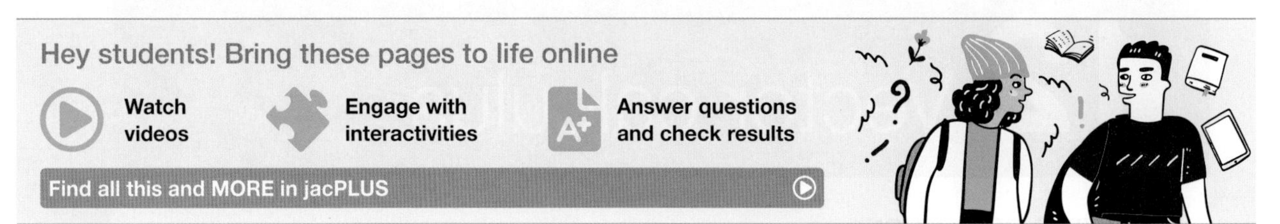

12.1.1 Introduction

The motion of a particle has been studied since ancient times and has intrigued both mathematicians and physicists. The motion of particles can take many forms, for example, a speck of dust sitting on the outer edge of a rotating vinyl record or a ball travelling toward the ground in a straight line. In this topic we will examine the motion of particles in both two and three dimensions and how differentiation and integration techniques can be used to analyse the motion. Consider the image below of the young athlete throwing a javelin. The initial speed and the angle at which the javelin is released will both have an effect on the distance the javelin will travel. These variables will be investigated in the study of projectile motion in this chapter.

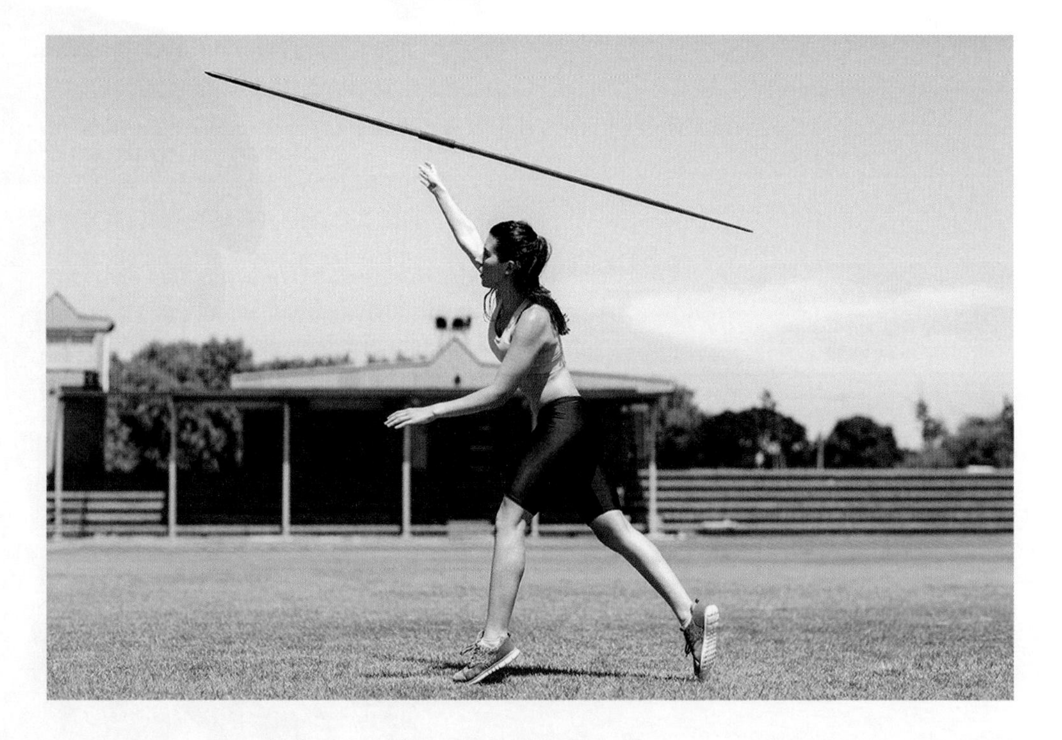

KEY CONCEPTS

This topic covers the following point from the VCE Mathematics Study Design:

- position vector as a function of time and sketching the corresponding path given the function, including circles, ellipses and hyperbolas in Cartesian or parametric forms
- the positions of two particles each described as a vector function of time, and whether their paths cross or if the particles meet
- differentiation and anti-differentiation of a vector function with respect to time and applying vector calculus to motion in a plane and in three dimensions.

Source: VCE Mathematics Study Design (2023–2027) extracts © VCAA; reproduced by permission.

12.2 Position vectors as functions of time

LEARNING INTENTION

At the end of this subtopic you should be able to:
- evaluate the position of a particle at a time t from its position vector
- determine the distance of a particle from the origin at a time t
- determine the time and position at which a particle is closest to the origin.

12.2.1 Parametric equations

Consider a particle moving around in 2-dimensional space so that its position with respect to the x- and y-coordinates are described by the following functions of the parameter, t.

$$x = x(t) \quad (1)$$
$$y = y(t) \quad (2)$$

The position vector of the particle is given by $\underset{\sim}{r}(t) = x\underset{\sim}{i} + y\underset{\sim}{j}$, where $\underset{\sim}{i}$ and $\underset{\sim}{j}$ are unit vectors in the x and y directions. This is also called a vector function of the scalar real variable t, where t is called the parameter, and often represents time.

If we can eliminate the parameter from the two parametric equations and obtain an equation of the form $y = f(x)$, this is called an explicit relationship or the equation of the path. It may not be possible to obtain an explicit relationship, but often an implicit relationship of the form $f(x, y) = 0$ can be formed. The relationship between x and y is called the Cartesian equation of the path. Careful consideration must be given to the possible values of t, which then specify the domain (the x values) and the range (the y values) of the equation of the path.

Closest approach

Given the position vector of a particle, $\underset{\sim}{r}(t) = x(t)\underset{\sim}{i} + y(t)\underset{\sim}{j}$, where $\underset{\sim}{i}$ and $\underset{\sim}{j}$ are unit vectors in the x and y directions, it is possible to find the position or coordinates of the particle at a given value of t. It is also possible to determine the value of t and the coordinates when the particle is closest to the origin.

The distance of the point from the origin is the magnitude of the position vector. Since both the x- and y-coordinates are expressed in terms of a parameter, t the distance can be expressed as a function of t, $|\underset{\sim}{r}(t)|$. To calculate the closest distance to the origin that the particle passes through, take the derivative of the distance function and determine the point at which the derivative is equal to zero.

WORKED EXAMPLE 1 Calculating the distance from the origin

A particle moves so that its vector equation is given by $\underset{\sim}{r}(t) = (3t - 4)\underset{\sim}{i} + (4t - 3)\underset{\sim}{j}$ for $t \geq 0$.
a. Calculate the distance of the particle from the origin when $t = 2$.
b. Determine the distance of the particle from the origin at any time t.
c. Calculate the closest distance of the particle from the origin.

THINK	WRITE
a. 1. Substitute the value of t.	a. Substitute $t = 2$: $\underset{\sim}{r}(2) = (6-4)\underset{\sim}{i} + (8-3)\underset{\sim}{j}$ $= 2\underset{\sim}{i} + 5\underset{\sim}{j}$ At the point (2, 5)

2. Calculate the magnitude of the vector, which represents the distance from the origin. $|\underset{\sim}{r}(2)| = \sqrt{2^2 + 5^2}$

3. State the distance from the origin at this time. $|\underset{\sim}{r}(2)| = \sqrt{29}$

b. 1. Determine the magnitude of the vector at time t.

b. $\underset{\sim}{r}(t) = (3t-4)\underset{\sim}{i} + (4t-3)\underset{\sim}{j}$

$$|\underset{\sim}{r}(t)| = \sqrt{(3t-4)^2 + (4t-3)^2}$$

2. Expand and simplify to state the distance in terms of t.

$$\begin{aligned}|\underset{\sim}{r}(t)| &= \sqrt{9t^2 - 24t + 16 + 16t^2 - 24t + 9}\\ &= \sqrt{25t^2 - 48t + 25}\end{aligned}$$

c. 1. For the closest or minimum distance, we need to use calculus.

c. The minimum distance occurs when $\frac{d}{dt}\left(|\underset{\sim}{r}(t)|\right) = 0$.

2. Use the chain rule.

$$|\underset{\sim}{r}(t)| = (25t^2 - 48t + 25)^{\frac{1}{2}}$$

$$\frac{d}{dt}\left(|\underset{\sim}{r}(t)|\right) = \frac{\frac{1}{2} \times (50t - 48)}{\sqrt{25t^2 - 48t + 25}} = 0$$

3. Solve for the value of t.

$$\begin{aligned}50t - 48 &= 0\\ t &= \frac{24}{25}\end{aligned}$$

4. Determine the position vector at this value.

Substitute $t = \frac{24}{25}$:

$$\begin{aligned}\underset{\sim}{r}\left(\frac{24}{25}\right) &= \left(3 \times \frac{24}{25} - 4\right)\underset{\sim}{i} + \left(4 \times \frac{24}{25} - 3\right)\underset{\sim}{j}\\ &= -\frac{28}{25}\underset{\sim}{i} + \frac{21}{25}\underset{\sim}{j}\end{aligned}$$

5. Calculate the magnitude of the vector at this time, which represents the closest distance of the particle from the origin.

$$\begin{aligned}|\underset{\sim}{r}(t)|_{\text{min}} &= \sqrt{\left(\frac{28}{25}\right)^2 + \left(\frac{21}{25}\right)^2}\\ &= \frac{1}{25}\sqrt{1225}\end{aligned}$$

6. State the final result.

$$|\underset{\sim}{r}(t)|_{\text{min}} = \frac{7}{5}$$

12.2.2 Collision problems

There are a number of problems that can be formulated around the motion of two moving particles on different curves in two dimensions.

1. Do the particles collide? For two particles to collide, they must be at exactly the same coordinates at exactly the same time.
2. Do the paths of the particles cross without colliding? This will happen when they are at the same coordinates but at different times.
3. What is the distance between the particles at a particular time? To determine this, the magnitude of the difference between their respective position vectors must be found.

WORKED EXAMPLE 2 Determining the point and time of collisions between two particles

Two particles move so that their position vectors are given by $\underset{\sim}{r}_A(t) = (3t - 8)\underset{\sim}{i} + \left(t^2 - 18t + 87\right)\underset{\sim}{j}$ and $\underset{\sim}{r}_B(t) = (20 - t)\underset{\sim}{i} + (2t - 4)\underset{\sim}{j}$ for $t \geq 0$. Determine:

a. **when and where the particles collide**
b. **where their paths cross**
c. **the distance between the particles when $t = 10$.**

THINK	WRITE
a. 1. Equate the $\underset{\sim}{i}$ components for each particle.	a. $3t - 8 = 20 - t$
2. Solve this equation for t.	$4t = 28$ $t = 7$
3. Equate the $\underset{\sim}{j}$ components for each particle.	$t^2 - 18t + 87 = 2t - 4$
4. Solve this equation for t.	$t^2 - 20t + 91 = 0$ $(t - 7)(t - 13) = 0$ $t = 7, 13$
5. Evaluate the position vectors at the common time.	The common solution is when $t = 7$. $\underset{\sim}{r}_A(7) = 13\underset{\sim}{i} + 10\underset{\sim}{j}$ $\underset{\sim}{r}_B(7) = 13\underset{\sim}{i} + 10\underset{\sim}{j}$
6. State the result for when the particles collide.	The particles collide when $t = 7$ at the point (13, 10).
b. 1. The particles paths cross at different values of t. Introduce a different parameter, s.	b. Let $\underset{\sim}{r}_A(s) = (3s - 8)\underset{\sim}{i} + \left(s^2 - 18s + 87\right)\underset{\sim}{j}$ and $\underset{\sim}{r}_B(t) = (20 - t)\underset{\sim}{i} + (2t - 4)\underset{\sim}{j}$.
2. Equate the $\underset{\sim}{i}$ components for each particle.	$3s - 8 = 20 - t$
3. Solve this equation for t.	$t = 28 - 3s$
4. Equate the $\underset{\sim}{j}$ components for each particle.	$s^2 - 18s + 87 = 2t - 4$
5. Solve this equation for s.	Substitute $t = 28 - 3s$: $s^2 - 18s + 87 = 2(28 - 3s) - 4$ $s^2 - 18s + 87 = 56 - 6s - 4$ $s^2 - 12s + 35 = 0$ $(s - 5)(s - 7) = 0$ $s = 5, 7$
6. Evaluate the position vectors at the required time. Note that when $s = 7$, the particles collide.	$\underset{\sim}{r}_A(5) = (15 - 8)\underset{\sim}{i} + (25 - 90 + 87)\underset{\sim}{j}$ $= 7\underset{\sim}{i} + 22\underset{\sim}{j}$ When $s = 5$, $t = 28 - 15 = 13$ $\underset{\sim}{r}_B(13) = (20 - 13)\underset{\sim}{i} + (26 - 4)\underset{\sim}{j}$ $= 7\underset{\sim}{i} + 22\underset{\sim}{j}$
7. State the required result.	The paths cross at the point (7, 22).
c. 1. Evaluate both the position vectors at the required time.	c. Substitute $t = 10$: $\underset{\sim}{r}_A(10) = (30 - 8)\underset{\sim}{i} + (100 - 180 + 87)\underset{\sim}{j}$ $= 22\underset{\sim}{i} + 7\underset{\sim}{j}$ $\underset{\sim}{r}_B(10) = (20 - 10)\underset{\sim}{i} + (20 - 4)\underset{\sim}{j}$ $= 10\underset{\sim}{i} + 16\underset{\sim}{j}$

2. Calculate the difference between these two vectors.	$\underset{\sim}{r}_A(10) - \underset{\sim}{r}_B(10) = 22\underset{\sim}{i} + 7\underset{\sim}{j} - (10\underset{\sim}{i} + 16\underset{\sim}{j})$ $= 12\underset{\sim}{i} - 9\underset{\sim}{j}$
3. The distance between the particles is the magnitude of the difference between these vectors.	$\left\|\underset{\sim}{r}_A(10) - \underset{\sim}{r}_B(10)\right\| = \left\|12\underset{\sim}{i} - 9\underset{\sim}{j}\right\|$ $= \sqrt{12^2 + (-9)^2}$ $= \sqrt{225}$
4. State the required distance.	$\left\|\underset{\sim}{r}_A(10) - \underset{\sim}{r}_B(10)\right\| = 15$

12.2 Exercise

Students, these questions are even better in jacPLUS

Receive immediate feedback and access sample responses

Access additional questions

Track your results and progress

Find all this and MORE in jacPLUS

Technology free

1. WE1 A particle moves so that its vector equation is given by $\underset{\sim}{r}(t) = (t-2)\underset{\sim}{i} + (3t-1)\underset{\sim}{j}$ for $t \geq 0$.
   a. Calculate the distance of the particle from the origin when $t = 5$.
   b. Determine the distance of the particle from the origin at any time t.
   c. Calculate the closest distance of the particle from the origin.

2. A particle moves so that its vector equation is given by $\underset{\sim}{r}(t) = \sqrt{t}\underset{\sim}{i} + (2t+3)\underset{\sim}{j}$ for $t \geq 0$.
   a. Calculate the distance of the particle from the origin when $t = 4$.
   b. Calculate the value t when the distance of the particle from the origin is $15\sqrt{2}$.

3. WE2 Two particles move so that their position vectors are given by $\underset{\sim}{r}_A(t) = (2t+6)\underset{\sim}{i} + \left(t^2 - 6t + 45\right)\underset{\sim}{j}$ and $\underset{\sim}{r}_B(t) = (t+11)\underset{\sim}{i} + (7t+5)\underset{\sim}{j}$ for $t \geq 0$. Determine:
   a. when and where the particles collide
   b. where their paths cross
   c. the distance between the particles when $t = 10$.

4. Two particles move so that their position vectors are given by $\underset{\sim}{r}_A(t) = (-t^2 + 12t - 22)\underset{\sim}{i} + (19 - 3t)\underset{\sim}{j}$ and $\underset{\sim}{r}_B(t) = (18 - 2t)\underset{\sim}{i} + (t+3)\underset{\sim}{j}$ for $t \geq 0$. Determine:
   a. when and where the particles collide
   b. where their paths cross
   c. the distance between the particles when $t = 5$.

5. A particle moves so that its vector equation is given by $\underset{\sim}{r}(t) = (2t-1)\underset{\sim}{i} + (t-3)\underset{\sim}{j}$ for $t \geq 0$.
   a. Calculate the distance of the particle from the origin when $t = 4$.
   b. Determine the distance of the particle from the origin at any time t.
   c. Calculate the closest distance of the particle from the origin.

6. A particle moves so that its vector equation is given by $\underset{\sim}{r}(t) = (4t - 3)\underset{\sim}{i} + (3t + 4)\underset{\sim}{j}$ for $t \geq 0$.
   a. Calculate the distance of the particle from the origin when $t = 2$.
   b. Determine the distance of the particle from the origin at any time t.
   c. Calculate the closest distance of the particle from the origin.

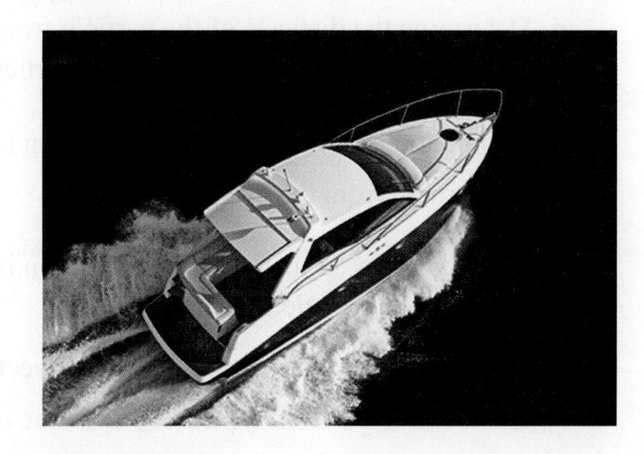

7. A boat moves so that its vector equation is given by $\underset{\sim}{r}(t) = (2t - 3)\underset{\sim}{i} + 2\sqrt{t}\underset{\sim}{j}$ for $t \geq 0$, where distance is measured in kilometres.
   a. Calculate the distance of the boat from the origin when $t = 4$.
   b. Determine the closest distance of the boat from the origin.
   c. Calculate the times when the boat is 3 kilometres from the origin.

8. A particle moves so that its vector equation is given by $\underset{\sim}{r}(t) = (2t - 7)\underset{\sim}{i} + (2t + 2)\underset{\sim}{j}$ for $t \geq 0$.

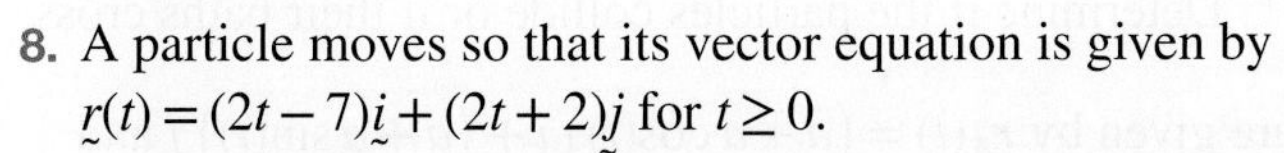

   a. Determine the closest distance of the particle from the origin.

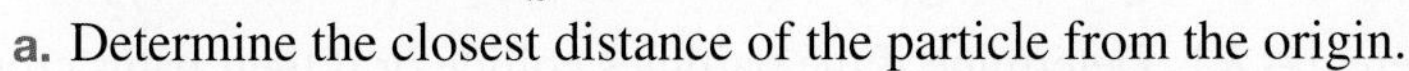

   b. Calculate the time when the particle is $9\sqrt{5}$ units from the origin.

9. A particle moves so that its vector equation is given by $\underset{\sim}{r}(t) = (at + b)\underset{\sim}{i} + \left(ct^2 + d\right)\underset{\sim}{j}$ for $t \geq 0$. If $\underset{\sim}{r}(2) = 5\underset{\sim}{i} + 7\underset{\sim}{j}$ and $\underset{\sim}{r}(4) = 13\underset{\sim}{i} + 19\underset{\sim}{j}$, determine the values of a, b, c and d.

10. A particle moves so that its vector equation is given by $\underset{\sim}{r}(t) = (at + b)\underset{\sim}{i} + \left(ct^2 + dt\right)\underset{\sim}{j}$ for $t \geq 0$. If $\underset{\sim}{r}(4) = 13\underset{\sim}{i} + 4\underset{\sim}{j}$ and $\underset{\sim}{r}(6) = 17\underset{\sim}{i} + 18\underset{\sim}{j}$, determine the values of a, b, c and d.

11. Two particles move so that their position vectors are given by $\underset{\sim}{r}_A(t) = (3t - 43)\underset{\sim}{i} + \left(-t^2 + 26t - 160\right)\underset{\sim}{j}$ and $\underset{\sim}{r}_B(t) = (17 - t)\underset{\sim}{i} + (2t - 25)\underset{\sim}{j}$ for $t \geq 0$. Determine:
    a. when and where the particles collide
    b. where their paths cross
    c. the distance between the particles when $t = 12$.

12. Two particles move so that their position vectors are given by $\underset{\sim}{r}_A(t) = \left(t^2 - 6\right)\underset{\sim}{i} + (2t + 2)\underset{\sim}{j}$ and $\underset{\sim}{r}_B(t) = (7t - 16)\underset{\sim}{i} + \frac{1}{5}\left(17t - t^2\right)\underset{\sim}{j}$ for $t \geq 0$. Determine:
    a. when and where the particles collide
    b. the distance between the particles when $t = 10$.

13. Two particles move so that their position vectors are given by $\underset{\sim}{r}_A(t) = \left(-t^2 + 12t + 53\right)\underset{\sim}{i} + (3t + 38)\underset{\sim}{j}$ and $\underset{\sim}{r}_B(t) = (2t + 29)\underset{\sim}{i} + (86 - t)\underset{\sim}{j}$ for $t \geq 0$. Determine:
    a. when and where the particles collide
    b. where their paths cross
    c. the distance between the particles when $t = 20$.

14. A toy train moves so that its vector equation is given by $\underset{\sim}{r}(t) = (3 - 3\cos(t))\underset{\sim}{i} + (3 + 3\sin(t))\underset{\sim}{j}$ for $t \geq 0$.
    a. Calculate the position of the toy train at the times $t = 0$, π and 2π.
    b. Determine and sketch the Cartesian equation of the path of the toy train, stating the domain and range.
    c. Determine the distance of the toy train from the origin at any time t.
    d. Calculate the closest distance of the toy train from the origin.

Technology active

15. A particle moves so that its vector equation is given by $\underset{\sim}{r}(t) = \left(2 + 4\cos(t)\right)\underset{\sim}{i} + \left(4 + 4\sin(t)\right)\underset{\sim}{j}$ for $t \geq 0$.
    a. Calculate the position of the particle at the times $t = 0$, π and 2π.
    b. Determine and sketch the Cartesian equation of the path, stating the domain and range.
    c. Determine the distance of the particle from the origin at any time t.
    d. Calculate the closest distance of the particle from the origin.

16. A particle moves so that its vector equation is given by $\underset{\sim}{r}(t) = t\underset{\sim}{i} + \frac{1}{t^2}\underset{\sim}{j}$ for $t > 0$. Determine the closest distance of the particle from the origin.

17. A particle moves so that its vector equation is given by $\underset{\sim}{r}(t) = e^{-t}\underset{\sim}{i} + e^{t}\underset{\sim}{j}$ for $t \in R$. Determine the closest distance of the particle from the origin.

18. Two particles move so that their position vectors are given by $\underset{\sim}{r}_A(t) = a\cos(t)\underset{\sim}{i} + a\sin(t)\underset{\sim}{j}$ and $\underset{\sim}{r}_B(t) = a\cos^2(t)\underset{\sim}{i} + a\sin^2(t)\underset{\sim}{j}$ for $t \geq 0$, where $a \in R^+$. Determine if the particles collide or if their paths cross.

19. Two particles move so that their position vectors are given by $\underset{\sim}{r}_A(t) = \left(a + a\cos(t)\right)\underset{\sim}{i} + \left(a + a\sin(t)\right)\underset{\sim}{j}$ and $\underset{\sim}{r}_B(t) = a\cos^2(t)\underset{\sim}{i} + a\sin^2(t)\underset{\sim}{j}$ for $t \geq 0$, where $a \in R^+$. Determine if the particles collide or if their paths cross.

20. A particle moves so that its vector equation is given by $\underset{\sim}{r}(t) = (5t - 2)\underset{\sim}{i} + (12t - 2)\underset{\sim}{j}$ for $t \geq 0$. Calculate the closest distance of the particle from the origin.

21. A particle moves so that its vector equation is given by $\underset{\sim}{r}(t) = (at + b)\underset{\sim}{i} + (ct + d)\underset{\sim}{j}$ for $t \geq 0$. Calculate the closest distance of the particle from the origin.

22. A particle moves so that its vector equation is given by $\underset{\sim}{r}(t) = (4t - 3)\underset{\sim}{i} + \left(t^2 + 3\right)\underset{\sim}{j}$ for $t \geq 0$. Calculate the closest distance of the particle from the origin, correct to 3 decimal places.

23. A particle moves so that its vector equation is given by $\underset{\sim}{r}(t) = (2t - 1)\underset{\sim}{i} + \left(t^2 + 3t\right)\underset{\sim}{j}$ for $t \geq 0$. Calculate the closest distance of the particle from the origin, correct to 3 decimal places.

12.2 Exam questions

Question 1 (3 marks) TECH-FREE

Source: VCE 2019, Specialist Mathematics Exam 1, Q4; © VCAA.

The position vectors of two particles A and B at time t seconds after they have started moving are given by $\underset{\sim}{r}_A(t) = \left(t^2 - 1\right)\underset{\sim}{i} + \left(a + \frac{t}{3}\right)\underset{\sim}{j}$ and $\underset{\sim}{r}_B(t) = \left(t^3 - t\right)\underset{\sim}{i} + \left(\arccos\left(\frac{t}{2}\right)\right)\underset{\sim}{j}$ respectively, where a is a real constant and $0 \leq t \leq 2$.

Find the value of a if the particles collide after they have started moving.

Question 2 (6 marks) TECH-ACTIVE

Source: VCE 2018, Specialist Mathematics Exam 2, Section B, Q4a-c; © VCAA.

Two yachts, A and B, are competing in a race and their position vectors on a certain section of the race after time t hours are given by

$$\underset{\sim}{r}_A(t) = (t + 1)\underset{\sim}{i} + \left(t^2 + 2t\right)\underset{\sim}{j} \quad \text{and} \quad \underset{\sim}{r}_B(t) = t^2\underset{\sim}{i} + \left(t^2 + 3\right)\underset{\sim}{j},\ t \geq 0$$

where displacement components are measured in kilometres from a given reference buoy at origin O.

a. Find the Cartesian equation of the path for each yacht. **(2 marks)**
b. Show that the two yachts will not collide if they follow these paths. **(2 marks)**
c. Find the coordinates of the point where the paths of the two yachts cross. Give your coordinates correct to three decimal places. **(2 marks)**

Question 3 (1 mark) TECH-ACTIVE

Source: *VCE 2015, Specialist Mathematics Exam 2, Section 1, Q18; © VCAA.*

The position vectors of two moving particles are given by
$\underset{\sim}{r}_1(t) = \left(2+4t^2\right)\underset{\sim}{i} + (3t+2)\underset{\sim}{j}$ and $\underset{\sim}{r}_2(t) = (6t)\underset{\sim}{i} + (4+t)\underset{\sim}{j}$, where $t \geq 0$.

The particles will collide at

A. $3\underset{\sim}{i} + 3.5\underset{\sim}{j}$ **B.** $6\underset{\sim}{i} + 5\underset{\sim}{j}$ **C.** $3\underset{\sim}{i} + 4.5\underset{\sim}{j}$ **D.** $0.5\underset{\sim}{i} + \underset{\sim}{j}$ **E.** $5\underset{\sim}{i} + 6\underset{\sim}{j}$

More exam questions are available online.

12.3 Differentiation of vectors

LEARNING INTENTION

At the end of this subtopic you should be able to:

- differentiate vectors
- determine velocity vectors by differentiating position vectors
- determine acceleration vectors by differentiating velocity vectors.

12.3.1 Vector functions

The function $\underset{\sim}{r}(t) = x(t)\underset{\sim}{i} + y(t)\underset{\sim}{j}$ is called a vector function. It represents the position vector of a particle at time t in two dimensions. As t changes, both the x- and y-coordinates change; thus, the particle is moving along a curve. The equation of the curve is called the Cartesian equation of the path.

The derivative of a vector function

Consider an origin, O, and let P be the position of the particle at time t, so that $\overrightarrow{OP} = \underset{\sim}{r}(t)$.

Suppose that Q is a neighbouring point close to P at time $t + \delta t$, so that $\overrightarrow{OQ} = \underset{\sim}{r}(t+\delta t)$.

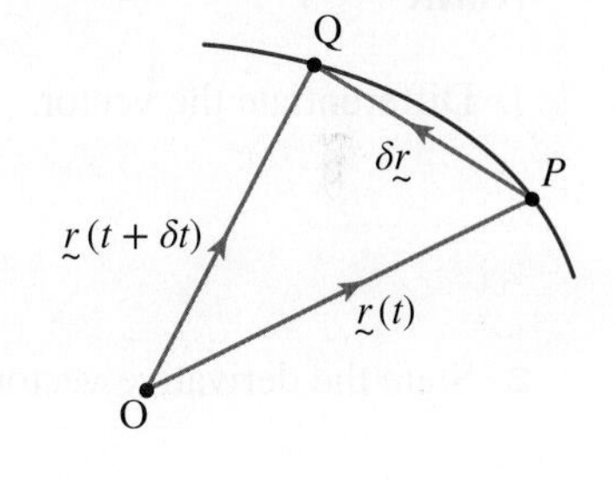

$$\begin{aligned}\overrightarrow{OQ} &= \overrightarrow{OP} + \overrightarrow{PQ}\\ \overrightarrow{PQ} &= \overrightarrow{OQ} - \overrightarrow{OP}\\ \delta\underset{\sim}{r} &= \underset{\sim}{r}(t+\delta t) - \underset{\sim}{r}(t)\end{aligned}$$

Consider $\dfrac{\delta\underset{\sim}{r}}{\delta t} = \dfrac{\underset{\sim}{r}(t+\delta t) - \underset{\sim}{r}(t)}{\delta t}$, where $\delta t \neq 0$. Because $\delta\underset{\sim}{r}$ is a vector and δt is a scalar, $\dfrac{\delta\underset{\sim}{r}}{\delta t}$ is a vector parallel to $\delta\underset{\sim}{r}$ or the vector $\overrightarrow{PQ}$. As $\delta t \to 0$, provided the limit exists, we define

$$\begin{aligned}\frac{d\underset{\sim}{r}}{dt} &= \lim_{\delta t\to 0}\frac{\underset{\sim}{r}(t+\delta t) - \underset{\sim}{r}(t)}{\delta t}\\ \frac{d\underset{\sim}{r}}{dt} &= \lim_{\delta t\to 0}\frac{x(t+\delta t)\underset{\sim}{i} + y(t+\delta t)\underset{\sim}{j} - (x(t)\underset{\sim}{i} + y(t)\underset{\sim}{j})}{\delta t}\\ \frac{d\underset{\sim}{r}}{dt} &= \lim_{\delta t\to 0}\frac{x(t+\delta t) - x(t)}{\delta t}\underset{\sim}{i} + \lim_{\delta t\to 0}\frac{y(t+\delta t) - y(t)}{\delta t}\underset{\sim}{j}\\ \frac{d\underset{\sim}{r}}{dt} &= \frac{dx}{dt}\underset{\sim}{i} + \frac{dy}{dt}\underset{\sim}{j}\end{aligned}$$

The vector $\frac{d\underset{\sim}{r}}{dt} = \dot{\underset{\sim}{r}}$, where the dot indicates the derivative with respect to t, is a vector parallel to the tangent T to the curve at the point P.

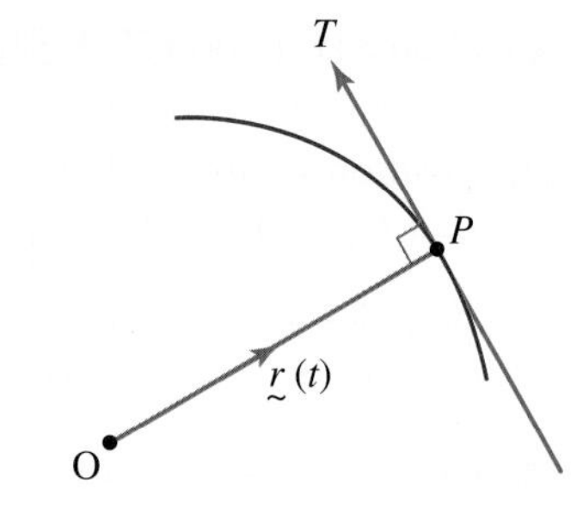

A unit tangent vector at $t = a$ is denoted by $\hat{\underset{\sim}{s}} = \frac{\dot{\underset{\sim}{r}}(a)}{|\dot{\underset{\sim}{r}}(a)|}$.

We do not need to use first principles to find the derivatives of vectors.

Derivative of a constant vector

If $\underset{\sim}{c}$ is a constant vector, that is a vector which does not change and is independent of t, then $\frac{d\underset{\sim}{c}}{dt} = \underset{\sim}{0}$. Note that $\frac{d\underset{\sim}{i}}{dt} = \frac{d\underset{\sim}{j}}{dt} = \frac{d\underset{\sim}{k}}{dt} = \underset{\sim}{0}$.

Derivative of a sum or difference of vectors

The sum or difference of two vectors can be differentiated as the sum or difference of the individual derivatives. That is,

$$\frac{d}{dt}(\underset{\sim}{a} + \underset{\sim}{b}) = \frac{d\underset{\sim}{a}}{dt} + \frac{d\underset{\sim}{b}}{dt} \text{ and } \frac{d}{dt}(\underset{\sim}{a} - \underset{\sim}{b}) = \frac{d\underset{\sim}{a}}{dt} - \frac{d\underset{\sim}{b}}{dt}.$$

Using these rules, if $\underset{\sim}{r}(t) = x(t)\underset{\sim}{i} + y(t)\underset{\sim}{j}$, then $\frac{d\underset{\sim}{r}}{dt} = \frac{dx}{dt}\underset{\sim}{i} + \frac{dy}{dt}\underset{\sim}{j}$. Simply put, to differentiate a vector we merely differentiate each component using the rules for differentiation.

WORKED EXAMPLE 3 Determining a unit tangent vector to a position vector

Determine a unit tangent vector to $\underset{\sim}{r}(t) = e^{3t}\underset{\sim}{i} + \sin(2t)\underset{\sim}{j}$ at the point where $t = 0$.

THINK	WRITE
1. Differentiate the vector.	$\underset{\sim}{r}(t) = e^{3t}\underset{\sim}{i} + \sin(2t)\underset{\sim}{j}$ $\frac{d\underset{\sim}{r}}{dt} = \frac{d}{dt}\left(e^{3t}\right)\underset{\sim}{i} + \frac{d}{dt}\left(\sin(2t)\right)\underset{\sim}{j}$
2. State the derivative vector.	$\frac{d\underset{\sim}{r}}{dt} = 3e^{3t}\underset{\sim}{i} + 2\cos(2t)\underset{\sim}{j}$
3. Evaluate the derivative vector at the required value.	$\left.\frac{d\underset{\sim}{r}}{dt}\right\|_{t=0} = 3\underset{\sim}{i} + 2\underset{\sim}{j}$
4. Determine the magnitude of the derivative vector.	$\left\|\frac{d\underset{\sim}{r}(0)}{dt}\right\| = \sqrt{3^2 + 2^2}$ $= \sqrt{13}$
5. State the required result, which is a unit vector.	$\hat{\underset{\sim}{r}} = \frac{1}{\sqrt{13}}(3\underset{\sim}{i} + 2\underset{\sim}{j})$

12.3.2 Velocity and acceleration vectors

Velocity vector

Because $\underset{\sim}{r}(t)$ represents the position vector, $\underset{\sim}{v}(t) = \frac{d\underset{\sim}{r}}{dt} = \dot{\underset{\sim}{r}}(t)$ represents the velocity vector. Note the single dot above $\underset{\sim}{r}$ indicates the derivative with respect to time. Furthermore, if $\underset{\sim}{r}(t) = x(t)\underset{\sim}{i} + y(t)\underset{\sim}{j}$, then $\dot{\underset{\sim}{r}}(t) = \dot{x}(t)\underset{\sim}{i} + \dot{y}(t)\underset{\sim}{j}$.

Speed

The speed of a moving particle is the magnitude of the velocity vector. The speed at time t is given by $|\underset{\sim}{v}(t)| = |\dot{\underset{\sim}{r}}(t)| = \sqrt{\dot{x}^2 + \dot{y}^2}$.

Acceleration vector

Since $\underset{\sim}{v}(t) = \frac{d\underset{\sim}{r}}{dt} = \dot{\underset{\sim}{r}}(t)$ represents the velocity vector, differentiating again with respect to t gives the acceleration vector. The acceleration vector is given by $\underset{\sim}{a}(t) = \frac{d}{dt}\left(\dot{r}(t)\right) = \ddot{\underset{\sim}{r}}(t) = \ddot{x}(t)\underset{\sim}{i} + \ddot{y}(t)\underset{\sim}{j}$. Note that the two dots above the variables indicate the second derivative with respect to time.

Position, velocity and acceleration vectors

Given a position vector $\underset{\sim}{r}(t)$:

The velocity vector is $\underset{\sim}{v}(t) = \frac{d\underset{\sim}{r}}{dt} = \dot{\underset{\sim}{r}}(t)$

The acceleration vector is $\underset{\sim}{a}(t) = \frac{d\underset{\sim}{v}}{dt} = \frac{d}{dt}\left(\dot{\underset{\sim}{r}}(t)\right) = \ddot{\underset{\sim}{r}}(t)$

WORKED EXAMPLE 4 Calculating velocity, speed and acceleration from a position vector

A particle spirals outwards so that its position vector is given by $\underset{\sim}{r}(t) = t\cos(t)\underset{\sim}{i} + t\sin(t)\underset{\sim}{j}$ for $t \geq 0$.

a. **Determine the velocity vector.**

b. **Determine the speed of the particle at time t and hence calculate the speed when $t = \frac{3\pi}{4}$.**

c. **Determine the acceleration vector.**

THINK	WRITE
a. 1. State the parametric equations.	a. $\underset{\sim}{r}(t) = t\cos(t)\underset{\sim}{i} + t\sin(t)\underset{\sim}{j}$ Then $x(t) = t\cos(t)$ and $y(t) = t\sin(t)$.
2. Differentiate x with respect to t. The dot notation is used for the derivative with respect to t. Use the product rule.	$\frac{dx}{dt} = \dot{x} = \frac{d}{dt}\left(t\cos(t)\right)$ $= \cos(t)\frac{d}{dt}(t) + t\frac{d}{dt}\left(\cos(t)\right)$ $= \cos(t) - t\sin(t)$
3. Differentiate y with respect to t. Use the product rule.	$\frac{dy}{dt} = \dot{y} = \frac{d}{dt}\left(t\sin(t)\right)$ $= \sin(t)\frac{d}{dt}(t) + t\frac{d}{dt}\left(\sin(t)\right)$ $= \sin(t) + t\cos(t)$
4. State the velocity vector.	$\dot{\underset{\sim}{r}}(t) = \left(\cos(t) - t\sin(t)\right)\underset{\sim}{i} + \left(\sin(t) + t\cos(t)\right)\underset{\sim}{j}$

b. 1. Determine the speed at time t. Substitute for the derivatives and expand.

b. $\left|\dot{\underset{\sim}{r}}(t)\right| = \sqrt{\dot{x}^2 + \dot{y}^2}$

$$\dot{x}^2 = (\cos(t) - t\sin(t))^2$$
$$= \cos^2(t) - 2t\cos(t)\sin(t) + t^2\sin^2(t)$$
$$\dot{y}^2 = (\sin(t) + t\cos(t))^2$$
$$= \sin^2(t) + 2t\sin(t)\cos(t) + t^2\cos^2(t)$$

2. Simplify using trigonometry, since $\sin^2(t) + \cos^2(t) = 1$.

$$\dot{x}^2 + \dot{y}^2 = 1 + t^2$$

3. State the speed at time t.

$$\left|\dot{\underset{\sim}{r}}(t)\right| = \sqrt{1 + t^2}$$

4. Calculate the speed for the required value of t.

Substitute $t = \dfrac{3\pi}{4}$:

$$\left|\dot{\underset{\sim}{r}}\left(\frac{3\pi}{4}\right)\right| = \sqrt{1 + \left(\frac{3\pi}{4}\right)^2}$$
$$= \sqrt{\frac{16 + 9\pi^2}{16}}$$

5. State the speed in simplified form.

$$\left|\dot{\underset{\sim}{r}}\left(\frac{3\pi}{4}\right)\right| = \frac{1}{4}\sqrt{16 + 9\pi^2}$$

c. 1. Determine the $\underset{\sim}{i}$ component of the acceleration vector.

c. $\dfrac{d^2x}{dt^2} = \ddot{x} = \dfrac{d}{dt}(\cos(t) - t\sin(t))$

$$= -\sin(t) - \frac{d}{dt}(t\sin(t))$$
$$= -\sin(t) - \sin(t) - t\cos(t)$$
$$= -2\sin(t) - t\cos(t)$$

2. Determine the $\underset{\sim}{j}$ component of the acceleration vector.

$$\frac{d^2y}{dt^2} = \ddot{y} = \frac{d}{dt}(\sin(t) - t\cos(t))$$
$$= \cos(t) - \frac{d}{dt}(t\cos(t))$$
$$= \cos(t) + \cos(t) - t\sin(t)$$
$$= 2\cos(t) - t\sin(t)$$

3. State the acceleration vector in terms of t.

$$\ddot{\underset{\sim}{r}}(t) = \ddot{x}(t)\underset{\sim}{i} + \ddot{y}(t)\underset{\sim}{j}$$
$$\ddot{\underset{\sim}{r}}(t) = -(2\sin(t) + t\cos(t))\,\underset{\sim}{i} + (2\cos(t) - t\sin(t))\,\underset{\sim}{j}$$

Extension to three dimensions

If $\underset{\sim}{r}(t) = x(t)\underset{\sim}{i} + y(t)\underset{\sim}{j} + z(t)\underset{\sim}{k}$ is the position vector of a particle moving in three dimensions, then the velocity vector is given by $\underset{\sim}{v}(t) = \dot{\underset{\sim}{r}}(t) = \dot{x}(t)\underset{\sim}{i} + \dot{y}(t)\underset{\sim}{j} + \dot{z}(t)\underset{\sim}{k}$. The speed is given by $\left|\dot{\underset{\sim}{r}}(t)\right| = \sqrt{\dot{x}^2 + \dot{y}^2 + \dot{z}^2}$ and the acceleration vector is given by $\underset{\sim}{a}(t) = \dfrac{d\underset{\sim}{v}(t)}{dt} = \ddot{\underset{\sim}{r}}(t) = \ddot{x}(t)\underset{\sim}{i} + \ddot{y}(t)\underset{\sim}{j} + \ddot{z}(t)\underset{\sim}{k}$. Note that alternative notations may be used.

WORKED EXAMPLE 5 Calculating velocity and acceleration from a position vector

A particle has a position vector given by $\underset{\sim}{r}(t) = t^3\underset{\sim}{i} + 6\sin(3t)\underset{\sim}{j} + 12e^{-\frac{t}{2}}\underset{\sim}{k}$ for $t \geq 0$. Determine:

a. **the velocity vector**
b. **the acceleration vector.**

THINK	WRITE
a. 1. Differentiate the position vector.	a. $\underset{\sim}{r}(t) = t^3\underset{\sim}{i} + 6\sin(3t)\underset{\sim}{j} + 12e^{-\frac{t}{2}}\underset{\sim}{k}$ $\dot{\underset{\sim}{r}}(t) = \frac{d}{dt}\left(t^3\right)\underset{\sim}{i} + \frac{d}{dt}\left(6\sin(3t)\right)\underset{\sim}{j} + \frac{d}{dt}\left(12e^{-\frac{t}{2}}\right)\underset{\sim}{k}$
2. State the derivative or velocity vector.	$\dot{\underset{\sim}{r}}(t) = 3t^2\underset{\sim}{i} + 18\cos(3t)\underset{\sim}{j} - 6e^{-\frac{t}{2}}\underset{\sim}{k}$
b. 1. Differentiate the velocity vector.	b. $\ddot{\underset{\sim}{r}}(t) = \frac{d}{dt}\left(3t^2\right)\underset{\sim}{i} + \frac{d}{dt}\left(18\cos(3t)\right)\underset{\sim}{j} - \frac{d}{dt}\left(6e^{-\frac{t}{2}}\right)\underset{\sim}{k}$
2. State the acceleration vector.	$\dot{\underset{\sim}{r}}(t) = 6t\underset{\sim}{i} - 54\sin(3t)\underset{\sim}{j} + 3e^{-\frac{t}{2}}\underset{\sim}{k}$

12.3.3 The gradient of the curve

Because we can determine the Cartesian equation of the curve as either an explicit relationship, $y = f(x)$, or an implicit relationship, $f(x, y) = c$, we can calculate the gradient of the curve using either explicit or implicit differentiation. Alternatively, we can calculate the gradient of the curve using parametric differentiation, since $\frac{dy}{dx} = \frac{dy}{dt} \cdot \frac{dt}{dx} = \frac{\dot{y}}{\dot{x}}$. Techniques such as these have been studied in earlier topics. It may be necessary to solve equations to determine the maximum or minimum speeds of a particle or the maximum or minimum values of the acceleration.

The gradient of a curve

Given a curve represented by a position vector $\underset{\sim}{r}(t) = \underset{\sim}{x}(t)\underset{\sim}{i} + \underset{\sim}{y}(t)\underset{\sim}{j}$:

$$\frac{dy}{dx} = \frac{dy}{dt} \cdot \frac{dt}{dx} = \frac{\dot{y}}{\dot{x}}$$

WORKED EXAMPLE 6 Sketching the Cartesian equation

A particle moves so that its position vector is given by $\underset{\sim}{r}(t) = \left(3 - 2\cos(2t)\right)\underset{\sim}{i} + \left(2 + 3\sin(2t)\right)\underset{\sim}{j}$ for $0 \leq t \leq \pi$.

a. **Determine the coordinates where the *gradient* of the curve is parallel to the x-axis.**
b. **Determine and sketch the Cartesian equation of the path, stating its domain and range.**
c. **Calculate the maximum and minimum values of the speed.**

THINK	WRITE/DRAW
a. 1. State the parametric equations.	a. $\underset{\sim}{r}(t) = \left(3 - 2\cos(2t)\right)\underset{\sim}{i} + \left(2 + 3\sin(2t)\right)\underset{\sim}{j}$ Then $x(t) = 3 - 2\cos(2t)$ and $y(t) = 2 + 3\sin(2t)$.

2. Differentiate x with respect to t. The dot notation is used for the derivative with respect to t.	$x(t) = 3 - 2\cos(t)$ $\frac{dx}{dt} = \dot{x} = 4\sin(2t)$
3. Differentiate y with respect to t.	$y(t) = 2 + 3\sin(2t)$ $\frac{dy}{dt} = \dot{y} = 6\cos(2t)$
4. Use the chain rule to calculate $\frac{dy}{dx}$.	$\frac{dy}{dx} = \frac{dy}{dt}\frac{dt}{dx} = \frac{\dot{y}}{\dot{x}}$
5. Substitute for the derivatives.	$\frac{dy}{dx} = \frac{6\cos(2t)}{4\sin(2t)}$ $= \frac{3}{2\tan(2t)}$
6. Determine where the *gradient* is parallel to the x-axis or where the gradient is zero.	$\frac{dy}{dx} = 0 \Rightarrow \cos(2t) = 0$, but $\sin(2t) \neq 0$.
7. Solve for the values of t.	Since $0 \le t \le \pi$, $2t = \frac{\pi}{2}, \frac{3\pi}{2}$ $t = \frac{\pi}{4}, \frac{3\pi}{4}$
8. Determine the coordinates of one point.	When $t = \frac{\pi}{4}$, $x\left(\frac{\pi}{4}\right) = 3 - 2\cos\left(\frac{\pi}{2}\right) = 3$ and $y\left(\frac{\pi}{4}\right) = 2 + 3\sin\left(\frac{\pi}{2}\right) = 5$. At (3, 5), the gradient is zero.
9. Calculate the other coordinate.	When $t = \frac{3\pi}{4}$, $x\left(\frac{3\pi}{4}\right) = 3 - 2\cos\left(\frac{3\pi}{2}\right) = 3$ and $y\left(\frac{3\pi}{4}\right) = 2 + 3\sin\left(\frac{3\pi}{2}\right) = -1$. At (3, −1), the gradient is zero.
b. 1. Use the parametric equations to eliminate the parameter.	**b.** $x = 3 - 2\cos(2t) \Rightarrow \cos(2t) = \frac{x-3}{-2}$ $y = 2 + 3\sin(2t) \Rightarrow \sin(2t) = \frac{y-2}{3}$
2. State the Cartesian equation of the path.	Since $\cos^2(2t) + \sin^2(2t) = 1$, $\frac{(x-3)^2}{4} + \frac{(y-2)^2}{9} = 1$.
3. State the curve and its domain and range.	The curve is an ellipse with centre at (3, 2) and semi-major and minor axes of 2 and 3. The domain is 3 ± 2 or [1, 5] and the range is 2 ± 3 or [−1, 5].

4. Sketch the curve.

Note that at the points (3, 5) and (3, −1) the gradient is zero.

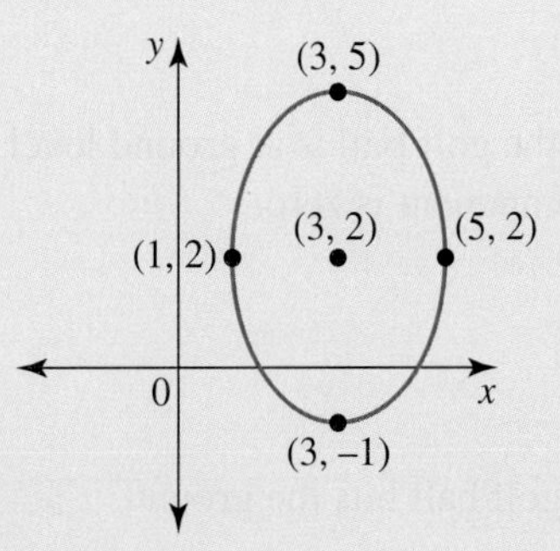

c. 1. Determine the speed at time t. Substitute for the derivatives.

c. $|\dot{\underset{\sim}{r}}(t)| = \sqrt{\dot{x}^2 + \dot{y}^2}$

$= \sqrt{(4\sin(2t))^2 + (6\cos(2t))^2}$

$= \sqrt{16\sin^2(2t) + 36\cos^2(2t)}$

2. Use $\sin^2(2t) = 1 - \cos^2(2t)$ to express the speed in terms of $\cos(2t)$ only.

$|\dot{\underset{\sim}{r}}(t)| = \sqrt{16\,(1 - \cos^2(2t)) + 36\cos^2(2t)}$

$= \sqrt{16 + 20\cos^2(2t)}$

$= \sqrt{4\,(4 + 5\cos^2(2t))}$

3. State the speed at time t.

$|\dot{\underset{\sim}{r}}(t)| = 2\sqrt{4 + 5\cos^2(2t)}$

4. Determine the maximum value of the speed will occur.

The maximum value of the speed occurs when $\cos^2(2t) = 1$; that is,

$|\dot{\underset{\sim}{r}}(t)|_{max} = 2\sqrt{4 + 5}$

5. State the maximum value of the speed.

$|\dot{\underset{\sim}{r}}(t)|_{max} = 6$

6. Determine the minimum value of the speed.

The minimum value of the speed occurs when $\cos^2(2t) = 0$; that is,

$|\dot{\underset{\sim}{r}}(t)|_{min} = 2\sqrt{4}$

7. State the minimum value of the speed.

$|\dot{\underset{\sim}{r}}(t)|_{min} = 4$

Applications of vector calculus

Vector equations can be used to model moving objects. The motion of the object can therefore be analysed using the techniques from above.

WORKED EXAMPLE 7 Applying vector calculus to a real life scenario

The position vector, $\underset{\sim}{r}(t)$, of a golf ball at a time t seconds is given by $\underset{\sim}{r}(t) = 15t\underset{\sim}{i} + (20t - 4.9t^2)\underset{\sim}{j}$ for $t \geq 0$, where the distance is in metres, $\underset{\sim}{i}$ is a unit vector horizontally forward and $\underset{\sim}{j}$ is a unit vector vertically upwards above ground level.

a. Calculate when the golf ball hits the ground, in seconds to 2 decimal places.

b. Calculate where the golf ball hits the ground, in seconds to 1 decimal place.

c. Determine the initial speed, in m/s, and angle of projection, in degrees to 2 decimal places.

d. **Calculate the maximum height reached, in metres to 2 decimal places.**
e. **Show that the golf ball travels in a parabolic path.**

THINK	WRITE
a. **1.** The time when the golf ball is at ground level is when the $\underset{\sim}{j}$ component is zero.	**a.** $y(t) = 20t - 4.9t^2 = 0$ $t(20 - 4.9t) = 0$ $t = 0$ or $20 - 4.9t = 0$ $\frac{20}{4.9} = 4.08$ Since $t \geq 0$, ignore the initial time when $t = 0$.
2. State when the golf ball hits the ground.	The golf ball hits the ground after 4.08 seconds.
b. **1.** The distance R travelled when the golf ball hits the ground is the value of the $\underset{\sim}{i}$ component at this time.	**b.** $R = x(4.08)$ $= 15 \times 4.08$
2. State where the golf ball hits the ground.	The golf ball hits the ground at a distance of 61.2 metres from the initial point.
c. **1.** Differentiate the position vector.	**c.** $\underset{\sim}{r}(t) = 15t\underset{\sim}{i} + (20t - 4.9t^2)\underset{\sim}{j}$ $\dot{\underset{\sim}{r}}(t) = 15\underset{\sim}{i} + (20 - 9.8t)\underset{\sim}{j}$
2. Calculate the initial velocity vector.	$\dot{\underset{\sim}{r}}(0) = 15\underset{\sim}{i} + 20\underset{\sim}{j}$
3. Calculate the initial speed.	The initial speed is the magnitude of the initial velocity vector. $\lvert\dot{\underset{\sim}{r}}(0)\rvert = \sqrt{15^2 + 20^2} = 25$
4. Calculate the initial angle of projection.	The angle the initial velocity vector makes with the $\underset{\sim}{i}$ axis is $\cos^{-1}\left(\frac{15}{25}\right) = 53.13°$.
5. State the required results.	The golf ball is struck with an initial speed of 25 m/s at an angle of 53.13°.
d. **1.** The golf ball will rise until the vertical component of its velocity is zero.	**d.** $\dot{y}(t) = 20 - 9.8t = 0$ $t = \frac{20}{9.8} = 2.04$
2. The maximum height reached, H, is the value of the $\underset{\sim}{j}$ component at this time.	$H = y(2.04) = 20 \times 2.04 - 4.9 \times 2.04^2$
3. State the maximum height reached.	The golf ball reaches a maximum height of 20.41 metres.
e. **1.** Write the parametric equations.	**e.** $x = 15t \Rightarrow t = \frac{x}{15}$ $y = 20t - 4.9t^2$
2. Substitute the value of t into the equation for y.	$y = 20\left(\frac{x}{15}\right) - 4.9\left(\frac{x}{15}\right)^2$
3. Simplify and form common denominators.	$y = -\frac{x}{2250}(49x - 3000)$
4. State the result.	The parametric equation is of the form of a parabola, $y = ax(x - b)$ with $a < 0$. Therefore, the golf ball travels in a parabolic path.

12.3 Exercise

Students, these questions are even better in jacPLUS

Receive immediate feedback and access sample responses

Access additional questions

Track your results and progress

Find all this and MORE in jacPLUS

Technology free

1. WE3 Determine a unit tangent vector to $\underset{\sim}{r}(t) = (e^{2t} + e^{-2t})\underset{\sim}{i} + (e^{2t} - e^{-2t})\underset{\sim}{j}$ at the point where $t = 0$.

2. Determine a unit tangent vector to $\underset{\sim}{r}(t) = \cos(2t)\underset{\sim}{i} + \sin(2t)\underset{\sim}{j}$ at the point where $t = \dfrac{\pi}{6}$.

3. WE4 A particle moves so that its position vector is given by $\underset{\sim}{r}(t) = te^{-2t}\underset{\sim}{i} + te^{2t}\underset{\sim}{j}$ for $t \geq 0$.
   a. Determine the velocity vector.
   b. Determine the speed of the particle at time t and hence calculate its speed when $t = \dfrac{1}{2}$.
   c. Determine the acceleration vector.

4. A particle moves so that its position vector is given by $\underset{\sim}{r}(t) = \cos^2(t)\underset{\sim}{i} + \sin^2(t)\underset{\sim}{j}$ for $t \geq 0$, where t is measured in seconds and the distance is in metres. Calculate the speed of the particle at time $t = \dfrac{3\pi}{8}$.

5. WE5 A particle moves so that its position vector is given by $\underset{\sim}{r}(t) = 2t^4\underset{\sim}{i} + 4\cos(2t)\underset{\sim}{j} + 6e^{-2t}\underset{\sim}{k}$ for $t \geq 0$. Determine:
   a. the velocity vector
   b. the acceleration vector.

6. A particle moves so that its position vector is given by $\underset{\sim}{r}(t) = 8\cos\left(\dfrac{\pi t}{4}\right)\underset{\sim}{i} + 8\sin\left(\dfrac{\pi t}{4}\right)\underset{\sim}{j} + 4e^{-2t}\underset{\sim}{k}$ for $t \geq 0$. Calculate the magnitude of the acceleration vector when $t = 1$.

7. WE6 A particle moves so that its position vector is given by $\underset{\sim}{r}(t) = \left(4 + 3\cos(2t)\right)\underset{\sim}{i} + \left(3 - 2\sin(2t)\right)\underset{\sim}{j}$ for $0 \leq t \leq \pi$.
   a. Determine the coordinates where the gradient of the curve is parallel to the x-axis.
   b. Determine and sketch the Cartesian equation of the path, stating its domain and range.
   c. Calculate the maximum and minimum values of the speed.

8. A particle moves so that its position vector is given by $\underset{\sim}{r}(t) = 3\sec(t)\underset{\sim}{i} + 2\tan(t)\underset{\sim}{j}$ for $0 \leq t \leq 2\pi$.
   a. Determine and sketch the Cartesian equation of the path, stating its domain and range.
   b. Determine the coordinates where the gradient is $\dfrac{4}{3}$.

Technology active

9. WE7 The position vector $\underset{\sim}{r}(t)$ of a soccer ball at a time $t \geq 0$ seconds is given by $\underset{\sim}{r}(t) = 5t\underset{\sim}{i} + (12t - 4.9t^2)\underset{\sim}{j}$, where the distance is in metres, $\underset{\sim}{i}$ is a unit vector horizontally forward and $\underset{\sim}{j}$ is a unit vector vertically upwards above ground level.
   a. Calculate when the soccer ball hits the ground, in seconds to 2 decimal places.
   b. Calculate where the soccer ball hits the ground, in metres to 2 decimal places.
   c. Determine the initial speed, in m/s, and angle of projection, in degrees to 2 decimal places.
   d. Calculate the maximum height reached, in metres to 2 decimal places.
   e. Show that the soccer ball travels in a parabolic path.

10. Determine a unit tangent vector to each of the following at the point indicated.

a. $\underset{\sim}{r}(t) = 2t\underset{\sim}{i} + 4t^2\underset{\sim}{j},\ t \geq 0$ at $t = 1$

b. $\underset{\sim}{r}(t) = 2t\underset{\sim}{i} + 8t^3\underset{\sim}{j},\ t \geq 0$ at $t = 1$

c. $\underset{\sim}{r}(t) = 3t^2\underset{\sim}{i} + (t^2 - 4t)\underset{\sim}{j},\ t \geq 0$ at $t = 3$

d. $\underset{\sim}{r}(t) = \left(t + \dfrac{1}{t}\right)\underset{\sim}{i} + \left(t - \dfrac{1}{t}\right)\underset{\sim}{j},\ t \geq 0$ at $t = 2$

11. Determine a unit tangent vector to each of the following at the point indicated.

a. $\underset{\sim}{r}(t) = e^{-2t}\underset{\sim}{i} + e^{2t}\underset{\sim}{j},\ t \geq 0$ at $t = 0$

b. $\underset{\sim}{r}(t) = \cos^2(t)\underset{\sim}{i} + \cos(2t)\underset{\sim}{j},\ t \geq 0$ at $t = \dfrac{\pi}{3}$

12. For each of the following position vectors, determine the velocity vector and the acceleration vector.

a. $\underset{\sim}{r}(t) = (t^2 + 9)\underset{\sim}{i} + \left(\dfrac{1}{1 + t}\right)\underset{\sim}{j}$

b. $\underset{\sim}{r}(t) = \log_e(3t)\underset{\sim}{i} + (5t^2 + 4t)\underset{\sim}{j}$

c. $\underset{\sim}{r}(t) = 3\cos(2t)\underset{\sim}{i} - 4\sin(2t)\underset{\sim}{j} + (12t - 5t^2)\underset{\sim}{k}$

13. A particle moves so that its position vector is given by $\underset{\sim}{r}(t) = 3\cos(2t)\underset{\sim}{i} + 3\sin(2t)\underset{\sim}{j},\ t \geq 0$.

a. Determine the Cartesian equation of the path.

b. Show that the speed is constant.

c. Show that the acceleration is directed inwards.

d. Show that the velocity vector is perpendicular to the acceleration vector.

14. A particle moves so that its position vector is given by $\underset{\sim}{r}(t) = 4\cos(3t)\underset{\sim}{i} + 2\sqrt{2}\sin(3t)(\underset{\sim}{j} - \underset{\sim}{k}),\ t \geq 0$.

a. Show that the speed is constant.

b. Show that the acceleration vector is perpendicular to the position vector.

15. A particle moves so that its position vector is given by $\underset{\sim}{r}(t) = a\cos(nt)\underset{\sim}{i} + a\sin(nt)\underset{\sim}{j}$ for $t \geq 0$, where a and n are positive constants.

a. Determine the Cartesian equation of the path.

b. Show that the speed is constant.

c. Show that the acceleration is directed inwards.

d. Show that the velocity vector is perpendicular to the position vector.

e. Describe the motion.

16. A particle moves so that its position vector is given by $\underset{\sim}{r}(t) = 2\sec(t)\underset{\sim}{i} + 3\tan(t)\underset{\sim}{j}$ for $0 \leq t \leq 2\pi$.

a. Determine and sketch the Cartesian equation of the path, stating its domain and range.

b. Calculate the gradient at the point where $t = \dfrac{\pi}{4}$.

17. A particle moves so that its position vector is given by $\underset{\sim}{r}(t) = 2\sec^2(t)\underset{\sim}{i} + 3\tan^2(t)\underset{\sim}{j}$ for $0 \leq t \leq 2\pi$.

a. Determine and sketch the Cartesian equation of the path, stating its domain and range.

b. Determine the gradient at any point.

c. Calculate the minimum value of the speed.

18. A particle moves so that its position vector is given by $\underset{\sim}{r}(t) = 3\,\text{cosec}(t)\underset{\sim}{i} + 4\cot(t)\underset{\sim}{j}$ for $0 \leq t \leq 2\pi$.

a. Determine and sketch the Cartesian equation of the path, stating its domain and range.

b. Calculate the gradient when $t = \dfrac{\pi}{3}$.

c. Calculate the minimum value of the speed.

19. A particle moves so that its position vector is given by $\underset{\sim}{r}(t) = (1 + 2\operatorname{cosec}(t))\,\underset{\sim}{i} + (4 - 3\cot(t))\,\underset{\sim}{j}$ for $0 \le t \le 2\pi$.

a. Determine and sketch the Cartesian equation of the path, stating its domain and range.
b. Determine the values of t for which the gradient of the curve is 3.

20. A boy throws a tennis ball. The position vector $\underset{\sim}{r}(t)$ of the tennis ball at a time $t \ge 0$ seconds is given by $\underset{\sim}{r}(t) = 10t\underset{\sim}{i} + (10t - 4.9t^2)\underset{\sim}{j}$ where the distance is in metres, $\underset{\sim}{i}$ is a unit vector horizontally forward and $\underset{\sim}{j}$ is a unit vector vertically upwards above ground level.

a. Calculate the time taken to reach the ground, in seconds to 2 decimal places.
b. Calculate the horizontal distance covered, in metres to 2 decimal places.
c. Determine the initial speed, in m/s to 2 decimal places, and angle of projection in degrees.
d. Calculate the maximum height reached, in metres to 1 decimal place.
e. Show that the tennis ball travels in a parabolic path.

21. A javelin is thrown by an athlete. After the javelin is thrown, its position vector is given by $\underset{\sim}{r}(t) = 35t\underset{\sim}{i} + (1.8 + 9t - 4.9t^2)\underset{\sim}{j}$ at a time $t \ge 0$ seconds, where the distance is in metres, t is the time in seconds after the javelin is released, $\underset{\sim}{i}$ is a unit vector horizontally forward and $\underset{\sim}{j}$ is a unit vector vertically upwards above ground level.

a. Determine the initial height above the ground when the javelin was released.
b. Calculate the time taken to reach the ground, in seconds to 3 decimal places.
c. Calculate the horizontal distance covered, in metres to 2 decimal places.
d. Determine the initial speed, in m/s to 2 decimal places, and angle of projection, in degrees to 2 decimal places.
e. Calculate the maximum height reached, in metres to 2 decimal places.
f. Show that the javelin travels in a parabolic path.

22. A soccer ball is kicked off the ground. Its position vector is given by $\underset{\sim}{r}(t) = 23t\underset{\sim}{i} + 5t\underset{\sim}{j} + 4\sqrt{2}\sin\left(\frac{\pi t}{2}\right)\underset{\sim}{k}$, where $\underset{\sim}{i}$ is a unit vector horizontally forward, $\underset{\sim}{j}$ is a unit vector to the right and $\underset{\sim}{k}$ is a unit vector vertically upwards above ground level. Measurements are in metres. Determine:

a. the time taken for the soccer ball to hit the ground
b. the distance the soccer ball lands from the initial point, in metres to 1 decimal place
c. the initial speed at which the soccer ball was kicked, in m/s to 2 decimal places
d. the maximum height above ground level reached by the soccer ball, in metres to 2 decimal places.

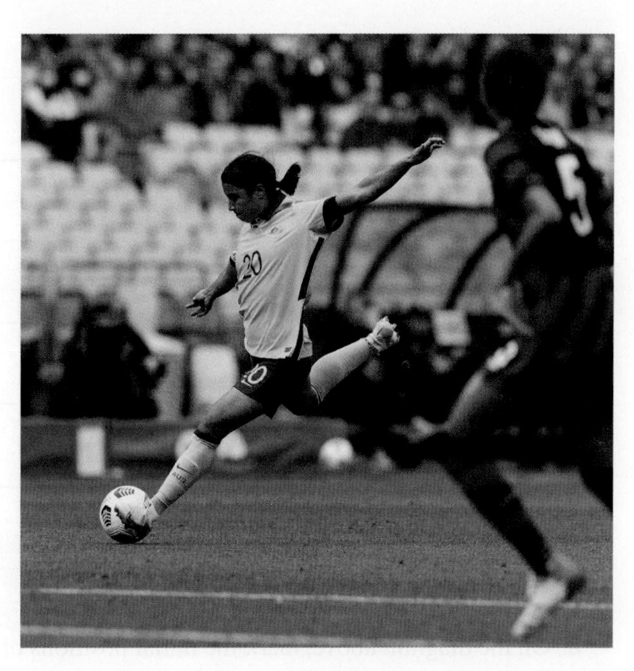

12.3 Exam questions

Question 1 (10 marks) TECH-ACTIVE

Source: *VCE 2020, Specialist Mathematics Exam 2, Section B, Q1 a-c; © VCAA.*

A particle moves in the x-y plane such that its position in terms of x and y metres at t seconds is given by the parametric equations

$$x = 2\sin(2t)$$
$$y = 3\cos(t)$$

where $t \geq 0$.

a. Find the distance, in metres, of the particle from the origin when $t = \dfrac{\pi}{6}$. **(2 marks)**

b. i. Express $\dfrac{dy}{dx}$ in terms of t and, hence, find the equation of the tangent to the path of the particle at $t = \pi$ seconds. **(3 marks)**

ii. Find the velocity, v, in ms^{-1}, of the particle when $t = \pi$. **(2 marks)**

iii. Find the magnitude of the acceleration, in ms^{-2}, when $t = \pi$. **(2 marks)**

c. Find the time, in seconds, when the particle first passes through the origin. **(1 mark)**

Question 2 (7 marks) TECH-ACTIVE

Source: *VCE 2017, Specialist Mathematics Exam 2, Section B, Q5 a-c; © VCAA.*

On a particular morning, the position vectors of a boat and a jet ski on a lake t minutes after they have started moving are given by $\underset{\sim}{r}_B(t) = \left(1 - 2\cos(t)\right)\underset{\sim}{i} + \left(3 + \sin(t)\right)\underset{\sim}{j}$ and $\underset{\sim}{r}_J(t) = \left(1 - \sin(t)\right)\underset{\sim}{i} + \left(2 - \cos(t)\right)\underset{\sim}{j}$ respectively for $t \geq 0$, where distances are measured in kilometres. The boat and the jet ski start moving at the same time. The graphs of their paths are shown below.

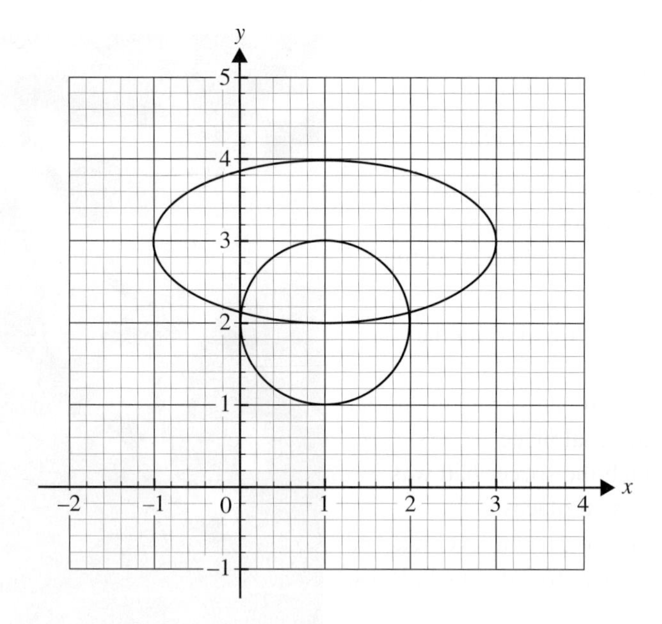

a. On the diagram above, mark the initial positions of the boat and the jet ski, clearly identifying each of them. Use arrows to show the directions in which they move. **(2 marks)**

b. i. Find the first time for $t > 0$ when the speeds of the boat and the jet ski are the same. **(2 marks)**

ii. State the coordinates of the boat at this time. **(1 mark)**

c. i. Write down an expression for the distance between the jet ski and the boat at any time t. **(1 mark)**

ii. Find the minimum distance separating the boat and the jet ski. Give your answer in kilometres, correct to two decimal places. **(1 mark)**

Question 3 (3 marks) TECH-FREE

Source: *VCE 2016, Specialist Mathematics Exam 1, Q8a,b; © VCAA.*

The position of a body with mass 3 kg from a fixed origin at time t seconds, $t \geq 0$, is given by $\underset{\sim}{r} = \left(3\sin(2t) - 2\right)\underset{\sim}{i} + \left(3 - 2\cos(2t)\right)\underset{\sim}{j}$, where components are in metres.

a. Find an expression for the speed, in metres per second, of the body at time t. **(2 marks)**

b. Find the speed of the body, in metres per second, when $t = \frac{\pi}{12}$. **(1 mark)**

More exam questions are available online.

12.4 Special parametric curves

LEARNING INTENTION

At the end of this subtopic you should be able to:

- calculate the gradients of curves defined by vector equations
- sketch the graphs of curves defined by vector equations
- calculate areas of curves defined by vector equations.

12.4.1 Plane curves

To sketch the graphs of curves defined by parametric equations, determine important features such as the turning points and axis intercepts. Alternatively, use a calculator.

Calculating areas using parametric forms

Areas bounded by curves can be calculated using the parametric equation of the curve and symmetrical properties. Consider the area bounded by a non-negative curve, which is a curve that lies entirely above the x-axis. The area bounded by the curve and the x-axis is given by $A = \int_a^b y\,dx$. However, we can substitute for $y = y(t)$ and $\frac{d}{dt}\left(x(t)\right)$ to obtain an integral in terms of t for the area, $\int_{t=t_0}^{t=t_1} y(t)\frac{d}{dt}\left(x(t)\right)dt$.

Calculating the length of curves

The length of a curve between the values of t_0 and t_1 is given by $\int_{t_0}^{t_1} \left|\underset{\sim}{v}(t)\right| dt$. Sometimes the definite integrals obtained cannot be integrated; however, we can use a calculator to determine the numerical value of a definite integral.

Arc length of a curve

The length s of the parametric curve $x = x(t)$, $y = y(t)$ from $t = a$ to $t = b$ is given by:

$$s = \int_a^b \sqrt{\left(\frac{dx}{dt}\right)^2 + \left(\frac{dy}{dt}\right)^2}\,dt$$

WORKED EXAMPLE 8 Solving problems involving curves defined by vector equations (1)

A particle moves along a curve defined by the vector equation $\underset{\sim}{r}(t) = 2\cos(t)\underset{\sim}{i} + \sin(2t)\underset{\sim}{j}$ for $t \in [0, 2\pi]$.

a. Determine the gradient of the curve in terms of t.

b. Determine the values of t when the tangent to the curve is parallel to the x-axis, and hence determine the turning points on the curve.

c. Sketch the graph of the curve defined by the parametric equations $x = 2\cos(t)$ and $y = \sin(2t)$ for $[0, 2\pi]$.

d. Calculate the speed of the particle when $t = \frac{\pi}{4}$ and $t = \frac{\pi}{2}$.

e. The area bounded by the curve and the x-axis can be expressed as $\int_{t=t_0}^{t=t_1} y(t)\frac{d}{dt}(x(t))\,dt$. Obtain a definite integral in terms of t for the area and show using calculus that the total area bounded by the curve and the x-axis is $\frac{16}{3}$ units^2.

f. Set up a definite integral that gives the total length of the curve and calculate the total length of the curve, correct to 4 decimal places.

g. Show that particle moves along the curve defined by the implicit equation $y^2 = \frac{x^2}{4}(4 - x^2)$.

h. Hence, verify that the total area bounded by the curve and x-axis is given by $\frac{16}{3}$ units^2.

THINK	WRITE/DRAW
a. 1. State the parametric equations.	a. $x = 2\cos(t)$ and $y = \sin(2t)$
2. Differentiate x with respect to t.	$\frac{dx}{dt} = \dot{x} = -2\sin(t)$
3. Differentiate y with respect to t.	$\frac{dy}{dt} = \dot{y} = 2\cos(2t)$
4. Use the chain rule to determine $\frac{dy}{dx}$.	$\frac{dy}{dx} = \frac{dy}{dt}\frac{dt}{dx} = \frac{\dot{y}}{\dot{x}}$
5. Substitute for the derivatives and state the gradient in terms of t.	$\frac{dy}{dx} = -\frac{2\cos(2t)}{2\sin(t)} = -\frac{\cos(2t)}{\sin(t)}$
b. 1. The tangent to the curve is parallel to the x-axis when the gradient is zero. Equate the numerator to zero but not the denominator.	b. $\frac{dy}{dx} = 0 \Rightarrow \cos(2t) = 0$, but $\sin(t) \neq 0$.
2. Solve and determine the values of t when the gradient is zero.	$2t = \frac{\pi}{2}, \frac{3\pi}{2}, \frac{5\pi}{2}, \frac{7\pi}{2}$ since $t \in [0, 2\pi]$, $2t \in [0, 4\pi]$ $t = \frac{\pi}{4}, \frac{3\pi}{4}, \frac{5\pi}{4}, \frac{7\pi}{4}$
3. Determine the coordinates of the turning points. Substitute $t = \frac{\pi}{4}$ into the position vector.	$\underset{\sim}{r}\left(\frac{\pi}{4}\right) = 2\cos\left(\frac{\pi}{4}\right)\underset{\sim}{i} + \sin\left(\frac{\pi}{2}\right)\underset{\sim}{j} = \sqrt{2}\underset{\sim}{i} + \underset{\sim}{j}$ The coordinates are $\left(\sqrt{2}, 1\right)$.

4. Substitute $t = \frac{3\pi}{4}$ into the position vector.

$$\underset{\sim}{r}\left(\frac{3\pi}{4}\right) = 2\cos\left(\frac{3\pi}{4}\right)\underset{\sim}{i} + \sin\left(\frac{3\pi}{2}\right)\underset{\sim}{j}$$
$$= -\sqrt{2}\underset{\sim}{i} - \underset{\sim}{j}$$

The coordinates are $\left(-\sqrt{2}, -1\right)$.

5. Substitute $t = \frac{5\pi}{4}$ into the position vector.

$$\underset{\sim}{r}\left(\frac{5\pi}{4}\right) = 2\cos\left(\frac{5\pi}{4}\right)\underset{\sim}{i} + \sin\left(\frac{5\pi}{2}\right)\underset{\sim}{j}$$
$$= -\sqrt{2}\underset{\sim}{i} + \underset{\sim}{j}$$

The coordinates are $\left(-\sqrt{2}, 1\right)$.

6. Substitute $t = \frac{7\pi}{4}$ into the position vector.

$$\underset{\sim}{r}\left(\frac{7\pi}{4}\right) = 2\cos\left(\frac{7\pi}{4}\right)\underset{\sim}{i} + \sin\left(\frac{7\pi}{2}\right)\underset{\sim}{j}$$
$$= \sqrt{2}\underset{\sim}{i} - \underset{\sim}{j}$$

The coordinates are $\left(\sqrt{2}, -1\right)$.

7. State the coordinates of the turning points.

There are maximum turning points at $\left(\pm\sqrt{2}, 1\right)$ and minimum turning points at $\left(\pm\sqrt{2}, -1\right)$.

c. 1. Use a calculator to sketch the graph of the parametric equations.

c.

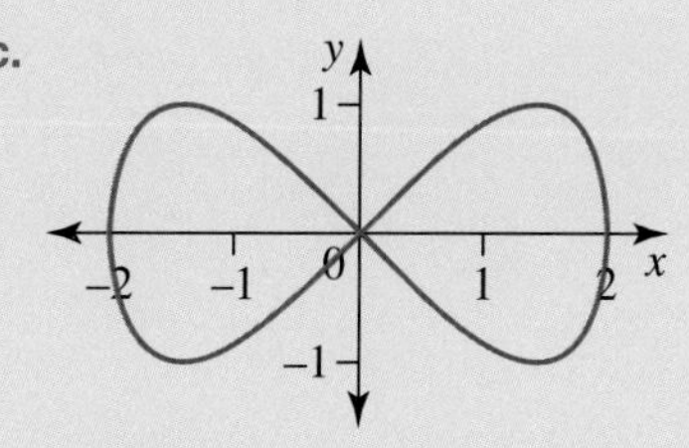

2. When the gradient of the denominator is zero, we have vertical asymptotes.

$$\sin(t) = 0$$
$$t = 0, \pi, 2\pi$$

3. State the coordinates on the graph where the tangent lines are vertical.

$$\underset{\sim}{r}(0) = 2\cos(0)\underset{\sim}{i} + \sin(0)\underset{\sim}{j}$$
$$= 2\underset{\sim}{i}$$

The coordinates are (2, 0).

$$\underset{\sim}{r}(\pi) = 2\cos(\pi)\underset{\sim}{i} + \sin(2\pi)\underset{\sim}{j}$$
$$= -2\underset{\sim}{i}$$

The coordinates are (–2, 0).

d. 1. Determine the velocity vector. Differentiate the position vector with respect to t.

d. $\underset{\sim}{r}(t) = 2\cos(t)\underset{\sim}{i} + \sin(2t)\underset{\sim}{j}$

$\dot{\underset{\sim}{r}}(t) = -2\sin(t)\underset{\sim}{i} + 2\cos(2t)\underset{\sim}{j}$

2. Determine the speed at time t by calculating the magnitude of the velocity vector.

$$\left|\dot{\underset{\sim}{r}}(t)\right| = \sqrt{\dot{x}^2 + \dot{y}^2}$$
$$= \sqrt{\left(-2\sin(t)\right)^2 + \left(2\cos(2t)\right)^2}$$
$$= \sqrt{4\left(\sin^2(t) + \cos^2(2t)\right)}$$

3. Calculate the speed at $t = \frac{\pi}{4}$.

Substitute $t = \frac{\pi}{4}$:

$$\left|\dot{\underset{\sim}{r}}\left(\frac{\pi}{4}\right)\right| = \sqrt{4\left(\sin^2\left(\frac{\pi}{4}\right) + \cos^2\left(\frac{\pi}{2}\right)\right)}$$
$$= \sqrt{4\left(\frac{1}{2} + 0\right)}$$
$$= \sqrt{2}$$

4. Calculate the speed at $t = \frac{\pi}{2}$.

Substitute $t = \frac{\pi}{2}$:

$$\left|\dot{\underset{\sim}{r}}\left(\frac{\pi}{2}\right)\right| = \sqrt{4\left(\sin^2\left(\frac{\pi}{2}\right) + \cos^2(\pi)\right)}$$
$$= \sqrt{4\left(1 + (-1)^2\right)}$$
$$= \sqrt{8} = 2\sqrt{2}$$

e. 1. Consider the movement of the particle.

e. When $t = 0$,

$$\underset{\sim}{r}(0) = 2\cos(0)\underset{\sim}{i} + \sin(0)\underset{\sim}{j}$$
$$= 2\underset{\sim}{i}$$

The coordinates are (2, 0).

When $t = \frac{\pi}{2}$,

$$\underset{\sim}{r}(0) = 2\cos\left(\frac{\pi}{2}\right)\underset{\sim}{i} + \sin\left(2 \times \frac{\pi}{2}\right)\underset{\sim}{j}$$
$$= \underset{\sim}{0}$$

The coordinates are (0, 0).

2. Use the symmetry of the curve to state a definite integral that gives the total area bounded by the curve.

$$\int_{t=t_0}^{t=t_1} y(t)\frac{d}{dt}\left(x(t)\right)dt = \int_{t=t_0}^{t=t_1} y(t)\dot{x}(t)dt$$

Where $y = \sin(2t)$ and $\frac{dx}{dt} = \dot{x} = 2\sin(t)$

The total area A is 4 times the area from $t = \frac{\pi}{2}$ to $t = 0$.

3. Write the definite integral that gives the area.

$$A = 4\int_{\frac{\pi}{2}}^{0} \sin(2t) \times -2\sin(t)dt$$

4. Simplify the integrand using the double-angle formula $\sin(2t) = 2\sin(t)\cos(t)$.

$$A = -16\int_{\frac{\pi}{2}}^{0} \sin^2(t)\cos(t)dt$$

5. Perform the integration.

$$A = -\frac{16}{3}\left[\sin^3(t)\right]_{\frac{\pi}{2}}^{0}$$

6. Evaluate the area.

$$A = -\frac{16}{3}\left[\left(\sin^3(0) - \sin^3\left(\frac{\pi}{2}\right)\right)\right]$$

7. State the value of the required area.

$$A = \frac{16}{3} \text{ units}^2$$

f. 1. The total length of a curve s between the values of t_0 and t_1 is given by $\int_{t_0}^{t_1} |\underset{\sim}{v}(t)|\, dt$.

f. Since $|\underset{\sim}{v}(t)| = |\dot{\underset{\sim}{r}}(t)| = 2\sqrt{(\sin^2(t) + \cos^2(2t))}$, by symmetry,

$$s = 4\int_{\frac{\pi}{2}}^{0} 2\sqrt{(\sin^2(t) + \cos^2(2t))}\,dt$$

2. This integral cannot be evaluated by our techniques of calculus, so we resort to using a calculator to obtain a numerical answer.

Using calculator, $s = 12.1944$ units.

g. 1. Substitute the given parametric equations into the implicit equation. Consider the left-hand side. Substitute $y = \sin(2t)$.

g. The implicit equation is $y^2 = \frac{x^2}{4}(4 - x^2)$.

LHS:

$$\begin{aligned} y^2 &= \sin^2(2t) \\ &= (2\sin(t)\cos(t))^2 \\ &= 4\sin^2(t)\cos^2(t) \end{aligned}$$

2. Consider the right-hand side. Substitute $x = \cos(t)$.

RHS:

$$\begin{aligned} \frac{x^2}{4}\left(4 - x^2\right) &= \frac{(2\cos(t))^2}{4}\left(4 - 4\cos^2(t)\right) \\ &= \frac{4}{4}\cos^2(t) \times 4\left(1 - \cos^2(t)\right) \\ &= 4\cos^2(t)\sin^2(t) \end{aligned}$$

h. 1. The implicit equation is a relation, not a function.

h.

$$\begin{aligned} y^2 &= \frac{x^2}{4}(4 - x^2) \\ y &= \pm\frac{x}{2}\sqrt{4 - x^2} \end{aligned}$$

Consider the graph of $y = \frac{x}{2}\sqrt{4 - x^2}$, which is a function.

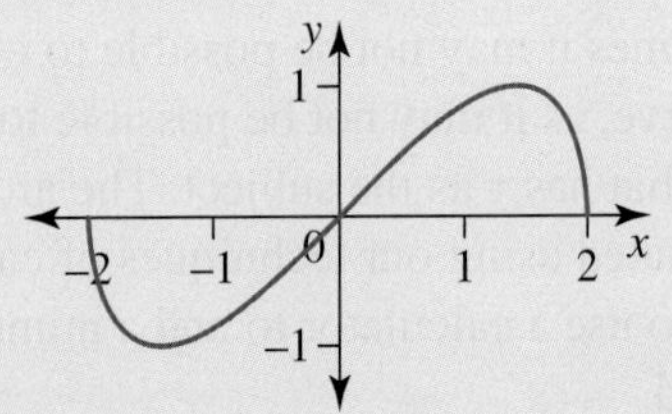

2. The total area bounded by the parametric curves can be found by symmetry.

$$A = \int_a^b y\,dx$$

$$A = 4\int_0^2 \frac{x}{2}\sqrt{4 - x^2}\,dx$$

3. Use a non-linear substitution.

Let $u = 4 - x^2$.

$$\frac{du}{dx} = -2x \text{ or } dx = -\frac{1}{2x}du$$

4. Change the terminals to the new variable. When $x = 0 \Rightarrow u = 4$, and when $x = 2 \Rightarrow u = 0$.

5. Substitute into the definite integral and use the properties of the definite integral. Note that the x terms cancel.

$$A = 4\int_4^0 \frac{x}{2}u^{\frac{1}{2}} \times -\frac{1}{2x}du$$

$$= -\int_4^0 u^{\frac{1}{2}}\,du$$

$$= \int_0^4 u^{\frac{1}{2}}\,du$$

6. Perform the integration.

$$= \frac{2}{3}\left[u^{\frac{3}{2}}\right]_0^4$$

7. Evaluate the definite integral.

$$= \frac{2}{3}\left[4^{\frac{3}{2}} - 0\right]$$

8. State the final result, which agrees with the alternative method.

$$A = \frac{16}{3}$$

12.4.2 Elegant curves

We can use calculators to sketch the graphs of parametric curves. The parametric form of some curves can appear to be quite complicated, but their graphs have interesting mathematical properties. For example, the types of curves in Worked example 8 are called Lissajous figures. Graphs of these types occur in electronics and appear on oscilloscopes. These graphs were first investigated by Nathaniel Bowditch in 1815, and were later explored in greater depth by Jules Lissajous in 1857.

From the parametric equations we can often but not always determine an explicit or implicit relation for the equation of the curve. Sometimes, by using symmetry and the parametric form, we can obtain the area bounded by the curve, but sometimes it may not be possible to obtain an area from an implicitly defined curve, as it may not be possible to rearrange the implicit curve into one that has y as the subject. The arc length of a curve can sometimes be calculated using our techniques of calculus, but often, as above, we will need to use a calculator to and a numerical value for the definite integral obtained.

As seen in many examples, we will often need to use trigonometric addition theorems and double-angle formulas to simplify the results.

WORKED EXAMPLE 9 Solving problems involving curves defined by vector equations (2)

A particle moves so that its position vector is given by $\underset{\sim}{r}(t) = (4\cos(t) + \cos(4t))\underset{\sim}{i} + (4\sin(t) - \sin(4t))\underset{\sim}{j}$ for $0 \le t \le 2\pi$.

a. **Determine the gradient of the curve.**
b. **Determine the speed at time t.**
c. **Calculate $\dot{\underset{\sim}{r}}(t) \cdot \underset{\sim}{r}(t)$.**
d. **Calculate the maximum and minimum values of the acceleration.**
e. **Sketch the path of the particle.**

THINK	WRITE/DRAW
a. 1. State the parametric equations.	a. $\underset{\sim}{r}(t) = (4\cos(t) + \cos(4t))\underset{\sim}{i} + (4\sin(t) - \sin(4t))\underset{\sim}{j}$ Then $x(t) = 4\cos(t) + \cos(4t)$ and $y(t) = 4\sin(t) - \sin(4t)$
2. Differentiate x with respect to t.	$x = 4\cos(t) + \cos(4t)$ $\frac{dx}{dt} = \dot{x} = -4\sin(t) - 4\sin(4t)$
3. Differentiate y with respect to t.	$y = 4\sin(t) - \sin(4t)$ $\frac{dy}{dt} = \dot{y} = 4\cos(t) - 4\cos(4t)$
4. Use the chain rule to determine $\frac{dy}{dx}$.	$\frac{dy}{dx} = \frac{dy}{dt}\frac{dt}{dx} = \frac{\dot{y}}{\dot{x}}$
5. Substitute for the derivatives.	$\frac{dy}{dx} = \frac{4\cos(t) - \cos(4t)}{-4\sin(t) - \sin(4t)}$
6. Simplify by cancelling common factors and state the gradient in terms of t.	$\frac{dy}{dx} = \frac{\cos(4t) - \cos(t)}{\sin(4t) + \sin(t)}$
b. 1. Determine the speed at time t. Substitute for the derivatives and expand.	b. $\lvert\dot{\underset{\sim}{r}}(t)\rvert = \sqrt{\dot{x}^2 + \dot{y}^2}$ $\dot{x}^2 = [-4\sin(t) - 4\sin(4t)]^2$ $= 16\sin^2(t) + 32\sin(t)\sin(4t) + 16\sin^2(4t)$ $\dot{y}^2 = [4(\cos(t) - 4\cos(4t))]^2$ $= 16\cos^2(t) + 32\cos(t)\cos(4t) + 16\cos^2(4t)$
2. Simplify using addition theorems and trigonometry. $\sin^2(A) + \cos^2(A) = 1$ and $\cos(A+B) = \cos(A)\cos(B) - \sin(A)\sin(B)$	$\dot{x}^2 + \dot{y}^2$ $= 16(\sin^2(t) + \cos^2(t)) + 16(\sin^2(4t) + \cos^2(4t)) + 32(\sin(t)\sin(4t) - \cos(t)\cos(4t))$ $= 32 - 32\cos(5t)$
3. Simplify using the double-angle formula $2\cos^2(A) = 1 - \cos(2A)$.	$\lvert\dot{\underset{\sim}{r}}(t)\rvert = \sqrt{32(1 - \cos(5t))}$ $= \sqrt{32 \times 2\cos^2\left(\frac{5t}{2}\right)}$
4. State the speed in simplified form at time t.	$\lvert\dot{\underset{\sim}{r}}(t)\rvert = 8\cos\left(\frac{5t}{2}\right)$

c. 1. Calculate the dot product of the position and velocity vectors.

c. $\dot{\underset{\sim}{r}}(t) \cdot \underset{\sim}{r}(t) = \dot{x}(t) \times x(t) + \dot{y}(t) \times y(t)$

$$= -4\left(\sin(t) + \sin(4t)\right) \times \left(4\cos(t) + \cos(4t)\right) + 4\left(\cos(t) - \cos(4t)\right) \times \left(4\sin(t) - \sin(4t)\right)$$

2. Expand the brackets.

$$= -16\sin(t)\cos(t) - 4\sin(t)\cos(4t) - 16\sin(4t)\cos(t) - 4\sin(4t)\cos(4t) + 16\sin(t)\cos(t) - 4\sin(4t)\cos(t) - 16\sin(t)\cos(4t) + 4\sin(4t)\cos(4t)$$

3. Cancel factors and group like terms.

$$= -20\left(\sin(4t)\cos(t) + \cos(4t)\sin(t)\right)$$

4. Simplify using addition theorems and trigonometry. $\sin(A + B) = \sin(A)\cos(B) + \cos(A)\sin(B)$

$$\dot{\underset{\sim}{r}}(t) \cdot \underset{\sim}{r}(t) = -20\sin(5t)$$

d. 1. Differentiate the velocity vector to determine the acceleration vector.

d. $\dot{\underset{\sim}{r}}(t) = -4\left(\sin(t) + \sin(4t)\right)\underset{\sim}{i} + 4\left(\cos(t) - \cos(4t)\right)\underset{\sim}{j}$

$$\ddot{\underset{\sim}{r}}(t) = -4\left(\cos(t) + 4\cos(4t)\right)\underset{\sim}{i} + 4\left(-\sin(t) - 4\sin(4t)\right)\underset{\sim}{j}$$

2. Calculate the magnitude of the acceleration vector at time t. Substitute for the derivatives and expand.

$$\left|\ddot{\underset{\sim}{r}}(t)\right| = \sqrt{\ddot{x}^2 + \ddot{y}^2}$$

$$\ddot{x}^2 = \left[-4\left(\cos(t) + 4\cos(4t)\right)\right]^2 = 16\left(\cos^2(t) + 8\cos(t)\cos(4t) + 16\cos^2(4t)\right)$$

$$\ddot{y}^2 = \left[4\left(-\sin(t) + 4\sin(4t)\right)\right]^2 = 16\left(\sin^2(t) + 8\sin(t)\sin(4t) + 16\sin^2(4t)\right)$$

3. Simplify using addition theorems and trigonometry. $\sin^2(A) + \cos^2(A) = 1$ and $\cos(A + B) = \cos(A)\cos(B) - \sin(A)\sin(B)$

$$\ddot{x}^2 + \ddot{y}^2$$

$$= 16\left(\cos^2(t) + \sin^2(t) + 16 \times 16(\cos^2(4t) + \sin^2(4t)\right) + 16 \times 8\left(\cos(t)\cos(4t) - \sin(t)\sin(4t)\right)$$

$$= 16\left(1 + 16 + 8\cos(5t)\right)$$

4. Simplify and state the magnitude of the acceleration in simplified form at time t.

$$\left|\ddot{\underset{\sim}{r}}(t)\right| = \sqrt{16\left(17 + 8\cos(5t)\right)}$$

$$= 4\sqrt{17 + 8\cos(5t)}$$

5. Determine when the maximum value of the acceleration will occur.

The maximum value of the acceleration occurs when $\cos(5t) = 1$; that is $\left|\ddot{\underset{\sim}{r}}(t)\right|_{max} = 4\sqrt{17 + 8}$.

6. State the maximum value of the acceleration.

$$\left|\ddot{\underset{\sim}{r}}(t)\right|_{max} = 20$$

7. Determine when the minimum value of the acceleration will occur.

The minimum value of the acceleration occurs when $\cos(5t) = -1$; that is $\left|\ddot{\underset{\sim}{r}}(t)\right|_{min} = 4\sqrt{17 - 8}$.

8. State the minimum value of the acceleration.

$$\left|\ddot{\underset{\sim}{r}}(t)\right|_{min} = 12$$

e. Use a calculator in parametric mode to sketch the graph of the parametric equations.

e. $x = 4\cos(t) + \cos(4t)$ and $y = 4\sin(t) - \sin(4t)$

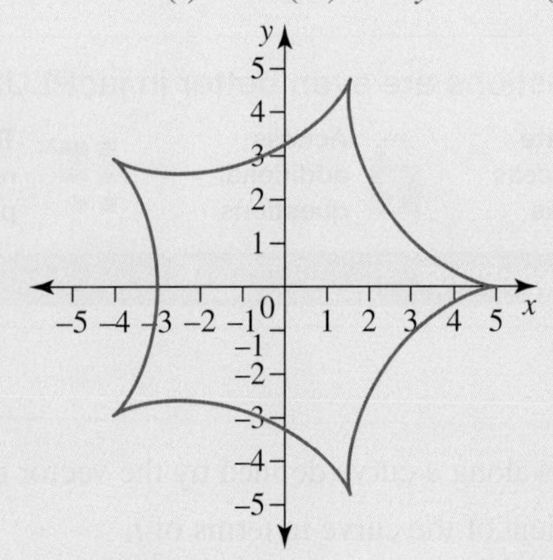

TI | THINK	DISPLAY/WRITE	CASIO | THINK	DISPLAY/WRITE
On a Notes page, select MENU 3: Insert 1: Math Box Click Enter, keep adding in Math Boxes and complete as shown.	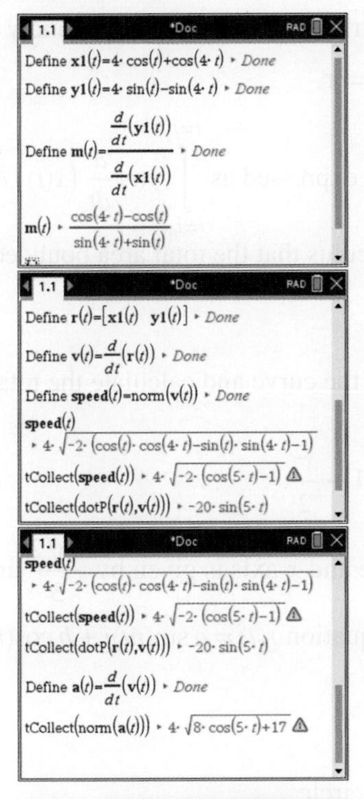	On the Main screen, complete the entry as shown.	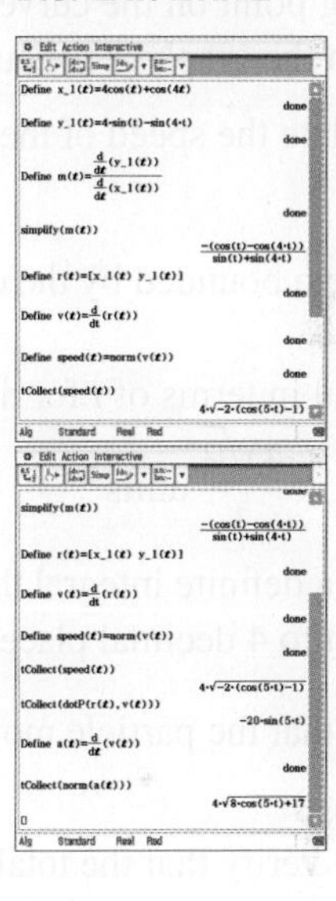
Open a Graphs page, select MENU 3: Graph/Entry Edit 3: Parametric Select $x1(t)$, $y1(t)$	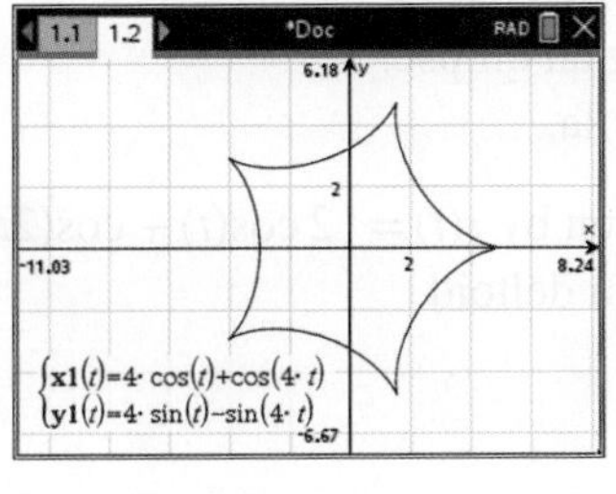	In a Graph Editor screen, choose the 'Parametric' form of equation entry.	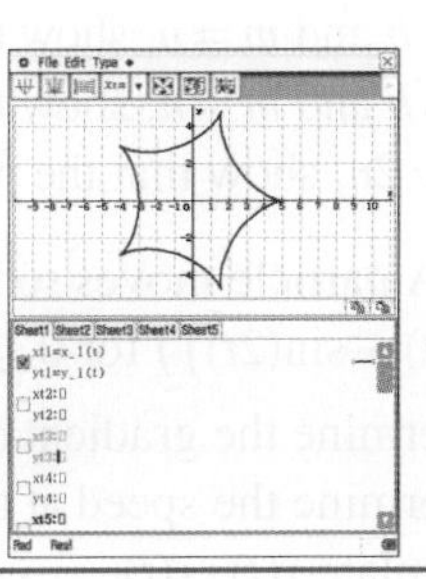

12.4 Exercise

Students, these questions are even better in jacPLUS

Receive immediate feedback and access sample responses

Access additional questions

Track your results and progress

Find all this and MORE in jacPLUS

Technology active

1. WE8 A particle moves along a curve defined by the vector equation $\underset{\sim}{r}(t) = 2\sin(t)\underset{\sim}{i} + \cos(2t)\underset{\sim}{j}$ for $t \in [0, 2\pi]$.
   a. Determine the gradient of the curve in terms of t.
   b. Determine the value of t when the tangent to the curve is parallel to the x-axis, and hence determine the turning point on the curve.
   c. Sketch the graph of the curve defined by the parametric equations $x = 2\sin(t)$ and $y = \cos(2t)$ for $[0, 2\pi]$.
   d. Calculate the speed of the particle when $t = \frac{\pi}{4}$ and $t = \pi$.
   e. The area bounded by the curve and the x-axis can be expressed as $\int_{t=t_0}^{t=t_1} y(t)\frac{d}{dt}\left(x(t)\right)dt$. Obtain a definite integral in terms of t for the area and show using calculus that the total area bounded by the curve and the x-axis is $\frac{4\sqrt{2}}{3}$ units^2.
   f. Write a definite integral that gives the total length of the curve and calculate the total length of the curve, correct to 4 decimal places.
   g. Show that the particle moves along the parabola $y = 1 - \frac{x^2}{2}$.
   h. Hence verify that the total area bounded by the curve and x-axis is given by $\frac{4\sqrt{2}}{3}$ unit^2.
2. A particle moves along a curve defined by the vector equation $\underset{\sim}{r}(t) = a\sin(nt)\underset{\sim}{i} + b\cos(mt)\underset{\sim}{j}$ for $t \geq 0$, where a, b, n and m are positive real numbers.
   a. Determine the gradient of the curve in terms of t.
   b. Determine the speed in terms of t.
   c. If $a = b$ and $m = n$, show that the particle moves in a circle.
   d. If $a \neq b$ and $m = n$, show that the particle moves along an ellipse.
   e. If $m = 2n$, show that the particle moves along a parabola.
3. WE9 A particle moves so that its position vector is given by $\underset{\sim}{r}(t) = \left(2\cos(t) + \cos(2t)\right)\underset{\sim}{i} + \left(2\sin(t) - \sin(2t)\right)\underset{\sim}{j}$ for $0 \leq t \leq 2\pi$. This curve is called a deltoid.
   a. Determine the gradient of the curve.
   b. Determine the speed at time t.
   c. Calculate $\dot{\underset{\sim}{r}}(t) \cdot \underset{\sim}{r}(t)$.
   d. Calculate the maximum and minimum values of the acceleration.
   e. Sketch the path of the particle.
4. A particle moves so that its position vector is given by $\underset{\sim}{r}(t) = \left(3\cos(t) - \cos(2t)\right)\underset{\sim}{i} + \left(3\sin(t) + \sin(2t)\right)\underset{\sim}{j}$ for $0 \leq t \leq 2\pi$. This curve is called a hypocycloid.
   a. Calculate the gradient of the curve at the point where $t = \frac{3\pi}{4}$.
   b. Calculate the maximum and minimum values of the speed.

c. Calculate the maximum and minimum values of the acceleration.
d. Sketch the path of the particle.

5. A ball rolls along a curve so that its position vector is given by $\underset{\sim}{r}(t) = (t - \sin(t))\,\underset{\sim}{i} + (1 + \cos(t))\,\underset{\sim}{j}$ for $t \in [0, 4\pi]$. The path of the curve is called a cycloid.
   a. Show that the gradient of the curve is given by $-\cot\left(\frac{t}{2}\right)$.
   b. Determine the coordinates on the curve where the gradient is parallel to the x-axis.
   c. Sketch the path of the particle.
   d. Show that the speed at time t is given by $2\sin\left(\frac{t}{2}\right)$.
   e. Show using calculus that the total length of curve is 16 units.
   f. Show that the total area bounded by the cycloid and the x-axis is given by 2π units^2.

6. A particle moves so that its position vector is given by $\underset{\sim}{r}(t) = (\cos(t) + t\sin(t))\,\underset{\sim}{i} + (\sin(t) - t\cos(t))\,\underset{\sim}{j}$ for $t \geq 0$. The path of the curve is called an involute of a circle.
   a. Show that the gradient of the curve is given by $\tan(t)$.
   b. Show that the speed at time t is given by t.
   c. Show that the magnitude of the acceleration at time t is given by $\sqrt{1+t^2}$.
   d. Show using calculus that the length of the curve between $t = t_0$ and $t = t_1$ is given by $\frac{1}{2}\left(t_1^2 - t_0^2\right)$.
   e. Sketch the path of the particle.

7. A particle moves so that its position vector is given by $\underset{\sim}{r}(t) = at\underset{\sim}{i} + \frac{a}{1+t^2}\underset{\sim}{j}$ for $t \in R$, where a is a positive real constant. The path of the curve is called the witch of Agnesi. Maria Agnesi (1718–1799) was an Italian mathematician and philosopher. She is often considered to be the first woman to have achieved a reputation in mathematics.

   a. Show that the gradient of the curve is given by $\frac{-2t}{\left(1+t^2\right)^2}$ and that the curve has a turning point at $(0, a)$.
   b. Show that $\frac{d^2y}{dx^2} = \frac{2\left(3t^2 - 1\right)}{a\left(1+t^2\right)^2}$ and hence show that the points of inflection on the curve are given by $\left(\pm\frac{\sqrt{3}a}{3}, \frac{3a}{4}\right)$.
   c. Sketch the graph of the witch of Agnesi.
   d. Show that the area bounded by the curve, the coordinate axes and the point $x = a$ is given by $\frac{\pi a^2}{4}$.
   e. Show that the Cartesian equation of the curve is given by $y = \frac{a^3}{a^2 + x^2}$ and hence verify that the area bounded by the curve, the coordinate axes and the point $x = a$ is given by $\frac{\pi a^2}{4}$.

8. A particle moves so that its position vector is given by $\underset{\sim}{r}(t) = \frac{3at}{1+t^3}\underset{\sim}{i} + \frac{3at^2}{1+t^3}\underset{\sim}{j}$ for $t \in R$, where a is a positive real constant. This curve is called the folium of Descartes.
   a. Show that the Cartesian equation of the curve is given by $x^3 + y^3 = 3axy$.
   b. Determine the gradient of the tangent to the curve in terms t.

c. Determine the gradient of the curve in terms of both x and y, and show that the curve has a turning point at $\left(a\sqrt[3]{2}, a\sqrt[3]{4}\right)$.

d. If $a = 3$:

  i. sketch the graph of the folium of Descartes
  ii. determine the position vector of the particle at the point where $t = 2$
  iii. calculate the speed of the particle at the point where $t = 2$
  iv. determine the equation of the tangent to the curve at the point where $t = 2$.

9. A curve given by the Cartesian equation $y = \dfrac{abx}{x^2 + a^2}$, where a and b are positive constants, is called the serpentine curve.

  a. Show that $\dfrac{dy}{dx} = \dfrac{ab(a - x)(a + x)}{\left(x^2 + a^2\right)^2}$ and that the graph of the serpentine curve has turning points at $\left(a, \dfrac{b}{2}\right)$ and $\left(-a, -\dfrac{b}{2}\right)$.

  b. Show that $\dfrac{d^2y}{dx^2} = \dfrac{2abx\left(x^2 - 3a^2\right)}{\left(x^2 + a^2\right)^3}$ and hence show that the points of inflection on the serpentine curve are given by $\left(\pm\sqrt{3}a, \pm\dfrac{\sqrt{3}b}{4}\right)$.

  c. Show that the vector equation of the serpentine curve is given by $\underset{\sim}{r}(t) = a\cot(t)\underset{\sim}{i} + \dfrac{b}{2}\sin(2t)\underset{\sim}{j}$ for $t \geq 0$.

  d. Sketch the graph of the serpentine curve.

  e. Show that the area bounded by the serpentine curve $y = \dfrac{abx}{x^2 + a^2}$, the x-axis, the origin and the point $x = a$ is given by $\dfrac{ab}{2}\log_e(2)$. Verify this result using another method.

10. A particle moves along the vector equation $\underset{\sim}{r}(t) = 8\cos^3(t)\underset{\sim}{i} + 8\sin^3(t)\underset{\sim}{j}$ for $t \in [0, 2\pi]$. This curve is called an astroid.

  a. Show that the Cartesian equation of the curve is given by $x^{\frac{2}{3}} + y^{\frac{2}{3}} = 4$.
  b. Show that the gradient at t is given by $-\tan(t)$.
  c. Show that $\dfrac{dy}{dx} = -\sqrt[3]{\dfrac{y}{x}}$ and deduce that the curve has no turning points.
  d. Sketch the graph of the astroid.
  e. Show that the speed of the particle at a time t is given by $12\sin(2t)$.
  f. Show that the length of the astroid is 48 units.
  g. Show that the total area inside the astroid is given by $768\displaystyle\int_0^{\frac{\pi}{2}} \sin^4(t)\cos^2(t)dt$, and, using technology, show that this area is equal to 24π units^2.

  Consider the general astroid with parametric equations $x = a\cos^3(t)$ and $y = a\sin^3(t)$, where a is a positive real constant.

  h. Let $P\left(a\cos^3(t), a\sin^3(t)\right)$ be a general point on the astroid. Determine, in terms of t, the equation of the tangent to the astroid at the point P.
  i. If the tangent to the astroid crosses the x-axis at Q and crosses the y-axis at R, show that distance between the points Q and R is always equal to a.

11. A curve given by the Cartesian equation $y^2(x^2+y^2)=a^2x^2$, where a is a positive constant, is called the kappa curve.

  a. Show that $x^2 = \dfrac{y^4}{a^2-y^2}$.

  b. Show that the curve can be represented by the parametric equations $x = a\cos(t)\cot(t)$ and $y = a\cos(t)$ for $t \in [0, 2\pi]$.

  c. Show that the gradient at t is given by $\dfrac{\sin^3(t)}{\cos(t)\left(\sin^2(t)+1\right)}$ and hence deduce that the kappa curve has a vertical tangent at the origin, no turning points and horizontal asymptotes at $y = \pm a$.

  d. Sketch the graph of the kappa curve.

  e. The area bounded by the kappa curve, the origin, the y-axis and the line $y = \dfrac{a}{2}$ is rotated about the y-axis to form a solid of revolution. Use calculus and partial fractions to show that the volume obtained is given by $\pi a^3\left(\log_e\left(\sqrt{3}\right) - \dfrac{13}{24}\right)$.

12. The function $f: D \to R, f(x) = x\sqrt{1-x^2}$ is called an eight curve.

  a. Determine the coordinates where the curve crosses the x-axis and hence state the maximal domain, D, of the function.

  b. Show that $f'(x) = \dfrac{1-2x^2}{\sqrt{1-x^2}}$ and that the graph of the function f has a maximum at $\left(\dfrac{\sqrt{2}}{2}, \dfrac{1}{2}\right)$ and a minimum at $\left(-\dfrac{\sqrt{2}}{2}, -\dfrac{1}{2}\right)$.

  c. Show that $f''(x) = \dfrac{2x^3-3x}{\sqrt{(1-x^2)^3}}$ and hence show that the only point of inflection is at the origin.

  d. Discuss what happens to the gradient of the curve as x approaches ± 1.

  e. Consider the curve defined by the implicit relationship $y^2 = x^2 - x^4$. Using implicit differentiation, determine $\dfrac{dy}{dx}$ and hence verify the result in **b**.

  The position vector $\underset{\sim}{r}(t)$ of a particle moving on a curve is given by $\underset{\sim}{r}(t) = \sin(t)\underset{\sim}{i} + \dfrac{1}{2}\sin(2t)\underset{\sim}{j}$, for $t \in [0, 2\pi]$.

  f. Show that the particle moves on the eight curve $y^2 = x^2 - x^4$.

  g. Sketch the graph of the eight curve $y^2 = x^2 - x^4$.

  h. Show that the total area inside the eight curve is given by $\dfrac{4}{3}$ units^2.

  i. If the eight curve is rotated about the x-axis, it forms a solid of revolution. Show that the total volume formed is $\dfrac{4\pi}{15}$.

  j. Determine the total length of the eight curve, correct to 4 decimal places.

13. Consider the functions $f: D \to R, f(x) = x^{\frac{3}{2}}\sqrt{4-x}$ and $g: D \to R, g(x) = -\sqrt{x^3(4-x)}$.

  a. If the maximal domain of both functions f and g is $D = [a, b]$, state the values of a and b.

  b. Explain how the graph of g is obtained from the graph of f.

  c. If $y^2 = x^3(4-x)$:

    i. use implicit differentiation to obtain an expression for $\dfrac{dy}{dx}$ in terms of both x and y

ii. show that $\frac{d^2y}{dx^2} = \frac{2(6-6x+x^2)}{\sqrt{x(4-x)^3}}$

iii. determine the exact coordinates of any turning points on the graph of f and use an appropriate test to verify its nature

iv. show that the point of inflection on the graph of f is $\left(3-\sqrt{3}, 2.36\right)$

v. discuss what happens to the gradient of both curves f and g as x approaches the value of b

vi. sketch the graphs of f and g.

d. The position vector $\underset{\sim}{r}(t)$ of a particle moving on a curve is given by

$\underset{\sim}{r}(t) = 4\sin^2\left(\frac{t}{2}\right)\underset{\sim}{i} + 16\cos\left(\frac{t}{2}\right)\sin^3\left(\frac{t}{2}\right)\underset{\sim}{j}$ for $t \geq 0$.

i. Show that the particle moves on the curve $y^2 = x^3(4-x)$.

ii. Calculate the first time the particle passes through the maximum turning point.

iii. Calculate the speed of the particle at the first time it passes through the maximum turning point.

14. A particle moves so that its position vector is given by $\underset{\sim}{r}(t) = \left(3\cos(t) - \cos(3t)\right)\underset{\sim}{i} + \left(3\sin(t) - \sin(3t)\right)\underset{\sim}{j}$ for $0 \leq t \leq 2\pi$. The graph of this curve is called a nephroid. The nephroid is a plane curve whose name means 'kidney-shaped'.

a. Show that the speed of the particle at time t is given by $6\sin(t)$ and hence show that the maximum and minimum values of the speed are 6 and –6.

b. Calculate the times when the gradient is zero and hence determine the turning points on the graph of the nephroid.

c. Sketch the path of the particle.

d. Show that the nephroid can be expressed in the implicit form $(x^2+y^2-4)^3 = 108y^2$.

e. Show that the total length of the nephroid is 24 units.

f. Show that the total area inside the nephroid is 12π units^2.

15. A particle moves so that its position vector is given by $\underset{\sim}{r}(t) = 2\cos(t)\left(1+\cos(t)\right)\underset{\sim}{i} + 2\sin(t)\left(1+\cos(t)\right)\underset{\sim}{j}$ for $0 \leq t \leq 2\pi$. The graph of this curve is called a cardioid. The cardioid is a plane curve that is heart-shaped.

a. Determine the distance of the particle from the origin at time t.

b. Determine the speed of the particle at time t.

c. Calculate the times when the position vector is perpendicular to the velocity vector.

d. Calculate the maximum and minimum values of the acceleration.

e. Sketch the path of the particle.

f. Show using calculus that the total length of the cardioid is 16 units.

g. Show that the total area inside the cardioid is 6π units^2.

h. Show that the cardioid can be expressed in the implicit form $(x^2+y^2-2x)^2 = 4(x^2+y^2)$.

16. A particle moves so that its position vector is given by $\underset{\sim}{r}(t) = \left(n\cos(t) + \cos(nt)\right)\underset{\sim}{i} + \left(n\sin(t) - \sin(nt)\right)\underset{\sim}{j}$, where $n > 1$.

a. Show that the gradient of the curve at time t is given by $\frac{\cos(nt) - \cos(t)}{\sin(nt) + \sin(t)}$.

b. Show that the speed at time t is given by $2n\cos\left(\left(\frac{n+1}{2}\right)t\right)$.

c. Show that $\dot{\underset{\sim}{r}}(t) \cdot \underset{\sim}{r}(t) = -(n^2+n)\sin\left((n+1)t\right)$.

d. Show that the maximum and minimum values of the acceleration are given by $n(n+1)$ and $n(n-1)$ respectively.

12.4 Exam questions

Question 1 (1 mark) TECH-ACTIVE

Source: VCE 2019, Specialist Mathematics Exam 2, Section A, Q7; © VCAA.

MC The length of the curve defined by the parametric equations $x = 3\sin(t)$ and $y = 4\cos(t)$ for $0 \le t \le \pi$ is given by

A. $\int_0^{\pi} \sqrt{9\cos^2(t) - 16\sin^2(t)}\, dt$

B. $\int_0^{\pi} \sqrt{9 + 7\sin^2(t)}\, dt$

C. $\int_0^{\pi} \sqrt{1 + 16\sin^2(t)}\, dt$

D. $\int_0^{\pi} \left(3\cos(t) - 4\sin(t)\right)\, dt$

E. $\int_0^{\pi} \sqrt{3\cos^2(t) + 4\sin^2(t)}\, dt$

Question 2 (5 marks) TECH-FREE

Source: VCE 2018, Specialist Mathematics Exam 1, Q10; © VCAA.

The position vector of a particle moving along a curve at time t seconds is given by

$\underset{\sim}{r}(t) = \frac{t^3}{3}\underset{\sim}{i} + \left(\arcsin(t) + t\sqrt{1 - t^2}\right)\underset{\sim}{j},\ 0 \le t \le 1$, where distances are measured in metres.

The distance d metres that the particle travels along the curve in three-quarters of a second is given by

$$d = \int_0^{\frac{3}{4}} \left(at^2 + bt + c\right)\, dt$$

Find a, b and c, where $a, b, c \in Z$.

Question 3 (4 marks) TECH-FREE

Source: VCE 2017, Specialist Mathematics Exam 1, Q7; © VCAA.

The position vector of a particle moving along a curve at time t is given by $\underset{\sim}{r}(t) = \cos^3(t)\underset{\sim}{i} + \sin^3(t)\underset{\sim}{j},\ 0 \le t \le \frac{\pi}{4}$.

Find the length of the path that the particle travels along the curve from $t = 0$ to $t = \frac{\pi}{4}$.

More exam questions are available online.

12.5 Integration of vectors

LEARNING INTENTION

At the end of this subtopic you should be able to:

- integrate vector functions
- determine velocity vectors by integrating acceleration vectors
- determine position vectors by integrating velocity vectors.

12.5.1 Rules for integrating vectors

The constant vector

When integrating a function, always remember to include the constant of integration, which is a scalar. When integrating a vector function with respect to a scalar, the constant of integration is a vector. This follows since if $\underset{\sim}{c}$ is a constant vector, then $\frac{d}{dt}(\underset{\sim}{c}) = \underset{\sim}{0}$.

Rules for integrating vectors

When differentiating a vector, we differentiate its components, so when we integrate a vector, we integrate each component using the usual rules for calculating antiderivatives. If $\underset{\sim}{q}(t) = x(t)\underset{\sim}{i} + y(t)\underset{\sim}{j}$ is a vector function, then we define $\int \underset{\sim}{q}(t)\,dt = \int x(t)\,dt\underset{\sim}{i} + \int y(t)\,dt\underset{\sim}{j} + \underset{\sim}{c}$, where $\underset{\sim}{c}$ is a constant vector.

Note that in the two-dimensional case, $\underset{\sim}{c} = c_1\underset{\sim}{i} + c_2\underset{\sim}{j}$, where c_1 and c_2 are real numbers.

Velocity vector to position vector

Because differentiating the position vector with respect to time gives the velocity vector, if we integrate the velocity vector with respect to time, we will obtain the position vector. Thus, given the velocity vector $\underset{\sim}{v}(t) = \dot{\underset{\sim}{r}}(t) = \dot{x}(t)\underset{\sim}{i} + \dot{y}(t)\underset{\sim}{j}$, we can obtain the position vector $\underset{\sim}{r}(t) = \int \dot{x}(t)dt\underset{\sim}{i} + \int \dot{y}(t)dt\underset{\sim}{j} = x(t)\underset{\sim}{i} + y(t)\underset{\sim}{j} + \underset{\sim}{c}$. Note that an initial condition must be given in order for us to be able to determine the constant vector of integration.

Integrating a velocity vector to obtain a position vector

Given a velocity vector $\underset{\sim}{v}(t) = \dot{x}(t)\underset{\sim}{i} + \dot{y}(t)\underset{\sim}{j}$, the position vector is:

$$\underset{\sim}{r}(t) = \int \underset{\sim}{v}(t)dt = \int \dot{x}(t)dt\underset{\sim}{i} + \int \dot{y}(t)dt\underset{\sim}{j} + \underset{\sim}{c}$$

WORKED EXAMPLE 10 Integrating a velocity vector to determine the position vector

The velocity vector of a particle is given by $\dot{\underset{\sim}{r}}(t) = 2\underset{\sim}{i} + 6t\underset{\sim}{j}$ for $t \geq 0$. If $\underset{\sim}{r}(1) = 3\underset{\sim}{i} + \underset{\sim}{j}$, determine the position vector at a time t.

THINK	WRITE
1. Integrate the velocity vector to obtain the position vector using the given rules.	$\dot{\underset{\sim}{r}}(t) = 2\underset{\sim}{i} + 6t\underset{\sim}{j}$ $\underset{\sim}{r}(t) = \int 2dt\,\underset{\sim}{i} + \int 6tdt\,\underset{\sim}{j}$

2. Perform the integration. Do not forget to add in a constant vector. | $\underset{\sim}{r}(t) = 2t\underset{\sim}{i} + 3t^2\underset{\sim}{j} + \underset{\sim}{c}$

3. Substitute to determine the value of the constant vector. | Substitute $t = 1$ and use the given condition.
$\underset{\sim}{r}(1) = 2\underset{\sim}{i} + 3\underset{\sim}{j} + \underset{\sim}{c} = 3\underset{\sim}{i} + \underset{\sim}{j}$

4. Solve for the constant vector. | $\underset{\sim}{c} = (3\underset{\sim}{i} + \underset{\sim}{j}) - (2\underset{\sim}{i} + 3\underset{\sim}{j})$
$= \underset{\sim}{i} - 2\underset{\sim}{j}$

5. Substitute back for the constant vector. | $\underset{\sim}{r}(t) = 2t\underset{\sim}{i} + 3t^2\underset{\sim}{j} + \underset{\sim}{c}$
$= (2t\underset{\sim}{i} + 3t^2\underset{\sim}{j}) + (\underset{\sim}{i} - 2\underset{\sim}{j})$

6. Simplify the position vector to give the final result. | $\underset{\sim}{r}(t) = (2t + 1)\underset{\sim}{i} + (3t^2 - 2)\underset{\sim}{j}$

Acceleration vector to position vector

Because differentiating the velocity vector with respect to time gives the acceleration vector, if we integrate the acceleration vector with respect to time, we will obtain the velocity vector. Given the acceleration vector $\underset{\sim}{a}(t) = \frac{d\underset{\sim}{v}(t)}{dt} = \ddot{\underset{\sim}{r}}(t) = \ddot{x}(t)\underset{\sim}{i} + \ddot{y}(t)\underset{\sim}{j}$, the velocity vector $\underset{\sim}{v}(t) = \dot{\underset{\sim}{r}}(t) = \int \ddot{x}(t)dt\underset{\sim}{i} + \int \ddot{y}(t)dt\underset{\sim}{j} = \dot{x}(t)\underset{\sim}{i} + \dot{y}(t)\underset{\sim}{j} + \underset{\sim}{c}_1$, where $\underset{\sim}{c}_1$ is a constant vector. By integrating again, as above, we can determine the position vector.

Note that two sets of information must be given to find the two constant vectors of integration. This process is a generalisation of the techniques used in earlier topics when we integrated the acceleration to determine the displacement.

Integrating an acceleration vector to obtain a velocity vector

Given an acceleration vector $\underset{\sim}{a}(t) = \ddot{x}(t)\underset{\sim}{i} + \ddot{y}(t)\underset{\sim}{j}$, the velocity vector is:

$$\underset{\sim}{v}(t) = \int \underset{\sim}{a}(t)dt = \int \ddot{x}(t)dt\underset{\sim}{i} + \int \ddot{y}(t)dt\underset{\sim}{j} + \underset{\sim}{c}$$

WORKED EXAMPLE 11 Integrating an acceleration vector to determine the velocity vector

The acceleration vector of a particle is given by $\ddot{\underset{\sim}{r}}(t) = 6t\underset{\sim}{i}$, where $t \geq 0$. Given that $\dot{\underset{\sim}{r}}(2) = 6\underset{\sim}{i} - 3\underset{\sim}{j}$ and $\underset{\sim}{r}(2) = 4\underset{\sim}{i} - 2\underset{\sim}{j}$, determine the position vector at time t.

THINK | **WRITE**

1. Integrate the acceleration vector to obtain the velocity vector using the given rules. | $\ddot{\underset{\sim}{r}}(t) = 6t\underset{\sim}{i}$
$\dot{\underset{\sim}{r}}(t) = \int 6tdt\underset{\sim}{i}$

2. Perform the integration. Do not forget to add in the first constant vector. | $\dot{\underset{\sim}{r}}(t) = 3t^2\underset{\sim}{i} + \underset{\sim}{c}_1$

3. Substitute to calculate the first constant vector. | Substitute $t = 2$ and use the first given condition.
$\dot{\underset{\sim}{r}}(2) = 12\underset{\sim}{i} + \underset{\sim}{c}_1 = 6\underset{\sim}{i} - 3\underset{\sim}{j}$

4. Solve for the first constant vector. | $\underset{\sim}{c}_1 = (6\underset{\sim}{i} - 3\underset{\sim}{j}) - 12\underset{\sim}{i}$
$= -6\underset{\sim}{i} - 3\underset{\sim}{j}$

5. Substitute back for the first constant vector.	$\dot{\underset{\sim}{r}}(t) = 3t^2\underset{\sim}{i} + (-6\underset{\sim}{i} - 3\underset{\sim}{j})$
6. Simplify to give the velocity vector.	$\dot{\underset{\sim}{r}}(t) = (3t^2 - 6)\underset{\sim}{i} - 3\underset{\sim}{j}$
7. Integrate the velocity vector to obtain the position vector using the given rules.	$\underset{\sim}{r}(t) = \int (3t^2 - 6)dt\underset{\sim}{i} - \int 3dt\underset{\sim}{j}$
8. Perform the integration. Do not forget to add in a second constant vector.	$\underset{\sim}{r}(t) = (t^3 - 6t)\underset{\sim}{i} - 3t\underset{\sim}{j} + \underset{\sim}{c}_2$
9. Substitute to calculate the value of the second constant vector.	Substitute $t = 2$ and use the second given condition. $\underset{\sim}{r}(2) = (8 - 12)\underset{\sim}{i} - 6\underset{\sim}{j} + \underset{\sim}{c}_2 = 4\underset{\sim}{i} - 2\underset{\sim}{j}$
10. Solve for the second constant vector.	$\underset{\sim}{c}_2 = (4\underset{\sim}{i} - 2\underset{\sim}{j}) - (-4\underset{\sim}{i} - 6\underset{\sim}{j})$ $= 8\underset{\sim}{i} + 4\underset{\sim}{j}$
11. Substitute back for the second constant vector.	$\underset{\sim}{r}(t) = (t^3 - 6t)\underset{\sim}{i} - 3t\underset{\sim}{j} + (8\underset{\sim}{i} + 4\underset{\sim}{j})$
12. Simplify the position vector to give the final result.	$\underset{\sim}{r}(t) = (t^3 - 6t + 8)\underset{\sim}{i} + (4 + 3t)\underset{\sim}{j}$

Determining the Cartesian equation

As previously, when we are given the position vector, we can determine the parametric equations of the path of the particle, and by eliminating the parameter, we can determine the Cartesian equation of the curve along which the particle moves.

WORKED EXAMPLE 12 Determining the Cartesian equation of the path of an object

The acceleration vector of a moving particle is given by $\cos\left(\frac{t}{2}\right)\underset{\sim}{i} - \sin\left(\frac{t}{2}\right)\underset{\sim}{j}$ for $0 \le t \le 4\pi$, where t is the time. The initial velocity is $2\underset{\sim}{j}$ and the initial position is $\underset{\sim}{i} - 3\underset{\sim}{j}$. Determine the Cartesian equation of the path.

THINK	WRITE
1. State and integrate the acceleration vector to obtain the velocity vector using the given rules.	$\ddot{\underset{\sim}{r}}(t) = \cos\left(\frac{t}{2}\right)\underset{\sim}{i} - \sin\left(\frac{t}{2}\right)\underset{\sim}{j}$ $\dot{\underset{\sim}{r}}(t) = \int \cos\left(\frac{t}{2}\right) dt\,\underset{\sim}{i} - \int \sin\left(\frac{t}{2}\right) dt\underset{\sim}{j}$
2. Perform the integration. Do not forget to add in the first constant vector.	Since $\int \cos(kx)dx = \frac{1}{k}\sin(kx)$ and $\int \sin(kx)dx = -\frac{1}{k}\cos(kx)$ with $k = \frac{1}{2}$, $\dot{\underset{\sim}{r}}(t) = 2\sin\left(\frac{t}{2}\right)\underset{\sim}{i} + 2\cos\left(\frac{t}{2}\right)\underset{\sim}{j} + \underset{\sim}{c}_1$.
3. Substitute to solve for the first constant vector.	Initially means when $t = 0$. Substitute $t = 0$ and use the first given condition. $\dot{\underset{\sim}{r}}(0) = 2\sin(0)\underset{\sim}{i} + 2\cos(0)\underset{\sim}{j} + \underset{\sim}{c}_1 = 2\underset{\sim}{j}$ $2\underset{\sim}{j} + \underset{\sim}{c}_1 = 2\underset{\sim}{j}$ $\underset{\sim}{c}_1 = \underset{\sim}{0}$

4. Substitute back for the first constant vector and integrate the velocity vector to obtain the position vector, using the given rules.	$\dot{\underset{\sim}{r}}(t) = 2\sin\left(\frac{t}{2}\right)\underset{\sim}{i} + 2\cos\left(\frac{t}{2}\right)\underset{\sim}{j}$ $\dot{\underset{\sim}{r}}(t) = \int 2\sin\left(\frac{t}{2}\right)dt\underset{\sim}{i} + \int 2\cos\left(\frac{t}{2}\right)dt\underset{\sim}{j}$
5. Perform the integration. Do not forget to add in a second constant vector.	$\underset{\sim}{r}(t) = -4\cos\left(\frac{t}{2}\right)\underset{\sim}{i} + 4\sin\left(\frac{t}{2}\right)\underset{\sim}{j} + \underset{\sim}{c}_2$
6. Substitute to solve for the second constant vector.	Initially means when $t = 0$. Substitute $t = 0$ and use the second given condition. $\underset{\sim}{r}(0) = -4\cos(0)\underset{\sim}{i} + 4\sin(0)\underset{\sim}{j} + \underset{\sim}{c}_2 = \underset{\sim}{i} - 3\underset{\sim}{j}$ $-4\underset{\sim}{i} + \underset{\sim}{c}_2 = \underset{\sim}{i} - 3\underset{\sim}{j}$ $\underset{\sim}{c}_2 = 5\underset{\sim}{i} - 3\underset{\sim}{j}$
7. Substitute back for the second constant vector and state the position vector.	$\underset{\sim}{r}(t) = -4\cos\left(\frac{t}{2}\right)\underset{\sim}{i} + 4\sin\left(\frac{t}{2}\right)\underset{\sim}{j} + \left(5\underset{\sim}{i} - 3\underset{\sim}{j}\right)$ $\underset{\sim}{r}(t) = \left(5 - 4\cos\left(\frac{t}{2}\right)\right)\underset{\sim}{i} + \left(4\sin\left(\frac{t}{2}\right) - 3\right)\underset{\sim}{j}$
8. State the parametric equations.	$x = 5 - 4\cos\left(\frac{t}{2}\right),\ y = 4\sin\left(\frac{t}{2}\right) - 3$
9. Eliminate the parameter.	$\cos\left(\frac{t}{2}\right) = \frac{5-x}{4},\ \sin\left(\frac{t}{2}\right) = \frac{y+3}{4}$ Since $\cos^2\left(\frac{t}{2}\right) + \sin^2\left(\frac{t}{2}\right) = 1$, $\left(\frac{5-x}{4}\right)^2 + \left(\frac{y+3}{4}\right)^2 = 1$
10. State the Cartesian equation in the implicit form.	$(x-5)^2 + (y+3)^2 = 16$ This is a circle with centre at (5, –3) and radius 4.

12.5 Exercise

Technology free

1. **WE10** The velocity vector of a particle is given by $\dot{\underset{\sim}{r}}(t) = (4t - 4)\underset{\sim}{i} - 3\underset{\sim}{j}$ for $t \geq 0$. If $\underset{\sim}{r}(1) = 3\underset{\sim}{i} + \underset{\sim}{j}$, determine the position vector at time t.
2. The initial position of a particle is given by $3\underset{\sim}{i} + \underset{\sim}{j}$. If the velocity vector of the particle is given by $\underset{\sim}{v}(t) = 6\sin(3t)\underset{\sim}{i} + 4e^{-2t}\underset{\sim}{j}$, determine the position vector.
3. a. The velocity vector of a particle is given by $\dot{\underset{\sim}{r}}(t) = e^{-\frac{t}{3}}\underset{\sim}{i} + 4t^3\underset{\sim}{j}$, where $t \geq 0$ is the time. If initially the particle is at the origin, determine the position vector.

b. The velocity vector of a particle is given by $\dot{\underset{\sim}{r}}(t) = 2t\underset{\sim}{i} + 6\sin(2t)\underset{\sim}{j}$, where $t \geq 0$ is the time. If initially the particle is at the origin, determine the position vector.

4. a. The velocity vector of a particle is given by $\dot{\underset{\sim}{r}}(t) = \frac{1}{\sqrt{16 - t^2}}\underset{\sim}{i} - \frac{t}{\sqrt{t^2 + 9}}\underset{\sim}{j}$, where $t \geq 0$ is the time. If $\underset{\sim}{r}(0) = 3\underset{\sim}{i} + 2\underset{\sim}{j}$, determine the position vector.

b. The velocity vector of a particle is given by $\dot{\underset{\sim}{r}}(t) = \frac{2}{2t+1}\underset{\sim}{i} + \frac{72}{(3t+2)^2}\underset{\sim}{j}$, where $t \geq 0$ is the time. $\underset{\sim}{r}(0) = 5\underset{\sim}{i} + \underset{\sim}{j}$, determine the position vector at time t.

5. **WE11** The acceleration vector of a particle is given by $\ddot{\underset{\sim}{r}}(t) = -12t^2\underset{\sim}{j}$, where $t \geq 0$ is the time. Given that $\dot{\underset{\sim}{r}}(2) = -2\underset{\sim}{i} - 16\underset{\sim}{j}$ and $\underset{\sim}{r}(2) = \underset{\sim}{i} + 6\underset{\sim}{j}$, determine the position vector at time t.

6. A particle is moving such that $\underset{\sim}{r}(1) = -2\underset{\sim}{i} + 7\underset{\sim}{j}$ and $\dot{\underset{\sim}{r}}(1) = 6\underset{\sim}{i} + 10\underset{\sim}{j}$. The acceleration vector of the particle is given by $\ddot{\underset{\sim}{r}}(t) = 6\underset{\sim}{i} + 2\underset{\sim}{j}$, where $t \geq 0$ is the time. Determine the position vector at time t.

7. a. A moving particle starts at position $\underset{\sim}{r}(0) = 3\underset{\sim}{i} - 2\underset{\sim}{j}$ with an initial velocity of zero. The acceleration vector of the particle is given by $\ddot{\underset{\sim}{r}}(t) = 8\underset{\sim}{i} + 6\underset{\sim}{j}$, where $t \geq 0$ is the time. Determine the position vector at time t.

b. A moving particle starts at position $\underset{\sim}{r}(0) = 3\underset{\sim}{i} + 4\underset{\sim}{j}$ with an initial velocity $\dot{\underset{\sim}{r}}(0) = 8\underset{\sim}{j}$. The acceleration vector of the particle is given by $\ddot{\underset{\sim}{r}}(t) = 4\underset{\sim}{i} + 2\underset{\sim}{j}$, where $t \geq 0$ is the time. Determine the position vector at time t.

8. a. The acceleration vector of a particle is given by $\ddot{\underset{\sim}{r}}(t) = \frac{-9}{(3t+1)^2}\underset{\sim}{i} + \frac{32}{(2t+1)^3}\underset{\sim}{j}$, where $t \geq 0$ is the time. If initially the velocity vector is $\dot{\underset{\sim}{r}}(0) = 3\underset{\sim}{i} - 8\underset{\sim}{j}$ and the initial position is $r(0) = 4\underset{\sim}{i} + 3\underset{\sim}{j}$, determine the position vector at time t.

b. The acceleration vector of a particle is given by $\ddot{\underset{\sim}{r}}(t) = \frac{-9}{(3t+1)^2}\underset{\sim}{i} - \frac{24}{(2t+1)^4}\underset{\sim}{j}$, where $t \geq 0$ is the time. If initially the velocity vector is $\dot{\underset{\sim}{r}}(0) = 2\underset{\sim}{i} - \underset{\sim}{j}$ and the initial position is $\underset{\sim}{r}(0) = 6\underset{\sim}{i} + 8\underset{\sim}{j}$, determine the position vector at time t.

9. **WE12** The acceleration vector of a moving particle is given by $-45\cos(3t)\underset{\sim}{i} + 45\sin(3t)\underset{\sim}{j}$, where $t \geq 0$ is the time. The initial velocity is $-15\underset{\sim}{j}$ and the initial position is $3\underset{\sim}{i} + 4\underset{\sim}{j}$. Determine the Cartesian equation of the path.

10. A particle is moving such that its initial position is $2\underset{\sim}{i} - 2\underset{\sim}{j}$ and its initial velocity is $10\underset{\sim}{j}$. The acceleration vector of the particle is given by $12\cos(2t)\underset{\sim}{i} - 20\sin(2t)\underset{\sim}{j}$, where $0 \leq t \leq \pi$ and t is the time. Determine the Cartesian equation of the path.

11. a. When a ball is thrown, its acceleration vector is given by $\ddot{\underset{\sim}{r}}(t) = -10\underset{\sim}{j}$, where $t \geq 0$ is the time. If initially the velocity vector is $15\underset{\sim}{i} + 20\underset{\sim}{j}$ and the initial position is $2\underset{\sim}{j}$, determine the Cartesian equation of the path.

b. When a ball is thrown, its acceleration vector is given by $\ddot{\underset{\sim}{r}}(t) = -9.8\underset{\sim}{j}$, where $t \geq 0$ is the time. If initially the velocity vector is $5\underset{\sim}{i} + 10\underset{\sim}{j}$ and the initial position is $\underset{\sim}{j}$, determine the Cartesian equation of the path.

12. a. The acceleration vector of a moving particle is given by $\ddot{\underset{\sim}{r}}(t) = -e^{-\frac{t}{2}}\underset{\sim}{i} + 2e^{\frac{t}{2}}\underset{\sim}{j}$, where $t \in R$. If $\dot{\underset{\sim}{r}}(0) = 2\underset{\sim}{i} + 4\underset{\sim}{j}$ and $\underset{\sim}{r}(0) = -2\underset{\sim}{i} + 3\underset{\sim}{j}$, determine the Cartesian equation of the path.

b. A particle is moving such that $\underset{\sim}{r}(0) = \underset{\sim}{i} + 5\underset{\sim}{j}$ and $\dot{\underset{\sim}{r}}(0) = 4\underset{\sim}{j}$. The acceleration vector of the particle is given by $\ddot{\underset{\sim}{r}}(t) = 8\cos(2t)\underset{\sim}{i} - 8\sin(2t)\underset{\sim}{j}$, where t is the time and $0 \leq t \leq 2\pi$. Determine the Cartesian equation of the path.

13. a. A particle is moving such that its initial position is $3\underset{\sim}{i} + 5\underset{\sim}{j}$ and its initial velocity $-6\underset{\sim}{j}$. The acceleration vector of the particle is given by $\underset{\sim}{a}(t) = 9\cos(3t)\underset{\sim}{i} + 18\sin(3t)\underset{\sim}{j}$, where t is the time and $0 \leq t \leq 2\pi$. Determine the Cartesian equation of the path.

b. The acceleration vector of a moving particle is given by $-3\cos\left(\frac{t}{2}\right)\underset{\sim}{i} + \sin\left(\frac{t}{2}\right)\underset{\sim}{j}$, where $0 \leq t \leq 4\pi$. If $\underset{\sim}{v}(0) = -2\underset{\sim}{j}$ and $\underset{\sim}{r}(0) = 5\underset{\sim}{i} + 3\underset{\sim}{j}$, determine the Cartesian equation of the path.

14. a. The acceleration vector of a moving particle is given by $-2\cos(t)\underset{\sim}{i} - 8\cos(2t)\underset{\sim}{j}$, where $t \geq 0$. If initially the velocity vector is zero and the initial position is $2\underset{\sim}{i}$, determine the Cartesian equation of the path, stating the domain and range.
    b. A moving particle is such that its initial position is $\underset{\sim}{i} + 3\underset{\sim}{j}$ and its initial velocity $4\underset{\sim}{i}$. The acceleration vector of the particle is given by $-8\sin(2t)\underset{\sim}{i} - 96\cos(4t)\underset{\sim}{j}$, where t is the time and $0 \leq t \leq \pi$. Determine the Cartesian equation of the path, stating the domain and range.

15. a. Particle A has an acceleration of $2\underset{\sim}{i} + 4\underset{\sim}{j}$, an initial velocity of $-6\underset{\sim}{i} + 5\underset{\sim}{j}$ and an initial position of $13\underset{\sim}{i} - 17\underset{\sim}{j}$. Particle B has an acceleration of $6\underset{\sim}{i} + 8\underset{\sim}{j}$, an initial velocity of $-8\underset{\sim}{i} - 20\underset{\sim}{j}$ and an initial position of $\underset{\sim}{i} + 40\underset{\sim}{j}$. Show that the two particles collide, and determine the time and point of collision.

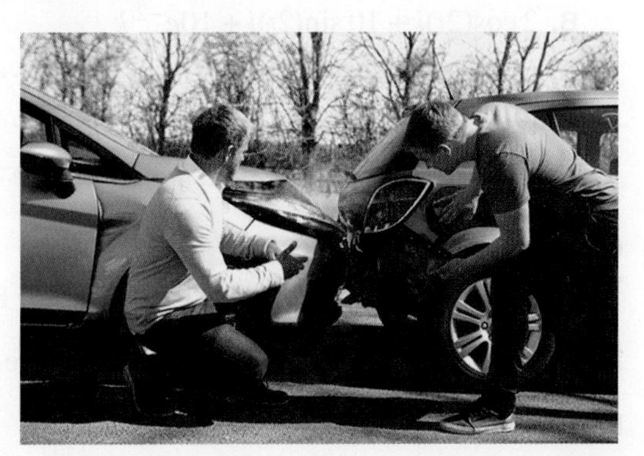

    b. Car A has an acceleration of $2\underset{\sim}{i} - 2\underset{\sim}{j}$ at time t, and after 1 second its velocity is $-5\underset{\sim}{i} + 6\underset{\sim}{j}$ and its position is $-3\underset{\sim}{i} + 2\underset{\sim}{j}$. Car B has an acceleration of $2\underset{\sim}{i} - 6\underset{\sim}{j}$ at time t, and after 1 second its velocity is $-2\underset{\sim}{i} + 6\underset{\sim}{j}$ and its position is $-15\underset{\sim}{i} + 34\underset{\sim}{j}$. Show that the two cars collide, and determine the time and point of collision.

16. The acceleration vector of a particle is given by $-n^2 r\cos(nt)\underset{\sim}{i} - n^2 r\sin(nt)\underset{\sim}{j}$, at time t where a, b and r are all real constants. If $\underset{\sim}{v}(0) = nr\underset{\sim}{j}$ and $\underset{\sim}{r}(0) = (a + r)\underset{\sim}{i} + b\underset{\sim}{j}$, determine the Cartesian equation of the path.

17. A particle is moving such that its initial position is $(h + a)\underset{\sim}{i} + k\underset{\sim}{j}$ and its initial velocity is $bn\underset{\sim}{j}$, where a, b, n, h and k, are all real constants. The acceleration vector of the particle is given by $-n^2 a\cos(nt)\underset{\sim}{i} - n^2 b\sin(nt)\underset{\sim}{j}$, where t is the time. Determine the Cartesian equation of the path.

Technology active

18. The acceleration vector of a moving particle is given by $\left(4\cos(2t) - 2\cos(t)\right)\underset{\sim}{i} + \left(4\sin(2t) - 2\sin(t)\right)\underset{\sim}{j}$, where $0 \leq t \leq 2\pi$, $\dot{\underset{\sim}{r}}(\pi) = -4\underset{\sim}{j}$ and $\underset{\sim}{r}(\pi) = -3\underset{\sim}{i}$. Determine the position vector and sketch the equation of the path.

19. A particle is moving such that $\underset{\sim}{r}\left(\frac{\pi}{2}\right) = \underset{\sim}{i} + 4\underset{\sim}{j}$ and $\dot{\underset{\sim}{r}}\left(\frac{\pi}{2}\right) = -4\underset{\sim}{i} - 4\underset{\sim}{j}$. The acceleration vector of the particle is given by $-\left(16\cos(4t) + 4\cos(t)\right)\underset{\sim}{i} + \left(16\sin(4t) - 4\sin(t)\right)\underset{\sim}{j}$. Determine the position vector and sketch the equation of the path.

20. The acceleration vector of a moving particle is given by $-8\cos(2t)\underset{\sim}{i} - 108\sin(6t)\underset{\sim}{j}$, where $0 \leq t \leq 2\pi$, $\dot{\underset{\sim}{r}}\left(\frac{\pi}{4}\right) = -4\underset{\sim}{i}$ and $\underset{\sim}{r}\left(\frac{\pi}{4}\right) = -3\underset{\sim}{j}$. Determine the position vector and sketch the equation of the path.

12.5 Exam questions

Question 1 (4 marks) TECH-FREE

Source: VCE 2015, Specialist Mathematics Exam 1, Q3; © VCAA.

The velocity of a particle at time t seconds is given by $\dot{\underset{\sim}{r}}(t) = (4t - 3)\underset{\sim}{i} + 2t\underset{\sim}{j} - 5\underset{\sim}{k}$, where components are measured in metres per second.

Find the distance of the particle from the origin in metres when $t = 2$, given that $r(0) = \underset{\sim}{i} - 2\underset{\sim}{k}$.

Question 2 (1 mark) TECH-ACTIVE

Source: *VCE 2014, Specialist Mathematics Exam 2, Section A, Q17; © VCAA.*

MC The acceleration vector of a particle that starts from rest is given by $\underset{\sim}{a}(t) = -4\sin(2t)\underset{\sim}{i} + 20\cos(2t)\underset{\sim}{j} - 20e^{-2t}\underset{\sim}{k}$, where $t \geq 0$.

The velocity vector of the particle, v(t), is given by

A. $-8\cos(2t)\underset{\sim}{i} - 40\sin(2t)\underset{\sim}{j} + 40e^{-2t}\underset{\sim}{k}$

B. $2\cos(2t)\underset{\sim}{i} + 10\sin(2t)\underset{\sim}{j} + 10e^{-2t}\underset{\sim}{k}$

C. $(8 - 8\cos(2t))\underset{\sim}{i} - 40\sin(2t)\underset{\sim}{j} + \left(40e^{-2t} - 40\right)\underset{\sim}{k}$

D. $(2\cos(2t) - 2)\underset{\sim}{i} + 10\sin(2t)\underset{\sim}{j} + \left(10e^{-2t} - 10\right)\underset{\sim}{k}$

E. $(4\cos(2t) - 4)\underset{\sim}{i} + 20\sin(2t)\underset{\sim}{j} + \left(20 - 20e^{-2t}\right)\underset{\sim}{k}$

Question 3 (2 marks) TECH FREE

The acceleration vector of a moving particle is given by $\ddot{\underset{\sim}{r}}(t) = 4e^{-2t}\underset{\sim}{i} + 2e^{2t}\underset{\sim}{j}$, where $t \in R$.

If $\dot{\underset{\sim}{r}}(0) = -2\underset{\sim}{i} + 2\underset{\sim}{j}$ and $\underset{\sim}{r}(0) = 5\underset{\sim}{i} - 2\underset{\sim}{j}$, determine the Cartesian equation of the path.

More exam questions are available online.

12.6 Projectile motion

LEARNING INTENTION

At the end of this subtopic you should be able to:

- model projectiles fired from the origin or a point above the origin
- determine the time of flight, range, maximum height, angle of projection and equation of path of a projectile.

12.6.1 Motion of projectiles on Earth

The motion of a particle when acted upon by gravity and air resistance is called projectile motion. The object or projectile considered can be a ball of any type, as considered in sport, or it can be a car or bullet; in fact, it can be any object which moves.

In our first modelling approach to projectile motion, certain assumptions are made. The first is to assume that the projectile, no matter how big, is treated as a point particle. Further assumptions are to ignore air resistance, assume that the Earth is flat, ignore the rotation of the Earth and ignore the variations in gravity due to height. For projectiles moving close to the Earth's surface at heights of no more than approximately two hundred metres, these assumptions are generally valid.

12.6.2 Projectiles fired from the origin

Consider a projectile fired in a vertical two-dimensional plane from the origin, O, with an initial speed of V m/s at an angle of α degrees to the horizontal. T is the time of flight, H is the maximum height reached by the projectile on its motion, and R is the range on the horizontal plane, that is, the horizontal distance travelled.

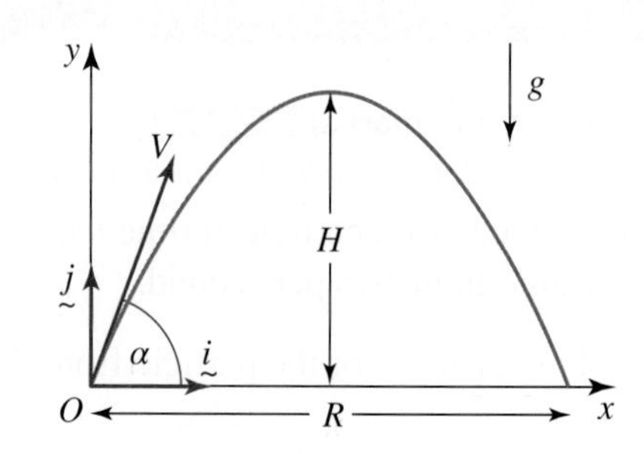

As the motion is in two dimensions, $x = x(t)$ is the horizontal displacement and $y = y(t)$ is the vertical displacement at time t seconds, where $0 \leq t \leq T$.

Taking $\underset{\sim}{i}$ as a unit vector of 1 metre in the x direction and $\underset{\sim}{j}$ as a unit vector of 1 metre in the positive upwards y direction,

$\underset{\sim}{r}(t) = x(t)\underset{\sim}{i} + y(t)\underset{\sim}{j}$ is the position vector,

$\dot{\underset{\sim}{r}}(t) = \dot{x}(t)\underset{\sim}{i} + \dot{y}(t)\underset{\sim}{j}$ is the velocity vector and

$\ddot{\underset{\sim}{r}}(t) = \ddot{x}(t)\underset{\sim}{i} + \ddot{y}(t)\underset{\sim}{j}$ is the acceleration vector.

The only force acting on the projectile is the weight force, which acts downwards, so that $\underset{\sim}{a} = -g\underset{\sim}{j}$, where $g = 9.8\ \text{m/s}^2$.

Using vectors:

$\ddot{\underset{\sim}{r}}(t) = -g\underset{\sim}{j}$ and $\dot{\underset{\sim}{r}}(t) = -gt\underset{\sim}{j} + \underset{\sim}{c}_1$ (integrating the acceleration vector to get the velocity vector).
But $\dot{\underset{\sim}{r}}(0) = V\cos(\alpha)\underset{\sim}{i} + V\sin(\alpha)\underset{\sim}{j}$ (the initial velocity vector).

Therefore, $\underset{\sim}{c}_1 = V\cos(\alpha)\underset{\sim}{i} + V\sin(\alpha)\underset{\sim}{j}$.

$\dot{\underset{\sim}{r}}(t) = V\cos(\alpha)\underset{\sim}{i} + \left(V\sin(\alpha) - gt\right)\underset{\sim}{j}$ (substituting for the first constant vector)

$\underset{\sim}{r}(t) = Vt\cos(\alpha)\underset{\sim}{i} + \left(Vt\sin(\alpha) - \frac{1}{2}gt^2\right)\underset{\sim}{j} + \underset{\sim}{c}_2$ (integrating the velocity vector to get the position vector)

As the projectile is fired from the origin, O, $\underset{\sim}{r}(0) = \underset{\sim}{0}$, so $\underset{\sim}{c}_2 = \underset{\sim}{0}$; thus, $\underset{\sim}{r}(t) = Vt\cos(\alpha)\underset{\sim}{i} + \left(Vt\sin(\alpha) - \frac{1}{2}gt^2\right)\underset{\sim}{j}$.

Projectile motion

A projectile fired from the origin with an initial velocity of V m/s and launch angle α can be described by the equations:

Cartesian equation:

$$\underset{\sim}{r}(t) = Vt\cos(\alpha)\underset{\sim}{i} + \left(Vt\sin(\alpha) - \frac{1}{2}gt^2\right)\underset{\sim}{j}$$

Parametric equations:

$$x(t) = Vt\cos(\alpha) \quad \text{and} \quad y(t) = Vt\sin(\alpha) - \frac{1}{2}gt^2$$

Time of flight

The time of flight is the time that the projectile takes to go up and come down again, or the time at which it returns to ground level and hits the ground. To determine the time of flight, solve $y = 0$ for t.

$$Vt\sin(\alpha) - \frac{1}{2}gt^2 = 0$$

$$t\left(V\sin(\alpha) - \frac{1}{2}gt\right) = 0$$

The result $t = 0$ represents the time when the projectile was fired, so $T = \dfrac{2V\sin(\alpha)}{g}$ represents the time of flight.

Time of flight

$$\boldsymbol{T = \frac{2V\sin(\alpha)}{g}}$$

The range

The range is the horizontal distance travelled in the total flight time, T.

$$R = x(T) = VT\cos(\alpha)$$

Substitute $T = \dfrac{2V\sin(\alpha)}{g}$:

$$= V\cos(\alpha)\left(\frac{2V\sin(\alpha)}{g}\right)$$

Expand:

$$= \frac{V^2 2\sin(\alpha)\cos(\alpha)}{g}$$

Using the double-angle formula $\sin(2A) = 2\sin(A)\cos(A)$,

$$R = \frac{V^2\sin(2\alpha)}{g}$$

Note that for maximum range, $\sin(2\alpha) = 1$, so $2\alpha = 90°$ or $\alpha = 45°$. This applies only for a projectile fired from ground level.

Range

$$\boldsymbol{R = \frac{V^2\sin(2\alpha)}{g}}$$

Maximum height

The maximum height occurs when the particle is no longer rising. This occurs when the vertical component of the velocity is zero, that is, $\dot{y}(t) = 0$.

When the projectile is fired from ground level the total flight time can be determined by solving $\dot{y} = V\sin(\alpha) - gt = 0$ which gives $t = \dfrac{V\sin(\alpha)}{g} = \dfrac{T}{2}$, which is half the time of flight.

This applies only for a projectile fired from ground level.

Substituting for t into the y component gives

$$\begin{aligned} H = y\left(\frac{T}{2}\right) &= V\sin(\alpha)\left(\frac{V\sin(\alpha)}{g}\right) - \frac{g}{2}\left(\frac{V\sin(\alpha)}{g}\right)^2 \\ &= \frac{V^2\sin^2(\alpha)}{g} - \frac{V^2\sin^2(\alpha)}{2g} \\ &= \frac{V^2\sin^2(\alpha)}{2g} \end{aligned}$$

Maximum height

$$H = \frac{V^2 \sin^2(\alpha)}{2g}$$

WORKED EXAMPLE 13 Determining the time of flight, range and max height of a golf ball

A golf ball is hit off the ground at an angle of 53.13° with an initial speed of 25 m/s. Determine:

a. **the time of flight, in seconds to 2 decimal places**

b. **the range, in metres to 2 decimal places**

c. **the maximum height reached, in metres to 2 decimal places.**

THINK	WRITE
a. 1. State the value of the parameters.	a. $V = 25$ and $\alpha = 53.13°$
2. Use the general results to determine the time of flight.	$T = \frac{2V\sin(\alpha)}{g}$ $T = \frac{2 \times 25\sin(53.13)°}{9.8}$
3. State the time of flight.	$T = 4.08$ seconds.
b. 1. Use the general results to determine the range.	b. $R = \frac{V^2\sin(2\alpha)}{g}$ $R = \frac{25^2\sin(2 \times 53.13°)}{9.8}$
2. State the range.	$R = 61.22$ metres.
c. 1. Use the general results to determine the maximum height reached.	c. $H = \frac{V^2\sin^2(\alpha)}{2g}$ $H = \frac{25^2\sin^2(53.13°)}{2 \times 9.8}$
2. State the maximum height reached.	$H = 20.41$ metres.

The equation of the path

It can be shown that the path of the projectile is a parabola. Transposing $x = Vt\cos(\alpha)$ for t gives $t = \frac{x}{V\cos(\alpha)}$.

Substituting this into $y(t) = Vt\sin(\alpha) - \frac{1}{2}gt^2$ gives

$$y = V\sin(\alpha)\left(\frac{x}{V\cos(\alpha)}\right) - \frac{g}{2}\left(\frac{x}{V\cos(\alpha)}\right)^2.$$

Simplifying this gives

$$y = x\tan(\alpha) - \frac{gx^2}{2V^2\cos^2(\alpha)} \text{ or } y = x\tan(\alpha) - \frac{gx^2\sec^2(\alpha)}{2V^2}.$$

This is of the form $y = ax + bx^2$, so the path of the projectile is a parabola.

Calculating the angle of projection

Using $\sec^2(\alpha) = 1 + \tan^2(\alpha)$, the Cartesian equation of a projectile can be written in the form $y = x\tan(\alpha) - \dfrac{gx^2}{2V^2}\left(1 + \tan^2(\alpha)\right)$. Alternatively, this can be rearranged into a quadratic in $\tan(\alpha)$ as $\dfrac{gx^2}{2V^2}\tan^2(\alpha) - x\tan(\alpha) + \left(y + \dfrac{gx^2}{2V^2}\right) = 0$.

This equation is useful for calculating the angle of projection, α, if we are given the coordinates of a point (x, y) through which the particle passes and the initial speed of projection, V. Because this equation is a quadratic in $\tan(\alpha)$, it is possible that we can obtain two values for α.

Angle of projection

To determine the angle of projection to the horizontal, solve the following equation for α:

$$\frac{gx^2}{2V^2}\tan^2(\alpha) - x\tan(\alpha) + \left(y + \frac{gx^2}{2V^2}\right) = 0$$

WORKED EXAMPLE 14 Modelling a basketball shot

A basketballer shoots for goal from the three-point line. She throws the ball with an initial speed of 15 m/s and the ball leaves her hands at a height of 2.1 metres above the ground. Determine the possible angles of projection if she is to score a goal.
Data: Distance from goal to three-point line: 6.25 m
Height of ring: 3.05 m

THINK	WRITE/DRAW
1. State the value of the parameters. Drawing a diagram will help.	$V = 15$ but α is unknown and required to be found.

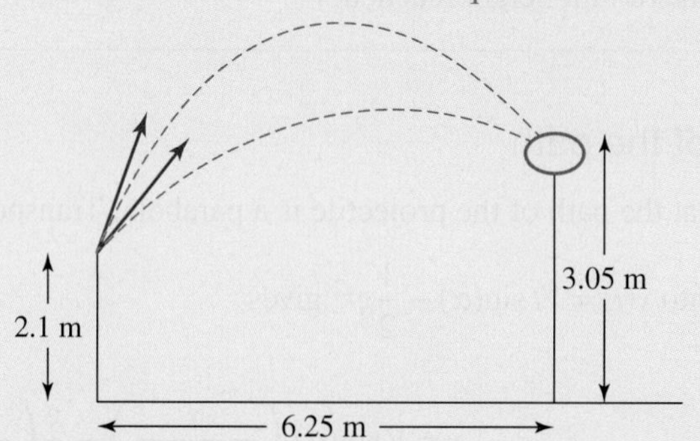

THINK	WRITE/DRAW
2. State where the projectile will pass through.	The basketball must pass through the point where $x = 6.25$ and $y = 3.05 - 2.1 = 0.95$.

3. Substitute in the appropriate values to obtain a quadratic in tan(α).	Substitute $V = 15$, $x = 6.25$ and $y = 0.95$: $$\frac{gx^2}{2V^2}\tan(\alpha) - x\tan(\alpha) + \left(y + \frac{gx^2}{2V^2}\right) = 0$$ $$\frac{9.8 \times 6.25^2}{2 \times 15^2}\tan^2(\alpha) - 6.25\tan(\alpha) + \left(0.95 + \frac{9.8 \times 6.25^2}{2 \times 15^2}\right) = 0$$
4. Simplify the quadratic to be solved.	$0.8507\tan^2(\alpha) - 6.25\tan(\alpha) + 1.8007 = 0$
5. First calculate the discriminant of the quadratic equation.	$a = 0.8507$, $b = -6.25$, $c = 1.8007$ $$\begin{aligned}\Delta &= b^2 - 4ac \\ &= (-6.25)^2 - 4 \times 0.8507 \times 1.8007 \\ &= 32.935\end{aligned}$$ $$\sqrt{\Delta} = 5.739$$
6. Use the quadratic formula to solve for tan(α).	$$\begin{aligned}\tan(\alpha) &= \frac{-b \pm \sqrt{\Delta}}{2a} \\ &= \frac{6.25 \pm 5.739}{2 \times 0.8507} \\ &= 7.0465,\ 0.3004\end{aligned}$$
7. Determine the values of α.	$\alpha = \tan^{-1}(7.0465)$, $\tan^{-1}(0.3004)$
8. State the final result.	There are two possible angles of projection: $\alpha = 81.92°$ or $16.92°$.

12.6.3 Projectiles fired from above the origin

The sections above apply to situations where the projectile is fired from the origin. We can also analyse situations where the projectile is fired from a height h above the origin.

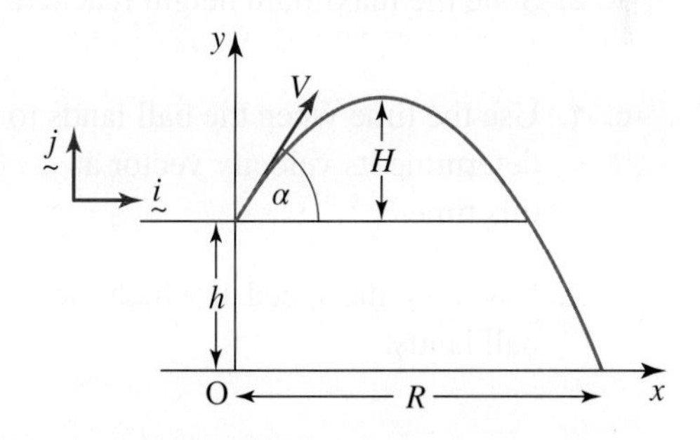

In these situations we need to solve the appropriate parametric equations $x(t) = Vt\cos(\alpha)$ and $y(t) = h + Vt\sin(\alpha) - \frac{1}{2}gt^2$ to determine the time of flight and the range. The result for the maximum height, $H = \frac{V^2\sin^2(\alpha)}{2g}$, is still valid, but it gives only the height above h.

WORKED EXAMPLE 15 Modeling a ball thrown from the top of a hill

A boy throws a ball from the top of a hill 2 metres above ground level with an initial speed of 10 m/s at an angle of 40° to the horizontal.

a. Calculate the time of flight, in seconds to 2 decimal places, and the horizontal distance travelled, in metres to 2 decimal places.

b. Calculate the maximum height reached, in metres to 2 decimal places.

c. Determine the speed, in m/s to 1 decimal place, and angle at which the ball lands, in degrees to 1 decimal place.

d. Show that the ball travels in a parabolic path.

THINK	WRITE/DRAW
a. 1. State the value of the parameters.	**a.** $V = 10$ m/s and $\alpha = 40°$

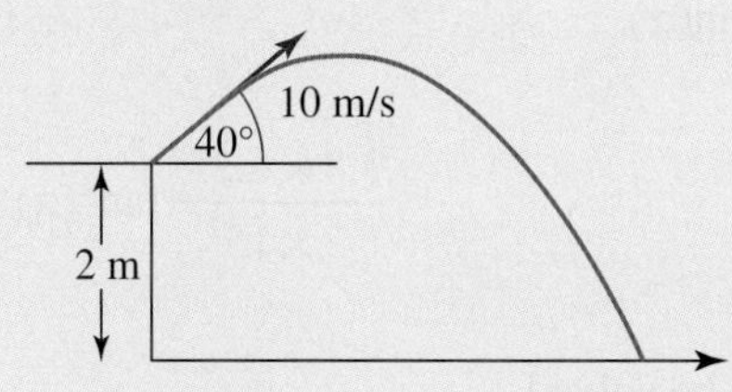

THINK	WRITE/DRAW
2. Start with the acceleration vector and the initial velocity vector to determine the velocity vector at time t.	$\ddot{r}(t) = -9.8\underset{\sim}{j}$ $\dot{r}(0) = 10\cos(40°)\underset{\sim}{i} + 10\sin(40°)\underset{\sim}{j}$ $\dot{r}(t) = 10\cos(40°)\underset{\sim}{i} + \left(10\sin(40°) - 9.8t\right)\underset{\sim}{j}$
3. State the position vector at time t.	Since the initial position is 2 metres above the ground, $\underset{\sim}{r}(0) = 2\underset{\sim}{j}$, $r(t) = 10t\cos(40°)\underset{\sim}{i} + \left(2 + 10t\sin(40°) - 4.9t^2\right)\underset{\sim}{j}$
4. The ball hits the ground when the vertical component is zero.	Solve $y = 2 + 10t\sin(40°) - 4.9t^2 = 0$ for t. Solving this quadratic gives $t = -0.26,\ 1.57$. Since $t \geq 0$, the time of flight is 1.57 seconds. $x(1.57) = 10 \times 1.57\cos(40°)$ $= 12.04$ The horizontal distance travelled is 12.04 metres.
b. 1. The ball will rise and reach maximum height when the vertical component of its velocity is zero.	**b.** Solve $\dot{y} = 10\sin(40°) - 9.8t = 0$ for t. $t = \dfrac{10\sin(40°)}{9.8}$ $= 0.656$ Note that this is not half the time of flight.
2. State the maximum height reached.	$y(0.656) = 2 + 10 \times 0.656\sin(40°) - 4.9 \times 0.656^2$ The maximum height reached is 4.11 metres.
c. 1. Use the time when the ball lands to determine its velocity vector at this time.	**c.** $\dot{\underset{\sim}{r}}(1.57) = 10\cos(40°)\underset{\sim}{i} + (10\sin(40°) - 9.8 \times 1.57)\underset{\sim}{j}$ $= 7.66\underset{\sim}{i} - 8.96\underset{\sim}{j}$
2. Calculate the speed at which the ball lands.	$\lvert\dot{\underset{\sim}{r}}(1.57)\rvert = \sqrt{\dot{x}^2 + \dot{y}^2}$ $= \sqrt{7.66^2 + (-8.96)^2}$ The ball lands with a speed of 11.8 m/s.
3. Determine the angle that the velocity vector makes with the ground, when the ball lands.	Let ψ be the angle with the ground when the ball lands. $\tan(\psi) = \dfrac{\dot{y}}{\dot{x}}$ $\psi = \tan^{-1}\left(\dfrac{8.96}{7.66}\right)$ $= 49.5°$
d. 1. State the parametric equations and eliminate the parameter t.	**d.** $x = 10t\cos(40°)$, $y = 2 + 10t\sin(40°) - 4.9t^2$ $t = \dfrac{x}{10\cos(40°)}$ $y = 2 + \dfrac{10x\sin(40°)}{10\cos(40°)} - 4.9\left(\dfrac{x}{10\cos(40°)}\right)^2$

2. The Cartesian equation is parabolic as it is of the form $y = ax^2 + bx + c$.	$y = 2 + x\tan(40°) - \dfrac{49x^2}{1000}\sec^2(40°)$

12.6.4 Proofs involving projectile motion

Often projectile motion problems involve parameters rather than specific given values. In these types of problems, we are required to prove or show that a certain equation is valid. To do this, we can use the general equations and mathematically manipulate these to show the desired result. Note that we can only give results in terms of the given parameters.

WORKED EXAMPLE 16 Proving general formulas for projectile motion

A ball is projected at an angle α from a point on a horizontal plane. If the ball reaches a maximum height of H, show that the time interval between the instants when the ball is at heights of $H\sin^2(\alpha)$ is $2\cos(\alpha)\sqrt{\dfrac{2H}{g}}$.

THINK

WRITE/DRAW

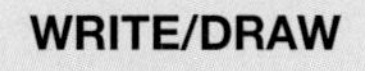

1. Draw a diagram.

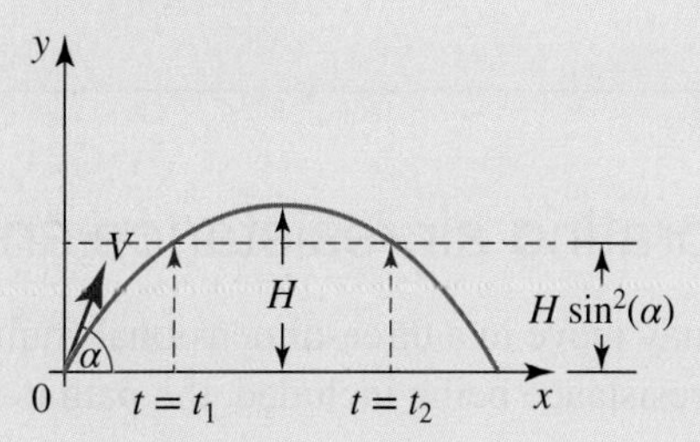

2. Use the result for the maximum height.

$$H = \frac{V^2\sin^2(\alpha)}{2g}$$

3. Since the initial speed of projection is unknown, express V in terms of the given values.

$$V^2 = \frac{2gH}{\sin^2(\alpha)}$$

4. Use the given equation for y.

$$y = Vt\sin(\alpha) - \frac{1}{2}gt^2$$

5. We need to determine the values of t at the given heights.

Let $y = H\sin^2(\alpha)$.

$$H\sin^2(\alpha) = Vt\sin(\alpha) - \frac{1}{2}gt^2$$

$$\frac{1}{2}gt^2 - Vt\sin(\alpha) + H\sin^2(\alpha) = 0$$

6. Evaluate the discriminant of the quadratic.

$a = \frac{1}{2}g,\ b = -V\sin(\alpha),\ c = H\sin^2(\alpha)$

$$\begin{aligned}\Delta &= b^2 - 4ac \\ &= (-V)^2\sin^2(\alpha) - 4 \times \frac{1}{2}g \times H\sin^2(\alpha) \\ &= V^2\sin^2(\alpha) - 2gH\sin^2(\alpha)\end{aligned}$$

7. Use previous results to eliminate the unknown, V.	Substitute $V^2\sin^2(\alpha) = 2gH$ into the discriminant: $\Delta = 2gH - 2gH\sin^2(\alpha)$ $= 2gH(1-\sin^2(\alpha))$ $= 2gH\cos^2(\alpha)$
8. Use the quadratic formula to determine the two times.	$t = \dfrac{V\sin(\alpha) + \sqrt{\Delta}}{g}$, so $t_1 = \dfrac{V\sin(\alpha) - \sqrt{\Delta}}{g}$ and $t_2 = \dfrac{V\sin(\alpha) + \sqrt{\Delta}}{g}$, where $t_2 > t_1$.
9. Determine the time interval and simplify.	$t_2 - t_1 = \dfrac{2\sqrt{\Delta}}{g}$ $= \dfrac{2\sqrt{2gH\cos^2(\alpha)}}{g}$ $= \dfrac{2\cos(\alpha)}{g}\sqrt{2gH}$
10. State the required result.	$t_2 - t_1 = 2\cos(\alpha)\sqrt{\dfrac{2H}{g}}$

12.6.5 Incorporating air resistance and three-dimensional motion

In reality, projectiles may move in a three-dimensional framework rather than a two-dimensional plane. Also, with the effects of air resistance being included, the path is not necessarily a parabola.

WORKED EXAMPLE 17 Projectile motion involving air resistance

A shot is thrown by a shot-put competitor on level ground. At a time t in seconds measured from the point of release of the shot, the position vector $\underset{\sim}{r}(t)$ of the shot is given by

$$\underset{\sim}{r}(t) = 7t\underset{\sim}{i} + \left(5t + 6\left(e^{-\frac{t}{2}} - 1\right)\right)\underset{\sim}{j} + (2 + 12t - 5t^2)\underset{\sim}{k}$$

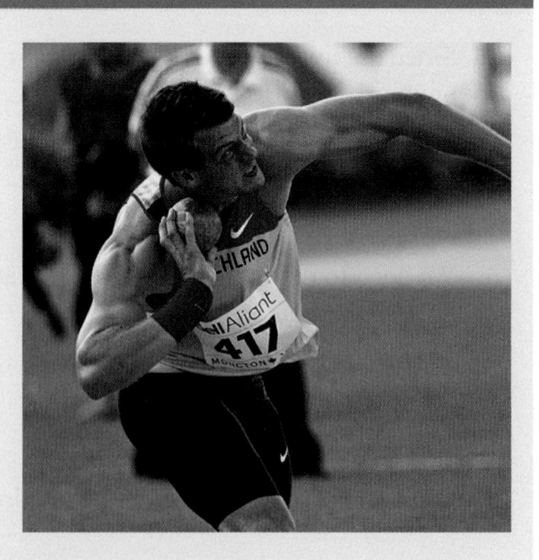

where $\underset{\sim}{i}$ is a unit vector in the east direction, $\underset{\sim}{j}$ is a unit vector in the north direction and $\underset{\sim}{k}$ is a unit vector vertically up. The origin, O, of the coordinate system is at ground level and displacements are measured in metres. Let P be the point where the shot hits the ground.

a. Calculate the time taken for the shot to hit the ground, in seconds to 2 decimal places.
b. Calculate how far from O the shot hits the ground, in metres to 2 decimal places.
c. Determine the initial speed of projection, in m/s to 2 decimal places.
d. Determine the speed, in m/s to 2 decimal places, and angle at which the shot hits the ground, in degrees to 2 decimal places.

THINK	WRITE/DRAW
a. 1. Determine when the shot hits the ground.	**a.** The shot hits the ground when the $\underset{\sim}{k}$ component is zero, that is, when $2 + 12t - 5t^2 = 0$.
2. Solve for the values of t.	Solving using the quadratic formula gives $t = -0.1565$ or $t = 2.5565$. Since $t \geq 0$, the shot hits the ground at 2.56 seconds.
b. 1. Determine the position vector where the shot hits the ground.	**b.** Substitute $t = 2.56$ into the position vector. $\underset{\sim}{r}(2.56) = 7 \times 2.56\underset{\sim}{i} + \left(5 \times 2.56 + 6\left(e^{-\frac{2.56}{2}} - 1\right)\right)\underset{\sim}{j}$ $+(2 + 12 \times 2.56 - 5 \times 2.56^2)\underset{\sim}{k}$ $\underset{\sim}{r}(2.56) = 17.896\underset{\sim}{i} + 8.453\underset{\sim}{j}$
2. Determine the distance where the shot hits the ground.	$\lvert\underset{\sim}{r}(2.56)\rvert = \sqrt{17.895^2 + 8.453^2}$ $= 19.79$ metres
c. 1. Determine the velocity vector.	**c.** $\dot{\underset{\sim}{r}}(t) = 7\underset{\sim}{i} + \left(5 - 3e^{-\frac{t}{2}}\right)\underset{\sim}{j} + (12 - 10t)\underset{\sim}{k}$
2. Determine the initial velocity vector.	Substitute $t = 0$ into the velocity vector. $\dot{\underset{\sim}{r}}(0) = 7\underset{\sim}{i} + (5 - 3e^0)\underset{\sim}{j} + 12\underset{\sim}{k}$ $= 7\underset{\sim}{i} + 2\underset{\sim}{j} + 12\underset{\sim}{k}$
3. Calculate the initial speed of projection.	$\lvert\dot{\underset{\sim}{r}}(0)\rvert = \sqrt{7^2 + 2^2 + 12^2}$ $= \sqrt{197}$ The initial speed of projection is 14.04 m/s.
d. 1. Determine the velocity vector when the shot hits the ground.	**d.** Substitute $t = 2.56$ into the velocity vector. $\dot{\underset{\sim}{r}}(2.56) = 7\underset{\sim}{i} + \left(5 - 3e^{-\frac{2.56}{2}}\right)\underset{\sim}{j} + (12 - 10 \times 2.56)\underset{\sim}{k}$ $= 7\underset{\sim}{i} + 4.16\underset{\sim}{j} - 13.56\underset{\sim}{k}$
2. Calculate the magnitude of the velocity vector.	$\lvert\dot{\underset{\sim}{r}}(2.56)\rvert = \sqrt{7^2 + 4.16^2 + (-13.56)}$ $= \sqrt{250.34}$
3. Determine the angle at which the shot hits the ground.	The required angle, ψ, is the angle between the downwards component and the combined east and north components. $\tan(\psi) = \frac{\dot{z}}{\sqrt{\dot{x} + \dot{y}^2}}$

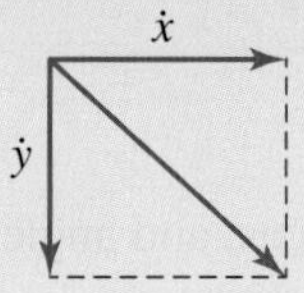

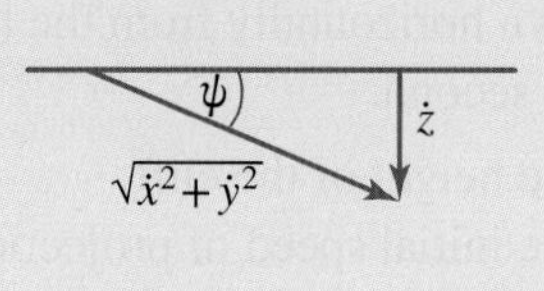

4. Calculate the angle.	$\tan(\psi) = \dfrac{13.56}{\sqrt{7^2 + 4.16^2}}$ $\psi = \tan^{-1}(1.665)$
5. State the final result.	The shot hits the ground with a speed of 15.82 m/s at an angle of 59.02°.

12.6 Exercise

Technology active

1. WE13 A soccer ball is kicked off the ground at an angle of 67.38° with an initial speed of 13 m/s. Calculate:
   a. the time of flight, in seconds to 2 decimal places
   b. the range, in metres to 2 decimal places
   c. the maximum height reached, in metres to 2 decimal places.
2. A ball is thrown so that its time of flight is $\dfrac{10\sqrt{3}}{g}$ and the ratio of the range to the maximum height reached is $\dfrac{4\sqrt{3}}{3}$. Determine the initial speed and angle of projection.
3. A cricket ball is hit by a batsman off his toes with an initial speed of 20 m/s at an angle of 30° with the horizontal. Calculate:
   a. the time of flight, in seconds to 2 decimal places
   b. the range on the horizontal plane, in metres to 2 decimal places
   c. the greatest height reached, in metres to 2 decimal places.
4. A golf ball is hit at an angle of 20° and its range is 150 metres. Determine:
   i. the initial speed of projection, in m/s to 2 decimal places
   ii. the greatest height reached, in metres to 2 decimal places
   iii. the time of flight, in seconds to 2 decimal places.
5. A rock is thrown horizontally from the top of a cliff face and strikes the ground 30 metres from the base of the cliff after 1 second.
   a. Calculate the height of the cliff.
   b. Calculate the initial speed of projection.
   c. Determine the speed, in m/s to 2 decimal places, and angle at which the rock strikes the ground, in degrees to 1 decimal place.
   d. Show that the rock travels in a parabolic path.

6. A motorcycle is driven at 150 km/h horizontally off the top of a cliff face and strikes the ground after 2 seconds. Determine:
   a. the height of the cliff
   b. how far from the base of the cliff the motorcycle strikes the ground, in metres to 3 decimal places
   c. the speed, in m/s to 2 decimal places, and angle at which it strikes the ground, in degrees to 1 decimal place.
7. WE15 A cricketer smashes a cricket ball at an angle of 35° to the horizontal from a point 0.5 m above the ground with an initial speed of 30 m/s.
   a. Determine the time of flight, in seconds to 2 decimal places, and the horizontal distance travelled, correct to the nearest metre.
   b. Calculate the maximum height reached, in metres to 2 decimal places.
   c. Determine the speed, in m/s to 2 decimal places, and angle at which the ball lands, in degrees to 2 decimal places.
   d. Show that the ball travels in a parabolic path.
8. A cricket outfielder attempts to throw the cricket ball back towards the stumps. He throws the ball from a height of 1.8 m above the ground with an initial speed of 20 m/s at an angle of 25° to the horizontal.
   a. Determine the time of flight, in seconds to 2 decimal places, and the horizontal distance travelled, in metres to 1 decimal place.
   b. Calculate the maximum height reached, in metres to 2 decimal places.
   c. Determine the speed, in m/s to 2 decimal places and angle at which the ball lands, in degrees to 2 decimal places.
   d. Show that the ball travels in a parabolic path.
9. An object is projected from the top of a building 100 metres high at an angle of 45° with a speed of 10 m/s. Determine:
   a. the time of flight to reach ground level, in seconds to 3 decimal places
   b. how far from the edge of the building the object strikes the ground, in metres to 2 decimal places
   c. the greatest height the object reaches above the building, in metres to 1 decimal place
   d. the speed, in m/s to 2 decimal places and angle at which the object strikes the ground, correct to the nearest degree.
10. A stone is thrown with a speed of 15 m/s at an angle of 20° from a cliff face and strikes the ground after 3.053 seconds. Determine:
    a. the height of the cliff, correct to the nearest metre
    b. how far from the edge of the cliff the stone strikes the ground, in metres to 2 decimal places
    c. the greatest height the stone reaches above the ground, in metres to 2 decimal places
    d. the speed, in m/s to 2 decimal places and angle at which the stone strikes the ground, in degrees to 1 decimal place.
11. WE14 A tennis player hits the ball 4 feet above the baseline of a tennis court with a speed of 100 feet per second. If the ball travels in a vertical plane towards the centre of the net and just grazes the top of the net, calculate the angles at which it could have been hit, in degrees to 1 decimal place.
    Data: Tennis court dimensions 78×27 feet
    Net height at the centre 3 feet
    Use $g = 32$ ft/s^2.

12. A rugby player uses a place kick to kick the ball off the ground. To score a goal, the ball must pass through a point 40 metres out and 8 metres vertically above the point of release. If he kicks the ball at a speed of 30 m/s, determine the possible angles of projection.

13. a. A shell is fired with an initial speed of 147 m/s for maximum range. There is a target 2 km away on the same horizontal level.
    i. Calculate how far above the target the shell passes, in metres to 1 decimal place.
    ii. Calculate how far beyond the target the shell strikes the ground.
    iii. Determine the minimum speed of projection to just reach the target.
 b. A projectile falls a metres short of its target when fired at an angle of α, and a metres beyond the target when fired at the same muzzle velocity but at an angle of β. If θ is the angle required for a direct hit on the target, show that $\sin(2\theta) = \frac{1}{2}\left(\sin(2\alpha) + \sin(2\beta)\right)$.

14. A baseball is initially hit from a distance of 1.5 metres above the ground at an angle of 35°. The ball reaches a maximum height of 8 metres above the ground. Determine:
 a. the initial speed of projection, in m/s to 2 decimal places
 b. the distance of the outfielder from the hitter if the outfielder just catches the ball at ground level, in metres to 3 decimal places
 c. the speed, in m/s to 2 decimal places, and angle at which the baseball strikes the fielder's hands, in degrees to 1 decimal place.

15. **WE16** A particle is projected at a given angle α from a point on a horizontal plane. If the particle reaches a maximum height of H, show that the time interval between the instants when the particle is at heights of $\frac{H}{2}$ is $2\sqrt{\frac{H}{g}}$.

16. A particle is projected with an initial speed of $3g$ m/s. The particle reaches heights of g metres at times t_1 and t_2. Given that $t_2 - t_1 = 1$, show that the range of the particle is given by $\frac{9g\sqrt{3}}{2}$.

17. A projectile is fired with an initial speed of $\sqrt{2ga}$ to hit a target at a horizontal distance of a from the point of projection and a vertical distance of $\frac{a}{2}$ above it. Show that there are two possible angles of projection, that the ratio of the two times to hit the target is $\sqrt{5}$, and that the ratio of the maximum heights reached in each case is $\frac{9}{5}$.

18. a. A particle is projected to just clear two walls. The walls are both of height 6 metres and are at distances of 5 metres and 10 metres from the point of projection. Show that if α is the angle of projection, then $\tan(\alpha) = \frac{9}{5}$.
 b. A particle is projected to just clear two walls. The walls are both h metres high and are at distances of a metres and b metres from the point of projection. Show that the angle of projection is $\tan^{-1}\left(\frac{h(a+b)}{ab}\right)$.
 c. A particle is projected to just clear two walls. The first wall is h_1 metres high and at a distance of a metres from the point of projection; the second wall is h_2 metres high and b metres from the point of projection, where $b > a > 0$. Show that the angle of projection is $\tan^{-1}\left(\frac{b^2h_1 - a^2h_2}{ab(b-a)}\right)$.

19. **WE17** A football is kicked by a footballer. At a time t in seconds measured from the point of impact, the position vector $\underset{\sim}{r}(t)$ of the tip of the football is given by

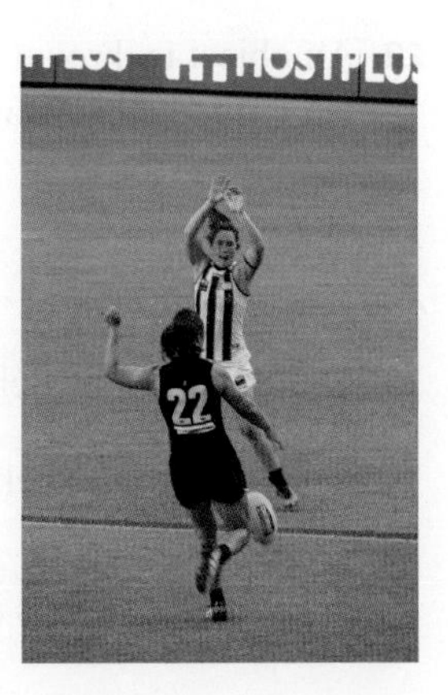

$$\underset{\sim}{r}(t) = 60\left(1 - e^{-\frac{t}{2}}\right)\underset{\sim}{i} + 2t\underset{\sim}{j} + (1 + 12t - 4.9t^2)\underset{\sim}{k}$$

where $\underset{\sim}{i}$ is a unit vector horizontally forward, $\underset{\sim}{j}$ is a unit vector to the right and $\underset{\sim}{k}$ is a unit vector vertically up.
The origin, O, of the coordinate system is at ground level and all displacements are measured in metres.

a. Calculate how far from the origin the football hits the ground, in metres to 2 decimal places.
b. Determine the speed, in m/s to 2 decimal places, and angle at which the football hits the ground, in degrees to 2 decimal places.

20. A javelin is thrown by an athlete on level ground. The time t is in seconds, measured from the release of the javelin. The position vector $\underset{\sim}{r}(t)$ of the tip of the javelin is given by
$\underset{\sim}{r}(t) = 20t\underset{\sim}{i} + \left(2\pi t - 3\sin\left(\frac{\pi t}{6}\right)\right)\underset{\sim}{j} + (1.8 + 14.4t - 5t^2)\underset{\sim}{k}$ where $\underset{\sim}{i}$ is a unit vector in the east direction, $\underset{\sim}{j}$ is a unit vector in the north direction and $\underset{\sim}{k}$ is a unit vector vertically up. The origin, O, of the coordinate system is at ground level, and all displacements are measured in metres.

a. Calculate how long will it take for the javelin to strike the ground.
b. Calculate how far from O the tip of the javelin hits the ground, in metres to 2 decimal places.
c. Determine the speed, in m/s to 2 decimal places, and angle at which the javelin's tip strikes the ground, in degrees to 2 decimal places.

21. A soccer ball is kicked off the ground. Its position vector is given by $\underset{\sim}{r}(t) = 6t\underset{\sim}{i} + 28t\underset{\sim}{j} + \frac{12\sqrt{2}}{5}\sin\left(\frac{\pi t}{2}\right)\underset{\sim}{k}$, where $\underset{\sim}{i}$ and $\underset{\sim}{j}$ are unit vectors in the horizontal plane at right angles to each other, and $\underset{\sim}{k}$ is a unit vector in the vertical direction. Displacements are measured in metres, and t is the time in seconds after the ball is kicked.

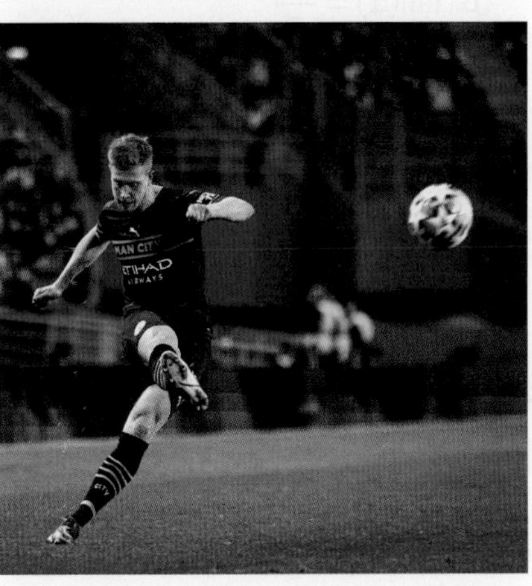

a. Calculate the initial speed at which the soccer ball is kicked, in m/s to 2 decimal places.

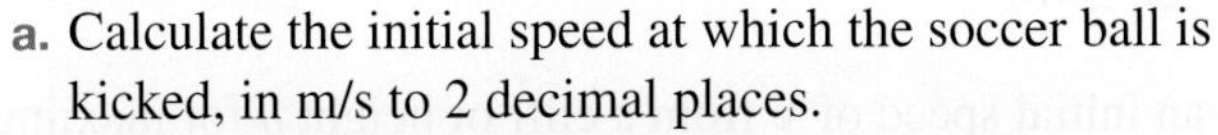

b. Calculate the angle from the ground that the soccer ball was kicked, in degrees to 1 decimal place.
c. Determine the maximum height reached by soccer ball, in metres to 1 decimal place.
d. After the soccer ball has reached its maximum height and is on its downwards trajectory, a player jumps and heads the soccer ball when it is 2.4 metres above the ground.

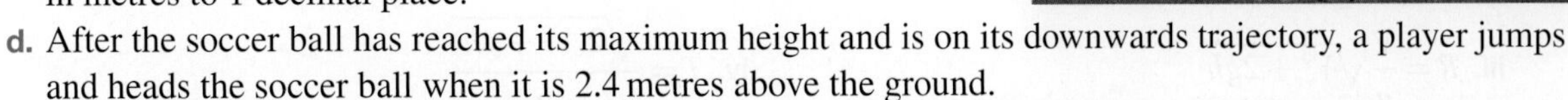

i. Calculate the time that has elapsed from the instant that the soccer ball is kicked until the player heads the ball, in seconds to 1 decimal place.
ii. Calculate how far, measured along the ground, the player was from where the ball was kicked, in metres to 2 decimal places.
iii. Determine the speed, in m/s, of the ball at the instant when it strikes the player's head. Give your answer correct to 2 decimal places.

22. An object is thrown horizontally with a speed of V m/s from the top of a cliff face h metres high and strikes the ground at a distance of R metres from the base of the cliff after a time of T seconds. Show that:

a. $T = \sqrt{\dfrac{2h}{g}}$

b. $R = VT$

c. the speed at which the object hits the ground is $\sqrt{V^2 + 2gh}$

d. the angle at which the object strikes the ground is $\tan^{-1}\left(\dfrac{2h}{R}\right)$

e. the object travels in a parabolic path, $y = h\left(1 - \dfrac{x^2}{R^2}\right)$.

23. A stone is projected horizontally from the top of a cliff of height H metres with a speed of U m/s. At the same instant another object is fired from the base of the cliff with a speed of V m/s at an angle of α. Given that the two objects collide after a time of T seconds, show that:

$$\alpha = \cos^{-1}\left(\frac{U}{V}\right) \text{ and } V^2 = U^2 + \frac{H^2}{T^2}.$$

24. A projectile is fired at an angle of α with an initial speed of V. It reaches a maximum height of H and has a horizontal range of R. Show that:

a. $R\tan(\alpha) = \dfrac{1}{2}gT^2$

b. $\tan(\alpha) = \dfrac{4H}{R}$

c. $T^2 = \dfrac{8H}{g}$

d. $V = \sqrt{2g\left(H + \dfrac{R^2}{16H}\right)}$

e. the equation of the path is $y = \dfrac{4Hx}{R^2}(R - x)$ for $0 \le x \le R$.

25. a. When a projectile is fired at an angle of α with an initial speed of V from a cliff of height h for maximum range R, provided that $V < \sqrt{gR}$, show that:

i. $\tan(\alpha) = \dfrac{V^2}{gR}$

ii. $h = \dfrac{g^2R^2 - V^4}{2V^2g}$

iii. $R = \dfrac{V}{g}\sqrt{V^2 + 2gh}$

iv. $T = \dfrac{\sqrt{2(V^2 + gh)}}{g}$.

b. A shot-putter can throw the shot with a release speed of 20 m/s. The shot leaves his hand 2 metres above the ground. Determine the maximum range, in metres to 2 decimal places, the angle of projection, in degrees to 1 decimal place, and the time of flight, in seconds to 2 decimal places.

12.6 Exam questions

Question 1 (1 mark) TECH-ACTIVE

Source: VCE 2016, Specialist Mathematics Exam 2, Section A, Q16; © VCAA.

MC A cricket ball is hit from the ground at an angle of 30° to the horizontal with a velocity of $20\,\text{ms}^{-1}$. The ball is subject only to gravity and air resistance is negligible.

Given that the field is level, the horizontal distance travelled by the ball, in metres, to the point of impact is

A. $\dfrac{10\sqrt{3}}{g}$

B. $\dfrac{20}{g}$

C. $\dfrac{100\sqrt{3}}{g}$

D. $\dfrac{200\sqrt{3}}{g}$

E. $\dfrac{400}{g}$

Question 2 (2 marks) TECH-ACTIVE

A catapult throws a stone with a speed of 7 m/s from a point at the top of a cliff face 40 metres high to hit a ship which is at a horizontal distance of 20 metres from the base of the cliff. Determine the possible angles of projection, in degrees to 1 decimal place.

Question 3 (4 marks) TECH-ACTIVE

A projectile is fired with an initial speed of $\sqrt{2ga}$ at an angle of α to hit a target at a horizontal distance of a from the point of projection and a vertical distance of b.

a. Show that $a\tan^2(\alpha) - 4a\tan(\alpha) + (4b + a) = 0$. **(2 marks)**

b. Show that it is impossible to hit the target if $4b > 3a$. **(1 mark)**

c. Show that if $4b = 3a$, then $\alpha = \tan^{-1}(2)$. **(1 mark)**

More exam questions are available online.

12.7 Review

 12.7.1 Summary

doc-37625

Hey students! Now that it's time to revise this topic, go online to:

Access the topic summary  Review your results Watch teacher-led videos  Practise exam questions

Find all this and MORE in jacPLUS

12.7 Exercise

Technology free: short answer

1. A particle moves so that at time t its position vector is given by $\underset{\sim}{r}(t) = 6\cos(t)\underset{\sim}{i} + 6\sin(t)\underset{\sim}{j}$, for $0 \le t \le 2\pi$. The particle moves along a curve C.
   a. Determine the coordinates of the point A when $t = \dfrac{\pi}{4}$.
   b. Determine the coordinates of the point B when $t = \dfrac{5\pi}{6}$.
   c. Determine the Cartesian equation of the curve C.
   d. Determine the distance between the points A and B, give your answer in exact form.
   e. If O is the origin, using vectors calculate the angle between OA and OB and hence show that $\cos\left(\dfrac{7\pi}{12}\right) = \dfrac{1}{4}\left(\sqrt{2} - \sqrt{6}\right)$.
   f. Show using calculus, that the distance moved by the particle along the curve C between the points A and B is $\dfrac{7\pi}{2}$.
2. A particle moves so that at time t its position vector is given by $\underset{\sim}{r}(t) = 8\cos(t)\underset{\sim}{i} + 6\sin(t)\underset{\sim}{j}$, for $0 \le t \le 2\pi$. The particle moves along a curve C.
   a. Determine the coordinates of the point A when $t = \dfrac{\pi}{4}$.
   b. Determine the coordinates of the point B when $t = \dfrac{5\pi}{6}$.
   c. Determine the Cartesian equation of the curve C.
3. A particle moves so that at time t its position vector is given by $\underset{\sim}{r}(t) = 8\cos(2t)\underset{\sim}{i} + 6\sin(2t)\underset{\sim}{j}$, for $0 \le t \le \pi$.
   a. Determine the position vector when $t = \dfrac{\pi}{6}$.
   b. Determine the velocity vector when $t = \dfrac{\pi}{6}$.
   c. Determine the times when the position vector is perpendicular to the velocity vector.
   d. Determine if there are any times when the position vector is parallel to the velocity vector.
4. When a kangaroo hops forwards and leaps into the air it is found that the ratio of its range to its maximum height reached is 4. If the kangaroo leaves the ground with a speed of 8 m/s, determine:
   a. the range for its jump
   b. the time intervals between touching the ground.

5. Two particles, P and Q, have position vectors
$\underset{\sim}{p} = 3\cos(2t)\underset{\sim}{i} + 4\sin(2t)\underset{\sim}{j}$ and $\underset{\sim}{q} = 4\sin(nt)\underset{\sim}{i} + 3\cos(nt)\underset{\sim}{j}$ respectively at a time, t seconds, $t \geq 0$.
   a. Determine the Cartesian equation of the paths for both P and Q.
   b. Determine the coordinates where the paths of P and Q cross.

6. A particle moves so that its acceleration vector is given by $\ddot{\underset{\sim}{r}}(t) = -\cos(t)\underset{\sim}{i} - 9\cos(3t)\underset{\sim}{j}$, for $0 \leq t \leq 2\pi$. The initial velocity vector $\dot{\underset{\sim}{r}}(0) = \underset{\sim}{0}$ and the initial position $\underset{\sim}{r}(0) = \underset{\sim}{i} + \underset{\sim}{j}$. Show that the particle moves on part of the curve $y = 4x^3 - 3x$.

Technology active: multiple choice

7. **MC** Two particles, P and Q have position vectors $\underset{\sim}{p} = \left(t^2 - 8t + 15\right)\underset{\sim}{i} + \left(t^2 - 4t + 3\right)\underset{\sim}{j}$ and $\underset{\sim}{q} = \left(t^2 - 5t + 6\right)\underset{\sim}{i} + \left(t^2 - 6t + 8\right)\underset{\sim}{j}$ respectively at a time t seconds, $t \geq 0$. Then
   A. P and Q are never in the same position
   B. both particles move on parabolic paths
   C. the particles paths intersect exactly once
   D. P and Q are in the same position when $t = 2$
   E. P and Q are in the same position when $t = 3$.

8. **MC** A particle moves so that its position vector at a time t, is given by $\underset{\sim}{r}(t) = 2t\underset{\sim}{i} + 2\sqrt{t}\underset{\sim}{j}$ for $t \geq 0$. The distance of the particle from the origin after 4 seconds is equal to
   A. $4\sqrt{5}$　　B. $\frac{3}{2}$　　C. $\frac{5}{2}$　　D. 8　　E. 12

9. **MC** A particle moves so that its position vector at a time t is given by $\underset{\sim}{r}(t) = a\cos(nt)\underset{\sim}{i} + b\sin(mt)\underset{\sim}{j}$ where m and n are positive integers, and a and b are real positive constants. A unit vector in the initial direction of motion of the particle is
   A. $bm\underset{\sim}{j}$　　B. $-\underset{\sim}{i}$　　C. $\underset{\sim}{i}$　　D. $\underset{\sim}{j}$　　E. $-\underset{\sim}{j}$

10. **MC** The position vector of a particle at a time $t \geq 0$ is given by $\underset{\sim}{r}(t) = \left(\cos(t) - \sin(t)\right)\underset{\sim}{i} + \left(\cos(t) + \sin(t)\right)\underset{\sim}{j}$. The speed of the particle at time t is equal to
   A. t^2　　B. t　　C. 1　　D. $\sqrt{2}$　　E. 2

11. **MC** A particle moves on a curve, so that its position at time $t \geq 0$ is given by $\underset{\sim}{r}(t) = 3\cos(2t)\underset{\sim}{i} + 4\sin(2t)\underset{\sim}{j}$. The statement which is false is
   A. The gradient of the curve is $\dfrac{-4}{3\tan(2t)}$.
   B. The domain of the curve is $[-3, 3]$.
   C. The maximum distance of the particle from the origin is 3.
   D. The minimum value of the speed is 6.
   E. The maximum value of the speed is 8.

12. **MC** A particle moves so that its position vector at time t is given by $\underset{\sim}{r}(t) = 4\cos(2t)\underset{\sim}{i} + 2\sin(2t)\underset{\sim}{j}$ for $t \geq 0$. The statement which is true is
   A. The velocity vector is perpendicular to the position vector.
   B. The velocity vector is perpendicular to the acceleration vector.
   C. The speed is constant.
   D. The particle moves in a circle.
   E. The particle moves on an ellipse.

13. **MC** A particle moves in such a way that its velocity vector at a time t is given by $4e^{\frac{t}{2}}\underset{\sim}{i} - 2\sin\left(\frac{t}{2}\right)\underset{\sim}{j}$.
Initially the position vector of the particle is $3\underset{\sim}{i} - 3\underset{\sim}{j}$. The position vector of the particle at a time t is given by

A. $\left(8e^{\frac{t}{2}} - 5\right)\underset{\sim}{i} - \left(4\cos\left(\frac{t}{2}\right) + 3\right)\underset{\sim}{j}$

B. $\left(8e^{\frac{t}{2}} - 5\right)\underset{\sim}{i} - \left(2 + \cos\left(\frac{t}{2}\right)\right)\underset{\sim}{j}$

C. $\left(8e^{\frac{t}{2}} - 5\right)\underset{\sim}{i} + \left(4\cos\left(\frac{t}{2}\right) - 7\right)\underset{\sim}{j}$

D. $\left(1 + 2e^{\frac{t}{2}}\right)\underset{\sim}{i} - \left(\cos\left(\frac{t}{2}\right) - 4\right)\underset{\sim}{j}$

E. $\left(1 + 2e^{\frac{t}{2}}\right)\underset{\sim}{i} + \left(4\cos\left(\frac{t}{2}\right) - 7\right)\underset{\sim}{j}$

14. **MC** A particle moves so that its position vector is given by $\underset{\sim}{r}$ and its acceleration vector is $\ddot{\underset{\sim}{r}}$. If $\underset{\sim}{r} = -4\ddot{\underset{\sim}{r}}$ then

A. the particle moves part of a straight line
B. the particle moves in a circle
C. the particle moves on an ellipse
D. the acceleration vector is perpendicular to the position vector
E. the distance of the particle from the origin is not constant.

15. **MC** A soccer ball is kicked off the ground with an initial velocity of 16 ms^{-1} at an angle of 30°.
The maximum height reached in metres is equal to

A. $\frac{16}{g}$ **B.** $\frac{32}{g}$ **C.** $\frac{64}{g}$ **D.** $\frac{32\sqrt{3}}{g}$ **E.** $\frac{128\sqrt{3}}{g}$

16. **MC** A golf ball is hit off the ground, its position vector, at a time t seconds after being hit, is given by $\underset{\sim}{r}(t) = 15\sqrt{2}\,t\,\underset{\sim}{i} + \left(15\sqrt{2}\,t - \frac{1}{2}gt^2\right)\underset{\sim}{k}$ for $t \geq 0$, where $\underset{\sim}{i}$ is a unit vector in metres horizontally forward, and $\underset{\sim}{k}$ is a unit vector in metres vertically upwards. Students, when analysing the motion of the golf ball, stated some propositions:

- Alicia stated that the golf ball is hit with an initial velocity of 30 m/s for maximum range.
- Betty stated that the golf ball hits the ground again after a time of 4.33 seconds.
- Colin stated that the golf ball reaches a maximum height of 22.96 metres.
- David stated that the golf ball first hits the ground at a distance of 91.84 metres from where it was hit.
- Edward stated that the golf ball travels in a parabolic path.

Then

A. only Alicia, Betty and Edward are correct
B. only Alicia, Betty and Colin are correct
C. only Alicia, Betty, Colin and David are correct
D. only Betty, Colin, David and Edward are correct
E. all of Alicia, Betty, Colin, David and Edward are correct.

Technology active: extended response

17. A particle moves so that its position vector is given by $\underset{\sim}{r}(t) = (2\cos(t) - \cos(3t))\underset{\sim}{i} + (2\sin(t) + \sin(3t))\underset{\sim}{j}$, for $0 \leq t \leq 2\pi$.

a. Determine the gradient of the curve when $t = \frac{\pi}{6}$.
b. Calculate the maximum and minimum values of the speed.
c. Sketch the path of the particle.

18. A shell is fired at 70 m/s and strikes a target on the same horizontal level as the gun at a distance of 480 metres.

a. Determine the two possible angles of projection at which it could have been fired, in degrees to 2 decimal places.

b. If T_1 and T_2 are the respective times of flight, evaluate the ratio $T_1 : T_2$.

c. If H_1 and H_2 are the respective greatest heights reached on each path, evaluate the ratio $H_1 : H_2$.

19. Complete the following.

a. Two bodies are projected at the same time from the same point, in the same vertical plane, with the same speed of projection but at different angles of projection. If the times of flight are 1 and 2 seconds, determine the range in each case if it is the same, in metres to 1 decimal place.

b. Two bodies are projected at the same time from the same point, in the same vertical plane, with the same speed of projection but at different angles of projection. If the times of flight are T_1 and T_2 show that the range is given by $\frac{1}{2} g T_1 T_2$.

20. A particle moves so that its position vector at time t is given by $\underset{\sim}{r}(t) = \sqrt{t} \cos(t)\underset{\sim}{i} + \sqrt{t} \sin(t)\underset{\sim}{j}$, for $0 \le t \le 9$.

a. Determine the distance of the particle from the origin at time t.

b. Determine the speed of the particle at time t.

c. Calculate the distance of the particle from the origin when $t = 4$.

d. Calculate, correct to 2 decimal places, the distance moved along the curve from $t = 0$ to $t = 4$.

e. Sketch the path of the particle.

f. Show that $\underset{\sim}{r}(t) \cdot \dot{\underset{\sim}{r}}(t) = \frac{1}{2}$.

g. Show that the gradient of the curve at time t is given by $\frac{\tan(t) + 2t}{1 - 2t\tan(t)}$.

h. Determine the times, correct to 2 decimal places, when the gradient is parallel to the x-axis.

i. Determine the times, correct to 2 decimal places, when the curve has vertical tangents.

12.7 Exam questions

Question 1 (8 marks) TECH-FREE

Source: *VCE 2021, Specialist Mathematics Exam 1, Q9; © VCAA.*

Let $\underset{\sim}{r}(t) = \left(-1 + 4\cos(t)\right)\underset{\sim}{i} + \frac{2}{\sqrt{3}}\sin(t)\underset{\sim}{j}$ and $\underset{\sim}{s}(t) = \left(3\sec(t) - 1\right)\underset{\sim}{i} + \tan(t)\underset{\sim}{j}$ be the position vectors relative to a fixed point O of particle A and particle B respectively for $0 \le t \le c$, where c is a positive real constant.

a. i. Show that the Cartesian equation of the path of particle A is $\frac{(x+1)^2}{16} + \frac{3y^2}{4} = 1$. **(1 mark)**

ii. Show that the Cartesian equation of the path of particle A in the first quadrant can be written as $y = \frac{\sqrt{3}}{6}\sqrt{-x^2 - 2x + 15}$. **(1 mark)**

b. i. Show that the particles A and B will collide. **(1 mark)**

ii. Hence, find the coordinates of the point of collision of the two particles. **(1 mark)**

c. i. Show that $\frac{d}{dx}\left(8\arcsin\left(\frac{x+1}{4}\right)+\frac{(x+1)\sqrt{-x^2-2x+15}}{2}\right)=\sqrt{-x^2-2x+15}$. **(2 marks)**

ii.

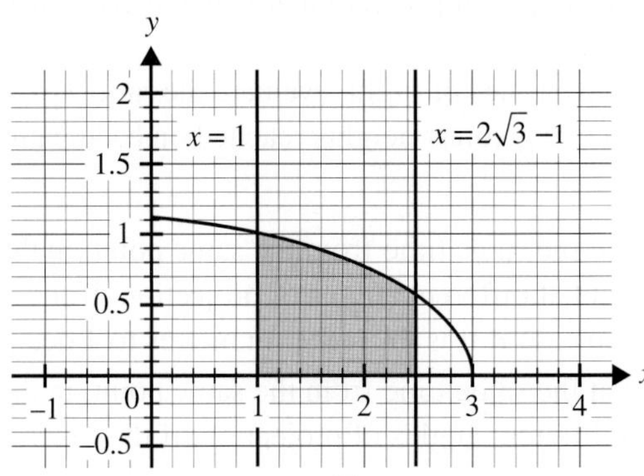

Hence, find the area bounded by the graph of $y=\frac{\sqrt{3}}{6}\sqrt{-x^2-2x+15}$, the x-axis and the lines $x=1$ and $x=2\sqrt{3}-1$, as shown in the diagram above. Give your answer in the form $\frac{a\sqrt{3}\pi}{b}$, where a and b are positive integers. **(2 marks)**

Question 2 (11 marks) TECH-ACTIVE

Source: *VCE 2021, Specialist Mathematics Exam 2, Section B, Q4; © VCAA.*

A car that performs stunts moves along a track, as shown in the diagram below. The car accelerates from rest at point A, is launched into the air by the ramp BO and lands on a second section of track at or beyond point C. This second section of track is inclined at 10° to the horizontal.

Due to a tailwind, the effect of air resistance is negligible. Point O is taken as the origin of a Cartesian coordinate system and all displacements are measured in metres. Point C has the coordinates (16, 4).

At point O, the speed of the car is u ms^{-1} and it takes off at an angle of θ to the horizontal direction.

After the car passes point O, it follows a trajectory where the position of the car's rear wheels relative to point O, at time t seconds after passing point O, is given by

$$\underset{\sim}{r}(t)=ut\cos(\theta)\underset{\sim}{i}+\left(ut\sin(\theta)-\frac{1}{2}gt^2\right)\underset{\sim}{j}$$

until the car lands on the second section of track that starts at point C.

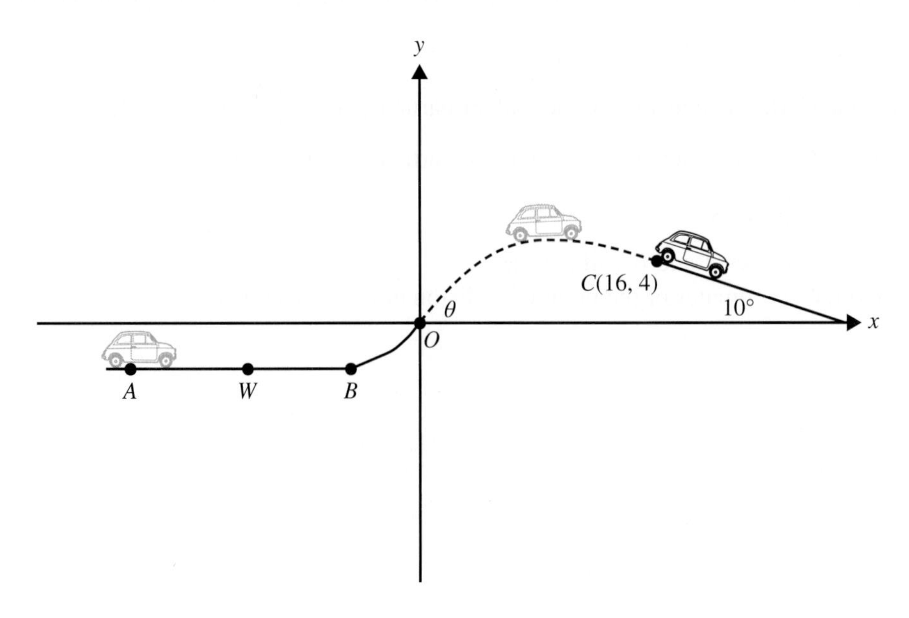

a. Show that the path of the rear wheels of the car, while in the air, is given in Cartesian form by $y = x\tan(\theta) - \dfrac{4.9x^2}{u^2\cos^2(\theta)}$. **(1 mark)**

b. If $\theta = 30°$, find the minimum speed, in ms^{-1}, that the car must reach at the point O for the rear wheels to land on the second section of track at or beyond point C. Give your answer correct to two decimal places. **(2 marks)**

c. The ramp BO is constructed so that the angle θ can be varied.
For what values of θ and u will the path of the rear wheels of the car join up **smoothly** with the beginning of the second section of track at point C? Give your answer for θ in degrees, correct to the nearest degree, and give your answer for u in ms^{-1}, correct to one decimal place. **(3 marks)**

The car accelerates from rest along the horizontal section of track AB, where its acceleration, $a\ \text{ms}^{-2}$ after it has travelled s metres from point A, is given by $a = \dfrac{60}{v}$, where v is its speed at s metres.

d. Show that v in terms of s is given by $v = (180s)^{\frac{1}{3}}$. **(2 marks)**

e. After the car leaves point A, it accelerates to reach a speed of $20\ \text{ms}^{-1}$ at point B. However, if the stunt is called off, the car immediately brakes and reduces its speed at a rate of $9\ \text{ms}^{-2}$. It is only safe to call off the stunt if the car can come to rest at or before point B. Point W is the furthest point along the section AB at which the stunt can be called off.
How far is point W from point B? Give your answer in metres, correct to one decimal place. **(3 marks)**

Question 3 (14 marks) TECH-ACTIVE

Source: VCE 2020, Specialist Mathematics Exam 2, Section B, Q4; © VCAA.

A pilot is performing at an air show. The position of her aeroplane at time t relative to a fixed origin O is given by $\underset{\sim}{r}_A(t) = \left(450 - 150\sin\left(\dfrac{\pi t}{6}\right)\right)\underset{\sim}{i} + \left(400 - 200\cos\left(\dfrac{\pi t}{6}\right)\right)\underset{\sim}{j}$, where $\underset{\sim}{i}$ is a unit vector in a horizontal direction, $\underset{\sim}{j}$ is a unit vector vertically up, displacement components are measured in metres and time t is measured in seconds where $t \geq 0$.

a. Find the maximum speed of the aeroplane. Give your answer in ms^{-1}. **(3 marks)**

b. i. Use $\underset{\sim}{r}_A(t)$ to show that the Cartesian equation of the path of the aeroplane is given by
$\dfrac{(x-450)^2}{22500} + \dfrac{(y-400)^2}{40\,000} = 1.$ **(2 marks)**

ii. Sketch the path of the aeroplane on the axes provided below. Label the position of the aeroplane when $t = 0$, using coordinates, and use an arrow to show the direction of motion of the aeroplane. **(3 marks)**

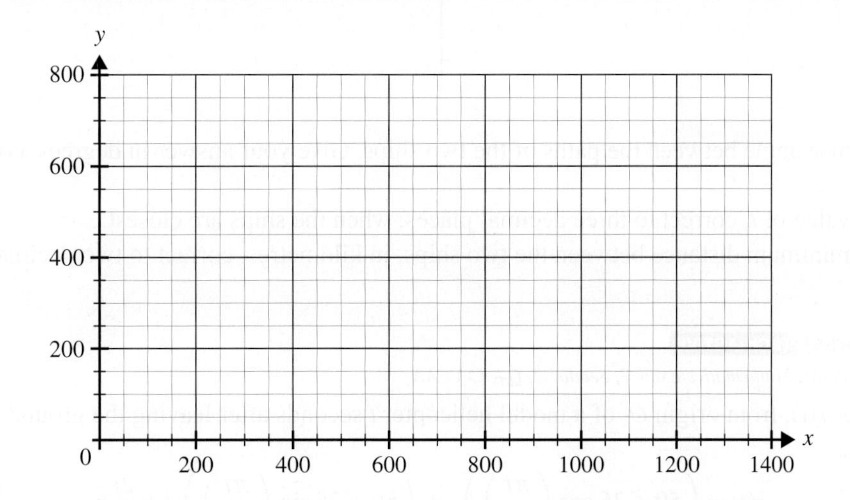

A friend of the pilot launches an experimental jet-powered drone to take photographs of the air show. The position of the drone at time t relative to the fixed origin is given by $\underset{\sim}{r}_D(t) = (30t)\underset{\sim}{i} + \left(-t^2 + 40t\right)\underset{\sim}{j}$ where t is in seconds and $0 \leq t \leq 40$, $\underset{\sim}{i}$ is a unit vector in the same horizontal direction, $\underset{\sim}{j}$ is a unit vector vertically up, and displacement components are measured in metres.

c. Sketch the path of the drone on the axes provided in part **b.ii**. Using coordinates, label the point where the path of the drone crosses the path of the aeroplane, correct to the nearest metre. **(3 marks)**

d. Determine whether the drone will make contact with the aeroplane. Give reasons for your answer. **(3 marks)**

Question 4 (10 marks) TECH-ACTIVE

Source: *VCE 2016, Specialist Mathematics Exam 2, Section B, Q4; © VCAA.*

Two ships, A and B, are observed from a lighthouse at origin O. Relative to O, their position vectors at time t hours after midday are given by

$$\underset{\sim}{r}_A = 5(1-t)\underset{\sim}{i} + 3(1+t)\underset{\sim}{j}$$
$$\underset{\sim}{r}_B = 4(t-2)\underset{\sim}{i} + (5t-2)\underset{\sim}{j}$$

where displacements are measured in kilometres.

a. Show that the two ships will not collide, clearly stating your reason. **(2 marks)**

b. Sketch and label the path of each ship on the axes below. Show the direction of motion of each ship with an arrow. **(3 marks)**

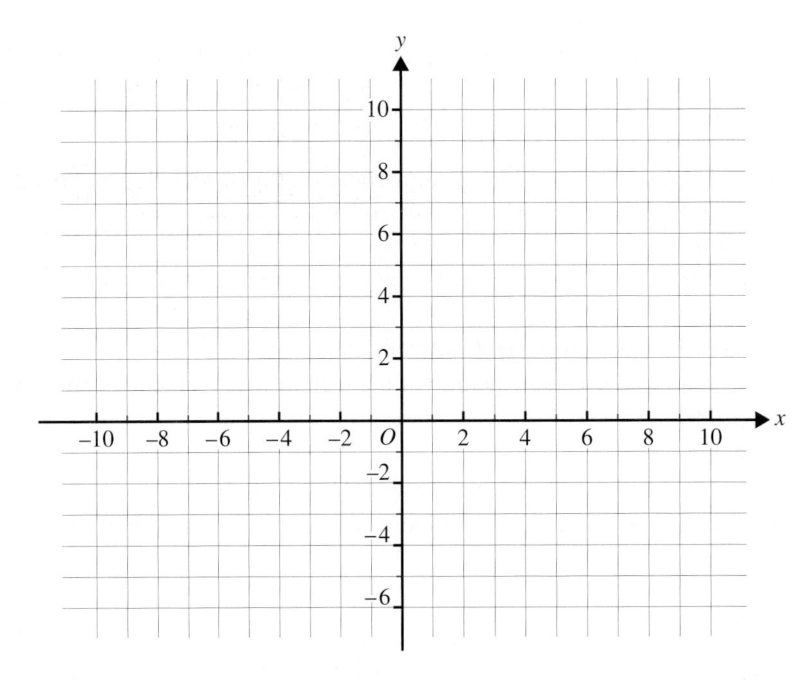

c. Find the obtuse angle between the paths of the two ships. Give your answer in degrees, correct to one decimal place. **(2 marks)**

d. i. Find the value of t, correct to three decimal places, when the ships are closest. **(2 marks)**

ii. Find the minimum distance between the two ships, in kilometres, correct to two decimal places. **(1 mark)**

Question 5 (12 marks) TECH-ACTIVE

Source: *VCE 2015, Specialist Mathematics Exam 2, Section 2, Q4; © VCAA.*

The position vector $\underset{\sim}{r}(t)$, from origin O, of a model helicopter t seconds after leaving the ground is given by

$$\underset{\sim}{r}(t) = \left(50 + 25\cos\left(\frac{\pi t}{30}\right)\right)\underset{\sim}{i} + \left(50 + 25\sin\left(\frac{\pi t}{30}\right)\right)\underset{\sim}{j} + \frac{2t}{5}\underset{\sim}{k}$$

where $\underset{\sim}{i}$ is a unit vector to the east, $\underset{\sim}{j}$ is a unit vector to the north and $\underset{\sim}{k}$ is a unit vector vertically up. Displacement components are measured in metres.

a. i. Find the time, in seconds, required for the helicopter to gain an altitude of 60 m. **(1 mark)**
   ii. Find the angle of elevation from O of the helicopter when it is at an altitude of 60 m. Give your answer in degrees, correct to the nearest degree. **(2 marks)**

b. After how many seconds will the helicopter first be directly above the point of take-off? **(1 mark)**

c. Show that the velocity of the helicopter is perpendicular to its acceleration. **(3 marks)**

d. Find the speed of the helicopter in ms^{-1}, giving your answer correct to two decimal places. **(2 marks)**

e. A treetop has position vector $\underset{\sim}{r} = 60\underset{\sim}{i} + 40\underset{\sim}{j} + 8\underset{\sim}{k}$.
   Find the distance of the helicopter from the treetop after it has been travelling for 45 seconds.
   Give your answer in metres, correct to one decimal place. **(3 marks)**

More exam questions are available online.

Answers

Topic 12 Vector calculus

12.2 Position vectors as functions of time

12.2 Exercise

1. a. $\sqrt{205}$
   b. $\sqrt{10t^2 - 10t + 5}$
   c. $\dfrac{\sqrt{10}}{2}$
2. a. $5\sqrt{5}$ b. 9
3. a. $t = 5, (16, 40)$ b. $(36, 180)$ c. $5\sqrt{5}$
4. a. $t = 4, (10, 7)$ b. $(-2, 13)$ c. $\sqrt{41}$
5. a. $5\sqrt{2}$
   b. $\sqrt{5t^2 - 10t + 10}$
   c. $\sqrt{5}$
6. a. $5\sqrt{5}$ b. $5\sqrt{t^2 + 1}$ c. 5
7. a. $\sqrt{41}$ b. $\sqrt{5}$ c. $0, 2$
8. a. $\dfrac{9\sqrt{2}}{2}$ b. 8
9. $a = 4, b = -3, c = 1, d = 3$
10. $a = 2, b = 5, c = 1, d = -3$
11. a. $t = 15, (2, 5)$ b. $(8, -7)$ c. 15
12. a. $t = 2, (-2, 6); t = 5, (19, 12)$
    b. $8\sqrt{26}$
13. a. $t = 12, (53, 74)$
    b. $(89, 56)$
    c. $80\sqrt{5}$
14. a. $3\underset{\sim}{j}, 6\underset{\sim}{i} + 3\underset{\sim}{j}, 3\underset{\sim}{j}$
    b. $(x-3)^2 + (y-3)^2 = 9$; domain $[0, 6]$, range $[0, 6]$

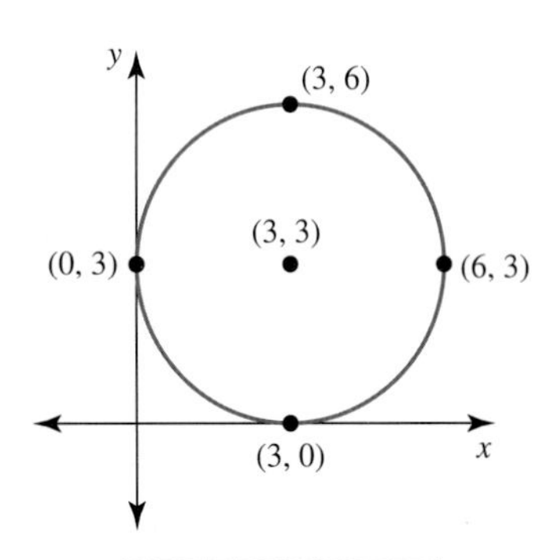

    c. $3\sqrt{3 + 2\sin(t) - 2\cos(t)}$
    d. $3\sqrt{2} - 3$
15. a. $6\underset{\sim}{i} + 4\underset{\sim}{j}, -2\underset{\sim}{i} + 4\underset{\sim}{j}, 6\underset{\sim}{i} + 4\underset{\sim}{j}$
    b. $(x-2)^2 + (y-4)^2 = 16$; domain $[-2, 6]$, range $[0, 8]$

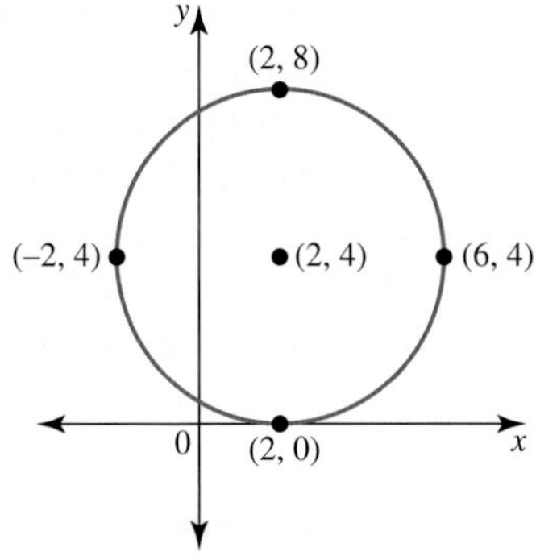

    c. $2\sqrt{9 + 4\cos(t) + 8\sin(t)}$
    d. $2\sqrt{5} - 4$
16. $\sqrt{2}$
17. $\sqrt{2}$
18. The particles collide at $t = 0, 2\pi, (a, 0)$ and $t = \dfrac{\pi}{2}, (0, a)$.
19. The paths intersect at $(a, 0)$ and $(0, a)$.
20. $\dfrac{14}{13}$
21. $\dfrac{ad - bc}{\sqrt{a^2 + c^2}}$
22. 3.397
23. 0.844

12.2 Exam questions

Note: Mark allocations are available with the fully worked solutions online.

1. $a = \dfrac{\pi - 1}{3}$
2. a. $y_A = x^2 - 1,\ x \geq 1$
   $y_B = x + 3,\ x \geq 0$
   b. Sample responses can be found in the worked solutions in the online resources.
   c. $(2.562, 5.562)$
3. B

12.3 Differentiation of vectors

12.3 Exercise

1. $\underset{\sim}{j}$
2. $\dfrac{1}{2}\left(-\sqrt{3}\underset{\sim}{i} + \underset{\sim}{j}\right)$
3. a. $e^{-2t}(1 - 2t)\underset{\sim}{i} + e^{2t}(1 + 2t)\underset{\sim}{j}$
   b. $2e$
   c. $4(t-1)e^{-2t}\underset{\sim}{i} + 4(t+1)e^{2t}\underset{\sim}{j}$
4. 1 m/s
5. a. $8t^3\underset{\sim}{i} - 8\sin(2t)\underset{\sim}{j} - 12e^{-2t}\underset{\sim}{k}$
   b. $24t^2\underset{\sim}{i} - 16\cos(2t)\underset{\sim}{j} + 24e^{-2t}\underset{\sim}{k}$
6. $\dfrac{1}{2}\sqrt{\pi^4 + 1024e^{-4}}$
7. a. $(4, 1), (4, 5)$

b. $\dfrac{(x-4)^2}{9}+\dfrac{(y-3)^2}{4}=1$; domain $[1, 7]$, range $[1, 5]$

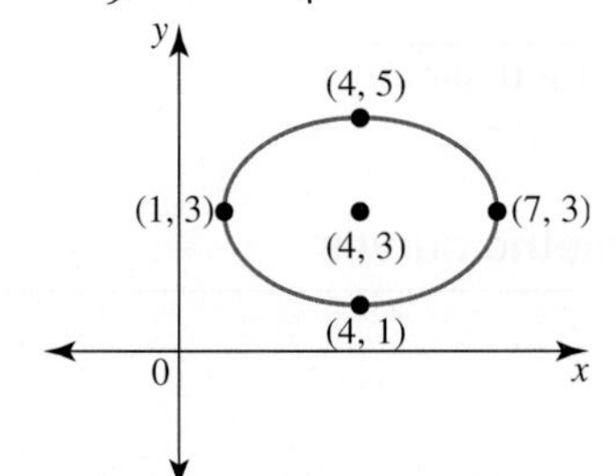

c. $\left|\dot{\underset{\sim}{r}}(t)\right|_{\max}=4$, $\left|\dot{\underset{\sim}{r}}(t)\right|_{\min}=3$

8. a. $\dfrac{x^2}{9}-\dfrac{y^2}{4}=1$; domain $|x|\geq 3$, range R,

asymptotes $y=\pm\dfrac{2x}{3}$

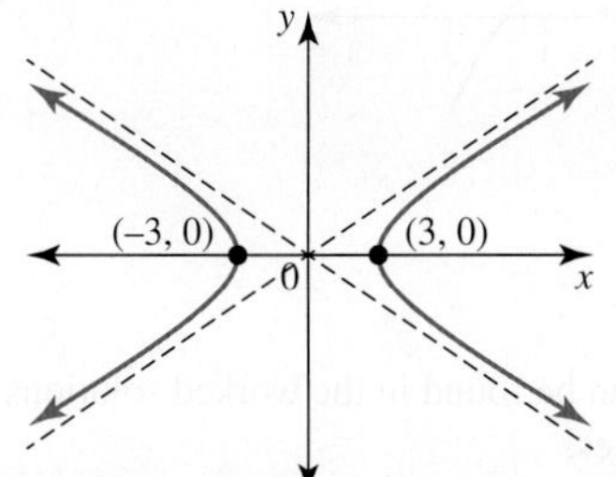

b. $\left(2\sqrt{3}, \dfrac{2\sqrt{3}}{3}\right), \left(-2\sqrt{3}, -\dfrac{2\sqrt{3}}{3}\right)$

9. a. 2.45 s b. 12.25 m

c. 13 m/s, 67.38° d. 7.35 m

e. $y=-\dfrac{x}{250}(49x-600)$

10. a. $\dfrac{1}{\sqrt{17}}(\underset{\sim}{i}+4\underset{\sim}{j})$ b. $\dfrac{1}{\sqrt{145}}(\underset{\sim}{i}+12\underset{\sim}{j})$

c. $\dfrac{1}{\sqrt{82}}(9\underset{\sim}{i}+\underset{\sim}{j})$ d. $\dfrac{1}{\sqrt{34}}(3\underset{\sim}{i}+5\underset{\sim}{j})$

11. a. $\dfrac{1}{\sqrt{2}}(-\underset{\sim}{i}+\underset{\sim}{j})$ b. $-\dfrac{1}{\sqrt{5}}(\underset{\sim}{i}+2\underset{\sim}{j})$

12. a. $\dot{\underset{\sim}{r}}(t)=2t\underset{\sim}{i}-\dfrac{1}{(1+t)^2}\underset{\sim}{j}$, $\ddot{\underset{\sim}{r}}(t)=2\underset{\sim}{i}+\dfrac{2}{(1+t)^3}\underset{\sim}{j}$

b. $\dot{\underset{\sim}{r}}(t)=\dfrac{1}{t}\underset{\sim}{i}+(10t+4)\underset{\sim}{j}$, $\ddot{\underset{\sim}{r}}(t)=\dfrac{-1}{t^2}\underset{\sim}{i}+10\underset{\sim}{j}$

c. $\dot{\underset{\sim}{r}}(t)=-6\sin(2t)\underset{\sim}{i}-8\cos(2t)\underset{\sim}{j}+(12-10t)\underset{\sim}{k}$,

$\ddot{\underset{\sim}{r}}(t)=-12\cos(2t)\underset{\sim}{i}+16\sin(2t)\underset{\sim}{j}-10\underset{\sim}{k}$

13. a. Circle, $x^2+y^2=9$

b. The speed is 6.

c. $\ddot{\underset{\sim}{r}}(t)=-4\underset{\sim}{r}(t)$

d. Sample responses can be found in the worked solutions in the online resources.

14. a. The speed is 12.

b. Sample responses can be found in the worked solutions in the online resources.

15. a. Circle, $x^2+y^2=a^2$

b. Sample responses can be found in the worked solutions in the online resources.

c. Sample responses can be found in the worked solutions in the online resources.

d. Sample responses can be found in the worked solutions in the online resources.

e. The motion is circular.

16. a. Hyperbola, $\dfrac{x^2}{4}-\dfrac{y^2}{9}=1$; domain $|x|\geq 2$, range R,

asymptotes $y=\pm\dfrac{3x}{2}$

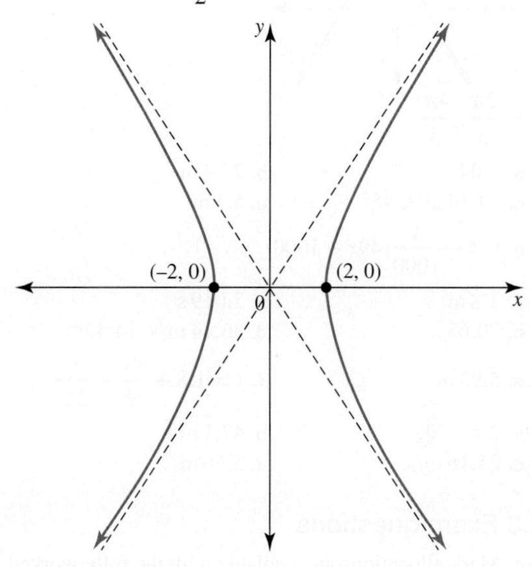

b. $\dfrac{3\sqrt{2}}{2}$

17. a. Straight line, $\dfrac{x}{2}-\dfrac{y}{3}=1$; domain $[2, \infty)$, range $[0, \infty)$

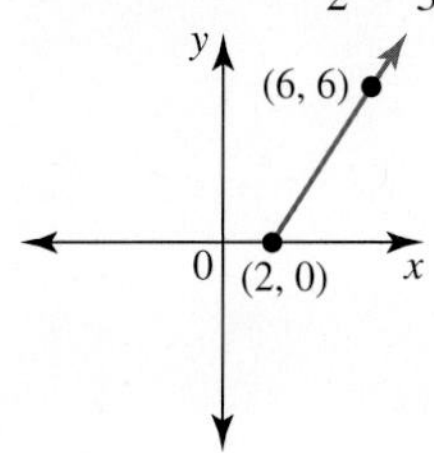

b. $\dfrac{3}{2}$

c. 3

18. a. Hyperbola, $\dfrac{x^2}{9}-\dfrac{y^2}{16}=1$; domain $|x|\geq 3$, range R;

asymptotes $y=\pm\dfrac{4x}{3}$

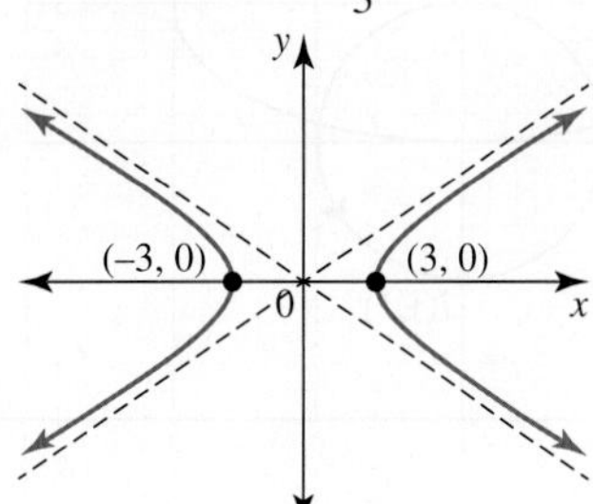

b. $\dfrac{8}{3}$

c. 4

19. a. Hyperbola, $\frac{(x-1)^2}{4} - \frac{(y-4)^2}{9} = 1$; domain $(-\infty, -1] \cup [3, \infty)$, range R, asymptotes $y = -\frac{3x}{2} + \frac{11}{2}, y = \frac{3x}{2} + \frac{5}{2}$

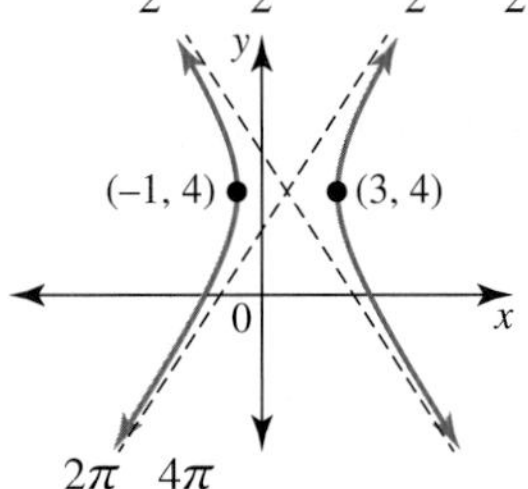

b. $\frac{2\pi}{3}, \frac{4\pi}{3}$

20. a. 2.04 s
b. 20.4 m
c. 14.14 m/s, 45°
d. 5.1 m
e. $y = -\frac{x}{1000}(49x - 1000)$

21. a. 1.8 m
b. 2.019 s
c. 70.65 m
d. 36.14 m/s, 14.42°
e. 5.93 m
f. $y = 1.8 + \frac{9x}{35} - \frac{x^2}{250}$

22. a. 2 s
b. 47.1 m
c. 25.16 m/s
d. 5.66 m

12.3 Exam questions

Note: Mark allocations are available with the fully worked solutions online.

1. a. $\frac{\sqrt{39}}{2}$ m
b. i. $\frac{dy}{dx} = \frac{-3\sin(t)}{4\cos(2t)}$
$\underset{\sim}{r}(\pi) = -3\underset{\sim}{j}$
ii. $\dot{\underset{\sim}{r}}(\pi) = 4\underset{\sim}{i}$
iii. $|\ddot{\underset{\sim}{r}}(\pi)| = 3$
c. $\frac{\pi}{2}$ s

2. a.

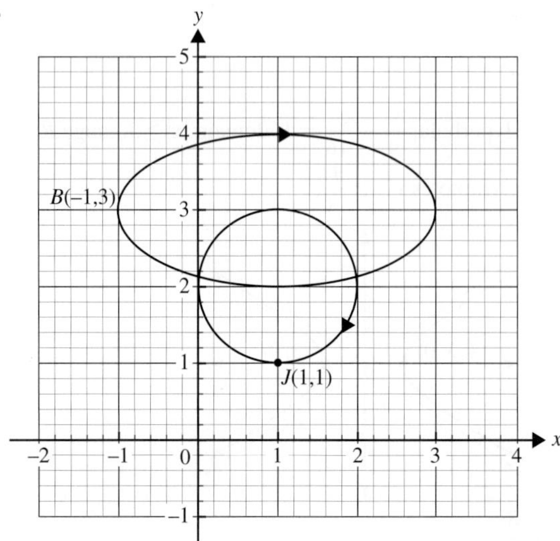

b. i. $t = \pi$ s
ii. $(3, 3)$
c. i. $d(t) = \sqrt{(\sin(t) - 2\cos(t))^2 + (1 + \sin(t) + \cos(t))^2}$
ii. 0.33 km

3. a. $|\dot{\underset{\sim}{r}}(t)| = \sqrt{36\cos^2(2t) + 16\sin^2(2t)}$
b. $\sqrt{31}$ m/s

12.4 Special parametric curves

12.4 Exercise

1. a. $-\frac{\sin(2t)}{\cos(t)}$
b. $t = 0, \pi, 2\pi$; maximum at $(0, 1)$
c.

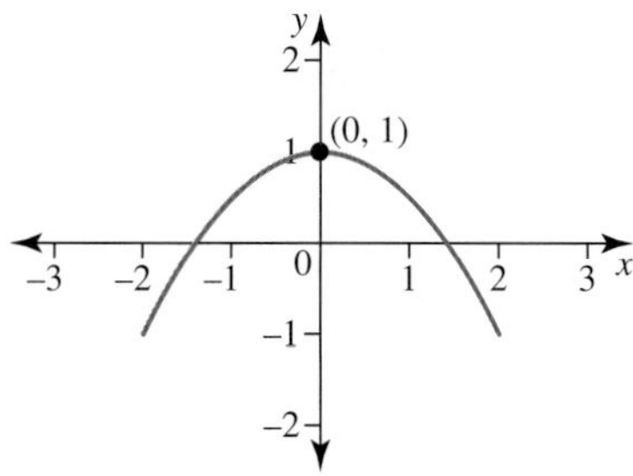

d. $\sqrt{6}, 2$
e. Sample responses can be found in the worked solutions in the online resources.
f. 5.9158
g. Sample responses can be found in the worked solutions in the online resources.
h. Sample responses can be found in the worked solutions in the online resources.

2. a. $-\frac{bm\sin(mt)}{an\cos(nt)}$
b. $\sqrt{a^2n^2\cos^2(nt) + b^2m^2\sin^2(mt)}$
c. Sample responses can be found in the worked solutions in the online resources.
d. Sample responses can be found in the worked solutions in the online resources.
e. Sample responses can be found in the worked solutions in the online resources.

3. a. $\frac{\cos(2t) - \cos(t)}{\sin(2t) + \sin(t)}$
b. $4\cos\left(\frac{3t}{2}\right)$
c. $-6\sin(3t)$
d. $|\dot{\underset{\sim}{r}}(t)|_{max} = 6, |\dot{\underset{\sim}{r}}(t)|_{min} = 2,$
e.

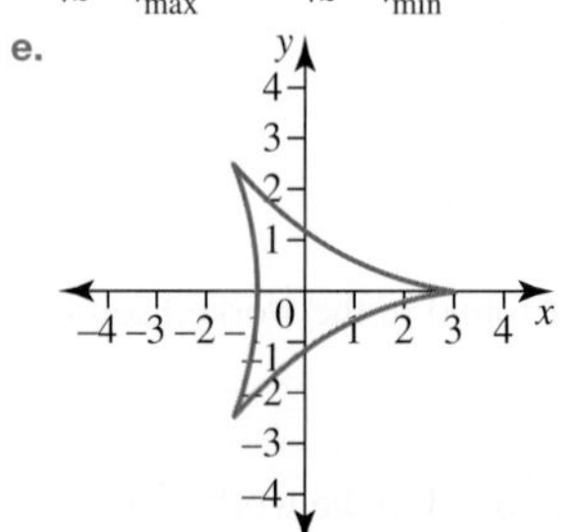

4. a. $9-6\sqrt{2}$
   b. $|\dot{\underset{\sim}{r}}(t)|_{max}=5, |\dot{\underset{\sim}{r}}(t)|_{min}=1$
   c. $|\ddot{\underset{\sim}{r}}(t)|_{max}=7, |\ddot{\underset{\sim}{r}}(t)|_{min}=1$
   d.

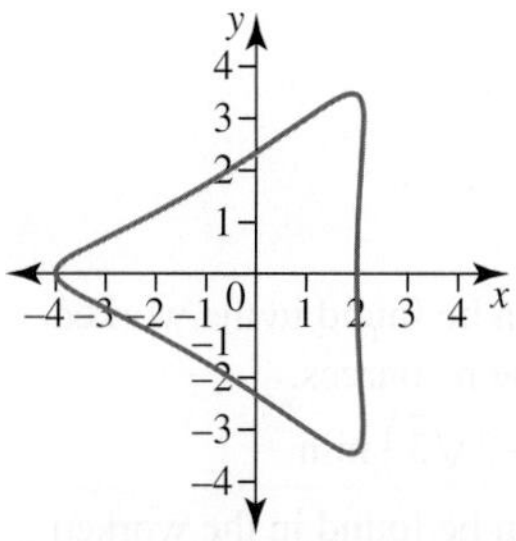

5. a. Sample responses can be found in the worked solutions in the online resources.
   b. $(\pi, 0), (3\pi, 0)$
   c.

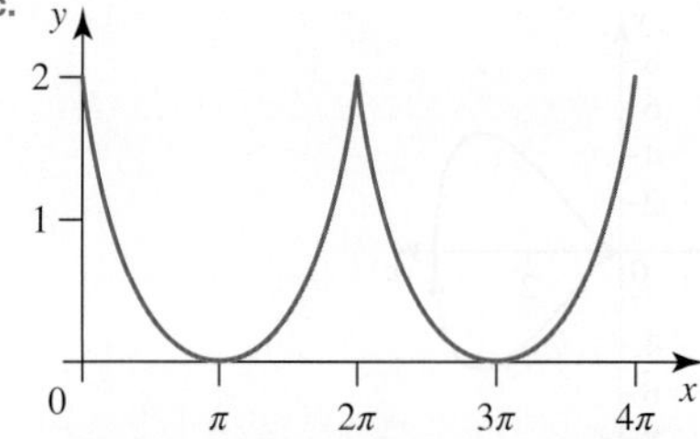

   d. Sample responses can be found in the worked solutions in the online resources.
   e. Sample responses can be found in the worked solutions in the online resources.
   f. Sample responses can be found in the worked solutions in the online resources.
6. a. Sample responses can be found in the worked solutions in the online resources.
   b. Sample responses can be found in the worked solutions in the online resources.
   c. Sample responses can be found in the worked solutions in the online resources.
   d. Sample responses can be found in the worked solutions in the online resources.
   e.

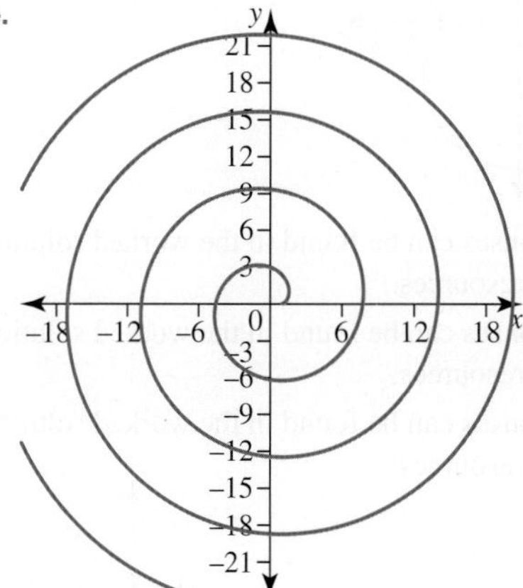

7. a. Sample responses can be found in the worked solutions in the online resources.
   b. Sample responses can be found in the worked solutions in the online resources.
   c.

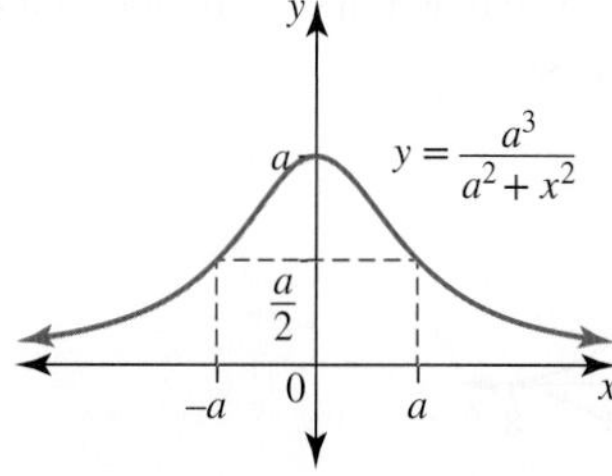

   d. Sample responses can be found in the worked solutions in the online resources.
   e. Sample responses can be found in the worked solutions in the online resources.
8. a. Sample responses can be found in the worked solutions in the online resources.
   b. $\frac{t(2-t^3)}{1-2at^3}$
   c. $\frac{ay-x^2}{y^2-ax}$
   d. i.

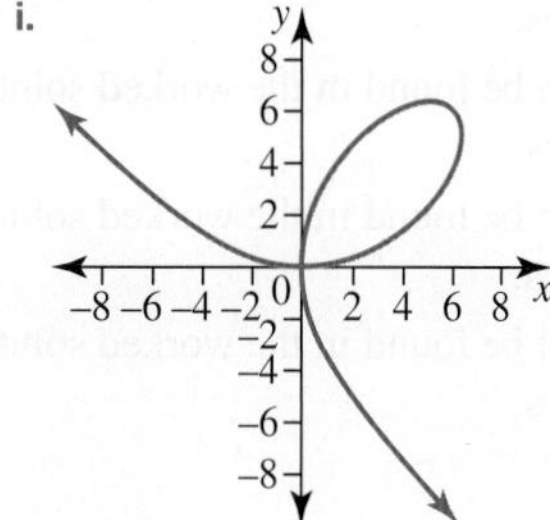

   ii. $2\underset{\sim}{i}+4\underset{\sim}{j}$
   iii. $\frac{\sqrt{41}}{3}$
   iv. $5y-4x-12=0$
9. a. Sample responses can be found in the worked solutions in the online resources.
   b. Sample responses can be found in the worked solutions in the online resources.
   c. Sample responses can be found in the worked solutions in the online resources.
   d.

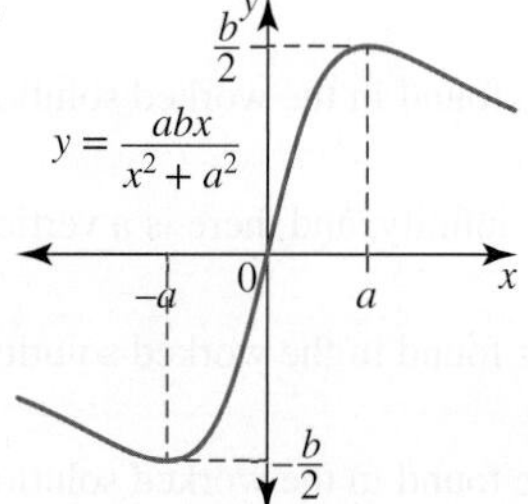

   Sample responses can be found in the worked solutions in the online resources.
   e. Sample responses can be found in the worked solutions in the online resources.
10. a. Sample responses can be found in the worked solutions in the online resources.
   b. Sample responses can be found in the worked solutions in the online resources.

c. Sample responses can be found in the worked solutions in the online resources.

d.

e. Sample responses can be found in the worked solutions in the online resources.

f. Sample responses can be found in the worked solutions in the online resources.

g. Sample responses can be found in the worked solutions in the online resources.

h. $y = a^3 \sin^3(t) - \tan(t)\left(x - a\cos^3(t)\right)$

i. Sample responses can be found in the worked solutions in the online resources.

11. a. Sample responses can be found in the worked solutions in the online resources.

b. Sample responses can be found in the worked solutions in the online resources.

c. Sample responses can be found in the worked solutions in the online resources.

d.

e. Sample responses can be found in the worked solutions in the online resources.

12. a. $(-1, 0), (0, 0), (1, 0)$, domain $[-1, 1]$

b. Sample responses can be found in the worked solutions in the online resources.

c. Sample responses can be found in the worked solutions in the online resources.

d. The gradient approaches infinity, and there is a vertical tangent at $x = \pm 1$.

e. Sample responses can be found in the worked solutions in the online resources.

f. Sample responses can be found in the worked solutions in the online resources.

g.

h. Sample responses can be found in the worked solutions in the online resources.

i. Sample responses can be found in the worked solutions in the online resources.

j. 6.0972

13. a. $a = 0, b = 4$

b. Reflection in the x-axis.

c. i. $\dfrac{2\sqrt{x}(3 - x)}{\sqrt{4 - x}}$

ii. Sample responses can be found in the worked solutions in the online resources.

iii. $(3, 3\sqrt{3})$ max, $\left(3, -3\sqrt{3}\right)$ min

iv. Sample responses can be found in the worked solutions in the online resources.

v. The gradient becomes infinite, and there is a vertical tangent.

vi.

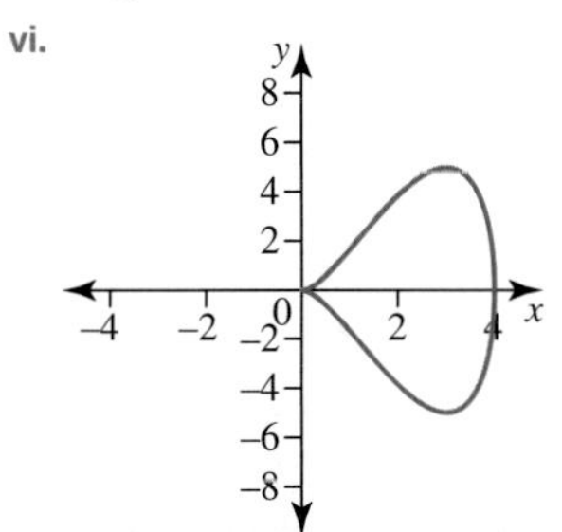

d. i. Sample responses can be found in the worked solutions in the online resources.

ii. $\dfrac{2\pi}{3}$

iii. $\sqrt{3}$

14. a. Sample responses can be found in the worked solutions in the online resources.

b. $t = \dfrac{\pi}{2}, \dfrac{3\pi}{2}$, $(0, 4)$ and $(0, -4)$

c.

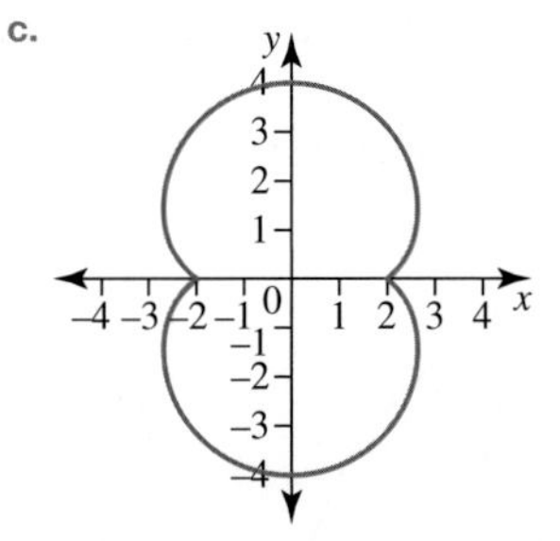

d. Sample responses can be found in the worked solutions in the online resources.

e. Sample responses can be found in the worked solutions in the online resources.

f. Sample responses can be found in the worked solutions in the online resources.

15. a. $4\cos^2\left(\dfrac{t}{2}\right)$

b. $4\cos\left(\dfrac{t}{2}\right)$

c. $0, \pi, 2\pi$

d. $\left|\ddot{\underset{\sim}{r}}(t)\right|_{\max} = 6$, $\left|\ddot{\underset{\sim}{r}}(t)\right|_{\min} = 2$

e. 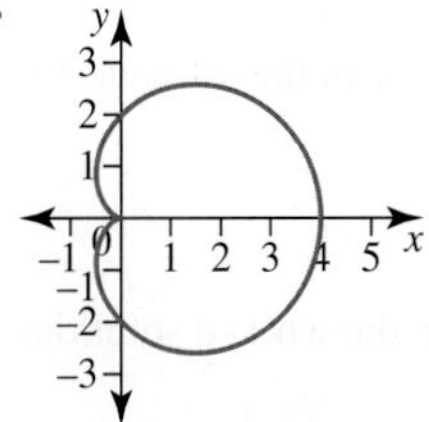

f. Sample responses can be found in the worked solutions in the online resources.

g. Sample responses can be found in the worked solutions in the online resources.

h. Sample responses can be found in the worked solutions in the online resources.

16. Sample responses can be found in the worked solutions in the online resources.

12.4 Exam questions

Note: Mark allocations are available with the fully worked solutions online.

1. B
1. $a = -1,\ b = 0,\ c = 2$
3. $s = \frac{3}{4}$ units

12.5 Integration of vectors

12.5 Exercise

1. $(2t^2 - 4t + 5)\underset{\sim}{i} + (4 - 3t)\underset{\sim}{j}$
2. $\left(5 - 2\cos(3t)\right)\underset{\sim}{i} + (3 - 2e^{-2t})\underset{\sim}{j}$
3. a. $3(1 - e^{-\frac{t}{3}})\underset{\sim}{i} + t^4\underset{\sim}{j}$
   b. $t^2\underset{\sim}{i} + 3\left(1 - \cos(2t)\right)\underset{\sim}{j}$
4. a. $\left(3 + \sin^{-1}\left(\frac{t}{4}\right)\right)\underset{\sim}{i} + \left(5 - \sqrt{t^2 + 9}\right)\underset{\sim}{j}$
   b. $\left(5 + \log_e(2t + 1)\right)\underset{\sim}{i} + \left(13 - \frac{24}{(3t + 2)}\right)\underset{\sim}{j}$
5. $(5 - 2t)\underset{\sim}{i} + (16t - t^4 - 10)\underset{\sim}{j}$
6. $(3t^2 - 5)\underset{\sim}{i} + (t^2 + 8t - 2)\underset{\sim}{j}$
7. a. $(4t^2 + 3)\underset{\sim}{i} + (3t^2 - 2)\underset{\sim}{j}$
   b. $(2t^2 + 3)\underset{\sim}{i} + (t^2 + 8t + 4)\underset{\sim}{j}$
8. a. $\left(\log_e(3t + 1) + 4\right)\underset{\sim}{i} + \left(\frac{4}{2t + 1} - 1\right)\underset{\sim}{j}$
   b. $\left(\log_e(3t + 1) + 3t + 2\right)\underset{\sim}{i} + \left(4t - \frac{1}{(2t + 1)^2}\right)\underset{\sim}{j}$
9. $(x + 2)^2 + (y - 4)^2 = 25$; circle with centre (–2, 4), radius 5
10. $\frac{(x - 5)^2}{9} + \frac{(y + 2)^2}{25} = 1$; ellipse with centre (5, –2), semi-major and minor axes 3, 5
11. a. $y = -\frac{x^2}{45} + \frac{4x}{3} + 2, x \geq 0$
    b. $y = -\frac{49x^2}{250} + 2x + 1, x \geq 0$
12. a. $y = \frac{32}{2 - x} - 5$
    b. $(x - 3)^2 + (y - 5)^2 = 4$
13. a. $(x - 4)^2 + \frac{(y - 5)^2}{4} = 1$
    b. $\frac{(x + 7)^2}{144} + \frac{(y - 3)^2}{16} = 1$
14. a. $y = x^2 - 4$, [–2, 2], [–4, 0]
    b. $y = -3x(x - 2)$, [–1, 3], [–9, 3]
15. a. $t = 3$, (4, 16)
    b. $t = 5$, (–7, 10)
16. $(x - a)^2 + \left(y - b\right)^2 = r^2$
17. $\frac{(x - h)^2}{a^2} + \frac{(y - k)^2}{b^2} = 1$
18. $\underset{\sim}{r}(t) = \left(2\cos(t) - \cos(2t)\right)\underset{\sim}{i} + \left(2\sin(t) - \sin(2t)\right)\underset{\sim}{j}$

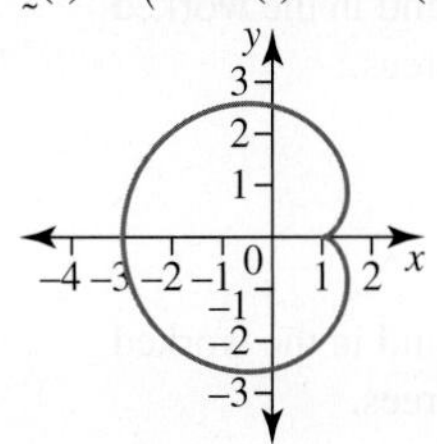

19. $\underset{\sim}{r}(t) = \left(4\cos(t) + \cos(4t)\right)\underset{\sim}{i} + \left(4\sin(t) - \sin(4t)\right)\underset{\sim}{j}$

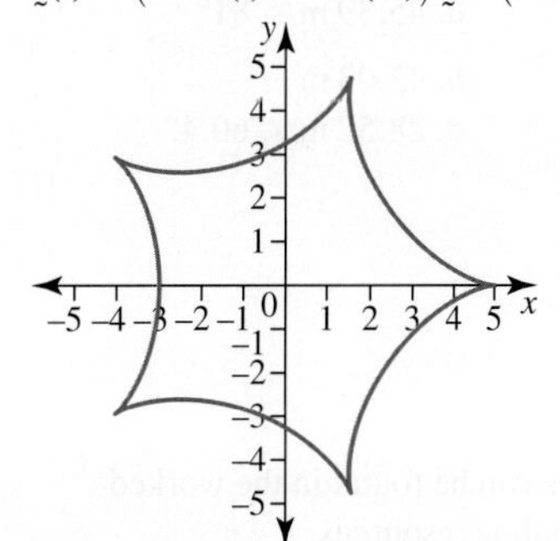

20. $\underset{\sim}{r}(t) = 2\cos(2t)\underset{\sim}{i} + 3\sin(6t)\underset{\sim}{j}$

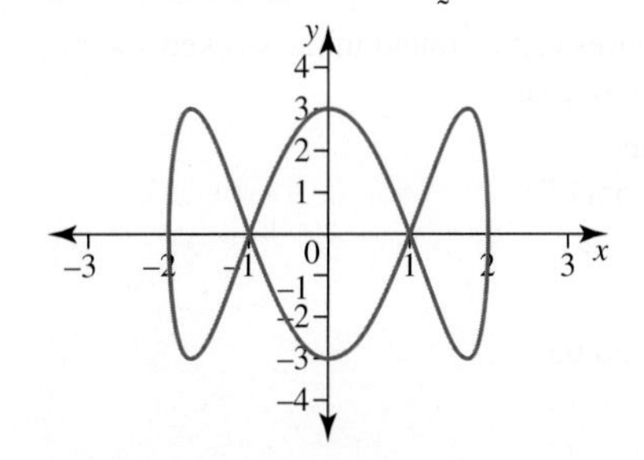

12.5 Exam questions

Note: Mark allocations are available with the fully worked solutions online.

1. 13 m
2. D
3. $y = \frac{1}{x - 4} + 3$

12.6 Projectile motion

12.6 Exercise

1. a. 2.45 s b. 12.24 m c. 7.35 m
2. 10 m/s, 60°
3. i. 2.04 s ii. 35.35 m iii. 5.10 m
4. i. 47.82 m/s ii. 13.65 m iii. 3.34 s
5. a. 4.9 m
   b. 30 m/s
   c. 31.56 m/s, 18.1°
   d. Sample responses can be found in the worked solutions in the online resources.
6. a. 19.6 m b. 83.333 m c. 46.05 m/s, 25.2°
7. a. 3.54 s, 87 metres
   b. 15.61 metres
   c. 30.16 m/s, 35.44°
   d. Sample responses can be found in the worked solutions in the online resources.
8. a. 1.92 s, 34.7 metres
   b. 5.45 metres
   c. 20.86 m/s, 29.76°
   d. Sample responses can be found in the worked solutions in the online resources.
9. a. 5.296 s b. 37.45 m
   c. 2.6 m d. 45.39 m/s, 81°
10. a. 30 m b. 43.03 m
    c. 1.34 m d. 28.52 m/s, 60.4°
11. 86.4°, 2.1°
12. 76.4°, 25°
13. a. i. 185.9 m
       ii. 205 m
       iii. 140 m/s
    b. Sample responses can be found in the worked solutions in the online resources.
14. a. 19.68 m/s b. 39.163 m c. 20.41 m/s, 37.8°

15–18. Sample responses can be found in the worked solutions in the online resources.

19. a. 43.36 metres
    b. 15.47 m/s, 55.77°
20. a. 3 s
    b. 62.06 metres
    c. 26.13 m/s, 36.65°
21. a. 29.13 m/s
    b. 10.5°
    c. 3.4 metres
    d. i. 1.5 seconds
       ii. 42.95 metres
       iii. 28.88 m/s

22–24. Sample responses can be found in the worked solutions in the online resources

25. a. Sample responses can be found in the worked solutions in the online resources.
    b. 42.77 m, 43.7°, 2.96 s

12.6 Exam questions

Note: Mark allocations are available with the fully worked solutions online.

1. D
2. 0° and 26.6°
3. Sample responses can be found in the worked solutions in the online resources

12.7 Review

12.7 Exercise

Technology free: short answer

1. a. $A\left(3\sqrt{2}, 3\sqrt{2}\right)$
   b. $B\left(-3\sqrt{3}, 3\right)$
   c. $x^2 + y^2 = 36$
   d. $3\sqrt{2(\sqrt{6} - \sqrt{2} + 4)}$
   e. $\dfrac{7\pi}{2}$
   f. Sample responses can be found in the worked solutions in the online resources.
2. a. $A\left(4\sqrt{2}, 3\sqrt{2}\right)$
   b. $B\left(-4\sqrt{3}, 3\right)$
   c. $\dfrac{x^2}{64} + \dfrac{y^2}{36} = 1$
3. a. $4\underset{\sim}{i} + 3\sqrt{3}\underset{\sim}{j}$ b. $-8\sqrt{3}\underset{\sim}{i} + 6\underset{\sim}{j}$
   c. $0, \dfrac{\pi}{4}, \dfrac{\pi}{2}, \dfrac{3\pi}{2}, \pi$ d. No
4. a. $\dfrac{64}{g}$ m b. $\dfrac{8\sqrt{2}}{g}$ s
5. a. P: $\dfrac{x^2}{9} + \dfrac{y^2}{16} = 1$, Q: $\dfrac{x^2}{16} + \dfrac{y^2}{9} = 1$
   b. $\left(\dfrac{12}{5}, \dfrac{12}{5}\right), \left(-\dfrac{12}{5}, \dfrac{12}{5}\right), \left(-\dfrac{12}{5}, -\dfrac{12}{5}\right), \left(\dfrac{12}{5}, -\dfrac{12}{5}\right)$
6. Sample responses can be found in the worked solutions in the online resources.

Technology active: multiple choice

7. A
8. A
9. D
10. D
11. C
12. E
13. C
14. B
15. B
16. E

Technology active: extended response

17. a. $\frac{\sqrt{3}}{2}$

 b. $\left|\dot{\underset{\sim}{r}}(t)\right|_{max} = 5$, $\left|\dot{\underset{\sim}{r}}(t)\right|_{min} = 1$

 c.

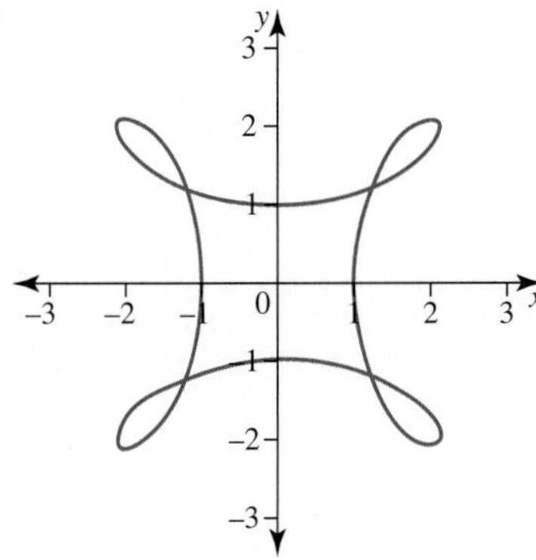

18. a. 36.87°, 53.13°

 b. 3:4

 c. 9:16

19. a. 9.8 m

 b. Sample responses can be found in the worked solutions in the online resources.

20. a. $\sqrt{t}$

 b. $\sqrt{\frac{1+4t^2}{4t}}$

 c. 2

 d. 6.08

 e.

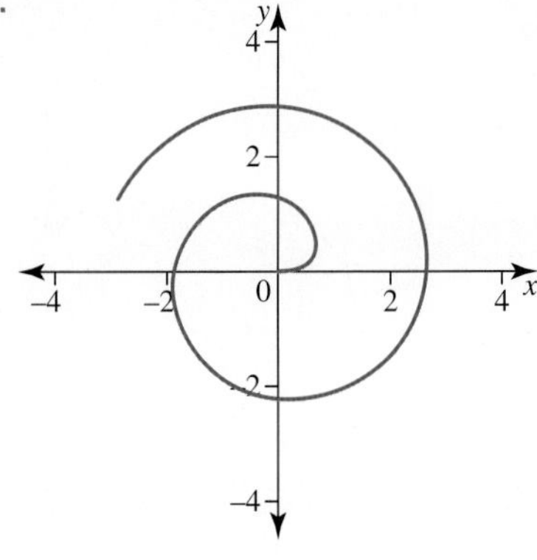

 f. Sample responses can be found in the worked solutions in the online resources.

 g. Sample responses can be found in the worked solutions in the online resources.

 h. 1.84, 4.82, 7.92

 i. 0.65, 3.29, 6.36

12.7 Exam questions

Note: Mark allocations are available with the fully worked solutions online.

1. a. Sample responses can be found in the worked solutions in the online resources.

 b. i. Sample responses can be found in the worked solutions in the online resources.

  ii. $\left(2\sqrt{3}-1, \frac{\sqrt{3}}{3}\right)$

 c. i. Sample responses can be found in the worked solutions in the online resources.

  ii. $A = \frac{2\sqrt{3}\pi}{9}$

2. a. Sample responses can be found in the worked solutions in the online resources.

 b. 17.87 m/s

 c. $u = 16.4$ m/s, $\theta = 34°$

 d. Sample responses can be found in the worked solutions in the online resources.

 e. 16.4 m

3. a. $\frac{100\pi}{3}$ ms^{-1}

 b. i. Sample responses can be found in the worked solutions in the online resources.

  ii. See graph at the bottom of the page*

---

*3 b ii.

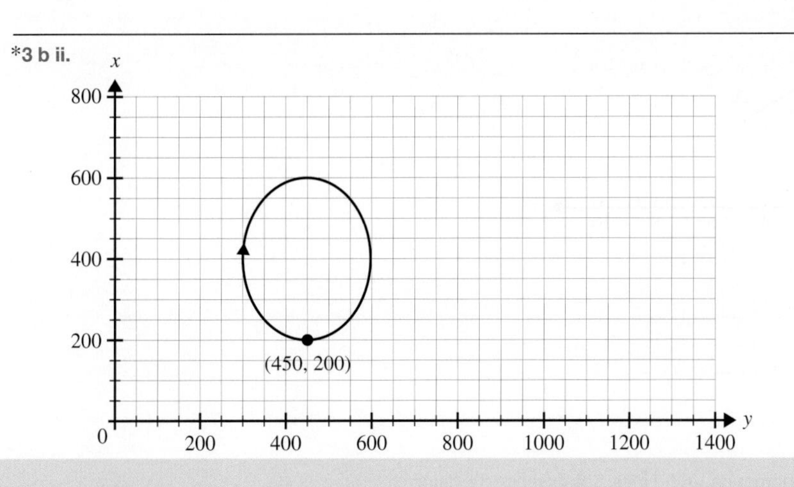

c. See graph at the bottom of the page*

d. The drone and aeroplane's paths cross, but at different times, therefore they will not collide.

4. a. Sample responses can be found in the worked solutions in the online resources.

b. See graph at the bottom of the page*

c. $\theta = 97.7°$

d. i. $t = 1.494$ hours ii. 2.06 km

5. a. i. 150 s ii. 47°

b. 60 s

c. Sample responses can be found in the worked solutions in the online resources.

d. $2.65\ \text{ms}^{-1}$

e. 20.6 m

*3 c.

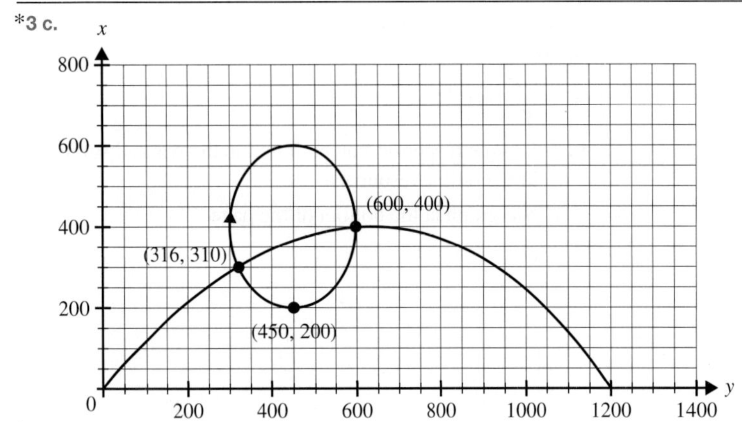

*4 b.

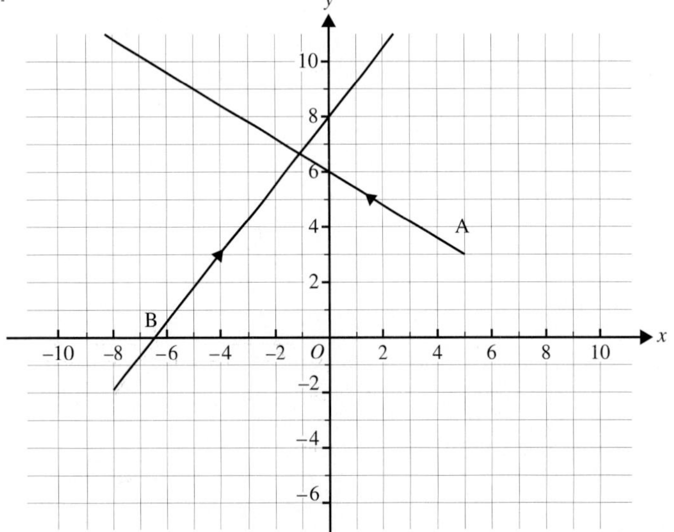

13 Probability and statistics

LEARNING SEQUENCE

Fully worked solutions for this topic are available online.

13.1 Overview

Hey students! Bring these pages to life online

Watch videos

Engage with interactivities

Answer questions and check results

Find all this and MORE in jacPLUS

13.1.1 Introduction

Confidence intervals, first introduced by the Polish mathematician Jerzy Neyman (1894–1981) in a paper published in 1937, give an indication of the likelihood that a particular sample contains a population parameter. His work on experiments and statistics eventually formed the basis for hypothesis testing, and the method adopted by the Federal Drug Authority (FDA) to test new medicines today. In this topic we will explore the uses of confidence intervals, with a particular focus on their application in hypothesis testing.

KEY CONCEPTS

This topic covers the following point from the VCE Mathematics Study Design:

- for n independent identically distributed random variables $X_1, X_2 \dots X_n$ each with mean μ and variance σ^2:
  - $\mathrm{E}(X_1 + X_2 + \dots + X_n) = n\mu$
  - $\mathrm{Var}(X_1 + X_2 + \dots + X_n) = n\sigma^2$
- for n independent random variables $X_1, X_2 \dots X_n$ and real numbers $a_1, a_2 \dots a_n$:
  - $\mathrm{E}(a_1X_1 + a_2X_2 + \dots + a_nX_n) = a_1\mathrm{E}(X_1) + a_2\mathrm{E}(X_2) + \dots + a_n\mathrm{E}(X_n)$
  - $\mathrm{Var}(a_1X_1 + a_2X_2 + \dots + a_nX_n) = a_1^2\mathrm{Var}(X_1) + a_2^2\mathrm{Var}(X_2) + \dots + a_n^2\mathrm{Var}(X_n)$
- for n normally distributed independent random variables $X_1, X_2 \dots X_n$ and real numbers $a_1, a_2 \dots a_n$ the random variable $a_1X_1 + a_2X_2 + \dots + a_nX_n$ is also normally distributed.
- the concept of the sample mean $\overline{X}$ as a random variable whose value varies between samples where X is a random variable with mean μ and the standard deviation σ
- simulation of repeated random sampling, from a variety of distributions and a range of sample sizes, to illustrate properties of the distribution of $\overline{X}$ across samples of a fixed size n including its mean μ its standard deviation $\frac{\sigma}{\sqrt{n}}$ (where μ and σ are the mean and standard deviation of X respectively) and its approximate normality if n is large
- determination of confidence intervals for means and the use of simulation to illustrate variations in confidence intervals between samples and to show that the likelihood of a confidence interval containing μ depends on the level of confidence chosen in the determination of the interval
- construction of an approximate confidence interval, $\left(\overline{x} - z\frac{\sigma}{\sqrt{n}},\ \overline{x} + z\frac{\sigma}{\sqrt{n}}\right)$ where σ is the population standard deviation and z is the appropriate quantile for the standard normal distribution or construction of an approximate confidence interval $\left(\overline{x} - z\frac{s}{\sqrt{n}},\ \overline{x} + z\frac{s}{\sqrt{n}}\right)$ where s is the sample standard deviation and z is the appropriate quantile for the standard normal distribution, and n is large ($n \geq 30$ in many practical contexts)

- concepts of null hypothesis, H_0, and alternative hypotheses, H_1, test statistic
- level of significance and p-value
- formulation of hypotheses and making a decision concerning a population mean based on:
  - a random sample from a normal population of known variance
  - a large random sample from any population
- 1-tail and 2-tail tests
- interpretation of the results of a hypothesis test in the context of the problem
- hypothesis test, relating the formulation, conduct, errors and results in terms of conditional probability.

13.2 Linear combinations of random variables

LEARNING INTENTION

At the end of this subtopic you should be able to:

- calculate the mean and variance of linear combinations of random variables
- calculate probabilities from a linear combination of independent normally distributed variables.

13.2.1 Discrete and continuous random variables

In Mathematical Methods, you have studied discrete and continuous random variables. In this topic, we will be expanding on this foundation.

Discrete random variables

Discrete random variables have a discrete set of possible values. The possible outcomes can therefore be listed and counted. The sum of the probabilities of all of the outcomes is 1.

Recall that the mean and variance of a discrete random variable X are defined as follows.

Mean and variance of discrete random variables

For a discrete random variable X, with outcomes $x_1, x_2, \ldots, x_n$:

$$\mathrm{E}(X) = \mu = \sum_{1}^{n} x_n \Pr(X = x_n)$$

and

$$\mathrm{Var}(X) = \sigma^2 = \mathrm{E}\left(X^2\right) - [\mathrm{E}(X)]^2$$

where:

$$\mathrm{E}(X^2) = \sum_{1}^{n} x_n^2 \Pr(X = x_n)$$

Continuous random variables

Recall that a random variable X can be called continuous if its set of possible values is an entire interval of numbers. That is, for some $A < B$, any number x between A and B is possible.

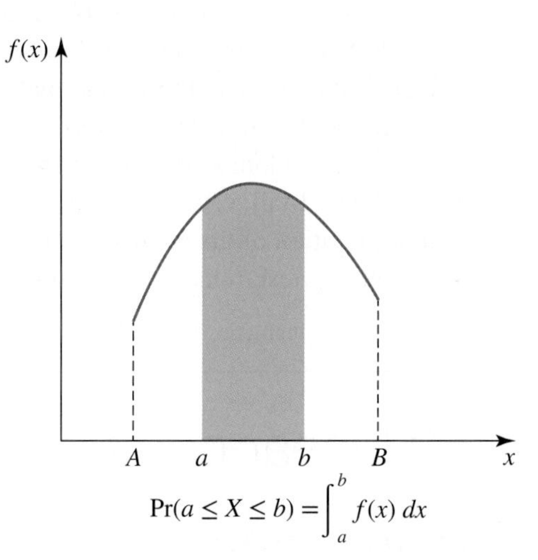

If the measurement scale used for X can be subdivided depending on the accuracy of the equipment available, then X is a continuous variable. For example, length may be measured in kilometres, metres, centimetres, millimeters and so on, so the variable length will be a continuous variable. The smallest unit for price of an object is cents, so the variable price will be a discrete variable. When deciding if a variable is discrete or continuous, a useful rule of thumb is: variables that can be counted are discrete and variables that are measured are continuous.

If X is a continuous random variable and X can be any value in the domain $[A, B]$, then $\Pr(A \leq X \leq B) = 1$. It is possible to find a function, called a probability density function $f(x)$ so that

$$\Pr(a \leq X \leq b) = \int_a^b f(x)\,dx.$$

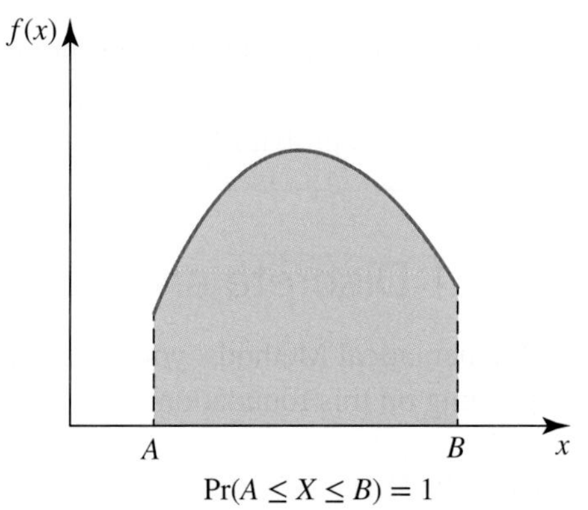

Note that as $\Pr(A \leq X \leq B) = 1$, then $\int_A^B f(x)\,dx = 1$.

Recall that the mean and variance of continuous random variables defined by probability density functions $f(x)$ are given by the following.

Mean and variance of continuous random variables

For a continuous random variable X, defined by the probability density function $f(x)$:

$$\mathrm{E}(X) = \mu = \int_{-\infty}^{\infty} x\,f(x)\,dx$$

and

$$\mathrm{Var}(X) = \sigma^2 = \int_{-\infty}^{\infty} (x-\mu)^2 f(x)\,dx$$

$$= \mathrm{E}(X^2) - [\mathrm{E}(X)]^2$$

where:

$$\mathrm{E}(X^2) = \int_{-\infty}^{\infty} x^2\,f(x)\,dx$$

The normal distribution

The normal distribution is a particular type of continuous random variable which appears frequently in nature. It is defined by a very complex probability density function which is not able to be integrated using the integration techniques covered in VCE mathematics. Fortunately, this does not matter as many calculators have specific functions which do it for you. On a CAS calculator this can be done using the Normal Cdf or Inverse Normal functions.

13.2.2 Transformations of distributions

In this section, we will look at how a distribution can undergo linear transformations, what those transformations mean, and how they affect the mean and variance of the distribution.

To bring this theory to life, consider the daily maximum and minimum temperatures in Melbourne over the month of June in a particular year as shown in the table.

	Temperatures	
	Min.	**Max.**
Date	**(°C)**	**(°C)**
1	11.4	15.1
2	11.5	17.4
3	12.6	16.6
4	10.8	18.6
5	11.9	17.1
6	8.5	16.6
7	10.5	17.5
8	11.2	15.9
9	8.6	15.6
10	9.7	19.0
11	6.9	16.6
12	9.2	16.3
13	12.2	17.0
14	10.3	15.8
15	9.9	16.4
16	10.4	17.0
17	10.1	16.3
18	11.1	14.3
19	8.0	12.3
20	9.3	15.5
21	11.3	15.7
22	7.7	15.8
23	8.5	14.7
24	7.1	14.5
25	8.6	16.0
26	11.2	16.8
27	11.0	14.7
28	9.6	13.0
29	8.8	11.3
30	6.5	13.3

Consider the distribution of minimum temperatures to be X and the distribution of maximum temperatures to be Y. The following summary statistics can be calculated (correct to 2 decimal places) from the data:

Minimum temperature, X	Maximum temperature, Y
$E(X) = \text{mean} = 9.81$ $\sigma(X) = \text{standard deviation} = 1.60$ $\text{Var}(X) = \text{variance} = 2.57$	$E(Y) = \text{mean} = 15.76$ $\sigma(Y) = \text{standard deviation} = 1.69$ $\text{Var}(Y) = \text{variance} = 2.85$

The data can be displayed on a scatterplot as shown in the following figure.

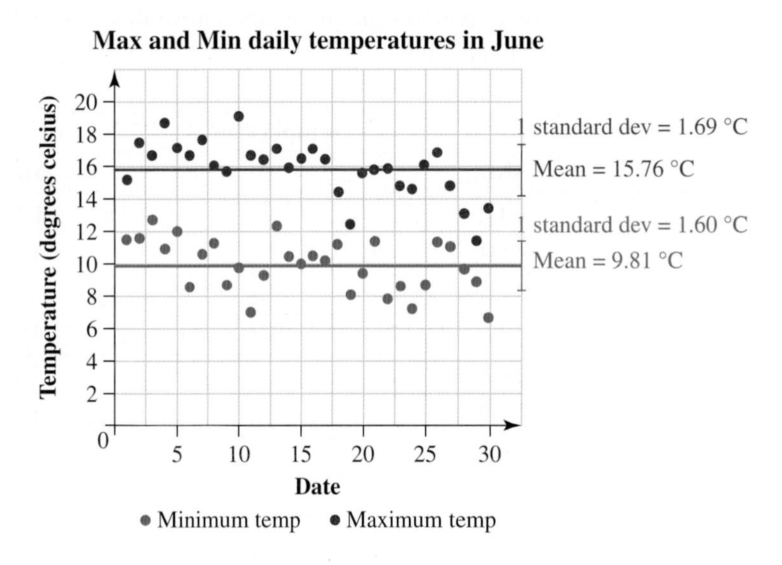

Change of origin

If, for some reason, we want to look at the minimum temperatures in Kelvin instead of degrees Celsius, it is a simple matter to add 273.15 to each measurement. A scatter plot of these temperatures is shown in the following figure.

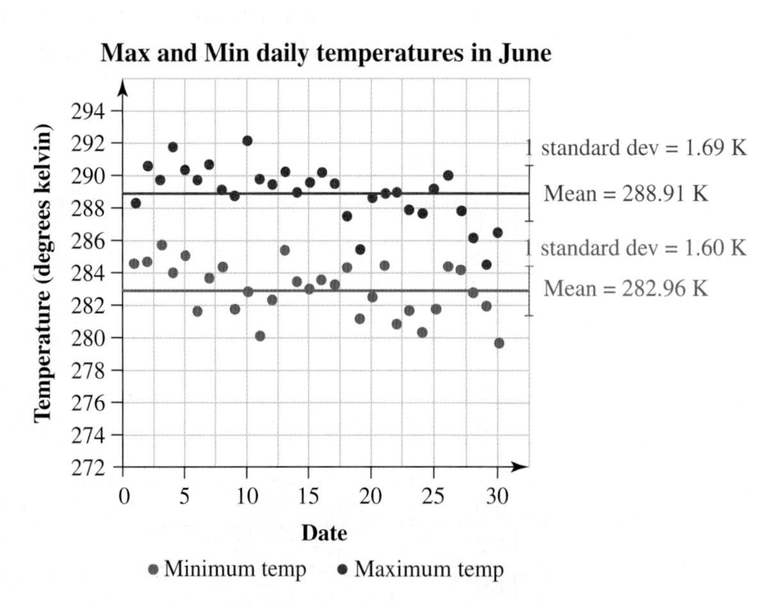

As you can see, the distribution has been moved up by 273.15 units. This repositioning is known as a change of origin. As all of the scores are increased by 273.15 units, the mean is also increased by 273.15. As the spread of the data is the same, the standard deviation and therefore the variance remain unchanged. This can be summarised as:

$$\begin{aligned} E(X+273.15) &= E(X)+273.15 \\ &= 9.81+273.15 \\ &= 282.96 \\ Var(X+273.15) &= Var(X) \\ &= 2.57 \end{aligned}$$

$$\begin{aligned} E(Y+273.15) &= E(Y)+273.15 \\ &= 15.76+273.15 \\ &= 288.91 \\ Var(Y+273.15) &= Var(Y) \\ &= 2.85 \end{aligned}$$

Change of scale

A distribution can be scaled by any real number by multiplying all values by a constant. This is called a change of scale.

For example, if the temperatures in the distributions X and Y were all multiplied by 2 the distributions would become $2X$ and $2Y$.

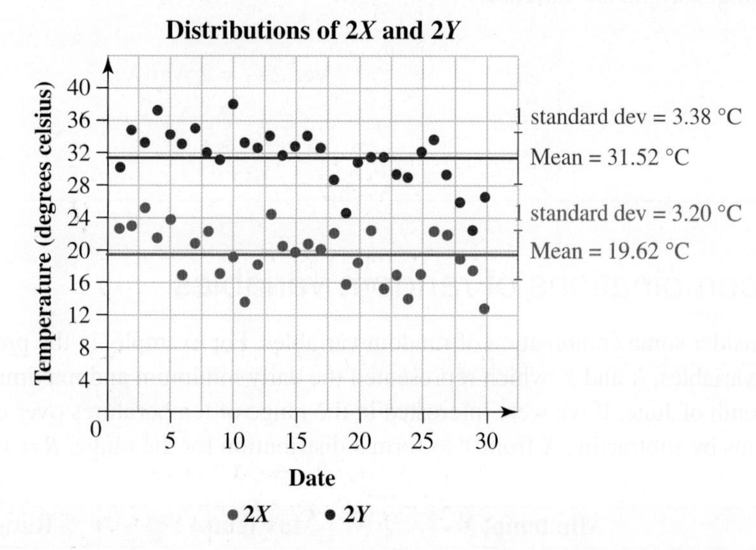

As you can see in the above scatter plot, the effect of multiplying the distributions by 2 is to multiply both the mean and the standard deviation by 2. Since the variance is the square of the standard deviation, the summary statistics become:

$$\begin{aligned} E(2X) &= 2E(X) \\ &= 2\times 9.81 \\ &= 19.62 \end{aligned}$$

$$\begin{aligned} E(2Y) &= 2E(Y) \\ &= 2\times 15.76 \\ &= 31.52 \end{aligned}$$

$$\begin{aligned} Var(2X) &= 2^2 Var(X) \\ &= 4\times 2.57 \\ &= 10.28 \end{aligned}$$

$$\begin{aligned} Var(2Y) &= 2^2 Var(Y) \\ &= 4\times 2.85 \\ &= 11.40 \end{aligned}$$

Change of origin and scale

In general, the combined effects of a transformation which changes origin and scale can be summarised as follows.

Change of origin and scale

$$\mathbf{E(aX+b) = aE(X)+b}$$

and

$$\mathbf{Var(aX+b) = a^2 Var(X)}$$

WORKED EXAMPLE 1 Change of scale and origin

For a random variable X, the expected value of the distribution is 3 and the variance is 2.2.

a. If 10 is added to each score in the distribution, calculate the new expected value and variance.

b. If each score in the distribution is doubled, calculate the new expected value and variance.

THINK	WRITE
Write the information using correct notation.	$E(X) = 3$, $Var(X) = 2.2$
a. Each score has 10 added to it. This will increase the expected value by 10 but will not change the variance.	a. $E(X + 10) = E(X) + 10$ $= 3 + 10$ $= 13$ $Var(X + 10) = Var(X)$ $= 2.2$
b. Each score is doubled. This will double the expected value and quadruple the variance.	b. $E(2X) = 2E(X)$ $= 2 \times 3$ $= 6$ $Var(2X) = 2^2 Var(X)$ $= 4 \times 2.2$ $= 8.8$

13.2.3 Linear combinations of random variables

Often it is useful to consider some combination of random variables. For example, in the previous section we looked at two random variables, X and Y, which represented the daily minimum and maximum temperatures in Melbourne over the month of June. If we were interested in the range of temperatures over each day we could combine the distributions by subtracting X from Y to form a distribution for the range, $R = Y - X$.

Date	Min temp, X	Max temp, Y	Range, $R = Y - X$
1	11.4	15.1	3.7
2	11.5	17.4	5.9
3	12.6	16.6	4.0
4	10.8	18.6	7.8
5	11.9	17.1	5.2
6	8.5	16.6	8.1
7	10.5	17.5	7.0
8	11.2	15.9	4.7
9	8.6	15.6	7.0
10	9.7	19.0	9.3
11	6.9	16.6	9.7
12	9.2	16.3	7.1
13	12.2	17.0	4.8
14	10.3	15.8	5.5
15	9.9	16.4	6.5
16	10.4	17.0	6.6
17	10.1	16.3	6.2

Date	Min temp, X	Max temp, Y	Range, $R = Y - X$
18	11.1	14.3	3.2
19	8.0	12.3	4.3
20	9.3	15.5	6.2
21	11.3	15.7	4.4
22	7.7	15.8	8.1
23	8.5	14.7	6.2
24	7.1	14.5	7.4
25	8.6	16.0	7.4
26	11.2	16.8	5.6
27	11.0	14.7	3.7
28	9.6	13.0	3.4
29	8.8	11.3	2.5
30	6.5	13.3	6.8
Mean	9.81	15.76	5.95
Standard deviation	1.60	1.69	1.82
Variance	2.57	2.85	3.31

Notice that the mean of the combined distribution is a function of each distribution's means:

$$\begin{aligned} \text{E}(Y - X) &= \text{E}(Y) - \text{E}(X) \\ &= 15.76 - 9.81 \\ &= 5.95 \end{aligned}$$

In general, the mean of a linear combination of random variables is:

$$\text{E}(aX + bY) = a\text{E}(X) + b\text{E}(Y)$$

If the distributions are independent of each other it is possible to calculate the variance of the linear combination as follows:

$$\text{Var}(aX + bY) = a^2\text{Var}(X) + b^2\text{Var}(Y)$$

In our example above, it is reasonable to assume that there is some correlation between the minimum and maximum temperatures for the day. Therefore, the variables are not independent and the relationship does not hold.

WORKED EXAMPLE 2 Linear combinations of two independent random variables

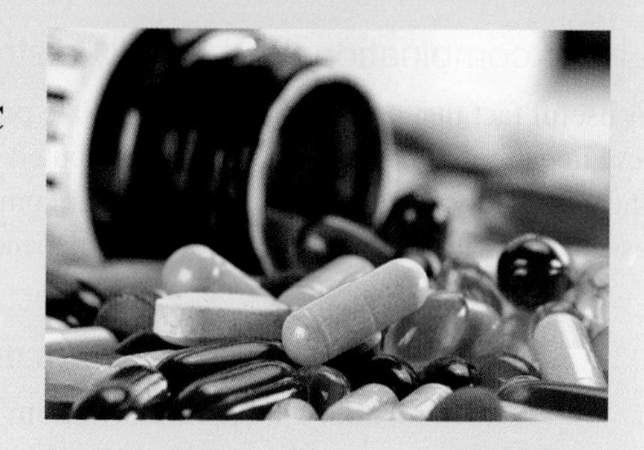

Ten patients in a hospital ward are taking vitamin C and vitamin D. Their doses of the two vitamins are shown in the table below. It is believed that the consumption of vitamin C and vitamin D by patients are independent random events. As vitamin C(X) is recorded in mg and vitamin $D(Y)$ is recorded in μg, the formula $T = X + 0.001\,Y$ is used to find the total amount of vitamins taken, in mg.

If $\text{E}(X) = 3.9032, \text{Var}(X) = 9.898\,976$, $\text{E}(Y) = 121.71$ and $\text{Var}(Y) = 24\,499.7129$, calculate the expected value and variance of T.
Note: **One microgram (μg) is equal to 10^{-6} g.**

Patient number	Vitamin C (mg), X	Vitamin D (μg), Y	Total vitamins (mg), $X + 0.001\,Y$
1	0.909	94.5	1.0035
2	2.201	52.1	2.2531
3	8.945	73.4	9.0184
4	6.262	88	6.35
5	1.866	79.9	1.9459
6	0.697	60.4	0.7574
7	3.114	95.3	3.2093
8	9.835	19.4	9.8544
9	3.723	67.3	3.7903
10	1.48	586.8	2.0668

THINK	WRITE
1. Calculate the expected value of T using $E(aX + bY) = aE(X) + bE(Y)$.	$T = X + 0.001Y$ $\text{E}(X + 0.001Y) = \text{E}(X) + 0.001\text{E}(Y)$ $= 3.9032 + 0.001 \times 121.71$ $= 4.02491$
2. Calculate the variance of T using $\text{Var}(aX + bY) = a^2\text{Var}(X) + b^2\text{Var}(Y)$.	$\text{Var}(X + 0.001Y) = \text{Var}(X) + 0.001^2\text{Var}(Y)$ $= 9.898\,976 + 0.000\,001 \times 24\,499.7129$ $= 9.923\,476$

Linear combinations of n independent random variables

The formulas above shows how to calculate the mean and variance of linear combinations of two random variables. More generally, these formulas can be extended to any number of independent random variables as follows.

Linear combinations of independent random variables

For n independent random variables $X_1, X_2, \ldots, X_n$ and real numbers $a_1, a_2, \ldots, a_n$:

$$\text{E}(a_1X_1 + a_2X_2 + \ldots + a_nX_n) = a_1\text{E}(X_1) + a_2\text{E}(X_2) + \ldots + a_n\text{E}(X_n)$$

and

$$\text{Var}(a_1X_1 + a_2X_2 + \ldots + a_nX_n) = a_1^2\text{Var}(X_1) + a_2^2\text{Var}(X_2) + \ldots + a_n^2\text{Var}(X_n)$$

Linear combinations of normally distributed random variables

A useful fact that can be applied to calculate probabilities associated with normally distributed variables is that any linear combination of normally distributed variables is normally distributed itself. Since we can determine the expected value and variance of a linear combination of independent variables, we can determine probabilities associated with linear combinations of independent normal distributions.

Linear combinations of normally distributed random variables

For n normally distributed random variables $X_1, X_2, \ldots, X_n$ and real numbers $a_1, a_2, \ldots, a_n$, the random variable $a_1X_1 + a_2X_2 + \ldots + a_nX_n$ is normally distributed.

WORKED EXAMPLE 3 Applications of linear combinations of random variables

Let X and Y be independent normally distributed random variables with means, 10 and 18 and variances 9 and 36 respectively. Determine the value of $\Pr(2X > Y)$. Give your answer rounded to 4 decimal places.

THINK	WRITE
1. State the mean and variance of X and Y. The notation $X \sim N\left(\mu, \sigma^2\right)$ means X is normally distributed with mean μ and variance σ^2.	$\text{E}(X)=10,\ \ \text{Var}(X)=9,$ $\text{E}(Y)=20,\ \ \text{Var}(Y)=36$ $X \sim N(10,9)$ $Y \sim N(18,36)$
2. Let D be the difference between the variables $2X$ and Y.	$D=2X-Y$
3. Calculate the mean of D.	$\text{E}(D)=2\text{E}(X)-\text{E}(Y)$ $=2\times 10-18=2$
4. Calculate the variance of D.	$\text{Var}(D)=2^2\times \text{Var}(X)+1^2\times \text{Var}(Y)$ $=4\times 9+36=72$
5. State the distribution of D.	$D \sim N(2,72)$
6. Determine the required probability.	$\Pr(2X>Y)=\Pr(2X-Y>0)$ $=\Pr(D>0)$
7. Use normcdf on the calculator.	$\Pr(D>0)=0.5932$

TI | THINK	DISPLAY/WRITE	CASIO | THINK	DISPLAY/WRITE
On a Calculator page, navigate to: Menu 5: Probability 5: Distributions 2: NormalCdf and complete the entry as shown.	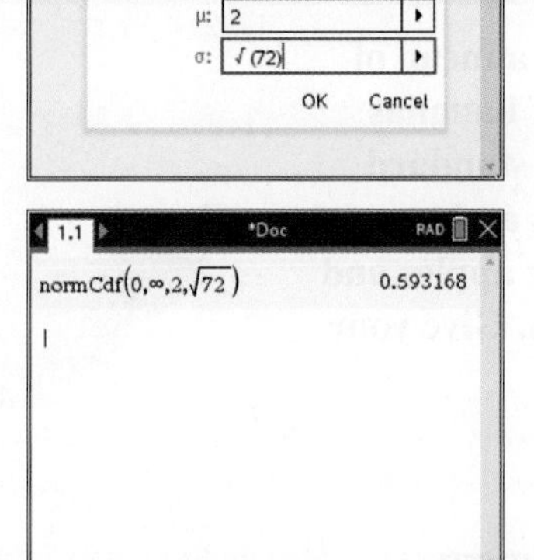	On a Main screen, navigate to: Action, Distribution/Inv.Dist, Continuous, normCDf and complete the entry as shown.	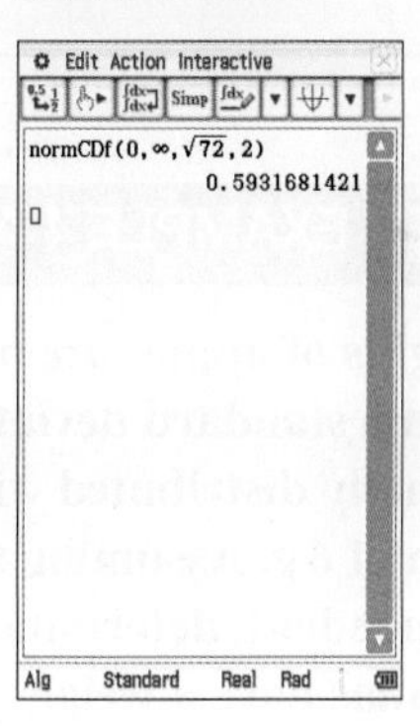

Independent identically distributed random variables

In some cases you may want to take multiple samples from a single distribution. For example, take the distribution X, which represents the weights/masses of potatoes in a market in grams, with $\text{E}(X)=250$ and $\text{Var}(X)=3000$. You may want to describe the distribution which represents the total weight of three potatoes taken from the market. This distribution is $X_1+X_2+X_3$, where X_1, X_2, X_3 are identical distributions which represent the weights of each potato respectively. Using the formulas in the above pink boxes and noting that $a_1 = a_2 = a_3 = 1$, we get:

$$\begin{aligned}\text{E}(X_1 + X_2 + X_3) &= 1 \times \text{E}(X_1) + 1 \times \text{E}(X_2) + 1 \times \text{E}(X_3)\\ &= \text{E}(X) + \text{E}(X) + \text{E}(X)\\ &= 3\text{E}(X)\\ &= 3 \times 250\\ &= 750\end{aligned}$$

$$\begin{aligned}\text{Var}(X_1 + X_2 + X_3) &= 1^2 \times \text{Var}(X_1) + 1^2 \times \text{Var}(X_2) + 1^2 \times \text{Var}(X_3)\\ &= \text{Var}(X) + \text{Var}(X) + \text{Var}(X)\\ &= 3\text{Var}(X)\\ &= 3 \times 3000\\ &= 9000\end{aligned}$$

Note that this is different to the distribution $3X$ which has a variance of $3^2 \times \text{Var}(X) = 9 \times \text{Var}(X) = 27000$.

Independent identically distributed random variables

For n independent identically distributed random variables $X_1, X_2, \ldots, X_n$ each with mean μ and variance σ^2:

$$\text{E}(X_1 + X_2 + \ldots + X_n) = \text{E}(X_1) + \text{E}(X_2) + \ldots + \text{E}(X_n) = n\mu$$

and

$$\text{Var}(X_1 + X_2 + \ldots + X_n) = \text{Var}(X_1) + \text{Var}(X_2) + \ldots + \text{Var}(X_n) = n\sigma^2$$

WORKED EXAMPLE 4 Applications of linear combinations of random variables (2)

The weights of apples are normally distributed with a mean of 180 g and a standard deviation of 7 g. The weights of bananas are normally distributed with a mean of 120 g and a standard deviation of 6 g. Assuming that the weights of apples and bananas are independent, determine the probability that four apples and two bananas have a weight in excess of one kilogram. Give your answer rounded to 4 decimal places.

THINK	WRITE
1. State the mean and variance in grams for apples A and bananas B.	$A \sim N(180, 7^2)$, $B \sim N(120, 6^2)$
2. Let T be the distribution representing the total weight of four apples and two bananas	$T = A_1 + A_2 + A_3 + A_4 + B_1 + B_2$
3. Calculate the expected value of T.	$\text{E}(T) = 4\text{E}(A) + 2\text{E}(B)$ $= 4 \times 180 + 2 \times 120$ $= 960$

4. Calculate the variance of T.	$\text{Var}(T) = 4\text{Var}(A) + 2\text{Var}(B)$ $= 4 \times 7^2 + 2 \times 6^2$ $= 268$
5. State the distribution of T.	$T \sim N(960, 268)$
6. Determine the required probability, using normcdf on a calculator.	$\Pr(T > 1000) = 0.0073$

13.2 Exercise

Students, these questions are even better in jacPLUS

Receive immediate feedback and access sample responses

Access additional questions

Track your results and progress

Find all this and MORE in jacPLUS

Technology free

1. **WE1** For a random variable X, the expected value of the distribution is 15.3 and the variance is 1.8.
   a. If 10 is subtracted from each score in the distribution, calculate the new expected value and variance.
   b. If each score in the distribution is doubled, calculate the new expected value and variance.

2. For a random variable X, the expected value of the distribution is 13 and the variance is 3.2.
   a. If 4 is added to each score in the distribution, calculate the new expected value and variance.
   b. If each score in the distribution is tripled, calculate the new expected value and variance.

3. **WE2** If $\text{E}(X) = 3.5$, $\text{Var}(X) = 1.1$, $\text{E}(Y) = 5.4$ and $\text{Var}(Y) = 2.44$, calculate the expected value and variance of $2X + 3Y$.

4. If $\text{E}(X) = 23.43$, $\text{Var}(X) = 5.89$, $\text{E}(Y) = 12.43$ and $\text{Var}(Y) = 9.7$, calculate the expected value and variance of $\frac{1}{2}X + Y$.

5. X and Y are two random variables with the same variance. Let $D = X - Y$ and $S = X + Y$. Given that $E(D) = 18$, $\text{E}(S) = 10$ and $\text{Var}(D) = 18$, calculate $\text{E}(X)$, $\text{E}(Y)$ and $\text{Var}(Y)$.

6. Let X be a binomially distributed random variables with $n = 25$ and $p = \frac{1}{5}$.

   Let Y be a binomially distributed random variable with $n = 16$ and $p = \frac{1}{4}$.

   X and Y are independent random variables. Let T be the random variable defined by $T = 3X - 2Y$. Calculate the mean and standard deviation of T.

7. If X is the random variable representing the weights of puppies, explain the difference between the distributions $2X$ and $X_1 + X_2$.

Technology active

8. **WE3** Let X and Y be independent normally distributed random variables with means 56 and 13 and variances 64 and 25 respectively. Calculate $\Pr\left(Y > \frac{X}{4}\right)$. Give your answer rounded to 3 decimal places.

9. **WE4** A fisherman goes fishing and catches 4 snapper and 2 flathead. The lengths of snapper are normally distributed with a mean of 43 cm and a standard deviation of 18 cm. The lengths of flathead are normally distributed with a mean of 35 cm and a standard deviation of 11 cm. Assuming that the lengths of each fish are independent, determine the probability that the total length of the fish caught is greater than 300 cm. Give your answer rounded to 3 decimal places.

10. Let X and Y be discrete random variables defined by

x	0	1	2
$\Pr(X=x)$	0.5	0.3	a

y	0	1	2
$\Pr(Y=y)$	0.4	b	0.5

Let T be the discrete random variable defined by $T=X+Y$

t	0	1	2	3	4
$\Pr(T=t)$	0.2	c	d	0.17	0.1

a. Determine the values of a, b, c and d.
b. Verify that $\mathrm{E}(T)=\mathrm{E}(X)+\mathrm{E}(Y)$ and that $\mathrm{Var}(T)=\mathrm{Var}(X)+\mathrm{Var}(Y)$.

11. Let X be a discrete random variable defined by $f(x)=cx^2, \;\; x=1, 2, 3$.
a. Determine the value of c.
b. Let $T=7X-3$, determine $\mathrm{E}(T)$ and $\mathrm{Var}(T)$.

12. Let Y be continuous random variable with a probability density function

$$f(y)=\begin{cases} ky^2 & 1\le y\le 3 \\ 0 & y<1 \text{ or } y>3 \end{cases}$$

a. Determine the value of k.
b. Let $T=13Y+8$. Calculate the mean of T and variance of T.

13. If X and Y are independent normally distributed random variables with means 10 and 11 and variances 4 and 5 respectively, calculate correct to 4 decimal places:
a. $\Pr(Y > X)$
b. $\Pr(X > Y)$
c. $\Pr(3X-2Y>5)$.

14. A child's toy contains cylindrical pegs which are fitted into circular holes. The diameters of the pegs and holes are independent and normally distributed with a mean of 0.48 cm and 0.50 cm, with standard deviations of 0.007 cm and 0.007 respectively.

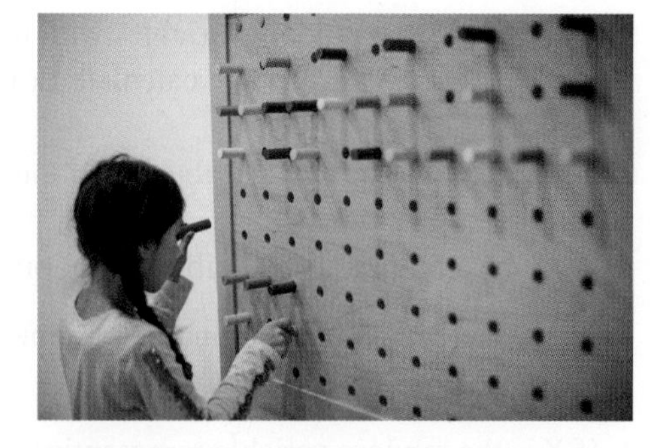

a. Determine the probability that a peg chosen at random will not fit into a hole chosen at random, correct to 4 decimal places.
b. Calculate the mean diameter of pegs, correct to three decimal places, which would give a probability of fit as 99%.

15. Let $X_1 \sim N(15, 25)$, $X_2 \sim N(12, 9)$ and $X_3 \sim N(10, 4)$ assuming that X_1, X_2 and X_3 are all independently distributed determine, correct to 4 decimal places:
a. $\Pr(4X_1+3X_2-5X_3 > 40)$
b. $\Pr(4X_1-3X_2-5X_3 > 0)$

16. A section of a wooden fence is to be constructed from twelve pieces of fencing posts. The width of each of these posts are normally distributed with a mean of 12 cm and a standard deviation of 0.5 cm. The fence also has two wooden end pieces, these end pieces have widths which are normally distributed with a mean of 75 cm and a standard deviation of 10 cm. Determine the probability that the total length of the section of fence is more than 300 cm, correct to 4 decimal places. Assume that the all the pieces of wood have widths that are independent.

17. The weights of medium sized dogs are normally distributed with a mean of 12 kg and a standard deviation of 3 kg. The weights of cats are normally distributed with a mean of 4 kg and a standard deviation of 1.5 kg. Assume that the weights of medium sized dogs and cats are independent. Calculate the probability that three cats and one medium sized dog weigh more than 25 kg combined, correct to 4 decimal places.

18. It has been found that the weights of males are normally distributed with a mean of 85 kg and a standard deviation of 15 kg. The weights of females are normally distributed with a mean of 65 kg and a standard deviation of 20 kg. Twelve males and eight females enter a lift. The lift will be overloaded if the total weight exceeds 1500 kg. Assuming that the weights of females and males are independent, determine the probability that the lift is overloaded. Give your answer correct to 4 decimal places.

13.2 Exam questions

Question 1 (1 mark) TECH-ACTIVE

Source: *VCE 2021, Specialist Mathematics Exam 2, Section A, Q19; © VCAA.*

MC The mean unscaled score for a certain assessment task is 25 and the variance is 36. The scores are scaled so that the mean score is 30 and the variance is 49. Let S be the scaled scores, to the nearest integer, and let X be the unscaled scores.

If the scaling function takes the form $S = mX + n$, where $m \in R^+$ and $n \in R$, then a score of 32 would be scaled to

A. 22 **B.** 34 **C.** 36 **D.** 38 **E.** 40

Question 2 (1 mark) TECH-ACTIVE

Source: *VCE 2019, Specialist Mathematics Exam 2, Section A, Q19; © VCAA.*

MC X and Y are independent random variables where each has a mean of 4 and a variance of 9.

If the random variable $Z = aX + bY$ has a mean of 8 and a variance of 90, possible values of a and b are

A. $a = 1, b = 1$
B. $a = 4, b = -2$
C. $a = 3, b = -1$
D. $a = 1, b = 3$
E. $a = -2, b = 4$

Question 3 (1 mark) TECH-ACTIVE

Source: *VCE 2018, Specialist Mathematics Exam 2, Section A, Q20; © VCAA.*

MC The scores on the Mathematics and Statistics tests, expressed as percentages, in a particular year were both normally distributed. The mean and the standard deviation of the Mathematics test scores were 71 and 10 respectively, while the mean and the standard deviation of the Statistics test scores were 75 and 7 respectively.

Assuming the sets of test scores were independent of each other, the probability, correct to four decimal places, that a randomly chosen Mathematics score is higher than a randomly chosen Statistics score is

A. 0.2877 **B.** 0.3716 **C.** 0.4070 **D.** 0.7123 **E.** 0.9088

More exam questions are available online.

13.3 Sample means and simulations

LEARNING INTENTION

At the end of this subtopic you should be able to:
- calculate the sample mean from a sample taken from a population and use it to estimate the population mean.

13.3.1 Estimating population parameters

In statistics, the entire underlying set of individuals of a group is called the **population**. In mathematics a population refers not just to people, but to any group of objects. The distribution of a population can be summarised by specific, fixed, values known as **parameters**. If the distribution of data is X, the parameters include:

1. The expected value of the distribution, E(X), which corresponds to the **mean**. That is, $E(X) = \mu$.
2. The **standard deviation**, σ, of the distribution. This parameter gives information about how the data are spread out from the mean. The standard deviation and mean have the same units making them excellent parameters for analysing the distribution.
3. The **variance**, σ^2, of the distribution can also be used to describe the spread of a distribution. The variance is the square of the standard deviation and therefore has different units compared to the mean and standard deviation. The variance, also denoted by Var(X), plays a central role in many areas of statistics.

The parameters of mean and standard deviation or variance are most frequently chosen to describe the population.

For very large populations, or where data for the entire population is very difficult and/or expensive to obtain, **samples** can be used to estimate the population parameters. A sample is a set of individuals or events selected from a population for analysis to give estimates of parameters of the whole population. In this section, we are concerned with determining the mean of different samples.

To estimate the mean of a large population we can take a random sample and then calculate the mean of the sample. The mean of the sample is called the **sample mean** and is denoted by $\bar{x}$. As the sample size increases, the sample mean provides a more accurate estimate of the population mean. In most cases a sample size of 30 or greater is considered large enough.

The notation that is used for the mean and standard deviation of populations and samples is summarised below.

Means and standard deviations of populations and samples

	mean	standard deviation
population	μ	σ
sample	$\bar{x}$	s

Calculating sample means

The sample mean is defined by:

$$\bar{x} = \frac{\sum x_i}{n}$$

When dealing with large sample sizes it is often easier to break the calculation up into multiple smaller samples of the same size and then calculate the average of the smaller sample's means.

WORKED EXAMPLE 5 Calculating the sample mean

Dylan is interested in calculating the average length of television shows on Netflix. He collects 12 samples and records the length of 10 different shows. All times are recorded in minutes. Use the sample means to estimate mean program length of the population.

Sample	Show 1	Show 2	Show 3	Show 4	Show 5	Show 6	Show 7	Show 8	Show 9	Show 10
1	60	55	60	60	60	5	55	30	30	30
2	60	30	30	60	35	130	35	60	30	35
3	85	30	50	55	60	60	55	25	55	30
4	30	30	30	30	30	30	30	30	30	30
5	30	30	30	30	30	25	25	25	25	25
6	60	60	60	60	60	60	60	30	30	60
7	30	60	60	60	90	60	35	35	35	30
8	55	55	60	35	35	55	55	55	50	55
9	85	60	60	30	30	60	60	60	60	60
10	30	30	30	45	45	45	45	45	45	45
11	70	70	130	35	35	70	70	35	35	70
12	60	105	60	120	60	60	60	60	60	105

THINK

1. Recall the formula for mean.

2. Determine the mean for each sample.

WRITE

$$\bar{x} = \frac{\sum x_i}{n}$$

$$\text{Sample 1: } \bar{x} = \frac{445}{10} = 44.5$$

$$\text{Sample 2: } \bar{x} = \frac{505}{10} = 50.5$$

$$\text{Sample 3: } \bar{x} = \frac{505}{10} = 50.5$$

$$\text{Sample 4: } \bar{x} = \frac{300}{10} = 30$$

$$\text{Sample 5: } \bar{x} = \frac{275}{10} = 27.5$$

$$\text{Sample 6: } \bar{x} = \frac{540}{10} = 54$$

$$\text{Sample 7: } \bar{x} = \frac{495}{10} = 49.5$$

$$\text{Sample 8: } \bar{x} = \frac{510}{10} = 51$$

$$\text{Sample 9: } \bar{x} = \frac{565}{10} = 56.5$$

$$\text{Sample 10: } \bar{x} = \frac{405}{10} = 40.5$$

$$\text{Sample 11: } \bar{x} = \frac{620}{10} = 62$$

$$\text{Sample 12: } \bar{x} = \frac{750}{10} = 75$$

3. Calculate the average $\bar{x}$.

Average $\bar{x}$

$$= \frac{44.5 + 50.5 + 50.5 + ... + 75}{12}$$

$$= \frac{591.5}{12}$$

$$\approx 49.3$$

4. Answer the question.

An estimate of the mean program length is 49.3 minutes.

TI | THINK — **DISPLAY/WRITE**

1. On a Home page, select: 1 New
4 Add Lists & Spreadsheet.

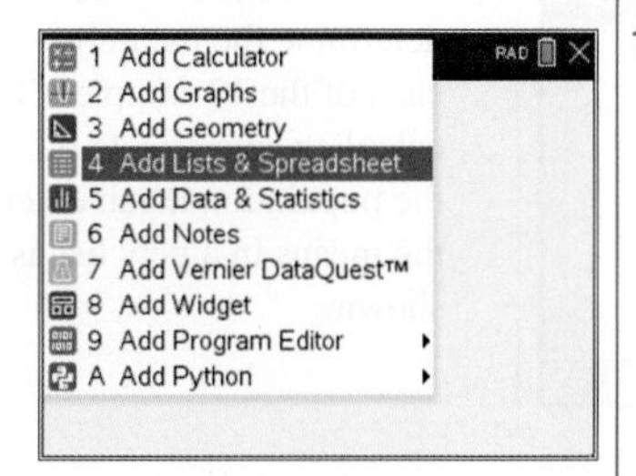

2. Enter the List 1 data by completing the entry lines
60
55
60
.
.
30
Continue entering the entire list.

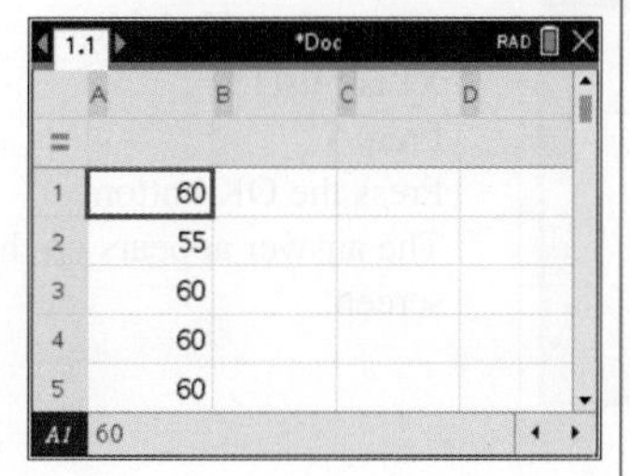

3. Select MENU
4 Statistics
1 Stat Calculations
1 One-Variable Statistics…

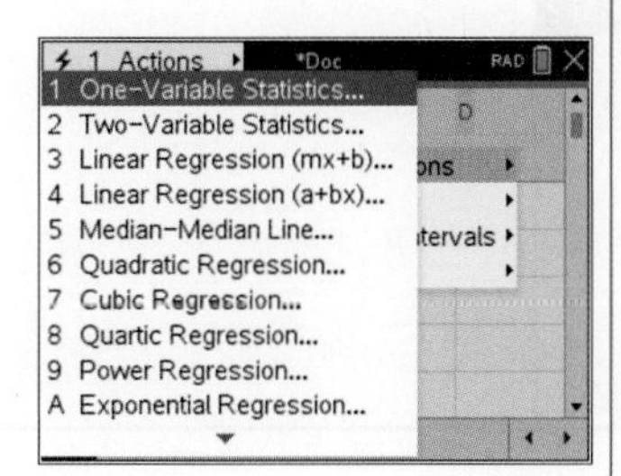

4. Set the number of lists as 1 by pressing the OK button.

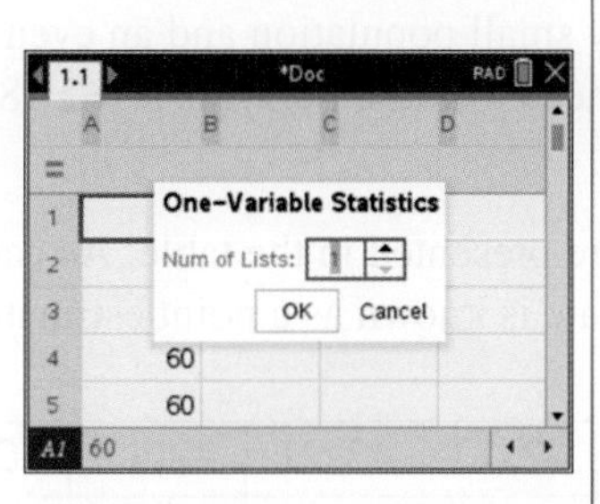

5. Complete the entry lines
X1 List: a[]
Frequency List: 1
1st Result Column: b[]

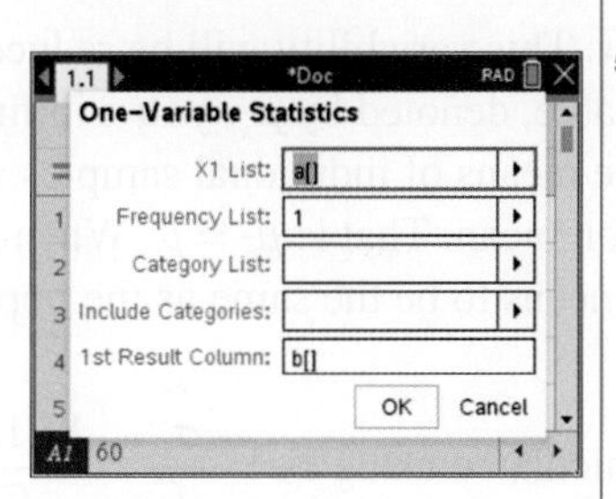

CASIO | THINK — **DISPLAY/WRITE**

1. From a Main screen, select the Statistics application.

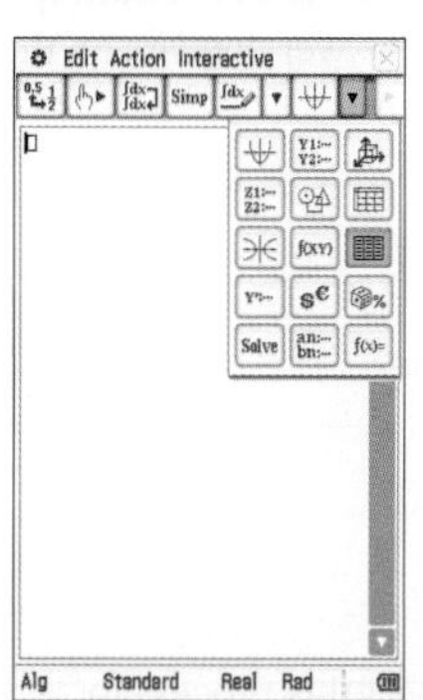

2. Enter the Sample 1 data by completing the entry lines
60
55
60
.
.
30
Continue entering the entire list.

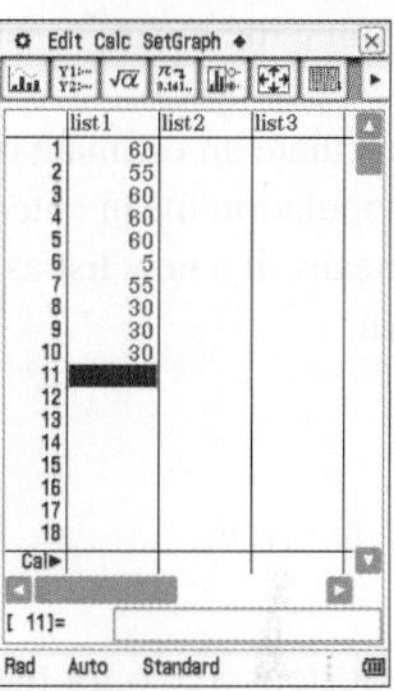

3. Select
Calc > One-Variable

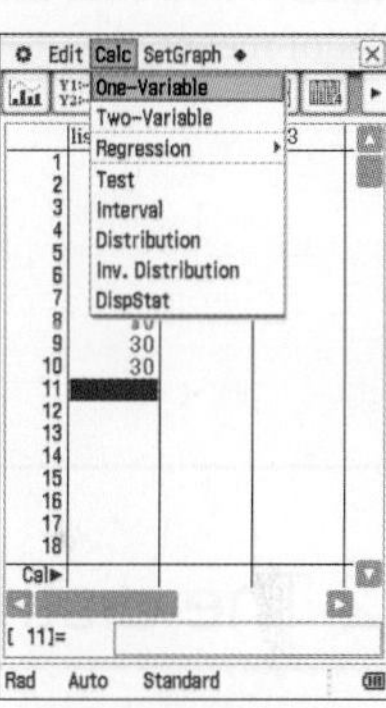

4. XList: list1
Freq: 1
Press the OK button

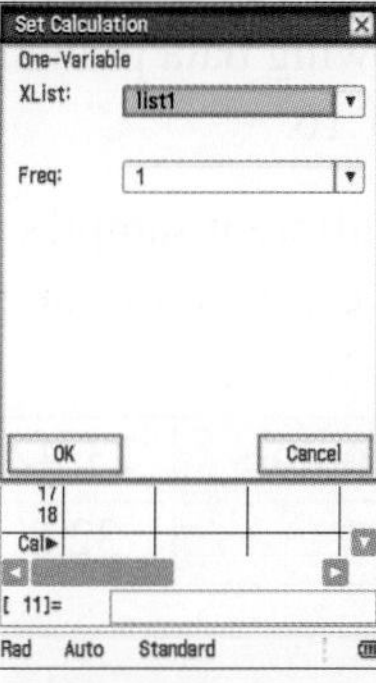

5. The summary statistics appear on the screen.

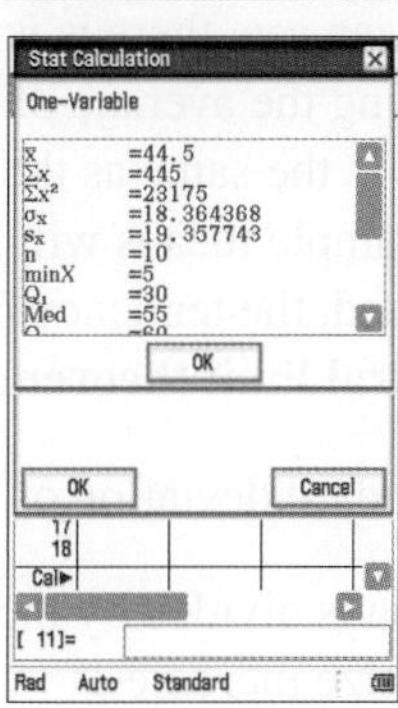

6. Press the ENTER button. The answer appears on the screen.

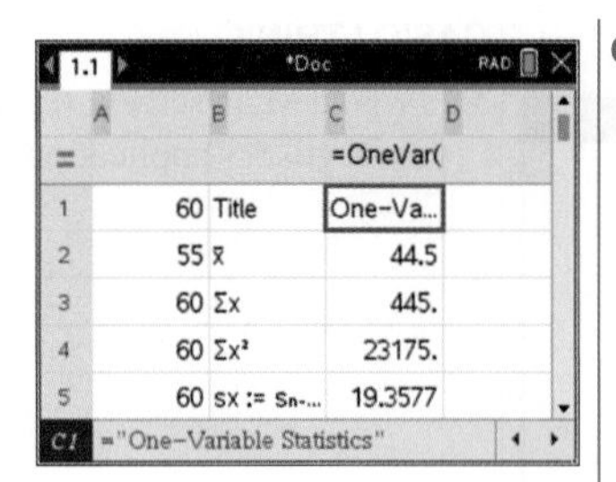

7. Repeat this process to determine the mean for each of the 12 samples.
To calculate an estimate of the population mean enter the means in a new list as shown.

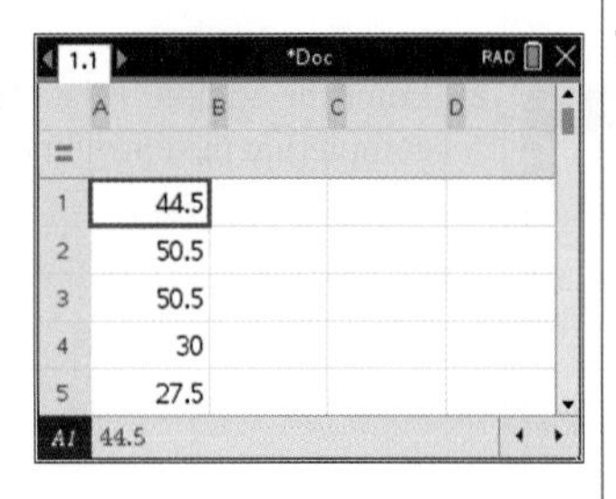

8. Repeat steps 3 to 6 for the new list of means.

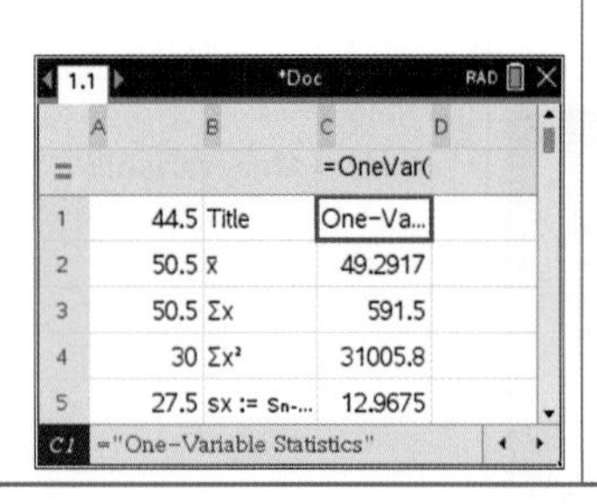

6. Repeat this process to determine the mean for each of the 12 samples. To calculate an estimate of the population mean enter the means in a new list as shown.

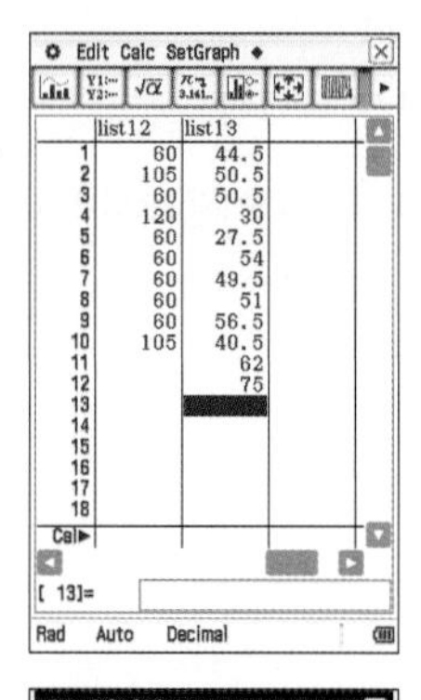

7. Select
Calc > One-Variable
XList: list13
Freq: 1
Press the OK button.
The answer appears on the screen.

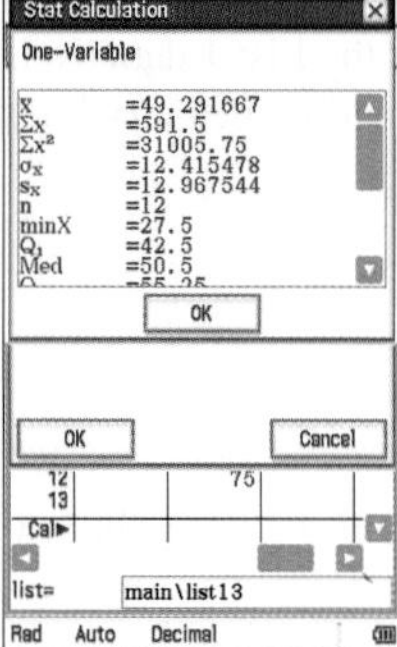

13.3.2 The distribution of sample means

For purposes of understanding we will examine a very small population and an even smaller sample. Consider the following data points: 23, 42, 12, 21 and 11. The mean of these points is $\mu = 21.8$ and the standard deviation is $\sigma \approx 11.16$.

All the different samples of size 2 from this data set are presented in the table. As each sample provides one possible estimate for the population mean, each estimate is known as a point estimate.

Data points	23, 42	23, 12	23, 21	23, 11	42, 12	42, 21	42, 11	12, 21	12, 11	21, 11
$\bar{x}$	32.5	17.5	22	17	27	31.5	26.5	16.5	11.5	16

As you can see, there is wide variety in the values of $\bar{x}$. This variability will be reduced in larger samples. Examining the average of the sample $\bar{x}$ values in the table, denoted by $\mu_{\bar{x}}$, you will find $\mu_{\bar{x}} = 21.8$. Notice that this is the same as the population mean. While the means of individual samples will vary, the mean of all the sample means will be the same as the population mean. That is $\mu_{\bar{x}} = \mu$. When all possible samples are considered, the tendency for the mean of the sample means to be the same as the population mean is known as the **central limit theorem**.

The standard deviation of the sample means can be calculated using $\sigma_{\bar{x}} = \dfrac{\sigma}{\sqrt{n}} = \dfrac{11.16}{\sqrt{2}} \approx 7.89$. As the standard deviation is divided by the square root of the sample size, this measure of variability becomes smaller as the sample size increases.

If the population is normally distributed, then the sample means will also be normally distributed with a mean of μ and a standard deviation of $\frac{\sigma}{\sqrt{n}}$. If the population is not normal, but the samples are sufficiently large (usually $n \geq 30$), then the sample means will also be normally distributed with a mean of μ and a standard deviation of $\frac{\sigma}{\sqrt{n}}$. This will be explored further in the next exercise.

13.3.3 Verifying the formulae

The original population X has an expected value $E(X) = \mu$ and a variance of $Var(X) = \sigma^2$. In this instance, we are concerned with the distribution of sample means taken from this population. Each sample has a mean $\bar{x}_i$, and the distribution of these means can be referred to as $\bar{X}$, known as the distribution of sample means (or sample mean distribution).

We observed that $E(\bar{X}) = \mu$. As each sample comes from the original population, the expected value for the mean of each sample is the same as the population mean, μ. That is $E(X_i) = E(X) = \mu$. (In reality the actual value is $\bar{x}_i$, but we can't expect a random value, we expect it to be μ).

$$\begin{aligned}
E(\bar{X}) &= E\left(\frac{X_1 + X_2 + ... + X_n}{n}\right) \\
&= \frac{1}{n}E(X_1 + X_2 + ... X_n) \\
&= \frac{1}{n}(E(X_1) + E(X_2) + ... E(X_n)) \text{ since } E(aX + bY) = aE(X) + bE(Y) \\
&= \frac{1}{n}(n\mu) \\
&= \mu
\end{aligned}$$

In a similar fashion, the variance of the distribution can be proven. It is assumed that the random samples are mutually independent and as each sample comes from the population, it will have the same variance as the population. That is $Var(X_i) = Var(X) = \sigma^2$.

$$\begin{aligned}
Var(\bar{X}) &= Var\left(\frac{X_1 + X_2 + ... + X_n}{n}\right) \\
&= \frac{1}{n^2}Var(X_1 + X_2 + ... + X_n) \text{ since } Var(aX + b) = a^2 Var(X) \\
&= \frac{1}{n^2}\left(n\sigma^2\right) \\
&= \frac{\sigma^2}{n} \\
sd(\bar{X}) &= \frac{\sigma}{\sqrt{n}}, \text{ recalling that } sd(X) = \sqrt{Var(X)}
\end{aligned}$$

WORKED EXAMPLE 6 The mean and standard deviation under different sample sizes

500 students are studying a Mathematics course at the University of the Sunshine Coast. On a recent exam, the mean score was 67.2 with a standard deviation of 17.

a. Samples of size 5 are selected and the means determined. Calculate the mean and standard deviation of the distribution of $\bar{X}$.

b. If the sample size was increased to 30, determine the effect this would have on the mean and standard deviation of the distribution of sample means.

THINK	WRITE
a. 1. The mean of the distribution of sample means is the same as the population mean.	**a.** $E(\overline{X}) = \mu$ $= 67.2$
2. Write down the formula for standard deviation of the distribution of sample means.	$sd(\overline{X}) = \dfrac{\sigma}{\sqrt{n}}$
3. Calculate the standard deviation.	$sd(\overline{X}) = \dfrac{17}{\sqrt{5}}$ ≈ 7.6
b. 1. The mean is not dependent on sample size.	**b.** $E(\overline{X}) = \mu$ $= 67.2$
2. Calculate the standard deviation using $n = 30$.	$sd(X^{-}) = \dfrac{\sigma}{\sqrt{n}}$ $= \dfrac{17}{\sqrt{30}}$ ≈ 3.1
3. Write your conclusions.	Increasing the sample size does not change the mean of the distribution, but the standard deviation of the distribution is reduced.

Calculating probabilities of sample means

WORKED EXAMPLE 7 Calculating probabilities relating to sample means

The gestation period of Indian elephants is normally distributed with a mean 645 days with a standard deviation of 5 days. From a random sample of nine Indian elephants, calculate the probability that the average gestation period is greater than 648 days, correct to 4 decimal places.

THINK	WRITE
1. Write down the required values of the distribution.	Let G be the gestation period of Indian elephants in days, $\mu_G = 645,\ \sigma_G = 5,\ G \sim N(645, 25)$
2. Write down the mean and standard deviation of the sample means.	$n = 9,\ \mu_{\overline{G}} = 645,\ \sigma_{\overline{G}} = \dfrac{5}{\sqrt{9}},\ \overline{G} \sim N\left(645, \dfrac{25}{9}\right)$
3. Determine the required probability using the normalCdf function on your CAS calculator.	$\Pr\left(\overline{G} > 648\right) = 0.0359$

13.3 Exercise

Students, these questions are even better in jacPLUS

Receive immediate feedback and access sample responses

Access additional questions

Track your results and progress

Find all this and MORE in jacPLUS

Technology active

1. **WE5** Ronit decides to measure his mean travel time to school. He records the time taken every day for 7 weeks and records the time in minutes, resulting in 7 samples. Use the sample means to estimate mean travel time of the population, correct to 1 decimal place.

Week	Monday	Tuesday	Wednesday	Thursday	Friday
1	92	43	41	39	35
2	118	81	46	51	38
3	62	48	46	41	49
4	82	48	42	43	41
5	78	51	42	41	38
6	63	62	41	43	44
7	55	41	46	41	32

2. Leesa wants to know the mean movie length for her favourite movie channel. Each day for a week she records the lengths of 8 movies. Her results are recorded in minutes. Use the sample means from each day to estimate mean movie length of the population, correct to 1 decimal place.

Day	Movie 1	Movie 2	Movie 3	Movie 4	Movie 5	Movie 6	Movie 7	Movie 8
Monday	115	95	105	95	115	100	90	95
Tuesday	95	85	90	90	105	95	75	95
Wednesday	110	95	80	110	95	90	105	80
Thursday	95	100	90	95	105	100	90	85
Friday	105	95	90	100	105	100	90	105
Saturday	90	85	90	110	80	100	90	90
Sunday	105	100	100	95	90	90	90	110

3. **WE6** Every year 500 students apply for a place at Monotreme University. The average enrolment test score is 600 with a standard deviation of $\sqrt{300}$.
   a. Samples of size 10 are selected and the means determined. Calculate the mean and standard deviation, correct to 2 decimal places, of the distribution of $\overline{X}$.
   b. If the sample size was increased to 20, determine the effect this would have on the mean and standard deviation, correct to 2 decimal places, of the distribution of sample means.

4. 1000 students are studying a statistics course at Echidna University. On a recent exam, the mean score is 90.5 with a standard deviation of 10.
   a. Samples of size 15 are selected and the means determined. Calculate the mean and standard deviation, correct to 2 decimal places, of the distribution of $\overline{X}$.
   b. If the sample size was increased to 30, determine the effect this would have on the mean and standard deviation, correct to 2 decimal places, of the distribution of sample means.

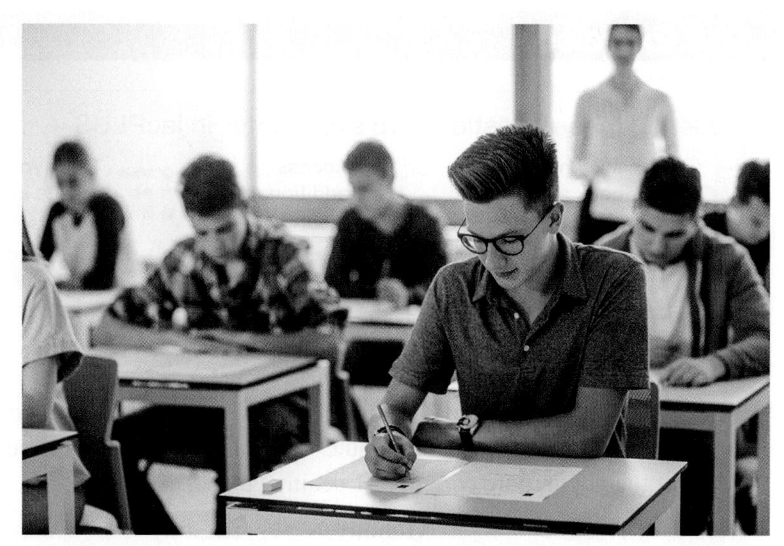

5. Use a random number generator to generate random numbers between 0 and 100. Perform the simulation 40 times and calculate the sample mean.
   a. Compare your results to those of your classmates. State how close they were to the expected value of 50.
   b. Calculate the average of $\overline{X}$ for your class. State how close it is to 50.

The following information relates to questions 6 and 7. A simulation of the total obtained when a pair of dice were tossed. There were 8 people who each tossed the dice 10 times.

	Player 1	Player 2	Player 3	Player 4	Player 5	Player 6	Player 7	Player 8
Toss 1	7	6	10	7	11	2	10	8
Toss 2	8	9	8	6	6	11	4	10
Toss 3	2	4	3	10	6	8	8	6
Toss 4	8	5	6	5	11	7	6	2
Toss 5	10	12	6	8	10	8	3	4
Toss 6	4	10	9	5	3	6	5	5
Toss 7	7	6	5	6	9	10	4	2
Toss 8	11	8	9	8	9	9	6	10
Toss 9	4	7	10	10	7	4	12	8
Toss 10	7	8	3	8	10	4	7	11

6. Calculate an estimate of the population mean, correct to 2 decimal places, and compare it to the theoretical mean.

7. Repeat the experiment with your own simulation of dice tosses. Use your results to estimate the population mean.

The following information relates to questions 8–10. Consider a population with a mean of 73 and a standard deviation of 12.

8. If samples of size 20 are selected, calculate the mean and standard deviation, correct to 2 decimal places, of the distribution of sample means.

9. If the sample size is increased to 30, calculate the mean and standard deviation, correct to 2 decimal places, of the sample means.

10. Determine the sample size that would be needed to reduce the standard deviation of the distribution of sample means to less than 2.

***The following information relates to questions* 11–13.** Consider a population with a mean of 123 and a standard deviation of 43.

11. If samples of size 25 are selected, calculate the mean and standard deviation, correct to 1 decimal place, of the distribution of sample means.

12. If samples of size 40 are selected, calculate the mean and standard deviation, correct to 1 decimal place, of the distribution of sample means.

13. Determine the sample size that would be needed to reduce the standard deviation of the distribution of sample means to less than 5.

14. **WE7** The amount of vegemite in a jar is normally distributed with a mean of 500 g with a standard deviation of 4 g. From a random sample of four jars of this type, calculate the probability that the average amount of vegemite is less than 498 g, correct to 4 decimal places.

15. A typical orange has a mass which is normally distributed with a mean of 131 g with a standard deviation of 17 g. From a random sample of three oranges of this type, calculate the probability that the average mass is more than 140 g, correct to 4 decimal places.

16. The amount of beer in a bottle is normally distributed with a mean of 330 mL with a standard deviation of 7 mL. From a random sample of a six pack of beer of this type, calculate the probability, correct to 4 decimal places, that:
    a. the average amount of beer in the bottles is less than 325 mL
    b. the total amount of beer in the six pack is more than 2 L.

17. The mass of medium size eggs are normally distributed with a mean of 50 g with a standard deviation of 6 g. From a random sample of one dozen medium eggs, calculate the probability, correct to 4 decimal places, that:
    a. the average mass of the sample is more than 52 g
    b. the total mass is less than 580 g.

18. The masses of baby girls are normally distributed with a mean of 3.2 kg and a standard deviation of 0.4 kg. A random sample of n baby girls was chosen and it was found that 2.3% had a mean weight of less of 3.0 kg, determine the value of n.

19. The heights of 17-year-old boys are normally distributed with a mean of 175.5 cm and a standard deviation of 5 cm. A random sample of n boys was chosen and it was found that 31% had a mean height in excess of 176 cm, determine the value of n.

20. We wish to create data points from a skewed population. Use Excel or another similar package to create the data points.
    a. Use a random number generator to generate 70 numbers between 1 and 20. Use the generator to generate another 30 numbers between 15 and 20. The distribution should now be skewed to the right. You should now have 100 numbers.
    b. Use technology to sketch your distribution.
    We now want to take samples from the distribution to see what the distribution of sample means might look like when the original population was skewed.
    c. Select random samples of size 10 from your distribution and calculate the sample means.
    d. Use technology to sketch the distribution of sample means. Compare the two sketches and comment on the similarities and difference.

13.3 Exam questions

Question 1 (1 mark) TECH-ACTIVE

Source: *VCE 2018, Specialist Mathematics Exam 2, Section A, Q19; © VCAA.*

MC The gestation period of cats is normally distributed with mean $\mu = 66$ days and variance $\sigma^2 = \frac{16}{9}$.

The probability that a sample of five cats chosen at random has an average gestation period greater than 65 days is closest to

A. 0.5000
B. 0.7131
C. 0.7734
D. 0.8958
E. 0.9532

Question 2 (7 marks) TECH-ACTIVE

Source: *VCE 2017, Specialist Mathematics Exam 2, Section B, Q6a-d; © VCAA.*

A dairy factory produces milk in bottles with a nominal volume of 2 L per bottle. To ensure most bottles contain at least the nominal value, the machine that fills the bottles dispenses volumes that are normally distributed with a mean of 2005 mL and a standard deviation of 6 mL.

a. Find the percentage of bottles that contain at least the nominal volume of milk, correct to one decimal place. **(1 mark)**

Bottles of milk are packed in crates of 10 bottles, where the nominal total volume per crate is 20 L.

b. Show that the total volume of milk contained in each crate varies with a mean of 20 050 mL and a standard deviation of $6\sqrt{10}$ mL. **(2 marks)**

c. Find the percentage, correct to one decimal place, of crates that contain at least the nominal volume of 20 L. **(1 mark)**

d. Regulations require at least 99.9% of crates to contain at least the nominal volume of 20 L. Assuming the mean volume dispensed by the machine remains 2005 mL, find the maximum allowable standard deviation of the bottle-filling machine needed to achieve this outcome. Give your answer in millimetres, correct to one decimal place. **(3 marks)**

Question 3 (1 mark) TECH-ACTIVE

Source: *VCE 2016, Specialist Mathematics Exam 2, Section A, Q20; © VCAA.*

MC The lifetime of a certain brand of batteries is normally distributed with a mean lifetime of 20 hours and a standard deviation of two hours. A random sample of 25 batteries is selected.

The probability that the mean lifetime of this sample of 25 batteries exceeds 19.3 hours is

A. 0.0401
B. 0.1368
C. 0.6103
D. 0.8632
E. 0.9599

More exam questions are available online.

13.4 Confidence intervals

LEARNING INTENTION

At the end of this subtopic you should be able to:
- construct confidence intervals for the mean of a population given information obtained by sampling
- determine the sample size required to reduce the margin of error to a certain amount for a given confidence level
- use a confidence interval to construct a new confidence interval with a different level of confidence.

13.4.1 Confidence intervals

We have seen that different samples from one population may have different means. As a sample is used to give an indication of the actual population parameters by making a point estimate, we need to be more specific than *'the population mean is approximately equal to the sample mean'*. Confidence intervals allow us to quantify the interval that the population mean might lie in.

For example, in the data used to explore sample means earlier, the first sample was $\{23, 42\}$ which had a mean of 32.5. The sample mean provides a **point estimate** for the actual population mean of 21.8. In this section, it will be shown that is possible to use this sample to calculate a 90% confidence interval of $(16.9, 48.1)$. Notice that the sample mean of 32.5 is in the middle of this interval. The lower limit of 16.9 is 15.6 units below the sample mean and the upper limit of 48.1 is 15.6 units above the sample mean. The 90% confidence interval means that from the possible samples, 90% of the calculated confidence intervals will contain the population mean. This will be demonstrated later.

Note: This is not the same as saying that there is a 90% chance that the population mean is in the confidence interval. The population mean is an actual value, in this case 21.8. Therefore, for any given sample, the confidence interval either contains the mean (which it does in this case) or it doesn't contain the mean.

13.4.2 Calculation of confidence intervals

A **confidence interval** for μ can be calculated from the sample mean, sample standard deviation and sample size as shown in the formula below. The confidence interval will be centered on the sample mean, and the size of the interval is determined by the sample standard deviation, sample size and the confidence level.

Formula for a confidence interval

The confidence interval for the population mean, μ, is:

$$\left(\bar{x} - z\frac{s}{\sqrt{n}},\ \bar{x} + z\frac{s}{\sqrt{n}}\right)$$

where: $\bar{x}$ is the sample mean

s is the sample standard deviation

n is the sample size

z is the z score for the level of confidence we are interested in.

Some important z scores to memorise are the ones which correspond to 90%, 95% and 99% confidence levels. The following graphs represent these values visually on the standard normal distribution.

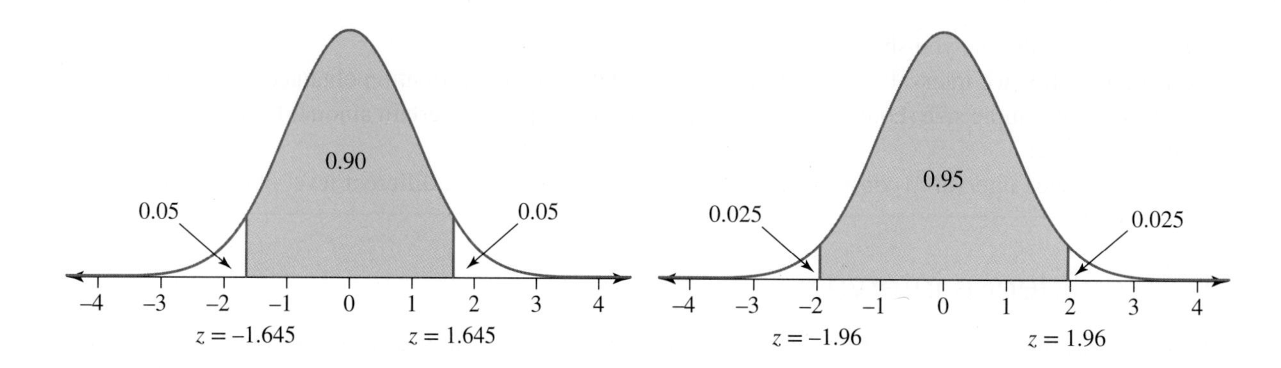

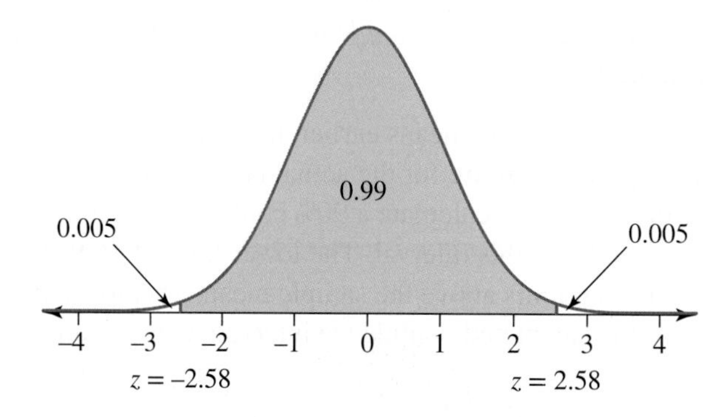

Confidence level z scores

Confidence level	z score
90%	$z = 1.645$
95%	$z = 1.96$
99%	$z = 2.58$

Returning to the sample data that we used for investigating sample means. Remember this population has a mean of 21.8. For each sample it is possible to calculate a 90% confidence interval (although not very accurately because of the small sample). As it is a 90% confidence interval, we use $z = 1.645$. For the first sample $\{23, 42\}$, $\bar{x} = 32.5$, $s = 13.435$ and $n = 2$. To calculate the lower limit:

$$\begin{aligned} \bar{x} - z\frac{s}{\sqrt{n}} &= 32.5 - 1.645 \times \frac{13.435}{\sqrt{2}} \\ &= 32.5 - 15.6 \\ &= 16.9 \end{aligned}$$

To calculate the upper limit:

$$\begin{aligned} \bar{x} + z\frac{s}{\sqrt{n}} &= 32.5 + 1.645 \times \frac{13.435}{\sqrt{2}} \\ &= 32.5 + 15.6 \\ &= 48.1 \end{aligned}$$

This sample estimates that the population mean will lie in the interval (16.9, 48.1). Notice that this interval does contain the population mean of 21.8. In a similar fashion, the 90% confidence interval for each of the 10 possible samples can be calculated. They are recorded in the table shown.

Data points	$\overline{x}$	s	90% confidence interval	Contains Population Mean
23, 42	32.5	13.435	(16.9, 48.1)	Yes
23, 12	17.5	7.778	(8.5, 26.5)	Yes
23, 21	22	1.414	(20.4, 23.6)	Yes
23, 11	17	8.485	(7.1, 26.9)	Yes
42, 12	27	21.213	(2.3, 51.7)	Yes
42, 21	31.5	14.849	(14.2, 48.8)	Yes
42, 11	26.5	21.920	(1, 52)	Yes
12, 21	16.5	6.364	(9.1, 23.9)	Yes
12, 11	11.5	0.707	(10.7, 12.3)	No
21, 11	16	7.071	(7.8, 24.2)	Yes

As you can see, 90% (9 out of 10) of the calculated confidence intervals contain the population mean. This is the meaning of the 90% confidence interval.

Notice that the size of the confidence intervals varies between samples. The smaller the sample standard deviation, the smaller the confidence interval for the sample.

13.4.3 Verifying the formulae

We have learnt that when the sample size is large, that is $n \geq 30$, the distribution of $\overline{x}$ is normally distributed with a mean of $\mu_{\overline{x}} = \mu$ and a standard deviation of $\frac{\sigma}{\sqrt{n}}$. As a sample is normally selected in order the estimate the population mean and standard deviation, the best estimates for μ and σ are $\overline{x}$ and s respectively. We know that for normal distributions, $z = \frac{x-\mu}{\sigma}$. This means that to find the upper and lower values of z,

$$z = \frac{x \pm \overline{x}}{\frac{s}{\sqrt{n}}}$$

$$x \pm \overline{x} = \frac{s}{\sqrt{n}} z$$

$$x = \overline{x} \pm \frac{s}{\sqrt{n}} z$$

Rearranging gives $x = \overline{x} \pm z\frac{s}{\sqrt{n}}$.

For a 95% confidence interval, 95% of the distribution is in the middle area of the distribution. This means that the tails combined contain 5% of the distribution (2.5% each). The z score for this distribution is 1.96.

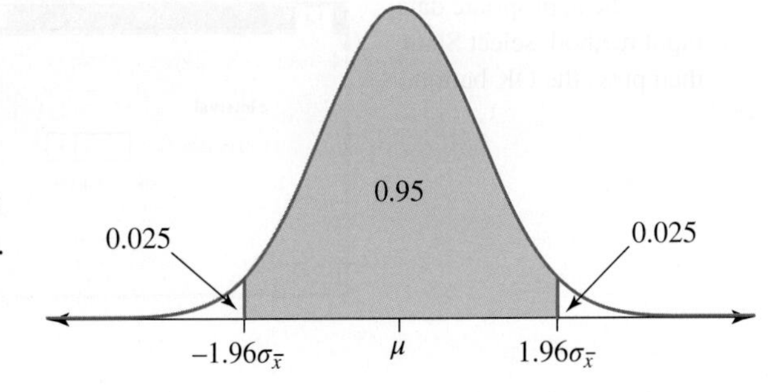

WORKED EXAMPLE 8 Constructing confidence intervals

After surveying the 20 people in your class, you find that they plan to spend an average of \$4.53 on their lunch today. The standard deviation for your sample was \$0.23. Use these statistics to state a 95% confidence interval for the average amount that will be spent on lunch today.

THINK	WRITE
1. There are 20 people on the class. This is the sample size. \$4.53 is spent on lunch, this is the sample mean. The sample standard deviation is \$0.23.	$n = 20$ $\bar{x} = 4.53$ $s = 0.23$
2. For a 95% confidence interval, $z = 1.96$.	$z = 1.96$
3. The confidence interval is $\left(\bar{x} - z\frac{s}{\sqrt{n}},\ \bar{x} + z\frac{s}{\sqrt{n}}\right)$. Use a calculator to calculate the values of $\bar{x} - z\frac{s}{\sqrt{n}}$ and $\bar{x} + z\frac{s}{\sqrt{n}}$.	$\bar{x} - z\frac{s}{\sqrt{n}} = 4.53 - 1.96 \times \frac{0.23}{\sqrt{20}} = 4.43$ $\bar{x} + z\frac{s}{\sqrt{n}} = 4.53 + 1.96 \times \frac{0.23}{\sqrt{20}} = 4.63$
4. Identify the 95% confidence interval.	$\left(\bar{x} - z\frac{s}{\sqrt{n}},\ \bar{x} + z\frac{s}{\sqrt{n}}\right) = (4.43, 4.63)$
5. State the answer.	The 95% confidence interval for the amount planned to be spent by a student on their lunch is (\$4.43, \$4.63).

TI | THINK	DISPLAY/WRITE	CASIO | THINK	DISPLAY/WRITE
1. On a Calculator page, press MENU then select 6 Statistics 6 Confidence intervals 1 z Interval…	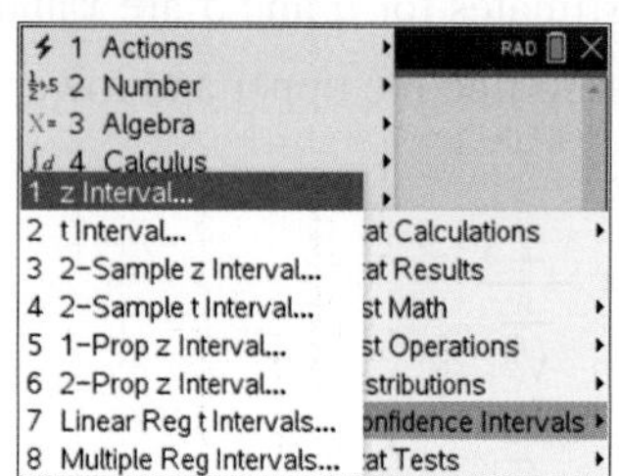	1. From a Main screen, select the Statistics application.	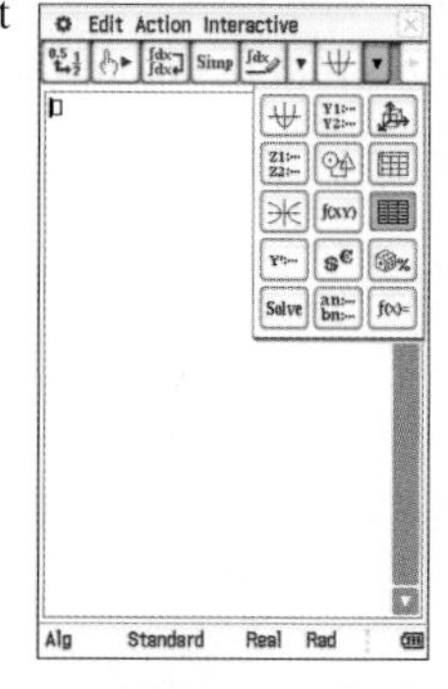
2. To set the appropriate data input method, select Stats then press the OK button.	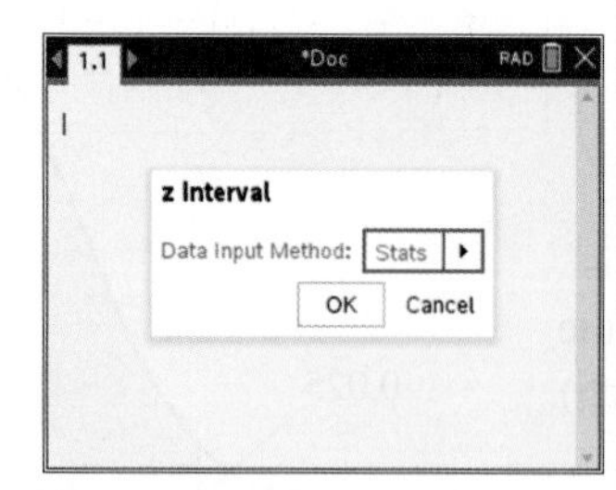	2. Select Calc > Interval	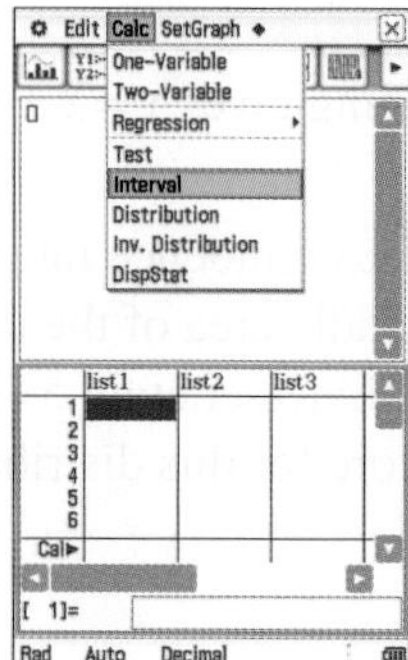

3. Complete the entry lines.
σ: 0.23
$\bar{x}$: 4.53
n: 20
C Level: 0.95

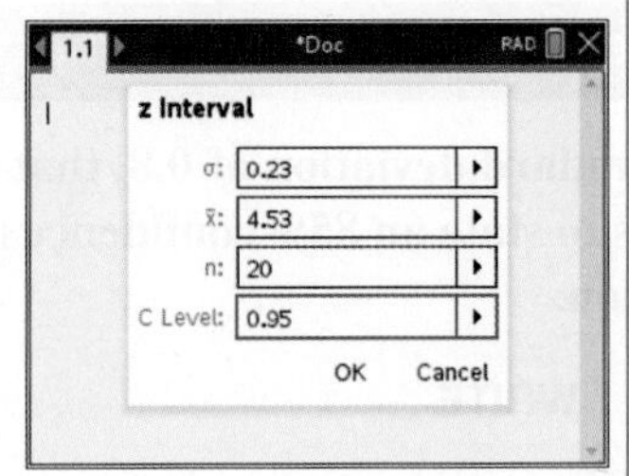

Press the OK button.
The answer will appear on the screen.

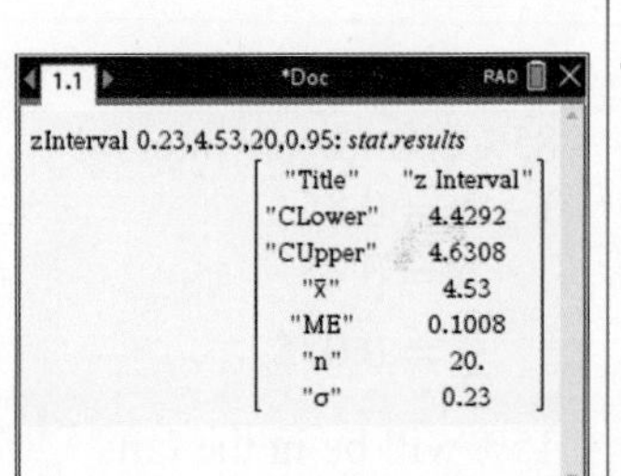

3. Select
One-Sample Z Int and the variable button

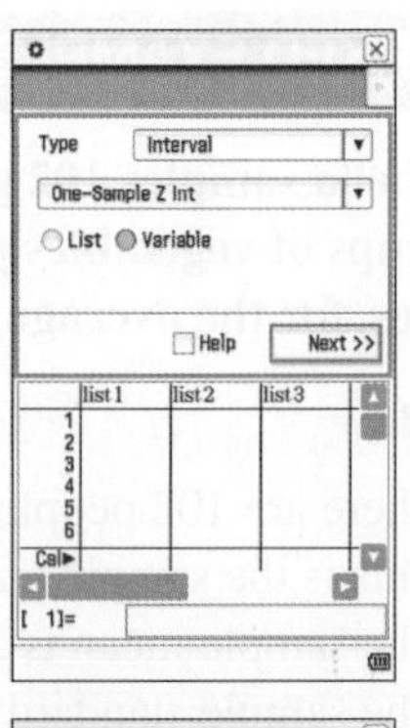

4. Complete the entry lines
0.95
0.23
4.53
20

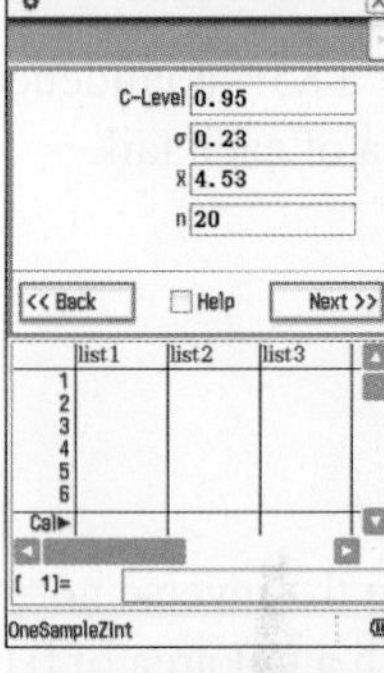

5. Select Next.
The answer appears on the screen.

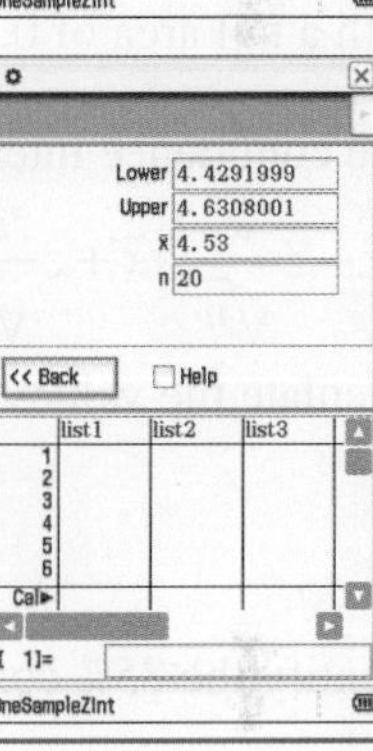

We have previously learnt that for a 90% or 99% confidence interval a z scores of 1.645 or 2.58 are used respectively.

In general, the inverse normal can be used to determine the z score to be used in the confidence interval calculation. Remember that the area under the curve is 1. The confidence interval is located in the middle of the distribution and the area of the right hand tail is used to determine the number of standard deviations from the mean. Because the curve is symmetric, the area of each tail will be equal. If the total tail area is α, then the middle area would be $1 - \alpha$ and the area of each tail is $\frac{\alpha}{2}$. As the middle area is $1 - \alpha$, we can talk about a $1 - \alpha$ confidence interval. As the area of each tail is $\frac{\alpha}{2}$, the z score that has a tail area of $\frac{\alpha}{2}$ is used.

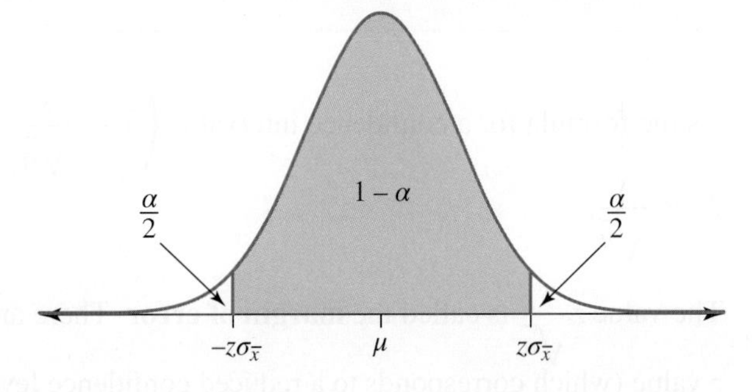

To calculate the z score that corresponds to a particular tail area use the Inverse Normal function on your CAS calculator. Remember to convert the area from a percentage into a decimal.

WORKED EXAMPLE 9 Constructing confidence intervals with other confidence levels

Arabella samples 102 people and finds, with a standard deviation of 0.8, that on average they eat 5.2 cups of vegetables per day. Use these statistics to state an 85% confidence interval, to 2 decimal places, for the average daily vegetable consumption.

THINK	WRITE
1. There are 102 people in the sample. This is the sample size. The sample mean is 5.2. The sample standard deviation is 0.8.	$n = 102$ $\bar{x} = 5.2$ $s = 0.8$
2. For a 85% confidence interval, determine the area of each tail.	85% confidence interval $1 - \alpha = 0.85$ $\alpha = 0.15$ $\frac{\alpha}{2} = 0.075$ 15% will be in the tails 7.5% in each tail
3. Use the inverse normal to determine the z score with a tail area of 0.075	$z_{0.075} = 1.44$
4. The confidence interval is $\left(\bar{x} - z\frac{s}{\sqrt{n}},\ \bar{x} + z\frac{s}{\sqrt{n}}\right)$. Use a calculator to calculate the values of $\bar{x} - z\frac{s}{\sqrt{n}}$ and $\bar{x} + z\frac{s}{\sqrt{n}}$.	$\bar{x} - z\frac{s}{\sqrt{n}} = 5.2 - 1.44 \times \frac{0.8}{\sqrt{102}}$ $= 5.09$ $\bar{x} + z\frac{s}{\sqrt{n}} = 5.2 + 1.44 \times \frac{0.8}{\sqrt{102}}$ $= 5.31$
5. Identify the 85% confidence interval.	$\left(\bar{x} - z\frac{s}{\sqrt{n}},\ \bar{x} + z\frac{s}{\sqrt{n}}\right) = (5.09, 5.31)$
6. State the answer.	An 85% confidence interval for the number of cups of vegetables consumed per day is (5.09, 5.31).

As the formula for a confidence interval is $\left(\bar{x} - z\frac{s}{\sqrt{n}},\ \bar{x} + z\frac{s}{\sqrt{n}}\right)$, the size of the interval is given by $2z \times \frac{s}{\sqrt{n}}$.

The value $z\frac{s}{\sqrt{n}}$ is called the **margin of error**. There are 2 ways to decrease the margin of error: decrease the z value (which corresponds to a reduced confidence level) or increase the sample size, n (which has no effect on the confidence level). Therefore, if you wish to reduce the margin of error for a confidence interval at a particular confidence level, you must increase the sample size.

Margin of error

The margin of error in a confidence interval is equal to:

$$z\frac{s}{\sqrt{n}}$$

WORKED EXAMPLE 10 Calculating the sample size

In Worked example 8, the average amount spent on lunch was between \$4.43 and \$4.63. Determine the sample size needed to reduce the interval to ± \$0.05 for a 95% confidence interval.

THINK / **WRITE**

1. The confidence interval formula is $\left(\bar{x}-z\frac{s}{\sqrt{n}},\ \bar{x}+z\frac{s}{\sqrt{n}}\right)$, this means that we need $z\frac{s}{\sqrt{n}}=0.05$.

   $z\frac{s}{\sqrt{n}}=0.05$

2. Although the sample standard deviation will change with a larger sample, the current value is the best estimate that we have.

   $s=0.23$

3. For a 95% confidence interval, $z=1.96$

   $z=1.96$

4. Solve $z\frac{s}{\sqrt{n}}=0.05$ for n.

$$z\frac{s}{\sqrt{n}}=0.05$$
$$1.96\times\frac{0.23}{\sqrt{n}}=0.05$$
$$1.96\times\frac{0.23}{0.05}=\sqrt{n}$$
$$\sqrt{n}=9.016$$
$$n=(9.016)^2$$
$$n=81.288256$$

At least 82 people would need to be surveyed.

13.4.4 Using confidence intervals to estimate other confidence intervals

If a sample's confidence interval for a population parameter is known, it is possible to estimate the confidence interval at a different level of significance.

WORKED EXAMPLE 11 Calculating a confidence interval from a given confidence interval

If we know that a 99% confidence interval for the average blog length is between 1200 and 2000 words, calculate a 95% confidence interval, to the nearest whole number, for the average blog length.

THINK / **WRITE**

1. As we are talking about the 99% confidence interval and the 95% confidence interval, 2 different z values will be used. To differentiate between them, call them $z_{0.005}$ and $z_{0.025}$ respectively. The number 0.005 represents the area of one tail for the 99% confidence interval. The number 0.025 represents the area of one tail for the 95% confidence interval.

   $z_{0.005}=2.58$
   $z_{0.025}=1.96$

2. The confidence interval formula is $\left(\bar{x}-z\frac{s}{\sqrt{n}},\ \bar{x}+z\frac{s}{\sqrt{n}}\right)$. The mean will be in the centre of the distribution.

$$\bar{x}=\frac{1200+2000}{2}$$
$$=1600$$

3. Use the upper limit of the confidence interval, $\bar{x}+z\frac{s}{\sqrt{n}}$, to calculate $z\frac{s}{\sqrt{n}}$.

$$2000=\bar{x}+z_{0.005}\frac{s}{\sqrt{n}}$$
$$2000=1600+z_{0.005}\frac{s}{\sqrt{n}}$$
$$z_{0.005}\frac{s}{\sqrt{n}}=400$$

4. As this is a 99% confidence interval, $z_{0.005}=2.58$, solve for $\frac{s}{\sqrt{n}}$.

$$2.58\times\frac{s}{\sqrt{n}}=400$$
$$\frac{s}{\sqrt{n}}=\frac{400}{2.58}$$

5. For a 95% confidence interval, $z_{0.025}=1.96$. Use this value to calculate the confidence interval $\left(\bar{x}-z\frac{s}{\sqrt{n}},\ \bar{x}+z\frac{s}{\sqrt{n}}\right)$.

$$\bar{x}-z\frac{s}{\sqrt{n}}=1600-1.96\times\frac{400}{2.58}$$
$$=1600-304$$
$$=1296$$
$$\bar{x}+z\frac{s}{\sqrt{n}}=1600+1.96\times\frac{400}{2.58}$$
$$=1600+304$$
$$=1904$$

6. State the 95% confidence interval.

A 95% confidence interval for the average blog length is (1296, 1904).

13.4 Exercise

Students, these questions are even better in jacPLUS

Receive immediate feedback and access sample responses

Access additional questions

Track your results and progress

Find all this and MORE in jacPLUS

Technology active

1. **WE8** Of 75 Victorians that were surveyed, the average amount spent on holidays this year was \$2314 with a standard deviation of \$567. Use these statistics to state a 95% confidence interval, correct to the nearest dollar, for the average amount spent on holidays by Victorians.

2. After surveying 30 swimmers as they entered the local swimming complex, it was found that the average distance that they intended to swim was 1.2 km. The sample standard deviation was 0.5 km. Use these statistics to state a 95% confidence interval, correct to 2 decimal places, for the average distance people were intending to swim that day.

3. **WE9** James samples 116 people and finds, with a standard deviation of \$537, that their average car value is \$23 456. Use these statistics to state an 80% confidence interval, correct to the nearest dollar, for the average car value for the population.

4. Charles samples 95 people and finds, with a standard deviation of 5.8 g, that on average, 25.7 g of chocolate is consumed per person per day. Use these statistics to determine a 75% confidence interval, correct to 2 decimal places, for an estimate for the amount of chocolate consumed per person per day.

5. **WE10** Determine the sample size needed to reduce the interval from question **1** to $\pm\$100$ for a 95% confidence interval.

6. Determine the sample size needed to reduce the interval from question **2** to ± 0.1 km for a 95% confidence interval.

7. You are auditing a bank. From a sample of 50 cash deposits, the mean and standard deviation are calculated to be \$203.45 and \$43.32 respectively.
   a. Determine a 95% confidence interval for the average cash deposit amount.
   b. Determine a 90% confidence interval for the average cash deposit amount.
   c. Calculate the sample size needed to reduce the interval to $\pm\$2$ for a 95% confidence interval.

8. A sample of 40 AA batteries were tested to determine their average lifetime. It was found that they lasted an average of 2314 minutes with a standard deviation of 243 minutes.
   a. Determine a 95% confidence interval, correct to the nearest minute, for the average battery life.
   b. Determine a 99% confidence interval, correct to the nearest minute, for the average battery life.
   c. Calculate the sample size needed to reduce the interval to ± 50 minutes for a 95% confidence interval.

9. For her class assignment, Holly records the time to travel to school. She records the times for 30 days and finds the average trip time is 24.6 minutes with a standard deviation of 7.6 minutes.
   a. Determine a 95% confidence interval, correct to 1 decimal place, for the average time that Holly will take to travel to school.
   b. Determine a 90% confidence interval, correct to 1 decimal place, for the average time that Holly will take to travel to school.
   c. Calculate the sample size needed to reduce the interval to ± 2 minutes for a 95% confidence interval.
   d. Calculate the sample size needed to reduce the interval to ± 2 minutes for a 90% confidence interval.

10. **WE11** If we know that a 99% confidence interval for the average television commercial break length is between 2 and 3 minutes, calculate a 95% confidence interval, correct to 2 decimal places, for the average television commercial break length.

11. A sample of feature movies has a 90% confidence level for the average running time of $(85, 90)$ minutes. Calculate a 99% confidence interval, correct to 1 decimal place, for the average running time length.

12. If the average song length of the top 100 songs in an online music store has a 95% confidence interval of $(215.171, 238.695)$ seconds, calculate a 99% confidence interval, correct to 3 decimal places.

13.4 Exam questions

Question 1 (1 mark) TECH-ACTIVE

Source: *VCE 2021, Specialist Mathematics Exam 2, Section A, Q18; © VCAA.*

MC A scientist investigates the distribution of the masses of fish in a particular river. A 95% confidence interval for the mean mass of a fish, in grams, calculated from a random sample of 100 fish is (70.2, 75.8).

The sample mean divided by the population standard deviation is closest to

A. 1.3 **B.** 2.6 **C.** 5.1 **D.** 10.2 **E.** 13.0

Question 2 (1 mark) TECH-ACTIVE

Source: *VCE 2019, Specialist Mathematics Exam 2, Section A, Q18; © VCAA.*

MC The masses of a random sample of 36 track athletes have a mean of 65 kg. The standard deviation of the masses of all track athletes is known to be 4 kg.

A 98% confidence interval for the mean of the masses of all track athletes, correct to one decimal place, would be closest to

A. (51.0, 79.0) **B.** (63.6, 66.7) **C.** (63.3, 66.7) **D.** (63.4, 66.6) **E.** (64.3, 65.7)

Question 3 (1 mark) TECH-ACTIVE

Source: *VCE 2018, Specialist Mathematics Exam 2, Section A, Q18; © VCAA.*

MC A 95% confidence interval for the mean height μ in centimetres, of a random sample of 36 Irish setter dogs is $58.42 < \mu < 67.31$

The standard deviation of the height of the population of Irish setter dogs, in centimetres, correct to two decimal places, is

A. 2.26 **B.** 2.27 **C.** 13.60 **D.** 13.61 **E.** 62.87

More exam questions are available online.

13.5 Hypothesis testing

LEARNING INTENTION

At the end of this subtopic you should be able to:

- set up and evaluate hypothesis tests to determine whether there is sufficient evidence to state that the mean of a population is different to the supposed mean.

Calculating confidence intervals involves finding an interval in which the population mean is likely to lie. Different samples can have different means. If a sample mean is different to what might have been expected, we need to determine if the difference is large enough to state that the sample is from a different population, or if the difference is just due to chance.

A parameter is a numerical property of a population. A statistic is a numerical property of a sample.

13.5.1 Null and alternative hypotheses

A hypothesis is a claim about a population that requires formal investigation before the claim can be determined. For example, when developing a new drug, it needs to undergo some testing to determine whether it is more, or less effective than the existing drug, or whether it has no perceptible difference in effect from the existing drug. To formalise the process of this testing we must make a couple of hypotheses. The first hypothesis is called the null hypothesis and is denoted by H_0. It states that the subject of the test has no significant effect on the outcome of the test. The alternative hypothesis is denoted by H_1, and states that the subject of the test does have a significant effect on the outcome. There are three different forms of alternative hypotheses. The alternative hypothesis can claim that the result of the test is different to the existing value (less than or greater than) in which case it is called a two sided or two-tailed test. It can also claim that the result is less than the existing value, or that it is greater than the existing value. In these cases the alternative hypothesis is called a one sided or one-tailed test.

In the example about the new drug, we could propose the following hypotheses:

H_0: The new drug has no difference in effectiveness to the existing drug.

H_1: The new drug is more effective than the existing drug.

Often hypothesis tests work by comparing a mean μ, to a known population mean μ_0. The null and alternative hypotheses can therefore be explained as follows.

Null and alternative hypotheses

- **The null hypothesis makes the claim that there is no difference, H_0: $\mu = \mu_0$.**
- **The alternative hypothesis makes the claim that the resulting value is different to the existing value, H_1: $\mu \neq \mu_0$. This is a two-tailed test.**
- **The alternative hypothesis makes the claim that the resulting value is less than the existing value, H_1: $\mu < \mu_0$. This is a one-tailed test.**
- **The alternative hypothesis makes the claim that the resulting value is greater than the existing value, H_1: $\mu > \mu_0$. This is a one-tailed-test.**

WORKED EXAMPLE 12 Identifying null and alternative hypotheses

Riley has decided to participate in the school's Maths tutoring process. He decides to go for an extra two hours per week, hoping to improve his grades. With reference to his exam results, state the null and alternative hypotheses.

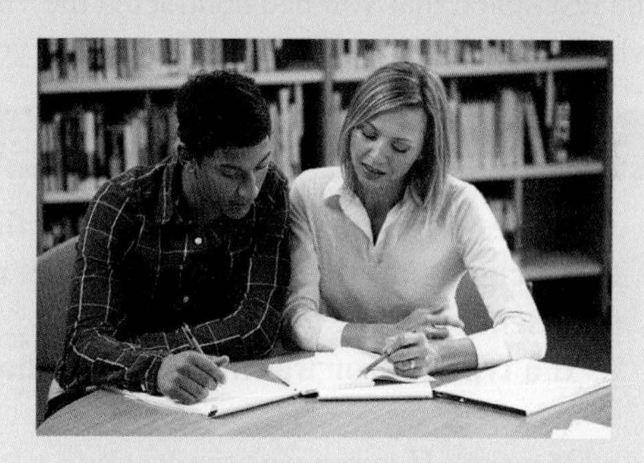

THINK	WRITE
1. The null hypothesis states there will be no difference.	H_0: The extra tutoring will have no effect on Riley's Maths results.
2. The alternative hypothesis states that there will be an effect. In this case we are hoping that there is an improvement.	H_1: The extra tutoring will improve Riley's Maths results.

13.5.2 Type I and Type II errors

In performing a hypothesis test we assume that the null hypothesis is true and then, after analysing the data, we make a decision regarding the alternative hypothesis that we are trying to prove. If there is sufficient evidence to support the alternative hypothesis, we reject the null hypothesis and accept the alternative hypothesis as true. Note that we never prove the null hypothesis, we just accept that it is true unless we have sufficient evidence to say otherwise. The details regarding *sufficient evidence* will be discussed later, but in short, we choose the level of significance we would like to test to, and if the statistics tell us with that level of significance (or higher) we reject the null hypothesis and accept the alternative hypothesis. A 5% level of significance is commonly used.

The fact that no hypothesis test is 100% certain means that there are some errors that can be concluded from the results of a statistical analysis. There are two types of errors; Type I and Type II, and they are defined as follows.

A Type I error is rejecting the null hypothesis when it is actually true.

A Type II error is accepting the null hypothesis when it is actually false.

Type I and Type II errors

- **A Type I error is rejecting the null hypothesis when it is actually true (false positive).**
- **A Type II error is accepting the null hypothesis when it is actually false (false negative).**

		Reality	
		H_0 is true	**H_0 is false**
Decision	**H_0 is accepted**	**Correct**	**Type II error**
	H_0 is rejected	**Type I error**	**Correct**

As an example of Type I and Type II errors consider the possible outcomes of a real-life test which is designed to detect a virus (such as the flu or COVID). The null hypothesis is that you do not have the virus and the alternative hypothesis is that you do have the virus. A Type I error is a false positive, which would happen if the test stated that you had contracted the virus when in reality you had not. A Type II error is a false negative, which would happen if the test stated that you had not contracted the virus when in reality you had contracted it.

WORKED EXAMPLE 13 Identifying Type I and Type II errors

Suppose you are testing a new medication. You are going to conclude that the drug has no effect unless there is sufficient evidence to say otherwise.

a. If a type I error was made, state what conclusions were reached.

b. If a type II error was made, state what conclusions were reached.

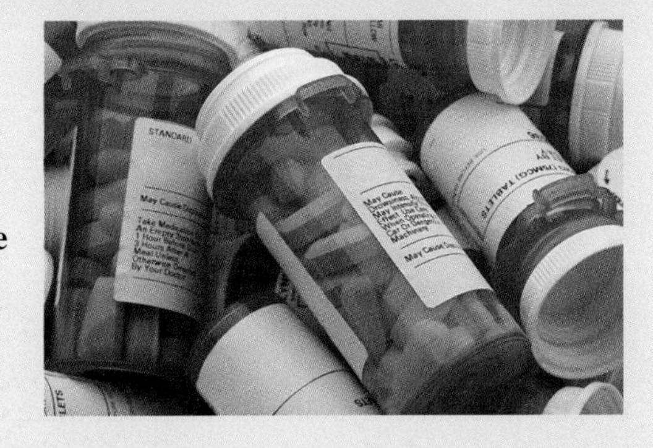

THINK	WRITE
Create the null and alternative hypotheses.	H_0: The new drug has the same effect as the normal drug. H_1: The new drug has a different effect.
a. A type I error means that the null hypothesis was rejected when it was actually true.	a. The null hypothesis was rejected. The new drug works differently to the normal drug.
b. A type II error means that the null hypothesis is accepted when it is actually false.	b. The null hypothesis was accepted. There is insufficient evidence to support the claim that the new drug works differently.

13.5.3 The level of significance, α

When hypothesis testing, we need to determine the point at which our result is significant enough to reject the null hypothesis and accept the alternative hypothesis. The level of significance, α, is what we use to set this point. α is the probability of rejecting the null hypothesis when it is true (obtaining a Type I error). Note that since α represents a probability, $0 < \alpha < 1$. Usually $\alpha = 0.05$ or $\alpha = 0.10$, which equate to probabilities of 5% and 10% respectively.

For two-sided hypothesis tests (H_1: $\mu \neq \mu_0$), α is the combined area of the tails as can be seen in the diagram below. Note that each tail has an area of $\frac{\alpha}{2}$.

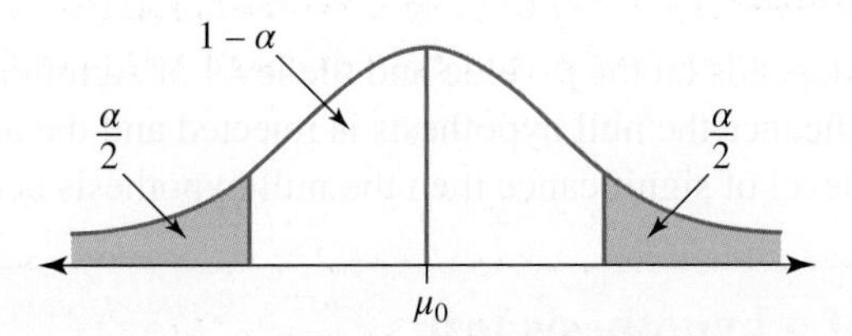

For one-sided hypothesis tests, α is the area of the tail. Depending on the alternative hypothesis, this tail will be on the left or right side of the distribution.

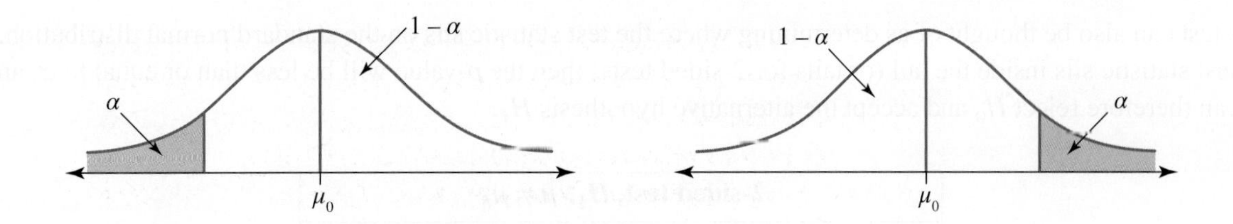

13.5.4 Hypothesis testing

The test statistic, z_c

The test statistic is used to determine whether the null hypothesis can be rejected or not. It takes the value of the sample mean, $\bar{x}$, which is calculated in the statistical test and transposes it onto the standard normal distribution, Z.

There are two formulas for the test statistic, one which is used when the population standard deviation, σ, is known, and one which is used when the population standard deviation is not known and the sample size, n, is greater than or equal to 30, in which case we use the sample standard deviation, s.

The test statistic, z_c

When the population standard deviation is known:

$$z_c = \frac{\bar{x} - \mu_0}{\left(\frac{\sigma}{\sqrt{n}}\right)}$$

When the population standard deviation is not known, but $n \geq 30$:

$$z_c = \frac{\bar{x} - \mu_0}{\left(\frac{s}{\sqrt{n}}\right)}$$

The p-value

Once you have calculated the test statistic it is time to determine the p-value, p, that it corresponds to. The p-value is the probability of obtaining results at least as extreme as the test statistic, assuming that the null hypothesis is true. The way that the p-value is calculated depends on whether the test being conducted is a one- or two-sided test.

For a two-tailed test: $p = \Pr(Z > |z_c|) = \Pr(Z > z_c) + \Pr(Z < -z_c) = 2\Pr(Z > z_c)$

For a one-tailed left test with $H_1:\ \mu < \mu_0$, $p = \Pr(Z < z_c)$

For a one-tailed right test with $H_1:\ \mu > \mu_0$, $p = \Pr(Z > z_c)$

The result of a hypothesis test

The result of the hypothesis test depends on the p-value and the level of significance, α. If the p-value is less than or equal to the level of significance the null hypothesis is rejected and the alternative hypothesis is accepted. If the p-value is greater than the level of significance then the null hypothesis is not rejected.

The result of a hypothesis test

If $p \leq \alpha$, reject H_0 and accept H_1.

If $p > \alpha$, do not reject H_0.

This test can also be thought of as determining where the test statistic sits on the standard normal distribution. If the test statistic sits inside the tail (or tails for 2-sided tests) then the p-value will be less than or equal to α, and we can therefore reject H_0 and accept the alternative hypothesis H_1.

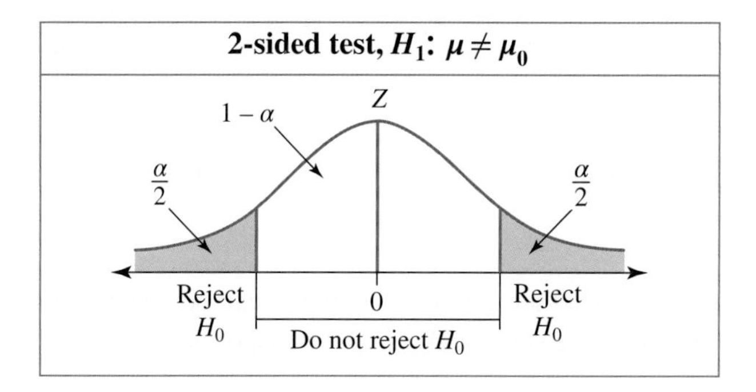

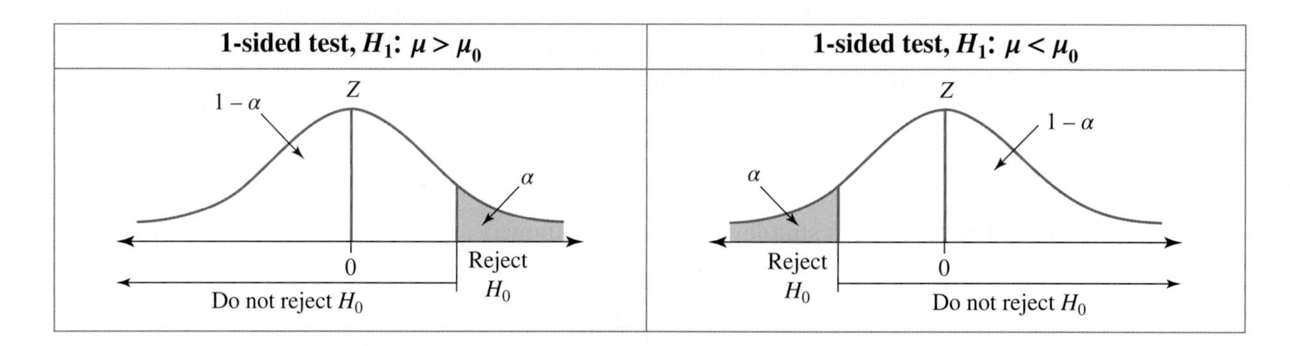

Hypothesis testing step-by-step

1. **Write down the null hypothesis, H_0, and alternative hypothesis, H_1.**
2. **Write down the parameters and statistics.**
3. **Calculate the test statistic $z_c = \dfrac{\overline{x} - \mu_0}{\sigma / \sqrt{n}}$**
4. **Calculate the p-value.**
5. **State the level of significance, α.**
6. **Compare the p-value to α.**
7. **State the result.**
   If $p \leq \alpha$, reject H_0 and accept H_1.
   If $p > \alpha$, do not reject H_0.

WORKED EXAMPLE 14 Evaluating data using hypothesis tests (1)

Bi More supermarkets' sales records show that the average monthly spending expenditure per person on a certain product was \$11 with a standard deviation of \$1.80. The company ran a promotion campaign and would like to know if sales have changed. They sampled 30 families and found their average expenditure to be \$11.50.

a. **Write down suitable hypotheses to test whether the promotion campaign has changed sales figures.**

b. **Determine the p-value for this test.**

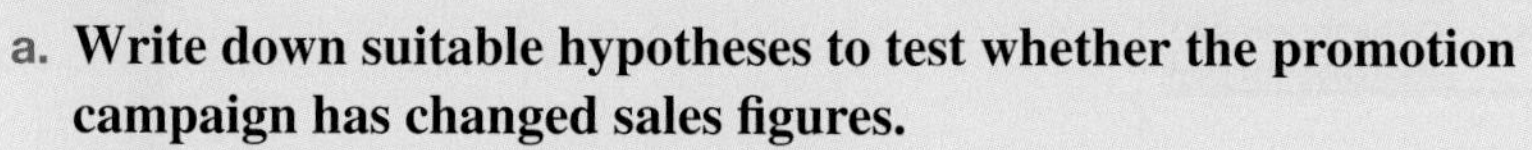

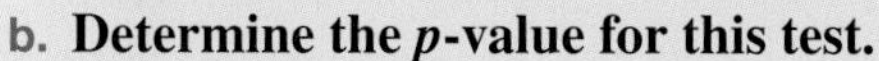

c. **State with reasons whether the null hypothesis should be rejected or not rejected at the 5% level of significance.**

THINK	WRITE/DRAW
a. State the null hypothesis. State the alternative hypothesis, which is what we are trying to prove. We are not sure if the sales have improved or not, so we are just testing if the promotion has changed the sales.	a. H_0: $\mu = 11$ H_1: $\mu \neq 11$ This is a two-sided test.
b. 1. Identify the required parameter values.	b. $\overline{x} = 11.5$, $\mu_0 = 11$, $\sigma = 1.8$, $n = 30$
2. Calculate the value of the test statistic.	$z_c = \dfrac{\overline{x} - \mu_0}{\left(\frac{\sigma}{\sqrt{n}}\right)} = \dfrac{11.5 - 11}{\left(\frac{1.8}{\sqrt{30}}\right)} = 1.5215$
3. Calculate the p-value	$p = \Pr(\lvert Z \rvert > 1.52) = 2\Pr(Z > 1.52) = 0.1281$
c. 1. Determine the value of α which corresponds to a level of significance of 5%.	c. $\alpha = 0.05$

2. Compare the p-value with the value of α.

$p = 0.1281 > 0.05 = \alpha$

3. State the result.

Since $p > \alpha$, there is insufficient evidence to accept H_1.
Do not reject the null hypothesis H_0.
This can also be shown by plotting the value of z_c on the standard normal distribution and noting that the z_c is **not** in the tails.

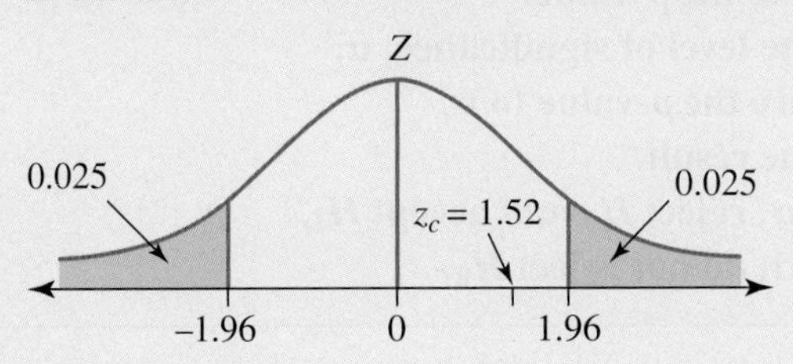

TI | THINK

1. On a Calculator page, press MENU then select
6 Statistics
7 Stat Tests
1 z Test…

DISPLAY/WRITE

2. Change the Data Input Method to 'Stats', click OK and then enter the required values and test type as shown.

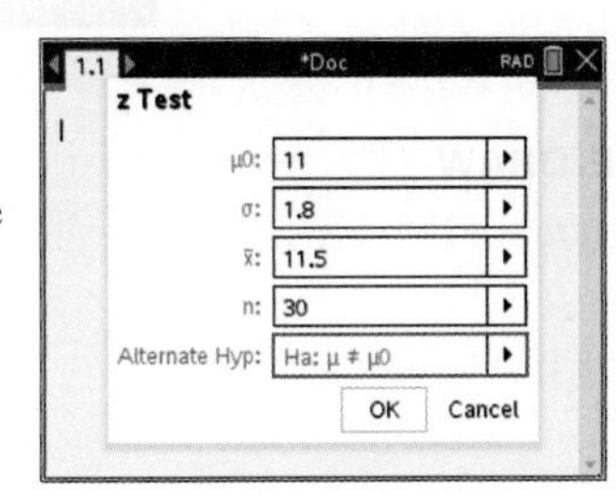

3. Click OK and the results will appear.

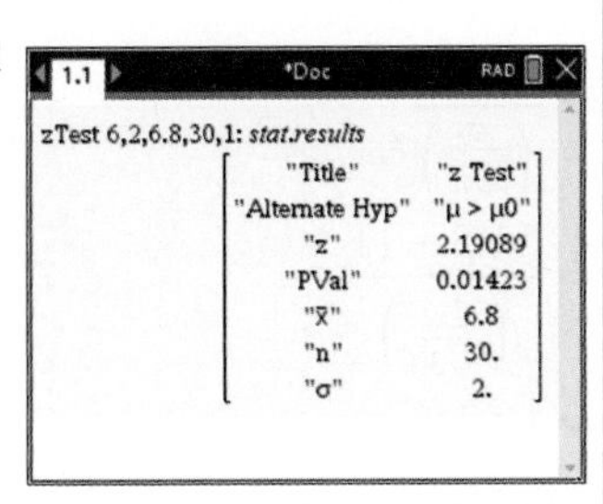

CASIO | THINK

1. From a Main screen, select the Statistics application.

DISPLAY/WRITE

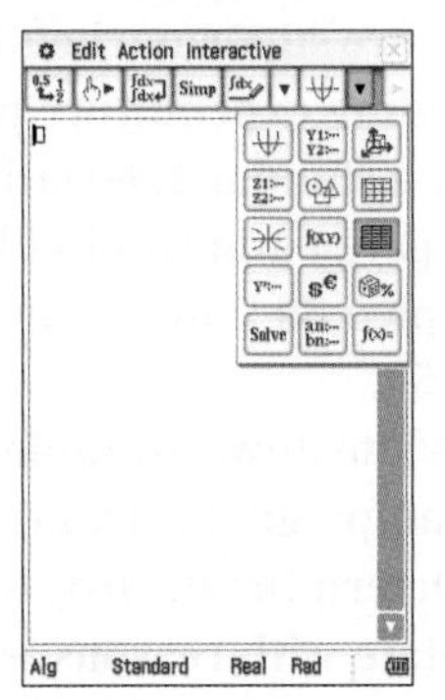

2. Select Calc > Test > One-Sample Z-Test > Variable, then click Next.

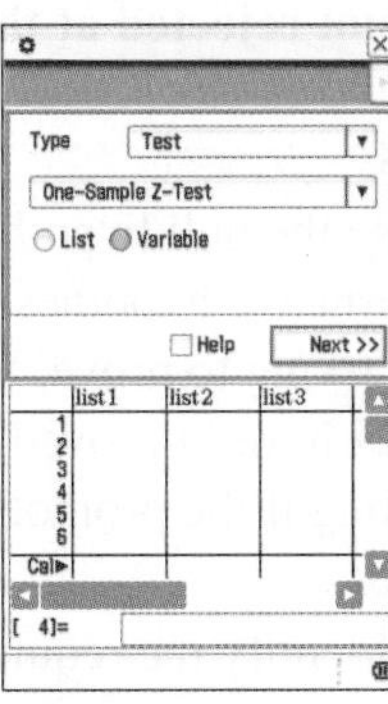

3. Enter the required condition and values as shown.

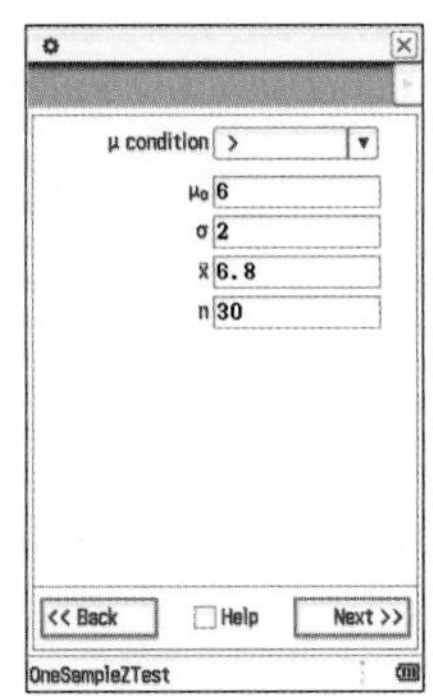

4. Click Next and the results will appear.

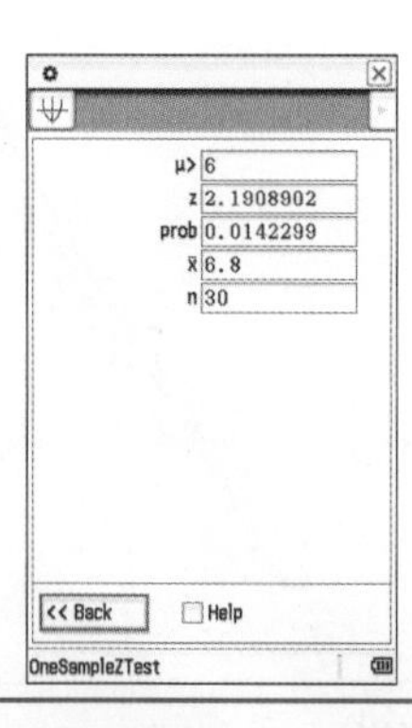

WORKED EXAMPLE 15 Evaluating data using hypothesis tests (2)

It is known that average number of houses sold by real estate agents from a certain company is 6 houses per month, with a standard deviation of 2 houses per month. 30 real estate agents from this company went on a sales improvement course, and after the course, they sold on average 6.8 houses per month.

a. **Write down suitable hypotheses to test whether the course has improved sales.**
b. **Determine the *p*-value for this test.**
c. **State with reasons whether the null hypothesis should be rejected or not rejected at the 1% level of significance.**

THINK	WRITE/DRAW
a. State the null hypothesis. State the alternative hypothesis, which is what we are trying to prove. Since the test is to see whether the sales numbers increased this is a one-sided test.	a. H_0: $\mu = 6$ H_1: $\mu > 6$ This is a one-sided test.
b. 1. Identify the required parameter values.	b. $\bar{x} = 6.8, \quad \mu_0 = 6, \quad \sigma = 2, \quad n = 30$
2. Calculate the value of the test statistic.	$z_c = \dfrac{\bar{x} - \mu_0}{\left(\frac{\sigma}{\sqrt{n}}\right)} = \dfrac{6.8 - 6}{\left(\frac{2}{\sqrt{30}}\right)} = 2.1909$
3. Calculate the *p*-value	$p = \Pr(Z > 2.1909)$ $= 0.0142$
c. 1. Determine the value of α which corresponds to a level of significance of 1%.	c. $\alpha = 0.01$
2. Compare the *p*-value with the value of α.	$p = 0.0142 > 0.01 = \alpha$
3. State the result.	Since $p > \alpha$, there is insufficient evidence to accept H_1. Do not reject the null hypothesis H_0.

This can also be shown by plotting the value of z_c on the standard normal distribution and noting that the z_c is **not** in the tail.

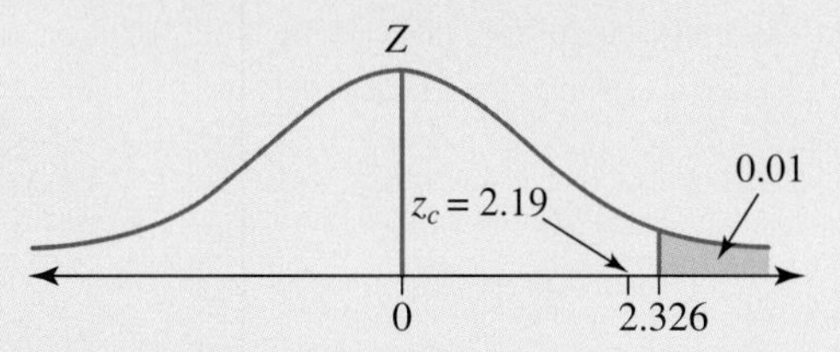

TI | THINK	DISPLAY/WRITE	CASIO | THINK	DISPLAY/WRITE
1. On a Calculator page, navigate to z Test… using the steps outlined in worked example 14. Enter the required values and test type as shown.	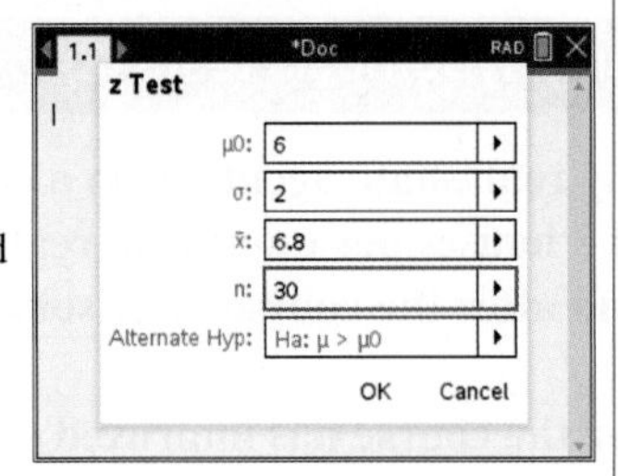	1. On a Main page, navigate to One Sample Z-Test using the steps outlined in worked example 14. Enter the required condition and values as shown.	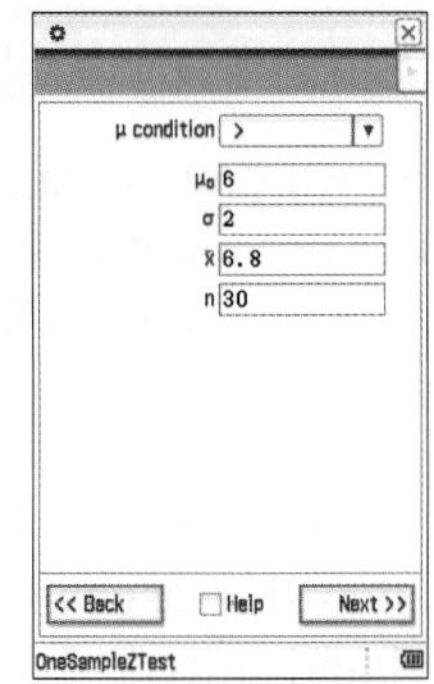
2. Click OK and the results will appear.	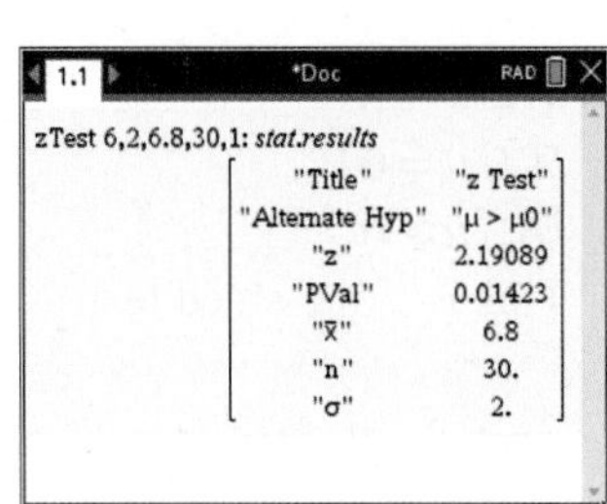	2. Click Next and the results will appear.	

WORKED EXAMPLE 16 Evaluating manufacturer claims using hypothesis tests

Apple manufacturers claim the new iPhone has a battery life of 12 hours. Julia thinks that the battery life is less than the 12 hours that the manufacturers claim. She sampled 36 iPhones and found that the average battery life was 11.6 hours, with a standard deviation of 90 minutes.

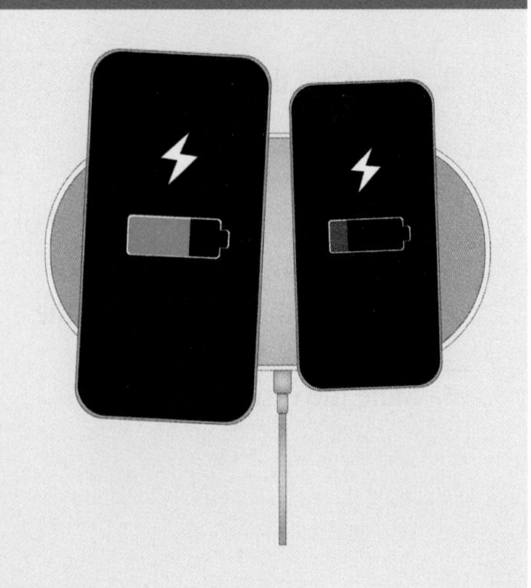

a. Write down suitable hypotheses to test whether the mean lifetime is less than that claimed by the manufacturer.

b. Determine the p-value for this test.

c. State with reasons whether the null hypothesis should be rejected or not rejected at the 10% level of significance.

d. Let $\overline{X}$ be the lifetime of a random sample of 36 such iPhones, assuming $\mu = 12$, determine the value I^* such that $\Pr\left(\overline{X} < I^*\right) = 0.1$, and interpret the meaning of the value of I^*.

THINK	WRITE/DRAW
a. State the null hypothesis. State the alternative hypothesis, which is what we are trying to prove. Since the test is investigating whether the battery life is lower than the stated value, this is a one-sided test.	**a.** $H_0\text{: } \mu = 12$ $H_1\text{: } \mu < 12$ This is a one-sided test.
b. 1. Identify the required parameter values.	**b.** $\bar{x} = 11.6,\ \ \mu_0 = 12,\ \ s = 1.5,\ \ n = 36$
2. Calculate the value of the test statistic.	We can substitute s for σ, since $n \geq 30$ $z_c = \dfrac{\bar{x} - \mu_0}{s/\sqrt{n}} = \dfrac{11.6 - 12}{1.5/\sqrt{36}}$ $z_c = -1.6$ $z_c = \dfrac{\bar{x} - \mu_0}{\left(\frac{s}{\sqrt{n}}\right)}$ $= \dfrac{11.6 - 12}{\left(\frac{1.5}{\sqrt{36}}\right)}$ $= -1.6$
3. Calculate the p-value	$p = \Pr(Z < -1.6)$ $= 0.0548$
c. 1. Determine the value of α which corresponds to a level of significance of 10%.	**c.** $\alpha = 0.1$
2. Compare the p-value with the value of α.	$p = 0.0548 < 0.1 = \alpha$
3. State the result.	Since $p < \alpha$, there is sufficient evidence to accept H_1. Reject the null hypothesis H_0 and accept the alternative hypothesis H_1. This can also be shown by plotting the value of z_c on the standard normal distribution and noting that the z_c **is** in the tail.
	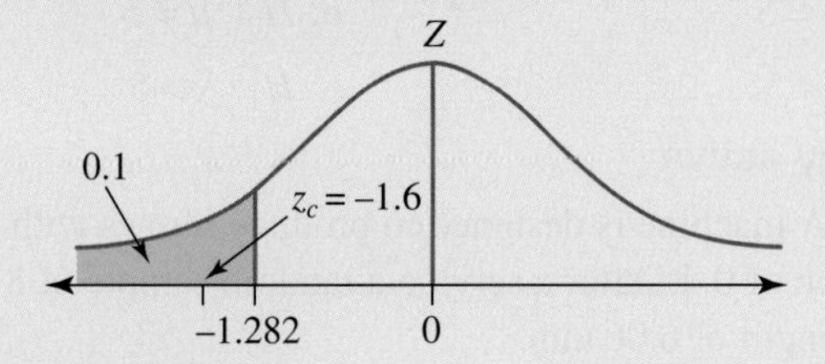
d. 1. Use the formula to convert the value of I^* into the test statistic. Recall that the $z_{0.1} = -1.282$.	**d.** $\Pr\left(\bar{X} < I^*\right) = 0.1$ $\dfrac{I^* - 12}{\left(\frac{1.5}{\sqrt{36}}\right)} = -1.282$
2. Solve for I^*.	$I^* = \dfrac{-1.282 \times 1.5}{6} + 12$ $= 11.68$
3. Give an interpretation of the results obtained.	If the mean of a sample of 36 iPhones is less than 11.68, we would have sufficient evidence to reject the null hypothesis, H_0.

13.5 Exercise

Students, these questions are even better in jacPLUS

Receive immediate feedback and access sample responses

Access additional questions

Track your results and progress

Find all this and MORE in jacPLUS

Technology free

1. WE12 Des has decided to increase his mountain bike training by adding an extra three 1-hour sessions to his weekly training schedule. He is hoping to improve his best race time. With reference to his race times, state the null and alternative hypotheses.

2. Gabby attempts to improve her cross country time by adding an extra training session per week. With reference to race times, state the null and alternative hypotheses.

3. WE13 Ezy Pet Food believes that it has a contaminated batch of pet food. The company decides to test some of the tins. It will assume that the batch is contaminated unless there is evidence to the contrary.
   a. If a type I error is made, state what conclusions are reached.
   b. If a type II error is made, state what conclusions are reached.

4. The probability of low birth weight in Australia is about 6%. Out of 200 babies born at Bundaberg during a period when aerial spraying of sugar cane was common, 18 had low birth weight. The local paper claimed that this proved that aerial spraying was dangerous because the rate of birth problems was 150% of the national average. Aerial spraying will only be stopped if there is sufficient evidence to do so.
   a. If a type I error was made, state what conclusions were reached.
   b. If a type II error was made, state what conclusions were reached.

5. Determine whether each of the following are valid null and alternative hypotheses.
   a. H_0: $\mu > 5$
      H_1: $\mu < 5$
   b. H_0: $\bar{x} \neq 5$
      H_1: $\bar{x} < 5$
   c. H_0: $\bar{x} = 5$
      H_1: $\bar{x} \neq 5$
   d. H_0: $\mu = 5$
      H_1: $\bar{x} < 5$
   e. H_0: $\mu \neq 5$
      H_1: $\bar{x} > 5$
   f. H_0: $\bar{x} = 5$
      H_1: $\mu \neq 5$

Technology active

6. WE14 A machine is designed to produce screws with a stated mean length of 6 mm, with a standard deviation of 0.4. After a service a random sample of 81 screws produced by the machine was found to have a mean length of 6.06 mm.
   a. Write down suitable hypothesis to test whether there is a change in the mean length of the screws produced by the machine.
   b. Determine the p-value for this test, correct to 3 decimal places.
   c. State with reasons whether the null hypothesis should be rejected or not rejected at the 5% level of significance.

7. Steel rods for car components produced by a machine are normally distributed with a mean of length of 4 cm in diameter and standard deviation 0.3 cm. The machine is serviced, after which a random sample of 49 steel rods gives a mean length of 4.12 cm. Perform a hypothesis test, stating the p-value, correct to 4 decimal places, and at the 1% level of significance, to determine if there is change in the mean lengths of the steel rods produced by the machine.

8. **WE15** Car tyres come in many different brands. Even on a car, the tyres can wear out after travelling different distances, due to cornering, braking and wheel alignment. Manufacturers of car tyres are interested in the average distance in kilometres travelled by tyres, or the lifetime of tyres until they are considered un-roadworthy and need to be replaced. For car tyres, of a certain car, records show that the lifetime is normally distributed with a mean of 40 000 km and a standard deviation of 9000 km. A random sample of 32 car tyres of a newer brand were driven and it was found that the average lifetime was 43 250 km.
   a. Write down suitable hypothesis to test whether the newer brand of tyres have a longer lifetime.
   b. Determine the p-value for this test, correct to 4 decimal places.
   c. State with reasons whether the null hypothesis should be rejected or not rejected at the 1% level of significance.

9. On average a ballpoint pen writes for a linear distance which is normally distributed with a mean of 2500 metres with a standard deviation of 750 metres. A newer brand of pen has been developed and is considered to last longer. To test this claim 35 pens were sampled and found to have an average linear distance of 2750 metres. Perform a hypothesis test, state the p-value, correct to 4 decimal places, and determine if these pens last longer than average using the following levels of significance.
   a. 10% b. 5% c. 1%

10. **WE16** The manufactures of a certain brand of TV, have claimed that the lifetime of a particular TV is normally distributed with a mean of 35 000 hours and a standard deviation of 12 000 hours. 25 such TVs were tested and failed after 32 100 hours.
   a. Write down suitable hypothesis to test whether the mean lifetime is less than that claimed by the manufacturer.
   b. Determine the p-value for this test, correct to 4 decimal places.
   c. State with reasons whether the null hypothesis should be rejected or not rejected at the 5% level of significance.
   d. Let $\overline{X}$ be the lifetime of a random sample of 25 such TV's, assuming $\mu = 35\,000$ hours, and the standard deviation of 12 000 calculate the value I^*, correct to 2 decimal places, such that $\Pr\left(\overline{X} < I^* | \mu = 35\,000\right) = 0.05$, and interpret the meaning of the value of I^*.

11. The time for an employee to complete a particular task is 10 minutes. In a random sample of 36 attempts to complete this task, the time taken was found to be 9.4 minutes with a standard deviation of 2 minutes. At the 5% level of significance, determine if there is evidence to suggest that the time taken to complete the task is less than 10 minutes.

12. A coffee dispensing machine is meant to fill cups to 200 ml. The amount of fluid in the cups is known to be normally distributed with a standard deviation of 20 ml. A random sample of 30 cups had a mean of 208 ml. Perform a hypothesis test, to decide if the machine needs servicing and by calculating the p-value, correct to 4 decimal places, state with reasons whether the null hypothesis should be rejected or not rejected at the following levels of significance.
   a. 10% b. 5% c. 1%

13. The IQ (intelligence quotient) scores are normally distributed with a mean of 100 and a standard deviation of 15.
   a. A teacher at school believes her class, 12A, have an above average IQ. To test this she samples the 25 students in her class and finds their average IQ to be 105. Set up a hypothesis test and by calculating the p-value determine if there is evidence at the 5% level of significance to suggest that the students in her class are above average.
   b. A teacher at school believes their class, 12B, have a below average IQ. To test this they samples the 36 students in the class and finds their average IQ to be 94. Set up a hypothesis test and by calculating the p-value determine if there is evidence at the 1% level of significance to suggest the students in class 12B are below average.

c. A teacher at school believes his class, 12C, to have an average IQ. To test this he samples the 40 students in his class and finds their average IQ to be 98. Set up a hypothesis test and by calculating the p-value determine if there is evidence at the 5% level of significance to suggest the students in his class have an average IQ.

14. The management at a hospital assume that the mean age of the patients in the hospital is 50 years old. To test this they took a random sample of 40 patients from the hospital and found the mean age of the sample was 54.5 years with a standard deviation of 17 years.

a. Determine if there is evidence at the 5% level of significant to suggest that the mean age is different.
b. Determine if there is evidence at the 5% level of significant to suggest that the mean age more than 50 years old.

15. The management at a hospital believe that the mean number of days that patients stay in the hospital is 10 days. To test this claim they took a random sample of 40 patients from the hospital and found the mean stay was 9 days with a standard deviation of 3 days.

a. Determine if there is evidence at the 5% level of significant to suggest that the average number of days that patients stay in the hospital is not 10 days.
b. Determine if there is evidence at the 5% level of significant to suggest that the average number of days that patients stay in the hospital is less than 10 days.

16. A normal variate has a mean of 50 and a standard deviation of 17. A two-sided hypothesis test is performed. A random sample of size 30 is taken, calculate a range of values for $\bar{x}$, correct to 2 decimal places, for each of the following situations.

a. H_0 is not rejected at the 10% level of significance. b. H_0 is rejected at the 5% level of significance.
c. H_0 is rejected at the 1% level of significance.

17. A normal variate has a mean of 30 and a standard deviation of 7. A one-sided hypothesis test is performed, the alternative hypothesis is $H_1: \mu < 30$. A random sample of size 36 is taken, calculate a range of values for $\bar{x}$, correct to 2 decimal places, for each of the following situations.

a. H_0 is rejected at the 10% level of significance. b. H_0 is rejected at the 5% level of significance.
c. H_0 is not rejected at the 1% level of significance.

18. A normal variate has a mean of 70 and a standard deviation of 17. A one-sided hypothesis test is performed, the alternative hypothesis is $H_1: \mu > 70$. A random sample of size n has found to have a mean of 75, calculate a range of values for n for each of the following situations.

a. H_0 is rejected at the 5% level of significance. b. H_0 is rejected at the 1% level of significance.

13.5 Exam questions

Question 1 (5 marks) TECH-FREE

Source: *VCE 2021, Specialist Mathematics Exam 1, Q3; © VCAA.*

A company produces a particular type of light called Shiny. The company claims that the lifetime of these globes is normally distributed with a mean of 200 weeks and it is known that the standard deviation of the lifetime of Shiny globes is 10 weeks. Customers have complained, saying Shiny globes were lasting less than the claimed 200 weeks. It was decided to investigate the complaints. A random sample of 36 Shiny globes was tested and it was found that the mean lifetime of the sample was 195 weeks.

Use $\Pr(-1.96 < Z < 1.96) = 0.95$ and $\Pr(-3 < Z < 3) = 0.9973$ to answer the following questions.

a. Write down the null and alternative hypotheses for the one-tailed test that was conducted to investigate the complaints. **(1 mark)**

b. i. Determine the p-value, correct to three decimal places, for the test. **(2 marks)**

ii What should the company be told if the test was carried out at the 1% level of significance? **(1 mark)**

c. The company decided to produce a new type of light globe called Globeplus.
Find an approximate 95% confidence interval for the mean lifetime of the new globes if a random sample of 25 Globeplus globes is tested and the sample mean is found to be 250 weeks. Assume that the standard deviation of the population is 10 weeks. Give your answer correct to two decimal places. **(1 mark)**

Question 2 (1 mark) TECH-ACTIVE

Source: *VCE 2019, Specialist Mathematics Exam 2, Section A, Q20; © VCAA.*

MC The random number function of a calculator is designed to generate random numbers that are uniformly distributed from 0 to 1. When working properly, a calculator, generates random numbers from a population where $\mu = 0.5$ and $\sigma = 0.2887$.

When checking the random number function of a particular calculator, a sample of 100 random numbers was generated and was found to have a mean of $\bar{x} = 0.4725$.

Assuming $H_0: \mu = 0.5$ and $H_1: \mu < 0.5$ and $\sigma = 0.2887$ the p-value for a one-sided test is

A. 0.0953 **B.** 0.1704 **C.** 0.4621 **D.** 0.8296 **E.** 0.9525

Question 3 (9 marks) TECH-ACTIVE

Source: *VCE 2019, Specialist Mathematics Exam 2, Section B, Q6; © VCAA.*

A company produces packets of noodles. It is known from past experience that the mass of a packet of noodles produced by one of the company's machines is normally distributed with a mean of 375 grams and a standard deviation of 15 grams.

To check the operation of the machine after some repairs, the company's quality control employees select two independent random samples of 50 packets and calculate the mean mass of the 50 packets for each random sample.

a. Assume that the machine is working properly. Find the probability that at least one random sample will have a mean mass between 370 grams and 375 grams. Give your answer correct to three decimal places. **(2 marks)**

b. Assume that the machine is working properly. Find the probability that the means of the two random samples differ by less than 2 grams. Give your answer correct to three decimal places. **(3 marks)**

To test whether the machine is working properly after the repairs and is still producing packets with a mean mass of 372 grams, the two random samples are combined and the mean mass of the 100 packets is found to be 375 grams. Assume that the standard deviation of the mass of the packets produced is still 15 grams. A two-tailed test at the 5% level of significance is to be carried out.

c. Write down suitable hypotheses H_0 and H_1 for this test. **(1 mark)**

d. Find the p-value for the test, correct to three decimal places. **(1 mark)**

e. Does the mean mass of the sample of 100 packets suggest that the machine is working properly at the 5% level of significance for a two-tailed test? Justify your answer. **(1 mark)**

f. What is the smallest value of the mean mass of the sample of 100 packets for H_0 to be **not** rejected? Give your answer correct to one decimal place. **(1 mark)**

More exam questions are available online.

13.6 Review

13.6.1 Summary

doc-37626

Hey students! Now that it's time to revise this topic, go online to:

Access the topic summary

Review your results

Watch teacher-led videos

Practise VCAA exam questions

Find all this and MORE in jacPLUS

13.6 Exercise

Technology free: short answer

1. $E(X) = 434$ and $Var(X) = 64$. If $Y = \frac{X}{4} + 10$, calculate:

   a. $E(Y)$

   b. $Var(Y)$

2. Given that $T = 2X + 3Y$, the mean and variance of T are 41 and 72. Let $D = X - Y$, the mean and variance of D are 3 and 13. Determine the mean and variance of both X and Y.

3. Throughout this question use an integer multiple of the standard deviation in calculations.

   a. From the results of a random sample of n students, a 95% confidence interval for the mean score of all students was calculated to be $(24, 36)$. If the standard deviation is 18, calculate the mean score and the size, n, of this random sample.

   b. Determine the size of another random sample for which the endpoints of the 95% confidence interval is 2 either side of the sample mean.

4. A population has a mean of 58.9 and a standard deviation of 5.2.

   a. If samples of size 20 are selected, determine the mean and variance for the distribution of $\overline{X}$.

   b. Determine the sample size that would be needed to reduce the standard deviation to less than 1.

5. Consider a continuous random variable X, with a probability density function

$$f(x) = \begin{cases} k(4-x) & 0 \le x \le 4 \\ 0 & x < 0 \text{ or } x > 4 \end{cases}$$

   a. Determine the value of k.

   b. Calculate $E(X)$.

   c. Calculate $Var(X)$.

   Let Y be a binomially distributed random variable with $n = 16$ and $p = \frac{1}{4}$.

   X and Y are independent random variables. Let Z be the random variable defined by $Z = 9X + Y$.

   d. Calculate the mean of Z.

   e. Calculate the standard deviation of Z.

Technology active: multiple choice

6 **MC** X and Y are independent normally distributed random variables, both with means of 12 and variances of 4. The random variable T is defined by $T = 2X - Y$. Let Z be the standard normal variable, then $\Pr(T > 18)$ is equal to

A. $\Pr\left(Z > \frac{3}{5}\right)$

B. $\Pr\left(Z > \frac{3}{2}\right)$

C. $\Pr(Z > 3)$

D. $\Pr\left(Z > \frac{3}{2\sqrt{3}}\right)$

E. $\Pr\left(Z > \frac{6}{\sqrt{20}}\right)$

7 **MC** The heights of 14-year-old boys in a given population are normally distributed, with a standard deviation 5 cm. A 93% confidence interval for the mean heights of all 14-year-old boys from a sample of n 14-year-old boys was found to be $(162.82, 165.18)$. Then n is closest to

A. 39

B. 40

C. 58

D. 59

E. 69

8 **MC** After sampling his local farm's tomatoes, Giovanni finds that a 90% confidence interval for the mean weight is 113–117 g. This means that

A. 90% of the tomatoes weigh between 113 g and 117 g.
B. the average tomato weighs 115 g.
C. the population mean weight of tomatoes could be between 113 g and 117 g.
D. 90% of the samples will have a mean between 113 g and 117 g.
E. 10% of the tomatoes will weigh less than 113 g.

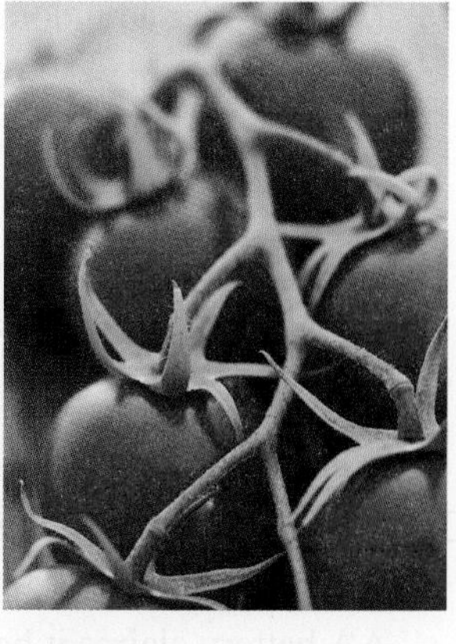

9 **MC** Our Local Grocer has invested in an advertising campaign and wish to see if it has changed the average number of daily customers entering the store. Prior to the campaign, the average number of customers is 342 per day. Identify H_0 and H_1.

A. $H_0: \mu = 342$, $H_1: \mu \neq 342$
B. $H_0: \mu = 342$, $H_1: \mu > 342$
C. $H_0: \mu = 342$, $H_1: \mu < 342$
D. $H_0: \mu \neq 342$, $H_1: \mu = 342$
E. $H_0: \mu \neq 342$, $H_1: \mu > 342$

10 **MC** A school principal must decide whether or not to cancel the school sports day due to a threatening rain day. The Type I and Type II errors for the null hypothesis, that the weather will remain dry are

A. Type I error: Don't cancel the sports day, but it rains.
Type II error: The weather remains dry, but the sports day is needlessly cancelled.
B. Type I error: The weather remains dry, but the sports day is needlessly cancelled.
Type II error: Don't cancel the sports day, and it rains.
C. Type I error: Cancel the sports day, and it rains.
Type II error: Don't cancel the sports day and it rains.
D. Type I error: Don't cancel the sports day, and it rains.
Type II error: Don't cancel the sports day, and the weather remains dry.
E. Type I error: Don't cancel the sports day, but it rains.
Type II error: Cancel the sports day, and it rains.

11 **MC** If a 99% confidence interval for the average cost of a tray of mangoes is (\$17.75, \$19.95), then a 97% confidence interval for the average cost of a tray of mangoes is closest to

A. (\$17.39, \$19.55) **B.** (\$17.88, \$19.81) **C.** (\$18.49, \$19.21) **D.** (\$17.92, \$19.78) **E.** (\$18.11, \$19.59)

12 **MC** In a two-sided test at the 8% level of significance, then

A. H_0 should be rejected if $p = 0.5$
B. H_0 should be rejected if $p = 0.25$
C. H_0 should be rejected if $p = 0.2$
D. H_0 should not be rejected if $p = 0.1$
E. H_0 should not be rejected if $p = 0.01$

13 **MC** A confidence interval is used to estimate the population mean μ based on a sample mean $\bar{x}$. To decrease the width of a confidence interval by 80%, the sample size must be multiplied by

A. 65 **B.** 25 **C.** 8 **D.** 4 **E.** 2

14 **MC** The volume of beer in a can varies normally with a mean of 376 mL and a standard deviation of 3 mL. The beer is sold in a 6 pack of cans. The probability that the mean volume of beer in a randomly selected 6 pack is less than 373 mL is closest to

A. 0.1402
B. 0.1587
C. 0.0668
D. 0.0228
E. 0.0072

15 **MC** Consider a random variable X, with a probability density function

$$f(x) = \begin{cases} 3x^2 & 0 \le x \le 1 \\ 0 & x < 0 \text{ or } x > 1 \end{cases}$$

If a large number of samples, each of size 75, is taken from this distribution, then the sample means $\overline{X}$, will be approximately normal. Then the standard deviation of $\overline{X}$ is equal to

A. $\dfrac{\sqrt{5}}{100}$ B. $\dfrac{45}{16}$ C. $\dfrac{3\sqrt{5}}{4}$ D. $\dfrac{\sqrt{15}}{20}$ E. $\dfrac{\sqrt{3}}{400}$

Technology active: extended response

16. The lifetime of AA batteries is stated to be 50 hours, with a standard deviation of 10 hours. A new long-life AA battery claims it has a longer lifetime, and a random sample of 25 new long-life batteries has a mean lifetime of 54 hours.

a. State the appropriate null and alternative hypothesis for the lifetime of these long life batteries.
b. The p value for this test, is given by $\Pr(Z \ge a)$, where Z has the standard normal distribution. Determine the value of a and p correct to 3 decimal places.
c. Determine whether or not the null hypothesis should be rejected at the 5% level of significance.
d. Determine whether or not the null hypothesis should be rejected at the 1% level of significance.
e. Calculate a 95% confidence interval, correct to 2 decimal places, for the lifetime of the long-life batteries.
f. Calculate a 99% confidence interval, correct to 2 decimal places, for the lifetime of the long-life batteries.

17. A farmer grows brown onions and potatoes. The brown onions have a mean mass of 180 grams with a standard deviation of 30 grams and the potatoes have a mean mass of 200 grams with a standard deviation of 50 grams. Assume that the masses of both brown onions and potatoes grown on the farm are normally distributed and independent.

a. Calculate the probability, correct to 4 decimal places, that 5 brown onions weigh more than4 potatoes.

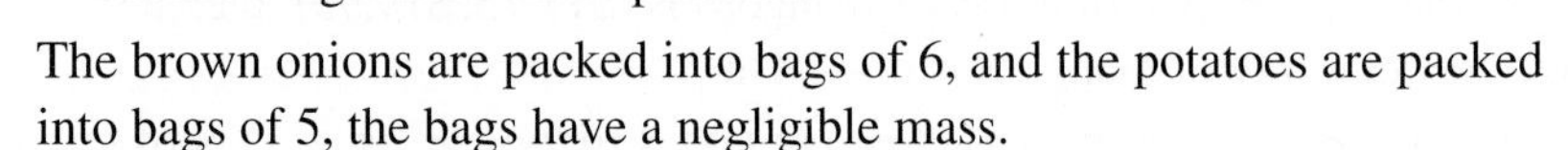

The brown onions are packed into bags of 6, and the potatoes are packed into bags of 5, the bags have a negligible mass.

b. Calculate the probability, correct to 4 decimal places, that one bag of brown onions has a mass of more than one kilogram.
c. Calculate the probability that one bag of potatoes has a mass of more than one kilogram.
d. Calculate the probability, correct to 4 decimal places, that one bag of brown onions and one bag of potatoes have a mass of more than 2 kilograms.

18. A factory makes blocks of chocolate and claim that the weights of their 200 g chocolate blocks, are normally distributed with a mean of 201 g, with a standard deviation of 1.0 g.

a. A box contains 5 blocks of chocolate, calculate the probability, correct to 4 decimal places, of boxes that contain at least the stated weight of one kilogram.
b. If regulations require 95% of blocks of chocolate to contain at least a weight of 200 g, determine the maximum allowable standard deviation to achieve this, correct to 2 decimal places.
c. Authorities visit the factory and check a random sample 25 blocks of chocolate and find their mean weight to be 200.5 g. Assuming the standard deviation of 1 g is correct, perform a hypothesis test and determine the p value correct to 4 decimal places. State with reasons whether the factories claim should be accepted at the 1% level of significance.
d. Determine the minimum weight, correct to 3 decimal places, of a random sample of 25 blocks of chocolate so that the claim can be accepted at the 1% level of significance.

19. a. Marina records the masses of a number of newborn babies during her shifts in January. She uses the sample standard deviation of 1.23 kg, to calculate a 95% confidence interval in kg to be (3.20, 3.62). Determine the number of babies in her sample.

b. Another sample of 25 newborn baby girls has a mean of 3.41 kg and a standard deviation of 1.23 kg. A $C\%$ confidence interval was calculated to be in kg (2.964, 3.856). Determine the value of C, correct to the nearest integer.

c. Assuming that newborn baby girls have a mean mass of 3.41 kg with a standard deviation of 1.23 kg. Calculate the probability, correct to 4 decimal places, that a random selection of 10 newborn baby girls have a combined mass of more than 34 kg.

20. Consider a random variable X, with a probability density function

$$f(x) = \begin{cases} ke^{-4x} & x \geq 0 \\ 0 & x < 0 \end{cases}$$

a. Determine the value of k.

b. Use integration by parts to show that $E(X) = \frac{1}{4}$.

c. Show that $\text{Var}(X) = \frac{1}{16}$.

Let Y be a binomially distributed random variable with $n = 18$ and $p = \frac{1}{3}$.
X and Y are independent random variables. Let Z be the random variable defined by $Z = 4X + 3Y$.

d. Determine the mean of Z.

e. Determine the exact standard deviation of Z.

13.6 Exam questions

Question 1 (10 marks) TECH-ACTIVE

Source: *VCE 2021, Specialist Mathematics Exam 2, Section B, Q6; © VCAA.*

The maximum load of a lift in a chocolate company's office building is 1000 kg. The masses of the employees who use the lift are normally distributed with a mean of 75 kg and a standard deviation of 8 kg. On a particular morning there are n employees about to use the lift.

a. What is the maximum possible value of n for there to be less than a 1% chance of the lift exceeding the maximum load? **(2 marks)**

Clare, who is one of the employees, likes to have a hot drink after she exits the lift. The time taken for the drink machine to dispense a hot drink is normally distributed with a mean of 2 minutes and a standard deviation of 0.5 minutes. Times taken to dispense successive hot drinks are independent.

b. Clare has a meeting at 9.00 am and at 8.52 am she is fourth in the queue for a hot drink. Assume that the waiting time between hot drinks dispensed is negligible and that it takes Clare 0.5 minutes to get from the drink machine to the meeting room.
What is the probability, correct to four decimal places, that Clare will get to her meeting on time? **(2 marks)**

Clare is a statistician for the chocolate company. The number of chocolate bars sold daily is normally distributed with a mean of 60 000 and a standard deviation of 5000. To increase sales, the company decides to run an advertising campaign. After the campaign, the mean daily sales from 14 randomly selected days was found to be 63 500.

Clare has been asked to investigate whether the advertising campaign was effective, so she decides to perform a one-sided statistical test at the 1% level of significance.

c. **i.** Write down suitable null and alternative hypotheses for this test. **(1 mark)**

ii. Determine the p value, correct to four decimal places, for this test. **(1 mark)**

iii. Giving a reason, state whether there is any evidence for the success of the advertising campaign. **(1 mark)**

d. Find the range of values for the mean daily sales of another 14 randomly selected days that would lead to the null hypothesis being rejected when tested at the 1% level of significance. Give your answer correct to the nearest integer. **(1 mark)**

e. The advertising campaign has been successful to the extent that the mean daily sales is now 63 000. A statistical test is applied at the 5% level of significance.
Find the probability that the null hypothesis would be incorrectly accepted, based on the sales of another 14 randomly selected days and assuming a standard deviation of 5000. Give your answer correct to three decimal places. **(2 marks)**

Question 2 (1 mark) TECH-ACTIVE

Source: *VCE 2021, Specialist Mathematics Exam 2, Section A, Q20; © VCAA.*

MC An office has two coffee machines that operate independently of each other. The time taken for each machine to produce a cup of coffee is normally distributed with a mean of 30 seconds and a standard deviation of 5 seconds. On a particular morning, a cup of coffee is produced from each machine.

The probability that the time taken by each coffee machine to produce one cup of coffee will differ by less than 3 seconds is closest to

A. 0.164
B. 0.236
C. 0.329
D. 0.451
E. 0.671

Question 3 (3 marks) TECH-FREE

Source: *VCE 2019, Specialist Mathematics Exam 1, Q3; © VCAA.*

A machine produces chocolate in the form of a continuous cylinder of radius 0.5 cm. Smaller cylindrical pieces are cut parallel to its end, as shown in the diagram below.

The lengths of the pieces vary with a mean of 3 cm and a standard deviation of 0.1 cm.

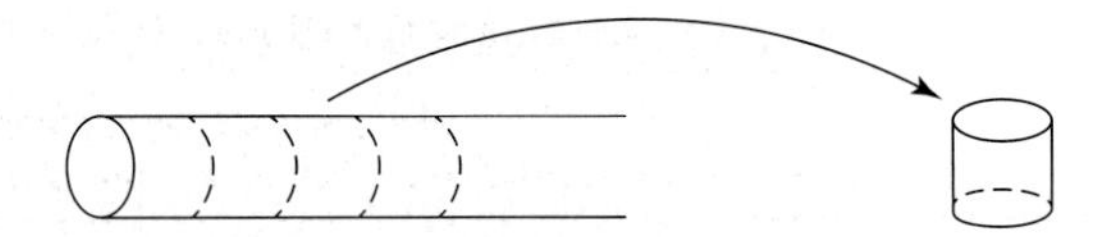

a. Find the expected volume of a piece of chocolate in cm^3. **(1 mark)**

b. Find the variance of the volume of a piece of chocolate in cm^6. **(1 mark)**

c. Find the expected surface area of a piece of chocolate in cm^2. **(1 mark)**

Question 4 (4 marks) TECH-FREE

Source: *VCE 2018, Specialist Mathematics Exam 1, Q4; © VCAA.*

X and Y are independent random variables. The mean and the variance of X are both 2, while the mean and the variance of Y are 2 and 4 respectively.

Given that a and b are integers, find the values of a and b if the mean and the variance of $aX + bY$ are 10 and 44 respectively.

Question 5 (9 marks) TECH-ACTIVE

***Source:** VCE 2016, Specialist Mathematics Exam 2, Section B, Q6; © VCAA.*

The mean level of pollutant in a river is known to be 1.1 mg/L with a standard deviation of 0.16 mg/L.

a. Let the random variable $\overline{X}$ represent the mean level of pollutant in the measurements from a random sample of 25 sites along the river.
Write down the mean and standard deviation of $\overline{X}$. **(2 marks)**

After a chemical spill, the mean level of pollutant from a random sample of 25 sites is found to be 1.2 mg/L.

To determine whether this sample provides evidence that the mean level of pollutant has increased, a statistical test is carried out.

b. Write down suitable hypotheses H_0 and H_1 to test whether the mean level of pollutant has increased. **(2 marks)**

c. i. Find the p-value for this test, correct to four decimal places. **(2 marks)**

ii. State with a reason whether the sample supports the contention that there has been an increase in the mean level of pollutant after the spill. Test at the 5% level of significance. **(1 mark)**

d. For this test, what is the smallest value of the sample mean that would provide evidence that the mean level of pollutant has increased? That is, find $\overline{x}_c$ such that $\Pr(\overline{X} > \overline{x}_c \mid \mu = 1.1) = 0.05$. Give your answer correct to three decimal places. **(1 mark)**

e. Suppose that for a level of significance of 2.5%, we find that $\overline{x}_c = 1.163$. That is,
$\Pr(\overline{X} > 1.163 \mid \mu = 1.1) = 0.025$.
If the mean level of pollutant in the river, μ is in fact 1.2 mg/L after the spill, find
$\Pr(\overline{X} < 1.163 \mid \mu = 1.2) = 1.2$. Give your answer correct to three decimal places. **(1 mark)**

More exam questions are available online.

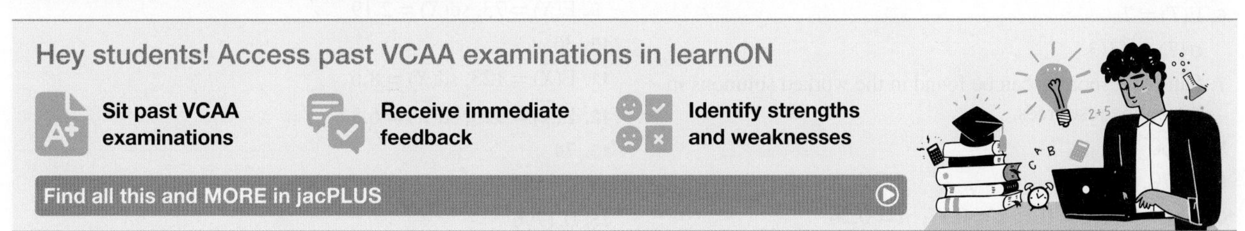

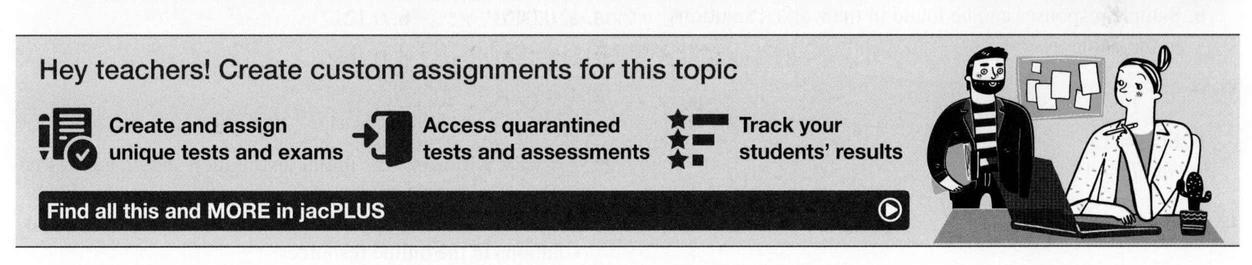

Answers

Topic 13 Probability and statistics

13.2 Linear combinations of random variables

13.2 Exercise

1. a. $E(X - 10) = 5.3$
   $Var(X - 10) = 1.8$

   b. $E(2X) = 30.6$
   $Var(2X) = 7.2$

2. a. $E(X + 4) = 17$
   $Var(X + 4) = 3.2$

   b. $E(3X) = 39$
   $Var(3X) = 28.8$

3. $E(2X + 3Y) = 23.2$
   $Var(2X + 3Y) = 26.36$

4. $E\left(\frac{1}{2}X + Y\right) = 24.145$

   $Var\left(\frac{1}{2}X + Y\right) = 11.1725$

5. $E(X) = 14$
   $E(Y) = -4$
   $Var(Y) = 9$

6. $E(T) = 7$
   $\sigma(T) = 4\sqrt{3}$

7. Sample responses can be found in the worked solutions in the online resources.

8. 0.460

9. 0.070

10. a. $a = 0.2,\ b = 0.1,\ c = 0.17,\ d = 0.36$

    b. Sample responses can be found in the worked solutions in the online resources.

11. a. $c = \frac{1}{14}$

    b. $E(T) = 15$
    $Var(T) = 19$

12. a. $k = \frac{3}{26}$

    b. $E(T) = 38$
    $Var(T) = 43.8$

13. a. $Pr(Y > X) = 0.6306$

    b. $Pr(X > Y) = 0.3694$

    c. $Pr(3X - 2Y > 5) = 0.6558$

14. a. 0.0217 b. 0.477 cm

15. a. 0.5983 b. 0.1404

16. 0.3368

17. 0.4005

18. 0.6987

13.2 Exam questions

Note: Mark allocations are available with the fully worked solutions online.

1. D
2. C
3. B

13.3 Sample means and simulations

13.3 Exercise

1. 51.5 minutes

2. 95.9 minutes

3. a. $\mu_{\bar{x}} = 600,\ \sigma_{\bar{x}} = 5.48$

   b. $\mu_{\bar{x}} = 600,\ \sigma_{\bar{x}} = 3.87$

4. a. $\mu_{\bar{x}} = 90.5,\ \sigma_{\bar{x}} = 2.58$

   b. $\mu_{\bar{x}} = 90.5,\ \sigma_{\bar{x}} = 1.83$

5. Answers will vary. The expected average is 50 but it is likely that results will be different to this. Sample responses can be found in the worked solutions in the online resources.

6. $E(\bar{X}) = 7.09$. Theoretical mean is 7

7. Sample responses can be found the worked solutions in the online resources.

8. $E(\bar{X}) = 73,\ sd(\bar{X}) = 2.68$

9. $E(\bar{X}) = 73,\ sd(\bar{X}) = 2.19$

10. 36

11. $E(\bar{X}) = 123,\ sd(\bar{X}) = 8.6$

12. $E(\bar{X}) = 123,\ sd(\bar{X}) = 6.8$

13. 74

14. 0.1587

15. 0.1796

16. a. 0.0401 b. 0.1217

17. a. 0.1241 b. 0.1680

18. 16

19. 25

20. Answers will vary. The initial distribution should be skewed to the right and the distribution of sample means should be symmetrical. Sample responses can be found in the worked solutions in the online resources.

13.3 Exam questions

Note: Mark allocations are available with the fully worked solutions online.

1. E

2. a. 79.8%

   b. Sample responses can be found in the worked solutions in the online resources.

   c. 99.6%

   d. 5.1 mL

3. E

13.4 Confidence intervals

13.4 Exercise

1. (\$2186, \$2442)
2. (1.02 km, 1.38 km)
3. (\$23 392, \$23 520)
4. (25.02 g, 26.38 g)
5. 124
6. 97
7. a. (\$191.44, \$215.46)
   b. (\$193.37, \$213.53)
   c. 1803
8. a. (2239, 2389) minutes
   b. (2215, 2413) minutes
   c. 91
9. a. (21.9, 27.3) minutes
   b. (22.3, 26.9) minutes
   c. 56
   d. 40
10. (2.12, 2.88) minutes
11. (83.6, 91.4) minutes
12. (211.450, 242.416) minutes

13.4 Exam questions

Note: Mark allocations are available with the fully worked solutions online.

1. C
2. D
3. D

13.5 Hypothesis testing

13.5 Exercise

1. H_0: The extra training sessions do not change his race time.
   H_1: The extra training sessions improve his race time.
2. H_0: The extra training sessions do not change her race time.
   H_1: The extra training sessions improve her race time.
3. a. The pet food is safe.
   b. The pet food is contaminated.
4. a. Aerial spraying is not safe.
   b. There is insufficient evidence to conclude that aerial spraying is safe.
5. None of the hypotheses are valid.
6. a. $H_0: \mu = 6$
   $H_1: \mu \neq 6$
   b. $p = 0.177$
   c. $p = 0.177 > 0.05 = \alpha$, do not reject H_0.
7. $p = 0.0051 < 0.01 = \alpha$, reject H_0.
8. a. $H_0: \mu = 40\,000$
   $H_1: \mu > 40\,000$
   b. $p = 0.0205$
   c. $p = 0.0205 > 0.01 = \alpha$, do not reject H_0.
9. a. $p = 0.0292 < 0.1 = \alpha$, reject H_0.
   b. $p = 0.0292 < 0.05 = \alpha$, reject H_0.
   c. $p = 0.0292 > 0.01 = \alpha$, do not reject H_0.
10. a. $H_0: \mu = 35000$
    $H_1: \mu < 35000$
    b. $p = 0.1135$
    c. $p = 0.1135 > 0.05 = \alpha$, do not reject H_0.
    d. $I^* = 31\ 052.35$ and is the maximum value at which H_0 can be rejected.
11. Yes
12. a. $p = 0.0285 < 0.1 = \alpha$, reject H_0.
    b. $p = 0.0285 < 0.05 = \alpha$, reject H_0.
    c. $p = 0.0285 > 0.01 = \alpha$, do not reject H_0.
13. a. Yes b. Yes c. No
14. a. No b. Yes
15. a. Yes b. Yes
16. a. $44.89 < \bar{x} < 55.11$
    b. $\bar{x} < 43.92$ or $\bar{x} > 56.08$
    c. $\bar{x} < 42.00$ or $\bar{x} > 58.00$
17. a. $\bar{x} < 28.50$ b. $\bar{x} < 28.08$ c. $\bar{x} \geq 27.29$
18. a. $n \geq 32$ b. $n \geq 63$

13.5 Exam questions

Note: Mark allocations are available with the fully worked solutions online.

1. a. $H_0: \mu = 200$
   $H_1: \mu < 200$
   b. i. $p = 0.001$
   ii. Reject H_0, and accept H_1.
   c. (246.08, 253.92)
2. B
3. a. 0.741
   b. 0.495
   c. $H_0: \mu = 375$
   $H_1: \mu \neq 375$
   d. $p = 0.046$
   e. The machine is not working properly.
   f. 372.1 g

13.6 Review

13.6 Exercise

Technology free: short answer

1. a. 118.5 b. 4
2. $\text{E}(X) = 10$, $\text{E}(Y) = 7$, $\text{Var}(X) = 9$, $\text{Var}(Y) = 4$
3. a. 36 b. 324
4. a. $\text{E}(\overline{X}) = 58.9$
   $\text{Var}(\overline{X}) = 1.352$
   b. The sample size needs to be at least 28.
5. a. $\frac{1}{8}$ b. $\frac{4}{3}$ c. $\frac{8}{9}$ d. 16 e. $5\sqrt{3}$

Technology active: multiple choice

6. E
7. D
8. C
9. A
10. B
11. D
12. D
13. B
14. E
15. A

Technology active: extended response

16. a. H_0: $\mu = 50$

    H_1: $\mu > 50$

    b. $a = 2,\ \ p = 0.023$

    c. Reject H_0

    d. Do not reject H_0

    e. $(50.08, 57.92)$

    f. $(48.85, 59.15)$

17. a. 0.7969 b. 0.8618

    c. 0.5 d. 0.7251

18. a. 0.9873

    b. 0.61

    c. Reject H_0, $p = 0.0062 < 0.01 = \alpha$

    d. 200.535

19. a. 132 b. 93% c. 0.5103

20. a. $k = 4$

    b. Sample responses can be found in the worked solutions in the online resources.

    c. Sample responses can be found in the worked solutions in the online resources.

    d. 19

    e. $\sqrt{37}$

13.6 Exam questions

Note: Mark allocations are available with the fully worked solutions online.

1. a. 12

   b. 0.3085

   c. i. H_0: $\mu = 60\,000$

      H_1: $\mu > 60\,000$

      ii. $p = 0.0044$

      iii. Since $p = 0.0044 < 0.01 = \alpha$, the advertising campaign was a success.

   d. $\bar{x} \geq 63109$

   e. 0.274

2. C

3. a. $\dfrac{3\pi}{4}$ cm^3 b. $\dfrac{\pi^2}{1600}$ cm^6 c. $\dfrac{7\pi}{2}$ cm^2

4. $a = 2,\ \ b = 3$

5. a. $E(\overline{X}) = 1.1$

      $sd(\overline{X}) = 0.032$

   b. H_0: $\mu = 1.1$

      H_1: $\mu > 1.1$

   c. i. $p = 0.0009$

      ii. Since $p = 0.0009 < 0.05 = \alpha$, there is evidence to support the alternative hypothesis that the level of pollutants has increased.

   d. $\bar{x} = 1.153$

   e. 0.124

GLOSSARY

Argand plane a plane used to represent complex numbers using a horizontal real axis and a vertical imaginary axis
argument of z the angle that a line segment of a complex number on an Argand plane makes with the positive real axis
axiom a statement that is considered to be true or a fact based on logic and does not need to be proven
Cartesian equation the relationship between x and y in a parametric equation
central limit theorem the theory that as all possible samples of a population are considered, the sample mean tends to be the same as that of the population
complex number system an extension of the real number system that includes numbers that involve i, the imaginary unit
concave down when the gradient of the curve decreases as x increases
concave up when the gradient of the curve increases as x increases
confidence interval an interval of values within which the true value of the population mean may lie between
conjecture a statement that seems to be true, but needs to be proven
contrapositive of $p \rightarrow q$ is $\neg q \rightarrow \neg p$
converse of $p \rightarrow q$ is $q \rightarrow p$
corollary the truth of these can be deduced from other statements
definition a statement that we accept as true, we cannot question it, it must be unambiguous and we accept it as fact
de Moivre's theorem states that if $z = r\,\text{cis}(\theta)$ then $z^n = r^n\,\text{cis}(n\theta)$ for $n \in N$
degree the highest power of the highest derivative present in the equation
derivative is the rate of change of a function with respect to a variable
direction the course along which an object moves
general solution contains arbitrary constants of integration and satisfies the differential equation
half-life the time it takes for half the original mass of a radioactive material to disintegrate
implications compound statements of the form if p, then q denoted by $p \rightarrow q$
implicit differentiation the computation of the derivative of an implicit function without explicitly determining the function
inverse of $p \rightarrow q$ is $\neg p \rightarrow \neg q$
lemma a statement used to help prove other theorems
linearly dependent a set of vectors is said to be linearly dependent if there is a non-trivial linear combination of the vectors that equals the zero vector
linearly independent a set of vectors is said to be linearly independent if no such linear combination of the vectors equals the zero vector
magnitude the size of an object
margin of error the distance between the sample estimate and the endpoints of the confidence interval
maximum turning points Points on a graph where the gradient is equal to zero and is the greatest value of the graph within the local domain. The gradient is positive to the left of the point and negative to the right of the point.
mean The average value of a set of numbers. Also corresponds to the expected value, E(X), of the distribution of a population.
minimum turning points Points on a graph where the gradient is equal to zero and is the lowest value of the graph within the local domain. The gradient is negative to the left of the point and positive to the right of the point.
negation the direct opposite of a statement or proposition, denoted by placing the symbol ¬ in front of the statement or proposition

Newton's law of cooling The law states that the rate of change of the temperature of a body is proportional to the difference between the temperature of the body and the temperature of the medium surrounding the body. It can be written as, $\frac{dT}{dt} \propto (T - T_m)$ or $\frac{dT}{dt} = k(T - T_m)$.

order the highest derivative present in the differential equation.

parallelogram a quadrilateral with two pairs of parallel sides

parameters varying constants used to describe a family of polynomials or relates two other variables

parametric equations equations written to model the path of a particle, involving a parameter

particular solution satisfies the differential equation and some initial value condition, known as the boundary value, that enables the constant(s) of integration to be calculated

point estimate A specific value that is an estimate of a parameter of a population based on sampling. For example, if the value of the sample mean is used as an estimate of the population mean, then it is called a point estimate.

point of discontinuity a point which is not in the domain of a graph, but which the graph approaches from both the positive and negative direction

point of inflection A point in which the concavity of the function changes. It means that the function changes from concave down to concave up or vice versa.

polar form a way to represent complex numbers using a modulus and an argument

population the entire group that you want to obtain information about in a statistical analysis

predicate logic a formal language in which propositions are expressed in terms of predicates, variables and quantifiers

principal value the angle $\text{Arg}(z)$ where $\text{Arg}(z) \in (-\pi, \pi]$

propositions statements which are either true of false

quadrilateral a four sided shape consisting of four straight lines, four angles and four vertices

rectangle a quadrilateral with four right angles

removable discontinuity a point which is not in the domain of a graph, but which the graph approaches from both the positive and negative direction

rhombus a quadrilateral whose four sides all have the same length

sample mean the mean of a sample

samples the group within the population that the information is being collected from

scalar a number that has no direction, and when multiplied with a vector, it produces another vector

second derivative the derivative of the derivative of a function

slope fields a graphical view of the solutions of a first-order differential equation, where the derivative at a certain point is represented by the midpoint of a line segment of the corresponding slope

solid of revolution a solid figure created by the 360° rotation of a function with defined boundaries around an axis

square a quadrilateral with four equal straight sides and four right angles

standard deviation A measure of spread. $\text{SD}(X) = \sigma = \sqrt{\text{Var}(X)}$. The larger the standard deviation, the more spread out the data. If the standard deviation is small, the data is clumped about the mean.

stationary point a point where the derivative of the function is equal to zero.

stationary point of inflection a point where the derivative of the function is zero but the derivative does not change sign on either side of the point

statistical inference the process of drawing conclusions about an underlying population based on a sample or subset of the data.

tautology a formulae or assertion that is true in every possible interpretation

theorem A statement that can be shown to be true using a proof technique. They can then be used as facts in proofs of other theorems.

trapezium a quadrilateral with exactly one pair of opposite sides parallel to each other

triangle a three sided figure, consisting of three straight lines three vertices and three angles

variance A measure of spread. $\text{Var}(X) = \sigma^2 = \text{E}(X^2) - \text{E}[(X)]^2$.

vector used to describe a quantity that has both a magnitude and a direction

INDEX

E

F

G

H

I

Q

R

S

T

U

V

W

Z